Volume 6

GENETIC RESOURCES, CHROMOSOME ENGINEERING, AND CROP IMPROVEMENT

Medicinal Plants

GENETIC RESOURCES, CHROMOSOME ENGINEERING, AND CROP IMPROVEMENT SERIES

Series Editor, Ram J. Singh

Genetic Resources, Chromosome Engineering, and Crop Improvement
Volume 1: Grain Legumes
edited by Ram J. Singh and Prem P. Jauhar

Genetic Resources, Chromosome Engineering, and Crop Improvement
Volume 2: Cereals
edited by Ram J. Singh and Prem P. Jauhar

Genetic Resources, Chromosome Engineering, and Crop Improvement
Volume 3: Vegetable Crops
edited by Ram J. Singh

Genetic Resources, Chromosome Engineering, and Crop Improvement
Volume 4: Oilseed Crops
edited by Ram J. Singh

Genetic Resources, Chromosome Engineering, and Crop Improvement
Volume 5: Forage Crops
edited by Ram J. Singh

Genetic Resources, Chromosome Engineering, and Crop Improvement
Volume 6: Medicinal Plants
edited by Ram J. Singh

GENETIC RESOURCES, CHROMOSOME ENGINEERING, AND CROP IMPROVEMENT SERIES

Series Editor, Ram J. Singh

Volume 6

GENETIC RESOURCES, CHROMOSOME ENGINEERING, AND CROP IMPROVEMENT

Medicinal Plants

EDITED BY

RAM J. SINGH

CRC Press
Taylor & Francis Group
Boca Raton London New York

CRC Press is an imprint of the
Taylor & Francis Group, an **informa** business

CRC Press
Taylor & Francis Group
6000 Broken Sound Parkway NW, Suite 300
Boca Raton, FL 33487-2742

First issued in paperback 2019

© 2012 by Taylor & Francis Group, LLC
CRC Press is an imprint of Taylor & Francis Group, an Informa business

No claim to original U.S. Government works

ISBN-13: 978-1-4200-7384-3 (hbk)
ISBN-13: 978-0-367-38240-7 (pbk)

Library of Congress Cataloging-in-Publication Data

Genetic resources, chromosome engineering, and crop improvement. Medicinal plants / editor, Ram J. Singh.
        p. cm. -- (Genetic resources, chromosome engineering, and crop improvement: medicinal plants, volume 6)
    Includes bibliographical references and index.
    ISBN 978-1-4200-7384-3 (alk. paper)
    1. Medicinal plants--Genetic engineering. 2. Medicinal plants--Breeding. 3. Crop improvement. I. Singh, Ram J. II. Title. III. Series.

SB293.G46 2012
581.6'34--dc22

2011001944

Visit the Taylor & Francis Web site at
http://www.taylorandfrancis.com

and the CRC Press Web site at
http://www.crcpress.com

Dedication

Homo Botanicus
Pietro Andrea Gregorio Matthioli (Mattioli, Matthiolus)

(1501, Siena, Italy–1577, Trident/Trento, Italy)

Pietro Andrea Gregorio Matthioli was a Renaissance physician and botanist of Italian origin. He spent his youth in Siena and Venezia studying medicine at the University of Padova. As a physician, he was active in Siena, Rome, Trident, and Gorizia and translated the Latin book *De Materia Medica* from P. Dioskuridos. He lived in Prague (Bohemia) from 1554 to 1566 as a physician on the court of Emperors Ferdinand I and Maxmilian II. The first edition of his famous book, *Herbarium,* was published in 1544 (in Italian) with description of over 1,000 plant species. This book contained the most comprehensive treatment of medicinal plants (botany, medicinal effect, preparation of different medicines, practical usage) grown in Europe in the sixteenth century. Matthioli brilliantly summarized the results of scientific knowledge of his époque and empirical knowledge related to the medical effects of plants. *Herbarium* was considered the "bible" of medicinal plants of the sixteenth century and was translated into Latin (Venice, 1554), Czech (Prague, 1562), German (Prague, 1563), French,

and many other languages until the twentieth century. P. Matthioli was considered a personality of European dimension because of his contribution to botany and knowledge of medicinal plants. His name was used to describe some new plant species, such as *Cortusa matthioli* and *Matthiola incana.*

Dedication

Homo Botanicus,
Pietro Andrea Gregorio Mattioli (Matthiolus)

(1501, Siena, Italy – 1577, Trento/Trento, Italy)

Pietro Andrea Gregorio Mattioli was a Renaissance physician and botanist of Italian origin. He spent his youth in Siena and Venezia, studying medicine at the University of Padova. As a physician, he was active in Siena, Roma, Trieste, and Gorizia and translated the famous book De Materia Medica from P. Dioskurides. He lived in Prague (Bohemia) from 1551 to 1562 as a physician on the court of Emperors Ferdinand I and Maximilian II. The first edition of his famous book, Herbarius, was published in 1544 (in Italian) with description of over 1,000 plant species. The book contained the most comprehensive treatment of medicinal plants thereby including effect, preparation of different medicines, practical usage, etc. grown in Europe in the sixteenth century. Mattioli brilliantly summarized the results of scientific knowledge of his epoque and comprised knowledge related to the medical effect of plants. Herbarium was considered the "bible" of medicinal plants of the sixteenth century and was translated into Latin (Venice, 1554), Czech (Prague, 1562), German (Prague, 1563), French, and many other languages until the twentieth century. P. Mattioli was considered a personal physician of European dimension because of his contribution to botany and knowledge of medicinal plants. His name was used to describe some new plant species, such as German wallflower and Matthiola incana.

Preface

Since time immemorial, people from all cultures of the world have selected plants for food, shelter, clothing, and medicine. Plants were identified according to their therapeutic properties by the priests, shamans, herbalists, spiritual leaders, and medicine men and women through trial and error; this knowledge was passed on through word of mouth. Thus, prehistoric knowledge of medicinal plants is either lost or is depicted on the walls of caves.

The science of medicinal plants developed independently on all the continents and the exchange of knowledge was limited because of geographical isolation. The result was Ayurveda in India and differing traditional plant-based medicines in China, the Middle East, Africa, Europe, Australia (Aboriginals), and the Americas (Native American tribes). Therefore, the widespread use of medicinal plants for curing and preventing diseases has been described in the ancient texts of the Vedas, Bible, Qur´an, and Ahadith, as well as in Chinese texts.

During the progression of time, our ancestors recognized the value of plants and initiated identification, domestication, and cultivation of medicinal plants, as well as plants for food and feed, such as cereals, grains, legumes, oilseeds, vegetable crops, forage crops, fruit trees, and flowering plants. The rapid explosion of human and animal populations, migration, civilization, urbanization, and deforestation created many new diseases for humans and domesticated animals, as well as for plants. Plant-based medicines were the only source for remedies to overcome catastrophic diseases. For example, the bark of the cinchona tree from the Andes of South America was the only source for treating malaria among the early European settlers in the sixteenth century.

The development of the sciences of botany, chemistry, pharmacology, anthropology, archeology, linguistics, history, sociology, comparative religion, and numerous other specialties contributed appreciably to the search for new biodynamic drug plants. The knowledge needed to determine the chemical basis of medicinal plants for treating diseases developed during the nineteenth century. Chemical compounds were isolated and identified, leading to the science of phytochemistry and the establishment of homeopathic institutes and drug industries worldwide. Medicinal properties of "nature's chest" have been or are being disseminated through many books, peer-reviewed research papers published in national and international journals of high repute, technical bulletins, and popular articles. Also, information technology has helped in assessing the rich knowledge of medicinal plants by making it possible to "surf" online resources (see the Bibliography).

In some Asian and African countries, 80% of the population depends on traditional medicine for primary health care, and in many developed countries, 70–80% of the population has used some form of alternative or complementary medicine. Along with the production of modern drugs, medicinal plants are used for producing beverages, cosmeceutical products, and spices for culinary uses.

Pharmaceutical companies in developed countries take genetic resources and traditional knowledge from developing and underdeveloped countries to create commercial products. In the United States, of the top 150 prescription drugs, at least 118 are based on natural sources. The plant-derived anticancer drug, Taxol, was first isolated from the Pacific yew (*Taxus brevifolia* Peattie). Madagascar periwinkle (*Catharanthus roseus* (L.) G. Don.) is an herbal medicine that contains vinblastine and vincristine, which are used for treating leukemia. Countries of natural diversity with traditional knowledge are protecting their heritage and treasures from biopiracy.

Unfortunately, progressive civilization and population growth, urbanization, industrialization, domestication of crops, deforestation, overharvesting, and warfare have led to air and water pollution and the extinction of many plants, including medicinal ones. National and international organizations are collecting, maintaining, characterizing, and identifying endangered species.

Medicinal Plants, volume 6 of the series *Genetic Resources, Chromosome Engineering, and Crop Improvement*, includes 30 unique chapters. Several medicinal plants that were covered in *Grain Legumes* (volume 1), *Cereals* (volume 2), *Vegetable Crops* (volume 3), *Oilseed Crops* (volume 4), and *Forage Crops* (volume 5) were excluded from this book. Chapter 1 summarizes landmark research that has been done on medicinal plants. Chapter 2 describes medicinal plants as "nature's pharmacy" and provides suggestions to conserve, maintain, examine, and protect plants, including medicinal ones, from deforestation, urbanization, and biopiracy. Successive chapters (3–12) provide a comprehensive account, including the history of medicinal plants used by our ancestors prior to civilization. Chapters 13–29 report on medicinal plants, including the history, origin, genetic resources, cytogenetics, varietal improvement through conventional and modern methods, and their use in producing pharmaceutical drugs, cosmeceutical products, nutrition and food products, and beverages. Chapter 30 elaborates on the use of molecular technology for maintaining authenticity and quality of plant-based products.

Each chapter is authored by world-renowned scientists who are conducting research in the field of medicinal plants. I am extremely grateful to all the authors for their outstanding contributions and to reviewers of all the chapters. I am fortunate to know them both professionally and personally, and our communication has been very cordial and friendly. I am particularly indebted to Ales Lebeda, Arthur Tucker, Brad Morris, Chamchalow Narong, David Walker, Govindjee, Joseph Nicholas, Melinda Anderson, Munir Ozturk, S. K. Malhotra, and Zohara Yaniv for their comments and suggestions.

This book is intended for scientists, professionals, and students whose interests center upon collection, maintenance, conservation, genetic improvement, utilization, and quality control of medicinal plants. This is a reference book for historians, germplasm explorers, geneticists, plant breeders, health professionals, food technologists, consumers, environmentalists, and plant-based pharmaceutical, cosmeceutical, nutrition, and beverage industries. I sincerely hope that the information assembled here will guide us to protect plants, particularly those with medicinal properties, from deforestation, urbanization, overgrazing, pollution, overharvesting, and, finally, biopiracy.

Ram J. Singh
Urbana-Champaign, Illinois
ramsingh@illinois.edu

The Editor

Ram J. Singh, MSc, PhD, is an agronomist and plant cytogeneticist in the Department of Crop Sciences at the University of Illinois at Urbana-Champaign. He received his PhD degree in plant cytogenetics under the guidance of the late Professor Takumi Tsuchiya from Colorado State University, Fort Collins, Colorado. He has benefited greatly from the expertise of and working association with Drs. T. Tsuchiya, G. Röbbelen, and G. S. Khush.

Dr. Singh has conceived, planned, and conducted pioneering research related to cytogenetic problems in barley, rice, rye, oat, wheat, and soybean. Thus, he has isolated monotelotrisomics and acrotrisomics in barley, identified them by Giemsa C- and N-banding techniques, and determined chromosome arm-linkage group relationships. By using pachytene chromosome analysis, Singh identified 12 possible primary trisomics in rice and chromosome-linkage group relationships. In soybean (*Glycine max* (L.) Merr.), he established genomic relationships among species of the genus *Glycine* and assigned genome symbols to all species. Singh constructed, for the first time, a soybean chromosome map based on pachytene chromosome analysis that laid the foundation for creating a global soybean map. By using fluorescent genomic *in situ* hybridization, he confirmed the tetraploid origin of the soybean.

Singh developed a methodology to produce fertile plants with $2n = 40$ chromosomes from an intersubgeneric cross between soybean and a perennial wild species, *Glycine tomentella Hayata* ($2n = 78$). This invaluable invention has been awarded a patent (US 7,842,850). Singh has published 71 research papers in reputable international journals, including the *American Journal of Botany*, *Chromosoma*, *Crop Science*, *Euphytica*, *Genetics*, *Genome*, *Journal of Heredity*, *Plant Breeding*, *The Nucleus*, and *Theoretical and Applied Genetics*. In addition, he has summarized his research results by writing 18 book chapters. His book, *Plant Cytogenetics* (first edition, 1993; second edition, 2003), is widely used to teach graduate students. He has presented research findings as an invited speaker at national and international meetings.

Singh is the chief editor of the *International Journal of Applied Agricultural Research* (IJAAR) and editor of *Plant Breeding*. In 2000 and 2007, he received the Academic Professional Award for Excellence: Innovation and Creativity from the Department of Crop Sciences at the University of Illinois at Urbana-Champaign. He is the recipient of the College of ACES Professional Staff Award for Excellence–Research in 2009 and the 2010 Illinois Soybean Association's Excellence in Soybean Research award. He was invited as a visiting professor by Dr. K. Fukui, Osaka University, Osaka, Japan and by Dr. G. Chung, Chonnam National University, Yeosu, Korea. Singh is an editor for the series entitled Genetic Resources, Chromosome Engineering, and Crop Improvement and has published *Grain Legumes* (volume 1), *Cereals* (volume 2), *Vegetable Crops* (volume 3), *Oilseed Crops* (volume 4), and *Forage Crops* (volume 5).

Contributors

Ricardo Acuña
National Coffee Research
  Center—CENICAFE
Chinchina (Caldas), Colombia

P. Addo-Fordjour
Department of Theoretical and Applied
  Biology
Kwame Nkrumah University of Science
  and Technology
Kumasi, Ghana

W. G. Akanwariwiak
Department of Theoretical and Applied
  Biology
Kwame Nkrumah University of Science
  and Technology
Kumasi, Ghana

Eren Akcicek
Faculty of Medicine
Department of Gastroenterology
Ege University
Bornova, Izmir, Turkey

Ernaz Altundag
Faculty of Arts and Sciences
Department of Biology
Duzce University
Duzce, Turkey

A. K. Anning
Department of Theoretical and Applied
  Biology
Kwame Nkrumah University of Science
  and Technology
Kumasi, Ghana

François Anthony
Institut de Recherche pour le
  Développement—IRD
Montpellier, France

Mehri Ashyraliyeva
Turkmen State Medical Institute
Ashgabat, Turkmenistan

Mamedova Gurbanbibi Atayevna
Turkmen State Medical Institute
Ashgabat, Turkmenistan

K. Nirmal Babu
Indian Institute of Spices Research
Marikunnu, Calicut
Kerala, India

S. Babu
Bayer Bio Sciences
Hyderabad, Andhra Pradesh, India

Uriel Bachrach
Department of Molecular Biology
Hebrew University-Hadassah Medical School
Jerusalem, Israel

E. J. D. Belford
Department of Theoretical and Applied
  Biology
Kwame Nkrumah University of Science
  and Technology
Kumasi, Ghana

Jenő Bernáth
Department of Medicinal and Aromatic Plants
Corvinus University of Budapest
Budapest, Hungary

Carlos Roberto Carvalho
Laboratório de Citogenética e Citometria
Departamento de Biologia Geral
Universidade Federal de Viçosa
Viçosa-MG, Brazil

Preeti Chavan-Gautam
Interactive Research School for Health Affairs
Bharati Vidyapeeth University
Pune, Maharashtra, India

Ali Celik
Faculty of Science and Arts
Biology Department
Pamukkale University
Denizli, Turkey

Sezgin Celik
Faculty of Education
Department of Primary Education
Kirikkale University
Yahsihan, Kirikkale, Turkey

Hubiao Chen
School of Chinese Medicine
Hong Kong Baptist University
Kowloon Tong, Hong Kong

Kwang-Tae Choi
Korea Institute of Science and Technology
  Information
Seocho-Gu, Seoul, Korea

Gyuhwa Chung
Department of Biotechnology
Chonnam National University
Yeosu, Chonnam, Korea

John Connett
USDA, ARS, Western Regional Plant
  Introduction
Pullman, Washington, U.S.A.

Hernando A. Cortina
National Coffee Research
  Center—CENICAFE
Chinchina (Caldas), Colombia

Marco A. Cristancho
National Coffee Research
  Center—CENICAFE
Chinchina (Caldas), Colombia

Manoj K. Dhar
Plant Genomics Laboratory
School of Biotechnology
University of Jammu
Jammu, India

Minoo Divakaran
Department of Botany
Providence Women's College
Calicut, Kerala, India

I. Doležalová
Faculty of Science
Department of Botany
Palacký University in Olomouc
Olomouc-Holice, Czech Republic

Nativ Dudai
The Unit of Medicinal and Aromatic Plants
Newe Ya'ar Research Center,
ARO, The Volcani Center
Ramat Yishay, Israel

E. Fernández
Institute of Tropics and Subtropics
Czech University of Life Sciences in Prague
Praha-Suchdol, Czech Republic

C. K. Firempong
Department of Biochemistry and
  Biotechnology
Kwame Nkrumah University of Science
  and Technology
Kumasi, Ghana

Jitendra Gaikwad
Indigenous Bioresources Research Group
Faculty of Science, Macquarie University
Sydney, Australia

Alvaro L. Gaitán
National Coffee Research
  Center—CENICAFE
Chinchina (Caldas), Colombia

Shahina A. Ghazanfar
Royal Botanic Gardens
Kew, United Kingdom

Cigdem Gork
Faculty of Science and Arts
Biology Department
Mugla University
Mugla, Turkey

Guven Gork
Faculty of Science and Arts
Biology Department
Mugla University
Mugla, Turkey

Salih Gucel
Institute of Environmental Sciences
Near East University
Nicosia, Cyprus

Ping Guo
School of Chinese Medicine
Hong Kong Baptist University
Kowloon Tong, Hong Kong

Mehak Gupta
Plant Genomics Laboratory
School of Biotechnology
University of Jammu
Jammu, India

David Harrington
Indigenous Bioresources Research Group
Faculty of Science, Macquarie University
Sydney, Australia

Barbara Hellier
USDA, ARS, Western Regional Plant
 Introduction
Pullman, Washington, USA

Juan Carlos Herrera
National Coffee Research
 Center—CENICAFE
Chinchina (Caldas), Colombia

Hossein Hosseinzadeh
Pharmaceutical Research Center
School of Pharmacy
Mashhad University of Medical Sciences
Mashhad, Iran

José Imery-Buiza
Laboratorio de Genética Vegetal
Departamento de Biología
Universidad de Oriente
Núcleo de Sucre, Cumaná, Venezuela

Vikas Jaitak
Institute of Himalayan Bioresource
 Technology
Council of Scientific and Industrial
 Research
Palampur, Himanchal Pradesh, India

Joanne Jamie
Indigenous Bioresources Research Group
Faculty of Science, Macquarie University
Sydney, Australia

Kalpana Joshi
Department of Biotechnology
Symbiosis International University
Pune, Maharashtra, India

Kiran Kaul
Institute of Himalayan Bioresource
 Technology
Council of Scientific and Industrial
 Research
Palampur, Himanchal Pradesh, India

Sanjana Kaul
Plant Genomics Laboratory
School of Biotechnology
University of Jammu
Jammu, India

V. K. Kaul
Institute of Himalayan Bioresource
 Technology
Council of Scientific and Industrial Research
Palampur, Himanchal Pradesh, India

Sherry Kitto
Department of Plant and Soil Sciences
University of Delaware
Newark, Delaware, U.S.A.

Philippe Lashermes
Institut de Recherche pour le
 Développement—IRD
Montpellier, France

Umesh C. Lavania
Central Institute of Medicinal and Aromatic
 Plants
Lucknow, Uttar Pradesh, India

Ales Lebeda
Faculty of Science
Department of Botany
Palacký University in Olomouc
Olomouc-Holice, Czech Republic

Zhitao Liang
School of Chinese Medicine
Hong Kong Baptist University
Kowloon Tong, Hong Kong

Darci R. Lima
Department of Clinical Pharmacology and
   History of Medicine
Federal University of Rio de Janeiro
Rio de Janeiro, Brazil

S. K. Malhotra
Indian Council of Agricultural Research
Horticulture Division, Krishi Anusandhan
   Bhawan—II
Pusa, New Delhi, India

A. K. A. Mandal
School of Bio Sciences and Technology
Plant Biotechnology Division
Vellore Institute of Technology University
Tamil Nadu, India

Tuba Mert
Faculty of Pharmacy
Ege University
Bornova, Izmir, Turkey

J. Bradley Morris
USDA-ARS Plant Genetic Resources
   Conservation Unit
Griffin, Georgia, U.S.A.

Marjan Nassiri-Asl
Faculty of Medicine
Department of Pharmacology
Qazvin University of Medical Sciences
Qazvin, Iran

Éva Németh
Department of Medicinal and Aromatic Plants
Corvinus University of Budapest
Budapest, Hungary

Luiz Orlando de Oliveira
Departamento de Bioquimica/Bioagro
Universidade Federal de Viçosa
Viçosa-MG, Brazil

Munir Ozturk
Faculty of Science, Botany Department
Ege University
Bornova-Izmir, Turkey

Joanne Packer
Indigenous Bioresources Research Group
Faculty of Science, Macquarie University
Sydney, Australia

Enzo A. Palombo
Environment and Biotechnology Centre
   British English
Faculty of Life and Social Sciences
Swinburne University of Technology
Hawthorn, Victoria, Australia

Bhushan Patwardhan
Symbiosis International University
Pune, Maharashtra, India

Svetlana Aleksandrovna Pleskanovskaya
Turkmen State Medical Institute
Ashgabat, Turkmenistan

Nayani Surya Prakash
Central Coffee Research Institute
Chikmagalur, Karnataka, India

Shoba Ranganathan
Indigenous Bioresources Research Group
Faculty of Science, Macquarie University
Sydney, Australia

P. N. Ravindran
Manasom, West Nadakkavu
Calicut, Kerala, India

M. Sabu
Department of Botany
University of Calicut
Kerala, India

R. S. Senthil Kumar
Harrisons Malayalam Ltd.
Coimbatore, Tamil Nadu, India

Pooja Sharma
Plant Genomics Laboratory
School of Biotechnology
University of Jammu
Jammu, India

K. N. Shiva
National Research Centre for Banana,
  Tiruchirapalli
Tamil Nadu, India

Bikram Singh
Institute of Himalayan Bioresource
  Technology
Council of Scientific and Industrial Research
Palampur, Himanchal Pradesh, India

Janardan Singh
Central Institute of Medicinal and Aromatic
  Plants
Lucknow, Uttar Pradesh, India

Ram J. Singh
Department of Crop Sciences
National Soybean Research Laboratory
University of Illinois
Urbana-Champaign, Illinois, USA

Suman Singh
Central Institute of Medicinal and Aromatic
  Plants
Lucknow, Uttar Pradesh, India

Virendra Singh
Institute of Himalayan Bioresource Technology
Council of Scientific and Industrial Research
Palampur, Himanchal Pradesh, India

Arthur O. Tucker
Department of Agriculture and Natural
  Resources
Delaware State University
Dover, Delaware, USA

Subramanyam Vemulpad
Indigenous Bioresources Research Group
  Faculty of Science, Macquarie University
Sydney, Australia

I. Viehmannová
Institute of Tropics and Subtropics
Czech University of Life Sciences in Prague
Praha-Suchdol, Czech Republic

Christophe Wiart
Department of Pharmacy
University of Nottingham
Selangor, Malaysia

Zohara Yaniv
Department of Genetic Resources
Institute of Plant Sciences
ARO, The Volcani Center
Bet-Dagan, Israel

Zhongzhen Zhao
School of Chinese Medicine
Hong Kong Baptist University
Kowloon Tong, Hong Kong

AEM	Ancient Egyptian Medicine
AFLP	Amplification Fragment Length Polymorphism
AGM	Ancient Greek Medicine
AICRPE	All India Coordinated Research Project on Ethnobiology
APOX	Ascorbate Peroxidase
AP-PCR	Arbitrarily Primed-Polymerase Chain Reaction
ARMS	Amplification Refractory Mutation System
AYUSH	Ayurveda, Yoga, Unani, Siddha, and Homeopathy
BAP	Benzylaminopurine
BCGC	Brazilian Coffee Genomics Consortium
BHA	Butylated Hydroxyanisole
BHT	Butylated Hydroxytoluene
BLAST	Basic Local Alignment Search Tool
BPH	Benign Prostatic Hyperplasia
CAD	Coronary Artery Diseases
CAM	Complementary and Alternative Medicines
CAPS	Cleaved Amplified Polymorphic Sequence
CBD	Convention on Biological Diversity
CDP	Copalyl Diphosphate
CDR	Clinical Dementia Rating
CFQA	Caffeoyferuloylquinic Acids
CHS	Chalcone Synthase
CIFC	Coffee Rust Research Centre
CIMAP	Central Institute of Medicinal and Aromatic Plants
CIP	Centro Internacional de la Papa
CIRAD	Centre de Coopération Internationale en Recherche Agronomique pour le Développement
CMV	Cytomegalovirus
CNS	Central Nervous System
CPVO	European Variety List
CSIR	Council of Scientific and Industrial Research
CTC	Crush, Tear, and Curl
DAD	Diode Array Detector
DAF	DNA Amplification Fingerprinting
DAMD	Directed Amplification of Minisatellite Region DNA
DFR	Dihydroflavonol 4-Reductase
DGETF	Digallate Equivalent of Theaflavins
DH	Doubled Haploid
DNA	Amplification Fingerprinting
DUS	Distinctness, Uniformity, and Stability
EBV	Epstein - Barr Virus
EGCG	Epigallocatechingallate
ELISA	Enzyme-Linked Immuno Sorbant Assay
ELSD	Evaporative Light Scattering Detector

EMBRAPA	Empresa Brasileira de Pesquisa Agropecuária
EMS	Ethyl Methane Sulphonate
EPO	European Patent Office
ESCOP	European Scientific Cooperative On Phototherapy
ESTs	Expressed Sequence Tags
FAO	Food and Agriculture Organization
FDA	US Food and Drug Administration
FISH	Fluorescence *in situ* Hybridization
FRD	Fumarate Reductase
FRF	Fiber Rich Fraction
FYM	Farm Yard Manure
GA3	Gibberellic Acid
GACP	Good Agricultural and Collection Practices
GAP	Good Agricultural Practice
GCP	Good Collecting Practice
GIS	Geographic Information System
GISH	Genomic *in situ* Hybridization
GMP	Good Manufacturing Process
GR	Glutathione Reductase
GSH	Glutathione Peroxidase
HAV	Hepatitis A Virus
HBV	Hepatitis B Virus
HCV	Hepatitis C Virus
HDL	High Density Lipoprotein
HIV	Human Immunodeficiency Virus
HIV/AID	Human Immunodeficiency Virus/Acquired Immunodeficiency Syndrome
HPLC	High Performance Liquid Chromatography
HPS	Highly Polymerized Substances
HSV	Herpes Simplex Virus
IAA	Indole-3-Acetic Acid
IBPGR	International Board for Plant Genetic Resources
IBRG	Indigenous Bioresources Research Group
ICFR	International Council for Fitogenetic Resources
ICMR	Indian Council of Medical Research
IISR	Indian Institute of Spices Research
INCB	International Narcotics Control Board
IPR	Intellectual Property Rights
IRD	Institut de Recherche pour le Développement
ISR	Inverse Sequence-tagged Repeat
ISSR	Inter Simple Sequence Repeat
ITS	Internal Transcribed Spacer
JECFA	Joint Expert Committee of Food Additives
KS	Kaurene Synthase
KSHV	Kaposi Sarcoma-Associated Herpesvirus
LDL	Low-Density Lipoprotein
LVM	Left Ventricular Mass

MAIDS	LP-BM5 Murine Leukemia Virus
MALDI-MS	Matrix-Assisted Laser Desorption/Ionization Mass Spectrometry
MAS	Marker Assisted Selection
MCAO	Middle Cerebral Artery Occlusion
MICs	Minimum Inhibitory Concentrations
MMS	Methyl Methane Sulphonate
MMSE	Mini-Mental State Examination
NAA	Naphthalene Acetic Acid
NBPGR	National Bureau of Plant Genetic Resources
NMITLI	New Millennium Indian Technology Leadership Initiative
NMPB	National Medicinal Plants Board
NMR	Nuclear Magnetic Resonance
NOR	Nucleolar Organizer Region
NPGS	National Plant Germplasm System
ORSTOM	Office de la Recherche Scientifique et Technique d'Outre-Mer
PAL	Phenylalanine Ammonia Lyase
PCNB	Pentachloronitrobenzene
PCR	Polymerase Chain Reaction
PDGF	Platelet-Derived Growth Factor
PEL	Primary Effusion Lymphoma
PGPR	Polyglycerol Polyricinoleate
PGs	Polygalacturonases
PHP	Practically Healthy Persons
PL	Pectin Lyase
PMV	Poliomyelitis Virus
PPO	Peroxidase and Polyphenol Oxidase
PTO	US Patents and Trademarks Office
PTZ	Pentylentetrazole
QTLs	Quantitative Trait Loci
RACE	Rapid Amplification of Complementary DNA
RAPD	Randomly Amplified Polymorphic DNA
RAPD-SCAR	Random Amplification of Polymorphic DNA- Sequence Characterized Amplified Region
rDNA	Ribosomal DNA
RETM	Research and Evaluation of Traditional Medicine
RFLP	Restriction Fragment Length Polymorphism
RuBisCo	Ribulose- 1, 5-Bisphosphate Carboxylase/oxygenase
rVSMC	Rat Vascular Smooth Muscle Cell
sscDL	Sanger Sequencing of cDNA Libraries
SARS	Severe Acute Respiratory Syndrome
SCAR	Sequence Characterized Amplified Region
SCFA	Short Chain Fatty Acids
SDF	Soluble Dietary Fibers
SFDA	State Food and Drug Administration
SNPs	Single Nucleotide Polymorphisms

SOD	Superoxide Dismutase
SSR	Simple Sequence Repeats
STS	Sequence-Tagged-Site
TCM	Traditional Chinese Medicine
TDZ	Thidiazuron
TGA	Therapeutic Goods Administration
TIM	Traditional Indian Medicine
TKDL	Traditional Knowledge Digital Library
TLC	Thin Layer Chromatography
TMV	Tobacco Mosaic Virus
TSH	Thyroid Stimulating Hormone
TSM	Traditional Systems of Medicine
TTO	Tea Tree Oil
UGTs	UDP-Glycosyltransferases
UNESCO	United Nations Educational, Scientific, and Cultural Organization
UNIDO	United Nations Industrial Development Organization
UPOV	Union for the Protection of New Varieties of Plants
USDA	United States Department of Agriculture
USFDA	United State Food and Drug Administration
USI	Urinary Stress Incontinence
USPTO	US Patent and Trademark Office
VFC	Volatile Flavor Compounds
VNTR	Variable Number Tandem Repeats
VZV	Varicella Zoster Virus
WHO	World Health Organization
ZOA	*Z.ingiber officinale* Agglutinin

Contents

CHAPTER 1

Landmark Research in Medicinal Plants

Ram J. Singh

CONTENTS

1.1 INTRODUCTION

Plants are an integral part of all the living organisms of planet Earth because they provide food, clean air, medicines, clothing, shade, and shelter. Further, they preserve our ecosystem by moderating natural calamities such as floods, droughts, and soil erosion. Since time immemorial, all living creatures coevolved in a symbiotic manner. Our early ancestors selected and initiated cultivation of plants based on their needs. As population grew, diseases evolved. Our early ancestors identified certain plants, by trial and error, for the healing of various ailments. These individuals were considered medicine men or medicine women. The knowledge was transferred from generation to generation through the word of mouth or preserved in epics of poems or stories. These medicines are now known as prehistoric medicines because they were identified prior to recorded history.

From wild plants growing in nature, our prehistoric ancestors identified and selected plants based on their use. This knowledge subsequently directed the path of domestication and cultivation. Progressive evolution by selection from the wild plants created domestication of many plants (legumes, cereals, vegetables, fruits, spices, oilseeds, forages, fibers, ornamentals, and medicines) and all have medicinal properties. Marinelli (2005) estimated 422,000 plant species worldwide. This includes 50,000–80,000 flowering plants being used medicinally. These plants are potentially rich sources of medicinal compounds, curing everything from the common cold to cancer and even HIV/AIDS, and are known as "nature's pharmacy" (Chapter 2).

Plant biodiversity from planet Earth is disappearing at a rapid pace due to deforestation, urbanization, overharvesting, and overgrazing. The objective of this chapter is to summarize knowledge of nature's pharmacy (Chapter 2) and history and use of medicinal plants of major continents (Chapters 3–11), and genetic resources and their taxonomy, diversity, collection, conservation, evaluation, and utilization in breeding for 16 medicinal plants (Chapters 13–29). Medicinal properties of legumes are covered in Chapter 12, and Chapter 30 describes molecular methods for determining the quality and authenticity of plant-based medicines.

1.2 HISTORY OF TRADITIONAL MEDICINES OF THE WORLD

Plants are the gift of our earth to nurture all the creatures, and medicinal plants (trees to herbs) have been the only source of remedy for all ailments since the inception of our species. The science of traditional medicine using medicinal plants developed independently in all the continents; exchange of knowledge was limited because of geographical isolation.

In 2003, the World Health Organization (WHO) defined traditional medicine as a combination of knowledge, skills, and practices based on the theories, beliefs, and experiences indigenous to different cultures that are used to maintain health, as well as to prevent, diagnose, improve, or treat physical and mental illnesses. Traditional medicine that has been adopted by other populations (outside its indigenous culture) is often termed complementary or alternative medicine (CAM). In some Asian and African countries, 80% of the population depends on traditional medicine for primary health care; in many developed countries, 70–80% of the population has used some form of CAM (http://www.who.int/mediacentre/factsheets/fs134/en/). The practice of traditional medicine, described next, has been adopted in different countries or continents since ancient times without the knowledge of those in other regions of the world:

Ayurveda: This system (science of life) of traditional medicine is native to the Indian subcontinent and originated around 2000 BC. It is still being used in combination with the modern system of medicine (allopathy) for total health care. Ayurveda, *Yoga* and naturopathy, *Unani, Siddha,* and *Homeopathy* (AYUSH: www.indianmedicine.nic.in) are the six nationally recognized systems of medicine in India. In India, over 4,000 terrestrial plants have been studied for their biological activities and 20% have exhibited promising new activities (Chapter 3).

Chinese medicine: This system of medicine originated in China about 3000 BC and is depicted in China's ancient poem collection and records more than 50 medicinal plant species. Compendia on plant-based medicines have been recorded in more than 400 books during several dynasties (Chapter 5). Southeast Asian countries followed both Indian and Chinese traditional medicines. China has about 7,000 species of medicinal plants (http://whqlibdoc.who.int/wpro/-1993/9290611022.pdf).

Australia: Aboriginal Australians have lived on the Australian continent for at least 40,000–50,000 years and have the longest continuous heritage of any human culture on the planet. Australia is one of the world's 17 megadiverse countries. It is home to over 20,000 vascular and 14,000 nonvascular plants: ~250,000 species of fungi and over 3,000 lichens and 85% of all the vascular plant species and nine plant families are endemic to the continent (Chapter 11).

Middle East: Medicinal use of plants by the Sumerians in southern Mesopotamia (modern-day Iraq) dates back to 3000 BC. Herbal and traditional medicine developed over a period of time through the contributions of the early Greeks, Romans, and Arabs (Chapter 6). A number of plant remedies have been described on the clay tablets that have survived from the Mesopotamian civilizations like Sumerians, Assyrians, Akkadians, and Hettites (Chapter 7). Several medicinal plants are described in the Bible and Qur'an and Ahadith (Chapter 2). Artemisia is mentioned eight times in the Bible—always as an example for evil, wrongdoing, and idol worshipping, or as an example for suffering of destruction and exile. This means that wormwood (*Artemisia*) was a common herb of the era and that its acrid taste was known as a drinkable preparation applied for specific reasons (Chapter 13). The narcotic properties of the poppy plant and the nutritive values of its seeds were recognized by the Greeks, the Egyptians, and the Romans. Hippocrates (460–377 BC) was one of the first who mentioned the medical application of the poppy and its preparations. Traditional Unani medicine originated in ancient Greece around 400 BC. Hippocrates, also known as the founder of allopathic medicine, is considered to be the first Unani physician. This medicine is practiced today in many Middle Eastern and Asian countries, including India, Pakistan, and Sri Lanka.

Europe: European (Western) traditional herbal medicine has a long history, with the roots of its practice found in the writings of the Greek physicians such as Hippocrates and Dioscorides (who was born in Anazarba, present-day Turkey) and in later Roman works, such as those of Galen, who contributed greatly to the understanding of numerous scientific disciplines including anatomy, physiology, pathology, pharmacology (http://en.wikipedia.org/wiki/Galen; see note 6), neurology, philosophy, and logic (Chapter 2). Pietro Andrea Gregorio Matthioli composed *Herbarium*, which was published in 1544 with the description of over 1,000 plant species. This was the compendium about medicinal plants in medieval Europe, continuing until the twenty-first century, and has had substantial impact on previous and recent plant-based modern medicine (see the dedication at the beginning of this book). Poppy was known and utilized by the cavemen in 4000–5000 BC in what is now known as Spain, France, Germany, and Hungary (Chapter 14).

Africa: African traditional medicine has a rich and diverse history. It is plant based and incorporates holistic beliefs, systems, and societies for combating various ailments. It was probably started around 1500 BC (www.greenthumbarticles.com). Systematic documentation of medicinal properties of plants (trees to herbs) of Africa did not develop because of vast continent size and the diversity in culture and language (Chapter 2).

Medicinal plants of the Americas: Native American nations or tribes have long used medicinal plants—in some cases for at least 10,000 years. These plants are linked to philosophy, religion, and spirituality, and treatments aim to balance the physical, emotional, mental, and spiritual components of a person. This involves a tribal healer and may involve the patient's family or the entire community (Chapter 20). Early people used almost 3,000 different plants as medicine. Black cohosh (*Actaea racemosa* (Nutt.) L.), a staple of Cherokee medicine, served many purposes ranging from use as a diuretic to a cure for rheumatic pains. Bloodroot (*Sanguinaria canadensis* L.) provided the Cherokee with medicine to cure coughs and lung inflammations. Blue cohosh (*Caulophyllum thalictroides* (L.) Michaux), another eastern woodland plant, helped cure toothaches for the Cherokee, while the Chippewa used its root to treat cramps (http://www.nps.gov/plants/medicinal/plants.htm). Duke (2009) lists 758 medicinal plants of Latin America.

1.3 IMPORTANCE OF PLANT-BASED MODERN MEDICINES

Plant-based natural products are collected from the wild and from cultivated plants. Our ancestors depended only on plants for treating ailments because they did not develop knowledge and technology to identify active compounds of plants being used for curing the diseases. The progressive civilization–population growth, urbanization, industrialization, and domestication of crops, worldwide, caused deforestation and overharvesting, leading to air and water pollution and extinction of many plants including medicinal plants. The early settlers in Peru (1638) used Cinchona bark for treating malaria. This tree originated from the slopes of the Andes in South America. The overharvesting of the bark of the Cinchona tree during the sixteenth and seventeenth centuries caused total extinction in South America; however, this species was saved by a large plantation in Java. An alkaloid, quinine, was isolated from the bark for treating malaria (Chapter 2).

The loss of a major biodiversity is being seen with the construction of Three Gorges Dam (TGD) on the Yangtze River in the middle of a biodiversity hot spot in central China. Approximately 90,000 plants will be submerged, including a medicinal plant, *Myricaria laxiflora* (Franch.) P.Y. Zhang & Y.J. Zhang (Liu, Wang, and Huang 2006). This species lost its entire habitat but it was rescued, kept alive in gardens for a decade, and replanted in a new habitat (Pennisi 2010).

All plants have medicinal properties because they tend to produce biologically active chemicals to survive in nature against pests and pathogens and also with abiotic stresses. Grain legumes supply proteins and antioxidants, cereals are major source of carbohydrates and fiber, and oilseeds are an excellent source of dietary fatty acids; vegetable crops are valuable sources of nutrition, including mineral nutrients, antioxidants, and vitamins. All these major crops have been domesticated from the wild resources, improved through genetics and breeding, and are being cultivated.

Food, cosmeceutical, and pharmaceutical industries have been using butylated hydroxyanisole (BHA) and butylated hydroxytoluene (BHT), synthesized antioxidants, to prevent oxidation (undesirable off-flavors and potentially toxic chemicals). These chemicals possess several harmful health effects. Plant-based antioxidants are healthy alternatives to synthetic antioxidants. Many medicinal and aromatic plants (anise, fennel, basil, mint, tarragon, marjoram, rosemary, thyme, parsley, juniper, laurel, and black pepper) are rich sources of phenolic compounds with strong antioxidant activity (Miguel 2010).

1.3.1 Modern Medicine

Pharmaceutical industries and national institutes are examining plants for compounds that may cure everything from the common cold to cancer and HIV/AIDS. Table 2.4, in Chapter 2, has a partial list of the names of the drugs isolated from the plants. However, the chemical basis for medicinal plants curing ailments, illustrated in old texts such as in Ayurveda (Chapter 3), Chinese compendiums (Chapter 5), and the Bible and the Qur'an, is now being examined and documented (Chapter 2). Chapters 2–29 describe the discovery of modern medicines used either in homoeopathy or allopathy. Products of these plants are sold worldwide as medicines, spices, health foods, nutrition supplements, food flavors, beverages, and cosmeceutical products (Figure 1.1). Presently, most drugstores, grocery stores, and even supermarkets have aisles of nature-made products; they are available to consumers without prescriptions from the doctors practicing allopathic medicines. Boundless information is freely reachable to masses having access to Internet technology.

Ginger is a stimulant of the gastrointestinal tracts, rubifacient, diaphoretic, diuretic, anti-inflammatory, antiemetic, sialagogic emmangogus, abortifacient, and vermifuge. Ginger relieves

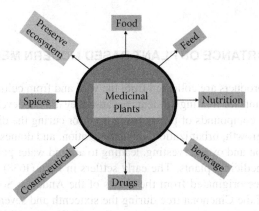

Figure 1.1   A diagram showing use of medicinal plants.

flatulence, stimulates the gastrointestinal tract, and acts as counterirritant. Alcoholic extracts of ginger have been found to stimulate the heart, increase fibrinolytic activity, and thereby protect against coronary artery disease. In veterinary practice, ginger is used as a stimulant and carminative in tonics for indigestion of horses and cattle (Chapter 15).

Curcumin from turmeric has anti-inflammatory, hypocholestremic, choleratic, antimicrobial, antirheumatic, antifibrotic, antivenomous, antiviral, antiamyloid, antidiabetic, antihepatotoxic, antioxidant, and anticancerous properties as well as insect repellent activity (Chapter 16).

In the United States, of the top 150 prescription drugs, at least 118 are based on natural sources. The plant-derived anticancer drug Taxol was first isolated from the Pacific yew (*Taxus brevifolia* Peattie). Madagascar periwinkle (*Catharanthus roseus* (L.) G. Don.) is an herbal medicine that contains Vinblastine and Vincristine, used in treating leukemia (Roberson 2008).

A topical anti-inflammatory drug, Acheflan®, has been developed from a Brazilian medicinal plant, *Cordia verbenacea* DC. An herbal drug, Catuama®, is a tonic medicine obtained from a mixture of *Trichilia catigua* A. Juss. (28.23%), *Paullinia cupana* Kunth (40.31%), *Ptychopetalum olacoides* Bentham (28.23%), and *Zingiber officinale* Roscoe (3.26%) and has an antidepressant-like effect. Vimang® is an extract from the bark of *Mangifera indica* L. and has anti-inflammatory, analgesic, and antioxidant properties (Calixto 2005)

1.3.2 Beverages

Medicinal plants are used as beverages. Mango (Chapter 4), ginger (Chapter 15), ginseng (Chapter 17), tea (Chapter 18), coffee (Chapter 19), mint (Chapter 21), licorice (Chapter 29), and several other plants grown in the Middle East (Chapters 6 and 7) are routinely used in beverages. Tea and coffee are the most commonly consumed plants, now cultivated as crops, worldwide. Both plants are rich in caffeine, and they are rich in antioxidants.

Herbal teas without caffeine are popular these days. Synthetic artificial sweeteners—sugar substitutes (saccharine, aspartame, sucralose [Splenda])—are in food and beverages worldwide for consumers with problems related to obesity and diabetes. The stevia plant, found in South America (300 times as sweet as sugar; Chapter 26), and licorice (50 times as sweet as sugar; Chapter 29) produce noncarbohydrate sweeteners and their products have additional medicinal properties. The U.S. Food and Drug Administration (USFDA) has granted approval to Coke and Pepsi for use of a new artificial sweetener, rebaudioside A (known as Reb-A), derived from stevia, in soft drinks (http://www.bevreview.com/2009/01/01/coke-pepsi-introduce-drinks-with-stevia-artificial-sweetener/). Based on the syrup of black elderberry (*Sambucus nigra* L.), Fanta has marketed a soft drink. Kava (*Piper methysticum* G. Forst.) is used in making a soft drink, known as Kava Cola, in Vanuatu (Figure 1.2) and Hawaiian-based Rzo.

Herb-based tea with medicinal values has been identified and developed, and is being marketed (http://en.wikipedia.org/wiki/Herbal_tea). Herbal teas are derived from many parts of a plant, such as roots, leaves, stems, flowers, and seeds. Some of the plants used in herbal teas are anise, artichoke, roasted barley, boldo, burdock root, cannabis (Bhang in India), caraway, cinnamon, chrysanthemum, red clover, citrus peel, dill tea, echinacea, elderberry, fennel, ginger root, ginseng, wolfberry, hawthorn, hibiscus, honeybush, hydrangea, kapor, labrador, lapacho, lemon balm, lemongrass, licorice root, tilia, mint, mountain tea, corn tea (Oksusu cha; Korean tea), St. John's wort, thyme, holy basil (Tulsi), rooibos, and erva mate.

1.3.3 Spices

Plants with medicinal properties also have culinary uses, known as spice, for flavor and taste. Ginger (Chapter 15), turmeric (Chapter 16), fenugreek (Chapter 24), and *Aloe vera* (L.) *Burum. f.* (Chapter 25) are being cultivated and have been covered extensively in this book. These crops have

Figure 1.2   Beverage from kava (*Piper methysticum*), known as Kava Cola in Vanuatu. (Courtesy of Cameron McLeod; www.vanuatukavastore.com; November 18, 2010.)

rich histories and are described in other chapters. *Artemisia dracunculus* L. (tarragon) is widely used as an herb and is particularly important in French cuisine (Chapter 13). Many vegetable crops, such as *Allium* species (onion and garlic) and *Capsicum* species (chili peppers), are used as vegetables and spices but have many medicinal compounds as described in *Vegetable Crops*, which is volume 3 of this series. Rapeseed and mustard are oilseed crops, described in *Oilseed Crops* (Volume 4), and are considered medicinal crops. Many spices (cinnamon, nutmeg and mace, bay leaf, clove, cardamom, black pepper, anise, coriander, cumin, turmeric, ginger, dill, fennel, saffron, vanilla) were initially used for disease remedy and food preservatives. These medicinal plants were main ingredients in the cuisine of India, China, Southeast Asia, and the Persians and Arabs and they are now used worldwide. The Western European countries' (Portugal and Spain) search for spices aided in the discovery of the New World.

1.3.4 Cosmeceutical Products

Chemicals found in medicinal and aromatic plants are in great demand by the cosmeceutical industries, and these products are preferred by many consumers. Turmeric (Chapter 16) is used as a cosmetic as well as a natural dye. *Aloe vera* (Chapter 25), described even in the Bible, has pharmaceutical properties related to cancer and gastrointestinal, skin, cardiovascular, respiratory, and metabolic disorders. *Aloe vera* leaves contain anthraquinones, saccharides, vitamins, amino acids, minerals, enzymes, fatty acids and other emollients, healing, clotting, moisturizing, antiallergic, disinfectant, anti-inflammatory, astringent, choleretic, laxative, and other compounds (Table 25.1 in Chapter 25). *Aloe* species are used in gel, lotion, cream, soap, and shampoo and in many other cosmeceutical products worldwide. Tea tree oil (*Melaleuca alternifolia* (Maiden et Betche) Cheel) is a tree used by the Australian Aboriginal people for centuries to treat coughs, colds, wounds, sore throats, and skin aliments. It produces oil that has antiseptic properties; other uses are for the treatment of acne. Antifungal gels are for treatment of infections of the nails and ringworm, and feminine hygiene products are for the relief of vaginal and anal itching, burning, and discomfort; the hair care products and shampoos are for general scalp health and the treatment of lice. Mouthwashes and toothpastes for oral care and other products such as hand and body lotions, deodorants, and cold sore creams are for personal care (Chapter 10). Fenugreek (Chapter 24) is a source of natural

Figure 1.3 (See color insert.) A cosmeceutical product VICCO produced from turmeric in India and marketed worldwide. This product was a gift from Shyamala Balgopal. (With permission from VICCO Laboratories; www.viccolabs.com; December 15, 2010.)

steroid sapogenin compounds and polysaccharide galactomannans and has unique application for control of diabetes.

Neem (*Azadirachta indica* A. Juss.) oil is used in the Ayurvedic medicine (Chapter 3) to treat acne, fever, leprosy, malaria ophthalmia, and tuberculosis and is also used as a biofungicide and insect repellent and in making soap, shampoo, balms, and creams. Chapter 16 covers extensively turmeric, which is traditionally used for medicinal and culinary purposes and also as a cosmetic (Figure 1.3) and a natural dye. Soybean is rich in protein (40%) and oil (20%) and is used for food and feed. It is also used for producing medicinal and cosmeceutical products (Chung and Singh 2008).

The global industries on plant-based medicines, spices, beverages, and cosmeceutical products have been estimated to about US$65 billion per year and are expected to grow at the rate of 15–20% per year (Ariyawardana, Govindasamy, and Simon 2009).

1.3.5 Illicit Use of Medicinal Plants

Mankind has been using hallucinogenic plants for centuries (Cunningham 2008). The use of hallucinogenic plants was associated with religion, black magic, rituals, spirituals, and medicine. Hallucinogenic plants include Jimson weed (*Datura stramonium* L.), henbane (*Hyoscyamus niger* L.), ayahuasca (*Banisteriopsis caapi* (Spruce ex Griseb.) C.V. Morton), marijuana (*Cannabis sativa* L.), poppy (*Papaver Somniferum* L.) (Chapter 14), and peyote (*Lophophora Williamsii* [Lem. ex Salm-Dyck] J. M. Coult.).

Marijuana has been used extensively worldwide, particularly in Asia, for millennia as an illicit drug, but the plant is also a source of fiber and seed oil. In India, *bhang* is a weak variety from dried green leaves while *ganja* is the stringer, dried form of flowering tops without the resin from the female plant. The active ingredients of marijuana are cannabinoids (cannabinol, cannabigerol, cannabichromene, tetrahydrocannabivarin). The most potent form of marijuana is hashish oil; it is a concentrated, reddish brown oil with a THC (delta-9-tetrahydrocannabinol) content of 15–60%, and one or two drops on a cigarette equal a single joint of marijuana (Cunningham 2008). Cannabis has been legalized for medicinal use in several countries (United States, Germany, Canada, Mexico, the United Kingdom, the Netherlands, Spain, and Austria). Marijuana has antitumor properties (Liang et al. 2009).

Certain medicinal plant species—*D. stramonium* (Chapter 3), *Artemisia arborescens* (Chapter 13), and poppy (Chapter 14)—are also hallucinogenic and lethal. Chapter 14 describes extensively the alkaloids (morphine, codeine, thebaine, narcotine, and papaverine) of poppy. Illicit use of morphine and heroin becomes addiction and overdose is lethal. In spite of the great pharmaceutical and nutraceutical values of the poppy, a large amount is cultivated in the world for illicit purposes; around 8,000 tons of opium is produced worldwide for abuse; this surpasses by about 10 times the official quantity of the opium used for medical purposes. The illicit cultivation area has been estimated to be about 200,000–250,000 ha in the first decade of the twenty-first century (Chapter 14).

1.3.6 Quality Control

Plant products are being used by pharmaceutical, nutraceutical, and cosmeceutical industries worldwide. Global trends in favor of botanicals have brought concerns over the quality, safety, and efficacy of these products. There are various pharmacopoeias and regulatory agencies maintaining and controlling the quality and authenticity of plant-based products:

(WHO) World Health Organization Guidelines on Good Agricultural and Collection Practices (GACP) and Research and Evaluation of Traditional Medicine (http://www.who.int/en/)
(USFDA) U.S. Food and Drug Administration (http://www.fda.gov/)
European Scientific Cooperative on Phototherapy (ESCOP) (www.escop.com)
(AYUSH) Ayurveda, Yoga, Unani, Siddha and Homeopathy, (http://www.ayush.com/) (government of India)
Therapeutic Goods Administration (TGA), Australia (www.tga.gov.au)
Complementary Healthcare Council of Australia (CHCA) (www.chc.org.au)
National Center for Complementary and Alternative Medicine, (NCCAM) United States (www.nccam. nih.gov)
Health Canada Natural Health Products Regulations (HCNHPR) (www.hc-sc.gc.ca)

These emphasize the need for establishing and verifying botanical identity of raw materials as a first step in ensuring quality, safety, and efficacy. Chapter 30 describes quality control of the plant-based products by using molecular approaches. Contamination of botanical products with pesticide, pathogens, packaging, and heavy metals, particularly lead and cadmium, should be monitored. For example, one of the five Ayurvedic herbal medicine products produced in south Asia, available in south Asian grocery stores in Boston, contains potentially harmful levels of lead, mercury, and/or arsenic (Chapter 2).

1.4 ESTABLISHMENT OF NATIONAL AND INTERNATIONAL PROGRAMS

Prior to urbanization, the earth was rich with plants, including medicinal plants. Plants acclimatized to certain climatic zones remained at the native place, and the dissemination was minimal. Population pressure, civilization, migration of humans and animals, and, finally, urbanization caused deforestation. Humans cleared forests for cultivating crops and grazing domesticated animals. During this process, many plant species, including medicinal plants, disappeared before they were identified and recorded and long before their value could be assessed. Medicinal plants were the only pharmacy for treating diseases before the dawn of drug discovery from plants (plant-based medicine—homoeopathy) and man-made synthetic medicines (allopathy).

Table 2.1 (Chapter 2) lists international organizations and national organizations that are listed in the chapters included in this book. The WHO (http://www.who.int/en) documents diseases and solutions through online resources, publications, and meetings. Conservation, plant exploration, maintenance, varietal improvement, characterization of medicinal properties, dissemination of knowledge, and regulation are active at the national centers (Table 2.1). National medicinal institutes worldwide collect, maintain, and examine medicinal properties of plants (Chapters 2–29). Only a fraction of plants have been examined for their medicinal values. About 250 species of fruit trees in Southeast Asia are wild, cultivated in villages and in some instances cultivated in large scale, but medicinal properties have been completely unstudied for pharmacology (Chapter 4).

The Central Institute of Medicinal and Aromatic Plants, Lucknow, India, maintains 303 pure compounds and 461 plant extracts from more than 80 medicinal and aromatic plant species. More than 182 accessions of 118 plant species are being maintained in the live gene bank (http://www.cimap.res.in/gene_bank.html#bck). Most of the medicinal plants are harvested from the wild, and if we do not follow "give-and-take" policy, some of the wild plants will become extinct.

The International Center for Trade and Sustainable Development (http://ictsd.org/i/news) estimates 400,000 tons of medicinal and aromatic plants are traded every year, with around 80% of the species harvested from the wild. About 15,000 species (21%) of all medicinal and aromatic plant species used in traditional and modern medicine in the world are at risk. The National Institute of Biomedical Innovation, Osaka, Japan, maintains, evaluates chemical and biological components, and cultivates more than 4,000 species and groups of the medicinal plants (http://www.nibio.go.jp/english/part/medicinal_plant/).

1.5 GENE POOLS OF MEDICINAL PLANTS

The discovery of the New World opened the door for migration and domestication of plants from the Old World to the New World and vice versa. Medicinal plants include herbs, shrubs, and trees, annual, perennial, cultivated, semiwild and wild, and they grow in a wide range of agro-eco climatic conditions (tropical rain forest to the arctic regions) of the world. Centers of origin of medicinal plants, herbs and spices, vegetables, fruits, and agronomic crops were initially established by Vavilov (1987; Duke 2002). Legume species range from large tropical trees to small herbs found in temperate areas, humid tropics, arid zones, mountains, prairies, and lowlands (Chapter 12). The gene pool concept for the medicinal plants is difficult to establish because study of hybridization is limited with plants grown as crops.

1.6 GERMPLASM ENHANCEMENT OF MEDICINAL PLANTS

In comparison with large-scale agricultural and horticultural crops (see previous Volumes 1–5), gene banking (maintenance, reproduction, characterization, and evaluation of germplasm) for medicinal plants has not been well developed. This area is for future international cooperation and development.

Varietal improvement of medicinal plants has lagged far behind that of the cultivated crops. Indigenous people in the modern era still depend on medicinal plants for their health needs; about 60–70% of Ghanaians rely on traditional plant-based medical systems (Chapter 9). Varietal improvement through selection, hybridization, induced mutation, polyploidy, and biotechnology is routine in poppy (Chapter 14), ginger (Chapter 15), turmeric (Chapter 16), ginseng (Chapter 17), tea (Chapter 18), coffee (Chapter 19), *Mentha* (Chapter 21), *Plantago* (Chapter 22), fenugreek (Chapter 24), and *Aloe vera* (Chapter 25). Micropropagation is routinely used in ginger, turmeric, tea, coffee, and *Aloe vera*. Genetic transformation in mint, tea, coffee, and *Aloe vera* is in progress but (GM) genetically modified crops have not yet been produced. Cytogenetics of poppy, ginger, turmeric, coffee, *Plantago ovata*, Forssk., mint, *Aloe vera*, ipecac, and fenugreek helped establish basic chromosome numbers and species relationships. Fluorescence *in situ* hybridization (FISH) and genomic *in situ* hybridization (GISH) are being used in *Plantago ovata* and coffee. Molecular linkage maps have been established in tea, coffee, and *Plantago ovata*.

1.7 BIOPIRACY

Plant exploration in countries of centers of diversity and also in the habitats of the indigenous people is not an open door presently because of biopiracy by the developed countries. Pharmaceutical companies in developed countries take genetic resources and traditional knowledge from developing and underdeveloped countries to create products for commercialization. This is known as biopiracy. A patent on the chemical ingredients of turmeric, for healing wounds, was withdrawn by the U.S. Patent and Trademark Office (USPTO) (Jayaraman 1997). Over 61 patents on turmeric were registered in India and many more are in the process (Chapter 16). A patent on Neem-based biopesticide including "Neemix" granted to W. R. Grace was revoked by the European Patent Office. Neem is a native plant of India that is used in Ayurvedic medicine (http://news.bbc.co.uk/2/hi/science/nature/4333627.stm).

Seven American and four Japanese companies have filed or have been granted patents on Ashwagandha (*Withania somnifera* (L.) Dunal; http://www.i-sis.org). Ashwagandha has been used in the Ayurvedic system as an aphrodisiac and diuretic and for restoring memory loss (Chapter 3). India and the United Sates have signed two intergovernmental agreements on intellectual property rights (IPRs) to help prevent misappropriation of traditional knowledge through mistaken issuance of patents. The first agreement is the Traditional Knowledge Digital Library (TKDL) Access Agreement signed between the Council of Scientific and Industrial Research (CSIR) and the USPTO. The CSIR will provide training to the USPTO examiners and staff to help them use TKDL tools for search and examination (www.patentbaristas.com).

Series of patents on the extraction of the fat from the cupuaçu (*Theobroma grandiflorum* (Willd. ex Spreng.) K. Schum.*)* seeds and the production of cupuaçu chocolate have been registered by the company ASAHI Foods Co., Ltd., from Kyoto, Japan. Cupuaçu is a native plant of Brazil and Peru. ASAHI has registered the plant name "Cupuaçu" as a trademark for various product classes (including chocolate) in Japan, the European Union, and the United States (www.amazonlink.org). The Brazilian government released a list containing the scientific names of around 3,000 species of Brazilian flora to prevent foreign companies from registering the names popularly used in Brazil to refer to these plants (www.brazzilmag.com).

1.8 SUMMARY

Plants have been an integral part of all living organisms of the earth since time immemorial. However, population pressure, urbanization, pollution, and climate change are causing loss of plant biodiversity worldwide. Plants are not keeping up the pace to clean the air with the rate of pollution we are generating. This causes extinction of plants before their values are assessed. Of the 422,000 plant species estimated worldwide, 50,000–80,000 flowering plants are being used medicinally. These plants are potentially rich sources of medicinal compounds curing everything from the common cold to cancer and even HIV/AIDS and are known as nature's pharmacy. National and international organizations are collecting, documenting, and conserving *in situ* or *ex situ* the biodiversity of medicinal plants. Pharmaceutical industries from developed countries are taking genetic resources and traditional knowledge from developing countries and patenting compounds for their financial gains.

REFERENCES

Ariyawardana, A., R. Govindasamy, and J. E. Simon. 2009. The natural products industry: A global and African economic perspective. In *African plant products: New discoveries and challenges in chemistry and quality*, ed. H. R. Juliani, J. E. Simon, and C.-T. Ho, ACS Symposium Series, American Chemical Society, Washington, D.C.

Calixto, J. B. 2005. Twenty-five years of research on medicinal plants in Latin America: A personal view. *Journal of Ethnopharmacology* 100:131–134.

Chung, G., and R. J. Singh. 2008. Broadening the genetic base of soybean: A multidisciplinary approach. *Critical Reviews in Plant Sciences* 27:295–341.

Cunningham, N. 2008. Hallucinogenic plants of abuse. *Emergency Medicine Australasia* 20:167–174.

Duke, J. A. 2002. *Handbook of medicinal herbs*, 2nd ed. Boca Raton, FL: CRC Press.

———. 2009. *Duke's handbook of medicinal plants of Latin America.* Boca Raton, FL: CRC Press, Taylor & Francis Group.

Jayaraman, K. S. 1997. United States withdraws patent on Indian herb. *Journal of Alternative and Complementary Medicine* 3:417–418.

Liang, C., M. D. McClean, C. Marsit, B. Christensen, E. Peters, H. H. Nelson, and K. T. Kelsey. 2009. A population-based case-control study of marijuana use and head and neck squamous cell carcinoma. *Cancer Prevention Research* 2:759–768.

Liu, Y., Y. Wang, and H. Huang. 2006. High interpopulation genetic differentiation and unidirectional linear migration patterns in *Myricaria laxiflora* (Tamariacaceae), an endemic riparian plant in the Three Gorges Valley of the Yangtze River. *American Journal of Botany* 93:206–215.

Marinelli, J., ed. 2005. *Plant: The ultimate visual reference to plants and flowers of the world.* New York: Dorling Kindersley Publishers, LTD.

Miguel, M. G. 2010. Antioxidant activity of medicinal and aromatic plants. A review. *Flavor and Fragrance Journal* 25:291–312.

Pennisi, E. 2010. Despite progress, biodiversity declines. *Science* 329:1272–1277.

Roberson, E. 2008. http://www.biologicaldiversity.org/publications/papers/Medicinal_Plants_042008_lores.pdf

Vavilov, N. I. 1987. *Origin and geography of cultivated plants.* Leningrad, USSR: Nauka (in Russian).

World Health Organization. 2003. http://www.who.int/en

CHAPTER **2**

Medicinal Plants—Nature's Pharmacy*

Ram J. Singh, Ales Lebeda, and Arthur O. Tucker

CONTENTS

2.1 INTRODUCTION

Plants are an integral part of all living organisms of the earth, and medicinal plants are widely distributed worldwide (Figure 2.1). Since time immemorial, humans from all the cultures of the world have independently selected plants for food, shelter, clothing, and medicine. Plants were identified, according to their therapeutic properties and through trial and error, by the priests, shamans, herbalists, spiritual leaders and medicine men, and this practice is still a routine in many countries of Asia, Africa, the Middle East, Australia, and Latin America. Indeed, the widespread use of natural herbs and medicinal plants for curing and preventing diseases (nature's pharmacy) has been described in the ancient texts of the Vedas and the Bible (Hoareau and DaSilva 1999) and the Qur'an and the Ahadith (Ahmad et al. 2009). Duke, Duke, and duCellier (2008) even wrote a book entitled *Duke's Handbook* of *Medicinal Plants of the Bible* and cataloged "faith-based farmaceuticals."

---

* This chapter is dedicated to the memory of the late Professor Edgar J. DaSilva, who suggested the inclusion of the topic of this chapter in this book.

Figure 2.1 Centers of origin for some medicinal plants. Medicinal plants include some herbs, spices, vegetables, fruits, and agronomic crops—legumes and oilseed crops—from herbs to trees. (Duke, J. A. 2002. *Handbook of medicinal herbs*, 2nd ed. Boca Raton, FL: CRC Press. With permission.)

The rapid explosion of human population, migration, urbanization, and deforestation created many diseases for humans as well as for domesticated animals during the nineteenth century. During this time, the systematic study of plants with medical properties was begun worldwide with advances in chemistry.

Some believe that natural medicines have been overly studied, but, in fact, we have hardly begun to tap the reservoir of potential therapeutics from plants. One study estimates that around 422,000 plant species exist worldwide (Marinelli 2005). These plants are a potentially rich source of medicinal compounds, curing everything from the common cold to cancer and HIV/AIDS. Schultes (1972) described chemical compounds of medicinal properties that are scattered throughout the plant kingdom—250,000–350,000 species: 18,000 algaes; 90,000 fungi (including bacteria); 15,000 lichens; 14,000–20,000 bryophytes; 6,000–9,000 pteridophytes; 675 gymnosperms; and 200,000 species of angiosperms in some 300 families. These are an untapped reservoir for future wonder drugs. Medicinal treasures from the wild may be available for future generations, but today only 50,000–80,000 flowering plants are used medicinally worldwide (Marinelli 2005).

Different medicinal plants have antibiotic, antidiabetic, antihyperglycemic, and hyperlipidemic properties (Mentreddy 2007). We present here some recent studies and reviews on the role of medicinal plants in treating human diseases and promoting healing:

Alzheimer's: Tucci (2008) reviewed herbal treatments for slowing the onset of Alzheimer's and remarked that "herbal medicines...have been used for centuries as cognitive activators in traditional cultures, and their consideration in the Western medical regime may eventually lead to greater options in the treatment of Alzheimer's." Zhang (2005) listed *on-compound multiple-target* strategies to combat Alzheimer's, both from natural and synthetic sources.

Cancer: Graham et al. (2000) listed 350 plant species used to fight cancer. These plants are native or domesticates from Asia, Europe, South America, the Caribbean, Africa, and Australia. Ionkova (2008) looked at several anticancer compounds from *in vitro* cultures of rare medicinal plants and focused upon two genera: *Astragalus* and *Linum*. Mahady et al. (2003) found that the isoquinoline alkaloids from *Sanguinaria canadensis* L. and *Hydrastis canadensis* L. were effective against *Helicobacter pylori*, the cause of stomach ulcers and linked to stomach cancer. O'Sullivan-Coyne et al. (2009) found that curcumin (the principal component in turmeric, *Curcuma longa* L.) induces apoptosis-independent death in esophageal cancer cells. In regard to the role of dietary herbs and spices in cancer prevention, Kaefer and Milner (2008) remarked:

> While culinary herbs and spices present intriguing possibilities for health promotion, more complete information is needed about the actual exposures to dietary compounds that are needed to bring about a response and the molecular agent(s) for specific herbs and spices. Only after this information is obtained will it be possible to define appropriate intervention strategies to achieve maximum benefits from herbs and spices without eliciting ill consequences.

Conjunctivitis: Sharma and Singh (2002) listed 169 species of the 66 families used during 1933–2000 for treating conjunctivitis; the maximum number (17) of species was from the family Asteraceae, followed by Euphorbiaceae (11) and Mimosaceae (8). These plant species are from Asia, Africa, the Middle East, Europe, the South Pacific Islands, and South America.

Diabetes: Yadav et al. (2007) listed 17 species of commonly used plants for treating diabetes (Table 2.1). Garg and Garg (2008) listed 113 species used for treating diabetes worldwide but emphasized that these must be used in conjunction with diet and exercise.

Immunomodulation: Sagrawat and Khan (2007) listed 18 plant species from India with immunomodulatory activity. *Withania somnifera* (L.) Dunal has immunomodulatory effects and could be useful in the treatment of colon cancer (Muralikrishnan, Donda, and Shakeel 2010).

Sexual dysfunction: Yakubu, Akanji, and Oladiji (2007) reviewed the 10 leading plants used to alleviate male sexual dysfunction. Yohimbe bark (*Pausinystalia yohimbe* Pierre ex Beille) contains tryptamine alkaloid yohimbine and is widely distributed over the counter as an herbal aphrodisiac

Table 2.1 Medicinal Plants with Antidiabetic Properties

Botanical Name	General Indian Name	Family
Momordica charantia L.	Karela	Cucurbitaceae
Swertia chirayita (Roxb.) H. Karst.	Chirayata	Gentianaceae
Gymnema sylvestre (Retz.) Schult.	Gudmar patra	Plumbaginaceae
Trigonella foenum-graecum L.	Methi dana	Fabaceae
Plumbago zeylanica L.	Chitrak mool	Plumbaginaceae
Syzygium cumini (L.) Skeels (*Eugena jambolana* Lam.)	Jamun	Myrtaceae
Aegle marmelos (L.) Corrêa	Bilva patra	Rutaceae
Terminalia chebula Retz.	Harad	Combretaceae
Terminalia bellirica (Gaertn.) Roxb.	Baheda	Combretaceae
Phyllanthus emblica L. (*Emblica officinalis* Gaertn.)	Amla	Euphorbiaceae
Curcuma longa L.	Haridra	Zingiberaceae
Pterocarpus marsupium Roxb.	Vijaysar	Fabaceae
Berberis aristata DC.	Daru harida	Berberidaceae
Citrullus colocynthis (L.) Schrad.	Indrayan mool	Cucurbitaceae
Cyperus rotundus L.	Nagarmotha	Cyperaceae
Piper longum L.	Pippali	Piperaceae
Zingiber officinale Roscoe	Adrak	Zingiberaceae

Source: Adapted from Yadav, H. et al. 2007. *Journal of Pharmacological Sciences* 105:12–21.

(Adeniyi et al. 2007). Yohimbe is native to western Africa, including Nigeria, Cameroon, Congo, and Gabon (Barceloux 2008).

Wound healing: Habbu, Joshi, and Patil (2007) reviewed potential wound healers of plant origin. They listed 81 plant species with potential to promote wound healing and stated that they "could be of enormous help in managing and treating various types of wounds."

The last detailed survey of the medicinal plants industry was 20 years ago (Wijesekera 1991). The main objective of this chapter is to update briefly the story of medicinal plants, nature's pharmacy, from time immemorial to the present day, where scientists have been and are desperate to find cures for illnesses such as Alzheimer's, cancer, cardiovascular diseases, diabetes, HIV/AIDS, malaria, and many others. The popularity of and demand for plant-based health food and cosmetics have been increasing over that for allopathic medicines. The collection, preservation, and maintenance of germplasm resources of the few medicinal plants described here will set the stage for the conservation of plants from the unexplored regions of the world before they become extinct.

2.2 MEDICINAL PLANTS, HERBS, AND CROPS: CONCEPT AND ISSUE

Not only botany, chemistry, and pharmacology, but also anthropology, archeology, linguistics, history, sociology, comparative religion, and numerous other specialties have contributed appreciably to the search for new biodynamic plants (ethnobotany, or with respect to drug plants, ethnopharmacology or pharmacognosy; see, for example, Lewis and Elvin-Lewis 2003; Samuelsson and Bohlin 2010; Tyler, Brady, and Robbers 1988).

Medicinal plants are mostly wild, and few have been domesticated or are currently being cultivated. Based on reviewing studies from eight countries (Mexico, Guatemala, Spain, Bulgaria, Ethiopia, Uganda, the Philippines, and Nepal) in four regions of the world, Aguilar-Støen and

Moe (2007) concluded that medicinal plants have wide distribution across countries and continents. Most plants are found wild (40.5%) or naturalized (33.3%), while only a small number (3.3%) are cultivated. Aguilar-Støen and Moe (2007) suggested that conservation and management interventions through collaboration among host countries should be introduced. The authors identified 611 plant species belonging to 132 families; the highest percentage of species reported were from Fabaceae (10.1%), followed by Asteraceae (9.9%), Poaceae (5.5%), Solanaceae (4.6%), and Euphorbiaceae (3.5%).

Why do plants have medicinal values? Phytochemists usually identify plant chemicals as either primary or secondary products; drugs generally belong to the class of secondary metabolites. In summary, primary compounds reflect the DNA sequence (DNA, RNA, and proteins), while secondary products are synthesized along metabolic pathways. Plants have a very limited immune response (often consisting of induced proteins), and they move on an extremely slow time scale in relation to that of animals. Thus, diseases, herbivores, and competition for resources from other plants must be avoided while simultaneously promoting dissemination of the species with pollinators and seed dispersal agents. Secondary constituents are not normally found in animals in any appreciable quantities except in corals and sponges, both of which harbor endophytic algae. A good introduction to secondary metabolites and their modes of action is provided by van Wyk and Wink (2004).

Plants may produce herbicides to inhibit the growth of competing plants, encompassing a vast literature termed allelopathy. For example, salicin, a salicyl alcohol glycoside from which salicylic acid is derived, is a water-soluble phytotoxin (plant poison) that washes off from the leaves of willows (*Salix*) and other plants to the ground below, inhibiting the growth of competing plants (Raskin 1992). Aspirin, or acetylsalicylic acid, is a semisynthetic drug (a modification of the natural precursors). The natural salicylates have the advantage of being readily absorbed through the skin, and for that reason they are often used as the active ingredients in creams for stopping muscle pain, especially by athletes. Salicylates inhibit the production of prostaglandins, important chemicals involved in the processes of temperature regulation, inflammation, and pain; they also help in maintaining the mucous layer in the gut, preventing the stomach from digesting itself (Moerman 2009).

Plants also produce toxic or repellent chemicals that deter browsing of insects and other herbivores. Nicotine is isolated from tobacco, and pyrethrins are isolated from chrysanthemums (Dev and Koul 1997). Monarch larvae feed on milkweeds (*Asclepias*), which are very toxic to other animals. The larvae ingest and sequester various cardiac glycosides throughout their bodies, deterring birds from feeding on them (Brower, Brower, and Corvine 1967). In addition, the activity of herbivores induces increases in protective chemicals (Karban and Baldwin 1997).

Flavonoids, coumarins, terpenoids, alkaloids, polylphenols, polysaccharides, and proteins all aid plants in their survival but may also be sources of drugs. For example, naturally derived anti-HIV compounds are being isolated and identified from plants (Asres et al. 2005). Biodiversity provides us not only with potential natural pharmaceuticals, but also with sources for biosynthesis and genetic engineering (Wrigley et al. 2000).

We emphasize that we should avoid the fallacy of equating "natural" with "safe." Poison ivy (*Toxicodendron radicans* (L.) Kuntze) is a perfect example. Comfrey (*Symphytum* spp.) contains documented hepatocarcinogens (pyrrolizidine alkaloids) that can also be absorbed through topical applications, yet comfrey still has very vocal adherents (Awang et al. 1993; Betz et al. 1994; Brauchli et al. 1982; Culvenor et al. 1980a, 1980b; Furuya and Hikichi 1971; Gomes et al. 2007; Hirono et al. 1978, 1979; Jaarsma et al. 1989; Westendorf 1992). Culinary herbs in North America must be designated GRAS (generally recognized as safe) by the FDA (Food and Drug Administration); GRAS protects both the consumer and distributor. Yet, Tucker and Maciarello (1998) identified five potentially toxic culinary herbs without GRAS status and freely sold in the United States: *hoja santa/yerba santa/acuyo* (*Piper auritum* Humb., Bonpl. & Kunth.), California bay (*Umbellularia californica* (Hook. & Arnott) Nutt), perilla/*shiso* (*Perilla frutescens* (L.) Britton), pink/red peppercorns (*Schinus terebinthifolius* Raddi and *S. molle* L.), and *epazote*/chenopodium (*Dysphania ambrosioides* (L.) Mosyakni & Clements).

To this list, we may add sassafras root (*Sassafras albidum* (Nutt.) Nees) with 74–80% safrole, a hepatocarcinogen (Tucker, Maciarello, and Broderick 1994).

We also warn readers about natural materials that come from improper GAP (good agricultural practice), GCP (good collecting practice), and GMP (good manufacturing process). Pesticide, pathogenic, and packaging contamination have all been recorded (Cho et al. 2007; Tucker and Maciarello 1993). Heavy metals, particularly lead and cadmium, are also a concern (Guédon et al. 2008; Krejpcio, Król, and Sionkowski 2007; Liu et al. 2008; Ozkutlu 2008; Saper et al. 2004, 2008; Sekeroglu et al. 2008). For example, Saper et al. (2004, 2008) reported that one of the five Ayurvedic herbal medicine products produced in southern Asia and available in Boston south Asian grocery stores contains potentially harmful levels of lead, mercury, and/or arsenic. This study recommended mandatory monitoring of Ayurvedic medicines because users may be at risk of heavy metal toxicity. While recommendations by individuals, states, and countries have been issued (Forschungsvereinigung der Arzneimittel-Herstell 2003; Harnischfeger 2000; Mathé and Franz 1999), the legislation for both GAP and GMP is still actively being pursued and advanced on an international level by FAO (http://www.fao.org/prods/gap/index_en.htm) and WHO (http://www.who.int/bloodproducts/gmp/en/).

2.3 MEDICINAL PLANTS OF ASIA

Although traditional Chinese medicine (TCM), traditional Indian medicine (TIM), ancient Egyptian medicine (AEM), and ancient Greek medicine (AGM) have developed independently, they have many similarities (Chapman and Chomchalow 2005; Patwardhan 2005).

In TIM, two major traditional indigenous systems of medicine are common: Ayurveda in the north and Siddha in the south, particularly in Tamil Nadu (Krishnan et al. 2008). The Athar Veda, one of the four Vedas, is the first and the earliest text dealing with medicines having antibiotics in India, probably started in the tenth to the twelfth centuries BCE (Chapter 3). Puranic text suggests a germ is the cause of leprosy, treatable with Ausadhi (medicine). From the description of Ausadhi as a black, branching entity with dusky patches, it was very likely lichen with antibiotic properties. Ayurveda (*ayu* = life; *veda* = the science) is a scholarly system of medicine that originated in India. The earliest foundations of Ayurveda were built on a synthesis of selected ancient herbal practices, together with a massive addition of theoretical discourse and conceptualization, new nosologies, and new therapies dating from about 400 BCE. The most famous text belongs to the school of Charaka (Chapter 3). Ayurvedic treatment involves diet; recommendations are individualized to a person's *dosha* (a unique mix of body and mind in Ayurveda) and the season. Cleansing and detoxification are done through fasting, enemas, diets, and body treatments. Ayurvedic herbs are *triphala, ashwaghandha, gotu kola, guggul,* and *boswellia*.

India has 15 agroclimatic zones, 47,000 plant species, and 15,000 medicinal plants; these include 7,000 plants used in Ayurveda, 700 in Unani medicine, and 600 in Siddha medicine. Traditional Siddha medicine is practiced mostly in Tamil Nadu (southeastern India) and is at least 1,000 years old and uses a large number of herbs (Subbarayappa 1997). Indian systems of medicine have identified 1,500 medicinal plants, of which 500 species are mostly used in the preparation of drugs. More than 150 species have been categorized as endangered.

Traditional Unani medicine originated in ancient Greece around 400 BCE. Hippocrates, also known as the founder of allopathic medicine, is considered to be the first Unani physician. This medicine is practiced today in many Middle Eastern and Asian countries, including India, Pakistan, and Sri Lanka. The literature on TIM is voluminous, but several texts stand out: Ajmal and Ajmal (2005); Bhattacharjee (2001); Husain et al. (1992); Jain and DeFilips (1991); Kapoor (1990); Nadkarni (1996); Sarin (1996); Singh, Pandey, and Kumar (2000); Sivarajan and Balachandran (1994); and Thakur, Puri, and Husain (1989).

TCM diagnosis looks for "patterns of disharmony" or imbalances rather than treating specific diseases. Restoring balance and harmony usually involves prescribing herbal tea decoctions, acupuncture, specific diet counseling, massage, and other therapies including cupping, moxibustion, exercise (*taichi* and *qi gong*), and meditation.

Chen et al. (2007) evaluated complete data sets of TCM outpatients' reimbursement claims from 1996 to 2001, including the use of Chinese herbal remedies, acupuncture, and traumatology, a manipulative therapy in Taiwan. They found 62.5% of patients had used TCM at least once during the whole 6-year period.

Anderson (1986) reported 121 plant species being used for medicinal use among the Akha tribe of the remote region of northern Thailand. De Boer and Lamxay (2009) summarized the use of 55 medicinal plants among Brou, Saek, and Kry ethnic groups of Lao People's Democratic Republic to facilitate childbirth, alleviate menstruation problems, assist recovery after miscarriage, mitigate postpartum hemorrhage, and aid postpartum recovery and infant care.

Amarasingham et al. (1964) screened 542 species of plants belonging to 295 genera from 89 families from the Malayan flora for the presence of alkaloids and saponins. Of the examined species, 101 showed positive alkaloid tests and 77 were found to contain saponin. Yeung et al. (2009) isolated seven antimycobacterial compounds from *Gentianopsis padulosa* (Hook. f.) Ma (Gentianaceae), which is a perennial herb that grows in the Tibetan plateau. Three compounds showed modest inhibitory effect against the growth of *Mycobacterium smegmatis* (Trevisan) Lehmann & Neumann and *M. tuberculosis* Lehman & Neumann.

The literature on TCM and medicinal plants in Southeast Asia is voluminous, but the following texts stand out as important references: Anderson (1993), Bensky and Gamble (1993), Chin and Keng (1992), Chuakul et al. (1997), Duke and Ayensu (1985), Farnsworth and Bunyapraphatsara (1992), Hsu (1986), Huang (1999), Keys (1976), Kletter and Kriechbaum (2001), Li (2006), National Institute for the Control of Pharmaceutical and Biological Products (1987), Perry (1980), Su (2003), Tierra and Tierra (1998), Van Duong (1993), World Health Organization (1997), Wu (2005), and Yen (1992).

2.4 MEDICINAL PLANTS OF AUSTRALIA AND NEW ZEALAND

Australian Aboriginal people have been using medicinal plants for treating ailments for thousands of years (Brouwer et al. 2005; Gaikwad et al. 2008; Chapter 11, this volume). In addition to these native uses, Australian medicinal plants also include those used by European settlers and other migrant groups, those plants used overseas (but also occurring in the native state in Australia), and plants with potential for the pharmaceutical industry (Collins et al. 1990; Lassak and McCarthy 2001).

Traditional medicines of the Maori people have been documented by Macdonald et al. (1973), while Brooker, Cambie, and Cooper (1981) have conducted a broader survey of medicinal plants in New Zealand.

2.5 MEDICINAL PLANTS OF OCEANIA

The reports on medicinal plants in Oceania are scattered in original research papers. Several books, however, are available; the authors are Krauss (2001), Staples and Kristiansen (1999), Whistler (1992a, 1992b, 1996), and Woodley (1991).

Bradacs, Maes, and Heilmann (2010) examined cytotoxic, antiprotozoal, and antimicrobial activities of 18 plants traditionally used in the South Pacific archipelago Vanuatu. Among 15 plant extracts with strong cytotoxic effects, one was specific for one cancer cell line. Leaf extracts of

Acalypha grandis Benth. significantly affected *Plasmodium falciparum* Welch. without showing obvious effects against the other protozoa tested. The leaf extracts of *Gyrocarpus americanus* Jacq. exhibited significant activity against *Trypanosoma brucei* Plimmer & Bradford. The extracts of leaves of *Tabernaemontana pandacaqui* Lam. and stems of *Macropiper latifolium* (L. f.) Miq. were active against *Trypanosoma cruzi* (Chagas).

2.6 MEDICINAL PLANTS OF EUROPE

Western traditional herbal medicine has a long history, with the roots of its practice found in the writings of the Greek physicians, such as Hippocrates and Dioscorides, as well as later in the works of the Romans, such as Galen. In Western Europe, with the collapse of Roman imperial authority, medicine became localized; folk medicine supplemented what remained of the preserved practices in many monastic institutions, which often had a hospital attached. Only a few effective drugs existed beyond opium and quinine. Folklore cures and potentially poisonous metal-based compounds were popular treatments. Knowledge passed from herbalist to herbalist has accumulated into a significant body of medicinal knowledge. For example, the use of foxglove in treating heart failure was developed from the observation of successful treatments of dropsy by a village wise-woman in 1775 (Norman 1985).

Probably the most comprehensive treatment of medicinal plants in medieval Europe (since the sixteenth century) was published by P. A. Matthioli (Matthioli 1544 (1954); see the dedication of this volume), who wrote the book *Herbarium*, where more than 1,000 plant species (mostly native in Europe) are described in detail from the viewpoint of botany, medicinal effects of different plant organs, preparation of medicines, and methods of application. This book was considered a medieval bestseller and was published repeatedly in many European languages.

The literature on European herbs is voluminous and beyond a brief summary here. However, to indicate the breadth of current use of European herbs, Lange (1998) provides the use, trade, and conservation, while Bisset and Wichtl (2001) provide a handbook of practice with European herbs.

Lange (1998) found that at least 2,000 medicinal and aromatic plant taxa are used on a commercial basis in Europe, of which two thirds (1200–1300 species) are native to Europe. Wild collection remains prominent in Albania, Turkey, Hungary, and Spain, and about 90% of the 1200–1300 European plant species are still wild collected. The overall volume of wild-collected material in Europe is estimated to be at least 20,000–30,000 tons, annually. In the European Union, about 130–140 plant species are cultivated, and the germplasm originates from either wild or cultivated stock. Europe, as a whole, imported an average of 120,000 tons of medicinal and aromatic plant material while exporting 70,000 tons, with Germany as a leading country in both import and export. Germany also appointed a Special Expert Committee of the German Federal Institute for Drugs and Medical Devices to produce the Commission E Monographs (Blumenthal et al. 1998, 2000).

2.7 MEDICINAL PLANTS OF AFRICA

Traditional medicine practitioners (TMPs) in Africa are based on holistic beliefs, systems, societies, and ancestors. Table 2.2 shows the relative ratios of traditional practitioners and medically trained doctors in relation to the whole population in some African countries (see Busia 2005). Herbal medicines are widely used and are the integral part of African traditional medical practices because they are cheaper than allopathic medicines. Incorporation of traditional medicines and Western medicines was suggested for the African countries as being followed in India and Southeast Asian countries including China and Vietnam (Busia 2005). Africa is extremely diverse and thus the literature on medicinal herbs is scattered; however, the following texts and articles

Table 2.2   Doctor–Patient Ratios in a Few Countries of Africa

Country	Doctor:Patient	TMP:Patient
Ghana	1:400 (Kwahu district)	1:12,000
Kenya	1:833 (Urban—Mathare)	1:897 (Urban—Mathare)
Malawi	1:50,000	1:138
Mozambique	1:50,000	1:200
Nigeria	1:110	1:16,400
South Africa	1:1639 (overall)	1:700–1200 (Venda);
	1:17,400	
Swaziland	1:10,000	1:100
Tanzania	1:33,000	1:350-450
Uganda	1:25,000	1:708
Zimbabwe	1:6,250	1:234 (urban); 1:956 (rural)

Source: Busia, K. 2005. *Phytotherapy Research* 19:919–923.

stand out: Ayensu (1978), De Smet (1999), Dokosi (1998), Fortin, Lô, and Maynart (1990), Iwu (1993), Kokwaro (1976), Neuwinger (1996), Simon et al. (2007), van Wyk, van Oudtschoorn, and Gericke (1997), Visser (1975), and von Koenen (1996).

Tabuti, Kukunda, and Waako (2010) documented 88 plant species used by traditional TMPs in the treatment of tuberculosis in Uganda. Seven species (*Eucalyptus* spp., *Warburgia salutaris* (G. Bertrol.) Chiov., *Ocimum gratissimum* L. (*O. suave* Willd.), *Zanthoxylum chalybeum* Engl., *Momordica foetida* Schumach., *Persea americana* Mill., and *Acacia hockii* De Wild.) were recommended for treatment by three or more TMPs. These authors suggested that TMPs in Uganda are playing a significant role in the primary health care and the authorities are integrating the allopathic and traditional medicine systems.

Persinos, Quimby, and Schermerhorn (1964) examined medicinal properties of 10 plant species from Nigeria; however, some of the species were growing not only in Nigeria but also in other parts of Africa. These species' descriptions follow:

- *Annona senegalensis* Pers.: The growing range for *A. senegalensis* extends from Cape Verde Islands to Gambia, northern Nigeria, and Sudan. The root and bark are used as a vermifuge, as an antidote for snakebites, as an insecticide, and as a treatment for sleeping sickness and dysentery. The powdered leaves are used for treating Guinea-worm sores.
- *Boswellia odorata* Hutch.: The growing range for *B. odorata* is from the Ivory Coast to the Cameroons and Ubangi-Shari. In Adamawa, eating of the fresh bark causes vomiting within a few hours and relieves the symptoms of giddiness and palpitations. A mixture of the root and bark is used as an antidote for arrow poison. The gum has a dual use as a stomachic and as a treatment for syphilis.
- *Canarium scweinfurthii* Engl.: The growing range for *C. scweinfurthii* is from Senegal to Angola, the Sudan, and eastern Africa. In Nigeria, the oleoresin is used at night by hunters as "bush candle" flares. In Liberia, the exudate is used as pitch or is burned black, and the carbon is used in tattooing. The bark is also used for colic, hemorrhoids, jaundice, dysentery, cough, chest pain, and cancers.
- *Jatropha curcas* L.: *Jatropha curcas* has an American origin and was introduced by the Portuguese; it is now cultivated in West Africa and extends as far as Rhodesia. Many uses are recorded for the leaves; an infusion mixed with lime juice is employed to reduce fever when taken internally and used externally as a wash. The Bakwiri people of the Cameroon Mountains drink an infusion of leaves with beer as a diuretic for rheumatism. In southern Nigeria, a decoction of leaves is used rectally as remedy for jaundice. Leaves are crushed and mixed with hot water, or they are burned and their ashes are applied to the sores.
- *Daniellia oliveri* (Rolf.) Hutch. & Dalziel.: In Sierra Leone and Guinea, the bark of *D. oliveri* is used in making beehives. In other parts of Africa, the resin is used for polishing furniture. An infusion of the powdered bark and buds is taken for headaches, migraines, and feverish pains. A decoction of

the leaves and bark relieves colic and toothache. The Yorubas of western Nigeria use gum for treating gonorrhea by chewing and swallowing it to produce a purgative action.

- *Dichrostachys cinerea* (L.) Wight. & Arn. (*D. nutans* (Pers.) Benth.): *Dichrostachys cinerea* is found in the savanna and the transition forest throughout tropical Africa, in Rhodesia, and in parts of South Africa. In Sudan, a decoction of the root is used to treat syphilis and leprosy. In Senegal and Guinea, the same decoction is employed as a purgative and in Sierra Leone the bark has an application as a vermifuge in treating elephantiasis.
- *Pterocarpus erinaceus* Lam.: *Pterocarpus erinaceus* is found from Senegal to Chad and Gabon. A decoction of the bark and resin makes a good astringent for use in severe diarrhea and dysentery, while a decoction from the terminal leaf of a stem is used in the Ivory Coast for fever.
- *Phoenix dactylifera* L.: *Phoenix dactylifera* grows widely in the tropics and subtropics. Fruit is mixed with capsicum pepper and added to beer to make it less intoxicating. In Morocco, the plant is used for tanning.
- *Crossopteryx febrifuga* (G. Don.) Benth.: *Crossopteryx febrifuga* is widely distributed in tropical Africa from Senegal to the Sudan and from eastern Africa to Rhodesia. In Sierra Leone, the bark is used as a cough medicine. In northern Nigeria, it is used against gonorrhea and worms. In Guinea, after being pulverized and mixed with rice, it is used as an astringent in treating diarrhea, dysentery, and fever.
- *Sarcocephalus latifolius* (Sm.) Bruce (*S. esculentus* Afzel. ex Sabine): An infusion of leaves and roots of *S. latifolius* is given internally to children for fever. In Liberia, the leaves are mixed with Guinea grains and taken for diarrhea and dysentery. In Sierra Leone, a decoction of the leaves is used against constipation.

As outlined by Eltohami (1997), the climate of Sudan ranges from completely arid to tropical zones, with a wide range of bioclimatic regions, from almost barren deserts in the north to the tropical rain forests in the extreme south of the country. Thus, this country is rich in floras. *Aloe crassipes* Baker, found in northern and eastern Sudan, is used in small doses as a laxative. *Balanites aegyptiacus* (L.) Delile is native to Africa; the roots contain steroidal sapogenins, whereas the bulb contains sugars and saponins. The leaves and fruits contain disogenin; kernels are high (30–40%) in oil and have valuable protein content. Extracts of fruit and bark are lethal to snails and water fleas. *Cymbopogon schoenanthus* (L.) Spreng. subsp. *proximus* (A. Rich) Maire & Weiller (*C. proximus* (Hochst. ex A. Rich.) Stapf.) is native to southern Egypt and northern Sudan and contains a bitter oleo resin, a toxic volatile oil and a saponin used extensively in indigenous medicine as a diuretic, colic painkiller, and antipyretic in fever.

Boswellia papyrifera (Delile ex Caill.) Hochst, a dry-land tree species native to Ethiopia, Eritrea, and Sudan, is used as incense that is burned in many churches worldwide; it is also used as a gum or oil in a number of applications, such as modern perfumery, traditional medicine, pharmaceuticals, fumigation powders, fabrication of varnishes, adhesives, painting, and chewing gum industries. It is also used for flavor in the food industry (e.g., bakery, milk products, and different alcoholic and soft drinks). *Acacia nilotica* (L.) Delile is native to Africa and the Indian subcontinent; it is used as a demulcent or for conditions such as gonorrhea, leucorrhoea, diarrhea, dysentery, or diabetes. It is also styptic and astringent. In Siddha medicine, the gum is used to consolidate otherwise watery semen.

Addo-Fordjour et al. (2008) examined medicinal properties of 52 plant species belonging to 47 genera and 22 families. The Fabaceae contained the most species. The medicinal plants were grouped into five growth forms: trees (63.5%), climbers (15.4%), herbs (11.5%), shrubs (9.6%), and mainly leaves (40.3%). Most of the traditional practitioners (75%) did not have any professional training; 56% of herbalists did not replant after harvesting. The majority of the herbalists did not keep records on diseases treated, the plants used, or the cultivation of the medicinal plants, and harvesting methods were destructive.

Kamatenesi-Mugisha et al. (2008) examined plants used in the treatment of fungal and bacterial infections in and around Queen Elizabeth Biosphere Reserve, western Uganda. They studied 67 medicinal plants distributed among 27 families and 51 genera that were used for treating fungal and bacterial infections. The medicinal plants used for the treatment of fungal and bacterial infections were harvested from wild/natural ecosystems (67.2%). The medicinal plants used for treating herpes zoster were obtained through cultivation (55.6%); the domesticated medicinal plants for treating herpes zoster are mainly food crops and are in high demand as nutritional supplements, such as *Cinnamomum verum* J. Presl, *Capsicum frutescens* L., *Avena sativa* L., *Aloe vera* (L.) Burm. f., and *Carica papaya* L. The highest number of species was from the Lamiaceae (13) and Asteraceae (11) families. The most commonly used plant parts were leaves (88.1%), roots (23.9%), and barks (10.5%).

Okigbo, Eme, and Ogbogu (2008) reviewed biodiversity and conservation of medicinal and aromatic plants of Africa. They estimated about 215,634,000 hectares of closed forest areas in Africa; the annual loss of forest is reported to be 1% per year due to deforestation. We believe that many medicinal and other genetic resources will become extinct before they can be collected and documented. Okigbo et al. (2008) estimated that 30% of the plants will be lost by the year 2040. However, about 70% of the wild plants in North Africa are known to have potential value in the fields of medicine, biotechnology and crop improvement.

According to Okigbo et al. (2008), phytomedicine used in Africa has demonstrated reduction in excessive mortality, morbidity, and disability due to diseases such as HIV/AIDS, malaria, tuberculosis, sickle-cell anemia, diabetes, and mental disorders. Ali, Wabel, and Blunden (2005) described the use of *Hibiscus sabdariffa* L., used for making cold and hot drinks (tea), to treat high blood pressure, liver diseases, cancer, and fever; it also has antifungal, antibacterial, and antiparasitic properties. It is native to the Old World tropics, is commonly known as roselle or red sorrel, and is widely grown in Central and West Africa and Southeast Asia.

Dièye et al. (2008) surveyed the use of medicinal plants of Senegal for the treatment of diabetes; they interviewed 220 patients and found that the most frequently used plants were *Moringa oleifera* Lam. (65.9%) and *Sclerocarya birrea* (A. Rich.) Hochst. (43.2%). The principal suppliers of plants were tradesmen in the market (66.8%) and traditional therapists (5%).

2.8 MEDICINAL PLANTS OF THE MIDDLE EAST AND NORTH AFRICA

Ancient Egyptian medicine, first recorded from 3100 BCE, was intimately tied to the rites of magic (Halioua and Ziskind 2005; Manniche 1989; Nunn 1996). Assyrian texts, dating from ca. 2000 BCE, are well documented by Thompson (1924, 1949) and include the concepts of diagnosis, prognosis, physical examination, and prescriptions. Greek and Roman medicines were influenced by the Assyrian and Egyptian medicinal traditions, as well as by local folklore, well represented by what became the *vade mecum* of the later physicians, the herbal of Dioscorides (Beck 2005). Jewish medicine was, at first, an amalgamation of Egyptian and Mesopotamian medicines, but it was later overshadowed by Hellenistic medicine (Jacob and Jacob 1993). Persian medicine provided a link of Europe and the Middle East with China and India (Laufer 1919).

Muslim physicians contributed significantly to the field of medicine, including anatomy, ophthalmology, pharmacology, pharmacy, physiology, surgery, and the pharmaceutical sciences. The translation of the Greek and Roman medical practices into Arabic set the template for Islamic medicine. Quantification of medicines by mathematics developed the strength of drugs and a system that could allow a doctor to determine, in advance, the most critical days of a patient's illness. The literature in English on medicinal plants from this region is scattered, but several texts are available by the following authors: Ahmed, Honda, and Miki (1979); Bellakhdar et al. (1982, 1991); Boulos (1983); Ghazanfar (1994); Miki (1976); and Rizk and El-Ghazaly (1995).

Ahmad et al. (2009) listed 32 medicinal plants belonging to 30 genera of 23 families described in the Qur'an and Ahadith. Health is defined as a state of the "body" (made up of the four elements: earth, air, water, and fire); there is an equilibrium in the four humors (blood, phlegm, yellow bile, and black bile), and the functions of the body are normal in accordance to its own "temperament" (cold, hot, wet, dry) and the environment. When the equilibrium of the humors is disturbed and functions of the body are abnormal, that state is disease.

Ghazanfar and Al-Sabahi (1993) described the use and chemical composition of 35 native and 21 cultivated plants in Oman used for curing a wide spectrum of diseases, from the common cold and fever to paralysis and diabetes. The flora of northern and central Oman consists of 600 plant species, the majority of which occur in the mountains and foothills. Of the native flora, Ghazanfar and Al-Sabahi recorded 35 species (ca. 6%) that are used medicinally. A large number of cultivated species are used for medicinal purposes. These species are used in parts of southwest Asia, especially India and Pakistan. There has been a long history of sea trade between Oman and India, and it is likely that many of the medicinal cultivated plants, garlic, lemon, and henna were imported from India for their medicinal values.

2.9 MEDICINAL PLANTS OF THE AMERICAS

Traditional medicine, a general term for the system of healing used by all Native American nations or tribes that has been practiced in some cases for at least 10,000 years, is linked to philosophy, religion, and spirituality, and treatments aim to balance the physical, emotional, mental, and spiritual components of a person. This involves a tribal healer, also known as a medicine man or medicine woman, and may involve the patient's family or the entire community. Treatments include prescribing medicinal herbal preparations, ritual purification or purging, and traditional smudging or burning of certain herbs, as well as chanting and prayers. Craker and Gardner (2006) evaluated an American perspective on medicinal plants and tomorrow's pharmacy and concluded that "American pharmacies of the future may well support both conventional and alternative systems enabling the consumer and medical practitioner to choose the best medicine for the medical condition." Craker, Gardner, and Etter (2003) summarized herb and medicinal plant cultivation in the United States from 1903 to 2003 and remarked: "Standardization of bioactive constituents and quality processing will undoubtedly be necessary before herbal remedies become mainstream medicines in the U.S."

Moerman (1977) listed 1,288 different medicinal plant species used by the Native North Americans, from 531 different genera from 118 different families. These plants were used in 48 different cultures in 4,869 different ways. Outside North America, the literature is scattered and not summarized, in part reflecting the diversity of the peoples. However, books by the following authors are available: Ayensu (1981); Balick, Elisabetsky, and Laird (1996); Breedlove and Laughlin (1993); Cáceres (1996); de Montellano, (1990); Lans (2003); Lorenzi and Abreu Matos (2000); Meléndez (1945); Mors, Rizzini, and Pereira (2000); Roersch (1994); Schultes and Raffauf (1990); and Seaforth, Adams, and Sylvester (1983).

Coe and Anderson (1996) examined the presence of bioactive compounds (alkaloids and glycosides) in 229 species, representing 177 genera and 72 families, used for medicinal purposes by the Garifuna people of Eastern Nicaragua. Of those examined, 113 species contained at least one of the bioactive compounds. The remaining 116 species were tested for alkaloids, and 51 contained alkaloids. Thus, 72% of species used by Garifuna natives have at least one medicinal compound.

Martín (1983) examined 131 species from 10 sites in Central Chile that have been used in popular medicine; the list included pteridophytes (2 families, 2 species) and angiosperms (52 families and 122 species of dicots; 3 families and 7 species of monocots). The number of native plants is equal to the number of exotic plants. Cultivated plants are as commonly used as wild plants. The families with greater number of species are Asteraceae (17), Rosaceae (12), and Labiatae (9). Some species, such as *Senecio pycnanthus* Phil., *Quinchamalium majus* Brongn., *Calceolaria thyrsiflora* Graham, and *Myoschilos oblongum* Ruiz & Pav., are extensively collected by the inhabitants and are thus in danger of extinction.

Mitchell and Ahmad (2006) reviewed medicinal plant research from 1948 to 2001 at the University of the West Indies, Jamaica. At least 334 plant species growing in Jamaica contained medicinal properties, and 193 were tested for their bioactivity. Only 80 possessed reasonable bioactivity; natural products were identified from 44 plants, and 29 of these natural products were bioactive. Patents have been obtained and drugs have been developed.

Darshan and Doreswamy (2004) reported patents obtained on anti-inflammatory drugs from 38 plants. These plants contain polysaccharides, terpenes, cucurminoids, and alkaloids and have potential in alleviating inflammatory diseases, including arthritis, rheumatism, acne, skin allergy, and ulcers. Chemicals that alleviate swelling are derived from plants including grape, boswellia, turmeric, devil's claw, and essential oils from clove, eucalyptus, rosemary, lavender, mint, myrrh, millefolia, and pine. Plants with polysaccharides have been reported to be the most potent in curing inflammatory diseases.

Medicinal plants from South America, particularly Brazil (Mans, da Rocha, and Schwartsmann 2000), have been screened for curing cancer. dos Santos Júnior et al. (2010) examined 51 plant species from Brazil against four tumor cell lines: B16 (murine skin), HL-60 (human leukemia), MCF-7 (human breast), and HCT-8 (human colon). The most active extracts against the tumor cells were those obtained from *Lantana fuscata* K. Koch. *Copaifera longsdorffii* Desf., and *Momordica charantia* L. Endringer et al. (2010) reported that several Brazilian plants are promising sources as cancer chemoprevention agents, particularly *Hancornia speciosa* Gomes, *Jacaranda caroba* (Vell.) DC., *Mansoa hirsuta* DC., and *Solanum paniculatum* L. De Mesquita et al. (2009) examined the cytotoxic potential of 412 extracts (from Brazilian cerrado plants used in traditional medicine that belonged to 21 families) against tumor cell lines in culture. Twenty-eight of 412 extracts demonstrated a substantial antiproliferative effect; 85% inhibition of cell proliferation, at 50 μg/mL, against one or more cell lines was observed. Extracts were obtained from plants belonging to several families: Anacardiaceae, Annonaceae, Apocynaceae, Clusiaceae, Flacourtiaceae, Sapindaceae, Sapotaceae, Simaroubaceae, and Zingiberaceae.

Plant-based antimalaria medicines are being developed at many places, but malaria-transferring mosquitoes are mutating at a much faster rate than identification of the drugs, and sometimes drug discovery takes about 10–15 years. The global resistance prevailing against the two most widely used antimalaria drugs—chloroquine and the antifolate sulphadoxine/pyrimethamine—is a major dilemma (Ridley 2002). Quinine from *Cinchona calisaya* Wedd. (*C. officinalis* Auct., non L.), a native of the Amazon rain forest, was the first European medicine used to treat malaria; artemisin from *Artemisia annua* L. later supplanted it but is rapidly becoming ineffective.

2.10 MODERN PLANT-BASED MEDICINE

Table 2.3 provides an historical perspective of important pharmaceutical drug discoveries since 1785. A number of compounds isolated and identified from plants have been summarized in books, bulletins, peer-reviewed papers, and popular articles (Table 2.4). Modern medicine is a multidisciplinary field, but drug discovery was originally plant based (Drews 2000; Fabricant and Farnsworth 2001; Sneader 2005). Drug research was revolutionized in the nineteenth century and beyond by advances in chemistry and laboratory techniques and equipment; old ideas of infectious disease epidemiology were replaced with newer ideas from bacteriology, biochemistry, and virology. The modern era of medicine really began with Robert Koch's postulates in 1890 concerning the transmission of disease by bacteria, and then with the discovery of penicillin by Alexander Fleming in 1928.

Table 2.5 estimates 287,685 species belonging to 14,066 genera compiled from several sources. This table suggests that Fabaceae (Leguminosae) contains 19,400 species while Poaceae (Gramineae) includes 10,000 species. In his book *Plant Alkaloids*, Raffauf (1996) estimates 260,145 species, including ferns, belonging to 13,209 genera worldwide. He tested 19,000 species and found

Table 2.3   Some Pharmaceutical Discoveries, 1785–1987

Year	Therapeutic	Drug Type	Drug
1785	Inotropic agent	Cardioglycoside	Digitoxin
1796	Smallpox vaccine	—	—
1803	Analgesic	Narcotic	Morphine
1867	Antiseptic	Phenol	Carbolic acid
1884	Analgesic	Alkaloid	Phenazone
1910	Antisyphillitic	Arsenical	Salversan
1935	Bactericide	Sulfonamides	Sulfamidochrysoidine
1942	Bactericide	Antibiotic	Penicillin
1987	Recombinant DNA	Hormone	Humulin

Source: Craker, L. E., and Z. E. Gardner. 2006. In *Medicinal and aromatic plants*, ed. R. J. Bogers, L. E. Craker, and D. Lange, 29–41. Dordrecht, Netherlands: Springer.

alkaloids in 3,660 species (19%) from 315 families. Plants of the family Papaveraceae are very high in alkaloids; by contrast Rosaceae and Lamiaceae lack appreciable levels of alkaloids.

Cardiac glycosides and sapogenins are found in the Apocynaceae, Asclepiadaceae, Liliaceae, Moraceae, Ranunculaceae, Scrophulariaceae (*s.l.*), and other families. Cardenolides have been used in primitive societies as arrow and dart poisons. Steroidal sapogenins are also known for the arrow poisons, and the most conspicuous genera are *Agave*, *Yucca*, and *Dioscorea*.

Anthraquinones, phenolic compounds, essential oils, and many other organic constituents of angiosperms are most certain to increase greatly in number and novelty when sophisticated methodologies are applied (Schultes 1972).

Busia (2005) reviewed the literature and reported over 15,000 compounds of significant therapeutic benefits, such as antibiotics, antimalarial drugs, analgesics, anti-inflammatory steroids, cardiotonics, hypotensives, tranquilizers, and sedatives. These compounds were isolated from about 200 plant species and most of the plants were of tropical origin. Plant-based medicines are still being used by up to 80% of the population in Africa, and nearly 25% of modern-medicine prescription drugs are derived from plants first used in traditional medicine (Fowler 2006).

Once science had a grasp on the interactions between drugs and the body, it became possible to synthesize natural compounds through chemistry. Then science and tradition diverged. In the first edition of the *American Pharmacopoeia*, published in 1820, 70% of drugs were plant based. In the 1960 edition, 5.3% were of plant origin and the science of ethnobotany developed—the winning combination of drugs and plants (Turner 1973).

2.11 LOOKING FORWARD

A large number of plant resources worldwide are unexplored, and only a fraction is being maintained in germplasm banks. International organizations, national institutes, and private drug industries worldwide are exploring and collecting plants, particularly from the tropical rain forests (Table 2.6). For example, the International Development Research Center in Ottawa, Canada (http://www.idrc.ca/en/ev-1-201-1-DO_TOPIC.html), is heavily engaged in preserving biodiversity of medicinal plants of Southeast Asia, Africa, and Latin America. These institutes are collecting medicinal plants and are examining chemical components and testing ingredients for curing diseases in humans and animals, as well as developing plant-based (bio) insecticides and pesticides for controlling pests and pathogens, respectively, of economically important crops.

We are alarmed over the loss of plant diversity, particularly endemic plants, which in turn will reduce the potential sources of plant-derived drugs. The highest endemic number of plants is in

Table 2.4 Plant-Based Drug Discovery

Botanical Name	Medicine	Family	Use	Native
Acacia nilotica (L.) Delile	Catechin, epicatechin, d-pinitol, dicatechin, quercetin, gallic acid, procyanidin	Fabaceae	Demulcent, gonorrhea, antimosquito larvae, leucorrhoea, diarrhea, dysentery, or diabetes	Africa and Indian subcontinent
Achillea millefolium L.	Isovaleric acid, salicylic acid, coumarin, tannin	Asteraceae	Diaphoretic, astringent, tonic, thrombosis, ulcers, measles	Eurasia
Aframomum melegueta K. Schum.	(6)-Paradol, (6)-gingerol, (6)-shogaol	Zingiberaceae	Body pains, rheumatism	Tropical West Africa
Ammi majus L.	Xanthotoxin	Umbelliferae	Vitiligo and psoriasis	Egypt
Ananas comosus (L.) Merr.	Bromelain	Bromeliaceae	Nematicidal	South America
Annona senegalensis Pers.	ent-kaurenoids	Annonaceae	Vermifuge, snakebites, insecticide, sleeping sickness, dysentery	Cape Verde Islands to Gambia, northern Nigeria, and Sudan
Artemisia annua L.	Artemisinin	Asteraceae	Malaria	Temperate Asia–China
Aspalathus linearis (Burm. f.) R. Dahlgren	Antioxidant aspalatin, nothofagin	Fabaceae	Rooibos tea—no caffeine antispasmodic, insomnia	South Africa
Atropa belladonna L.	Atropine	Solanaceae	Pupil dilator	Europe, North Africa, and western Asia
Avena sativa L.	Oatmeat extract	Poaceae	Lower cholesterol	Near East
Azadirachta indica A. Juss.	Azadirachtin	Meliaceae	Herpes, insecticide, antibacterial, antiviral, contraceptive, sedative, cosmeceutical	India
Balanites aegyptiaca (L.) Delile	Sapogenins, diosgenin	Zygophyllaceae	Lethal to snails and water fleas, headache	Africa
Boswellia papyrifera (Delile ex Caill.) Hochtst.	Oil, incense	Burseraceae	Stimulant	Ethiopia, Eritrea, Sudan
Cannabis sativa L.	Cannabinoids (cannabinol, cannabigerol, cannabichromene, tetrahydrocannabivarin)	Cannabaceae	Illicit drugs, stimulant, pain reliever, anticancer and HIV/AIDS	South and central Asia

(continued)

Table 2.4   Plant-Based Drug Discovery (Continued)

Botanical Name	Medicine	Family	Use	Native
Castanospermum australe A. Cunn & C. Fraser ex Hook.	Castanospermine, celgosivir, combretastatin	Fabaceae	Chronic HCV infection	Australia, Vanuatu, and New Caledonia
Catharanthus roseus (L.) G. Don	Vinblastine and vincristine	Apocynaceae	Diabetes, malaria, Hodgkin's leukemia, antineoplastic	Madagascar periwinkle is a native of Madagascar, in India, known as Nithyakalyani
Cephalotaxus harringtonia Knight ex J. Forbes) K. Koch.	Homoharringtonine	Taxaceae	Chronic myeloid leukemia	Japan
Chondrodendron tomentosum Ruiz & Pav.	(+)−Tubocurarine	Menispermaceae	Arrow poison	Central and South America
Chrysopogon zizanioides (L.) Roberty	Complex oil	Poaceae	Essential oil—perfumes	India
Cinchona pubescens Vahl.	Quinine	Rubiaceae	Antimalaria	Costa Rica, Venezuela, Ecuador, Peru
Citrullus colocynthis (L.) Schrad.	Phytosterol, glycoside	Cucurbitaceae	Purgative, laxative, diabetes	North Africa, Iran to India and tropical Asia
Coffea arabica L.	Caffeine	Rubiaceae	Stimulant	Yemen, Arabian Peninsula, Ethiopia, Sudan
Colchicum autumnale L.	Colchicine	Colchicaceae	Gout	South and Center of Europe
Cordia verbenacea DC.	Artemetin, beta-sitosterol, alpha-humulene, and beta-caryophyllene	Boraginaceae	Anti-inflammatory, antiulcer, and antirheumatic	Brazil
Crataegus laevigata (Poir.) DC.	Arterio-K	Rosaceae	Cardiac drug	Northern Europe, Asia, and North America
Curcuma longa L.	Curcumin	Zingiberaceae	Antioxidants, antiparasitic, antispasmodic, anticancer, anti-inflammatory, gastrointestinal	India, China
Cymbopogon proximus (Hochst. ex A. Rich.) Stapf	Piperitone, volatile oil	Poaceae	Neurotic disease	North Africa

Species	Compound	Family	Use	Region
Datura stramonium L.	Toxiferine	Solanaceae	Relaxant in surgery	South America
Digitalis purpurea L.	Digitoxin	Plantaginaceae	Cardiac drug	Africa
	Digoxin		Arterial fibrillation	Native to most of Europe
Dioscorea composita Hemsl.	Diosgenin	Dioscoreaceae	Steroid hormones	Asia and South America
Duboisia myoporoides R. Br.	Atropine, hyoscyamine, scopolamine	Solanaceae	Stimulant, euphoric, antispasmodic, analgesic	Australia
Erythroxylum coca Lam.	Cocaine	Erythroxylaceae	Local anesthetic	Northwestern South America
Eucalyptus globulus Labill.	Oil	Myrtaceae	Antimicrobial, biopesticide	Australia
Euphorbia peplus L.	Ingenol 3-angelate	Euphorbiaceae	Warts, actinic keratosis	Europe, northern Africa, western Asia
Filipendula ulmaria (L.) Maxim.	Aspirin–salicin	Rosaceae	Common cold and fever	Europe and western Asia
Foeniculum vulgare Mill.	Anethole	Apiaceae	Used in gripe water, digestive disorders, glaucoma	Mediterranean
Galega officinalis L.	Guanidine	Fabaceae	Type II diabetes	Middle East
Ginkgo biloba L.	Quercetin, kaempferol isorhamnetin, terpenoid lactones, ginkgolides A, B, C, bilobalide	Ginkgoraceae	Alzheimer's	China
Glycine max (L.) Merr.	Sitosterol, isoflavones, tamoxifen, genistein	Fabaceae	Cancers, estrogen	China
Glycyrrhiza glabra L.	Glycyrrhizin, saponin, asparagine	Fabaceae	Asthma, bronchitis, coughs, cancers	Mediterranean, central though southwest Asia
Guarea guidonia (L.) Sleumer	Diterpenoid, labdane, clerodane	Meliaceae	Antimalaria, insecticide	South America
Hamamelis virginiana L.	Optrex	Hamamelidaceae	Anti-inflammatory, antioxidant	North America
Harpagophytum procumbens (Burch.) DC. ex Meisn.	Harpagoside, beta-sitosterol	Pedaliaceae	Analgesic, sedative, diuretic, arthritis	South Africa
Hibiscus sabdariffa L.	Protocatechuic acid	Malvaceae	Antihypertensive	Old World tropics

(continued)

Table 2.4 Plant-Based Drug Discovery (Continued)

Botanical Name	Medicine	Family	Use	Native
Hoodia currorii (Hook.) Decne.	Chlorogenic acid, sterol glycoside	Asclepiadaceae	Hypertension, diabetes, stomachache	Kalahari, South Africa, Namibia, Angola, Botswana
Huperzia serrata (Thunb.) Trevis.	Huperzine A	Lycopodiaceae	Nutraceutical, Alzheimer's	India and Southeast Asia
Hydrastis canadensis L.	Gingivitol	Ranunculaceae	Cardiac drug, antihemorrhagic, anti-inflammatory	North America
Hyoscyamus niger L.	Hyoscyamine	Solanaceae	Anticholinergic	Eurasia
Hypericum perforatum L.	Hyperforin	Clusiaceae	Antidepressant, positive effect to control Parkinson's; many side effects	Europe
Ilex paraguariensis A. St.-Hil.	Xanthines	Aquifoliaceae	Antiobesity, anticarcinogenic	South America
Laburnum anagyroides Medik.	Cytisine	Fabaceae	Tobacco dependence	Central and southern Europe
Leucojum aestivum L.	Galanthamine	Amaryllidaceae	Cholinesterase inhibitor	Central and southern Europe
Linum usitatissimum L.	Linseed oil	Linaceae	Nutritional supplement, anticancer, purgative	Central Asia and Mediterranean
Lophophora williamsii (Lem.) J. Coult.	Mescaline	Cactaceae	Neurasthenia, asthma, hysteria; illegal drugs	South America
Mangifera indica L.	Mangiferin (Vimang®)	Anacardiaceae	Antioxidant, antiviral, chemopreventive, hemochromatosis	India
Melissa officinalis L.	Rosmarinic acid	Lamiaceae	Mosquito repellent, antibacterial, antiviral	Southern Europe, Mediterranean
Mentha canadensis L.	Menthol, methanolic	Lamiaceae	Anesthetic, antispasmodic, antiseptic, aromatic	Temperate regions of Europe and western and central Asia, east to the Himalayas and eastern Siberia, and North America
Momordica charantia Descourt.	Momordicin, lectin	Cucurbitaceae	Dyspepsia and constipation, antimalarial; controls diabetes	Indian subcontinent
Mondia whitei (Hook. f.) Skeels	Alkaloid, 5-chloropacin	Apocynaceae	Appetite stimulation, gonorrhea, postpartum bleeding, pediatric asthma	Tropical Africa from Guinea through Cameroon to eastern Africa

Species	Active compounds	Family	Uses	Distribution
Ocimum tenuiforum L.	Oleanolic acid, ursolic acid, rosmarinic acid, eugenol, carvacrol, linalool, β-caryophyllene	Lamiaceae	Common colds, headaches, stomach disorders, inflammation, heart disease, malaria, diabetes, antibacterial	India
Panax ginseng C.A. Mey.	Protopanaxadiol	Araliaceae	Cancer	China and Korea
Panax quinquefolius L.	Ginsenosides	Araliaceae	Diabetes, decreases fever, stomach pain, hemorrhage	North America
Papaver somniferum L.	Codeine, morphine	Papaveraceae	Analgesic	Southeastern Europe, western Asia
Pausinystalia johimbe (K. Schum.) Pierre ex Beille	Yohimbine	Rubiaceae	Depression, high blood pressure, high blood sugar	Nigeria, Cameroon, and the Congo
Pilocarpus jaborandi Vahl.	Pilocarpine	Rutaceae	Glaucoma	Neotropics of South America
Pimpinella anisum L.	Anethole, phytoestrogen	Apiaceae	Epilepsy, insomnia, lice, scabies, intestinal bacteria	Eastern Mediterranean and Southwest Asia
Piper methysticum G. Forst.	Kavain	Piperaceae	Sedative, soar throat, soft drink	Polynesia
Podophyllum peltatum L.	Podophyllin	Berberidaceae	Cathartic, insecticide, rheumatism, cancer	Eastern North America
Polygala senega L.	Senega fluid extract, saponin glycosides, polygalic acid, senegin	Polygalaceae	Anti-inflammatory	North America
Prunus africana (Hook. f.) Kalkman	Phytosterol, triterpene esters, pentacyclic acids, aliphatic alcohols	Rosaceae	Chest and stomach pain, gonorrhea, inflammations, kidney diseases, benign prostrate therapy	South Africa, tropical Africa, eastern Africa
Rauwolfia serpentina (L.) Benth. ex. Kurzo	Reserpine	Apocynaceae	Antihypertensive	Tropical regions (commonly known as Indian snakeroot or Sarpagandha)
Ricinus communis L.	Ricin	Euphorbiaceae	Laxative, purgative, cathartic	Southeastern Mediterranean basin, eastern Africa, and India
Salix alba L.	Salicin	Salicaceae	Aches, pains; reduces fevers	Europe, western and central Asia
Sanguinaria canadensis L.	Sanguinarine, chelerythrine	Papaveraceae	Gastrointestinal ailments	United States of America

(continued)

Table 2.4 Plant-Based Drug Discovery (Continued)

Botanical Name	Medicine	Family	Use	Native
Santalum album L.	Sandalwood	Santalaceae	Chronic bronchitis, gonorrhea, oil against *Eberthella typhosa* and *Escherichia coli*, incense, cosmeceutical—wood paste	Indian peninsula, South Pacific, northern Australia
Scutellaria baicalensis Georgi	Flavocoxid	Lamiaceae	Osteoarthritis	China
Senna alexandrina Mill.	Glycoside	Fabaceae	Laxative	Egypt
Stevia rebaudiana (Bertoni) Bertoni	Steviosides	Asteraceae	Natural sweetener	Paraguay
Styrax tonkinensis (Pierre) Craib ex Hartwich	Benzoin	Styracaceae	Oral disinfectant	Southeast Asia
Syzygium aromaticum (L.) Merr. & L. M. Perry	Eugenol	Myrtaceae	Toothache, Ayurvedic and Chinese medicines; spice	Moluccas (Spice) Islands, Indonesia;
Tabebuia impetiginosa (Mart. ex DC.) Standl.	Lapachol	Bignoniaceae	Anticancer, antibiotic, disinfectant; higher dose is toxic	Trinidad and Tobago, Mexico, Argentina
Taxus brevifolia Nutt.	Taxol	Taxaceae	Antineoplastic	Pacific Northwest of North America
Terminalia arjuna (Roxb. ex DC.) Wight & Arn.	Glycosides, antioxidant	Combretaceae	Coronary artery disease, anti-inflammatory	India
Theobroma cacao L.	Theobromine	Sterculiaceae	Heart healthy	Deep tropical region of America
Thymus vulgaris L.	Thymol	Lamiaceae	Antibacteria and antifungal	Southern Europe
Uncaria tomentosa (Willd. ex Schult.) DC.	Rhynchophylline	Rubiaceae	Anti-inflammatory, antioxidant, anticancer	South and Central America
Viscum album L.	Iscador	Viscaceae	Cancer	Europe and western, southern Asia
Voacanga africana Stapf	Vincamine, voacangine, vobasine, ibogaine, vobtusine	Apocynaceae	Leprosy, diarrhea, edema, madness, diuretic, asthma, anticancer	West Africa from Senegal to the Sudan and south to Angola
Zingiber officinale Roscoe	Zingerone, shogaol, gingerol	Zingiberaceae	Arthritis, lowering blood thinning and cholesterol, treating chemotherapy, diarrhea, nausea	South Asia

Source: Farnswoth and Soejarto, 1985; Fowler, 2006; Aftab and Vieira, 2010; da Silva et al., 2000; Noble, 1990; O'Sullivan-Coyne, 2009; Salim et al., 2008; Simon et al., 2009; Subbarayappa, 1997; Chaubal et al., 2005; Padro-Andreu et al., 2006; http://en.wikipedia.org, http://plants.usda.gov/, http://www.ars-grin.gov/

Table 2.5 Distribution of plants in the world

Family	Genus	Species	Distribution
Acanthaceae	346	4,300	Amazon, Central America, Africa, Indo-Malaysia
Aceraceae	2	113	Northern North America
Achariaceae	3	3	Africa
Actinidiaceae	3	355	Eastern Asia, Northern Australia, tropical America
Agavaceae	20	670	Arid America
Aizoaceae	100	2,400	Tropical Africa, Asia, Australia, South America, USA
Alangiaceae	1	17	Tropics and semitropics of the Old World
Alismataceae	11	95	Temperate and tropical regions of the Northern hemisphere
Aloeaceae	7	400	Arabia, South Africa, Madagascar
Alstroemeriaceae	4	200	Central and South America
Amaranthaceae	71	800	Tropical America, Africa
Amaryllidaceae	85	1,100	Worldwide
Anacardiaceae	73	850	North temperate regions of Eurasia, North America
Annonaceae	128	2,050	Old World Tropics
Apocynaceae	215	2,100	Tropical with some in the temperate zones
Aquifoliaceae	4	420	South America, North America and South Pacific
Araceae	106	2,950	Tropical and subtropical but extends into temperate regions
Araliaceae	57	800	Indo-Malaysia, tropical America
Araucariaceae	2	32	Southern hemisphere except in Africa and southeast Asia
Aristolochiaceae	7	410	Tropical and temperate zone
Asclepiadaceae	347	2,850	Pantropical, South America
Balanitaceae	1	25	Tropical Asia and Africa
Balanopacaceae	1	9	Southwest Pacific including Queensland, Australia
Balsaminaceae	2	850	Asian and African tropics
Basellaceae	4	15	Tropical America and West Indies, one species native to Asia
Bataceae	1	2	Tropical and subtropics of the New World and Hawaii
Begoniaceae	2	900	Throughout tropics, especially in South America
Berberidaceae	15	570	South America
Betulaceae	6	150	Northern hemisphere
Bignoniaceae	112	725	Northern South America
Bixaceae	3	16	American tropics
Bombacaceae	30	250	American tropics
Boraginaceae	54	2,500	Wide distribution
Bromeliaceae	46	2,100	Native to tropical and warm America
Bruniaceae	11	69	South Africa
Burseraceae	8	540	Tropical America, northeastern regions of Africa
Buxaceae	5	60	Tropics and subtropics of the Old World
Cabombaceae	2	8	Warm temperate areas
Cactaceae	130	1,650	New World
Callitrichaceae	1	17	New World and Old World
Calycanthaceae	3	9	South China, Southeast Asia, Pacific Islands, Australia, New Zealand, Chile
Calyceraceae	6	55	South America
Campanulaceae	87	1,950	Temperate and subtropical regions

(continued)

Table 2.5　Distribution of Plants in the World (Continued)

Family	Genus	Species	Distribution
Canellaceae	5	16	Caribbean, Madagascar, Africa
Cannaceae	1	25	New World tropics
Capparidaceae	45	675	Paleotropic
Caprifoliaceae	16	365	Asia, North America
Caricaceae	4	31	Tropical America, and tropical Africa
Carpinaceae	3	47	East Asia
Caryocaraceae	2	24	Tropical America
Caryophyllaceae	89	1,070	Northern temperate zone
Casuarinaceae	4	70	Australia
Celastraceae	49	1,300	Wide distribution except in the arctic regions
Centrolepidaceae	3	28	Southeast Asia to Australia
Cephalotaxaceae	1	4	Asia
Ceratophylaceae	1	2	Worldwide
Cercidiphyllaceae	1	1	China and Japan
Chenopodiaceae	120	1,300	Worldwide
Chloranthaceae	4	56	New World; tropics and semitropics of the Old World
Cistaceae	7	135	Mediterranean region, Europe, North America
Clethraceae	1	64	Tropical Asia and America
Cochlospermaceae	2	25	Tropical regions of the World
Combretaceae	20	500	Pantropical
Commelinaceae	42	620	Old and New World tropics and subtropics
Compositae (Asteraceae)	1,314	21,000	Worldwide except Antarctica
Connaraceae	20	380	Tropical
Convolvulaceae	58	1,650	Tropical and subtropical
Coriariaceae	1	30	Temperate regions of the world
Cornaceae	12	100	Temperate North America and Asia
Corylaceae (>Betulaceae)	35	1,500	Temperate Northern and South Hemisphere
Crassulaceae	35	1,500	Worldwide
Cruciferae (Brassicaceae)	300	3,700	Cool areas of the northern hemisphere
Crypteroniaceae	5	11	Asia, South Africa, South America
Cucurbitaceae	125	960	Tropical and semitropical regions of the Old and New Worlds
Cunoniaceae	26	350	Australia, New Zealand, southern South America, and southern Africa
Cupressaceae	30	140	Worldwide
Cycadaceae	1	95	East Africa to Japan and Australia
Cyclanthaceae	12	190	West Indies and South America
Cyperaceae	115	1,600	Worldwide
Cyrillaceae	3	14	Southeastern USA, Cuba, Brazil and Colombia
Daphniphyllaceae	1	25	Eastern Asia from China through Malaysia to tropical Australia
Datiscaceae	3	4	Malaysia to Australia; Western North America
Diapensiaceae	5	12	New World (Arctic regions of the northern hemisphere)
Dichapetalaceae	3	165	Tropical and subtropical regions of the world
Dilleniaceae	12	300	Australia
Dioscoreaceae	8	750	Africa, Asia, Latin America, Oceania

Table 2.5   Distribution of Plants in the World (Continued)

Family	Genus	Species	Distribution
Dipsacaceae	8	350	Mediterranean, Eurasia, and Africa
Dipterocarpaceae	16	530	Pantropical, northern south America to Africa
Droseraceae	3	200	Tropical and subtropical regions of the both hemispheres
Ebenaceae	2	500	Africa, Comoro Island, and Arabia
Elaeagnaceae	3	45	Southern Asia, Europe, and North America, Australia
Elaeocarpaceae	12	605	Madagascar, South East Asia, Eastern Australia, South America
Elatinaceae	2	50	Worldwide except Arctic region
Ephedraceae	1	40	Northern Hemisphere, Asia, South America
Equisetaceae	1	29	Cosmopolitan except for Austalia and New Zealand
Ericaceae	145	3,343	Temperate zones; some tropical region
Eriocaulaceae	10	1,200	America
Erythroxylaceae	4	240	Tropical america
Eucommiaceae	1	1	China
Euphorbiaceae	321	7,950	Tropical America, Mediterranean basin, Middle East and Africa
Eupomatiaceae	1	2	Australia, New Guinea
Fagaceae	7	1,050	Northern Hemisphere from temperate to subtropical, Pantropical
Filicopsida (Ferns)			
Adiantaceae	56	1,150	Tropical and subtropical regions of the Old World
Aspleniaceae	78	2,200	Cosmopolitan distribution epiphytes or rock plants, Hawaii
Blechnaceae	10	260	New and Old world
Cyatheaceae	2	625	Tropical regions
Davalliaceae	13	220	Warm and tropical regions-epiphytic
Dennstaedtiaceae	24	410	Pantropicals
Dicksoniaceae	5	26	Tropical America, the southwest Pacific and the island of St. Helena
Gleicheniaceae	6	125	Paleotropics and warmer regions of the New World
Graimitidiaceae	11	500	Tropical and warm south temperate zones
Hymenophyllaceae	33	460	Tropical rain forest, temperate rain forest (New Zealand)
Isoetaceae	2	150	Cosmopolitan aquatics except in the islands of the Pacific
Marattiaceae	7	100	East-Asian regions, tropical and warm regions
Marsileaceae	3	70	Worldwide
Ophioglossaceae	4	65	Tropical and temperate regions
Osmundaceae	4	25	Tropical and temperate regions-ornamentals
Parkeiaceae (=Ceratopteridaceae)	2	4	Tropical, subtropical and warm temperate regions-worldwide
Plagiogyriace	1	37	Asia and Americas (tropical and subtropical regions)
Polypodiaceae	60	1,000	Tropical regions
Psilotaceae	2	9	Tropical and subtropical-Hawaii
Schizaceae	4	150	Warm and tropical areas
Thelypteridaceae	30	900	Tropical and temperate regions
Thrysopteridaceae	3	20	Macronesia

(continued)

Table 2.5 Distribution of Plants in the World (Continued)

Family	Genus	Species	Distribution
Flagellariaceae	2	4	Tropical and subtropical regions of the Old World and Australia
Fouquieriaceae	1	11	Mexico
Frankeniaceae	1	100	Worldwide primarily Mediterranean
Fumariaceae	20	575	Northern hemisphere, South Africa
Garryaceae	2	28	Western North and Central America, eastern Asia
Gentianaceae	87	1,500	Worldwide distribution
Geraniaceae	14	800	Temperate and tropical regions of both hemispheres
Gesneriaceae	150	3,200	Tropical and subtropical regions-Old and New World
Ginkgoaceae	1	1	China
Globulariaceae	10	250	Mediterranean
Gnetaceae	1	35	Tropical regions of Old and New World
Goodeniaceae	16	430	Australasia
Gramineae (Poaceae)	635	10,000	Widely distributed
Grossulariaceae	23	340	Temperate regions of the Northern Hemisphere
Guttiferae (=Clusiaceae)	37	1,610	Tropical Central and South America
Haemodoraceae	16	85	Tropical North and South America, Australia, New Guinea, South Africa
Haloragaceae	8	145	Australia, Southern Hemisphere
Hamamelidaceae	28	90	Subtropics of Southeast Asia
Hernandiaceae	4	68	Pantropical
Himantandraceae	1	2	Eastern Malaysia to northern Australia
Hippocastanaceae	2	15	North and South America
Humiriaceae	8	50	Neotropical; one species in tropical West Africa
Hydrocharitaceae	16	90	Warm fresh and salt waters of the world
Hydrophyllaceae	20	300	Western United States
Hypoxidaceae	7	120	America, Asia, Africa, Australia
Icacinaceae	55	400	Tropical to temperate regions
Iridaceae	92	2,000	South and Central Africa, the eastern Mediterranean, Eurasia, North America
Juglandaceae	8	59	Central and South America, Temperate Asia to Java and New Guinea
Julianaceae	2	5	Peru north to Central America
Juncaceae	8	400	Europe, South America
Juncaginaceae	4	18	Temperate and cold regions-worldwide
Labiatae (Lamiaceae)	236	7,200	Cosmopolitan distribution
Lacistemataceae	2	14	Mesoamerica and South America (excluding Chile and Argentina)
Lardizabalaceae	8	50	Japan to China, Chile, western Argentina
Lauraceae	55	4,000	Tropical southeast Asia, Africa, Mediterranean, central Chile
Lecythidaceae	20	280	South America, Madagascar
Leguminosae (Fabaceae)	730	19,400	Worldwide distribution
Lemnaceae	6	30	Worldwide distribution
Lentibulariaceae	4	245	Worldwide distribution
Liliaceae	294	4,500	Warm temperate and tropical regions

Table 2.5   Distribution of Plants in the World (Continued)

Family	Genus	Species	Distribution
Limnanthaceae	2	8	North America
Linaceae	15	300	Cosmopolitan distribution
Loasaceae	20	260	Western South America, Africa, Arabia
Loganiaceae	29	600	South America
Loranthaceae	75	1,000	Tropical to temperate regions (Australia, Central to South America)
Lycopodiaceae	5	300	Cosmopolitan distribution
Lythraceae	31	580	American tropics
Magnoliaceae	7	225	Tropical to warm temperate regions (Old to New Worlds)
Malpighiaceae	75	1,300	Old World to New World
Malvaceae	200	3,300	Cosmopolitan distribution
Marantaceae	31	550	American tropics
Marcgraviaceae	5	108	Tropical America, Australasia, Sub-Saharan Africa
Martyniaceae	18	95	New World Tropical region
Mayacaceae	1	4	South America, West Africa
Melastomataceae	215	4,750	South America-tropics
Meliaceae	51	575	Tropical and subtropical regions-temperate China, Southeast Australia
Melianthaceae	2	8	Tropical and Southern Africa, Chile
Mendonciaceae	2	80	Tropical America and Africa
Menispermaceae	78	570	Tropical to temperate-North America and eastern Asia
Menyanthaceae	5	70	Worldwide distribution
Monimiaceae	35	450	Warm and tropical regions-Madagascar
Moraceae	48	1,200	Tropical and warm to some temperate regions
Moringaceae	1	14	Asia and Africa (Madagascar, Arabia to India)
Musaceae	2	42	Old World tropics
Myoporaceae	5	220	Australia and the Indian ocean
Myricaceae	4	50	Cosmopolitan distribution
Myristicaceae	19	440	Northern Amazon
Myrothamnaceae	1	2	South Africa and Madagascar
Myrsinaceae	37	1,250	Old to New World
Myrtaceae	150	5,650	Australia, Tropical to warm temperate regions
Najadaceae	1	35	Cosmopolitan distribution
Nelumbonaceae	1	2	Eastern Asia, North America
Nepenthaceae	1	130	Seychelles and Madagascar to Australia and New Caledonia
Nyctaginaceae	34	350	Tropical and subtropical of both hemispheres
Nymphaeaceae	8	70	Worldwide distribution
Nyssaceae	5	10	North America, China, southeast Asia
Ochnaceae	53	600	South America-Brazil
Olacaceae	14	100	Tropical regions-America, Africa and Asia
Oleaceae	29	900	Temperate and tropical Asia
Oliniaceae	1	8	Southern Africa
Onagraceae	24	650	Warm and temperate America
Opiliaceae	10	30	Tropical woody family
Orchidaceae	700	30,000	Cosmopolitan-Asia, South America, Africa, Oceania, Europe

(continued)

Table 2.5   Distribution of Plants in the World (Continued)

Family	Genus	Species	Distribution
Orobanchaceae	90	2,000	Northern hemisphere, subtropicals of the Old and New World
Oxalidaceae	8	900	Tropical and subtropical regions of the world
Palmae (=Arecaceae)	202	2,600	Tropical and subtropical regions of the world
Pandanaceae	3	700	Old Word tropics, New Zealand
Papaveraceae	26	250	Cosmopolitan family-temperate and subtropical regions
Passifloraceae	18	530	Tropical and warm temperate regions of America
Pedaliaceae	18	95	Warm Tropical areas-Old World
Penaeaceae	7	25	Southern Africa
Philydraceae	4	5	Southeast Asia to Australia
Phytolaccaceae	18	65	Tropical and warm areas of the America
Pinaceae	11	250	South to Central America, West Indies, Southwest China to Sumatra and Java
Piperaceae	14	3,610	Southeast Asia, Pacific Islands, Central and South America
Pittosporaceae	9	240	Old World family, especially Australia, Malaysia
Plantaginaceae	90	1,700	Cosmopolitan family
Platanaceae	1	7	Northern hemisphere
Plumbaginaceae	24	800	Cosmopolitan distribution
Podocarpaceae	19	156	Asia, tropical Africa, Central Africa
Podostemaceae	50	275	Tropical Asia, America
Polemoniaceae	25	400	Western and northern America, Eurasia
Polygalaceae	17	1,000	Cosmopolitan family but absent in the western Pacific
Polygonaceae	50	1,200	Cosmopolitan family
Pontederiaceae	9	31	North America
Portulacaceae	29	500	Southern hemisphere in Africa, Australia, and South America
Potamogetonaceae	6	120	Sub-cosmopolitan family
Primulaceae	24	800	Northern hemisphere
Proteaceae	80	2,000	Australia and South Africa
Punicaceae (>Lythraceae)	1	2	Southeastern Europe to the Himalayas
Pyrolaceae (>Ericaceae)	4	42	North temperate zone to Sumatra
Quinaceae	4	44	Amazonia
Rafflesiaceae	8	50	Tropical family but few temperate zones
Ranunculaceae	60	1,750	North temperate family-Worldwide
Resedaceae	6	75	Old World especially of the temperate zones
Restionaceae	50	520	Australia and South Africa
Rhabdodendraceae	1	6	South American trees
Rhamnaceae	60	900	Cosmopolitan-tropical and warm regions
Rhizophoraceae	16	130	Old World tropics
Rosaceae	107	3,100	Subcosmopolitan-warm temperate
Rubiaceae	611	13,000	Pantropical and subtropical
Rutaceae	161	1,700	Cosmopolitan family-tropical
Sabiaceae	3	160	Warm temperate regions of Southeast Asia, tropical Africa, and America
Salicaceae	52	453	Subcosmopolitan-northern hemisphere
Salvadoraceae	3	12	Southeast Asia, Africa-Madagascar

Table 2.5 Distribution of Plants in the World (Continued)

Family	Genus	Species	Distribution
Santalaceae	36	1,000	Warm and tropical regions-cosmopolitan
Sapindaceae	150	2,000	Temperate regions
Sapotaceae	107	1,000	Tropical and few temperate-Pantropicals
Sarraceniaceae	3	15	Eastern and western North America, northeastern South America
Saururaceae	5	7	North America and eastern and southern Asia
Saxifragaceae	36	475	Subcosmopolitan distribution
Scrophulariaceae	275	5,000	Cosmopolitan distribution
Selaginellaceae	1	700	Cosmopolitan distribution
Simaroubaceae	22	170	Tropical to temperate Asia
Smilacaceae	10	225	Tropical and warm zones-Southern hemisphere
Solanaceae	90	2,800	Subcosmopolitan - Andean South America
Sonneratiaceae	2	7	Old World species
Sparangiaceae	1	12	North temperate zone and south to Australia and New Zealand
Sphenocleaceae	1	2	Tropical family, Indonesia, areas of Amazonia
Stachyuraceae	1	6	Asian family extends to Japan to the Himalayas
Stackhousiaceae	3	28	Australasia-Australia, New Zealand, the islands of the Pacific
Staphyleaceae	5	27	Northern Hemisphere and South America
Stemonaceae	4	32	Eastern Asia, Indomalaysia, south to tropical Australia, eastern North America
Sterculiaceae	72	1,500	Warm and tropical regions-few temperate zones
Stylidaceae	5	150	Southern and southeastern Asia, Australasia, South America
Styracaceae	12	165	Warm and tropical areas of Americas, Mediterranean, Southeast Asia
Symplocaceae	2	320	Tropical and warm America, eastern Old World, Australia
Taccaceae	1	31	Tropical family
Tamaricaceae	5	78	Eurasian, Africa, Asia
Taxaceae	12	30	Temperate regions-New Caledonia
Taxodiaceae	10	14	North America, Central Asia, Tasmania
Theaceae	40	520	Tropical family
Theophrastaceae	7	95	Tropical regions of the Americas
Thymelaeaceae	50	898	Australia, tropical Africa
Tiliaceae	50	725	Subcosmopolitan family
Tremandraceae	3	43	Australia
Trigoniaceae	4	35	Madagascar, Tropical America, Western Malaysia
Trochodendraceae	2	2	Asia-Korea, Japan to Taiwan
Tropaeolaceae	3	88	Central and South America
Turneraceae	10	120	Warm and tropical areas of America and Africa
Typhaceae	1	12	Cosmopolitan marsh plants
Ulmaceae	18	150	Tropical and north tmperate areas
Umbelliferae (=Apiaceae)	418	3,100	Cosmopolitan-north temperate zone and tropical mountains
Urticaceae	79	1,050	Tropical to temperate species
Valerianaceae	17	400	Cosmopolitan family-north temperate and the Andes of South America

(continued)

Table 2.5 Distribution of Plants in the World (Continued)

Family	Genus	Species	Distribution
Velloziaceae	6	252	South America, Africa, Madagascar, southern Arabia
Verbenaceae	91	1,900	Tropical regions
Violaceae	28	830	Cosmopolitan species
Vitaceae	13	800	Tropical to warm regions
Vochysiaceae	8	210	Tropical America, West Africa
Winteracea	9	120	South America, Australia, New Guinea, Southwestern Pacific, Madagascar
Xanthophyllaceae	1	39	Indo-Malaysian
Xanthorrhoeaceae	1	66	Australia to New Guinea and New Caledonia
Xyridaceae	5	300	Tropical and warm area, temperate regions
Zamiaceae	8	150	America, Africa, Australia
Zingiberaceae	53	1,300	Tropical Africa, Asia, and Americas
Zygophyllaceae	27	285	Tropical and warm regions
Totals	14,066	287,685	

Source: Lloyd, R. M. 1974. Systematics of the genus *Ceratopteris* Brongn. (Parkeriaceae) II. Taxonomy. Brittonia 26(2): 139–160; Benedict, R. C. 1909. The *genus Ceratopteris*: A Preliminary revision. Bull. Torrey Botanical Club 36: 463–476.; http://www.efloras.org/, http://en.wikipedia.org/, http://www.britannica.com/, http://www.mobot.org/, http://www.botany.hawaii.edu/, http://deltaintkey.com/, http://www.ncbi.nlm.nih. gov/; Kornaš, J., J. Dzwonka, K. Harmata, and A. Pacyna. 1982. Biometrics and numerical taxonomy of the genus Actiniopteric (Adiantaceae, Fillicopsida) in Zambia. Bulletin du Jardin Botanique National de Belgique 52: 265–309.; Conant, D. S., L. A., Raubeson, D. K. Attwood, and D. B. Stein. 1995. The relationships of Papuasian Cyatheaceae to new world tree ferns. American Fern Journal 85: 328–340.; Raffauf, R. F. 1996. *Plant Alkaloids: A Guide to their Discovery and Distribution*. New York: Haworth Press.

the moist and the dry forests of Madagascar, New Caledonia, and Hawaii; Brazil's Atlantic coastal moist forest; and the fynbos of South Africa's cape floral kingdom. The United Nations Educational, Scientific, and Cultural Organization (UNESCO) provides long-term protection of landscapes, ecosystems, and species. According to a list compiled in 2003 by the United Nations, 102,102 areas are protected, worldwide, covering 18.8 million km^2. This amounts to 12.65% of the earth's land surface—an area equivalent to South America (Chape et al. 2003). Thus, we must balance utilization and conservation (Akerele, Heywood, and Synge 1991; Alexiades 1996; Freese 1998; Sheldon, Balick, and Laird 1997).

Overharvesting of plants has placed many medicinal plant species at risk of extinction. Commercial exploitation has also sometimes led to traditional medicines becoming unavailable to the indigenous people who have depended on them for centuries or millennia. Already, about 15,000 medicinal plant species may be threatened with extinction worldwide (Roberson 2008). In 1995, Mendelsohn and Balick (1995) estimated that "each new drug is worth an average of $94 million to a private drug company and $449 million to society as a whole" and that "screening of all tropical species should be worth about $3–4 billion to a private pharmaceutical company and as much as $147 billion to society as a whole." Various guides on conservation of medicinal plants have been published (Freese 1998; Fuller 1991; Given 1994; Jenkins and Oldfield 1992; Lewington 1993; Marshall 1998). In addition, organizations such as United Plant Savers have been formed to promote not only conservation but also ecologically conscious cultivation of at-risk medicinal plants (Cech 2002; Gladstar and Hirsch 2000).

In order to reduce overharvesting of wild plants, more selection and breeding must be done for both increased and stable yields of pharmacologically active constituents under cultivation. Johnson and Franz (2002) are cited as a good example of research in this direction. As compendia of guidelines for cultivation and harvesting of medicinal plants, the books by Hay and Waterman (1993), Hornok (1992), Kumar et al. (1993), Peter (2001, 2004, 2006), and Weiss (1997, 2002) should be studied.

Table 2.6 Activities of the Organizations Responsible for Coordinating Medicinal Plants

Organization	Activities
ASCOPAP African Scientific Co-operation on Phytomedicine and Aromatic Plants http://www.nextdaysite.net/bioresourses/index.htm/	This institution was initiated in October, 1995 at Douala, Cameroon. The mission of ASCOPAP is to promote and facilitate the research, development and commercialization of safe, efficient, affordable, standardized phytomedicines based on sustainable utilization of African plant resources. It is an internet home for the African pharmacopoeia.
EFMC European Federation for Medicinal Chemistry http://www.efmc.info/	An independent association founded in 1970, organizes biennial international symposium in medicinal chemistry.
ESCOP European Scientific Co-operation on Phytotherapy http://www.escop.com/	This organization was founded in 1989. The objectives are: • To develop a coordinated scientific framework to assess phytomedicines • To promote the acceptance of phytomedicine, especially within general medicinal practice • To support and initiate clinical and experimental research in phytotherapy • To produce reference monographs on the therapeutic use of plant drugs
FAO Food and Agricultural Organization of the United Nations http://www.fao.org/	This organization was founded in 1945. The objectives are: • Defeat hunger • FAO coordinates with other medicinal and aromatic plants research institutes • Commissions report on use of medicinal plants
FIADREP International Federation of Associations of Defense in Phytotherapy Research and Training www.phyto2000.org/fiadrepus.html/	• To gather doctors, chemists and other members of equivalent scientific disciplines who, according to the medicinal tradition, carry on with researches and practices on Phytotherapy and Aromatherapy and on all natural techniques, biological and physical which will help health to be ameliorated • To promote research, practice, and education worldwide in Phytotherapy • to improve and extend the international accumulation of scientific and practical knowledge in the field of Phytotherapy • to support all appropriate measures that will secure optimum protection for those who use herbal medicinal products • To unify and coordinate, in each countries member, in Europe and all over the world, the defense of the use of medicinal plants of a pharmacomedicinal quality • To pursue research and practices in Phytotherapy and to spread all over the world an education in accordance with the various points of the present statutes. • To bring up to date yearly, the criteria of the pre-established clinical consensus which are necessary to join the FIADREP, and to publish an opuscule which describes those criteria

(continued)

Table 2.6   Activities of the Organizations Responsible for Coordinating Medicinal Plants

Organization	Activities
GIFTS The Global Initiative for Traditional Systems of Health http://www.giftsofhealth.org/html/history.html/	This organization was founded in 1993. • Has outreach programs in Asia and Africa It focuses on policy, education and research in the are of Malaria, HIV/AIDS and TB as priority diseases, traditional medicine in managing common ailments, especially skin conditions, refugees and their utilization of traditional health services and intellectual property rights, traditional resource rights and conservation and sustainable use of medicinal plant resources
IDRC International Development Research Centre http://www.idrc.ca/en/ev-1-201-1-DO_TOPIC.html/	This organization was founded in 1970 in Canada. • Activities in 60 countries and has several projects: • Medicinal and Aromatic Plants Program in Asia (www.medplant.net) • Preserve medicinal plants of Rwanda • Latin America
ICMAP International Council for Medicinal and Aromatic Plants http://www.icmap.org/	This organization was established in 1993. • Promoting international understanding and cooperation between national and international organizations on the role of medicinal and aromatic plants in science, medicine, and industry • Improve relation of exchange of knowledge
IOCD International Organization for Chemical Sciences in Development http://www.iocd.org/index.shtml/	This organization was established in 1981. • Synthesized of the first steroid oral contraceptive; now well-known birth control pil • Promoting identification of natural products • Collaborating with the responsible groups in South Africa, Kenya, Uganda, and Guatemala, in making plans to, eventually, implement bio-prospecting
IUCN International Union for the Conservation of Nature http://www.iucn.org/	This organization was founded in 1948. • The largest professional global conservation network; about 15,000 species of medicinal plants are globally threatened due to loss of habitat, overexploitation, invasive species and pollution. • Manage natural products • Conservation of medicinal plants and traditional knowledge
TRAFFIC Joint wildlife monitoring program of the World Wildlife Fund and International Union for the Conservation of Nature http://www.traffic.org/	This organization was established in 1976. • Mission is the wildlife trade monitoring network, works to ensure that trade in wild plants and animals is not a threat to the conservation of nature • Nine regional programs (Central Africa, East Asia, East/Southern Africa, Europe, North America, Oceana, South America, South Asia, Southeast Asia) • Promote collection and conservation of medicinal plants threatened by over-collection, and deforestation

UNESCO
United Nations, Educational, Scientific and Cultural Organizations
http://www.unesco.org/new/en/unesco/

This organization was created in 1945.
• Develop guidelines and training on conservation of medicinal plants worldwide
• Organize international symposium on medicinal plants and spices, and other natural products
• Preservation and publication of traditional knowledge of medicinal plants of African countries

UNIDO
United Nations Industrial Development Organization
http://www.unido.org/

This organization was established in 1966.
• To establish 2020 vision for Andean medicinal plants
• Expansion and upgrading of small sized pharmaceuticals enterprises in selected developing countries of Asia and Africa.
• Manufacturing generic drugs affordable to poor people

WHO
World Health Organization
http://www.who.int/en/

This organization was founded in 1948.
• Publication of books, monographs, bulletins, flyers on medicinal plants
• Formulation of national policies on traditional medicine
• Combating communicable diseases
• Building healthy communities and populations
• Education

WWF
world Wildlife Fund
http://www.worldwildlife.org/

This organization was founded in 1961.
• The mandate is conservation of nature and natural resources including medicinal plants
• Develop guidelines for global trade on medicinal plants

The term "biopiracy" has been coined to describe the practice of private companies patenting traditional remedies from the wild and selling them at a vast profit, often allowing little or none of that profit to go back to the country or indigenous and local communities of origin. Neem tree (*Azadirachta indica* A. Juss.), a south Asian relative to mahogany, has yielded extremely effective and profitable germicide, fungicide, and other compounds (see, for example, Tiwari et al. 2010). By 1995, at least 50 patents had been granted to U.S. and international corporations for products extracted from this species (Torrance 2000).

Man-made disasters in Iraq and Afghanistan, such as war, may cause many medicinal plants to become extinct because of lack of plant exploration, preservation, and maintenance. Younos et al. (1987) identified 215 medicinal plants in Afghanistan whose current status is unknown.

The quality and efficacy of the Chinese traditional medicines, Ayurvedic medicines, and natural health products should be monitored through private and public agencies after extensive clinical trials. Molecular technologies can help quality control of plant-based drugs and extracts (Chavan, Joshi, and Patwardhan 2006). Bioactive compounds of the Ayurvedic plants need to be determined (Khan and Balick 2001; Samy, Pushparaj, and Gopanakrishnakone 2008).

ACKNOWLEDGMENTS

The work of A. Lebeda in this chapter was supported by the project MSM 6198959215 (Ministry of Education, Youth and Sports of the Czech Republic).

REFERENCES

Addo-Fordjour, P., A. K. Anning, E. J. D. Belford, et al. 2008. Diversity and conservation of medicinal plants in the Bomaa community of the Brong Ahafo region, Ghana. *Journal of Medicinal Plants Research* 2:226–233.

Adeniyi, A. A., G. S. Brindley, J. P. Pryor, and D. J. Ralph. 2007. Yohimbine in the treatment of orgasmic dysfunction. *Asian Journal of Andrology* 9:403–407.

Aftab, N., and A. Vieira. 2010. Antioxidant activities of curcumin and combinations of this curcuminoid with other phytochemicals. *Phytotherapy Research* 24: 500–502.

Aguilar-Støen, M., and S. R. Moe. 2007. Medicinal plant conservation and management: Distribution of wild and cultivated species in eight countries. *Biodiversity Conservation* 16:1973–1981.

Ahmad, M., M. A. Khan, S. K. Marwat, et al. 2009. Useful medicinal flora enlisted in Holy Qur'an and Ahadith. *American-Eurasian Journal of Agricultural and Environmental Science* 5:126–140.

Ahmed, M. S., G. Honda, and W. Miki. 1979. *Herb drugs and herbalists in the Middle East.* Tokyo: Institute for Study of Language and Culture of Asia and Africa.

Ajmal, B., and S. Ajmal. 2005. *Handbook of medicinal & aromatic plants of North East India.* Woodstock, GA: Spectrum Publishers.

Akerele, O., V. Heywood, and H. Synge, eds. 1991. *The conservation of medicinal plants.* Cambridge, England: Cambridge University Press.

Alexiades, M. N., ed. 1996. *Selected guidelines for ethnobotanical research: A field manual.* New York: New York Botanical Garden.

Ali, B. H., N. A. Wabel, and G. Blunden. 2005. Phytochemical, pharmacological and toxicological aspects of *Hibiscus sabdariffa* L.: A review. *Phytotherapy Research* 19:369–375.

Amarasingham, R. D., N. G. Bisset, A. H. Millard, et al. 1964. A phytochemical survey of Malaya. Part III. Alkaloids and saponins. *Economic Botany* 18 (3): 270–278.

Anderson, E. F. 1986. Ethnobotany of hill tribes of northern Thailand. I. Medicinal plants of Akha. *Economic Botany* 40:38–53.

———. 1993. *Plants and people of the Golden Triangle: Ethnobotany of the Hill Tribes of Northern Thailand.* Portland, OR: Dioscorides Press.

Asres, K., A. Seyoum, C. Veeresham, F. Bucar, and S. Gibbons. 2005. Naturally derived anti-HIV agents. *Phytotherapy Research* 19:557–581.

Awang, D. V. C., B. A. Dawson, J. Fillion, et al. 1993. Echimine content of commercial comfrey (*Symphytum* spp.—Boraginaceae). *Journal of Herbs Spices and Medicinal Plants* 2 (1): 21–34.

Ayensu, E. S. 1978. *Medicinal plants of West Africa*. Algonac, MI: Reference Publications.

————. 1981. *Medicinal plants of the West Indies*. Algonac, MI: Reference Publications.

Balick, M. J., E. Elisabetsky, and S. A. Laird, eds. 1996. *Medicinal resources of the tropical forest: Biodiversity and its importance to human health*. New York: Columbia Univ. Press.

Barceloux, D. G. 2008. *Medical toxicology of natural substances: Foods, fungi, medicinal herbs, plants, and venomous animals*. Hoboken, NJ: John Wiley & Sons Inc.

Beck, L. 2005. *De materia medica: Dioscorides Pedanius, of Anazarbos*. New York: Olms-Weidmann.

Bellakhdar, J., R. Claisse, J. Fleurentin, et al. 1991. Repertory of standard herbal drugs in the Moroccan pharmacopoeia. *Journal of Ethnopharmacology* 35:123–143.

Bellakhdar, J., G. Honda, and W. Miki. 1982. *Herb drugs and herbalists in the Maghrib*. Tokyo: Institute for Study of Language and Culture of Asia and Africa.

Bensky, D., and A. Gamble. 1993. *Chinese herbal medicine*, rev. ed. Seattle, WA: Eastland Press.

Betz, J. M., R. M. Eppley, W. C. Taylor, et al. 1994. Determination of pyrrolozidine alkaloids in commercial comfrey products (*Symphytum* sp.). *Journal of Pharmaceutical Sciences* 83:649–653.

Bhattacharjee, S. K. 2001. *Handbook of medicinal plants*, 3rd rev. ed. Jaipur, India: Pointer Publ.

Bisset, N. G., and M. Wichtl, eds. 2001. *Herbal drugs and phytopharmaceuticals*, 2nd ed. Boca Raton, FL: CRC Press.

Blumenthal, M., W. R. Busse, A. Goldberg, et al. 1998. *The complete German Commission E monographs: Therapeutic guide to herbal medicine*. Austin, TX: American Botanical Council.

Blumenthal, M., A. Goldberg, and J. Brinckmann. 2000. *Herbal medicine: Expanded Commission E monographs*. Austin, TX: American Botanical Council.

Boulos, L. 1983. *Medicinal plants of North Africa*. Algonac, MI: Reference Publications.

Bradacs, G., L. Maes, and J. Heilmann. 2010. *In vitro* cytotoxic, antiprotozoal and antimicrobial activities of medicinal plants from Vanuatu. *Phytotherapy Research* 24:800–809.

Brauchli, J., J. Lüthy, U. Zweifel, et al. 1982. Pyrrolozidine alkaloids from *Symphytum officinale* L. and their percutaneous absorption in rats. *Experientia* 38:1085–1087.

Breedlove, D. E., and R. M. Laughlin. 1993. *The flowering of man: A Tzotzil botany of Zinacantán*. Washington, D.C.: Smithsonian Inst. Press.

Brooker, S. G., R. C. Cambie, and R. C. Cooper. 1981. *New Zealand medicinal plants*. Auckland: Heinemann.

Brouwer, N., Q. Liu, D. Harrington, et al. 2005. An ethnopharmacological study of medicinal plants in New South Wales. *Molecules* 10:1252–1262.

Brower, L. P., J. V. Z. Brower, and J. M. Corvine. 1967. Plant poisons in a terrestrial food chain. *Proceedings of National Academies of Science* 57:893–898.

Busia, K. 2005. Medical provision in Africa—Past and present. *Phytotherapy Research* 19:919–923.

Cáceres, A. 1996. *Plantas de uso medicinal en Guatemala*. Univ. San Carlos Guatemala.

Cech, R. 2002. *Growing at-risk medicinal herbs: Cultivation, conservation and ecology*. Portland, OR: Horizon Herbs.

Chape, S., S. Blyth, L. Fish, et al. 2003. *2003 United Nations list of protected areas*. Cambridge: IUCN.

Chapman, K., and N. Chomchalow. 2005. Production of medicinal plants in Asia. *Acta Horticulturae* 679:45–59.

Chavan, P., K. Joshi, and B. Patwardhan. 2006. DNA microarrays in herbal drug research. *Evidence-Based Complementary Alternative Medicine* 3 (40):447–457.

Chen, F.-P., T.-J. Chen, Y.-Y. Kung, et al. 2007. Use frequency of traditional Chinese medicine in Taiwan. *BMC Health Services Research* 7:26.

Chin, W. Y., and H. Keng. 1992. *Chinese medicinal herbs*. Sebastopol, CA: CRCS Publ.

Chaubal, R., P. V. Pawar, G. D. Hebbalkar, V. B. Tungikar, V. G. Puranik, V. H. Deshpande, and N. R. Deshpande. 2005. Larvicidal activity of *Acacia nilotica* extracts and isolation of d-pinitol abioactive carbohydrate. *Chemistry and Biodiversity* 2:684–688.

Cho, S.-H., C. H. Lee, M.-R. Jang, et al. 2007. Aflatoxins contamination in spices and processed spice products commercialized in Korea. *Food Chemistry* 107:1283–1288.

Chuakul, W., P. Saralamp, W. Paonil, R. Temsiririrkkul, and T. Clayton. 1997. *Medicinal plants in Thailand.* Bangkok: Mahidol Univ.

Coe, F. G., and G. J. Anderson. 1996. Screening of medicinal plants used by the Garifuna of Eastern Nicaragua for bioactive compounds. *Journal of Ethnopharmacology* 53:29–50.

Collins, D. J., C. C. J. Culvenor, J. A. Lamberton, J. W. Loder, and J. R. Price. 1990. *Plants for medicine: A chemical and pharmacological survey of plants in the Australian region.* Melbourne: CSIRO.

Craker, L. E., and Z. E. Gardner. 2006. Medicinal plants and tomorrow's pharmacy. In *Medicinal and aromatic plants*, ed. R. J. Bogers, L. E. Craker, and D. Lange, 29–41. Dordrecht, Netherlands: Springer.

Craker, L. E., Z. Gardner, and S. C. Etter. 2003. Herbs in American fields: A horticultural perspective of herb and medicinal plant production in the United States, 1903 to 2003. *HortScience* 38:977–983.

Culvenor, C. C. J., M. Clarke, J. A. Edgar, et al. 1980a. Structure and toxicity of the alkaloids of Russian comfrey (*Symphytum* × *uplandicum* Nyman), a medicinal herb and item of human diet. *Experientia* 36:377–379.

———. 1980b. The alkaloids of *Symphytum* × *uplandicum* (Russian comfrey). *Australian Journal of Chemistry* 33:1105–1113.

Darshan, S., and R. Doreswamy. 2004. Patented anti-inflammatory plant drug development from traditional medicine. *Phytotherapy Research* 18:343–357.

de Silva, A. P., R. Rocha, C. M. L. Silva, L. Mira, M. Filomena Duarte, and M. H. Florêncio. 2000. Antioxidants in medicinal plant extracts: A research study of the antioxidant capacity of *Crataegus, Hamamelis* and *Hydrastis. Phytotherapy Research* 14: 612–616.

de Boer, H., and V. Lamxay. 2009. Plants used during pregnancy, childbirth and postpartum healthcare in Lao PDR: A comparative study of the Brou, Saek and Kry ethnic groups. *Journal of Ethnobiology and Ethnomedicine* 5:25.

De Mesquita, M. L., J. E. de Paula, C. Pessoa, et al. 2009. Cytotoxic activity of Brazilian cerrado plants used in traditional medicine against cancer cell lines. *Journal of Ethnopharmacology* 123:439–445.

De Montellano, B. R. O. 1990. *Aztec medicine, health, and nutrition.* New Brunswick, NJ: Rutgers University Press.

De Smet, P. A. G. M. 1999. *Herbs, health & healers: Africa as ethnopharmacological treasury.* Berg en Dal, Netherlands: Afrika Museum.

Dev, S., and O. Koul. 1997. *Insecticides of natural origin.* Amsterdam: Harwood Academy Publishers.

Dièye, A. M., A. Sarr, S. N. Diop, et al. 2008. Medicinal plants and the treatment of diabetes in Senegal: Survey with patients. *Fundamental and Clinical Pharmacology* 22:211–216.

Dokosi, O. B. 1998. *Herbs of Ghana.* Accra: Ghana University Press.

dos Santos Júnior, H. M., D. F. Oliveira, D. A. de Carvalho, et al. 2010. Evaluation of native and exotic Brazilian plants for anticancer activity. *Journal of Natural Medicine* 64:231–238.

Drews, J. 2000. Drug discovery: A historical prospective. *Science* 287:1960–1964.

Duke, J. A., and E. S. Ayensu. 1985. *Medicinal plants of China.* Algonac, MI: Reference Publications.

Duke, P., A. Duke, and J. L. duCellier. 2008. *Duke's handbook of medicinal plants of the Bible.* Boca Raton, FL: CRC Press, Taylor & Francis Group.

Eltohami, M. S. 1997. *Medicinal and aromatic plants in Sudan.* Proceedings of International Expert Meeting for Prod. Div. FAO Fore. Dept. FAO Reg. Office Near East, Cairo, Egypt.

Endringer, D. C., Y. M. Valadares, P. R. V. Campana, et al. 2010. Evaluation of Brazilian plants on cancer chemoprevention targets *in vitro. Phytotherapy Research* 24:928–933.

Fabricant, D. S., and N. R. Farnsworth. 2001. The value of plants used in traditional medicine for drug discovery. *Environmental Health Perspectives* 109 (supplement 1): 69–75.

Farnsworth, N. R., and N. Bunyapraphatsara. 1992. *Thai medicinal plants recommended for primary health care systems.* Bangkok: Prachadhon Co.

Farnsworth, N. R., and D. D. Soejarto. 1985. Potential consequence of plant extinction in the United States on the current and future availability of prescription drugs. *Economic Botany* 39:231–240.

Forschungsvereinigung der Arzneimittel-Herstell, E. V. 2003. Standard operating procedures for inspecting cultivated and wild crafted medicinal plants. *Journal of Herbs Spices and Medicinal Plants* 10 (3): 109–125.

Fortin, D., M. Lô, and G. Maynart. 1990. *Plantes médicinales du Sahel.* Montréal: CECI.

Fowler, M. W. 2006. Plants, medicines and man. *Journal of Science of Food and Agriculture* 86:1797–1804.

Freese, C. H., ed. 1998. *Harvesting wild species: Implications for biodiversity conservation.* Washington, D.C.: Island Press.

Fuller, D. O. 1991. *Medicine from the wild: An overview of the U.S. native medicinal plant trade and its conservation implications.* Baltimore, MD: WWF Publ.

Furuya, T., and M. Hikichi. 1971. Alkaloids and triterpenoids of *Symphytum officinale. Phytochemistry* 10:2217–2220.

Gaikwad, J., V. Khanna, S. Vemulpad, et al. 2008. CMKb: A Web-based prototype for integrating Australian aboriginal customary medicinal plant knowledge. *BMC Bioinformatics* 9 (suppl. 12): S25.

Garg, M., and C. Garg. 2008. Scientific alternative approach in diabetes—An overview. *Pharmacognosy Review* 2 (4): 284–301.

Ghazanfar, S. A. 1994. *Handbook of Arabian medicinal plants.* Boca Raton, FL: CRC Press.

Ghazanfar, S. A., and A. M. A. Al-Sabahi. 1993. Medicinal plants of Northern and Central Oman (Arabia). *Economic Botany* 47:89–98.

Given, D. R. 1994. *Principles and practice of plant conservation.* Portland, OR: Timber Press.

Gladstar, R., and P. Hirsch. 2000. *Planting our future: Saving our medicinal herbs.* Rochester, VT: Healing Arts Press.

Gomes, M. F. P. L., C. de Oliveira Massoco, J. G. Xavier, et al. 2007. Comfrey (*Symphytum officinale* L.) and experimental hepatic carcinogenesis: A short-term carcinogenesis model study. *eCAM* 7 (2): 197–202.

Graham, J. G., M. L. Quinn, D. S. Fabricant, et al. 2000. Plant used against cancer—An extension of the work of Jonathan Hartwell. *Journal of Ethnopharmacology* 73:347–377.

Guédon, D., M. Brum, J.-M. Seigneuret, et al. 2008. Impurities in herbal substances, herbal preparations and herbal medicinal products, IV. Heavy (toxic) metals. *Natural Product Communications* 12:2107–2122.

Habbu, P. V., H. Joshi, and B. S. Patil. 2007. Potential wound healers from plant origin. *Pharmacognosy Review* 1 (2): 271–282.

Halioua, B., and B. Ziskind. 2005. *Medicine in the days of the pharaohs.* Transl. M. B. DeBevoise. Cambridge, MA: Beknap Press.

Harnischfeger, G. 2000. Proposed guidelines for commercial collection of medicinal plants. *Journal of Herbs Spices and Medicinal Plants* 7 (1): 43–50.

Hay, R. K. M., and P. G. Waterman, eds. 1993. *Volatile oil crops: Their biology, biochemistry and production.* Essex, England: Longman Science & Technology.

Hirono, I., M. Haga, M. Fujii, et al. 1979. Induction of hepatic tumors in rats by senkirkine and symphytine. *Journal of National Cancer Institute* 63:469–471.

Hirono, I., H. Mori, and M. Haga. 1978. Carcinogenic study of *Symphytum officinale. Journal of National Cancer Institute* 61:865–869.

Hoareau, L., and E. J. DaSilva. 1999. Medicinal plants: A reemerging health aid. *EJB Electronic Journal of Biotechnology* 2:56–70.

Hornok, L., ed. 1992. *Cultivation and processing of medicinal plants.* New York: John Wiley & Sons Inc.

Hsu, H.-Y. 1986. *Oriental materia medica.* New Canaan, CT: Keats Publ.

Huang, K. C. 1999. *The pharmacology of Chinese herbs,* 2nd. ed. Boca Raton, FL: CRC Press.

Husain, A., O. P. Virmani, S. P. Popli, et al. 1992. *Dictionary of Indian medicinal plants.* Lucknow, India: Central Institute of Medicinal and Aromatic Plants.

Ionkova, I. 2008. Anticancer compounds from *in vitro* cultures of rare medicinal plants. *Pharmacognosy Review* 2 (4): 206–218.

Iwu, M. M. 1993. *Handbook of African medicinal plants.* Boca Raton, FL: CRC Press.

Jaarsma, T. A., E. Lohgmanns, T. W. J. Gadella, et al. 1989. Chemotaxonomy of the *Symphytum officinale* agg. (Boraginaceae). *Plant Systematics and Evolution* 167:113–127.

Jacob, I., and W. Jacob. 1993. *The healing past: Pharmaceuticals in the Biblical and Rabbinic world.* Leiden, Holland: E. J. Brill.

Jain, S. K., and R. A. DeFilips. 1991. *Medicinal plants of India.* Algonac, MI: Reference Publications.

Jenkins, M., and S. Oldfield. 1992. *Wild plants in trade.* Baltimore, MD: WWF Publ.

Johnson, C. B., and C. Franz, eds. 2002. *Breeding research on aromatic and medicinal plants.* New York: Haworth Herbal Press.

Kaefer, C. M., and J. A. Milner. 2008. The role of herbs and spices in cancer prevention. *Journal of Nutrition and Biochemistry* 19:347–361.

Kamatenesi-Mugisha, M., H. Oryem-Origa, O. Odyek, et al. 2008. Medicinal plants used in the treatment of fungal and bacterial infections in and around Queen Elizabeth Biosphere Reserve, western Uganda. *African Journal of Ecology* 46:90–97.

Kapoor, L. D. 1990. *CRC handbook of Ayurvedic medicinal plants.* Boca Raton, FL: CRC Press.

Karban, R., and I. T. Baldwin. 1997. *Induced responses to herbivory.* Chicago: Univ. Chicago Press.

Keys, J. D. 1976. *Chinese herbs: Their botany, chemistry, and pharmacodynamics.* Rutland, VT: Charles E. Tuttle Co.

Khan, S., and M. J. Balick. 2001. Therapeutic plants of Ayurveda: A review of selected clinical and other studies of 166 species. *Journal of Alternative Complementary Medicine* 7:405–515.

Kletter, C., and M. Kriechbaum. 2001. *Tibetan medicinal plants.* Boca Raton, FL: CRC Press.

Kokwaro, J. O. 1976. *Medicinal plants of East Africa.* Kampala, Africa: East African Literature Bureau.

Krauss, B. H. 2001. *Plants in Hawaiian medicine.* Honolulu: Bess Press.

Krejpcio, Z., E. Król, and S. Sionkowski. 2007. Evaluation of heavy metals contents in spices and herbs available on the Polish market. *Polish Journal of Environmental Studies* 16:97–100.

Krishnan, A., P. Bagyalakshimi, S. Ramya, et al. 2008. Revitalization of Siddha medicine in Tamilnadu, India— Changing trends in consumers' attitude: A survey. *Ethnobotanical Leaflets* 12:1246–1251.

Kumar, N., A. Khader, P. Rangaswami, et al. 1993. *Introduction to spices, plantation crops, medicinal and aromatic plants.* Nagercoil, India: Rajalakshmi Publ.

Lange, D. 1998. *Europe's medicinal and aromatic plants: Their use, trade and conservation.* TRAFFIC Intern.

Lans, C. 2003. Struggling over the direction of Caribbean medicinal plant research. *Futures* 35:473–491.

Lassak, E. V., and T. McCarthy. 2001. *Australian medicinal plants.* Auckland: JB Books.

Laufer, B. 1919. *Sino-Iranica.* Chicago: Field Museum.

Lewington, A. 1993. *Medicinal plants and plant extracts: A review of their importation into Europe.* TRAFFIC Intern.

Lewis, W. H., and M. P. F. Elvin-Lewis. 2003. *Medical botany: Plants affecting human health,* 2nd ed. New York: John Wiley & Sons Inc.

Li, T. S. C. 2006. *Taiwanese native medicinal plants: Phytopharmacology and therapeutic values.* Boca Raton, FL: CRC Press.

Liu, J., Y. Lu, Q. Wu, R. A. Goyer, et al. 2008. Mineral arsenicals in traditional medicines: Orpiment, realgar, and arsenolite. *Journal of Pharmacology and Experimental Therapeutics* 326:363–368.

Lorenzi, H., and F. J. de Abreu Matos. 2000. *Plantas medicinais no Brasil.* Avenida, Brazil: *Instituto Plantarum de Estudos da Flora.*

Macdonald, C. 1973. *Medicines of the Maori.* Auckland: Collins.

Mahady, G. B., S. L. Pendland, A. Stoia, et al. 2003. *In vitro* susceptibility of *Helicobacter pylori* to isoquinoline alkaloids from *Sanguinaria canadensis* and *Hydrastis canadensis. Phytotherapy Research* 17:217–221.

Manniche, L. 1989. *An ancient Egyptian herbal.* Austin: Univ. of Texas Press.

Mans, D. R. A., A. B. da Rocha, and G. Schwartsmann. 2000. Anticancer drug discovery and development in Brazil: Targeted plant collection as a rational strategy to acquire candidate anticancer compounds. *Oncologist* 5:185–198.

Marinelli, J., ed. 2005. *Plant: The ultimate visual reference to plants and flowers of the world.* New York: Dorling Kindersley Publishers Ltd.

Marshall, N. T. 1998. *Searching for a cure: Conservation of medicinal wildlife resources in east and southern Africa.* TRAFFIC Intern.

Martín, J. S. 1983. Medicinal plants in central Chile. *Economic Botany* 37:216–227.

Mathé, A., and C. Franz. 1999. Good agricultural practices and the quality of phytomedicines. *Journal of Herbs Spices and Medicinal Plants* 6 (3): 101–113.

Matthioli, P. A. 1544 (1554). *Di Pedacio Dioscoride Anazarbeo Libri cinque Della historia, et materia medicinale tradotti in lingua volgare italiana da M. Pietro Andrea Matthiolo Sanese Medico, con amplissimi discorsi, et comenti, et dottissime annotationi, et censure del medesimo interprete (Commentarrii in sex libros Pedacii Dioscoridis).* Vincento Valgrisio, Venecia, Italy.

Meléndez, E. N. 1945. *Plantas medicinales de possible cultivo en Puerto Rico.* Rio Piedras: Estacion Experimental Agricola.

Mendelsohn, R., and M. J. Balick. 1995. The value of undiscovered pharmaceuticals in tropical forests. *Economic Botany* 49:223–228.

Mentreddy, S. R. 2007. Medicinal plant species with potential antidiabetic properties. *Journal of Science of Food and Agriculture* 87:743–750.

Miki, W. 1976. *Index of the Arab herbalist's materials.* Tokyo: Institute for Study of Language and Culture of Asia and Africa.

Mitchell, S. A., and M. H. Ahmad. 2006. A review of medicinal plant research at the University of the West Indies, Jamaica, 1948–2001. *West Indian Medical Journal* 55:243–269.

Moerman, D. E. 1977. *American medical ethnobotany: A reference dictionary.* New York: Garland Publishers.

———. 2009. *Native American medicinal plants: An ethnobotanical dictionary.* Portland, OR: Timber Press.

Mors, W. B., C. T. Rizzini, and N. A. Pereira. 2000. *Medicinal plants of Brazil*, ed. R. A. De Filipps. Algonac, MI: Reference Publications.

Muralikrishnan, G., A. K. Donda, and F. Shakeel. 2010. Immunomodulatory effects of *Withania somnifera* on azoxymethane induced experimental colon cancer in mice. *Immunological Investigation* 39:688–698.

Nadkarni, A. K. and K. M. Nadkarni. 1996. Dr. K. M. Nadkarni's *Indian Materia Medica.* Vol.2. Popular Prakashan Private Limited.

National Institute for the Control of Pharmaceutical and Biological Products. 1987. *Color atlas of Chinese traditional drugs.* Beijing: Sci. Press.

Neuwinger, H. D. 1996. *African ethnobotany: Poisons and drugs.* London: Chapman & Hall.

Noble, R. L. 1990. The discovery of the vinca alkaloids—Chemotherapeutic agents against cancer. *Biochemistry and Cell Biology* 68:1344–1351.

Norman, J. M. 1985. William Withering and the purple foxglove: A bicentennial tribute. *Journal of Clinical Pharmacology* 25:479–483.

Nunn, J. F. 1996. *Ancient Egyptian medicine.* Norman: University of Oklahoma Press.

Okigbo, R. N., U. E. Eme, and S. Ogbogu. 2008. Biodiversity and conservation of medicinal and aromatic plants in Africa. *Biotechnology and Molecular Biology Reviews* 3:127–134.

O'Sullivan-Coyne, G., G. C. O'Sullivan, T. R. O'Donovan, et al. 2009. Curcumin induces apoptosis-independent death in esophageal cancer cells. *British Journal of Cancer* 101:1585–1595.

Ozkutlu, F. 2008. Determination of cadmium and trace elements in some spices cultivated in Turkey. *Asian Journal of Chemistry* 20:1081–1088.

Pardo-Andreu, G. L., C. Sánchez-Baldoquín, R. Ávila-González, et al. 2006. Interaction of vimang (*Mangifera indica* L.) extract with Fe(III) improves its antioxidant and cytoprotecting activity. *Pharmacological Research* 54:389–395.

Patwardhan, B. 2005. Ethnopharmacology and drug discovery. *Journal of Ethnopharmacology* 100:50–52.

Perry, L. M. 1980. *Medicinal plants of East and Southeast Asia: Attributed properties and uses.* Cambridge, MA: MIT Press.

Persinos, G. J., M. W. Quimby, and J. W. Schermerhorn. 1964. A preliminary pharmacognostical study of 10 Nigerian plants. *Economic Botany* 18:329–341.

Peter, K. V., ed. 2001. *Handbook of herbs and spices.* Boca Raton, FL: CRC Press.

———, ed. 2004. *Handbook of herbs and spices*, vol. 2. Boca Raton, FL: CRC Press.

———, ed. 2006. *Handbook of herbs and spices*, vol. 3. Boca Raton, FL: CRC Press.

Raffauf, R. F. 1996. *Plant alkaloids: A guide to their discovery and distribution.* New York: Haworth Press.

Raskin, I. 1992. Role of salicylic acid in plants. *Annual Review of Plant Physiology and Plant Molecular Biology* 43:439–463.

Ridley, R. G. 2002. Medical need, scientific opportunity and the drive for antimalaria drugs. *Nature* 415: 686–693.

Rizk, A. M., and G. A. El-Ghazaly. 1995. *Medicinal and poisonous plants of Qatar.* Doha: University of Qatar.

Roberson, E. 2008. *Medicinal plants at risk.* Tucson, AZ: Center for Biological Diversity. (www.biologicaldiversity.org)

Roersch, C. 1994. *Plantas medicinales en el sur Andino del Perú.* Königstein: Koeltz Scientific Books.

Sagrawat, H., and M. Y. Khan. 2007. Immunomodulatory plants: A phytopharmacological review. *Pharmacognosy Review* 1 (2):248–260.

Salim, A. A., Y.-W. Chin, and A. D. Kinghorn. 2008. Drug discovery from plants. In *Bioactive molecules and medicinal plants*, eds. K. G. Ramawat and J. M. Mérillon, 1–24. Springer (DOI: 10. 1007/978-3-540–7460304_1.)

Samuelsson, G., and L. Bohlin. 2010. *Drugs of natural origin: A treatise of pharmacognosy*, 6th ed. Stockholm: Swedish Pharm. Press.

Samy, R. P., P. N. Pushparaj, and P. Gopanakrishnakone. 2008. A compilation of bioactive compounds from Ayurveda. *Bioinformation* 3:100–110.

Saper, R. B., S. N. Kales, J. Paquin, et al. 2004. Heavy metal content of Ayurvedic herbal medicine products. *JAMA* 292:2868–2873.

Saper, R. B., R. S. Phillips, A. Sehgal, et al. 2008. Lead, mercury, and arsenic in U.S.- and Indian-manufactured Ayurvedic medicines sold via the Internet. *JAMA* 300:915–923.

Sarin, Y. K. 1996. *Illustrated manual of herbal drugs used in Ayurveda*. New Delhi: National Institute of Science Communications.

Schultes, R. E. 1972. The future of plants as source of new biodynamic compounds. In *Plants in the development of modern medicine*, ed. T. Swain, 103–124. Cambridge: Harvard Univ. Press.

Schultes, R. E., and R. F. Raffauf. 1990. *The healing forest: Medicinal and toxic plants of the Northwest Amazonia*. Portland, OR: Dioscorides Press.

Seaforth, C. E., C. D. Adams, and Y. Sylvester. 1983. *A guide to the medicinal plants of Trinidad & Tobago*. London: Commonwealth Secretariat.

Sekeroglu, N., F. Ozkutlu, S. M. Kara, et al. 2008. Determination of cadmium and selected micronutrients in commonly used and traded plants in Turkey. *Journal of Science of Food and Agriculture* 88:86–90.

Sharma, P., and G. Singh. 2002. A review of plant species used to treat conjunctivitis. *Phytotherapy Research* 16:1–22.

Sheldon, J. W., M. J. Balick, and S. A. Laird. 1997. *Medicinal plants: Can utilization and conservation coexist?* New York: New York Botanical Garden.

Simon, J. E., A. R. Koroch, D. Acquaye, et al. 2007. Medicinal crops of Africa. In *New crops and new uses*, ed. J. Janick and A. Whipkey, 322–331. Alexandria, VA: ASHS Press.

Singh, S., P. Pandey, and S. Kumar. 2000. *Traditional knowledge on the medicinal plants of Ayurveda*. Lucknow, India: Central Institute of Medicinal and Aromatic Plants.

Sivarajan, V. V., and I. Balachandran. 1994. *Ayurvedic drugs and their plant sources*. New Delhi-Bombay-Calcutta: India: Oxford & IBH Publishing Company.

Sneader, W. 2005. *Drug discovery: A history*. Hoboken, NJ: John Wiley & Sons Inc.

Staples, G. W., and M. S. Kristiansen. 1999. *Ethnic culinary herbs: A guide to identification and cultivation in Hawaii*. Honolulu: Univ. Hawaii Press.

Su, E. G. 2003. *Asian botanicals*. Carol Stream, IL: Allured.

Subbarayappa, B. V. 1997. Siddha medicine: An overview. *Lancet* 350:1841–1844.

Tabuti, J. R. S., C. B. Kukunda, and P. J. Waako. 2010. Medicinal plants used by traditional medicine practitioners in the treatment of tuberculosis and related ailments in Uganda. *Journal of Ethnopharmacology* 127:130–136.

Thakur, R. S., J. S. Puri, and A. Husain. 1989. *Major medicinal plants of India*. Lucknow, India: Central Institute of Medicinal and Aromatic Plants.

Thompson, R. C. 1924. *The Assyrian herbal*. London: Luzac and Co.

———. 1949. *A dictionary of Assyrian botany*. London: British Academy.

Tierra, M., and L. Tierra. 1998. *Chinese traditional herbal medicine*. Twin Lakes, WI: Lotus Press.

Tiwari, V., N. A. Darmani, B. Y. J. T. Yue, et al. 2010. *In vitro* antiviral activity of neem (*Azadirachta indica* L.) bark extract against herpes simplex virus type-1 infection. *Phytotherapy Research* 24:1132–1140.

Torrance, A. W. 2000. *Bioprospecting and the convention on biological diversity*. Harvard Law School document archive, http://leda.law.harvard.edu/leda/data/258/Torrence,_Andrew_00.pdf (accessed September 9, 2010).

Tucci, J. 2008. Herbs for Alzheimer's. *Journal of Complementary Medicine* 7 (6):32–35.

Tucker, A. O., and M. J. Maciarello. 1993. Shelf life of culinary herbs and spices. In *Shelf life studies of foods and beverages*, ed. G. Charalambous, 469–485. Amsterdam: Elsevier.

———. 1998. Some toxic culinary herbs in North America. In *Food flavors: Formation, analysis and packaging influences*, ed. E. T. Contis, C.-T. Ho, C. J. Mussinan, T. H. Parliment, F. Shahidi, and A. M. Spanier, 401–414. Amsterdam: Elsevier.

Tucker, A. O., M. J. Maciarello, and C. E. Broderick. 1994. File and the essential oils of the leaves, twigs, and commercial root teas of *Sassafras albidum* (Nutt.) Nees (Lauraceae). In *Spices, herbs and edible fungi*, ed. G. Charlambous, 595–604. Amsterdam: Elsevier.

Turner, P. 1973. Plants in the development of modern medicine. *Proceedings of Royal Society of Medicine* 66:490.

Tyler, V. E., J. R. Brady, and J. E. Robbers. 1988. *Pharmacognosy*. Philadelphia: Lea and Febiger.

Van Duong, N. 1993. *Medicinal plants of Vietnam, Cambodia and Laos*. Mekong, Vietnam: Mekong Printing.

van Wyk, B.-E., B. van Oudtschoorn, and N. Gericke. 1997. *Medicinal plants of South Africa*. Pretoria: Briza Publ.

van Wyk, B.-E., and M. Wink. 2004. *Medicinal plants of the world*. Portland, OR: Timber Press.

Visser, L. E. 1975. Plantes médicinales de la Côte d'Ivoire. *Meded. Landbouwhoogeschool* 75–15.

von Koenen, E. 1996. *Medicinal, poisonous, and edible plants in Namibia.* Göttingen: Klaus Hess Publ.

Weiss, E. A. 1997. *Essential oil crops.* New York: CAB Intern.

————. 2002. *Spice crops.* New York: CABI Publ.

Westendorf, J. 1992. Pyrrolizidine alkaloids—*Symphytum* species. *Adverse. Effects of Herbal Drugs* 1:219–222.

Whistler, W. A. 1992a. *Tongan herbal medicine.* Honolulu: Isle Botany.

————. 1992b. *Polynesian herbal medicine.* Lawai, Kauai, Hawaii: National Tropical Botantical Garden.

————. 1996. *Samoan herbal medicine.* Honolulu: Isle Botany.

Wijesekera, R. O. B. 1991. *The medicinal plant industry.* Boca Raton, FL: CRC Press.

Woodley, E., ed. 1991. *Medicinal plants of Papua New Guinea. Part I: Morobe Province.* Weikersheim, Germany: Verlag Josef Margraf.

World Health Organization. 1997. *Medicinal plants in China: A selection of 150 commonly used species.* Manila: WHO.

Wrigley, S. K., M. A. Hayes, R. Thomas, et al., eds. 2000. *Biodiversity: New leads for the pharmaceutical and agrochemical industries.* Cambridge, England: Royal Society of Chemists.

Wu, J.-N. 2005. *Chinese materia medica.* New York: Oxford Univ. Press.

Yadav, H., S. Jain, G. B. K. S. Prasad, et al. 2007. Preventive effect of diabegon, a polyherbal preparation, during progression of diabetes induced by high-fructose feeding in rats. *Journal of Pharmacological Sciences* 105:12–21.

Yakubu, M. T., M. A. Akanji, and A. T. Oladiji. 2007. Male sexual dysfunction and methods used in assessing medicinal plants with aphrodisiac potentials. *Pharmacognosy Review* 1 (1):49–56.

Yen, K.-Y. 1992. *The illustrated Chinese materia medica: Crude and prepared.* Taipei: SMC Publ.

Yeung, M-F, C. B. S. Lau, R. C. Y. Chan, et al. 2009. Search for antimycobacterial constituents from a Tibetan medicinal plant, *Gentianopsis paludosa.* *Phytotherapy Research* 23:123–125.

Younos, C., J. Fleurentin, D. Notter, et al. 1987. Repertory of drugs and medicinal plants used in traditional medicine in Afghanistan. *Journal of Ethnopharmacology* 20:245–290.

Zhang, H.-Y. 2005. One-compound multiple-targets strategy to combat Alzheimer's disease. *FEBS Letters* 579:5260–5264.

CHAPTER 3

Indian Traditional and Ethno Medicines from Antiquity to Modern Drug Development

Janardan Singh, Umesh C. Lavania, and Suman Singh

CONTENTS

3.1 INTRODUCTION

Traditional medicine (TM) is generally defined as health practices, knowledge, and beliefs incorporating plant-, animal-, and mineral-based medicines, spiritual therapies, manual techniques, and exercises applied singularly or in combination to treat, diagnose, and prevent illness or maintain well being (WHO 2008). TM is based on holistic principles laying emphasis on eradication of root cause of the disease rather than just treatment of the disease per se.

Over 100 types of traditional systems of medicine (TSM) identified by the World Health Organization (WHO) are in vogue worldwide today (Anonymous 1978, 2006). Some of the more important are Ayurveda, Yoga and Naturopathy, Unani, Siddha, and Homeopathy (AYUSH) (India); Traditional Chinese medicine, Acupuncture (China); Shiastu (Japan); Magnetic healing (France); Heilpraxis (Germany); Herbalism (Sweden); and So-wa rig-pa (Bhutan) (Anonymous 1978, 2006). These TSMs continue to be the tradition of every country, in spite of the advancement of modern medicine and the discovery of new and novel drugs in recent times.

Traditional medicines are the precious cultural heritage of the different societies of the world and continue to cater to the medicinal needs of 80% of the world's population (according to WHO estimates). In general, they have a strong doctrinaire base, well documented, and recognized or codified indigenous systems of medicine. The major TSMs in India are Ayurveda, Yoga and Naturopathy, and Siddha, which have their origins in India, and the Unani system, which comes from Greece and was introduced in India by the Arabs and Persians around the eleventh century. Homeopathy originated in Germany and is practiced in many parts of the world, including India; it is recognized as one of the national systems of medicine in India (AYUSH: www.indianmedicine.nic.in). These systems of medicine are heavily dependent on plant-based drugs, although drugs of mineral and animal origin are also utilized for total health care management. The majority of the population in developing economies still relies on TM for their health care needs, and the popularity of TM is fast spreading in the industrialized countries.

Another form of traditional medicament, variously termed as ethno, tribal, or folk medicine (Jain 2001, 2004), is used by tribals and aboriginals inhabiting rural or remote inaccessible areas. These forms of medicines are generally uncoded or unrecorded and perpetuated by word of mouth among the communities.

It is an underlying fact that most of the phytopharmaceuticals presently in use as therapeutic agents owe their origin primarily to the knowledge gained over thousands of years from traditional or ethno–medico–botanical usage. Many valuable drugs (e.g., atropine, ephedrine, tubocurarine, digoxin, reserpine, morphine, quinine, artemisinin, etc.) came through indigenous remedies. Such knowledge still continues to be the preferred choice for bioprospection toward development of new and novel drugs. Pharmaceutical chemists strive to use plant-derived drugs as a prototype (e.g., morphine, taxol, physostigmine, quinidine, and emetine) to facilitate development of more effective and less toxic drugs.

In view of this, the present chapter aims to provide

- an insight into the Indian traditional systems and ethnomedicines and their possible role in developing new remedies, especially for lifestyle diseases and degenerative and psychosomatic disorders, where these systems have special strength
- an overview of certain important and extensively used Indian medicinal plants with respect to their traditional uses, bioactive molecules, and preclinical (pharmacological) and clinical uses that promise new leads and clinically useful drugs

3.2 TRADITIONAL SYSTEMS OF MEDICINE IN INDIA

India has a rich cultural heritage and ancient medical wisdom and health knowledge where traditional and indigenous systems of medicine have been in use since time immemorial. Even today, such medicinal systems are extensively in use along with a modern system of medicine (allopathy) for total health care. Ayurveda, Yoga and Naturopathy, Unani, Siddha, and Homeopathy (AYUSH) (www.indianmedicine.nic.in) are the six nationally recognized systems of medicines in India. Their education, practice, drug manufacture, quality, and sale are regulated through various acts, rules, and regulations enacted by the Government of India.

3.3 AYURVEDIC SYSTEM OF MEDICINE

3.3.1 Origin and History

The word "Ayurveda" connotes *ayur* = life and *veda* = knowledge; its origin in India was long ago in the pre-Vedic period. *Rigveda* and *Ayurveda* (2000 BC), perhaps the oldest documented repository of human knowledge, represents the first mention of diseases and medicinal herbs in *Osadhi-Sukta* (Rv. 10.47, 123). Similarly, the earliest comprehensive descriptions of the drugs and diseases are found in *Atharvaveda* (1600–100 BC). In Vedic literature, about 200 drugs are mentioned that were in use for treatment of diseases (Singh et al. 2003), although during the Vedic period, treatment was mainly by recitation of prayers and incantations (*mantras*). Ayurveda is thus often considered as *Upa Veda* (subsidiary Veda) or even as *fifth Veda*. The growth of Ayurveda was a part of general philosophical development and many of its concepts were derived from various schools of philosophy, particularly the *Samakhya*, *Nyaya*, and *Vaiseshika* schools (Anand 1990; Dwarakanath 2003).

However, the scientific foundation of Ayurveda was laid during the *Samhita* (compilation) period by the three legendary and authoritative classical texts: *Charak Samhita* (500–200 BC), *Sushrata Samhita* (1000–200 BC), and *Astang Hridayam* (AD 700), collectively known as *Brihattrayi*. *Charak* emphasizes medicine and *Sushruta Samhita* decribes surgery. The *Astang Hridayam* deals with basic principles and practices of Ayurvedic medicine, primarily based on the information contained in *Charak* and *Shushruta Samhita*. These compendia were used for teaching of Ayurveda in the ancient universities of Takshashila and Nalanda. Based on Sanskrit names, the number of plants in these compendia include 1,270 in *Sushruta*, 1,100 in *Charak*, and 1,150 in *Astang Hridayam* (Singh and Chunekar 1972). Further elaboration of drugs and disease is found in subsequent classical treatises like *Madhava Nidan* (AD 1200), *Sarangdhar Samhita*, and *Bhava Prakash* (AD 1500) and in over 70 other lexicons (*Nighantu*, *Granths*) written between the sixth and seventh centuries.

During the early period BC, Ayurveda was the only overall health care system that enjoyed the patronage and support of the common people and rulers. However, progress of Ayurveda suffered a setback during the subjugation of India by the Mughals and British, who patronized their own systems of medicine: Unani and Allopathy, respectively. Nevertheless, the Ayurveda survived owing to its strong native roots and popularity among the people, especially those living in remote and inaccessible areas. After independence from British rule, interest in Ayurveda got a fillip on account of its official recognition by the government of India. Now, the Ayurvedic system enjoys the same status as Allopathy in terms of official health policies, national plans, and programs of the government of India.

Ayurveda is now going through the process of reorientation to modern rigorous scientific scrutiny and is poised to emerge globally for much greater and effective utilization in health care systems. At the present time, researchers of many disciplines are exploring Ayurveda to find remedies for lifestyle-related diseases and degenerative and psychosomatic health problems. The Walton committee appointed by the House of Lords (England) has included Ayurveda in the group-1 category of "professionally practiced complementary and alternative medicines" (AYUSH: www.indianmedicine.nic.in).

The Ayurvedic system of medicine manages health care through eight disciplines generally called *Astang Ayurved:*

- Internal medicine (*Kaya Chikitsa*)
- Pediatrics (*Kumarbhritya*)
- Psychiatry (*Bhut-Vidya*)
- Surgery (*Shalya-Tantra*)
- Otorhinolaryngology and ophthalmology (*Shalakya Tantra*)
- Toxicology (*Agada Tantra*)
- Geriatrics (*Rasayana Tantra*)
- Eugenics and aphrodisiacs (*Vaji-Karaao-Tantra*)

Of these, generally Rasayan and Svasthavrita (means and methods for living a healthy life) deal with the preservation and promotion of health of the healthy persons; the remaining disciplines deal with the diseased persons (Majumdar 1971; Anonymous 1978; Kapoor 2001).

3.3.2 The Concepts: Health, Sickness, and Treatment in Ayurveda

Ayurveda, the "science of life", not only encompasses the preventive and curative aspects of diseases but also provides a unique approach of health promotion, leading to a healthy, active, and long life span (Katiyar 2010). Conceptually, Ayurveda considers "life" as the union of body, senses, mind, and soul. The total body matrix of human beings comprises three humors (*Dosa—Vata, Pitta*, and *Kapha*) considering the biological units of the body—seven body tissues (*Rasa, Rakta, Mansa, Meda, Asthi, Majja*, and *Sukra*) and the waste products of the body, such as feces, urine, and sweat. In the normal state of equilibrium, they support body growth and decay of the body matrix and revolve around food, which is significantly affected by psychosomatic mechanisms as well as by biofire (*Agni*).

From the Indian philosophical point of view, the matters of the universe including the human body are composed of five basic elements (*Panchamahabhutas*): namely, *Akasa* (vacuum-ether), *Vayu* (air), *Tejas* (fire), *Jal* (water), and *Ksiti* (earth). There is a balanced condensation of these elements in different proportion to suit the needs and requirements of different structures and functions of the body matrix and its parts. Health and sickness depend upon the presence or absence of a balanced state of the total body matrix, including the balance between its different constituents. Both the intrinsic and extrinsic factors can cause disturbance in the natural equilibrium, giving rise to disease. The treatment of diseases consists of restoring the balance of the disturbed body–mind matrix through regulating diet, physical activity and regimen, and medicines (Kapoor 2001; www.indianmedicine.nic.in).

In the Ayurvedic system of medicine, the treatment of diseases can be broadly classified as *Sodhan Therapy* (purification treatment), *Shaman Therapy* (palliative treatment), *Nidan Parivarjan* (avoidance of disease-causing and aggravating factors), *Satvavajay* (physchotherapy), and *Rasayana Therapy* (use of immunomodulators and rejuvenation psychotherapy).

The two most important modes of treatments (i.e., Pancha-karma and Rasayana Therapy) are specialties of Ayurveda. The two are quite popular and have attracted world attention (www. indianmedicine.nic.in).

3.3.3 Panchakarma

The treatment planned for elimination of the increased and morbid dosa (vata, pitta, and kapha) biological unit of the living body responsible for its malfunctions is known as *Sodhan Therapy* (purification therapy) and technically termed as *Pancha-Karma* (fivefold therapy). It involves internal and external purification by medically induced emesis (*Vamana*), purgation (*Virechan*), oil enema (*Anuvasan Basti*), decoction enema therapy (*Asthapana* or *Niruha Basti*), and nasal insufflations therapy (*Sirovirechana*). Two necessary measures are carried out before (*Purva*) and after (*Pashchat*) performing the Pancha-Karma. These measures are considered as a part of this therapy consisting of *Snehana* and *Svedana* (external and internal oleation and induced sweating), before and after Pancha-Karma therapy. The patient is gradually allowed to do his normal routines of diet and duties (Anonymous 1978). This treatment is used especially for promotion of health and prevention and treatment of neurological disorders, musculoskeletal disease conditions, certain vascular or neurovascular states, respiratory diseases, and metabolic and degenerative disorders.

3.3.4 Rasayana Therapy (Health Promoters)

The whole drug regimens of Ayurveda are divided into two distinct groups: one for maintenance and promotion of good health and longevity, and another for treatment of diseases. The *Rasayana* group of drugs falls under the first category. The word "Rasayana" literally means the path that *rasa* takes (*Rasa:* plasma; *Ayana:* path). It is defined as a therapeutic measure that promotes longevity, prevents aging, provides positive health and mental faculties, increases memory, and imparts resistance and immunity against diseases (Anonymous 1978). *Rasayanas* are broadly classified into four groups:

Kamya Rasayan—dealing with health and vigor of a healthy person
Naimittic Rasayan—for increasing the strength of a diseased person, which may include drugs that can be used as an adjunct to the specific medical treatment for early cure
Medhya Rasayan—for improving mental faculties
Acharya Rasayan—religious way of living with high morals and virtues following established rules of conduct, practice, usage, precept, etc. (Anonymous 1978)

Thus, the rasayana group of drugs and recipes has a multidimensional mode of action through nutritional dynamics of tissues. The disease-specific concepts have similar ideas of adoptogens or antistress drugs, tonics, and "unstimmungs therapy" of other medical systems and traditions (Anand 1990). After a detailed study, Wagner and Proksch (1985) opined that Rasayana preparation, which acts as an herbal immunostimulant and as an adoptogen, regulates the immunological and endocrine systems with relatively low doses, without damaging the autoregulative functions of the organism.

The quintessence of all *Dhatus* (tissues) is known as *Ojas*. The expression of Ojas through *Bala* (resistance power) in turn is responsible for immunity or resistance against diseases. Ayurvedic approaches to the phenomenon of *Vyadhi-Kshamatva* (defense mechanism to resist diseases) are also fundamental and important factors to mediate through functioning of *Ojas*

(AYUSH: www.indianmedicine.nic.in). The concept of modulating immune response can be equated with the *Vyadhi-kshamatva* and *Oja vridhi* as postulated in the Ayurvedic texts. Consequent upon a range of structural and functional changes, the inevitable and progressive aging phenomenon of life takes place and, because of aging, the human system becomes prone to a variety of age-related diseases. Stress is another important factor that is directly related with body immunity. In Ayurvedic parlance, stress, adoption and immunity complex coexist and form the basis for better understanding of human body functions and cause of diseases. Moreover, rasayana drugs act inside the human body by modulating the neuro–endocrino–immune system and have been found to be a rich source of antioxidants (Brahma and Debnath 2003).

In recent times, extensive studies have been undertaken in the contexts of validating the age-old concepts from a modern point of view. Puri (1970a, 1970b, 1971, 2003) provided a detailed account of the herbs used in Rasayana preparations. The effects of Rasayana drugs on psychosomatic stress (Udupa 1973), epilepsy (Singh and Murthy 1989), and conclusive disorders (Diwedi and Singh 1992) have been studied. The role of free radicals in various diseases and 15 *Rasayana* plants with potent antioxidant activity has been reviewed for their traditional uses and mechanism of antioxidant action (Govindarajan et al. 2005).

3.4 UNANI SYSTEM OF MEDICINE

3.4.1 Origin and History

The Unani system of medicine was primarily developed in Greece. Its foundation was laid by Hippocrates. The system owes its present form to the Arabs. The system is based on well-defined scientific principles derived from experiences and documentation in the ancient cultural heritage of Egypt, Arabia, Iraq, China, Syria, and India (Kumar et al. 1997). The Unani or Greco-Arab system was introduced in India during the medieval period (eleventh century) by the Arabs. Ayurvedic practitioners enriched this system by their own experience and amalgamated it with Ayurveda to be known as "Unani-Tibb." During the reformative period, attention was paid to medicinal herbs found in India, and several books on the therapeutic properties of herbs were brought out by the Arab physicians. Later, several schools, colleges, and laboratories were established for imparting education, training, and research, along with several hospitals and dispensaries for the treatment.

Along with Ayurveda, this system also suffered a setback during the British rule, causing retardation of growth of education, research and development, practice, and popularization of this system in India. However, Hakim Ajamal Khan, a renowned physician, initiated efforts for revival of the Unani system in India. After independence on August 15, 1947, the Unani system, along with other systems of medicine, received necessary impetus for its all-round development in India with support from the government of India. Today, India is one of the leading countries that practices the Unani system and supports the largest number of Unani system-related educational, research and health care institutions (AYUSH: www.indianmedicine.nic.in).

3.4.2 Principles and Concepts

The basic principles and concepts of the Unani system of medicine are based upon the well-known four humors theory of Hippocrates as developed in the Arab world; they assess the patent's needs in the forms of temperament. The four humors (blood, phlegm, yellow and black bile) are the fluids that the human body obtains from food and include various types of hormones and enzymes. According to this system, whenever any disturbance occurs in the equilibrium of humors, it causes diseases. The treatment, therefore, aims at restoring the equilibrium of humors (AYUSH: www.indianmedicine.nic.in; Khanuja et al. 2006; Ahmad and Akhtar 2007).

The diagnostic process in the Unani system is dependent on observation and physical examination. Keeping all interrelated factors in view, the cause and nature of illness is determined and treatment is given. For diagnosis and treatment, physicians depend mainly on pulse reading and examination of urine and stool. For therapeutics, the entire personality of the patient is taken into account and various types of treatments comprising regimental diets and pharmacological and surgical therapies are provided.

3.4.3 Specialty

The type of treatment involves the use of naturally occurring drugs of plant, animal, and mineral origins. Since this system emphasizes the importance of temperament of the drugs as well as patients, the medicines prescribed are those that go well with the temperament of patients, thereby accelerating the process of recovery and eliminating the risk of reaction. This system is especially popular in the treatment of chronic diseases such as skin problems, leucoderma, eczema, filaria, liver disorders, metabolic disorders, arthritis, etc., well supported with research and development in concerned areas.

3.5 SIDDHA SYSTEM OF MEDICINE

3.5.1 Origin and History

This system of medicine is supposed to have been developed within the *Dravidian* culture and dates back to the pre-Vedic period. This is one of the oldest systems of medicine in India. The term "siddha" means achievement, and "siddhars" were saintly figures who seem to have contributed to the development of the Siddha system of medicine. This system of medicine is practiced predominantly in the state of Tamil Nadu in the southern part of India. It is largely therapeutic in nature (Kumar et al. 1997; Khanuja et al. 2006).

3.5.2 Principles and Concepts

The principle and doctrine of this system, both fundamental and applied, have a close similarity to those of Ayurveda, with specialization in Iatrochemistry. According to this system, the human body is the replica of the universe and the drugs are to be found in the universe irrespective of their origin. This system also considers the human body as a conglomeration of three humors, like Ayurveda, in *vatasu*, *pittam*, and *kapham*. The equilibrium of humans is considered to be health and its disturbance or imbalance leads to disease or sickness.

The diagnosis of disease involves identifying its cause through the examination of pulse reading, examination of urine and eyes, study of voice, color of the body, examination of the tongue, and status of digestive system (Kumar et al. 1997). The treatment is mainly directed toward restoration of equilibrium of the previously stated humors, for which one or more forms of cleansing procedure (vomiting, purgation, enema, and nasal drops) are adopted. This is followed by application of various forms of medicine.

3.5.3 Specialty

Siddha treatment covers all types of diseases and has achieved success in treating peptic ulcer, infective hepatitis, psoriasis, rheumatoid arthritis, and abnormal growth of tissues (AYUSH: www.indianmedicine.nic.in).

3.6 YOGA

3.6.1 Origin and History

The concept and practice of *Yoga* originated in India several thousand years ago. It is one among the six systems of Vedic philosophy. The founders were great saints and sages. Maharishi Patanjali, known as the "Father of *Yoga*," compiled and refined various aspects of yoga systematically and advocated the eightfold path of *Yoga*, popularly known as "*Astang Yoga*" for all-around development of human personality. These are *Yama, Niyam, Asana, Pranayama, Pratyahasa, Dharna, Dhyana,* and *Samadhi. Yoga* today is no longer restricted to hermits, saints, and sages but rather has taken its place in everyday life. At present it has attracted attention all over the world and its science and techniques now have been reoriented to suit modern sociological and physiological practices and proponents' needs and lifestyles (AYUSH: www.indianmedicine.nic.in; Khanuja et al. 2006).

3.6.2 Principles and Concepts

Yoga can be defined as a means for uniting the individual spirit with the universal spirit of God. It is promoted as complex therapy comprising eight components (*Astang Yoga*): restraint, observance of austerity, physical posture, breathing exercises, controlling the sense organs, contemplation, meditation, and *Samadhi*. These steps are believed to have potential for the improvement of physical health by encouraging better circulation of oxygenated blood in the body—restraining the sense organs and thereby inducing tranquility and serenity of mind. The practice of *Yoga* prevents psychosomatic disorders and diseases and improves an individual's resistance and ability to overcome stressful situations.

3.6.3 Specialty

Yoga is effective in the management of many disorders and indications: anxiety, neurosis, depression, arthritis, bronchial asthma, cervical spondolysis, diabetes, hypertension, obesity, back pain, sciatica, insomnia, and epilepsy (AYUSH: www.indianmedicine.nic.in).

3.7 NATUROPATHY

Naturopathy deals with the healing power of nature; it believes that all healing powers are within our body and work in harmony with constructive principles of nature on physical, mental, moral, and spiritual planes of life. Generally, it is considered that health is normal and constitutes a harmonious vibration of the elements and forces, whereas disease is abnormal and an unharmonious vibration of the elements and forces constituting human entity. Nature cure is a very old science of healing and method of living. It has great health promoting, curative, and rehabilitative potential (AYUSH: www.indianmedicine.nic.in; Khanuja et al. 2006).

3.7.1 Principles and Concepts

Naturopathy is projected not only as a system of treatment but also as a way of life. It is often referred to as drugless therapy. Attention is given to dietary combinations using natural, mostly uncooked food, fruits, and vegetables; adoption of purification practices; and use of hydrotherapy, cold pack, mild pack, bath massage, and a variety of methods/measures to tone up the body system and increase energy levels, with the aim of producing a state of good health and happiness.

3.7.2 Specialty

Naturopathy has many proponents among chronic patients who have found relief and cures where conventional treatment has failed. It has been found to be efficacious in disease conditions such as allergy and bronchial asthma, allergic skin diseases, chronic, nonhealing cancer, colitis, depression, anxiety, neurosis, cirrhosis of liver, and diabetes (AYUSH: www.indianmedicine.nic.in).

3.8 HOMEOPATHY

3.8.1 Origin and History

Historically, the principle of *Homeopathy* has been known since the time of Hippocrates, around 450 BC. But it was not until the late eighteenth century that *Homeopathy* as it is practiced today was evolved by the great German physician Dr. Samuel Hahnemann. He was appalled by the medical practice of that time and set about to develop a method of healing that would be safe, generic, and effective. He believed that human beings have a capacity for healing themselves and that the symptoms of disease reflect the individual's struggle to overcome his illness.

Homeopathy, in a simple way, signifies treatment of diseases with remedies, usually prescribed in minute doses, that are capable of producing symptoms similar to those of the disease when taken by healthy people (*Similia Similibus Curontur*). In India, it is recognized as one of the national systems of medicine. The system focuses on promotion of physical and mental values in health, disease, and treatment. Therefore, it is widely known as a haliotic system. It has blended into the traditions of India because of its close similarity in philosophy and principles to other Indian systems of medicine (Khanuja et al. 2006).

3.8 2 Principles and Concepts

The important demonstrable law and principles of *Homeopathy* are (a) the law of similar, (b) the law of single remedies, (c) the law of direction of cure, and (d) the law of minimum dose. Every law and principle signifies the use of *Homeopathic* medicines. A correct *Homeopathic* remedy or drug should be able to produce symptomatology in the patient. Cure is considered complete when there is a full restoration of the functioning of all the vital parts of the body. A single remedy can be chosen that is similar to the symptom complex of the sick person. The selected remedy for a sick person should be prescribed in a minimum dose. It treats a patient as a whole but considers each patient differently from others and aims not only to alleviate the patient's present symptoms but also to result in long-term well-being.

3.8.3 Specialty

It is believed that *Homeopathy* has definite and effective treatments for chronic diseases such as diabetes, arthritis, bronchial asthma, immunological disorders, and behavioral and mental disorders. Recently, this system has attracted debate on the cure versus molecule in placebo and drug comparison. Therefore, extensive research is warranted to dispel doubt about the effectiveness of *Homeopathic* medicines for the treatment of various diseases and ailments.

3.9 ETHNOMEDICAL BOTANICAL RESEARCH IN INDIA

The well-being of living organisms including human beings is inextricably linked with biological resources. Traditional communities that live close to nature make judicious and sustainable use of natural resources for their existence to meet their food, feed, medicine, and all other necessities of life based on their unique knowledge acquired by necessities, instinct observations, trial and error, and lengthy experiences. In a few geographic regions of the world (e.g., remote areas in Africa, Asia, and South America) that are not yet affected by fast-encroaching modern civilization, there exists a wealth of information on the properties of plants built up by the people in primitive societies over millennia (Schultes 1962, 1991). The multidisciplinary science that deals with the traditional and natural relationship between human societies and plants has been duly recognized as "ethnobotany" (Jain 1989, 2001). In the same context, other terms (e.g., ethnobiology, ethnopharmacology) have been used from time to time by various researchers. In recent times, the interest in ethnobotany has greatly increased, leading to a large number of publications in primary scientific journals such as *Economic Botany* (New York), *Journal of Ethnopharmacology* (Ireland; United States), *Pharmaceutical Biology* (United Kingdom), *Fitoterapia* (the Netherlands), and *Phytotherapy Research* (United Kingdom).

The Indian subcontinent is inhabited by over 53 million tribal people belonging to over 550 tribal communities of 227 ethnic groups spread over varied geographic and climatic zones of the country. There are about 106 different languages and 227 subsidiary dialects spoken by tribals in India. The tribal communities inhabit about 5,000 forested villages in India and lead a nomadic life. They are generally populated by the four races: Negroid, Australoid, Europeoid, and Mongoloid. The Negroid type is represented by natives of the Andaman Islands; Australoid by Veddah of Kerala and Sri Lanka and people of the Rajmahal Hills; Europeoid by Santhals, Mundas, and others of south and central India; and Mongoloid by tribals of the Himalayan belt and Nicobar Islands. These tribals enjoy certain protection and privileges in the constitution of India.

Considerable ethnobotanical information has been collected in India through organized studies conducted by the Botanical Survey of India in the latter part of the twentieth century by the late Dr. E. K. Janaki Amal. These were further intensified by Dr. S. K. Jain and other groups (Jain 1991, 2001, 2004; Ignacimuthu et. al. 2006). Ethnobotanical work in India has been devoted to four major categories:

- Ethnically distinct primitive, preventive, or interesting human societies—Mikir of Assam, Hils of Rajasthan, Thanus of Uttar Pradesh and Uttrakhand
- Specific geographic regions—central India, Kumaon Garhwal, Santhal Pargana, Balphakarm Sanctuary, Meghalaya and Indian desert
- Particular utility of groups of plans—food, medicine, hallucinogens
- Specific ethnophamacological uses—ethnogynecology, ethnodermatology, rheumatism, snakebite, ethno-orthopedics, and contraceptives

An All India Coordinated Research Project on Ethnobiology (AICRPE) funded by the Ministry of the Environment and Forests, Government of India, was launched in 1982. This project adopted a multidisciplinary approach involving 24 research institutes, centers, and university departments at different stages of the project. Ethnobotanical investigations under this ambitious program led to the documentation of over 9,500 wild plants used by tribals for meeting their various requirements. Of these, 7,500 plant species were found to be used for medicinal purposes and 950 to be new claims worthy of scientific scrutiny; 3,900 plant species were used as edible products, 525 for fiber and cordage, 400 as fodder, and 300 plant species as piscicides and pesticides. Over 175 plant species show promise for development of safe biopesticides.

3.10 REVIVAL OF TRADITIONAL KNOWLEDGE ON MEDICINAL PLANTS IN INDIA

3.10.1 Traditional Knowledge Digital Library

The Traditional Knowledge Digital Library (TKDL) is a collaborative project between the Council of Scientific and Industrial Research (CSIR), Ministry of Science and Technology, and Department of AYUSH, Ministry of Health and Family Welfare, Government of India. TKDL is being created on the codified traditional knowledge existing in local languages on Indian systems of medicine, which are *Ayurveda*, *Unani*, *Siddha*, and *Yoga*. The project TKDL was initiated in 2001. By May 2010, the TKDL database had 85,500 formulations in Ayurveda, 120,200 formulations from Unani, and 13,470 formulations from Siddha. With respect to *Yoga*, 1,098 postures have been transcribed. So far a total of 220,268 medicinal formulations have been transcribed (www.csir.res.in).

The origin of establishment of TKDL goes back to the legal battle waged by CSIR for the revocation of turmeric and basmati patents granted by the US Patent and Trademark Office (USPTO) and a neem patent granted by the European Patent Office (EPO). The TKDL has been established to prevent the misappropriation of traditional knowledge at international patent offices, as well as to derive cues for bioprospection.

3.10.2 CSIR Coordinated Program on Discovery and Preclinical Studies of New Bioactive Molecules and Traditional Preparations

Under this program of universal importance, 21 institutes of CSIR and university departments are involved in bioresource prospecting toward development and commercialization of bioactive substances from plant and microbial sources, including drug formulation, and the combination of plants processed in a particular form from *Ayurveda*, *Unani*, and *Siddha* systems of medicine. This program aims at the following major objectives:

- Isolation and characterization of bioactive molecules from selected plants with established efficacy in Indian systems of medicine—namely, *Ayurveda*, *Unani*, and *Siddha*; folk; or tribal medicine, based on their preclinical evaluation (pharmacognosy/biological activities and clinical evidence)
- Development of internationally acceptable herbal formulations for treatment of various dysfunctional diseases manifested in the form of arthritis, diabetes, viral hepatitis, cancer, tuberculosis, and other immune disorders

Based on the leads obtained during the last decade, 25 discovery groups in 13 institutes have been instituted. Currently, many leads (not disclosed here) are at the advanced stage in the CSIR Network Project.

3.10.3 New Millennium Indian Technology Leadership Initiative (NMITLI) Herbal Project

To realize the vision of developed India through science and technology, the project was initiated by CSIR in the year 2002. The project envisages a fast-track innovation-driven development of herbals through contemporary science and technology. The aim of NMITLI is to place the ancient Indian system of medicine, the *Ayurveda*, with enough scientific evidence for its global acceptability. The targets covered under this project include diabetes mellitus, hepatitis, and arthritis. Under this project, several drugs (names are not disclosed) are under clinical trial using a reverse pharmacology approach.

3.10.4 Golden Triangle Partnership

The knowledge base of traditional systems such as *Ayurveda* combined with modern science could provide new functional leads to reduce time, money, and toxicity—the three main limitations in drug development. Keeping with such a prospective approach, three organizations of the government of India—AYUSH, ICMR (Indian Council of Medical Research), and CSIR—under a tripartite agreement are developing formulations/plant-based drugs for identified diseases. AYUSH is providing technical guidance regarding formulations to be used, CSIR is carrying the preclinical studies on the scientific formulations, and ICMR is conducting the clinical trials. The ultimate aim and objectives of the project are to

- Develop safe and effective standardized herbal products for the identified disease conditions
- Develop new Ayurvedic and plant-based products in disease conditions of national/global importance
- Develop mechanisms to make products affordable in the domestic market
- Use appropriate technology for development of single and bioherbal products to make it globally acceptable
- Develop patentable products

The work is in progress in 14 areas of this "Rasayan" partnership: joint disorders, memory disorders, menopausal syndrome, bronchial allergy, fertility and infertility, cardiac disorders, sleep disorders, irritable bowel syndrome, vision disorders, unilethias and benign prostrate hypertrophy (BPH), malaria/filaria/leishmaniasis, and diabetes.

3.10.5 National Medicinal Plant Board

Plants yield ingredients for some of the most advanced medications available today. They also provide remedies that have been used by indigenous communities for centuries. The very communities that survive on harvesting medicinal plants are unable to do so in a sustainable manner. With the primary mandate of coordinating all matters relating to medicinal plants and to support policies and programs for growth of trade, export, conservation, and cultivation, the government of India in November 2002 set up the National Medicinal Plants Board (NMPB). The NMPB aims to augment the efforts of communities to conserve and harvest plants in a sustainable manner and to bring the benefits of this knowledge to the world. At the same time, the board will ensure that communities involved in the cultivation, procurement, and intellectual property rights (IPR) get their true value in return. The board is located in the Department of AYUSH of the Ministry of Health and Family Welfare (http://nmpb.nic.in).

3.10.6 Traditional Indian Medicinal Plants Yielding Useful Drugs

The indigenous medicinal plants are not merely used in production of traditional formulations, but also serve as resource material for production of clinically useful prescription drugs. There are at least 121 major clinically useful prescription drugs derived from higher plants (Farnsworth and Soejarto 1991; Kumar et al. 1997; Lavania 2005) and their number is ever increasing. Production of plant-based modern drugs has become an important segment of the Indian pharmaceutical industry for the past three and a half decades. Notable single active molecules currently in use are morphine, codeine and papaverine, quinine, quinidine, hyoscine and hyocymine, colchcine, digoxin, lanatoside, berberine, vinblastin, vincrystin, reserpine, recinamine and deserpidin, diosgenin, sennosides, podophylotoxin, artemisinin, taxol, pilocarpine, and sylmarine (Handa 1992; Lavania and Lavania 1995; Kumar et al. 1997; Patwardhan, Vaidya, and Chorghade 2004). A selective list of important traditional Indian medicinal plants yielding useful drugs is provided in Table 3.1 (drawn from the important work and compilation by Chopra, Nayer, and Chopra 1956; Thakur, Puri, and Husain 1989; Farnsworth and Soejarto 1991; Husain et al. 1992; Husain 1993; Lavania and Lavania 1994, 1995; Evans 2009; Kumar et al. 1997; Singh et al. 1999, as well as authors' own understanding).

3.10.7 Promising/Emerging Indian Medicinal Plants for Bioprospection and Drug Development

Traditional and ethnomedicine have played a prominent role in the development of new therapeutic agents. A large number of plant-based traditional remedies for the treatment of various diseases and health care are recorded for Indian medicinal plants, but their global acceptance is limited for want of supporting scientific evidence. Approximately 60% of the antitumor and anti-infective agents currently in use or in the stage of advanced clinical trials are of natural product origin

Table 3.1 Traditional Indian Medicinal Plants Yielding Useful Drugs

S. No.	Species/Family	Common or Trade Names	Distribution	Clinical Use/Active Principle
1	*Adhatoda zeylanica* Medic; syn. *A vasica* Nees; *Justicia adhatoda* L. (Acanthaceae)	Vasaka, Malabar nut	Wild, plains and sub-Himalayan tracts, up to 1200 m	Leaf and root—in chronic bronchitis, expectorant; alkaloids—vasicine (peganine)—bronchodilator, oxytocic, vasicinone—bronchoconstrictor
2	*Anamirta cocculus* L. W & A; syn; *A. paniculata* Coleb. (Menispermaceae)	Kakmari, cocoulus, fish berries	Wild, northeast regions; Malabar coast	Berry fish and crow poison; a highly toxic substance—picrotoxin from berries analeptic; used in barbiturate and other narcotics, poisoning etc.
3	*Ananas comosus* L. Merr.; syn. *A sativus* Schult. & Schult. f. (Bromeliaceae)	Annanas, pineapple	Cultivated in Assam, West Bengal, and west coastal area	Leaf and green fruit—abortifacient; bromelain—therapeutically valued in modulating (a) tumor growth, (b) blood coagulation, (c) inflammatory changes, (d) debridement of third degree burn, and (d) enhancing absorption of the drug
4	*Andrographis paniculata* (Burn. F.) Wall. ex Nees (Acanthaceae)	Kalmegh	Wild in tropical, moist, deciduous forests; cultivated in limited scale	Plant—a bitter tonic, used as a febrifuge; in many ailments of liver and alimentary tract; andrographolide and neoandrographolide used in bacillary dysentery; andrographolide—antihepatotoxic
5	*Areca catechu* L. (Arecaceae)	Supari, areca nut	Cultivated in coastal and tropical regions of India and foothills of Assam	Nut—extensively used as masticatory, anthelmintic nervine, emmenagogue; alkaloid—arecoline—anthelmintic
6	*Atropa acuminata* Royle ex. Lindl.; *A. belladonna* auct. non L. (Solanaceae)	Suchi, luffah, Indian belladonna	Occurs naturally in northwestern Himalaya from Kashmir to Himachal Pradesh at 1500–3600 m	Tropane alkaloids—anticholinergic action, used in respiratory diseases, bronchial trouble, intestinal disorders, ulcers, colic pain, and as a source of atropine and substitute for true belladonna (*A. belladonna*)
7	*Berberis aristata* DC. (Berberidaceae)	Daruharidra, berberis	Grows in temperate Himalaya up to an altitude of 1800–3500 m	Tonic, astringent, febrifuge, hepatic dysfunction, laxative menorrhagia; alkaloid—berberin—antibacterial
8	*Brassica juncea* (L.) Czern. & Coss. (Brassicaceae)	Rai; Indian mustard	Cultivated in plains of north India	Seed—antirheumatic, emetic for children, stimulant to gastric mucosa and increases pancreatic secretion; allyl—isothiocyanate–rubefacient
9	*Camellia sinensis* (L.) Kuntze; syn. *C. thea* Link, *Thea sinensis* L. (Theaceae)	Chai, tea	Grown at high elevation as plantation crop in Himachal Pradesh, Uttarakhand, Assam, and west Bengal	Leaf—stimulant, anticephalalgic, cardiotonic, hypoglycemic, anti-inflammatory, caffeine—CNS (central nervous system) stimulant
10	*Cannabis sativa* L.; syn., *C. indica* L. (Cannabaceae)	Bhang, Ganja, Indian hemp, marijuana	Found wild in lower Himalaya from Kashmir to east of Assam	Inflorescence top—intoxicant, stomachic, analgesic, stimulant, aphrodisiac, THC (tetrahydrocannabinol)—antiemetic, decrease ocular tension

(continued)

Table 3.1 Traditional Indian Medicinal Plants Yielding Useful Drugs (Continued)

S. No.	Species/Family	Common or Trade Names	Distribution	Clinical Use/Active Principle
11	Carica papaya L. (Caricaceae)	Papita, papaya	Cultivated in most of the states of India	Fruits—antibacterial, digestive, diuretic, anthelmintic, source of proteolytic enzyme—papain, mucolytic
12	Cassia senna L. syn. C. angustifolia Vahl (Caesalpiniaceae)	Sonamukhi, senai; Indian senna, Tinnevelly senna	Cultivated in dry areas of Tirunelveli, Madurai and Triruchirapalli districts in Tamil Nadu, and in Rajasthan and Gujarat and other dry parts of India	Senna leaves and pods in form of powder and infusion used for either habitual or occasional constipation; sennoside A and B—used as laxative
13	Centella asiatica (L.) Urban; syn. Hydrocotyle asiatica L. (Hydrocotylaceae)	Mandukparni/Brahmi Gotukola	Common in moist, shady places all over India, up to 200 m	Whole herb is rejuvenating, improves memory and is used in skin diseases of varied nature; asiaticoside–vulnerary
14	Cephaelis ipecacuanha (Brot.) A. Rich; syn. Psychotria ipecacuanha Stokes (Rubianceae)	Ipecac (Brot.)	Cultivated in Mungpoo, near Darjeeling (W.B.) and in Nilgiris (Kallar) and Sikkim (Rungbee)	Root—emetic, expectorant, diaphoretic and in amoebic dysentery, emetine hydrochloride in the form of injection and emetine bismathiodide is orally used for amoebic dysentery
15	Cinchona spp. (Rubiaceae) (a) C. calisaya Wedd. (b) C. calisaya Wedd. var. Ledgerinana How. (c) C. officinalis L.		Yellow cinchona cultivated in Nilgiri (at 270 m) and Sikkim, crown bark cinchona cultivated in west Bengal, Khasi Hills and Tamil Nadu	The bark of Cinchona spp. has been used as antimalarial for a long time in oriental medicine; source of over 30 quinoline alkaloids. The salts of quinine are used as antimalarial and in tonic water and carbonated drinks as a bitter; quinidine sulfate is cardiac depressant and used to cure arterial fibrillation
16	Cissampelos pareira L. (Menispermaceae)	Laghu Patha, velvet leaf	Grows wild throughout India in tropical and subtropical zones	Intermittent fever, heat stroke, colic, diarrhea, cissampeline; skeletal muscle relaxant
17	Colchicum luteum Baker (Colchicaceae)	Hirantuliya, golden collyrium	Temperate Himalaya from Kashmir to Himachal Pradesh at 700–2800 m	Corm—anti-inflammatory, antirheumatic, used in gout, diseases of liver and spleen, colchicines analogs—DTC, DMC, and TMC—effective in gout; colchicine amide demecolcine—in cancer
18	Coptis teeta Wall (Ranunculaceae)	Mamira, coptis	Occurs in localized areas of Lohit, Dibang Valley, east and west Siang and upper Subansiri districts of Arunachal Pradesh; cultivated in Nagaland in limited scale	Rhizome, bitter tonic antiperiodic and alterative; alkaloid—palmatine—antipyretic, detoxicant
19	Coscinium fenestratum (Gaertn.) Coleb. (Menispermaceae)	Daru-haridra, false calumbo, tree turmeric	In most evergreen and semideciduous forests of western ghats, Kerala, Karnataka, and Tamil Nadu	Plant, stomachic, tonic, used in cholera and in tetanus, aqueous extract and alkaloid berberine—prophylactic agent against tetanus

No.	Botanical name (Family)	Common names	Distribution	Uses
20	*Curcuma domestica* Veleton; syn. *C. longa* L. (Zingiberaceae)	Haldi, haridra, turmeric	Cultivated in humid climatic regions of India	Rhizome—aromatic stomachic, diuretic, used in jaundice, hepatitis, wound healing and in respiratory diseases; curcumin, chloretic—strong antihepatotoxic action
21	*Datura metel* L.; syn. *D. fastulosa* L. (Solanaceae)	Datura, downy datura	Common weed growing in wastelands of the country	Used since ancient times as intoxicant and narcotic, whooping cough and asthma, etc. Scopolamine (hyoscine) — used as a sedative
22	*D. stramonium* L.; syn. *D. tatula* L. (Solanaceae)	Sada datura, stinkweed, thorn apple	Himalaya from Kashmir to Sikkim up to 2700 m; hilly districts of central and south India	Bechic, boils, leaf, and seeds—antispasmodic, narcotic; atropine and hyoscine used in ophthalmic practice, hyoscine—sedative
23	*Ephedra gerardiana* Wall. ex Stapf (Ephedraceae)	Som, ephedra	Drier regions of temperate Himalaya from Kashmir to Sikkim from 2350 to 5350 m	Root—alternative, stomachic, tonic, cardiac circulatory stimulant and antisudorific, etc.; ephedrine, pseudoephedrine, and norpseudoephedrine—bronchodilator, used in asthma and hay fever
24	*Gaultheria fragrantissima* Wall; syn. *G. leschenaultia* DC. (Ericaceae)	Gandhpura, Hemantari, Indian wintergreen	Hill regions of north and east India, from 2000 to 2700 m and in the western Ghats above 1700 m	Leaf oil used as a stimulant, carminative, and antiseptic, externally applied in sciatica, rheumatism, and neuralgia, internally for controlling hookworm infections; essential oil is identical with *G. procumbens* oil and used exclusively as a flavoring agent and insect repellent
25	*Gentiana kurroo* Royle (Gentianaceae)	Karu, Indian Gentian	In hilly regions of Jammu and Kashmir, Himachal Pradesh and Uttarakhand at an altitude of 2300–3500 m	Roots and rhizome—a bitter tonic, anthelmintic and blood purifier, substitute for true gentian (*G. lutea*)
26	*Glycyrrhiza glabra* L. (Fabaceae)	Mulethi, licorice	Grows in Punjab, Jammu Kashmir, Gujrat, Rajasthan to a limited extent	Crude, licorice is used as demulcent in peptic ulcer, expectorant, antitussive, laxative and sweetener; licorice extract and glycyrrhetinic acid—used in rheumatoid arthritis, Addison's disease and in various inflammatory conditions
27	*Hyoscyamus niger* L. (Solanaceae)	Khursani Ajawayan, Indian henbane, black henbane	Found in northwestern Himalaya, between 1700 and 3700 m	Extract used in diseases of respiratory and intestinal tracts and in urinary bladder pain to control spasms and secretion; hyoscyamine, atropine—anticholinergic
28	*Lobelia nicotianifolia* Roem & Schult. (Campanulaceae)	Jangali Tamaku, Nala, wild tobacco	Western ghats from Maharashtra to Kerala, from 1000 to 2000 m	Leaf antiseptic; root in scorpion sting, lobeline—tobacco deterrent and respiratory stimulant
29	*Mucuna pruriens* (L.) DC.; syn. *M. Prurita* Hook. (Fabaceae)	Kiwanch; Kapikachhu horse-eye bean, cowhage	Common in forest plains of India	Seeds aphrodisiac, good source of l-dopa used in Parkinsonism and other muscular disorders

(continued)

Table 3.1 Traditional Indian Medicinal Plants Yielding Useful Drugs (Continued)

S. No.	Species/Family	Common or Trade Names	Distribution	Clinical Use/Active Principle
30	*Panax pseudo-ginseng* Wall. (Araliaceae)	Ajambari (Nepali), Himalayan ginseng	Found in temperate Himalaya in Sikkim, Arunachal Pradesh, at the altitude of 2900a–4000 m	Considered allied to Korean and Japanese ginseng, undoubtedly useful as a tonic in increasing longevity, mental agility, and checking of hypertension; saponin fractions have adaptogenic immunostimulant and anti-inflammatory properties comparable to Korean ginseng (*P. ginseng*)
31	*Papaver somniferum* L. (Papaveraceae)	Afeem, poppy	Cultivated in India in the states of Rajasthan, Uttar Pradesh, Madhya Pradesh, under government control	Crude opium is used in traditional medicine as narcotic, sedative, anodyne, controlling dysentery and diarrhea; alkaloids—morphine, analgesic, codeine—antitussive, noscapine—antitussive, papaverine—smooth muscle relaxant
32	*Plantago ovata* Forsk.; syn. *P. ispaghula* Roxb. (Plantaginaceae)	Isabgol, blend psyllium, Ispaghula	Cultivated in north Gujarat, Haryana, and Rajasthan	The husk and seed in the form of powder, infusion and decoction used in chronic constipation; also used to control diarrhea and dysentery
33	*Rheum australe* D. Don; syn. *R. emodi* Wall. ex. Meisn. (Polygonaceae)	Revanchini, Himalayan rhubarb, Indian rhubarb	Found in subalpine and alpine Himalaya from Himachal Pradesh to Nepal at 3000–4200 m	The root and rhizomes are stomachic bitter and cathartic, safe for children and the elderly; anthraquinomes—emodine and rhein—purgative
34	*Ricinus communis* L. (Euhorbiaceae)	Erand, castor	Cultivated or found wild throughout India	Seed oil—antirheumatic, castor oil—purgative, castor oil gel—in dermatosis and eczema
35	*Silybum marianum* (L.) Gaertn. (Asteraceae)	Holy thistle, milk thistle	Northwestern Himalaya from 2000 to 2500 m	Leaf operient, sudorific; seed demulcent, antihemorrhagic; flavonolignans from fruits (seed) exert marked antihepatotoxic activity; commercially available silymarin is standardized mixture of compounds
36	*Strychnos nux-vomica* (Loganiaceae)	Kuchala, nux vomica	Trees wild in tropical semi-evergreen forests of India	Seed after purification—stimulant, used for chronic indigestion, diseases of nervous systems, diarrhea, fever, strychnine—circulatory stimulant, improves the appetite and digestion

No.	Botanical name	Common name	Distribution	Uses
37	*Swertia chirayita* (Roxb. ex Fleming) Karton; syn. *S. chirata* (Wall.) C.B. Clarke (Gentianaceae)	Chireta, Chirayita	Found wild in the temperate Himalaya at 1200–1300 m in Himachal Pradesh, Uttarakhand, Sikkim, Nagaland, Meghalaya	Whole plant is valued for its bitter and tonic properties and credited for febrifugal, laxative, stomachic, anthelmintic, and antidiarrheal properties in indigenous systems of medicine
38	*Taxus baccata* L. subsp. *wallichiana* (Zucc.) Pilger, syn. *T. Wallichiana* Zucc. (Taxaceae)	Talishpatra, Himalayan yew	Temparate Himalaya at 1800–3300 m, from Kashmir to Bhutan and in Khasi hills at 1500 m	In indigenous systems of medicine, considered as expectorant, emmengogue, sedative and antiseptic indicated in asthma, bronchitis, hiccough, and indigestion; Taxol and its related analogues as anticancer
39	*Urginia indica* (Roxb.) Kunth, syn. *Scilla indica* Roxb. Non (Wt.) Baker (Liliaceae)	Jangali Piyaz, Indian squill	Throughout India from north to south up to 1500 m	Bulb, cardiac stimulant, expectorant, digestive, diuretic, emetic, antirheumatic and used in skin diseases, hemorrhages in kidney and uterus, scillarin A and B—bronchial troubles
40	*Valeriana jatamansi* Jones; syn. *V. Wallichii* DC. (Valerianaceae)	Tagar, Indian valerian	Temperate Himalaya from Kashmir to Bhutan above 300 m	Whole plant—carminative, sedative and used as an antispasmodic in hysteria and other nervous disorders, and in perfumery industries; valepotriates—sedative and spasmolytic

(Cragg, Newman, and Snader 1997). Over 5,000 species and more than 120,000 phytochemicals from higher plants have been examined for their therapeutic potential worldwide. During 1991–1995 around 200 NCEs (new chemical entities) were introduced globally, of which 10 (2%) led to the drug development and commercialization (Dev 1997).

In India, over 4,000 terrestrial plants have been studied for their biological activities and 20% have exhibited promising new activities (Figure 3.1). As such, exhaustive studies on bioprospection of plant species have been undertaken in several laboratories in India, particularly in the national laboratories of CSIR, with the hope to discover new phytopharmaceuticals and clinically validated plant-based drugs in the near future. In this respect, Table 3.2 provides a list of select promising

Figure 3.1　Close-up of a few prospective Indian medicinal plants offering promising drug development. From left to right: upper row—*Withania somnifera, Andrographis paniculata;* lower row—*Phyllanthus amarus, Gloriosa superba.*

Table 3.2 Promising/Emerging Indian Medicinal Plants for Bioprospection and Drug Development

Plant Name/Brief Description	Traditional Uses	Therapeutic Applications	Major Chemical Constituents	Major Source Ref.
Acorus calamus L. (Acoraceae), commonly known as Vacha, is a semiaquatic, perennial, aromatic herb with creeping rhizomes. It is distributed throughout India at moist places. Since antiquity, calamus rhizome has been used in medicinal baths, in incense, and in tea.	In Ayurveda, the rhizome is used as an aromatic, stimulant, bitter, and tonic, carminative, antispasmodic, emetic, and expectorant, emmenagogue, aphrodisiac, laxative, and diuretic. Traditionally, the rhizome has been used in flatulent colic and chronic dyspepsia, and ethnobotanically it is used in asthma, bronchitis, body ache, cold, cough, and inflammation.	The ethyl acetate extract of *Acorus calamus* is found to be potent antioxidant by inhibition of 1,1-diphenyl-2-picrylhydrazyl (DPPH) free radicals. In clinical trial, chewing of 1–2 g fresh rhizome for 1–2 h showed excellent response in patients of Tamakswas (bronchial asthma).	The essential oil obtained from rhizome is constituted of α-asarone (up to 82%) and its β-isomer; other constituents of the oil are β-pinene, myrcene, camphene, *p*-cymene, camphor, and linalool.	Parotta (2001); Govindarajan, Agnihotri, et al. (2003)
Andrographis paniculata (Burm. f.) Wall. ex. Nees (Acanthaceae) is commonly known as "Kālmegh." The plant is widely distributed in moist, deciduous forests in India. It is used as a substitute for *Swertia chirayta* and is popular as green Chirayta.	In Ayurveda, the leaf juice is a household remedy for flatulence, loss of appetite, bowel complaints of children, diarrhea, dysentery, dyspepsia, and general debility; it is preferably given with the addition of aromatics. Decoction or infusion of the leaves gives good results in sluggish liver, neuralgia, general debility, convalescence after fevers, and advanced stages of dysentery.	*Andrographis paniculata* shows antioxidant and hepatoprotective activity; it increases kidney TBARS in normal rats along with increase in the activity of SOD and CAT but has no effect on GSH-px activity in diabetic rats, showing that it possesses an antihyperglycemic property and may also reduce oxidative stress in diabetic rats.	The herb contains diterpenoid and sesquiterpenoid compounds—paniculosides and andrographolides.	Mishra, Sangwan, and Sangwan (2007); Niranjan, Tewari, and Lehri (2010)
Asparagus racemosus Willd. (Asparagaceae), commonly known as Shatavari, is a tall climber undershrub found all over India.	Almost all parts of this plant are used by the Indian traditional system of medicine (Ayurveda and Unani) for the treatment of various ailments. In particular, the roots are used in dysentery, diarrhea, tuberculosis, leprosy, skin diseases, epilepsy, inflammations, and as an expectorant.	The aqueous extracts of fresh and dry roots were found to have amylase and lipase activities. Both the crude extract and an active fraction consisting of polysaccharides provide partial protection against radiation-induced loss of protein thiols and inactivation of SOD. The inhibitory effect of these active principles is comparable to that of the established antioxidants GSH and ascorbic acid. An antioxidant compound racemofuran isolated from *A. racemosus* shows activity against DPPH with IC 50 value of 130 µM. It possesses moderate antidiabetic activity but exhibits potent antioxidant potential in diabetic conditions.	Root contains saponins—Shatavarin I–IV.	Jain (1991); Govindarajan et al. (2004); Wiboonpun, Phuwapraisirisan, and Tip-pyang (2004); Velavan et al. (2007)

(continued)

Table 3.2 Promising/Emerging Indian Medicinal Plants for Bioprospection and Drug Development (Continued)

Plant Name/Brief Description	Traditional Uses	Therapeutic Applications	Major Chemical Constituents	Major Source Ref.
Bacopa monnieri (L.) Penn. (Scrophulariaceae), commonly known as Brahmi in Ayurvedic materia medica. It is a small prostrate herb, growing wild on damp places and marshy lands of India and ascending to an altitude of 1320 m. The drug Brahmi is elaborately described in classical texts of Ayurveda and is well known for increase of intellect and memory.	In the indigenous system of medicine it is prescribed for the treatment of nervous disorders such as insanity, epilepsy, neurasthenia, nervous breakdown, dermatosis, cough, fever, anorexia, anemia, diabetes, and loss of memory. In folk or tribal medicine it is used for the treatment of insanity, asthma, leucorrhoea, and liver diseases. Brahmi extract is also known to have anticancer and antioxidant properties.	The saponins present in the plant—namely, bacosides A and B—have been indicated for memory-enhancing properties. Clinical studies performed on *B. monnieri* show its usefulness in treatment of gastrointestinal disorders, hypothyroidism, and cancer. Treatment with *B. monnieri* extract significantly increased the antioxidant enzyme status, reduced the rate of lipid peroxidation, and markers of tumor progression in the fibrosarcoma-bearing rats. It is shown that Bacopa extracts modulate the expression of certain enzymes involved in generation and scavenging of reactive oxygen species in the brain. *In vitro* research has shown that *Bacopa* exerts a protective effect against DNA damage in astrocytes in human fibroblasts and broncho-vasodilatory activity in guinea pig trachea, pulmonary artery, and aorta. The bacosides A and B isolated from the plant have been found to improve the performance of rats in several learning tests as evidenced by better *acquisition*, improved retention, and delayed extinction of information.	Major chemical components are alkaloids—brahmine and herpestine, and saponins—bacoside A and B.	Singh et al. (1988); Russo et al. (2003); Rohini, Sabitha, and Devi (2004); Singh et al. (2005); Gohil and Patel (2010)
Boswellia serrata Roxb. ex Colebr. (Buseraceae) is a moderately sized gregarious tree found throughout India in the drier forests of western and central India from Bihar to Rajasthan, in northern India on dry hills, and southward into the Deccan Peninsula. Trunk exudes oleogum resin commonly known as Shallaki, Loban, or Salai Guggal.	Traditionally, it has been used for mitigating osteoarthritis, juvenile rheumatoid arthritis, soft tissue fibrosis, and spondilytis without any side effect and also as demulcent, aperient, to make it an alternative, and a purifier of blood. It is used in rheumatism, nervous diseases, scrofulous affections, urinary disorders, and skin diseases, generally combined with aromatics. It is also prescribed with clarified butter in syphilitic diseases	In addition to its sedative effect, the gum resin of *B. serrata* possesses marked analgesic activity in experimental animals; it produces reduction in the spontaneous motor activity and causes ptosis in rats. Also, it shows anti-inflammatory, antiarthritic, antiasthmatic, antimicrobial, antidiarrheal, and chronic toxicity activity. Alcoholic extract of Salai Guggal (AESG) causes hepatoprotection in galactosamine-	The essential oil of *B. serrata* comprises monoterpenoids, of which pinenes are a major constituent; others present in less than 2% each are *cis*-verbenol, *trans*-pinocarveol, borneol, myrcene, verbenone, limonene, α-thujene and *p*-cymene, boswellic acid, and several other triterpenoids have also	Anonymous (2008); Upaganlawar and Ghule (2009); Sharma et al. (2007, 2009)

Botanical name and description	Therapeutic/ethnomedical uses	Chemistry/pharmacology	References
and as a stimulant in pulmonary diseases and bronchitis. Herbal formulation containing *B. serrata* oleogum resin as one of the ingredients has been reported to produce significant antidiabetic activity on non-insulin-dependent diabetes mellitus in streptozocin-induced diabetic rat model where reduction in blood-glucose level was comparable to that of phenformin.	and endotoxin-induced liver damage in mice, and it exhibits anticarcinogenicity in mice with ehrlic ascites carcinoma and S-180 tumor, inhibition of tumor growth by inhibiting cell proliferation, and cell growth due to its interference with biosynthesis of DNA, RNA, and proteins. Water-soluble fraction of *B. serrata* extract decreases total cholesterol and increases HDL in rats fed an atherogenic diet.	been isolated from the gum resin.	Agarwal and Parks (1983); Dubey, Srimal, and Tandon (1997); Srivastava et al. (2002); Kavitha, Rajamani, and Vadivel (2010)
Coleus forskohlii (Willd.) Briq.; syn. *Plectranthus forshkohlii* Willd.: *Coleus barbatus* (Andrews) Benth.; *Plectranthus barbatus* Andrews (Lamiaceae); commonly known as Gandeer. Gandeer is an important drug used in indigenous systems of medicine as a remedy for heart diseases; abdominal, colic and respiratory disorders; painful micturition; and certain CNS disorders, such as insomnia and convulsions.	Root extract exhibits antihypertensive activity and antisenescence property. The diterpenes found in the roots are useful for curing different ailments, including cardiovascular diseases, hypertension, asthma, glaucoma, and Alzheimer's disease.	Chemical investigation led to the isolation of a number of diterpenoids: coleonol, barbatusin, and forskolin offers a new class of anti-glaucoma drug.	
Commiphora wightii (Arnott.) (Bursepaceae), commonly known as Indian Bdellium or Guggul, is an important medicinal plant of the herbal heritage of India. It provides Guggul, an oleogum resin of which medicinal and curative properties are mentioned in the classical Ayurvedic texts. *Commiphora* is widely distributed in tropical regions of Africa, Madagascar, and Asia. In India, it is found in wild state in Rajasthan, Gujarat, and Maharashtra. The Guggul is used in Ayurvedic system of medicine for a variety of ailments and is useful in several conditions—namely, urinary infections, ascites, piles, fistula, arthritis, rheumatism, swelling, ulcers, bone fractures, obesity, and disorders of lipid metabolism.	Guggul has been extensively investigated for its theraperutic applications. It is reported to be efficacious as an antiarthritic, antirheumatic, antiphlogistic, and hypolipidemic agent. Oral administration of standardized extract gugulipid exhibits dose-related lowering of serum cholesterol and triglycerides and also protects against cholesterol-induced atherosclerosis. The ethyl acetate-insoluble but water-soluble polysaccharide fraction, which constitutes 55% of the gum, was found to be toxic to rats and devoid of hypolipidemic activity, while ethylacetate-soluble fraction exhibited marked hypolipidemic activity.	Major chemical constituents are gugulsterones E and Z, guggulsterols I–V, guggulsterol VI, Z-guggulsterol, steroids, diterpenoids, aliphatic esters.	Satyavati, Dwarkanath, and Tripathi (1969); Purushothaman and Chandrashekharan (1976); Kakrani (1981); Jain and Gupta (2006)

(continued)

Table 3.2   Promising/Emerging Indian Medicinal Plants for Bioprospection and Drug Development (Continued)

Plant Name/Brief Description	Traditional Uses	Therapeutic Applications	Major Chemical Constituents	Major Source Ref.
Curcuma longa L. (Zingiberaceae), commonly called Haldi, is a perennial herb that measures up to 1 m high with a short stem; it is distributed throughout tropical and subtropical regions of the world, being widely cultivated in Asiatic countries, mainly in India and China. Its rhizomes are oblong, ovate, pyriform, and often short branched. Turmeric powder has been in continuous use for its flavoring digestive properties as a spice in both vegetarian and nonvegetarian food preparations.	Current traditional Indian medicine claims the use of its powder against biliary disorders, anorexia, coryza, cough, diabetic wounds, hepatic disorder, rheumatism, and sinusitis. In old Hindu medicine, it is extensively used for the treatment of sprains and swellings caused by injury. The traditional medicine in China uses *C. longa* in diseases associated with abdominal pains.	Curcumin is a good antioxidant and inhibits lipid peroxidation in rat liver microsomes, erythrocyte membranes, and brain homogenates. The lipid peroxidation has the main role in inflammation, heart diseases, and cancer. Curcumin has inhibitory effects on proliferation of blood mononuclear cells and vascular smooth muscle cells. Ethanolic extract from *C. longa* inhibits *in vitro* growth of *Plasmodium falciparum* and *Leishmania major* parasites, shows antimicrobial, anti-HIV, antitumor, antimutagenic, and nematocidal activity. Administration of turmeric powder provides relief in symptoms like dyspnea, cough, sputum or physical signs, rheumatoid arthritis. Turmeric, when administered orally with phenytoin, significantly prevents cognitive impairment and oxidative stress in rats.	Turmeric comprises a group of three curcuminoids: curcumin (diferuloylmethane), demethoxycurcumin, and bisdemethoxycurcumin, as well as volatile oils (tumerone, atlantone, and zingiberone), sugars, proteins, and resins.	Govindarajan (1980); Chattopadhyay et al. (2004); Sasikumar (2005); Jurenka (2009)
Gloriosa superba L. (Liliaceae) commonly known as "Kali hari," is a perennial, climbing or scrambling glabrous herb occurring in light porous soils of plains and low hills almost all over India. Rhizomes and seeds are the source of colchicine used in modern medicine.	Rhizomes are used in the indigenous systems of medicine. The paste of the drug is applied on the palms and soles for easy and painless delivery.	The water-soluble portion of the rhizome has been found to exhibit oxytocic effect on the isolated uterus of guinea pigs, rabbits, dogs, and human beings. The fresh juice showed strong ecobolic, spasmodic activity of varying degrees in isolated dog tracheal chain and CNS depressant action on isolated frog heart. Colchicine is used for the relief of pain in acute gout and prevents attacks of Mediterranean fever. It has antimitotic properties and its derivatives have been used in the control of neoplasm of skin.	Major chemical constituents include colchine, 3-demethyl colchicines and colchicoside from different parts of the plant.	Chaudhary (1993); Chaudhary and Thakur (1994); Ade and Rai (2009)

Glycyrrhiza glabra (L.) (Fabaceae), commonly known as *Vashtimadhu* or *Mulethi*, is native to the Mediterranean and certain areas of Asia. The plant is reported to be cultivated to a limited extent in India.	In Ayurveda, infusion, decoction, or extract of roots is used as a demulcent in inflammatory or irritable conditions of the bronchial tubes or bowels, such as cough, hoarseness, sore throat, asthma and dysuria, and catarrh of the genitourinary passage.	In modern medicine, licorice extracts are often used as a flavoring agent to mask bitter taste in preparations and used after preparations, as an expectorant in cough and cold preparations, for treatment of chronic hepatitis, and for therapeutic benefit against viruses, including human immunodeficiency virus (HIV), cytomegalovirus (CMV), and *Herpes simplex*. The isoflavones glabridin and hispaglabridins A and B have significant antioxidant activity, and both glabridin and glabrene possess estrogen-like activity. The licorice preparation reduces CAT activity in the peripheral blood and increases animal resistance to vibration stress.	Glycyrrhizin, a triterpenoid compound providing sweet taste to licorice root, represents a mixture of potassium–calcium–magnesium salts of glycyrrhizic acid. Other constituents are isoflavones hispaglabridin A, hispaglabridin B, glabridin, and methylglabridin; chalcones: isoprenylchalcone derivative and isoliquiritigenin, and the isoflavone: formononetin.	Vaya, Belinky, and Aviram (1997); Belinky et al. (1998); Oganesyan (2002); Biondi, Rocco, and Ruberto (2003)
Gymnema sylvestre (Retz.) R. Br. (fam: Asclepiadaceae), commonly called Gurmar, is a perennial, tomentose, woody climber found in wet areas of forests and distributed throughout the Indian plains and other countries such as Sri Lanka and South Africa	Sushruta describes Gurmar as a destroyer of Madhumeha (glycosuria) and other urinary disorders. On account of its property of abolishing the taste of sugar, it has been given the name of Gur-mar (meaning sugar destroying). It is believed that the drug neutralizes the excess sugar present in the body in diabetes mellitus. In cases of diabetes mellitus, 3–6 g of leaf powder is recommended to be taken with 5–10 g honey/water (preferably) thrice a day. The plant is also reported to be bitter, astringent, acrid, thermogenic, anti-inflammatory, anodyne, digestive, liver tonic, emetic, diuretic, stomachic, stimulant, anthelmintic, laxative, cardiotonic, expectorant, antipyretic, and uterine tonic. It is useful in dyspepsia, constipation, jaundice, hemorrhoids, renal and vesical calculi, cardiopathy, asthma, bronchitis, amenorrhea, conjunctivitis, and leucoderma.	Patients with insulin-dependent diabetes mellitus (IDDM) who were on insulin therapy were administered water-soluble extract of the leaves of *Gymnema sylvestre*; their insulin requirement came down with fasting blood glucose and glycosylated hemoglobin (HBA1c) and glycosylated plasma protein levels. While serum lipids returned to near normal level, glycosylated hemoglobin and glycosylated plasma protein levels remained higher than controls. IDDM patients on insulin therapy show no significant reduction in serum lipids, HBA1c, or glycosylated plasma proteins. *Gymnema sylvestre* therapy enhances endogenous insulin, possibly by regeneration/revitalization of the residual β-cells in IDDM.	The leaves contain antisweet compounds, which have been found to be a mixture of triterpene saponins designated as gymnemic acids A, B, C, and D.	Baskaran et. al. (1990); Masayuki et. al. (1997); Kanetkar, Singhal, and Kamat (2007); Saneja et al. (2010)

(continued)

Table 3.2   Promising/Emerging Indian Medicinal Plants for Bioprospection and Drug Development (Continued)

Plant Name/Brief Description	Traditional Uses	Therapeutic Applications	Major Chemical Constituents	Major Source Ref.
Holarrhena antidysenterica (G. Don) Wall. ex A. DC. (Apocynaceae), commonly known as Kutaja, is a glabrous or pubescent tree or large shrub found throughout the deciduous forest areas of India at low elevations and up to 1100 m in the tropical Himalayan tract. Most of the scholars of Ayurveda, like Bhava Mishra, Madanapala, Narahari, and PV Sharma, mention Kutaja as having Titka and Kashaya Rasa, Shita Virya, and Tridosha Hara, especially Kapha Pitta Shaman properties.	*Holarrhena antidysenterica* (stem and root bark) is a reputed traditional remedy for amoebic dysentery and other gastric disorders. Bark is used as an astringent, anthelmintic, antidontalgic, stomachic, febrifuge, antidropsical, diuretic, in piles, colic, dyspepsia, chest affections, and as a remedy in diseases of the skin and spleen.	Conessine from the bark killed free living amoebae and also kills *Entamoeba histolytica* in the dysenteric stools of experimentally infected kittens. It is markedly lethal to the flagellate protozoan. It is antitubercular also. Conkurchine and its hydrochloride are hypotensive and a vasodilator.	Around 30 alkaloids have been isolated from the plant, mostly from the bark. These include conessine, kurchine, kurchicine, holarrhimine, conarrhimine, conaine, conessimine, iso-conessimine, conimine, holacetin, and conkurchin.	Dwivedi and Sharma (1990); Daswani et al. (2002); Pal, Sharma, and Mukherjee (2009)
Genus *Phyllanthus* L. (Euphorbiaceae) contains over 500 species in 10 or 11 subgenera that are distributed in all tropical regions of the world from Africa to Asia, South America, and the West Indies. *Phyllanthus amarus* Schumach. & Thonn. commonly called Bhumyamalaki, is the most widespread and medicinally important species found along roads and valleys, on riverbanks, and near lakes in tropical areas.	The plant is bitter, astringent, cooling, diuretic, stomachic, febrifuge, and antiseptic. It is useful in dropsy, jaundice, diarrhea, dysentery, intermittent fevers, diseases of urinogenital system, scabies ulcers, and wounds.	*Phyllanthus amarus* is highly valued in the treatment of liver ailments and kidney stones and has been shown to posses antihepatitis-B virus surface antigen activity in both *in vivo* and *in vitro* studies. Phyllanthin and hypophyllanthin impart antihepatotoxic action against carbon tetrachloride- and galactosamine-induced hepatotoxicity in primary cultured rat hepatocytes. They are also found to show antiviral, antihyperglycemic activity.	Phyllanthin and hypophyllanthin are two major lignans found in the genus.	Bagchi, Srivastava, and Singh (1992); Sane et. al. (1995); Srividya and Periwal (1995); Lee et al. (1996); Ott, Thyagarajan, and Gupta (1997); Khatoon et al. (2006)

Picrorhiza kurooa Royle ex Benth. (Scrophulariaceae), commonly known as Kutki, is a small perennial herb found in the Himalayan region growing at an altitude of 3000–5000 m. It is a well-known herb in the Ayurvedic system of medicine and has traditionally been used to treat disorders of the liver and upper respiratory tract, reduce fever, and treat dyspepsia, chronic disorders, and scorpion stings.	The bitter tonic of the herb is antiperiodic, cholagogue, blood purifier, stomachic aperient, and cardiotonic; it is considered useful in treating dyspepsia, respiratory disorders, and diseases of spleen, jaundice, and anemia. In bilious fever, a compound decoction of its root, licorice, raisins, and Neem bark is very curative.	Picroliv, the active principle of *Picrorhiza kurooa* comprising a mixture of the iridoid glycosides, picroside-I, and kutkoside, possesses antioxidant properties that are mediated through SOD-like activities, metal ion chelation, and xanthine oxidase inhibition. The drug is reported to produce marked reduction in serum chlolesterol and coagulation time, and it also brings down serum bilirubin levels to normal range in patients with infective hepatitis with jaundice. The alcoholic extract of the drug imparts hepatoprotective effect against carbon tetrachloride, paracetamol, and aflatoxin-induced liver damage. A drug, "picroliv," containing irridoid glycoside fraction of roots and its rhizome, containing at least 60% of 1:15 mixture of picroside I and kutkoside, was developed for the treatment of acute and chronic hepatitis in healthy carriers. Root and rhizome contain the bitter principle kutkin, which comprises two iridoid glycosides: picroside and kutkoside. A novel cucurbitacin glycoside has also been isolated from roots.	Chander, Kapoor, and Dhawan (1992); Anonymous (2001); Govindarajan, Vijayakumar, et al. (2003); Govindarajan, Vijayakumar, and Pushpangadan (2005)
Psoralea corylifolia Linn. (Fabaceal), commonly known as Babchi, is extensively used in India for skin problems. The plant is found all over India, particularly in drier parts of Rajasthan and Uttar Pradesh.	Babchi is extensively used for mitigating skin problems, especially for leprosy and leucoderma. It is particularly useful as a laxative for bilious disorders.	The major active constituents found in this species are furocoumarin and psoralen; seeds and roots contain chalcones, flavones, isoflavones, furanocoumarins. Psoralen stimulates/induces the formation of melanin pigmentation. Psoralen's administration combined with UV light treatment (PUVA) is used to increase tolerance to sunlight and in the treatment of idiopathic vitiligo.	Anderson and Voorhees (1980); Haraguchi et al. (2002); Jiangninga et al. (2005); Govindrajan et al. (2005)

(continued)

Table 3.2   Promising/Emerging Indian Medicinal Plants for Bioprospection and Drug Development (Continued)

Plant Name/Brief Description	Traditional Uses	Therapeutic Applications	Major Chemical Constituents	Major Source Ref.
Rauvolfia serpentina (L.) Benth. ex Kurz. (Apocynaceae), commonly known as Sarpgandha, is one of the most important drugs of the Indian system of medicine. It is indigenous to moist, deciduous forests of Southeast Asia. In India, it is found wild in plains and hills in deep, fertile soils that are rich in organic matter in subtropical areas and is also cultivated to some extent.	Sarpagandha is one of those medicinal plants known in folk medicine even in the pre-Vedic period in India. Roots are used in the folk and traditional system of medicine and in homeopathy for insomnia, schizophrenia, epilepsy, stress, anxiety, and depression. In folk or traditional systems of medicine, 5–10 g fine powder of Sarpagandha roots along with 15 mL honey or 10–30 mL water decoction is taken twice daily for the treatment of schizophrenia, insanity, epilepsy, and blood pressure problems.	In modern medicine, alkaloid reserpine is used in the management of essential hypertension and certain neuropsychiatric disorders, ajmaline in the cardiac arrhythmias, and ajmalicine for the treatment of circulatory diseases. Rauvolfia in various forms of powder, dry or liquid extract, and tablets is official in different pharmacopoeias of the world. Many proprietary medicines prepared out of it are being used mainly as a hypotensive agent in different doses.	The roots contain numerous alkaloids, the most important being reserpine and rescinnamine; others include ajmaline (rauvolfine), ajmalcine, serpentine, rauvolfinine, and sarpagine.	Kurup, Ramdas, and Joshi (1979); Husain (1991); Husain et al. (1992); Kumar et al. (1997)
Terminalia arjuna W. & A. (Combretaceae), commonly known as Arjun, is an important drug well described in classical texts of the Indian system of medicine for its medicinal properties for treatment of heart diseases. The plant is found commonly occurring throughout the greater part of the Indian Peninsula along rivers. It extends northward to the sub-Himalayan tract, where it is found along the banks of streams.	Terminalia bark is prescribed in the form of powder and decoction in milk. The bark is acrid, styptic, cardiotonic, febrifuge, antidysenteric in fractures and contusions, with excessive echymosis. It also gives relief in symptomatic complaint in hypertension and is a diuretic tonic in cases of cirrhosis of liver.	The drug has been extensively investigated for its various pharmacological and clinical activities—namely, cardioprotective, antianginal, spasmogenic, oxytocic, antifertility, cytotoxic, antifungal, antibacterial, hepatoprotective, antimutagenic, and anticarcinogenic. Clinically, it has been found to reduce the left ventricular mass (LVM) in coronary artery disease (CAD). The drug also shows electrocardiographic improvement and is beneficial in modifying various known coronary risk factors like obesity, hypertension, and hyperglycemia. Animal studies suggest terminalia drug might reduce blood lipids.	Major chemical components are glucosides-arjunosides I, II, III, and IV identified as arjunic acid-3-0-β-d-galactoside, arjunic acids, arjunolic acid, terminic acid, oleanolic acid, hentriacontane, arjunolone, baicalein, arjunetin.	Singh et al. (1982); Ram et al. (1997); Kaur, Grover, and Kumar (2001); Anonymous (2002)

				References

Tinospora cordifolia (Willd.) Miers (Menispermaceae), commonly known as Amrita or Guduchi, is an important drug of the Ayurvedic system of medicine known for the treatment of diseases such as jaundice, fever, diabetes, and skin diseases. The plant is a large, glabrous, perennial, deciduous climbing shrub of weak and fleshy stem found throughout India and distributed in tropical regions up to 1200 m above sea level from Kumaon to Assam, extending through West Bengal, Bihar, Deccan, Konkan, Karnataka, and Kerala.

The drug is a prescribed for treatment in chronic diarrhea, leprosy, rheumatoid arthritis, and dysentery. The starch obtained from the stem, known as guduchi-satva, is highly nutritive and digestive. The drug is mentioned in various classical texts of Ayurveda—namely, Charak, Sushrut, and Astang Hridaya—and other treatises like Bhava Prakash and claimed to be useful in leprosy, fever, asthma, anorexia, and jaundice. It is reported to be a potent tonic and rejuvenator and is fortified into many classical preparations. The drug attracted the attention of European physicians in India as a specific tonic, antiperiodic, and for its diuretic properties. *T. cordifolia* finds special mention for its use in tribal or folk medicine in treatment of fever, jaundice, diarrhea, dysentery, and cancer in different parts of India.

Potential medicinal properties reported by scientific research include antidiabetic, antipyretic, antispasmodic, anti-inflammatory, antiarthritic, antioxidant, antiallergic, antistress, antileprotic, antimalarial, hepatoprotective, immunomodulatory, and antineoplastic activities. During the last two decades, utility of *T. cordifolia* has been demonstrated in various preclinical activities in animal models. Some of the notable findings are that administration of *T. cordifolia* extract to BALB/c mice increased total white blood cell count, reduced solid tumor growth, exhibited therapeutic effect against leishmania, and exerted significant hypoglycemic effect in normal and alloxan diabetic-induced rabbits. The root extract was found to lower hepatic glucose-6-phosphatase and serum acid phosphatase, alkaline phosphatase, and lactate dehydrogenase in diabetic rats. The decoction of *T. cordifolia* showed anti-inflammatory, antioxidant, antistress, and antiulcer activity. In clinical studies, it showed immunosuppression on obstructive jaundice patients and provided protection against cancer chemotherapy-induced leucopenia. It also plays an important role in treatment of hepatic disorders, postmenopausal syndrome, and antitubercular therapy. Arabinogalactan polysaccharide isolated from *T. cordifolia* showed protection against iron-mediated LPO of rat brain homogenate as revealed by the TBARS and lipid hydroperoxide assays and high reactivity toward DPPH, superoxide, and hydroxyl radicals. *T. cordifolia* is found to be

The chemical constituents reported from this shrub belong to different classes, such as alkaloids, diterpenoid lactones, glycosides, steroids, sesquiterpenoids, phenolics, aliphatic compounds, and polysaccharides. The main constituents are tinosporide, tinosporaside, ecdysterone, makisterone, cordifoliosides A, B, C, D, and E, tinocordifolioside, tinosporine, berberine, jatrorrhizine.

Wadood, Wadood, and Shah (1992); Goel, Premkumar, and Rana (2002); Sinha et al. (2004); Sharma et al. (1995); Upadhyay et al. (2010)

(continued)

Table 3.2 Promising/Emerging Indian Medicinal Plants for Bioprospection and Drug Development (Continued)

Plant Name/Brief Description	Traditional Uses	Therapeutic Applications	Major Chemical Constituents	Major Source Ref.
Withania somnifera (L.) Dunal. (Solanaceae; Ayurvedic name: ashwagandha) is an important medicinal plant mentioned in various texts of indigenous systems of medicine (Ayurveda, Siddha, and Unani). It is considered as a Rasayana for strength, vigor, and rejuvenation. Roots are the main plant part used therapeutically. The plant is found widely occurring in the drier regions all over India and is also cultivated in several states.	In Ayurveda, it is used as a nerve tonic, aphrodisiac, sedative, antirheumatic, and for immune strength. It relieves inflammation, pain, and backache and is also known to stimulate sexual impulses and increase sperm count. Roots are used as an application in obstinate ulcers and rheumatic swelling. It infuses fresh energy and vigor in a system worn out due to any constitutional diseases. Powdered roots are used with equal portions of Ghee and honey to treat impotence or seminal debility.	effective in elevating the GSH levels and expression of the γ-glutamylcysteine ligase and Cu–Zn SOD genes and also exhibits strong free radical scavenging properties against ROS and RNS. Much of Ashwagandha's pharmacological activity has been attributed to its two main withanolides: withaferin A and withanolide A. Oral intake of *Withania somnifera* root powder for 30 days leads to significant decrease in LPO and increase in both SOD and CAT, thus indicating that Asshwagandha root powder possesses free radical scavenging activity. Its glycowithanolides, administered orally 1 h prior to the stress procedure for 21 days, in doses of 10, 20, and 50 mg/kg body weight, induce dose-related reversal of stress effects. *Withania somnifera* glycowithanolides tend to normalize the augmented SOD and LPO activities and enhance activities of CAT and GSH-px, lending support to the clinical use of the plant as an antistress adaptogen. Lately, five discreate chemotypes of ashwagandha (rich in specific withanolides such as withaferin A) that possess different pharmacological activities (immunomodulation, antioxidant) have been developed by CSIR, India.	The major biochemical constituents of ashwagandha root are steroidal alkaloids and steroidal lactones in a class of constituents called withanolides. At present, 12 alkaloids, more than 35 withanolides, and several sitoindosides have been isolated and studied.	Singh and Kumar (1998); Gupta, Dua, and Vohra (2003); Puri (2003); Gupta and Rana (2007); Tuli and Sangwan (2009)

Note: Abbreviations: DPPH (1, 1-dipheyl-2-picrylhydrazyl); TBARS (ThioBarbituric Acid Reactive Substances); SOD (superoxide dismutase); CAT (catalase); GSG-px (erythrocyte glutathione peroxidase); AESG (Alcoholic extract of Salai Guggal); CNS (Central nervous system); HDL (High-density lipoprotein)
Sources: Thakur et. al. (1989), Husain et al. (1992), Singh et al. (2000), Kapoor (2001), Evans (2009).

Indian medicinal plants vis-à-vis their prospective potential; further research needs to be intensified to realize target drug molecules for treatment of specific ailments.

3.11 PERSPECTIVES

There is no denying the fact that most of the phytopharmaceuticals used as therapeutic agents throughout the world today have been developed based on cues obtained from ethnobotanical or traditional uses over thousands of years (Dev 1997). Such information still continues to be the obvious choice for bioprospection in the current research programs of many countries of the world. A good deal of research has lately been done to prove the concepts, therapeutic regimens, therapies, and other modalities pertaining to traditional medicine. As a result, certain excellent leads have emerged, including Guggul (gum resin of *Commiphora* wightii) for hypercholesterolemia, *Boswellia* for inflammatory disorders, Arjuna (*Terminalia arjuna*) for cardioprotection, turmeric (*Curcuma longa*) for wound healing and antioxidant and anticancer properties, Kutaki (*Picrorhiza kurroa*) for hepatoprotection, *Kshaara-Sutra* (a medicated thread coated with herbal alkaline drugs—ash of *Achyranthus ascera*, latex of *Euphorbia nerufolia*, and powder of *Curcuma longa* in a specific order) for anorectal disorders, and *Pancha Karma* for neurodegenerative disorders (Katiyar 2010).

In the twentieth century, Sir R. N. Chopra laid the foundation for pharmacological studies by providing taxonomic information on the medicinal plant wealth of India. Since then, the herbal drugs have occupied a central stage of Ayurvedic research in academia and industry in India and traditional medicine-inspired approaches to drug discovery (Patwardhan and Mashelkar 2009). Traditional knowledge-driven drug development can follow a reverse pharmacology path and reduce time and cost of development (Patwardhan et al. 2004). In their guest editorial on the launch of the *International Journal of Ayurvedic Research*, Valiathan and Thatte (2010) emphasize that the use of modern science as a research tool in *Ayurveda* has never been more necessary, more promising, or more compelling than it is in the present and that it is time to experiment.

As such, the following needs due consideration by researchers, practitioners, and policy makers:

- The Indian systems of medicine, Ayurveda, Siddha, and Unani, with their ancient origin, are of greater relevance today on account of widespread use and continued faith of the people. The classical texts of Ayurveda are a unique record of long experience. The doctrine base of Ayurveda reflects a concern of the philosopher physicians for a unified holistic approach to the maintenance of positive health. However, much of the knowledge contained in the texts needs reinterpretation in the light of present knowledge and modern science, since most of the drugs of Indian medicine lack standards of identity, purity, and safety (especially compound formulations). Thus, in order to make traditional remedies more acceptable in the modern era, it is desirable that efforts are made to validate the *Ayurvedic*, *Unani*, and *Siddha* pharmacopoeia scientifically by incorporating scientific data per WHO guidelines on use of herbal drugs. This will help revive the glory of ancient science of positive health to put forth the concepts and practice of traditional systems of medicine.
- Among the various established national traditional medicinal systems, *Ayurveda* is not only the oldest medicinal system consisting of treatises on some medicines or treatment of the diseases, but also a well-organized system of medicine that has enunciated general philosophical approaches, concepts, principles, and methods of diagnosis and treatments relating to promotion, maintenance of good health and long life, and cure or management of diseases and ailments. It also prescribes drugs, diet, and other regimens that include a code of conduct conducive to the maintenance and promotion of positive healthy status of body, mind, and spirit. As such, intensive efforts are needed to validate scientifically the established medicaments of this ancient and established system for global acceptance of known remedies as well as to search for further potential of Ayurvedic plants and formulations.
- TKDL contains vast information on plants/formulations and their traditional medicinal usage across the societies and communities. This could provide initial guidance and serve as a powerful search engine to short-list the target molecules for scientific validation of therapeutic claims as well as

bioprospection for potential molecules to combat hitherto incurable diseases/physiological disorders, and support health care benefits.

- A judicious comparison of various traditional systems, including ethnomedicinal claims across the ethnic groups, wherein the same plants help cure similar ailments across the communities could be another area for further research for scientific validation and drug discovery.
- Further, in order to make the traditional systems sustainable, it would be necessary to produce the target molecules in required quantity and desirable quality. This could be better done through organized cultivation and appropriate management of plant genetic resources, their genetic enhancement for value addition, and development of production strategies to get the molecule of choice or its prototype at will through microbial "pharming."

REFERENCES

Ade, R., and M. K. Rai. 2009. Current advances in *Gloriosa superba* L. *Biodiversitas* 10:210–214.

Agarwal, K. C., and R. E. Parks, Jr. 1983. Forskolin: A potential antimetastatic agent. *International Journal of Cancer* 32:801–804.

Ahmad, A., and J. Akhtar. 2007. Unani system of medicine. *Pharmacognosy Review* 1:210–214.

Anand, N. 1990. Contribution of ayurvedic medicines to medicinal chemistry. In *Comprehensive medicinal chemistry*, vol. 1, ed. P. D. Kennewell, 113–131. Oxford, England: Pergamon Press.

Anderson, T. F., and W. G. Voorhees. 1980. Psoralen photochemotherapy of cutaneous disorders. *Annual Review of Pharmacology and Toxicology* 20:235–257.

Anonymous. 1978. *Handbook of domestic medicine and common Ayurvedic remedies.* New Delhi, India: Central Council for Research in Indian Medicine and Homeopathy.

———. 2001. *Picrorhiza kurroa* monograph. *Alternative Medicine Review* 6:319–321.

———. 2002. *Terminalia arjuna* monograph. *Alternative Medicine Review* 7:411–415.

———. 2006. Overview of current scenario on traditional medicines. In *International conclave on traditional medicine*, New Delhi, India, p. 31. Background: http://www.niscair.res.in/conclave/

———. 2008. *Boswellia serrarta* monograph. *Alternative Medicine Review* 13:165–167.

AYUSH (www.indianmedicine.nic.in). *About the systems—An overview of Ayurveda, Yoga and Naturopathy, Unani, Siddha and Homeopathy.* New Delhi: Department of Ayurveda, Yoga and Naturopathy, Unani, Siddha and Homoeopathy (AYUSH), Ministry and Family Welfare, Government of India.

Bagchi, G. D., G. N. Srivastava, and S. C. Singh. 1992. Distinguishing features of medicinal herbaceous species of *Phyllanthus* occuring in Lucknow district (U.P), India. *International Journal of Pharmacognosy* 30:161–168.

Baskaran, K., B. K. Ahamath, K. R. Shanmugasundaram, and E. R. B. Shanmugasundaram. 1990. Antidiabetic effect of a leaf extract from *Gymnema sylvestre* in non-insulin-dependent diabetes mellitus patients. *Journal of Ethnopharmacology* 30:295–305.

Belinky, P. A., M. Aviram, B. Fuhrman, M. Rosenblat, and J. Vaya. 1998. The antioxidative effects of the isoflavaon glabridin on endogenous constituents of LDL during its oxidation. *Atherosclerosis* 137:49–61.

Biondi, D. M., C. Rocco, and G. Ruberto. 2003. New dihydrostilbene derivatives from the leaves of *Glycyrrhiza glabra* and evaluation of their antioxidant activity. *Journal of Natural Products* 66:477–480.

Brahma, S. K., and P. K. Debnath. 2003. Therapeutic importance of Rasayana drugs with special reference to their multidimensional actions. *Aryavaidyan* 16:160–163.

Chander, R., N. K. Kapoor, and B. N. Dhawan. 1992. Picroliv, picroside-I and kutkoside from *Picrorhiza kurroa* are scavengers of superoxide anions. *Biochemical Pharmacology* 44:180–183.

Chattopadhyay, I., K. Biswas, U. Bandyopadhyay, and R. K. Banerjee. 2004. Turmeric and curcumin: Biological actions and medicinal applications. *Current Science* 87:44–53.

Chaudhary, P. K. 1993.Colchicine from Indian *Gloriosa superba* L., a substitute of *Colchicum autumnale* L. *Indian Drugs* 30:529–530.

Chaudhary, P. K., and R. S. Thakur. 1994. *Glorioa superba* L—A review. *Current Research on Medicinal and Aromatic Plants* 16:51–64.

Chopra, R. N., S. L. Nayer, and I. C. Chopra. 1956. *Glossary of Indian medicinal plants.* New Delhi: Council of Scientific and Industrial Research.

Cragg, G. M., D. J. Newman, and K. M. Snader. 1997. Natural products in drug discovery and drug development. *Journal of Natural Products* 60:52–60.

Daswani, P. G., T. J. Birdi, D. S. Antarkar, and N. H. Antia. 2002. Investigation of the antidiarrheal activity of *Holarrhena antidysenterica*. *Indian Journal of Pharmaceutical Sciences* 64:164–167.

Dev, S. 1997. Ethnotherapeutics and modern drug development: The potential of Ayurveda. *Current Science* 73:909–928.

Diwedi, K. K., and R. H. Singh. 1992. A clinical study of Medhya Rasayan therapy in management of convulsive disorders. *Journal of Ayurveda and Siddha* 13:97–106.

Dubey, C. B., R. C. Srimal, and J. S. Tandon. 1997. Clinical evaluation of ethanolic extract of *Coleus forskohlii* in hypertensive patients. *Sachitra Ayurveda* 49:931–936.

Dwarakanath, C. 2003. *Fundamental principles of Ayurveda*. Part I. Introductory and outlines of Nyaya Vaisesika system of natural philosophy. Part II. Outlines of Samkhya Patajala system. Part III. Ayuskamiya and Dravyadi Vijnana (including Rasabhedia). Varanasi, India: Chowkhamaba Krishnadas Academy.

Dwivedi, R. K., and R. K. Sharma. 1990. Quantitative estimation of *Holarrhena antidysenterica* bark total alkaloids in crude drugs and in the body fluids of man and rat. *Journal of Ethnopharmacology* 30:75–89.

Evans, W. C. 2009. *Trease and Evans pharmacognosy*, 16th ed. Philadelphia, PA: Elsevier Saunders Ltd.

Farnsworth, N. R., and D. D. Soejarto. 1991. Global importance of medicinal plants. In *The conservation of medicinal plants*, ed. O. Akerele, V. Heywood, and H. Synge, 25–63. New York: Cambridge University Press.

Goel, H. C., I. Premkumar, and S. V. Rana. 2002. Free radical scavenging and metal chelation by *Tinospora cordifolia*, a possible role in radioprotection. *Indian Journal of Experimental Biology* 40:727–734.

Gohil, K. J., and J. Patel. 2010. A review on *Bacopa monniera*: Current research and future prospects. *International Journal of Green Pharmacy* 4:1–9.

Govindarajan, R., A. K. Agnihotri, S. Khatoon, A. K. S. Rawat, and S. Mehrotra. 2003. Pharmacognostical evaluation of an antioxidant plant—*Acorus calamus*. *Natural Product Sciences* 9:264–269.

Govindarajan, R., M. Vijayakumar, and P. Pushpangadan. 2005. Antioxidant approach to disease management and the role of "Rasayan" herbs of Ayurveda. *Journal of Ethnopharmacology* 99:165–178.

Govindarajan, R., M. Vijayakumar, Ch. V. Rao, V. Kumar, A. K. S. Rawat, and P. Pushpangadan. 2004. Action of *Asparagus racemosus* against streptozotocin-induced oxidative stress. *Natural Product Sciences* 10:177–181.

Govindarajan, R., M. Vijayakumar, A. K. S. Rawat, and S. Mehrotra. 2003. Free radical scavenging potential of *Picrorhiza kurroa* Royle ex Benth. *Indian Journal of Experimental Biology* 41:875–879.

Govindarajan, V. S. 1980. Turmeric—Chemistry, technology and quality. *CRC Critical Reviews in Food Science and Nutrition* 12:199–301.

Gupta, G. L., and A. C. Rana. 2007. *Withania somnifera* (Ashwagandha): A review. *Pharmacognosy Review* 1:129–136.

Gupta, S. K., A. Dua, and B. P. Vohra. 2003. *Withania somnifera* (Ashwagandha) attenuates antioxidant defense in aged spinal cord and inhibits copper induced lipid peroxidation and protein oxidative modifications. *Drug Metabolism and Drug Interaction* 19:211–222.

Handa, S. S. 1992. Medicinal plants-based drug industry and emerging plant drugs. *Current Research on Medicinal Aromatic Plants* 14:233–262.

Haraguchi, H., J. Inoue, Y. Tamura, and K. Mizutani. 2002. Antioxidative components of *Psoralea corylifolia* (Leguminosae). *Phytotherapy Research* 16:539–544.

Husain, A. 1991. Economic aspects of exploitation of medicinal plants. In *Conservation of medicinal plants*, ed. O. Akerele, V. Heywood, and H. Synge, 125–139. Cambridge, England: Cambridge University Press.

———. 1993. *Medicinal plants and their cultivation*. Lucknow, India: Central Institute of Medicinal and Aromatic Plants.

Husain, A., O. P. Virmani, S. P. Popli, L. N. Misra, M. M. Gupta, G. N. Srivastava, Z. Abraham, and A. K. Singh. 1992. *Dictionary of Indian medicinal plants*. Lucknow, India: Central Institute of Medicinal and Aromatic Plants.

Ignacimuthu, S., M. Ayyanar, and K. Sankarasivaraman. 2006. Ethnobotanical investigations among tribes in Madurai district of Tamil Nadu (India). *Journal of Ethnobiology and Ethnomedicine* 2:25: doi:10.1186/1746-4269-2-25.

Jain, A., and V. B. Gupta. 2006. Chemistry and pharmacological profile of Guggul—A review. *Indian Journal of Traditional Knowledge* 5:478–483.

Jain, S. K. 1989. *Methods and approaches in ethnobotany.* Lucknow, India: Society of Ethnobotanists, CDRI, Lucknow.

———. 1991. *Dictionary of folk medicine and ethnobotany.* New Delhi, India: Deep Publications.

———. 2001. Ethnobotany in modern India. *Phytomorphology (Golden Jubilee Issue: Trends in Plant Sciences)* 51:39–54.

———. 2004. Credibility of traditional knowledge: The criterion of multilocational and medicinal use. *Indian Journal of Traditional Knowledge* 3:137–153.

Jiangninga, G., W. Xinchu, W. Houb, L. Qinghuab, and B. Kaishuna. 2005. Antioxidants from a Chinese medicinal herb—*Psoralea corylifolia* L. *Food Chemistry* 91:287–292.

Jurenka, J. S. 2009. Anti-inflammatory properties of curcumin, a major constituent of *Curcuma longa:* A review of preclinical and clinical research. *Alternative Medicine Review* 14:141–153.

Kakrani, H. K. 1981. Guggul—A review. *Indian Drugs* 18:417–421.

Kanetkar, P., R. Singhal, and M. Kamat. 2007. *Gymnema sylvestre:* A memoir. *Journal of Clinical Biochemistry and Nutrition* 41:77–81.

Kapoor, L. D. 2001. *Handbook of Ayurvedic medicinal plants.* Boca Raton, FL: CRC Press.

Katiyar, C. K. 2010. Indian initiative in the revival of traditional systems of medicine: An overview and recent leads from Ayurveda. In *Science in India: Achievements and aspirations*, ed. H. Y. M. Ram and P. N. Tandon, 243–256. New Delhi, India: Indian National Science Academy.

Kaur, S., I. S. Grover, and S. Kumar. 2001. Antimutagenic potential of extracts isolated from *Terminalia arjuna. Journal of Environmental Pathology, Toxicology and Oncology* 20:9–14.

Kavitha, C., K. Rajamani, and E. Vadivel. 2010. *Coleus forskohlii:* A comprehensive review on morphology, phytochemistry and pharmacological aspects. *Journal of Medicinal Plants Research* 4:278–285.

Khanuja, S. P. S., J. Singh, A. K. Shasany, and A. Sharma. 2006. Inventorization and strategic utilization of medicinal and aromatic plants biodiversity in Asia. In *Perspectives on biodiversity and division of mega-diverse countries*, ed. D. D. Verma, S. Arora, and R. K. Rai, 197–242. New Delhi, India: Ministry of Environment and Forests, Government of India.

Khatoon, S., V. Rai, A. K. S. Rawat, and S. Mehrotra. 2006. Comparative pharmacognostic studies of three *Phyllanthus* species. *Journal of Ethnopharmacology* 104:79–86.

Kumar, S., J. Singh, N. C. Shah, and V. Ranjan. 1997. *Indian medicinal plants facing genetic erosion.* Lucknow, India: Central Institute of Medicinal and Aromatic Plants.

Kurup, P. N. V., V. N. K. Ramdas, and P. Joshi. 1979. *Handbook of medicinal plants.* New Delhi, India: Central Council for Research in Ayurveda and Siddha.

Lavania, S., and U.C. Lavania. 1995. Sustainable progress in the realization of high value medicinal principles of plant origin. *Journal of Indian Botanical Society* 74A (platinum jubilee volume): 227–232.

Lavania, U. C. 2005. "Pharming" plant genetic resources. *Plant Genetic Resources* 3:81–82.

Lavania, U. C., and S. Lavania. 1994. Plants for medicine. In *Botany in India: History and progress*, ed. B. M. Johri, 43–61. New Delhi, India: Oxford & IBH Publishing Co.

Lee, C. D., M. Ott, S. P. Thayagarajan, D. A. Shafritz, R. D. Burk, and S. Gupta. 1996. *Phyllanthus amarus* down regulates hepatitis B virus mRNA transcription and replication. *European Journal of Clinical Investigation* 26:1069–1076.

Majumdar, R. C. 1971. Medicine. In *A concise history of science in India*, ed. D. M. Bose, S. N. Sen, and B. V. Subbarayappa, 213–273. New Delhi, India: Indian National Science Academy.

Masayuki, Y., M. Toshiyuki, K. Masashi, L. Yuhao, M. Nubotoshi, Y. Johji, and M. Hisash. 1997. Medicinal foodstuffs (IX1) the inhibitors of glucose absorption from the leaves of *Gymnema sylvestre* R.Br. (Asclepiadaceae): Structures of gymneomosides A and B. *Chemical & Pharmaceutical Bulletin* 45:1671–1676.

Mishra, S. K., N. S. Sangwan, and R. S. Sangwan. 2007. *Andrographis paniculata* (Kalmegh): A review. *Pharmacognosy Review* 1:283–298.

Niranjan, A., S. K. Tewari, and A. Lehri. 2010. Biological activities of Kalmegh (*Andrographis paniculata* Nees) and its active principles—A review. *Indian Journal of Natural Products and Resources* 1:125–135.

Oganesyan, K. R. 2002. Antioxidant effect of licorice root on blood catalase activity in vibration stress. *Bulletin of Experimental Biology and Medicine* 134:135–136.

Ott, M., S. P. Thyagarajan, and S. Gupta. 1997. *Phyllanthus amarus* suppresses hepatitis B virus by interrupting interaction between HBV enhancer I and cellular transcription factors. *European Journal of Clinical Investigation* 27:908–915.

Pal, A., R. P. Sharma, and P. K. Mukherjee. 2009. A clinical study of Kutaja (*Holarrhena antidysenterica* Wall.) on Shonitarsha. *AYU* 30:369–372.

Parotta, J. A. 2001. *Healing plants of peninsular India.* Oxford, England: CABI Publishing.

Patwardhan, B., and R. A. Mashelkar. 2009. Traditional medicine-inspired approaches to drug discovery: can Ayurveda show the way forward? *Drug Discovery Today* 14: 804–811.

Patwardhan, B., A. D. B. Vaidya, and M. Chorghade. 2004. Ayurveda and natural products drug discovery. *Current Science* 86:789–799.

Puri, H. S. 1970a. Indian medicinal plants use in elixirs and tonics. *Quarterly Journal of Crude Drug Research* 10:1555–1566.

———. 1970b. Chavanprasha—An ancient Indian preparation for respiratory diseases. *Indian Drugs* 7:15–16.

———. 1971. Vegetable aphrodisiacs of India. *Quarterly Journal of Crude Drug Research* 11:1742–1748.

———. 2003. *Rasayana—Ayurvedic herbs for longevity and rejuvenation.* London: Taylor & Francis.

Purushothaman, K. K., and S. Chandrashekharan.1976. Guggulsterols from *C. mukul. Indian Journal of Chemistry* 14B:802–804.

Ram, A., P. Lauria, R. Gupta, P. Kumar, and V. N. Sharma. 1997. Hypocholesterolemic effects of *Terminalia arjuna* tree bark. *Journal of Ethnopharmacology* 55:165–169.

Rohini, G., K. E. Sabitha, and C. S. Devi. 2004. *Bacopa monniera* Linn. Extract modulates antioxidant and marker enzyme status in fibrosarcoma bearing rats. *Indian Journal of Experimental Biology* 42:776–780.

Russo, A., A. Izzo, F. Borrelli, et al. 2003. Free radical scavenging capacity and protective effect of *Bacopa monniera* L. on DNA damage. *Phytotherapy Research* 17:870–875.

Sane, R. T., V. V. Kuber, M. S. Chalissery, and S. Menon. 1995. Hepatoprotection by *Phyllanthus amarus* and *Phyllanthus debili* in Ccl4 induced liver dysfunction. *Current Science* 68:1243–1246.

Saneja, A., C. Sharma, K. R. Aneja, and R. Pahwa. 2010. *Gymnema sylvestre* (Gurmar): A review. *Der Pharmacia Lettre* 2 (1): 275–284.

Sasikumar, B. 2005. Genetic resources of curcuma: Diversity, characterization and utilization. *Plant Genetic Resources* 3:230–251.

Satyavati, G. V., C. Dwarkanath, and S. N. Tripathi. 1969. Experimental studies on the hypocholesterolemic effect of *Commiphora mukul* (guggul). *Indian Journal of Medical Research* 57:1950–1962.

Schultes, R. E. 1962. The role of ethnobotanists in search of new medicinal plants. *Lloydia* 25:257–266.

———. 1991. The reason for ethnobotanical conservation. In *Conservation of medicinal plants,* ed. O. Akerele, V. Heywood, and H. Synge, 65–75. New York: Cambridge University Press.

Sharma, A., S. Chhikara, S. N. Ghodekar, S. Bhatia, M. D. Kharya, V. Gajbhiye, A. S. Mann, A. G. Namdeo, and K. R. Mahadik. 2009. Phytochemical and pharmacological investigations on *Boswellia serrata. Pharmacognosy Review* 3:206–215.

Sharma, A., A. S. Mann, V. Gajbhiye, and M. D. Kharya. 2007. Phytochemical profile of *Boswellia serrata*: An overview. *Pharmacognosy Review* 1:137–142.

Sharma, D. N. K., R. L. Khosa, J. P. N. Chansauria, and M. Sahai. 1995. Antiulcer activity of *Tinospora cordifolia* Meirs and *Centella asiatica* Linn. extracts. *Phytotherapy Research* 9:589–590.

Singh, B., and K. C. Chunekar. 1972. Glossary of vegetable drugs in Brihatrayi. Varanasi, India: Chaukhambah Amarbharti Prakashan.

Singh, H. K., R. P. Rastogi, R. C. Srimal, and B. N. Dhawan. 1988. Effect of bacosides A and B on avoidance responses in rats. *Phytotherapy Research* 2:70–75.

Singh, J., G. D. Bagchi, and S. P. S. Khanuja. 2003. Manufacturing and quality control of Ayurvedic and herbal preparations. In *GMP for botanicals: Regulatory and quality issues on phytomedicine,* ed. P. K. Mukherjee and R. Verpoorte, 201–230. New Delhi, India: Business Horizons Pharmaceutical Publishers.

Singh, J., N. P. Mishra, G. Joshi, S. C. Singh, A. Sharma, and S. P. S. Khanuja. 2005. Traditional uses of *Bacopa monneiri* (Brahmi). *Journal of Medicinal and Aromatic Plant Sciences* 27:122–124.

Singh, J., A. Sharma, S. C. Singh, and S. Kumar. 1999. *Medicinal plants for bioprospection.* Lucknow, India: Central Institute of Medicinal and Aromatic Plants.

Singh, N., K. K. Kapoor, S. P. Singh, K. Shanker, J. N. Sinha, and P. P. Kohli. 1982. Mechanism of cardiovascular action of *T. arjuna. Planta Medica* 45:102–104.

Singh, R. H., and A. R. V. Murthy. 1989. Medhya Rasayana therapy in the management of apsmara vis-a-vis epilepsies. *Journal of Research and Education in Indian Medicine* 8:13–16.

Singh, S., and S. Kumar. 1998. *Withania somnifera: The Indian ginseng ashwagandha.* Lucknow, India: Central Institute of Medicinal and Aromatic Plants.

Singh, S., P. Pandey, and S. Kumar. 2000. *Traditional knowledge on the medicinal plants of Ayurveda.* Lucknow, India: Central Institute of Medicinal and Aromatic Plants.

Sinha, K., N. P. Mishra, J. Singh, and S. P. S. Khanuja. 2004. *Tinospora cordifolia* (Guduchi), a reservoir plant for therapeutic applications: A review. *Indian Journal of Traditional Knowledge* 3:257–270.

Srivastava, S. K., M. Chaubey, S. Khatoon, A. K. S. Rawat, and S. Mehrotra. 2002. Pharamcognostic evaluation of *Coleus forskohlii. Pharmaceutical Biology* 40:129–134.

Srividya, N., and S. Periwal. 1995. Diuretic, hypotensive and hypoglycaemic effect of *Phyllanthus amarus. Indian Journal of Experimental Biology* 33:861–864.

Thakur, R. S., H. S. Puri, and A. Husain. 1989. *Major medicinal plants of India.* Lucknow, India: Central Institute of Medicinal and Aromatic Plants.

Tuli, R., and R. S. Sangwan. 2009. *Ashwagandha (Withania somnifera): A model Indian medicinal plant.* New Delhi, India: Council of Scientific and Industrial Research.

Udupa, K. N. 1973. Psychosomatic stress in Rasayana. *Journal of Research in Indian Medicine* 8:1–2.

Upadhyay, A. K., K. Kumar, A. Kumar, and H. S. Mishra. 2010. *Tinospora cordifolia* (Willd.) Hook. f. and Thoms. (Guduchi)—Validation of the Ayurvedic pharmacology through experimental and clinical studies. *International Journal of Ayurveda Research* 1:112–121.

Upaganlawar, A., and B. Ghule. 2009. Pharmacological activities of *Boswellia serrata* Roxb.—Minireview. *Ethnobotanical Leaflets* 13:766–774.

Valiathan, M. S., and U. Thatte. 2010. Ayurveda: The Time to experiment. *International Journal of Ayurveda Research* 1:3.

Vaya, J., P. A. Belinky, and M. Aviram. 1997. Antioxidant constituents from licorice roots: Isolation, structure elucidation and antioxidative capacity toward LDL oxidation. *Free Radical Biology and Medicine* 23:302–313.

Velavan, S., K. R. Nagulendran, R. Mahesh, and V. H. Begum. 2007. The chemistry, pharmacological and therapeutic applications of *Asparagus racemosus*—A review. *Pharmacognosy Review* 1:350–360.

Wadood, N., A. Wadood, and S. A. W. Shah. 1992. Effect of *Tinospora cordifolia* on blood glucose and total lipid levels of normal and alloxan diabetic rabbits. *Planta Medica* 58:131–136.

Wagner, H., and A. Proksch. 1985. Immunostimulatory drugs of fungi and higher plants. In *Economic and medicinal plant research*, ed. N. Farnsworth and H. Wagner, 113–153. London: Academic Press.

Wiboonpun, N., P. Phuwapraisirisan, and S. Tip-pyang. 2004. Identification of antioxidant compound from *Asparagus racemosus. Phytotherapy Research* 18:771–773.

WHO (World Health Organization). 2008. http://www.who.int/mediacentre/factsheets/fs134/en/print.html

CHAPTER 4

Medicinal Fruit Trees of Southeast Asia

Christophe Wiart

CONTENTS

4.1 INTRODUCTION

The purpose of this chapter is not to provide information on the most common Southeast Asian fruit trees but rather to bring some light to some less known medicinal species. One can estimate that there are about 250 species of fruit trees in Southeast Asia, a number of which are medicinal and completely unstudied for pharmacology. The trees mentioned here are wild, cultivated in villages, or, in some instances, cultivated on a large scale. These species are potentially valuable as genetic resources, as new commercial products, or as sources of more proactive hybrids; they could hold some potential as sources of drugs. Therefore, the study of the medicinal trees of Southeast Asia is an important aspect of plant research, and the plants listed in this chapter should be considered as serious candidates for further research development. For each species listed, the reader is provided with Latin names, synonyms (if any), local names, main botanical features, and pharmacological and chemical status.

4.2 *GNETUM GNEMON* L.

Synonyms: *Gnetum acutatum* Miq., *Gnetum vinosum* Elm., *Gnetum gnemon* L. var. *gnetum*
The plant is known as gnetum (English); *voe*, *khalet* (Cambodia); *sukau motu* (Fiji); *melinjo* (Indonesia); *bago* (Philippines); *maninjau* (Malaysia); and *melindjo* (Singapore). It is a tree that grows wild in the rain forests of Southeast Asia. It can be recognized in field collection by its dark-green leaves—which are simple, opposite, about 15 cm long, lanceolate, and glossy—and particularly by its verticilate inflorescence and fruits, which are ovoid, about 2.5 cm in length, and completely filled with a single ovoid or ellipsoid seed (Figure 4.1). It is used as an antidote for poison in Cambodia, Laos, and Vietnam. The young leaves and the seeds are edible and used to make

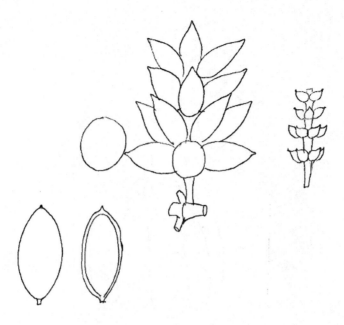

Figure 4.1 *Gnetum gnemon* L.

crackers. The plant produces a series of stilbenoids with antibacterial activities against vancomycin-resistant *Enterococci* (VRE) and methicillin-resistant *Staphylococcus aureus* (MRSA) (Iliya et al. 2002, 2003a, 2003b; Sakagami et al. 2007).

4.3 *GARCINIA ATROVIRIDIS* GRIFF.

In Malaysia, the plant is known as *asam gelugor* and it is used for earache, postpartum, and as a remedy for skin diseases. It is a tree that grows to a height of 20 m. The bark is smooth and exudes yellow latex if incised. The leaves are glossy and oblong; the flowers are dark red and the fruit is like a little pumpkin that is ribbed about 10 cm in diameter containing 12–16 ribs (Figure 4.2). The ripe fruits are sliced, dried, and used in curries. The fruit contains several organic acids, such as hydrocitrate, which rend it very sour. In Thailand, water-soluble calcium hydroxycitrate in *Garcinia atroviridis* Griff. is used to treat obesity (Roongpisuthipong, Kantawan, and Roongpisuthipong 2007; Jena, Jayaprakasha, Singh, et al. 2002). The plant contains atrovirinone, which is anti-inflammatory (Syahida et al. 2006). Esters from the fruits are antifungal (Mackeen et al. 2002). Extract of the plant exhibited antitumor properties (Mackeen et al. 2002; Permanaa et al. 2005).

4.4 *GARCINIA CAMBOGIA* (GAERTN.) DESR.

The plant is known as Malabar tamarind or bitter kola. This tree grows to about 10 m in height in the forests of Thailand, Laos, Cambodia, and Vietnam. The plant produces a yellow resin. The leaves are dark green, glossy, elliptic, and about 10 cm long. The fruits are ovoid, about 5 cm long, and yellow ripening to red, with six to eight grooves (Figure 4.3). The dried rind of the plant is used as a condiment for flavoring curries in India and Southeast Asia. The resin is used as a laxative in Cambodia, Laos, and Vietnam. *Garcinia cambogia*-derived (–)-hydroxycitric acid (HCA) is a safe, natural supplement for weight management. HCA is a competitive inhibitor of ATP citrate lyase, a

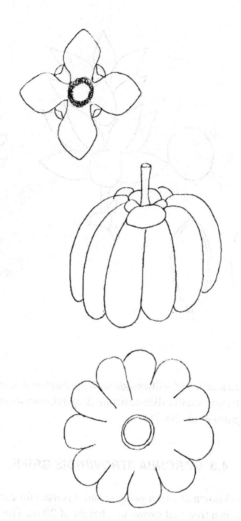

Figure 4.2 *Garcinia atroviridis* Griff.

key enzyme that facilitates the synthesis of fatty acids, cholesterol, and triglycerides (Preuss et al. 2004). Ethanolic extracts of *G. cambogia* seeds have hematologically enhancing and antiobesity effects. The decrease in the high-density lipoprotein level and an increase in the LDL level may play an important role in cardiovascular disease (Oluyemi et al. 2007). *Cambogia* was able to decrease the acidity and to increase the mucosal defenses in the gastric areas, thereby justifying its use as an antiulcerogenic agent (Mahendran, Vanisree, and Devi 2002).

4.5 *GARCINIA HOMBRONIANA* PIERRE

The plant is a tree that grows to a height of 20 m on seashores in a geographical area covering east India to Malaysia. Its common name is seashore mangosteen, *beruas*, *manggis hutan* in Malaysia and *waa* in Thailand. The plant produces a white latex. The leaves are oblong to elliptic, 14 cm × 7 cm with flowers that are rose-red outside and cream-yellow inside. The arillode around the seeds is palatable and sour (Figure 4.4). Malays use the plant after childbirth and to relieve itchiness. So

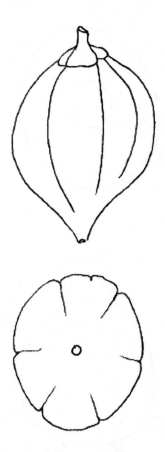

Figure 4.3 *Garcinia cambogia* (Gaertn.) Desr.

far the plant is known to produce friedolanostane and lanostane triterpenes (Rukachaisirikul et al. 2000, 2005).

4.6 *GARCINIA PRAINIANA* KING

The plant is known as button mangostin or *cherapu*. It is native to Thailand and Malaysia. It grows wild or cultivated to 10 m tall. Leaves are subsessile, the weakly heart-shaped base often slightly clasping the stem. The leaf blade is large and can reach 23 cm long. Flowers are yellowish to pink in dense, round, 2-cm-diameter clusters. There are five petals and sepals, 2.5 cm across. The fruit is about 3 cm across, bun-like, smooth, ripening golden yellow, with vestigial stigma at apex and containing a few seeds embedded in a sour pulp (Figure 4.5). In Malaysia, the fruits are used to stimulate appetite in children. The pharmacological property of the plant is unknown.

4.7 *GARCINIA COWA* ROXB.

This tree grows to a height of 30 m in the forests of India and Southeast Asia, called *kandis* by Malays and *bhava* in India. The plant exudes a pale yellow resin upon incision. The leaves are simple, about 4 cm × 12 cm, with straight secondary nerves and black dots underneath. The flowers are in axillary clusters, with four pinkish red petals and about 2 cm in diameter.

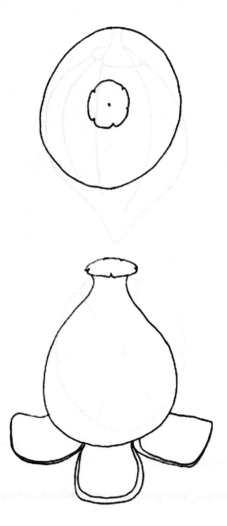

Figure 4.4 *Garcinia hombroniana* Pierre.

The fruit is dull orange-yellow, coarsely longitudinally grooved, and about 3 cm in diameter (Figure 4.6). The young leaves are edible. The plant is grown for its edible fruits, which are considered medicinal in Malaysia. The fruits contain (–)-hydroxycitric acid (Jena, Jayaprakasha, and Sakariah 2002). The plant contains flavonoids and xanthones, some with antimalarial properties (Likhitwitayawuid, Phadungcharoen, and Krungkrai 1998) and anticancer potential, such as dulxanthone A, which induced S-phase arrest and apoptosis in the most sensitive cell line HepG2 (Tian et al. 2008).

4.8 *GARCINIA DULCIS* (ROXB.) KURZ.

This is a tree that grows to 9 m tall in the rain forests of Southeast Asia. In Malaysia, it is called *mundu* or *munu*. The leaves are large, oblong, and to 25 cm long with a few hairs below. The flowers are about 0.6 cm across. The fruits are globose and pear shaped, to 6.5 cm long, ripening light yellow and pulpy and drying black, glossy, and wrinkled. It contains one to five seeds embedded in a sour pulp (Figure 4.7). In Malaysia, the fruits are made into a paste applied to glandular swelling.

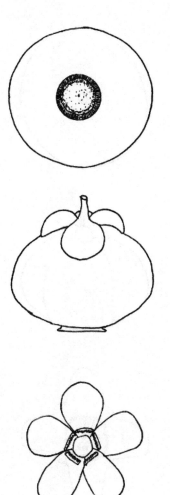

Figure 4.5 *Garcinia prainiana* King.

The plant contains xanthones with antimalarial activity (Likhitwitayawuid, Chanmahasathien, et al. 1998). A biflavonoid, morelloflavone, and prenylated xanthone, camboginol, isolated from the fruits exhibited strong antioxidant activity (Hutadilok-Towatana, Kongkachuay, and Mahabusarakam 2007).

4.9 *GARCINIA GRIFFITHII* T. ANDERS

This forest tree can reach 23 m tall in Malaysia, Singapore, and Sumatra. The leaves are very large, cabbagy, to 30 cm long. The flowers are packed in dense clusters. Each flower contains four petals, sepals, and numerous stamens. The fruits are apple-like, ribbed, subsessile, green turning browning yellow with a watery acid flesh, 5–9 cm across, drying jet black, and coarsely wrinkled. In Malaysia, the fruits are eaten to promote vomiting. A benzophenone, guttiferone I, together with 1,7-dihydroxyxanthone, 1,3,6,7-tetrahydroxyxanthone, and 1,3,5,6-tetrahydroxyxanthone, was isolated from the stem bark of *Garcinia griffithii* (Nguyen et al. 2005).

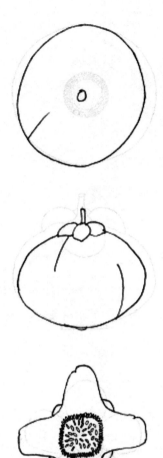

Figure 4.6   *Garcinia cowa* Roxb.

4.10 *GARCINIA FORBESI* KING

This is a forest tree that grows to 18 m tall in the forests of Malaysia and Sumatra. The leaves are ovate, about 20 cm long, thin, with faint and straight secondary nerves. The flowers show four sepals and petals. The fruits are apple-like, ripening red, juicy, subsessile, deeply depressed at the base, to 8 cm across, drying jet black, and showing at the apex a roundish and raised vestigial stigma. Slices of dried fruits are given to children to stimulate appetite. The pharmacological potential of this plant is yet unexplored.

4.11 *GARCINIA PARVIFOLIA* (MIQ.) MIQ.

Synonym: *Garcinia globulosa* Ridl.

This tree grows in the forests of Malaysia and Indonesia. A common name for this plant is Brunei cherry or *asam aur aur* in Malaysia. The plant reaches 30 m tall. The bark exudes yellow

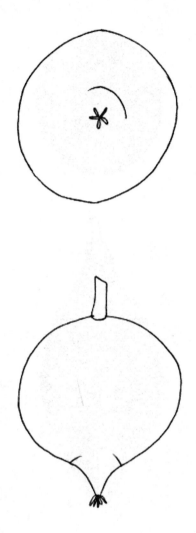

Figure 4.7 *Garcinia dulcis* (Roxb.) Kurz.

latex upon excision. The leaves are similar to *Garcinia cowa* but abruptly taper to the tip. The flowers show four sepals and petals. The petals are clear yellow to light pink and about 0.5 cm long. The fruit is roundish or elongate, light green, not grooved, and containing a few seeds. The latex is used in Malaysia to check bleeding of small cuts. The plant contains xanthones, such as rubraxanthone, which inhibit platelet activating factor receptor binding *in vitro* (Jantan et al. 2002). The plant produces cytotoxic prenylated depsidones as well (Xu et al. 2000).

4.12 *GARCINIA MADRUNO* (HBK) HAMMEL

Synonym: *Calophyllum madruno* HBK

This tree, native to Central America, grows to 20 m and is found cultivated in several tropical countries. The bark is blackish. Red latex exudes when the bark is incised. The leaves are simple, opposite, elliptic, and about 15 cm × 7 cm. The flowers are white. The fruit is ovoid, bright yellow, about 3 cm long, with 1–2 seeds, and covered with numerous conical formations (Figure 4.8).

Figure 4.8 *Garcinia madruno* (HBK) Hammel.

In tropical Asia, the pulp of the fruit is used to promote appetite. An extract of the plant showed antimycobacterial activity (Graham et al. 2003).

4.13 *BERTHOLLETIA EXCELSA* HUMB. & BONPL.

Synonym: *Bertholletia nobilis* Miers.

The plant is known as Brazil nut or para nut and is native to Amazonia. It is a tall tree that can reach 50 m in height. The bark is grayish and smooth. The leaves are simple and alternate. The leaf blade is oblong, about 30 cm × 12 cm, and entire or crenate at the margin. The inflorescences are panicles about 10 cm long. The calyx is bifid. The corolla comprises six unequal cream-colored petals and numerous stamens united into a mass. The fruit itself is a woody capsule that is about 12 cm and contains 8–25 triangular seeds, which are elongated and about 5 cm long (Figure 4.9). The seeds are edible, accumulate selenium, and are of economical value because the oil is used both in food and the cosmetic industry. It is not used medicinally in Southeast Asia, where it has

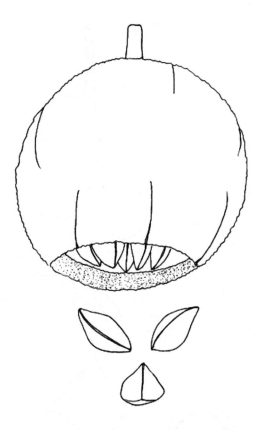

Figure 4.9 *Bertholletia excelsa* Humb. & Bonpl.

been introduced. Extracts of *Bertholletia excelsa* stem bark showed significant *in vitro* trypanocidal activity against the trypomastigote form of *Trypanosoma cruzi* (Campos et al. 2005). The high selenium content is of value for cancer prevention (Ip and Lisk 1994). The plant has some levels of antifungal activity (Mohamed et al. 1999).

4.14 *PITHECELLOBIUM JIRINGA* (JACK) PRAIN

Synonyms: *Albizia jiringa* (Jack) Kurz, *Feuilleea jiringa* (Jack) Kuntze, *Inga jiringa* (Jack) DC., *Inga kaeringa* (Roxb.) Voigt, *Mimosa jiringa* Jack, *Mimosa kaeringa* Roxb., *Pithecellobium lobatum* Benth., *Zygia jiringa* (Jack) Kosterm.

This is a tree that grows wild in secondary forests or is cultivated in Burma, Thailand, Malaysia, and Indonesia. The bark is smooth and pale gray. The leaves are compound with two pairs of folioles. The folioles are purple when young, 5–14.5 cm × 2.7–5.5 cm, lanceolate, papery, discretely acuminate at apex, and acute to asymmetrical at the base, with four to five pairs of secondary nerves. The inflorescence is a panicle. The flowers are minute. The fruit is a pod about 20 cm long, coiled, lobed, dull purplish brown, thickly leathery, and used as a source of dye for silk. The seeds are large, reddish brown, smelling of garlic, and edible (Figure 4.10). The plant is used against itchiness in Malaysia. The pod contains methyl gallate (Lajis and Khan 1994), which probably accounts for the toxicity against brine shrimp (Mackeen and Khan et al. 2000).

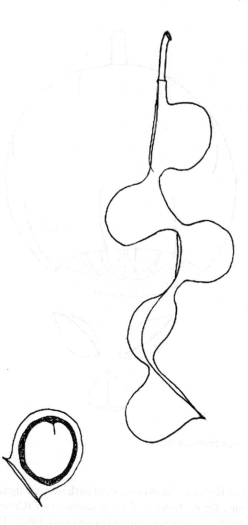

Figure 4.10   *Pithecellobium jiringa* (Jack) Prain.

4.15 *DIALIUM COCHINCHINENSE* PIERRE

This timber is known as *keranji kertas kechil* in Malaysia and *xoay* in Vietnam. The plant grows in Thailand, Cambodia, Laos, Vietnam, and Malaysia. The leaves are about 15 cm long with seven subopposite folioles that are oblong–elliptic, about 10 cm long and 2.5 cm in diameter, and hairy when young. The flowers are minute. The fruits are brittle, blackish pods that are roundish and about 2 cm in diameter (Figure 4.11). In Cambodia, Laos, and Vietnam, the plant is used to treat diarrhea, ringworm infection, and urticaria. It seems that the plant has been unexplored for pharmacology.

4.16 *DIALIUM INDUM* L.

This is a tree that grows in Thailand, Malaysia, Borneo, and Indonesia. It is called velvet tamarind, *keranji kertas besar* in Malaysia, and *yee* in Thailand. The leaves are compound, to 25 cm long, with five to nine subopposite leaflets that are ovate–oblong, about 10 cm × 3.5 cm, and

Figure 4.11 *Dialium cochinchinense* Pierre.

thin. The flowers are small. The pod is roundish, dark, about 2.5 cm in diameter, brittle, shortly apiculate hairy, and contains squarish seeds, which are about 1 cm in diameter and embedded in a pulp that is sweet and sour. A trade in the fruits from Indonesia and Malaysia occurred in the past. The pulp is used to stimulate appetite. The plant seems not to have been studied for pharmacology yet.

4.17 *PITHECELLOBIUM BUBALINUM* (JACK) BENTH.

Synonyms: *Archidendron bubalinum* (Jack) I.C. Nielsen, *Cylindrokelupha bubalina* (Jack) Kosterm., *Feuilleea bubalina* (Jack) Kuntze, *Inga bubalina* Jack, *Ortholobium bubalinum* (Jack) Kosterm., *Pithecellobium bigeminum* var. *bubalinum* (Jack) Benth.

The plant is called *keredas* or *tangki* in Malaysia. It is very much like *Pithecellobium jiringa* (Jack) Prain, from which it differs by the bark, which is closely fissured; the reddish brown young leaves; the terminal and axillary inflorescence; and the pod, which is 4–15 cm × 2.5 cm, plump, straight, or slightly coiled and not lobed. The seeds and young shoots are edible. The plant is used for fever in Indonesia. One might have some interest in studying the plant for pharmacology and especially anti-inflammatory, antipyretic properties.

4.18 *CYNOMETRA CAULIFLORA* L.

This shrub is cultivated in Southeast Asia and is native in east Indonesia. The plant is called *hima* in Thailand, *kopi anjing* in Indonesia, and *nam-nam* in Malaysia. The leaves are compound and short, with a pair of folioles that are about 8 cm × 4 cm, blunt, and notched at the tip. The inflorescence is about 10 cm, crowded on woody knobs on the trunk. The pods are short, flattish, about 10 cm long, brown, and edible (Figure 4.12). The flesh is white and includes a single large seed.

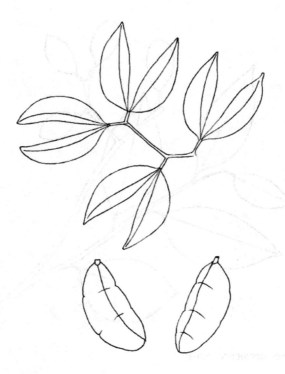

Figure 4.12   *Cynometra cauliflora* L.

In Burma, the plant is used as a purgative and in the Philippines it is used to treat herpes. The plant seems to be unexplored for pharmacology.

4.19 *MANGIFERA FOETIDA* LOUR.

Synonym: *Mangifera horsfieldii* Miq.

The plant is known as horse mango, *bacang* (Malaysia, Indonesia), and *ma chae* (Thailand). It is a medium-sized tree reaching 40 m tall with a gray-brown bark. The plant grows throughout Southeast Asia. The leaves are simple, on a petiole swollen at the base, which can reach 10 cm long. The blade is leathery, about 30 cm × 6 cm, blunt at the apex and cuneate at the base, with 11–30 pairs of secondary nerves. The inflorescence is deep red. The petals are whitish. The fruit is a drupe to 10 cm long with a strong smell of turpentine (Figure 4.13). In Indonesia, the plant is used for itchiness. The plant seems to have been unexplored for pharmacology.

4.20 *MANGIFERA PAJANG* KOSTERM.

This is a tree that grows to 35 m in Borneo, where it is known as *asem payang* or *bambangan*. The bark is brown. The leaves are simple, large, and glossy. The inflorescence is a large pyramidal raceme of little, fragrant flowers. The fruit is ovoid, 9.5–20 cm × 6.5–17 cm, and brown, containing a yellowish aromatic pulp and a seed that is 9 cm × 4.5 cm. The plant has antioxidant properties (Abu Bakar et al. 2009).

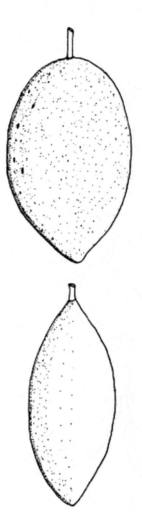

Figure 4.13 *Mangifera foetida* Lour.

4.21 *MANGIFERA CAESIA* JACK

Synonyms: *Mangifera verticillata* C.B. Rob., *Mangifera caesia* Jack var. *verticillata* (C.B. Rob.) Mukh., *Mangifera caesia* Jack var. wanji Kosterm.

This tree is cultivated in Southeast Asia and is occasionally found in the forests of Thailand, Malaysia, and Indonesia. In Malaysia, the plant is known as *binjai* and as *kemang* in Indonesia. The tree grows to 40 m tall and presents a shallowly domed crown. The bark is pinkish. The leaves are pinkish, spiral, and on a 0.5–0.25 cm long petiole. The blade is thickly leathery, elliptic–lanceolate, and 9–30 cm × 3.5–9 cm. The flowers are violet, with five sepals and petals, and small. The fruit is drupe, which is brownish yellow, ellipsoid, pear shaped, and 12–19 cm × 6–10 cm, with a yellowish white flesh that is sour to sweet and edible. The sap of this plant causes dermatitis. The plant is known to contain alkenylphenol and alkenylsalicylic acid (Masuda et al. 2002).

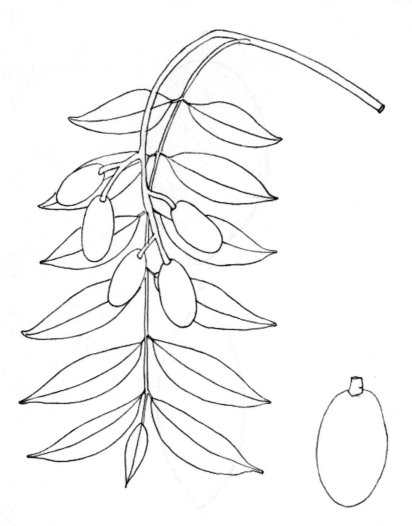

Figure 4.14   *Spondias cytherea* Sonnerat.

4.22 *SPONDIAS CYTHEREA* SONNERAT

Synonym: *Spondias dulcis* Soland. ex Forst. f.

The common names for this plant include ambarella, great hog plum, *kedondong* (Malaysia), and *makok-farang* (Thailand). It is a tree reaching 40 m tall. The plant grows throughout Southeast Asia, wild or cultivated. The bark is grayish and smooth to shallow fissured. The leaves are compound with 4–10 pairs of leaflets on a 10–20 cm long rachis. The foliole is ovate or lanceolate, 7.5–15 cm × 3–5 cm, pointed at the apex, and cuneate at the base and shows 14–24 pairs of secondary nerves forming an intramarginal nerve. The inflorescence is a terminal panicle to 35 cm long. The flowers are cream or white and small. The fruit is a drupe, which is edible, ellipsoid, 4–10 cm × 3–8 cm, bright orange, and five celled (Figure 4.14). In Cambodia, Laos, and Vietnam, the plant is used to stop diarrhea. In Indonesia, it is used after childbirth. The pulp of the fruit contains some polysaccharides, which activate peritoneal macrophages (Iacomini et al. 2005).

4.23 *MANGIFERA ODORATA* GRIFF.

This is a tree reaching 25 m tall in the forests of Southeast Asia. It is known in Malaysia as *kuini* or *kuining*. The bark is gray. The sap is irritating. The leaves are simple, with a stalk 2–5 cm long. The blade is leathery, 10–23 cm × 3–6.5 cm, pointed at the apex, cuneate at the base, and with 15–26 pairs of secondary nerves. The inflorescence is a terminal panicle of small, pinkish, fragrant flowers. The fruit is an orange drupe, which is oblong and 7.5–12.5 cm long, containing a light-yellow, sweet flesh. In Malaysia, the bark is used as a sedative. The plant contains mangiferin, which has significant neuraminidase inhibitory activity (Li et al. 2007).

4.24 *BOUEA MACROPHYLLA* GRIFF.

Synonym: *Bouea gandaria* Bl. ex Miq

The plant is globally known as marian plum; *kundang daun besar, kudang hutan* in Malaysia, where it is not so common; and *ma praang* in Thailand, where it grows wild, as well as Indonesia (*ramania*) and Cambodia. It is also found cultivated in villages for its fruits. It is a tree that grows to 35 m with a reddish-brown bark. The stems are angled. The leaves are simple, opposite. The petiole is 1–2.5 cm long and stout. The leaf blade is 10–30 cm × 2.5–11 cm and oblong–lanceolate, with 15–25 pairs of nerves. The flowers are small and light yellowish green. The fruit is a drupe, which is subglobose, 3–3.5 cm long, and yellow or orange when ripe, with a purple seed (Figure 4.15). The ripened

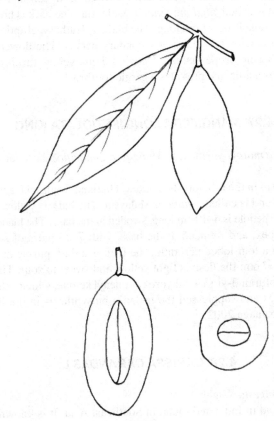

Figure 4.15 *Bouea macrophylla* Griff.

fruits are eaten fresh or cooked to make compote. Immature fruits are used in making spicy condiments. The young leaves are eaten. The plant is used externally to relieve headaches and to gargle for thrush. To date, the plant awaits further pharmacological research.

4.25 *MANGIFERA PENTANDRA* HOOK. F.

Synonym: *Mangifera lanceolata* Ridl.

This tree grows wild or cultivated in Malaysia, Indonesia, and Thailand. The Malay name for this plant is *mempelan bemban*. The leaves are simple and spiral. The petiole is 1.5–3.5 cm long and swollen at the base. The leaf blade is leathery, oblong or lanceolate, 12–30 cm × 4–11 cm, pointed at the apex, and round at the base, with 12–23 pairs of secondary nerves. The flowers are small, yellowish white. The fruit is a drupe that is oblong, 7.5–10 cm × 5–6 cm, pale orange, sweet, and juicy. The plant affords an astringent remedy and has not been studied for pharmacology yet.

4.26 *BOUEA MICROPHYLLA* GRIFF.

Synonyms: *Bouea oppositifolia* (Roxb.) Meisn., *Bouea burnamica* Griff., *Mangifera oppositifolia* Roxb, *Matania laotica* Gagnep.

This is a common tree of Southeast Asia that grows to 35 m tall and is generally known as Burmese plum. The plant is called *kundang Siam* in Malaysia, The bark is brownish. The leaves are simple and spiral. The petiole is 0.5–1 cm long. The blade is leathery, elliptic, 4–12 cm × 1–5.5 cm, and acuminate at the apex, with 7–14 pairs of secondary nerves. The flowers are yellowish white. The fruit is a broadly ellipsoid drupe that is 2.5 cm × 1.5 cm, yellow turning red. The fruit can be eaten raw or steamed; it is acidic and provides a gargle for thrush.

4.27 *MANGIFERA LONGIPETIOLATA* KING

Synonyms: *Mangifera maingayi* Hook. f., *Mangifera quadrifida* Jack var. *longipetiolata* (King) Kochummen.

This tree grows to 40 m in the rain forests of Laos, Thailand, Malaysia, and Indonesia. The plant is known as Asam wood and is called *sepam* in Malaysia. The bark is yellowish brown. The leaves are simple and spiral. The petiole is 4–10 cm long, swollen at the base. The blade is elliptic, 6.5–30 cm × 3–9 cm blunt at the apex, and rounded at the base, with 7–22 pairs of secondary nerves. The flowers are whitish, with a four-lobed perianth. The fruit is a dark-purple drupe with pale brownish spots, 8–10 cm × 5.5–7 cm; the flesh is light yellow and sweet to sour. The plant contains (4,6-dihydroxy)-dihydrobenzofuran-3-yl-(3,4-dihydroxy)-phenyl ketone, which exhibited strong tyrosine kinase inhibitor properties and suppressed the melanin biosynthesis in the B16 mouse melanoma cell line (Takagi and Mitsunaga 2002).

4.28 *CARISSA CARANDAS* L.

Synonym: *Carissa congesta* Wight

The plant is widespread in India and found in Southeast Asia. It is known as *karanda kerenda* in Malaysia, *karaunda* in India, *nam phrom* in Thailand, and *caramba* in the Philippines. It is a climbing shrub that grows to 5 m long. The plant produces white latex. The leaves are simple, opposite, 2.5–7.5 cm long, dark green, and elliptic. The white flowers are tubular, five lobed, contortate,

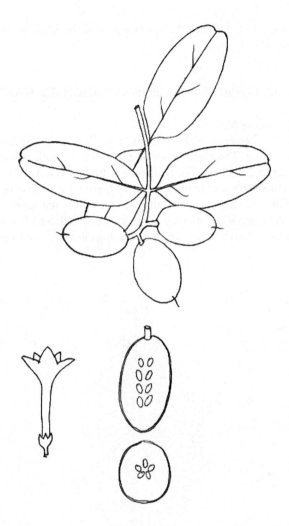

Figure 4.16   *Carissa carandas* L.

and fragrant; they are twisted to the left in the bud instead of to the right as in other species. The fruit is broad-ovoid or round, 1.25–2.5 cm in diameter, purplish red turning dark purple to black. The pulp is juicy, bittersweet, red or pink, and contains two to eight small, flat, brown seeds (Figure 4.16). The fruits are eaten raw and made into compote. In India, the plant is medicinal and used as cooling bitter stomachic itches, fevers, diarrhea, earache, soreness of the mouth and throat, and syphilis and as astringents. The plant has cardiotonic properties (Vohra and De 1963) and exhibited potent antiviral activity against sindbis virus, human poliovirus, and herpes simplex virus (Taylor et al. 1996).

4.29 *WILLUGHBEIA FIRMA* BL.

Synonyms: *Ancylocladus firmus* (Blume) Kuntze, *Willughbeia coriacea* Wall.

The name of the plant is Borneo rubber. It is a climber that produces white latex known as *akar getah gaharu* in Malaysia. The leaves are simple, opposite. The petiole is about 1 cm long. The leaf blade is elliptic, about 16 cm × 6 cm, with about 15 pairs of secondary nerves. The flowers are yellowish. The fruits are edible follicles. Malaysians spread the latex on yaws. The plant has not been studied for pharmacology.

4.30 *DURIO MALACCENSIS* PLANCH. EX MAST.

Synonym: *Durio oblongus* Mast.

This tree grows in the forests of Malaysia, Thailand, Burma, and Indonesia. In Malaysia, the plant is called *durian batang*. The plant grows to 25 m tall; the bark is brownish. The leaves are alternate, simple, and about 2 cm long. The leaf blade is 10–20 cm × 3–1.5 cm, pointed at the apex, rounded at the apex, and silvery brown underneath with numerous secondary nerves. The flowers are creamy white to pink with conspicuous stamens. The fruit is red, globular, 15 cm in diameter, with 1 cm long spines. The seeds are brown in an ivory-white aril, which is palatable (Figure 4.17). The plant has no particular medicinal use and the pharmacological properties are unknown.

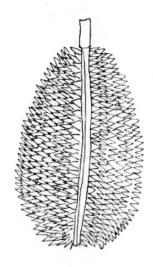

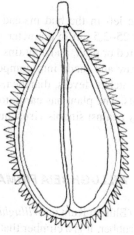

Figure 4.17 *Durio malaccensis* Planch. ex Mast.

4.31 *DURIO DULCIS* BECC.

Synonym: *Durio conicus* Becc.

Red durian, or *durian merah* (Malayan and Indonesian), is a tree native to Borneo that grows to a height of 40 m in Indonesia and Malaysia. The bark is reddish brown. The leaves are elliptical or obovate, 7–15 cm × 3.5–5 cm, and scaly below. The flowers are large, to 10 cm in diameter, with pink petals. The stamens are grouped in bundles. The fruit is 15 cm in diameter, dark red, and covered with 1.5–2 cm long spines. The fruit flesh is dark yellow and very much esteemed among all durian types.

4.32 *DURIO OXLEYANUS* GRIFF.

This is a tall tree that grows to 45 m in the forests of Malaysia, where it is known as *durian beludu*. The bark is dark brown. The leaves are simple and alternate. The petiole is about 2 cm long. The leaf blade is broadly elliptic, about 6–10 cm × 3–7 cm, rounded at the apex, and hairy and scaly below with 15–20 pairs of secondary nerves. The flowers are white to pale cream. The fruit is grayish, globular, and 15–20 cm in diameter with 3-cm spines. The seeds are glossy red in a fleshy, dark-yellow aril. In Indonesia, the plant is used for malaria, ulcers, and wounds. The plant has not been studied for pharmacology.

4.33 *DURIO LOWIANUS* SCORT. EX KING

Synonyms: *Durian wrayi* King, *Durian zibethinus* Murr. var. *roseiflorus* Corner.

The plant is known as *durian daun* or *durian sangka* in Malaysia, where it grows to a height of 55 m. The bark is purple brown. The leaves are simple and alternate. The petiole is about 1–2 cm long. The blade is oblong 8–18 cm × 2.5–7 cm, pointed at the apex and round at the base with looped secondary nerves. The flowers are red and pilose. The fruit is green to yellow, 25 cm × 20 cm, with 1 cm long spines. The seed aril is dark yellow and palatable. The plant is not medicinal and unexplored for pharmacology.

4.34 *DURIO GRAVEOLENS* BECC.

The red-fleshed durian is a tree that grows in the forest of Malaysia (*durian merah, durian kuning*) and Indonesia (*tinambela*) to a height of 55 m. The bark is reddish brown. The petiole is 1.3–2 cm. The leaves are simple and alternate. The blade is 10–15 cm × 5–7 cm, pointed at the apex, round at the base, and scaly below, with 10–13 pairs of secondary nerves. The flowers are white and large with conspicuous stamens in bundles. The fruits are orange-yellow, globose, and 10 cm in diameter, with 2 cm long spines. The seeds are ellipsoid and embedded in a dark-red aril that is palatable (Figure 4.18). The plant has not been studied for pharmacology.

4.35 *DURIO CARINATUS* MAST.

This tree grows in the peat swamp forests of Indonesia and Malaysia. The plant is called *durian paya* or *durian burung* in Malaysia. The bark is pink and lenticelled. The leaves are simple and alternate. The petiole is 1–1.5 cm long. The leaf blade is ovate-oblong, 8–12.5 cm × 3–5 cm, pointed at the apex, rounded at the base, and scaly underneath, with 8–14 pairs of secondary nerves. The flowers are yellow to pink. The fruits are pale, orange-yellow, ovoid, 13 cm × 10 cm, and covered with 1.5-cm spines. The seeds are black and glossy in a bright-red aril, which is edible. The plant has not been studied for pharmacology.

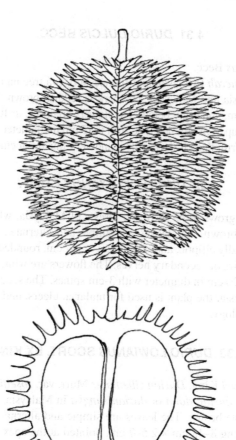

Figure 4.18   *Durio graveolens* Becc.

4.36 *DURIO KUTEJENSIS* (HASSK.) BECC.

Synonym: *Lahia kutejensis* Hassk.

The plant is known as lai or nyekak in Borneo, where it grows wild. The tree grows to 30 m tall. It is cultivated in Malaysia and other neighboring countries. The leaves are simple and alternate. The flowers are large and red. The fruit is globose or oblong, 20 cm × 12 cm, and dirty yellow with slightly curved soft spines 1–1.5 cm long; it contains a few dark-brown glossy seeds embedded in a yellow to orange flesh, which is palatable. The bark contains 3β-*O*-*trans*-caffeoyl-2α-hydroxyolean-12-en-28-oic acid, 3β-*O*-*trans*-caffeoyl-2α-hydroxytaraxest-12-en-28-oic acid, maslinic acid, arjunolic acid, 2,6-dimethoxy-*p*-benzoquinone, and fraxidin (Rudiyansyah and Garson 2006).

4.37 *DURIO TESTUDIRARIUM* BECC.

Synonym: *Durio macrophyllus* (King) Ridley

This is a medium-sized tree of the forest of Malaysia, where it is called *durian daun besar* or *durian kura*, and Borneo, where it is known as *lujian beramatai*. The bark is light gray-fawn. The leaves are simple and alternate. The petiole is about 1–3.5 cm long. The leaf blade is oblong,

8–10 cm × 23–28 cm, pointed at the apex, rounded at the base, and scaly underneath. The flowers are white to pink. The fruits are white to pink, 9 cm × 6 cm, blue green, with 1 cm long spines. The seeds are light yellow in a white palatable aril. The pharmacological property of the plant is unknown.

4.38 *CANARIUM ODONTOPHYLLUM* MIQ.

Synonyms: *Canarium beccarii* Engl. in DC., *Canarium multifidum* H.J. Lam, *Canarium palawanense* Elm.

The plant is known as sibu olive or kembayaui in Borneo, where it grows wild. It is also found in Malaysia (local name: *dabai*) and the Philippines. It grows wild in secondary forests to a height of 35 m. The leaves are alternate and compound, the folioles oblong and toothed. The flowers are 1 cm in diameter and white-yellow. The fruits are about 4 cm long and purple-black; fleshy drupes are edible (Figure 4.19). The plant has not been studied for pharmacology.

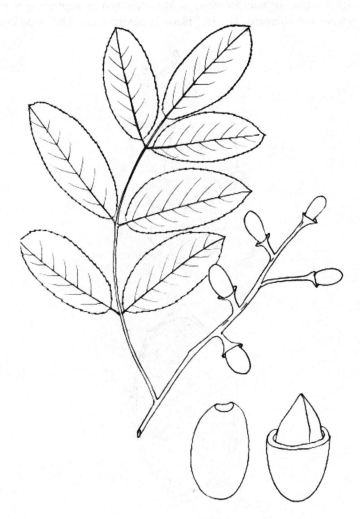

Figure 4.19 *Canarium odontophyllum* Miq.

4.39 *HYLOCEREUS UNDATUS* (HAW.) BRITTON & ROSE

Synonyms: *Cereus undatus* Haw., *Hylocereus guatemalensis* (Weing.) Britton & Rose

The plant is a cactus climber known as dragon fruit or night blooming cactus, *pitahaya*, and strawberry pear. The cactus is native to Brazil and cultivated in several tropical countries, including Southeast Asia, where it is know as *thanh long* (Vietnam), *kaew mung korn* (Thailand), *kaktus madu* (Malaysia), and *long guo* (China). The stems branch profusely and are about 1 m long and 10 cm in diameter, three ribbed, and spiny. The flowers are large to 30 cm × 5 cm and white. The fruit is oblong, about 10 cm × 8 cm, red with large bracteoles, and contains a whitish pulp containing several tiny black seeds (Figure 4.20). The pulp is edible. In China, the plant is used as a remedy for pulmonary ailments. The plant has shown healing properties on diabetic rodents (Perez, Vargas, and Ortiz 2005).

4.40 *DIOSPYROS DISCOLOR* WILLD.

This tree grows to 15 m tall. It is native to the Philippines and cultivated in Southeast Asia. It is known as velvet apple, as *buah mentega* in Malaysia, and as *kamagong* in the Philippines. The leaves are simple and alternate. The leaf blade is oblong lanceolate, 8–30 cm × 2.5–12 cm,

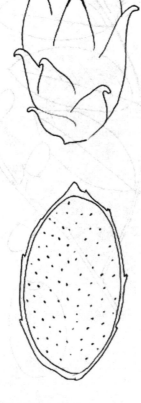

Figure 4.20 *Hylocereus undatus* (Haw.) Britton & Rose.

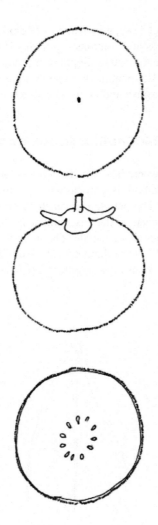

Figure 4.21 *Diospyros discolor* Willd.

acuminate at apex and hairy below, with 9–12 pairs of secondary nerves. The flowers are male or female. The male flowers contain four sepals, and 25 stamens; the female flower has 4–10 staminodes. The fruit is globose, red, 5–12 cm × 7–10 cm, hairy with persistent calyx and up to 10 seeds (Figure 4.21). The fruits are edible. No medicinal properties have been reported. An extract of the plant showed good antioxidant properties (M. H. Lee et al. 2006) probably on account of naphthoquinones (Ganapaty et al. 2005).

4.41 *DIOSPYROS DIGYNA* JACQ.

Synonyms: *Diospyros nigra* (J.F. Gmelin) Perrot., *Diospyros ebenaster* Hiern (non Retz.), *Diospyros obtusifolia* Humb. & Bonpl. ex Willd.

This tree grows to 25 m tall and is native to Central America and cultivated in Southeast Asia and other tropical regions. The common name for this plant is black sapote; in the Philippines it is called *zapote negro*. The bark is black. The leaves are simple and alternate. The leaf blade is elliptic–oblong, acute at the apex, leathery, glossy, and to 30 cm long. The flowers are white and

1–1.6 cm in diameter. The fruit is globose, green, and to 12 cm in diameter, with a prominent, four-lobed calyx, which is to 5 cm across and contains a pulp that is palatable and contains 1–10 seeds. The plant is used as a blistering agent in the Philippines. It contains a C-alkylglucoside, diospyrodin [β-1C-(1′S*,2′R*,3′R*,4′S*-1′,2′,3′,4′,5′-pentahydroxypentyl)-glucopyranoside], which showed antimicrobial activity against Gram-positive and Gram-negative bacteria (Dinda et al. 2006).

4.42 *BACCAUREA ANGULATA* MERR.

The plant is native to Borneo, where it is called *tampoi belimbing.* It is a tree that reaches 20 m tall and is profusely branched. The bark is gray-brown. The flowers are yellow and the fruits red. The leaves are simple and spiral. The petiole can reach 12 cm long. The leaf blade is elliptic to obovate, 12–39 cm × 4–14 cm, cuneate at the base, and acuminate at the apex, with 9–16 pairs of secondary nerves. The flowers are male or female. The male flowers are about 0.3 cm in diameter and the female flowers are about 0.5 cm in diameter. The fruits are angular, juicy, and sour–sweet, with edible white pulp berries, to 3 cm long (Figure 4.22). The pharmacological property of this plant is unknown.

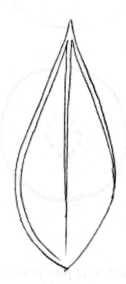

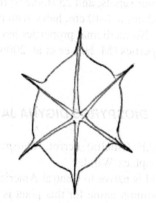

Figure 4.22 *Baccaurea angulata* Merr.

4.43 *PHYLLANTHUS EMBLICA* L.

Synonyms: *Emblica officinalis* Gaertn., *Emblica pectinata* (Hook. f.) Ridley

Indian gooseberry or emblic myrobalan is a tree that grows from India to Southeast Asia to China. In Malaysia, the plant is known as *pokok melaka;* in Indonesia, as *kimalaka;* in Burma, as *ta-sha-pen;* and as *ma-khaam pom* in Thailand. The bark is gray-fawn. The leaves are simple and spiral on 15-cm rachis. The blade is linear, 0.7–2.5 cm × 0.13–0.4 cm, and rounded at the apex and base. The flowers are minute and axillary. The fruits are globose, to 2 cm in diameter, and ripening to greenish yellow (Figure 4.23). The stem bark is used for tanning leather. The plant is astringent and used to treat inflammation in Burma; fever in Malaysia; diarrhea in Cambodia, Laos, and Vietnam; and headaches in Indonesia.

Putranjivain A, 1,6-di-*O*-galloyl-β-d-glucose, 1-*O*-galloyl-β-d-glucose, kaempferol-3-*O*-β-d-glucoside, quercetin-3-*O*-β-d-glucoside, and digallic acid from the plant inhibited the human immunodeficiency virus-1 reverse transcriptase (el-Mekkawy et al. 1995). The anti-inflammatory property is confirmed (Ihantola-Vormisto et al. 1997). Phenolic constituents isolated from this plant showed some anticancer properties (Rajeshkumar et al. 2003). The plant has antipyretic and

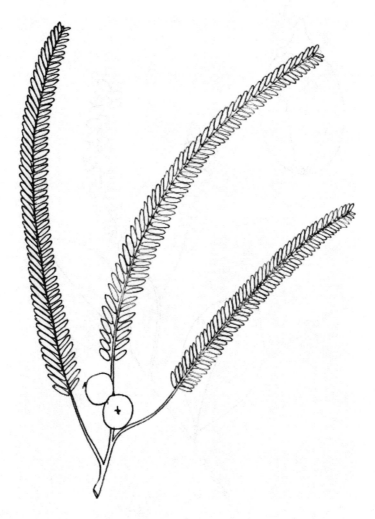

Figure 4.23 *Phyllanthus emblica* L.

analgesic properties (Perianayagam et al. 2004). The fruit extract has a chondroprotective property in osteoarthritis (Sumantran et al. 2008).

4.44 *ANTIDESMA BUNIUS* (L.) SPRENG.

Synonyms: *Stilago bunius* L., *Antidesma rumphii* Tul., *Antidesma dallachyanum* Baill.

The plant is known as salamander tree—*ma mao luang* in Thailand, *bignai* in the Philippines, *choi moi* in Vietnam, *hooni* in Indonesia, *berunai* in Malaysia, *kho lien tu* in Laos, and *moi-kin* and *chunka* in Australia. It is a tree that grows to 30 m, wild or cultivated, from India to Australia. The leaves are simple and alternate. The leaf blade is elliptic lanceolate to 23 cm long and 8 cm in diameter. The flowers are minute, reddish, and either male or female on different inflorescences. The fruits are globose, glossy, red turning black, about 1 cm in diameter, and contain one seed in a slightly sweet pulp which is edible (Figure 4.24). The fruits are eaten raw. The plant is medicinal in Cambodia, Laos, and Vietnam. In the Philippines, the plants are used to treat syphilis. The

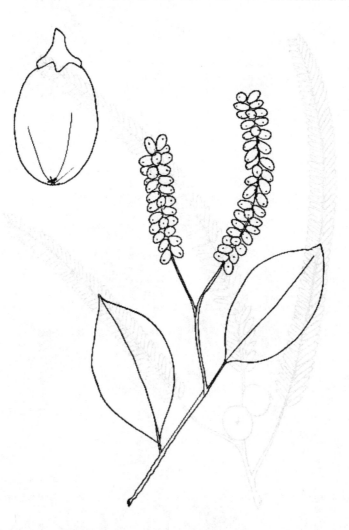

Figure 4.24 *Antidesma bunius* (L.) Spreng.

fruits contain flavonoids (Butkhup and Samappito 2008). Extract of the plant showed some toxicity against *Artemia salina* (Micor, Deocaris, and Mojica 2005).

4.45 *PHYLLANTHUS ACIDUS* (L.) SKEELS

Synonym: *Cicca acida* (L.) Merr., *Cicca disticha* L., *Phyllanthus distichus* Müll.-Arg.

The plant is known as otaheite gooseberry. It is a tree native to Brazil and is cultivated throughout Southeast Asia. In Malaysia, it is known as *chermai* and as *cheremoi* in Indonesia, *iba* in the Philippines, and *mayom* in Thailand. It grows to 8 m. The leaves are simple, and alternate. The leaf blade is 2–7.5 cm long and glaucous below. The flowers are minute in 15 cm long racemes. The fruit is edible, pulpy, six to eight lobed, 1–2.5 cm in diameter, acid, light green, and about 2 cm long and contains four to six seeds (Figure 4.25). In the Philippines, the plant is used to treat bronchitis and urticaria. In Indonesia, it is used to treat lumbago. In Malaysia, the plant is used for cough, headache, and psoriasis. In Cambodia, Laos, and Vietnam, it is used to treat smallpox and constipation. The plant contains some cytotoxic principles (Vongvanich et al. 2000), displays antibacterial properties (Meléndez and Capriles 2006) and hepatoprotective properties (C. Y. Lee et al. 2006), and may have some potential as a treatment for cystic fibrosis (Sousa et al. 2007).

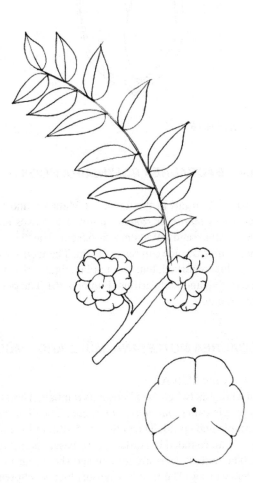

Figure 4.25 *Phyllanthus acidus* (L.) Skeels.

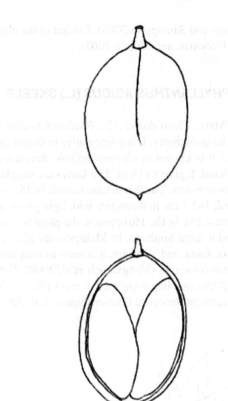

Figure 4.26 *Baccaurea polyneura* Hook. f.

4.46 *BACCAUREA POLYNEURA* HOOK. F.

This is a tree that grows to 25 m tall in the forests of Malaysia and Sumatra. It is endemic to Malaysia, where it is called *jintek merah* or *rambai hutan*. The leaves are simple and spiral. The blade is thick, oblong shaped, obtuse, and 10–14 cm × 5–6.5 cm. The blade is cordate at the base and acute at the apex. The female inflorescence can reach 30 cm. The fruit is an ovoid capsule, which is 2.5 cm long and reddish, with big, recurved, long, triangular stigmas. It opens to release a few seeds embedded in an edible, glossy, red, and juicy arillus (Figure 4.26). The plant is not medicinal. It has not been studied for pharmacology.

4.47 *BACCAUREA MOTLEYANA* (MÜLL ARG.) MÜLL ARG.

Synonym: *Pierardia motleyana* Müll Arg.

This tree grows wild in Malaysia (where it is known as rambai), Thailand, and Indonesia, and it is cultivated in China, where it is called *duo mai mu nai guo*. The plant grows to 10 m. The stems are hairy. The leaves are simple and spiral. The petiole is 5–10 cm long and hairy. The leaf blade is elliptic, 20–35 cm × 7.5–17 cm, rounded to cordate at the base, and acute at the apex, with 12–16 pairs of secondary nerves. The male flowers are minute and show four to six stamens. The female flowers are minute, with a hairy ovary. The fruit is a capsule that is yellowish, baccate, globose, and 2.5–3 cm and has three seeds. The white pulp is sweet and edible.

4.48 *BACCAUREA BREVIPES* HOOK. F.

This tree grows in Thailand, Malaysia, and Indonesia. In Thailand, it is known as *fai tao*. The plant grows to 10 m. The stems are hairy. The leaves are simple and spiral. The petiole is about 3 cm long. The blade is elliptic, 15–35 cm × 6–15 cm, and round at the base, with 11–15 pairs of secondary nerves. The flowers are male or female, minute, and yellowish green. The fruits are globose, fleshy, 1.5–2 cm × 1–2 cm, and white to red to pink and contain a few seeds with blue arillode. The fruits are edible. In Indonesia, the plant is used to promote menses. The plant has not been studied for pharmacology.

4.49 *BACCAUREA LANCEOLATA* (MIQ.) M.A.

Synonyms: *Hedycarpus lanceolatus* Miq., *Adenocrepis lanceolatus* (Miq.) Mull. Arg.
Green rambai is a small tree that grows in Thailand (known as *som huuk*), Malaysia (*rambai utan*, *asam pahong*), Indonesia (*lompayang*), and the Philippines (*limpahung*). The bark is gray-green. The leaves are simple and spiral, large, elliptic, and about 25 cm × 10 cm. The fruits are arranged in racemes about 18 cm long. The fruit is an ovoid capsule, up to 5 cm long, yellow-green, contains few seeds, and has a translucent, juicy, sour, edible pulp. The plant is not medicinal.

4.50 *BACCAUREA MACROCARPA* (MIQ.) MÜLL. ARG.

Synonyms: *Baccaurea borneensis* (Mull. Arg.) Mull. Arg., *Baccaurea griffithii* Hook. f., *Mappa borneensis* Müll. Arg., *Pierardia macrocarpa* Miq.
The plant grows in Malaysia and Indonesia. In Malaysia, it is called *tampoi*; in Indonesia, it is called *tampoi saya*. It is a tree that can reach 18 m. The leaves are alternate, simple, and crowded at the twig apex. The petiole is long and the leaf blade is broadly ovoid. The flowers are minute, white-yellow, and arranged in racemes. The fruits are depressed, globose, to 4 cm in diameter, yellow-orange-red, indehiscent, cauliflorous capsules; seeds have a white-yellow-orange aril, which is edible. The plant has not been studied for pharmacology.

4.51 *CASTANOPSIS INERMIS* (LINDL. EX WALL.) BENTH. & HOOK.

Synonyms: *Callaeocarpus sumatrana* Miq., *Castanopsis sumatrana* A. DC.
This is a big tree that grows to 30 m in the forests of Malaysia, Thailand, and Indonesia, where it is called *berangan*. The English name for the plant is braided chestnut. The bole is buttressed and the bark is flaky. The leaves are spiral and simple. The stalk is about 2 cm long. The blade is elliptic-ovate 7–23 cm × 3–9 cm, with 12–16 pairs of secondary nerves and tiny scales below. The flowers are minute and white, with a strong musty smell. The cupule is four angled with short blunt warts 5 cm × 3 cm, sessile, and splitting onto four valves. The cupule contains three fruits, which are brown, triangular, and pointed with a large basal scar. The fruit is cooked. The pharmacological properties of this plant are unknown.

4.52 *FLACOURTIA INERMIS* ROXB.

This tree grows to 15 m tall, either wild or cultivated, from India to Southeast Asia and Papua New Guinea. It is known as governor plum, *lobi-lobi* in Indonesia, *rukam masam* in Malaysia, and *lovi-lovi* in Thailand. The leaves are simple, ovate–oblong, and 7–20 cm × 3–12 cm. The fruit is a globose berry, which is about 2 cm in diameter, pink to red, and glossy (Figure 4.27). The fruits

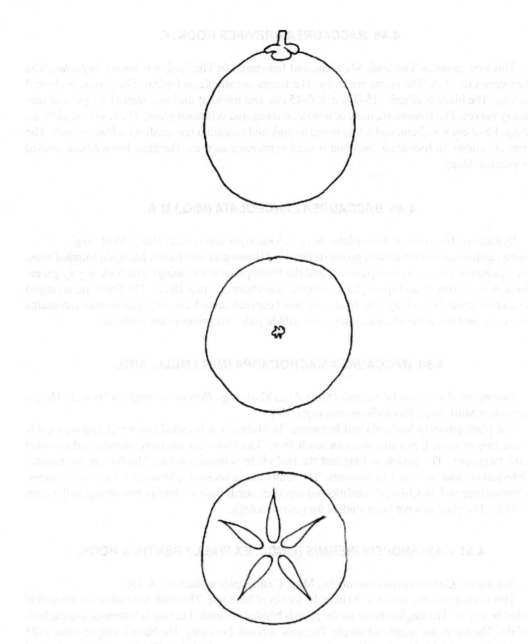

Figure 4.27 *Flacourtia inermis* Roxb.

are eaten raw and used to make jams. The fruit is an astringent. The plant has not been studied for pharmacology.

4.53 CONCLUSIONS

Southeast Asia is richly endowed with a diversity of tree species that can serve as useful feed and medicinal resources. The plants listed in this chapter could be more widely adopted for long-term benefits to the overall farming or pharmaceutical systems. While substantial work has been conducted on common species, a lot remains to be studied in wild medicinal species.

In this chapter, 52 species of medicinal fruit trees have been described. Most species belong to the Clusiaceae family, *Garcinia* and *Mangifera* species. Out of all the species, fewer than half have been studied for pharmacology or chemistry. Some species have shown interesting results, such as *Garcinia parvifolia* (Miq.) Miq., *Mangifera longipetiolata* King., *Phyllanthus emblica* L., and *Antidesma bunius* (L.) Spreng. This warrants further study of the medicinal fruit trees of Southeast Asia.

REFERENCES

Abu Bakar, M. F., Mohamed, M., Rahmat, R., and Fry, J. 2009. Phytochemicals and antioxidant activity of different parts of bambangan (*Mangifera pajang*) and tarap (*Artocarpus odoratissimus*). *Food Chemistry* 113:479–483.

Butkhup, L., and Samappito, S. 2008. An analysis on flavonoid contents in mao luang fruits of fifteen cultivars (*Antidesma bunius*), grown in northeast Thailand. *Pakistan Journal of Biological Sciences* 11 (7): 996–1002.

Campos, F. R., Januário, A. H., Rosas, L. V., Nascimento, S. K., Pereira, P. S., França, S. C., Cordeiro, M. S., Toldo, M. P., and Albuquerque, S. 2005. Trypanocidal activity of extracts and fractions of *Bertholletia excelsa*. *Fitoterapia* 76 (1): 26–29.

Dinda, B., Bhattacharya, A., De, U. C., Arima, S., Takayanagi, and Harigaya, Y. 2006. Antimicrobial C glucoside from aerial parts of *Diospyros nigra*. *Chemical and Pharmaceutical Bulletin (Tokyo)* 54 (5): 679–681.

el-Mekkawy, S., Meselhy, M. R., Kusumoto, I. T., Kadota, S., Hattori, M., and Namba, T. 1995. Inhibitory effects of Egyptian folk medicines on human immunodeficiency virus (HIV) reverse transcriptase. *Chemical & Pharmaceutical Bulletin (Tokyo)* 43 (4): 641–648.

Ganapaty, S., Thomas, P. S., Mallika, B. N., Balaji, S., Karagianis, G., and Waterman, P. G. 2005. Dimeric naphthoquinones from *Diospyros discolor*. *Biochemical Systematics and Ecology* 33 (3): 313–315.

Graham, J. G., Pendland, S. L., Prause, J. L., Danzinger, L. H., Schunke, J., Cabieses, F., and Farnsworth, N. R. 2003. Antimycobacterial evaluation of Peruvian plants. *Phytomedicine* 10 (6–7): 528–535.

Hutadilok-Towatana, N., Kongkachuay, S., and Mahabusarakam, W. 2007. Inhibition of human lipoprotein oxidation by morelloflavone and camboginol from *Garcinia dulcis*. *Natural Product Research* 21 (7): 655–662.

Iacomini, M., Serrato, R. V., Sassaki, G. L., Lopes, L., Buchi, D. F., and Gorin, P. A. 2005. Isolation and partial characterization of a pectic polysaccharide from the fruit pulp of *Spondias cytherea* and its effect on peritoneal macrophage activation. *Fitoterapia* 76 (7–8): 676–683.

Ihantola-Vormisto, A., Summanen, J., Kankaanranta, H., Vuorela, H., Asmawi, Z. M., and Moilanen, E. 1997. Anti-inflammatory activity of extracts from leaves of *Phyllanthus emblica*. *Planta Medica* 63 (6): 518–524.

Iliya, I., Ali, Z., Tanaka, T., Iinuma, M., Furusawa, M., Nakaya, K., Murata, J., Darnaedi, D., Matsuura, N., and Ubukata, M. 2003a. Stilbene derivatives from *Gnetum gnemon* Linn. *Phytochemistry* 62 (4): 601–606.

Iliya, I., Ali, Z., Tanaka, T., Iinuma, M., Furasawa, M., Nakaya, K., Shirataki, Y., Murata, J., Darnaedi, D., Matsuura, N., and Ubukata, M. 2003b. Three new trimeric stilbenes from *Gnetum gnemon*. *Chemical and Pharmaceutical Bulletin (Tokyo)* 51 (1): 85–88.

Iliya, I., Tanaka, T., Iinuma, M., Ali, Z., Furasawa, M., Nakaya, K., Shirataki, Y., Murata, J., and Darnaedi, D. 2002. Stilbene derivatives from two species of Gnetaceae. *Chemical and Pharmaceutical Bulletin (Tokyo)* 50 (6): 796–801.

Ip, C., and Lisk, D. J. 1994. Bioactivity of selenium from Brazil nut for cancer prevention and selenoenzyme maintenance. *Nutrition Cancer* 21 (3): 203–212.

Jantan, I., Pisar, M. M., Idris, M. S., Taher, M., and Ali, R. M. 2002. *In vitro* inhibitory effect of rubraxanthone isolated from *Garcinia parvifolia* on platelet-activating factor receptor binding. *Planta Medica* 68 (12): 33–34.

Jena, B. S., Jayaprakasha, G. K., and Sakariah, K. K. 2002. Organic acids from leaves, fruits, and rinds of *Garcinia cowa*. *Journal of Agricultural and Food Chemistry* 50 (12): 3431–3434.

Jena, B. S., Jayaprakasha, G. K., Singh, R. P., and Sakariah, K. K. 2002. Chemistry and biochemistry of (-) hydroxycitric acid from *Garcinia*. *Journal of Agricultural and Food Chemistry* 50 (1):10–22.

Lajis, N. H., and Khan, M. N. 1994. Extraction, identification and spectrophotometric determination of second ionization constant of methyl gallate, a constituent present in the fruit shells of *Pithecellobium jiringa*. *Indian Journal of Chemistry* Sect B 33 (6): 609–612.

Lee, C. Y., Peng, W. H., Cheng, H. Y., Chen, F. N., Lai, M. T., and Chiu, T. H. 2006. Hepatoprotective effect of *Phyllanthus* in Taiwan on acute liver damage induced by carbon tetrachloride. *American Journal of Chinese Medicine* 34 (3): 471–482.

Lee, M. H., Jiang, C. B., Juan, S. H., Lin, R. D., and Hou, W. C. 2006. Antioxidant and heme oxygenase-1 (HO-1)-induced effects of selected Taiwanese plants. *Fitoterapia* 77 (2): 109–115.

Li, X., Ohtsuki, T., Shindo, S., Sato, M., Koyano, T., Preeprame, S., Kowithayakorn, T., and Ishibashi, M. 2007. Mangiferin identified in a screening study guided by neuraminidase inhibitory activity. *Planta Medica* 73 (11): 1195–1196.

Likhitwitayawuid, K., Chanmahasathien, W., Ruangrungsi, N., and Krungkrai, J. 1998. Xanthones with antimalarial activity from *Garcinia dulcis*. *Planta Medica* 64 (3): 281–282.

Likhitwitayawuid, K., Phadungcharoen, T., and Krungkrai, J. 1998. Antimalarial xanthones from *Garcinia cowa*. *Planta Medica* 64 (1): 70–72.

Mackeen, M. M., Ali, A. M., Lajis, N. H., Kawazu, K., Hassan, Z., Amran, M., Habsah, M., Mooi, L. Y., and Mohamed, S. M. 2000. Antimicrobial, antioxidant, antitumor-promoting and cytotoxic activities of different plant part extracts of *Garcinia atroviridis* Griff. ex T. Anders. *Journal of Ethnopharmacology* 72 (3): 395–402.

Mackeen, M. M., Ali, A. M., Lajis, N. H., Kawazu, K., Kikuzaki, H., and Nakatani, N. 2002. Antifungal garcinia acid esters from the fruits of *Garcinia atroviridis*. *Zeitschrift für Naturforschung* 57 (3–4): 291–295.

Mackeen, M. M., Khan, M. N., Samadi, Z., and Lajis, N. H. 2000. Brine shrimp toxicity of fractionated extracts of Malaysian medicinal plants. *Natural Product Sciences* 6 (3): 131–134.

Mahendran, P., Vanisree, A. J., and Devi, C. S. 2002. The antiulcer activity of *Garcinia cambogia* extract against indomethacin-induced gastric ulcer in rats. *Phytotherapy Research* 16 (1): 80–83.

Masuda, D., Koyano, T., Fujimoto, H., Okuyama, E., Hayashi, M., Komiyama, K., and Ishibashi, M. 2002. Alkenylphenol and alkenylsalicylic acid from *Mangifera caesia*. *Biochemical Systematics and Ecology* 30 (5): 475–478.

Meléndez, P. A., and Capriles, V. A. 2006. Antibacterial properties of tropical plants from Puerto Rico. *Phytomedicine* 13 (4): 272–276.

Micor, J. R. L., Deocaris, C. C., and Mojica, E. R. E. 2005. Biological activity of bignay [*Antidesma bunius* (L.) Spreng] crude extract in *Artemia salina*. *Journal of Medical Sciences (Pakistan)* 5 (3): 195–198.

Mohamed, S., Saka, S., El-Sharkawy, S. H., Ali, A. M., and Muid, S. 1999. Antimycotic screening of 58 Malaysian plants against plant pathogens. *Pesticide Science* 47 (3): 259–264.

Nguyen, L. H., Venkatraman, G., Sim, K. Y., and Harrison, L. J. 2005. Xanthones and benzophenones from *Garcinia griffithii* and *Garcinia mangostana*. *Phytochemistry* 66 (14): 1718–1723.

Oluyemi, K. A., Omotuyi, I. O., Jimoh, O. R., Adesanya, O. A., Saalu, C. L., and Josiah, S. J. 2007. Erythropoietic and antiobesity effects of *Garcinia cambogia* (bitter kola) in Wistar rats. *Biotechnology & Applied Biochemistry* 46 (1): 69–72.

Perez, G. R. M., Vargas, S. R., and Ortiz, H. Y. D. 2005. Wound healing properties of *Hylocereus undatus* on diabetic rats. *Phytotherapy Research* 19 (8): 665–668.

Perianayagam, J. B., Sharma, S. K., Joseph, A., and Christina, A. J. 2004. Evaluation of anti-pyretic and analgesic activity of *Emblica officinalis* Gaertn. *Journal of Ethnopharmacology* 95 (1): 83–85.

Permanaa, D., Abas, F., Shaari, K., Stanslas, J., Ali, A. M., and Lajis, N. H. 2005. Atrovirisidone B, a new prenylated depsidone with cytotoxic property from the roots of *Garcinia atroviridis*. *Zeitschrift für Naturforschung* 60 (7–8): 523–526.

Preuss, H. G., Rao, C. V., Garis, R., Bramble, J. D., Ohia, S. E., Bagchi, M., and Bagchi, D. 2004. An overview of the safety and efficacy of a novel, natural (–)-hydroxycitric acid extract (HCA-SX) for weight management. *Journal of Medicine* 35 (1–6): 33–48.

Rajeshkumar, N. V., Pillai, M. R., and Kuttan, R. 2003. Induction of apoptosis in mouse and human carcinoma cell lines by *Emblica officinalis* polyphenols and its effect on chemical carcinogenesis. *Journal of Experimental and Clinical Cancer Research* 22 (2): 201–212.

Roongpisuthipong, C., Kantawan, R., and Roongpisuthipong, W. 2007. Reduction of adipose tissue and body weight: Effect of water-soluble calcium hydroxycitrate in *Garcinia atroviridis* on the short-term treatment of obese women in Thailand. *Asia Pacific Journal of Clinical Nutrition* 16 (1): 25–29.

Rudiyansyah, and Garson, M. J. 2006. Secondary metabolites from the wood bark of *Durio zibethinus* and *Durio kutejensis. Journal of Natural Products* 69 (8): 1218–1221.

Rukachaisirikul, V., Adair, A., Dampawan, P., Taylor, W. C., and Turner, P. C. 2000. Lanostanes and friedolanostanes from the pericarp of *Garcinia hombroniana. Phytochemistry* 55 (2): 183–188.

Rukachaisirikul, V., Saelim, S., Karnsomchoke, P., and Phongpaichit, S. 2005. Friedolanostanes and lanostanes from the leaves of *Garcinia hombroniana. Journal of Natural Products* 68 (8): 1222–1225.

Sakagami, Y., Sawabe, A., Komemushi, S., All, Z., Tanaka, T., Iliya, I., and Iinuma, M. 2007. Antibacterial activity of stilbene oligomers against vancomycin-resistant *Enterococci* (VRE) and methicillin-resistant *Staphylococcus aureus* (MRSA) and their synergism with antibiotics. *Biocontrol Science* 12 (1): 7–14.

Sousa, M., Ousingsawat, J., Seitz, R., Puntheeranurak, S., Regalado, A., Schmidt, A., Grego, T., Jansakul, C., Amaral, M. D., Schreiber, R., and Kunzelmann, K. 2007. An extract from the medicinal plant *Phyllanthus acidus* and its isolated compounds induce airway chloride secretion: A potential treatment for cystic fibrosis. *Molecular Pharmacology* 71 (1): 366–376.

Sumantran, V. N., Kulkarni, A., Chandwaskar, R., Harsulkar, A., Patwardhan, B., Chopra, A., and Wagh, U. V. 2008. Chondroprotective potential of fruit extracts of *Phyllanthus emblica* in osteoarthritis. *Evidence-Based Complementary and Alternative Medicine* 5 (3): 329–335.

Syahida, A., Israf, D. A., Permana, D., Lajis, N. H., Khozirah, S., Afiza, A. W., Khaizurin, T. A., Somchit, M. N., Sulaiman, M. R., and Nasaruddin, A. A. 2006. Atrovirinone inhibits pro-inflammatory mediator release from murine macrophages and human whole blood. *Immunology and Cell Biology* 84 (3): 250–258.

Takagi, K., and Mitsunaga, T. 2002. A tyrosinase inhibitor from the wood of Asam (*Mangifera quadrifida* J.). *Natural Medicines* 56 (3): 97–103.

Taylor, R. S. L., Hudson, J. B., Manandhar, N. P., and Towers, G. H. N. 1996. Antiviral activities of medicinal plants of southern Nepal. *Journal of Ethnopharmacology* 53 (2): 105–110.

Tian, Z., Shen, J., Moseman, A. P., Yang, Q., Yang, J., Xiao, P., Wu, E., and Kohane, I. S. 2008. Dulxanthone A induces cell cycle arrest and apoptosis via up-regulation of p53 through mitochondrial pathway in HepG2 cells. *International Journal of Cancer* 122 (1): 31–38.

Vohra, M. M., and De, N. N. 1963. Comparative cardiotonic activity of *Carissa caranda* L. and *Carissa spinarum* A. DC. *Indian Journal of Medical Research* 51:937–940.

Vongvanich, N., Kittakoop, P., Kramyu, J., Tanticharoen, M., and Thebtaranonth, Y. 2000. Phyllanthusols A and B, cytotoxic norbisabolane glycosides from *Phyllanthus acidus* Skeels. *Journal of Organic Chemistry* 65 (17): 5420–5423.

Xu, Y. J., Chiang, P. Y., Lai, Y. H., Vittal, J. J., Wu, X. H., Tan, B. K., Imiyabir, Z., and Goh, S. H. 2000. Cytotoxic prenylated depsidones from *Garcinia parvifolia. Journal of Natural Products* 63 (10): 1361–1363.

CHAPTER 5

Medicinal Plants of China

Zhongzhen Zhao, Zhitao Liang, Ping Guo, and Hubiao Chen

CONTENTS

5.1 INTRODUCTION

China is one of the countries that have the most abundant medicinal plant species and the longest medicinal plant application history. Under the guidance of the traditional Chinese medicine (TCM) theory, Chinese medicinals are mostly of botanical origin. There have been great achievements in the aspects of research and development, protection, and sustainable utilization of medicinal plants of China. Systematic quality evaluation of medicinal plants has been gradually established as well. Medicinal plants have been continuously making great contributions to the health of people at home and abroad.

5.1.1 History

The use of medicinal plants in China began more than 3000 years ago. *The Book of Poetry* (*Shi Jing*), China's ancient poem collection (ca. 3000 BC), records more than 50 species of medicinal plants (T. P. Zhu et al. 2007). *The Philosophers of Huainan* (Huai Nan Zi), an ancient book published in the early Western Han Dynasty (ca. 200 BC), records that "the Divine Husbandman [*shennong*] tasted hundreds of herbs and met 70 toxicities within a day." Ancient medical books excavated in 1973 from tombs of the Western Han Dynasty at Mawangdui list 169 herbal substances (Ma 1986).

In China, books that record the sources and applications of medicinal materials are commonly known as *ben cao* (materia medica). Because most medicinal materials are from botanical sources, the words ben cao imply that medicinal materials are primarily plant derived. There are more than 400 such books, from all the past dynasties of China. These various ben cao in documenting ancient people's experience represent centuries of accumulated wisdom in combating disease and preserving health. Some prominent representatives are introduced next (Association of Chinese Culture Research 1999):

The Divine Husbandman's Classic of Materia Medica (*Shen Nong Ben Cao Jing*): This is the earliest materia medica, compiled in the late Eastern Han Dynasty (ca. AD 200); it is known only from compiled versions dating from the Ming Dynasty (ca. AD 1400) because the original book has been lost. This book records 365 medicinals, including 237 herbal substances, and summarizes medicinal experiences as of the Han Dynasty. Medicinals are classified into three categories: high grade, medium grade, and low grade. Entries for each substance include definition, compatibilities, properties, harvesting details, processing methods, and medical applications. Production areas and macroscopic descriptions of some medicinal plants are also recorded briefly.

Collection of Commentaries on the Classic of the Materia Medica (*Ben Cao Jing Ji Zhu*): This book records 730 medicinals. It was compiled by Tao Hongjing, a physician of the Liang Dynasty, by revising *The Divine Husbandman's Classic of Materia Medica* and adding another 365 medicinals. In this book, Tao introduces new methods of medicinal classification based on the natural and therapeutic properties of the medicinal materials. The materials are classified into seven categories: herbs, trees, crops, insects or beasts, jades or stones, fruits or vegetables, and medicinals with names but without actual applications. Each entry includes morphological description, production area, harvesting time, processing, dosage, usage, and authentication.

Newly Revised Materia Medica (*Xin Xiu Ben Cao*): Commissioned by the government of the Tang Dynasty in AD 659, this 54-volume materia medica is considered the earliest national pharmacopoeia in China. It was compiled by experts and is divided into three parts: texts, drawings, and illustrations. It records 850 medicinals with excellent drawings and written descriptions. In the original book, all drawings were in color. Unfortunately, the original book has been lost, and the only extant versions were reconstructed by later authors.

Materia Medica Arranged according to Pattern (*Zheng Lei Ben Cao*): This book was finished in 1082 by a physician named Tang Shenwei and is the most praiseworthy materia medica of the Song Dynasty as it comprehensively summarizes herbal knowledge up to that time. There are three versions of this book (*Da Guan*, *Zheng He*, and *Shao Xing*) currently in circulation. In this book, medicinal materials are divided into 13 categories: jades and stones, herbs, trees, human beings, beasts, birds, insects and fishes, fruits, crops, vegetables, medicinals with names but without actual applications, herbs without illustrations, and trees and vines without illustrations. It records 1,746 medicinals including their synonyms, medicinal properties, indications, production areas, harvesting times, processing methods, differentiations, and prescriptions.

Compendium of Materia Medica (*Ben Cao Gang Mu*): Written by Li Shizhen, a physician of the Ming Dynasty, the *Compendium of Materia Medica* was first published in 1593. It records 1,892 medicinal materials including 1,095 herbal substances. Based on morphological characteristics, medicinal materials are classified into mineral substances, herbal substances, and zoological substances. Herbal substances are further classified into five categories—herbs, crops, vegetables, fruits, and trees—according to their properties, morphological characteristics, and growth environments.

Under each entry, the name, production area, nature, taste, morphological characteristics, and processing method have been documented. Thus, the *Compendium of Materia Medica* represents the highest academic achievement among all the ancient Chinese materia medica with novel plant classification methods and highly informative documentations.

Illustrated Reference of Botanical Nomenclature (*Zhi Wu Ming Shi Tu Kao*): Written by Wu Qijun, a botanist of the Qing Dynasty, this 38-volume book records 1,714 plant species. Plants are divided into 12 categories: crops, vegetables, herbs growing in mountains, herbs growing in wet places, herbs growing near stones, water weeds, herbs with creeping stems, toxic herbs, fragrant herbs, flowering herbs, fruits, and trees. Under each category, the name, morphological description, color, taste, variety, production area, growing habitat, and usage are recorded. This book includes 1,805 drawings and 1,500 sketches that are the most accurate drawings found among all the ancient Chinese materia medica. This book has served as a bridge linking ancient Chinese herbology to modern botany and agricultural sciences.

Since 1949, China has carried out much work on medicinal plants, including field surveys and studies on the development, utilization, introduction, and cultivation of medicinal herbs. China has also made great efforts in the determination, extraction, and isolation of chemical components, as well as in conducting pharmacological experiments with medicinal plants. Based on these efforts, many monographs have been published. The following are representative of ongoing, concerted efforts to document and promote the use of medicinal plants and their medicinal materials in the modern world:

The nine-volume *Medicinal Plants of China* (*Zhong Guo Yao Yong Zhi Wu Zhi*), published from 1955 to 1985, documents 450 species with illustrations (Pei and Zhou 1955–1985).

The *Chinese Materia Medica* (*Zhong Yao Zhi*), published from 1959 to 1961, records more than 500 species. The revised six-volume edition (1982–1998) records 637 herbal substances (including spores, volatile oils, and processed products), involving more than 2,100 medicinal plant species (Institute of Materia Medica, Chinese Academy of Medical Sciences, 1959, 1982–1998).

The two-volume *National Collection of Chinese Herbal Medicinals* (*Quan Guo Zhong Cao Yao Hui Bian*) records more than 4,000 species of medicinal plants (Editorial Committee 1975–1978, 1996).

The 30-volume *Materia Medica of China* (*Zhong Hua Ben Cao*), published in 1999, comprehensively and thoroughly documents knowledge of 7,815 herbal substances. Contents of this work include textual investigation, botanical origin, cultivation and harvesting notes, marketing, authentication, chemical composition, pharmacological activities, processing methods, medicinal properties, functions and indications, applications and compatibility, usage and dosage, precautions, prescriptions, preparations, modern clinic research, and references (Editorial Committee 1999).

The 80-volume *Flora of China*, published from 1959 to 2004, covers 31,142 plant species involving 301 families and 3,408 genera. The medicinal effects of plants are specifically mentioned in *Flora of China* (Editorial Committee 1959–2004).

The *Pharmacopoeia of the People's Republic of China* records 551 crude medicinal materials and decoction pieces (Chinese Pharmacopoeia Commission 2005).

Thus, since early history, the Chinese people have attempted to document and facilitate the use of natural medicinal materials. As medicine and agriculture have progressed, so has the cultivation of medicinal plants. Indeed, the cultivation of medicinal plants has a long history in China. In *Important Arts for the People's Welfare* (*Qi Min Yao Shu*), an agricultural encyclopedia of the late Northern Wei Dynasty compiled by Jia Sixie, cultivation methods of more than 20 medicinal plants are recorded. In the Sui Dynasty, the imperial medical department assigned officials specifically to take charge of the cultivation of medicinal plants. According to the *Index of Classics of the Sui Dynasty* (*Sui Shu Jing Ji Zhi*), monographs on the cultivation of medicinal plants such as *Cultivation Methods of Medicinal Plants* (*Zhong Zhi Yao Fa*) and *Plantation of the Divine Herbs* (*Zhong Shen Zhi*) had already been published (Chen and Xiao 2006). Written in the Ming Dynasty, the *Compendium of Materia Medica* records more than 180 species of medicinal plants with their cultivation methods.

Up to the present, medicinal plants such as *Angelica sinensis* (Oliv.) Diels, *Ligusticum chuanxiong* Hort., *Aconitum carmichaeli* Debx., *Coptis chinensis* Franch., *Carthamus tinctorius* L., *Lycium barbarum* L., *Panax ginseng* C.A. Mey., *Achyranthes bidentata* Bl., *Chrysanthemum morifolium* Ramat., *Dioscorea opposita* Thunb., *Rehmannia glutinosa* Libosch., *Fritillaria thunbergii* Miq., and *Ophiopogon japonicus* Ker-Gawl. have been cultivated for hundreds and thousands of years in China. Not all medicinal herbs have been traditionally cultivated. However, now, the global popularity of natural medicines is putting unprecedented pressure on limited resources of certain wild medicinal plants, and their cultivation is an inevitable trend to ensure their survival. Up to now, there are about 300 species of cultivated medicinal plants in China. Since June 1, 2002, Good agricultural practice (GAP) for Chinese crude medicinal materials has been carried out in China, which provides guidelines for the management and the cultivation of medicinal plants.

5.1.2 Distribution of Resources

China is vast in territory, stretching both north and south at about 50° latitude and from sea to midcontinent. It has diversified land forms and mountain ranges and hence diverse weather. Precisely because of the complexity and diversity of the climate, China has abundant botanical and zoological resources.

According to a national survey on Chinese medicinal materials (1985–1989), China has 11,118 species of medicinal plants (including 1,208 subspecies, varieties, and forms) belonging to 385 families and 2,313 genera. Among these species, angiosperms account for more than 90% and are the main body of Chinese medicinal plant resources (China Medicinal Materials Group, 1995).

According to the statistics, each of the following 32 families contains more than 100 medicinal plant species:

Asteraceae	Fabaceae
Lamiaceae	Ranunculaceae
Rosaceae	Apiaceae
Scrophulariaceae	Rubiaceae
Euphorbiaceae	Saxifragaceae
Papaveraceae	Ericaceae
Polygonaceae	Primulaceae
Berberidaceae	Urticaceae
Gesneriaceae	Lauraceae
Araliaceae	Campanulaceae
Gentianaceae	Caryophyllaceae
Vitaceae	Caprifoliaceae
Verbenaceae	Rutaceae
Liliaceae	Orchidaceae
Poaceae	Cyperaceae
Araceae	Zingiberaceae

Among these, the Asteraceae is the largest, with 778 species of medicinal plants.

In addition to families with many species, some genera also contain a large number of medicinal plants. The following each contain more than 50 species of medicinal plants: *Aconitum, Corydalis, Clematis, Polygonum, Artemisia, Berberis, Rhododendron, Rubus, Euonymus, Lysimachia, Salvia,* and *Gentiana.* Among them, 103 species in *Aconitum* are used medicinally, making it the biggest medicinal genus in the angiosperm. Commonly used species include *Aconitum carmichaeli* Debx., *Aconitum kusnezoffii* Reichb., and *Aconitum coreanum* (Levl.) Rapaics.

In some families with many medicinal species, the common medicinal plants are found in only a few genera. For example, medicinal plants of the Polygonaceae are mainly in the genus *Polygonum*, including *Polygonum multiflorum* Thunb., *Polygonum orientale* L., *Polygonum avicu-lare* L., *Polygonum tinctorium* Ait., and *Polygonum bistorta* L. Medicinal plants of Araliaceae are mostly distributed in *Panax* and *Acanthopanax*.

Medicinal parts of plants in some families also have similarities. For example, roots or rhi-zomes of common medicinal plants of Ranunculaceae are often medicinally used: Rhizomes of *Coptis chinensis* Franch., *Coptis deltoidea* C.Y. Cheng et Hsiao, and *Coptis teeta* Wall. are Chinese medicinal *huanglian* (Rhizoma Coptidis); the root of *Pulsatilla chinensis* (Bunge) Regel. is Chinese medicinal *baitouweng* (Radix Pulsatillae); and rhizomes of *Cimicifuga foetida* L., *Cimicifuga heracleifolia* Kom., and *Cimicifuga dahurica* (Turcz.) Maxim. are Chinese medicinal *shengma* (Rhizoma Cimicifugae). Fruits or seeds of medicinal plants in Rosaceae are often medicinally used: The fruit of *Rosa laevigata* Michx. is Chinese medicinal *jinyingzi* (Fructus Rosae Laevigatae); fruits of *Crataegus pinnatifida* Bge. var. *major* N.E. Br. and *Crataegus pinnatifida* Bge. are Chinese medicinal *shanzha* (Fructus Crataegi); and seeds of *Prunus armeniaca* L. var. *ansu* Maxim., *Prunus sibirica* L., *Prunus mandshurica* (Maxim.) Koehne, and *Prunus armeniaca* L. are Chinese medici-nal *kuxingren* (Semen Armeniacae Amarum).

According to climatic differences from northern to southern China, medicinal plant resources are divided into seven distribution areas: northeastern cold temperate zone, northern warm temper-ate zone, central subtropical zone, southern subtropical zone, southwestern subtropical zone, north-western arid zone, and Qinghai-Tibet plateau zone (Figure 5.1).

Figure 5.1 A geographical map of Chinese medicinal materials. (A): northeastern cold temperate zone; (B): northern warm temperate zone; (C): central subtropical zone; (D): southern subtropical zone; (E): southwestern subtropical zone; (F): northwestern arid zone; (G): Qinghai-Tibet plateau zone.

Northeastern cold temperate zone: There are over 1,000 species of medicinal plants, mostly wild, with relatively rich deposits, including *Gentiana manshurica* Kitag., *Cimicifuga dahurica* (Turcz.) Maxim., *Schisandra chinensis* (Turcz.) Baill., *Panax ginseng* C.A. Mey., *Asarum heterotropoldes* Fr. Schmidt var. *mandshuricum* (Maxim.) Kitag, *Atractylodes japonica* Koidz. ex Kitam., *Saposhnikovia divaricata* (Turcz.) Schischk., *Bupleurum chinense* DC., *Anemarrhena asphodeloides* Bunge, and *Epimedium grandiflorum* Morr.

Northern warm temperate zone: There are about 1,000–1,500 species of cultivated and/or wild medicinal plants. Among them, the annual output of 30 cultivated medicinal materials accounts for 50–70% of the national output. The following four famous medicinal materials of Henan are from this zone: *dihuang* (root of *Rehmannia glutinosa* Libosch.), *juhua* (capitulum of *Chrysanthemum morifolium* Ramat.), *shanyao* (rhizome of *Dioscorea opposita* Thunb.), and *niuxi* (root of *Achyranthes bidentata* Bl.).

Central subtropical zone: There are more than 3,000 species of cultivated and/or wild medicinal plants. Among them, the annual output of 60 cultivated medicinal materials accounts for 50–70% of the national output. The following eight famous medicinal materials of Zhejiang are from this zone: *baizhu* (rhizome of *Atractylodes macrocephala* Koidz.), *yanhusuo* (tuber of *Corydalis yanhusuo* W.T. Wang), *xuanshen* (root of *Scrophularia ningpoensis* Hemsl.), *zhebeimu* (bulb of *Fritillaria thunbergii* Miq.), *baishao* (root of *Paeonia lactiflora* Pall.), *juhua* (capitulum of *Chrysanthemum morifolium* Ramat.), *yujin* (root tube of *Curcuma wenyujin* Y.H. Chen et C. Ling), and *maidong* (root tuber of *Ophiopogon japonicus* (Thunb.) Ker-Gawl.). Additionally, other common medicinal plants include *Cornus officinalis* Sieb. et Zucc., *Belamcanda chinensis* (L.) DC., *Pinellia ternata* (Thunb.) Breit., *Magnolia officinalis* Rehd. et Wils., *Paeonia suffruticosa* Andr., *Changium smyrnioides* Wolff, *Pseudostellaria heterophylla* (Miq.) Pax ex Pax et Hoffm., and *Atracylodes lancea* (Thunb.) DC.

Southern subtropical zone: There are more than 4,500 species of medicinal plants. This is an area that assembles medicinal materials produced in southern China. Common medicinal plants include *Citrus medica* L. var. *sarcodactylis* (Noot.) Swingle, *Citrus reticulata* Blanco, *Citrus grandis* (L.) Osbeck 'Tomentosa,' *Aristolochia fangchi* Y.C. Wu ex L.D. Chow et S.M. Hwang, *Desmodium styracifolium* (Osbeck.) Merr., *Amomum villosum* Lour., *Morinda officinalis* How., *Polygonum multiflorum* Thunb., *Cinnamomum cassia* Presl, and *Aquilaria sinensis* (Lour.) Spreng.

Southwestern subtropical zone: Medicinal plants are abundant, with more than 5,000 species. Among them, natural deposits of 40 and 20 species of wild medicinal plants account for 50 and 80% of the nation's total natural deposits, respectively. The annual outputs of 30 and 10 cultivated medicinal materials account for 50 and 80% of the national outputs, respectively. Many reputed famous medicinal materials produced in Sichuan, Yunnan, and Guizhou Provinces are from this zone. Representative medicinal plants of Sichuan Province include *Ligusticum chuanxiong* Hort., *Fritillaria cirrhosa* D. Don, *Iris tectorum* Maxim., *Clematis armandii* Franch., *Cyathula officinalis* Kuan, *Dipsacus asperoides* C.Y. Cheng et T.M. Ai, *Coptis deltoidea* C.Y. Cheng et Hsiao, *Phellodendron chinense* Schneid., *Gastrodia elata* Bl., and *Aconitum carmichaeli* Debx. Representatives of Yunnan Province include *Panax notoginseng* (Burk.) F.H. Chen, *Paris polyphylla* Smith var. *yunnanensis* (Franch.) Hand.-Mazz., *Saussurea lappa* C.B. Clarke, *Dendrobium loddigesii* Rolfe., *Dendrobium nobile* Lindl., *Dendrobium candidum* Wall. ex Lindl., *Acacia catechu* (L.) Willd., *Picrorhiza scrophulariiflora* Pennell, *Amomum tsao-ko* Crevost et Lemaire, *Taxus yunnanensis* Cheng et L.K. Fu, and *Cyrtomium yunnanense* Ching et Shing. Representatives of Guizhou Province include *Evodia rutaecarpa* (Juss.) Benth., *Gentiana rigescens* Franch., *Eucommia ulmoides* Oliv., *Adenophora tetraphylla* (Thunb.) Fisch., *Bletilla striata* (Thunb. ex A. Murray) Rchb. f., *Platycodon grandiflorum* (Jacq.) A. DC., and *Tinospora capillipes* Gagnep.

Northwestern arid zone: There are more than 2,000 species of medicinal plants with a rich deposit of wild species but an uneven distribution. Common medicinal plants include *Prunus armeniaca* L., *Astragalus membranaceus* (Fisch.) Bge. var. *mongholicus* (Bge.) Hsiao, *Stellaria dichotoma* L. var. *lanceolata* Bge., *Ephedra intermedia* Schrenk ex C.A. Mey., *Ligusticum sinense* Oliv., *Gentiana macrophylla* Pall., *Rheum palmatum* L., *Rheum tanguticum* Maxim. ex Balf., *Rhodiola rosea* L., *Lycium barbarum* L., *Tribulus terrestris* L., *Artemisia capillaris* Thunb., *Paeonia lactiflora* Pall., *Fritillaria pallidiflora* Schrenk, *Fritillaria walujewii* Regel, *Cistanche deserticola* Ma,

Trollius altaicus C.A. Mey., *Cynomorium songaricum* Rupr., *Ferula sinkiangensis* K.M. Shen, and *Carthamus tinctorius* L.

Qinghai-Tibet plateau zone: There are more than 1,000 species, mostly wild. Common medicinal plants include *Saussurea involucrata* Kar. et Kir. et Maxim., *Ranunculus ternatus* Thunb., *Rheum palmatum* L., *Rheum tanguticum* Maxim. ex Balf., *Ephedra intermedia* Schrenk ex C.A. Mey., *Hippophae rhamnoides* L., *Stellera chamaejasma* L., *Meconopsis florindae* Kingdon-Ward, *Cimicifuga foetida* L., *Panax pseudo-ginseng* Wall. var. *bipinnatifidus* (Seem.) Li, and *Stachyurus himalaicus* Hook. f. et Thoms. ex Benth.

5.2 RESEARCH AND DEVELOPMENT OF MEDICINAL PLANTS

Over centuries of treating disease and attempting to preserve health, the Chinese have used their rich experiences to develop a unique and comprehensive medical system, known as TCM. Plant resources have been and are the material basis for the application of theories in preventing and treating disease. Today, these plant materials represent a huge potential resource for the development of new drugs and herbal products.

5.2.1 New Drugs

Developing new drugs from medicinal plants has a long history. In 1804, morphine, a famous analgesic, was isolated from *Papaver somniferum* L. (Huxtable and Schwarz 2001); in 1820, quinine, an antimalarial agent, was isolated from *Cinchona succirubra* Pav. (Dagani 2005); in 1838, salicylic acid, a precursor compound of antipyretic and analgesic aspirin, was isolated from the bark of *Salix alba* L. (Lagowski 1997); and in 1887, ephedrine was extracted from *Ephedra sinica* Stapf (Chen 1928). Inspired by these successes, researchers have adopted, adapted, and applied the strategy in seeking to develop even more drugs. At the same time, more and more people worldwide are turning to natural remedies for help in combating disease with fewer side effects than from chemical drugs. Large pharmaceutical companies such as GlaxoSmithKline, Merck, and DuPont are following this trend as they have begun to carry out systematic studies screening bioactive compounds from plants.

Chinese scientists have made great contributions to the research on bioactive compounds of medicinal plants and have created a series of new drugs, such as berberine, scopolamine, anisodamine, huperzine A, bulleyaconitine A, and artemisinin (Figure 5.2). Chemical drugs that originate from medicinal plants or via chemical modification of phytochemicals have been recorded in the *Pharmacopeia of the People's Republic of China* (2005 edition) (Table 5.1) (Chinese Pharmacopoeia Commission, 2005).

Thus, by conducting chemical isolation studies and pharmacological evaluations of medicinal plants, new active components can be found. By applying modern technologies, these can be developed into new drugs via chemical synthesis and structure modification.

5.2.2 Herbal Medicinal Products

In the past decades, the pharmaceutical industry of China has greatly expanded by taking herbal substances as raw materials. Over 1,300 pharmaceutical manufacturers have manufactured more than 400 phytochemicals and their derivative products, as well as more than 800 herbal preparations from bioactive fractions of medicinal plants. More than 200 bioactive components of medicinal plants have been made into more than 500 drugs of different dosage forms after chemical and pharmacological studies and clinical trials. In addition, a large number of proprietary TCM products are manufactured by pharmaceutical manufacturers all over China.

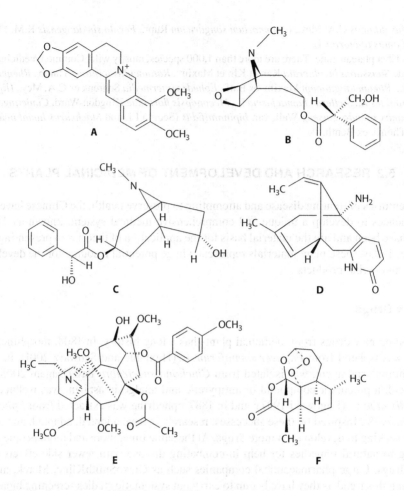

Figure 5.2 Chemical structures of (A) berberine; (B) scopolamine; (C) anisodamine; (D) huperzine A; (E) bulleyaconitine A; (F) artemisinin.

Proprietary TCM products are dosage forms made of crude medicinal materials with certain processing methods. Proprietary TCM products have fixed formulating ingredients and are dispensed by TCM practitioners under the guidance of TCM theory. Nonprescription proprietary TCM products are available as well. New proprietary TCM products can also be developed from medicinal plants. Table 5.2 shows the number of proprietary TCM products listed in Chinese pharmacopoeias since 1953.

According to Directive 2004/24/EC of the European Parliament and of the Council of March 31, 2004 amending, as regards traditional herbal medicinal products, Directive 2001/83/EC on the Community code relating to medicinal products for human use, a herbal medicinal product is defined as any medicinal product exclusively containing as active ingredients one or more herbal substances or one or more herbal preparations, or one or more such herbal substances in combination with one or more such herbal preparations. Herbal substances are defined as all mainly whole, fragmented, or cut plants; plant parts; algae; fungi; or lichen in an unprocessed, usually dried, form, but sometimes fresh. Herbal preparations are defined as preparations obtained by subjecting herbal substances to treatments such as extraction, distillation, expression, fractionation, purification, concentration, or fermentation. These include comminuted or powdered herbal substances, tinctures, extracts, essential oils, expressed juices, and processed exudates (European Parliament and the Council of the European Union 2004).

Table 5.1   Chemical Drugs Originated from Medicinal Plants or via Modification of Phytochemicals

No.	Chemical Drug	Category	Botanical Origin
1	Artemether	Antimalarial	*Artemisia annua*
2	Artemisinin	Antimalarial	*Artemisia annua*
3	Artesunate	Antimalarial	*Artemisia annua*
4	Berberine hydrochloride	Antibacterial	*Coptis chinensis, Thalictrum foliolosum, Mahonia japonica, Berberis julianae, Phellodendron amurense*
5	Bulleyaconitine A	Anti-inflammatory, analgesic	*Aconitum bulleyanum*
6	Camphor	Skin irritant	*Cinnamomum camphora*
7	Colchicine	Antipodagric, antineoplastic	*Iphigenia indica, Colchicum autumnale, Tulipa edulis, Gloriosa superba*
8	Deslanoside	Cardiotonic	*Digitalis lanata*
9	Digoxin	Cardiotonic	*Digitalis lanata*
10	Dihydroartemisinin	Antimalarial	*Artemisia annua*
11	Ephedrine hydrochloride	β_2 adrenergic agonist	*Ephedra sinica, Ephedra intermedia, Ephedra equisetina*
12	Homoharringtonine	Antineoplastic	*Cephalotaxus fortunei*
13	Huperzine A	Cholinesterase inhibitor	*Huperzia serrata*
14	Ligustrazine phosphate	Vasodilator	*Ligusticum chuanxiong*
15	Metildigoxin	Cardiotonic	*Digitalis lanata*
16	Morphine hydrochloride	Analgesic	*Papaver somniferum*
17	Papaverine hydrochloride	Vasodilator	*Papaver somniferum*
18	Pseudoephedrine hydrochloride	β_2 adrenergic agonist	*Ephedra sinica, Ephedra intermedia, Ephedra equisetina*
19	Puerarin	Vasodilator	*Pueraria lobata*
20	Raceanisodamine	Anticholinergic	*Scopolia tangutica*
21	Salicylic acid	Antiseptic, disinfectant	*Salix alba, Pterocarya stenoptera, Plantago major*
22	Salsalate	Anti-inflammatory, analgesic, nonsteroidal anti-inflammatory agent	*Salix alba, Pterocarya stenoptera, Plantago major*
23	Scopolamine butylbromide/ hydrobromide	Anticholinergic	*Atropa belladonna, Scopolia japonica, Hyoscyamus niger, Datura metel, Datura innoxia*
24	Strophanthin K	Cardiotonic	*Strophanthus kombe*
25	Theophylline	Smooth-muscle relaxant	*Camellia sinensis*

In June 2004, "Guidance for Industry—Botanical Drug Products" was issued by the U.S. Food and Drug Administration (USFDA) (Center for Drug Evaluation and Research 2004). On October 31, 2006, the USFDA approved an herbal extract as a prescription drug for the topical treatment of genital warts caused by the human papilloma virus (American Botanical Council 2007). This new drug, called Polyphenon® E (Veregen™) ointment, is composed of a proprietary mixture of phytochemicals of green tea leaves, and it is the first prescription botanical drug approved by the USFDA under the drug amendments of 1962. It is also the first proprietary drug of a German biotech company on the market. This breakthrough will encourage further research and development of proprietary

Table 5.2   Proprietary TCM Products Listed in Chinese Pharmacopoeias

Edition	1953	1963	1977	1985	1990	1995	2000	2005
Number	0	197	270	207	275	398	458	564

TCM products that are based on abundant medicinal plant resources. More botanical drugs or herbal medicinal products will be developed to meet European and American drug regulations.

5.2.3 New Resources from Ethnic and Folk Herbal Medicine

Application of medicinal plants for the prevention and treatment of diseases has a long history in China. A great amount of practical experience has been accumulated in various nationalities of different places in China. Chinese medicinals in a broad sense include traditional Chinese, ethnic, and folk medicinals. Traditional Chinese medicinal refers to natural medicinal substances and their processed products that are widely in clinical application under the guidance of TCM theory and are commercially circulated in the herbal markets. There are 500–600 commonly used traditional Chinese medicinal materials.

Ethnic medicinal refers to natural medicinal substances that are used in ethnic regions under the guidance of ethnic medical theory and experience. Ethnic medicinal materials are harvested and circulated within ethnic communities. In addition to the native Han nationality, China has 55 ethnic groups. More than 50% of ethnic groups have their own ethnic medicines. About 30% of ethnic medicines have their unique medical theories. Among them, the Tibetan, Mongolian, and Uygur medicinals are most representative. For example, *Jingzhu Materia Medica* (*Jing Zhu Ben Cao*), a classic of Tibetan medicine, has recorded 2,294 Tibetan medicinals. Tibetan medicinals mainly come from wild botanical resources, such as the Asteraceae, Fabaceae, Ranunculaceae, Papaveraceae, Apiaceae, and Gentianaceae. Representative medicinal plants include *Carum carvi* L., *Saussurea medusa* Maxim., *Scopolia tangutica* Maxim., *Onosma hookeri* C.B. Clarke, and *Onosma hookeri* Clarke. var. *longiforum* Duthie.

Folk medicinal refers to medicinal substances passed on by oral instructions. Without systematic theory, folk medicinals are limited to certain regions and are not commercially circulated in the herbal market. Treatment of diseases with folk medicinals is a simple and convenient way with a clear aim. Folk medicinal resources are abundant as well. According to statistical data, there are 5,000 folk medicinals (Zhang et al. 1995). For example, there are hundreds of folk medicinals bearing the name of *qi* in Taibai mountain regions of Shaanxi Province: *Sinopodophyllum emodi* (Wall.) Ying (*tao er qi*), *Herminium monorchis* (L.) R. Br. (*ren tou qi*), *Aconitum sinomontanum* Nakai (*ma bu qi*), *Polygonum ciliinerve* Ohwi (*zhu sha qi*), *Sedum aizoon* Li (*tu san qi*), *Aconitum szechenyianum* Gay. (*tie niu qi*), *Panax pseudoginseng* Wall var. *japonica* (G.K. Mey) G. Hoo et C.S. Tseng (*niu zi qi*), etc. (Hu and Xu 1997; Guo et al. 2006).

Influenced by historical, geographic regional, economical, and cultural factors, only a small part of ethnic and folk medicinal knowledge has been documented. The understanding of ethnic and folk medicinals is far less than that of traditional Chinese medicinal. Therefore, there are huge potentials in the research and development of medicinal plant resources and new medicinal substances from ethnic and folk medicinals. For examples, raceanisodamine, a chemical drug recorded in the *Pharmacopoeia of the People's Republic of China*, (2005). (Chinese Pharmacopoeia Commission 2005) has been developed from the Tibetan medicinal *Scopolia tangutica* Maxim. Tripterygium glycosides tablets have been developed from the folk medicinal *Tripterygium wilfordii* Hook. f. for the treatment of rheumatoid arthritis.

5.3 GOOD AGRICULTURAL PRACTICE

Along with the rapid development of the research on medicinal plants, the international demand for plant resources has increased greatly. Although China has abundant medicinal plant resources, germplasm resources have been damaged seriously due to the damage and degeneration of the ecological system in large areas. Deposits of a large number of medicinal plants are decreasing gradually; 168 medicinal plant species have been included in the list of rare and endangered plants of China. In recent years, the Chinese government has attached importance to the protection and sustainable utilization of medicinal plant resources. The sustainable resource utilization and industrial development is one of the basic principles in "The Development Outline of the Modernization of Chinese Medicinals (2002–2010)" proclaimed by eight governmental departments, including the Ministry of Science and Technology of the People's Republic of China. The protection and sustainable utilization of Chinese medicinal resources are made clear as one of the priority missions of this development outline. Its detailed contents include:

- To carry out the Chinese medicinal resource survey and to establish the early-warning mechanism of endangered wild resources; to protect Chinese medicinal germplasm and genetic resources and to strengthen the research on selective breeding and seed origins of medicinal plants, in order to prevent species degeneration and to solve the problem of botanical origin confusion
- To establish the Chinese medicinal database and the germplasm resource bank, in order to collect information on botanical origin, production area, and pharmacological effects and to preserve the germplasm resource of Chinese medicinal materials
- To reinforce research on the cultivation of wild species and the cultivation technology, in order to realize standardized plantation and industrialized production of Chinese medicinal materials; to strengthen research on plant protection technology in order to facilitate the development of green medicinal materials
- To reinforce research on the breeding of new species of medicinal materials and to carry out research on seeking substitutes of the rare and endangered species as well as to ensure the sustainable development

It can be concluded that protection and sustainable utilization of medicinal plant resources have become an unavoidable trend. Standardized plantation of some medicinal plants is one of the effective approaches to realize the sustainable utilization of medicinal plant resources. Standardized plantation of medicinal plants involves cultivation technologies of wild species and the construction of standardized plantation sites.

5.3.1 Objectives and Certification

Medicinal plants have been cultivated in China with a long history to yield precious and main Chinese medicinal materials, such as Radix Ginseng, (renshen) and Radix et Rhizoma Glycyrrhizae (*gancao*). A nationwide investigation indicates that about 800 plantation sites have been established, cultivating medicinal plants for the production of about 500 Chinese medicinal materials (Li and Wu 2006). Pictures of plantation sites of 10 common medicinal plants are shown in Figure 5.3(a)–5.3(j).

Along with the development of TCM modernization, more and more medicinal plants are cultivated to meet the increasing demand for Chinese medicinal materials. Standardized management for the cultivation of medicinal plants is gradually being establishing in China. The State Food and Drug Administration (SFDA) issued a trial edition of GAP for Chinese medicinal materials in March 2002; this has been enforced since June 2002. It provides guidelines for GAP of medicinal plants and animals. Its aim is to control factors that influence the quality of medicinal materials and to standardize all links, and even the entire production process of medicinal materials, in order to make medicinal materials "safe, with high quality, stable, and controllable." Its contents include 10 chapters with 57 articles, covering aspects of ecological environment of production sites, seeds and propagation material, management of cultivation and animal husbandry, harvesting and primary processing, packaging, transportation and storage, quality

Figure 5.3(a) *Panax ginseng* C.A. Mey. (Fusong, Jilin Province).

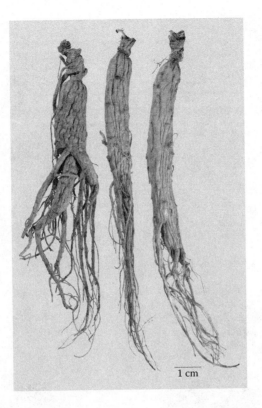

Figure 5.3(a) Continued

Figure 5.3(b) *Angelica sinensis* (Oliv.) Diels. (Min, Gansu Province).

Figure 5.3(b)   Continued

control, personnel and equipment, and documentation. A specific standard operating procedure (SOP) has been established to meet GAP requirements for each individual medicinal plant. The contents of SOPs include:

- Quality evaluation and dynamic change of the ecological environment of production sites
- Quality standard and operating procedure of seed and seedling
- Operating procedure of selective breeding
- Operating procedure of cultivation technology and measures
- Operating procedure of field management
- Operating procedure of pest control
- Operating procedure of pesticide and herbicide applications
- Operating procedure of the use of fertilizers and hygienic processing of organic fertilizers
- Operating procedure of quality control of medicinal materials
- Operating procedure of harvesting and primary processing at the production sites
- Operating procedure of packaging, transportation, and storage
- Operating procedure of personnel training and document management

1 cm

Figure 5.3(b)   Continued

In November 2003, the SFDA began to handle the certification of the GAP qualification of manufacturers of Chinese medicinal materials. Since October 14, 2004, the certification of GAP for Chinese medicinal materials has been included in items that need administrative licensing. This has established the legal status of GAP certification and has promoted the construction of GAP plantation sites and hence has fully facilitated GAP coming into force. Up to 2009, GAP plantation sites for 42 medicinal plants had been approved by the SFDA (Table 5.3).

5.3.2 Description (Selection of Plant Species/Selection of GAP Farm Sites)

The concept of "daodi medicinal material," also known as genuine medicinal material, has existed since ancient times. Daodi medicinal material refers to Chinese medicinal materials that are originating from specific ecological locations and are produced with specific cultivation techniques and processing methods. The medicinal properties of daodi medicinal material have been well proved by long-term practice of TCM. According to documentation, there are about 200 daodi medicinal materials in China (Xiao et al. 2009).

Superior plant species is the internal factor of the formation of daodi medicinal material. For example, Radix et Rhizoma Rhei (*dahuang*) originates from *Rheum palmatum* L., *R. tanguticum* Maxim. ex Balf., and *R. officinale* Baill. However, species with sinuate leaves from the same genus, *R. hotaoense* C.Y. Cheng et C.T. Kao and *R. franzenbachii* Munt., hardly contain any anthraquinones. These two species will never be the botanical origins of the daodi medicinal

Figure 5.3(c)   *Glycyrrhiza uralensis* Fisch. (Huhehaote, Inner Mongolia).

Figure 5.3(c)   Continued

material Radix et Rhizoma Rhei, no matter how the ecological environment changes. The ecological environment is the external factor of the formation of daodi medicinal material. Ecological factors such as humidity, soil, and climate affect not only the growth of plants, but also the formation and accumulation of active components in plants. The morphological character and chemical components of a plant species vary greatly under different ecological environments.

For example, *Pogostemon cablin* grown in Hainan Province yields more volatile oils but with a low content of the antibacterial pogostone in the volatile oil. However, the same species grown in

Figure 5.3(d)   **(See color insert.)** *Lycium barbarum* L. (Zhongning, Ningxia Province).

Figure 5.3(d) Continued

Guangdong Province yields less volatile oils but with a high content of pogostone in the volatile oil. In addition, the traditional cultivation techniques and harvesting and processing methods also influence the formation of daodi medicinal material. For example, Radix Aconiti Lateralis Praeparata (*fuzi*) produced in Jiangyou of Sichuan Province gains its reputation through complicated cultivation techniques and unique processing methods.

According to the environmental adjustability and population differentiation, daodi medicinal material can be divided into four types:

- One plant species with one specific production site, such as Radix Aconiti Lateralis Praeparata (*fuzi*) produced in Jiangyou of Sichuan Province
- One plant species with multiple production sites, such as Radix Angelicae Dahuricae (*baizhi*) produced in Sichuan, Zhejiang, Anhui, Henan, and Hebei Provinces, respectively
- Multiple plant species with one specific production site, such as all of the botanical origins of Bulbus Fritillariae (*chuanbeimu*), *Fritillaria cirrhosa* D. Don, *F. unibracteata* Hsiao et K.C. Hsia, *F. przewalskii* Maxim., and *F. delavayi* Franch. distributed in the northwestern plateau in Sichuan Province
- Multiple plant species with multiple production sites, such as botanical origins of Rhizoma Coptidis (*huanglian*) involving three species (*Coptis chinensis* Franch., *C. deltoidea* C.Y. Cheng et Hsiao, and *C. teeta*, respectively) and three production sites (Shizhu of Chongqing Province, Hongya of Sichuan Province, and Zhongdian of Yunnan Province, respectively).

Figure 5.3(e)   *Paeonia lactiflora* Pall. (Bozhou, Anhui Province).

1 cm

Figure 5.3(e) Continued

From this, it can be concluded that the selection of medicinal plant species and GAP plantation sites is closely related to the plant species and geographic and ecological environments. In addition, modern science and technology can be used to analyze the climate, soil, topography, geomorphology, ecology, and plant colonies in order to seek the dominating and limiting factors that influence active components of medicinal materials, thus ensuring that medicinal plant species are growing in suitable locations (Chen et al. 2006).

5.4 QUALITY CONTROL OF MEDICINAL PLANTS AND THEIR PRODUCTS

Traditional Chinese medicine is undergoing rapid globalization. At the same time, the safety and efficacy of Chinese medicinals are drawing international concerns as well. As the main source of Chinese medicinals, medicinal plants must undergo systematic quality evaluation, which includes authentication, purity tests, and assays.

Figure 5.3(f) *Pinellia ternata* (Thunb.) Breit. (Bijie, Guizhou Province).

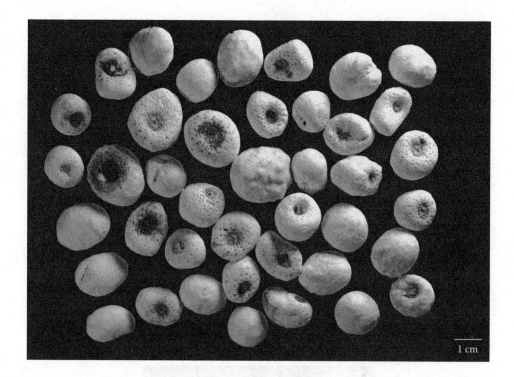

Figure 5.3(f)   Continued

5.4.1 Authentication

Authentication is a prerequisite and key procedure to ensure the quality of a medicinal plant. It includes the identification of the botanical origin by means of taxonomic methods and identification of medicinal parts by means of macroscopic, microscopic, physical–chemical, and molecular biological approaches (Zhao et al. 2006) (Figure 5.4).

5.4.1.1 Botanical Origin Identification

Botanical origin identification establishes the taxonomic position of a medicinal plant. The scientific name is identified by observing and verifying morphological characters of a medicinal plant with documentations (such as *Flora of China* and *Index Kewensis*) and authenticated specimens in herbariums such as in the Institute of Botany of Chinese Academy of Sciences, the Kew Gardens' Chinese Medicinal Plants Authentication and Conservation Center, and Bank of China (HK) Chinese Medicine Center of Hong Kong Baptist University.

5.4.1.2 Macroscopic Identification

Medicinal parts of a medicinal plant are known as medicinal materials. Macroscopic identification is a principal method to identify Chinese medicinal materials. This identification method is formed by the valuable experiences of previous TCM experts and the long-term practical applications. It depends largely on organoleptic properties of medicinal materials. By observing, touching, smelling, and tasting, such characteristics of medicinal materials as shape, size, color, surface, texture, cross-section, odor, and taste are used to identify their genuineness and quality. Some

Figure 5.3(g) *Morinda officinalis* How. (Deqing, Guangdong Province).

medicinal materials can also be tested with water and/or fire. Some medicinal materials sink, float, or dissolve in the water. They may also change their colors, transparencies, expansibilities, or viscosities when they are treated with water.

For example, when the flower of *Carthamus tinctorius* (Flos Carthami, *honghua*) is soaked in water, the water changes into a golden-yellow solution, and the color of the flower does not fade. When the stigma of *Crocus sativus* (Stigma Croci, *xihonghua*) is soaked in water, the stigma expands into the shape of a long trumpet, a yellow color strip forms, and the water gradually becomes a yellow solution (Figure 5.5). Testing with water helps to distinguish Flos Carthami (*honghua*) and Stigma Croci (*xihonghua*). Some medicinal materials produce specific odors, colors, smoke, or sounds, or melt when they are treated with fire. For example, spores of *Lygodium japonicum* (Spora Lygodii, *haijingsha*) produce sounds of explosion with bright flames, but without any ash left, when they are set on fire (Figure 5.6).

Modern research has indicated that macroscopic characteristics of medicinal materials are somewhat related to their phytochemicals. For example, the content of sinomenine changes with the diameter change of the stem of *Sinomenium acutum* (Caulis Sinomenii, *qingfengteng*). The content of sinomenine is in the following descending sequence: medicinal materials of small size (stem diameter < 1 cm) < middle size (stem diameter 1–3 cm) < large size (stem diameter > 3 cm)

1 cm

Figure 5.3(g) Continued

Figure 5.3(h) *Momordica grosvenori* Swingle (Nanning, Guangxi Province).

1 cm

Figure 5.3(h)   Continued

(Zhao et al. 2005). Further, analysis and spatial profiling of phytochemicals in Caulis Sinomenii *qingfengteng* by matrix-assisted laser desorption/ionization mass spectrometry (MALDI-MS) has been conducted. The results reveal that menisperine is mainly localized in the cortex and phloem tissue (outer part of the stem), magnoflorine in the pith region (central part of the stem), and stepharanine, sinomenine, and sinomendine in the xylem region (inner part of the stem) (Ng et al. 2007).

5.4.1.3 *Microscopic Identification*

Microscopic identification is a commonly used method in the authentication of medicinal materials. By means of microscopic technique, structural and cellular features of medicinal materials are analyzed to facilitate the identification of their botanical origins and qualities. This method is useful for the identification of broken or powdered medicinal materials, identifying plant species with similar morphological characters, and determining the ingredients in proprietary TCM products. Since the 1980s, microscopic identification has been widely used, and microscopic data of many medicinal materials have been established. In recent years, digital microscopic imaging technique has been applied in the microscopic identification and related monographs have been published.

Figure 5.3(i)   *Platycodon grandiflorum* (Jacq.) A. DC. (Chifeng, Inner Mongolia).

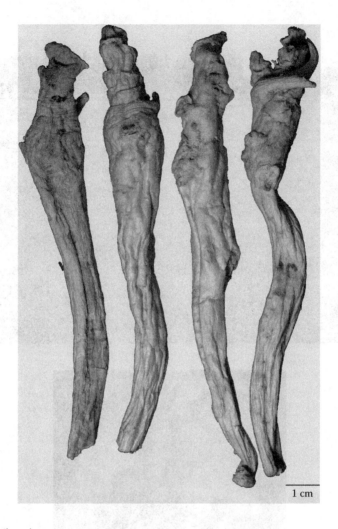

1 cm

Figure 5.3(i)   Continued

Representatives include *A Colored Atlas of Microscopic Identification of Chinese Materia Medica in Powdered Form as Specified in the Pharmacopoeia of the People's Republic of China* (Zhao 1999) and *An Illustrated Microscopic Identification of Chinese Materia Medica* (an instructional DVD on the standard operating procedure of microscopic identification is included with the book) (Zhao 2005).

A normal light microscope combined with polariscope and fluorescent microscopy has been used in the authentication of medicinal materials and proprietary TCM products. Structures and cells of medicinal plants have been found to demonstrate stable and specific polariscopic or fluorescent characteristics and have been successfully used in authenticating proprietary TCM products (Zhao et al. 1996, 1997, 1998) and different species of *Ilex* (Tam et al. 2006) and *Oldenlandia* (Liang et al. 2006). In addition, microscopic techniques combined with mathematic analysis have made great progress in the authentication of medicinal materials. Microscopic images based on granulometric operations are successfully used to identify medicinal materials that contain starch grains (Tong et al. 2007, 2008). A new method that combines histological and microscopic analysis of laticifers by "blob" analysis has been successfully applied to identify different species of *Ficus* (Au et al. 2009).

Figure 5.3(j) *Scutellaria baicalensis* Georgi (Chengde, Hebei Province).

1 cm

Figure 5.3(j)   Continued

5.4.1.4 Physical and Chemical Identification

This method is used to identify the authenticity and quality of medicinal materials by means of physical, chemical, and/or instrumental analytical ways. Traditionally, physical and chemical identification includes color reaction, precipitation reaction, fluorescence spectrometry, microsublimation, and spectrophotometry. With the further exploration into phytochemicals in medicinal plants and the rapid development of modern spectroscopic and chromatographic techniques, new methods and techniques, especially chemical fingerprint, have been applied in the identification of medicinal materials. Chemical fingerprint uses modern analytical techniques, including high-performance thin layer chromatography, high-performance liquid chromatography (HPLC), gas chromatography, capillary electrophoresis, mass spectrometry (MS), infrared spectrometry, and nuclear magnetic resonance spectrometry to analyze chemical information, which is then expressed and described in forms of figures for the authentication and quality evaluation of medicinal materials.

Recently, systematic monographs on chromatographic fingerprinting techniques, such as *Chromatographic Fingerprints of Chinese Medicinals*, have been published (Xie 2005). HPLC chromatographic fingerprinting techniques have been adopted in *Hong Kong Chinese Materia Medica*

Table 5.3   GAP Plantation Sites Approved by SFDA

No.	Medicinal Plants	Medicinal Materials	GAP Plantation Sites
1	*Aconitum carmichaeli* Debx.	Radix Aconiti Lateralis Praeparata	Jiangyou, Sichuan Province
2	*Alisma orientale* (Sam.) Juzep.	Rhizoma Alismatis	Jian'ou, Fujian Province
3	*Andrographis paniculata* (Burm. f.) Nees	Herba Andrographis	Qingyuan, Guangdong Province
4	*Angelica dahurica* Benth. et Hook	Radix Angelicae Dahuricae	Suining, Shehong, and Pengxi, Sichuan Province
5	*Angelica sinensis* (Oliv.) Diels	Radix Angelicae Sinensis	Dangchang and Min, Gansu Province
6	*Artemisia annua* L.	Herba Artemisiae Annuae	Youyang, Chongqing
7	*Astragalus membranaceus* (Fisch.) Bunge	Radix Astragali	Ulanqab, Inner Mongolia
8	*Codonopsis pilosula* (Franch.) Nannf.	Radix Codonopsis	Lingchuan, Shanxi Province
9	*Coix lacroyma-jobi* L. var. *ma-yuen* (Roman.) Stapf	Semen Coicis	Taishun, Zhejiang Province
10	*Coptis chinensis* Franch.	Rhizoma Coptidis	Shizhu, Chongqing
11	*Cornus officinalis* Sieb. et Zucc.	Fructus Corni	Xixia, Henan Province; Linan, Zhejiang Province; Foping, Shaanxi Province; Nanyang, Henan Province
12	*Corydalis bungeana* Turcz.	Herba Corydalis Bungeanae	Yutian, Hebei Province
13	*Corydalis yanhusuo* W.T. Wang	Rhizoma Corydalis	Fuzhou, Jiangxi Province
14	*Crocus sativus* L.	Stigma Croci	Shanghai
15	*Dendrobium candidum* Wall. ex Lindl.	Caulis Dendrobii	Tiantai, Zhejiang Province
16	*Dioscorea opposita* Thunb.	Rhizoma Dioscoreae	Wuzhi and Wen, Henan Province
17	*Erigeron breviscapus* (Vaniot) Hand.-Mazz.	Herba Erigerontis	Luxi, Yunnan Province
18	*Fritillaria ussuriensis* Maxim.	Bulbus Fritillariae Ussuriensis	Tieli and Yichun, Heilongjiang Province
19	*Gardenia jasminoides* Ellis	Fructus Gardeniae	Zhangshu, Jiangxi Province
20	*Gastrodia elata* Bl.	Rhizoma Gastrodiae	Lueyang, Shaanxi Province
21	*Gentiana manshurica* Kitag.	Radix Gentianae	Qingyuan, Liaoning Province
22	*Ginkgo biloba* L.	Folium Ginkgo	Pizhou, Jiangsu Province
23	*Gynostemma pentaphyllum* (Thunb.) Mak.	Herba Gynostemmatis Pentaphylli	Ankang, Shaanxi Province
24	*Houttuynia cordata* Thunb.	Herba Houttuyniae	Ya'an, Sichuan Province
25	*Isatis indigotica* Fort.	Radix Isatidis	Taihe, Anhui Province; Yutian, Hebei Province; Daqing, Liaoning Province
26	*Ligusticum chuanxiong* Hort.	Rhizoma Chuanxiong	Pengzhou and Wenchuan, Sichuan Province
27	*Ophiopogon japonicus* (Thunb.) Ker-Gawl.	Radix Ophiopogonis	Mianyang, Sichuan Province
28	*Panax ginseng* C.A. Mey.	Radix et Rhizoma Ginseng	Jingyu, Linjiang, Changbai, Ji'an, and Fusong, Jilin Province
29	*Panax notoginseng* (Burk.) F.H. Chen	Radix et Rhizoma Notoginseng	Wenshan, Yunnan Province
30	*Panax quinquefolium* L.	Radix Panacis Quinquefolii	Jingyu, Jilin Province
31	*Papaver somniferum* L.	Pericarpium Papaveris	Wuwei, Zhangye, Jinchang, and Baiyin, Gansu Province

(continued)

Table 5.3   GAP Plantation Sites Approved by SFDA (Continued)

No.	Medicinal Plants	Medicinal Materials	GAP Plantation Sites
32	*Platycodon grandiflorus* (Jacq.) A. DC.	Radix Platycodonis	Yiyuan, Shandong Province
33	*Pogostemon cablin* (Blanco) Benth.	Herba Pogostemonis	Guangzhou and Zhanjiang, Guangdong Province
34	*Polygonum capitatum* Buch.-Ham. ex D. Don	Herba Polygoni Capitati	Shibing and Guiyang, Guizhou Province
35	*Polygonum multiflorum* Thunb.	Radix Polygoni Multiflori	Shibing, Congjiang, Qingong, Jinping, and Kaili, Guizhou Province
36	*Pseudostellaria heterophylla* (Miq.) Pax ex Pax et Hoffm.	Radix Pesudostellariae	Shibing, Huangping, Leishan, and Kaili, Guizhou Province; Zherong, Fujian Province
37	*Rehmannia glatinosa* Libosch.	Radix Rehmanniae	Wuzhi, Wen, and Mengzhou, Henan Province
38	*Salvia miltiorrhiza* Bge.	Radix et Rhizoma Salviae Miltiorrhizae	Shangluo, Shaanxi Province
39	*Schisandra chinensis* (Turcz.) Baill.	Fructus Schisandrae Chinensis	Xinbin, Liaoning Province
40	*Schizonepeta tenuifolia* (Bench.) Briq.	Herba Schizonepetae	Yutian, Hebei Province
41	*Scrophularia ningpoensis* Hemsl.	Radix Scrophulariae	Enshi, Hubei Province
42	*Tussilago farfara* L.	Flos Farfarae	Chongqing

Standards (volumes 1 and 2) as well (Department of Health, Hong Kong Special Administrative Region, P.R.C. 2005, 2008).

Chemical fingerprinting data of medicinal parts of *Oldenlandia diffusa* (Liang et al. 2007), *Houttuynia cordata* (Meng et al. 2005), *Angelica sinensis* (Lu et al. 2005), *Salvia miltiorrhiza* (Hu et al. 2005), *Angelica dahurica* (Kang et al. 2008), *Cistanche deserticola* (Jiang et al. 2009), *Belamcanda chinensis* (Li et al. 2009), and *Psoralea corylifolia* (Qiao et al. 2007) have been established by means of HPLC or HPLC-MS approaches. For example, in the chromatographic fingerprints of roots of *Angelica dahurica*, nine characteristic peaks have been found in 13 batches of medicinal materials, and 20 furocoumarins have been identified by the HPLC/DAD/ESI-MS technique. Chemical fingerprints of *Oldenlandia diffusa* from different production sites are different. Chemical fingerprints of *O. diffusa*, *O. corymbosa*, and *O. tenelliflora* are also different. On the basis of these differences, authentication of *O. diffusa* and its related plant species has been achieved (Figures 5.7 and 5.8).

In addition to being used to identify medicinal plant species, these chromatographic fingerprints provide abundant chemical information that better reflects the intrinsic quality of medicinal materials. However, chemical components of a medicinal plant are somewhat affected by variables such as harvest time, production area, storage period, and processing method. Kindred plants have similar chemical components. Therefore, these factors should be considered in evaluating a chromatographic fingerprint for authentication and in other qualitative and quantitative analyses.

In addition to chromatographic fingerprints, spectrum fingerprints are also important in the authentication of medicinal materials. Spectrum fingerprints mainly include infrared spectrum, nuclear magnetic resonance spectrum, and x-ray diffraction fingerprint.

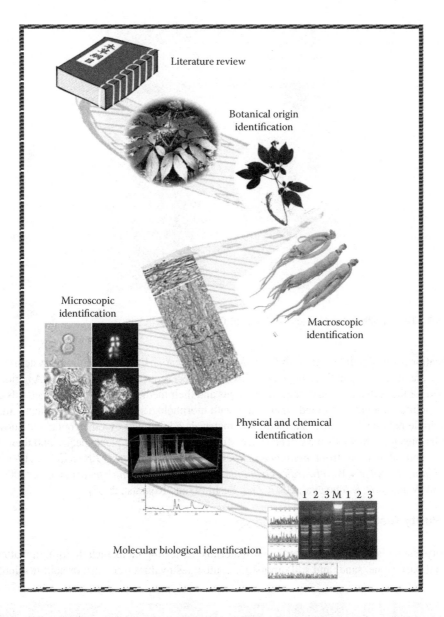

Literature review

Botanical origin
identification

Microscopic
identification

Macroscopic
identification

Physical and chemical
identification

1 2 3 M 1 2 3

Molecular biological identification

Figure 5.4   Authentication procedure of a medicinal plant (taking *Panax ginseng* as an example).

5.4.1.5 *Molecular Biological Identification*

Since the 1990s, with the rapid development of molecular biology, the uniqueness of genotype of plant species has been used for the authentication of medicinal plants and their medicinal parts. Because molecular genetic markers of plants are not influenced by their growing process and environment, this genotype-based method has proven to be accurate, specific, stable, and convenient.

In the past two decades, a number of molecular methods, including random amplified polymorphic DNA (RAPD), arbitrarily primed polymerase chain reaction (AP-PCR), restriction fragment length polymorphism (RFLP), PCR-RFLP, amplification fragment length polymorphism (AFLP), directed

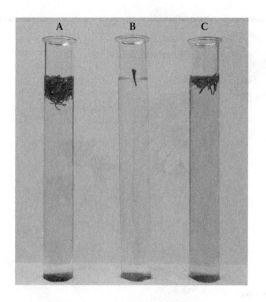

Figure 5.5 Testing with water: (A) Flos Carthami; (B,C) Stigma Croci.

amplification of minisatellite region DNA (DAMD), amplification refractory mutation system (ARMS), sequence-characterized amplified region (SCAR), and direct sequencing of certain DNA regions, have been used for the authentication of medicinal plants and their medicinal parts. These methods are particularly helpful in identifying medicinal plants with morphological similarities, medicinal materials with multiple botanical origins, and medicinal materials adulterated by other botanical substances. Molecular biological methods have been successfully used to identify plant species and their medicinal parts and adulterants from *Echinacea*, *Rheum*, *Panax*, *Fritillaria*, *Epimedium*, *Adenophora*, *Dendrobium*, *Dryopteris*, *Woodwardia*, *Osmunda*, *Ligusticum*, and *Cnidium* (Zhao et al. 2006; Zhao, Leng, and Wang 2007; S. Zhu et al. 2007; Zhu, Fushimi, and Komatsu 2008).

5.4.2 Purity Testing

Purity testing is applicable to medicinal materials. Its testing items include foreign matter (nonmedicinal part, stone, sand, and soil), moisture content, ash values (total ash or acid-insoluble ash),

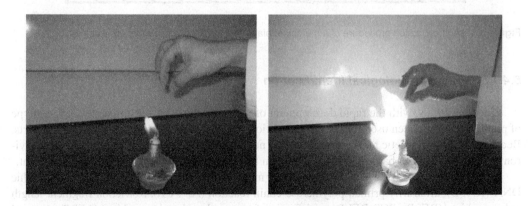

Figure 5.6 Testing with fire: Spora Lygodiis.

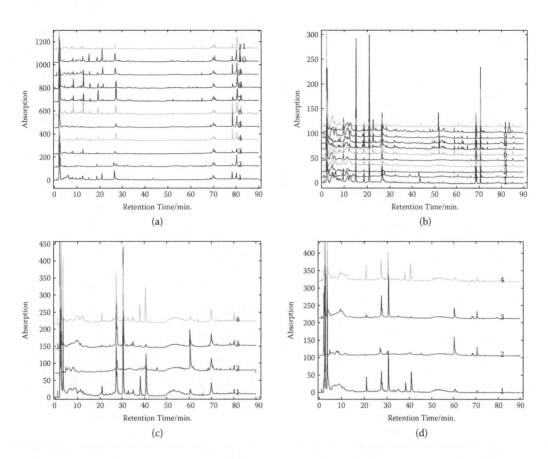

Figure 5.7 HPLC chromatograms of (a) *O. corymbosa*, detection wavelength 210 nm; (b) *O. corymbosa*, detection wavelength 238 nm; (c) *O. telleniflora*, detection wavelength 210 nm; (d) *O. telleniflora*, detection wavelength 238 nm.

extractive values (water-soluble or ethanol-soluble extractive), microbial contamination, and toxic residues (pesticides and heavy metals). Pharmacopoeias of different countries have specified limits for these items.

Residues of pesticides and heavy metals in herbal substances have been a constant international concern. The Chinese government also has attached importance to control the residues of pesticides and heavy metals in Chinese medicinals and proprietary TCM products.

Atomic absorption spectrophotometry and inductively coupled plasma mass spectrometry have been adopted in the *Pharmacopoeia of the People's Republic of China* (2005 edition) to determine heavy metals and other poisonous elements. The *Chinese Pharmacopoeia* specifies that the contents of lead, cadmium, mercury, arsenic, and copper shall be not more than 5.0, 0.3, 0.2, 2.0, and 20.0 mg/kg, respectively, in the following six herbal substances: Radix Astragali (*huangqi*), Radix Paeoniae Alba (*baishao*), Radix Panacis Quinquefolii (*xiyangshen*), Radix et Rhizoma Salviae Miltiorrhizae (*danshen*), Radix et Rhizoma Glycyrrhizae (*gancao*), and Flos Lonicerae (*jinyinhua*). The gas chromatographic method has been used for the determination of residual organochlorine, organophosphorous, and pyrethrin pesticides. The pharmacopoeia specifies that residual hexachloro-cyclohexane (BHC), dichloro-diphenyl-trichloroethane (DDT), and pentachloronitrobenzene (PCNB) in Radix Astragali (*huangqi*) and Radix et Rhizoma Glycyrrhizae (*gancao*) shall be not more than 0.2, 0.2, and 0.1 mg/kg, respectively (Chinese Pharmacopoeia Commission 2005).

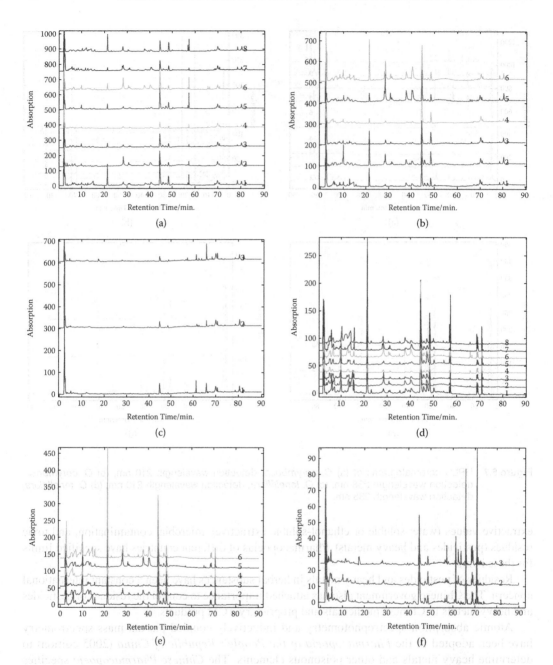

Figure 5.8 HPLC chromatograms of *O. diffusa* (OD): (a) pattern one of OD 1–8, detection wavelength 210 nm; (b) pattern two of OD 9–14, detection wavelength 210 nm; (c) pattern three of OD 15–17, detection wavelength 210 nm; (d) pattern one of OD 1–8, detection wavelength 238 nm; (e) pattern two of OD 9–14, detection wavelength 238 nm; (f) pattern three of OD 15–17, detection wavelength 238 nm.

5.4.3 Assays

A medicinal plant usually contains complicated chemical components. A particular group of components are selected as marker compounds and are assayed for quality evaluation. Choosing suitable marker compounds from a medicinal plant is of significance in the identification of plant species, selection of production site, determination of harvesting time and processing method, and quality evaluation of medicinal material and its products. Ideal marker compounds should be the active components of a medicinal plant. However, active components of most medicinal plants are not clear yet. Moreover, a medicinal plant may contain multiple components with diversified pharmacological activities. Therefore, in addition to active components, characteristic components, main components, and toxic components should be considered to be chosen as marker compounds for quality evaluation.

Modern chromatographic techniques, such as HPLC, HPLC coupled with MS, diode array detector (DAD), and evaporative light scattering detector (ELSD), have been applied for the determination of chemical components in medicinal plants. Different from that of the 1990s, simultaneous determination of multiple components has become a trend in the quality evaluation of medicinal materials. For example, 15 bioactive components, including danshensu; protocatechuic acid; protocatechuic aldehyde; caffeic acid; rosmarinic acid; lithospermic acid; salvianolic acids A, B, and C; dihydrotanshinone I; cryptotanshinone, tanshinones I and IIA; methylene tanshiqunone; and miltirone, have been simultaneously determined in Radix et Rhizoma Salviae Miltiorrhizae (*danshen*) (Cao et al. 2008).

Other examples include simultaneous determinations of iridoids, phenolic acids, flavonoids, and saponins in Flos Lonicerae (*jinyinhua*) (C. Y. Chen et al. 2007); phytoecdysones and the triterpenoids in Radix Achyranthis Bidentatae (*niuxi*) (Li et al. 2007); flavonoids and saponins in Radix Astragali (*huangqi*) (Yu et al. 2007); and naphthoquinone derivatives in plants of the Boraginaceous family (Hu et al. 2006) by the previously mentioned modern chromatographic techniques. These simultaneous qualification and quantification methods with multiple marker compounds have been proved to be effective and comprehensive in the quality evaluation of medicinal plants.

5.5 SUMMARY

China is one of the countries that have the most abundant medicinal plant species and longest medicinal plant application history. Mostly of botanical origins, Chinese medicinals have a long history of clinical application under the guidance of the systematic TCM theory. In this capacity, Chinese herbal medicinals are different from Western herbal substances and those used in other traditional medical systems. Comparative studies of global medicinal plants on their plant species, medicinal parts, medicinal properties, chemical constituents, and pharmacological activities will promote research and development and facilitate the protection and sustainable utilization of medicinal plant resources.

In view of the importance of global information exchange, the *Encyclopedia of Medicinal Plants* has been published based on a systematic investigation of contemporary medicinal plants in the world. Published in traditional Chinese, simplified Chinese, and English, 500 commonly used herbal substances all over the world, involving over 800 species of medicinal plants, are documented in this four-volume book (Zhao and Xiao 2009). The entire book consists of the eastern chapter (volumes 1 and 2: commonly used medicinal plants of traditional Oriental medical systems, such as those from China, Japan, the Korean Peninsula, and India), the western chapter (volume 3: commonly used American and European medicinal plants, such as those from Europe, Russia, and the United States), and the Lingnan chapter (volume 4: medicinal plants commonly used and produced in the Lingnan area, including those commercially circulated via this area). The contents are described in the sequence of names, overview, photos of original plants, photos of medicinal materials, chemical composition and chemical structures, pharmacological activities, applications, comments, and references. This encyclopedia serves as a tool that the world can use to understand more about Chinese medicinal plants. It also opens a gate for China to know medicinal plants of the international society.

REFERENCES

American Botanical Council. 2007. FDA approves special green tea extract as a new topical drug for genital warts. *HerbalGram* 74:62–65.

Association of Chinese Culture Research. 1999. *The complete collection of traditional texts on Chinese materia medica*, vol. 2, 5–7, 38–41, 127. Beijing: Cathay Press.

Au, D. T., Chen, H. B., Jiang, Z. H., and Zhao, Z. Z. 2009. A novel method to identify the Chinese herbal medicine wuzhimaotao by quantification of laticifers. *Microscopy Research and Technique* 72 (4): 293–298.

Cao, J., Wei, Y. J., Qi, L. W., Li, P., Qian, Z. M., Luo, H. W., Chen, J., and Zhao. J. 2008. Determination of fifteen bioactive components in Radix et Rhizoma Salviae Miltiorrhizae by high-performance liquid chromatography with ultraviolet and mass spectrometric detection. *Biomedical Chromatography* 22 (2): 164–172.

Center for Drug Evaluation and Research. 2004. *Guidance for industry—Botanical drug products*. Silver Spring, MD: U.S. Food and Drug Administration.

Chen, C. Y., Qi, L. W., Li, H. J., Li, P., Yi, L., Ma, H. L., and Tang, D. 2007. Simultaneous determination of iridoids, phenolic acids, flavonoids, and saponins in Flos Lonicerae and Flos Lonicerae Japonicae by HPLC-DAD-ELSD coupled with principal component analysis. *Journal of Separation Science* 30 (18): 3181–3192.

Chen, K. K. 1928. A comparative study of synthetic and natural ephedrines. *Journal of Pharmacology and Experimental Therapeutics* 33 (2): 237–258.

Chen, S. L., Wei, J. H., Sun, C. Z., Liu, Z. Q., Zhao, R. H., Wang, J. R., Zhou, Y. Q., and Xiao, X. H. 2006. Development of TCMGIS-I and its application in suitable producing area evaluation of *Astragalus membranaceus*. *World Science and Technology—Modernization of Traditional Chinese Medicine and Materia Medica* 8 (3): 47–53.

Chen, S. L., and Xiao, P. G. 2006. *Introduction to the sustainable utilization of Chinese herbal medicine resources*. Beijing: China Medical Science and Technology Press.

China Medicinal Materials Group. 1995. *Resources of Chinese medicinals in China*. Beijing: Science Press.

Chinese Pharmacopoeia Commission. 2005. *Pharmacopoeia of the People's Republic of China*. Beijing: People's Medical Publishing House.

Dagani, R. 2005. Quinine. The top pharmaceuticals that changed the world. *Chemical and Engineering News* 83 (25): 3–8.

Department of Health, Hong Kong Special Administrative Region, P. R. C. 2005, 2008. *Hong Kong Chinese materia medica standards*. Hong Kong: Department of Health.

Editorial Committee. 1959–2004. *Flora of China*. Beijing: Science Press.

———. 1975–1978, 1996. *National collection of Chinese herbal medicinals*. Beijing: People's Medical Publishing House.

———. 1999. *Materia medica of China*. Shanghai: Shanghai Science and Technology Publishing House.

European Parliament and the Council of the European Union. 2004. Directive 2004/24/EC of the European Parliament and of the Council of 31 March 2004 amending, as regards traditional herbal medicinal products, Directive 2001/83/EC on the Community code relating to medicinal products for human use. *Official Journal of European Union* L 136:85–90.

Guo, Z. J., Pu, Y. Q., Wang, J. X., and Lu, J. X. 2006. Research overview and plant resources of "*qi* (seven) medicinals" in Shanxi. *Chinese Journal of Ethnomedicine and Ethnopharmacy* 79:79–81, 36.

Hu, B. C., and Xu, W. Y. 1997. The classification and ecological distribution of herbal medicinals of Taibai mountain regions in Shaanxi. *Chinese Journal of Ethnomedicine and Ethnopharmacy* 25:34–36.

Hu, P., Luo, G. A., Zhao, Z. Z., and Jiang, Z. H. 2005. Multicomponent HPLC fingerprinting of Radix Salviae Miltiorrhizae and its LC-MS-MS identification. *Chemical & Pharmaceutical Bulletin* 53 (6): 677–683.

Hu, Y. N., Jiang, Z. H., Leung, K. S. Y., and Zhao, Z. Z. 2006. Simultaneous determination of naphthoquinone derivatives in Boraginaceous herbs by high-performance liquid chromatography. *Analytica Chimica Acta* 577 (1): 26–31.

Huxtable, R. J., and Schwarz, S. K. W. 2001. The isolation of morphine—First principles in science and ethics. *Molecular Interventions* 1:189–191.

Institute of Materia Medica, Chinese Academy of Medical Sciences, 1959 (1st ed.), 1982–1998 (2nd ed.). *Chinese materia medica*. Beijing: People's Medical Publishing House.

Jiang, Y., Li, S. P., Wang, Y. T., Chen, X. J., and Tu, P. F. 2009. Differentiation of Herba Cistanches by finger-print with high-performance liquid chromatography-diode array detection-mass spectrometry. *Journal of Chromatography A* 1216 (11): 2156–2162.

Kang, J., Zhou, L., Sun, J., Han, J., and Guo, D. A. 2008. Chromatographic fingerprint analysis and character-ization of furocoumarins in the roots of *Angelica dahurica* by HPLC/DAD/ESI-MS technique. *Journal of Pharmaceutical and Biomedical Analysis* 47 (4–5): 778–785.

Lagowski, J. J. 1997. *Macmillan encyclopedia of chemistry*, vol. 3. New York: Simon & Schuster/Macmillan.

Li, J., Li, W. Z., Huang, W., Cheung, A. W., Bi, C. W., Duan, R., Guo, A. J., Dong, T. T., and Tsim, K. W. 2009. Quality evaluation of Rhizoma Belamcandae (*Belamcanda chinensis* (L.) DC.) by using high-performance liquid chromatography coupled with diode array detector and mass spectrometry. *Journal of Chromatography A* 1216 (11): 2071–2078.

Li, J., Qi, H., Qi, L. W., Yi, L., and Li, P. 2007. Simultaneous determination of main phytoecdysones and triter-penoids in Radix Achyranthis Bidentatae by high-performance liquid chromatography with diode array-evaporative light scattering detectors and mass spectrometry. *Analytica Chimica Acta* 596 (2): 264–272.

Li, M., and Wu, R. 2006. *Practice and certification of GAP on Chinese medicinal materials.* Beijing: China Medical Science and Technology Press.

Liang, Z. T., Jiang, Z. H., Ho, H. M., and Zhao, Z. Z. 2007. Comparative analysis of *Oldenlandia diffusa* and its substitutes by high performance liquid chromatographic fingerprint and mass spectrometric analysis. *Planta Medica* 73:1502–1508.

Liang, Z. T., Jiang, Z. H., Leung, K. S. Y., Peng, Y., and Zhao, Z. Z. 2006. Distinguishing the medicinal herb *Oldenlandia diffusa* from similar species of the same genus using fluorescence microscopy. *Microscopy Research and Technique* 69 (4): 277–282.

Lu, G. H., Chan, K., Liang, Y. Z., Leung, K. S. Y., Chan, C. L., and Jiang, Z. H. 2005. Development of high performance liquid chromatographic fingerprints for distinguishing of Chinese *angelica* from related Umbelliferae herbs. *Journal of Chromatography A* 1073:383–392.

Ma, J. X. 1986. The pharmaceutical achievement of ancient medical books excavated from tombs of Han Dynasty at Mawangdui. *Journal of Traditional Chinese Medicine* 5:57–60.

Meng, J., Leung, K. S. Y., Jiang, Z. H., Dong, X. P., Zhao, Z. Z., and Xu, L. J. 2005. Establishment of HPLC-DAD-MS fingerprint of fresh *Houttuynia cordata*. *Chemical & Pharmaceutical Bulletin* 53 (12): 1604–1609.

Ng, K. M., Liang, Z. T., Lu, W., Tang, H. W., Zhao, Z. Z., Che, C. M., and Cheng, Y. C. 2007. *In vivo* analysis and spatial profiling of phytochemicals in herbal tissue by matrix-assisted laser desorption/ionization mass spectrometry. *Analytical Chemistry* 79 (7): 2745–2755.

Pei, J., and Zhou, T. Y. 1955–1985. *Medicinal plants of China.* Beijing: Science Press.

Qiao, C. F., Han, Q. B., Song, J. Z., Mo, S. F., Kong, L. D., Kung, H. F., and Xu, H. X. 2007. Chemical fin-gerprint and quantitative analysis of *Fructus Psoraleae* by high-performance liquid chromatography. *Journal of Separation Science* 30 (6): 813–818.

Tam, C. H., Peng, Y., Liang, Z. T., He, Z. D., and Zhao, Z. Z. 2006. Application of microscopic techniques in authentication of herbal tea—Ku-Ding-Cha. *Microscopy Research and Technique* 69 (11): 277–282.

Tong, C. S., Choy, S. K., Chiu, S. N., Zhao, Z. Z., and Liang, Z. T. 2008. Characterization of shapes for use in classification of starch grains images. *Microscopy Research and Technique* 71 (9): 651–658.

Tong, C. S., Choy, S. K., Zhao, Z. Z., Liang, Z. T., and Chen, H. 2007. Identification of starch grains in micro-scopic images based on granulometric operations. *Microscopy Research and Technique* 70 (8): 724–732.

Xiao, X. H., Chen, S. L., Huang, L. Q., and Xiao, P. G. 2009. Survey of investigations on daodi Chinese medici-nal materials in China since 1980s. *China Journal of Chinese Materia Medica* 34 (5): 519–523.

Xie, P. S. 2005. *Chromatographic fingerprints of Chinese medicinals.* Beijing: People's Medical Publishing House.

Yu, Q. T., Qi, L. W., Li, P., Yi, L., Zhao, J., and Bi, Z. 2007. Determination of seventeen main flavonoids and saponins in the medicinal plant Huang-qi (Radix Astragali) by HPLC-DAD-ELSD. *Journal of Separation Science* 30 (9): 1292–1299.

Zhang, H. Y., Zhao, R. H., Yuan, C. Q., Sun, C. Q., and Zhang, Z. Y. 1995. The variety of Chinese medicinal resources. *China Journal of Chinese Materia Medica* 20 (7): 387–390.

Zhao, Z. L., Leng, C. H., and Wang, Z. T. 2007. Identification of *Dryopteris crassirhizoma* and the adul-terant species based on cpDNA rbcL and translated amino acid sequences. *Planta Medica* 73 (11): 1230–1233.

Zhao, Z. Z. 1999. *A colored atlas of microscopic identification of Chinese materia medica in powdered form as specified in the Pharmacopoeia of the People's Republic of China.* Guangzhou: Guangdong Science and Technology Press.

———. 2005. *An illustrated microscopic identification of Chinese materia medica.* Macau: International Society for Chinese Medicine.

Zhao, Z. Z., Hu, Y. N., Liang, Z. T., Yuen, J. P. S., Jiang, Z. H., and Leung, K. S. Y. 2006. Authentication is fundamental for standardization of Chinese medicines. *Planta Medica* 72 (10): 865–874.

Zhao, Z. Z., Kazami, T., Suzuki, R., and Shimomura, H. 1998. Identification of traditional Chinese patent medicines by polariscope (3): Polariscopic characteristics of ZhiBao Sanbian Wan. *Natural Medicines* 6:485–490.

Zhao, Z. Z., Liang, Z. T., Zhou, H., Jiang, Z. H., Liu, Z. Q., Wong, Y. F., Xu, H. X., and Liu, L. 2005. Quantification of sinomenine in Caulis Sinomenii collected from different growing regions and wholesale herbal markets by a modified HPLC method. *Biological & Pharmaceutical Bulletin* 28 (1): 105–109.

Zhao, Z. Z., Shimomura, H., Sashida, Y., Ishikawa, R., Ohamoto, T., and Kazami, T. 1997. Identification of crude drugs in traditional Chinese patent medicines by means of microscope and polariscope (2): Polariscopic characteristics of stone cells, vessels and fibers. *Natural Medicines* 6:504–511.

Zhao, Z. Z., Shimomura, H., Sashida, Y., Tujino, R., Ohamoto, T., and Kazami, T. 1996. Identification of grains in traditional Chinese patent medicine by a polariscope (1): Polariscopic characteristics of starch and calcium oxalate crystals. *Natural Medicines* 6:389–398.

Zhao, Z. Z., and Xiao, P. G. 2009. *Encyclopedia of medicinal plants*, vols. 1–4. Shanghai: Shanghai World Publishing Corporation.

Zhu, S., Fushimi, H., Han, G., Tsuchida, T., Uno, T., Takano, A., and Komatsu, K. 2007. Molecular identification of "Chuanxiong" by nucleotide sequence and multiplex single base extension analysis on chloroplast trnK gene. *Biological & Pharmaceutical Bulletin* 30 (3): 527–531.

Zhu, S., Fushimi, H., and Komatsu, K. 2008. Development of a DNA microarray for authentication of ginseng drugs based on 18S rRNA gene sequence. *Journal of Agricultural and Food Chemistry* 56 (11): 3953–3959.

Zhu, T. P., Liu, L., and Zhu, M. 2007. *Plant resources of China.* Beijing: Science Press.

CHAPTER 6

Medicinal Plants of the Middle East

Shahina A. Ghazanfar

CONTENTS

6.1 INTRODUCTION

The recorded uses of plants as medicinals dates back over 5,000 years to the earliest known civilization: the Sumerians in southern Mesopotamia (modern-day Iraq). Uses of plants such as laurel, caraway, and thyme were described for treating diseases and ailments. Sumerians also developed the art of writing, the practice of organized agriculture, and irrigation; the wheel and the arch were developed during their time, which spanned over 3,000 years.

The use of castor oil, coriander, garlic, indigo, mint, and opium has been recorded by the Egyptians from about 1000 BC and the Chinese herbals from about 2700 BC (where *Ephedra* species that yields ephedrine is mentioned). The uses of turmeric were described in Indian Ayurvedic medicine as early as 1900 BC.

Plants (and animals and minerals) were used by the ancient Greeks and Romans, but it was Hippocrates (ca. 460–370 BC) and Galen (AD 129–199/217) who developed the principles of diagnosis and the use of plants that eventually became the basis of modern medicine. Hippocrates based his medical diagnosis on bodily humors that he assigned to a human body. Thus, according to him, the human body contains blood, phlegm, yellow bile, and black bile (Chadwick and Mann 1950):

> *These are the things which make up its constitution and cause its pains and health. Health is primarily that state in which these constituent substances are in correct proportion to each other, both in strength and quality and are mixed well. Pain occurs when one of the substances presents either a deficiency or an excess or is separated in the body and not mixed with the others.*

The Hippocratic system of the classification of bodily humors prevailed in the practice of medicine and was later adopted by the Muslims as the classical Greco-Arab medicine, or *Unani tibb*, still taught and practiced today.

In Unani tibb, four elements—earth, water, air, and fire—are related to the four bodily humors (blood, phlegm, yellow bile, and black bile). In addition, over a period of time, a concept of four "qualities"—hot, cold, dry, and wet—was integrated with the original Greek system (Foster and Anderson 1987). Thus, blood was hot and moist, phlegm cold and moist, yellow bile hot and dry, and black bile cold and dry. Certain equilibrium between the four humors was considered necessary for a healthy body, and an imbalance of any of the four resulted in sickness.

The concepts of Greek medicine were acquired by the Arabs through the works of Dioscorides (first century AD) and Galen (AD 131–201). During the fifth century, Greek and Roman medicinal texts were translated by the Nestorians and taken to Persia (Iran). They also translated Indian medicinal texts from Sanskrit. With the spread of Islam and during the ninth to thirteenth centuries, many medical works were translated into Arabic and developed and adapted by Muslim scholars into the Islamic system of medicine. The Arabs, being seafarers and traders, had access to medical knowledge and material from far countries such as India and China that they incorporated into their medical knowledge. *Bimaristan* (medical schools) appeared from the ninth century and all aspects of medicine were taught in the Islamic world.

The scholar Ibn Dawoud Dinawari (of Persia, AD 828–896) described more than 600 plants and their uses in his book, *Kitab al Nibat* (*Book of Plants*), for which he is considered the father of Arabic botany. Another scholar, Ibn Sina (Avicenna, Persian, AD 980–1037), a renowned physician, philosopher, physicist, and astronomer of his time, wrote about 450 texts on various subjects and 40 dealt with medicine. The Book of *Healing* and *Qanoon f'il tibb* (*The Canon of Medicine*) are some of his most famous and comprehensive works; the latter was a standard medical text in European universities up until the 17th century. He introduced the concept of contagious diseases and quarantine, testing, and efficacy of medicines, as well as the benefits of exercise for good health. He described several medicinal plants that were fully incorporated into the practice of herbal medicine.

Medicine developed further under Muslim physicians from Persia, Iraq, and Andalus (Spain). Al Razi (Razes) (AD 865–925) is credited with writing a treatise on small pox and measles, which had not been described earlier, and was the first to use animal gut for sutures and plaster of Paris for casts. Abu al Kassim Al Zahrawi (Albucasis, b. AD 936) developed surgical methods and is considered to be the father of modern surgery. Abn Rushd (Averroes, b. AD 1200), a scholar of many disciplines, wrote on rules of medicine. Ibn Al Nafis (AD 1210–1288) is famous for describing the pulmonary circulation. Ibn al-Baitar (of Malaga, AD 1197–1248) described more than 1,400 food and medicine plants. By the 13th century the experimental scientific method had become a part of medicine, taught and reinforced by Abu al Abbas al-Nabati (from Andalus and teacher of Ibn al Baitar). Al-Nabati taught and practiced verifiable techniques in identifying, describing, and testing the efficacy of medicinal plants, thus initiating the science of pharmacology.

Together with knowledge and scholarship, a strong belief in God and Prophet Mohammed are considered of foremost importance for any cure in *Unani tibb*. Hadith, volume 7, book 71, number 582, narrated by Abu Huraira, states: The Prophet said, '*There is no disease that Allah has created, except that He also has created its treatment.*' Based on the Hadith, several works were written on Mohammed's life and sayings relating to medicine; they became known as *Tibb-al-Nabbi (Medicine of the Prophet)*. Three of the better known works are those compiled by Abu Nu'aim (Persian, d. AD 1038), Ibn al-Qayyim al Jawziyya (from North Africa, d. AD 1350), and Jalaluddin ibn Abi Bakr as-Sayuti (from Egypt, b. AD 1445). The books describe the theory of medicine (Islamic medicine) based on the Hippocratic four elements that constitute man and what was in practice at that time, and the practice of medicine with references to the life and sayings of Prophet Mohammed and the Qur'an. *Tibb al Nabbi* also includes instructions on food intake, sleep, rest, body hygiene, etc. for a healthy body and gives the properties of food and remedies according to the sayings of the Prophet and the Qur'an.

In addition to the use of plants for medicine (*al tadawee bil a' ashaib*), bone setting (*al tajbeer*), cupping (*al hajamah*), and cauterization (*wasm, qai*) are also practiced in traditional medicine. In any of these forms of treatment, specific plants and food may be used as part of the treatment. A brief review is given here (adapted from Ghazanfar 1994 and references therein):

Al Tajbeer (bone setting): Bone fractures are described as "cracks" when the fracture is not right through the bone and as "real fractures" when the bone is broken right through. Once the bone fracture is corrected, a plaster made from the resin of *Acacia* spp., seeds of tamarind or lentils mixed with egg is applied. The fracture is then bound with cloth and the limb stabilized with wooden planks. If the setting is not done correctly, the bone setter may break the bone and set it again. Special plants are used for this procedure. During the period of recovery from bone fractures a special diet of honey (date or bee honey) is prescribed. In the Middle East, traditional bone setting is rarely carried out because access to modern hospitals and clinics is available to most people.

Al Hajamah (cupping): Cupping is the traditional method of treating specific diseases that are believed to have been caused by "bad blood" in the body. Chronic pain in legs, headaches, obesity, and paralysis of the lower limbs are often treated by cupping. Cupping involves making superficial cuts in the skin on the back of or below the knee, inverting a cup on it, creating a vacuum by making a small light, and sealing the edges of the cup to the skin. The blood (black blood) that oozes out is considered to cure or help with the cure. Cupping is recommended in *Tibb al Nabbi* as a cure for several diseases and conditions.

Wasm (cauterization): Cauterization and the use of heat therapy were known as early as 3000 BC, when Egyptians used it for treating various conditions. Cauterization has also been used by the ancient Greek and pre-Islamic Arabs. Among the Arab physicians, Abu al Qassim al Zahrawi (Abulcassis, AD 936–1013) greatly developed the use of cauterization for treating as many as 50 diseases. In his book, *Kitab al Tassrif*, he described various techniques and instruments that were used for cauterization and for surgery. Cauterized parts were treated with plants to help with the process of healing (Ghazanfar 1995).

It must be noted that in traditional medicine certain beliefs relate to having a healthy and a sound body. It is believed that too much food causes illness and dieting or abstinence from food is the main medicine. Ibn Khaldun (AD 1332–1406), Muslim philosopher and historian, wrote in his book *Al Muqaddimah* (Dawood 1967):

...the nourishment is boiled by natural heat, stage by stage, until it actually becomes a part of the body. Now, illnesses originate from fevers and most illnesses are fevers. The reason for fevers is that the natural heat is too weak to complete the process of boiling of food in the body. The nourishment thus does not assimilate. The reason for that, as a rule, is either that there is a great amount of food in the stomach that becomes too much for the natural heat, or that food is put into the stomach before the first food has been completely boiled. Consequently, the unassimilated nourishment becomes putrid. Anything in the process of putrefaction develops a strange heat. This heat is what, in the human body, is called a fever.

Though this belief is not commonly accepted, the concept of eating food and taking drugs with a "cool" quality for diseases and conditions that are "hot" is generally followed in traditional *Unani* medicine.

6.2 MEDICINAL PLANTS

The use of native plants for medicinal purposes is declining throughout the Middle East as development sees the establishment of hospitals and primary health clinics. Health care is available to most people, albeit with some constraints, but during the past 20 or so years, some sort of medical care has become available to even the most remote villages. With the ready availability of Western medicine, the use of herbal medicines and other forms of traditional medicine is becoming scare, and consequently the knowledge of medicinal plants is being forgotten and lost. Today the knowledge lies mostly with the elders, most of whom are women. In many instances, the lack of market has led to the closure of many local herb shops, or they are stocked with exotic imported herbal treatments, mainly Indian or Chinese (Ghazanfar 1994).

In the Middle East, herbal medicines are, to some extent, still used for common ailments such as common cold, cough, headache, sunburn, muscular pain, and some digestive problems (e.g., colic, indigestion, etc.). Herbal treatments are also commonly used during the periods before and after childbirth.

Tables 6.1–6.8 list the medicinal plants commonly used for the treatment of different diseases and conditions in the Middle East. The information has been taken from Abu-Rabia (2005), Al-Douri (2000), Batanouny et al. (1999), Duke (2007), Ghazanfar (1994 and references therein; 1996), Ghazanfar and Al-Sabahi (1993), Marshall et al. (1995), Miller and Morris (1988), Morris (1984), Schopen (1983), and Townsend and Guest (1966–1985).

6.2.1 Carminatives, Laxatives, and Antidiarrheals

Among the most commonly used plant-based treatments are those for the digestive system (Table 6.1). In Arabia and Egypt, the use of dill, coriander, cumin, and nutmeg as carminatives; senna and colocynth as laxatives; and the multiple uses of sumak from a tanning agent to providing relief from earache, toothache, dysentery, and gangrene were known as early as the second millennium BC (Simpson and Conner-Ogorzaly 1986). Volatile oils in the family Apiaceae, to which coriander, fennel, cumin, and dill belong, are responsible for the carminative properties and are now widely used in commercial pharmaceutical preparations.

6.2.2 Anthelmintics

About 16 plant species are known to be used for the treatment of stomach worms (Table 6.2). Several of these are also used as laxatives and for stomachache.

6.2.3 Muscular Pain and Swollen Joints

Muscular pain, inflammation of joints, and swelling are generally treated by massage with oils or by applying a poultice of medicinal leaves (Table 6.3). Gums and resins are also used for treatment.

Table 6.1   Plants Used as Carminatives, Laxatives, and Antidiarrheals

Species	Plant Parts	Medicinal Properties
Allium cepa, A. sativum	Bulb, cloves	Juice of fresh bulb and cloves used for colic and abdominal pain
Cadaba farinosa	Leaves	Infusion of leaves taken for colic
Carissa edulis	Berries	Eaten for abdominal colic and constipation
Citrullus colocynthis	Leaves, roots, and seeds	Infusion of crushed leaves or roots with goat's milk, or seeds taken with food as a laxative
Croton confertus	Leaves	Slightly crushed in water, and solution taken orally for constipation and stomachache
Cynomorium coccineum	Whole plant	Eaten as a laxative
Dorstenia foetida	Seeds	Eaten for flatulence and indigestion
Euclea schimperi	Leaves	Eaten for colic, diarrhea, and indigestion
Euphorbia hadramautica	Seeds	Eaten to cure flatulence, colic, and indigestion
Ficus carica	Fruit	Eaten as a laxative and as a general tonic
Heliotropium fartakense	Leaves	Soaked in water, taken for indigestion and colic
Ipomoea pes-caprae	Seeds	Eaten as a purgative
Jasminum grandiflorum	Flowers	Infusion of dried flowers taken for colic, dysentery, and abdominal pain
Lavandula spp.	Leaves and stems	Boiled in water and solution taken for colic and stomachache
Maerua crassifolia	Leaves	Infusion of leaves taken for constipation and colic
Matricaria aurea	Flowers	Infusion of flowers taken as tea for colic and stomach cramps
Moringa peregrina	Seed oil	Taken orally for constipation and stomach cramps
Nepeta deflersiana	Leaves	Infusion of leaves taken as tea for indigestion
Ocimum basilicum	Seeds	Infusion of leaves or crushed seeds taken as tea for diarrhea
Plantago coronopus, P. major	Whole plant, seeds	Infusion of plant taken as a laxative, and seeds of *P. major* crushed and swallowed with water for diarrhea
Reichardia tingitana	Leaves	Dried powdered leaves with water to treat colic and constipation
Rhazya sticta	Leaves	Mixed with senna and milk, taken for abdominal colic and constipation
Ricinus communis	Seed oil (castor oil)	Used as a purgative
Senna alexandrina, S. holosericea, S. italica	Leaves and seeds	Decoction of leaves, often mixed with other herbs, used as a laxative and for stomach cramps
Tamarindus indica	Fruit	Soaked in water, solution taken as a laxative, blood cleanser, and a general tonic
Teucrium polium, T. mascatensis	Leaves and stems	Fresh or dried, boiled in water for colic and stomach pains
Thymus vulgaris	Leaves	Infusion of dried, powdered leaves to which salt and cumin are added, taken for colic, coughs, and colds

6.2.4 Skin Disorders, Burns, Wounds, Bruises, Stings, and Bites

Over 50 plant species are used for treating skin problems (Table 6.4). General skin disorders include eczema, skin rash, and skin infections. Treatments are also available for removing freckles. Special cooling pastes are applied topically for burns, sunburn, and heat rash. Treatment for snakebite and scorpion stings includes both topical application at the site of the bite or sting and

Table 6.2   Anthelmintic Plants

Species	Plant Parts	Medicinal Properties
Artemisia sieber	Leaves	Crushed and an infusion made, taken orally (the plant contains essential oils, which are reported to be toxic against *Ascaris*)
Capparis decidua	Fruit	Eaten for expelling worms, also to treat constipation, hysteria, and other psychological problems
Capparis spinosa	Leaves, root, bark	Soaked in water, extract used as an anthelmintic, expectorant, and as a tonic
Carica papaya (cultivated)	Leaves and seeds	Pounded and eaten as an anthelmintic; also used to treat diarrhea; raw fruit is digestive and carminative
Cyperus rotundus	Tubers	Dried or eaten fresh to expel worms; also eaten to regulate menstruation; powdered and used as an insecticide
Fumaria parviflora	Whole plant	Extract as an anthelmintic, laxative, and for treating dyspepsia
Peganum harmala	Leaves and seeds	Tea made from leaves or powdered seeds mixed with water, taken orally as a vermifuge against tapeworms and for removing kidney stones; seeds mixed with senna and honey used for stomachache
Punica granatum (cultivated)	Fruit rind	Dried, mixed with thyme and flour, baked as bread and eaten as an anthelmintic and to cure diarrhea
Rhazya stricta	Seeds	Powdered, eaten to remove stomach worms; in central Oman dried leaves powdered with Halwa (sweets) and leaves of *Cassia senna* are mixed with milk and used for abdominal pain, colic, and constipation; the mixture is drunk on alternate days; also given to infants from about 3 months to make their joints strong
Vernonia cinerea	Roots and seeds	Seeds are eaten and decoction of roots used as an anthelmintic and diuretic

Table 6.3   Plants Used for Treating Muscular Pain and Swollen Joints

Species	Plant Parts	Medicinal Properties
Calotropis procera	Leaves	Oil is rubbed on the site of pain, then leaves are heated and placed on it
Cissus rotundifolia	Leaves	Heated and used as a poultice for backache
Conyza incana	Leaves	Heated and used as a poultice for muscular pain
Dracaena serrulata	Gum resin	Mixed with water and made into a paste, applied on legs and feet to relieve pain
Euphorbia amak	Latex	Mixed with oil and massaged on painful joints
Heliotropium fartakense	Leaves	Pounded with salt, turmeric, and ginger and made into a paste, applied to sprains and swellings
Juniperus excelsa subsp. *polycarpos*	Leaves	Soaked in oil, massaged over painful joints, muscles, and paralyzed limbs
Kleinia odora	Stems	Juice from the stems applied over painful muscles
Psiadia punctulata	Stems	Heated and placed on site of pain
Tephrosia apollinea	Leaves	Powdered with salt, heated, and placed on swellings or swollen joints

Table 6.4   Plants Used for Skin Disorders, Burns, Wounds, Bruises, Stings, and Bites

Species	Plant Parts	Medicinal Properties
Acacia gerardii, A. nilotica	Gum resin, leaves	Applied to soothe burns; leaves are pounded into a paste and used as a poultice on boils and swellings or applied around boils to draw out the pus
Achyranthes aspera	Roots	Crushed and applied on scorpion stings to reduce itching and swelling
Adenium obesum	Sap, crushed bark	Pounded and applied on wounds and skin infections
Allium cepa, A. sativum	Bulb	Juice rubbed to remove spots from face or treat skin rash
Allophylus rubifolius	Leaves	Pounded and placed on skin boils to bring out the pus
Aloe dhufarensis, A. tomentosa, A. vera	Leaves	Juice extracted from the leaves is used by itself or mixed with indigo and applied on skin rash, burns, and wounds; also used as disinfectant for burns and wounds; also applied around the eyes to relieve itching and pain
Amaranthus graecizans, A. viridis	Leaves	Crushed and applied on scorpion stings, snakebites, and itchy skin rash
Anagallis arvensis	Whole plant	Crushed and applied to soothe skin rashes, snakebites, and skin ulcers
Aristolochia bracteolata	Whole plant	Rubbed on place of snakebites or scorpion stings
Aerva javanica	Flowers	Mixed with water, made into a paste, and used as a wound dressing and to stop bleeding
Becium dhofarense	Whole plant	Boiled in water, infusion used to soothe skin sores, dry skin, itch, and insect bites
Calotropis procera	Leaves and latex	Latex from stems is dropped around a pus-filled wound to draw out the pus; also applied on scorpion stings to relieve pain and swelling
Capparis cartilaginea	Leaves and stems	Crushed in water and heated, the solution is strained and applied on bruises, snakebites, swellings, and itchy skin rash
Caralluma aucheriana	Stems	Extract from the succulent stems applied to soothe sunburned or itchy skin
Citrullus colocynthis	Leaves	Poultice made from crushed leaves and garlic, applied on bites and stings
Citrus aurantifolia	Fruit	Dried and crushed, mixed with salt and water, and heated to make a paste, applied on skin to extract thorns
Convolvulus arvensis	Roots and leaves	Crushed and applied on wounds and cuts to stop bleeding
Cyphostemma ternatum	Stems and leaves	Juice extracted, used to wash feet to remove fungal infections; heated with salt, is used as a poultice for skin infections
Eulophia petersii	Pseudobulbs	Juice extracted to treat eczema, ringworm, skin rash, and sores
Euphorbia cactus, E. larica, E. schimperiana	Whole plant	Extract applied around boils to draw out the pus; also applied on sores, ringworm, and skin ulcers; latex from *E. larica* is applied around pus-filled boils and that from *E. schimperiana* is used to loosen thorns for extraction
Euryops arabicus	Leaves and stems	Heated leaves and stems applied to sore feet
Ficus carica, F. cordata subsp. *salicifolia*	Latex, leaves	Applied to burn warts; dried leaves are crushed, mixed with honey, and applied to treat skin discoloration and freckles
Heliotropium fartakense, H. kotschyi	Leaves and stems	Pounded with salt, turmeric, and ginger, applied on burns and ulcers; paste of leaves used for ringworm, eczema, wounds, and skin sores
Impatiens balsamina	Leaves and stems	Extract used to soothe itchy and inflamed skin
Jatropha dhofarica	Sap	Applied over wounds and skin sores as an antiseptic and to stop bleeding

(continued)

Table 6.4   Plants Used for Skin Disorders, Burns, Wounds, Bruises, Stings, and Bites (Continued)

Species	Plant Parts	Medicinal Properties
Medicago sativa	Leaves	Mixed with leaves of tamarind and salt, applied on bruises
Melilotus indicus	Whole plant	Crushed and used to soothe skin rash
Myrtus communis	Leaves	Paste applied on wounds and scorpion stings; ashes of burned leaves applied on blisters and ulcer
Olea europaea	Leaves and bark	Pounded bark and leaves applied on blisters and skin ulcer
Pergularia tomemtosa	Latex	Latex applied on skin sores
Phoenix dactylifera	Fruit	Paste made from dates, mixed with salt, applied to bruises
Pluchea arabica	Whole plant	Pounded with water and boiled and solution applied to boils and swellings
Polycarpaea repens	Whole plant	Crushed and applied at site of snakebite
Psoralea corylifolia	Seeds	Mixed with thyme and made into a paste, for treating general skin disorders
Rhazya stricta	Leaves	Extract mixed with oil, *Nigella sativa*, and ginger, or leaves crushed in water, applied on skin rash
Rumex vesicarius	Leaves and seeds	Eaten as an antidote for scorpion stings
Salvadora persica	Leaves	Fresh or dried, powdered, applied on skin blisters and scorpion stings
Sansevieria ehrenbergii	Leaves	Juice applied to skin sores and eruptions
Tamarix aphylla	Leaves	Dried leaves applied on wounds to stop bleeding
Teucrium polium	Leaves	Fresh or dried, boiled in water and poured on bites and skin ulcers
Trichilia emetica	Seeds	Crushed with sulfur, used as a skin ointment
Vernonia cinerea	Leaves	Rubbed on scorpion stings to relieve pain and inflammation
Withania somnifera	Leaves	Pounded and applied as a poultice on burns and sunburned skin; mixed with garlic, applied on stings and bites

a plant extract taken orally. Special plants are used for the treatment of pus-filled boils and for removing thorns.

6.2.5 Diuretics and Urinary Disorders

Table 6.5 lists the species that have been identified for the treatment of disorders of the urinary tract. Remedies include boiling infusions, extraction of dry or fresh leaves, flowers, seeds, or whole plants (Abu-Rabia 2005).

6.2.6 Fertility, Childbirth, Pre- and Postnatal Care

Several plant species are used to improve lactation and strengthen back muscles after childbirth (Table 6.6). Suppositories and tampons are prepared with a selection of plants to relieve backache after childbirth. A special diet including honey, fenugreek, ginger, and garlic is given to mothers from 7 to 40 days after they give birth. Several herbal treatments are used for fertility, contraception, and abortion (Colfer 1990; Musallam 1983).

6.2.7 Cold, Coughs, Fever, and Headaches

Infusions of plants are commonly used for treating colds and coughs, and smoke from burning particular species is inhaled to reduce chest congestion and ease difficult breathing (Table 6.7). To

Table 6.5 Diuretics and Urinary Disorders

Species	Plant Parts	Method	Medicinal Properties
Acacia raddiana	Resin	Boil in water or milk and drink	Urinary retention and urinary tract infections, and prostate
Adiantum capillus-veneris	Leaves and roots	Boil in water and drink	Colds, diuretic, urinary tract infections
Adonis aestivalis	Leaves	Soak in water and drink	Urinary tract infections, prostate, impotency
Alhagi maurorum	Rhizome	Boil in water and drink	Urinary retention and urinary tract infections
Allium ampeloprasum	Fruit, leaves, and stalks	Mix with olive oil and eat	Urinary tract infections and retention, kidney pain
Althaea officinalis	Leaves	Eat as raw salad	Urinary tract infections and incontinence
Anastatica hierochuntica	Dried fruit	Crush the fruit, soak in water, and drink	Urinary retention and urinary tract infections, female sterility, male impotency
Anchusa strigosa	Leaves and roots	Crush, soak in water, and drink	Urinary tract infections
Anisum vulgare	Seeds	Soak or boil in water and drink	Diuretic, urinary tract infections, and retention
Apium graveolens	Leaves, roots	Boil in water and drink	Urinary tract infections and pain, incontinence and retention
Asparagus aphyllus	Leaves	Eat as raw salad	Diuretic, urinary retention, and urinary tract infections
Asphodelus aestivus	Roots	Boil in water and drink	Urinary tract infections
Beta Brassica	Leaves, roots, stems, seeds	Raw; eat or soak in water and drink	Urinary retention and urinary tract infections, stomachache
Capparis spinosa	Bulbs	Soak in water and drink	Diuretic, urinary tract infections, and retention
Capsella bursa-pastoris	Dried leaves	Boil in water and drink	Urinary retention and urinary tract infections
Cassia italica	Leaves	Boil in water and drink	Urinary retention
Cercis siliquastrum	Leaves and flowers	Boil in water and drink	Urinary retention and infections
Chenopodium album	Whole herb	Soak in water and drink	Urinary tract infections and retention
Cichorium pumilum	Green leaves	Eat, or soak in water and drink	Diuretic, urinary tract infections, and retention
Citrulus vulgaris	Fruit	Eat	Urinary tract infections
Citrus limon	Fruit	Drink with water or tea	Urinary tract infections and retention, stomachache, diarrhea, stomach upset, nausea
Cnicus benedictus	Leaves, stalks, and roots	Boil in water and drink	Urinary retention and urinary tract infections, prostate illness
Colchicum ritchii	Corms	Apply hot corms on aches, peel, and eat	Kidney infections, urination pains, and prostate
Conium maculatum	Dried leaves	Soak in water and drink	Urinary tract infections and prostate illnesses
Coridothymus capitatus	Leaves and flowers	Boil leaves and flowers in water and drink	Urinary tract infections and retention, prostate, testicle pains
Cucumis melo	Fruit	Eat	Urinary retention and urinary tract infections, kidney pains, constipation
Cucumis sativa	Fruit	Eat	Urinary retention and urinary tract infections
Ecballium elaterium	Fruit, seeds	Squeeze or soak in water and drink	Urinary retention and urinary tract infections, yellow fever, liver infections

(continued)

Table 6.5   Diuretics and Urinary Disorders (Continued)

Species	Plant Parts	Method	Medicinal Properties
Elymus repens	Leaves and roots	Boil in water and drink	Urinary tract infections and cystitis
Globularia arabica	Leaves, flowers	Soak in water and drink	Urinary incontinence and urinary tract infections, kidney pains
Hordeum sativum, Hordeum spontaneum	Seeds, bran	Eat as pita, boil crushed seeds in water and drink	Urinary tract infections and retention
Avena sterilis	Leaves, seeds	Eat bread or prepare porridge	Urinary retention and infections, diabetes, increasing breast milk, and strengthening mother's body after childbirth
Inula viscosa	Leaves	Boil in water and drink	Urinary retention and urinary tract infections
Iris pallida, Iris palaestina	Leaves, rhizome	Boil in water and drink	Urinary tract infections and retention
Juniperus communis, Juniperus oxycedrus, Juniperus phoenicea	Berries	Soak in water and drink	Diuretic, urinary retention, urinary tract infections, kidney pains, prostate, impotency, female sterility, vaginal infection
Lavatera trimestris	Leaves, seeds	Eat as raw salad or cook	Urinary tract infections, retention, and prostate illnesses
Lepidium sativum	Seeds	Boil in water and drink	Urinary retention and urinary tract infections, prostate illnesses
Linaria cymbalaria	Leaves	Boil in water and drink	Diuretic, urinary tract infections, retention, breast infection
Linum pubescens	Seeds	Boil in water and drink	Urinary tract infections, urinary pains, and prostate illnesses
Lupinus albus	Seeds, whole herb	Soak in water and drink	Diuretic, urinary tract infections, and retention
Malva sylvestris	Leaves	Eat as raw salad, cook, or boil in water and drink	Urinary tract infections, retention, prostate illnesses, vaginal infection, and skin irritation
Mandragora autumnalis	Ripe fruit	Eat	Urinary tract infections, sterility, vaginal infections, increasing breast milk, "evil eye"
Matricaria aurea	Leaves and flowers	Boil in water and drink	Urinary retention, urinary tract infections, kidney pains, testicle pains
Mentha longifolia	Leaves	Boil in water and drink	Blood in the urine, urinary tract infections and retention, vomiting, nausea
Micromeria myrtifolia	Leaves	Boil in water and drink	Urinary retention and urinary tract infections.
Nigella sativa	Seeds or extracted oil	Boil the seeds in water and drink, or drink the oil	Urinary tract infections and retention, prostate illnesses, blood in urine, diabetes, impotency, sterility
Olea europaea	Oil	Drink	Urinary retention and urinary tract infections, prostate, kidney pains, skin irritation, constipation
Ononis antiquorum	Leaves, flowers, and dried roots	Boil in water and drink	Diuretic, urinary tract infections, and retention
Oxalis corniculata	Whole plant	Soak in water and drink	Urinary retention and urinary tract infections
Parietaria judaica	Leaves	Boil in water and drink	Urinary tract infections, urinary retention, vaginal infection
Paronychia argentea	Leaves and flowers	Boil in water and drink	Diuretic, urinary retention, and urinary tract infections
Petroselinum crispum, Petroselinum sativum	Leaves	Soak in water and drink	Urinary tract infections and retention, and prostate illnesses

Table 6.5 Diuretics and Urinary Disorders (Continued)

Species	Plant Parts	Method	Medicinal Properties
Phagnalon rupestre	Leaves and stalks	Boil in water and drink	Urinary retention and urinary tract infections
Pinus halepensis	Leaves	Boil in water and drink	Urinary retention, cystitis, and prostate illnesses
Pituranthos tortuosus	Stems and flowers	Boil in water and drink	Blood in the urine, urinary retention
Plantago lagopus	Leaves	Boil in water and drink	Urinary tract infections
Polygonum equisetiforme	Leaves	Boil in water and drink	Urinary retention and urinary tract infections, blood in the urine
Portulaca oleracea	Leaves and stalks	Crush, soak in water, and drink, or eat as raw salad	Diuretic, urinary tract infections and retention, cystitis
Prosopis farcta	Leaves, fruit, and roots	Boil in water and drink	Urinary tract infections and retention, diabetes, high blood pressure
Raphanus raphanistrum	Leaves	Boil in water and drink	Urinary tract infections and urinary retention
Rheum palaestinum	Leaves and roots	Boil in water and drink	Urinary retention and urinary tract infections
Rhus coriaria	Fruit	Drink extracted liquid from the fruit	Urinary tract infections and urinary retention
Rosmarinus officinalis	Leaves	Boil in water and drink	Urinary tract infections and retention, vagina infection, sterility
Rubia tenuifolia	Roots	Boil in water and drink	Diuretic, urinary retention, skin irritation
Rubus sanctus	Berries or leaves	Eat the berries; soak leaves in water and drink	Urinary retention and prostate illnesses
Rumex cyprius	Leaves	Eat as raw salad	Urinary tract infections and retention, blood in the urine
Ruscus aculeatus	Roots, leaves	Boil in water and drink	Urinary tract infections and retention
Scolymus maculatus	Stems, stalks	Boil in water	Urinary tract infections
Scorpiurus muricatus	Leaves, stems, and stalks	Eat as raw salad	Urinary tract infections
Sisymbrium irio	Leaves and flowers	Eat as raw salad	Urinary retention and urinary tract infections
Taraxacum megalorrhizon	Flowers and leaves	Eat as raw salad	Urinary retention and urinary tract infections
Tribulus terrestris	Dried leaves	Boil in water and drink	Urinary tract infections and retention
Trigonella foenum-graecum	Seeds	Boil in water and drink	Urinary tract infections and retention, diabetes, and high blood pressure
Triticum aestivum	Leaves, seeds, bran	Eat as pita, soak seeds or bran in water and drink	Urinary retention and prostate illnesses, impotency, diabetes, high blood pressure
Urtica pilulifera, Urtica urens	Leaves	Boil in water and drink	Urinary tract infections and retention
Varthemia iphionoides	Leaves and stalks	Boil in water and drink	Urinary retention and urinary tract infections
Zea mays	Maize fruit	Soak in water and drink	Diuretic, urinary retention, urinary tract infections and cystitis, diabetes
Zizyphus spina-christi	Fruit or leaves	Eat the fruit; soak leaves in water and drink	Urinary tract infections and prostate

Table 6.6   Plants Used for Fertility, Childbirth, and Pre- and Postnatal Care

Species	Plant Parts	Methods and Medicinal Properties
Anastatica hierochuntica	Dried plant	Soaked in water and the water taken orally at the time of childbirth
Delonix elata	Leaves	Infusion given for prolonged or difficult labor
Haplophyllum tuberculatum	Leaves	An anal suppository is made with leaves of Azadirachta indica, resin of Commiphora spp., and thyme, given to the new mother after childbirth, to strengthen back muscles
Launaea nudicaulis	Leaves	Inserted in the vagina after childbirth to stop excessive bleeding
Moringa peregrina	Seed oil	Mixed with clove oil and cardamom oil, taken during labor
Morus nigra	Fruit	Infusion of fruits and berries of Salvadora persica, given as a general tonic to women and to regulate menstruation
Psoralea corylifolia	Seeds	Crushed with butter, applied to treat mastitis
Pulicaria jaubertii	Seeds	Infusion of seeds taken to improve lactation; a decoction of leaves taken to stimulate digestion
Rhazya stricta	Seeds	Crushed and added to milk, given to increase lactation
Trigonella foenum-graecum	Seeds	Boiled in water, mixed with egg, and given to the new mother for 1 week after childbirth

lower a fever, the body is bathed in water to which plants have been added. The body is cooled down by applying herbal pastes and by imbibing cooling drinks. Pastes and oils are applied or massaged on the forehead for relief from headaches.

6.2.8 Health Tonics

Approximately 15 plant species are used as general health tonics, of which seven are commonly used (Table 6.8). Concoctions and infusions of plants are used for the frail and elderly seven, for children, and during convalescence. Several species are used to "give strength" or to "make the body strong," especially after illness. Health tonics are also used to "purify blood" and to cleanse the blood system.

6.3 IMPORTANT MEDICINE-RELATED FOOD AND CULTURAL PLANTS

In the Middle East, several plants are used for both medicinal and cultural purposes. Among these, perhaps one of the most important is the date palm, which is used in traditional medicine, as building material, and for food and shade. Other plants are used as dyes, cosmetics, and perfumes. A summary of the most widely and commonly used plants is given here.

6.3.1 The Date Palm

That dates are more or less synonymous with the Middle East is not surprising since they have been a staple food in the Middle East for over 5000 years BC. The date palm (*Phoenix dactylifera*) is believed to have originated around the Persian Gulf and has been cultivated from Mesopotamia (Iraq) to Egypt, possibly as early as 4000 BC. Charred date stones have been found in eastern Arabia dating back to 6000 BC (Berthoud and Cleuziou 1983; Cleuziou and Constantini 1980).

The date palm is mentioned several times in the Qur'an; it is believed that the tree is blessed by God with gifts not found in other trees. All parts of the tree are useful and it is undoubtedly a versatile plant of great economic importance to the people of the Middle East; every part of the tree is used. Over 40 varieties (of mostly three main cultivar groups) exist: soft (e.g., *barhee, halawy, khadraw, medjool*), semidry (e.g., *dayri, deglet, deglet noor, zahidi*), and dry (e.g., *thoory*). Each

Table 6.7 Plants Used for Colds, Coughs, Fever, and Headache

Species	Plant Parts	Methods and Medicinal Properties
Acridocarpus orientalis	Seed oil	Massaged on the forehead to relieve headache
Alkanna orientalis	Leaves	Infusion of fresh leaves, mixed with thyme, taken for sore throat
Aloe dhufarensis, A. vera	Leaves	Juice/gel applied to forehead for headache and to lower fever; applied around eyes to soothe pain and redness
Ambrosia maritima	Whole plant	Crushed, burned, and the smoke inhaled to relieve difficult breathing
Arnebia hispidissima	Whole plant	Infusion taken as tea to lower fever
Cichorium intybus	Leaves	Eaten or boiled in water to lower fever
Citrus aurantifolia	Fruit juice	Taken for cold and fevers; also used for digestive problems
Cocculus hirsutus	Leaves and roots	Used for lowering fevers
Commiphora myrrha	Resin	For fever, smoke from burned resin inhaled or resin soaked in water and solution taken orally
Croton confertus	Shoots	Dipped in butter and sucked to cure coughs
Fagonia indica	Whole plant	Boiled in water and water used to bathe body to reduce fevers
Glycyrrhiza glabra	Roots	Soaked in water, or powdered and mixed with water, taken for coughs and as an expectorant
Launaea nudicaulis	Leaves	Infusion or paste made with water is applied to cool the forehead and body
Mentha longifolia	Leaves	Infusion with honey, taken for coughs, headaches, and as a carminative
Nigella sativa	Seeds	Eaten for chest congestion and difficult breathing
Phoenix dactylifera	Fruit	Paste made from dates, applied to forehead and eyes to relieve headache
Rhazya stricta	Leaves and stem	Burned and the smoke inhaled to ease chest pains; fresh ground leaves mixed with lemon or boiled in water to bathe body to reduce fever
Senecio asirensis	Leaves	Infusion taken to lower fever
Sisymbrium irio	Seeds	Boiled in water, solution taken as a drink to reduce fever
Solanum nigrum	Whole plant	Boiled in water and water used to bathe forehead to reduce fever
Sonchus oleraceus	Whole plant	Used as a coolant, diuretic, laxative, and general tonic
Trigonella foenum-graecum	Seeds	Boiled with dates and figs in water, taken as a drink for bronchitis and coughs
Vernonia cinerea	Leaves	Decoction taken to reduce fever

Table 6.8 Plants Used as Health Tonics

Species	Plant Part	Method of Use and Medicinal Properties
Caralluma arabica	Stems	Juice added to curdle milk, given to speed convalescence; also eaten raw to improve health
Citrus aurantifolia, C. limetoides	Fruit	Juice or fruit, eaten for general health
Ficus carica	Fruit	Juice or fruit, eaten for general health
Morus nigra	Fruit	Eaten for general health
Punica granatum	Fruit	Juice or fruit, eaten for general health and convalescence, especially when recovering from fever
Rhus somalensis	Berries	Eaten raw or roasted as a general tonic
Sarcostemma viminale	Whole plant	Inner tissue of the stem eaten as a general tonic and to cleanse the digestive system

cultivar variety has its distinct shape, color, taste, and fruiting time. At one time, date cultivation was so extensive in Bahrain that it was known as the land of a million date palms. Several products are made from the fruit, including date syrup (*dibs*)—often called date honey, a product unique to Arabia and the use of which dates back several thousand years. The trunk and leaves provide material for making thatch, screens, baskets, mats, brooms, ropes, and hives for traditional beekeeping.

Date palms are cultivated wherever water is sufficient for irrigation and the temperature is suitable. The plant requires a warm, dry climate for growth and a high temperature with low humidity for fruit-set and ripening. Date palms can grow in relatively saline soils and can withstand irrigation with brackish water. Young plants produce basal suckers used for vegetative propagation. Trees mature in 5 years and can yield up to 200 kg of fruit per tree. Pollination is by hand and in a date-palm grove one or two male trees are planted for approximately 40 female trees. Good-quality date plants are highly valued and can fetch high prices.

6.3.2 Dyes, Cosmetics, and Perfumes

The use of incense and perfumes in the Middle East dates from over 4,000 years ago, when perfumes and incense were used in religious rituals. Incense was also used in healing the sick, in exorcisms, and in festivities and other social events (Ghazanfar 1998).

6.3.2.1 Indigo

Indigo dye has been used in the Middle East as far back as 3000 BC, but the true identification of the source of the dye (whether from *Indigo* species (*nil*), woad, or shells) is not fully known (Balfour-Paul 1997). India is believed to be the oldest center of indigo dyeing in the Old World and a primary supplier of the dye. Cuneiform tablets from Mesopotamia from the seventh century give instructions for dyeing wool with indigo, and the Romans used it for medicinal and cosmetic purposes (Balfour-Paul 1997). During the Islamic era, Egypt and Palestine were the main source of indigo dyeing, and *Indigo tinctoria* was cultivated on a small scale in Arabia until about 50 years ago (Miles 1901; Stone 1985). Among the medicinal uses in the Middle East are its uses as an emetic, to reduce inflammation and fever, and to treat hemorrhoids and scorpion stings. *Indigo tinctoria* is not native to the Middle East but is often found as an escape from cultivation.

6.3.2.2 Frankincense

By the first millennium BC, with the domestication of the camel, overland trade routes became established between southern Arabia and Palestine. Frankincense, a product from southern Arabia and northeastern Africa, along with other items of trade, was traded from southern Arabia, Eritrea, and Somalia to Jordan and Palestine. Frankincense was highly regarded by ancient Romans and was in great demand. The earliest recorded account of the use of frankincense and myrrh from Arabia by the ancient Greeks comes from Herodotus (Groom 1981), suggesting that by 500 BC a well-established trade existed between southern Arabia and Greece. To ascertain the origin of frankincense, Alexander the Great (356–323 BC) sent Anaxicrates to southern Arabia and his account was recorded by Theophrastus (ca. 295 BC):

> The trees of frankincense and myrrh grow partly in the mountains, partly on private estates at the foot of the mountains; wherefore some are under cultivation, others not; the mountains, they say, are lofty, forest-covered and subject to snow, and rivers from them flow down into the plain. The frankincense tree, it is said, is not tall, about five cubits and is much branched...The myrrh tree is said to be still much smaller in stature and more bushy. (quoted in Groom 1981)

The incense trade expanded considerably in the first two centuries AD and Dhofar (southern Oman) became the main region for the production of Arabian frankincense. By the fourth century, with the economic decline of the Roman Empire and the spread of Christianity, the frankincense and myrrh trade declined.

Frankincense is obtained from *Boswellia sacra* (Burseraceae), which grows in the upper plateau on the northern face of the escarpment mountains of Dhofar. Classical and modern accounts of the harvesting and handling of frankincense in southern Arabia were given by Theophrastus (Hort 1916), Pliny, Bent (Groom 1981), and Thomas (1932). Pliny's account suggests that, by the first century, the demand for frankincense had increased so much that two harvests were being extracted each year. In his book on frankincense and myrrh, Groom (1981) gives detailed historical and modern accounts of the distribution, harvest, and trade of frankincense and other incenses from southern Arabia. The account of the harvesting of frankincense given by the explorer Bertram Thomas (1932) more or less describes the method still in use today:

> The tree begins to bear in its third or fourth year. The collectors, women as well as men, come to make slight incisions here and there in the low and stout branches with a special knife. A gum exudes at these points and hardens into large lozenge-shaped tears of resinous substance which is known as frankincense (liban). After ten days the drops are large enough for collection, and the tree will continue to yield from these old incisions opened as necessary at intervals of ten days for a further period of five months. After this the tree dries up and is left to recover, the period varying from six months to two years according to its condition. Collection of the "liban" is made chiefly during the monsoon months. It is stored in the mountain caves until the winter, when it is sent down to the ports for export, for no country craft is put to sea during the gales of the summer south-west monsoon. This delay enables the product to dry well, though normally it is ready for export from ten to twenty days after collection.

6.3.2.3 Rose Water

The fragrance of roses has always held a place of high esteem in Islamic culture. Rose water and rose oil are used in medicinal preparations, to sanctify mosques, on religious occasions, and as one of the most favored perfumes. The oldest historical evidence of rose cultivation comes from the Minoan civilization, from the remains of a fresco from the palace at Knossos, Crete (ca. 1500 BC). This was identified as *Rosa damascena*, the damask rose, which is one of the most important roses for the rose oil distillation industry. The parentage of *R. damascena* is not fully known, but it is believed to be of hybrid origin, and *R. gallica* is one of its most probable parents (Widrlechner 1981).

During the eighth and ninth centuries, Greek texts on medicine, botany, and alchemy were translated and became available to Arab herbalists and physicians. Arab scholars, such as Al Jawbari (thirteenth century) and Al Dimashqi (d. 1327), developed the techniques of distillation, and the production of perfumes and scented oils became a flourishing industry during the Islamic era (Al Hassan and Hill, 1986, and references therein). With the establishment of caravan and pilgrimage routes between Arabia, North Africa, and Asia, the knowledge of distillation techniques became more widely available, and by the thirteenth century Iran became one of the principal centers of rose oil production (Dawood 1967). The use of rose oil and rose water spread widely and was eventually introduced by a Turkish merchant into the Ottoman province of Eastern Roumelia (now part of Bulgaria) around the end of the seventeenth century. This area is now one of the world's largest rose oil producers (Widrlechner 1981).

In addition to the production of rose water, the Islamic perfume industry included the manufacture of preparations such as musk, *ban* (a perfume from *Moringa oleifera*), *ghalia* (a perfume from musk and ambergris), and others.

In addition to frankincense and rose water, myrtle and jasmine are also commonly used. The leaves of myrtle (*Myrtus communis*) and flowers of jasmine (*Jasminum* spp.) are often placed among clothes for their perfume. Other plants used in a similar way include leaves and flowers of *Lavandula* spp. and *Ocimum sanctum*.

6.3.2.4 Henna and Kohl

Since imported cosmetics have become cheap and freely available, the use of plants for adornment has declined. However, a few plants still retain their popularity. One of the most popular that is used throughout the Middle East is *henna* (*Lawsonia inermis*). The dried, powdered leaves are made into a paste with water to which oils, lime juice, and other ingredients may be added to enhance the color and perfume. The paste is then applied and left for some time to give an orange-red dye. Palms and backs of hands and feet are stained in intricate patterns for religious festivities, weddings, and birthdays. Henna is also popularly used as a hair color and for dyeing cloth. Henna plants are cultivated in most homes and villages, and henna powder is imported from India, Iran, and Pakistan.

Oils of certain plants are used as face lotions. In Oman and Yemen, the seed oil of *Moringa peregrina* is used as a skin lotion (Ghazanfar and Rechinger 1996). The powdered leaves of *Ziziphus spina-christi* are used as a shampoo and are believed to leave the hair clean and lustrous. Hair is often lacquered with frankincense. Roots and stems of *Salvadora persica* (Arabic: *rak*, *miswak*) are used widely throughout Arabia as teeth-cleaning sticks.

Among cosmetics, *kohl* is perhaps one of the best known Middle Eastern adornments, widely used by both men and women. The use of black eye and eyebrow paint goes back to the Bronze Age; it has been used across North Africa, the Middle East, and south Asia. Kohl is made from lampblack, though galena (lead sulfide) and stibnite (an antimony compound) have also been used for black, as well as copper compounds for blues and greens (Cartwright-Jones 2005; Hardy et al. 1995). In the Bronze Age, Sumerian women used eye shadow made of finely ground malachite, a green-blue mineral. Malachite was traded from Sinai into Egypt and the Middle East, and black cosmetic remnants found in tombs in Ur contained manganese dioxide, turquoise, and lead (probably galena) (Cartwright-Jones 2005). Pliny and Discorides describe the manufacture of black eye paint by ancient Egyptians; galena was pounded with frankincense and gum and then mixed with goose fat. It was put in dough or cow dung and burned. The burning drove sulfur out of the galena to form lead oxide. This was quenched with milk and then pounded in a mortar with rainwater. This was decanted several times and the finest powder was collected, dried, and divided into tablets, which were powdered for use (Cartwright-Jones 2005).

Facial decoration with black or red dyes is also common in some areas, especially rural ones. In Yemen, a black paste, *khidab*, used for facial decoration is made traditionally with oak galls and minerals such as copper oxide, potash, and sal ammoniac (Schönig 1995). In parts of the Arabian Peninsula, a paste made from sandalwood oil and powdered flowers of *Flemingia* or powdered turmeric, called *wers*, is applied on the face to give a yellow complexion.

6.4 CONCLUSIONS

Man has depended on plants for his livelihood for as long as he has been around. Early civilizations used plants for all purposes, from providing food to building materials and making everyday utilitarian objects to making carts and boats for travel. The recorded uses of plants for medicinal purposes in the Middle East go as far back as the earliest civilization, that of the Sumerians in Mesopotamia (modern-day Iraq), where uses for a few plants were recorded. Herbal and traditional medicine developed over a period of time and became an integral part of modern medicine through the contributions of the early Greeks, Romans, and Muslims.

Today, over 300 plants in the Middle East have been recorded for their medicinal value; however, due to the ready availability of modern medicines, traditional herbal medicine is fast losing popularity. The knowledge of medicinal plants and their uses now lies in isolated pockets, mostly with the older generations or within a few nomadic tribes. Many plants used in traditional medicine have not yet been investigated fully for their potent active ingredients, and not all plants used have been documented for their traditional uses. Documentation of uses of plants (be it for medicine, food, or any other use) is therefore of utter importance before this knowledge is lost forever. As well, it is important to protect and conserve plants before they are lost to development and before there is a chance to investigate and learn about their uses for the betterment of all.

REFERENCES

Abu-Rabia, A. 2005. Unirary diseases and ethnobotany among pastoral nomads in the Middle East. *Journal of Ethnobiology and Ethnomedicine* 1:4.

Al-Douri, N. A. 2000. A survey of medicinal plants and their traditional uses in Iraq. *Pharmaceutical Biology* 38 (1): 74–79.

Al Hassan, A. Y., and D. R. Hill. 1986. *Islamic technology, an illustrated history.* Cambridge, England: Cambridge University Press.

Balfour-Paul, J. 1997. *Indigo in the Arab world.* London: Curzon Press.

Batanouny, K. H., S. A. Tabl, H. Shabana, and F. Soliman. 1999. Wild medicinal plants in Egypt: An inventory to support conservation and sustainable use. Academy of Scientific Research and Technology (Egypt) and IUCN. Unpublished report.

Berthoud, T., and S. Cleuziou. 1983. Farming communities of the Oman Peninsula and the copper of Makkan. *Journal of Oman Studies* 6:239–246.

Cartwright-Jones, C. 2005. *Introduction to Harquus: Part 2: Kohl as traditional women's adornment in North Africa and the Middle East.* Tap Dancing Lizard Publications (www.harquus.com).

Chadwick, J., and W. N. Mann. 1950. *The medical works of Hippocrates: A new translation from the original Greek made especially for English readers.* Oxford, England: Blackwell Scientific Publications.

Cleuziou, S., and L. Constantini. 1980. Premiers éléments sur l'agriculture protohistorique de l'Arabie Orientale. *Paléorient* 6:245–251.

Colfer, C. 1990. Indigenous knowledge of midwives in Oman's interior: A preliminary account. Paper presented at the III International Congress on Traditional Asian Medicine, Bombay, January 4–7, 1990.

Dawood, N. J., ed. 1967. *The Muqaddimah: Ibn Khaldûn, an introduction to history,* translated from the Arabic by Franz Rosenthal. London: Routledge & Kegan Paul, in association with Secker & Warburg, London.

Duke, J. A. 2007. *Duke's medicinal plants of the Bible.* Boca Raton, FL: CRC Press.

Foster, G. M., and B. G. Anderson. 1987. *Medical anthropology.* New York: John Wiley & Sons.

Ghazanfar, S. A. 1994. *Handbook of Arabian medicinal plants.* Boca Raton, FL: CRC Press.

———. 1995. Wasm: A traditional method of healing by cauterization. *Journal of Ethnopharmacology* 47:125–128.

———. 1996. Traditional health plants of the Arabian Peninsula. Paper presented at the joint meeting of the Society for Economic Botany and International Society for Ethnopharmacology, Plants for Food and Medicine, London, July 1–7, 1996.

———. 1998. Plants of economic importance. In *Vegetation of the Arabian Peninsula,* ed. S. A. Ghazanfar and M. Fisher, 241–264. Dordrecht, the Netherlands: Kluwer Academic Press.

Ghazanfar, S. A., and B. Rechinger. 1996. Two multipurpose seed oils from Oman. Paper presented at the joint meeting of the Society for Economic Botany and International Society for Ethnopharmacology, Plants for Food and Medicine. London, July 1–7 1996.

Ghazanfar, S. A., and A. A. Al-Sabahi. 1993. Medicinal plants of northern and central Oman. *Economic Botany* 47:89–98.

Groom, N. 1981. *Frankincense and myrrh. A study of the Arabian incense trade.* London: Longman and Beirut: Libraire du Liban.

Hardy, A. D., H. H. Sutherland, R. Vaishnav, and M. A. Worthing. 1995. A report on the composition of mercurials used in traditional medicines in Oman. *Journal of Ethnopharmacology* 49:17–22.

Hort, A. 1916. *Theophastus: "Enquiry into plants" and "concerning odors"* (in one volume). Cambridge, MA: Loeb Classical Library, Heinemann and Harvard University Press.

Marshall, S. J., S. A. Ghazanfar, G. C. Kirby, and J. D. Phillipson. 1995. *In vitro* antimalarial activity of some Arabian medicinal plants. *Annals of Tropical Medicine and Parasitology* 89 (2):199.

Miles, S. B. 1901. On the border of the great desert: A journey in Oman. *Geographical Journal* 1901:405–425.

Miller, A. G., and M. Morris. 1988. *Plants of Dhofar, the southern region of Oman: Traditional, economic and medicinal uses.* The Office of the Adviser for Conservation of the Environment, Diwan of Royal Court, Oman.

Morris, E. T. 1984. *Fragrance: The story of perfume from Cleopatra to Chanel.* New York: Charles Scribner's Sons.

Musallam, B. F. 1983. *Sex and society in Islam: Birth control before the nineteenth century.* (Cambridge Studies in Islamic Civilization). Cambridge, England: Cambridge University Press.

Schönig, H. 1995. Traditional cosmetics. Paper presented at the seminar for Arabian Studies 1995, Cambridge, England.

Schopen, A. 1983. *Traditionelle Heilmittel in Jemen.* Wiesbaden, Germany: Franz Steiner Verlag.

Simpson, B. B., and M. Conner-Ogorzaly. 1986. *Economic botany: Plants in our world.* New York: McGraw–Hill.

Stone, F. 1985. *Studies on the Tihama. The report of the Tihama expedition, 1982, and related papers.* Harlow, England: Longman.

Thomas, B. 1932. *Arabia felix.* London: Cape.

Townsend, C., and E. Guest. 1966–1985. *Flora of Iraq,* vols. 1–9. Royal Botanic Gardens, Kew, and Baghdad University, Baghdad.

Widrlechner, M. P. 1981. History and utilization of *Rosa damascena. Economic Botany* 35:42–58.

CHAPTER 7

An Overview of the Medicinal Plants of Turkey

**Munir Ozturk, Salih Gucel, Ernaz Altundag, Tuba Mert,
Cigdem Gork, Guven Gork, and Eren Akcicek**

CONTENTS

7.1 INTRODUCTION

Plant species have different uses in different countries as well as different areas of the same country (Farnsworth and Soejarto 1991; Plotkin 2000; Hamilton 2004; Halberstein 2005). The relationship between humans and plants has existed since the existence of human beings, and the earliest documented record dates from the Paleolithic age (50,000 BC); plants were found in the grave of a Neanderthal man in the southern part of Hakkari (far southeast edge of Turkey) (Baytop 1984, 1999).

A number of plant remedies have been described on the clay tablets that have survived from Mesopotamian civilizations such as Sumerians, Assyrians and Akkadians, and Hittites. In fact, the study of medicinal botany began when plants were classified according to their uses, such as pain- and illness-healing plants and poisonous ones (Ozturk and Ozcelik 1991; Mert et al. 2008; Ozturk et al. 2008). It was during this period that the knowledge and manipulation of plant properties became associated with individuals.

Nature has bestowed upon plants active molecules with natural affinities with the human body. As such, it is not surprising that we find remedies for several diseases by using plants. People nowadays are using the powerful curative properties of the plant world to stimulate the functioning of the organism and improve their physical as well as mental well-being. The role of medicinal plants in the maintenance of health and treatment of diseases as therapeutic alternatives throughout the world is progressing at a fast speed (WHO 2002). Nature cures and herbal medicinal products are indispensable. These have played a dominating role in the development of human civilization. The search for lands of spices led to the discovery of continents but at the same time colonial invasions of other areas. Presently, more than 20,000 species of plants from our rich global plant diversity are used as herbal drugs. Out of these, more than 120 compounds from 90 plant species are available as prescription drugs.

Developing countries are slowly realizing that they do not have the means to provide comprehensive health care to their masses and they have started to become more interested in traditional medicines. There is now more acceptance for phytotherapy, and demand for plant-based medicines for age-related disease (autoimmune, degenerative diseases), preventive medicines (antioxidants, edible vaccines, nutritional therapy, etc.), and plant-based anti-infectious agents is becoming more important. Medicinal plants will continue to play this role as long as modern medicine continues to be unable to meet the health care needs of people of the developing world (Melo, Filho, and Guerra 2005).

Turkey is one of the industrializing countries and one among the important gene centers of plant diversity in the world. Among the countries in southwest Asia and the Mediterranean basin, as well as the whole of Europe, the richest flora has been reported for the Anatolian peninsula. The number of flowering plant taxa distributed in the country is estimated to be around 10,000, which is very near to the number recorded from the whole of Europe (Davis 1965–1985; Davis, Mill, and Tan 1988; Guner et al. 2000). These taxa are distributed in different phytogeographical regions and include nearly 3,300 endemics, which are mostly found in the Irano-Turanian region (Figures 7.1 and 7.2) (Ozgokce and Ozcelik 2004; Simsek et al. 2004).

The country includes a large part of the "fertile crescent" covering the valleys of the Tigris and Euphrates rivers (Figure 7.3), the southern slopes of the Taurus Mountains, and the eastern shores of the Mediterranean Sea. It was largely within these lands that many ancient civilizations flourished and domestication of many food and medicinal plants started (Baytop 1984). The plant wealth in Turkey has been evaluated as a source of medicines from ancient times. Most of the 600 plants documented in the book of Dioscorides, *Materia Medica*, originated from the Anatolian peninsula (Baytop 1984). A recent survey of traditional and folk medicine in Turkey has revealed that most of these plants are still in use by the local inhabitants (Yesilada and Sezik 2003). Therefore, *Materia Medica* may be assumed to be the oldest comprehensive document on Anatolian folk medicine. For example, *Ecballium elaterium* fruit juice is used widely even now to treat sinusitis in Turkey; this had been reported only by Dioscorides. Its activity has also been reported in *Materia Medica* for some other illnesses and these usages are also practiced in the folk medicine of Turkey (Baytop 1984).

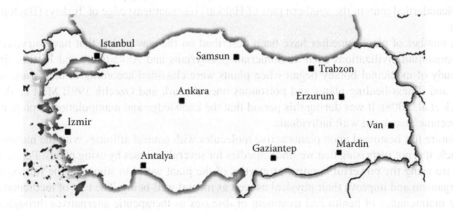

Figure 7.1a Geographical map of Turkey showing different states.

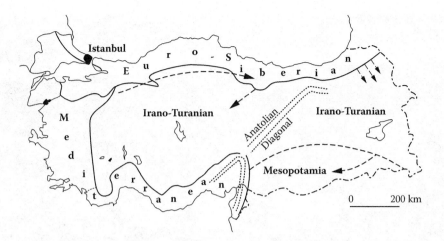

Figure 7.1b Phytogeographical divisions of Turkey.

However, abiotic pressures are posing a great threat for the traditional heritages in the country and a wealth of information is slowly vanishing. Extensive research has been conducted by several workers on the medicinal plants of Turkey during the last few decades (Baytop 1984, 1999; Ozturk and Ozcelik 1991; Sezik et al. 1991; Sezik, Zor, and Yesilada 1992; Yesilada et al. 1993; Tabata et al. 1994; Fujita et al. 1995; Surmeli et al. 2000; Orhan et al. 2003). The present chapter synthesizes the use of plants in folk medicine. An attempt has been made here to bring together the scattered information on the medicinal plants from the published records of Turkey for its availability to the researchers in the fields of pharmacy, ethnobotany, and phytotherapy.

Figure 7.2 (a, b, c) A general view of the traditional herbal drug dealers.

Figure 7.2 (a, b, c)   Continued

Figure 7.3 *Tilia europaea.*

7.2 PRESENT SITUATION OF MEDICINAL PLANTS

The number of plant species used in Turkey as folk remedies has recently been figured to lie around 1,500, although the book of Baytop (1999) previously estimated them at around 500. Although scarcely practiced in Turkey, the traditional Greco-Arabic (Unani) medicine is still being practiced widely in the south and southeast regions of the country. Some principles reflect important components of the ancient Mesopotamian traditions. Akhtars or attars are still popular in procuring such remedies (Baser, Honda, and Miki 1986; Bingol 1995). The rural inhabitants are applying different plant remedies around their villages as well as specific remedies prepared from animals or natural materials (Sezik and Tumen 1984; Sezik and Basaran 1989).

Presently, the trade of medicinal plants mainly takes place in three groups:

- Traditional shops, which are old, with stalls and shelves arranged simply, unpackaged herbs stored in large plastic bags or in bundles; the dealers are not professional healers
- Stalls in open-air markets, which run in different places once a week, with bunches of herbs; stall keepers are professional plant collectors with a low educational level and an empirical experience in herbal medicine based on tradition
- Modern shops scattered in the city but carefully decorated; the products sold are industrially manufactured, attractively packaged, and displayed on shelves; the dealers are traders with knowledge about herbal drugs mostly derived from modern textbooks and the information provided by the company suppliers

7.3 HERBAL TEAS

In Turkey, tea commonly known as *çay* is a form of black tea produced from the Black Sea region. Wherever people go in Turkey, tea or coffee will be offered as a sign of friendship and hospitality, at homes, bazaars, and restaurants, before or after a meal. The Turks evolved their own way of making and drinking the black tea that became a way of life for Turkish culture (Baser et al. 1986; Sezik 1990; Baser 1996; Atoui et al. 2005). It is prepared by using two stacked kettles called *caydanlik*. Turkish tea is full flavored and too strong to be served in large cups; thus, it is always offered in small tulip-shaped glasses to enjoy it hot in addition to showing its color. The glasses are usually held by the rim in order to save the drinker's fingertips from being burned. As a matter of fact, tea replaces both alcohol and coffee as the social beverage. Tea replaced coffee after the retreat of the Turks from Yemen.

Recently, in addition to black tea, other teas of plant origin—in particular, sage tea, linden flower tea, rose hip tea, and apple tea—have become popular in the country and are a buzzword of the day (Akcicek and Ozturk 1995). Herbal tea, also known as a tisane or ptisan, has been used for thousands of years, as is evident from the documents dating back to the ancient Egyptians, who discuss the enjoyment and uses of herbal tea. The word "tisane" originated from the Greek word *ptisane* (a drink made from pearl barley). It is any herbal infusion other than from the leaves of the tea bush (*Camellia sinensis*). Herbal teas are often consumed for their physical or medicinal effects, especially as stimulants, relaxants, or sedatives. Generally, the leaves, flowers, fruits, fruit peelings, dried fruits, dried flowers, roots, or herb are added to the boiling water (Ozturk and Ozcelik 1991; Zeybek 2003; Ozcan 2005). Almost all contain essential oils. Although the majority of the herbal teas are safe for regular consumption, some can have toxic or allergenic effects; in particular, different effects can be observed in different people. The problem is consumption of misidentified herbs as tea.

There are 53 natural herbal teas used in Turkey. The most important ones are

dried flowers of lime tree (*Tilia*, locally known as Ihlamur) (Figure 7.3)

sage (*Salvia* spp., locally known as adacay; Ulubelen 1964; Bayrak and Akgul 1987; Solmaz 1993a, 1993b; Haznedaroglu, Karabay, and Zeybek 2001; Bozan, Karabay, and Zeybek 2002; Sagdic 2003; Tepe, Donmez, and Unlu 2004; Tepe, Daferera, and Sokmen 2005; Tepe et al. 2006; Nickavar, Kamalinejad, and Hamidreza 2007; Tepe, Eminagaoglu, and Akpulat 2007; Akkol et al. 2008)

mint, especially peppermint (*Mentha* sps., locally known as nane)

rosehip (*Rosa canina*, locally known as kuşburnu)

thyme (*Thymus* spp., locally known as kekik)

oregano (*Origanum* spp, locally known as Izmir kekigi)

apple (*Malus sylvestris*, locally known as elma)

lavender (*Lavandula stoechas*, locally known as karabas otu

melisa (*Melissa officinalis*, locally known as ogul otu)

camomile (*Matricaria chamomilla*)

anise (*Pimpinella anisum*, locally known as anason)

licorice root (*Glycyrrhiza glabra*, locally known as meyan)

the species of *Sideritis, Chrysanthemum, Stachys, Phlomis, Rubus*, and *Citrus* tea, including bergamot, lemon, and orange peel (Ezer and Sezik 1988; Sezik 1990; Ezer et al. 1995; Isik et al. 1995; Sayar et al. 1995; Baser 1996; Surmeli et al. 2000; Dogan et al. 2005; Kukic, Petrovic, and Niketic 2006; Sagdic, Aksoy, and Ozkan 2006)

A special drink locally known as "salep" is prepared from the tubers of orchids (*Orchis* and *Ophrys*) during the winter season as a protection from a bad cold (Figure 7.4).

Figure 7.4 *Orchis collina.*

7.4 AROMATIC AND ESSENTIAL-OIL-BEARING PLANTS

A large number of plants from over 300 families are rich in aromatics (Miliauskas, Venskutonis, and van Beek 2004). The number of plants used industrially for the production of aromatic compounds is around 40. The family Lamiaceae (Labiatae), commonly known as the mint family, is one of the few plant families with various aromatic plants rich in essential oils—in particular, volatile oils, iridoids, flavonoids, and diterpenoids (Baytop 1984; Triantaphyllou 2001; Stefanini et al. 2006). It includes nearly 200 genera and more than 3,000 taxa; a large number of these are spices, medicinals, and odorous plants. Several pharmacological activities have been illustrated for these plants (Richardson 1992). Members of this family have been used in Turkish traditional and folk medicine since ancient times (Dogan, Bayrak, and Akgul 1985; Basar et al. 1986; Baytop 1999; Dogan et al. 2005; Everest and Ozturk 2005).

Lavandula officinalis, Lavandula angustifolia ssp. *angustifolia, Melissa officinalis, Mentha pulegium, M. longifolia, M. aquatica, M. sauveolens, M. viridis, M. spicata, Origanum dictamnus, O. majorana, O. vulgare* ssp. *hirtum, O. vulgare* ssp. *vulgare, Rosmarinus officinalis, Salvia fruticosa, Satureja thymbra, Sideritis* spp., *Teucrium chamaedrys, T. polium, Thymus* spp., *Glycyrrhiza glabra, Paliurus spina-christi,* and *Crataegus monogyna* are used as culinary herbs also in the toothpaste and chewing gum industries (Mimica-Dukic et al. 2003; Sagdic 2003). These species are also mixed with green tea to make mint tea.

The following plants are used for the following purposes:

Coridothymus spicatus for hyperglycemia, hypertension, and stomach ailments
Lavandula stoechas against cold, rheumatismal pains, and urinary diseases, and as a diuretic, pain
  killer, wound healer, antiseptic, and expectorant

Melissa officinalis as a sedative, in gastric disorders, as a degasifier, sudorific, and antiseptic

Mentha species for nervous gastric pains and nausea, as a degasifier, an antiseptic, and a spice

Origanum majorana for abdominal pain and gastric troubles; as a tranquilizer, diuretic, degasifier, and sudorific; for constipation; and as spice

O. riganlum onites for diabetes, abdominal pain, stomachache, and colds and as an immunostimulant (children)

Rosmarinus officinalis for constipation, as a stimulant of the digestive system, for bilious conditions, as a diuretic, and in wounds (external)

Salvia fruticosa as an antiflatulent, for vigor, as a stimulant and antiseptic (ear–throat) (Yildirim et al. 2000; Tepe et al. 2006)

Sideritis brevidens as a carminative and degasifier

Teucrium polium against eczema, hemorrhoids, and diabetes

Thymus species for gastric disorders, as a sedative and an antiseptic, for worm removal, for speeding blood circulation, for bilious conditions, against psoriasis (with honey), and as a spice tea (Sezik and Saracoglu 1988; Altanlar et al. 2006)

Tymbra spicata for colds

Lamiaceae has a great diversity and distribution in Turkey, being represented by 45 genera and 567 taxa. From these species, 256 taxa (45.3%) are endemic to Turkey (Davis 1965–1985; Hedge 1986; Davis et al. 1988; Guner et al. 2000; Triantaphyllou, Blekas, and Boskou 2001; Dorman et al. 2003; Kosar, Dorman, and Hiltunen 2005), which includes the endemic monotypic genus *Dorystoechas hastata* occurring in southern parts of Turkey. The most important genera are *Coridothymus, Lamium, Lavandula, Marrubium, Melissa, Mentha, Origanum, Phlomis, Rosmarinus, Salvia, Saturaeja, Sideritis, Stachys, Teucrium, Thymus,* and *Thymbra.* The genera of *Nepeta* (76 species), *Salvia* (88 species), *Stachys* (81 species), *Sideritis* (45 species), *Thymus* (39 species), *Phlomis* (34 species), *Nepeta* (34 species), *Lamium* (30 species), *Teucrium* (29 species), *Origanum* (24 species), *Marrubium* (19 species), *Scutellaria* (16 species), *Micromeria* (14 species), *Satureja* (14 species), *Ajuga* (13 species), and *Ballota* (11 species) are the largest genera in Turkey (Ozgen et al. 2006; Nakiboglu et al. 2007). A comparison of the total number of species, endemics percentages, and number of medicinal species is given in Figure 7.5.

7.5 ENDEMICS OF SOME IMPORTANT GENERA

Dorystoechas *hastata*

Origanum *boisseri, O. saccatum, O. solymicum, O. hypericifolium, O. simpyleum, O. haussknechtii, O. brevidens., O. leptocladum, O. amanum, O. bilgeri, O. micranthum, O. minutiflorum, O. husnubaseri, O. symes, O. munzurense*

Thymus *cilicicus, T. revolutus, T. pulvinatus, T. convolutus, T. argaeus, T. brachychilus, T. cappadocicus* var. *pruinosus* var. *T. cappadocicus* var. *globifer, T. haussknechtii, T. pectinatus* var. *pectinatus* var. *pallasicus, T. canoviridis, T. spathulifolius, T. cariensis, T. samius, T. zygioides* var. *lycaonicus, T. aznavouri, T. fedtschenkoi* var. *handelii, T. spipyleus* var. *sipyleus* var. *davisianus, T. bornmuelleri, T. praecox* var. *laniger, T. longicaulis* var. *antalyanus*

Salvia *divaricata, S. aucheri, S. tigrina, S. recognita, S. pilifera, S. reeseana, S. cedronella, S. adenophylla, S. rosifolia, S. huberi, S. wiedemannii, S. pisidica, S. freyniana, S. potentillifolia, S. albimaculata, S. tchihatcheffii, S. heldreichiana, S. caespitosa, S. haussknechtii, S. ballsiana, S. quezelii, S. cadmica, S. smyrnaea, S. blepharochlaena, S. euphratica, S. kronenburgii, S. sericeotomentosa, S. cryptantha, S. hypargeia, S. eriophora, S. chrysophylla, S. chionantha, S. longipedicellata, S. yosgadensis, S. modesta, S. tobeyi, S. odontochlamys, S. cyanescens, S. vermifolia, S. halophila, S. adenocaulon, S. dichroantha, S. aytachii, S. nydeggeri*

Sideritis *sipylea, S. hololeuca, S. phlomoides, S. erythrantha* var. *erythrantha* var. *cedretorum, S. brevidens, S. strica, S. vulcanica, S. condensata, S. tmolea, S. congesta, S. cilicica, S. niveotomentosa,*

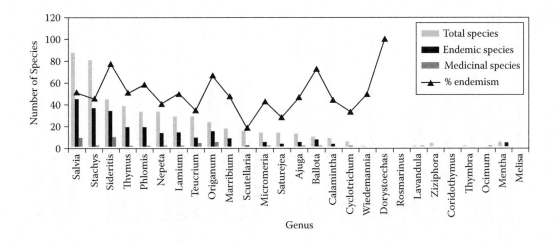

Figure 7.5   Total species, endemics, and percentage and number of medicinal species from Lamiaceae.

S. arguta, S. lycia, S. leptoclada, S. brevibracteata, S. albiflora, S. rubriflora, S. argyrea, S. bilgerana, S. hispida, S. dichotoma, S. trojana, S. phrygia, S. amasiaca, S. galatica, S. armeniaca, S. germanicopolitana subsp. *germanicopolitana* subsp. *viridis, S. libanotica* subsp. *linearis, S. serratifolia, S. pisidica, S. akmanii, S. gulendamiae, S. caesarea, S. vuralii, S. huber-morathii*

Family Lamiaceae is one of the major sources of culinary and medicinal plants all over the world (Marinova and Yanishlieva 1997). The rich spicy quality of basil, rosemary, lavender, oregano, thyme, sage, and mint makes the plants of this family useful in cooking. People in various cultures in different parts of the world in one way or another use these aromatic plants to season their food, perfume their bodies, and treat ailments (Dogan et al. 1985; Ozturk and Ozcelik 1991; Ozturk, Pirdal, and Uysal 1992; Gue´don and Pasquier 1994; Isik et al. 1995; Sayar et al. 1995; Atta and Alkofahi 1998; Basar et al. 1999; Barbour et al. 2004; Ghorbani and Motamed 2004; Capecka, Mareczek, and Leja 2005; Nickavar, Alinaghi, and Kamalinejad 2008). Species of *Mentha, Thymus, Lavandula, Ocimum, Origanum, Melissa,* and *Satureja* are used as culinary and flavoring plants in addition to their wider use as herbal teas. These species are available in local markets or in traditional medicinal plant stores.

Many species of the family are used in traditional and modern medicine. Endemics are especially used locally by indigenous people in different parts of Turkey (Table 7.1) (Sezik and Basaran 1989; Yucel and Ozturk 1998; Baser 2001; Elmastas et al. 2005; Gulluce et al. 2007). According to Baytop (1984), 70 species of Lamiaceae belonging to 23 genera are used for medicinal purposes in Turkey. If we add to this the thyme species, this number goes up to 110. However, based on our findings, more than 78 taxa have medicinal uses. Considering the total number of the mint family in Turkey, 12.4% of taxa are of medicinal value. They are used in the treatment of 30 different ailments (Tables 7.2–7.5).

7.6 COMMENTS ON TURKISH MINT SPECIES

Genus *Mentha* is represented by 15 taxa in Turkey; the most common are *M. longifolia, M. suaveolens, M. pulegium,* and *M. aquatica* (Ozturk, Secmen, and Pirdal 1986; Richardson 1992; Voirin and Bayet 1992; Voirin, Saunois, and Bayet 1994; Miura, Kikuzaki, and Nakatani 2002; Parejo et al. 2002; Damien Dorman et al. 2003). All are perennial herbs, flowering in late spring and early

Table 7.1 Most Important Plants Used as Medicine in Turkey

Family	Botanical Name	Local Name	Plant Part Used	Treatment
Amaryllidaceae	Galanthus nivalis L.	Kardelen, aktas	Bulb	Nervine, emmenagogue
Anacardiaceae	Pistacia lentiscus L. var. latifolius Coss.	Sakiz, damla sakizi, mezeke sakizi, mesteki	Gum	Antihalitosis, mucolytic, strengthen gums
Anacardiaceae	Pistacia vera L.	Fistik sakizi	Gum	Antihemorrhoidal, stomachic
Anacardiaceae	Pistacia terebinthus L.	Menengic sakizi	Gum	Antiseptic
Anacardiaceae	Cotinus coggyria Scop.	Boyaci somagi, sariboya, sarican	Leaf	Antiseptic, anti-inflammatory
Anacardiaceae	Rhus coriaria L.	Sumak, somak	Leaf	Antiseptic
Apocynaceae	Nerium oleander L.	Zakkum	Leaf	Diuretic, cardiac
Araceae	Acorus calamus L.	Egir, azakegeri, hazambel, hazambel	Root	Antispasmodic, carminative, diaphoretic
Araceae	Arum italicum Miller	Yilanyastigi	Tuber	Laxative, vesicant
Araliaceae	Hedera helix L.	Duvarsarmasigi	Leaf	Vesicant
Aristolochiaceae	Aristolochia sp.	Zeravent, lohusaotu	Root	Analgesic, laxative, antidermatosic
Aristolochiaceae	Asarum europaeum L.	Azaron, avsarotu, cetukotu, kediotu, meyhaneciotu	Root	Emetic, mucolytic, diuretic, anti-inflammatory, laxative, emmenagogue
Berberidaceae	Leontice leontopetalum L.	Kirkbas yavrusu, arslanayagi, arslankulagi, patlangac	Bulb	Emmenagogue
Boraginaceae	Alkanna orientalis (L.) Boiss.	Kamburuyan, sari havaciva, kurbagaotu	Root	Emmenagogue
Boraginaceae	Alkanna tinctoria (L.) Tausch	Havaciva	Root	Antidiarrheal, antidermatosic
Cannabaceae	Humulus lupulus (L.) Vill.	Serbetciotu	Flower	Stomachic, diuretic
Capparaceae	Capparis ovata Desf.	Kebere	Fruit	Diuretic, antidiarrheal, nervine
Caprifoliaceae	Viburnum opulus L.	Gilaburu	Fruit	Diuretic, sedative, laxative
Caryophyllaceae	Dianthus barbatus L.	Husnuyusuf	Flower	Diaphoretic, diuretic, sedative, cardiac
Caryophyllaceae	Gypsophila arrostii Guss. var. nebulosa (Boiss. et Heldr.) Bark.	Coven	Root	Diuretic, mucolytic
Caryophyllaceae	Silene vulgaris (Moench) Garcke	Givisganotu, ibis-gibis	Whole plant	Excretory or genitourinary system diseases
Caryophyllaceae	Stellaria media (L.) Vill.	Kusotu, sercedili, serceotu	Whole plant	Diuretic, mucolytic
Compositae	Achillea millefolium L.	Civanpercemi, akbasli, barsamaotu, beyaz civanpercemi, binbiryaprakotu, kandilcicegi, marsamaotu	Whole plant	Antidermatosic, antihemorrhoidal
Compositae	Artemisia absinthium L.	Pelinotu	Whole plant	Diuretic, stomachic, nervine, anthelmintic
Compositae	Carthamus lanatus L.	Yunlu aspir	Flower	Diaphoretic, anthelmintic, emmenagogue
Compositae	Centaurea cyanus L.	Peygambercicegi	Flower	Antidiarrheal, nervine, stomachic

Family	Scientific name	Turkish name	Part used	Uses
Compositae	Cichorium intybus L.	Hindiba	Leaf	Diuretic, laxative, diaphoretic, nervine, stomachic
Compositae	Cnicus benedictus L.	Mubarekdikeni	Whole plant	Diuretic, appetizer, anti-inflammatory, hypotensive, sedative
Compositae	Cynara scolymus L.	Enginar	Leaf	Stomachic, diuretic
Compositae	Gundelia tournefortii L.	Kenger, enger, kengel, kengiotu, kengir	Gum	Strengthen gums, stomachic
Compositae	Helianthus tuberosus L.	Yerelmasi	Tuber	Galactogogue, diuretic, aphrodisiac
Compositae	Helichrysum sp.	Olmezcicek, altinotu, altincicek, herdemtaze	Flower	Diuretic, cholagogue, lithontriptic
Compositae	Inula helenium L.	Andizotu koku	Root	Diuretic, antitussive, anthelmintic, nervine
Compositae	Inula viscosa (L.) nokta yok Aiton.	Yapiskan andiz	Leaf	Antidermatosic
Compositae	Lactuca sativa L.	Marul	Whole plant	Analgesic, galactogogue, diuretic, laxative
Compositae	Matricaria chamomilla L.	Papatya	Flower	Diuretic, carminative, sedative, stomachic, antidermatosic
Compositae	Scolymus hispanicus L.	Altindikeni, saridiken	Whole plant	Diuretic, lithontriptic
Compositae	Scorzonera latifolia (Fisch. et Mey.) DC.	Benis, cingenesakizi, kandilsakizi, karasakiz, markosakizi, selsepetsakizi, yerlemesakizi	Root	Analgesic, anthelmintic
Compositae	Silybum marianum (L.) Geartner	Devedikeni, akkiz, devekengeli, kengel, kibbun, meryem ana dikeni, sutlukengel, sevkulmeryem	Whole plant	Anti-inflammatory, appetizer, sedative, analgesic
Compositae	Tanacetum vulgare L.	Solucanotu	Flower	Nervine, stomachic, anthelmintic
Compositae	Taraxacum officinale Weber	Radika, arslandisi	Root	Diuretic, laxative
Compositae	Tragopogon porrifolius L.	Sari iskorcina, salsifi, tekesakali	Root	Analgesic, anthelmintic
Compositae	Tussilago farfara L.	Oksurukotu	Leaf	Antitussive
Compositae	Xanthium spinosum L.	Pitrak	Leaf	Diuretic, sedative, diaphoretic
Convolvulaceae	Convolvulus arvensis L.	Tarla sarmasigi, kuzu sarmasigi	Root	Laxative
Convolvulaceae	Convolvulus scammonia L.	Mahmude	Root	Laxative
Cornaceae	Cornus mas L.	Kizilcik	Fruit	Antidiarrheal
Corylaceae	Corylus maxima Miller	Findik	Leaf	Diuretic
Cruciferae	Brassica nigra (L.) Koch.	Hardal	Seed	Stomachic, analgesic
Cruciferae	Eruca sativa Miller	Roka	Leaf	Stimulant, nervine, antitussive
Cruciferae	Isatis tinctoria L.	Civiotu	Whole plant	Antidiarrheal, antidermatosic

(continued)

Table 7.1 Most Important Plants Used as Medicine in Turkey (Continued)

Family	Botanical Name	Local Name	Plant Part Used	Treatment
Cruciferae	*Nasturtium officinale* R. Br.	Suteresi	Whole plant	Nervine, nutritive, diuretic, stomachic
Cucurbitaceae	*Citrullus colocynthis* (L.) Schrader	Acikarpuz	Fruit	Diuretic, antidiarrheal
Cucurbitaceae	*Cucumis sativus* L.	Hiyar	Seed	Antihelminthic
Cucurbitaceae	*Ecballium elaterium* (L.) A. Richard	Esek hiyari, acidulek, aciduvelek, acikavun, cirtlak, cirtatan, hiyarcik, kargaduvelegi, seytankelegi, yabanihiyar	Fruit	Laxative, diuretic
Cucurbitaceae	*Lagenaria vulgaris* Ser.	Su kabagi, kantar kabagi, testi kabagi	Seed	Diuretic, laxative, anti-inflammatory
Cupressaceae	*Juniperus communis* L.	Ardic kozalagi	Cone	Diuretic, diaphoretic, antiseptic
Dioscoraceae	*Tamus communis* L.	Dovulmusavratotu, aciot, karaasma, sincan	Root	Diuretic, laxative, emetic
Eleagnaceae	*Eleagnus angustifolia* L.	Igde	Leaf, flower	Diuretic, anti-inflammatory
Ephedraceae	*Ephedra campylopoda* C.A. Mayer	Denizuzumu	Whole plant	Diaphoretic, analgesic
Ericaceae	*Arbutus andrachne* L.	Sandal	Leaf	Antidiarrheal, antiseptic
Ericaceae	*Arbutus unedo* L.	Kocayemis, dagcilegi	Leaf, fruit	Antidiarrheal, antiseptic, diuretic
Ericaceae	*Rhododendron ponticum* L.	Komar	Leaf	Analgesic, diuretic
Ericaceae	*Vaccinium vitis idaea* L.	Kirmizi meyvali ayi uzumu	Leaf	Diuretic, antiseptic
Fagaceae	*Castanea sativa* Miller	Kestane	Leaf, bark	Hypotensive, antidiarrheal
Gentianaceae	*Centaurium erythaea* Rafn.	Kantaron, kucuk kantaron, kantarion, kirmizi kantaron	Flower	Stomachic, digestive
Gentianaceae	*Gentiana lutea* L.	Centiyane	Root	Stomachic, antidermatosic
Gentianaceae	*Gentiana olivieri* Griseb.	Afat	Root	Stomachic, anti-inflammatory
Poaceae	*Arundo donax* L.	Kamis, masura kamisi	Root	Diuretic, diaphoretic, blood purifier
Poaceae	*Hordeum distichon* L.	ikisirali arpa	Seed	Diuretic, nervine
Poaceae	*Secale cereale* L.	Cavdar	Seed	Laxative
Guttiferae	*Hypericum perforatum* L.	Binbirdelikotu	Whole plant	Antiseptic, antidiarrheal, sedative, antihelminthic
Hamamelidaceae	*Liquidambar orientalis* Miller	Gunluk	Oil	Stomachic
Hippocastanaceae	*Aesculus hippocastanum* L.	Atkestanesi	Bark	Antidiarrheal, antifibrinolytic, anti-inflammatory, antihemorrhoidal
Iridaceae	*Crocus sativus* L.	Safran	Flower, stigma	Stimulant, stomachic, emmenagogue
Iridaceae	*Iris pseudacorus* L.	Bataklik suseni, kazip egir, sari susen, yalanci egir	Seed	Carminative, antidiarrheal, stomachic

Family	Species	Turkish name	Part	Uses
Juglandaceae	Juglans regia L.	Ceviz	Leaf	Stomachic, antidiarrheal, antidiabetes, antiseptic
Lauraceae	Laurus nobilis L.	Defne	Leaf	Diaphoretic, antiseptic
Fabaceae	Anagyris foetida L.	Zivircik	Seed	Laxative, anthelminthic, emetic
Fabaceae	Astragalus aureus Willd.	Kitre zamki	Gum	Vesicant
Fabaceae	Ceratonia siliqua L.	Keciboynuzu, harnup, harup	Fruit	Diuretic, laxative
Fabaceae	Genista lydia Boiss.	Katirtirnagi	Flower	Diuretic, laxative, diaphoretic
Fabaceae	Glycyrrhiza glabra L.	Meyan	Root	Diuretic, mucolytic
Fabaceae	Lupinus albus L. subsp. albus	Termiye, acibakla, delicebakla, lupen	Seed	Diuretic, anthelmintic, nervine, antidiabetes
Fabaceae	Lathyrus sativus L.	Mudurmuk, burcak, kulur	Seed	Sedative, diuretic, nervine, aphrodisiac
Fabaceae	Ononis spinosa L.	Kayiskiran, kayikcicegi, yandak, yantak	Root	Diuretic, lithontriptic
Fabaceae	Robinia pseudoacacia L.	Akasya	Flower	Sedative, antidiarrheal
Fabaceae	Spartium junceum L.	Katirtirnagi, adi katirtirnagi	Flower	Diuretic, anesthetic
Fabaceae	Trigonella foenum-graecum L.	Boyotu, buyotu, cemenotu	Seed	Mucolytic, laxative
Leguminosae	Vicia faba L.	Bakla	Flower	Diuretic, lithontriptic
Liliaceae	Asparagus acutifolius L.	Yabani kuskonmaz, tilkisen, aciot	Root	Diuretic, lithontriptic
Liliaceae	Asparagus officinalis L.	Kuskonmaz	Root	Diuretic, lithontriptic
Liliaceae	Asphodelus aestivus Brot.	Ciris	Root	Diuretic, antidermatosic
Liliaceae	Colchicum autumnale L.	Acicigdem tohumu	Seed, bulb	Diuretic, diaphoretic, purgative
Liliaceae	Polygonatum multiflorum (L.) All.	Muhrusuleyman, bogumlucaotu	Root	Antidiarrheal, analgesic, antidiabetes
Liliaceae	Ruscus aculeatus L.	Tavsanmemesi	Root	Diuretic, diaphoretic, stomachic, lithontriptic, anti-inflammatory
Liliaceae	Urginia maritima (L.) Baker	Adasogani	Bulb	Diuretic, cardiac
Linaceae	Linum usitatissimum L.	Keten	Seed	Analgesic, laxative
Loranthaceae	Viscum album L.	Cekem	Fruit	Hypotensive, analgesic,
Malvaceae	Alcea rosea L.	Gul hatmi	Flower	Diuretic, anti-inflammatory
Malvaceae	Althaea officinalis L.	Hatmi	Flower	Anti-inflammatory
Malvaceae	Hibiscus esculentus L.	Bamya	Flower	Laxative, vesicant
Malvaceae	Malva sylvestris L.	Buyuk ebegumeci, ebegumeci	Leaf	Gargle
Moraceae	Morus alba L.	Beyaz dut	Leaf	Diuretic, anti-inflammatory
Moraceae	Morus nigra L.	Kara dut	Fruit, bark	Gargle, laxative, anthelmintic
Myrtaceae	Eucalyptus globulus Labill.	Okaliptus	Leaf	Antidiarrheal, antiseptic, mucolytic

(continued)

Table 7.1 Most Important Plants Used as Medicine in Turkey (Continued)

Family	Botanical Name	Local Name	Plant Part Used	Treatment
Orchidaceae	Orchis sp., Ophrys sp., Dactylorhiza sp.	Salep	Tuber	Aphrodisiac, antidiarrheal
Paeoniaceae	Paeonia mascula (L.) Miller	Sakayik, Orman gulu	Root	Sedative, antidiarrheal
Papaveraceae	Chelidonium majus L.	Kirlangicotu, temereotu	Whole plant	Diuretic, laxative, sedative
Papaveraceae	Fumaria officinalis L.	Sahtereotu, tilki kisnisi	Whole plant	Diuretic, weight loss, sedative, hypotensive
Papaveraceae	Glaucium flavum Crantz	Boynuzlu hashas, gulfatma, sari boynuzlugelincik	Whole plant	Sedative, antitussive
Papaveraceae	Papaver rhoeas L.	Gelincik, asotu	Flower	Sedative, antitussive, sedative,
Phytolaccaceae	Phytolacca americana L.	Sekerciboyasi	Root	Emetic, laxative
Pinaceae	Pinus pinea L.	Cam fistigi	Seed	Nervine
Pinaceae	Pinus sp.	Cam terementisi	Terebinth	Respiratory and excretory or genitourinary system diseases
Plantaginaceae	Plantago coronopus L.	Sinirliot, kargaayagi	Leaf	Antidiarrheal, diuretic, mucolytic
Polygonaceae	Polygonum bistorta L.	Kurtpencesi	Root	Antidiarrheal, antiseptic, diuretic, mouth ulcers, antifibrinolytic
Polygonaceae	Polygonum cognatum Meissn.	Madimak	Root	Diuretic, antidiabetes
Polygonaceae	Rheum ribes L.	Ravent, ucgun	Root	Stomachic, laxative
Polygonaceae	Rumex acetosella L.	Kuzukulagi	Root/leaf	Diuretic, anti-inflammatory, cholagogue/vesicant
Primulaceae	Cyclamen coum Miller	Siklamen	Tuber	Emetic, laxative, stimulant
Primulaceae	Primula vulgaris L.	Cuhacicegi	Root	Diuretic, mucolytic, sedative
Ranunculaceae	Adonis aestivalis L	Keklikgozu otu	Whole plant	Cardiac, diuretic
Ranunculaceae	Delphinium staphisagria L.	Mevzek, kokarot, mezevek, muzudek	Whole plant	Lice treatment
Ranunculaceae	Nigella sativa L.	Corekotu	Whole plant	Diuretic, galactogogue, stomachic, emmenagogue
Ranunculaceae	Nigella arvensis L.	Yabani corekotu	Whole plant	Diuretic
Ranunculaceae	Ranunculus ficaria L	Basurotu, yaglicicek	Root	Antihemorrhoidal
Ranunculaceae	Thalictrum flavum L.	Cayirsedefi	Root	Laxative, diuretic
Rhamnaceae	Paliurus spina-christi Miller	Karacali, calidikeni, calitohumu, caltidikeni, cesmezen, isadikeni, karadiken, kunar, sincandikeni	Fruit	Antidiarrheal, diuretic, lithontriptic
Rhamnaceae	Frangula alnus Miller L.	Cehri, barut agaci, erkek akdiken	Bark	Laxative, stomachic

Family	Scientific name	Turkish name	Aboveground parts	Uses
Rhamnaceae	Ziziyphus jujuba Miller	Hunnap		Antidiarrheal, stomachic
Rosaceae	Amygdalus communis L.	Badem	Oil	Laxative
Rosaceae	Cerasus avium (L.) Moench	Kiraz	Bark, leaf, flower, gum	Antidiarrheal, anti-inflammatory, laxative, antihelminthic, antitussive
Rosaceae	Crataegus monogyna Jacq	Alic meyvasi	Fruit	Sedative, hypotensive, diuretic, antidiarrheal
Rosaceae	Cydonia oblonga Miller	Ayva	Seed	Antidiarrheal, gargle
Rosaceae	Geum urbanum L.	Sukaranfili, yalliceotu, zencefil	Root	Antidiarrheal, nervine, stomachic
Rosaceae	Mespilus germanica L.	Musmula	Fruit	Antidiarrheal
Rosaceae	Prunus domestica L.	Erik	Fruit	Laxative
Rosaceae	Laurocerasus officinalis Roemer	Taflan, lazkirazi, lazuzumu, karayemis	Leaf	Sedative, antitussive, antidiarrheal
Rosaceae	Prunus spinosa L.	Cakal erigi	Leaf, flower, fruit	Antidiarrheal, diuretic, antihelminthic, laxative
Rosaceae	Rosa canina L.	Yabani gul, kusburnu, kopek gulu, gulburnu, gulelmasi, itburnu, sillan	Fruit	Antidiarrheal, nervine, antidiabetes
Rosaceae	Rubus idaeus L	Bogurtlen	Leaf	Antidiarrheal, diuretic, nervine, antidiabetes
Rosaceae	Sarcopoterium spinosum (L.) Spach.	Aptesbozanotu	Bark	Antidiabetes
Rosaceae	Sorbus domestica L.	Uvez	Fruit	Antidiarrheal
Rubiaceae	Rubia tinctorum L.	Kokboya	Root	Urinary disorders, antibilious, laxative
Rutaceae	Ruta graveolens L.	Sedefotu	Whole plant	Sedative, diaphoretic, stomachic
Salicaceae	Salix alba L.	Sogut	Bark	Sedative, analgesic, nervine, anti-inflammatory, antidiarrheal
Scrophulariaceae	Digitalis purpurea L.	Yuksukotu	Leaf	Diuretic, cardiac
Solanaceae	Atropa belladonna L.	Guzelavratotu	Leaf	Analgesic, antispasmodic
Solanaceae	Hyoscyamus niger L.	Banotu	Leaf	Sedative, analgesic
Solanaceae	Mandragora autumnalis Bertol. nokta var sonunda	Adamotu	Root	Analgesic, sedative, aphrodisiac
Solanaceae	Solanum nigrum L.	Kopekuzumu, ituzumu,tilkiuzumu	Fruit	Analgesic, antidermatosic, antihemorrhoidal
Taxaceae	Taxus baccata L.	Porsuk, kadin agaci, puren agaci	Leaf	Sedative, carminative
Thymelaceae	Daphne mezereum L.	Defne	Bark	Diuretic, laxative, diaphoretic
Umbelliferae	Anethum graveolens L.	Dereotu, durakotu, tereotu, turakotu	Fruit	Carminative, digestive
Umbelliferae	Angelica sylvestris L.	Melekotu	Root	Antidiarrheal, sedative
Umbelliferae	Ammi visnaga (L.) Lam.	Disotu	Fruit	Diuretic, lithontriptic, carminative, antihelminthic

(continued)

Table 7.1 Most Important Plants Used as Medicine in Turkey (Continued)

Family	Botanical Name	Local Name	Plant Part Used	Treatment
Umbelliferae	*Apium graveolens* L.	Kereviz	Seed	Carminative, diuretic, stimulant
Umbelliferae	*Conium maculatum* L.	Baldiran otu	Fruit	Analgesic, sedative, aphrodisiac
Umbelliferae	*Coriandrum sativum* L.	Kisnis, kara kimyon	Seed	Stomachic, carminative, digestive
Umbelliferae	*Daucus carota* L. subsp. *sativus* (Hoffm.) Arc.	Havuc, yergecen, kesur, porcuklu	Fruit	Carminative, emmenagogue, diuretic, anthelminthic
Umbelliferae	*Eryngium campestre* L.	Bogadikeni, deveelmasi, devecidikeni, devedikeni, gozdikeni, tengeldikeni	Whole plant	Antitussive, diuretic, stomachic, aphrodisiac
Umbelliferae	*Ferula elaeochytris* Korovin	Caksirotu	Root	Aphrodisiac
Umbelliferae	*Foeniculum vulgare* Miller	Rezene, arapsaci	Seed	Carminative, galactogogue, stomachic
Umbelliferae	*Petroselinum crispum* (Miller) A.W. Hill	Maydanoz	Seed	Diuretic, antibilious, emmenagogue
Umbelliferae	*Pimpinella anisum* L.	Anason meyvasi, enison, nanahan, raziyanei-rumi	Fruit	Carminative, sedative, stomachic, galactogogue
Umbelliferae	*Prangos pabularia* Lindl.	Prangos	Fruit	Carminative, stimulant
Urticaceae	*Urtica dioica* L.	Isirgan, dizlegen	Leaf	Blood purifier, diuretic, stomachic
Valerianaceae	*Centranthus ruber* (L.) DC.	Kirmizi kantaron, kirmizi kediotu, kirmizi mahmuzcicegi	Root	Antispasmodic, sedative
Valerianaceae	*Valeriana officinalis* L.	Kediotu	Root	Antispasmodic, sedative
Verbenaceae	*Vitex agnus-castus* L.	Hayit, ayid, ayit, besparmakotu	Fruit	Diuretic, carminative, sedative
Violaceae	*Viola odorata* L.	Kokulu menekse	Whole plant	Diaphoretic, gargle
Violaceae	*Viola tricolor* L.	Menekse	Whole plant	Diuretic, blood purifier
Zygophyllaceae	*Peganum harmala* L.	Uzerlik	Seed	Antihelminthic, sedative, anesthetic, diaphoretic
Zygophyllaceae	*Tribulus terrestris* L.	Demirdikeni, carikdikeni, cobancokerten, devecokerten	Fruit	Lithontriptic, diuretic, nervine

Table 7.2   Some Lamiaceae Genera Used Medicinally in Turkey

Genera	Number of Species
Salvia	10
Sideritis	11
Lamium	2
Wiedemannia	1
Rosmarinus	1
Stachys	3
Phlomis	3
Thymbra	1
Lavandula	2
Dorystoechas	1
Ocimum	2
Thymus	3
Origanum	7
Coridothymus	1
Teucrium	5
Ajuga	2
Mentha	5
Cyclotrichium	1
Nepeta	2
Micromeria	2
Ziziphora	1
Melissa	1
Ballota	2

Table 7.3   Medicinal Uses of Some Important Taxa from the Mint Family

Scientific Name	Pharmacological Effects
Coridothymus spicatus	Hyperglycemia, hypertension, stomach ailments
Lavandula stoechas	Cold, diuretic, urinary diseases, pain killer, wound healer, antiseptic, rheumatismal pains, expectorant
Melissa officinalis	Sedative, gastric disorders, degasifier, sudorific, antiseptic
Mentha species	Nervous gastric pains and nausea, degasifier, antiseptic, spice
Origanum majorana	Abdominal pain, tranquilizer, gastric troubles, diuretic, degasifier, sudorific, constipation, spice
Origanum onites	Diabetes, abdominal pain, stomachache, cold, immunostimulant (children)
Rosmarinus officinalis	Constipation, stimulant of digestive system, antibilious, diuretic, wounds (external)
Salvia fruticosa	Removal of gas, increase vigor, stimulant, antiseptic (ear–throat)
Sideritis brevidens	Carminative, degasifier
Teucrium polium	Eczema, hemorrhoids, diabetes
Tymbra spicata	Cold
Thymus species	Gastric disorders, sedative, antiseptic, worm removal, speeding blood circulation, antibilious, psoriasis (with honey), spice, tea

Table 7.4   Disease Types and Genera of Lamiaceae Used for the Treatment of Diseases

Disease Type	Genera
Antiseptic	*Salvia, Mentha, Cyclotrichum, Micromeria, Nepeta, Ziziphora, Thymus, Origanum, Thymbra, Coridothymus, Lavandula, Melissa*
Constipation	*Lamium, Rosmarinus, Origanum, Ajuga*
Carminative	*Rosmarinus*
Stomachic	*Origanum, Dorystoechas, Mentha, Cyclotrichum, Ocimum, Micromeria, Nepeta, Teucrium, Ballota, Stachys Thymus, Origanum, Melissa Thymbra, Coridothymus*
Lack of appetite	*Sideritis, Phlomis, Thymbra, Stachys, Dorystoechas, Mentha, Cyclotrichum, Micromeria, Nepeta, Ziziphora, Teucrium*
Tranquilizer	*Sideritis, Phlomis, Thymbra, Stachys, Origanum, Ocimum, Mentha, Thymus, Origanum, Thymbra, Coridothymus, Lavandula, Melissa,*
Tonic	*Salvia, Lamium, Teucrium, Ajuga, Stachys, Lavandula*
Diabetes	*Salvia, Teucrium*
Psoriasis	*Thymus, Origanum, Thymbra, Coridothymus*
Stimulant	*Salvia, Sideritis, Phlomis, Thymbra, Stachys, Teucrium, Stachys, Rosmarinus, Thymus, Origanum, Thymbra, Coridothymus*
Wound healer	*Rosmarinus, Dorystoechas, Ajuga, Stachys, Lavandula, Salvia*
Diuretic	*Rosmarinus, Origanum, Ajuga, Ballota, Ocimum, Lavandula*
Urinary antiseptic	*Ocimum, Lavandula,*
Bilious	*Rosmarinus, Thymus, Origanum, Thymbra, Coridothymus, Lavandula, Mentha,*
Expectorant, balsamic	*Lavandula,*
Degasifier	*Salvia, Sideritis, Phlomis, Thymbra, Stachys, Origanum, Mentha, Cyclotrichum, Micromeria, Nepeta, Ziziphora, Ocimum, Melissa*
Sudorific	*Origanum, Ajuga, Melissa*
Menstruation	*Ajuga, Ballota*
Worm removal	*Ballota, Thymus, Origanum, Thymbra, Coridothymus, Mentha*
Asthma	*Stachys, Lavandula, Mentha*
Speeding blood circulation	*Thymus, Origanum, Thymbra, Coridothymus*
Aphrodisiac	*Salvia*
Rheumatism pains	*Lavandula, Mentha*
Hair tonic	*(Thymus, Origanum, Thymbra, Coridothymus)*
Itch	*Thymus, Origanum, Thymbra, Coridothymus,*
Hair loss, dandruff	*Thymus, Origanum, Thymbra, Coridothymus, Lavandula, Ocimum*
Forgetfulness	*Rosmarinus*
Eczema	*Lavandula*
Headache	*Mentha*
Stinging of poisonous insects	*Ajuga*
Spice	*Ocimum, Thymus, Origanum, Thymbra, Coridothymus*
Tea	*Thymus, Origanum, Thymbra, Coridothymus, Mentha*
Perfume	*Mentha, Lavandula*

summer and reproducing with the help of seeds as well as vegetatively. Tables 7.2–7.5 reveal that members of the mint family are widely used by locals as condiments as well as for the preparation of mint tea against flu, lung and stomach disorders, whooping cough, and as a diuretic. The plants flourish equally well in sunny and shady habitats; however, growth in moist habitats and alongside watercourses is more profuse, with dark-green, bigger leaves, short roots, longer shoots, and higher fruit production.

Table 7.5 Disease Types Where Lamiaceae Genera Are Mixed with Other Plants for Treatment

Disease/Condition Type	Genera Used and Numbers of Other Plants Mixed for Treatment
Mouth odor (M)	*Lavandula* + *Salvia* + other plants
Feet sweating (M)	*Salvia* + two other plants
Headache (M)	*Lavandula* + three other plants
Toothache (M)	*Mentha* + *Anthemis*
Degasifier (M)	*Mentha* + two, three, or five other plants
Indigestion (M)	*Salvia* + three other plants
Appetizer (M)	*Teucrium* + two other plants, *Mentha* + two other plants
Heart palpitation (M)	*Rosmarinus* + three other plants
Anemia (M)	*Salvia* + three other plants
Vomiting (children) (M)	*Mentha* + two other plants
Vigor (M)	*Mentha* + two other plants, *Salvia* + two other plants
Diarrhea (M)	*Mentha* + *Thymus* + three other plants
Cough (M)	*Salvia* + two other plants, *Thymus* + three other plants
Nerve sedative (M)	*Thymus* + *Rosmarinus* + *Lavandula* + *Mentha* + *Rosa*
Sedative (M)	*Lavandula* + *Ociumn* + four other plants
Insomnia (M)	*Mentha* + two other plants

Note: M = mixture.

M. longifolia and *M. suaveolens* grow in open as well as shady habitats, but best growth is seen in moist places. In moist habitats they are completely dominant or are one of the dominant species in the community. Sometimes these are seen to grow in the form of pure stands in moist habitats. *M. longifolia* is distributed between 0 and 1450 m, mostly in wet habitats. *M. longifolia* and *M. suaveolens* are found alongside roads, along wet sea coasts, among the stones in running waters, and in abandoned fields, calcareous habitats, woods, ponds, and on steep slopes. They also grow in open and dry habitats. Altitude affects the number of individuals in *M. suaveolens* and *M. aquatica*. These are found between 0 and 1000 m. *M. aquatica* is not resistant to dry habitats (Figures 7.6–7.8) (Ozturk and Gork 1978, 1979a, 1979b, 1979c, 1979d).

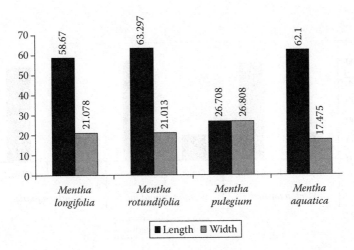

Figure 7.6 Length and width of wild mint plant species.

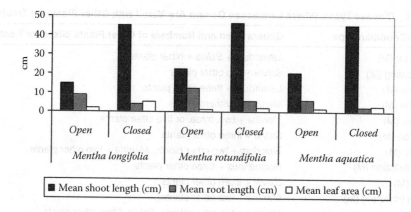

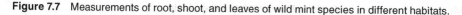

Figure 7.7   Measurements of root, shoot, and leaves of wild mint species in different habitats.

The soils are generally of sandy-loam texture, calcareous or noncalcareous in nature; pH varies between acidic to alkaline. Major constituents of the mint oil are menthone, isomenthone, pulegone, and menthofuran. *M. longifolia* plants are 40–120 cm long, with white or purplish white flowers; shoots contain 0.22% volatile oil, with 15–29% menthol content in dry matter, and terpinolene lies around 20 ppm. *M. suaveolens* plants are 40–100 cm tall, with white or lilac flowers and volatile oil content of 0.14%. *M. pulegium* is 10–40 cm tall, with lilac flowers; volatile oil content in shoots is 0.12% and 60% is pulegon. These plants contain 90 ppm terpinolene.

M. aquatica plants are 20–90 cm tall, with lilac flowers; shoots contain 0.42% volatile oil and 5–37% menthol in the dry matter. Only a small amount of oleum menthae is produced on a household basis in Turkey; a major amount consumed by the pharmaceutical industry is imported. Out of 10 tons of *M. pulegium* exported by some Mediterranean countries, nearly 2 tons are from Turkey. Wild mints are widely used by locals as condiments as well as for the preparation of mint tea, which is used against flu, lung and stomach disorders, and whooping cough and as a diuretic.

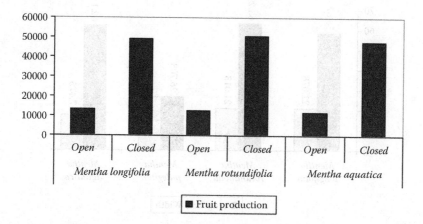

Figure 7.8   Fruit measurements of wild mint species in different habitats.

7.7 CONCLUSIONS

Nearly 80% of the world's population (approximately 5.3 billion humans) relies on plant-based med-icines and billions of dollars are spent on phytochemicals. The world trade of herbal raw materials is around $8 billion, which comes to over $60 billion of global consumer sales value. The annual growth rate for botanicals is 10% in Europe and 20% in the United States. Approximately 0.5 mil-lion tons of medicinal plants are shown in the world trade. In 2001 global sales of herbal drugs were around $62 billion and expectations for 2010 are US$3 trillion.

Several modern medicines of plant origin are sold on the market in huge quantities. The most important ones are ajmaline (*Rauwolfia serpentina*), atropine (*Atropa belladona*), codeine and noscapine (*Papaver somniferum*), colchicine (*Colchicum autumnale*), digoxin (*Digitalis purpurea*), ephedrine (*Ephedra sinica*), and monocrotaline (*Crotalaria sessilifora*)—used for the treatment of heart arrhythmia, traditionally for insanity, dilation of the pupil of the eye, analgesic-sedative, antitumor/gout, cardiotonic, chronic bronchitis, and antitumor/skin cancer, respectively (Choudhary et al. 2000). At present herbal products with top annual sales in Europe are tebonin ($200 million), ginsana ($50 million), kwai ($40 million), and efarmol/epogam ($30 million). The highest importer of medicinal plants in Europe is Germany (64,000 herbal drugs; 38%) followed by France (17%), Italy (9%), UK (6%), and Spain (6%). Germany alone is earning $230 million from *Ginkgo biloba*, $30 million from *Hypericum* species, $18 million from *Myrtus communis*, and $18 million from *Urtica* species. In India and Pakistan 75% of the population depend on medicinal plants and in Africa around 80% do.

The extraordinarily varied flora of Turkey includes many plants with medicinal value. The gen-eral abuse of the environment is affecting these species in the same way as it does others in Turkey's flora. The harvesting and export of certain plants for medicinal purposes or the extraction of oil is damaging the flora and vegetation and is seriously threatening certain species (Demiriz and Baytop

Figure 7.9 *Gypsophila* spp.

Figure 7.10   *Digitalis grandiflora.*

1985; Myers et al. 2000; Aguilar-Støen and Moe 2007). The widespread harvesting of as yet abundant species, such as *Laurus nobilis*, *Origanum heracleoticum*, and *O. onites*, for the export of their leaves and flowers suggests that soon they will eventually be threatened by extinction.

Some species of *Gypsophila* (Figure 7.9) (especially *G. paniculata* L. and *G. venusta* Fenzl.) are gathered for the export of their roots. Widespread harvesting of these species and deep ploughing by mechanical methods have greatly reduced their numbers in central Anatolia, so collectors are now concentrating on the plants found in eastern Anatolia. An alarming decrease in the numbers of *Gentiana lutea* is an example. It has been widely harvested for the medicinal value of its roots in the mountain localities of western Anatolia (Uludag, Bozdag, etc.), where it was abundant before 1940. Consequently, this species can now be found only in very isolated, precipitous localities—and, even then, infrequently. *Digitalis grandiflora* (Figure 7.10), widely distributed throughout eastern and central Europe, is represented in Turkey in Europe by one isolated colony, restricted to the locality of Kirklareli. This species was collected in large quantities for chemical and pharmacological research. Today, 10 years after that intensive harvesting, the population has been unable to recover its former strength and only a few individual specimens are to be encountered.

Some biennial species of *Papaver* are in a similar situation. The collection of practically whole populations of these species for chemical research is leading to their extinction. A report has been prepared about the most endangered plant species, which have been collected for traditional purposes. Out of 50 species mentioned in this endangered plant list, because of intensive exploitation, *Acorus calamus* (rhizome), *Ankyropetalum gypsophiloides* (root), *Ballota cristata* (whole plant), *Barlia robertiana* (tuber), *Gentiana lutea* (root) (Figure 7.11), *Gypsophila arrostii* var. *nebulosa* (root), *Lycopodium annotinum* (whole plant), *Origanum minutiflorum* (whole plant), *Paeonia mascula* (tuber), and *Ruscus aculeatus* (root) are the first 10 most endangered medicinal plant species in Turkey.

Figure 7.11 (See color insert.) *Gentiana lutea.*

REFERENCES

Aquillar-Støen, M. and Moe, S. 2007. Medicinal plant conservation and management: Distribution of wild and cultivated species in eight countries *Biodiversity and Conservation* 16: 1973–1981.

Akcicek, E., and Ozturk, M. 1995. Application of rose and rose products in folk medicine. Proc. of the 13th Int. Cong. of Flavors, fragrances, and essential oils. Baser K. H. C. (ed). AREP Publication. Istanbul, Turkey. 260–268.

Akkol, E. K., Yalcin, F. N. Kaya, D., Ihsan, C., Yesilada, E., and Ersoz, T. 2008. *In vivo* anti-inflammatory and antinoiceptive actions of some *Lamium* species. *Journal of Ethnopharmacology* 19:166–172.

Atoui, A. K., Mansouri, A., Boskou, G., Panagiotis, K. 2005. Tea and herbal infusions: Their antioxidant activity and phenolic profile. *Food Chemistry* 89: 27–36.

Yalcin, F .N., Akkol, E. K., Kaya, D., Ihsan, C., Yesilada, E., and Ersoz, T. 2008. *In vivo* anti-inflammatory and antinoiceptive actions of some *Lamium* opecies. *Journal of Ethnopharmacology* 19: 166–172.

Atoui, A. K., Mansouri, A., Boskou, G., Panagiotis, K. 2005. Tea and herbal infusions: their antioxidant activity and phenolic profile. *Food Chemis* 89:27–36.

Barbour, E. K., Sharif, M. A., Sagherian, V. K., Habre, A. N., Talhouk, R. S., and Talhouk, S. N. 2004. Screening of selected indigenous plants of Lebanon for antimicrobial activity. *Journal of Ethnopharmacology* 93:1–7.

Baser, K. H. C. 1996. Turkiye'de Bitkisel Cay Olarak Kullanilan Aromatik Bitkilerin Ucucu Yaglari. *GIDA* 11:18–21.

———. 2001. Her Derde Deva Bir Bitki-KEKİK. *Bilim ve Teknik Mayis* 74–77.

Baser, K. H. C., Honda, G., and Miki, W. 1986. Herb drugs and herbalists in Turkey. Institute for the Study of Languages and Cultures of Asia and Africa, no. 27. Tokyo.

Bayrak, A., and Akgul, A. 1987. Composition of essential oils from Turkish *Salvia* species. *Phytochemistry* 26:846–847.

Baytop, T. 1984. Turkiye'de Bitkilerle Tedavi. Istanbul Universitesi Yayinlari, 3255, Eczacilik Fakultesi No. 40, s. 235, Istanbul.

———. 1999. Treatment with plants in Turkey, past and present. Nobel Tip Kitapevleri.

Bingol, F. 1995. Some drug samples sold in the herbal markets of Ankara. *Sistematik Botanik* 2:83–110.

Bozan, B., Ozturk, N., and Kosar, M. 2002. Antioxidant and free radical scavenging activities of eight *Salvia* species. *Chemistry of Natural Compounds* 38:198–200.

Capecka, E., Mareczek, A., and Leja, M. 2005. Antioxidant activity of fresh and dry herbs of some Lamiaceae species. *Food Chemistry* 93:223–226.

Choudhary, M., Ahmed, S., Ali, A., Sher, H., and Malik S. 2000. Technical report. Market study of medicinal herbs in Malakand, Peshawar, Lahore, and Karachi. SDC-Inter co-operation, Peshawar.

Damien Dorman, D. H. J., Koar, M., Kahlos, K., Holm, Y., and Hiltunen, R. 2003. Antioxidant properties and composition of aqueous extracts from *Mentha* species, hybrids, varieties, and cultivars. *Journal of Agricultural and Food Chemistry* 51:4563–4569.

Davis, P. H. 1965–1985. *Flora of Turkey and the East Aegean Islands*, vol. 1–9. Edinburgh: Edinburgh Univ. Press.

Davis, P. H., Mill, R. R., and Tan, K. 1988. *Flora of Turkey and the East Aegean Islands*, vol. 10 (supplement 1). Edinburgh: Edinburgh Univ. Press.

Demiriz, H., and Baytop, T. 1985. The Anatolian peninsula. In *Plant conservation in the Mediterranean area*, ed C. Gomez-Campo, chap. 7, 113–121. Dodrecht, the Netherlands: Junk Publishers.

Dogan, A., Bayrak, A., and Akgul, A. 1985. Bazi Kekik Turlerinin Ucucu Yag Bilesimi Uzerinde Arastirma. *GIDA* 10 (4): 213–217.

Dogan, Y., Baslar, S., Ay, G., and Mert, H. H. 2005. The use of wild edible plants in western and central Anatolia (Turkey). *Economic Botany* 54:684–690.

Elmastas, M., Gucin, I., Ozturk, L., and Gokce, I. 2005. Investigation of antioxidant properties of spearmint (*Mentha spicata* L.). *Asian Journal of Chemistry* 17:137–148.

Everest, A., and Ozturk, E. 2005. Focusing on the ethnobotanical uses of plants in Mersin and Adana provinces (Turkey). *Journal of Ethnobiology and Ethnomedicine* 1:6, doi: 10.1186/1746-4269-1-6.

Ezer, N., Akcos, Y., Rodriguez, B., and Abbasoglu, U. 1995. *Sideritis libanotica* Labill., ssp. *linearis* (Bent.) Bornm., den elde edilen iridoit heteroziti ve antimikrobiyal aktivitesi. *Hacetepe University Eczacılık Dergisi* 15:15–21.

Ezer, N., and Sezik, E. 1988. Morphological and anatomical investigations on the plants used as folk medicine and herbal tea in Turkey VI. *Sideritis arguta* Boiss. et Heldr. *Doğa TUBITAK Tıp ve Eczacılık Dergisi* 12:136–142.

Farnsworth, N. R., and Soejarto, D. D. 1991. Global importance of medicinal plants. In *Conservation of medicinal plants*, ed. O. Akerele, V. Heywood, and H. Synge, 25–51. New York: Cambridge Univ. Press.

Fujita, T., Sezik, E., Tabata, M., Yesilada, E., Honda, G., Takeda, Y., Tanaka, T., and Takaishi, Y. 1995. Traditional medicine in Turkey VII. Folk medicine in middle and west Black Sea region. *Economic Botany* 49:406–422.

Ghorbani, A. B., and Motamed, S. M. 2004. A review on the ethnobotany of Labiatae family in Iran. *2nd International Congress on Traditional Medicine and Materia Medica*, October 4–7, 2004, Tehran, Iran.

Gue´don, D. J., and Pasquier, B. P. 1994. Analysis and distribution of flavonoid glycosides and rosmarinic acid in 40 *Mentha* × *Piperita* clones. *Journal of Agricultural and Food Chemistry* 42:679–684.

Gulluce, M., Sahin, F., Sökmen, M., Ozer, H., Daferera, D., Sökmen, A., Polissiou, M., Adiguzel, A., and Ozkan, H. 2007. Antimicrobial and antioxidant properties of the essential oils and methanol extract from *Mentha longifolia* L. ssp. *longifolia*. *Food Chemistry* 103:1449–1456.

Guner, A., Ozhatay, N., Ekim, T., and Baser, K. C. H. 2000. *Flora of Turkey and the East Aegean Islands*, vol. 11. Edinburgh: Edinburgh Univ. Press.

Halberstein, R. 2005. Medicinal plants: Historical and cross cultural usage patterns. *Annals of Epidemiology* 15:686–699.

Hamilton, A. C. 2004. Medicinal plants, conservation and livelihoods. *Biodiversity Conservation* 13:1477–1517.

Haznedaroglu, M. Z., Karabay, N. U., and Zeybek, U. 2001. Antibacterial activity of *Salvia tomentosa* essential oil. *Fitoterapia* 72:829–831.

Hedge, I. C. 1986. Labiatae of Southwest Asia: Diversity, distribution and endemism. *Proceedings of the Royal Society of Edinburgh* 89B:23–35.

Isik, S., Gonuz, A., Arslan, U., and Ozturk, M. 1995. Ethnobotanical studies in the state of Afyon. *OT Systematic Botany Journal* 2:161–166.

Kosar, M., Dorman, H. J. D., and Hiltunen, R. 2005. Effect of an acid treatment on the phytochemical and antioxidant characteristics of extracts from selected Lamiaceae species. *Food Chemistry* 91:525–533.

Kukic, J., Petrovic, S., and Niketic, M. 2006. Antioxidant activity of four endemic *Stachys* taxa. *Biological and Pharmaceutical Bulletin* 29:725–729.

Marinova, E. M., and Yanishlieva, N. V. 1997. Antioxidative activity of extracts from selected species of the family Lamiaceae in sunflower oil. *Food Chemistry* 58:245–248.

Melo, E. A., Filho, J. M., and Guerra, N. B. 2005. Characterization of antioxidant compounds in aqueous cori-ander extract (*Coriander sativum* L.). *Lebensmittel-Wissenschaft und Technologie* 38:15–19.

Mert, T., Akcicek, E., Celik, S., Uysal, I., and Ozturk, M. 2008. Ethnoecology of poisonous plants from West Anatolia in Turkey. *European Journal of Scientific Research* 19:828–834.

Miliauskas, G., Venskutonis, P. R., and van Beek, T. A. 2004. Screening of radical scavenging activity of some medicinal and aromatic plant extracts. *Food Chemistry* 85:231–237.

Mimica-Dukic, N., Bozin, B., Sokovic, M., Mihajlovic, B., and Matavulj, M. 2003. Antimicrobial and antioxi-dant activities of three *Mentha* species essential oils. *Planta Medica* 69:413–419.

Miura, K., Kikuzaki, H., and Nakatani, N. 2002. Antioxidant activity of chemical components from sage (*Salvia officinalis* L.) and thyme (*Thymus vulgaris* L.) measured by the oil stability index method. *Journal of Agricultural and Food Chemistry* 50:1845–1851.

Myers, N., Mittermeier, R. A., Mittermeier, C. G., da Fonseca, G. A. B., and Kent, J. 2000. Biodiversity hotspots for conservation priorities. *Nature* 403:853–858.

Nakiboglu, M., Urek, R. O., Kayali, H. A., and Tarhan, L. 2007. Antioxidant capacities of endemic *Sideritis sipylea* and *Origanum sipyleum* from Turkey. *Food Chemistry* 104:630–635.

Nickavar, B., Alinaghi, A., and Kamalinejad, M. 2008. Evaluation of the antioxidant properties of five *Mentha* species. *Iranian Journal of Pharmaceutical Research* 7:203–209.

Nickavar, B., Kamalinejad, M., and Hamidreza, I. 2007. *In vitro* free radical scavenging activity of five *Salvia* species. *Pakistan Journal of Pharmaceutical Science* 20:291–294.

Orhan, I., Aydin, A., Colkesen, A., Sener, B., and Isimer, A. I. 2003. Free radical scavenging activities of some edible fruit seeds. *Pharmaceutical Biology* 41:163–165.

Ozcan, M. 2005. Determination of mineral contents of Turkish herbal tea (*Salvia aucheri* var. *canescens*) at different infusion periods. *Journal of Medicinal Food* 8:110–112.

Ozgen, U., Mavi, A., Terzi, Z., Yildirim, A., Coskun, M., and Houghton, P. J. 2006. Antioxidant properties of some medicinal Lamiaceae (Labiatae) species. *Pharmaceutical Biology* 44:107–112.

Ozgokce, F., and Ozcelik, H. 2004. Ethnobotanical aspects of some taxa in East Anatolia, Turkey. *Economic Botany* 58:697–704.

Ozturk, M., and Gork, G. 1978. Studies on the chronology and economical evaluation of *Mentha* species in West Anatolia. *Ege University, Science Faculty Journal* 2:339–356.

———. 1979a. Ecological factors affecting the distribution and plasticity of *Mentha* species in West Anatolia. BITKI 6:39–51.

———. 1979b. Studies on the morphology and taxonomy of *Mentha* species in West Anatolia. *Ege University, Science Faculty Journal* 3:43–55.

———. 1979c. Ecology of *Mentha pulegium. Ege University, Science Faculty Journal* 3:57–72.

———. 1979d. Edaphic relations of *Mentha* species in West Anatolia. *Ege University, Science Faculty Journal* 3:95–110.

Ozturk, M., and Ozcelik, H. 1991. *Useful plants of East Anatolia*. Ankara: Siskav Press.

Ozturk, M., Pirdal, M., and Uysal, I. 1992. Ecology and importance of Turkish endemics. *Tarim and Koy* 74:20–21.

Ozturk, M., Secmen, O., and Pirdal, M. 1986. Mint farming in upper Euphrates. *Firat Basin Medical and Industrial Plants Symposium*, Elazig, 119–126.

Ozturk, M., Uysal, I., Gücel, S., Mert, T., Akcicek, E., and Celik, S. 2008. Ethnoecology of poisonous plants of Turkey and Northern Cyprus. *Pakistan Journal of Botany* 40:1359–1386.

Parejo, I., Viladomat, F., Bastida, J., Rosas-Romero, A., Flerlage, N., Burillo, J., and Codina, C. 2002. Comparison between the radical scavenging activity and antioxidant activity of six distilled and nondistilled Mediterranean herbs and aromatic plants. *Journal of Agricultural and Food Chemistry* 50:6882–6890.

Plotkin, M. J. 2000. *Medicine quest*. New York: Penguin Books Ltd.

Richardson, P. M. 1992. The chemistry of the Labiatae: An introduction and overview. In *Advances in Labiatae science*, ed. R. M. Harley, and T. Reynolds, 291–297. Kew, England: Royal Botanic Gardens.

Sagdic, O. 2003. Sensitivity of four pathogenic bacteria to Turkish thyme and oregano hydrosols. *Food Science and Technology* 36:467–473.

Sagdic, O., Aksoy, A., and Ozkan, G. 2006. Evaluation of the antibacterial and antioxidant potentials of gilab-uru (*Viburnum opulus* L.) fruit extract. *Acta Alimentaria* 35:487–492.

Sayar, A., Guvensen, A., Ozdemir, F., and Ozturk, M. 1995. Ethnobotanical studies in the state of Mugla. *OT Systematic Botany Journal* 2 (1): 151–160.

Sezik, E. 1990. Anadolu da Cay Olarak Kullanilan Yabani Bitkiler. *Bilim ve Teknik* 23:18–20.

Sezik, E., and Basaran, A. 1989. The volatile oil of *Origanum sipyleum* L. *Acta Pharmacologia Turcica* 31:129–133.

Sezik, E., and Saracoglu, I. 1988. Morphological and anatomical investigations on the plants used as folk medicine and herbal tea in Turkey.V. *Thymus eigii* (M. Zohary et P.H. Davis). *Jalas Doga TUBITAK Tip ve Eczacılık Dergisi* 12:32–37.

Sezik, E., Tabata, M., Yesilada, E., Honda, G., Goto, K., and Ikeshiro, Y. 1991. Traditional medicine in Turkey. Folk medicine in Northeast Anatolia. *Journal of Ethnopharmacology* 35:191–196.

Sezik, E., and Tumen, G. 1984. Morphological and anatomical investigations on the plants used as folk medicine and herbal tea in Turkey. II. *Ziziphora taurica* Bieb. ssp. *taurica. Doga Bilim Dergisi* 8:98–103.

Sezik, E., Zor, M., and Yesilada, E. 1992. Traditional medicine in Turkey II. Folk medicine in Kastamonu. *International Journal of Pharmacognosy* 30:233–239.

Simsek, I., Aytekin, F., Yesilada, E., and Yildirimli, S. 2004. An ethnobotanical survey of the Beypazari, Ayas and Gudul district towns of Ankara Province (Turkey). *Economic Botany* 58:705–720.

Solmaz, Z. 1993a. Ihlamur (*Tilia cordata*). Yesil Saglik. *Tarim ve Koy* 91:42–43.

———. Adacayi (*Salvia*). Yesil Saglik. *Tarim ve Koy* 92:42–43.

Stefanini, I., Piccaglia, R., Marotti, M., and Biavati, B. 2006. Characterization and biological activity of essential oils from fourteen Labiatae species. *Acta Horticultura* 723:221–226.

Surmeli, B., Sakcali, S., Ozturk, M., and Serin, M. 2000. Kilis ve Cevresinde Halk Hekimliginde Kullanilan Bitkiler. XIII. Plant Raw Materials Meeting, Istanbul, 211–220.

Tabata, M., Sezik, E., Honda, G., Yesilada, E., Fukui, H., Goto, K., and Ikeshiro, Y. 1994. Traditional medicine in Turkey III. Folk medicine in East Anatolia, Van and Bitlis provinces. *International Journal of Pharmacognosy* 32:3–12.

Tepe, B., Daferera, D., and Sokmen, A. 2005. Antimicrobial and antioxidant activities of the essential oil and various extracts of *Salvia tomentosa* Miller (Lamiaceae). *Food Chemistry* 90:333–340.

Tepe, B., Donmez, E., and Unlu, M. 2004. Antimicrobial and antioxidative activities of the essential oils and methanol extracts of *Salvia cryptantha* (Montbret et Aucher ex Benth.) and *Salvia multicaulis* (Vahl). *Food Chemistry* 84:519–525.

Tepe, B., Eminagaoglu, O., and Akpulat, H. A. 2007. Antioxidant potentials and rosmarinic acid levels of the methanolic extracts of *Salvia verticillata* (L.) subsp. *verticillata* and S. *verticillata* (L.) subsp.*amasiaca* (Freyn & Bornm.) Bornm. *Food Chemistry* 100:985–989.

Tepe, B., Sökmen, M., Akpulat, H. A., and Sökmen, A. 2006. Screening of the antioxidant potentials of six *Salvia* species from Turkey. *Food Chemistry* 95:200–204.

Triantaphyllou, K., Blekas, G., and Boskou, D. 2001. Antioxidative properties of water extracts obtained from herbs of the species *Lamiaceae. International Journal of Food Sciences and Nutrition* 52:313–317

Ulubelen, A. 1964. Cardioactive and antibacterial terpenoids from some *Salvia* species. *Phytochemistry* 64:395–399.

Voirin, B., and Bayet, C. 1992. Developmental variations in leaf flavonoid aglycones of *Mentha* × *Piperita. Phytochemistry* 31 (7): 2299–2304.

Voirin, B., Saunois, A., and Bayet, C. 1994. Free flavonoid aglycones from *Mentha* × *Piperita:* Developmental, chemotaxonomical and physiological aspects. *Biochemical Systematics and Ecology* 22:95–99.

WHO. 2002. *WHO traditional medicine strategy 2002–2005.* Geneva: WHO.

Yesilada, E., Honda, G., Sezik, E., Tabata, M., Goto, K., and Ikeshiro, Y. 1993. Traditional medicine in Turkey IV. Folk medicine in the Mediterranean subdivision. *Journal of Ethnopharmacology* 39:31–38.

Yesilada, E., and Sezik, E. 2003. A survey on the traditional medicine in Turkey: Semi-quantitative evaluation of the results. In *Recent Progress in Medicinal Plants.* Vol. VII. V.K. Singh, J. N Govil, S. H ashimi, G. Singh (eds.). Stadium Press, LLC: Houston, TX. 389–412.

Yildirim, A., Mavi, A., Oktay, M., Kara, A. A., Algur, O. F., and Bilaloglu, V. 2000. Comparison of antioxidant and antimicrobial activities of tilia (*Tilia argentea* Desf. ex. DC.), sage (*Salvia triloba* L.) and black tea (*Camellia sinensis* L.) extracts. *Journal of Agriculture and Food Chemistry* 48:5030–5034.

Yucel, E., and Ozturk, M. 1998. Studies on the autecology of *Origanum sipyleum* L. In *Plant life in southwest and central Asia, 5th International Symposium,* ed. Ashurmetov, O., Khassanov, F., and Salieva, Y., 201–204. Tashkent, Uzbekistan.

Zeybek, U. 2003. Dogal Yaglar, Bitkisel Proteinler, Organik Bitki ve Meyve Caylari. BUKAS Tarim Urunleri San. ve Tic. A. S., Izmir, Turkiye, pamphlet, 20 pp.

CHAPTER 8

An Overview of the Ethnobotany of Turkmenistan and Use of *Juniperus turcomanica* in Phytotherapy

Svetlana Aleksandrovna Pleskanovskaya, Mamedova Gurbanbibi
Atayevna, Munir Ozturk, Salih Gucel, and Mehri Ashyraliyeva

CONTENTS

8.1 INTRODUCTION

Turkmenistan is situated in the western part of central Asia, 35–43° north latitude and 53–67° east longitude. The elevation ranges between –110 and 2942 m. The country occupies an area of more than 488,000 km^2. Its territory stretches 1,100 km from west to east and 650 km from north to south, bordering Kazakhstan in the north, Uzbekistan in the east and northeast, Iran in the south, and Afghanistan in the southeast. The Caspian Sea, a landlocked saltwater lake, forms Turkmenistan's entire western border. The Akdzhakaya Depression, located in the north-central part of the country, is the lowest point in the republic at 110 m below sea level. The Karakum, one of the largest sand deserts in the world, covers 90% of the country, an area of 350,000 km^2. It lies in the entire central part of the country at an elevation of 500 m. The Kopetdag Mountains fringe the Karakum Desert along the country's southern border with Iran. The Lorne, Murgab, and Tecen rivers are located near the desert (Figure 8.1a and 8.1b). Along the mountain foothills is a belt of oases, which are fed by mountain streams. Turkmenistan oasis is to the east of the lake Karaboğaz along the River Amuderya, originating from the mountainous Pamirs region of Tajikistan in the east of Turkmenistan. It forms part of the country's border with Uzbekistan. Amuderya and the Murgap are the two largest permanent rivers.

The climate is continental subtropical, with cold winters and very hot summers. The summer temperatures exceed 50°C, but in winter temperatures fall to –30°C. For most of the country, the average daily temperature in January ranges from –6 to 5°C, while in July it is 27–32°C. The absolute minimum air temperature is 46–47°C, with a lower limit of 50°C in Repetek. Absolute

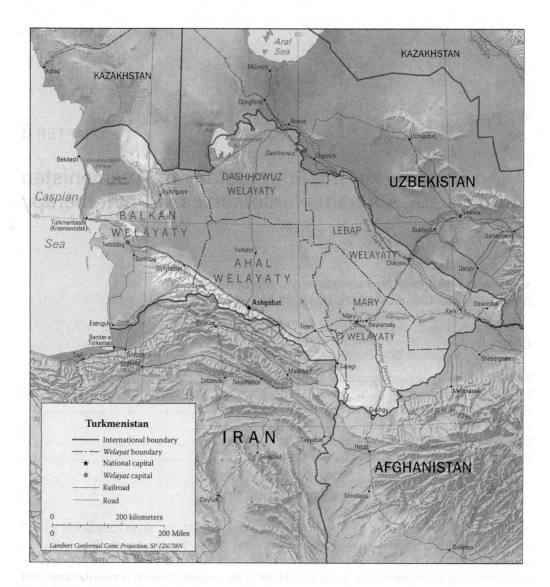

Figure 8.1(a) Map of Turkmenistan.

maximum on the soil surface in most areas exceeds 70°C. Average annual precipitation ranges from 80 to 400 mm, although two thirds of the country receives only 150 mm or less of rain.

For the most part, the lands are meadow and the soils are takyr, light sierozem, dark sierozems, meadow sierozems gray-brown, sandy desert, residual meadow, meadow-takyr, marsh, solonchaks, cinnamon-colored mountains, weakly fixed and shifting sands, steep precipices, and ledges (UN 2009).

Turkmenistan shows a rich plant diversity. Nearly 311 plant taxa are of medicinal value, mainly found at Kopetdag, Greater and Lesser Balkans, and the Turkmen part of the Kugitang-tau. This plant cover is squeezed into the 10% zone, whereas 90% of the country is desert. The most important medicinal plants are *Ephedra*, Turkmen juniper, common St. John's wort, *Acanthophyllum*, *Ziziphora*, and others. The Turkmen and Zarafshan junipers contain both food and medicinal components.

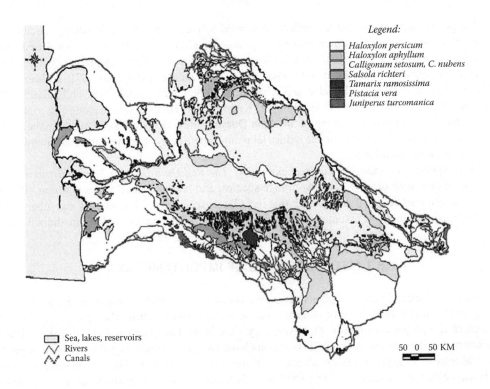

Legend:

☐ *Haloxylon persicum*
☐ *Haloxylon aphyllum*
☐ *Calligonum setosum, C. nubens*
☐ *Salsola richteri*
☐ *Tamarix ramosissima*
☐ *Pistacia vera*
☐ *Juniperus turcomanica*

☐ Sea, lakes, reservoirs
/\/ Rivers
/\/ Canals

50 0   50 KM

Figure 8.1(b)   Map showing plant diversity of Turkmenistan.

The aim of this review is to provide information about the important medicinal plants growing under harsh environmental conditions in Turkmenistan and to present newly obtained data on the use of *Juniperus turcomanica* in the treatment of hypothyrosis.

8.2 PLANT DIVERSITY

Plant life is sparse in the vast, arid desert, where only drought-resistant grasses and desert scrub grow. The mountain valleys in the south support wild grapevines, fig plants, and ancient forests of wild walnut trees. The mountain slopes are covered with forests of juniper and pistachio trees. Dense thickets called *tugai* grow along riverbanks. About four fifths of the country is *steppe* (semiarid grassy plain), which is part of the southern portion of the vast Turan lowland. The plant communities, with dominating xerophytic low semishrubs and halophytes and sparse groups of saxaul as well as shrub psammophytes, ephemers, and ephemeroids, define the landscape of the desert zone.

The flora of Turkmenistan includes over 2,900 species of higher plants belonging to 105 families. The dominant life forms are herbaceous plants (2,137 species), followed by 47 species of trees, 88 of shrubs, 44 of low shrubs, 46 of semishrubs, and 238 of low semishrubs. Nearly 1,700–1,800 species are found on Kopetdag, where the country's major diversity of valuable plants (729 species) and wild crop relatives (106 species) is concentrated. The distribution of other groups of plants is as follows: vegetables (five), tuber crops (six), textile fiber plants (three), glue-producing plants (two), resin and gum-producing plants (five), pigment-producing plants (12), tanning plants (five), fat- and oil-bearing plants (three), insecticide and repellent plants (three), plants used for mats (one), volatile oil-producing plants (seven), forage crops (12), ornamentals (trees and shrubs [eight], grasses [70]), and medicinal plants (172).

The perspective regions for the collection of wild medicinal plants are Kopetdag, Greater and Lesser Balkans, and the Turkmen part of Kugitangtau. Many valuable species are found there: *Ephedra*, common St. John's wort, *Acanthophyllum*, *Ziziphora*, and Turkmen juniper (Walker 1968; Goriaev and Ignatova 1969; Dembitsky 1969; Kitchens, Dorsky, and Kaiser 1971; Akimov et al. 1976; Lawrence 1985; Chavchanidze and Kharabava 1989; Clark, McChesney, and Adams 1990; Rafique, Hanif, and Chaudary 1993; Tunalier, Kirimer, and Baser 2004). Among the great variety of valuable wild plants growing in the Amu Darya Valley, licorice should be mentioned especially, together with the most valuable industrial plant, saponine-containing Turkestan soapwort, an endemic plant of central Asia.

Many species are under the threat of extinction. *The Red Data Book* of *Turkmenistan* includes 10 fruit species, nine ornamentals, two forage species, and two medicinal species. About 400 species are considered economically useful; this includes 41 fruit trees, 14 vegetables and tuber plants, five oil-bearing plants, 131 essential oil-bearing plants, 142 ornamentals, and 311 medicinal plants.

8.3 MEDICINAL USES OF IMPORTANT TAXA

The importance of medicinal plants has gained a great momentum in recent years (Pleskanovskaya et al. 2003). In this field, a lot of research has been done and scientific data put forward for the benefit of the global community. However, very little work has been carried out in this field in Turkmenistan. Severe environmental conditions have forced and encouraged the Turkmen people for centuries to struggle in taking advantage of the sparse vegetation. The information presented here was collected between 2000 and 2007 from Turkmen people as a result of direct conversations and has been compiled from several references (Ashuralyeva and Ozturk 2000; Adams 2004, 2008). The list presented here includes information on the most commonly used 78 taxa. These are presented here alphabetically in Table 8.1, together with the distribution, plant part used, and medicinal applications.

Out of these taxa, 24 are distributed in Kopetdag, 16 are cultivated, 7 occur on hilly areas, 7 are widespread, and 6 are distributed in Karakala. The most important dominating families that include the highest number of taxa among the medicinal plants are Asteraceae (9; 11.5%), Rosaceae (9; 11.5%), Fabaceae (8; 10.3%), Lamiaceae (6; 7.7%), and Solanaceae (4; 5.1%). Nearly 0.5% of the medicinal plants are applied internally and 29.5% externally. Most commonly used plant parts are leaves (31 preparations), fruits (22 preparations), and aboveground green parts (16 preparations). Medicinal plants have been used mainly as diuretics (20 applications), laxatives (13 applications), stomachics (13 applications), analgesics (10 applications), vasoconstrictors, and antidermatosics (8 applications).

8.4 APPLICATION OF *JUNIPERUS TURCOMANICA*

Out of these taxa, junipers have attracted much attention in recent years. *Juniperus turcomanica* B. Fedtsch. (Figure 8.2), a member of Cupressaceae family, is a medicinally important rare species endemic to the northern Kopetdag-Khorasan chains distributed at an altitude of 1100–2800 m (Atamuradov et al. 1999).

The immunomodulating and anti-inflammatory effects of the *J. turcomanica* (JT) water extract in the treatment of patients suffering from chronic tonsillitis has been demonstrated (Yeger 1986; Gofman and Smirnov 2000; Pleskanovskaya, Mamedova, and Allaberdiyev 2008). It is known that tonsillitis is a cause as well as a manifestation of chronic immunodeficiency. The water extract of JT fruits was given to 25 practically healthy persons (PHPs) and 21 hypothyrosis patients (HT-p) (volunteers aged from 19 to 45 years). The HT diagnosis was confirmed by using urinary stress

Table 8.1 List of Important Medicinal Plant Taxa from Turkmenistan and Their Characteristics

Botanical Name	Distribution	Plant Part Used	Medical Usage	Treatment
Allium cepa (Alliaceae)	Cultivated	Leaf	Externally, internally	Pectoral, antipruritic, antiulcer, antidermatosic, vesicant, antitussive
Allium sativum (Alliaceae)	Cultivated	Leaf	Internally	Stomachic, anticoagulant
Ungernia victoris (Amaryllidaceae)	Kopetdag	Bulbs	Externally	Vulnerary
Rhus coriaria (Anacardiaceae)	Kopetdag	Leaf	Internally	Laxative, styptic
Foeniculum vulgare (Apiaceae)	Cultivated	Whole plant	Internally	Expectorant, digestive, stomachic, demulcent
Anethum graveolens (Apiaceae)	Cultivated	Seed	Internally	Stomachic, decongestant, antiecchymotic, galactogogue, digestive
Carum carvi (Apiaceae)	Hilly areas	Fruit	Internally	Laxative, stomachic, carminative
Coriandrum sativum (Apiaceae)	Aşkabat, Merv	Fruit, aboveground green part	Internally	Stomachic, hypotensive, hypoglycemic diuretic, antitussive
Ferula assafoetida (Apiaceae)	Karakum	Juice, root	Externally, internally	Antihelminthic, carminative, antidermatosic
Achillea millefolium (Asteraceae)	Karakum	Aboveground green parts	Internally	Anti-inflammatory, antibacterial, antiallergy, appetizer, coagulant
Artemisia kopetdaghensis (Asteraceae)	Kopetdag	Leaf	Internally, externally	Stomachic, ophthalmic, antipyretic, inflammation of the pancreas
Artemisia turcomanica (Asteraceae)	Kopetdag	Leaf	Internally, externally	Stomachic, ophthalmic, antipyretic, inflammation of the pancreas
Bidens tripartita (Asteraceae)	Karakala, Kopetdag	Leaf	Externally	Antidermatosic, scabies, and rickets
Calendula officinalis (Asteraceae)	Cultivated	Flower	Internally	Diuretic, tonsillitis, mouthwash
Cichorium intybus (Asteraceae)	Widespread	Root	Externally	Antidermatosic
Matricaria recutita (Asteraceae)	Cultivated	Flower	Internally	Sedative, analgesic, antibilious, mouthwash, diuretic, hepatic
Taraxacum officinale (Asteraceae)	Türkmenbaşi, merv	Root	Externally,	Vasoconstrictor, appetizer, sedative, stimulant
Tussilago farfara (Asteraceae)	Nohur, Kara-kala	Leaf	Internally, externally	Anti-inflammation, vasodilator, hepatic, cardiac diseases
Berberis turcomanica (Berberidaceae)	Kurendag	Leaf, root, and fruit	Internally	Febrifuge, hepatic, antibilious, ophthalmic, stimulant
Brassica oleracea (Brassicaceae)	Widespread	Leaf, seed	Internally, externally	Appetizer, laxative, jaundice, gastric ulcer, analgesic, rheumatism, diuretic

(continued)

Table 8.1 List of Important Medicinal Plant Taxa from Turkmenistan and Their Characteristics (Continued)

Botanical Name	Distribution	Plant Part Used	Medical Usage	Treatment
Capsella bursa pastoris (Brassicaceae)	Widespread	Aboveground part	Externally	Vasodilator, lithontriptic, febrifuge, diuretic, hemostatic, vasoconstrictor
Capparis spinosa (Capparaceae)	Kara-kum	Root, fruit	Internally	Jaundice, antiperiodic, antihemorrhoidal, antiasthmatic, goiter, analgesic, vulnerary
Sambucus ebulus (Caprifoliaceae)	Kopetdag	Leaf	Internally	Diuretic, antihalitosis
Acanthophyllum gypsophiloides (Caryophyllaceae)		Root	Internally	Expectorant, anti-inflammatory, pectoral
Acanthophyllum kugitangum (Caryophyllaceae)		Root	Externally	Expectorant, anti-inflammatory, pectoral
Anabasis aphylla (Chenopodiaceae)	Taşoguz	Root	Externally, internally	TB, antiasthmatic, antidermatosic
Salsola richteri (Chenopodiaceae)	Karakum	Fruit and leaf	Internally	Vasoconstrictor
Juniperus turcomanica (Cupressaceae)	Kopetdag, Kurendag	Fruit	Internally	Diuretic, antibacterial, anti-inflammatory
Diospyros kaki (Ebenaceae)	Kara-kala	Fruit	Internally	Nutritive, anemia
Elaegnus angustifolia (Elaegnaceae)	Lebab	Fruit	Internally	Antiasthmatic, antilaxative, diarrhea, pectoral, stomachic, intestinal inflammation
Equisetum arvense (Equisetaceae)	Hilly areas	Aboveground part	Internally, externally	Rheumatism, colds, skin diseases, asthma, allergies
	Kopetdag	Aboveground green part	Internally	Antiasthmatic, antiemetic, dysentery, hepatic disorders, vascular congestion, throat inflammation, vasoconstrictor
Alhagi persarum (Fabaceae)	Kopetdag	Aboveground green part	Internally	Laxative, diuretic, appetizer, styptic against dysentery
Astragalus pulvinatus (Fabaceae)	Kopetdag	Aboveground part	Internally	Vasoconstrictor, diuretic, analgesic
Cassia angustifolia (Fabaceae)	Tedjen	Leaf	Internally	Laxative
Glycyrrhiza glabra (Fabaceae)	Kopetdag	Root	Internally	Antitussive, decongestant, diuretic, against intestine and stomach ulcer
Melilotus officinalis (Fabaceae)	Field edges	Leaf	Internally	Analgesic, rheumatism, hypnosis, anticoagulant, decongestant
Robinia pseudoacacia (Fabaceae)	Cultivated	Flower, leaf, seed	Externally, internally	Coagulant, laxative, decongestant

Species (Family)	Location	Plant part	Use	Properties/Indications
Sophora pachycarpa Schrenk (Fabaceae)	Kopetdag	Aboveground part	Externally, internally	Decongestant
Sphaerophysa salsola (Fabaceae)	Taşoguz	Aboveground part	Externally, internally	Coagulant, vasoconstrictor
Centaurium pulchellum (Gentianaceae)	Kopetdag	Aboveground part	Externally, internally	Anthelminthic, antidermatosic
Hypericum helianthomoides (Hypericaceae)	Kopetdag	Leaf	Externally, internally	Abdominal pains, tuberculosis, diuretic, decongestant, anti-inflammatory, nervine, coagulant, stimulant
Juglans regia (Juglandaceae)	Alongside streams	Leaf, fruit	Externally	Arthritis, styptic, tonsillitis, antipruritic, antihelminthic
Lagochilus inebrians (Lamiaceae)	Kugitangum	Flower	Internally, externally	Styptic, sedative, vasoconstrictor
Melissa officinalis (Lamiaceae)	Kopetdag	Leaf	Internally, externally	Sedative, decongestant, nervine
Mentha piperita (Lamiaceae)	Cultivated	Leaf	Internally	Sedative, digestive, stomachic
Salvia sclerea (Lamiaceae)	Kopetdag	Leaf	Internally	Decongestant, demulcent
Thymus transcarpicus (Lamiaceae)	Kopetdag	Leaf	Internally, externally	Diuretic, antibacterial
Ziziphora clinopodioides (Lamiaceae)	Kopetdag	Aboveground part	Internally	Diuretic, cardiotonic, bowel diseases
Althaea officinalis (Malvaceae)	Kopetdag	Root	Internally	Antitussive, decongestant, demulcent
Morus alba (Moraceae)		Leaf and bark	Internally	Cardiac, febrifuge
		Fruit resin		Stomach and intestine ulcers, cardiac, febrifuge
Ficus carica (Moraceae)		Leaf, Fruit	Internally	Lithontriptic, kidney, cardiac, digestive
Olea europaea (Oleaceae)	Kara-kala	Fruit, leaf	Internally, externally	Antibilious, vasoconstrictor, laxative, emollient
Sesamum indicum (Pedaliaceae)	Cultivated	Seed	Externally, internally	Laxative, styptic
Plantago major (Plantaginaceae)	Widely distributed in the hilly areas	Leaf	Internally	Decongestant
Polygonum aviculare (Polygonaceae)		Aboveground green part	Internally	Diuretic, laxative, styptic
Portulaca oleracea (Portulacaceae)		Aboveground green part	Internally	Styptic, antihemorrhoidal, hepatic, against dysentery, hypotensive

(continued)

Table 8.1   List of Important Medicinal Plant Taxa from Turkmenistan and Their Characteristics (Continued)

Botanical Name	Distribution	Plant Part Used	Medical Usage	Treatment
Punica granatum (Punicaceae)	Naturally growing in Karakala and Kiziletrik; cultivated at other places	Fruit	Internally	Diuretic, febrifuge
		Bark		Against dysentery
Zizyphus jujuba (Rhamnaceae)	Hilly areas	Fruit	Externally	Vasoconstrictor, sedative
Amygdalus communis (Rosaceae)	Kara-kala	Seed, fruit	Internally, externally	Stomachic, nervine, analgesic
Cerasus austera (Rosaceae)	Cultivated	Fruit	Internally	Anemia, diuretic, lithontriptic
Crataegus sanguinea (Rosaceae)	North and central Kopetdag	Fruit	Internally	Cardiac, vasodilator, analgesic
Crataegus curvisepala (Rosaceae)				
Cydonia oblonga (Rosaceae)	Cultivated	Seed	Internally	Stomachic, decongestant, anti-inflammatory
Persica vulgaris (Rosaceae)	Cultivated	Fruit	Internally	Diuretic, laxative, cardiac, anemia,
Potentilla reptans (Rosaceae)	Kopetdag	Root	Internally, externally	Antilaxative, coagulant, dysentery
Rosa canina (Rosaceae)	Kopetdag	Fruit	Externally, internally	
Rubus idaeus (Rosaceae)	Hilly areas	Fruit, leaf	Internally, externally	Coagulant, stomachic, antipyretic Digestive, Decongestant, anticoagulant, antidermatosic, antipyretic, hypotensive
Rubia tinctorum (Rubiaceae)	Wide distribution in central Kopetdag	Root	Internally	Lithontriptic, gout
Verbascum thapsus (Scrophulariaceae)	Caspian Sea	Leaf, root	Externally, internally	Liver inflammation, emollient, expectorant, diuretic
Ailanthus altissima (Simarubaceae)	Cultivated	Leaf	Externally	Antidermatosic and for dysentery

Capsicum annuum (Solanaceae)	Cultivated	Leaf	Internally, externally	Appetizer, antipyretic, drip disease, and for muscle inflammation
Datura stramonium (Solanaceae)	Ruderal	Leaf	Internally	Hypnotic, analgesic, antiasthmatic, and some respiratory diseases
Hyoscyamus niger (Solanaceae)	Fields and hilly areas	Leaf	Externally	Analgesic, antiasthmatic, stomachic, painkillers, asthma, antispasmodic
Solanum nigrum (Solanaceae)	Hilly areas	Aboveground part	Internally	Analgesic, laxative, sedative, decongestant
Urtica dioica (Urticaceae)	Widely distributed	Leaf	Internally	Diuretic, decongestant, antibilious, febrifuge
Vitis vinifera (Vitaceae)	Cultivated	Fruit	Internally	Tuberculosis, anemia, liver, lung, intestinal disorders
Peganum harmala (Zygophyllaceae)	Widely distributed	Aboveground green part	Internally	Febrifuge, nervine, bayfainting
Tribulus terrestris (Zygophyllaceae)	Widely distributed	Seed and root	Internally	Antihelminthic
		Aboveground green part Fruit	Internally	Diuretic, stimulant, anticoagulant, lithontriptic

Figure 8.2 A general view of *Juniperus turcomanica*.

incontinence (USI), the thyroid-stimulating hormone (TSH), thyroxine (T4), and thyroperoxydase antibody (TPA).

The three groups of patients (seven persons in each) were arranged. Group I received only the traditional l-thyroxine treatment HT-p, group II (not sensitive to JT) *in vitro* HT-p, and group III (sensitive to JT) *in vitro* HT-p. HT-p of groups II and III received 20.0 mL of the 10% water extract of JT per oral supply (os) during 15 days, each 2–3 hours at night in addition to the traditional hormonal treatment. All of HT-p experienced twice the positive effect of JT examined, before the treatment began and a month after it. The serum TPA, TSH, and T4 concentrations were determined by the ELISA (Enzyme-Linked Immunosorbent Assay) method; results were expressed as ME per milliliter. The character of the JT *in vitro* influence on the TG was determined by the patented original method (Pleskanovskaya et al. 2008).

It was found that JT in an equal number of cases has a positive, negative, or no effect on the PHP TG *in vitro* (Figure 8.3). At the same time, JT has a positive effect on the TG of 65.5% of HT-p and in equal cases negative or no effect at all. The serum thyroid hormone concentration of any group of HT-p before the treatment was changed in equal degree to the PHP for comparing the degree of its change. During the treatment, the serum TSH and TPA concentrations decreased in all HT-p-groups (Figures 8.4 and 8.5); simultaneously, the concentration increased to the initial level. The positive dynamics were more expressed in group III of HT-p, who were sensitive to the JT *in vitro* and had taken it per os. Thus, JT water extract *in vitro* has both positive and negative effects upon the PHP and HT-p TG tissues. However, more often (1.8 times), a positive effect of the JT was seen in the HT-p groups.

The hormonal status rehabilitation and TPA reduction of JT receiving HT-p was more significant in the group I taking only traditional l-thyroxine treatment. The JT *in vitro* sensitive HT-p

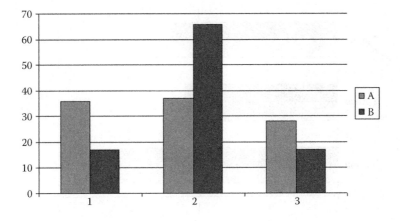

Figure 8.3 *Juniperus turcomanica* effects on the TG tissue of (A) PHP and (B) HT-p.

(group III) showed the best combined treatment results. The working hypothesis concerning the JT curative effect is that TG tissues have definite (background) "recognizing ability" (RA) (immunoattraction) for the mammalian immunocompetent cells (ICCs). This RA promotes the timely elimination of dead, apoptotic, or mutant cells of any functional organ. The RA thus ensures the organ's homeostasis.

In the case of disease, the "target" organ tissue renovation would be either suppressed or stimulated. In the first case, the tissue faces sclerosis and, in the second, it is altered; the RA (or immunoattraction) of the TG tissue of the HT-p for their own ICC changes. As a result, renovation of the TG tissue is suppressed. Possibly, there are some molecules (RA molecules; RA-m) in the JT water extract that increase the TG immune attraction; at the same time, TG tissue has suitable receptors for RA-m. Expression of these receptors is genetically determined, and this causes different individual sensitivity of mammalian tissues to the action of phytopreparations. RA-m of JT in an equal number of cases stimulate, block, or do not change the RA-m-receptors of the PHP-TG expression.

This study has shown that in the case of HT *in vitro*, the stimulation of the TG receptors by the JT-RA molecules was more often observed as compared to the PHP. This activity of RA-m possibly

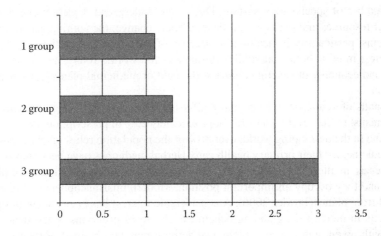

Figure 8.4 *Juniperus turcomanica* effects on the TG tissue of (a) PHP and (b) HT-p. Group I: traditional treatment; group II: traditional treatment + JT (patients who were not sensitive); 3 group III: traditional treatment + JT (sensitive patients).

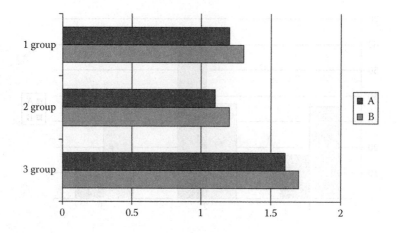

Figure 8.5   Serum TTG and T4 concentrations changing degree in dependents on the HT-p treatment. Group I: traditional treatment; group II: traditional treatment + JT (patients who were not sensitive); group III: traditional treatment + JT (sensitive patients).

is due to the curative effect of JT water extract. If it will be possible to select RA-m from JT water extract or RA-m-receptors from TG tissue, the approach to the HT treatment can be significantly changed. It will be possible to leave the HT hormone therapy to the management of the TG function through the "ligand-receptor" type of interaction. Undoubtedly, RA-m are presenting some glyco-proteides containing phytoextract. They are not exclusive components of JT. Future investigation in this direction is needed.

8.5 FUTURE DIRECTION

Within the next 50 years, the human population on our planet will probably pass nine billion, leading toward scarcities of renewable resources. The increase in the residential and agricultural areas will lead to a decrease in the total area of productive land and forests and a concomitant drop in the number of species they sustain. Due to the widespread depletion and degradation of land and water resources and perhaps significant climate change, the coming generations will face greater problems, particularly in health care because of a decrease in medicinal plants. This has triggered interest in ethnobotanical studies throughout the world. Today, due to its contributions to the health and economy of several countries, the field of medicinal plants has started attracting larger attention.

For thousands of years, one of our many gifts from the earth has been medicines to treat a myriad of ailments. In the developed world, approximately 25% of prescription products are derived from plants, and in the developing world, over 80% of the population relies on plants, in the form of traditional medicine, for their primary health care. Herbal medicines have been improved in developing countries as an alternative solution to health problems due to the high costs of pharmaceutical products, and they occupy an important position. Modern pharmacopoeia still contain 25% of drugs derived from plants (Fowler 2006). A considerable interest has been shown in the screening of plant extracts in modern drug and agrochemical discovery programs, since structurally novel chemotypes with potent and selective biological activity may be obtained of relevance to human disease treatment or improved crop production. Many semisynthetic derivatives of plant natural products also offer promise as useful drugs (Kinghorn 2000). The screening for biological activity of plant extracts, chromatographic fractions, and pure isolates requires the close cooperation of phytochemists and biologists.

The interest in herbal medicine in Turkmenistan has progressed parallel to the increased interest in other countries. The traditional herbal medicines are important for the life of people. For centuries, Turkmen have been using herbal medicine for the treatment of some diseases. Folk medicine is still commonly practiced in the country because of the geographical remoteness of many villages from modern medical facilities. Herbal treatment is largely based on traditional knowledge, which has survived since ancient times.

Relatively few medicinal plants are cultivated. Thus, the conservation of medicinal plant species requires that efforts be focused on key habitats. Many species are wild, cultivated, and naturalized in several places and, accordingly, their conservation and management would be better served by increased cooperation among the local people in this field. This gives an added sense of urgency to the task of recording their identity and uses, as well as initiating a program of preservation of the genetic resources of medicinal plants of the country.

Recently, some studies have been conducted in Turkmenistan to prevent the knowledge on folk medicine from disappearing. The conservation and sustainable use of plant diversity in the country is an essential element of any strategy to adapt to climate change. The response to these challenges needs to move much more rapidly and with more determination at all levels.

This information underlines the ethnobotanical richness of the region and the need to broaden this study. Furthermore, this constitutes a base for future phytochemical and pharmacological studies, which could lead to new therapeutic products. Turkmenistan has a great potential due to existing plant diversity in its natural flora. For a better look at our future, we need to understand and look deeply at our past. If this problem is not tackled now, we will regret not examining our existing natural resources for their medicinal potential. At the same time, we are ethically bound to think about the health care issues that we are leaving for our descendants as a result of our choices today.

REFERENCES

Adams, R. P. 2004. *Juniperus deltoides*, a new species and nomenclatural notes on *Juniperus poljcarpos* and *J. turcomanica* (Cupressaceae). *Phytologia* 86:49–53.

———. 2008. *Junipers of the world: The genus* Juniperus, 2nd ed., 402 pp. Vancouver: Trafford Publishing Co.

Akimov, A. P., Kuznetsov, S. I., Nilov, G. I., Chirkina, N. N., Krylova, A. P., and Litvinenko, R. M. 1976. Essential oils of junipers from ancient Mediterranean region. Composition, properties and prospectives of use. *Nikitskij Botanicheskij Sad Nikitsk Botan Sad* 69:79–93.

Ashuralyeva, M., and Ozturk, M. 2000. Plants used in the folk medicine of Turkmenia. *XIII. Plant Raw Materials Meeting*, Marmara Univ., Goztepe-Istanbul.

Atamuradov, K. I., Karyeva, O., Shammakov, S., and Yazkulyev, A. 1999. *The red data book of Turkmenistan*, vol. 2, 272 pp. Ministry of Nature Protection of Turkmenistan.

Chavchanidze, V. Y., and Kharabava, L. G. 1989. Juniper essential oils. *Subtropical Culture* 4:131–143.

Clark, A. M., McChesney, J. D., and Adams, R. P. 1990. Antimicrobial properties of the heartwood, bark/sapwood and leaves of *Juniperus* species. *Phytotherapy Research* 4:15–19.

Dembitsky, A. D. 1969. *Chemical content of steam oils of some wild growing plants from southern USSR*. Uzbek Academy of Science, Tashkent, Uzbekistan.

Fowler, M. W. 2006. Plants, medicines and men. *Journal of the Science of Food and Agriculture* 86:1797–1804.

Gofman, V. R., and Smirnov, V. S. 2000. The immune systems condition in the acute and chronic tonsillitis immunodeficits. *Foliant* 7:163–184.

Goriaev, M. I., and Ignatova, L. A. 1969. Chemistry of the junipers. *Science, Kazak SSR*, 1969:1–80.

Kinghorn, A. D. 2000. Plant secondary metabolites as potential anticancer agents and cancer chemopreventives. *Molecules* 2000 (5): 285–288.

Kitchens, G. D., Dorsky, J., and Kaiser, K. 1971. Cedarwood oil and derivatives. *Givaudanian* 1:3–9.

Lawrence, B. M. 1985. A review of the world production of essential oils. *Perfumer and Flavorist* 10:1–16.

Pleskanovskaya, S. A., Gurbandurdyyev, A., Konstantinova, T. G., and Berdiyew, M. 2003. On the possibility of using the *Juniperus turcomanica* decoction in the chronic tonsillitis treatment. *Turkmen Health Care Journal* 4:18–20.

Pleskanovskaya, S. A., Mamedova, K. A., and Allaberdiyev, A. A. 2008. Some immunological characteristics of autoimmune thyroditis. *Allergy and Immunology Journal* 9:132.

Rafique, M., Hanif, M., and Chaudary, F. M. 1993. Evaluation and commercial exploitation of essential oil of juniper berries of Pakistan. *Pakistan Journal of Scientific and Industrial Research* 36:107–109.

Tunalier, Z., Kirimer, N., and Baser, K. H. C. 2004. A potential new source of cedarwood oil: *Juniperus foetidissima* Willd. *Journal of Essential Oil Research* 16:233–235.

UN. 2009. Human development report, Turkmenistan. hhtp://hdrstats.undp.org/en/countries/country_fact_sheets/cty_fs_TKM.html

Walker, G. T. 1968. Cedarwood oil. *Perfume and Essential Oil Research* 59:346–350.

Yeger, L. 1986. Clinical immunology and allergology: Effect of *Juniperus turcomanica* decoction treatment on the functional activity of the TG depends on patients' sensitivity to phytopreparation *in vitro*. *Medicina* 3:262–289.

CHAPTER 9

Medicinal Plants of Ghana

P. Addo-Fordjour, A. K. Anning, W. G. Akanwariwiak,
E. J. D. Belford, and C. K. Firempong

CONTENTS

9.1 INTRODUCTION

The use of plants as medicines started many hundreds of years ago but it was not until the eighteenth and nineteenth centuries that medicinal plants gained recognition for their role in providing for the health care needs of many people in the world (Ohja 2000; Ticktin et al. 2002). Traditional medicine over the years has played a vital role in providing healing and contributing toward the discovery of most pharmaceutically active substances in plants used in the commercial production of drugs (Principe 1991; Pearce and Puroshothaman 1992; Robbers, Speedie, and Tyler 1996).

The use of medicinal plants in treating diseases is widespread in Ghana (located in Western Africa, latitude 8°00'N and longitude 00'W; Figure 9.1), and this often involves a wide variety of species. This popularity is partly attributed to the fact that traditional medical practices are less expensive and more accessible to a majority of the rural communities, which constitute the largest proportion of the country's population. Available estimates show that between 60 and 70% of Ghanaians rely on traditional medical systems for their health needs (Sarpong 2000).

In addition to providing for the health care needs of the people, medicinal plants also have the potential to increase the economic status of traditional healers and people who trade in them in many communities throughout Ghana. Although Ghana's tropical vegetation is rich in medicinal plants, knowledge of their distribution and uses appear to be the preserve of the elderly and, particularly the herbalists (Abel and Busia 2005), most of whom acquired this knowledge through oral transmission or as a result of using the plants in traditional medicine preparation. Unfortunately, many people in Ghana, especially the younger generation, have barely any knowledge of medicinal plants.

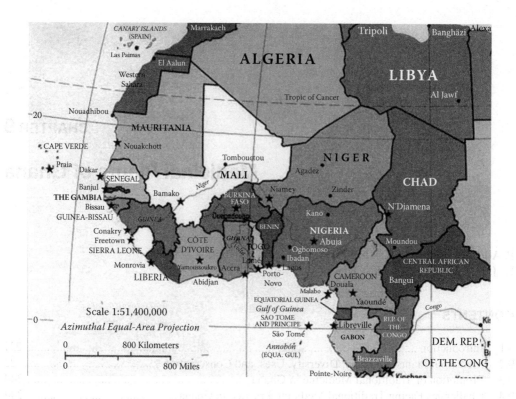

Figure 9.1   A section of a map of Africa showing Ghana.

In general, information on the diversity and uses of medicinal plants in Ghana has only been made known through ethnobotanical studies by some individual researchers (Abbiw 1990; Sarpong 2000; Agbovie et al. 2002; Abel and Busia 2005; Asase et al. 2005; Addo-Fordjour et al. 2008). Such studies, beside being few in number and isolated, have typically focused on the efficacy of plant extracts as well as their safety assessment. Consequently, scant literature exists on the diversity of medicinal plants and their uses in Ghana. This chapter lists some plants commonly used by herbal practitioners or cited for their medicinal values in Ghana, and also provides an overview of their uses and conservation status. Efforts aimed at promoting traditional medicinal plants to complement orthodox medicine as well as some associated challenges are also discussed. This compilation, sourced from various publications, is intended to serve as a reference source for any future studies on medicinal plants in Ghana and other countries of the world.

9.2 MEDICINAL PLANTS IN GHANA: DIVERSITY, USES, AND CONSERVATION

The tropical forest and savannah ecosystems in Ghana are noted for the richness of their plant species, many of which have, at one point or the other, been used traditionally in treating various disease conditions. Estimates by the National Biodiversity Council (Ministry of Environment and Science 2002) suggest there may be a total of 3,600 species of vascular plants in Ghana, out of which the tropical forests of Ghanas alone are home to about 21,000 species (Agyeman et al. 1999). While a comprehensive database is still lacking in Ghana, isolated documentations of medicinal plant species used in different parts of the country exist. These have been possible through the efforts of a few individuals or researchers (Abbiw 1990; Agbovie et al. 2002; Abel and Busia 2005; Ameyaw et al. 2005; Asase et al. 2005; Addo-Fordjour et al. 2008). Although the majority of medicinal plants are derived from the forests, some species are also obtained from other ecosystems (Table 9.1).

Table 9.1 Diversity of Medicinal Plant Species and Their Uses in Ghana

Scientific Name	Family	Local Name	Habit	Part Used	Purpose	Sources
Abrus precatorius	Fabaceae	Obirekuaiura	Liana	Stem, bark	Oliguria, rheumatism	Agbovie et al. (2002)
				Leaf	Asthma	Abbiw (1990)
				Root	Aphrodisiac	Abbiw (1990)
Acacia kamerunensis	Fabaceae	Nnwere	Liana	Leaf	Measles, fever, skin diseases, ringworm infection, wounds	Abbiw (1990), Addo-Fordjour et al. (2008)
				Root	Aphrodisiac, toothache	Abbiw (1990)
Acanthospermum hispidum	Asteraceae	Sharaha-nsoe	Herb	Whole plant	Malaria, stomach disorder	Asase et al. (2005), Ayensu (1978)
Adenia cissampeliodes	Passifloraceae	Hamakyem	Liana	Bark, seed, stem, whole leaf, plant	Hypertension, numbness, wound	Agbovie et al. (2002)
Afromomum latifolium	Zingiberaceae	Sensan	Herb	Root	Cough, fibroid	Agbovie et al. (2002)
Afromomum melegueta	Zingiberaceae	Fomwisa		Seed	Chicken pox, flatulence, dysentery, earache, cough, female infertility	Abbiw (1990), Addae-Mensah and Aryee (1992), Addo-Fordjour et al. (2008)
				Root	Antihelminthic	Abbiw (1990)
				Leaf	Anemia, rheumatism	Addae-Mensah and Aryee (1992)
				Fruit	Dysentery	Abbiw (1990), Addo-Fordjour et al. (2008)
Afzelia africana	Fabaceae	Papao	Tree	Bark	Piles, pneumonia, diuretic, fever	Asase et al. (2005)
				Leaf	Malaria	Abbiw (1990), Agbovie et al. (2002)
Ageratum conyzoides	Asteraceae	Oboakro	Herb	Whole plant	Menstrual disorder, diuretic, cough, dysentery, female infertility, craw-craw	Agbovie et al. (2002)
				Leaf	Dysentery, skin ulcer	Ameyaw et al. (2005)
				Aerial part	Female infertility	Addo-Fordjour et al. (2008), Agbovie et al. (2002)
Albizia zygia	Fabaceae	Okuro	Tree	Bark	Stomach upset, infertility, arthritis, appetizer	
				Leaf	Aphrodisiac	
				Root	Cough	Abbiw (1990), Agbovie et al. (2002)
Alchornea cordifolia	Euphorbiaceae	Gyama	Shrub	Root, leaf	Whitlow, stomachache, jaundice, yaws, fever, ringworms, wounds	Abbiw (1990)
				Root	Abortion	

(continued)

Table 9.1　Diversity of Medicinal Plant Species and Their Uses in Ghana (Continued)

Scientific Name	Family	Local Name	Habit	Part Used	Purpose	Sources
				Bark	Appetizer	Abbiw (1990)
Allophylus africanus	Sapindaceae	Odwendwena	Shrub	Bark	Lactogenic, hemorrhoids	Agbovie et al. (2002)
Alstonia boonei	Apocynaceae	Onyamedua	Tree	Bark, root	Stomachache	Agbovie et al. (2002)
				Stem, root, leaf	Malaria, hypertension, wound	Agbovie et al. (2002)
				Bark	Fever	Agbovie et al. (2002)
				Leaf	Placenta retention	Agbovie et al. (2002)
				Root	Measles	Agbovie et al. (2002)
Alternanthera pungens	Amaranthaceae	Nsoesoe	Herb	Whole plant	Dysentery, catarrh, purgative	Abbiw (1990), Agbovie et al. (2002)
				Leaf	Dysentery, yaws, diarrhea, lactogenic, neuralgia	Abbiw (1990), Agbovie et al. (2002)
Anacardium occidentale	Anacardiaceae	Atea	Tree	Leaf, root	Diarrhea, yaws	Abbiw (1990), Agbovie et al. (2002)
				Fruit	Diuretic	Abbiw (1990)
				Leaf	Pile	Abbiw (1990)
				Leaf, bark	Sore mouth/gums, toothache	Abbiw (1990)
				Seed	Vesicant	Abbiw (1990)
Ananas comosus	Bromeliaceae	Aborobe	Crop	Root, fruit	Jaundice	Abbiw (1990), Agbovie et al. (2002)
				Fruit	Antihelminthic, bladder trouble, diuretic, dropsy	Abbiw (1990)
Anthocleista nobilis	Loganiaceae	Owudifo kete	Tree	Root, stem, bark	Catarrh, hemorrhoids, laxative, constipation, hepatitis, syphilis,	Agbovie et al. (2002)
				Bark	Antihelminthic, colic, wounds, dysmenorrhea, stomachache	Abbiw (1990)
				Root	Piles	Abbiw (1990)
				Young shoot	Ulcers	Abbiw (1990)
Antiaris toxicaria	Moraceae	Kyenkyen	Tree	Bark	Stomach upset, anemia, bronchial trouble, leprosy, labor facilitation	Abbiw (1990), Addo-Fordjour et al. (2008)
				Stem, bark	Epilepsy	Agbovie et al. (2002)
Antrocaryon micraster	Anacardiaceae	Aprokuma	Tree	Bark	Chicken pox	Addo-Fordjour et al. (2008)
				Fruit	Stomachache	Abbiw (1990)
Azadirachta indica	Meliaceae	Nyeedua	Tree	Bark	Ringworm, boil, malaria	Agbovie et al. (2002)

Scientific name	Family	Local name	Habit	Part used	Uses	References
Balanites aegyptica	Zygophyllaceae	Ohwirem	Tree	Leaf, seed	Fever, hepatitis, ringworm,	Abbiw (1990), Agbovie et al. (2002)
				Root	Fever	Abbiw (1990)
				Seed	Wound, antihelminthic, antidote to snake venom	Abbiw (1990)
Baphia nitida	Fabaceae	Odwen	Tree	Root, fruit	Skin diseases	Agbovie et al. (2002)
				Bark	Skin diseases	Ameyaw et al. (2005)
				Leaf	Fever, high blood pressure, flatulence, enteritis, jaundice, purgative	Abbiw (1990), Addo-Fordjour et al. (2008)
Bauhinia thonningii	Fabaceae	Otokotaka	Shrub	Leaf, bark, seed	Diarrhea, yaws	Agbovie et al. (2002)
				Leaf, root	Diarrhea, migraine	Agbovie et al. (2002)
				Stem, bark	Arthritis, snakebite	Agbovie et al. (2002)
Berlina confusa	Fabaceae	Nseduansehoma	Shrub	Leaf, root	Menstrual pain, purgative	Abbiw (1990), Addo-Fordjour et al. (2008)
				Leaf	Gastrointestinal pain	Abbiw (1990)
				Bark	Dysmenorrhea, dyspepsia, earache	Abbiw (1990)
Bidens pilosa	Asteraceae	Gyinantwi	Herb	Whole plant	Jaundice, hypertension	Agbovie et al. (2002)
				Leaf	Hypertension, anemia, earache, styptic	Abbiw (1990), Addo-Fordjour et al. (2008)
Blighia sapida	Sapindaceae	Akye	Tree	Leaf, bark	Stomach upset	Addo-Fordjour et al. (2008)
				Bark	Conjunctivitis, consumption, convulsion, fits, epilepsy, hernia, intercostal pains	Abbiw (1990)
				Root	Sexual weakness, cancer, burns, bruises, bronchitis	Abbiw (1990)
				Leaf, fruit	Diarrhea, migraine, yaws	Agbovie et al. (2002)
				Leaf	Sore eyes, conjunctivitis, iritis, trachoma, yaws	Abbiw (1990)
				Pulp of leafy twig	Hemoptysis, hemorrhoids, headache	Abbiw (1990)
				Seed	Smallpox	Abbiw (1990)
Blighia welwitschii	Sapindaceae	Akyekobri	Tree	Bark	Measles	Addo-Fordjour et al. (2008)
Boerhavia diffusa	Nyctaginaceae	Nkokodwe	Herb	Whole plant	Asthma, boils, dropsy	Abbiw (1990), Agbovie et al. (2002)

(continued)

Table 9.1   Diversity of Medicinal Plant Species and Their Uses in Ghana (Continued)

Scientific Name	Family	Local Name	Habit	Part Used	Purpose	Sources
				Root	Anemia, boil, genitourinary troubles, giddiness, gonorrhea, griping, guinea worm sores	Abbiw (1990)
				Root, leaf	Asthma	Abbiw (1990)
				Leaf	Emetic	Abbiw (1990)
Bombax buonopozense	Bombaceae	Akatanini	Tree	Leaf	Candidiasis	Addo-Fordjour et al. (2008)
				Bark	Stomachache, purification of blood, lactogenic, amenorrhea	Abbiw (1990), Agbovie et al. (2002)
				Gum resin	Craw-craw	Abbiw (1990)
Caesalpina benthamiana	Fabaceae	Akoobowerew	Liana	Leaf, root	Sexual weakness, yaws	Agbovie et al. (2002)
Calliandra portoricensis	Fabaceae	Nhwatenhurat	Shrub	Whole plant	Headache, lumbago	Agbovie et al. (2002)
				Leaf	Headache	Abbiw (1990)
				Root	Lumbago	Abbiw (1990)
Capparis erythrocarpus	Capparidaceae	Pitipiti	Shrub	Root	Aphrodisiac	Agbovie et al. (2002)
				Root bark	Arthritis	Ameyaw et al. (2005)
Carapa procera	Meliaceae	Kwaebese	Tree	Bark	Body pains, syphilis, sinusitis, chest pains, sore eyes, iritis, conjunctivitis, trachoma, fever	Abbiw (1990), Agbovie et al. (2002), Addo-Fordjour et al. (2008)
				Dry leaf	Hypertension	Abbiw (1990)
				Stem, bark	Tuberculosis, anemia	Agbovie et al. (2002)
				Root	Hematuria, lactogenic, bronchial trouble	Abbiw (1990)
				Seed	Emetic	Abbiw (1990)
				Root bark	Headache	Abbiw (1990)
				Leaf, bark	Bladder trouble, biliousness	Abbiw (1990)
Carica papaya	Cariaceae	Brofere	Tree	Seed	Jaundice, skin ulcer, cough	Agbovie et al. (2002)
				Leaf, root	Malaria	Addo-Fordjour et al. (2008)
				Dry leaf	Abortion	Abbiw (1990)
				Root of male	Headache	Abbiw (1990)
Cassia occidentalis	Fabaceae	Mmofrabrode	Shrub	Leaf, seed	Stomachache, toothache, hypertension	Agbovie et al. (2002)
				Root	Antihelminthic, asthenia, bladder trouble	Abbiw (1990)

Species	Family	Local name	Habit	Part used	Use	Reference
Cassia podocarpa	Fabaceae	Mumuaha	Shrub	Seed	Cataract	Abbiw (1990)
				Root, leaf, fruit	Fainting recovery	Abbiw (1990)
				Leaf	Malaria, constipation, Guinea worm sores, gonorrhea	Abbiw (1990), Agbovie et al. (2002)
Cassia rotundifolia	Fabaceae	Asase neobo	Herb	Fruit	Purgative	Abbiw (1990)
Cassia sieberiana	Fabaceae	Nkokowu	Tree	Whole plant	Stomachache	Agbovie et al. (2002)
				Root	Stomachache	Agbovie et al. (2002)
				Root bark	Aphrodisiac	Abbiw (1990)
				Root, fruit	Apoplexy, appetizer, dizziness	Abbiw (1990)
Ceiba pentandra	Bombaceae	Onyina	Tree	Bark	Hernia, emetic, fever	Abbiw (1990), Addo-Fordjour et al. (2008)
				Root	Diarrhea, leprosy	Abbiw (1990)
				Leaf	Colic	Abbiw (1990)
				Leaf, fruit	Hemoptysis, hemorrhoids, hemostatic, headache	
				Oil	Rheumatism	Abbiw (1990)
Celtis milbraedii	Ulmaceae	Esa	Tree	Bark	Hernia, pneumonia	Addo-Fordjour et al. (2008)
				Root	Arthritis	Agbovie et al. (2002)
Cercestis afzelii	Araceae	Batatwene	Vine	Stem	Gonorrhea	Addo-Fordjour et al. (2008)
Chromolaena odorata	Asteraceae	Acheampong		Leaf	Wound healing, styptic	Addo-Fordjour et al. (2008) Agbovie et al. (2002)
Citrus aurantifolia	Rutaceae	Ankaa		Fruit juice, leaf, root	Urinary retention, yaws	Agbovie et al. (2002)
Clausena anisata	Rutaceae	Sesadua	Tree	Root	Arthritis, cough	Agbovie et al. (2002)
				Root bark	Rheumatoid arthritis	Ameyaw et al. (2005)
				Leaf, root	Asthma, dysentery, abdominal pains, antihelminthic	Agbovie et al. (2002)
Cleistopholis patens	Annonaceae	Abotokuradua	Tree	Bark	Measles, improper growth in children, hepatitis	Abbiw (1990), Addo-Fordjour et al. (2008)
				Leaf	Sleeping sickness, fever, antihelminthic	Abbiw (1990)
Cnestis ferruginea	Connaraceae	Apoosen	Shrub	Leaf	Dysentery, cold, sore eyes, iritis, trachoma, conjunctivitis	Abbiw (1990), Addo-Fordjour et al. (2008)

(continued)

Table 9.1 Diversity of Medicinal Plant Species and Their Uses in Ghana (Continued)

Scientific Name	Family	Local Name	Habit	Part Used	Purpose	Sources
				Leafy twig	Fever	Abbiw (1990)
				Leaf, stem	Cough, anemia	Agbovie et al. (2002)
				Root	Cough, abortion, headache, mental trouble	Abbiw (1990), Agbovie et al. (2002)
				Fruit	Oral hygiene	Addo-Fordjour et al. (2008)
Cocos nucifera	Arecaceae	Kube	Tree	Fruit	Herpes	Agbovie et al. (2002)
				Bark	Earache	Abbiw (1990)
Cola gigantea	Sterculiaceae	Awapuo	Tree	Bark	Waist pain, syphilis	Abbiw (1990), Addo-Fordjour et al. (2008)
				Stem	Toothache	Abbiw (1990)
				Dry leaf	Stomach ulcer	
Cola nitida	Sterculiaceae	Bese	Tree	Bark, fruit	Fracture, herpes, dystocia	Agbovie et al. (2002)
Combretum fibribundum	Combretaceae	Ohwienba	Shrub	Leaf, root	Guinea worm eradication	Agbovie et al. (2002)
Corynanthe pachyceras	Rubiaceae	Duagya	Tree	Stem, bark	Aphrodisiac	Agbovie et al. (2002)
Cryptolepis sanguinolenta	Asclepiadaceae	Nibima		Root	Malaria	Ameyaw et al. (2005)
Cucumis melo	Curcurbitaceae	Kuradonton	Herb	Root	Cough, blood tonic, stomach ache	Agbovie et al. (2002)
Datura suaveolens	Solanaceae	Korantema	Shrub	Leaf	Tooth cleaning	Agbovie et al. (2002)
Desmodium adscendens	Fabaceae	Nkatenkate	Herb	Leaf, stem	Asthma	Addo-Fordjour et al. (2008)
				Whole plant	Asthma, pneumonia	Agbovie et al. (2002)
				Leaf	Antihelminthic, convulsion, cough, diarrhea	Abbiw (1990)
				Dry leaf	Asthma	Abbiw (1990)
				Leafy twig, root	Chest pain	Abbiw (1990)
				Leafy stem	Purgative	Abbiw (1990)
Dialium guineense	Fabaceae	Osenafo	Tree	Bark, root	Stomatitis, toothache, hemorrhoids	Agbovie et al. (2002)
				Leaf	Asthma, catarrh, chest pain, labor facilitation, kidney disease, edema, palpitations	Abbiw (1990)
				Fruit	Diarrhea	Abbiw (1990)
				Shoot	Fever	Abbiw (1990)
				Bark	Sore throat, stomachache	Abbiw (1990)

Species	Family	Local name	Habit	Part used	Uses	Reference
Dichapetalum toxicarium	Dichapetalaceae	Ofoabiri	Shrub	Leaf	Malaria	Agbovie et al. (2002)
				Bark	Swellings, fainting recovery	Abbiw (1990)
Dioclea reflexa	Fabaceae	Ntewhama	Shrub	Seed	Asthma	Agbovie et al. (2002)
Diodia scandens	Rubiaceae	Apaproyem	Herb	Whole plant	Cough, arthritis	Agbovie et al. (2002)
Dioscorea dumentorum	Dioscoreaceae	Nkanto hama	Vine	Tuber	Antihelminthic	Agbovie et al. (2002)
Dissotis rotundifolia	Melastomaceae	Boreakete	Herb	Leaf, root	Abdominal pain, diarrhea	Agbovie et al. (2002)
				Leaf	Iritis, conjunctivitis, trachoma, eyes sore	Abbiw (1990)
				Dry leaf	Cold, cough	Abbiw (1990)
				Whole plant	Sinusitis	Abbiw (1990)
Dracaena arborea	Dracaenaceae	Nsomme	Shrub	Dry leaf	Stomach ulcer	Addo-Fordjour et al. (2008)
Elaeis guineensis	Arecaceae	Abe	Tree	Fruit (oil)	Boil, dracontiasis, wound, filaries, craw-craw	Abbiw (1990), Addo-Fordjour et al. (2008)
				Root	Headache, labor facilitation	Abbiw (1990)
				Young petioles	Wound	Abbiw (1990)
Elaeophorbia grandifolia	Euphorbiaceae	Kanne	Tree	Leaf, root	Boils, contraceptives	Agbovie et al. (2002)
Euadenia eminens	Capparaceae	Dinsinkro	Tree	Root	Aphrodisiac, tuberculosis, earache	Abbiw (1990), Addo-Fordjour et al. (2008)
			Shrub	Root, bark	Otalgia, rectal prolapse, aphrodisiac	Agbovie et al. (2002)
				Fruit	Aphrodisiac, eyes sore, iritis, conjunctivitis, trachoma	Abbiw (1990)
Eugenia aromatica	Myrtaceae	Pepra	Shrub	Flower bud	Stomach pains	Agbovie et al. (2002)
Ficus asperifolia	Moraceae		Tree	Bark	Cancer	Abbiw (1990)
Ficus capensis	Moraceae	Oketewanfro	Tree	Bark	Stomachache	Agbovie et al. (2002)
				Stem, bark	Lactogenic, wound, diarrhea	Agbovie et al. (2002)
				Leaf, seed	Lactogenic	Addo-Fordjour et al. (2008)
				Bark	Convulsion, fits, epilepsy	Abbiw (1990)
				Leafy stem	Diarrhea	Abbiw (1990)
Ficus exasperata	Moraceae	Nyankyerene	Shrub	Leaf	Asthma, cataract, cough, cold, antihelminthic	Abbiw (1990), Addo-Fordjour et al. (2008)
				Leaf, latex	Eyes sore, conjunctivitis, iritis, trachoma	Abbiw (1990)
Flacourita flavescens	Flacourtiaceae	Pitipiti	Shrub	Root	Toothache	Agbovie et al. (2002)

(continued)

Table 9.1 Diversity of Medicinal Plant Species and Their Uses in Ghana (Continued)

Scientific Name	Family	Local Name	Habit	Part Used	Purpose	Sources
Flagellaria guineensis	Flagellariaceae	Mmirebia	Tree	Stem	Diarrhea	Abbiw (1990)
Funtumia elastica	Apocynaceae	Ofuntum	Tree	Root	Piles	Agbovie et al. (2002)
				Root	Edema, incontinence, in urine	Abbiw (1990), Addo-Fordjour et al. (2008)
Garcinia kola	Guttiferae	Tweapea	Tree	Bark	Piles	Abbiw (1990)
				Root, stem	Tooth cleaning	Agbovie et al. (2002)
				Fruit	Antihelminthic	Abbiw (1990)
				Seed	Bronchial trouble, chest pain, colic	Abbiw (1990)
				Bark	Purgative	Abbiw (1990)
Gardenia tenuifolia	Rubiaceae	Petebiri	Shrub	Bark, root, stem	Hypertension	Agbovie et al. (2002)
				Leaf, root	Hypertension, skin disease	Agbovie et al. (2002)
Gongronema latifolium	Asclepiadaceae	Ansrogya	Liana	Stem	Pneumonia	Addo-Fordjour et al. (2008)
				Leafy stem	Colic	Abbiw (1990)
				Leaf	Antihelminthic, aids infants to walk early, cough	Abbiw (1990), Addo-Fordjour et al. (2008)
				Fruit	Purgative	Abbiw (1990)
				Stem	Stomachache	Agbovie et al. (2002)
Gossypium arboreum	Malvaceae	Asaawa	Shrub	Leaf, root	Dysentery, malaria, vomiting	Agbovie et al. (2002)
				Root	Abortion, amenorrhea	Abbiw (1990)
				Leaf	Blood purifier, diarrhea	Abbiw (1990)
				Seed	Headache	Abbiw (1990)
				Leaf-seed kernel	Swellings	Abbiw (1990)
Griffonia simplicifolia	Fabaceae	Kagya	Liana	Root	Impotence	Addo-Fordjour et al. (2008)
				Leaf	Headache, bladder trouble	Abbiw (1990), Addo-Fordjour et al. (2008)
				Leafy stem, leaf	Aphrodisiac	Abbiw (1990)
				Leafy stem	Purgative	Abbiw (1990)
				Leaf, root	Congestion, fracture	Abbiw (1990)
Guarea cedrata	Meliaceae	Kwabohoro	Tree	Stem, bark	Anemia, stomach ulcer	Agbovie et al. (2002)
Heritiera utilis	Sterculiaceae	Kwakuoaduaba	Tree	Dry leaf	Kwashiokor	Addo-Fordjour et al. (2008)
				Leaf	Styptic	Abbiw (1990)

Species	Family	Local name	Habit	Part	Uses	References
Hildegardia barteri	Sterculiaceae		Tree	Leaf, bark	Aphrodisiac	Abbiw (1990)
Hilleria latifolia	Phytolaccaceae	Ofosow	Herb	Bark	Epilepsy	Agbovie et al. (2002)
Holarrhena floribunda	Apocynaceae	Anakranaku	Tree	Whole plant	Skin diseases	Agbovie et al. (2002)
				Bark	Stomachache, fever, jaundice, ringworm, skin diseases	Abbiw (1990), Addo-Fordjour et al. (2008)
Hoslundia opposita	Lamiaceae	Abrewa aninso	Tree	Leaf, bark	Diarrhea	Abbiw (1990)
				Leaf	Malaria, styptic	Agbovie et al. (2002)
				Root	Antiseptic, cold	Abbiw (1990)
				Bark	Convulsion	Abbiw (1990)
			Shrub	Whole plant	Gonorrhea, jaundice, diabetes	Agbovie et al. (2002)
Hyptis pectinata	Lamiaceae	Opeabaa	Herb	Leaf	Hypermesis gravidarum, ulcer, fever, labor facilitation	Abbiw (1990), Agbovie et al. (2002)
Jatropha curcas	Euphorbiaceae	Nkanyadua	Shrub	Leaf	Wound, convulsion, diarrhea, guinea worm, candidiasis	Abbiw (1990), Addo-Fordjour et al. (2008), Musa (2009)
				Root, leaf	Measles, styptic, impotence, wounds, jaundice, yellow fever	Agbovie et al. (2002)
				Bark	Antihelminthic	Abbiw (1990)
				Oil	Craw-craw	Abbiw (1990)
				Seed	Dropsy	Abbiw (1990)
				Root	Incontinence of urine	Abbiw (1990)
Justicia flava	Acanthaceae	Ntumunum	Herb	Leaf	Hemorrhoids, stomachache, fever	Abbiw (1990), Addo-Fordjour et al. (2008)
				Bark	Diarrhea	Abbiw (1990)
				Whole plant	Yaws	Abbiw (1990)
				Flower	Hemorrhoids	Abbiw (1990)
Kalanchoe pinnata	Crassulaceae	Afare	Herb	Leaf	Cough, eye drop	Agbovie et al. (2002)
Khaya senegalensis	Meliaceae	Odupong	Tree	Bark	Blood purifier, blood tonic, piles, female infertility, tuberculosis, abortion, antihelminthic, boils, emetic, fever, appetizer, gastrointestinal disorders, dysmenorrhea	Abbiw (1990), Agbovie et al. (2002)
				Stem, bark	Headache, arthritis, convulsion, anemia	Agbovie et al. (2002)

(continued)

Table 9.1   Diversity of Medicinal Plant Species and Their Uses in Ghana (Continued)

Scientific Name	Family	Local Name	Habit	Part Used	Purpose	Sources
Kigelia africana	Bignoniaceae	Nufuten	Tree	Bark	Infertility, lactogenesis	Abbiw (1990), Addo-Fordjour et al. (2008)
				Bark, seed	Amenorrhea	Abbiw (1990)
				Fruit	Lumbago	Abbiw (1990)
				Root bark	Piles	Abbiw (1990)
				Root	Anthelminthic	Abbiw (1990)
				Leaf	Diarrhea	Abbiw (1990)
				Bark, leaf	Bladder trouble	Abbiw (1990)
				Whole plant	Piles, wounds, anemia	Agbovie et al. (2002)
				Leaf, fruit, root	Constipation, tapeworm	Agbovie et al. (2002)
				Bark, seed, root	Stomachache	Agbovie et al. (2002)
Landolphia dulcis	Apocynaceae	Hama-fufu	Liana	Root	Chest pains, aphrodisiac	Agbovie et al. (2002)
Landolphia owariensis	Apocynaceae	Kooko ahoma	Liana	Leaf, root	Malaria	Addo-Fordjour et al. (2008)
Lannea welwithschii	Anacardiaceae	Okumanini	Tree	Bark, stem	Abdominal pain, skin ulcer	
Lantana camara	Verbenaceae	Anansedokon	Shrub	Whole plant	Wound, jaundice, fever	Agbovie et al. (2002)
				Latex	Lactogenic	Abbiw (1990)
Lecaniodiscus cupanioides	Sapindaceae	Odwindwera	Tree	Leaf, stem, bark	Cough, wound	Addo-Fordjour et al. (2008)
Lippia multiflora	Saanunum			Leaves	Hypertension, laxative, febrifuge	Ameyaw et al. (2005)
Mallotus oppositifolius	Euphorbiaceae	Satadua	Shrub	Leaf, root	Wounds, dysentery, lumbago, anemia	Abbiw (1990), Addo-Fordjour et al. (2008)
				Root	Lumbago	Abbiw (1990)
				Leaf	Headache, styptic, ulcers	Abbiw (1990)
				Leaf, seed	Styptic, measles	Agbovie et al. (2002)
				Whole plant	Migraine, lumbago	Agbovie et al. (2002)
Mangifera indica	Anacardiaceae	Mango	Tree	Stem, bark, root	Cough, diarrhea, toothache, jaundice, fever	Abbiw (1990), Agbovie et al. (2002)
				Leaf	Diuretic, toothache	Abbiw (1990)
				Leaf, bud, root	Fever	Abbiw (1990)
				Leaf, bark	Sore gums	Abbiw (1990)
Manihot esculenta	Euphorbiaceae	Bankye	Shrub	Leaf	Wound, hemorrhage, snakebite	Agbovie et al. (2002)
				Leaf, root	Blood, eye trouble	Agbovie et al. (2002)
Mansonia altissima	Sterculiaceae	Oprono	Tree	Dry leaf	Body pains	Addo-Fordjour et al. (2008)

Species	Family	Local name	Habit	Part used	Uses	References
Marantochloa leucantha	Marantaceae	Sibrie	Herb	Root	Aphrodisiac	Abbiw (1990)
				Seed	Boil	Abbiw (1990), Addo-Fordjour et al. (2008)
				Flower	Stomachache	Abbiw (1990)
Margaritaria discoides	Euphorbiaceae	Opapea	Tree	Leaf	Wound, earache, eyes sore, iritis, conjunctivitis, trachoma	Abbiw (1990), Agbovie et al. (2002)
				Root	Aphrodisiac, diarrhea	Abbiw (1990)
				Bark	Boil	Abbiw (1990)
Maytenus senegalensis	Celastraceae	Okumapaafo	Tree	Bark, root	Dyspepsia, wound	Agbovie et al. (2002)
				Root	Colic	Abbiw (1990)
				Leaf	Antihelminthic, boil	Abbiw (1990)
Milicia excelsa	Moraceae	Odum	Tree	Bark	Headache, cough	Abbiw (1990), Addo-Fordjour et al. (2008)
				Leaf	Fever	Abbiw (1990)
				Latex	Craw-craw	Abbiw (1990)
				Bark, stem	Cough	Agbovie et al. (2002)
				Bark	Toothache, stomachache	Agbovie et al. (2002)
				Leaf, root	Stomachache	Agbovie et al. (2002)
Momordica charantia	Cucurbitaceae	Nyanya	Vine	Leaf, seed	Boil	Agbovie et al. (2002)
				Whole plant	Diabetes, hypertension, appetizer	Abbiw (1990), Agbovie et al. (2002)
				Fruit	Blood purifier	Abbiw (1990)
				Fruit, oil	Swellings	Abbiw (1990)
Momordica foetida	Cucurbitaceae	Seprepe	Vine	Leaf	Wound, fever, counterirritant	Abbiw (1990), Agbovie et al. (2002)
Monodora aegyptica	Annonaceae	Tree	Tree	Young leaf	Stomachache	Abbiw (1990)
				Root, seed	Stomachache	Agbovie et al. (2002)
Monodora myristica	Annonaceae	Wedeaba	Tree	Seed	Stomachache, candidiasis, numbness, guinea worm, headache, purgative, hemorrhoids	Abbiw (1990), Addo-Fordjour et al. (2008), Agbovie et al. (2002), Ameyaw et al. (2005)
Morinda lucida	Rubiaceae	Konkroma	Tree	Root	Anemia, wound	Agbovie et al. (2002)
				Leaf	Fever, malaria, blood purifier, typhoid fever	Abbiw (1990), Addo-Fordjour et al. (2008), Agbovie et al. (2002), Ameyaw et al. (2005)
				Bark	Candidiasis	
				Bark, root	Chest pain, jaundice	Abbiw (1990)
				Leaf, root	Colic	Abbiw (1990)

(continued)

Table 9.1 Diversity of Medicinal Plant Species and Their Uses in Ghana (Continued)

Scientific Name	Family	Local Name	Habit	Part Used	Purpose	Sources
Murraya micrantha	Rutaceace	Dubrafo	Tree	Leaf	Purgative	Agbovie et al. (2002)
Musa paradisiaca	Musaceae	Brode	Herb	Leaf, root	Goiter, wound, palpitation	Agbovie et al. (2002)
				Leaf	Cancer	Abbiw (1990)
Musanga ceropiodes	Cecropiaceae	OdwumaTree		Bark	Chest pain, intoxicant	Abbiw (1990)
				Trunk sap	Lactogenic	Abbiw (1990)
Nauclea diderrichii	Rubiaceae	Kusia	Tree	Root	Sexual weakness	Addo-Fordjour et al. (2008)
				Bark	Stomachache, jaundice	Abbiw (1990)
Nauclea latifolia	Rubiaceae	Peyarediasa	Tree	Root, stem, bark	Arthritis, malaria	Agbovie et al. (2002), Ameyaw et al. (2005)
				Root, leaves	Malaria	Asase et al. (2005)
Nesogordonia papaverifera	Sterculiaceae	Odanta	Tree	Bark	Chest pain, peptic ulcer	Agbovie et al. (2002)
Newbouldia laevis	Bignoniaceae	Anansent!entan	Tree	Leaf	Bone fracture, fever	Abbiw (1990), Addo-Fordjour et al. (2008)
				Root, leaf	Epilepsy, convulsion, skin ulcer, peptic ulcer, malaria, anemia, aphrodisiac	Abbiw (1990), Agbovie et al. (2002)
				Bark	Antihelminthic	Abbiw (1990)
				Root, stem	Syphilis, antihelminthic, stomachache	Abbiw (1990)
Nicotiana tabacum	Solanaceae	Numuaha	Shrub	Leaf	Wound, toothache	Agbovie et al. (2002)
Ocimum canum	Lamiaceae	Mme	Shrub	Leaf	Treatment of poisoning, malaria	Agbovie et al. (2002), Asase et al. (2005)
				Whole plant	Malaria	Asase et al. (2005)
Ocimum gratissimum	Lamiaceae	Onunum	Shrub	Leaf	Wound healing, appetizer, catarrh, colds, cough, diarrhea	Abbiw (1990), Addo-Fordjour et al. (2008)
				Leaf, root	Fever, stomachache, snakebite, dysentery, malaria, catarrh	Agbovie et al. (2002)
				Whole plant	Diaphoretic	Abbiw (1990)
Oncoba spinosa	Flacourtiaceae	Astrotoa	Tree	Leaf, root	Cough, wounds	Agbovie et al. (2002)
				Seed	Fever, leprosy	Abbiw (1990)
				Root	Diarrhea, bladder trouble	Abbiw (1990)

Species	Family	Local name	Habit	Plant part	Uses	References
Pachypodanthium staudtii	Annonaceae	Duawusa	Tree	Stem, bark	Abdominal pain, cough, arthritis tumor	Abbiw (1990)
				Bark	Chest pain, gastrointestinal disorder, tumor	Abbiw (1990)
Palisota hirsuta	Commelinaceae	Mpentem	Herb	Root	Anemia, dysentery, labor facilitation, stomachache	Abbiw (1990), Agbovie et al. (2002)
				Whole plant	Analgesic, convulsion	Abbiw (1990)
				Leaf	Colic, antiseptic, earache	Abbiw (1990), Ayensu (1978)
Parkia biglobosa	Fabaceae	Dawadawa	Tree	Bark, seed	Hemorrhoid, malaria	Agbovie et al. (2002)
				Fruit	Diuretic, strains, burns, bruises	Abbiw (1990)
				Bark	Fever, rickets, malaria	Abbiw (1990), Asase et al. (2005)
				Leaf	Malaria	Asase et al. (2005)
Parquetina nigrescens	Asclepiadaceae	Abakamo	Liana	Leaf	Boils, craw-craw, eyes sore, iritis, conjunctivitis, trachoma, lactogenic	Abbiw (1990), Addo-Fordjour et al. (2008)
				Leaf latex	Diarrhea	Abbiw (1990)
				Latex	Abortion	Abbiw (1990)
				Whole plant	Backache	Abbiw (1990)
				Root	Waist pains	Addo-Fordjour et al. (2008)
				Whole plant	Asthma, jaundice, lumbago, malaria, stomachache	Agbovie et al. (2002)
Paullinia pinnata	Sapindaceae	Twentini	Liana	Root	Sexual weakness, rheumatism, stroke	Abbiw (1990), Addo-Fordjour et al. (2008)
				Leaf, root	Prevents miscarriage, aphrodisiac, fracture, cough	Abbiw (1990), Agbovie et al. (2002)
				Stem	Boil	Abbiw (1990)
				Whole plant	Asthenia	Abbiw (1990)
				Leaf	Malaria	Asase et al. (2005)
Pennisetum pedicellatum	Poaceae	Akokonisuo	Herb	Leaf	Cutaneous	Agbovie et al. (2002)
Pericopsis laxiflora	Fabaceae	Obonsamdua	Tree	Bark	Edema	Agbovie et al. (2002)
				Leaf	Dentition/teething	Abbiw (1990)
				Root, bark	Diarrhea	Abbiw (1990)

(continued)

Table 9.1   Diversity of Medicinal Plant Species and Their Uses in Ghana (Continued)

Scientific Name	Family	Local Name	Habit	Part Used	Purpose	Sources
Persea americana	Lauraceae	Pear	Tree	Leaf	Hypertension	Agbovie et al. (2002)
				Dry leaf	Hypertension	Abbiw (1990)
Petersianthus macrocarpus	Lecythidaceae	Esia	Tree	Leaf, stem, bark	Fibroids, headache, lumbago	Agbovie et al. (2002)
Phyllanthus amarus	Euphorbiaceae	Bomagueakire	Herb	Bark	Bronchial trouble, lumbago	Abbiw (1990)
				Whole plant	Typhoid fever, malaria	Addo-Fordjour et al. (2008) Agbovie et al. (2002)
Physalis angulata	Solanaceae	Tutotuto	Herb	Shoot, leaf	Edema, female infertility	Addo-Fordjour et al. (2008)
Piper guineense	Piperaceae	Nsesaa	Liana	Root	Aphrodisiac, bronchial trouble	Abbiw (1990), Agbovie et al. (2002)
				Seed	Appetizer, flatulence, cold, catarrh, fibroids, joint pain	Abbiw (1990), Ameyaw et al. (2005)
				Seed, stem, fruit	Rheumatism, cough, bronchitis	Agbovie et al. (2002), Addae-Mensah and Aryee (1992)
Piper umbellatum	Piperaceae	Amumuaha	Liana	Leaf	Skin diseases, catarrh	Agbovie et al. (2002)
Piptadeniastrum africanum	Fabaceae	Odahoma	Tree	Bark, stem	Hernia	Addo-Fordjour et al. (2008)
Pleicarpa pycnantha	Apocynaceae	Okanwen	Shrub	Bark	Toothache, abortion	Abbiw (1990)
				Whole plant	Angina pectoris	Agbovie et al. (2002)
Portulaca oleracea	Portulacaceae	Adwera	Shrub	Whole plant	Dermatitis, whitlow, palpitation	Agbovie et al. (2002)
Psidium guajava	Myrtaceae	Guava	Tree	Leaf	Typhoid fever, measles, cough	Abbiw (1990), Agbovie et al. (2002)
				Root	Diarrhea	Abbiw (1990)
				Ripe fruit	Purgative	Abbiw (1990)
				Leaf, root	Fever, blood tonic	Agbovie et al. (2002)
Psychotria calva	Rubiaceae	Nkonkonua	Shrub	Root	Pregnancy booster, stroke	Agbovie et al. (2002)
Pycnanthus angolensis	Myristicaceae	Otie	Tree	Bark, leaf, root,	Anemia, chest pains, headache	Agbovie et al. (2002)
				Stem	Ulcer	
				Root	Antihelminthic	Abbiw (1990)
				Bark	Appetizer	Abbiw (1990)
Raphia hookeri	Arecaceae	Adobe	Tree	Leaf, juice	Laryngitis, lactogenic	Agbovie et al. (2002)
				Root	Earache	Abbiw (1990)

Species	Family	Local name	Habit	Part used	Uses	References
Rauvolfia vomitoria	Apocynaceae	Akakapenpen	Tree	Bark	Measles, stomachache, malaria, lumbago	Agbovie et al. (2002)
				Leaf	Rheumatism	Abbiw (1990)
				Leaf, latex	Ringworm, skin diseases, itch	Abbiw (1990)
				Root	Malaria, lumbago, yaws, snakebite	Agbovie et al. (2002)
Ricinodendron heudelotii	Euphorbiaceae	Wama	Tree	Leaf, stem, bark	Infertility, anemia	Addo-Fordjour et al. (2008) Agbovie et al. (2002)
				Bark	Anemia, stomachache, elephantiasis	Abbiw (1990), Addo-Fordjour et al. (2008), Agbovie et al. (2002)
				Root, bark	Diarrhea	Abbiw (1990)
Ricinus communis	Euphorbiaceae	Adedenkuma	Shrub	Leaf, root, seed	Lumbago, constipation	Agbovie et al. (2002)
				Leaf, root, fruit	Lumbago, headache, dermatitis	Agbovie et al. (2002)
				Root, seed	Hypertension	Agbovie et al. (2002)
				Seed	Asthma	Abbiw (1990)
				Oil	Craw-craw	Abbiw (1990)
				Leaf	Purgative, eyes sore, conjunctivitis, iritis, trachoma, malaria	Abbiw (1990)
Ritchiea reflexa	Capparaceae	Alevo	Shrub	Root	Migraine	Agbovie et al. (2002)
Rourea coccinea	Connaraceae	Awendade	Shrub	Whole plant	Wound dressing, jaundice, poison	Agbovie et al. (2002)
Secamone afzelii	Asclepiadaceae	Ahaban kroratima	Herb	Whole plant	Sore throat, edema	Agbovie et al. (2002)
				Leaf	Purgative	Abbiw (1990)
				Latex	Blood purifier	Abbiw (1990)
Schwenkia americana	Solanaceae	Agengyansu	Herb	Root, whole plant	Cough, boils, yellow fever	Abbiw (1990)
Smilax kraussiana	Smilacaceae	Sawoma	Liana	Root	Impotence, diuretic, eyes sore, conjunctivitis, iritis, trachoma, fainting recovery, fever	Abbiw (1990), Addo-Fordjour et al. (2008)
				Bark	Piles	
Spathodia campanulata	Bignoniaceace	Osisiriw	Tree	Root, bark, seed	Stomach ulcer	Agbovie et al. (2002)
				Bark	Toothache, stomachache, appetizer, backache, bladder trouble	Agbovie et al. (2002)
				Stem, bark, leaf	Dyspepsia, peptic ulcer	Agbovie et al. (2002)
				Leaf, bark, fruit	Arthritis, dyspepsia, fracture	Agbovie et al. (2002)

(continued)

Table 9.1 Diversity of Medicinal Plant Species and Their Uses in Ghana (Continued)

Scientific Name	Family	Local Name	Habit	Part Used	Purpose	Sources
Sterculia tragacantha	Sterculiaceae		Tree	Bark, leaf	Dysentery, whitlow, syphilis, antihelminthic	Agbovie et al. (2002)
Strophantus hispidus	Apocynaceae	Omaatura	Liana	Root	Arthritis, heart failure, stroke, rheumatism	Agbovie et al. (2002)
Synedrella notifolia	Asteraceae	Mamponfo	Herb	Leaf	Epilepsy, purgative	Agbovie et al. (2002)
Synsepalum dulcificum	Sapotaceace	Asaa	Tree	Seed	Prolapsed rectum	Agbovie et al. (2002)
Talbotiella gentii	Fabaceae	Takurowanua	Tree	Root	Cancer	Addo-Fordjour et al. (2008)
Terminalia catappa	Combretaceae		Tree	Leaf	Typhoid fever	Addo-Fordjour et al. (2008)
Terminalia ivorensis	Combretaceae	Emire	Tree	Leaf	Fever	Addo-Fordjour et al. (2008)
				Bark	Stomachache, ulcers, wounds, skin ulcers	Abbiw (1990), Addo-Fordjour et al. (2008)
Terminalia superba	Combretaceae	Framo	Tree	Dry leaf	Stomach ulcer	Addo-Fordjour et al. (2008)
Tetracera affinis	Dilleniaceae	Atwehama	Shrub	Root	Yaws	Abbiw (1990), Agbovie et al. (2002)
Tetrapleura tetraptera	Fabaceae	Prekese	Tree	Bark, fruit	Hypertension, stomachache	Agbovie et al. (2002)
				Bark	Emetic, purgative	Abbiw (1990)
				Fruit	Fever, intoxicant, rheumatism	Abbiw (1990)
Thaumatococcus daniellie	Marantaceae	Anwonomoo		Leaf	Dewormer	Addo-Fordjour et al. (2008)
Theobroma cacao	Sterculiaceae	Kookoo (Cocoa)		Root	Cough, chest pains	Addo-Fordjour et al. (2008), Agbovie et al. (2002)
				Seed	Diuretic	Abbiw (1990)
Teclea verdoorniana	Rutaceae	Owebiribi	Tree	Bark	Cough	Agbovie et al. (2002)
Treculia africana	Moraceae	Ototim	Tree	Bark, stem	Abdominal pain, cough, anemia, arthritis	Agbovie et al. (2002)
				Bark	Cough	Abbiw (1990)
Trema orientalis	Ulmaceae	Osesea	Tree	Leaf	Diabetes, jaundice, oliguria	Agbovie et al. (2002)
Trichilia martineaui	Meliaceae	Tanuronua	Tree	Bark	Candidiasis	Addo-Fordjour et al. (2008)
Trichilia monadelpha	Meliaceae	Tanuro	Tree	Bark	Blood purification, arthritis, dysentery, skin ulcers, stomach disorders, waist pains, candidiasis, yaws, wounds, gonorrhea, ulcers, nausea, stomach pain	Abbiw (1990), Addo-Fordjour et al. (2008), Agbovie et al. (2002), Ameyaw et al. (2005)

Species	Family	Local name	Habit	Part used	Uses	References
Trilepsium madagascariense	Moraceae	Okure	Tree	Stem, bark	Anemia	Agbovie et al. (2002)
				Bark, root	Cough, arthritis, skin ulcer, dyspepsia, dysentery	Agbovie et al. (2002)
				Bark	Candidiasis, stomach ulcer	Addo-Fordjour et al. (2008), Agbovie et al. (2002)
Triplochiton scleroxylon	Sterculiaceae	Wawa	Tree	Bark, stem	Rheumatism, anemia	Agbovie et al. (2002)
				Root	Proper positioning of a baby in the womb	Addo-Fordjour et al. (2008)
Turraea heterophylla	Meliaceae	Ahunayankura	Shrub	Leaf, root	Whooping cough, male impotence	Agbovie et al. (2002)
				Root, bark, seed	Aphrodisiac	Agbovie et al. (2002)
Vernonia amygdalina	Asteraceae	Anwonwen	Shrub	Leaf, root	Asthma, cough, skin diseases, cough, hypertension, fever	Agbovie et al. (2002)
				Seed	Cough, hypertension, fever	Agbovie et al. (2002)
				Root, stem	Cataract, asthma, cough	Agbovie et al. (2002)
Voacanga africana	Apocynaceae	Papaku	Tree	Stem	Dental caries	Addo-Fordjour et al. (2008)
Wedelia africana	Asteraceae	Mfofo	Herb	Leaf	Asthma/cataract, styptic, ulcer	Addae-Mensah and Aryee (1992), Agbovie et al. (2002)
Xylopia aethiopica	Annonaceae	Hwenetia		Seed	Chicken pox, stomachache, bladder trouble	Addo-Fordjour et al. (2008), Agbovie et al. (2002)
				Fruit	Malaria, wound, arthritis, anemia, joint pain, anorexia	Addae-Mensah and Aryee (1992), Ameyaw et al. (2005)
				Bark	Dizziness	Addae-Mensah and Aryee (1992)
Xylopia villosa	Annonaceae	Obaafufuo		Bark	Blood purifier	Addo-Fordjour et al. (2008)
				Seeds	Amenorrhea	Abbiw (1990)
Zanthoxylum xanthoxyloides	Rutaceae	Okanto	Tree	Root, bark	Hypertension	Agbovie et al. (2002)
				Stem, bark	Cough, abdominal pain, toothache	Agbovie et al. (2002)
				Root	Toothache, whitlow	Agbovie et al. (2002)
				Root bark	Joint pain, skin diseases	Ameyaw et al. (2005)
				Stem	Stomachache	Agbovie et al. (2002)
Zingiber officinale	Zingiberaceae	Akakaduro	Herb	Rhizome	Cough, chest pains, stomachache, dyspepsia	Agbovie et al. (2002)

In addition to the diversity of their ecological zones, medicinal plants are often contributed by many plant taxa. Though medicinal plant species in Ghana belong to a wide variety of families, most of them include Fabaceae, Euphorbiaceae, Apocynaceae and Sterculiaceae. The medicinal plants belong to different growth for forms with trees, herbs, and climbers being the most notable ones.

Medicinal plants are often used in a variety of ways among forms and practitioners. (Table 9.1). Depending on practitioners and towns or villages, the same plant may be used to treat a multitude of diseases. As in other African countries (Okello and Ssegawa 2007; Kamatenesi-Mugisha et al. 2008), different parts of the same plant are used to treat the same or different diseases in Ghana (Figures 9.2–9.8). Common usage medicinal plants in Ghana are used to treat malaria, infertility, sexually transmitted diseases, headache, stomachache, and fever (Table 9.1). In recent times, the use of herbal medicine for treating infertility has gained ground among youth and the older generation.

A major barrier to understanding the diversity and uses of medicinal plants in Ghana has been the lack of research. Fortunately, research on traditional uses of plants in Ghana seems to be gathering momentum lately. For instance, an ethnobotanical survey by Agbovie et al. (2002) recorded 339 plant species used as medicines by traditional healers in the eastern region of the country. Asase et al. (2005), working on the ethnobotany of antimalaria plants, identified 41 species of plants used in treating malaria in the Wechiau Community Hippopotamus Sanctuary area in Ghana. A recent study conducted by Addo-Fordjour et al. (2008) in the Bomaa community of the Brong Ahafo region of Ghana recorded 52 plant species as medicinal plants. Other botanical surveys have concentrated on determining the quality standards and harvesting procedures of some medicinal plant parts used in herbal preparations in Ghana (Ameyaw et al. 2005).

Concerns about sustainable utilization and conservation of medicinal plants in Ghana seem to be growing in view of the continuous exploitation of species and the destruction of the country's remaining forests at unprecedented rates. Medicinal plants form an important component of biodiversity in many ecosystems and contribute significantly to services provided by these ecosystems. Conservation of medicinal plants and their habitats (especially in the forests where most of them occur) in Ghana is critical if present and future generations are to obtain maximum benefits from them. This realization has led to a few initiatives to conserve medicinal plants in Ghana.

Figure 9.2 *Ageratum conyzoides.*

Figure 9.3 *Elaeis guineensis* (plant).

Figure 9.4 **(See color insert.)** *Elaeis guineensis* (fruit).

Figure 9.5 *Antiaris toxicaria* (bark).

The Aburi Botanic Garden was engaged in a program funded by the National Lottery, the United Kingdom, and the Darwin Initiative for the Survival of Species to cultivate medicinal plant species in a 20.23-hectare medicinal plant garden at Aburi. The program, which was a 3-year capacity building project to support the conservation and sustainable use of medicinal plants in Ghana, was undertaken from 1999 to 2002. In addition, a nursery with about 5,000 medicinal plant seedlings was set up. From these projects, seedlings are provided to communities for cultivation in their private gardens. Another project by the Africa First Limited Liability Company, aimed at cultivating important plant

Figure 9.6 *Griffonia simplicifolia* (leaves).

Figure 9.7 *Marantochloa leucantha.*

species and trees of medicinal and commercial value that are threatened around the world by exploitation and other forces such as deforestation and bushfires, is presently underway. These measures would go a long way to ensure that not only are medicinal plant species protected from total extinction but also communities continue to benefit from these plants.

The efforts to conserve medicinal plants in Ghana, it must be stated, have not been made without some practical challenges. The forests are a major source of livelihood for many people in Ghana, particularly the forest-fringed communities (Anning and Grant 2008; Opoku 2006). These communities derive many products from the forests, including most of the plants used in traditional medicine (Addo-Fordjour et al. 2008), with only a few cultivating medicinal plants in

Figure 9.8 *Smilax kraussiana* (leaves).

gardens to supplement those in the forest. In many cases, the kinds of harvesting methods adopted by herbal practitioners and the forest-dependent communities from different parts of the country pose a potential threat to biodiversity in the forest, such as those plants with known medicinal values. Addo-Fordjour et al. (2008) reported the use by some herbalists of unsustainable harvesting techniques in the Brong Ahafo region of Ghana.

9.3 PROMOTION OF TRADITIONAL MEDICINE IN GHANA

The government of Ghana, like those of many other African countries, encourages the practice of traditional medicine alongside conventional medicine in law and promotes their coexistence in order to reach the greatest number of citizens (Agbovie et al. 2002). The use of medicinal plants in Ghana is strengthened by the existence of some governmental institutions involved in research in medicinal plants. One such research body is the Center for Scientific Research into Plant Medicine (CSRPM), which was established in 1773 and given legal backing two years later. The main aim of the center, located at Mampong–Akuapim in the eastern region of Ghana, is to screen and evaluate the efficacy of plant materials reported by some herbalists to have medicinal properties, as well as their preparation as tinctures, extracts, and decoctions for treating certain diseases in humans.

Some public universities in Ghana are contributing immensely toward the promotion of medicinal plant use in the country through the inclusion of relevant courses or programs in their curricula. At the Kwame Nkrumah University of Science and Technology, for example, a herbal medicine department has been established to train students, among other things, in the use of medicinal plants. This demonstrates the recognition of the role of herbal medicine in health care delivery in Ghana. The existence of these institutions has resulted in the availability of enormous information about medicinal plants and their efficacy in a scientific way.

9.4 CHALLENGES FACING TRADITIONAL MEDICINE PRACTICE IN GHANA

Despite the essential roles of traditional medicine in the health delivery system of Ghana and the enormous strides made in its promotion, a section of the Ghanaian populace are yet to come to terms with its acceptance, considering it as primitive. It is common knowledge that some orthodox practitioners still look down upon herbal medicine and therefore forbid their patients to patronize them (Busia 2005). Interestingly, however, a few medical doctors recommend herbal medicines to their patients.

Another major challenge to effective practice of traditional medicine is the difficulty in medicinal plant identification. Accurate identification of plants by herbalists is very vital in sustaining their profession and ensuring the safety of their clients. Different species of plants may appear similar or plants of the same species may exhibit morphological differences under different environmental conditions due to phenotypic plasticity, posing identification difficult for traditional herbalists. There have been a few reported cases of death resulting from wrong identification and, consequently, wrong use of plant materials for herbal preparation.

The success of traditional medicine practice in Ghana, like many other businesses, to a large extent can be influenced by availability of funds. The herbal products of some herbalists do not go beyond the boundaries of their respective communities due to the inability to raise adequate capital to increase production and transport them to other parts of the country. In view of the financial constraints, some herbalists find it difficult to package their products for both the local and international markets, resulting in the low patronage of these products.

In a country where oral transmission of knowledge still constitutes an essential means of acquisition and preservation (Abel and Busia 2005), Ghana stands a high risk of losing vital information on traditional uses of plants in treating diseases. This situation is likely to arise because not only are the herbalists a repository of information on medicinal plants, but the younger generation also does not appear ready to learn anything from them. A recent report indicates that the country's youth do not show much interest in learning traditional medicine practice (Addo-Fordjour et al. 2008) since they see the profession as outmoded. Consequently, some traditional herbalists do not have trainees in the profession. There is an urgent need to move toward effective documentation of the country's medicinal plant resources.

9.5 CONCLUSION

Traditional medicine remains an important feature of Ghana's health care system. Several parts of plants are used for treating various diseases. However, it is important to note that the part used and the disease conditions for which they are applied vary among herbalists and communities in Ghana. In view of the lack of extensive literature on the diversity of medicinal resources, this compilation is timely to provide useful information on some of the plants used by herbalists in Ghana. It is hoped that this chapter will serve as a good reference for those researchers and practitioners who might be interested in further study on medicinal plants in Ghana and other tropical countries. We do not recommend under any circumstances, the application of any of the plants listed here for treatment of any ailment, as their efficacy and potential side effects are not known. While it is necessary to promote the use of medicinal plants to complement conventional medicine, it is equally important to ensure that these natural resources are conserved in view of the threats posed to them by exploitation and destruction of their habitats.

REFERENCES

Abbiw, D. 1990. *Useful plants of Ghana.* London: Intermediate Technology Publications Ltd. and the Royal Botanic Gardens, Kew.

Addae-Mensah, I., and Aryee, G. 1992. *Ghana herbal pharmacopoeia.* Osu, Accra, Ghana: Advent Press Ltd.

Addo-Fordjour, P., Anning, A. K., Belford, E. J. D., and Akonnor, D. 2008. Diversity and conservation of medicinal plants in the Bomaa community of the Brong Ahafo region, Ghana. *Journal of Medicinal Plants Research* 2 (9): 226–233.

Agbovie, T., Amponsah, K., Crentsil, O. R., Dennis, F., Odamtten, G. T., and Ofusohene-Djan, W. 2002. Conservation and sustainable use of medicinal plants in Ghana: Ethnobotanical survey. http://www.unep-wcmc.org/species/plants/ghana

Agyeman, V. K., Appiah, S. K., Siisi-Wilson, E., Ortsin, G., and Patience, D. 1999. Timber harvesting in Ghana: A review of ecological impacts and potentials for increased harvesting of timber resources. *Ghana Journal of Forestry* 7:1–14.

Ameyaw, Y., Aboagye, F. A., Appiah, A. A., and Blagogee, H. R. 2005. Quality and harvesting specifications of some medicinal plant parts set up by some herbalists in the eastern region of Ghana. *Ethnobotanical Leaflets.* http://www.ethnoleaflets.com/index2005.htm

Anning, A. K., and Grant, N. T. 2008. An evaluation of local community participation in forest resources conservation in some selected districts of Ghana. *The African Journal of Plant Science and Biotechnology* 2 (2): 39–45.

Asase, A., Oteng-Yeboah, A. A., Odamtten, G. T., and Simmonds, M. S. 2005. Ethnobotanical study of some Ghanaian antimalarial plants. *Journal of Ethnopharmacology* 99 (2): 273–279.

Ayensu, E. S. 1978. *Medicinal plants in West Africa.* Michigan: Reference Publications Inc.

Busia, K. 2005. Medical provision in Africa—Past and present. *Phytotherapy Research* 19:919–923.

Kamatenesi-Mugisha, M., Oryem-Origa, H., Odyek, O., and Makawiti, D. W. 2008. Medicinal plants used in the treatment of fungal and bacterial infections in and around Queen Elizabeth Biosphere Reserve, Western Uganda. *African Journal of Ecology* 46:90–97.

Ministry of Environment and Science. 2002. National biodiversity strategy for Ghana. Ministry of Environment and Science document, pp. 1–55.

Musa, A. A. 2009. Effects of crude ethanol extract on *Jatropha curcas* leaves on the antimicrobial activity of some antibiotics against selected microorganisms. BSc thesis submitted to the Department of Theoretical and Applied Biology, Kwame Nkrumah University of Science and Technology, Ghana.

Ohja, H. R. 2000. Current policy issues in NTFP development in Nepal Asia Network for small-scale biore-sources, Kathmandu, Nepal.

Okello, J., and Ssegawa, P. 2007. Medicinal plants used by communities of Ngai subcounty, Apac district, Northern Uganda. *African Journal of Ecology* 45:76–83.

Opoku, K. 2006. Forest governance in Ghana: An NGO perspective. A report produced by Forest Watch Ghana for FERN. Zuidam Uithof, Utrecht, the Netherlands, 32 pp.

Pearce, D. W., and Puroshothaman, S. 1992. Protecting biological diversity: The economic value of phar-maceutical plants. CSERGE global environmental change working paper 92-27, Center for Social and Economic Research on the Global Environment, University College.

Principe, P. P. 1991. Valuing the biodiversity of medicinal plants. In *The conservation of medicinal plants. Proceedings of an International Consultation*, ed. Akerele et al., 79–124, March 21–27, 1988, Chiang Mai, Thailand. Cambridge, England: Cambridge University Press.

Robbers, J. M., Speedie, M. K., and Tyler, V. 1996. *Pharmacognosy and pharmacobiotechnology*, 1–14. Baltimore, MD: Williams and Wilkins.

Sarpong, K. 2000. Traditional medicine for the 21st century in Ghana—The role of the scientist/researcher. Commonwealth Lecture Theatre, University of Ghana, as part of traditional medicine week celebrations for the year 2000.

Ticktin, T., Nantel, P., Ramirez, F., and Johns, T. 2002. Effects of variation of harvest limits for nontimber forest species in Mexico. *Conservation Biology* 16:691–705.

CHAPTER 10

Medicinal Plants of Australia
Melaleuca alternifolia for the Production of Tea Tree Oil

Enzo A. Palombo

CONTENTS

10.1 INTRODUCTION

Melaleuca alternifolia (Maiden et Betche) Cheel (Myrtaceae), also known as *Melaleuca linariifolia* var. *alternifolia*, is a small shrub or tree of up to 5 m in height with a papery bark. The species is unique to Australia and is usually found in swampy or wet ground on the northern coastal strip of New South Wales and southern Queensland (Lassak and McCarthy 2001). The plant was used medicinally by the Australian aboriginal people for centuries to treat coughs, colds, wounds, sore throats, and skin aliments. However, the antiseptic properties of the essential oils obtained by steam distillation of the plant foliage were only discovered in the 1920s (Penfold and Morrison 1946). At that time, the major component of the oil, terpinen-4-ol, was also found to be associated with the activity of the oil, commonly referred to as "tea tree oil."

Numerous medicinal applications have now been described for tea tree oil (TTO), including antibacterial, antifungal, antiviral, and anti-inflammatory activities. *Melaleuca alternifolia* has been intensively cultivated only in the past 20 years and most commercial production of TTO is located in the areas where the plant is naturally distributed (Figure 10.1). There are currently about 3000 ha

Figure 10.1 Tea tree oil plantation and harvesting equipment, Coraki, New South Wales, Australia. (Photo: John Moss.)

of cultivated *M. alternifolia* and about 100 producers of TTO. Plants are first grown in greenhouses prior to high-density field planting; then, they are harvested after 1–3 years by cutting the whole plant close to the ground and chipping it into small pieces before oil extraction (Carson, Hammer, and Riley 2006). More than 80% of the world's TTO is produced in Australia, although about 90% is exported.

10.2 PHYSICAL CHARACTERISTICS AND COMPONENTS OF TEA TREE OIL

Commercial TTO is extracted from the foliage and terminal branchlets of *M. alternifolia*, although it can also be produced from *M. dissitiflora* F. Mueller and *M. linariifolia* Smith. Other species of *Melaleuca* can be used, provided the oil conforms to the international standard (see later discussion). The extraction of TTO involves boiling the leaves and branchlets with water and separation of the oil from the aqueous steam distillate. One kilogram of foliage yields 20–25 g of oil (Lassak and McCarthy 2001).

The properties and chemical composition of TTO are defined by the international standard ISO 4730 (2004) and the identical Australian standard AS 2782-2009 ("Oil of Melaleuca, Terpinen-4-ol Type"), which specifies levels of 15 of the more than 100 components in pure Australian TTO. These are listed in Table 10.1 and a typical chromatogram is presented in Figure 10.2. TTO is composed of terpene hydrocarbons (mainly monoterpenes) and their associated alcohols. Terpinen-4-ol is the most abundant component and constitutes at least 30% of TTO. This component is also the major contributor to the antimicrobial properties of the oil. The oil is a clear, mobile liquid that is

Table 10.1 Components of Tea Tree Oil as Specified under ISO 4730

Component	Percentage Composition (Range)
Terpinen-4-ol	30–48
γ-Terpinene	10–28
α-Terpinene	5–13
1,8-cineole	Trace–15
α-Terpineol	1.5–8
p-Cymene	0.5–8
α-Pinene	1–6
Terpinolene	1.5–5
Sabinene	Trace–3.5
Aromadendrene	Trace–3
Ledene	Trace–3
δ-Cadenine	Trace–3
Limonene	0.5–1.5
Globulol	Trace–1
Viridiflorol	Trace–1

colorless to pale yellow with a characteristic odor. At 20°C, the relative density is 0.885–0.906, the refractive index is 1.475–1.482, and the optical rotation is between +5 and +15°.

The composition of TTO changes over time, particularly when it is exposed to the air, light, and high temperatures. A 12-month study supported by the Rural Industries Research and Development Corporation of Australia showed that TTO components were relatively unchanged over the first 6 months. However, changes were seen after that time—in particular, increased levels of p-cymene and peroxidase, which are indicators of oxidation. However, levels of terpinen-4-ol, α-terpinene, and γ-terpinene were largely unchanged. Nevertheless, to reduce oxidation, TTO products should be kept in a tightly sealed bottle and stored away from light and heat.

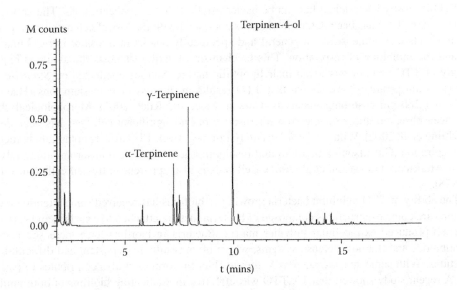

Figure 10.2 Typical chromatogram of tea tree oil, with major peaks labeled. Analysis was performed using a Varian CP-3800 gas chromatograph coupled with a Varian Saturn 2000 GC/MS/MS instrument, with a DB-5.625 column.

10.3 MEDICINAL PROPERTIES OF TEA TREE OIL

TTO has a long history of medicinal use and many products containing the oil are listed as antiseptics by the Therapeutic Goods Administration of Australia, although TTO is not listed as a pharmaceutical. TTO-containing products available commercially include 100% pure oil and aqueous solutions used as topical antiseptics. Other products include face washes for the treatment of acne; antifungal gels for the treatment of infections of the nails and ringworm; lozenges for the relief of sore throats; feminine hygiene products to relieve vaginal and anal itching, burning, and discomfort; hair care products and shampoos for general scalp health and the treatment of lice; insect repellents; mouthwashes; and toothpastes for oral care; and personal care products such as hand and body lotions, deodorants, and cold-sore creams. The reported medicinal properties of TTO are described next.

10.3.1 Antibacterial Activity

The antibacterial properties of TTO are the most widely investigated and many reports have appeared in the scientific literature over the past 20 years. The antibacterial activity has been assessed against a range of bacteria and most species are susceptible to concentrations of 1% or less. However, minimum inhibitory concentrations (MICs) of some bacteria are in excess of 2%, with *Enterococcus faecalis* and *Pseudomonas aeruginosa* having high levels of tolerance (Table 10.2). TTO appears to be equally effective against Gram-positive and Gram-negative bacteria. The activity of TTO against antibiotic-resistant bacteria has received some attention; studies have shown that the MICs are similar to antibiotic-susceptible strains (Carson et al. 1995; Nelson 1997; Chan and Louden 1998; Elsom and Hide 1999; May et al. 2000; Hada et al. 2001). The main antibacterial activity is associated with terpinen-4-ol and α-terpineol, although some degree of activity is also associated with minor components, such as α-pinene, β-pinene, and linalool (Carson and Riley 1995a). Other components are less active. No significant synergistic relationships between different components have been identified (Cox, Mann, and Markham 2001).

TTO is mostly bactericidal but can be bacteriostatic at lower concentrations. The antibacterial mechanism of TTO has been determined and is consistent with the loss of cell membrane integrity, which affects cell homeostasis in general and specifically results in a loss of intracellular material and the inhibition of respiration. This mechanism of action supports the assumed biological activity of TTO components given their lipophilic nature. Studies conducted on *Staphylococcus aureus* (Gram positive) have shown that TTO results in the loss of potassium ions (Hada et al. 2003) and 260 nm absorbing materials (Carson, Mee, and Riley 2002). Morphological changes have been observed under the electron microscope but no significant cell lysis has been detected (Reichling et al. 2002). With *Escherichia coli* (Gram negative), TTO affects potassium homeostasis and respiration (Gustafson et al. 1998) and results in a loss of 260 nm absorbing materials (Cox, Mann, Markham, Gustafson, et al. 2001). Cell lysis is also apparent in treated cells (Gustafson et al. 1998).

The ability of TTO to inhibit bacteria growing as biofilms has received some attention recently. Biofilms are complex structures composed of bacteria that are attached to a surface and surrounded by a self-produced extracellular polymer matrix. Bacteria in biofilms have increased resistance to antibiotics and the host defense systems, which often results in persistent and difficult-to-treat infections. With some species, such as *S. aureus*, biofilm formation also contributes to pathogenesis. A recent study showed that 1% TTO was effective in eradicating biofilms of both antibiotic-susceptible and -resistant *S. aureus* (Kwieciński, Eick, and Wójcik 2009). Effective killing of cells in biofilms was achieved within 30–60 minutes of exposure to TTO, resulting is an 85% reduction of viable cells. TTO has also been shown to affect the viability of *S. epidermidis* biofilms, although

Table 10.2 Susceptibility of Selected Bacteria and Fungi to Tea Tree Oil

Species	Percent (v/v)	
	MIC	MBC/MFC
Bacteria		
Acinetobacter baumannii	1	1
Actinomyces spp.	1	1
Bacillus cereus	0.3	
Bacteroides spp.	0.06–0.5	0.06–0.12
Corynebacterium sp.	0.2–2	2
Enterococcus faecalis	0.5 to >8	>8
E. faecium (vancomycin resistant)	0.5–1	0.5–1
Escherichia coli	0.08–2	0.25–4
Fusobacterium nucleatum	0.6 to >6	0.25
Klebsiella pneumoniae	0.25–0.3	0.25
Lactobacillus spp.	1–2	2
Micrococcus luteus	0.06–0.5	0.25–6
Porphyromonas gingivalis	0.11–0.25	0.13 to >0.6
Prevotella intermedia	0.003–0.1	0.003–0.1
Propionibacterium acnes	0.05–0.63	0.5
Proteus vulgaris	0.08–2	4
Pseudomonas aeruginosa	1–8	2 to >8
Staphylococcus aureus	0.5–1.25	1–2
S. aureus (methicillin resistant)	0.04–0.53	0.5
S. epidermidis	0.45–1.25	4
Streptococcus pyogenes	0.12–2	0.25–4
Fungi		
Alternaria spp.	0.016–0.12	0.06–2
Aspergillus niger	0.016–0.4	2–8
Candida albicans	0.06–8	0.12–1
Cryptococcus neoformans	0.015–0.06	
Epidermophyton floccosum	0.008–0.7	0.12–0.25
Fusarium spp.	0.008–0.25	0.25–2
Penicillium spp.	0.03–0.06	0.5–2
Saccharomyces cerevisiae	0.25	0.5
Trichophyton mentagrophytes	0.11–0.44	0.25–0.5
Trichosporon spp.	0.12–0.22	0.12

Source: Data are taken from Carson, C. F. et al. 2006. *Clinical Microbiology Reviews* 19:50–62.

Notes: MIC = minimum inhibitory concentration; MBC = minimum bactericidal concentration; MFC = minimum fungicidal concentration.

higher concentrations were required to inhibit or kill biofilm bacteria compared to planktonic bacteria (Karpanen et al. 2008).

Recent studies have indicated that TTO is also effective as a sporicidal agent and is able to reduce the number of viable *Bacillus* spores at concentrations of >1% (Messager et al. 2006; Lawrence and Palombo 2009). Scanning electron microscopic analysis of TTO-treated spores suggested that the likely mode of action was leakage of spore contents; interestingly, individual TTO components did not exert the same sporicidal effect as the whole oil (Lawrence and Palombo 2009). TTO thus may be useful in applications where bacterial spore reduction is desired.

Other specific antibacterial applications where TTO may prove effective (and, indeed, many TTO-containing products are commercially available for these conditions) are in the treatment of acne and oral diseases and in the maintenance of hand hygiene. A clinical trial evaluated the effectiveness of 5% TTO gel in treating mild to moderate acne compared with 5% benzoyl peroxide (BP) lotion. Both the TTO gel and BP lotion had a significant effect in reducing acne (lower numbers of inflamed and noninflamed lesions). However, fewer side effects (skin dryness, itching, stinging, burning, and redness) were experienced by participants treated with TTO (Bassett, Pannowitz, and Barnetson 1990). With respect to the prevention of oral diseases, 0.5% TTO has been shown to rapidly kill two bacterial species, *Streptococcus mutans* and *Lactobacillus rhamnosus*, which are associated with the development of dental caries (Riley 2003).

Hospital-acquired bacterial infections are a significant problem worldwide, and most infections are thought to be caused by bacteria that are passed from the hands of hospital staff to patients or from patient to patient. The effectiveness of a 5% TTO hygienic skin wash, a 5% TTO alcohol hygienic skin wash, and a 3% TTO alcohol hand rub against four bacterial species has been compared to povidone iodine, a commonly used hospital hand wash. The study showed that some TTO formulations could be useful in reducing hospital-acquired infections such as those caused by *S. aureus* (Messager et al. 2005). Also, hospital staff with access to TTO hand washes have shown better compliance to hand washing (Carson and Riley 1995b).

There have been few studies addressing bacterial resistance to TTO and these have been largely limited to Gram-positive species, although one report suggested that some *E. coli* strains may have decreased susceptibility to TTO (Gustafson et al. 2001). Recently, a systematic study conducted using a wider range of bacteria (*S. aureus*, *S. epidermidis*, and *Enterococcus faecalis*, including clinical and antibiotic-resistant isolates) indicated that TTO resistance frequencies were very low ($<10^{-9}$), suggesting that that it is unlikely that microorganisms will acquire single-step mutations rendering them resistant to TTO (Hammer, Carson, and Riley 2008). Overall, there is little evidence of resistance to TTO; this is not unexpected, given that multiple simultaneous mutations would be required to overcome the many antibacterial components found in TTO that target the bacterial cell membrane.

10.3.2 Antifungal Activity

A number of fungi, including yeasts, dermatophytes, and other filamentous fungi, have been shown to be susceptible to TTO (Table 10.2). While most species are inhibited by TTO concentrations of 0.03–0.5%, *Aspergillus niger* is notable as having greater tolerance (MFC up to 8%), although germinated conidia are more susceptible than intact conidia (Hammer, Carson, and Riley 2002). TTO vapors can inhibit fungal growth and sporulation (Inouye et al. 1998, 2000; Inouye, Uchida, and Yamaguchi 2001). Similar to bacterial cells, TTO appears to alter the permeability of fungal cells by affecting the properties of the cell membrane (Hammer, Carson, and Riley 2004). Additionally, TTO has been shown to inhibit germ tube formation in *Candida albicans* (Hammer, Carson, and Riley 2000; D'Auria et al. 2001).

10.3.3 Antiviral Activity

There have been few studies of the antiviral properties of TTO. An early study investigated the inhibitory activity of TTO against tobacco mosaic virus on tobacco plant leaves and indicated that fewer lesions were present on plants treated with 100, 250, or 500 ppm TTO compared to controls (Bishop 1995). More recent studies have focused on herpes simplex virus (HSV) and have shown that TTO reduced viral plaque formation (*in vitro*) by 98.2% for HSV-1 and 93% for HSV-2 at a concentration of 0.003% (Schnitzler, Schön, and Reichling 2001). At 1% concentration, TTO was found to inhibit HSV-1 plaque formation completely.

Recently, the antiviral activity of TTO and its main components was evaluated against poliovirus type 1, echovirus 9, coxsackievirus B1, adenovirus type 2, HSV-1 and -2, and influenza virus A (H1N1) (Garozzo et al. 2009). TTO and some of its components (terpinen-4-ol, terpinolene, and α-terpineol) showed inhibitory effects on influenza virus H1N1 replication at doses (ID_{50} = 0.0006%) below the cytotoxic dose (CD_{50} = 0.025%). While none of the compounds or TTO was effective against the other viruses, TTO at 0.125% had some virucidal activity against HSV-1 and -2. The authors concluded that TTO has antiviral activity against influenza A subtype H1N1, with terpinen-4-ol being the main active component, and that TTO could be useful in the treatment of influenza virus infection.

10.3.4 Antiprotozoal Activity

TTO is able to reduce the growth of *Leishmania major* and *Trypanosoma brucei* by 50% at concentrations of 403 mg/mL and 0.5 mg/mL, respectively; terpinen-4-ol is the main active component (Mikus et al. 2000). At 300 mg/mL, TTO is able to kill all cells of *Trichomonas vaginalis* (Viollon, Mandin, and Chaumont 1996).

10.3.5 Anti-inflammatory Activity

In addition to antimicrobial activity, TTO and various components have been shown to affect the immune response and the inflammatory response in particular. TTO applied topically suppresses edema associated with contact hypersensitivity in mice (Brand et al. 2002). This activity is supported by human studies of histamine-induced wheal and flare, where topically applied TTO has been shown to reduce wheal volume, but not flare area (Koh et al. 2002). In addition, terpinen-4-ol controls the vasodilation and plasma extravasation of histamine-induced inflammation (Klimmek et al. 2002).

In vitro studies have shown that the water-soluble fraction of TTO (TTO-H_2O) reduces the production of inflammatory mediators (TNF-α, IL-8β, IL-8, IL-10, and prostaglandin E_2) induced by lipopolysaccharide and this activity is linked to terpinen-4-ol (Hart et al. 2000). TTO-H_2O, terpinen-4-ol, and α-terpineol also suppress superoxide by agonist-stimulated monocyte production (Brand et al. 2001). However, TTO does not suppress the adherence reaction of neutrophils induced by TNF-α stimulation (Abe et al. 2003) or casein-induced recruitment of neutrophils (Abe et al. 2004), suggesting specific modes of anti-inflammatory action.

10.4 CLINICAL EFFICACY OF TEA TREE OIL

Clinical studies investigating the efficacy of TTO have been reviewed previously (Carson et al. 2006), so only the findings of randomized control studies are presented here. As mentioned earlier, a study of the efficacy of a 5% TTO gel compared to 5% benzoyl peroxide (BP) in participants with mild to moderate acne found that both treatments significantly reduced inflamed lesions (Bassett et al. 1990). However, BP was better at reducing oiliness, while TTO produced less scaling, purities, and dryness. TTO also produced fewer adverse events, such as dryness, stinging, burning, and redness.

A 6% TTO gel did not produce significantly better results (measured as time to re-epithelization) in participants with recurrent herpes cold sores compared to a placebo gel (Carson et al. 2001), while a 5% TTO shampoo was shown to be more effective than a placebo shampoo in a study of participants with moderate dandruff (Satchell et al. 2002a).

In a study of hospital patients colonized or infected with methicillin-resistant *S. aureus* (MRSA), use of a 4% TTO nasal ointment plus 5% TTO body wash showed no significant difference when compared to using a mupirocin nasal ointment plus triclosan body wash in terms of patients who cleared infection or became chronically infected (Caelli et al. 2000). Similarly, a study of patients with culture-positive onychomycosis found no significant difference in full or partial resolution of infection between patients treated with 100% TTO or 1% clotrimazole (Buck, Nidorf, and Addino 1994). However, in another study of outpatients with onychomycosis, 80% of participants treated with a cream containing 2% butenafine hydrochloride plus 5% TTO were cured, while none of the participants treated with 5% TTO only were cured—a highly significant difference (Syed et al. 1999).

Clinical improvement and mycological cure were observed in participants diagnosed with tinea pedis when treated with 10% TTO in sorbolene and 1% tolnafate, but not in participants treated with placebo (sorbolene) only (Tong, Altman, and Barnetson 1992). A similar study showed that 25% TTO and 50% TTO treatments were significantly better than placebo in effective curing in participants with culture-positive tinea pedis (Satchell et al. 2002b).

While most of the studies described before demonstrate the promising clinical efficacy of TTO in treating bacterial, fungal and viral conditions, Carson et al. (2006) have pointed out that many studies have limitations, such as low participant numbers. The characteristic odor of TTO also reportedly compromised the participant blinding in several studies. In addition, most of the studies have not been independently replicated. Clearly, further work in this area is required to demonstrate the clinical efficacy of TTO unequivocally.

10.5 SAFETY AND TOXICITY CONSIDERATIONS

TTO has been used for over 80 years and anecdotal evidence suggests that it is safe to use with few and minor adverse effects (Carson et al. 2006). Nevertheless, there is limited scientific information about the oral and dermal toxicity of TTO. Animal studies and cases of human poisoning indicate that TTO can be toxic if ingested, although there are no reports of human deaths after ingestion if supportive treatment is given. Apart from some rare reports of human poisoning, there is no evidence of systemic toxicity. Even so, TTO is classified as a schedule 6 poison in Australia as a substance with "a moderate potential for causing harm, the extent of which can be reduced through the use of distinctive packaging with strong warnings and safety directions on the label" (Hammer, Carson, and Riley 2006). The LD_{50} in rats is 1.9–2.6 mg/kg (Russell 1999). TTO can be an irritant and allergen. Although studies of irritancy are contradictory, they indicate that irritant reactions can be avoided if lower concentrations of TTO are used. The allergic properties of TTO are purported to be due to oxidation products or aged or improperly stored oil (Hausen, Reichling, and Harkenthal 1999).

TTO has been shown to exhibit toxicity to human cells *in vitro* with IC_{50} values of 20–2700 μg/mL or toxic doses of ≥0.004–0.016%, depending on the cells used (Carson et al. 2006). Terpinen-4-ol, α-terpineol, terpinolene, α-phellandrene, aromandendrene, sabinene, α-pinene, and β-pinene are more active than the whole oil, while 1,8-cineole is less active (Hayes, Leach, and Markham 1997; Mikus et al. 2000). However, such data may not reflect the *in vivo* effects of a topically applied agent like TTO where dermal penetration and metabolism issues are relevant (Carson et al. 2006). The cytotoxic activity of TTO against human lung cancer, human breast cancer, and human prostate cancer (Liu et al. 2009); cervix carcinoma; and liver carcinoma (Hayes et al. 1997; Mikus et al. 2000) cell lines has been demonstrated with IC_{50} values of 20–2700 μg/mL or 0.012–0.037%, depending on the study.

The available data on mutagenicity indicate that TTO and its individual components have low mutagenic potential as assessed using both bacterial and mammalian systems (Carson et al. 2006). There are limited data on exotoxicity, although it has been observed that TTO is toxic to whiteflies (Choi et al. 2003), mites (Sammataro et al. 1998), and lice (Lee et al. 2001); no effect on the rice weevil is apparent (Lee et al. 2001). In aquatic environments, TTO is not lethal to rainbow trout eggs at 1500 ppm, yet it is toxic to brine shrimp at 500 ppm (McCage et al. 2002).

10.6 CONCLUSIONS

There is increasing interest in complementary and alternative therapies, especially those using products derived from plants. While the therapeutic properties of *Melaleuca alternifolia* and its volatile oil, known popularly as tea tree oil, have been known for many years, and the availability of TTO as an "over-the-counter" medicine in Australia, Europe, and North America has increased, detailed scientific evaluation of its medicinal properties has only occurred in the past 20 years. Such studies have validated the use of TTO primarily as an antiseptic (with broad-spectrum activity against bacteria, fungi, and, to a lesser extent, viruses) and as an anti-inflammatory.

With increasing rates of resistance to antimicrobial agents, there is an urgent need for new products to control microbial growth and a number of nonantibiotic approaches are currently under investigation. *In vitro* studies clearly show the efficacy of TTO, and some recent studies have demonstrated the clinical efficacy of TTO and TTO-containing products. However, further clinical investigations (including larger clinical trials) are needed to corroborate these findings. Similarly, *in vitro* studies suggesting that TTO may have anticancer potential need to be supported by clinical studies. Nonetheless, the number of commercially available products containing TTO indicates that this product will remain a popular treatment for minor ailments such as skin infections.

REFERENCES

Abe, S., N. Maruyama, K. Hayama, S. Inouye, H. Oshima, and H. Yamaguchi. 2004. Suppression of neutrophil recruitment in mice by geranium essential oil. *Mediators of Inflammation* 13:21–24.

Abe, S., N. Maruyama, K. Hayama, et al. 2003. Suppression of tumor necrosis factor-alpha-induced neutrophil adherence responses by essential oils. *Mediators of Inflammation* 12:323–328.

Bassett, I. B., D. L. Pannowitz, and R. S. Barnetson. 1990. A comparative study of tea tree oil versus benzoyl peroxide in the treatment of acne. *Medical Journal of Australia* 153:455–458.

Bishop, C. D. 1995. Antiviral activity of the essential oil of *Melaleuca alternifolia* (Maiden & Betche) Cheel (tea tree) against tobacco mosaic virus. *Journal of Essential Oil Research* 7:641–644.

Brand, C., A. Ferrante, R. H. Prager, et al. 2001. The water soluble-components of the essential oil of *Melaleuca alternifolia* (tea tree oil) suppress the production of superoxide by human monocytes, but not neutrophils, activated *in vitro*. *Inflammation Research* 50:213–219.

Brand, C., M. A. Grimbaldeston, J. R. Gamble, J. Drew, J. J. Finlay-Jones, and P. H. Hart. 2002. Tea tree oil reduces the swelling associated with the efferent phase of a contact hypersensitivity response. *Inflammation Research* 51:236–244.

Buck, D. S., D. M. Nidorf, and J. G. Addino. 1994. Comparison of two topical preparations for the treatment of onychomycosis: *Melaleuca alternifolia* (tea tree) oil and clotrimazole. *Journal of Family Practice* 38:601–605.

Caelli, M., J. Porteous, C. F. Carson, R. Heller, and T. V. Riley. 2000. Tea tree oil as an alternative topical decolonization agent for methicillin-resistant *Staphylococcus aureus*. *Journal of Hospital Infection* 46:236–237.

Carson, C. F., L. Ashton, L. Dry, D. W. Smith, and T. V. Riley. 2001. *Melaleuca alternifolia* (tea tree) oil gel (6%) for the treatment of recurrent herpes labialis. *Journal of Antimicrobial Chemotherapy* 48:450–451.

Carson, C. F., B. D. Cookson, H. D. Farrelly, and T. V. Riley. 1995. Susceptibility of methicillin-resistant *Staphylococcus aureus* to the essential oil of *Melaleuca alternifolia*. *Journal of Antimicrobial Chemotherapy* 35:421–424.

Carson, C. F., K. A. Hammer, and T. V. Riley. 2006. *Melaleuca alternifolia* (tea tree) oil: A review of antimicrobial and other medicinal properties. *Clinical Microbiology Reviews* 19:50–62.

Carson, C. F., B. J. Mee, and T. V. Riley. 2002. Mechanism of action of *Melaleuca alternifolia* (tea tree) oil on *Staphylococcus aureus* determined by time-kill, lysis, leakage, and salt tolerance assays and electron microscopy. *Antimicrobial Agents and Chemotherapy* 48:1914–1920.

Carson, C. F., and T. V. Riley. 1995a. Antimicrobial activity of the major components of the essential oil of *Melaleuca alternifolia*. *Journal of Applied Bacteriology* 78:264–269.

———. 1995b. Toxicity of the essential oil of *Melaleuca alternifolia* or tea tree oil. *Journal of Toxicology–Clinical Toxicology* 33:193–195.

Chan, C. H., and K. W. Loudon. 1998. Activity of tea tree oil on methicillin-resistant *Staphylococcus aureus* (MRSA). *Journal of Hospital Infection* 39:244–245.

Choi, W. I., E. H. Lee, B. R. Choi, H. M. Park, and Y. J. Ahn. 2003. Toxicity of plant essential oils to *Trialeurodes vaporariorum* (Homoptera: Aleyrodidae). *Journal of Economic Entomology* 96:1479–1484.

Cox, S. D., C. M. Mann, and J. L. Markham. 2001. Interactions between components of the essential oil of *Melaleuca alternifolia*. *Journal of Applied Microbiology* 91:492–497.

Cox, S. D., C. M. Mann, J. L. Markham, J. E. Gustafson, J. R. Warmington, and S. G. Wyllie. 2001. Determining the antimicrobial actions of tea tree oil. *Molecules* 6:87–91.

D'Auria, F. D., L. Laino, V. Strippoli, et al. 2001. *In vitro* activity of tea tree oil against *Candida albicans* mycelial conversion and other pathogenic fungi. *Journal of Chemotherapy* 13:377–383.

Elsom, G. K. F., and D. Hide. 1999. Susceptibility of methicillin-resistant *Staphylococcus aureus* to tea tree oil and mupirocin. *Journal of Antimicrobial Chemotherapy* 43:427–428.

Garozzo, A., R. Timpanaro, B. Bisignano, P. M. Furneri, G. Bisignano, and A. Castro. 2009. *In vitro* antiviral activity of *Melaleuca alternifolia* essential oil. *Letters in Applied Microbiology* 49:806–808.

Gustafson, J. E., S. D. Cox, Y. C. Liew, S. G. Wyllie, and J. R. Warmington. 2001. The bacterial multiple antibiotic resistant (Mar) phenotype leads to increased tolerance to tea tree oil. *Pathology* 33:211–215.

Gustafson, J. E., Y. C. Liew, S. Chew, J. Markham, H. C. Bell, S. G. Wyllie, and J. R. Warmington. 1998. Effects of tea tree oil on *Escherichia coli*. *Letters in Applied Microbiology* 26:194–198.

Hada, T., S. Furuse, Y. Matsumoto, et al. 2001. Comparison of the effects *in vitro* of tea tree oil and plaunotol on methicillin-susceptible and methicillin-resistant strains of *Staphylococcus aureus*. *Microbios* 106 (Suppl. 2): 133–141.

Hada, T., Y. Inoue, A. Shiraishi, and H. Hamashima. 2003. Leakage of K$^+$ ions from *Staphylococcus aureus* in response to tea tree oil. *Journal of Microbiological Methods* 53:309–312.

Hammer, K. A., C. F. Carson, and T. V. Riley. 2000. *Melaleuca alternifolia* (tea tree) oil inhibits germ tube formation by *Candida albicans*. *Medical Mycology* 38:355–362.

———. 2002. *In vitro* activity of *Melaleuca alternifolia* (tea tree) oil against dermatophytes and other filamentous fungi. *Journal of Antimicrobial Chemotherapy* 50:195–199.

———. 2004. Antifungal effects of *Melaleuca alternifolia* (tea tree) oil and its components on *Candida albicans*, *Candida glabrata* and *Saccharomyces cerevisiae*. *Journal of Antimicrobial Chemotherapy* 53:1081–1085.

———. 2006. A review of the toxicity of *Melaleuca alternifolia* (tea tree) oil. *Food and Chemical Toxicology* 44:616–625.

———. 2008. Frequencies of resistance to *Melaleuca alternifolia* (tea tree) oil and rifampicin in *Staphylococcus aureus*, *Staphylococcus epidermidis* and *Enterococcus faecalis*. *International Journal of Antimicrobial Agents* 32:170–173.

Hart, P. H., C. Brand, C. F. Carson, T. V. Riley, R. H. Prager, and J. J. Finlay-Jones. 2000. Terpinen-4-ol, the main component of the essential oil of *Melaleuca alternifolia* (tea tree oil), suppresses inflammatory mediator production by activated human monocytes. *Inflammation Research* 49:619–626.

Hausen, B. M., J. Reichling, and M. Harkenthal. 1999. Degradation products of monoterpenes are the sensitizing agents in tea tree oil. *American Journal of Contact Dermatitis* 10:68–77.

Hayes, A. J., D. N. Leach, and J. L. Markham. 1997. *In vitro* cytotoxicity of Australian tea tree oil using human cell lines. *Journal of Essential Oil Research* 9:575–582.

Inouye, S., T. Tsuruoka, M. Watanabe, et al. 2000. Inhibitory effect of essential oils on apical growth of *Aspergillus fumigatus* by vapor contact. *Mycoses* 43:17–23.

Inouye, S., K. Uchida, and H. Yamaguchi. 2001. *In vitro* and *in vivo* anti-*Trichophyton* activity of essential oils by vapor contact. *Mycoses* 44:99–107.

Inouye, S., M. Watanabe, Y. Nishiyama, K. Takeo, M. Akao, and H. Yamaguchi. 1998. Antisporulating and respiration-inhibitory effects of essential oils on filamentous fungi. *Mycoses* 41:403–410.

Karpanen, T. J., T. Worthington, E. R. Hendry, B. R. Conway, and P. A. Lambert. 2008. Antimicrobial efficacy of chlorhexidine digluconate alone and in combination with eucalyptus oil, tea tree oil and thymol against planktonic and biofilm cultures of *Staphylococcus epidermidis*. *Journal of Antimicrobial Chemotherapy* 62:1031–1036.

Klimmek, J. K., R. Nowicki, K. Szendzielorz, et al. 2002. Application of tea tree oil and its preparations in combined treatment of dermatomycoses. *Mikologia Lekarska* 9:93–96.

Koh, K. J., A. L. Pearce, G. Marshman, J. J. Finlay-Jones, and P. H. Hart. 2002. Tea tree oil reduces histamine-induced skin inflammation. *British Journal of Dermatology* 147:1212–1217.

Kwieciński, J., S. Eick, and K. Wójcik. 2009. Effects of tea tree (*Melaleuca alternifolia*) oil on *Staphylococcus aureus* in biofilms and stationary growth phase. *International Journal of Antimicrobial Agents* 33:343–347.

Lassak, E. V., and T. McCarthy. 2001. *Australian medicinal plants*. Sydney: Reed New Holland.

Lawrence, H. A., and E. A. Palombo. 2009. Activity of essential oils against *Bacillus subtilis* spores. *Journal of Microbiology and Biotechnology* 19:1590–1595.

Lee, B. H., W. S. Choi, S. E. Lee, and B. S. Park. 2001. Fumigant toxicity of essential oils and their constituent compounds towards the rice weevil, *Sitophilus oryzae* (L.). *Crop Protection* 20:317–320.

Liu, X., Y. Zu, Y. Fu, et al. 2009. Antimicrobial activity and cytotoxicity towards cancer cells of *Melaleuca alternifolia* (tea tree) oil. *European Food Research and Technology* 229:247–253.

May, J., C. H. Chan, A. King, L. Williams, and G. L. French. 2000. Time-kill studies of tea tree oils on clinical isolates. *Journal of Antimicrobial Chemotherapy* 45:639–643.

McCage, C. M., S. M. Ward, C. A. Paling, D. A. Fisher, P. J. Flynn, and J. L. McLaughlin. 2002. Development of a paw paw herbal shampoo for the removal of head lice. *Phytomedicine* 9:743–748.

Messager, S., K. A. Hammer, C. F. Carson, and T. V. Riley. 2005. Assessment of the antibacterial activity of tea tree oil using the European EN 1276 and EN 12054 standard suspension tests. *Journal of Hospital Infection* 59:113–125.

———. 2006. Sporicidal activity of tea tree oil. *Australian Infection Control* 11:112–121.

Mikus, J., M. Harkenthal, D. Steverding, and J. Reichling. 2000. *In vitro* effect of essential oils and isolated mono- and sesquiterpenes on *Leishmania major* and *Trypanosoma brucei*. *Planta Medica* 66:366–368.

Nelson, R. R. S. 1997. In-vitro activities of five plant essential oils against methicillin-resistant *Staphylococcus aureus* and vancomycin-resistant *Enterococcus faecium*. *Journal of Antimicrobial Chemotherapy* 40:305–306.

Penfold, A. R., and Morrison, F. R. 1946. Australian tea trees of economic value, part I (Third edition). Technological Museum. *Sydney Bulletin* No. 14.

Reichling, J., A. Weseler, U. Landvatter, and R. Saller. 2002. Bioactive essential oils used in phytomedicine as anti-infective agents: Australian tea tree oil and manuka oil. *Acta Phytotherapeutica* 1:26–32.

Riley, T. V. 2003. Antimicrobial activity of tea tree oil against oral microorganisms. RIRDC report #03/019.

Russell, M. 1999. Toxicology of tea tree oil. In *Tea tree: The genus* Melaleuca, vol. 9, ed. I. Southwell and R. Lowe, 191–201. Amsterdam: Harwood Academic Publishers.

Sammataro, D., G. Degrandihoffman, G. Needham, and G. Wardell. 1998. Some volatile plant oils as potential control agents for varroa mites (Acari: Varroidae) in honey bee colonies (Hymenoptera: Apidae). *American Bee Journal* 138:681–685.

Satchell, A. C., A. Saurajen, C. Bell, and R. S. Barnetson. 2002a. Treatment of dandruff with 5% tea tree oil shampoo. *Journal of the American Academy of Dermatology* 47:852–855.

———. 2002b. Treatment of interdigital tinea pedis with 25% and 50% tea tree oil solution: A randomized, placebo controlled, blinded study. *Australasian Journal of Dermatology* 43:175–178.

Schnitzler, P., K. Schön, and J. Reichling. 2001. Antiviral activity of Australian tea tree oil and eucalyptus oil against herpes simplex virus in cell culture. *Pharmazie* 56:343–347.

Syed, T. A., Z. A. Qureshi, S. M. Ali, S. Ahmad, and S. A. Ahmad. 1999. Treatment of toenail onychomycosis with 2% butenafine and 5% *Melaleuca alternifolia* (tea tree) oil in cream. *Tropical Medicine and International Health* 4:284–287.

Tong, M. M., P. M. Altman, and R. S. Barnetson. 1992. Tea tree oil in the treatment of tinea pedis. *Australasian Journal of Dermatology* 33:145–149.

Viollon, C., D. Mandin, and J. P. Chaumont. 1996. Activite´s antagonistes, *in vitro*, de quelques huiles essentielles et de compose´s naturels volatils vis-a-vis de la croissance de *Trichomonas vaginalis*. *Fitoterapia* 67:279–281.

CHAPTER 11

Medicinal Plants of New South Wales, Australia

Joanne Packer, Jitendra Gaikwad,* David Harrington, Shoba Ranganathan,
Joanne Jamie, and Subramanyam Vemulpad

CONTENTS

11.1 INTRODUCTION

11.1.1 Plant Biodiversity of Australia and New South Wales

Australia is one of the world's 17 megadiverse countries (Williams et al. 2001). It is home to over 20,000 vascular and 14,000 nonvascular plants, ~250,000 species of fungi, and over 3,000 lichens. Of all the vascular plant species and nine plant families, 85% are endemic to the continent (Orchard 1999). This biodiversity is due to the continent's long isolation after the

---

* Joanne Packer and Jitendra Gaikwad are equal first authors.

early Cretaceous breakup of the Gondwana supercontinent, and consequent speciation across its large latitudinal and climatic range (White and Frazier 1986; White 1994). The Australian flora is highly adapted to drought, low soil nutrients, and fire through various physiological characteristics, such as hard seededness, scleromorphy, ligonotuber development, and epicormic buds (Benson 1999).

The Australian state of New South Wales (NSW) has a land mass of approximately 800,000 km² and occupies around 10° of latitudinal range. Its vegetation represents a microcosm of the Australian continent with tropical savannas and rain forests, temperate broadleaf forests, alpine and subalpine, semiarid and arid rangelands and deserts (Department of the Environment 2010). The major vegetation associations of NSW are woodlands and forests, of varying heights and canopy cover, which are dominated by many species of *Eucalyptus*. The type of forest cover varies according to rainfall, hydrology, and soil fertility, with taller, moister associations such as rain forest limited to small pockets of high-nutrient soils of the coast and ranges. Dependent on rainfall, the vegetation becomes lower and sparser following a westerly gradient along the western slopes and plains, progressing to semiarid rangelands and grasslands (Thackway et al. 1995).

Some authors claim that between 600 and 1,000 distinct vegetation associations are present in NSW (Specht 1981; Walker and Hopkins 1990), with an estimated 5,300 native plants (Harden 1990) and 1,000 species that are either exotic or have been naturalized (Benson 1999). This enormous diversity, across such a wide range of ecological biota, constitutes a significant source of medicinal resources as well as a source of phytochemicals and potentially novel compounds (Wickens and Pennacchio 2002; PMSEIC 2005).

11.1.2 Aboriginal Australians and Customary Medicines

Aboriginal Australians have lived on the Australian continent for at least 40 to 50 thousand years (Kohen 1995) and have the longest continuous heritage of any human culture on the planet (Crotty et al. 1995; Pearn 2004). Aboriginal people have occupied every possible habitat, from the arid central deserts to the alpine areas of NSW, Victoria and Tasmania. Correspondingly, a wide range of different food and medicinal resources, particularly plants, has been exploited by different groups. Consistent with many Indigenous cultures, Australian Aboriginal people, including those throughout NSW, have used plants as medicines for thousands of years and have a vast knowledge of Australia's unique flora. In fact, contemporary Aboriginal societies still cite medicinal uses for species from over 100 plant genera (Crotty et al. 1995; Pearn 2004).

Indigenous medicinal plant knowledge is usually referred to as traditional knowledge. In this chapter, we would like to distinguish between "traditional knowledge" and the related term, "customary knowledge." Customary knowledge incorporates knowledge associated with a contemporary Indigenous society, including innovations that may have occurred using new resources (for example, introduced plant species), technologies, and practices. As such, customary knowledge connotes a modern, evolving knowledge system. Traditional knowledge could imply a static body of knowledge, untouched by changing times.

The customary knowledge of medicinal plants held by Indigenous peoples is a significant medicinal, cultural, and economic resource. It is estimated that ~80% of the population in developing countries rely predominantly on customary medicines, mostly from plants, for their primary healthcare (WHO 2002). Furthermore, approximately 80% of all plant-derived drugs that are currently in use globally were discovered as a direct result of the study of the medicinal use of plants by humans (ethnomedicine), and are still used for the same or similar ailments as the original medicinal plant was (Fabricant and Farnsworth 2001). Natural products from plants used medicinally by Indigenous people are the most consistently successful source of structurally diverse and novel compounds and drugs (Harvey 2000; Newman and Cragg 2007).

11.2 CONTEXT OF THE REVIEW

The systematic documentation of ethnomedicinal data is an important avenue for the identification of potential resources and an understanding of biodiversity and its sustainability. There are excellent publications that provide information on medicinal plants of Australia (Cribb and Cribb 1984; Kneale and Johnson 1984; Leiper and Hauser 1984; Low 1990; Hiddins 1999; Everard et al. 2002; Lassak and McCarthy 2008), including some firsthand information from Aboriginal communities (Kneale and Johnson 1984; Leiper and Hauser 1984; Hiddins 1999; Everard, Kalotas et al. 2002). These resources, however, provide information only up to the 1990s (Lassak and McCarthy, 2008, is a reprint of a 1983 edition). The literature available on NSW customary medicines has a botanical focus with a deficiency of literature on phytochemistry or biological activity and only a fraction of the information collated on NSW plants of ethnomedicinal importance.

In Australia, as is occurring in other parts of the world (Cox 2000; Cordell 2002), Indigenous knowledge of medicinal plants is being lost as elderly people possessing this knowledge pass away and communities are dislocated and Westernized (Horstman and Wightman 2001). It is essential that this knowledge be conserved for both its cultural and scientific value (Fabricant and Farnsworth 2001; Cordell 2002; Heinrich 2003).

This review aims to consolidate information presented in existing literature of plants found in NSW that have been used as customary medicines by Australian Aboriginal people. It will cover distribution and habitat, documented customary use, biological activity of the plant extracts, and phytochemistry of secondary metabolites. An additional focus of this review is to address contemporary issues of working with Indigenous people on customary medicines and to introduce the work of our group, the Indigenous Bioresources Research Group (IBRG). The work of this group is aimed at the systematic documentation and preservation of customary medicinal plant knowledge of Indigenous people as an exemplar of best practice in working with custodians of customary knowledge.

11.3 DISCUSSION

This review focuses on plant species that grow in NSW and have been documented as Australian Aboriginal customary medicines in reference sources well regarded in this field (Cribb and Cribb 1984; Kneale and Johnson 1984; Leiper and Hauser 1984; Low 1990; Hiddins 1999; Everard et al. 2002; Lassak and McCarthy 2008). This information has been enhanced in this review by drawing from additional resources for information on the distribution, habitat, biological activity, and phytochemistry of these medicinal plants. Systematic searches of the available literature were performed, focusing on currently accepted scientific names, recent synonyms, plant families and genera, and key words, using online literature databases including the Australian Plant Name Index, Australian National Herbarium Specimen Information Register, Integrated Botanical Information System, Flora of Australia Online, Western Australian Herbarium, PLANTNET, Scifinder Scholar, and Google Scholar. The information is summarized in Tables 11.1–11.3 and key features from this review are highlighted in the following sections.

Table 11.1 Names and Australian Distribution of NSW Medicinal Plants

Scientific Name	Family	Common Name	Habitat	Distribution
Acacia bivenosa	Fabaceae	Small cooba, umbrella bush, wattle	Arid zones, coastal sands, rocky hillsides, gullies, shrubland and open shrubland[1]	NSW, NT, QLD, SA, VIC, WA
Acacia falcata	Fabaceae	Hickory, lignum vitae, Sally	Coastal regions, on the eastern slopes of the Great Divide, often in shallow stony soil[1]	NSW, QLD
Acacia implexa	Fabaceae	Bastard myall, black wattle, broad leaf wattle, fish wattle, hickory, hickory wattle, lightwood, lignum vitae, Sally wattle, screw-pod wattle, scrub wattle, wattle	Shallow soil on hills in open forest[1]	ACT, NSW, QLD, TAS, VIC
Acacia melanoxylon	Fabaceae	Black Sally, black wattle, blackwood, hickory, lightwood, mudgerabah, paluma blackwood, silver wattle, Tasmanian blackwood, wattle	Widespread; fertile soils in valleys and on flats in mountainous areas, forested seasonal swamps[1]	ACT, NSW, QLD, SA, TAS, VIC, WA
Acacia tetragonophylla	Fabaceae	Curara, dead finish, kurara, wattle	Arid and semiarid areas, near watercourses, in Mulga communities[1]	NSW, NT, QLD, SA, VIC, WA,
Aegiceras corniculatum	Primulaceae	Black mangrove, river mangrove	Brackish water[1]	NSW, NT, QLD
Ailanthus triphysa	Simaroubaceae	Ferntop ash, white bean, white siris	Dry, littoral, and subtropical rain forests; not common[2]	NSW, QLD, WA
Ajuga australis	Lamiaceae	Australian bugle, bugle	Rocky areas of forests and woodlands, common in waste areas and sandy to clay rich soils[2]	NSW, QLD, SA, TAS, VIC
Aleurites moluccana	Euphorbiaceae	Buah keras, candlenut, candleberry, candlenut siris, candlenut tree, carie nut, Indian walnut	Coastal areas[2]	NSW, QLD
Alphitonia excelsa	Rhamnaceae	Coopers wood, humbug, leather jacket, mountain ash, pink almond, red almond, red ash, red tweedie, sarsaparilla, soap tree, white leaf, white myrtle	Margin of warm rain forests on the coast, scrub, open forests, sheltered gullies, steep slopes[2]	NSW, NT, QLD, VIC, WA
Alstonia constricta	Apocynaceae	Bitter bark, fever bark, Peruvian bark, quinine bark, quinine bush, quinine tree, whitewood	Red sandy soils, open forests, gully banks[3]	NSW, QLD
Amyema maidenii	Loranthaceae	Maidens mistletoe, mulka wertibi	Dry riverbeds, riverine woodlands, sand plain, open shrublands, bunchgrass paddocks, shallow red sand[3]	NSW, NT, SA, QLD, VIC, WA

Atriplex nummularia	Amaranthaceae	Cabbage saltbush, giant saltbush, old man saltbush	Heavy soils in or around seasonally inundated areas or floodways, drier areas of all mainland states[2]	NSW, NT QLD, SA, VIC, WA
Avicennia marina	Acanthaceae	Gray mangrove, mangrove, white mangrove	Saltwater swamps and estuaries, most common and widespread mangrove in NSW[2]	NSW, QLD, SA, VIC, WA
Canarium australianum	Burseraceae	Brown cudgerie, canarium, island white beech, mango bark, Melville Island white beech, scrub turpentine	Savannahs, gravelly soil along edge of creeks, stabilized dunes of shell and lateritic pebbles at edge of beaches, rain forests, forest on headlands, moist sandy scrub, monsoon forests[3]	NSW, NT, QLD, WA
Canavalia rosea	Fabaceae	Beach bean, coastal canavalia, coastal jack bean, fire bean, McKenzie bean, Norfolk Island bean, wild jack bean	Pantropical and subtropical areas, on ocean strands, coralline sands[3]	NSW, NT, WA, QLD
Cassia odorata	Fabaceae	Australian senna	Wet sclerophyll or subtropical rain forests[1]	NSW, NT, QLD
Cassytha glabella	Lauraceae	Devil's twine, dodder laurel, slender devil's twine, slender dodder-laurel, smooth cassytha	Heath, sandstone derived soil, dry sclerophyll forests[3]	NSW, QLD, SA, TAS, VIC, WA
Casuarina equisetifolia	Casuarinaceae	Beach casuarina, beach she-oak, beachoak, beefwood, casuarina, coast oak, coast she-oak, horsetail she-oak, swampoak, whistling tree	Tropical and subtropical coastlines of northern and northeastern Australia[1]	NSW, NT, QLD, WA
Centaurium spicatum	Gentianaceae	Australian centaury, bushman's headache-cure, centaury, native centaury, spike centaury	Damp flood areas, saline flats, occasionally on grassy swards near roadside ditches, banks of brackish lagoons behind coastal sand dunes[3]	NSW, NT, QLD, SA, TAS, VIC, WA
Centipeda cunninghamii	Asteraceae	Common sneezeweed, old man weed, scentwood	Damp places, dry sclerophyll forests, muddy sandbanks, low-lying paddocks, sandy shores of arid zone lakes[3]	NSW, NT, QLD, SA, VIC
Centipeda minima	Asteraceae	Scentweed, sneezeweed, spreading sneezeweed	River banks, sandy banks, scattered in damp swampy soils[3]	NSW, NT, QLD, SA, TAS, VIC, WA
Centipeda thespidioides	Asteraceae	Desert sneezeweed	Shrublands, edge of swamps, damp red loamy sand in areas subject to temporary, slight inundation, wet banks of dams[3]	NSW, QLD, SA, VIC
Chamaesyce alsiniflora	Euphorbiaceae		Coastal dunes[3]	NSW, NT, QLD
Chamaesyce drummondii	Euphorbiaceae	Caustic creeper, caustic weed, creeping spurge, flat spurge, mat spurge, milkweed, spurgewort	Woodland habitat, arid shrub-steppes and woodlands, mature Mallee, low areas between rolling sandhills[3]	NSW, NT, QLD, SA, VIC, WA

(continued)

Table 11.1　Names and Australian Distribution of NSW Medicinal Plants (Continued)

Scientific Name	Family	Common Name	Habitat	Distribution
Clematis glycinoides	Ranunculaceae	Headache vine, traveler's joy, traveler's vine	Wet sclerophyll forests, rain forests, dry sclerophyll forests[3]	NSW, NT, QLD, VIC
Clematis microphylla	Ranunculaceae	Oldman's beard, small clematis, small leaf clematis	Mallee scrub, exposed headlands, riparian shrubland, semievergreen vine thicket remnants[3]	NSW, QLD, SA, TAS, VIC, WA
Cleome viscosa	Cleomaceae	Mustard bush, spider flower, tick-weed, wild caia, yellow spiderflower	Sandy soil such as coral beaches, tropical, hummock grasslands, adjacent to permanent creeks, along rocky creek banks[3]	NSW, NT, SA, QLD, WA
Clerodendrum floribundum	Lamiaceae	Abundant clerodendron, clerodendron, lolly bush, smooth clerodendrum, thurkoo	Savannah woodlands, along the creek banks in sandy soil, eucalypt woodlands, wet sclerophyll forests, coastal headlands, strand rain forests, dune scrub, semievergreen vine thicket, rain forests, semideciduous monsoon forests[3]	NSW, NT, QLD, WA
Clerodendrum inerme	Lamiaceae	Harmless clerodendron, scrambling clerodendron	Tropical coastal areas on margins of mangrove communities[1,2]	NSW, NT, QLD, WA
Convolvulus erubescens	Convolvulaceae	Australian bindweed, blushing bindweed, pink bindweed	Coastal and subtropical areas, wetter eucalypt forests and in rain forest margins[2]	NSW, NT, QLD, SA, TAS, VIC, WA
Corymbia dichromophloia	Myrtaceae	Bloodwood, gum topped bloodwood, Mount Cooper bloodwood, red topped bloodwood, variable barked bloodwood	Mixed eucalypt grasslands, scree slopes of sandstone plateaus, rocky laterite tablelands, flat gravelly laterite, skeletal soil, low savannah woodland on flats[3]	NSW, NT, SA, WA
Corymbia gummifera	Myrtaceae	Bloodwood, pale bloodwood, red bloodwood	Sandstone mesas, shallow sandy soil, mixed woodlands, over heathy understorey[3]	NSW, VIC, QLD
Corymbia maculata	Myrtaceae	Spotted gum, spotted iron gum	Tall open forests with sparse understorey of *Acacia* and *Persoonia*, skeletal clay-loam on metamorphic rock, old sand dunes, shrub woodlands[3]	NSW, VIC, QLD
Corymbia polycarpa	Myrtaceae	Grey bloodwood, kulcha, long fruited bloodwood, Mallee bloodwood, pale bloodwood, red bloodwood, small flowered bloodwood	Low open woodlands, sandy alluvial terrace at base of sandstone scarp, with sparse mixed woodlands[3]	NSW, NT, QLD

Species	Family	Common names	Habitat	States
Corymbia terminalis	Myrtaceae	Bloodwood, long fruited bloodwood, pale bloodwood	Pale sandy clay, open savannah woodlands with scattered *Eucalyptus* spp. and low shrubs, open woodlands, red sandy soil[3]	NSW, QLD, SA, WA
Corymbia tessellaris	Myrtaceae	Carbeen, manna gum, Moreton Bay ash, ribbon gum, rough barked manna gum, roughbark ribbon gum, white gum carbeen bloodwood	Widespread and abundant, in grassy woodlands or forest on fertile loamy soils,[2] low sand dunes, sandy-surfaced soils	NSW, QLD, SA, TAS, VIC
Crinum pedunculatum	Alliaceae	Crinum lily, river lily, spider lily, swamp lily	Grows in swamps and along stream banks in coastal districts and on offshore islands[2]	NSW, NT, QLD
Crotalaria cunninghamii	Fabaceae	Green bird-flower, parrot pea, rattlepod	In Mulga communities, unstable sand dunes[2]	NSW, NT, QLD, SA, WA
Croton insularis	Euphorbiaceae	Cascarilla bark, native cascarilla bark, Queensland cascarilla, Queensland cascarilla bark, silver croton, warrel, white croton	Margins of or in dry rain forests, escarpment ranges[2]	NSW, QLD
Cyathea australis	Cyatheaceae	Rough tree fern	Rain forests, open forests, gullies, on hillsides in moist, shady areas[2]	ACT, NSW, QLD, TAS, VIC
Cymbidium canaliculatum	Orchidaceae	Black orchid, rock orchid, tree orchid	Hollows of trees in dry sclerophyll forest or woodlands[2]	NSW, NT, SA, QLD, VIC
Cymbopogon obtectus	Poaceae	Cottongrass, lemon scentgrass, silky-heads, turpentine grass	Near watercourses, riverine cliffs, in mixed high, open shrubland, interspersed with abundant perennial grasslands and areas of woody shrubs and dense Mulga/Box woodland along drainage lines and fringing ephemeral swamps[3]	NSW, NT, QLD, SA, VIC, WA
Daviesia latifolia	Fabaceae	Hop bitter-pea	Widespread in dry sclerophyll communities[2]	NSW, VIC, TAS
Dendrocnide excelsa	Urticaceae	Fiberwood, giant nettle tree, giant stinging tree, gympi gympi, stinging bush, stinging tree	Warm rain forests on the coast, drier rain forests in gullies[2]	NSW, QLD, VIC
Dioscorea transversa	Dioscoreaceae	Long yam, native yam, yam	Chiefly in warmer rain forests and moist sclerophyll forests[2]	NSW, NT, QLD, WA
Dodonaea viscosa	Sapindaceae	Broad-leaf hop bush, broadleaf hopbush, candlewood, giant hop bush, hop bush, narrow leaf hopbush, native hop, native hop bush, soapwood, sticky dodonea, sticky hop bush, switchsorrel, wedge leaf hopbush, wild hops	Riverbanks, dry sclerophyll forests, in open Mallee communities, tall disturbed shrubland, creek terraces[3]	NSW, NT, QLD, SA, TAS, VIC, WA

(continued)

Table 11.1 Names and Australian Distribution of NSW Medicinal Plants (Continued)

Scientific Name	Family	Common Name	Habitat	Distribution
Doryphora sassafras	Atherospermataceae	Canary sassafras, golden deal, golden sassafras, NSW sassafras, sassafras, yellow sassafras	Rain forests, disturbed wet sclerophyll forests, montane rain forests[3]	NSW, QLD
Eremophila debilis	Scrophulariaceae	Winter apple	Box and white cypress communities, on a variety of soils[2]	NSW, QLD
Eremophila freelingii	Scrophulariaceae	Emu bush, fuchsia bush, limestone fuchsia, native fuchsia, rock fuchsia bush	Mostly in Mulga communities, in rocky areas in hilly and ridge country[2]	NSW, NT, QLD, SA, WA
Eremophila gilesii	Scrophulariaceae	Charleville turkey bush, desert fuchsia bush, green turkey bush, turkey bush	In Mulga communities on sand plains and stony ridge areas[2]	NSW, NT, QLD, SA, WA
Eremophila longifolia	Scrophulariaceae	Berrigan, dogwood, emu bush, long leaf emu bush, native plum, native plum tree	Variety of communities on most soil types,[2] open Mulga shrubland, arid shrub-steppe and woodlands, floodplains[3]	NSW, NT, QLD, SA, VIC, WA
Eremophila maculata	Scrophulariaceae	Fuchsia bush, native fuchsia, spotted emu bush, spotted fuchsia, wild fuchsia	Variety of communities, mainly on heavy clay to clay loams of river and creek floodplains[2]	NSW, NT, QLD, SA, VIC, WA
Eremophila sturtii	Scrophulariaceae	Narrow-leaved emu bush, turpentine bush	Usually in Mallee, Mulga, and Bimble Box communities on sandy and loamy red earth[2]	NSW, NT, SA, QLD, VIC
Ervatamia angustisepala	Apocynaceae	Banana bush, bitterbark, iodine plant, native gardenia, windmill bush	Closed rain forests[3]	NSW, QLD
Eucalyptus camaldulensis	Myrtaceae	Blue gum, murray red gum, red gum, river gum, river red gum, yarrow	Community dominant, in grassy woodland or forests on deep, rich alluvial soils adjacent to large, permanent water bodies[2]	NSW, NT, QLD, SA, VIC, WA
Eucalyptus haemastoma	Myrtaceae	Scribbly gum, snappy gum, white gum	Abundant, in dry sclerophyll woodlands, on shallow, infertile sandy soil, on sandstone[2]	NSW
Eucalyptus mannifera	Myrtaceae	Brittle gum, broad leaved manna gum, capertee brittle gum, manna gum, mountain spotted gum, red spotted gum, white brittle gum	Widespread and abundant, in open dry sclerophyll woodlands on usually shallow and rocky, somewhat infertile soils[2]	ACT, NSW, VIC
Eucalyptus microtheca	Myrtaceae	Coolibah, flooded box, western coolibah	Box woodlands, scrubby deserts, margins of seasonally dry lagoons, floodways in open forest, clay of dry creek bed through grasslands, margins of swamp in heavy textured soil, with panicum and sedges[3]	NSW, NT, QLD, SA, WA

Species	Family	Common names	Habitat	Distribution
Eucalyptus pilularis	Myrtaceae	Blackbutt, great blackbutt	Widespread, in wet sclerophyll or grassy coastal forests, on lighter soils of medium fertility[2]	NSW, QLD
Eucalyptus piperita	Myrtaceae	Sydney peppermint, urn fruited peppermint, white stringybark	Locally frequent, in dry sclerophyll forest or woodlands on moderately fertile, often alluvial, sandy soil[2]	NSW
Eucalyptus pruinosa	Myrtaceae	Kullingal, silver box, silver leaved box	Open woodlands, undulating plains, red skeletal soil, tall open shrublands with tall grass stratum, stony red sand, open scrub[3]	NSW, NT, QLD, SA, VIC, WA
Eucalyptus racemosa	Myrtaceae	Scribbly gum, snappy gum, white gum	Abundant in dry sclerophyll woodlands on shallow, infertile sandy soil on sandstone[2]	NSW
Eucalyptus resinifera	Myrtaceae	Red mahogany, red messmate, red stringybark	Abundant in wet or dry sclerophyll forests on deeper soils of medium to high fertility[2]	NSW, QLD
Evolvulus alsinoides	Convolvulaceae	Speedwell	Sandy plains to rocky outcrops, often in grassy *Eucalyptus* and *Acacia* woodlands, widespread but not common[2]	NSW, NT, SA, QLD, WA
Excoecaria agallocha	Euphorbiaceae	Blind your eye, blind your eye mangrove, blinding tree, blind-your-eyes-tree, milky mangrove, river poison tree, scrub poison tree	Chiefly adjacent to or in mangrove communities and swamp rain forest of palms and *Elaeocarpus obovatus*[2]	NSW, NT, QLD, WA
Excoecaria dallachyana	Euphorbiaceae	Blind your eye, brown birch, brush poison tree, milkwood, milky birch, scrub poison tree	Dry or riverine rain forests[2]	NSW, QLD
Exocarpos aphyllus	Santalaceae	Leafless ballart	Woodland communities, widespread in various habitats[2]	NSW, QLD
Ficus coronata	Moraceae	Creek fig, figwood, sandpaper fig	Often along creeks, in rain forests and open country, occasionally in sheltered rocky areas on western slopes, widespread on coast and tablelands[2]	NSW, NT, QLD, VIC
Flagellaria indica	Flagellariaceae	Bush cane, flagellaria, lawyer vine, supple jack, whip vine	In or near warmer coastal rain forests, frequently along streams or in gullies and often forming dense thickets[2]	NSW, QLD, NT, WA
Flindersia maculosa	Rutaceae	Leopard tree, leopardwood	Sandplains and stony hilly areas[2]	NSW, QLD
Geijera parviflora	Rutaceae	Australian willow, dogbush, greenheart, lavender bush, sheep bus, tree wilga, wilga	Mixed woodland communities,[2] creek bank, red soil, semievergreen vine thicket, in low shrub woodlands[3]	NSW, NT, QLD, SA, VIC, WA

(continued)

Table 11.1 Names and Australian Distribution of NSW Medicinal Plants (Continued)

Scientific Name	Family	Common Name	Habitat	Distribution
Goodenia varia	Goodeniaceae	Sticky goodenia	Triodia grasslands and Mallee communities[2]	NSW, SA, VIC, WA
Gratiola pedunculata	Scrophulariaceae	Heartsease, stalked brooklime	River or lagoon banks and other damp areas, widespread[2]	NSW, QLD, SA, VIC, WA
Grevillea striata	Proteaceae	Beef oak, beef silky oak, beefwood, silvery honeysuckle, western beefwood	Woodlands, shrublands or Triodia communities, on red sand (sometimes on dunes) or black soil or clay[2]	NSW, NT, QLD, SA, WA
Grewia retusifolia	Malvaceae	Diddle diddle, dog's balls, dog's nuts, dysentery bush, dysentery plant, emu-berry, turkey bush	Sandy soil, open woodlands, grassy forests, sandy scrub bordering creeks[3]	NSW, NT, QLD
Helichrysum luteoalbum	Asteraceae	Cud-weed, flannel-leaf, Jersey cudweed	Near river flats and margins, swamps[3]	NSW, QLD, SA, TAS, VIC, WA
Hibiscus tiliaceus	Malvaceae	Beach hibiscus, coast cottonwood, coast hibiscus, cotton tree, cottonwood, cottonwood hibiscus, gatapa, green cottonwood, mahoe, majagua, native hibiscus, native rosella, sea hibiscus, yellow hibiscus	Margins of rain forests or swamp forests near the coast, banks of tidal rivers[2]	NSW, NT, QLD
Hybanthus enneaspermus	Violaceae	Spade flower, yellow spade	Mixed woodlands, stream banks, open savanna woodlands, swampy plains[3]	NSW, NT, QLD, SA, WA,
Ipomoea pes-caprae	Convolvulaceae	Bay hops, beach convolvulus, coast morning glory, convolvulus, goat's foot, morning glory	Coastal sand dunes,[2] sand above high-tide mark[3]	NSW, NT, QLD, WA
Isopogon ceratophyllus	Proteaceae	Horny conebush	Mostly in sclerophyll forests, woodlands or heathlands, in sand or sandy soils[1]	NSW, SA, TAS, VIC
Lythrum salicaria	Lythraceae	Loosestrife, purple loosestrife, spiked loosestrife, willow-like loosestrife	Widespread in moist places or near water, often in swamps[2]	ACT, NSW, QLD,SA, TAS, VIC
Melaleuca alternifolia	Myrtaceae	Medicinal tea tree, narrow leaved tea tree, narrow leaved paperbark	Along streams and on swampy flats, on the coast and adjacent ranges[2]	NSW, QLD
Melaleuca hypericifolia	Myrtaceae	Red honey myrtle	Wet places in sandy heath and sclerophyll woodlands and coastal headlands[2]	NSW, QLD
Melaleuca quinquenervia	Myrtaceae	Belbowrie, broad leaved tea tree, broadleaf paperbark, broad-leaved tree, five veined paperbark, paperbark, paperbarked tea tree, tea tree	Widespread in swamps and around lake margins[2]	NSW, QLD

Species	Family	Common names	Habitat	Distribution
Melaleuca uncinata	Myrtaceae	Broom honey myrtle, broombush	In Mallee, woodlands and dry sclerophyll forests, usually on sandy soils, widespread, often forming dense stands[2]	NSW, SA, QLD, VIC
Melicope vitiflora	Rutaceae	Coast euodia, fishpoison wood, leatherjacket, leatherwood, north evodia	Grows in subtropical and littoral rain forests[2]	NSW, QLD
Mentha diemenica	Lamiaceae	Native pennyroyal, slender mint	Communities similar to that of *M. satureioides* but also on sandy soils[2]	NSW, NT, SA, TAS, VIC
Mentha satureioides	Lamiaceae	Brisbane pennyroyal, creeping mint, native pennyroyal	Sandy clay to clay-rich soils, frequently in grassy areas and in open woodland communities, widespread[2]	NSW, NT, QLD, SA, VIC, WA
Mucuna gigantea	Fabaceae	Cow-itch, velvet bean	Coastal rain forests[2]	NSW, NT, QLD
Myoporum platycarpum	Scrophulariaceae	Dogwood, false sandalwood, red sandalwood, sandalwood, sugar tree, sugar wood, sugarwood	Woodlands, especially Mallee, Belah and Belah–rosewood communities on red and red-brown earth[2]	NSW, SA, VIC, WA
Nymphaea gigantea	Nymphaeaceae	Blue waterlily, giant waterlily	Permanent water with deep muddy substrate in tropical and subtropical areas[2]	NSW, QLD, NT, WA
Omalanthus populifolius	Euphorbiaceae	Native bleedingheart, native poplar, Queensland poplar	Rain forests[3]	NSW, QLD
Owenia acidula	Meliaceae	Colane, emu apple, gooya, gruie, mooley apple, native nectarine, native peach, sour apple, sour plum	Low open woodlands and tall shrubland, sparsely vegetated dry plains, light-textured alluvial deserts[3]	NSW, NT, QLD, SA, WA
Petalostigma pubescens	Euphorbiaceae	Bitter bark, downy cracker bush, forest quinine, native quince, quinine berry, quinine tree, red hearted forest quinine, strychnine tree, wild quinine	Dry rain forests or open forests[2]	NSW, NT, QLD, WA
Pimelea microcephala	Thymelaeaceae	Mallee riceflower, scrub kurrajong, small head rice flower	Sandy soils in open forests and Mallee, widespread[2]	NSW, NT, QLD, SA, VIC, WA
Piper hederaceum	Piperaceae	Australian pepper vine, climbing pepper, curtain vine, native pepper, native pepper vine, pepper vine	Widespread in warmer coastal rain forests[2]	NSW, NT, QLD
Pittosporum angustifolium	Pittosporaceae	Bitter bush, butterbush, cattle bush, native willow, poison berry tree, weeping pittosporum, willow	Open woodlands[3]	NSW, NT, QLD, SA, VIC, WA
Plectranthus parviflorus	Lamiaceae	Cockspur flower, native coleus	Widespread, frequently in rocky areas and associated with creeks and rivers[2]	NSW, QLD, VIC
Plumbago zeylanica	Plumbaginaceae	Leadwort, native plumbago	Salt-tolerant species that can survive in habitats subject to sea spray[1]	NSW, NT, QLD, WA

(continued)

Table 11.1 Names and Australian Distribution of NSW Medicinal Plants (Continued)

Scientific Name	Family	Common Name	Habitat	Distribution
Pouteria pohlmaniana	Sapotaceae	Black apple, engraver's wood, yellow boxwood, Pohlmanns jungle plum, Queensland yellow box	Dry rain forests[2]	NSW, QLD
Pratia purpurascens	Campanulaceae	White root	Shady wet areas in wet sclerophyll forests, woodlands and grasslands, widespread[2]	NSW, QLD, SA, VIC
Prunella vulgaris	Lamiaceae	Selfheal	Disturbed areas, particularly along roadsides, especially in moist sites[2]	NSW, QLD, SA, TAS, VIC
Pteridium esculentum	Dennstaedtiaceae	Austral bracken, bracken fern, common bracken	Open forests or on cleared land where it can form extensive colonies[2]	ACT, NSW, NT, QLD, SA, TAS, VIC, WA
Pterocaulon sphacelatum	Asteraceae	Fruit salad plant	Open woodlands, red clay mudflats, open shrublands, sandstone escarpments[3]	NSW, NT, QLD, SA, VIC, WA
Rhizophora mucronata	Rhizophoraceae	Black mangrove, loop root mangrove, red mangrove, upriver stilt mangrove	Lower tidal reaches, mangrove swamps[3]	NSW, NT, QLD, WA
Rubus hillii	Rosaceae	Native raspberry, Queensland bramble, wild raspberry	Damp gullies with rain forest elements, creek flood plains[3]	NSW, NT, QLD
Rubus rosifolius	Rosaceae	Forest bramble, rain forest raspberry, raspberry, roseleaf raspberry, wild raspberry	Widespread on margins of rain forests and tall open forests[3]	NSW, QLD, VIC
Santalum lanceolatum	Santalaceae	Blue bush, bush plum, cherry bush, native plum, northern sandalbox, northern sandalwood, plum bush, plumwood, quandong, Queensland sandalwood, sandalwood, wild plum	Wide range of woodland communities, from sandy sites to rocky hillsides, widespread but scattered[2]	NSW, NT, QLD, SA, VIC, WA
Santalum obtusifolium	Santalaceae	NSW sandalwood	Sclerophyll forests, often along creek banks[2]	NSW, QLD, VIC
Santalum spicatum	Santalaceae	Australian sandalwood, fragrant sandalwood, sandalwood, West Australian sandalwood	Open low shrubs and Mallees, red loamy sand over clay[3]	NSW, QLD, SA, WA
Sarcostemma australe	Apocynaceae	Caustic bush, caustic vine, milk bush, milk vine, pencil caustic	Widespread, in woodlands and scrub, often in rocky outcrops, also in coastal sites in drier rain forests and on cliffs near sea[2]	NSW, QLD, SA, WA
Scaevola spinescens	Goodeniaceae	Currant bush, maroon bush, prickly fan-flower	Drier areas usually on hillsides or stony areas[2]	NSW, NT, QLD, SA, VIC, WA
Sebaea ovata	Gentianaceae	Yellow centaury, yellow sebaea	Widespread, damp sites in woodlands[2]	ACT, NSW, NT, QLD, SA, TAS, VIC, WA

Species	Family	Common names	Habitat	Distribution
Sida rhombifolia	Malvaceae	Big Jack, common sida, jelly leaf, paddys lucerne, Queensland hemp, sida-retusa, sidratusa	Weed of wastelands, now pantropical[1]	NSW, NT, QLD, SA, VIC, WA
Smilax australis	Smilacaceae	Austral sarsaparilla, barbwire vine, lawyer vine, sweet sarsaparilla, wait-a-while	Widespread and common in rain forests, sclerophyll forests, woodlands and heath, often forming dense thickets[2]	NSW, NT, QLD, VIC, WA
Smilax glyciphylla	Smilacaceae	Native sarsaparilla, sarsaparilla, sweet sarsaparilla, sweet tea, wild licorice	Widespread in rain forests, sclerophyll forests and woodlands, chiefly in coastal districts[2]	NSW, QLD
Sophora tomentosa	Fabaceae	Golden chain, sea coast laburnum	Seashores[2]	NSW, QLD
Spartothamnella juncea	Lamiaceae		Sclerophyll forests and in dry rain forests and vine thickets[2]	NSW, QLD
Sterculia quadrifida	Sterculiaceae	Kuman, monkey nut tree, orange fruited kurrajong, orange fruited sterculia, peanut tree, red fruited kurrajong, redfruit kurrajong, red-fruited kurrajong, small flowered kurrajong, smooth seeded kurrajong, white crowsfoot	Chiefly in littoral and riverine rain forests[2]	NSW, NT, QLD
Swainsona galegifolia	Fabaceae	Darling pea, indigo plant, smooth darling pea	Widespread in a variety of habitats,[2] tall open woodlands with grass and herb understorey, wet and dry sclerophyll forests[3]	NSW, QLD
Tinospora smilacina	Menispermaceae	Snakevine	In dry rain forests[2]	NSW, NT, QLD, WA
Toona ciliata	Meliaceae	Australian red cedar, cedar, red cedar, Suren or Indian mahogany, toon	Warmer rain forests on the coast and coastal ranges[2]	NSW, QLD
Trichodesma zeylanicum	Boraginaceae	Camel bush, cattle bush	Open communities, often on sand dunes or rocky hills[2]	NSW, NT, QLD, SA, WA
Trichosanthes palmata	Cucurbitaceae	Redball snakegourd, thowan	In moist open forest or rain forests[4]	NSW, QLD
Verbena officinalis	Verbenaceae	Common verbena, common vervain, vervain	Widespread weed of disturbed ground and wastelands[2]	NSW, NT, QLD, SA, TAS, VIC, WA
Vigna vexillata	Fabaceae	Cow pea, native cowpea, pea vine	Moist sites, often along roadsides[2]	NSW, NT, QLD
Zieria smithii	Rutaceae	Lanoline bush, sandfly bush, sandfly zieria	Widespread on the coast and ranges[2]	NSW, QLD, VIC

Notes: ACT—Australian Capital Territory; NSW—New South Wales; NT—Northern Territory; QLD—Queensland; SA—South Australia; TAS—Tasmania; VIC—Victoria; WA—Western Australia.

Table 11.2 Australian Customary Uses of NSW Medicinal Plants

Medical Condition	Medicinally Used Plants			
Skin Conditions				
Sores and skin complaints	Acacia falcata	Cleome viscosa	Ervatamia angustisepala	Plumbago zeylanica
	Acacia implexa	Clerodendrum floribundum	Eucalyptus camaldulensis	Pterocaulon sphacelatum
	Acacia tetragonophylla	Clerodendrum inerme	Eucalyptus haemastoma	Santalum lanceolatum
	Ailanthus triphysa	Corymbia tessellaris	Eucalyptus microtheca	Sarcostemma australe
	Ajuga australis	Cymbidium canaliculatum	Excoecaria agallocha	Scaevola spinescens
	Alphitonia excelsa	Dendrocnide excelsa	Exocarpos aphyllus	Sterculia quadrifida
	Alstonia constricta	Dioscorea transversa	Grewia retusifolia	Trichodesma zeylanicum
	Avicennia marina	Eremophila freelingii	Hibiscus tiliaceus	Trichosanthes palmata
	Centaurium spicatum	Eremophila gilesii	Ipomoea pes-caprae	
	Centipeda cunninghamii	Eremophila longifolia	Melaleuca alternifolia	
	Clematis microphylla	Eremophila sturtii	Pittosporum angustifolium	
Wounds	Cleome viscosa	Dodonaea viscosa	Grevillea striata	Prunella vulgaris
	Clerodendrum inerme	Eremophila freelingii	Hibiscus tiliaceus	Sarcostemma australe
	Corymbia dichromophloia	Eucalyptus camaldulensis	Melaleuca alternifolia	Sterculia quadrifida
	Cymbidium canaliculatum	Flagellaria indica	Pouteria pohlmaniana	Verbena officinalis
Swellings	Aegiceras corniculatum	Crotalaria cunninghamii	Ipomoea pes-caprae	Swainsona galegifolia
	Avicennia marina	Dodonaea viscosa	Prunella vulgaris	Tinospora smilacina
	Cleome viscosa	Eremophila maculata	Santalum spicatum	
	Clerodendrum floribundum	Hybanthus enneaspermus	Sarcostemma australe	
Astringent	Casuarina equisetifolia	Eucalyptus pilularis	Pteridium esculentum	Rhizophora mucronata
	Corymbia gummifera	Petalostigma pubescens		
Antiseptic	Corymbia tessellaris	Corymbia gummifera	Melaleuca alternifolia	Petalostigma pubescens
Coagulant/antihemorrhagic	Lythrum salicaria	Omalanthus populifolius	Pittosporum angustifolium	Swainsona galegifolia
Burns	Cymbidium canaliculatum	Corymbia tessellaris	Excoecaria dallachyana	
Body Aches and Pains				
Aches and pains/rheumatism	Aegiceras corniculatum	Dendrocnide excelsa	Ipomoea pes-caprae	Santalum lanceolatum
	Aleurites moluccana	Eremophila freelingii	Melaleuca quinquenervia	Santalum spicatum
	Alphitonia excelsa	Eremophila sturtii	Melicope vitiflora	Sida rhombifolia
	Acacia melanoxylon	Eucalyptus pruinosa	Mentha satureioides	Smilax glyciphylla

Sore eyes	Canavalia rosea	Excoecaria agallocha	Mucuna gigantea	Tinospora smilacina
	Cassytha glabella	Excoecaria dallachyana	Pimelea microcephala	Trichosanthes palmate
	Clematis glycinoides	Flindersia maculosa	Pittosporum angustifolium	Verbena officinalis
	Cleome viscosa	Geijera parviflora	Prunella vulgaris	
	Clerodendrum floribundum	Hybanthus enneaspermus	Pteridium esculentum	
Toothache	Alphitonia excelsa	Corymbia tessellaris	Eremophila sturtii	Sarcostemma australe
	Centipeda cunninghamii	Crotalaria cunninghamii	Grewia retusifolia	Sterculia quadrifida
	Centipeda minima	Cymbopogon obtectus	Owenia acidula	Tinospora smilacina
	Centipeda thespidioides	Eremophila freelingii	Petalostigma pubescens	
	Clerodendrum floribundum	Eremophila longifolia	Pterocaulon sphacelatum	
Sore throat	Alphitonia excelsa	Corymbia polycarpa	Flagellaria indica	Melicope vitiflora
	Casuarina equisetifolia	Corymbia tessellaris	Flindersia maculosa	Petalostigma pubescens
	Corymbia dichromophloia	Dodonaea viscosa	Geijera parviflora	Santalum lanceolatum
	Centipeda thespidioides	Eucalyptus camaldulensis	Flagellaria indica	
	Corymbia tessellaris	Eucalyptus microtheca	Pimelea microcephala	
Muskuloskeletal complaints	Canavalia rosea	Pittosporum angustifolium	Tinospora smilacina	Trichosanthes palmata
	Centipeda thespidioides	Sarcostemma australe		
Sore ears	Aegiceras corniculatum	Cleome viscosa	Crotalaria cunninghamii	
General Malaise and Lethargy				
Blood cleanser/tonic, sickness	Ailanthus triphysa	Croton insularis	Eucalyptus camaldulensis	Mentha satureioides
	Alphitonia excelsa	Cyathea australis	Excoecaria agallocha	Petalostigma pubescens
	Alstonia constricta	Daviesia latifolia	Gratiola pedunculata	Piper hederaceum
	Atriplex nummularia	Dodonaea viscosa	Helichrysum luteoalbum	Sebaea ovata
	Centipeda cunninghamii	Doryphora sassafras	Hybanthus enneaspermus	Smilax glyciphylla
	Corymbia dichromophloia	Eremophila freelingii	Isopogon ceratophyllus	Smilax australis
	Corymbia tessellaris	Eremophila sturtii	Melaleuca quinquenervia	
Fevers	Ailanthus triphysa	Clerodendrum inerme	Evolvulus alsinoides	Toona ciliata
	Alstonia constricta	Daviesia latifolia	Petalostigma pubescens	Verbena officinalis
	Cassytha glabella	Eremophila freelingii	Prunella vulgaris	
	Chamaesyce alsiniflora	Ervatamia angustisepala	Pterocaulon sphacelatum	
	Cleome viscosa	Eucalyptus camaldulensis	Santalum obtusifolium	

(continued)

Table 11.2 Australian Customary Uses of NSW Medicinal Plants (Continued)

Medical Condition	Medicinally Used Plants			
	General Malaise and Lethargy			
Headache	Alphitonia excelsa	Clerodendrum floribundum	Eremophila maculata	Melaleuca quinquenervia
	Centaurium spicatum	Crotalaria cunninghamii	Eucalyptus pruinosa	Pimelea microcephala
	Clematis glycinoides	Eremophila freelingii	Ipomoea pes-caprae	Pterocaulon sphacelatum
	Cleome viscosa	Eremophila longifolia	Melaleuca hypericifolia	Zieria smithii
	Santalum lanceolatum			
	Eremophila longifolia			
	Goodenia varia			
Fatigue				
Insomnia				
Sedative				
	Stomach and Intestinal Conditions			
Diarrhea/dysentery	Acacia tetragonophylla	Cleome viscosa	Eucalyptus camaldulensis	Hybanthus enneaspermus
	Ailanthus triphysa	Clerodendrum floribundum	Eucalyptus haemastoma	Lythrum salicaria
	Alstonia constricta	Convolvulus erubescens	Eucalyptus microtheca	Rubus rosifolius
	Canarium australianum	Corymbia polycarpa	Eucalyptus racemosa	Sebaea ovata
	Casuarina equisetifolia	Corymbia tessellaris	Evolvulus alsinoides	Sida rhombifolia
	Centaurium spicatum	Cymbidium canaliculatum	Flindersia maculosa	Tinospora smilacina
	Chamaesyce alsinifora	Eremophila sturtii	Grewia retusifolia	Toona ciliata
Stomachache	Ailanthus triphysa	Clematis microphylla	Eucalyptus microtheca	Rubus rosifolius
	Alphitonia excelsa	Clerodendrum inerme	Eucalyptus piperita	Scaevola spinescens
	Alstonia constricta	Convolvulus erubescens	Mentha diemenica	Sida rhombifolia
	Canarium australianum	Croton insularis	Mentha satureioides	Verbena officinalis
	Centaurium spicatum	Eremophila freelingii	Rubus hillii	
Constipation	Aleurites moluccana	Eucalyptus mannifera	Myoporum platycarpum	Santalum obtusifolium
	Cassia odorata	Ipomoea pes-caprae	Santalum lanceolatum	Vigna vexillata
	Corymbia tessellaris			
Emetic	Gratiola pedunculata	Santalum lanceolatum		
Wasting disease	Exocarpos aphyllus			
Induce appetite	Croton insularis			

Respiratory Conditions

Condition				
Coughs, cold, and congestion	Acacia bivenosa	Clerodendrum floribundum	Eucalyptus microtheca	Pittosporum angustifolium
	Acacia tetragonophylla	Corymbia dichromophloia	Exocarpos aphyllus	Pterocaulon sphacelatum
	Ailanthus triphysa	Cymbopogon obtectus	Grewia retusifolia	Santalum spicatum
	Alstonia constricta	Eremophila freelingii	Hybanthus enneaspermus	Scaevola spinescens
	Canavalia rosea	Eremophila gilesii	Melaleuca alternifolia	Smilax glyciphylla
	Centipeda minima	Eremophila longifolia	Melaleuca quinquenervia	Spartothamnella juncea
	Centipeda thespidioides	Eremophila maculate	Melaleuca uncinata	Tinospora smilacina
	Clematis glycinoides	Eremophila sturtii	Mentha satureoides	
	Cleome viscosa	Eucalyptus camaldulensis	Pimelea microcephala	
Respiratory disease	Corymbia dichromophloia	Eremophila sturtii	Melaleuca uncinata	Spartothamnella juncea
	Cymbopogon obtectus	Eucalyptus microtheca	Pterocaulon sphacelatum	Trichosanthes palmata
	Eremophila longifolia	Helichrysum luteoalbum	Sida rhombifolia	
Asthma	Aegiceras corniculatum	Ailanthus triphysa	Hybanthus enneaspermus	Trichosanthes palmata
Tuberculosis	Centipeda cunninghamii	Exocarpos aphyllus	Ipomoea pes-caprae	

Genitourinary Conditions

Venereal disease	Corymbia gummifera	Ipomoea pes-caprae	Plectranthus parviflorus	Verbena officinalis
	Eremophila debilis	Melaleuca alternifolia	Santalum lanceolatum	
	Eucalyptus resinifera	Piper hederaceum	Smilax glyciphylla	
Kidney complaints, urine retention	Hybanthus enneaspermus	Mentha diemenica	Verbena officinalis	
Liver complaints	Centaurium spicatum	Hybanthus enneaspermus	Sophora tomentosa	Toona ciliata
	Gratiola pedunculata			
Bladder inflammation/urinary tract infection	Corymbia maculata	Hybanthus enneaspermus	Santalum lanceolatum	Scaevola spinescens
Hemorrhoids	Centaurium spicatum	Lythrum salicaria		
Inflammation of the genital regions	Amyema maidenii	Centaurium spicatum		

Women's Health and Fertility

Condition				
Pre-/postnatal	Eremophila longifolia	Mentha satureoides	Rubus rosifolius	Spartothamnella juncea
	Eucalyptus camaldulensis			

(continued)

Table 11.2 Australian Customary Uses of NSW Medicinal Plants (Continued)

Medical Condition	Medicinally Used Plants			
Women's Health and Fertility				
Increase lactation	Atriplex nummularia	Eremophila longifolia	Pittosporum angustifolium	Sarcostemma australe
Promote female sterility	Alstonia constricta	Flagellaria indica	Petalostigma pubescens	Toona ciliata
Stomach cramps	Mentha diemenica	Mentha satureioides	Pittosporum angustifolium	Rubus rosifolius
Regulate menstruation	Lythrum salicaria	Mentha diemenica	Mentha satureioides	Toona ciliata
Abortion	Mentha satureioides			
Treat male sterility	Hybanthus enneaspermus			
Aphrodisiac	Aleurites moluccana			
Infectious Diseases				
Leprosy	Canavalia rosea	Excoecaria agallocha		
Malaria	Alstonia constricta	Owenia acidula	Tinospora smilacina	
Infectious diseases	Sarcostemma australe			
Typhoid	Alstonia constricta			
Stings and Bites				
Stings	Avicennia marina	Dodonaea viscosa	Ipomoea pes-caprae	Sterculia quadrifida
Envenomation/poison	Crinum pedunculatum	Excoecaria agallocha	Pteridium esculentum	Tinospora smilacina
	Eucalyptus microtheca	Hybanthus enneaspermus	Pratia purpurascens	Trichodesma zeylanicum
	Excoecaria agallocha	Ipomoea pes-caprae	Sida rhombifolia	
Infections and Infestations				
Parasitic infestation	Centipeda minima	Daviesia latifolia	Melaleuca quinquenervia	Pteridium esculentum
	Cleome viscosa	Dendrocnide excelsa	Petalostigma pubescens	
Fungal infections	Aleurites moluccana	Corymbia gummifera	Ipomoea pes-caprae	Melaleuca alternifolia
Nervous System Conditions				
Nervous system "disease"	Melaleuca quinquenervia	Verbena officinalis		
Convulsions	Hybanthus enneaspermus			

			Other	
Heart disease, chest pain	Atherosperma moschatum	Corymbia tessellaris	Flagellaria indica	Santalum lanceolatum
	Corymbia dichromophloia	Eremophila freelingii	Pimelea microcephala	
Diabetes	Aegiceras corniculatum	Alstonia constricta	Lythrum salicaria	
Scurvy	Atriplex nummularia	Sarcostemma australe	Smilax glyciphylla	
Cancer	Dioscorea transversa	Scaevola spinescens		
Insecticide/repellent	Aegiceras corniculatum	Nymphaea gigantea		
Circumcision	Acacia tetragonophylla			
Excessive sweating	Mentha diemenica			
Fistulas	Ajuga australis			
Narcotic	Geijera parviflora			
Opium antidote	Petalostigma pubescens			
Sore/bleeding gums	Piper hederaceum			

Note: References 4–29 were used for Table 11.2.

Table 11.3   Customary Uses, Biological Activities, and Phytochemistry of NSW Medicinal Plants

Plant Name	Australian Customary Medicinal Use	Major Biological Activities of Extracts	Relevant/Major Phytochemicals
Acacia bivenosa	Colds, cough[4,7]	Antibacterial[30]	No secondary metabolites reported
Acacia melanoxylon	Rheumatism[4]	Antibacterial[31]	*n*-Alkyl caffeates; flavonoids (e.g., quercetin 3-*O*-β-D-galactoside, isomelacacidin, melacacidin, melanoxetin; sterols (e.g., β-sitosterol, spinasterol, stigmasterol; quinones (e.g., acamelin)[32–38]
Acacia tetragonophylla	Circumcision wounds, colds, cough, dysentery, warts[4,6,7,27]	Antibacterial[27]	No secondary metabolites reported
Aegiceras corniculatum	Earache[4]	Antioxidant, cytotoxic (to cancer cell lines)[29,39]	Flavonoids (e.g., isorhamnetin); phenols (e.g., gallic acid); quinones (e.g., 5-*O*-ethylembelin, 5-*O*-methylembelin, rapanone); sterols (e.g., α-spinasterol, stigmasterol, triterpenoids (e.g., aegicerin, embellinone)[40–45]
Ailanthus triphysa	Asthma, bronchitis, dysentery, fever, ulcers[4]	None reported	Coumarins (e.g., 4,7-dimethoxy-4-methylcoumarin, 4,6,7-trimethoxy-5-methylcoumarin); kaempferol glycosides; phenols (e.g., 2,3-dihydroxybenzoic acid, methyl 3,4-dihydroxybenzoate); sterols (e.g., β-sitosterol, daucosterol); triterpenoids (e.g., β-amyrin, malabaricol)[46–48]
Ajuga australis	Boils, fistulas, gangrene, sores, ulcers[4,6–8]	None reported	Neoclerodane diterpenoids (e.g., ajugapitin, ajugorientin)[49]
Aleurites moluccana	Fungal infections of the skin[9]	Antiinflammatory, antinociceptive, antipyretic, hypolipidemic[50–52]	Flavonoids (e.g., 2″-*O*-rhamnosylswertisin, swertisin); sterols (e.g., stigmasterol, β-sitosterol, campesterol); terpenoids (e.g., α-amyrin, β-amyrin, moluccanic acid, spruceanol)[50,53–57]
Alphitonia excelsa	Body pain, headache, liniment, skin diseases, sore eyes, stomach upset, tonic, toothache[4,7,9]	Inhibits platelet aggregation[58]	Benzofuranone alphitonin; salicylic acid; triterpenoids (e.g., alphitexolide, alphitolic acid, betulin, betulinic acid, ceanothic acid)[59–61]
Alstonia constricta	Fevers[4]	None reported	Alkaloids (e.g., alstonine, quebrachidine, reserpine, vincamajine)[62–64]
Atriplex nummularia	Increase lactation[4,6]	Antibacterial[31]	Phytoecdysteroids 20-hydroxyecdysone and polypodine B); triterpenoids (e.g., calenduloside E, echinocystic acid, oleanolic acid)[65–67]
Centipeda cunninghamii	Colds, cough, eye inflammation, sandy blight (trachoma), sickness, skin infections, tuberculosis[4,7]	Antiinflammatory, antioxidant[68]	*cis*-Chrysanthenyl acetate; flavonoids (e.g., 5,7,2′,4′-tetrahydroxy-6-methoxyflavone-3-*O*-β-D-glucoside)[68,69]

Species	Traditional uses	Pharmacological activities	Secondary metabolites
Centipeda minima	Colds, intestinal worms, eye inflammation, sandy blight (trachoma)[4]	Antiallergic, antibacterial, anticancer, platelet inhibition[70–73]	Flavonoids (e.g., apigenin, quercetin 3,3'-dimethylether, quercetin 3-methylether); terpenoids (e.g., 6-O-angeloylplenolin, arnicolide C, 6-O-isobutyroylplenolin, minimaoside A and B, 6-O-senesioylplenolin, thymol-β-D-glucoside)[72,74,75]
Chamaesyce drummondii	Diarrhea, dysentery, fever, genital diseases, rheumatism, snakebite, sore eyes, warts[6,7,10]	Anticancer, antiinflammatory[26,76]	No secondary metabolites reported
Clematis glycinoides	Colds, headache, nasal congestion, pain, sinus[4,6,8,9]	Antiinflammatory[77]	Protoanemonin[78]
Clematis microphylla	Skin irritations[4]	Antiinflammatory[77]	Protoanemonin[78]
Cleome viscosa	Colds, diarrhea, fever, headache, intestinal worms, painful joints, rheumatism, sores, swelling[4,6,7]	Analgesic, anthelmintic, antibacterial, antidiarrheal, antiinflammatory, antipyretic, hepatoprotective, immunomodulatory[79]	Coumarinolignoids (e.g., cleomiscosin A–D); flavonoids (e.g., naringenin 4'-galactoside, dihydrokaempferol 4'-xyloside); glucosinolates (e.g., glucocapparin, glucocleomin); terpenoids (e.g., β-amyrin, cleomeolide, lupeol)[79]
Clerodendrum floribundum	Aches, bronchial congestion, cough, headache, pain[4,6,7,9]	Xanthine oxidase inhibition[80]	No secondary metabolites reported
Clerodendrum inerme	Sores[4]	Antibacterial, anticancer, antifungal, antioxidant, hepatoprotective[81–84]	Iridoids (e.g., melittoside, monomelittoside); phenylethanoid glycosides (e.g., isoverbascoside, verbascoside); terpenoids (e.g., betulinic acid, clerodermic acid, inermes A and B); 4-α-methylsterols[85–88]
Corymbia dichromophloia	Bronchial disease, cough, heart disease, influenza, lung disease, tonic, toothache, wounds[4,6,7]	None reported	*cis*- and *trans*-3,5,4'-Trihydroxystilbene, *cis*- and *trans*- 3,5,4'-trihydroxystilbene-3-O-β-D-glucoside[89]
Corymbia gummifera	Antiseptic, astringent, ringworm, venereal sores[4,7]	None reported	Terpenoids (e.g., α-pinene, β-pinene, γ-terpinene, terpinene-4-ol)[90]
Corymbia maculata	Bladder inflammation[4]	Antioxidant[91]	*p*-Coumaric acid; phenylpropanoid glucosides (e.g., 6-O-cinnamoyl 1-O-*p*-coumaroyl β-D-glucopyranoside); terpenoids (e.g., aromadendrene, 1,8-cineole, *p*-cymene, globulol, limonene, α-pinene, spathulenol)[92,93]
Corymbia polycarpa	Dysentery, toothache[4,94]	None reported	Terpenoids (e.g., aromadendrene, globulol, β-pinene, γ-terpinene, viridiflorol); *cis*- and *trans*-3,5,4'-trihydroxystilbene, *cis*- and *trans*-3,5,4'-trihydroxystilbene-3-O-β-D-glucoside[89,95]
Corymbia terminalis	Antiseptic, burns, chest pain, constipation, diarrhea, heart pain, sickness, sore eyes, sore lips, sore throat, sores, toothache[4,6–8,94]	None reported	*cis*- and *trans*-3,5,4'-Trihydroxystilbene-3-O-β-D-glucoside[89]

(continued)

Table 11.3 Customary Uses, Biological Activities, and Phytochemistry of NSW Medicinal Plants (Continued)

Plant Name	Australian Customary Medicinal Use	Major Biological Activities of Extracts	Relevant/Major Phytochemicals
Corymbia tessellaris	Diarrhea, dysentery, laxative, sore eyes[4,6,7,9]	Antibacterial, antifungal, toxic to mosquito (*Aedes aegypti* adults)[96,97]	Terpenoids (e.g., aromadendrene, 1,8-cineole, α-, β-, γ-eudesmol, globulol, limonene, α-pinene, *trans*-pinocarveol, spathulenol)[92,98,99]
Crotalaria cunninghamii	Earache, eyewash (sore eyes), headache, swellings[4,7]	None reported	Pyrrolizidine alkaloid monocrotaline[100]
Croton insularis	Inducing appetite, intestinal catarrh, tonic[4]	None reported	Aromatic ethers and their glycosides (e.g., ferulic acid, kelampayoside A, vanilic acid); flavonoids (e.g., isokaempferide); terpenoids (e.g., crotinsulactone, crotinsularin, furocrotinsulolide A and B)[101,102]
Daviesia latifolia	Fever, hydatids, tonic[4]	None reported	2-O-β-D-Apiosyl-D-glucose Iβ,5'-dibenzoate[103]
Dodonaea viscosa	Cuts, open wounds, sickness, stings, toothache[4,6,7,9,12]	Antibacterial, antifungal, antioxidant, antiproliferative, antiulcer[104–109]	Flavonoids (e.g., 3,5,7-trihydroxy-4'-methoxyflavone, 5,7,4'-trihydroxy-3,6-dimethoxyflavone, 5,7-dihydroxy-3,6,4'-trimethoxyflavone (santin), kaempferol); sterols (e.g., β-sitosterol, stigmasterol); triterpenoids (e.g., dodoneasides A and B, methyl dodovisate A and B)[46,105,109–111]
Doryphora sassafras	Tonic[4]	Antimalarial[112]	Alkaloids (e.g., anonaine, isocorydine, 1-(4-hydroxybenzyl)-6,7-methylenedioxy-2-methylisoquinolinium trifluoroacetate, liriodenine, 2-methyl-1-(p-methoxybenzyl)-6,7-methylenedioxyisoquinolinium chloride); aromatic ethers (e.g., methyl eugenol, safrole); terpenoids (e.g., camphor)[112–115]
Eremophila freelingii	Body pain, chest pain, colds, cough, diarrhea, headache, sickness, skin cuts, sores, wounds[4,6,7,12]	Antibacterial, inhibits platelet aggregation, inhibits platelet serotonin release[27,58]	Sesquiterpenes (e.g., freelingnite, freelingyne)[116–118]
Eremophila gilesii	Colds, sores[6,7]	Inhibits platelet serotonin release[119]	Phenylethanoid glycosides verbascoside and poliumoside[120]
Eremophila longifolia	Boils, colds, headache, increase lactation, insomnia, respiratory tract infection, skin sores, smoking babies, sore eyes (eye wash)[4,6,7,27]	Cardioactive, inhibits platelet serotonin release[119,121]	Aromatic ethers (e.g., methyl eugenol, safrole); geniposidic acid; terpenoids (e.g., cineole, fenchone)[116,121,122]
Eremophila maculata	Antiinflammatory, colds, headache[4,6,7]	Antibacterial, xanthine oxidase inhibition cardioactive properties[30,80]	Cyanogenetic glycosides (e.g., prunasin); iridoid glycosides (e.g., catalpol, melampyroside, verminoside); lignans (e.g., pinoresinol)[123,124]
Eremophila sturtii	Backache, cough, diarrhea, fly repellent, general sickness, skin infection, sore eyes[6,12]	Antibacterial, antiinflammatory[125,126]	Serrulatane diterpenes 3,8-dihydroxyserrulatic acid and serrulatic acid[126]

Species	Uses	Activity	Secondary metabolites
Ervatamia angustisepala	Fever, sores[4,7]	None reported	Alkaloids epiervatamine and ervatamine[127]
Eucalyptus camaldulensis	Colds, diarrhea, fever, making babies strong, sickness, sore throat, sores, wounds[4,6]	Antibacterial, antifungal, antioxidant, toxic to mosquitoes (*Aedes aegypti* adults)[97,128,129]	Aldehydes cuminal and phellandral; terpenoids (e.g., aromadendrene, betulinic acid, 1,8-cineole, *p*-cymene, α-, β-, γ-eudesmol, spathulenol, terpinen-4-ol, α-terpineol, ursolic acid)[129–133]
Eucalyptus haemastoma	Cuts, dysentery, ulcers, wounds[4]	None reported	Terpenoids (e.g., aromadendrene, 1,8-cineole, *p*-cymene, α-, β-, γ-eudesmol, α-pinene)[134]
Eucalyptus microtheca	Colds, diarrhea, sore throat, sores, snakebite, stomach complaints, whooping cough[4,6-10]	Antibacterial, antifungal[96]	Terpenoids (e.g., 1,8-cineole, *p*-cymene, α-pinene, α-terpineol, ursolic acid)[135,136]
Eucalyptus pilularis	Astringent[4]	None reported	Terpenoids (e.g., *p*-cymene, α-pinene, spathulenol, terpinen-4-ol)[90]
Eucalyptus piperita	Colicky complaints, stomach upset[4]	None reported	Terpenoids (e.g., 1,8-cineole)[90]
Eucalyptus pruinosa	Headache, pain, rheumatism[4,7]	Antiinflammatory[26]	Terpenoids (e.g., aromadendrene, 1,8-cineole, *p*-cymene, α-pinene, *trans*-pinocarveol, pinocarvone)[92]
Eucalyptus resinifera	Syphilitic sores[4]	Antibacterial[137]	Flavonoids (e.g., resinoside A and B); terpenoids (e.g., aromadendrene, 1,8-cineole, *p*-cymene, α-, β-, γ-eudesmol, globulol, limonene, α-pinene, *trans*-pinocarveol, spathulenol)[92,138]
Evolvulus alsinoides	Dysentery, fever[4]	Antiamnesic, antioxidant, antistress, immunomodulatory[139-144]	Alkaloids (e.g., betaine, evolvine); coumarins (e.g., scoppletin, umbelliferone); flavonoid glycosides of apigenin, kaempferol, and quercetin[143,145,146]
Excoecaria agallocha	Leprosy, marine stings, pain, sickness[4,7]	Antioxidant, antifilarial (cytotoxic), antihistamine, anti-HIV[147-149]	Diterpenoids (e.g., agallochaexcoerins A–C, agallochins A–L); flavonoid glycoside 3,5,7,3',5'-pentahydroxy-2R,3R-flavanonol 3-*O*-α-L-rhamnoside; phorbol ester 12-deoxyphorbol 13-(3*E*,5*E*)-decadienoate; triterpenoids (e.g., β-amyrin acetate, taraxerone)[147,150-156]
Excoecaria dallachyana	Burns, pain[7]	None reported	Terpenoids (e.g., β-amyrin acetate, 3-epilupeol, 3-epitaraxerol, 16-β-hydroxy-ent-atisan-3-one, taraxerone, taraxerol)[154,157]
Exocarpos aphyllus	Colds, sores, tuberculosis[4,7]	Antibacterial[31]	No secondary metabolites reported
Flagellaria indica	Chest complaints, to produce female sterility, toothache, sore throat, wounds[4,6,8,9]	None reported	Kaempferol 3-glycosides[158]
Flindersia maculosa	Diarrhea, rheumatism[4]	None reported	Alkaloids (e.g., flindersiamine, kokusaginine, maculosidine, maculosine); C12 hydrocarbons geijerene, pregeijerene; coumarin collinin; terpenes and terpenoids (e.g., 1,8-cineole, α-pinene, β-carophyllene, bicyclogermacrene, flindissol)[159,160]

(continued)

Table 11.3 Customary Uses, Biological Activities, and Phytochemistry of NSW Medicinal Plants (Continued)

Plant Name	Australian Customary Medicinal Use	Major Biological Activities of Extracts	Relevant/Major Phytochemicals
Geijera parviflora	Pain, smoked as narcotic, toothache[4]	None reported	Alkaloid flindersine; coumarins (e.g., 6′-dehydromarmin, 2′,3′-dihydrogeiparvarin, geiparvarin, 7-geranyloxycoumarin)[161,162]
Grevillea striata	Wounds[4]	None reported	Phenols (e.g., striatol, striatol-B, *bis*-norstriatol)[163,164]
Helichrysum luteoalbum	Sickness[4]	Antiinflammatory[165]	Aldehydes (e.g., decanal, terpenes (e.g., β-caryophyllene, α-gurjunene))[165]
Hibiscus tiliaceus	Ulcers, wounds[4,6,8]	Antioxidant, anticancer, cytotoxic[166–171]	Coumarins (e.g., hibiscusin); flavonoids (e.g., kaempferol-3-O-β-D-galactoside, quercetin-3-O-β-D-galactoside); hibiscusamide; sterols (e.g., stigmastadienol, stigmasterol); triterpenoids (e.g., friedelin, hibiscones A–D)[166,169–173]
Hybanthus enneaspermus	Colds, skin problems, urinary tract infections[4]	Antibacterial, antimalarial[18,22]	Alkaloids (e.g., aurantiamide acetate); terpenoids (e.g., β-sitosterol, isoarborinol)[18,174]
Ipomoea pes-caprae	Boils; colic; diuretic; hemorrhoids; headache; inflammation; laxative; pain; ringworm; skin irritations/infections; snakebite; sores; stings of green tree ants, stingrays, and stonefish; swellings; venereal disease[4,6–9]	Antihistamine, antinociceptive, neutralize jellyfish venom[58,175–177]	Aromatic ethers (e.g., eugenol, 4-vinylguaiacol); flavanoids and glycosides (e.g., isoquercetin, kaempferol, myricetin quercetin); terpenoids (e.g., betulinic acid, α-cadinol, 8-cedren-13-ol, *E*-nerolidol, guaiol, limonene, *E*-phytol)[178–180]
Lythrum salicaria	Bleeding of gums, diarrhea, dysentery, menstrual flow regulation[4,6]	Antibacterial, antifungal, anti-inflammatory, antioxidant, hypoglycemic, lipase inhibition[5,21,181–183]	Flavonoids (e.g., isoorientin, isovitexin, orientin, vitexin); phenols ellagic acid and gallic acid; triterpenoids (e.g., oleanolic acid, ursolic acid)[21,184]
Melaleuca alternifolia	Abrasions, abscesses, antibacterial, antifungal (ringworm), boils, colds, cuts, gonorrhea, sores[4]	Acetylcholine esterase inhibition, antibacterial, antifungal, anti-inflammatory, antioxidant, antiviral[185–190]	Terpenes and terpenoids (e.g., 1,8-cineole, *p*-cymene, α-terpinene, γ-terpinene, terpinen-4-ol)[191]
Melaleuca hypericifolia	Headache[4]	None reported	Terpenes and terpenoids (e.g., 1,8-cineole, limonene, α-pinene, α-terpineol)[192]
Melaleuca quinquenervia	Colds, cough, headache, intestinal worms, neuralgia, rheumatism, sickness[4,7]	Antibacterial, antifungal, insecticidal[193–195]	Flavonoids (e.g., luteolin, myricetin, quercetin, tricetin); phenols (e.g., ellagic acid, gallic acid); terpenes and terpenoids (e.g., 1,8-cineole, citronellol, geraniol, α-pinene, β-pinene, α-terpineol, viridiflorol)[196–199]
Melaleuca uncinata	Catarrh, respiratory complaints[4,7]	Insecticidal[200]	Terpenes and terpenoids (e.g., 1,8-cineole, α-pinene, terpinen-4-ol, α-terpinene, γ-terpinene)[201–203]

Plant	Traditional uses	Reported activity	Chemical constituents
Melicope vitiflora	Leprosy, marine stings, pain, sickness, toothache, ulcers[4,7]	Antibacterial[204]	Benzopyran evodionol; coumarins (e.g., 7-(3'-carboxybutoxy)coumarin, 7-(3,3'-dimethylallyloxy) coumarin; terpenes (e.g., limonene, ocimene, α-pinene, sabinene, γ-terpinene)[205,206]
Mentha diemenica	Diaphoretic, diuretic, indigestion, intestinal cramps, irregular menstruation[4,6]	Antispasmodic, carminative[207]	Terpenes and terpenoids (e.g., isomenthone, menthol, menthone, menthyl acetate, neomenthol, neomenthyl acetate, pulegone)[208]
Mentha satureioides	Stomach complaints[4,6]	None reported	Terpenoids (e.g., menthol, menthone, menthyl acetate, pulegone)[4]
Owenia acidula	Malaria, sore eyes[4,7]	None reported	Liminoids 6α-acetoxydeoxyhavanensin and 3-isobutyryl-7-deaacetylglabretal[209]
Petalostigma pubescens	Antiseptic, astringent, contraceptive, fever, malaria, opium antidote, sore eyes, tonic, toothache[4,6]	None reported	Diterpenoids (e.g., petalostigmone A–C; pubescenone, sonderianol)[210]
Piper hederaceum	Gonorrhea, sore gums, tonic[4,7,9]	Anticancer[211]	Amides (e.g., Δα and Δβ-dihydropiperine, 3,4-methylene dioxycinnamoylpiperidide, fagaramide, piperine); dillapiole, ω-hydroxyisodillapiole[211,212]
Pittosporum angustifolium	Colds, cramps, eczema, pain, pruritus, sprains[4,6,7,9]	Antibacterial, antiviral[13,31]	Triterpenoids (e.g., R_1-barrigenol, barringtogenol C, dihydropriverogenin A, phillyrigenin)[213]
Plectranthus parviflorus	Syphilitic sores[7]	None reported	Quinonemethanes parviflorone A–F[214,215]
Plumbago zeylanica	Blisters[4]	Antibacterial, anticancer, anti-inflammatory, antinociceptive, antioxidant[216–219]	Coumarins (e.g., seselin, suberosin, xanthoxyletin, xanthyletin); napthoquinones, (e.g., maritinone, elliptinone, plumbagin); sterols (e.g., β-sitosterol, β-sitosteryl-3β-glucoside-6'-O-palmitate)[219–221]
Pouteria pohlmaniana	Boils, sores[4,6]	None reported	Saponin aglycones bayogenin and hederagenin[222]
Prunella vulgaris	Antispasmodic, cuts, expectorant, fever, rheumatism, wounds[4]	Antibacterial, antiinflammatory, antioxidant, antiviral[223–226]	Flavonoids (e.g., hesperidin, luteolin, quercetin, quercetin-3-O-β-D-glucoside, rutin); phenols (e.g., rosmarinic acid); sterols (e.g., β-sitosterol, stigmasterol); triterpenoids (e.g., betulinic acid, 2α-hydroxyursolic acid, ursolic acid)[226–230]
Pteridium esculentum	Astringent, insect stings, intestinal worms, rheumatism[4,6,8]	Carcinogenic[231]	Norsesquiterpenes (e.g., caudatoside, ptaquiloside, ptesculentoside); phenols (e.g., chlorogenic acid, kaempferol 3-O-β-D-glucoside)[232,233]
Pterocaulon sphacelatum	Colds, cough, respiratory infections, sinusitis[6]	Antiviral (polio, rhinoviruses)[13,23]	Chrysosplenol C (3,7,3'-trimethoxy-5,6,4'-trihydroxyflavone), 6,7-dimethoxycoumarin, 6,7,8-trimethoxycoumarin[23,234]
Rhizophora mucronata	Astringent[4]	Antioxidant[235]	Diterpenoids (e.g., rhizophorins A–D); flavonoids (e.g., myricetin, quercetin); triterpenoids (e.g., 3β-O-(E)-(4-methoxy)cinnamoyl-15α-hydroxy-β-amyrin, lupeol, simiarenol)[46,236,237]

(continued)

Table 11.3 Customary Uses, Biological Activities, and Phytochemistry of NSW Medicinal Plants (Continued)

Plant Name	Australian Customary Medicinal Use	Major Biological Activities of Extracts	Relevant/Major Phytochemicals
Rubus rosifolius	Diarrhea, labor pain, menstrual pain, morning sickness, stomach problems[6,8,9]	Analgesic[238]	Monoterpenes (e.g., α-amorphene, T-muurolol); sesquiterpenes and sesquiterpenoids (e.g., β-caryophyllene, humulene, rosifoliol); triterpenoids (e.g., euscaphic acid, 28-methoxytormentic acid)[197,238–240]
Santalum lanceolatum	Boils, chest complaints, colds, gonorrhea, purgative, rheumatism, sores, tiredness[4,6–8]	Antibacterial[27]	Fatty acids (e.g., oleic acid, ximenynic acid); sesequiterpene alcohols (e.g., lanceol)[241,242]
Santalum obtusifolium	Aches, constipation, pain[4,7]	None reported	Fatty acids (e.g., oleic acid, ximenynic acid)[242,243]
Santalum spicatum	Colds, cough, stiffness[7]	Antibacterial, antifungal[244]	Sesquiterpene alcohols (e.g., α-acorenol, β-acorenol, *epi*-α-bisabolol, *epi*-β-santalol, *E,E*-farnesol, Z-lanceol, Z-nuciferol, (Z)-α-santalol, (Z)-β-santalol)[244–249]
Sarcostemma australe	Corns, eye complaints, increase lactation, rashes, skin sores, small-pox, sprains, sickness, warts, wounds[4,6,7,12]	None reported	Sarcostin (aglycone)[250]
Scaevola spinescens	Alimentary ulcers, boils, colds, rashes, sores, stomachache, urinary trouble[4,7]	Antiviral (human cytomegalovirus)[13]	Coumarins (e.g., ammirin, seselin, xanthyletin); taraxerenes (e.g., myricadiol); terpenes (e.g., α-bisabolol)[15,251]
Sida rhombifolia	Diarrhea, indigestion[6]	Antiarthritic, antiamoebic, antibacterial, antiinflammatory, antioxidant, cytotoxic, xanthine oxidase inhibition[252–255]	Alkaloids (e.g., vasicinol, vasicinone, vasicine); ecdysteroids (e.g., 20-hydroxyecdysone, 5,20-dihydroxyecdysterone)[253,256,257]
Smilax glyciphylla	Blood tonic, chest complaints, congestion, cough, scurvy (including prevention)[4,6]	Antioxidant[24]	Phenols (e.g., glycyphyllin (phloretin 2′-α-L-rhamnoside)), mangiferin (1,3,6,7-tetrahydroxyxanthone-C2-β-D-glucoside)[24,258]
Sophora tomentosa	Bilious sickness[4]	Antioxidant[259]	Alkaloids (e.g., anagyrine, baptifoline, cytisine, matrine, sophocarpine *N*-oxide); flavonoids (e.g., isosophoranone, sophoraflavanone A–E, sophoraisoflavanone A, sophoronol, tomentosanols A–E); pterocarpans (e.g., sophoracarpans A and B)[260–262]
Spilanthes grandiflora	Toothache[4]	None reported	May contain spilanthol[4]
Sterculia quadrifida	Eye infections, wounds[4,6–9]	Antibacterial[263]	No secondary metabolites reported
Swainsona galegifolia	Bruises, swelling[4,9]	Lysosomal α-D-mannosidase inhibition[264]	Sphaerophysin, swainsoninechen[264,265]

Species	Traditional use	Activity	Secondary metabolites
Tinospora smilacina	Body pain, headache, inflammatory disorders, rheumatoid arthritis, snakebite, stonefish stings[4,7]	Antiinflammatory[266]	Alkaloids (e.g., berberine, magnoflorine, palmatine); lignans (e.g., cubebin, isolariciresinol); columbin; dihydrosyringenin[267-270]
Toona ciliata	Regulating menstrual flow[4]	Antibacterial[271]	Coumarins (e.g., siderin); triterpenoids (e.g., 6-acetoxytoonacilin, cedrelone, 21- and 23-hydroxytoonacilide, toonacilin, toonafolin)[209,272-274]
Trichodesma zeylanicum	Snakebite, sores[4,6,7]	None reported	Alkaloid supinine; malvalic acid, ricinoleic acid, sterculic acid[275,276]
Trichosanthes palmata	Asthma[4]	None reported	Saponins (e.g., santholin, trichonin); sterols (e.g., cycloeucalenol, β-sitosterol, α-spinosterol[277,278]
Verbena officinalis	Venereal disease[4,6]	Analgesic, antibacterial, antifungal, antiinflammatory, antioxidant, apopotic inducing, gastroprotective, neuroprotective, wound healing[182,279-284]	Flavonoids (e.g., apigenin, 6-hydroxyluteolin, luteolin and their glycosides); iridoid glucosides (e.g., hastatoside, verbenalin); phenylethanoid glycosides (e.g., isoverbascoside, verbascoside); terpenes and terpenoids (e.g., citral, limonene, isobornyl formate, ursolic acid)[279,285-287]
Vigna vexillata	Constipation[4,6,7]	Antifungal[288]	No secondary metabolites reported
Wikstroemia indica	Cough[4]	Antibacterial, anticancer, antifungal, antiinflammatory, antimalarial, antioxidant, antiviral[289-293]	Coumarins (e.g., daphnoretin); flavonoids (e.g., kaempferol-3-O-β-D-glucoside, sikokianin B and C, 5,7,4'-trinydroxy-3,5'-dimethoxyflavone (tricin); wikstrol A, wikstrol B); indicanone (guaiane type sesquiterpene); lignans (e.g., lariciresinol, lirioresinol B, nortrachelogenin)[289,290,292,294-298]
Zieria smithii	Headache[4]	None reported	Aromatic ethers (e.g., elimicin, eugenol, methyl eugenol, safrole); (3,4-dimethoxycinnamaldehyde); terpenoids (e.g., chrysanthenone, linalool, α-pinene)[46,299,300]

11.3.1 NSW Medicinal Plants—Major Classes, Habitat, and Distribution

Table 11.1 documents 128 medicinal plants of NSW that have been used as important Australian Aboriginal medicines, along with their common names, distribution, and habitat. Due to the varied geography of NSW, a wide variety of families (53) and genera (93) found in diverse habitats are represented. Some genera, such as *Acacia, Corymbia, Eucalyptus, Eremophila,* and *Melaleuca,* and families, including Asteraceae, Euphorbiaceae, Fabaceae, Lamiaceae, Malvaceae, Myrtaceae, Rutaceae, Santalaceae, and Scrophulariaceae, are particularly well represented. As seen by the wide distribution of the plants, NSW is truly a microcosm of the rest of Australia, with almost all plants being found in other Australian states and territories. Only three plants, *Eucalyptus haemastoma, Eucalyptus piperita,* and *Eucalyptus racemosa,* are unique to NSW.

11.3.2 Customary Preparation and Uses of Medicinal Plants

This literature survey highlights that, unlike for many Indigenous people in other parts of the world, fresh plant material was (and is) the most common method for the customary preparation of Australian Aboriginal medicines and often all of the plant material is utilized, including roots and bark, wood, leaves, and flowers.

Plant-based Aboriginal medicinal preparations generally fall into several main groups: plant material that is crushed, often heated, and employed as topical poultices; plants that are infused or decocted in hot or boiling water or macerated in cold water and then applied topically or rubbed into the skin, inhaled, or, less commonly, drunk as decoctions; plants that are used directly, including applying the sap topically, by chewing leaves and twigs, or eating fruits; plants that are burned and used for smoking ceremonies or whole-body fumigation; and plants that are macerated or decocted and combined with animal-based emollients for topical application.

For NSW Aboriginal people and Australian Aboriginal people as a whole, the knowledge of plants and their healing properties is often integrated into a unique world view, where the physical and spiritual worlds are unified. As such, healing, culture, and spirituality are often indistinguishable and Aboriginal healers concentrate on treating both the physical and spiritual symptoms of an illness (Pearn 2004). The Aboriginal people have used their medicinal preparations to address symptoms or groups of symptoms, rather than diseases, and the various groups of remedies reflect this focus.

Some common symptom groups include sores and skin complaints, toothache, coughs and congestion, sore ears and eyes, stomachaches, diarrhea and constipation, fevers, and aches and pains. Medicinal plants and other techniques, including bleeding, blistering, and whole-body fumigation (or smoking), have also been used for stings, burns, envenomation, wounds, and parasitic infestation. Preventative treatments are also very important and many fruits, leaves, and decoctions are consumed to confer disease resistance or, applied to the skin, to prevent insect bites and parasitic infestation. Table 11.2 summarizes the customary medicinal uses of the plants listed in Table 11.1.

11.3.3 Bioactivity and Phytochemistry of Selected NSW Medicinal Plants

Of the 128 plants documented in Table 11.1, 93 have been further investigated for their biological activities and/or chemical constituents. Table 11.3 summarizes the confirmed biological activities reported for extracts of these plants and major secondary metabolites. It also provides the documented customary uses of the plants by Australian Aboriginal communities.

From Table 11.3 it can be seen that these plants collectively have a wide range of biological activities, including those of great importance for treating human diseases and infections. The most frequently reported are antibacterial, antiinflammatory, and antioxidant activities. Table 11.3 also shows that there is a great diversity in the types of natural products that have so far been isolated, with bioactive flavonoids, terpenoids, and sterols being particularly well represented.

For many of the plants, the biological activities and phytochemistry are aligned with their ethnomedicinal use. For example, *Melaleuca alternifolia* (tea tree), which is customarily used for treatment of abrasions, cuts, and sores, is well known for its antibacterial and antiinflammatory properties, which are attributed especially to terpenes and terpenoids present in its essential oil. Many of the plants documented have also been used as traditional medicines in various parts of the world for similar ailments.

In spite of documentation of their use in traditional medicine, 66 of the plants have not been investigated for their biological activities and 42 have not been investigated for their secondary metabolites. As already noted, 35 plants have not been reported for their biological activities or their phytochemistry. This highlights gaps in the current literature and scope for further biological and chemical investigations. Interestingly, there is a clear resurgence of interest in chemical and biological investigations of traditional medicinal plants. This is highlighted by the increased number of publications over the last decade, predominantly from China and India.

11.3.4 Contemporary Issues in Customary Medicines Research

Over the last decade, governments, private industry, and scientists have expressed an unprecedented interest in Indigenous medicinal knowledge, especially in the quest for new medicines. The World Health Organization (WHO 2002) and governments, including the Australian federal government (http://www.environment.gov.au/indigenous/ipa/index.html), acknowledge the value of traditional medicinal knowledge in terms of its cultural and scientific significance; the protection and preservation of these knowledge systems is an objective of many organizations, such as Foundation for Revitalization of Local Health Traditions (http://www.frlht.org), Royal Botanic Gardens, Kew (http://www.kew.org/), and Plant Resources of Tropical Africa (http://www.prota.org/uk/About+PROTA/Home.htm).

11.3.5 Ethical Engagement with Indigenous People

The increased interest in traditional medicinal knowledge has resulted in cases where Indigenous people have been treated as a resource to be exploited. Australia is a signatory to the Convention of Biological Diversity (CBD) (http://www.cbd.int/), which recognizes the importance of fair and equitable sharing of benefits arising from the use of genetic resources and the resultant knowledge. Ethnobotanical research therefore requires the research community to engage with Indigenous people in a spirit of cooperation, consultation, and support. For this to occur, the researcher must follow best practices of conforming to rigorous ethical principles and developing relationships of trust with individual Indigenous communities.

To abide by these international guidelines, researchers must have sufficient time for and commitment to negotiations with Indigenous people and the development of benefit-sharing strategies. Despite the required commitment of time, the involvement of Indigenous people in all areas of this research is crucial. Indigenous people can provide vital details to direct the research, such as harvest times for optimal efficacy, details about therapeutic variation within particular species, and information about a remedy's therapeutic index (Li et al. 2003; Pearn 2004). The benefits of such cooperative research with Indigenous people can often extend beyond the original context of the project.

11.3.6 Conservation and Sustainability of Biodiversity

In Australia, loss of biodiversity is almost as pressing an issue as the loss of Indigenous knowledge—nowhere more so than in NSW, Australia's most populous state. Compared to pre-European forest cover, it is estimated that between 50 and 60% of NSW land has been cleared (Benson 1999). In 1995 Australia's rate of land clearance was the fifth highest in the world (Glanznig 1995). Over 700 of the native plant taxa in NSW are listed as either threatened or rare (Briggs and Leigh 1995). The most disturbed areas of the state are in the western regions, where upwards of 70% of the pre-European vegetation cover has been cleared or altered (Benson 1999).

The higher nutrient soils have been the most affected, with some areas having less than 15% of the original vegetation cover remaining. In these areas, native vegetation has been replaced with intensive cropping or mixed cropping and improved pasture (Thackway et al. 1995). Even in areas where cropping and land clearance have been minimal, the vegetation has been altered by the impact of grazing and introduced plant and animal species (Benson 1999). The conservation status of NSW vegetation also reflects this pattern, with less than 7% of NSW overall and less than 5% of NSW western regions, respectively, protected in conservation reserves. This conservation status and high level of degradation of the native vegetation of NSW represent a critical threat to the sustainability of the state's natural products industry.

11.3.7 The Macquarie University Indigenous Bioresources Research Group

Our research group, the Indigenous Bioresources Research Group (IBRG), is a multidisciplinary team with the main objectives of systematically documenting, firsthand, customary medicinal plant knowledge of Indigenous people, using best ethical practices, and applying this knowledge to identify medicinally important compounds following targeted chemical and biological studies. Additional benefits of this research include a greater understanding of biodiversity and its conservation and providing capacity-building opportunities for Indigenous people.

The IBRG follows the principles of the CBD, along with the stepwise participatory action research methodology (PAR) of UNESCO (Tuxill and Nabhan 2001) and the ethical guidelines of the National Health and Medical Research Council (NHMRC 2003) for conduct of research with Indigenous people. We have developed collaborative research partnership agreements with Indigenous people of Australia (northern NSW) and India (Nagaland). These agreements recognize ownership of the customary knowledge by the Indigenous partners, joint ownership of any subsequent intellectual property from the research and coauthorship of published materials, and the commitment by us to provide benefits back to the Indigenous stakeholders.

In line with the CBD directives mentioned under articles 7, 8j, and 9, for systematic documentation and conservation of medicinal plant knowledge, we have developed a web-based prototype, a customary medicinal plant knowledge database (CMKb) (Gaikwad et al. 2008). CMKb contains information on customary medicinal plants from published ethnobotanical literature and firsthand information from our Indigenous research partners. The information gathered from interviews and augmented by literature searches and bioactivity assays is transferred back to the Indigenous groups in the form of a relevant database that is easy to update. This database also serves a vital role as a repository of cumulative knowledge and an educational aid for Indigenous schoolchildren and future generations. Since the documented information reflects the customary knowledge of the Indigenous people, it is password protected and access privileges are determined in consultation with the Indigenous stakeholders.

Tangible benefits of the collaborative partnerships have included employment and training opportunities for Indigenous people (for example, in field work and documentation), development of bush food and medicine gardens, and the establishment of the Indigenous Science Education Program (http://web.science.mq.edu.au/groups/isep/). The latter was developed in direct response to specific requests from elders to increase the motivation of Indigenous youth to complete their high school studies.

11.4 CONCLUSION

This review has extended the information, presented in previous literature, of plants found in NSW that have been used as customary medicines by Australian Aboriginal people. Information on the distribution, habitat, and known customary uses of 128 plants has been documented. The biological activities and phytochemistry of 93 plants have also been described. This review identifies scope for further biological and chemical investigations of medicinal plants of NSW to add to the growing understanding of this resource. It also highlights certain contemporary issues of working in partnership with Indigenous people on customary knowledge.

REFERENCES

Benson, J. S. 1999. Setting the scene: The native vegetation of New South Wales. Background paper no. 1, National Vegetation Advisory Council, Australia.

Briggs, J. D., and J. H. Leigh.1995. *Rare or threatened Australian plants*. Collingwood, Vic., Australia: [Darwin, N.T.], CSIRO, Australia; Australian Nature Conservation Agency.

Cordell, G. A. 2002. Natural products in drug discovery—Creating a new vision. *Phytochemistry Reviews* 1:261–273.

Cox, P. A. 2000. Will tribal knowledge survive the millennium? *Science* 287 (5450): 44–45.

Cribb, A. B., and J. W. Cribb.1984. *Wild medicine in Australia*. Sydney, Australia: Collins.

Crotty, M., R. H. Crotty, et al. 1995. *Finding a way—The religious worlds of today*. Melbourne, Australia: Collins Dove.

Department of the Environment, W., Heritage and the Arts. 2010. IBRA—Interim Biogeographic Regionalisation for Australia V6.1 (http://www.environment.gov.au/parks/nrs/science/ibra.html).

Everard, P., A. C. Kalotas, et al. 2002. *Punu: Yankunytjatjara plant use—Traditional methods of preparing foods, medicines, utensils and weapons from native plants*. Alice Springs, Australia: Jukurrpa Books.

Fabricant, D. S., and N. R. Farnsworth. 2001. The value of plants used in traditional medicine for drug discovery. *Environmental Health Perspectives* 109 (Suppl 1): 69–75.

Gaikwad, J., V. Khanna, et al. 2008. CMKb: A web-based prototype for integrating Australian aboriginal customary medicinal plant knowledge. *BMC Bioinformatics* 9 (Suppl 12): S25.

Glanznig, A. 1995. Native vegetation clearance, habitat loss and biodiversity decline: An overview of recent native vegetation clearance in Australia and its implications for biodiversity. Canberra, Australia: ACT, Dept. of the Environment, Sport and Territories.

Harden, G. J. 1990. *Flora of New South Wales*. Kensington, New South Wales: New South Wales University Press.

Harvey, A. 2000. Strategies for discovering drugs from previously unexplored natural products. *Drug Discovery Today* 5 (7): 294–300.

Heinrich, M. 2003. Ethnobotany and natural products: The search for new molecules, new treatments of old diseases or a better understanding of indigenous cultures? *Current Topics in Medicinal Chemistry* 3:141–154.

Hiddins, L. J. 1999. *Explore wild Australia with the bush tucker man*. Ringwood, Vic., Australia: Penguin.

Horstman, M., and G. Wightman. 2001. *Karparti* ecology: Recognition of aboriginal ecological knowledge and its application to management in northwestern Australia. *Ecological Management & Restoration* 2 (2): 99–109.

Kneale, K. E., and A. S. Johnson.1984. *A mee mee's memories*. Inverell, New South Wales: Kneale KE.

Kohen, J. L. 1995. *Aboriginal environmental impacts*. Sydney, Australia: University of New South Wales Press.

Lassak, E. V., and T. McCarthy. 2008. *Australian medicinal plants*. Sydney, Australia: New Holland Publishers.

Leiper, G., and J. Hauser. 1984. *Mutooroo: Plant use by Australian aboriginal people*. Eagleby, Qld., Australia: Eagleby South State School.

Li, R. W., S. P. Myers, et al. 2003. A cross-cultural study: Antiinflammatory activity of Australian and Chinese plants. *Journal of Ethnopharmacology* 85 (1): 25–32.

Low, T. 1990. *Bush medicine: A pharmacopoeia of natural remedies.* North Ryde, New South Wales: Angus & Robertson.

Newman, D. J., and G. M. Cragg. 2007. Natural products as sources of new drugs over the last 25 years. *Journal of Natural Products* 70 (3): 461–477.

NHMRC. 2003. Values and ethics: Guidelines for ethical conduct in aboriginal and Torres Strait Islander health research. Canberra: Commonwealth of Australia.

Orchard, A. E. 1999. Introduction. In *Flora of Australia.* Canberra, Australian Biological Resources Study; CSIRO. 1:1–9.

Pearn, J. 2004. *Medical ethnobotany of Australia—Past and present.* Piccadilly, London: Linnean Society.

PMSEIC (Prime Minister's Science, Engineering and Innovation Council). 2005. Biodiscovery. I. Department of Innovation, Science and Research.

Specht, R. 1981. *Major vegetation formations in Australia.* The Hague: Dr. W. Junk B.V. Publishers.

Thackway, R., I. Cresswell, et al. 1995. An interim biogeographic regionalization for Australia: A framework for setting priorities in the National Reserves System Cooperative Program. R. Thackway and I. Cresswell. Canberra, Australian Nature Conservation Agency, Reserve Systems Unit. 4.0.

Tuxill, J., and G. P. Nabhan. 2001. *People, plants and protected areas: A guide to in situ management.* London: Earthscan Publications Ltd.

Walker, J., and M. S. Hopkins. 1990. Vegetation. In *Australian soil and land survey field handbook,* ed. R. C. McDonald, R. F. Isbell, J. G. Speight, J. Walker, and M. S. Hopkins. Melbourne, Australia: Inkata Press.

White, M. E. 1994. *After the greening: The browning of Australia.* Kenthurst, New South Wales: Kangaroo Press.

White, M. E., and J. Frazier. 1986. *The greening of Gondwana.* Frenchs Forest, New South Wales: Reed.

Wickens, K., and M. Pennacchio. 2002. A search for novel biologically active compounds in the phyllodes of *Acacia* species. *Conservation Science Western Australia* 4 (3): 139–144.

Williams, D. J., M. C. Read, et al. 2001. Biodiversity. *Australia state of the environment report.* Canberra, RMIT University, Melbourne: 227.

WHO (World Health Organization). 2002. *WHO—Traditional medicine strategy 2002–2005.* Geneva: World Health Organization.

REFERENCES FOR TABLES

1. Flora of Australia Online. 2010. Australian Biological Resources Study. http://www.anbg.gov.au/abrs/online-resources/flora/main-query-styles.html (Department of the Environment Water Heritage and the Arts).

2. PLANTNET. 2010. Plant Net—The Plant Information Network System of Botanic Gardens Trust, Sydney, Australia. http://plantnet.rbgsyd.nsw.gov.au (Botanic Gardens Trust).

3. ANHSIR. 2010. Australian National Herbarium Specimen Information Register. http://www.cpbr.gov.au/cgi-bin/anhsir (Integrated Botanical Information System, IBIS, Australian National Botanic Gardens).

4. Lassak, E. V., and T. McCarthy. 2008. *Australian medicinal plants.* Sydney Australia: New Holland Publishers.

5. Tunalier, Z. et al. 2007. *Journal of Ethnopharmacology* 110 (3): 539–547.

6. Low, T. 1990. *Bush medicine: A pharmacopoeia of natural remedies.* North Ryde, New South Wales: Angus & Robertson.

7. Cribb, A. B., and J. W. Cribb. 1984. *Wild medicine in Australia.* Sydney, Australia: Collins.

8. Hiddins, L. J. 1999. *Explore wild Australia with the bush tucker man.* Ringwood, Vic., Australia: Penguin.

9. Leiper, G., and J. Hauser. 1984. *Mutooroo: Plant use by Australian aboriginal people.* Eagleby, Qld., Australia: Eagleby South State School.

10. Kneale, K. E., and A. S. Johnson. 1984. *A mee mee's memories.* Inverell, New South Wales: Kneale KE.

11. Sudhakar, M et al. 2006. *Fitoterapia* 77 (1): 47–49.
12. Everard, P. et al. 2002. *Punu. Yankunytjatjara plant use* A. C. Kaktas *Traditional methods of preparing foods, medicines, utensils and weapons from native plants*, ed. A. K. Cliff Goddard, 121. Alice Springs, Australia: Jukurrpa Books.
13. Semple, S. J. et al. 1998. *Journal of Ethnopharmacology* 60 (2): 163–172.
14. Paddy, E. et al. 1988. *Boonja bardak korn = All trees are good for something.* Perth, Western Australia: Anthropology Dept., Western Australian Museum.
15. Ghisalberti, E. L. 2004. *Fitoterapia* 75 (5): 429–446.
16. Devanesen, D. D. 2000. Traditional aboriginal medicine practice in northern territory. *International Symposium on Traditional Medicine—Better Science, Policy and Services for Health Developments.* Awaji Island, Japan.
17. Boreham, P. F. L. 1995. *International Journal for Parasitology* 25 (9): 1009–1022.
18. Weniger, B. et al. 2004. *Journal of Ethnopharmacology* 90 (2–3): 279–284.
19. Boominathan, R. et al. 2003. *Phytotherapy Research* 17 (7): 838–839.
20. Boominathan, R. et al. 2004. *Journal of Ethnopharmacology* 91 (2–3): 367–370.
21. Becker, H. et al. 2005. *Fitoterapia* 76 (6): 580–584.
22. Sahoo, S. et al. 2006. *Indian Journal of Pharmaceutical Sciences* 68 (5): 653–655.
23. Semple, S. J. et al. 1999. *Journal of Ethnopharmacology* 68 (1–3): 283–288.
24. Cox, S. D. et al. 2005. *Journal of Ethnopharmacology* 101 (1–3): 162–168.
25. Beck, W., and J. Balme. 2003. *Australian Aboriginal Studies* 2003 (2): 4–20.
26. Li, R. W. et al. 2003. *Journal of Ethnopharmacology* 85 (1): 25–32.
27. Palombo, E. A., and S. J. Semple. 2001. *Journal of Ethnopharmacology* 77 (2–3): 151–157.
28. Bandaranayake, W. M. 1998. *Mangroves and Salt Marshes* 2:133–148.
29. Roome, T. et al. 2008. *Journal of Ethnopharmacology* 118 (3): 514–521.
30. Pennacchio, M. et al. 2005. *Journal of Ethnopharmacology* 96 (3): 597–601.
31. Durmic, Z., et al. 2008. *Animal Feed Science and Technology* 145 (1–4): 271–284.
32. Falco, M. R., and J. X. Vries. 1964. *Naturwissenschaften* 51 (19): 462–463.
33. Foo, L. Y. 1987. *Phytochemistry* 26 (3): 813–817.
34. Foo, L. Y., and H. Wong. 1986. *Phytochemistry* 25 (8): 1961–1965.
35. Tindale, M. D., and D. G. Roux. 1969. *Phytochemistry* 8 (9): 1713–1727.
36. Freire, C. et al. 2005. *Lipids* 40 (3): 317–322.
37. Freire, C. S. R. et al. 2007. *Phytochemical Analysis* 18 (2): 151–156.
38. Schmalle, H., and B. Hausen. 1980. *Tetrahedron Letters* 21:149–151.
39. Uddin, S. J. et al. 2009. *eCAM* 1–6.
40. Rao, K. V. 1964. *Tetrahedron* 20 (4): 973–977.
41. Rao, K. V., and P. K. Bose. 1961. *Annals of Biochemistry and Experimental Medicine* 21:354–358.
42. Gomez, E. et al. 1989. *Journal of Natural Products* 52 (3): 649–651.
43. Hensens, O. D., and K. G. Lewis. 1966. *Australian Journal of Chemistry* 19 (1): 169–174.
44. Xu, M. et al. 2004. *Journal of Natural Products* 67 (5): 762–766.
45. Zhang, D. et al. 2005. *Fitoterapia* 76 (1): 131–133.
46. Hegnauer, R. 1973. *Chemotaxonomie der Pflanzen.* Basel and Stuttgart: Birkhäuser Verlag.
47. Qi, S.-H. et al. 2003. *Chinese Journal of Chemistry* 21 (2): 200–203.
48. Qi, S. W. et al. 2003. *Zhongcaoyao* 34 (7): 590–592.
49. de la Torre, M. C. et al. 1997. *Phytochemistry* 45 (1): 121–123.
50. Meyre-Silva, C. M. et al. 1998. *Phytomedicine* 5 (2): 109–113.
51. Niazi, J. et al. 2010. *Asian Journal of Pharmaceutical and Clinical Research* 3 (1): 35–37.
52. Pedrosa, R. et al. 2002. *Phytotherapy Research* 16 (8): 765–768.
53. Girardi, L. G. J. et al. 2003. *Pharmazie* 58 (9): 629–630.
54. Liu, H. et al. 2008. *Tetrahedron Letters* 49 (35): 5150–5151.
55. Liu, H.-Y. et al. 2007. *Helvetica Chimica Acta* 90 (10): 2017–2023.
56. Meyre-Silva, C. et al. 1999. *Planta Medica* 65 (03): 293–294.
57. Satyanarayana, P. et al. 2001. *Fitoterapia* 72 (3): 304–306.
58. Rogers, K. L. et al. 2000. *European Journal of Pharmaceutical Sciences* 9 (4): 355–363.
59. Birch, A. J., and R. N. Speake. 1960. *Journal of the Chemical Society* 1960: 3593–3599.
60. Branch, G. B. et al. 1972. *Australian Journal of Chemistry* 25 (10): 2209–2216.

61. Guise, G. B. et al. 1962. *Australian Journal of Chemistry* 15 (2): 314–321.
62. Allam, K. et al. 1987. *Journal of Natural Products* 50 (4): 623–625.
63. Crow, W. D. et al. 1970. *Australian Journal of Chemistry* 23 (12): 2489–2501.
64. Crow, W. D. G., and M. Yolande. 1955. *Australian Journal of Chemistry* 8:460–463.
65. Christensen, S. B., and A. A. Omar. 1985. *Journal of Natural Products* 48 (1): 161.
66. Keckeis, K. et al. 2000. *Fitoterapia* 71 (4): 456–458.
67. Omar, A. A. et al. 1984. *Fitoterapia* 55 (1): 59.
68. Leach, D. et al. 2007. In *Organization WIP*, ed. Bio-actives Export Pty Ltd: Australia.
69. Pinhey, J. T., and I. A. Southwell. 1971. *Australian Journal of Chemistry* 24 (6): 1311–1313.
70. Iwakami, S. et al. 1992. *Chemical and Pharmaceutical Bulletin* (Tokyo) 40 (5): 1196–1198.
71. Taylor, R. S. et al. 1995. *Journal of Ethnopharmacology* 46 (3): 153–159.
72. Wu, J. et al. 1991. *Chemical and Pharmaceutical Bulletin* (Tokyo) 39 (12): 3272–3275.
73. Su, M. et al. 2010. *Natural Product Communications* 5 (1): 151–156.
74. Ding, L. F. et al. 2009. *Journal of Asian Natural Product Research* 11 (8): 732–736.
75. Wu, J. B. et al. 1985. *Chemical and Pharmaceutical Bulletin* 33 (9): 4091–4094.
76. Aylward, J. H. 1999. In *Organization WIP*, ed. Peplin Biotech Pty, Ltd.
77. Li, R. W. et al. 2006. *Journal of Ethnopharmacology* 104 (1–2): 138–143.
78. Southwell, I. A., and D. J. Tucker. 1993. *Phytochemistry* 33 (5): 1099–1102.
79. Mali, R. G. 2010. *Pharmaceutical Biology* 48:105–112.
80. Sweeney, A. P. et al. 2001. *Journal of Ethnopharmacology* 75 (2–3): 273–277.
81. Gopal, N., and S. Sengottuvelu. 2008. *Fitoterapia* 79 (1): 24–26.
82. Rajalingam, K. et al. 2008. *Asian Journal of Scientific Research* 1 (3): 246–255.
83. Yln, M. N. et al. 2009. *Journal of Pharmacy and Chemistry* 3 (2): 51–56.
84. Gurudeeban, S. et al. 2010. *World of Fish and Marine Sciences* 2 (1): 66–69.
85. Kanchanapoom, T. et al. 2001. *Phytochemistry* 58 (2): 333–336.
86. Nan, H. et al. 2005. *Pharmazie* 60 (10): 798–799.
87. Pandey, R. et al. 2005. *Phytochemistry* 66 (6): 643–648.
88. Pandey, R. et al. 2003. *Phytochemistry* 63 (4): 415–420.
89. Hathaway, D. E. 1962. *Biochemical Journal* 83:80–84.
90. Molangui, T. et al. 1997. *Flavor and Fragrance Journal* 12 (6): 433–437.
91. Mohamed, A. et al. 2005. *Medical Science Monitor* 11 (11): BR426–431.
92. Bignell, C. M. et al. 1997. *Flavor and Fragrance Journal* 12 (1): 19–27.
93. Rashwan, O. A. 2002. *Molecules* 7:75–80.
94. Paddy, E. et al. 1987. *Community Report* 87 (1).
95. Silou, T. L. et al. 2009. *Journal of Essential Oil Research* 21 (3): 203–11..
96. Takahashi, T. et al. 2004. *Letters in Applied Microbiology* 39 (1): 60–64.
97. Lucia, A. et al. 2009. *Bioresource Technology* 100 (23): 6083–6087.
98. Elaissi, A. et al. 2010. *Chemistry & Biodiversity* 7 (4): 909–921.
99. Brophy, J. J., and I. A. Southwell. 2002. Eucalyptus: *The genus* Eucalyptus, ed. J. J. W. Coppen. London: Taylor & Francis.
100. Pilbeam, D. J. et al. 1983. *Journal of Natural Products* 46 (5): 601–605.
101. Graikou, K. et al. 2005. *Helvetica Chimica Acta* 88 (10): 2654–2660.
102. Graikou, K. et al. 2004. *Journal of Natural Products* 67 (4): 685–688.
103. Bowden, B. F., and D. J. Collins. 1988. *Journal of Natural Products* 51 (2): 311–313.
104. Getie, M. et al. 2003. *Fitoterapia* 74 (1–2): 139–143.
105. Veerapur, V. P. et al. 2004. *Indian Journal of Pharmaceutical Sciences* 66 (4): 407–411.
106. Arun, M., and V. V. Asha. 2008. *Journal of Ethnopharmacology* 118 (3): 460–465.
107. Patel, M., and M. M. Coogan. 2008. *Journal of Ethnopharmacology* 118 (1): 173–176.
108. Khurram, M. et al. 2009. *Molecules* 14 (3): 1332–1341.
109. Teffo, L. S. et al. 2010. *South African Journal of Botany* 76 (1): 25–29.
110. Cao, S. et al. 2009. *Journal of Natural Products* 72 (9): 1705–1707.
111. Niu, H.-M. et al. 2010. *Journal of Asian Natural Products Research* 12 (1): 7–14.
112. Buchanan, M. S. et al. 2009. *Journal of Natural Products* 72 (8): 1541–1543.
113. Brophy, J. J. et al. 1993. *Journal of Essential Oil Research* 5 (6): 581–586.
114. Carroll, A. R. et al. 2001. *Journal of Natural Products* 64 (12): 1572–1573.

115. Chen, C. R. et al. 1974. *Lloydia* 37 (3): 493–500.
116. Ghisalberti, E. L. 1994. *Journal of Ethnopharmacology* 44 (1): 1–9.
117. Knight, D. W., and G. Pattenden. 1975. *Tetrahedron Letters* 16 (13): 1115–1116.
118. Massy-Westropp, R. A. et al. 1966. *Tetrahedron Letters* 7 (18): 1939–1946.
119. Rogers, K. et al. 2001. *Life Sciences* 69 (15): 1817–1829.
120. Grice, I. D. et al. 2003. *Journal of Ethnopharmacology* 86 (1): 123–125.
121. Pennacchio, M. et al. 1996. *Journal of Ethnopharmacology* 53 (1): 21–27.
122. Della, E., and P. Jefferies. 1961. *Australian Journal of Chemistry* 14 (4): 663–664.
123. Pennacchio, M. et al. 1997. *Phytomedicine* 4 (4): 325–330.
124. Syah, Y. M., and E. L. Ghisalberti. 1996. *Fitoterapia* 67 (5): 447–451.
125. Palombo, E. A., and S. J. Semple. 2002. *Journal of Basic Microbiology* 42 (6): 444–448.
126. Liu, Q. et al. (2006). *Phytochemistry* 67(12):1256–1261.
127. Bick, I. R. C. 1996. In *Alkaloids: Chemical and biological perspectives*, ed. S. W. Pelletier. Oxford, England: Pergamon Press.
128. Ayepola, O. O., and B. A. Adeniyi. 2008. *Journal of Applied Sciences Research* 4 (11): 1410–1413.
129. Barra, A. C. et al. 2010. *Natural Product Communications* 5 (2): 329–335.
130. Cheng, S.-S. et al. 2009. *Bioresource Technology* 100 (1): 452–456.
131. Rasooli, I. et al. 2009. *International Journal of Dental Hygiene* 7 (3): 196–203.
132. Siddiqui, B. S. et al. 2000. *Phytochemistry* 54 (8): 861–865.
133. Tsiri, D. et al. 2008. *Helvetica Chimica Acta* 91 (11): 2110–2114.
134. Burchfield, E. et al. 2006. *Australian Journal of Zoology* 53 (6): 395–402.
135. Dayal, R. 1982. *Current Science* 51 (20): 997–998.
136. Zrira, S. et al. 2004. *Flavor and Fragrance Journal* 19 (2): 172–175.
137. El-Shafae, A. M. 1997. *Zagazig Journal of Pharmaceutical Sciences* 6 (2): 30–34.
138. Hyodo, S. et al. 1992. *Bioscience, Biotechnology, and Biochemistry* 56 (1).
139. Auddy, B. et al. 2003. *Journal of Ethnopharmacology* 84 (2–3): 131–138.
140. Ganju, L. et al. 2003. *Biomedicine and Pharmacotherapy* 57 (7): 296–300.
141. Tharan, N. T. et al. 2003. *Antibacterial activity of* Evolvulus alsinoides, vol. 40. Bombay, India: Indian Drug Manufacturers' Association.
142. Siripurapu, K. B. et al. 2005. *Pharmacology Biochemistry and Behavior* 81 (3): 424–432.
143. Kumar, M. et al. 2010. *Fitoterapia* 81 (4): 234–242.
144. Nahata, A. et al. 2010. *Phytotherapy Research* 24 (4): 486–493.
145. Baveja, S. K., and R. D. Singla. 1969. *Indian Journal of Pharmacy* 31 (4): 108–110.
146. Cervenka, F. et al. 2008. *Journal of Enzyme Inhibition and Medicinal Chemistry* 23 (4): 574–578.
147. Erickson, K. et al. 1995. *Journal of Natural Products* 58 (5): 769–772.
148. Hossain, S. J. et al. 2009. *Pharmacology Online* 2:927–936.
149. Patra, J. K. et al. 2009. *International Journal of Integrative Biology* 7 (1): 9–15.
150. Gowri, P. M. et al. 2009. *Helvetica Chimica Acta* 92 (7): 1419–1427.
151. Konishi, T. et al. 2003. *Chemical and Pharmaceutical Bulletin* 51 (10): 1142–1146.
152. Konishi, T. et al. 2003. *Phytochemistry* 64 (4): 835–840.
153. Wang, Z. et al. 2009. *Molecules* 16 (14): 414–422.
154. Zou, J.-H. et al. 2006. *Chemical and Pharmaceutical Bulletin* 54 (6): 920–921.
155. Anjaneyulu, A. et al. 2003. *Natural Product Research* 17 (1): 27–32.
156. Anjaneyulu, A., and V. Rao. 2000. *Phytochemistry* 55 (8): 891–901.
157. Kang, J. et al. 2005. *Journal of Asian Natural Products Research* 7 (5): 729–734.
158. Williams, C. A. et al. 1971. *Phytochemistry* 10 (5): 1059–1063.
159. Brophy, J. J. et al. 2005. *Journal of Essential Oil Research* 17 (4): 388.
160. Brown, R. et al. 1954. *Australian Journal of Chemistry* 7 (4): 348–377.
161. Dreyer, D. L., and A. Lee. 1972. *Phytochemistry* 11 (2): 763–767.
162. Lahey, F., and J. Macleod. 1967. *Australian Journal of Chemistry* 20 (9): 1943–1955.
163. Rasmussen, M. et al. 1968. *Australian Journal of Chemistry* 21 (12): 11.
164. Ridley, D. et al. 1970. *Australian Journal of Chemistry* 23 (1): 37.
165. Demirci, B. et al. 2009. *Chemistry of Natural Compounds* 45 (3): 2.
166. Chen, J. et al. 2006. *Planta Medica* 72 (10): 935–938.
167. Rosa, R. M. et al. 2006. *Journal of Agricultural and Food Chemistry* 54 (19): 7324–7330.

168. Rosa, R. M. et al. 2007. *Toxicology in Vitro* 21 (8): 1442–1452.
169. Feng, C. et al. 2008. *Helvetica Chimica Acta* 91 (5): 850–855.
170. Maganha, E. G. et al. 2010. *Food Chemistry* 118 (1): 1–10.
171. Li, X.-J., and H.-Y. Zhang. 2008. *Trends in Molecular Medicine* 14 (1): 1–2.
172. Li, L. et al. 2008. *Phytochemistry* 69 (2): 511–517.
173. Subramanian, S., and A. G. R. Nair. 1973. *Current Science* 42 (21): 2.
174. Narayanaswamy, V. B. K. et al. 2006. *Natural Product Sciences* 12 (2): 5.
175. Wasuwat, S. 1970. *Nature* 225 (5234): 758–758.
176. Pongprayoon, U. et al. 1991. *Journal of Ethnopharmacology* 35 (1): 65–69.
177. De Souza, M. et al. 2000. *Journal of Ethnopharmacology* 69 (1): 85–90.
178. Marie, D. et al. 2007. *Natural Product Communications* 2 (12): 1225–1228.
179. Wang, Q. et al. 2008. *Zhongguo Yaoxue Zazhi (Beijing, China)* 43 (1): 20–22.
180. Tao, H. et al. 2008. *Journal of Natural Products* 71 (12): 1998–2003.
181. Lamela, M., I. Cadavid, and J. M. Calleja. 1986. *Journal of Ethnopharmacology* 15 (2): 153–160.
182. Lopez, V. et al. 2008. *Pharmaceutical Biology* 46 (9): 602–609.
183. Slanc, P. et al. 2009. *Phytotherapy Research* 23 (6): 874–877.
184. Humadi, S., and S. I. Viorica. 2009. *Farmacia* 57 (2): 192–200.
185. Kim, H. J. et al. 2004. *Journal of Agricultural and Food Chemistry* 52 (10): 2849–2854.
186. Miyazawa, M., and C. Yamafuji. 2005. *Journal of Agricultural and Food Chemistry* 53 (5): 1765–1768.
187. Carson, C. F. et al. 2006. *Clinical and Microbiology Reviews* 19 (1): 50–62.
188. Furneri, P. M. et al. 2006. *Journal of Antimicrobial Chemotherapy* 58 (3): 706–707.
189. Garozzo, A. et al. 2009. *Letters in Applied Microbiology* 49 (6): 806–808.
190. Liu, X. et al. 2009. *European Food Research and Technology* 229 (2): 247–253.
191. Swords, G. H., and G. L. K. Hunter. 1978. *Journal of Agricultural and Food Chemistry* 26 (3): 734–737.
192. Silva, C. J. et al. 2010. *Química Nova* 33:104–108.
193. Chao, S. et al. 2008. *Flavor and Fragrance Journal* 23 (6): 444–449.
194. Pavela, R. 2008. *Phytotherapy Research* 22 (2): 274–278.
195. Lee, Y.-S. et al. 2008. *Flavor and Fragrance Journal* 23 (1): 23–28.
196. Moharram, F. A. et al. 2003. *Phytotherapy Research* 17 (7): 767–773.
197. Sardans, J. et al. 2010. *Journal of Chemical Ecology* 36 (2): 210–226.
198. Wheeler, D. L. et al. 2000. *Nucleic Acids Research* 28 (1): 10–14.
199. Yao, L. et al. 2004. *Food Research International* 37 (2): 166–174.
200. Park, I.-K., and S.-C. Shin. 2005. *Journal of Agricultural and Food Chemistry* 53 (11): 4388–4392.
201. Jones, G. P. et al. 1987. *Phytochemistry* 26 (12): 3343–3344.
202. Brophy, J. J. et al. 2006. *Journal of Essential Oil Research* 18 (6): 591–599.
203. Lassak, E. V., and J. J. Brophy. 2004. *Flavor and Fragrance Journal* 19 (1): 12–16.
204. O'Donnell, F. et al. 2009. *Analytica Chimica Acta* 634 (1): 115–120.
205. Lassak, E., and I. Southwell. 1972. *Australian Journal of Chemistry* 25 (11): 2491–2496.
206. Brophy, J. J. et al. 2004. *Journal of Essential Oil Research* 16 (4): 286.
207. Abbaszadeh, B. et al. 2009. *African Journal of Plant Science* 3 (10): 217–221.
208. Brophy, J. J. et al. 1996. *Journal of Essential Oil Research* 8 (2): 179–181.
209. Mulholland, D. A., and D. A. H. Taylor. 1992. *Phytochemistry* 31 (12): 4163–4166.
210. Grace, M. H. et al. 2006. *Phytochemistry* 67 (16): 1708–1715.
211. Loder, J. et al. 1969. *Australian Journal of Chemistry* 22 (7): 1531–1538.
212. Falkiner, M. et al. 1972. *Australian Journal of Chemistry* 25 (11): 2417–2420.
213. Errington, S. G., and P. R. Jefferies. *Phytochemistry* 27 (2): 543–545.
214. Abdel-Mogib, M. et al. 2002. *Molecules* 7:271–301.
215. Rüedi, P., and C. H. Eugster. 1978. *Helvetica Chimica Acta* 61 (2): 709–715.
216. Ahmad, I., and F. Aqil. 2007. *Microbiological Research* 162 (3): 264–275.
217. Sheeja, E. et al. 2010. *Pharmaceutical Biology* 48 (4): 381–387.
218. Zahin, M. et al. 2009. *International Journal of Pharmacy and Pharmaceutical Sciences* 1 (Suppl. 1): 88–95.
219. Nguyen, A. T. et al. 2004. *Fitoterapia* 75 (5): 500–504.
220. Gunaherath, G. M. K. B., and A. A. L. Gunatilaka. 1988. *Journal of the Chemical Society* 2:407–410.

221. Lin, L.-C. et al. 2003. *Phytochemistry* 62 (4): 619–622.
222. Eade, R. et al. 1969. *Australian Journal of Chemistry* 22 (12): 2703–2707.
223. Psotová, J. et al. 2003. *Phytotherapy Research* 17 (9): 1082–1087.
224. Sárosi, S., and J. Bernáth. 2008. *Acta Alimentaria* 37 (2): 293–300.
225. Brindley, M. et al. 2009. *Virology Journal* 6 (1): 8.
226. Huang, N. et al. 2009. *Journal of Agricultural and Food Chemistry* 57 (22): 10579–10589.
227. Kojima, H., and H. Ogura. 1986. *Phytochemistry* 25 (3): 729–733.
228. Kojima, H. et al. 1990. *Phytochemistry* 29 (7): 2351–2355.
229. Ryu, S. Y. et al. 2000. *Planta Medica* 66 (4): 358–360.
230. Wang, Z.-J. et al. 2008. *Lishizhen Medicine and Materia Medica Research* 8.
231. Smith, B. L. et al. 1988. *New Zealand Veterinary Journal* 36 (2): 56–58.
232. Fletcher, M. T. et al. 2010. *Tetrahedron Letters* 51 (15): 1997–1999.
233. Tanaka, N. et al. 1993. *Phytochemistry* 32 (4): 1037–3080.
234. Johns, S. R. et al. 1968. *Australian Journal of Chemistry* 21 (12): 3079–3080.
235. Agoramoorthy, G. et al. 2008. *Asian Journal of Chemistry* 20 (2): 1311–1322.
236. Anjaneyulu, A. S. R. et al. *Journal of Asian Natural Products Research* 4 (1): 53–61.
237. Rohini, R. M., and A. K. Das. 2010. *Natural Product Research* 24 (2): 197–202.
238. Kanegusuku, M. et al. 2007. *Biological and Pharmaceutical Bulletin* 30 (5): 999–1002.
239. Bowen-Forbes, C. S. et al. 2009. *Food Chemistry* 116 (3): 633–637.
240. Southwell, I. 1978. *Australian Journal of Chemistry* 31:2527–2538.
241. Bradfield, A. E. et al. 1936. *Journal of the Chemical Society* 1936:1619–1625.
242. Butaud, J. F. et al. 2008. *Journal of the American Oil Chemists' Society* 85 (4): 353–356.
243. Vickery, J. R. et al. 1984. *Journal of the American Oil Chemists' Society* 61 (5): 890–891.
244. Jirovetz, L. et al. 2006. *Flavor and Fragrance Journal* 21 (3): 465–468.
245. Brophy, J. J. et al. 1991. *Journal of Essential Oil Research* 3 (6): 381–385.
246. Piggott, M. J. et al. 1997. *Flavor and Fragrance Journal* 12 (1): 43–46.
247. Braun, N. A. et al. 2003. *Journal of Essential Oil Research* 15 (6): 381–386.
248. Valder, C. N. et al. 2003. *Journal of Essential Oil Research* 15 (3): 178–186.
249. Shellie, R. et al. 2004. *Journal of Chromatographic Science* 42 (8): 417–422.
250. Cornforth, J. W. E., and J. Campbell. 1939. *Journal of the Chemical Society* 1939:737–742.
251. Kerr, P. G. et al. 1996. *Planta Medica* 62 (06): 519–522.
252. Dhalwal, K. et al. 2007. *Journal of Medicinal Food* 10 (4): 683–688.
253. Gupta, S. R. et al. 2009. *Natural Product Research* 23 (8): 689–695.
254. Iswantini, D. D. et al. 2009. *Journal of Biological Sciences* 9 (5): 504–508.
255. Tona, G. L. et al. 2009. *Recent Progress in Medicinal Plants* 25:209–224.
256. Wang, Y.-H. et al. 2008. *Rapid Communications in Mass Spectrometry* 22 (16): 2413–2422.
257. Islam, M. E. et al. 2003. *Phytotherapy Research* 17 (8): 973–975.
258. Williams, A. H. 1967. *Phytochemistry* 6 (11): 1583–1584.
259. Leu, T. et al. 2008. *ACGC Chemical Research Communications* 22:22–29.
260. Murakoshi, I. et al. 1981. *Phytochemistry* 20 (7): 1725–1730.
261. Kinoshita, T. et al. 1990. *Chemical and Pharmaceutical Bulletin* 38 (10): 2756–2759.
262. Tanaka, T. et al. 1997. *Phytochemistry* 46 (8): 1431–1437.
263. Smyth, T. et al. 2009. *Journal of Pharmacognosy and Phytotherapy* 1 (6): 82–86.
264. Huxtable, C. R., and P. R. Dorling. 1982. *American Journal of Pathology* 107 (1): 124–126.
265. Steiniger, J. R. 1970. *Pharmazie* 28 (10): 682–683.
266. Li, R. W. et al. 2004. *Phytotherapy Research* 18 (1): 78–83.
267. Hungerford, N. L. et al. 1998. *Australian Journal of Chemistry* 51 (12): 1103–1112.
268. Bisset, N. G., and J. Nwaiwu. 1983. *Planta Medica* 48 (08): 275–279.
269. Umezawa, T. 2003. *Phytochemistry Reviews* 2 (3): 371–390.
270. Umezawa, T. 2003. *Wood Research* 90:27–110.
271. Chowdhury, R. et al. 2003. *Fitoterapia* 74 (1–2): 155–158.
272. Kraus, W. G., and W. Grimminger. 1980. *Nouveau Journal de Chimie* 4 (11): 651–655.
273. Kraus, W. G., and W. Grimminger. 1981. *Liebigs Annalen der Chemie* 10:1838–1843.
274. Da Silva, M. F.d. G. F. et al. 1999. *Pure and Applied Chemistry* 71 (6): 1083–1087.
275. Hosamani, K. M. 1994. *Phytochemistry* 37 (6): 1621–1624.

276. O'Kelly, J., and K. Sargeant. 1961. *Journal of the Chemical Society* 484.
277. Bhandari, P., and R. P. Rastogi. 1983. *Indian Journal of Chemistry, Section B: Organic Chemistry Including Medicinal Chemistry* 22B (7): 624–626.
278. Bhandari, P., and R. P. Rastogi. 1983. *Indian Journal of Chemistry, Section B: Organic Chemistry Including Medicinal Chemistry* 22B (3): 252–256.
279. Calvo, M. I. 2006. *Journal of Ethnopharmacology* 107 (3): 380–382.
280. Lai, S.-W. et al. 2006. *Neuropharmacology* 50 (6): 641–650.
281. Speroni, E. et al. 2007. *Planta Medica* 73 (3): 227–235.
282. Casanova, E. et al. 2008. *Plant Foods for Human Nutrition* 63 (3): 93–97.
283. De Martino, L. et al. 2008. *Pharmacology Online* 2:170–175.
284. De Martino, L. et al. 2009. *International Journal of Immunopathology and Pharmacology* 22 (4): 1097–1104.
285. Deepak, M., and S. S. Handa. 2000. *Phytotherapy Research* 14 (6): 463–465.
286. Deepak, M., and S. S. Handa. 2000. *Phytochemical Analysis* 11 (6): 351–355.
287. De Martino, L. et al. 2009. *Natural Product Communications* 4 (12): 1741–1750.
288. Carvalho, A. O. et al. 2001. *Plant Physiology and Biochemistry* 39 (2): 137–146.
289. Hu, K. et al. 2000. *Planta Medica* 66 (6): 564–567.
290. Nunome, S. et al. 2004. *Planta Medica* 70 (1): 76–78.
291. Wang, L.-Y. et al. 2005. *Chemical and Pharmaceutical Bulletin* 53 (1): 137–139.
292. Wang, L.-Y. et al. 2005. *Chemical and Pharmaceutical Bulletin* 53 (10): 1348–1351.
293. Ho, W. S. et al. 2010. *Phytotherapy Research* 24 (5): 657–661.
294. Chen, Y. et al. 2009. *Chinese Chemical Letters* 20 (5): 592–594.
295. Kato, A. et al. 1979. *Journal of Natural Products* 42 (2): 159–162.
296. Wang, H.-K. et al. 1998. *Advances in Experimental Medicine and Biology* 439:191–225.
297. Lee, K.-H. et al. 1981. *Journal of Natural Products* 44 (5): 530–535.
298. Li, Y. M. et al. 2009. *Current Pharmaceutical Biotechnology* 10 (8): 743–752.
299. Flynn, T. M., and I. A. Southwell. 1987. *Phytochemistry* 26 (6): 1673–1686.
300. Islam, S. K. N., and A. Monira. 1997. *Phytotherapy Research* 11 (1): 64–66.

CHAPTER 12

Medicinal Properties of Legumes

J. Bradley Morris, Barbara Hellier, and John Connett

CONTENTS

12.1 INTRODUCTION

Legumes, or beans, are members of a family of flowering plants known as Fabaceae (Leguminosae). It is the third largest family of flowering plants, with approximately 650 genera and nearly 20,000 species (Doyle 1994). Legume species range from large tropical trees to small herbs found in temperate areas, humid tropics, arid zones, mountains, prairies, and lowlands. The legume family consists of three subfamilies including the Papilionoideae, Caesalpinioideae, and Mimosoideae, which are identified based on their flower characteristics. The Papilionoideae consists of about 476 genera and 14,000 species, and Caesalpinioideae contains 162 genera and about 3,000 species, and Mimosoideae consists of 77 genera and approximately 3,000 species (Doyle and Luckow 2003).

Many legumes possess unique phytochemicals commonly referred to as primary or secondary metabolites. Secondary metabolites are derived from primary metabolites (Balandrin et al. 1985). Considerable research is focused on identifying plants, including legumes, with valuable secondary products or metabolites for pharmaceutical use (Morris 2003) as well as legume organs with medicinal qualities.

The U.S. National Plant Germplasm System, USDA, Agricultural Research Service (ARS), has collections of many medicinal legumes. The majority of these are curated at the Plant Genetic Resources Conservation Unit, Griffin, Georgia—for example, hyacinth bean (*Lablab purpureus* L.

Sweet), butterfly pea (*Clitoria ternatea* L.), guar (*Cyamopsis tetragonoloba* (L.) Taub), horse gram (*Macrotyloma uniflorum* Lam. Verdc.), and sunn hemp (*Crotalaria juncea* L.)—and the Western Regional Plant Introduction Station, Pullman, Washington—for example, membranous milk vetch (*Astragalus embranaceous* Fisch. ex Link Bunge), kidney vetch (*Anthyllis vulneraria* L.), and common licorice (*Glycyrrhiza glabra* L.).

12.2 GENETIC RESOURCES

The U.S. collection of medicinal legumes consists of more than 8,800 accessions, representing about 43 species collected or donated from countries worldwide (NPGS 2009; www.ars-grin.gov/ npgs/searchgrin.html). A list of worldwide collections for several of these medicinal legumes can be found in Table 12.1. The most commonly used medicinal legumes are listed in Table 12.2. Of these legumes, only peanut (*Arachis hypogaea* L.), orchid tree (*Bauhinia variegata* L.), swordbean (*Canavalia ensiformis* (L.) D.C.), butterfly pea (*Clitoria ternatea* L.), annual lespedeza (*Kummerowia striata* (Thunb.) Schindl.), shrub lespedeza (*Lespedeza bicolor* Turcz.), virgata lespedeza (*L. virgata* (L.) P. Beauv.), and red clover (*Trifolium pratense* L.) have been used in cultivar development (NPGS 2009) for use as food and forages.

Most of the legumes reported for medicinal research consist of wild species or plants grown in cultivation but not improved through breeding. Genetically improved accessions for *Anthyllis vulneraria* L., *Glycyrrhiza echninata* L. (Figure 12.1), *G. glabra* L., and *G. uralensis* Fisch. are listed in the GRIN database (NPGS 2009; www.ars-grin.gov/npgs/searchgrin.html). Since most of these medicinal legumes consist of limited numbers of accessions, additional acquisitions of these valuable species must be obtained through collection trips in the countries of origin as well as donations from participating agencies.

12.3 MAINTENANCE

Quality maintenance of medicinal legume germplasm is required for long-term preservation and future utilization. Seed viability is determined through standard germination testing (Association of Official Seed Analysts 2000). Medicinal legume seeds are cleaned and dried at 21°C, 25% relative humidity (RH), for 3–7 days prior to storage in sealed bags at –18°C at the Griffin, Georgia, repository. Seeds are dried and cleaned at ambient temperature and RH and stored at 4°C, 30% RH, at the Pullman, Washington, repository.

Regeneration is a component of maintenance and is an important curation responsibility. All legumes at Griffin and all but *A. vulneraria* at Pullman are scarified prior to planting. Scarified seeds are planted in jiffy pots containing potting soil or vermiculite, during early spring, in a greenhouse. Vermiculite-produced seedlings are transplanted to soil-less potting medium or Q-plugs™(International Horticulture Technologies, LLC, Hollister, California). The self-pollinated species are regenerated in normal field conditions at a rate of 50 plants per accession and planted in 6 m rows at the USDA, ARS location in Griffin. Similar field transplanting efforts are conducted at USDA, ARS, in Pullman (Figure 12.2).

If available for the particular legume species, *Rhizobium* is applied to the seed at greenhouse planting or field transplanting. Mature seeds from each accession are harvested from dry pods or field plants. Cross-pollinated species are planted in two rows consisting of 100 plants within pollination cages. Honeybee hives are placed in each cage at 50% bloom. Some self-pollinating legumes, such as *Clitoria ternatea*, require regeneration inside cages without honeybees because of fairly high outcrossing rates. Harvest begins at seed maturity and continues until about 3,000 seeds have been obtained in Griffin or at least 10,000 seeds have been collected in Pullman.

Table 12.1 Worldwide Genetic Resource Collections of Medicinal Legumes

Species	Type	No.	Germplasm Collection Site
Acacia nilotica	Wild, weedy species	2	Institut National de la Recherche Agronomique d'Algerie
	Wild, weedy species	25	Centre National de Semences Forestieres (CNSF) Burkina Faso, cnsf@fasonet.bf
	Wild, weedy species	6	International Livestock Research Institute (ILRI), ILRI-ETHIOPIA@cgiar.org
	Wild, weedy species	4	National Genebank of Kenya, Crop Plant Genetic Resources Centre, KARI ngbk@wananchi.com
	Wild, weedy species	1	Ministry of Agriculture and Fisheries, Oman
	Landrace	1	Grassland Research Centre, Department of Agricultural Development, http://www.arc.agric.za/
	Wild, weedy species	1	El-Kod Agricultural Research Centre, Republic of Yemen
Alhagi maurorum	Wild, weedy species	2	Dept. of E.S.E., Institute of Life Science, Hebrew Univ. of Jerusalem, ilanahs@vms.huji.ac.il
	Wild, weedy species	1	Royal Botanic Gardens, Kew, r.smith@rbgkew.org.uk
	Wild species	2	USDA, ARS, Western Regional Plant Introduction, http://www.ars-grin.gov/npgs/
Amorpha canescens	Unknown	1	National Genebank of Kenya, Crop Plant Genetic Resources Centre, KARI ngbk@wananchi.com
	Wild species	13	USDA, ARS, Western Regional Plant Introduction, http://www.ars-grin.gov/npgs/
A. nana	Wild species	1	USDA, ARS, Western Regional Plant Introduction, http://www.ars-grin.gov/npgs/
Anthyllis vulneraria	Cultivar	3	USDA, ARS, Western Regional Plant Introduction, http://www.ars-grin.gov/npgs/
	Wild species	42	
	Wild, weedy species	51	Australian Medicago Genetic Resources Centre, SARDI, http://www.sardi.sa.gov.au
	Advanced cultivar	6	Research Institute for Fodder Crops Ltd., Czech Republic, http://www.vupt.cz
	Wild, weedy species	1	OSEVA PRO Ltd. Grassland Research Station, Czech Republic, http://www.oseva.cz
	Landrace	27	Institute for Agrobotany, Hungary, http://www.rcat.hu
	Advanced cultivars wild, weedy species		Dip. di Biologia Veg. e Biotecnologie Agroambientali e Zootecniche, Universit, http://www.agr.unipg.it/dbvba/
	Unknown	1	Institute of Medicinal Plants, Poland, jl_iripz@man.poznan.pl
	Landrace	4	Plant Breeding and Acclimatization Institute (IHAR), http://www.ihar.edu.pl
	Wild, weedy species	2	Banca de Resurse Genetice Vegetale Suceava, Romania, genebank@assist.ro
	Advanced cultivar	1	Research Institute of Plant Production Piestany, Slovakia, benedikova@vurv.sk
	Wild, weedy species	34	Kmetijski Institut Slovenije, Slovenia, vladimir.meglic@kis-h2.si
	Wild, weedy species	2	Nordic Gene Bank, Sweden, http://www.nordgen.org/ngb/
	Wild, weedy species	2	
	Wild, weedy species	22	Seed Bank, Seed Conservation Sect. Royal Botanic Gardens, Kew, r.smith@rbgkew.org.uk

(continued)

Table 12.1 Worldwide Genetic Resource Collections of Medicinal Legumes (Continued)

Species	Type	No.	Germplasm Collection Site
Astragalus canadensis	Wild, weedy species	1	Australian Medicago Genetic Resources Centre, SARDI, http://www.sardi.sa.gov.au
	Cultivar	1	USDA, ARS, Western Regional Plant Introduction, http://www.ars-grin.gov/npgs/
	Unknown	10	
	Wild, weedy species	1	Seed Bank, Seed Conservation Sect. Royal Botanic Gardens, Kew, r.smith@rbgkew.org.uk
A. chinensis	Unknown	2	USDA, ARS, Western Regional Plant Introduction, http://www.ars-grin.gov/npgs/
	Wild, weedy species	3	Australian Medicago Genetic Resources Centre, SARDI, http://www.sardi.sa.gov.au
A. complanatus	Wild, weedy species	1	Australian Medicago Genetic Resources Centre, SARDI, http://www.sardi.sa.gov.au
A. crassicarpus	Unknown	2	USDA, ARS, Western Regional Plant Introduction, http://www.ars-grin.gov/npgs/
A. exscapus	Wild, weedy species	1	Research Institute for Fodder Crops, Czech Republic, http://www.vupt.cz
A. hamosus	Unknown	51	USDA, ARS, Western Regional Plant Introduction, http://www.ars-grin.gov/npgs/
	Advanced cultivar	1	Australian Medicago Genetic Resources Centre, SARDI, http://www.sardi.sa.gov.au
	Breeding, inbred lines	3	
	Wild, weedy species	292	
	Wild, weedy species	42	Dept. of E.S.E., Institute of Life Science, Hebrew Univ. of Jerusalem, Israel, ilanahs@vms.huji.ac.il
	Landrace	1	Grassland Research Centre, Department of Agricultural Development, South Africa, http://www.arc.agric.za/
	Landrace	1	Int. Center for Agricultural Research in the Dry Areas (ICARDA), Syrian Arab Republic, http://www.icarda.cgiar.org
	Wild, weedy species	356	
	Wild, weedy species	4	Seed Bank, Seed Conservation Sect. Royal Botanic Gardens, Kew, r.smith@rbgkew.org.uk
A. membranaceus	Unknown	3	USDA, ARS, Western Regional Plant Introduction, http://www.ars-grin.gov/npgs/
	Wild, weedy species	1	Federal Centre for Breeding Research on Cultivated Plants (BAZ), Research and Coordination Centre for Plant Genetic Resources (FKZPGR), Germany, http://www.bafz.de
A. mongholicus	Unknown	5	USDA, ARS, Western Regional Plant Introduction, http://www.ars-grin.gov/npgs/
A. sinicus	Cultivar	30	USDA, ARS, Western Regional Plant Introduction, http://www.ars-grin.gov/npgs/
	Wild, weedy species	1	Australian Medicago Genetic Resources Centre, SARDI, http://www.sardi.sa.gov.au
	Landrace	95	Institute of Crop Science (CAAS), Beijing, China, http://lcgr.caas.net.cn/cgrisngb.html
	Unknown	159	National Institute of Livestock and Grassland Science, Japan, http://nilgs.naro.affrc.go.jp
	Unknown	10	Hokuriku National Agricultural Experimental Station, Niigata-ken, Japan
	Unknown	33	Genetic Resources Management Section, NIAR (MAFF), http://www.gene.affrc.go.jp
Baphia nitida	Wild, weedy species	1	International Livestock Research Institute (ILRI), ILRI-Ethiopia@cgiar.org

Species	Type	No.	Source
Bituminaria bituminosa	Wild, weedy species	1	International Institute of Tropical Agriculture, http://www.cgiar.org/iita/
	Wild species	15	USDA, ARS, Plant Genetic Resources Conservation Unit, http://www.ars-grin.gov/npgs/
	Wild, weedy species	7	Australian Medicago Genetic Resources Centre, SARDI, http://www.sardi.sa.gov.au
Canavalia brasiliensis	Unknown	1	USDA, ARS, Plant Genetic Resources Conservation Unit, http://www.ars-grin.gov/npgs/
	Wild, weedy species	1	Australian Tropical Crops & Forages Genetic Resources, http://www.dpi.qld.gov.au/auspgris
	Unknown	4	Cenargen/Embrapa, http://www.cenargen.embrapa.br
	Wild, weedy species	55	Centro Internacional de Agricultura Tropical (CIAT), http://www.ciat.cgiar.org
	Wild, weedy species	2	International Livestock Research Institute (ILRI), ILRI-Ethiopia@cgiar.org
C. ensiformis	Unknown	20	USDA, ARS, Plant Genetic Resources Conservation Unit,
	Cultivated	2	http://www.ars.usda.gov/main/site_main.htm?modecode=66-07-00-00
	Breeding material	1	Australian Tropical Crops & Forages Genetic Resources,
	Landrace	6	http://www.dpi.qld.gov.au/auspgris
	Wild, weedy species	23	
	Cultivated	1	National Botanical Garden of Belgium, http://www.br.fgov.be/
	Advanced cultivar	1	Centro de Pesquisas para Pequenas Propiedades, EPAGRI-CPPP, http://www.epagri.rct-sc.br
	Unknown	12	Cenargen/Embrapa, http://www.cenargen.embrapa.br
	Unknown	4	Embrapa Gado de Corte (CNPGC), http://www.cnpgc.embrapa.br
	Unknown	2	Empressa Brasileira de Pesquisa Agropecuaria, Embrapa Gado de Leite, http://www.cnpgl.embrapa.br
	Unknown	1	Centro Nacional de Pesquisa de Caprinos (CNPC), http://www.cnpc.embrapa.br
	Wild, weedy species	14	Centro Internacional de Agricultura Tropical (CIAT), http://www.ciat.cgiar.org
	Unknown	1	Corpoica, C.I. La Selva, pnegv@epm.net.co
	Unknown	1	Asociacion ANAI, anaicr@sol.racsa.co.cr
	Landrace	1	Banco de Germoplasma, Instituto de Investigaciones, inifat@ceniai.inf.cu
	Wild, weedy species	6	International Livestock Research Institute (ILRI), ILRI-Ethiopia@cgiar.org
	Unknown	8	National Genebank of Kenya, Crop Plant Genetic Resources, ngbk@wananchi.com
	Unknown	7	International Institute of Tropical Agriculture, http://www.cgiar.org/iita/
	Unknown	1	Instituto de Investigacion Agropecuaria de Panama, IDIAP, idiap@sinfo.net
	Unknown	1	Facultad de Ciencias Agropecuarias, Universidad de Panama, jgaonab@hotmail.com
	Unknown	9	Institute of Plant Breeding, College of Agriculture UPLB, http://www.uplb.edu.ph/ca/ipb/main/
	Unknown	2	Division of Plant and Seed Control, Department of Agriculture, http://www.agris.agric.za
	Landrace	7	Grassland Research Centre, Department of Agricultural Development, http://www.arc.agric.za/
Chamaecrista nictitans	Breeding material	1	USDA, ARS, Plant Genetic Resources Conservation Unit

(continued)

Table 12.1 Worldwide Genetic Resource Collections of Medicinal Legumes (Continued)

Species	Type	No.	Germplasm Collection Site
	Unknown	5	http://www.ars.usda.gov/main/site_main.htm?modecode=66-07-00-00
	Wild, weedy species	21	Australian Tropical Crops & Forages Genetic Resources http://www.dpi.qld.gov.au/auspgris
	Unknown	3	Cenargen/Embrapa, http://www.cenargen.embrapa.br
	Wild, weedy species	98	Centro Internacional de Agricultura Tropical (CIAT), http://www.ciat.cgiar.org
Clitoria ternatea	Wild species	1	USDA, ARS, Plant Genetic Resources Conservation Unit
	Unknown	25	http://www.ars.usda.gov/main/site_main.htm?modecode=66-07-00-00
	Landrace	5	Australian Tropical Crops & Forages Genetic Resources
	Unknown	121	http://www.dpi.qld.gov.au/auspgris
	Wild, weedy species; advanced cultivars	109	CSIRO Townsville Division of Tropical Crops and Pastures, mike.foale@tag.csiro.au
	Wild, weedy species	4	National Botanical Garden of Belgium, http://www.br.fgov.be/
	Unknown	19	Cenargen/Embrapa, http://www.cenargen.embrapa.br
	Unknown	10	Embrapa Gado de Corte (CNPGC), http://cnpgc.embrapa.br
	Unknown	4	Empresa Catarinense de Pesquisa Agropecuaria e Extensao
	Unknown	14	Empresa Brasileira de Pesquisa Agropecuaria, Embrapa Gado, http://www.cnpgl.embrapa.br
	Unknown	2	Centro Nacional de Pesquisa de Caprinos (CNPC), http://www.cnpc.embrapa.br
	Wild, weedy species	64	Centro Internacional de Agricultura Tropical (CIAT), http://www.ciat.cgiar.org
	Advanced cultivars	15	Banco de Germoplasma, Instituto de Investigaciones, inifat@ceniai.inf.cu
	Advanced cultivars; wild, weedy species	6	Estacion Experimental de Pastos y Forrajes INDIO Hatuey, interaca@indio.atenas.inf.cu
	Wild, weedy species	23	International Livestock Research Institute (ILRI), ILRI-Ethiopia@cgiar.org
	Advanced cultivars; landraces; wild, weedy species	96	Indian Grassland and Fodder Research Institute (IGFRI), http://www.icar.org.in/igfri/index.html
	Advanced cultivars; wild, weedy species	34	Central Arid Zone Research Institute, Div. of Plant Studies, http://www.icar.org.in/cazri.htm
	Unknown	1	Genetic Resources Management Section, NIAR (MAFF), http://www.gene.affrc.go.jp
	Wild, weedy species	9	Kenya Agricultural Research Institute, National Agricultural Research Centre, Kitale, http://www.kari.org
	Landraces	35	National Genebank of Kenya, Crop Plant Genetic Resources, ngbk@wananchi.com
	Wild, weedy species	10	
	Unknown	343	

Species	Status	No.	Source
	Wild, weedy species	1	Instituto Nacional de Investigaciones Forestales, Agricolas y Pecuarias (INIFAP), http://www.inifap.conacyt.mx
	Unknown	1	Institute of Plant Breeding, College of Agriculture UPLB, http://www.uplb.edu.ph/ca/ipb/main/
	Landraces	5	Grassland Research Centre, Department of Agricultural Development, http://www.arc.agric.za/
Colutea arborescens	Unknown	3	USDA, ARS, Western Regional Plant Introduction, http://www.ars-grin.gov/npgs/
	Wild, weedy species	2	Federal Centre for Breeding Research on Cultivated Plants (BAZ), Research and Coordination Centre for Plant Genetic Resources (FKZPGR), Germany, http://www.bafz.de
	Wild, weedy species	1	Greek Genebank, Agric. Res. Center of Macedonia and Thraki, NAGREF, kgeggb@otenet.gr
	Unknown	1	National Genebank of Kenya, Crop Plant Genetic Resources Centre, KARI, ngbk@wananchi.com
	Wild, weedy species	1	Seed Bank, Seed Conservation Sect. Royal Botanic Gardens, Kew, r.smith@rbgkew.org.uk
Daniellia oliveri	Unknown	1	USDA, ARS, Plant Genetic Resources Conservation Unit, http://www.ars.usda.gov/main/site_main.htm?modecode=66-07-00-00
	Wild, weedy species	3	Centre National de Semences Forestieres (CNSF), cnsf@fasonet.bf
	Wild, weedy species	1	International Livestock Research Institute (ILRI), ILRI-Ethiopia@cgiar.org
	Wild, weedy species	1	International Institute of Tropical Agriculture, http://www.cgiar.org/iita/
Desmodium adscendens	Unknown	5	USDA, ARS, Plant Genetic Resources Conservation Unit, http://www.ars.usda.gov/main/site_main.htm?modecode=66-07-00-00
	Unknown	10	Australian Tropical Crops & Forages Genetic Resources, http://www.dpi.qld.gov.au/auspgris
	Unknown	6	Cenargen/Embrapa, http://www.cenargen.embrapa.br
	Unknown	2	Embrapa Gado de Corte (CNPGC), http://www.cnpgc.embrapa.br
	Unknown	3	Empresa Brasileira de Pesquisa Agropecuaria, Embrapa Gado de Leite, http://www.cnpgl.embrapa.br
	Wild, weedy species	112	Centro Internacional de Agricultura Tropical (CIAT), http://www.ciat.cgiar.org
	Wild, weedy species	30	International Livestock Research Institute (ILRI), ILRI-Ethiopia@cgiar.org
	Unknown	1	Institute of Plant Breeding, College of Agriculture UPLB, http://www.uplb.edu.ph/ca/ipb/main/
	Wild, weedy species	1	Seed Bank, Seed Conservation Section Royal Botanic Gardens, Kew, r.smith@rbgkew.org.uk
D. gangeticum	Unknown	1	USDA, ARS, Plant Genetic Resources Conservation Unit, http://www.ars.usda.gov/main/site_main.htm?modecode=66-07-00-00
	Unknown	53	Australian Tropical Crops & Forages Genetic Resources, http://www.dpi.qld.gov.au/auspgris
	Unknown	1	Empresa Brasileira de Pesquisa Agropecuaria, Embrapa Gado de Leite, http://www.cnpgl.embrapa.br
	Wild, weedy species	238	Centro Internacional de Agricultura Tropical (CIAT), http://www.ciat.cgiar.org
	Wild, weedy species	1	Seed Bank, Seed Conservation Section Royal Botanic Gardens, Kew, r.smith@rbgkew.org.uk
Dorycnium rectum	Unknown	3	USDA, ARS, Plant Genetic Resources Conservation Unit, http://www.ars.usda.gov/main/site_main.htm?modecode=66-07-00-00

(continued)

Table 12.1   Worldwide Genetic Resource Collections of Medicinal Legumes (Continued)

Species	Type	No.	Germplasm Collection Site
	Wild, weedy species	4	Australian Medicago Genetic Resources Centre, SARDI, http://www.sardi.sa.gov.au
	Wild, weedy species	1	Department of E.S.E., Institute of life Science, Hebrew University of Jerusalem, ilanahs@vms.huji.ac.il
	Wild, weedy species	1	Dip. Di Biologia Veg. e Biotecnologie Agroambientali e, Zootecniche, Universit, http://www.agr.unipg.it/dbvba/
	Wild, weedy species	1	Seed Bank, Seed Conservation Section Royal Botanic Gardens, Kew, r.smith@rbgkew.org.uk
Genista tinctoria	Unknown	7	USDA, ARS, Western Regional Plant Introduction, http://www.ars-grin.gov/npgs/
	Wild, weedy species	13	Research Institute for Fodder Crops Ltd., Czech Republic, http://www.vupt.cz
	Wild, weedy species	1	OSEVA PRO Ltd. Grassland Research Station, Czech Republic, http://www.oseva.cz
Glycyrrhiza echinata	Unknown	2	USDA, ARS, Western Regional Plant Introduction, http://www.ars-grin.gov/npgs/
	Landrace	1	Faculty of Horticulture, Mendel Agricultural and Forestry University, Czech Republic, http://www.zf.mendelu.cz/
	Wild, weedy species	1	Federal Centre for Breeding Research on Cultivated Plants (BAZ), Research and Coordination Centre for Plant Genetic Resources (FKZPGR), Germany, http://www.bafz.de
	Landrace	1	Medicinal and Aromatic Plants Research Station, Fundulea, Romania
G. glabra	Unknown	22	USDA, ARS, Western Regional Plant Introduction, http://www.ars-grin.gov/npgs/
	Advanced cultivar	1	Australian Tropical Crops & Forages Genetic Resources Centre, http://www.dpi.qld.gov.au/auspgris
	Landrace	5	Faculty of Horticulture, Mendel Agricultural and Forestry University, Czech Republic, http://www.zf.mendelu.cz/
	Landrace	1	Medicinal and Aromatic Plants Research Station, Fundulea, Romania
	Wild, weedy species	1	Centro de Investigacion y Tecnologia Agraria, Spain, carravedo@mizar.csic.es
	Wild, weedy species	8	Int. Center for Agricultural Research in the Dry Areas (ICARDA), http://www.icarda.cgiar.org
G. lepidota	Unknown	8	USDA, ARS, Western Regional Plant Introduction, http://www.ars-grin.gov/npgs/
G. uralensis	Unknown	10	USDA, ARS, Western Regional Plant Introduction, http://www.ars-grin.gov/npgs/
	Landrace; wild, weedy species	6	Grassland Research Institute Chinese Academy of Agric. Sciences, cgi@public.hh.nm.cn
	Wild, weedy species	14	Int. Center for Agricultural Research in the Dry Areas (ICARDA), http://www.icarda.cgiar.org
Haematoxylum brasiletto	Unknown	5	USDA, ARS, Plant Genetic Resources Conservation Unit, http://www.ars.usda.gov/main/site_main.htm?modecode=66-07-00-00
Indigofera suffruticosa	Unknown	11	USDA, ARS, Plant Genetic Resources Conservation Unit, http://www.ars.usda.gov/main/site_main.htm?modecode=66-07-00-00
	Unknown	23	Australian Tropical Crops & Forages Genetic Resources, http://www.dpi.qld.gov.au/auspgris
	Unknown	6	Cenargen/Embrapa, http://www.cenargen.embrapa.br

Species	Category	No.	Source
	Unknown	1	Empresa Catarinense de Pesquisa Agropecuaria e Extensao, unknown
	Unknown	1	Centro Nacional de Pesquisa de Caprinos, http://www.cnpc.embrapa.br
	Wild, weedy species	50	Centro Internacional de Agricultura Tropical (CIAT), http://www.ciat.cgiar.org
	Wild, weedy species	1	International Livestock Research Institute (ILRI), ILRI-Ethiopia@cgiar.org
	Unknown	1	National Genebank of Kenya, Crop Plant Genetic Resources, ngbk@wananchi.com
	Landrace	7	Grassland Research Centre, Department of Agricultural Development, http://www.arc.agric.za/
	Wild, weedy species	9	Seed Bank, Seed Conservation Section Royal Botanic Gardens, Kew, r.smith@rbgkew.org.uk
I. tinctoria	Unknown	9	USDA, ARS, Plant Genetic Resources Conservation Unit, http://www.ars.usda.gov/main/site_main.htm?modecode=66-07-00-00
Kummerowia striata	Cultivated	1	USDA, ARS, Plant Genetic Resources Conservation Unit
	Unknown	26	http://www.ars.usda.gov/main/site_main.htm?modecode=66-07-00-00
	Advanced cultivar	1	Australian Medicago Genetic Resources Centre, SARDI, http://www.sardi.sa.gov.au
	Wild, weedy species	3	
	Wild, weedy species	6	Grassland Research Institute Chinese Academy of Agricultural Sciences, cgi@public.hh.nm.cn
	Advanced cultivar	1	Grassland Research Centre, Department of Agricultural Development
	Landrace	2	http://www.arc.agric.za/
Lespedeza bicolor	Cultivated	2	USDA, ARS, Plant Genetic Resources Conservation Unit
	Wild, weedy species	1	http://www.ars.usda.gov/main/site_main.htm?modecode=66-07-00-00
	Landrace		
L. cyrtobotrya	Unknown	4	USDA, ARS, Plant Genetic Resources Conservation Unit, http://www.ars.usda.gov/main/site_main.htm?modecode=66-07-00-00
L. virgata	Cultivar	1	USDA, ARS, Plant Genetic Resources Conservation Unit
	Unknown	1	http://www.ars.usda.gov/main/site_main.htm?modecode=66-07-00-00
Mucuna pruriens	Landrace	1	USDA, ARS, Plant Genetic Resources Conservation Unit
	Unknown	1	http://www.ars.usda.gov/main/site_main.htm?modecode=66-07-00-00
	Breeding material	6	Australian Tropical Crops & Forages Genetic Resources
	Landrace	10	http://www.dpi.qld.gov.au/auspgris
	Unknown	4	Cenargen/embrapa, http://www.cenargen.embrapa.br
	Landrace	44	Institute of Crop Science (CAAS), http://icgr.caas.net.cn/cgrisngb.html
	Wild, weedy species	22	Centro Internacional de Agricultura Tropical (CIAT), http://www.ciat.cgiar.org
	Wild, weedy species	3	International Livestock Research Institute (ILRI), ILRI-Ethiopia@cgiar.org

(continued)

Table 12.1 Worldwide Genetic Resource Collections of Medicinal Legumes (Continued)

Species	Type	No.	Germplasm Collection Site
	Unknown	55	Centre for Plant Conservation, Bogor Botanic Gardens, unknown
	Breeding material	1	National Dryland Farming Research Station, Kenya, karikat@kari.org
	Unknown	2	National Genebank of Kenya, Crop Plant Genetic Resources, ngbk@wananchi.com
	Landrace	5	
	Wild, weedy species	1	
	Wild, weedy species	1	International Institute of Tropical Agriculture, http://www.cgiar.org/iita/
Ononis spinosa	Wild, weedy species	1	Australian Medicago Genetic Resources Centre, SARDI, http://www.sardi.sa.gov.au
	Wild, weedy species	1	Faculty of Science, Palacky University, http://genbank.vurv.cz/genetic resources/
	Wild, weedy species	2	Federal Centre for Breeding Research on Cultivated Plants, http://www.bafz.de
	Wild, weedy species	1	Seed Bank, Seed Conservation Section Royal Botanic Gardens, Kew, r.smith@rbgkew.org.uk
O. spinosa ssp. spinosa	Unknown	1	USDA, ARS, Western Regional Plant Introduction, http://www.ars-grin.gov/npgs/
Senna alata	Unknown	2	USDA, ARS, Plant Genetic Resources Conservation Unit, http://www.ars.usda.gov/main/site_main.htm?modecode=66-07-00-00
	Landrace	1	Australian Tropical Crops & Forages Genetic Resources
	Wild, weedy species	1	http://www.dpi.qld.gov.au/auspgris
	Wild, weedy species	1	Centro Internacional de Agricultura Tropical (CIAT), http://www.ciat.cgiar.org
	Wild, weedy species	1	Seed Bank, Seed Conservation Section Royal Botanic Gardens, Kew, r.smith@rbgkew.org.uk
Sesbania gradiflora	Cultivar	1	USDA, ARS, Plant Genetic Resources Conservation Unit, http://www.ars.usda.gov/main/site_main.htm?modecode=66-07-00-00
	Wild, weedy species	3	Centro Internacional de Agricultura Tropical (CIAT), http://www.ciat.cgiar.org
	Wild, weedy species	16	International Livestock Research Institute (ILRI), ILRI-Ethiopia@cgiar.org
	Unknown	4	National Genebank of Kenya, Crop Plant Genetic Resources, ngbk@wananchi.com
	Wild, weedy species	1	International Institute of Tropical Agriculture, http://www.cgiar.org/iita/
	Advanced cultivars	8	Institute of Plant Breeding, College of Agriculture UPLB, http://www.uplb.edu.ph/ca/ipb/main/
Sophora flavescens	Unknown	4	USDA, ARS, National Arboretum, http://www.ars-grin.gov/npgs/
Spartium junceum	Unknown	2	USDA, ARS, Miami, FI repository, http://www.ars-grin.gov/npgs/
	Unknown	1	Banco Base Nacional de Germoplasma, Instituto de Recursos Biologicos, INTA, Argentina, http://cirn2.inta.gov.ar
	Wild, weedy species	1	International Livestock Research Institute (ILRI), ILRI-Ethiopia@cgiar.org
	Wild, weedy species	2	Dept. of E.S.E., Institute of Life Science, Hebrew Univ. of Jerusalem, ilanahs@vms.huji.ac.il

Species	Status	No.	Source
	Unknown	1	National Genebank of Kenya, Crop Plant Genetic Resources Centre, KARI, ngbk@wananchi.com
	Wild, weedy species	1	Oddelek za agronomijo University of Ljubljana, Slovenia, zlata.luthar@uni-lj.si
	Wild, weedy species	3	Seed Bank, Seed Conservation Sect. Royal Botanic Gardens, Kew, r.smith@rbgkew.org.uk
Tephrosia purpurea	Unknown	2	USDA, ARS, Plant Genetic Resources Conservation Unit, http://www.ars.usda.gov/main/site_main.htm?modecode=66-07-00-00
	Unknown	29	Australian Tropical Crops & Forages Genetic Resources,
	Wild, weedy species	1	http://www.dpi.qld.gov.au/auspgris
	Wild, weedy species	13	Centro Internacional de Agricultura Tropical (CIAT), http://www.ciat.cgiar.org
	Wild, weedy species	8	International Livestock Research Institute (ILRI), ILRI-Ethiopia@cgiar.org
	Landrace	2	National Genebank of Kenya, Crop Plant Genetic Resources, ngbk@wananchi.com
	Landrace	2	Grassland Research Centre, Department of Agricultural Development, http://www.arc.agric.za/
	Wild, weedy species	9	Seed Bank, Seed Conservation Section Royal Botanic Gardens, Kew, r.smith@rbgkew.org.uk
Thermopsis lanceolata	Unknown	2	USDA, ARS, Western Regional Plant Introduction, http://www.ars-grin.gov/npgs/
Zornia brasiliensis	Unknown	2	USDA, ARS, Plant Genetic Resources Conservation Unit, http://www.ars.usda.gov/main/site_main.htm?modecode=66-07-00-00
	Unknown	3	Australian Tropical Crops & Forages Genetic Resources
	Wild, weedy species	5	http://www.dpi.qld.gov.au/auspgris
	Unknown	1	Agricultural Research Centre for Pasture Species, unknown
	Unknown	9	Cenargen/Embrapa, http://www.cenargen.embrapa.br
	Wild, weedy species	16	Centro Internacional de Agricultura Tropical (CIAT), http://www.ciat.cgiar.org
	Wild, weedy species	2	International Livestock Research Institute (ILRI), ILRI-Ethiopia@cgiar.org
	Landrace	1	National Genebank of Kenya, Crop Plant Genetic Resources, ngbk@wananchi.com
Z. diphylla	Unknown	10	USDA, ARS, Plant Genetic Resources Conservation Unit
	Wild	1	http://www.ars.usda.gov/main/site_main.htm?modecode=66-07-00-00
	Unknown	15	Australian Tropical Crops & Forages Genetic Resources
	Wild, weedy species	1	http://www.dpi.qld.gov.au/auspgris
	Unknown	6	Cenargen/Embrapa, http://www.cenargen.embrapa.br
	Wild, weedy species	16	Centro Internacional de Agricultura Tropical (CIAT), http://www.ciat.cgiar.org
	Wild, weedy species	12	International Livestock Research Institute (ILRI), ILRI-Ethiopia@cgiar.org
	Unknown	1	Institut Penyelidikan dan Kemajuan Pertanian, Malaysia, http://www.mardi.my

Table 12.2 Legumes Used as Medicine

Species	Common Name	Extract or Organ	Potential Use	Ref.
Alhagi graecorum	Alhagi mana	Sugar secretion extract	Purgative, stomach ache, fever reduction	Lev and Amor (2000)
A. maurorum	Cammal thorn	Leaf, flower extract	Diaphoretic, diuretic, expectorant, laxative	Uphof (1959)
Acacia nilotica (L.) Delile	Egyptian acacia	Extract	Antiplasmodial against *Plasmodium falciparum*	El-Tahir, Satti, and Khalid (1999)
		Kaempferol	Antioxidant	Singh et al. (2008)
Amorpha canescens	Leadplant	Gum, leaf, flower extract	Chemopreventive and antimutagenic	Meena et al. (2006)
		Leaf, root, twig extract	Eczema, wounds, stomach pain, rheumatism	Moerman (1998)
A. nana	Fragile false indigo, dwarf indigo bush	Plant	Expectorant	Moerman (1998)
Anthyllis vulneraria L.	Kidney vetch	Extract	Inhibit poliovirus	Suganda et al. (1983)
Arachis hypogaea L.	Peanut	Products	Reduced postprandial glycemia	Johnston and Buller (2005)
		Oil	Controls iodine deficiency	Untoro et al. (2006)
Astragalus canadensis	Canadian milk vetch	Root	Analgesic, antihemorrhagic, chest and back pain	Moerman (1998)
A. chinensis	Hua huang qi	Seed	Kidney diseases	Yeung (1985)
A. complanatus	Bei bian huang qi	Seed	Kidney diseases	Yeung (1985)
A. crassicarpus	Ground plum	Root	Wounds, stimulant	Moerman (1998)
A. exscapus		Plant	Venereal disease	Weiner (1980)
A. floridus	Duo hua huang qi	Whole plant	Diabetes, sores, diarrhea	Tsarong (1994)
A. gummifer	Tragacanth	Root stem	Stimulant, tumor suppression	Bown (1995)
A. hamosus		Whole plant	Mucous membrane irritation	Chopra et al. (1986)
A. hoantchy	Wu la te huang qi	Root	Diuretic, tonic	Stuart (1979)
A. membranaceus	Membranous milk vetch	Root	Immune system, lowers blood pressure	Bown (1995)
A. mongholicus		Root	Cardiotonic, diuretic, vasodilator	Yeung (1985)
A. multiceps		Seed	Colic	Chopra et al. (1986)
A. sinicus	Chinese milk vetch	Plant	Burns	Duke and Ayensu (1985)
Baphia nitida Lodd. et al.	African sandalwood	Extract	Antidiarrhea	Adeyemi and Akindele (2008)
		Leaf extract	Sedative, anxiolytic and skeletal muscle relaxant	Adeyemi, Yemitan, and Taiwo (2006)
		Leaves	Anti-inflammatory	Onwukaeme (1995)
Bauhinia purpurea L.	Butterfly orchid tree	Leaf extract	Anti-inflammatory	Zakaria et al. (2007)
		Root extract	Antimycobacterial, antimalarial, antifungal, anti-inflammatory	Boonphong et al. (2007)
B. variegata L.	Orchid tree	Stem bark extract	Hepatoprotection	Bodakhe and Ram (2007)

Species	Common name	Plant part/compound	Medicinal property	Reference
Bituminaria bituminosa		Erybraedin C	Anticlastogenic	Maurich et al. (2006)
			Antineoplactic on human colon adenocarcinoma cells	Maurich, Pistelli, and Turchi (2004)
Canavalia brasiliensis Mart. ex Benth.	Barbicou bean	Bitucarpin A	Anticlastogenic	
C. ensiformis L. (DC.)	Swordbean	Lectin	Antiadhesion of streptococci on teeth	Teixeira et al. (2006)
Chamaecrista nictitans L. Moench.	Partridge pea	Extract	Antiviral	Herrero Uribe, Chaves Olarte, and Tamayo Castillo (2004)
Clitoria ternatea L.	Butterfly pea	Extract	Antidepressant, anticonvulsant, antistress	Jain et al. (2003)
		Root extract	Antipyretic	Parimaladevi, Boominathan, and Mandal (2004)
		Extract	Mosquito larvicide	Mathew et al. (2008)
Colutea arborescens L.	Bladder senna	Leaf, seed	Diuretic, emetic, purgative	Grieve (1984)
		Isoflavonoids from roots	Antibacterial, antifungal	Erturk (2006)
Daniellia oliveri (Rofe) Hutch. & Dalziel	African copaiba balsalm tree	Leaf extract	Antimicrobial against *Staphylococcus aureus*	Ahmadu et al. (2004)
Desmodium adscendens (Sw.) (DC.)	Tick clover	Extract	Antiasthma, anti-inflammatory	Barreto (2002)
Desmodium gangeticum (L.) DC.	Sarivan	Extract	Free radical scavenging	Kurian et al. (2008)
		Glycosphingolipid	Antileishmanial	Mishra et al. (2005)
		Extract	Improves memory	Joshi and Parle (2006)
		Extract	Antidiabetes	Govindarajan et al. (2007)
Dorycnium rectum (L.) Ser.			Reduces *Trichostrongylus colubriformis* in lambs	Niezen et al. (2002)
Genista hispanica L.	Spanish broom, Spanish gorse	Flowers	Diuretic	Grieve (1984)
G. tinctoria L.	Dyer's broom, greenwood	Leaf, seed, twig	Cathartic, diaphoretic, diuretic, emetic, stimulant	Grieve (1984); Launert (1981); Lust (1983); Uphof (1959)
Gleditsia sinensis Lam.	Zao jia	Lupane acid and derivatives	Anti-HIV	Li et al. (2007)
		Fruit extract	Anticancer	Pak et al. (2009)
Glycine max (L.) Merr.	Soybean	Extract	Breast cancer inhibition	Hyeona, KyuShik, and YooKyeong (2008)
		Extract	Radical scavenger	Chung et al. (2008)

(continued)

Table 12.2 Legumes Used as Medicine (Continued)

Species	Common Name	Extract or Organ	Potential Use	Ref.
		Lunasin from defatted soybean flour	Suppress inflammation	Dia et al. (2009)
	Black soybean	Glyceollin	Prostate cancer preventive	Payton-Stewart et al. (2009)
		Seed coats	Protect against UVB-induced skin aging	Tsoyi et al. (2008)
Glycyrrhiza echinata L.	Chinese licorice	Root	Expectorant, tonic	Stuart (1979)
G. glabra L.	Licorice	Glycyrrhizin	Antiviral	Fiore et al. (2008)
		Glabridin from roots	Antituberculosis	Gupta et al. (2008)
		Spice extract	Radical scavenging	Yadav and Bhatnagar (2007)
G. lepidota Pursh	American licorice	Leaf, root	Cough, diarrhea, chest pains, fever, stomach ache, toothache, sore throats, sores	Coffey (1993); Moerman (1998)
G. uralensis Fisch. ex DC.	Licorice	Glycyrrhizin	Antiprostate cancer	Thirugnanam et al. (2008)
Gymnocladus chinensis Baill.		Seeds	Antiproliferative activity on murine leukemia, a hepatoma, and inhibited HIV-1 reverse transcriptase	Wong and Ng (2003)
		Triterpenoid saponin extracted from fruit	Cancer inhibitor	Ma et al. (2007)
Haematoxylum brasiletto H. Karst.	Peachwood	Extract	Highly active against Staphylococcus aureus	Yasunaka et al. (2005)
Hedysarum polybotrys Hand.-Mazz		Formononetin and calycosin from roots	Antitumor	Li et al. (2008)
Indigofera suffruticosa Mill.	Anil indigo	Leaf extract	Treat skin disease caused by dermatophytes	Leite et al. (2006)
		Leaf extract	Anticancer	Vieira et al. (2007)
Indigofera tinctoria L.	Indigo	Extract	Antioxidant	Bakasso et al. (2008)
		Trans-tetracos-15-enoic acid	Hepatoprotection	Singh et al. (2006)
Kummerowia striata (Thunb.) Schindl.	Annual lespedeza	Extract	Anti-inflammatory	Tao et al. (2008)
Lespedeza bicolor Turcz.	Shrub lespedeza	Extract	Estrogenic	Yoo et al. (2005)
Lespedeza cyrtobotrya		Extract	Estrogenic	Yoo et al. (2005)
		Haginin A	Inhibits hyperpigmentation caused by UV irradiation	Kim et al. (2008)
L. virgata (Thunb.) DC.		Flavonoid	Antioxidant	Tan et al. (2007)

Species	Common name	Part/Extract	Property	Reference
Lonchocarpus sericeus (Poir.) Kunth ex DC.	Savonette	Lectin from seeds	Inhibits inflammatory response and bacterial colonization of infectious peritonitis	Alencar et al. (2005)
Lupinus angustifolius L.	Blue lupine	Kernel fiber	Improvement of bowel function and reduction of colon cancer risk in men	Johnson et al. (2006)
Mucuna pruriens (L.) DC.	Velvet bean		Antioxidant	Dhanasekaran, Tharakan, and Manyam (2008)
		Seed powder	Anti-Parkinson's disease	Katzenschlager et al. (2004)
Ononis repens L.	Common restharrow	Whole plant	Bladder stones, delirium	Grieve (1984)
O. spinosa L.	Restharrow	Extract	Analgesic	Yilmaz et al. (2006)
Peltophorum pterocarpum (DC.) Backer ex K. Heyne	Copper pod	Extract	Antimicrobial	Duraipandiyan, Ayyanar, and Ignacimuthu (2006)
Pterocarpus marsupium Roxb.	East Indian kino		Antidiabetes	Hariharan et al. (2005)
Senna alata (L.) Roxb.	Candlebush	Extract	Antiacne	Chomnawang et al. (2005)
Sesbania grandiflora (L.)	Scarlet wisteria tree	Leaf extract	Protect against erythromycin estolate-induced hepatotoxicity	Pari and Uma (2003)
Sophora flavescens Aiton	Shrubby sophora	Aqueous suspension	Heart protection, antioxidant	Ramesh et al. (2008)
		Kangke injection	Regressed viral myocarditis	Chen, Mei, and Wang (1997)
		Kurorinone	Antihepatitis B	Chen, Guo, and Liu (2000)
		Herbal extract	Antiasthma	Hoang et al. (2007)
Spartium junceum L.	Spanish broom	Flower extract	Analgesic, antiulcer	Menghini et al. (2006)
Tamarindus indica L.	Indian tamarind	Pulp fruit extract	Reduces risk of atherosclerosis	Martinello et al. (2006)
Tephrosia purpurea (L.) Pers.	Purple tephrosia	Aerial parts extract	Wound healing	Lodhi et al. (2006)
Tetrapleura tetraptera (Schumach.) Taub.		Fruit extract	Hepatoprotection	Khatri, Garg, and Agrawal (2008)
			Antiarthritis, anti-inflammatory, antidiabetes	Ojewole and Adewunmi (2004)
			Antiplasmodial	Okokon, Udokpoh, and Antia (2007)
Trifolium pratense L.	Red clover	Biochanin	Lowers LDL cholesterol in males	Nestel et al. (2004)
Trigonella foenum graecum L.	Fenugreek	Seed extract	Analgesic, anti-inflammatory	Vyas et al. (2008)
Zornia brasiliensis Vogel		Extracts	Radical scavenging and antioxidant	David et al. (2007)
Z. diphylla (L.) Pers.		Extracts	Treatment of gastrointestinal disorders	Rojas et al. (1999)

Figure 12.1 *Glycyrrhiza echinata*, PI 477120.

Figure 12.2 (See color insert.) Regeneration nursery, Pullman, Washington.

12.4 CHARACTERIZATION AND EVALUATION

During regeneration of medicinal legume accessions in the field or greenhouse, scientists observe and record characterization data. Characterization is the observation and documentation of plant morphological, phenological, and reproductive traits. The traits measured are plant height, plant width, plant maturity, relative number of branches and relative amount of foliage, winter hardiness, seed quantity, and harvest date. These traits are assigned common terms and scored using agreed-upon conventions. These data provide a variety of information for use in plant development and breeding (NPGS 2009). These traits are usually highly heritable and observed in any environment where the accession is grown.

Additionally, molecular markers are being used to characterize medicinal legume accessions. Genetic markers, including expressed sequence tag-simple sequence repeats (EST-SSR), have been identified in *Lespedeza bicolor*, *L. cyrtobotrya*, *L. virgata*, and *Kummerowia striata*. These markers are used for identifying genetic diversity, clarifying phylogenetic relationships, and correcting misidentified accessions within the genus *Lespedeza* (Wang et al. 2009).

Evaluation data are typically recorded from all accessions of a particular species and can include disease ratings, agronomic traits, or chemical constituents in the plants. Historically, legumes such as *T. pratense* have been evaluated primarily for uses such as soil enrichment and livestock herbage (Taylor, Gibson, and Knight 1977). Many legumes have been studied for various pharmaceutical traits. Seeds of hyacinth bean accessions differed significantly for flavonoid content (Wang et al. 2007). In another study, butterfly pea (Morris and Wang 2007) and horse gram (Morris 2008) accessions differed for anthocyanin content. Numerous phytochemicals in butterfly pea plants have potential for nutraceutical and pharmaceutical uses (Morris 2009).

Both guar (Kays, Morris, and Kim 2006) and sunn hemp (Morris and Kays 2005) accessions varied in total dietary fiber, while guar also varied for soluble dietary fiber. Interestingly, Wang and Morris (2007) also identified several guar accessions that varied in flavonoid content. High-performance liquid chromatography coupled with diode array and evaporative light scattering detectors has been developed to evaluate the quality of *Astragalus membranaceus* (Fisch.) Bunge through a simultaneous determination of six major active isoflavonoids and four main saponins (Lian-Wen et al. 2006).

12.5 UTILIZATION

Table 12.2 lists several legumes that have been used as medicinal remedies. Several medicinal legumes have been tested in clinical trials, including peanut, blue lupin, velvet bean, shrubby sophora, and red clover. In 8- to 10-year-old children, peanut oil was more efficacious in controlling iodine deficiency than iodized poppy seed oil containing similar amounts of iodine (Untoro et al. 2006). Peanut products significantly reduced postprandial glycemia in healthy human subjects (Johnston and Buller 2005). Addition of blue lupin kernel fiber to the human diet improved some markers of healthy bowel function and reduced colon cancer risk in men (Johnson et al. 2006). Velvet bean powder was found to possess advantages over conventional levodopa preparations in the long-term management of Parkinson's disease (Katzenschlager et al. 2004).

An extract from shrubby sophora was proven to be a safe and potentially effective alternative treatment for refractory chronic asthma (Hoang et al. 2007). Interestingly, an effective ingredient extracted from shrubby sophora was also shown to regress the pathologic status of viral myocarditis in humans (Chen, Mei, and Wang 1997). Another extract from shrubby sophora was shown to inhibit hepatitis B virus replication and improve disease remission in patients with chronic hepatitis B (Chen, Guo, and Liu 2000).

Various isoflavones from red clover lowered low density lipoprotein cholesterol in men (Nestel et al. 2004). *Bituminaria bituminosa* contains the phytochemical psoralen with potential use as a pharmaceutical agent. Psoralen combined with ultraviolet A is a widely accepted treatment of cutaneous T-cell lymphomas (Geskin 2007). In fact, a recent clinical trial revealed that this method of treatment resulted in longer remissions of chronic plaque psoriasis (Yones et al. 2006). *Astraglus membranaceus* can be used as an adjuvant chemotherapeutic agent in gastric cancer therapy (Na et al. 2009).

Several legumes contain useful phytochemicals identified through literature surveys with medicinal qualities, including common indigo (*Indigofera tinctoria* L.), wing bean (*Psophocarpus tetragonolobus* (L.) DC.), kudzu (*Pueraria montana* var. *lobata* (Willd.) Maesen & S. Almeida), coffee senna (*Senna occidentalis* (L.) Link), white tephrosia (*Tephrosia candida* DC.), and fish poison bean (*T. vogelii* Hook. f.) (Morris 1997). Kudzu contains both genistein and daidzein and it has recently been demonstrated that dietary supplementation with the isoflavones, genistein, daidzein, and equol caused a decline in the rate of recurrence of prostate cancer (Pendleton et al. 2008).

Positive results have been reported for the immune cell activation in humans by *Glycyrrhiza glabra* (Brush et al. 2006). Recent research has focused on the identification of medicinal traits based on similar literature surveys in various legumes, including *Canavalia ensiformis* (Morris 2007) and *Mucuna pruriens* (Morris, Moore, and Eitzen 2004). Based on these studies, variable anthocyanin indexes have been discovered in *Clitoria ternatea* and *Desmodium adscendens* (Morris and Wang 2007). Anthocyanin indexes are relative values recorded from a modified chlorophyll meter with a 520 nm LED diode, which measures the absorbance near the wavelength at which free anthocyanin aglycones in beans, cyanidin, and pelargonidin monoglucosides absorb (Macz-Pop et al. 2004). The licorice root (*G. glabra*) isolate licochalcone A has been reported to have anticancer and chemopreventive properties (Fu et al. 2004). Soybean consists of numerous phytochemicals with medicinal qualities as well (see Table 12.2 for a brief list).

12.6 CYTOGENETICS

Most of the cytogenetics research conducted on medicinal legumes was aimed at chromosome number determination (Table 12.3). Butterfly pea and velvetbean have been determined to contain diploid chromosome numbers of $2n = 16$ and $2n = 22$, respectively (Frahm-Leliveld 1953). Karyological studies carried out in scarlet wisteria tree resulted in a tetraploid chromosome number of $2n = 4x = 24$. Abou-El-Enain, El-Shazly, and El-Kholy (1998) determined that scarlet wisteria tree belongs to a primitive genus in its tribe, *Robinieae*, from which additional genera may have evolved through aneuploid and polyploidy changes. A tetraploid chromosome number of $2n = 4x = 36$ has been determined in shrubby sophora as well (Kodama 1977). *In vitro* tetraploid induction and generation of tetraploids from mixoploid populations of *Astragalus membranaceus* have been accomplished (Chen and Gao 2006). Earlier reports indicated *Indigofera suffruticosa* had a chromosome count of $2n = 32$. However, Ram Singh (University of Illinois collaborator) recently proved that *I. suffruticosa* actually has $2n = 16$ (Figure 12.3).

12.7 BREEDING

Breeding of medicinal legumes has been limited also; however, some progress has been made in some species, including velvetbean, scarlet wisteria tree, and purple tephrosia. Random amplified polymorphic DNA (RAPD) has been used to estimate velvet bean genetic diversity (Padmesh et al. 2006) as well as genetic relationships in purple tephrosia (Acharya, Mukherjee, and Panda 2004).

Table 12.3 Taxonomic Classification for Medicinal Legumes

Species	Chromosome No. (2n)	Life Form	Distribution
Alhagi graecorum Boiss		P	North Africa, Middle East, Greece
A. maurorum Medik.	8	P	Middle East, Caucasus, central Asia, Mongolia, China, India, Pakistan
Acacia nilotica (L.) Delile	52, 104, 208	P	Algeria, Egypt, Libya, Ethiopia, Somalia, Sudan, Kenya, Tanzania, Uganda, Gambia, Ghana, Guinea-Bissau, Mali, Niger, Nigeria, Senegal, Togo, Malawi, Mozambique, Zambia, Zimbabwe, Botswana, South Africa, Oman, Saudi Arabia, Yemen, Iran, Iraq, Israel, Syria, Bangladesh, India, Nepal, Pakistan, Sri Lanka
Amorpha canescens Pursh	20	P	Canada, United States
A. nana Nutt.	20	P	Canada, United States
Anthyllis vulneraria L.	12	P	North Africa, Ethiopia, Iran, Syria, Turkey, Caucasus, Europe
Arachis hypogaea L.	40	A	Only cultivated
Astragalus canadensis	16	P	Central and eastern N. America, Quebec to Saskatchewan, New York, Louisiana, Nebraska, Utah
A. chinensis	16	P	Eastern Asia, Mongolia, far eastern Russia
A. complanatus	16	P	Eastern Asia, China
A. crassicarpus	22	P	Western N. America, eastern Rocky Mountains and eastward to Nebraska
A. exscapus	16	P	Moldavia, Ukraine
A. floridus	16, 32		Himalayas
A. gummifer	16	P	Iraq, Kurdistan
A. hamosus	8, 22, 32, 42, 48	A	Europe, Mediterranean to Armenia, Ukraine, Caucasus
A. hoantchy	16		China, Manchuria
A. membranaceus	16	P	Siberia, Russian Federation, China, Korea
A. mongholicus	16	P	China, Mongolia
A. multiceps	8	P	Western Himalayas
A. sinicus	16, 32	P	China
Baphia nitida Lodd. et al.	44	P	Cameroon, Equatorial Guinea, Gabon, Benin, Cote D'Ivoire, Ghana, Liberia, Nigeria, Senegal, Sierra Leone, Togo
Bituminaria bituminosa	20	P	Portugal, Spain, Algeria, Libya, Morocco, Tunisia, temperate Asia, Cyprus, Egypt, Israel, Jordan, Lebanon, Syria, Turkey, Georgia, Russian Federation, Ukraine, Albania, Bulgaria, former Yugoslavia, Greece, Italy, Romania, France
Canavalia brasiliensis Mart. ex Benth.	22	A	Florida, Mexico, Belize, Costa Rica, El Salvador, Guatemala, Honduras, Nicaragua, Panama, Antigua, Barbuda, Barbados, Cuba, Haiti, Martinique, St. Vincent, Grenadines, Trinidad, Tobago, U.S. Virgin Islands, Venezuela, Brazil, Colombia, Ecuador, Argentina, Paraguay
C. ensiformis L. (DC.)	22	A	Cultivated and naturalized worldwide, origin from cult. of American Indians

(continued)

Table 12.3 Taxonomic Classification for Medicinal Legumes (Continued)

Species	Chromosome No. (2n)	Life Form	Distribution
Chamaecrista nictitans L. Moench.		P	Connecticut, Indiana, Massachusetts, Michigan, New Hampshire, New York, Ohio, Pennsylvania, Rhode Island, Vermont, West Virginia, Illinois, Kansas, Missouri, Oklahoma, Wisconsin, Alabama, Arkansas, Florida, Georgia, Kentucky, Louisiana, Maryland, Mississippi, North Carolina, South Carolina, Tennessee, Virginia, New Mexico, Texas, Arizona, Mexico, Belize, Costa Rica, El Salvador, Guatemala, Honduras, Nicaragua, Panama, Aruba, Bahamas, Cuba, Grenada, Guadeloupe, Hispaniola, Jamaica, Martinique, the Netherlands Antilles, Puerto Rico, St. Kitts, Nevis, St. Vincent, Grenadines, Trinidad, Tobago, British Virgin Islands, U.S. Virgin Islands, French Guiana, Guyana, Suriname, Venezuela, Brazil, Bolivia, Colombia, Ecuador, Peru, Argentina, Paraguay
Clitoria ternatea L.	16	A	Paleotropics, widely cultivated, native range obscure
Colutea arborescens L.	16	P	Algeria, Morocco, Europe
Daniellia oliveri (Rofe) Hutch. & Dalziel		P	Chad, Sudan, Uganda, Cameroon, Central African Republic, Zaire, Benin, Cote D'Ivoire, Gambia, Ghana, Guinea, Guinea-Bissau, Mali, Nigeria, Senegal, Sierra Leone, Togo
Desmodium adscendens (Sw.) (DC.)		A	Ethiopia, Kenya, Tanzania, Uganda, Burundi, Cameroon, Central African Republic, Equatorial Guinea, Gabon, Rwanda, Sao Tome, Principe, Zaire, Cote D'Ivoire, Ghana, Guinea, Guinea-Bissau, Liberia, Nigeria, Senegal, Sierra Leone, Togo, Angola, Malawi, Mozambique, Zambia, Zimbabwe, South Africa, Transvaal, Swaziland, Madagascar, India, Sri Lanka, Thailand, Malaysia, Papua New Guinea, Mexico, Belize, Costa Rica, Guatemala, Honduras, Nicaragua, Panama, Antigua, Barbuda, Cuba, Dominica, Grenada, Guadeloupe, Hispaniola, Jamaica, Martinique, Montserrat, Puerto Rico, St. Kitts, Nevis, St. Lucia, St. Vincent, Grenadines, French Guiana, Guyana, Suriname, Venezuela, Brazil, Bolivia, Colombia, Ecuador, Peru
D. gangeticum (L.) DC.		A	Tropical Africa, China, Japan, Taiwan, Bhutan, India, Nepal, Pakistan, Sri Lanka, Cambodia, Laos, Myanmar, Thailand, Vietnam, Malaysia, Papua New Guinea, Philippines, Australia
Dorycnium rectum (L.) Ser.		P	Algeria, Libya, Morocco, Tunisia, Cyprus, Israel, Lebanon, Syria, Turkey, Albania, Greece, Crete, Italy, Sardinia, Sicily, France, Portugal, Spain
Genista hispanica L.	36	P	France, Spain
G. tinctoria L.	48	P	Afghanistan, Iran, Turkey, Russian Federation, Kazakhstan, Europe
Glycyrrhiza echinata L.	16	P	Iran, Turkey, Armenia, Azerbaijan, Georgia, Russian Federation, Kazakhstan, Hungary, Moldova, Ukraine, Bulgaria, former Yugoslavia, Greece, Italy
G. glabra L.	16	P	North Africa, central Asia, Turkey, Russian Federation, Mongolia, India, Pakistan, eastern and southern Europe
G. inflata Batalin	16	P	Kazakhstan, Kyrgyzstan, Tajikistan, Turkmenistan, Uzbekistan, Mongolia, China
G. lepidota Pursh	16	P	Canada, United States
G. uralensis Fisch. ex DC.	16	P	Afghanistan, Russian Federation, Kazakhstan, Kyrgyzstan, Tajikistan, Mongolia, China, Pakistan
Haematoxylum brasiletto H. Karst.		P	Mexico, Belize, Costa Rica, Guatemala, Honduras, Nicaragua, Colombia
Hedysarum neglectum Ledeb.		P	Russian Federation, Kazakhstan, Kyrgyzstan, Mongolia

Species	No.	P/A	Distribution
H. polybotrys Hand.-Mazz.	14	P	China
Indigofera suffruticosa Mill.	16,32	P	Florida, Louisiana, Mississippi, Texas, Mexico, Belize, Costa Rica, El Salvador, Guatemala, Honduras, Nicaragua, Panama, Anguilla, Antigua, Barbuda, Bahamas, Barbados, Cayman Islands, Cuba, Dominica, Dominican Republic, Grenada, Guadeloupe, Haiti, Jamaica, Martinique, Montserrat, Puerto Rico, St. Lucia, St. Vincent, Grenadines, British Virgin Islands, U.S. Virgin Islands, French Guiana, Guyana, Suriname, Venezuela, Brazil, Bolivia, Colombia, Ecuador, Peru, Argentina, Paraguay
I. tinctoria L.	16	A	Cape Verde, Chad, Ethiopia, Somalia, Sudan, Yemen, Kenya, Tanzania, Uganda, Cameroon, Central African Republic, Gabon, Sao Tome, Principe, Benin, Cote D'Ivoire, Gambia, Ghana, Guinea, Guinea-Bissau, Mali, Niger, Nigeria, Senegal, Togo, Angola, Malawi, Mozambique, Zambia, Zimbabwe, Botswana, Madagascar, Yemen, China, Taiwan, Bangladesh, India, Pakistan, Sri Lanka, Maldives, Cambodia, Myanmar, Thailand, Vietnam, Indonesia, Malaysia, Papua New Guinea, Australia
Kummerowia striata (Thunb.) Schindl.		A	Russian Federation, China, Japan, Korea, Taiwan, Vietnam
Lespedeza bicolor Turcz.	18, 20, 22	P	Russian Federation, Mongolia, China, Japan, Korea
L. cyrtobotrya Miq.	22	P	Russian Federation, China, Japan, Korea
L. virgata (Thunb.) DC.		P	China, Japan, Korea, Taiwan
Mucuna pruriens (L.) DC.	22	A	Chad, Ethiopia, Somalia, Sudan, Kenya, Tanzania, Uganda, Burundi, Cameroon, Central African Republic, Equatorial Guinea, Sao Tome, Principe, Zaire, Ghana, Guinea, Guinea-Bissau, Liberia, Nigeria, Senegal, Sierra Leone, Togo, Angola, Malawi, Zambia, Zimbabwe, South Africa, Madagascar, Bangladesh, Bhutan, India, Nepal, Sri Lanka, Cambodia, Myanmar, Thailand, Vietnam, Indonesia, Malaysia, Papua New Guinea, Philippines
Ononis repens L.	30	P	Europe
O. spinosa L.	30	P	Algeria, Libya, Morocco, Tunisia, Afghanistan, Iran, Iraq, Israel, Jordan, Lebanon, Syria, Turkey, India, Pakistan, Denmark, Norway, Sweden, United Kingdom, Austria, Belgium, Czechoslovakia, Germany, Hungary, Netherlands, Poland, Switzerland, Lithuania, Russian Federation, Ukraine, Albania, Bulgaria, former Yugoslavia, Greece, Crete, Italy, Romania, France, Corsica, Portugal, Spain
O. spinosa L. ssp. spinosa	30, 32	P	Europe
Senna alata (L.) Roxb.	28	A	French Guiana, Guyana, Suriname, Venezuela, Brazil, Colombia

(continued)

Table 12.3 Taxonomic Classification for Medicinal Legumes (Continued)

Species	Chromosome No. (2n)	Life Form	Distribution
Sesbania grandiflora (L.) Pers	14	A	Indonesia
Sophora flavescens Aiton	36	P	Russian Federation, Mongolia, China, Japan, Korea, Taiwan
Spartium junceum L.		P	Portugal, Algeria, Libya, Morocco, Tunisia, Israel, Lebanon, Syria, Turkey, Armenia, Azerbaijan, Georgia, Albania, former Yugoslavia, Greece, Crete, Italy, Sardinia, Sicily, France, Portugal, Spain
Tephrosia purpurea (L.) Pers.	22, 44	A	Algeria, Egypt, Chad, Djibouti, Ethiopia, Somalia, Sudan, Yemen, Kenya, Tanzania, Uganda, Burundi, Cameroon, Central African Republic, Zaire, Benin, Cote D'Ivoire, Gambia, Ghana, Guinea-Bissau, Mali, Mauritania, Niger, Nigeria, Senegal, Togo, Angola, Malawi, Mozambique, Zimbabwe, Botswana, Namibia, South Africa, Transvaal, Madagascar, Mauritius, Seychelles, Oman, Saudi Arabia, Yemen, Egypt, Iran, China, Taiwan, Bhutan, India, Nepal, Pakistan, Sri Lanka, Indonesia, Western Australia
Tetrapleura tetraptera (Schumach.) Taub.		P	Sudan, Kenya, Tanzania, Uganda, Cameroon, Central African Republic, Gabon, Zaire, Cote D'Ivoire, Ghana, Guinea-Bissau, Liberia, Nigeria, Senegal, Sierra Leone, Togo, Angola
Thermopsis lanceolata R. Br.	18	P	Russian Federation, Kazakhstan, Mongolia, China, Nepal
Trifolium pretense L.	14	P	Algeria, Morocco, Tunisia, Afghanistan, Cyprus, Iran, Iraq, Lebanon, Turkey, Armenia, Azerbaijan, Georgia, Russian Federation, Siberia, Kazakhstan, Kyrgyzstan, Tajikistan, Turkmenistan, India, Pakistan, Denmark, Finland, Ireland, Norway, Sweden, United Kingdom, Austria, Belgium, Czechoslovakia, Germany, Hungary, Netherlands, Poland, Switzerland, Albania, Bulgaria, former Yugoslavia, Greece, Italy, Sardinia, Sicily, Romania, France, Portugal, Spain
Zornia brasiliensis Vogel		A	Venezuela, Brazil
Z. diphylla (L.) Pers.		A	India, Sri Lanka

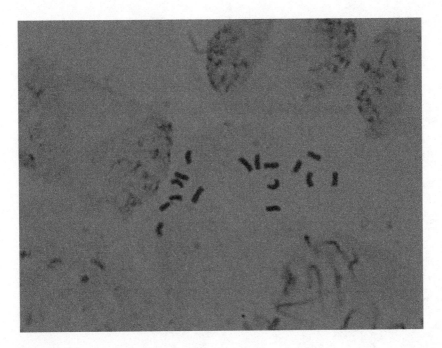

Figure 12.3 *Indigofera suffruticosa*, PI 331110 chromosomes.

Cluster analysis indicated two major clusters of accessions according to their geographic locations, and levodopa content was uniformly distributed among accessions. Velvet bean breeding for optimum levodopa and seed yield has been conducted with favorable results (Krishnamurthy et al. 2002).

Scarlet wisteria tree has been characterized for genetic variability using isozyme analysis (Veasey et al. 2002), where a high incidence of intraspecific monomorphism has been detected. *Astragalus membranaceus* and *A. mongholicus* have been screened for morphological differences, growth habit, physiological characteristics, and powdery mildew resistance for identifying superior genotypes for use in breeding programs (Cao et al. 2008).

12.8 CELL AND TISSUE CULTURE

In vitro research has been conducted on a number of medicinal legume species both for developing cultural techniques and enhancing phytochemical production. Effective and efficient shoot regeneration protocols for butterfly pea establishment have been developed (Shahzad, Faisal, and Anis 2007; Barik et al. 2007). Cryopreservation of butterfly pea embryogenic callus for 8 weeks has been accomplished (Malabadi and Nataraja 2004). Progress has occurred for encapsulated butterfly pea somatic embryo storage (Malabadi and Nataraja 2002). An optimum micropropagation system was developed for velvet bean with normal appearance when plantlets were transplanted to plastic pots containing potting soil (Faisal, Siddique, and Anis 2006). In fact, successful cell suspension cultures of velvet bean accumulating up to 6% levodopa (dry weight) have been accomplished (Pras et al. 1993).

The dynamics for a cell suspension culture system in *Glycyrrhiza inflata* producing flavonoids has been reported (Yang, He, and Yu 2008). Phenolic compounds have been successfully produced from *G. glabra* hairy root cultures (Toivonen and Rosenqvist 1995). Shoot cultures of *Genista tinctoria* produced high concentrations of the phytoestrogens, genistin malonate, and genistin acetate (Luczkiewicz and Glod 2005). In addition, *G. tinctoria* (Figure 12.4) coculture of shoots and hairy

Figure 12.4   *Genista tinctoria*, Ames 25524.

roots produced large amounts of isoflavone phytoestrogens (Luczkiewicz and Kokotkiewicz 2005). Callus cultures of *G. tinctoria* produced high isoflavone content as well (Luczkiewicz and Glod 2003). A clinical trial has been conducted for the phenol licochalcone A found in *Glycyrrhiza inflata*. *In vitro* research proved that licochalcone A has therapeutic skin care benefits when applied to sensitive or irritated skin (Kolbe et al. 2006). Several flavonoid constituents have been isolated and identified from hairy root cultures of *G. glabra* (Wei, Asada, and Yoshikawa 2000).

12.9 SUMMARY

Many legumes contain useful phytochemicals that can provide numerous medicinal uses for humans. The USDA, ARS, National Plant Germplasm System conserves a broad range of leguminous species that offer considerable opportunities for the study of medicinal properties. These legumes have been characterized for their cultivation as well as optimum plant and seed regeneration, and seed storage requirements have been determined. While these characterizations provide breeders with new germplasm for cultivar development, they also offer a new realm of possibilities in the area of medicinal properties from these species.

These legumes have been studied for medicinal purposes, and some have actually entered the market as supplements or nutraceuticals. Cytogenetic studies have been conducted on only a few of these species, opening up another area in need of research consideration. Few breeding studies have been conducted on these species outside of velvet bean and scarlet wisteria tree. Screening studies have resulted in the agronomic characterizations and some disease resistance and more evaluation is needed. Cell and tissue culture has successfully been used to produce flavonoids in licorice species adequately but, again, more research is needed to encompass the entire group of legumes discussed here. Most of these underutilized legumes require additional agronomic, botanical, and cytogenetic screening of valuable compounds, as well as other research to sample adequately the broad diversity of medicinal phytochemicals found in these species.

REFERENCES

Abou-El-Enain, M. M., H. H. El-Shazly, and M. A. El-Kholy. 1998. Karyological studies in some African species of the genus *Sesbania* (Fabaceae). *Cytologia* 63:1–8.

Acharya, L., A. K. Mukherjee, and P. C. Panda. 2004. Genome relationship among nine species of millettieae (Leguminosae: Papilionoideae) based on random amplified polymorphic DNA (RAPD). *Journal of Biosciences* 59:868–873.

Adeyemi, O. O., and A. J. Akindele. 2008. Antidiarrheal activity of the ethyl acetate extract of *Baphia nitida* (Papilionaceae). *Journal of Ethnopharmacology* 116:407–412.

Adeyemi, O. O., O. K. Yemitan, and A. E. Taiwo. 2006. Neurosedative and muscle-relaxant activities of ethyl acetate extract of *Baphia nitida* AFZEL. *Journal of Ethnopharmacology* 106:312–316.

Ahmadu, A., A. K. Haruna, M. Garba, J. O. Ehinmidu, and S. D. Sarker. 2004. Phytochemical and antimicrobial activities of the *Daniellia oliveri* leaves. *Fitoterapia* 75:729–732.

Alencar, N. M., C. F. Cavalcante, M. P. Vasconcelos, K. B. Leite, K. S. Aragao, A. M. Assreuy, N. A. Nogueira, B. S. Cavada, and M. R. Vale. 2005. Anti-inflammatory and antimicrobial effect of lectin from *Lonchocarpus sericeus* seeds in an experimental rat model of infectious peritonitis. *Journal of Pharmacy and Pharmacology* 57:919–922.

Association of Official Seed Analysts. 2000. Rules for testing seeds. AOSA.

Bakasso, S., A. Lamien-Meda, C. E. Lamien, M. Kiendrebeogo, J. Millogo, A. G. Ouedraogo, and O. G. Nacoulma. 2008. Polyphenol contents and antioxidant activities of five *Indigofera* species (Fabaceae) from Burkina Faso. *Pakistan Journal of Biological Sciences* 11:1429–1435.

Balandrin, M. F., J. A. Klocke, E. S. Wurtele, and W. H. Bollinger. 1985. Natural plant chemicals: Sources of industrial and medicinal materials. *Science* 228:1154–1160.

Barik, D. P., S. K. Naik, A. Mudgal, and P. K. Chand. 2007. Rapid plant regeneration through *in vitro* axillary shoot proliferation of butterfly pea (*Clitoria ternatea* L.)—A twinning legume. *In Vitro Cellular and Developmental Biology—Plant* 43:144–148.

Barreto, G. S. 2002. Effect of butanolic fraction of *Desmodium adscendens* on the anococcygeus of the rat. *Brazilian Journal of Biology* 62:223–230.

Bodakhe, S. H., and A. Ram. 2007. Hepatoprotective properties of *Bauhinia* variegate bark extract. *Yakugaku Zasshi* 127:1503–1507.

Boonphong, S., P. Puangsombat, A. Baramee, C. Mahidol, and S. Ruchirawat. 2007. Bioactive compounds from *Bauhinia purpurea* possessing antimalarial, anti-mycobacterial, antifungal, anti-inflammatory and cytotoxic activities. *Journal of Natural Products* 70:795–801.

Bown, D. 1995. *Encyclopedia of herbs and their uses.* London: Dorling Kindersley.

Brush, J., E. Mendenhall, A. Guggenheim, T. Chan, E. Connelly, A. Soumyanath, R. Buresh, R. Barrett, and H. Zwickey. 2006. The effect of *Echinacea purpurea, Astragalus membranaceus* and *Glycyrrhiza glabra* on CD69 expression and immune cell activation in humans. *Phytotherapy Research* 20:687–695.

Cao, J. J., Z. S. Liang, W. L. Wang, and Q. M. Duan. 2008. Study on difference of biological characteristics and resistance to powdery mildew of different *Astragalus* populations. *Zhongguo Zhong Yao Za Zhi* 9:992–996.

Chen, C., S. M. Guo, and B. Liu. 2000. A randomized controlled trial of kurorinone versus interferon-alpha2a treatment in patients with chronic hepatitis B. *Journal of Viral Hepatitis* 7:225–229.

Chen, L. L., and S. L. Gao. 2006. *In vitro* tetrapolid induction and generation of tetraploids from mixoploids in *Astragalus membranaceus. Scientia Horticulturae* 3:339–344.

Chen, S. X., S. W. Mei, and P. Q. Wang. 1997. Therapeutic effect of kangke injection on viral myocarditis and its anticoxsackie virus mechanism. *Zhongguo Zhong Xi Yi Jie He Za Zhi* 7:207–209.

Chomnawang, M. T., S. Surassmo, V. S. Nukoolkarn, and W. Gritsanapan. 2005. Antimicrobial effects of Thai medicinal plants against acne-inducing bacteria. *Journal of Ethnopharmacology* 101:330–333.

Chopra, R. N. et al. 1986. *Glossary of Indian medicinal plants (including the supplement).* New Delhi: Council of Scientific and Industrial Research.

Chung, H., S. Hogan, L. Zhang, K. Rainey, and K. Q. Zhou. 2008. Characterization and comparison of antioxidant properties and bioactive components of Virginia soybeans. *Journal of Agricultural and Food Chemistry* 56:11515–11519.

Coffey, T. 1993. *The history and folklore of North American wild flowers.* New York: Facts on File.

David, J. P., M. Meira, J. M. David, H. N. Brandao, A. Branco, A. M. de Fatima, M. R. Barbosa, L. P. de Queiroz, and A. M. Giulietti. 2007. Radical scavenging, antioxidant and cytotoxic activity of Brazillian caatinga plants. *Fitoterapia* 78:215–218.

Dhanasekaran, M., B. Tharakan, and B. V. Manyam. 2008. Antiparkinson drug—*Mucuna puriens* shows antioxidant and metal chelating activity. *Phytotherapy Research* 22:6–11.

Dia, V. P., W. Wang, V. L. Oh, B. O. Lumen, and E. G. Mejia. 2009. Isolation, purification and characterization of lunasin from defatted soybean flour and *in vitro* evaluation of its anti-inflammatory activity. *Food Chemistry* 114:108–115.

Doyle, J. J. 1994. Phylogeny of the legume family: An approach to understanding the origins of nodulation. *Annual Review of Ecology and Systematics* 25:325–349.

Doyle, J. J., and M. A. Luckow. 2003. The rest of the iceberg. Legume diversity and evolution in a phylogenetic context. *Plant Physiology* 131:900–910.

Duke, J. A., and E. S. Ayensu. 1985. *Medicinal plants of China.* Algonac, MI: Reference Publications, Inc.

Duraipandiyan, V., M. Ayyanar, and S. Ignacimuthu. 2006. Antimicrobial activity of some ethnomedicinal plants used by Paliyar tribe from Tamil Nadu, India. *BMC Complementary and Alternative Medicine* 6:35.

El-Tahir, A., G. M. Satti, and S. A. Khalid. 1999. Antiplasmodial activity of selected Sudanese medicinal plants with emphasis on *Acacia nilotica*. *Phytotherapy Research* 13:474–478.

Erturk, O. 2006. Antibacterial and antifungal activity of ethanolic extracts from eleven spice plants. *Biologia* 61:275–278.

Faisal, M., I. Siddique, and M. Anis. 2006. An efficient plant regeneration system for *Mucuna pruriens* L. (DC.) using cotyledonary node explants. *In Vitro Cellular and Developmental Biology—Plant* 42:59–64.

Fiore, C., M. Eisenhut, R. Krausse, E. Ragazzi, D. Pellati, D. Armanini, and J. Bielenberg. 2008. Antiviral effects of *Glycyrrhiza* species. *Phytotherapy Research* 22:141–148.

Frahm-Leliveld, J. A. 1953. Some chromosome numbers in tropical leguminous plants. *Euphytica* 2:46–48.

Fu, Y., T. C. Hsieh, J. Guo, J. Kunicki, M. Y. W. T. Lee, Z. Darzynkiewicz, and J. M. Wu. 2004. Licochalcone-A, a novel flavonoid isolated from licorice root (*Glycyrrhizaglabra*), causes G2 and late-G1 arrests in androgen-independent PC-3 prostate cancer cells. *Biochemical and Biophysical Research Communications* 322:263–270.

Geskin, L. 2007. ECP versus PUVA for the treatment of cutaneous T-cell lymphoma. *Skin Therapy Letters* 12:1–4.

Govindarajan, R., H. Asare-Anane, S. Persaud, P. Jones, and P. J. Houghton. 2007. Effect of *Desmodium gangeticum* extract on blood glucose in rats and on insulin secretion *in vitro*. *Planta Medica* 73:427–432.

Grieve, M. 1984. *A modern herbal.* New York: Penguin Books.

Gupta, V. K., A. Fatima, U. Faridi, A. S. Negi, K. Shanker, J. K. Kumar, N. Rahuja, S. Luqman, B. S. Sisodia, D. Saikia, et al. 2008. Antimicrobial potential of *Glycyrrhiza glabra* roots. *Journal of Ethnopharmacology* 116:377–380.

Hariharan, R. S., S. Venkataraman, P. Sunitha, S. Rajalakshmi, K. C. Samal, B. M. Routray, R. V. Jayakumar, K. Baiju, G. V. Satyavati, V. Muthuswamy, et al. 2005. Efficacy of vijayasar (*Pterocarpus marsupium*) in the treatment of newly diagnosed patients with type 2 diabetes mellitus: A flexible dose double-blind multicenter randomized controlled trial. *Diabetologia Croatica* 34:13–20.

Herrero Uribe, L., E. Chaves Olarte, and G. Tamayo Castillo. 2004. *In vitro* antiviral activity of *Chamaecrista nictitans* (Fabaceae) against herpes simplex virus: Biological characterization of mechanisms of action. *Revista de Biologia Tropical* 52:807–816.

Hoang, B. X., D. G. Shaw, S. Levine, C. Hoang, and P. Pham. 2007. New approach in asthma treatment using excitatory modulator. *Phytotherapy Research* 21:554–557.

Hyeona, K, J. KyuShik, and K. YooKyeong. 2008. Soy extract is more potent than genistein on tumor growth inhibition. *Anticancer Research* 28:2837–2842.

Jain, N. N., C. C. Ohal, S. K. Shroff, R. H. Bhutada, R. S. Somani, V. S. Kasture, and S. B. Kasture. 2003. *Clitoria ternatea* and the CNS. *Pharmacology Biochemistry and Behavior* 75:529–536.

Johnson, S. K., V. Chua, R. S. Hall, and A. L. Baxter. 2006. Lupin kernel fiber foods improve bowel function and beneficially modify some putative fecal risk factors for colon cancer in men. *British Journal of Nutrition* 95:372–378.

Johnston, C. S., and A. J. Buller. 2005. Vinegar and peanut products as complementary foods to reduce postprandial glycemia. *Journal of the American Dietetic Association* 105:1939–1942.

Joshi, H., and M. Parle. 2006. Antiamnesic effects of *Desmodium gangeticum* in mice. *Yakugaku Zasshi* 126:795–804.

Katzenschlager, R., A. Evans, A. Manson, P. N. Patsalos, N. Ratnaraj, H. Watt, L. Timmermann, R. Van der Giessen, and A. J. Lees. 2004. *Mucuna pruriens* in Parkinson's disease: A double blind clinical and pharmacological study. *Journal of Neurology, Neurosurgery and Psychiatry* 75:1672–1677.

Kays, S. E., J. B. Morris, and Y. Kim. 2006. Total and soluble dietary fiber variation in *Cyamopsis tetragonoloba* (L.) Taub. genotypes. *Journal of Food Quality* 29:383–392.

Khatri, A., A. Garg, and S. S. Agrawal. 2008. Evaluation of hepatoprotective activity of aerial parts of *Tephrosia purpurea* L. and stem bark of *Tecomella undulate*. *Journal of Ethnopharmacology* Nov. 18. (Epub ahead of print).

Kim, J. H., S. H. Baek, D. H. Kim, T. Y. Choi, T. J. Yoon, J. S. Hwang, M. R. Kim, H. J. Kwon, and C. H. Lee. 2008. Downregulation of melanin synthesis by haginin A and its application to *in vivo* lightening model. *Journal of Investigative Dermatology* 128:1227–1235.

Kodama, A. 1977. Karyological and morphological observations on root nodules of some woody and herbaceous leguminous plants. *Bulletin of the Hiroshima Agricultural College* 5:389–394.

Kolbe L., J. Immeyer, J. Batzer, U. Wensorra, K. tom Dieck, C. Mundt, R. Wolber, F. Stab, U. Schonrock, R. I. Ceilley, and H. Wenck. 2006. Anti-inflammatory efficacy of licochalcone A: Correlation of clinical potency and *in vitro* effects. *Archives of Dermatological Research* 298:23–30.

Krishnamurthy, R., M. S. Chandorkar, M. R. Palsuledesai, J. M. Pathak, and R. Gupta. 2002. Breeding in velvetbean (*Mucuna pruriens*) for improvement in seed yield and quality traits. *Indian Journal of Agricultural Sciences* 72:709–715.

Kurian, G. A., N. Yagnesh, R. S. Kishan, and J. Paddikkala. 2008. Methanol extract of *Desmodium gangeticum* roots preserves mitochondrial respiratory enzymes, protecting rat heart against oxidative stress induced by reperfusion injury. *Journal of Pharmacy and Pharmacology* 60:523–530.

Launert, E. 1981. *Edible and medicinal plants*. London: Hamlyn.

Leite, S. P., J. R. Vieira, P. L. de Medeiros, R. M. Leite, V. L. de Menezes Lima, H. S. Xavier, and E. de Oliveira Lima. 2006. Antimicrobial activity of *Indigofera suffruticosa*. *Evidence-Based Complementary Alternative Medicine* 3:261–265.

Lev, E., and Z. Amor. 2000. Ethnopharmacological survey of traditional drugs sold in Israel at the end of the 20th century. *Journal of Ethnopharmacology* 72:191–205.

Li, S., D. Wang, W. Tian, X. Wang, J. Zhao, Z. Liu, and R. Chen. 2008. Characterization and anti-tumor activity of a polysaccharide from *Hedysarum polybotrys* Hand.-Mazz. *Carbohydrate Polymers* 73:344–350.

Li, W. H., X. M. Zhang, R. R. Tian, Y. T. Zheng, W. M. Zhao, and M. H. Oiu. 2007. A new anti-HIV lupine acid from *Gleditsia sinensis* Lam. *Journal of Asian Natural Products Research* 9:551–555.

Lian-Wen, Q., Y. Qing-Tao, L. Ping, L. Song-Lin, W. Yu-Xia, S. Ling-Hong, and Y. Ling. 2006. Quality evaluation of Radix Astragali through a simultaneous determination of six major active isoflavonoids and four main saponins by high-performance liquid chromatography coupled with diode array and evaporative light scattering detectors. *Journal of Chromatography* 2:162–169.

Lodhi, S., R. S. Pawar, A. P. Jain, and A. K. Singhai. 2006. Wound healing potential of *Tephrosia purpurea* (Linn.) Pers. in rats. *Journal of Ethnopharmacology* 108:204–210.

Luczkiewicz, M., and D. Glod. 2003. Callus cultures of *Genista* plants—*In vitro* material producing high amounts of isoflavones of phytoestrogenic activity. *Plant Science* 165:1101–1108.

———. 2005. Morphogenesis-dependent accumulation of phytoestrogens in *Genista tinctoria in vitro* cultures. *Plant Science* 168:967–979.

Luczkiewicz, M., and A. Kokotkiewicz. 2005. Co-cultures of shoots and hairy roots of *Genista tinctoria* L. for synthesis and biotransformation of large amounts of phytoestrogens. *Plant Science* 169:862–871.

Lust, J. 1983. *The herb book*. New York: Bantam Books.

Ma, Y. X., H. Z. Fu, M. Li, W. Sun, B. Xu, and J. R. Cui. 2007. An anticancer effect of a new saponin component from *Gymnocladus chinensis* Baillon through inactivation of nuclear factor kappa B. *Anticancer Drugs* 18:41–46.

Macz-Pop, G. A., J. C. Rivas-Gonzalo, J. J. Perez-Alonso, and A. M. Gonzalez-Paramas. 2004. Natural occurrence of free anthocyanin aglycones in beans (*Phaseolus vulgaris* L.). *Food Chemistry* 94:448–456.

Malabadi, R. B., and K. Nataraja. 2002. *In vitro* storage of synthetic seeds in *Clitoria ternatea* Linn. *Phytomorphology: An International Journal of Plant Morphology* 52:231–237.

———. 2004. Cryopreservation and plant regeneration via somatic embryogenesis in *Clitoria ternatea* Linn. *Phytomorphology: An International Journal of Plant Morphology* 54:7–17.

Martinello, F., S. M. Soares, J. J. Franco, A. C. Santos, A. Sugohara, S. B. Garcia, C. Curti, and S. A. Uyemura. 2006. Hypolipemic and antioxidant activities from *Tamarindus indica* L. pulp fruit extract in hypercholesterolemic hamsters. *Food and Chemical Toxicology* 44:810–818.

Mathew, N., M. G. Anitha, T. S. Bala, S. M. Sivakumar, R. Narmadha, and M. Kalyanasundaram. 2008. Larvicidal activity of *Saraca indica*, *Nyctanthes arbor-tristis*, and *Clitoria ternatea* extracts against three mosquito vector species. *Parasitology Research* Nov. 28. (Epub ahead of print).

Maurich, T., L. Pistelli, and G. Turchi. 2004. Anticlastogenic activity of two structurally related pterocarpans purified from *Bituminaria bituminosa* in cultured human lymphocytes. *Mutation Research* 56:75–81.

Maurich, T. et al. 2006. Erybraedin C and bitucarpin A, two structurally related pterocarpans purified from *Bituminaria bituminosa*, induced apoptosis in human colon adenocarcinoma cell lines MMR- and p53-proficient and -deficient in a dose-, time-, and structure-dependent fashion. *Chemico-Biological Interactions* 159:104–116.

Meena, P. D., P. Kaushik, S. Shukla, A. K. Soni, M. Kumar, and A. Kumar. 2006. Anticancer and antimutagenic properties of *Acacia nilotica* (Linn.) on 7,12-dimethylbenz(a)anthracene-induced skin papillomagenesis in Swiss albino mice. *Asian Pacific Journal of Cancer Prevention* 7:627–632.

Menghini, L., P. Massarelli, G. Bruni, and R. Pagiotti. 2006. Anti-inflammatory and analgesic effects of *Spartium junceum* L. flower extracts: A preliminary study. *Journal of Medicinal Food* 9:386–390.

Mishra, P. K., N. Singh, G. Ahmad, A. Dube, and R. Maurya. 2005. Glycolipids and other constituents from *Desmodium gangeticum* with antileishmanial and immunomodulatory activities. *Bioorganic & Medicinal Chemistry Letters* 15:4543–4546.

Moerman, D. 1998. *Native American ethnobotany.* Portland, Oregon: Timber Press.

Morris, J. B. 1997. Special-purpose legume genetic resources conserved for agricultural, industrial, and pharmaceutical use. *Economic Botany* 51:251–263.

———. 2003. Bio-functional legumes with nutraceutical, pharmaceutical, and industrial uses. *Economic Botany* 57:254–261.

———. 2007. Swordbean (*Canavalia ensiformis* (L.) DC.) genetic resources regenerated for potential medical, nutraceutical and agricultural traits. *Genetic Resources and Crop Evolution* 54:585–592.

———. 2008. *Macrotyloma axillare* and *M. uniflorum*: Descriptor analysis, anthocyanin indexes, and potential uses. *Genetic Resources and Crop Evolution* 55:5–8.

———. 2009. Characterization of butterfly pea (*Clitoria ternatea* L.) accessions for morphology, phenology, reproduction and potential nutraceutical, pharmaceutical trait utilization. *Genetic Resources and Crop Evolution* DOI 10.1007/s10722-008-9376-0.

Morris, J. B., and S. E. Kays. 2005. Total dietary fiber variability in a cross section of *Crotalaria juncea* genetic resources. *Crop Science* 45:1826–1829.

Morris, J. B., K. M. Moore, and J. B. Eitzen. 2004. Nutraceuticals and potential sources of phytopharmaceuticals from guar and velvetbean genetic resources regenerated in Georgia, USA. In *Current topics in phytochemistry*, vol. 6, 125–130. Trivandrum, India: Research Trends.

Morris, J. B., and M. L. Wang. 2007. Anthocyanin and potential therapeutic traits in *Clitoria*, *Desmodium*, *Corchorus*, *Catharanthus* and *Hibiscus* species. In *Medicinal and nutraceutical plants*, ed. A. K. Yadav, 381–388. *Proceedings of the International Symposium*. ACTA Horticulturae, Ft. Valley, GA., Mar. 19–23, 2007.

Na, D., F. N. Liu, Z. F. Miao, Z. M. Du, and H. M. Xu. 2009. *Astragalus* extract inhibits destruction of gastric cancer cells to mesothelial cells by anti-apoptosis. *World Journal of Gastroenterology* 15:570–577.

NPGS (National Plant Germplasm System). 2009. Germplasm Resources Information Network (GRIN). Database management unit (DBMU), National Plant Germplasm System, U.S. Dep. Agric., Beltsville, MD.

Nestel, P., M. Cehun, A. Chronopoulos, L. DaSilva, H. Teede, and B. McGrath. 2004. A biochanin-enriched isoflavone from red clover lowers LDL cholesterol in men. *European Journal of Clinical Nutrition* 58:403–408.

Niezen, J. H., G. C. Waghorn, T. Graham, J. L. Carter, and D. M. Leathwick. 2002. The effect of diet fed to lambs on subsequent development of *Trichostrongylus colubriformis* larvae *in vitro* and on pasture. *Veterinary Parasitology* 105:269–283.

Ojewole, J. A., and C. O. Adewunmi. 2004. Anti-inflammatory and hypoglycemic effects of *Tetrapleura tetraptera* (Taub) [Fabaceae] fruit aqueous extract in rats. *Journal of Ethnopharmacology* 95:177–182.

Okokon, J. E., A. E. Udokpoh, and B. S. Antia. 2007. Antimalaria activity of ethanolic extract of *Tetrapleura tetraptera* fruit. *Journal of Ethnopharmacology* 111:537–540.

Onwukaeme, N. D. 1995. Anti-inflammatory activities of flavonoids of *Baphia nitida* Lodd. (Leguminosae) on mice and rats. *Journal of Ethnopharmacology* 46:121–124.

Padmesh, P., J. V. Reji, M. Jinish Dhar, and S. Seeni. 2006. Estimation of genetic diversity in varieties of *Mucuna pruriens* using RAPD. *Biologia Plantarum* 50:367–372.

Pak, K. C., K. Y. Lam, S. Law, and J. C. Tang. 2009. The inhibitory effect of *Gleditsia sinensis* on cyclooxygenase-2 expression in human esophageal squamous cell carcinoma. *International Journal of Molecular Medicine* 23:121–129.

Pari, L., and A. Uma. 2003. Protective effect of *Sesbanis grandiflora* against erythromycin estolate-induced hepatotoxicity. *Therapie* 58:439–443.

Parimaladevi, B., R. Boominathan, and S. C. Mandal. 2004. Evaluation of antipyretic potential of *Clitoria ternatea* L. extract in rats. *Phytomedicine* 11:323–326.

Payton-Stewart, F., N. W. Schoene, Y. S. Kim, M. E. Burow, T. E. Cleveland, S. M. Boue, and T. T. Wang. 2009. Molecular effects of soy phytoalexin glyceollins in human prostate cancer cells. *Molecular Carcinogenesis* 48:862–871.

Pendleton, J. M., W. W. Tan, S. Anai, M. Chang, W. Hou, K. T. Shiverick, and C. J. Rosser. 2008. Phase II trial of isoflavone in prostate-specific antigen recurrent prostate cancer after previous local therapy. *BMC Cancer* 8:132.

Pras, N., H. J. Woerdenbag, S. Batterman, J. F. Visser, and W. Van Uden. 1993. *Mucuna pruriens:* Improvement of the biotechnological production of the anti-Parkinson drug l-dopa by plant cell selection. *Pharmacy World and Science* 15:263–268.

Ramesh, T., R. Mahesh, C. Sureka, and V. H. Begum. 2008. Cardioprotective effects of *Sesbania grandiflora* in cigarette smoke-exposed rats. *Journal of Cardiovascular Pharmacology* 52:338–343.

Rojas, A., M. Bah, J. I. Rojas, V. Serrano, and S. Pacheco. 1999. Spasmolytic activity of some plants used by the Otomi Indians of Queretaro (Mexico) for the treatment of gastrointestinal disorders. *Phytomedicine* 6:367–371.

Shahzad, A., M. Faisal, and M. Anis. 2007. Micropropagation through excised root culture of *Clitoria ternatea* and comparison between *in vitro*-regenerated plants and seedlings. *Annals of Applied Biology* 150:341–349.

Singh, B., B. K. Chandan, N. Sharma, V. Bhardwaj, N. K. Satti, V. N. Gupta, B. D. Gupta, K. A. Suri, and O. P. Suri. 2006. Isolation, structure elucidation and *in vivo* hepatoprotective potential of *trans*-tetracos-15-enoic acid from *Indigofera tinctoria* Linn. *Phytotherapy Research* 20:831–839.

Singh, R., B. Singh, N. Kumar, S. Kumar, and S. Arora. 2008. Anti-free radical activities of kaempferol isolated from *Acacia nilotica* (L.) Willd. ex. Del. *Toxicology in Vitro* 22:1965–1970.

Stuart, G. A. 1979. *Chinese material medica.* Taipei: Southern Materials Center.

Suganda, A. G., M. Amoros, L. Girre, and B. Fauconnier. 1983. Inhibitory effects of some crude and semipurified extracts of indigenous French plants on the multiplication of human herpes virus 1 and poliovirus 2 in cell culture. *Journal of Natural Products* 46:626–632.

Tan, L., X. F. Zhang, B. Z. Yan, H. M. Shi, L. B. Du, Y. Z. Zhang, L. F. Wang, Y. L. Tang, and Y. Liu. 2007. A novel flavonoid from *Lespedeza virgata* (Thunb.) DC.: Structural elucidation and antioxidative activity. *Bioorganic and Medicinal Chemistry Letters* 17:6311–6315.

Tao, J. Y., L. Zhao, Z. J. Huang, X. Y. Zhang, S. L. Zhang, Q. G. Zhang, X. Fei, B. H. Zhang, Q. L. Feng, and G. H. Zheng. 2008. Anti-inflammatory effects of ethanol extract from *Kummerowia striata* (Thunb.) Schindl on lps-stimulated RAW 264.7 cell. *Inflammation* 31:154–166.

Taylor, N. L., P. B. Gibson, and W. E. Knight. 1977. Genetic vulnerability and germplasm resources of the true clovers. *Crop Science* 17:632–634.

Teixeira, E. H., M. H. Napimoga, V. A. Carneiro, T. M. de Oliveira, R. M. Cunha, A. Havt, J. L. Martins, V. P. Pinto, R. B. Goncalves, and B. S. Cavada. 2006. *In vitro* inhibition of Streptococci binding to enamel acquired pellicle by plant lectins. *Journal of Applied Microbiology* 101:111–116.

Thirugnanam, S., L. Xu, K. Ramaswamy, and M. Gnanasekar. 2008. Glycyrrhizin induces apoptosis in prostate cancer cell lines DU-145 and LNCaP. *Oncology Reports* 20:1387–1392.

Toivonen, L., and H. Rosenqvist. 1995. Establishment and growth of *Glycyrrhiza glabra* hairy root cultures. *Plant Cell, Tissue and Organ Culture* 41:249–258.

Tsarong, T. 1994. *Tibetan medicinal plants.* India: Tibetan Medical Publications.

Tsoyi, K., H. B. Park, Y. M. Kim, J. I. Chung, S. C. Shin, H. J. Shim, W. S. Lee, H. G. Seo, J. H. Lee, K. C. Chang, and H. J. Kim. 2008. Protective effect of anthocyanins from black soybean seed coats on UVB-induced apoptotic cell death *in vitro* and *in vivo*. *Journal of Agricultural and Food Chemistry* 56:10600–10605.

Untoro, J., W. Schultink, C. E. West, R. Gross, and J. G. Hautvast. 2006. Efficacy of oral iodized peanut oil is greater than that of iodized poppy seed oil among Indonesian schoolchildren. *American Journal of Clinical Nutrition* 84:1208–1214.

Uphof, J. C. 1959. *Dictionary of economic plants.* New York: Hafner Publishing Co.

Veasey, E. A., R. Vencovsky, P. S. Martins, and G. Bandel. 2002. Germplasm characterization of *Sesbania* accessions based on isozyme analysis: Isozyme electrophoresis in *Sesbania* species. *Genetic Resources and Crop Evolution* 49:449–462.

Vieira, J. R., I. A. de Souza, S. C. do Nascimento, and S. P. Leite. 2007. *Indigofera suffruticosa:* An alternative anticancer therapy. *Evidence-Based Complementary Alternative Medicine* 4:355–359.

Vyas, S., R. P. Agrawal, P. Solanki, and P. Trivedi. 2008. Analgesic and anti-inflammatory activities of *Trigonella foenum-graecum* (seed) extract. *Acta Poloniae Pharmaceutica* 65:473–476.

Wang, M. L., A. G. Gillaspie, J. B. Morris, R. N. Pittman, J. Davis, and G. A. Pederson. 2007. Variability of flavonoid content and seed-coat color in different legumes. *Plant Genetic Resources Newsletter* 6:300–325.

Wang, M. L., and J. B. Morris. 2007. Flavonoid content in seeds of guar germplasm using HPLC. *Plant Genetic Resources: Characterization and Utilization* 5:96–99.

Wang, M. L., J. A. Mosjidis, J. B. Morris, Z. B. Chen, N. A. Barkley, and G. A. Pederson. 2009. Evaluation of *Lespedeza* germplasm genetic diversity and its phylogenetic relationship with the genus *Kummerowia*. *Conservation Genetics* 10:79–85.

Wei, L., Y. Asada, and T. Yoshikawa. 2000. Flavonoid constituents from *Glycyrrhiza glabra* hairy root cultures. *Phytochemistry* 55:447–456.

Weiner, M. A. 1980. *Earth medicine, earth food.* New York: Ballantine Books.

Wong, J. H., and T. B. Ng. 2003. Gymnin, a potent defensin-like antifungal peptide from the Yunnan bean (*Glymnocladus chinensis* Baill). *Peptides* 24:963–968.

Yadav, A. S., and D. Bhatnagar. 2007. Free radical scavenging activity, metal chelation and antioxidant power of some of the Indian spices. *Biofactors* 31:219–227.

Yang, Y., F. He, and L. J. Yu. 2008. Dyanmics analyses of nutrients consumption and flavonoids accumulation in cell suspension culture of *Glycyrrhiza inflata*. *Biologia Plantarum* 52:732–734.

Yasunaka, K., F. Abe, A. Nagayama, H. Okabe, L. Lozada-Pérez, E. López-Villafranco, E. Estrada Muñiz, A. Aguilar, and R. Reyes-Chilpa. 2005. Antibacterial activity of crude extracts from Mexican medicinal plants and purified coumarins and xanthones. *Journal of Ethnopharmacology* 97:293–299.

Yeung, H. C. 1985. *Handbook of Chinese herbs and formulas.* Los Angeles: Institute of Chinese Medicine.

Yilmaz, B. S., H. Ozbek, G. S. Citoglu, S. Ugras, I. Bayram, and E. Erdogan. 2006. Analgesic and hepatotoxic effects of *Ononis spinosa* L. *Phytotherapy Research* 20:500–503.

Yones, S. S., R. A. Palmer, T. T. Garibaldinos, and J. L. Hawk. 2006. Randomized double-blind trial of the treatment of chronic plaque psoriasis: Efficacy of psoralen-UVA therapy vs. narrowband UVB therapy. *Archives of Dermatology* 142:836–842.

Yoo, H. H., T. Kim, S. Ahn, Y. J. Kim, H. Y. Kim, X. L. Piao, and J. H. Park. 2005. Evaluation of the estrogenic activity of Leguminosae plants. *Biological and Pharmaceutical Bulletin* 28:538–540.

Zakaria, Z. A., L. Y. Wen, N. I. Abdul Rahman, A. H. Abdul Ayub, M. R. Sulaiman, and H. K. Gopalan. 2007. Antinociceptive, anti-inflammatory and antipyretic properties of the aqueous extract of *Bauhinia purpurea* leaves in experimental animals. *Medical Principles and Practice* 16:443–449.

CHAPTER 13

Artemisia spp.

Zohara Yaniv, Nativ Dudai, and Uriel Bachrach

CONTENTS

13.1 THE GENUS *ARTEMISIA*

13.1.1 Botany and Distribution

Artemisia is a large, diverse genus of plants with between 200 and 400 species belonging to the daisy family Asteraceae. It comprises hardy herbs and shrubs known for their volatile oils. They grow in temperate climates of the Northern and Southern hemispheres, usually in dry or semidry habitats. The fern-like leaves of many species are covered with white hairs. Some botanists split the genus into several genera, but DNA analysis (Watson et al. 2002) does not support the existence of the genera *Crossostephium, Filifolium, Neopallasia, Seriphidium,* and *Sphaeromeria;* three other segregated genera—*Stilnolepis, Elachanthemum,* and *Kaschgaria*—are maintained by this evidence.

Common names used for several species include wormwood, mugwort, sagebrush, and sagewort; a few species have unique names, notably tarragon (*A. dracunculus*) and southernwood (*A. abrotanum*). Occasionally, some of the species are called "sages," causing confusion with the *Salvia* sages in the family Lamiaceae.

The aromatic leaves of many species of *Artemisia* are medicinal, and some are used for flavoring. Most species have an extremely bitter taste. *Artemisia dracunculus* (tarragon) is widely used as an herb and is particularly important in French cuisine.

Artemisia absinthium L. (absinth wormwood) was used to repel fleas and moths and in brewing (wormwood beer, wormwood wine). The aperitif *vermouth* (derived from the German word *Wermut*, wormwood) is a wine flavored with aromatic herbs, but originally with wormwood. The highly potent spirit absinthe also contains wormwood. Wormwood has been used medicinally as a tonic, stomachic, febrifuge, and anthelmintic.

Artemisia arborescens L. (tree wormwood, or *sheeba* in Arabic) is a very bitter herb indigenous to the Middle East, which is used in tea, usually with mint. It may have some hallucinogenic properties (Palevitch and Yaniv 1991).

Within such religious practices as Wicca, both wormwood and sheeba are believed to have multiple effects on the psychic abilities of the practitioner. Because of the power believed to be inherent in certain herbs of the genus *Artemisia*, many believers cultivate the plants in a "moon garden." The beliefs surrounding this genus are founded upon the strong association between the herbs of the genus *Artemisia* and the moon goddess *Artemis*, who is believed to hold these powers (Dafni, Yaniv, and Palevitch 1984).

It has been speculated that the genus *Artemisia* is named after an ancient botanist. Artemisia was the wife and sister of the Greek/Persian King Mausolus from the name of whose tomb we get the word "mausoleum." Artemisia, who ruled for 3 years after the king's death, was a botanist and medical researcher and died in 350 BC (Wright 2002).

A few species are grown as ornamental plants; the fine-textured ones are used for clipped borders. All grow best in free-draining sandy soil, unfertilized, and in full sun.

13.1.2 History and Traditional Medicine

Artemisia is mentioned eight times in the Bible—always as an example for evil, wrongdoing, and idol worshipping, or as an example for suffering of destruction and exile. This means that wormwood was a common herb of the era and that its awful taste was known as a drinkable preparation applied for specific reasons.

Wormwood (*apsinthos* in the Greek text) is the "name of the star" in the Book of Revelation (8:11) (*kai to onoma tou asteros legetai ho apsinthos*) that John the Evangelist envisions as cast by the angel and falling into the waters, making them undrinkably bitter.

Many species of *Artemisia* are mentioned in the earliest sources, but the more detailed descriptions are cited in the classical literature (Pliny XXV 73, in Jones 1956; Dioscorides I, 127, in Günther 1959). However, even from these descriptions it is difficult to distinguish clearly among the different species. Tree wormwood (*A. arborescens*) is considered by many researchers to be the plant known as "*Artemisia leptophullos*" referred to by Dioscorides (I, 128; Günther 1959), although in our opinion there is no absolute certainty of this identification. It is possible that the species mentioned in the ancient sources is the absinthe wormwood (*A. absinthium*), or perhaps it is a collective term referring to several species. It is worth noticing the similar morphology of these two species (see Figures 13.1 and 13.2). Later descriptions, written during the Middle Ages, also do not allow an unequivocal identification. *Assaf Harofe*, which is the earliest Hebrew medical text still extant (written in the tenth century at the latest; Muntner 1967–1969), describes a type of wormwood that may correspond to tree wormwood, but could also fit other species.

Figure 13.1 *Artemisia absinthium.*

The bitterness of the plant led to its use by wet nurses for weaning infants from the breast, as is indicated in Shakespeare's *Romeo and Juliet.*

In Russian culture, *Artemisia* species are commonly used in medicine. Their bitter taste is associated with medicinal properties. These facts have caused wormwood to be seen as a symbol for a "bitter truth" that must be accepted by a deluded (often self-deluded) person. This symbol has

Figure 13.2   (See color insert.) *Artemisia arborescens.*

acquired a particular poignancy in modern Russian poetry, which often deals with the loss of illusory beliefs in various ideologies (Krispil 1996).

13.1.3 Chemistry

The essential oil components of the genus *Artemisia* include two groups of distinct biosynthetic origin. The main group is composed of terpenes and terpenenoids. The other includes aromatic and aliphatic constituents (Bakkali et al. 2008): myrcene (20–25%), sabinene (9–14%), I-limonene (1–13%), β-phellandrene (3%), ketones (9–18%), α-thujone (4–8%), β-thujone (4–8%), camphor (1–2%), oxides (1–2%), 1,8 cineole (1–2%), lactones (0–2%), vulgarin (0–1%), and pilotachyin (0–1%) (Stewart 2005).

Thujone, a terpene present in the essential oil of *A. absinthium*, is probably responsible for its toxicity. This component in large doses is poisonous and the FDA classifies it as an unsafe herb containing "a volatile oil which is an active narcotic poison" (Duke 1985).

Artemisinins are derived from extracts of sweet wormwood (*Artemisia annua*) and are well established for the treatment of malaria, including highly drug-resistant strains. Their efficacy also extends to phylogenetically unrelated parasitic infections such as schistosomiasis. More recently, they have also shown potent and broad anticancer properties in cell lines and animal models (Krishna et al. 2008).

The essential oil of three species—*A. sieberi*, *A. arborescens*, and *A. judaica*—were analyzed in our laboratories and their essential oil profile is shown in Table 13.1. The plant samples were collected in their typical natural habitats: *Artemisia judaica* in Nahal Paran, in the Negev Desert, Israel; *Artemisia sieberi* near Yrucham in the Negev Desert, Israel; and *Artemisia arborescens* at the Montfort Castle, Nahal Kziv, in the Upper Galilee, Israel.

Samples of fresh plant material weighing 200 g were hydrodistilled for 1.5 h in a modified Clevenger apparatus. The obtained essential oil was cooled and separated from the water as described by Dudai et al. (1992). In order to prepare the samples for GC-MS analysis, 20 µL of each sample were diluted in 1 mL petroleum ether (~1:30,000). Essential oil was determined by GC-MS analysis as described by Dudai et al. (1992). It may be seen that *Artemisia sieberi* is rich in 1,8-cineole (23.1%), while *Artemisia arborescens* contains high amounts of β-thujone (29.2%). *A. judaica*, on the other hand, contains 21.8% of *Artemisia* ketone.

13.1.4 Biological Activities

Artemisia species, widespread in nature, are frequently utilized for the treatment of diseases such as malaria, hepatitis, cancer, inflammation, and infections by fungi, bacteria, and viruses. Furthermore, some *Artemisia* constituents were found to be potential insecticides and allelochemicals. This genus is receiving growing attention presumably due to (1) the diversified biology and chemistry of the constituents, (2) the frequent application in traditional medical practice, and (3) the rich source of the plant material (Tan, Zheng, and Tang 1998).

Only a few selected species were chosen to be included in this review. Most of them are native to the Mediterranean area and are important medicinal plants. These species could be evaluated as having a potential to become useful crops for human and therapeutic uses, based on new scientific research cited in this review. Our review deals with the following species:

Artemisia absinthium
Artemisia arborescens (sheeba)
Artemisia dracunculus (tarragon)
Artemisia judaica
Artemisia sieberi confused with *herba-alba*

Table 13.1   Content[a] and Composition[b] of the Essential Oils Obtained by Hydrodistillation of Three Local *Artemisia* Species in Israel

Compound	Artemisia. Sieberi	Artemisia Judaica	Artemisia Arborescens
Santolina triene	3.0	0.2	
α-Pinene	0.5	0.1	1.5
Camphene	4.6	1.8	1.2
Sabinene	0.7	0.1	3.7
Myrcene			2.7
Yomogi alcohol	4.1	1.6	
α-Terpinene	1.0		1.8
para-Cymene	1.3	0.7	0.4
Limonene		0.1	0.5
1,8-Cineole	23.1		0.4
Artemisia ketone		21.8	
γ-Terpinene	1.6		3.2
cis-Sabinene hydrate	0.4	0.1	1.3
Artemisia alcohol	8.5	2.9	
β-Thujone	2.2		29.2
Chrysanthenone		14.4	
Camphor	12.6	6.8	13.2
Santolinyl acetate	4.7		
Artemisyl acetate	1.0	4.6	
Borneol	2.5	1.5	1.8
Terpinen-4-ol	2.2	0.2	8.5
α-Terpineol	1.1		0.7
Piperitone		8.8	
cis-Chrysanthenyl acetate	0.7	0.5	
cis-Verbenyl acetate	1.8		
Bornyl acetate	1.4	1.4	
trans-Sabinyl acetate	1.6		
(Z)-Ethyl cinnamate		5.6	
(E)-Methyl cinnamate		0.8	
(E)-Caryophyllene	0.3	0.1	1.5
(E)-Ethyl cinnamate		9.5	
Germacrene D	4.7	0.6	1.7
β-Selinene		0.5	
Bicyclogermacrene	0.7	0.9	
(E)-Nerolidol	1.0		
Davanone B		1.5	
Spathulenol	0.5		
Caryophyllene oxide		0.1	0.7
Chamazulene			14.3
Total %	87.7	87.3	88.3

[a] Percent in fresh weight.
[b] Percent of total.

An ethnobotanical survey was conduced in Israel on native plants that are known as medicinal plants by different ethnic groups practicing traditional medicine (Dafni et al. 1984; Palevitch and Yaniv 1991). *Artemisia* species, which are used as medicinal plants in traditional medicine, are included and will be presented in the review. The literature on *Artemisia* was previously summarized in a book edited by C. Wright, in 2002, entitled *The Genus* Artemisia, *Medicinal and Aromatic Plants—Industrial Profiles* (vol. 18). Therefore, the present review does not include information about *A. annua* and the important metabolite artemisinin, but rather focuses on Mediterranean species and includes data obtained from recent publications and new information added in recent years.

13.2 *ARTEMISIA ABSINTHIUM* L.

13.2.1 Botany

A. absinthium, or common wormwood, is a very aromatic herbaceous plant. The root is perennial, and from it arise branched, firm, leafy stems, sometimes almost woody at the base. The flowering stem is 60–120 cm high and whitish and closely covered with fine silky hairs (Figure 13.1). The leaves, having a characteristic silvery color, are cut deeply and repeatedly; the segments are narrow (linear) and blunt. The leaf stalks are slightly winged at the margin. The small, nearly globular flower heads are arranged in an erect, leafy panicle, the leaves on the flower stalks are reduced to three or even one linear segment, and the little flowers themselves are pendulous and of a greenish yellow tint. They bloom from May to June in Israel and from July to October in Europe. The ripe fruits are not crowned by a tuft of hairs, or pappus, as in the majority of the Compositae family (Grieve 1971).

The leaves and flowers are very bitter, with a characteristic odor resembling that of thujone. The root has a warm and aromatic taste (Palevitch and Yaniv 1991).

13.2.2 Distribution

The common wormwood grows on roadsides and waste places and is found over the greater part of Europe and Siberia, having been formerly cultivated widely for its qualities. In Great Britain, it appears to be truly indigenous near the sea and locally in many other parts of England and Scotland. In Ireland, it is probably a native. It has become naturalized in the United States (Grieve 1971). The whole herb—leaves and tops—is used for preparation of traditional remedies.

13.2.3 History and Traditional Medicine

Artemisia absinthium (wormwood, absinthe) was named after the Greek goddess Artemis, who discovered its virtues. According to legend and commemorated on the Selinus coin, in the fifth century BC, the people of the city of Selinus in Sicily were suffering from a feverous plague (thought to be malaria). Artemis, accompanied by her brother Apollo, the god of healing, used wormwood to heal the plague. The oil leaves and inflorescences were used in ancient Egypt, the Mediterranean basin, and Europe as a digestive remedy, anthelmintic, antiseptic, sedative, stimulant, and tonic. According to the Ancients, wormwood counteracted the effects of poisoning by hemlock and toadstools and the biting of the sea dragon (Grieve 1971).

The plant was of great importance among the Mexicans, who celebrated their great festival of the goddess of salt by a ceremonial dance of women who wore garlands of wormwood on their heads (Grieve 1971).

In Europe, wormwood was used as a folk remedy against colds, rheumatism, fevers, diabetes, and arthritis—and could be taken by drinking beer, which contained wormwood as a flavoring and preservative before hops were used. Wormwood is still used to flavor alcoholic beverages such as vermouth, bitters, tonics, and absinthe, a highly addictive, stupor-inducing, licorice-tasting Swiss liqueur that became the national drink of France in the 1890s (Johnson 1999).

Among the most popular and intriguing intoxicants of the Victorians, absinthe did not disappear after it was banned in nearly all developed countries in the early 1900s. A number of great artists and writers from the late 1800s used absinthe as a social drink, including van Gogh and Toulouse-Lautrec. Wormwood and similar liquors were banned in France in 1915, clearing the way for anise-flavored substitutes (Bruneton 1993). A recent resurgence of absinthe use has occurred in Europe and is rapidly spreading to the United States. Despite its increasing popularity, limited information exists on the mechanism of action and neurotoxicity of absinthe, which is linked to the presence of thujone (Holstege, Saylor, and Rusyniak 2002).

13.2.3.1 Traditional Uses among American Indians

Moerman (2000), in his anthology, listed the traditional uses of *A. absinthium* by the American Indians:

- Boiled plant top is used as warm compress for sprain or strained muscles.
- Infusion of leaves is taken as a vermifuge.
- Decoction or infusion of twigs is taken for head colds.
- Poultice of pounded leaves is applied for chest colds.
- Infusion of split roots is taken for stomach ailments.
- Plant is used as a sanitary napkin to heal the mother's insides after a baby's birth.
- Poultice of mashed, boiled plant is applied or decoction of plant is used as a wash for broken limbs.
- Decoction or infusion of twigs is taken for tuberculosis and for venereal disease.
- Branches are used under mattresses as a repellent for bedbugs and other pests.

13.2.3.2 Uses of Artemisia absinthium in the Middle East

In the Old World, it was customary to drink wine mixed with leaves or juice from *A. absinthium*. The wine was bitter and intoxicating and referred to as absinthe wine (Lev 2002). According to Lev (2002), the plant called absinthe was common in Syria and Lebanon. It was used as a diuretic and for treating drunkenness and skin diseases. In medieval times, it was used for toothaches and tooth decay and for improving the smell of breath. The medieval doctor and philosopher Maimonides recommended using wine with leaves of *A. absinthium* for treating stomach problems. In the Middle East, it is used to strengthen the body, stimulate digestion, reduce fever, and expel worms (Palevitch and Yaniv 1991).

13.2.4 Chemistry

Upon distillation, the herb yields between 0.5 and 1.0% volatile oil. The oil is usually dark green or sometimes blue in color, and it has a strong odor and bitter, acrid taste. The oil contains thujone (absinthol or tenacetone), thujyl alcohol (both free and combined with acetic, isovaleri-anic, succine, and malic acids; their concentration increases after blooming), monoterpenoid hydro-carbons, azulene, cadinene, phellandrene, and pinene. The herb also contains the bitter glucoside *absinthin* and absinthic acid, together with tannin, resin, starch, potassium nitrate, and other salts (Bruneton 1993).

13.2.5 Biological Activities

13.2.5.1 Senescence–Promoting Substance

A senescence-promoting substance was detected in *A. absinthium* by the oat leaf assay and was identified as (–)-methyl jasmonate (Ueda and Kato 1980). Its senescence-promoting effect was much stronger than that of abscisic acid; even at such a low concentration as 1–2.5 μg/mL, it could completely eliminate the antisenescence action of 2 μg/mL kinetin.

13.2.5.2 Hepatoprotective Activity

The effect of aqueous-methanolic extract of *A. absinthium* against hepatic damage in mice was investigated by a group of scientists in Pakistan. The damage was induced by acetaminophen (dose of 1 g/kg produced 100% mortality) and CCl_4. Pretreatment of animals with plant extract (500 mg/kg) reduced the death rate to 20%. Pretreatment of rats with plant extract (500 mg/kg, orally, twice daily for 2 days) prevented both toxic compounds to increase in serum transaminases (Gilani and Janbaz 1995). The authors concluded that the crude extract of *A. absinthium* exhibits hepatoprotective action partly through MDME (microsomal drug metabolizing enzymes) inhibitory action and validates the traditional use of the plant in hepatic damage.

13.2.5.3 Antiparasitic Effects

The antiparasitic activity of native medicinal herbs in Iran was studied (Esfandiari et al. 2007). The authors examined direct effects of *Artemisia absinthium* extract in removing *Syphacia* parasite in mice. Experimental mice were treated with *A. absinthium* extract 10 days after infection by oral inoculation with *Syphacia* ova. Mice were examined by observation of *Syphacia* ova in their feces. Three groups of mice were treated with *A. absinthium* extract at concentrations of 2.5, 5, and 10%, respectively; the fifth group was administrated with Pyrantel Pamoate. Microscopic examination of feces indicated no *Syphacia* ova in all experimental groups, which was verified later by histopathological study of target organs at the terminal stage of the experiment. It was concluded that pharmacological application of *Artemisia absinthium* extract was able to decrease the number of *Syphacia* parasite ova in mice with fewer pathophysiological side effects. The authors concluded that these native plants could serve as a source of novel antiparasitic drugs (Esfandiari et al. 2007).

13.2.5.4 Anthelmintic Activity

A. absinthium, which is called "tethwen" in India, is used traditionally by people as a vermifuge (Tariq et al. 2009). The objective of this study was to evaluate the anthelmintic efficacy of crude aqueous extracts (CAEs) and crude ethanol extracts (CEEs) of the aerial parts of *A. absinthium* in comparison to albendazole against the gastrointestinal (GI) nematodes of sheep. Significant anthelmintic effects of CAEs and CEEs on live adult *Haemonchus contortus* worms ($P < 0.005$) were observed in terms of the paralysis and/or death of the worms at different hours after treatment. However, CEEs were more efficacious than CAEs. The oral administration of the extracts in sheep was associated with significant reduction in fecal egg output by the GI nematodes. Dosage had a significant ($P < 0.05$) influence on the anthelmintic efficacy of *A. absinthium*. The better activity of CEEs can be attributed to the greater concentration of alcohol-soluble active anthelmintic principles and more rapid transcuticular absorption of the CEEs into the body of the worms when compared with the CAEs. The results suggest that *A. absinthium* extracts are a promising alternative to the commercially available anthelmintics for the treatment of GI nematodes of sheep (Tariq et al. 2009).

13.2.5.5 Artemisinin Production in Callus

An effort was made to produce artemisinin, the important antimalarial compound extracted usually from *A. annua*, in the callus of *A. absinthium* (Zia, Mannan, and Chaudhary 2007). Leaves contain 223 μg/g artemisinin. Callus cultures initiated from leaf extract on MS medium without any growth regulator failed to show the presence of artemisinin. Addition of different growth regulators such as valine and cystine and manipulation of the medium by different hormones enhanced the production of artemisinin up to 3 μg/g (Zia et al. 2007). This is a remarkable increase!

13.3 *ARTEMISIA ARBORESCENS* L.

13.3.1 Botany

Artemisia arborescens, known in the Middle East as *sheeba*, is a perennial shrub and morphologically highly variable species with gray-green to silver leaves. The leaves are highly aromatic and quite bitter. It flowers in Israel from April to October, with yellow inflorescence (Figure 13.2).

13.3.2 Distribution

Artemisia arborescens is native to the various habitats of the Mediterranean region. According to popular folklore, the plant was spread by Moorish invaders and Knights Templars during the times leading up to the Crusades. Colonies of *A. arborescens* found on the European shores and the Mediterranean islands may have originated in North Africa. In Israel the plant is usually found as a native plant close to old Crusader sites, so it is reasonable to assume that it was brought by the Crusaders. Zohary (1955) was probably the first to speculate that sheeba was introduced by the Crusaders as a medicinal plant and had since then become a dominant plant near Crusader fortresses and monasteries. This view is supported by other workers (Feinbrun-Dothan 1978; Dafni 1983).

In a study by Dudai and Amar (2005), the composition and chemical variation of the essential oil in wild populations of *A. arborescens* grown in Israel was compared with a genetic collection from cultivated plants and plants from wild populations abroad. The main compounds identified in the essential oil were α-thujone, β-thujone, camphor, and chamazulene. There were wide variations among individual plants, and three main chemotypes were identified according to the levels of their main compounds:

     camphor and thujone: a camphor type (52% of the individuals)
     a β-thujone type (14%)
     a type containing both compounds (34%)

It follows that *A. arborescens* plants growing in Israel did not all originate from the same place. The origin of the camphor type is probably Europe, while the origin of the β-thujone type is probably North Africa.

This examination may thus offer clues as to the relations and origins of *A. arborescens* in Israel. Moreover, the exploration of this issue may assist the historical research following the routes of the Crusaders.

13.3.3 History and Traditional Uses

Artemisia arborescens (sheeba) was popular as a tea in ancient Egypt. The Romans called it "the seed of Herod." The Jews of Morocco use sheeba tea in the same way as mint tea: for stimulating the appetite, as a diuretic, and for stomachaches (Palevitch and Yaniv 1991). The Yemenite Jews prepare from the leaves of sheeba a cream to treat infected wounds. Hair is washed with a decoction made from leaves to prevent a loss of hair. The Jewish communities of Tunis and Algeria use sheeba tea to treat digestive problems such as anthelmintic and kidney stones and to prevent weakness and vomiting (Krispil 1996).

Typical aromatherapy uses of *A. arborescens* include anti-inflammatory, antiallergenic, antihistamine, anticatarrh, choleretic, and mucolytic (Sheppard-Harrger 1995).

13.3.4 Chemistry

The main components of *A. arborescens* essential oil in plants growing in Europe are usually chamazulene, β-thujone, and camphor. The chemotypes have been mainly differentiated by the dominance of one of these compounds (Sacco, Frattini, and Bicchi 1983; Biondi et al. 1993; Cotroneo et al. 2001). Biondi et al. (1993) suggested that the essential oil composition is affected by ecological factors such as climate and characteristics of the soil. In other studies *Artemisia arborescens* has been shown to contain various nonvolatile secondary metabolites: sesquiterpene lactones (Appendino and Gariboldi 1982; Grandolini et al. 1988; Marco et al. 1997), lignans, flavonoids, and diterpenes (Marco et al. 1997). Phytochemical analysis of the aerial parts of *A. arborescens* growing in Jordan resulted in isolation from the ethanolic extract of the compounds: artemisinin, arborescin, sesamin, lirioresinol, β-dimetyl ether, chrysoeriol, apigenin, β-sitosteryl glucoside, dihydroridentin, and chrysoeriol 4-glucoside. The last six compounds were isolated from this plant for the first time by Zarga et al. (1995). The same ethanolic fraction also yielded a new eudesmanolide: jordanolide.

A nor-caryophyllane derivative, artarborol, has been isolated from *A. arborescens* growing in Italy (Fattorusso et al. 2007). The authors established the stereo structure by using a combination of chemical derivatization, NMR data, molecular modeling, and quantum mechanical calculations.

13.3.5 Biological Activities

13.3.5.1 Antiherpetic Activity

The antiviral activity of the essential oil obtained from leaves of *Artemisia arborescens* was studied using HSV-1 and HSV-2. Cytotoxicity was observed in Vero cells using the MTT (dye) reduction method (Saddi et al. 2007). The mode of action of this oil as an anti-herpes-virus agent seems to be particularly interesting in consideration of its ability to inactivate the virus and to inhibit the cell-to-cell virus diffusion.

The effect of liposomal inclusion on the *in vitro* antiherpetic activity of *A. arborescence* essential oil was investigated (Sinico et al. 2005). Results show that *Artemisia* essential oil can be incorporated in good amounts in the prepared vesicular dispersions. Antiviral assays demonstrated that the liposomal incorporation of *A. arborescens* essential oil enhanced its *in vitro* antiherpetic activity (Sinico et al. 2005).

13.3.5.2 Antimycoplasmal Activity

The *in vitro* effect of methanol extracts of *A. herba-alba* and *A. arborescens* were tested against 32 isolates of *Mycoplasma* species. All were isolated from sheep and goats in different regions in Jordan (Al-Momani et al. 2007). All *Mycoplasma* species showed susceptibility to both species of *Artemisia*. The authors concluded that the species of *Artemisia* tested could be used for the treatment of *Mycoplasma* infections.

13.3.5.3 Antioxidant Activity

Extracts of four plant species indigenous to Sardinia were tested on lipid peroxidation in simple *in vitro* systems. One of these plants was *A. arborescens*. All extracts were active. Methanol extracts were more active than the essential oils (Dessi et al. 2001).

13.3.5.4 Effect on Smooth Muscles

The effect of an aqueous extract (AE) of *A. arborescens* was studied on rat isolated ileum, uterus, and urinary bladder. The extract caused a concentration-dependent reduction in the amplitude of the phasic contractions and in the tone of the ileum. On the other hand, AE caused a significant increase in the frequency as well as the amplitude of the phasic contractions and increased the tone of the isolated uterus and the urinary bladder strips. On the uterus, quinacrine, an inhibitor of the release of arachidonic acid and its metabolites, and indomethacin, a cyclo-oxygenase inhibitor, potentiated rather than inhibited the effects of AE on this tissue (Zarga et al. 1995.)

13.3.5.5 Anticancer Activity

Recently a case has been reported about a cancer patient diagnosed with CLL (chronic lymphocytic leukemia). He was given an extract from Chinese *A. annua*. However, he prepared a tea of *A. arborescens* from his garden and had it twice daily. Within 6 months, he had complete remission of his disease, as was confirmed by his doctors (Dr. Zvi Orlan, personal communication). Due to the anticancer properties related to artemisinin (Efferth 2006), this effect could be related to the content of artemisinin in *A. arborescence*. However, no data are yet available about the presence of artemisinin in this species.

13.4 ARTEMISIA DRACUNCULUS L.

13.4.1 Botany

Tarragon or dragon's wort (*Artemisia dracunculus*) is a perennial shrub, 80–120 cm tall, with slender branched stems. The leaves are thin, elongated, glossy, and dark green. Inflorescence, in yellow florets, is on top of branches. Flowering is in summer (Figure 13.3). Two varieties are known as French and Russian tarragon. The French tarragon has a better taste, but seed production is rare. The Russian tarragon can be cultivated from seeds, but its taste is considered inferior (Palevitch and Yaniv 1991; Werker et al. 1994).

13.4.2 Distribution

Tarragon is native to a wide area of the Northern Hemisphere, from easternmost Europe across central and eastern Asia to India, western North America, and south to northern Mexico. The North American populations may, however, be naturalized from early human introduction. Although tarragon is most closely associated with French and European cuisine, it was not cultivated in Europe until the late 1500s, when the Tudor family introduced it into the royal gardens from its origins in Siberia. Later, when the colonists settled in America, they brought along tarragon for their kitchen gardens, along with burnett to flavor ale, horehound for cough syrup, and chamomile for soothing tea and insect repellent (Lev 2002).

13.4.3 History and Traditional Uses

The name tarragon is a corruption of the French *esdragon*, derived from the Latin *dracunculus* (a little dragon), which also serves as its specific name. It was sometimes called little dragon mugwort and in French also has the name *herbe au dragon*. The common name, tarragon, is thought to be a corruption of the Arabic word *tarkhum*, meaning little dragon. The name is practically the same in most countries.

Figure 13.3 *Artemisia dracunculus.*

13.4.3.1 Medicinal Uses in Medieval Times

It is not known whether the plant was used in the Hellenistic and Roman periods. The first evidence for its use is provided by Assaf Harofeh, a medieval doctor (Lev 2002). According to him, the plant named tarragon in Greek was used to cure eye and kidney problems, kidney stones, as a diuretic, as a remedy for insect bites, and for many other pains. Among his other medieval medicinal uses are treatment of mouth wounds, improving the ability of taste and the appetite, preventing diarrhea, and curing toothaches (Lev 2002). This could be attributed to the presence of eugenol, an anesthetic compound, which is the major constituent of anesthetic clove oil. Clove oil is used even today as an anesthetic for toothaches.

13.4.3.2 Traditional Uses of the Present Time

The drug (aerial parts) is traditionally used to treat the symptoms of various digestive ailments and as an adjunctive therapy for the painful components of spasmodic colitis (Bruneton 1993). The following uses were common among the American Indian population for many years (Moerman 2000):

- Crushed plant was mixed with water and used on bed clothing as a bedbug repellent.
- Decoction of roots was used for infants with colic, for urinary problems, and for dysentery. Infusion of stems and leaves was used as eyewash for snow blindness.
- Infusion of foliage was used lukewarm for swollen feet and legs.
- Foliage was dried, powdered, and used for open sores.
- Infusion of leaves was used as a wash for rheumatism.
- Poultice of mashed, dampened leaves was applied to the forehead for headaches.
- Leaves were used in diapers or used as a diaper for diaper rash and skin rawness.
- Plant was burned to keep away mosquitoes.

- Decoction of leaves and roots was used as a bath for tiredness.
- Plants were used as a wash for colds, especially for babies, and for chicken pox to help the itching.
- Leaves and young stems were boiled to make a nonintoxicating beverage.
- Infusion of leaf and flower was taken or fresh leaf chewed for heart palpitations.
- Strong decoction of root was used "for steaming old people to make them stronger." (Moerman 2000).

13.4.4 Chemistry

There are differences in the essential oil composition of French and the Russian tarragon. Table 13.2 shows a typical essential oil profile of both varieties. The French tarragon contains an essential oil rich in estragole (68–80%), *cis*- and *trans*-ocimene (6–12%), and limonene (2–6%). Due to the presence of estragole, this variety has an aromatic property reminiscent of anise. Estragole is a known carcinogen and teratogen in mice (Surburg and Panten 2006). However, the main components of the essential oil in Russian tarragon are sabinene, methyl eugenol, and elemicin (Werker et al. 1994). The essential oil accumulates in two secretary structures: glandular hairs on the epidermis and secretary cavities in the mesophyll. The main components, methyl chavicol in French tarragon

Table 13.2 Content[a] and Composition[b] of the Essential Oils Obtained by Hydrodistillation of French and Russian Tarragon (*Artemisia dracunculus* L.)

Sample Name	*A. dracunculus* (French Type)	*A. dracunculus* (Russian Type)
α-Pinene		0.3
Sabinene		12.4
β-Pinene		1.4
Myrcene		2.7
α-Terpinene		0.3
Limonene	1.1	0.2
(Z)-β-Ocimene	3.6	2.8
(E)-β-Ocimene	3.5	6.9
γ-Terpinene		0.5
cis-Sabinene hydrate		0.2
Linalool		0.1
Terpinen-4-ol		1.6
Methyl chavicol	74.1	0.7
Bornyl acetate	0.5	
Citronellyl acetate		3.5
Eugenol	0.9	
(E)-Methyl cinnamate	5.4	
Geranyl acetate		1.3
Methyl eugenol	3.8	16.1
(E)-Caryophyllene	0.4	0.1
Germacrene D	0.7	1.0
Bicyclogermacrene	0.6	1.2
Elemicin		43.3
Spathulenol	3.3	
Caryophyllene oxide	0.4	
Total %	98.2	96.8

[a] Percent in fresh weight.
[b] Percent of total.

and elemicin and methyl eugenol in the Russian tarragon, accumulate in the mesophyll. Most of the other compounds were detected in the epidermis (Werker et al. 1994).

cis-Pellitorin, an isobutylamide eliciting a pungent taste, was isolated from tarragon plant (Gatfield et al. 2004; Ley et al. 2004). From the aerial parts of *Artemisia dracunculus*, one known alkamide, pellitorine; two new alkamides, neopellitorine A and neopellitorine B; and one known coumarin, herniarine, were isolated. Structures were elucidated by means of UV, IR, MS, ^{1}H and ^{13}C NMR. These compounds showed insecticidal activity against *Sitophilus oryzae* and *Rhyzopertha dominica* at 200 μg/mL concentrations (Saadali et al. 2001).

The plant samples were collected in the living aromatic plants collection at Newe Ya'ar Research Center, Israel. The essential oil distillation and analysis are the same as described in the legend of Table 13.1.

13.4.5 Culinary Use

Tarragon is used as a seasoning for food. As such, it is mentioned in the French "Herbal Remedies, Notice to Applicants for Marketing Authorization" of 1990 (Bruneton 1993). It is one of the four *fines herbes* of French cooking and is particularly suitable for chicken, fish, and egg dishes. French tarragon is the variety generally considered best for the kitchen, but cannot be grown from seed. Russian tarragon (*A. dracunculoides*) can be grown from seed but is much weaker in flavor (Deans and Simpson 2002).

13.4.6 Biological Activities

13.4.6.1 Diabetes

An alcoholic extract of *Artemisia dracunculus* (PM1 5011) has been shown to decrease glucose and improve insulin levels in animal models, suggesting an ability to enhance insulin sensitivity. In order to assess the cellular mechanism, some parameters were measured, such as basal and insulin-stimulated glucose uptake, glycogen accumulation, phosphoinositide-3 kinase activity, and Akt phosphorylation in primary skeletal muscle culture from subjects with type 2 diabetes incubated with or without various concentrations of PM1 5011. Glucose uptake was significantly increased and glycogen accumulation was partially restored. It was concluded that an alcoholic extract of *A. dracunculus* improves carbohydrate metabolism by enhancing insulin receptor signaling and modulation levels of a specific protein, tyrosine phosphatase (Wang et al. 2008).

This ethanolic extract was found to contain at least six bioactive compounds responsible for its antidiabetic properties. With the help of the bioenhancer Labrasol, the activity of the extract was enhanced three- to fivefold, making it comparable to the activity of the antidiabetic drug Metformin (Ribnicky et al. 2009). The same group of scientists (Govorko et al. 2007) was able to isolate and identify two polyphenolic compounds that inhibited PEPCK (phosphoenolpyruvate carboxykinase) mRNA expression in diabetic rats. These two compounds are responsible for much of the glucose-lowering activity of the extract. The compounds were identified as 6-demethoxycapillarisin and 2'4'-dihydroxy-4-methoxydihydrochalcone (Govorko et al. 2007).

13.4.6.2 Blood Platelet Adhesion

A. dracunculus is used as blood anticoagulator in Iranian folk medicine (Yazdanparast and Shahriyary 2008). In order to investigate this claim, the inhibitory effect of a methanol extract on adhesion of activated platelet to laminin-coated plates was studied. Based on observations, the methanol extract of *A. dracunculus*, at a concentration of 200 μg/mL, inhibited platelet adhesion by 51%! It inhibited aggregation and secretion as well (Shahriyary and Yazdanparast 2007). These observations provide the explanation for the traditional use of this herb in treatments of cardiovascular diseases and thrombosis.

13.5 *ARTEMISIA JUDAICA* L.

13.5.1 Botany and Distribution

Artemisia judaica, or Judean wormwood, is a perennial small shrub with pubescent leaves. It grows in desert riverbeds in the eastern Sahara (Libya and Egypt) and in southwestern parts of the Middle East (Sinai Peninsula, Israel, Jordan, and Saudi Arabia) (Ravid et al. 1992). Similarly to other desert *Artemisia* species, this species is strongly aromatic. The plant flowers in the early spring with light yellow flowers (Figure 13.4).

13.5.2 History and Traditional Uses

Artemisia judaica is commonly used by the local population for various purposes. Leaves are added to flavor tea and milk products in the local milk and cheese production, and they are used in the local cosmetic industry (Krispil 1996).

Medicinal uses are partly similar to those of *A. sieberi*. The plant is cardiotonic, anthelmintic, antispasmodic, stomachic, expectorant, and analgesic (Dafni et al. 1984; Palevitch and Yaniv 1991). Crushed leaves are used against snake and insect bites. Crushed leaves with onion and olive oil are used to treat skin wounds. Drops of juice from this preparation are used in infected ears and as an aid for other infections. The Bedouins in Sinai eat the flowers with sugar in cases of severe constipation. Sometimes, after a heavy meal, the whole branch is eaten with the purpose of causing vomiting. Spring is the best time to use the leaves and branches. In autumn, the branches become woody and the smell is too strong (Palevitch and Yaniv 1991).

Figure 13.4 *Artemisia judaica.*

13.5.3 Chemistry

Artemisia judaica contains, in addition to volatile compounds, sesquiterpene lactones, eudesmanolodes, seco-eudesmanolides, and glaucolide-like lactones (Ravid et al. 1992). A detailed study was performed in Israel comparing the chemistry of some desert chemotypes. It was found that *Artemisia* ketone and *Artemisia* alcohol were predominant in two different Negev oil types, while piperitone was the major constituent of oils of *A. judaica* from Sinai. The main difference between the chemotypes is the absence of most of the artemisyl-skeleton type compounds from the Sinai populations (Putievsky et al. 1992; Ravid et al. 1992).

No santonin was found in *A. judaica*. However, studies performed in Egypt describe the isolation and identification of the unique isoflavone judaicin (Khafagy and Tosson 1968). A colorimetric method for the estimation of judaicin was described by Khafagy, Abdel Salam, and El Ghazooly (1976). Studies have shown that judaicin has a cardiotonic effect somewhat similar to the effect of digitoxin (Galal et al. 1974). Judaicin is a very promising compound and more research is needed to evaluate its therapeutic potential.

13.5.4 Micropropagation and Tissue Culture

An *in vitro* propagation system for *A. judaica* has been developed (Liu et al. 2003). The successful regeneration protocol is an important achievement in this difficult field of micropropagation since it provides a basis for germplasm conservation and for further investigation of medicinally active constituents of *A. judaica*. Another study was performed in China (Liu et al. 2004). *A. judaica* was mass propagated and grown using solid, paper-bridge-support liquid, liquid-flask, and bioreactor cultures. Assays of antioxidant activity and total flavonoid content of *in vitro* and *in vivo* grown tissues were evaluated as gross parameters of medicinal efficacy.

Significantly higher antioxidant activity and flavonoid contents were observed in the tissues of mature greenhouse-grown plants. The efficient *in vitro* production systems provided sterile, consistent tissues for investigating bioactivity and germplasm conservation of *A. judaica*.

13.5.5 Biological Activities

13.5.5.1 Antioxidant Activity

Methanol extracts of *A. judaica* shoots possess antioxidant activity. This activity is in agreement with its potential use by the Jordanian population as a traditional antidiabetic agent (Al-Mustafa and Al-Thunibat 2008).

13.5.5.2 Insecticidal and Antifungal Activities

Few studies have been reported on the biological activity of two major constituents of *A. judaica* oil: piperitone and *trans*-ethyl cinnamate. Piperitone showed insecticidal activity against *Callosobruchus maculates* and antifungal activity against human (Saleh, Belai, and el-Baroty 2006) and plant pathogen fungi. In addition, *trans*-ethyl cinnamate displayed antifeedant activity and antifungal activity against human pathogen fungi (Abdelgaleil et al. 2008).

Recent studies demonstrated that both compounds showed pronounced insecticidal and antifeedant activity against the third instar larvae of *Spodoptera littoralis*; *trans*-ethyl cinnamate was more toxic. The two isolated compounds revealed antifeedant as well as antifungal activity against four plant pathogenic fungi: *Rhizocotonia solani*, *Pythium debaryanum*, *Botrytis fabae*, and *Fusarium oxysporum* (Abdelgaleil et al. 2008).

13.6 *ARTEMISIA SIEBERI* L.

Misapplied name: *Artemisia herba-alba* Asso

13.6.1 Botany

There is a long-time confusion concerning the identity of *A. herba-alba* and *A. sieberi* in the Middle East. For many years the Eastern species (reaching westward to Turkey and Egypt and eastward to Iran and Russia) was called by both names. However, according to Greuter, Burdet, and Long (2009), *A. herba-alba* is an endemic species of Spain, Morocco, and France in the West Mediterranean. Application of the name *A. herba-alba* to the Eastern species by some authors is an error (Greuter et al. 2009). In this publication we use both names, according to the citations by the authors. However, it is considered as one species: *A. sieberi*.

The white wormwood (*A. sieberi*) is a dwarf shrub 40 cm tall, heavily branched from the base, with gray, densely haired, much dissected leaves shed at the end of the rainy season and replaced by small, scale-like summer leaves. This is the mechanism for adaptation to dry conditions. The stems and branches develop small flowers in autumn. They are arranged in heads, each comprising two to four florets. They produce minute fruits with a tuft of hairs that facilitates dispersal (Figure 13.5). The whole plant is strongly aromatic and very bitter. However, the leaves are eaten by desert goats.

13.6.2 Distribution and Folk Medicine

Artemisia sieberi is a typical desert plant that grows in Iran, Palestine, Syria, Iraq, Turkey, Afghanistan, and Central Asia.

Iran: *Artemisia sieberi* is an endemic medicinal herb widely used by rural healers in the north of Iran (Arab et al. 2006; Behmanesh et al. 2007; Ghasemi et al. 2007; Negahban, Moharramipour, and

Figure 13.5 *Artemisia sieberi = A. herba-alba.*

Figure 13.6 *Artemisia sieberi* in habitat.

Sefidkon 2007; Mahboubi, Mehdi Feizabadi, and Safra 2008; Rad et al. 2008; Yaghmaie, Soltani, and Khodagholi 2008).

Israel: In the Negev Desert of Israel, 580 plant species were described; 66 of them are considered to be of medicinal use (Palevitch and Yaniv 1991). The Negev plants have a long history of use in traditional medicine by the Bedouin nomads. Desert plants of the Negev and Beer Sheva Bedouin market provided a rich source of biologically active natural compounds. *Artemisia sieberi* was one of the 66 desert plants screened (Figure 13.6). An ethnobotanical survey was also conducted in the northern part of Israel (Dafni et al. 1984). In that study 43 species, which had medicinal importance, were described.

Iraq: *Artemisia herba-alba* is widely used in Iraq folk medicine for the treatment of diabetes mellitus. Extensive studies were carried out to test this therapeutic activity in animals and human patients (Al-waili 1986; Twaij and Al-Badr 1988; al-Khazraji, al-Shamaony, and Twaij 1993; al-Shamaony, al-Khazraji, and Twaij 1994).

Libya: Aqueous extracts of *Artemisia herba-alba* were also tested as hypoglycemic agents in Libya (Marrif, Ali, and Hassan 1995).

Morocco: In Morocco a survey was undertaken to select the main medicinal plants used in folk medicine (Ziyyat et al. 1997; Tahraoui et al. 2007; Zeggwagh et al. 2008). Out of more than 600 patients, about 70% regularly used medicinal plants, which included *Artemisia herba-alba*.

Extensive studies concerning the biological activities of *Artemisia herba-alba* were also carried out in other countries in the Middle East (Saad et al. 2006). These countries included Jordan (Almasad, Qazan, and Daradka 2007; Al-Momani et al. 2007; Al-Mustafa and Al-Thunibat 2008; Hudaib et al. 2008), Algeria (Gharzouli et al. 1999), Lebanon (Salah and Jager 2005), Turkey (Tastekin et al. 2006), and Tunisia (Abid et al. 2007).

13.6.3 Biblical History

Therefore thus says the lord of hosts concerning the prophets: Behold, I will feed them with wormwood and give them poisoned water to drink. (Jeremiah 23:15)

O you who turn justice to wormwood, and cast down righteousness to the earth. (Amos 5:7)

The Hebrew word *laanah* appears eight times in the Bible. Laanah, according to most commentators, is identified with the word "wormwood." The name *herba-alba*, white herb, refers to the whitish color of the leaves.

13.6.4 Traditional Uses

The Bedouin of the Sinai and the Negev (southern part of the Israeli desert) prepare a tea from the dried leaves of *Artemisia herba-alba* for a number of ailments, such as colds, coughs, infections, and even heart problems. In cases of eye infections, a strong tea is prepared and drops are applied to the eyes. To ease chest pains and breathing problems, a steam bath is employed. Cradles of newborn babies are decorated during the first week of their life with fresh branches of *Artemisia herba-alba* as a sign of health and long life. Fresh branches hanging from the tent or house are put there against the evil eye (Palevitch and Yaniv 1991). Infusion of shoots was also used by the Bedouins from the Negev Desert in Israel to treat stomach disorders (Friedman et al. 1986).

13.6.5 Chemistry

Chemical analyses revealed that different conditions yield different percentages of constituents (Ghasemi et al. 2007). The effect of climatic factors on the distribution of *Artemisia sieberi* and the content of its essential oils was studied in Isfahan Province in Iran. It has been shown that rainfall, temperature, and light affect growth (Yaghmaie et al. 2008). Similarly, differences in the content of artemisinin were noticed when plants were collected in summer and autumn (Arab et al. 2006). It is therefore not surprising that different data may be found in the literature if conditions of growth and harvesting were not identical in the reported studies. The use of different analytical and extraction procedures also led to different results.

Essential oils hydrodistilled from the aerial parts of *Artemisia herba-alba* grown in Jordan were analyzed by GC and GC-MS technologies. Dried leaves, stems, and flowers contained thujones (24.7%) and santolina alcohol (13.0%). Other analyses, based on GC and GC-MS technologies, indicated the presence of ketones (48.5%) and 1, 8-cineole (19.7%) (Behmanesh et al. 2007).

Essential oils obtained by hydrodistillation contained thujone (19.55%), camphor (19.55%), verbenol (9.69%), *p*-mentha-1,5-dien-8-ol (6.39%), and davanone (5.79%) (Farzaneh et al. 2006). Other purifications of essential oil by hydrodistillation showed camphor (54.7%), camphene (11.7%), 1,8-cinrol (9.9%), thujone (5.6%), and pinene (2.5%) (Negahban et al. 2007).

Similar purifications of essential oils by hydrodistillation showed carvacrol (39.8%), citronellol (45.2%), pinene (23.7%), 1, 8-cineol (30.2%), and thujone (38.8%) (Mahboubi et al. 2008). Carbon dioxide extractions showed the presence of camphor (54.7%), camphebe (11.8%), 1,8-cineol (9.9%), thujone (5.7%), and pinene (2.5%) (Ghasemi et al. 2007). Earlier studies showed that santolina alcohol was the active antibacterial component of *Artemisia* (Yashphe et al. 1979).

13.6.6 Biological Activities

13.6.6.1 Antimicrobial Activity

Antibacterial: Essential oil, isolated from *Artemisia*, was effective against Gram-positive and Gram-negative bacteria (Yashphe et al. 1979; Behmanesh et al. 2007).

Antimycoplasmal: The antimycoplasmic effect of methanolic extracts of *Artemisia herba-alba* was tested against 32 *Mycoplasma* species isolated from either nasal swabs or milk from sheep or goats in different regions of Jordan. All *Mycoplasma* species showed susceptibility to *Artemisia herba-alba* extracts. It has been concluded that *Artemisia herba-alba* may be useful for the treatment of *Mycoplasma* infections (Al-Momani et al. 2007).

13.6.6.2 Antifungal Activity

Antifungal activity was evaluated *in vitro* against four soil-borne phytopathogenic fungi. Essential oil of *Artemisia sieberi* was slightly effective against *Tiarosporella phaseolina, Fusairum moniliforme,* and *Fusarium solani. Rhizoctonia solan* was highly sensitive (Farzaneh et al. 2006). Extracts of *Artemisia sieberi* showed a moderate antifungal effect when tested against *Candida albicans* (Mahboubi et al. 2008). Patients suffering from fungal infection caused by *Pityriasis versicolor* were treated with an *Artemisia sieberi* 5% lotion. This treatment was more effective than that obtained with clotrimazole (Rad et al. 2008).

The antifungal activity of *Artemisia herba-alba* was found to associate with two major volatile compounds isolated from fresh leaves of the plant: carvone and piperitone. This inhibited the growth of *Penicillum citrinum* (Saleh et al. 2006).

13.6.6.3 Insecticide Activity

Essential oils from *Artemisia sieberi* were used as fumigants against *Callosobruchus maculates, Sitophilus oryzae,* and *Tribolium castaneum* (Negahban et al. 2007). Artemisinin, which was extracted from *Artemisia sieberi,* was as active as artemisinin isolated from other species, including *Artemisia annua.* Artemisinin extracted from *Artemisia sieberi* with petrol ether was able to reduce the severity of coccidial infection induced by *Eimeria tenella* in broiler chickens (Arab et al. 2006).

13.6.6.4 Anthelmintic Activity

Anthelmintic activity of powdered shoots of *Artemisia herba-alba* was studied. The clinical signs of caprine hemonchosis were not observed in Nubian goats after treatment with 2–20 g of *Artemisia* shoots (Idris, Adam, and Tartour 1982).

13.6.6.5 Therapeutic Activities

Artemisia species have been known as a folk medicine for anti-inflammatory remedies (Strzelecka et al. 2005). Extracts of *Artemesia sieberi* showed cytotoxicity and inhibited the growth of drug-sensitive human melanoma cells (Golan-Goldhirsh et al. 2000).

Diabetes mellitus: Aqueous extracts of *Artemesia sieberi* produced a significant reduction of blood glucose levels in diabetic rats (Palevitch and Yaniv 1991). Extensive studies were carried out in Iraq to test the hypoglycemic effect of *Artemisia sieberi* extracts. When 15 patients with diabetes mellitus were treated with aqueous plant extracts, 14 patients had good remission of diabetic symptoms (Al-waili 1986). Similar hypoglycemic effects were also observed in alloxan-diabetic rabbits (Twaij and Al-Badr 1988) and rats (al-Shamaony et al. 1994). It was also reported that aqueous extracts of plant leaves or barks produced a significant reduction in blood sugar, while aqueous extracts of roots and methanolic extract of the aerial parts of the plants did not reduce blood glucose levels (al-Khazraji et al. 1993).

In Libya, extracts of *Artemisia herba-alba* were tested in alloxan-treated rabbits and mice (Marrif et al. 1995). In northeastern Morocco, diabetic patients were treated with similar plant extracts (Ziyyat et al. 1997). The same treatment was used in Turkey (Tastekin et al. 2006). *Artemisia sieberi* was also used in Jordan for diabetes treatment. This use was explained by its antioxidant activity (Al-Mustafa and Al-Thunibat 2008).

Cardiovascular effects: Aqueous extracts of *Artemisia herba-alba* showed antihypertensive activity in hypertensive rats. Those extracts increased urine and electrolyte output in rats and were therefore regarded as antihypertensive agents (Zeggwagh et al. 2008).

Digestive problems: *Artemisia sieberi* was used for treatment of digestive problems in Jordan (Hudaib et al. 2008). Infusion of the roots of *Artemisia* were used by the Bedouins of the Negev Desert of Israel to relieve stomach disorders (Friedman et al. 1986). In Algeria, *Artemisia herba-alba* leaves were used to cure ethanol-induced gastric damage in rats (Gharzouli et al. 1999).

13.6.6.6 Neurological Activities

Aqueous, ethanol, and ethyl acetate extracts of plants, including *Artemisia herba-alba*, were used traditionally in Lebanon for neurological disorders such as Alzheimer's disease and epilepsy (Salah and Jager 2005).

13.6.6.7 Antioxidant Activity

The antioxidant activity of *Artemisia herba-alba* extracts was compared with that of green tea. After 9 weeks, *Artemisia* or green tea decoctions decreased weight gain. The beneficial antioxidant effects were in descending order: *Artemisia* ≥ green tea decoction > black tea (Abid et al. 2007).

13.6.6.8 Toxicity

The toxic effect of *Artemisia herba-alba* on the reproductive system after administration to female Sprague-Dawley rats was tested in Jordan. Exposure to *Artemisia herba-alba* for 12 weeks resulted in a reduction in the percentage of pregnancies. It has been concluded that ingestion of *Artemisia herba-alba* by adult female rats causes adverse effects on fertility and reproduction (Al-Momani et al. 2007).

13.6.6.9 Mode of Action

Ethanolic extracts of *Artemisia herba-alba* had good affinity to the GABA-benzodiazepine receptor site. This may explain its activity in treatment of neurological disorders (Salah and Jager 2005). Ethyl acetate extracts of *Artemisia herba-alba* also showed high affinity to the GABA-benzodiazepine receptor site.

13.7 GENERAL CONCLUSIONS

Mankind has relied on plants as food, medicine, and much of its clothing and shelter. Since the earliest beginnings, higher plants have served as sources of drugs (Balandrin, Kinghorn, and Farnsworth 1993). *Artemisia* species were widely used in the Middle East since Biblical times. Traditional uses include tea and decoctions prepared from the aerial parts of the plants, including leaves and branches. One of the most common therapeutic uses of many *Artemisia* species, both in past and present times, is the treatment of diabetes. In recent years, various purification methods were used to prepare extracts that were more active. This fact permitted the identification and characterization of active components. These components were tested both *in vivo* and *in vitro*. Studies concerning the mode of action of the active extracts revealed that, among others, they possess antioxidant activity.

Artemisinin extracted from *Artemisia annua* is a proven antimalarial agent, very important to mankind. There are indications that other species of Artemisia, such as *A. absinthium*, produce artemisinin in lower quantities. More studies are needed to verify the extent of these findings, in order to evaluate the potential of these species as antimalarial and anticancer future drugs.

The traditional treatment of various diseases by *Artemisia* extracts is only one example of the wide potential of herbal medicine. At the present time, when many bacteria and parasites have developed resistance to antibiotics and other drugs, the use of herbal medicine should be explored.

REFERENCES

Abdelgaleil, S. A., Abbassy, M. A., Belal, A. S., and Abdel Rasoul, M. A. 2008. Bioactivity of two major constituents isolated from the essential oil of *Artemisia judaica* L. *Bioresource Technology* 99:5947–5950.

Abid, Z. B., Feki, M., Hedhili, A., and Hamdaoui, M. H. 2007. *Artemisia herba-alba* Assoc (Asteraceae) has equivalent effects to green and black tea decoctions on antioxidant processes and some metabolic parameters in rats. *Annals of Nutrition and Metabolism* 51:216–222.

al-Khazraji, S. M., al-Shamaony, L. A., and Twaij, H. A. 1993. Hypoglycemic effect of *Artemisia herba alba*. I. Effect of different parts and influence of the solvent on hypoglycemic activity. *Journal of Ethnopharmacology* 40:163–166.

Almasad, M. M., Qazan, W. S., and Daradka, H. 2007. Reproductive toxic effects of *Artemisia herba alba* ingestion in female Spague-Dawley rats. *Pakistan Journal of Biological Science* 10:3158–3161.

Al-Momani, W., Abu-Basha, E., Janakat, S., Nicholas, R. A., and Ayling, R. D. 2007. *In vitro* antimycoplasmal activity of six Jordanian medicinal plants against three *Mycoplasma* species. *Tropical Animal Health and Production* 39:515–559.

Al-Mustafa, A. H., and Al-Thunibat, O. Y. 2008. Antioxidant activity of some Jordanian medicinal plants used traditionally for treatment of diabetes. *Pakistan Journal of Biological Science* 11:351–358.

al-Shamaony, L. A., al-Khazraji, S. M., and Twaij, H. A. 1994. Hypoglycemic effect of *Artemisia herba alba*. II. Effect of valuable extract on some blood parameters in diabetic animals. *Journal of Ethnopharmacology* 43:167–171.

Al-waili, N. S. 1986. Treatment of diabetes mellitus by *Artemisia herba-alba* extract. Preliminary study. *Clinical Experimental Pharmacology and Physiology* 13:569–572.

Appendino, G., and Gariboldi, P. 1982. The stereochemistry of matricin and 4-epimatricin, proazulene sesquiterpene lactones from *Artemisia arborescens*. *Photochemistry* 21:2555–2557.

Arab, H. A., Rahbari, S., Rassouli, A., Moslemi, M. H., and Khosravirad, F. 2006. Determination of artemisinin in *Artemisia sieberi* and anticoccidial effects of the plant extract in broiler chickens. *Tropical Animal Health and Production* 38:497–503.

Bakkali, F., Averbeck, S., Averbeck, D., and Idaomar, M. 2008. Biological effects of essential oils—A review. *Food and Chemical Toxicology* 46:446–475.

Balandrin, M. F., Kinghorn, A. D., and Farnsworth, W. R. 1993. Plant-derived natural products drug development. In *Human medicinal agents*, ed. A. D. Kinghorn and M. F. Balandrin, 2–12. Washington, D.C.: American Chemical Society.

Behmanesh, B., Heshmati. G. A., Mazandrani, M., et al. 2007. Chemical composition and antibacterial activity from essential oil of *Artemisia sieberi Besser* subsp. *Sieberi* in North Iran. *Asian Journal of Plant Science* 6:562–564.

Biondi, E., Valentini, G., Bellomaria, B., and Zuccarello, V. 1993. Composition of essential oil in *Artemisia arborescens* L. from Italy. *Acta Horticulturae* (ISHS) 344:123–131.

Bruneton, J. 1993. *Pharmacognosy, phytochemistry, medicinal plants.* Paris: Lavoisier.

Cotroneo, A., Dugo, P., Verzera, A., Previti, P., and Rupisarda, A. 2001. L'olio essenziale della foglie di piante tipiche della flora mediterranea. Nota IV. *Artemisia arborescens* L. (Asteraceae). *Essenze Derivati Agrumari* 71:15–18.

Dafni, A. 1983. *Artemisia arborescens*. In *Plants and animals of the land of Israel: A practical illustrated encyclopedia* 11, 154. ed. A. Alon, 154. Tel Aviv: Massada (Hebrew).

Dafni, A., Yaniv, Z., and Palevitch, D. 1984. Ethnobotanical survey of medicinal plants in northern Israel. *Journal of Ethnopharmacology* 10:295–310.

Deans, S. G., and Simpson, E. J. M. 2002. *Artemisia dracunculus*. In *The genus* Artemisia. *Medicinal and aromatic plants—Industrial profiles*, vol. 18, ed. C. Wright. London: Taylor & Francis.

Dessi, M. A., Deiana, M., Rosa, A., et al. 2001. Antioxidant activity of extracts from plants growing in Sardinia. *Phytotherapy Research* 15:511–518.

Dudai, N., and Amar, Z. 2005. *Artemisia arborescens*, a medicinal plant which arrived in Israel from Europe with the Crusaders. *Yaar* 7:35–40 (in Hebrew).

Dudai, N., Putievsky, E., Ravid, U., Palevitch, D., and Halevy, A. H. 1992. Monoterpene content in *Origanum syriacum* as affected by environmental conditions and flowering. *Phsyologia Plantarum* 84:453–459.

Duke, J. A. 1985. *The handbook of medicinal herbs*. Boca Raton, FL: CRC Press.

Efferth, T. 2006. Molecular pharmacology and pharmacogenomics of artemisinin and its derivatives in cancer cells. *Current Drug Targets* 7:407–421.

Esfandiari, B., Youssefi, M., Keighobadi, M., and Nahrevanian, H. 2007. *In vivo* evaluation of antiparasitic effects of *Artemisia absinthium* extracts on Syphacia parasite. *International Journal of Parasitic Diseases* 2:1–6.

Farzaneh, M., Ghorbani-Ghouzhdi, H., Ghorbani, M., and Hadian, J. 2006. Composition and antifungal activity of essential oil of *Artemisia sieberi* Bess. on soil-borne phytopathogens. *Pakistan Journal of Biological Science* 9:1979–1982.

Fattorusso, C., Stendardo, E., Appendino, G., et al. 2007. Artarborol, a nor-caryophyllane sesquiterpene alcohol from *Artemisia arborescens*, stereostructure assignment through concurrence of NMR data and computational analysis. *Organic Letters* 9:2377–2380.

Feinbrun-Dothan, N. 1978. *Flora Palestina*, III. The Israel Academy of Science and Humanities, Jerusalem.

Friedman, J., Yaniv, Z., Dafni, A., and Palevitch, D. 1986. A preliminary classification of the healing potential of medicinal plants, based on a rational analysis of an ethnopharmacological field survey among Bedouins in the Negev Desert. *Israel Journal of Ethnopharmacology* 16:275–287.

Galal, E. E., Kandil, A., Abdel-Latif, M., Khedr, T., and Khafagy, M. S. 1974. Cardiac pharmaco-toxicological studies of judaicin, isolated from *Artemisia judaica*. *Planta Medica* 25:88–91.

Gatfield, I. L., Ley, J. P., Foerstner, J., Krammer, G., and Machinek, A. 2004. Production of *cis*-pellitorin and use as a flavoring. World patent WO2004000787 A2.

Gharzouli, K., Khennouf, S., Amira, S., and Gharzouli, A. 1999. Effect of aqueous extracts from *Quercus ilex* L. root bark, *Punica granatum* L. fruit peel and *Artemisia herba-alba* Asso leaves on ethanol-induced gastric damage in rats. *Phytotherapy Research* 13:42–45.

Ghasemi, E., Yamini, Y., Bahramifar, N., and Sefidkon, F. 2007. Comparative analysis of the oil and supercritical CO_2 extract of *Artemisia sieberi*. *Journal of Food Engineering* 79:306–311.

Gilani, A. H., and Janbaz, K. H. 1995. Preventive and curative effects of *Artemisia absinthium* on acetaminophen and CCL_4-induced hepatotoxicity. *General Pharmacology* 26:309–315.

Golan-Goldhirsh, A., Lugasi-Evgi, H., Sathiyamoorthy, P., Pollack, Y., and Gopas, J. 2000. Biotechnological potential of Israeli desert plants of the Negev. *Acta Horticulturae* 25:29–33.

Govorko, D., Logrndra, S., Wang, Y., et al. 2007. Polyphenolic compounds from *Artemisia dracunculus* L. inhibit PEPCK gene expression and gluconeogenesis in an H411E hepatome cell line. *American Journal of Physiology—Endocrinology and Metabolism* 293:E1503–1510.

Grandolini, G., Casinovi, C., Betto, P., et al. 1988. A sesquiterpene lactone from *Artemisia arborescens*. *Phytochemistry* 27:3670–3672.

Greuter, W., Burdet, H. M., and Long, G. 2009. *Med-Checklist: A critical inventory of vascular plants of the circum-Mediterranean countries*, vol. 2. Conservatoire Botanique de Geneve.

Grieve, M. 1971. *A modern herbal*. New York: Dover Press.

Günther, T. R. 1959. *The Greek herbal of Dioscorides*. New York: Hafner Publishing.

Holstege, C. P., Saylor, M. R., and Rusyniak, D. H. 2002. Absinthe: Return of the green fairy. *Seminars in Neurology* 22:89–93.

Hudaib, M., Mohammad, M., Bustanji, Y., et al. 2008. Ethnopharmacological survey of medicinal plants in Jordan, Mujib nature reserve and surrounding area. *Journal of Ethnopharmacology* 120:63–71.

Idris, U. E., Adam, S. E., and Tartour, G. 1982. The antihelminic efficacy of *Artemisia herba-alba* against *Haemonchus contortus* infection in goats. *National Institute of Animal Health Quarterly* 22:138–143.

Johnson, T., ed. 1999. *CRC ethnobotany desk reference*. Boca Raton, FL: CRC Press.

Jones, W. H., ed. 1956. *Pliny natural history*. London: Loeb Classical Library.

Khafagy, S. M., Abdel Salam, M. A., and El Ghazooly, M. G. 1976. A colorimetric method for the estimation of judaicin, bitter principle of *Artemisia judaica* L. *Planta Medica* 30:21–24.

Khafagy, S. M., and Tosson, S. 1968. Crystallographic, optical and chromatographic studies of judaicin, bitter principle of *Artemisia judaica* L. *Planta Medica* 16:446–449.

Krishna, S., Bustamante, L., Haynes, R. K., and Staines, H. M. 2008. Artemisinins: Their growing importance in medicine. *Trends in Pharmacological Sciences* 29 (10): 520–527.

Krispil, N. 1996. *Taste of life*, 60–65. Or Yehuda, Israel: Maariv Book Guild (in Hebrew).

Lev, E. 2002. *Medicinal substances of the medieval Levant*. Eretz Publications. 181 pp.

Ley, J. P., Hilmer, J. M., Weber, B., Krammer, G., Gatfield, I. L., and Bertram, H. J. 2004. Stereoselective enzymatic synthesis of *cis*-pellitorine, a taste active alkamide naturally occurring in tarragon. *European Journal of Organic Chemistry* 24:5135–5140.

Liu, C. Z., Murch, S. J., El-Demerdash, M., and Saxena, P. K. 2003. Regeneration of the Egyptian medicinal plant *Artemisia judaica* L. *Plant Cell Reports* 21:525–530.

———. 2004. *Artemisia judaica* L.: Micropropagation and antioxidant activity. *Journal of Biotechnology* 13:63–71.

Mahboubi, M., Mehdi Feizabadi, M., and Safra, M. 2008. Antifungal activity of essential oils from *Zataria multiflora*, *Rosmarinus officinalis*, *Lavandula stoechas*, *Artemisia sieberi* Besser and *Pelargonium graveolens* against clinical isolates of *Candida albicans*. *Pharmacognosy Magazine* 4:S15–18.

Marco, J. A., Sanz-Cervera, J. F., García-Lliso, V., and Vallès-Xirau, J. 1997 Sesquiterpene lactones and lignans from *Artemisia arborescens*. *Phytochemistry* 44:1133–1137

Marrif, H. I., Ali, B. H., and Hassan, K. M. 1995. Pharmacological studies on *Artemisia herba-alba* (Asso.) in rabbits and mice. *Journal of Ethnopharmacology* 49:51–55.

Moerman, D., 2000. *Native American ethnobotany*. Portland, OR: Timber Press.

Muntner, Z. 1967–1969. Sefer Assaf Harofe. *Korot* 4:403 (Hebrew).

Negahban, M., Moharramipour, S., and Sefidkon, F. 2007. Fumigant toxicity of essential oil from *Artemisia sieberi* Besser against three stored-product insects. *Journal of Stored Products Research* 43:123–128.

Palevitch, D., and Yaniv, Z. 1991. *Medicinal plants of the Holy Land*. Tel Aviv: Tamuz Modan (in Hebrew).

Putievsky, E., Ravid, U., Dudai, N., Katzir, I., Carmeli, D., and Eshel, A. 1992. Variations in the essential oil of *A. judaica* L. Chemotypes related to phonological and environmental factors. *Flavor and Fragrance Journal* 7:253–257.

Rad, F., Aala, F., Reshadmanesh, N., and Yaghmaie, R. 2008. Randomized comparative clinical trial of *Artemisia sieberi* 5% lotion and clorimazole 1% lotion for the treatment of pityriasis versicolor. *Indian Journal of Dermatology* 53:115–118.

Ravid, U., Putievsky, E., Katzir, I., Carmeli, D., Eshel, A., and Schenk, H. P. 1992. The essential oil of *Artemisia judaica* L. chemotypes. *Flavor and Fragrance Journal* 7:69–72.

Ribnicky, D. M., Kuhn, P., Poulev, A., et al. 2009. Improved absorption and bioactivity of active compounds from an antidiabetic extract of *Artemisia dracunculus* L. *International Journal of Pharmacy* 370:87–92.

Saad, B., Azaizeh, H., Abu-Hijleh, G., and Said, O. 2006. Safety of traditional Arab herbal medicine. *Annals of Oncology* 3:433–439.

Saadali, B., Boriky, D., Blaghen, M., Vanhaelen, M., and Talbi, M. 2001. Alkamides from *Artemisia dracunculus*. *Phytochemistry* 58:1083–1086.

Sacco, T., Frattini, C., and Bicchi, C. 1983. Constituents of essential oil of *Artemisia arborescens*. *Planta Medica* 47:49–51.

Saddi, M., Sanna, A., Cottiglia, F., et al. 2007. Antiherpevirus activity of *Artemisia arborescens* essential oil and inhibition of lateral diffusion in Vero cells. *Annals of Clinical Microbiology and Antimicrobials* 6:10.

Salah, S. M., and Jager, A. K. 2005. Screening of traditionally used Lebanese herbs for neurological activities. *Journal of Ethnopharmacology* 97:145–149.

Saleh, M. A., Belai, M. H., and el-Baroty, G. 2006. Fungicidal activity of *Artemisia herba alba* Assoc (Asteraceae). *Journal of Environmental Science and Health B* 41:237–244.

Shahriyary, L., and Yazdanparast, R. 2007. Inhibition of blood platelet adhesion, aggregation and secretion by *Artemisia dracunculus* leaves extracts. *Journal of Ethnopharmacology* 114:194–198.

Sheppard-Harrger, S. 1995. *The aromatherapy practitioner reference manual*. Tampa, FL: Atlantic Institute of Aromatherapy.

Sinico, C., De Logu, A., Lai, F., et al. 2005. Liposomal incorporation of *Artemisia arborescens* essential oil and *in vitro* antiviral activity. *European Journal of Pharmaceutics and Biopharmaceutics* 59:161–168.

Stewart, D. 2005. *The chemistry of essential oil*. Marble Hill, MO: Care Publications.

Strzelecka, M., Bzowska, M., Koziel, J., et al. 2005. Anti-inflammatory effects of extracts from some traditional Mediterranean diet plants. *Journal of Physiology and Pharmacology* 56:139–156.

Surburg, H., and Panten, J. 2006. *Common fragrance and flavor materials: Preparation, properties and uses*, 233. New York: Wiley-VCH.

Tahraoui, A., El-Hilaly, J., Israili, Z. H., and Lyoussi, S. 2007. Ethnopharmacological survey of plants used in traditional treatment of hypertension and diabetes in southeastern Morocco. *Journal of Ethnopharmacology* 110:105–117.

Tan, R. X., Zheng, W. F., and Tang, H. Q. 1998. Biologically active substances from the genus *Artemisia*. *Planta Medica* 64 (4): 295–302.

Tariq, K. A., Chishti, M. Z., Ahmad, F., and Shaw, A. S. 2009. Anthelmintic activity of extracts of *Artemisia absinthium* against ovine nematodes. *Veterinary Parasitology* 160:83–88.

Tastekin, D. M., Atasever, M., Adiguzel, G., Keles, M., and Tastekin, A. 2006. Hypoglycemic effect of *Artemisia herba-alba* in experimental hyperglycemic rats. *Bulletin of the Veterinary Institute in Pulawy* 50:23–38.

Twaij, H. A., and Al-Badr, A. A. 1988. Hypoglycemic activity of *Artemisia herba alba*. *Journal of Ethnopharmacology* 24:123–126.

Ueda, J., and Kato, J. 1980. Isolation and identification of a senescence-promoting substance from wormwood (*Artemisia absinthium* L.). *Plant Physiology* 66:246–249.

Wang, Z. Q., Ribnicky, D., Zhang, X. H., Raskin, I., Yu, Y., and Cefalu, W. T. 2008. Bioactives of *Artemisia dracunculus* L. enhances cellular insulin signaling in primary human skeletal muscle culture. *Metabolism* 57:s58–64.

Watson, L. E., Bates, P. L., Evans, T. M., Unwin, M. M., and Estes, J. R. 2002. Molecular phylogeny of subtribe *Artemisiinae* (Asteraceae), including *Artemisia* and its allied and segregate genera. *Biomed Central Evolutionary Biology* 2:17.

Werker, E., Putievsky, E., Ravid, U., Dudai, N., and Katzir, I. 1994. Glandular hairs, secretory cavities and the essential oils in leaves of tarragon (*Artemisia dracunculus* L.). *Journal of Herbs, Spices and Medicinal Plants* 2:19–31.

Wright, C., ed., 2002. *The genus* Artemisia, *medicinal and aromatic plants. Industrial profiles*, vol. 18, series ed. R. Hardman. London: Taylor & Francis.

Yaghmaie, L., Soltani, S., and Khodagholi, M. 2008. Effect of climatic factors on distribution of *Artemisia sieberi* and *Artemisia aucheri* in Isfahan Province using multivariate statistical methods. *Journal of Science and Technology of Agriculture and Natural Resources* 12:371.

Yashphe, J., Segal, R., Breuer, A., and Erdreich-Naftali, G. 1979. Antibacterial activity of *Artemisia herba-alba*. *Journal of Pharmaceutical Science* 68:924–925.

Yazdanparast, R., and Shahriyary, L. 2008. Comparative effects of *Artemisia dracunculus*, *Satureja hortensis* and *Origanum majorana* on inhibition of blood platelet adhesion, aggregation and secretion. *Vascular Pharmacology* 48:32–37.

Zarga, A. M., Qauasmeh, R., Sabri, S., Munsoor, M., and Abdalla, S. 1995. Chemical constituents of *Artemisia arborescens* and the effect of the aqueous extract on rat isolated smooth muscle. *Planta Medica* 61:242–245.

Zeggwagh, N. A., Farid, O., Michel, J. B., and Eddouks, M. 2008. Cardiovascular effect of *Artemisia herba alba* aqueous extract in spontaneously hypertensive rats. *Methods & Findings in Experimental Clinical Pharmacology* 30:375–381.

Zia, M., Mannan, A., and Chaudhary, M. F. 2007. Effect of growth regulators and amino acids on artemisinin production in the callus of *Artemisia absinthium*. *Pakistan Journal of Botany* 39:799–805.

Ziyyat, A., Legssyer, A., Mekhfi, H., et al. 1997. Phytotherapy of hypertension and diabetes in oriental Morocco. *Journal of Ethnopharmacology* 58:45–54.

Zohary, M. 1955. *Geobotany*. Tel Aviv: Merhavia.

CHAPTER 14

Poppy
Utilization and Genetic Resources

Jenő Bernáth and Éva Németh

CONTENTS

14.1 INTRODUCTION

There is historical evidence that poppy (*Papaver somniferum* L.) has been utilized and cultivated since prehistoric times (Tétényi 1997). The narcotic properties of the plant and the nutritive values of the seeds were recognized by Greeks, Egyptians, and Romans. Hippocrates (460–377 BC) was one of the first to mention the medical application of the poppy and its preparations. The utilization of poppy was introduced into the Roman provinces, where its nutritive value was admitted, too. After the Roman period, the cultivation of the plant spread in both Europe and Asia. However, great differences arose in the profile of the production: the opium became the main product in the East and the seed and the seed oil in Europe.

The opium production in Southeast Asia ("Golden Triangle"), west Asia ("Golden Crescent"), and some other geographical regions (Central and South America) is going on even now. At present, the politics of international organizations (WHO, FAO, UNIDO, INCB) are oriented to gain control of production in order to reduce the illicit trade and consumption.

In Europe, the first epoch of expansion of poppy cultivation can be dated back to the eighteenth century. At that time, a large amount of poppy oil was produced as a result of the increasing food and industrial demand. This prosperity of seed production and oil processing lasted to the end of the first half of the nineteenth century. Worldwide known producers came into the existence in Provence and Alsace, France, and some former German states. The economical importance of the poppy seed and oil declined after that and a new age of poppy utilization appeared when its high medicinal value was discovered (Bernáth 1998).

The second epoch of enlarging poppy cultivation was initiated by the recognition and scientific affirmation of the outstanding medical value of the secondary compounds of the plant. The manufacture of morphine started in the nineteenth century in small European pharmaceutical companies. The production was based on opium as a raw material imported from Turkey and Persia. Macfarlane and Smith was one of the first companies that specialized in opium processing in 1837 in the UK. Only 25 tonnes of opium were processed at that time. Realizing the economical and medical importance of the products, the processing of opium developed in many other companies later—for example, Francopia (France 1847) and Mallinckrodt (United States 1898). The number of countries and the cultivation area of licit opium production have been decreasing. At present, India is the only country in which the licit cultivation of opium is going on at an industrial scale. Some other countries, such as the Republic of Korea, China, and Japan, produce restricted amounts of opium for utilization in their traditional medication systems. However, the total amount of cultivation area in the world for opium production is less than 10,000 ha (INCB 2009).

The real cultivation of poppy started with the development of processing technology. In the first half of the twentieth century, a remarkable advance was achieved by the Hungarian pharmacist Kabay in 1928 that opened a new perspective for the plant and for its industrial utilization (Bernáth 1998). Based on his method, the alkaloids, especially morphine, were extracted from the dry capsule of the poppy. The poppy straw had been known as a waste that had to be separated from the seed in the final step of the commercial poppy cultivation process. Using this method, high seed quality and valuable raw material for the pharmaceutical industry were available at the same time. This invention and the increasing demand for opiates resulted in a continuous increase of the poppy cultivation area in Europe and, most recently, in Australia.

Considering the licit production area and quantity of harvested poppy straw and seed, the main European producers are Ukraine, Czech Republic, Hungary, Spain, France, Romania, Slovakia, and Poland (INCB 2009). Germany, Austria, and the Netherlands also produce smaller amounts of poppy seeds, almost exclusively for local utilization. In the Balkan region, poppy production is on about 40,000 ha in Turkey (INCB 2009). In the late 1960s, Australia appeared as a new producer in the poppy business. The cultivation area in Tasmania increased up to 21,000 ha until the first years of the twenty-first century, but as a result of intensification of production systems, some reduction in the cultivation area has occurred since then. According to the data of INCB (2009), recent worldwide production of poppy straw has been about 150,000–200,000 ha.

In spite of the great pharmaceutical and nutraceutical values of the poppy, we have to take into consideration that a large amount of the plant is cultivated in the world for illicit purposes. Referring to the data of INCB (2009), around 8000 tonnes of opium is produced worldwide for abuse, which is about 10 times the official quantity of opium used for medical purposes. The illicit cultivation area was estimated to be about 200,000–250,000 ha in the first decade of the twenty-first century.

According to estimates, the licit production of poppy will increase in the future, in parallel with reduction of illegal activity. This can be explained by the well known narcotic, pharmacological, and nutritive value of poppy. The utilization of opiate alkaloids (either taken from the opium or extracted from the dry capsule) is important. Analgesics of morphine origin are used mainly to control severe pain and for their antidiarrhea and sedative effects. Codeine and, to a lesser extent, pholcodine, ethyl morphine, and narcotine are utilized as cough depressant agents. Apomorphyne hydrochloride can be used as an emetic in small quantities. Moreover, its anti-Parkinson efficacy has been recognized and tested.

According to the Technical Report of INCB (2008) the medical consumption of opioids in the world doubled in the past 20-year period. While the global consumption of opioids expressed in defined daily doses had been about 2,300 million in 1987, more than 5,500 million were applied in 2006 for medications. The demand for opiates is expected to increase steadily in the future. From the point of view of raw material production, continuous improvement of the biological background, modernization of cultivation practices, and development of processing methods are needed. Simultaneously, the limitation of illicit cultivation activity should be an important international goal, which needs new antidrug policy and strict cooperation between INCB and national agencies.

14.2 GENE CENTER OF POPPY

14.2.1 Geographical Origin

The origin of poppy shows a lot of uncertainty in spite of the fact that it belongs to the few species that have been utilized, even cultivated, since prehistoric times. Neither the geographical nor botanical origin of poppy has been cleared up yet with full confidence. Based on archaeological investigations, it became clear that some sort of poppy was known and utilized by cavemen living in the territory of Spain, France, Germany, and Hungary in 4000–5000 BC (Tétényi 1997). Recent archaeological findings (Herbig 2009) indicate that poppy was cultivated in the early phase of the late Neolithic period in the upper Swabia region by the inhabitants of lakeside settlements (4000–2400 BC). Kohler-Schneider and Caneppele (2009) came up with similar findings. Their archaeological investigations proved that farmers in the Jevisovice culture, around 3000 BC, cultivated poppy in the territory of eastern Austria.

The first written historical records hypothesize that the gene center of poppy could be located in western Asia (Simmonds 1976). The early historical data justify that the plant and its extracts were applied in ritual ceremonies for its therapeutic action. The name of the plant appears in classic literature, such as *The Odyssey* and *The Iliad* of Homer. Greeks portrayed their divinities Hypnos, Nyx, and Thanatos with poppies (Kritikos and Papadaki 1967). In the opinions of some scientists, the name of the plant comes from the name of the town of Mekone (in the Corinth region), where the extensive cultivation of the poppy was going on at that time; others think that it was a place of the first discovery of the plant. In the early records, the plant was mentioned as a tool for getting easy and painless death. Hippocrates (460–377 BC) was one of the first who emphasized the medical usage of poppy and its preparations. The nutritive property of the seeds was first mentioned by him. Herakleides (340 BC) reported the use of this plant as a tool for euthanasia in some Greek islands. People, especially women, took poppy to shorten the time that is left for them until natural death.

The poppy was also known to the ancient Egyptians. Some scientists state that the plant was introduced from abroad, particularly from Greece and Babylon. The time of the introduction coincides with the period shortly before the Roman time. By the data of Gabra (1956), an oleaginous ointment was found in the tomb of the XVIII Dynasty. The author added that the flower and the pieces of *Papaver rhoeas* were frequently found on the tombs and monuments from the time of this dynasty. There is other evidence in the Ebers papyrus (1500 BC). In the opinion of many scientists, the poppy is mentioned in this document under the name of *seter-seref*. However, the first data on large-scale cultivation and preparation of opium were registered only in Egyptian Thebes.

There is no doubt about the evidence that the poppy was cultivated by Sumerians, Babylonians, and Assyrians as well around 3000–6000 BC. On the clay tables of Sumerians the production method of poppy juice was detailed: It was collected very early in the morning. They called it *gil* and used it for curing.

The name and preparations of poppy appear in the Bible and in the Talmud. Probably the plant head "rosch" refers to the capsule of *Papaver setigerum*. In Europe, the utilization of poppy was spread by the growth of the Roman Empire. The cultivation of the plant for food and medicines might have started at that time in each province. The cultivation of poppy has been continued by local inhabitants and adapted by invasive nations as well. Based on archeobotanical investigations, the cultivation of poppy was adapted, for instance, by the German tribe of the Alamanni in southwestern Germany from the third to the sixth centuries (Rosch 2008).

The exact time of the introduction of poppy into India is under discussion. Probably it was introduced at the time of the invasion of Alexander the Great (fourth century BC). The Persians took it with them for the needs of their army. There is no evidence of application of poppy in India until the seventh century AD. However, it is hard to believe that opium was unknown by the Indian physicians until that time. It is clear that, from the sixteenth century, the Europeans played a significant impact in the development of the large-scale production and consumption of opium. Portuguese and later Dutch merchants carried Indian opium thorough Macao into China. Under the Mughal Empire, the poppy was extensively grown and became an important article of trade with China and other Eastern countries. Fine opium was derived from the poppies cultivated in the fertile alluvial plains of the Ganges. In 1757, the British East India Company began to control the production and trade, which became the British government's right in 1874.

At present, the distribution and cultivation area of poppy can be distinguished according to the legality of the production and form of final product. This type of diversification is demonstrated in Figure 14.1.

14.2.2 Domestication

The large diversity of poppy was recognized in early times by Dioscorides (first century AD). Several kinds of poppy were distinguished by him. The "cultivated" or "garden" poppy was used in baking bread. Two types were known in this category: plants with elongated capsules and white seeds and plants forming involuted and elongated capsules with black seeds (Kritikos and Papadaki 1967). From the present botanical point of view, the previously mentioned taxa could have been *P. somniferum*. The next group, named "flowering" poppy, showed high hypnotic properties and may

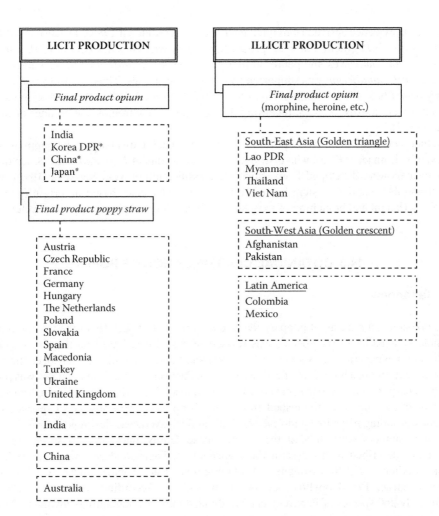

*Traditional medication

Figure 14.1 Countries involved in the licit and illicit cultivation of poppy. (INCB. 2008. *World drug report*. New York (Vienna): UN Office on Drugs and Crime, 2007.)

refer to the species of *P. hybridum*. The "wild" poppy group may be equivalent with species of *P. orientalis*. Later, Pliny mentioned an "intermediate" type between the wild and cultivated poppy, which could have been *P. rhoeas*.

The definition of the botanical origin of the poppy caused a lot of problems for scientists from the beginning of investigations. The phylogenetic origin of the opium poppy was approached from different points of view. Some botanists (De Candolle 1883; Fedde 1909; Soó 1968; Hammer and Fritsch 1977) assumed that opium poppy originated from *P. setigerum*. Based on traditional genetic investigations, large differences between the *P. setigerum* and the cultivated poppy have been proved (Hrishi 1960). The gene center of the two species was distinguished by the results of the botanical expeditions. It seems to be acceptable that the gene center of cultivated poppy is outside Europe, located in middle Asia, especially in the territories of Iran and Afghanistan.

Another hypothesis was that *P. glaucum* is the ancestor of opium poppy (Rothmaler 1949); others assumed the parallel evolution of *P. somniferum* and *P. aculeatum* (Reckin 1971). The triploid hybrid origin of *P. somniferum* ($2n = 22$) was assumed as well, which was supported by morphological evidence derived from interspecific crosses (Kadereit 1986). The crosses of *P. glaucum* (2*n* = 14) with *P. gracile* (2*n* = 14) produced progeny with capsules similar to *P. somniferum* subsp. *setigerum*, but leaves and petals similar to subspecies *somniferum*. Nevertheless, the triploid origin of *P. somniferum* was supported by cytogenetic studies. Thus, the F_1 hybrid of subspecies *setigerum* (2*n* = 44) and *somniferum* (2*n* = 22) has 2*n* = 22. When these were selfed, the F_2 descendants were mostly 2*n* = 28, regaining the original basic chromosome number of the genus *Papaver* (*x* = 7).

The theoretical and practical importance of the intraspecific diversity of the poppy was recognized even by Linnaeus (1753), who distinguished five variants of *P. somniferum*. Since that time, several systems were developed from which the classification of Vesselovskaya (1975) seems to be a considerable one. Her system has three levels: the subspecies based on origin (geography), morphology (height and branching of axis, shape of capsule), and physiology (life cycle) of poppy plants.

14.3 BOTANICAL CLASSIFICATION OF POPPY

14.3.1 Taxonomy

The taxonomical ranking of poppies changed a lot in the last decades of the twentieth century. The whole systematics of the species was revised several times. Based on the chemosystematic evaluation of angiosperms by Gottlieb et al. (1993) and the revision of their phylogenetic relationships (Kubitzky 1993), a high level of affinity between Papaverales and Ranunculales was proved. It led to the incorporation of order Papaverales into Ranunculales at the family level (Papaveraceae). The Papaveraceae can be distinguished from other families by the presence of special chemical compounds including benzylisoquinoline alkaloids and Papaveraceae flavonoids.

Different authors have divided the Papaveraceae family into three to seven subfamilies. Fedde (1936) described in his system the Hypecoideae, Papaveroideae, and Fumaroideae subfamilies. Takhtajan (1959) distinguished Platystemonoideae, Papaveroideae, Chelidonioideae, Eschscholzioideae, Pteridophylloideae, and Hypecoideae. According to Kadereit (1993), the approximately 260 species of Papaveraceae are divided into four subfamilies involving 23 genera. The *Papaver* genus belongs to the subfamily Papaveroideae.

Based on high variability, the genus *Papaver* was divided into 5–11 sections (Elkan 1839; Fedde 1936; Günther 1975). The latest revision has been made by Kadereit (1988). He organized 11 sections of the genus into four groups based on morphological traits, particularly characters of capsules:

Group 1: *Meconella, Meconidium, Californicum*
Group 2: *Argemonidium*
Group 3: *Pilosa, Pseudopilosa, Horrida*
Group 4: *Carinatae, Rhoeadium, Macrantha, Papaver*

The members of section *Meconella* (Spach.) belonging to the first group of sections are characterized by their scapose growth habit. All of them are perennial. This section contains about 30 species (Table 14.1). In the high mountains of central Europe, *P. alpinum* L. is the most characteristic species. Six of its subspecies are spread in the Alps; one subspecies is indigenous in the eastern and southern Carpathian regions and one of them in Macedonia. However, the main area of distribution of *Meconella* section is in central, inner, and eastern Asia and the northeastern part of the former Soviet Union (northern Siberia, Kamchatka). It is supposed (Rändel 1974) that the area of Asian diploid taxa is the center of origin. Based on the change of ploidy level, it could be hypothesized that distribution of these taxa follows two directions: northeast and west-west-north. The former direction results in the species of the Behringian region and western North America. The great

Table 14.1 First Group of Sections of Genus *Papaver*

Meconella	Meconidium	Californicum
Papaver alpinum L.	*P. acrochaetum* Born.	*P. californicum* A. Gray
P. nudicaule L.	*P. armeniacum* (L.) DC	
P. croceum Ledeb.	*P. curviscapum* Nab.	
P. radicatum Rottb.	*P. cylindricum* Cullen	
P. canescens Tolm.	*P. persicum* Lindl	
P. tianshanicum Popov	*P. triniifolium* Boiss.	
P. angrenicum Pazij.	*P. fugax* Poir	
P. lapponicum Tolm.	*P. libanoticum* Boiss.	
P. indigirkense Jurc.	*P. polychaetum* Schott et Kotschy	
P. minutifolium Tolm.	*P. arcochaetum* Born.	
P. leucotrichum Tolm.	*P. armeniacum* (L.) DC	
P. laestadianum Nordh.		
P. pulvinatum Tolm.		
P. nivale Tolm.		
P. microcarpum DC		
P. alboroseum Hult.		
P. walpolei Pors.		
P. relictum (Lundstr.) Nordh.		
P. macounii Greene		
P. miyabeanum Tatew.		
P. pygmaeum Rydb.		
P. czekanowskii Tolm.		

Source: Kadereit, J. W. 1988. *Beitrage zur Biologie der Pflanzen* 63:139–156.

morphological similarity of northern European and eastern North American species could be a consequence of a transatlantic transport of this group involving also Iceland and Greenland.

The characteristic alkaloids in these groups are amurensine, amurensinine, amurine, nudaurine, alpinigenine, alpinine, papaverrubine, and muramine (Boit and Flentje 1960; Maturová, Pavlásková, and Šantavý 1966; Rändel 1974). Tétényi (1993) describes the accumulation of isopavine, retroprotoberberine, and protomorphinane alkaloids. It seems that the lack of aporphine alkaloids is the main chemical characteristic of this section (Preininger 1986).

Meconidium (Spach.) is the least known section in the genus (Table 14.1). It involves biennial and perennial species. Species of this section are diploid ($2n = 40$). The distribution of this section includes northwest Iran, the Caucasus, and the Middle East. One species exists in Lebanon.

The major specific chemical constituents in this section are rhoeadanes and armepavine as 1-benzyltetrahydroisoquinoline type alkaloids (Phillipson, Thomas, et al. 1981; Phillipson 1983). In *P. curviscapum*, 1-methoxiallocryptopine was detected (Sariyar, Baytop, and Phillipson 1989). This section is heterogenous not only in the morphology of its species, but in chemical characters as well. At least three chemical races exist: benzylisoquinoline-proaporphine type, morphinane type, and rhoeadane type (Preininger 1986).

The section *Californicum* (Kadereit) has only a single annual species of North American origin (*P. californicum* A. Gray). Chemically, it is characterized by the accumulation of rhoeadine and protopine (Santavý et al. 1960), muramine, and benzylisoquinoline latericine (Table 14.1).

The section *Argemonidium* (Spach.) includes annual species (Table 14.2). They are called halfrosette plants. The peculiarity of their branch system is the few and very long internodes. These species occurr from the southern Alps and Adriatic Sea to the western Himalayas; *P. hybridum* can be found in some Pacific Ocean islands.

Table 14.2 Second Group of Sections of Genus *Papaver*

Argemonidium
P. apulum Ten.
P. argemone L.
P. pavonium Fisch et Mey
P. hybridum L.

Source: Kadereit, J. W. 1988. *Beitrage zur Biologie der Pflanzen* 63:139–156.

The basic chromosome number (x) is 6 (*P. hybridum*) or 7 (*P. argemone*); the maximum level of ploidy is $6x$ for *P. argemone* subsp. *argemone* ($2n = 40$ or 42) (Kadereit 1990). Although *P. apulum* is morphologically between *P. hybridum* and *P. argemone*, cytologically *P. apulum* and *P. hybridum* are diploids while *P. argemone* is hexaploid. In *P. argemone*, rhoeadine and isorhoeadine are present; *P. hybridum* accumulates glaudine and pahybrine, and *P. pavonium* contains roemeridine and its carbonyl derivative (Tétényi 1993). Characteristic alkaloid groups are protoberberine, benzophenantridine, protopine, rhoeadine, and papaverrubine (Preininger 1986).

The section *Pilosa* (Prantl) contains perennial half-rosette subscapose plants with racemose inflorescence (Table 14.3). The level of maximum ploidy in this group is $4x$, chromosome numbers are $2n = 14$ or 28. The occurrence of the section is limited to western Anatolia. Species of this group accumulate morphinane amurine and aporphine, glaucine, and roemerine alkaloids (Öztekin et al. 1985). Rhoeadine, papaverrubine, and protopine alkaloids are found. The presence or absence of amurine (promorphinane alkaloid) means the chemical difference between sections *Pilosa* and *Pseudopilosa* (Preininger 1986).

The habit of the species of section *Pseudopilosa* (Pop. et Günther) shows subscapose form (Table 14.3). These species are separated into two areas of distribution: *P. lateritium* is endemic in northeastern Turkey and occurs also in the Caucasus; *P. rupifragum* and *P. atlanticum* are species of northwestern Africa and southern Spain. The most characteristic chemical compounds that accumulate in the plants are protopine and rhoeadine lacking amurine and glaucine (Tétényi 1993).

The section *Horrida* (Elkan) is represented by a single species, *P. aculeatum* Thunb. (Table 14.3). The habit of this species is usually annual, but in some circumstances it may be biennial (Günther 1975). Its chromosome number is $2n = 22$ and it occurs in eastern South Africa as a species of temperate mixed grasslands. It is also found in Australia, as a result of human activity (Kadereit 1988). The species accumulates salutaridine and aculeatine (Maturová et al. 1966).

The section *Carinatae* Fedde contains only a single species: *P. macrostomum* Boiss et Huet. with three varieties: *bornmülleri* (Fedde) Kadereit, *macrostomum* Kadereit, and *dalechianum* (Fedde) Kadereit (Kadereit 1987). It is characterized by an annual habit (Table 14.4). The chromosome number is diploid ($2n = 14$). The species occurs in Iraq, Iran, and Turkey and in the southern area of the former Soviet Union. A characteristic alkaloid of this species is the macrostomine (Hegnauer 1990). In the natural habitat (Central Asia), macrostomine and sevanine accumulation have been detected.

Table 14.3 Third Group of Sections in Genus *Papaver*

Pilosa	Pseudopilosa	Horrida
P. pilosum Sibth. et Sm.	*P. atlanticum* Ball et Cross.	*P. aculeatum* Thunb.
P. apokrinomenon Fedde	*P. lateritium* Koch.	
P. strictum Boiss. et Bal.	*P. oreophilum* Rupr.	
P. feddei Schwz.	*P. rupifragum* Boiss. et Reut.	
P. spicatum Boiss. et Bal.	*P. montanum* Trautv.	

Source: Kadereit, J. W. 1988. *Beitrage zur Biologie der Pflanzen* 63:139–156.

Table 14.4   Fourth Group of Sections in Genus *Papaver*

Carinatae	Rhoeadium	Macrantha	Papaver
P. macrostomum Boiss et Huet.	*P. albiflorum* Pacz.	*Papaver bracteatum* Lindl.	*P. glaucum* Boiss. and Hausskn.
	P. atlanticum Ball.	*P. orientale* L.	*P. gracile* Boiss.
	P. arenarium Marsch.et Bich.	*P. pseudo-orientale* (Fedde) Medv.	*P. decaisnei* Hochst. Steud. ex Elkan
	P. commutatum Fisch et Mey		*P. somniferum* L.
	P. guerlekense Stapf		
	P. stylatum Boiss. et Bal. ex Boiss.		
	P. carmeli Feinbrun		
	P. clavatum Boiss. and Hausskn. ex Boiss.		
	P. dubium L.		
	P. dubium Kadereit		
	P. lacerum Pop.		
	P. litwinowii Fedde		
	P. rhoeas L.		
	P. rechingeri Kadereit		
	P. pinnatifidum Moris		
	P. purpureomarginatum Kadereit		
	P. umbonatum Boiss.		
	P. postii Fedde		
	P. rumelicum Velen		
	P. syriacum		
	P. humile Fedde		

Source: Kadereit, J. W. 1988. *Beitrage zur Biologie der Pflanzen* 63:139–156.

In the same species, cultivated in Europe, rhoeadine, papaverrubine, and protopine alkaloids are found, showing the existence of chemical races (Preininger 1986).

The characteristic habit of the species belonging to the section *Rhoeadium* Spach. is annual, although some species may be biennial in special circumstances (Günther 1975; Markgraf 1986) (Table 14.4). Level of ploidy is mostly diploid (*P. rhoeas*, $2n = 14$), tetraploid (*P. dubium*, $2n = 28$), or hexaploid (*P. dubium* subsp. *dubium*, $2n = 42$) (Kadereit 1990). Hybridization often occurs among *P. rhoeas*, *P. carmeli*, *P. humile*, and *P. umbonatum*. Spontaneous hybrids occur quite often, such as *P. rhoeas* × *P. dubium* = *P. exspectatum* Fedde and *P. rhoeas* var. *rhoeas* × var. *strigosum* Boenn = *P. feddeanum* K. Wein recognized in Harz (Markgraf 1986; Kadereit 1990). The native geographical distribution area of this group is a bicentral one—southwestern Europe and northwestern Africa—and the other is from the eastern Mediterranean to central Asia. According to Kadereit (1988, 1990), *P. rhoeas* and *P. dubium* have been introduced to Europe together with cereals.

This group is chemically quite unclear. Rhoeadine content (Kalav and Sariyar 1989) and the various latex colors in *P. rhoeas* allies and the existence of different phtalide, morphinoide, and aporphine alkaloids in *P. arenarium* illustrate the great chemical diversity of these species. Thebaine has been detected from a tetraploid cytotype of *P. dubium* (Espinasse 1981). Studies of this section have shown the occurrence of different chemotypes (Preininger 1986). One chemotype contains rhoeadines, and other contains aporphines and proaporphines. Interesting is the occurrence of latericine in this section (*P. californicum*) because this compound has been detected formerly in the section *Pilosa*.

Species belonging to the section *Macrantha* (Elk.) = Oxytona Bernh. are characterized by sca-pose perennial habit (Table 14.4). The ploidy level provides clear delimitation of species in this sec-tion. According to the systematic study of Goldblatt (1974), the hexaploid *P. pseudo-orientale* ($2n =$ 42) is regarded as the transient form between the diploid *P. bracteatum* ($2n = 14$) and the tetraploid *P. orientale* ($2n = 28$). This group contains only three species, which occur in different areas of the Caucasus, Iran, and Turkey. *Papaver bracteatum* tolerates dry habitats, *P. orientale* can be found in alpine conditions, and *P. pseudo-orientale* prefers moist climates.

Papaver bracteatum accumulates thebaine as a dominant alkaloid and, to a much lesser extent, alpinigenin codeine, protopine, and some other alkaloids (Nyman and Bruhn 1979; Meshulam and Lavie 1980). In *P. orientale*, isothebaine (Vágújfalvi 1970) and mecambridine (Phillipson, Scutt, et al. 1981) are characteristic traits. Main alkaloids in different populations of *P. pseudo-orientale* are isothebaine (Tétényi 1993), marcantaline, orientalidine, and salutaridine (Sariyar and Baytop 1980). The most important and intensively studied species of this section is *P. bracteatum* for its thebaine content and the lack of morphine.

According to Kadereit (1986), four species can be found in the section *Papaver* (Tourn) L. They are annuals (Table 14.4) with more or less branched flower axis. The ploidy level in this section is between $2x$ and $8x$. Species are self-compatible with the exception of *P. glaucum*. In the case of *P. somniferum*, two subspecies (*somniferum* Kadereit and *setigerum* (DC) Corb.) are described. The geographical distribution of this group is quite clear. *Papaver somniferum* subsp. *setigerum* occurs in the western Mediterranean region; *P. glaucum* is the species of the Irano-Turanian region; *P. decaisnei* is spread from Sudan to Iran, Pakistan, and Afghanistan; and *P. gracile* is the species of the eastern Mediterranean area.

Based on the presence or absence of morphinanes, the species in this section can be divided into two groups: *P. somniferum* and *P. decaisnei*, which are characterized by the presence of mor-phinane alkaloids, thebaine, codeine, and morphine with narceine. *Papaver glaucum* and *P. gracile* lack morphinanes and rhoeadine but papaverrubine occurs at a higher concentration (Preininger, Novák, and Santavý 1981; Preininger 1986). In *P. somniferum* subsp. *setigerum*, both morphinanes and rhoeadines have been detected (Fairbairn and Williamson 1978). *Papaver somniferum* also accumulates oripavine (Nielsen, Röe, and Brochmann-Hanssen 1983).

Geographical distribution was used for further intraspecific classification of *P. somniferum* by Basilevskaja (1928). She separated eight subspecies. Seven of them (*songoricum*, *tarbagaticum*, *turcicum*, *austro-asiaticum*, *chinense*, *tianchanicum*, and *eurasiaticum*) comprise the varieties of a distinct area, and the last one, *subspontaneum*, involves the cultivated taxa provar. *oleiferum* and provar. *opiiferum*.

More recently, discrimination among species of the *Papaver* genus has been carried out in Japan by molecular tools, comparing the nucleotide sequences of the plasmid rpl16 gene and the rp 116–rpl14 spacer region. Five species were clearly distinguishable, but *P. somniferum* and *P. setigerum* proved to be identical (Hosokawa et al. 2004). Although the authors mention the utiliza-tion of this method against illegal cultivation of *P. bracteatum* in gardens, the results seem to have mainly theoretical importance.

14.3.2 Classification of Domesticated *Papaver somniferum*

The classification of the domesticated *P. somniferum* is rather difficult and results in many contra-dictions. Using the system of Danert (1958), Hammer (1981), and Maas (1986), Hanelt and Hammer (1987) divided the domesticated races into three subspecies: *somniferum* and *songaricum* com-prise the cultivated races (Table 14.5) and the third subspecies, *setigerum*, represents the ances-tral wild one. These subspecies differ in their geographical distribution and the form of stigmatic lobes. In Europe the sulcate predominate and in Asia the flat stigmatic lobes predominate. Within a subspecies, the varieties differ in dehiscence or indehiscence of capsules and in seed color. This

Table 14.5   Classification of Cultivated *Papaver somniferum* L.

	Papaver somniferum		
	subsp. somniferum	subsp. songaricum	subsp. setigerum
Seed color: white, yellowish, pink	var. *somniferum*[a]	var. *albescens* Vess.[a]	
	var. *candidum* Vess.[a]	var. *rubicundum* Vess.[a]	
	var. *roseolum* Vess.[a]	var. *rhodanthum* Vess.[a]	
	var. *paeonifolium* Alef.[a]	var. *igneum* Danert[a]	
	var. *macrocarpum* Coss.[a]	var. *parmulatum* Danert[a]	
	var. *papyrinum* Danert[a]	var. *apiatum* Vess.[a]	
	var. *clausum* Danert[a]	var. *limboflorum* Danert[a]	
	var. *haageanum* Alef.[a]	var. *mundum* Danert[a]	
	var. *coerulescens* Rothm.[a]	var. *livens* Vess.[a]	
	var. *dinocarpum* Alef.[b]	var. *orientale* Danert[b]	
	var. *rubrospermum* Vess.[a]	var. *apertum* Danert[b]	
	var. *sanguineum* Vess.[b]	var. *foratum* Danert[b]	
	var. *rutilum* Danert[b]	var. *fulgidum* Danert[b]	
	var. *hussenotii* Alef.[b]	var. *maculosum* Danert[b]	
	var. *pictiflorum* Danert[b]	var. *hapalanthum* Danert[b]	
	var. *tenerum* Danert[b]	var. *palleolum* Danert[b]	
	var. *contrasticum* Danert[b]	var. *leucomelum* Danert[b]	
	var. *pallidum* Rothm.[b]	var. *gaucescens* (Rothm) Danert[b]	
Seed color: light gray, light blue, dark blue	var. *oculatum* Danert[a]	var. *holonatum* Danert[a]	
	var. *nigrum* Hayne[a]	var. *rubidum* Vess.[a]	
	var. *serenum* Danert[a]	var. *sigillatum* Danert[a]	
	var. *subgriseum* Vess.[a]	var. *nubeculosum* Danert[a]	
	var. *madritense* Vess.[b]	var. *praetextum* Danert[b]	
	var. *quassandrum* Alef.[b]	var. *rotundilobum* Danert[b]	
	var. *spilanthum* Danert[b]	var. *ocellatum* Danert[b]	
	var. *subviolaceum* Vess.[b]	var. *poriferum* Danert[b]	

Source: Hanelt, P., and Hammer, K. 1987. *Feddes Report* 98:553–555.
[a] Indehiscent capsule.
[b] Dehiscent capsule.

classification is mostly artificial because the characters used may be influenced by growing site and are variable, so it is difficult to establish their exact taxonomic status. In modern cultivars, the capsule is usually indehiscent and the seed color is light.

However, none of these intraspecific classifications dealt with the diversification of alkaloid biosynthesis, which can be considered one of the most important characteristics of cultivated poppy. Some attempts were made by Tétényi (1963) based on the diversity of alkaloid synthesis and accumulation. He distinguished chemoconvars morphinan and isoquinoline as well as their chemoprovars. This classification was developed afterward, taking into consideration the two demethylation pathways and the response to photoperiod (Tétényi 1989).

More recently, because of practical considerations (including industrial and narcotic control requirements), the chemical characteristics of the intraspecific taxa have become increasingly important. As a result of the intensification of the worldwide selection work, new cultivars representing diverse chemical taxa are under development. The appearance of the new cultivars, both in Europe and in other parts of the world, made a revision of the old systems for evaluation of poppy cultivars necessary. With the help of Hungarian experts, including specialists of the Hungarian

Figure 14.2   (See color insert.) Cultivar evaluation at the research station of the Hungarian Agricultural Quality Control Organization (Tordas) according to the new DUS evaluation method. (Köck et al. 2001. In *A mák (*Papaver somniferum *L.) Magyarország Kultúrflórája*, ed. Sárkány, S., Bernáth, J., and Tétényi, pp. 244–259. Budapest: Akadémiai Kiadó.)

Agricultural Quality Control Organization, a new distinctness, uniformity, and stability (DUS) evaluation method has been developed (Köck, Bernáth, and Sárkány 2001) and is accepted by the European agency International Union for the Protection of New Varieties of Plants (UPOV). This evaluation process takes into consideration both the morphophenological and chemical characteristics (Figure 14.2). From the chemical point of view, the cultivars represent at least five main chemical groups (chemocultivars):

cultivar group: low alkaloid content (less than 0.01%)
cultivar group: medium alkaloid content (0.03–0.07%),
cultivar group: high morphine content (1.5–3.0%)
cultivar group: high thebaine content (1.0–2.5%)
cultivar group: high narcotine content (1.5–2.5%)

To establish the correlation between the alkaloid formation and other metabolic pathways, H-1 nuclear magnetic resonance metabolic profiling was applied by Hagel et al. (2008). Based on their analysis of latex extracts, a low-alkaloid variety and high-thebaine, low-morphine cultivar were distinguished. Distinction was also made between pharmaceutical-grade poppy cultivars and a condiment variety. Reduced alkaloid levels in the condiment variety were associated with reduced abundance of transcripts encoding several alkaloid biosynthetic enzymes.

Neither morphophenological nor chemotaxonomical classification provides a good and precise distinction of cultivated races. In the Gatersleben gene bank, a collection of 300 accessions from nearly all over the world was analyzed by the method of amplified fragment length polymorphism

(AFLP) (Dittbrenner et al. 2008). The preliminary results showed that all tetraploid *P. somniferum* subsp. *setigerum* accessions form one cluster, while the diploid accessions form another group including subspecies *somniferum* and subspecies *songaricum*. However, concerning the amount and composition of alkaloids, a large and independent variability was found.

14.4 GENETIC RESOURCES

During the thousands of years of poppy cultivation, different populations were used depending on region and the main goal of production. Numerous, more or less selected populations differed in morphological characteristics, life form, seed yield and color, opium yield, alkaloid content, etc.

There is only a little information, even nowadays, on plant populations cultivated in the illicit cultivation areas (the Golden Triangle, the Golden Crescent, Central and South America). Based on available data, rather inhomogeneous populations are applied in both morphophenological and chemical points of view. The selection of populations is motivated by two main considerations: The plant material should adapt well to the local ecological conditions and should give a high opium yield.

In several European and first of all Asian areas, even today, the majority of cultivated materials are landraces, populations without any former consequent breeding. They are well adapted to the local conditions, but usually are far from fulfilling the requirements of intensive agriculture. In Romania, a great diversity was found concerning the investigated morphological characteristics of local poppy populations (Handrea 1996). In Turkey, the seed market offers populations characterized by flowers of mixed color and shape (Anonymous 2007a). Because of industrial demand, the cultivated populations have been revised recently (Gumuscu, Arslan, and Sarihan 2008). Based on chemical analysis of 99 poppy lines, 15 lines with high morphine, 3 lines with high noscapine, and 1 line with high papaverine content were selected.

Indian landraces consist of 20–25 basic types, which have given rise to a wide diversity by intermixing and hybridization during long years of cultivation (Singh, Shukla, and Khanna 1997). The plant material cultivated in India is commonly known as opium poppy. It is a plant of about 120 cm height, with brightly colored flowers. The form of the capsule is oblong to globose and it is filled with white, flat seeds. Singh (1982) attempted to describe three main varieties:

- *P. somniferum* var. *nigrum* DC is a seminaturalized form with purple-red flowers and roundish oblong, dehiscent capsules. The color of the seeds is greenish black and it is a good source for oil production.
- *P. somniferum* var. *album* DC has white flowers. The capsule is roundish ovate and has no opening pores after ripening. It is widely used for opium production.
- *P. somniferum* var. *normale* is characterized by small flowers, which are streaked with green and red; the petals are crumpled and never expand fully. The capsule is oblong roundish, opening by pores. It is an ornamental type.

Some of the cultivated populations of higher importance and those of the newly selected races in India are the following: Telia-1, Bhakua, Aphuri, Kantia Pink, Galania, Mandraj, and Kasuha. Differences include plant height, flower color, capsule shape, morphine content, and vegetation length, among others. Today, an intensive analysis of genetic resources is carried out in India. The local populations and strains may ensure a wide and prosperous genetic basis for developing new cultivars. Shukla et al. (2006) screened 1,470 individual plant samples of 98 germplasm lines for five major economic alkaloids. The chemical diversity was rather high. The morphine ratio ranged from 92.0 to 20.8%, the codeine from 1.69 to 6.48%, thebaine from 0.52 to 7.95%, and narcotine from 8.79 to 17.9%; the papaverine content was from 0.00 to 6.07%. Dubey, Dhawan, and Khanuja (2009) evaluated 35 selected germplasm accessions of opium poppy and found that two genotypes (I-14 and Pps-1) were stable sources for downy mildew resistance. Significant reciprocal differences were found due to maternal transmission of resistance. This indicates the involvement of cytoplasmic genes under nuclear control. They concluded that these genotypes can be used for improving culti-vated populations.

The production of poppy in Europe is by using selection from local populations and landraces. The selection of poppy in European countries interested in production of both seed and industrial straw started in the first half of the twentieth century. It was mentioned by Heeger (1956) that five cultivars of poppy were known at that time in Germany, but he emphasized the presence and dominance of landraces in the commercial production. In Hungary, science-based breeding started around 1930 (Köck et al. 2001). Today, the varietal background of the production shows character-istic differences among cultivation areas depending on the main goal of production.

In areas of highly developed industrial production (western Europe, Australia) and high poten-tial, patented strains are used without cultivar registration by the varietal authority. Because of industrial interest, detailed information on poppy is almost fully missing. These materials are homogenous, developed by different breeding methods and selected for special production charac-teristics. Consequently, their narrow—presumably homozygote—genetic potential can be used in further breeding only after enhancement of variability.

In some middle European countries, the selected cultivars are registered by the national vari-ety offices, similarly to the varieties of many other agricultural crops. The registration authority investigates the candidates according to the valid DUS guideline for *Papaver somniferum*. The official EU cultivar trials for poppy are running in Hungary. Seed of these cultivars and basic information on them are available. However, in cases of industrial varieties, because of the strict regulation of their production the propagation material is distributed only by the processing fac-tories inside their agricultural producers. For the breeders' interest, the best varieties can also be patented. Table 14.6 shows the registered varieties present in the European variety list (Community Plant Variety Office).

Varieties can, most practically, be grouped on the basis of primary utilization (Anonymous 2007b). Thus, industrial cultivars of high alkaloid contents are used for extraction and processing of pharmaceuticals. The culinary varieties are primarily used by households and food industries. For human consumption, either the seeds or the oil extracted from the seed can be used. Currently, no variety is known where the primary aim of breeding would have been to increase seed oil content or to create a special composition.

Table 14.6 Cultivars of Poppy (*Papaver somniferum* L.) on the European List

Industrial (Capsule)	Double Use (Capsule and Seed)	Culinary (Seed)
A-1: Hungary	Bergam: Slovakia	Albin: Slovakia
Alfa: Hungary	Edel-weiss: Austria	Agat: Poland
Botond: Hungary	Gerlach: Slovakia	Albakomp: Hungary
Buddha: Hungary	Kék Duna: Hungary	Ametiszt: Hungary
Csiki kék: Hungary	Major: Slovakia	Aristo: Austria
Extaz: Romania	Malsar: Slovakia	Florian: Austria
Evelin: Hungary	Marathon: Slovakia	Kozmosz: Hungary
Kék Gemona: Hungary	Marianne: the Netherlands	Josef: Austria
Lazur: Poland	Sokol: Czech Rep.	Michalko: Poland
Medea: Hungary	Opal: Slovakia	Mieszko: Poland
Minoan: Hungary	Parmo: Denmark	Przemko: Poland
Monaco: Hungary	Rubin: Poland	Zeno: Austria
Nigra: Hungary	Rosemarie: the Netherlands	Zeno 2002: Austria
Riesenmohn: Germany		Zeta: Austria
Tebona: Hungary		

Source: Anonymous 2007c.

14.5 PRODUCTION OF POPPY

14.5.1 Straw and Seed Production

Traditionally, poppy is cultivated in several countries of the temperate zone (Figure 14.3). It is cultivated for two purposes: for straw (Figure 14.4), which is an important pharmaceutical raw material, and for seeds and fatty oils, which are for alimentary and industrial production. According to the official data of the International Narcotics Control Board (INCB 2007; http://www.incb.org/), the total cultivation area of poppy in the temperate zone is about 100,000–200,000 hectare. However, it is changing year by year (Figure 14.5). The most important countries and their share in poppy straw and seed production are shown in Table 14.7. Considering the production area and quantity of harvested poppy straw as well as seed, the main European producers are Czech Republic, Turkey, Ukraine, France, Hungary, Spain, Slovakia, and Austria. Furthermore, Australian poppy production is increasing continuously.

Poppy can be successfully grown under various ecological conditions. The required heat amount is 2000–2200°C, which is ensured in the majority of the previously mentioned cultivation areas. Using the proper cultivars, the photoperiod (duration of sunshine) might not be a limiting factor either. However, the amount and distribution of precipitation may limit the cultivated area.

Figure 14.3 Large-scale cultivation of poppy in Hungary producing industrial raw material and seed for culinary use. (Dunaföldvár)

Figure 14.4   Poppy straw, which is an important pharmaceutical raw material for industrial production of opiates.

The highest demand for water appears at the time of shoot elongation and capsule development. In some cases, artificial irrigation may be necessary. On the other hand, too much rain can be unfavorable during flowering and capsule-ripening stages. This decreases dry matter and alkaloid production. In humid regions of the temperate zone, damage by plant pathogenic fungi also increases.

Poppy grows in clay-loam soil of high fertility with neutral or slightly alkaline pH. The most suitable soil types are chernozem and forest- and washland soils of neutral character (Földesi 1995). As the poppy has very small seeds, it requires careful soil preparation before sowing.

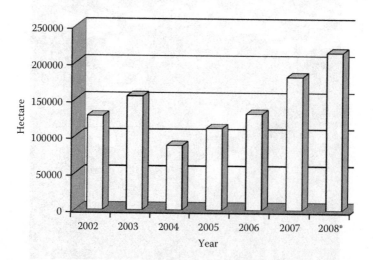

Figure 14.5   Changes of cultivation area of poppy in the temperate zone between 2002 and 2008 according to the data of INCB. (INCB. 2008. *World drug report.* New York (Vienna): UN Office on Drugs and Crime, 2007.) (*Estimated data.)

Table 14.7 Countries Involved in Cultivation of Poppy for Production of Straw and Seed

Country	Area of Cultivation (ha)	Yield of Poppy Straw (kg)	Yield/ha (kg)
Australia	3,457	399,4000	1,155.3
Austria	3,858	—	—
China	1,400	1,575,000	1,125.0
Czech Republic	53,290	36,695,200	688.6
France	6,632	5,040,420	760.0
Germany	14	—	—
Hungary	4,322	2,005,064	463.9
Poland	1,510	—	—
Slovakia	1,326	178,000	134.2
Spain	2,146	1,442,000	671.9
Macedonia	245	73,551	300.0
Turkey	42,023	27,443,000	653.0
Ukraine	10,387	—	—

Source: INCB data for 2006.

Fertilization requirements depend on soil type, previous crop, etc. Because of the short vegetation period of poppy, nutrients prepared in easily taken-up forms can be utilized. Farmyard manure is generally avoided because of its favorable effect on growing of weeds and lowering the efficacy of herbicides in weed control. Manure is better recommended for the previous crop. Because of the interaction of many factors, the optimum dose of fertilizer changes from site to site and has to be characterized by broad intervals. Generally, the main macroelements are added in the following dosages (calculated in active agents): nitrogen 140–160 kg/ha, phosphorous 70–110 kg/ha, and potassium 80–100 kg/ha.

Nitrogen is the most effective nutrient. It was observed that in certain cases the high dosage of nitrogen may harm plants and is not recommended above 60–80 kg/ha (Ruminska 1973). However, higher doses are required to get economical yields under intensive cultivation. Under irrigated conditions, an increase of the nitrogen amount (200–240 kg/ha) may be effective (Földesi 1994). The contradictory data concerning the effectiveness of nitrogen might be the result of the differences found in both soil types and nitrogen reserves (Dachler 1990). A crop producing 3500 kg/ha vegetative mass and 1200 kg/ha capsules with seeds uses 103 kg nitrogen, 192 kg phosphorous, and 11 kg potassium (Földesi 1992). Boron is the most important among the microelements in the practice. Generally, an amount of 25–30 kg/ha ensures the proper development of poppy plantation. According to Bulgarian results, application of boron with Zn, Mn, and Cu increases yields by 28% (Popov et al. 1971).

One of the preconditions of successful poppy production is the optimal time of sowing. In southern Europe, both autumn and spring sowing is applicable (Földesi 1995). In general, poppy should be sown as early as possible in spring, not later than early grain crops of the given region. Practically, the depth of sowing is between 0.1 and 1.5 cm. Under optimal soil conditions, 0.5–1.0 cm gives the best results. In very loose soils, 2–3 cm depth is applied; this is a practice mainly in southeastern Europe (Shulgjin 1969). On average, 300,000–400,000 plants per hectare give optimal development and production (Földesi 1992). In Australia, about four to five times more plants per hectare are cultivated (Laughlin, Chung, and Beattie 1998).

Poppy seeds germinate within 10–14 days. During this period, cracking of the soil may severely harm or even stop germination. The surface of the soil that is capable for cracking has to be loosened regularly.

If the plantation is settled by traditional methods, the poppy stand should be thinned at the phase of four to six leaves. In large-scale cultivation, the thinning is avoided by special methods like sowing an irradiated seed mixture (Földesi 1994) or applying precise seed drills.

Irrigation may be necessary on very loose soils or in a region where the precipitation is extremely low during sprouting, at the time of development of the leaf rosette and budding. According to Shulgjin (1969), the usual quantity of water is 800 m³/ha, which should drain out when the soil moisture content falls below 70%. In southern European conditions, poppy sown in spring is regularly irrigated and three to four irrigations are recommended. However, the majority of poppy cultivation in temperate zones is managed without irrigation or irrigation is applied only in extremely dry years. In Australia, the intensive cultivation technique is combined by regular irrigation and at least three irrigations are applied during the vegetation period (Laughlin et al. 1998).

In large-scale culture, proper weed control is one of the most important elements of cultivation technology and is responsible for success. Poppy may be attacked by a number of parasites and pests during the vegetation period. The most dangerous insect in poppy plantations is the weevil (*Ceutorrhynchus macula-alba* or *C. denticulatus* in southern regions). Poppy fly (*Dasyneura papaveris*) and poppy gnat (*Perrisia papaveris*) are parasites that use the hole made by the weevil for laying the eggs into the capsule. In certain years, plant lice (e.g., *Aphis fabae*) and soil parasites may also cause higher damage.

The diseases of poppy cause increasing damage in rainy weather. Seed treatment against some fungi (*Fusarium* spp., *Erwinia* spp., *Helminthosporium* spp.) before sowing is useful. It is also part of the protection against peronospore (*Peronospora arborescens*), which is the most harmful disease of poppy. Powdery mildew (*Erisiphe communis*) first damages the stem and then attacks the leaves, which then begin to get white. As a secondary pest, smut mold (*Apiosporium* spp. or *Microsphaerella tulasnei*) may often appear in the form of a black layer on the damaged capsules. In certain cases, other parasites and pests (e.g., *Ceutorrynchus denticulatus, Stenocarus fuliginosus, Clinodiplosis papaveris, Timaspis papaveris, Opatrum sabulosum, Ervinia carotovora,* and *Entyloma fuscum*) may appear in the poppy field, which necessitates protection.

In the temperate zone, poppy is usually produced for both straw and seed; therefore, it has to be harvested when the seeds are completely ripe and the capsules are "straw yellow" or greenish yellow and crack when pressed (Figure 14.6). The optimum condition for harvesting is dry weather, after dew. After harvesting and separating both seeds and capsule, straw has to be stored properly. Capsules are stored at first in 5-cm-thick layers and turned over every day. Generally, no artificial dryers are necessary, but natural air ventilation is advised under wet weather conditions. After some days, the product may be packed in bigger sacks. The required water content of the capsules is 10–12% and it cannot be stored when content is over 14%. The water content of the seeds, which is a precondition for long-term storage, is about 9% (Hörömpöli 1995). The yield of poppy is changing to a large extent, varying between 0.3 and 2.0 t/ha for seeds and 0.5 and 4.0 t/ha for the capsule.

14.5.2 Production of Opium

14.5.2.1 Licit Production of Opium

In tropical countries, the plant is mainly cultivated for the opium (Singh 1982). The seed, which is considered to be a by-product, is utilized in culinary activities and is said to be a good source of oil and protein. The main product is opium, which is the sun-dried latex of the unripe capsule taken by lancing and formulated afterward by a simple technological procedure. From the practical point of view, five alkaloids of main importance are distinguished in opium: morphine, codeine, thebaine, narcotine, and papaverine. The opium, on dry basis, contains 9–14% morphine, 0.7–2.5% codeine, 0.3–1.5% thebaine, 5.5–11.0% narcotine, and up to 1% papaverine (Madyastha and Bhatnagar 1982).

Figure 14.6 Optimal development stage of capsules in the temperate zone for harvesting stand producing both straw and seed.

About 200–300 tonnes of opium have been produced recently (INCB 2008); only eleven countries are involved into the illicit production (Table 14.8). The importance of licit opium production is decreasing continuously (INCB 2007). World production of opium decreased from 820 tonnes to the third between 2002 and 2008 (Figure 14.7).

From a practical point of view, India is the only country that plays an important role in the licit production of opium. The first records on cultivation of opium poppy in India date back to the fifteenth century (Kohli 1966). At first, it was cultivated along the seacoast and penetrated into the peninsula later. During the Mughul Empire, the production of opium had a great importance and it became a valuable object of trade with China and other countries. In the second half of the eighteenth century, the East India Company took the right for controlling opium production, especially in Bengal and Bihar, which went later into the hands of the British governor. From that time up to the independence of India, the British authorities controlled all the fields of production, processing, distribution, and sale of opium. Since April 1, 1950, the Indian government has had control of poppy cultivation and the opium business. A central organization, the Narcotics Commission, was established to unify and rationalize the control system throughout the country.

Based on practical and scientific experiences, the opium poppy can be cultivated in deep, clay-loam soil of high fertility in India. At the same time, a good water supply is required that can be obtained using irrigation facilities. The climate of the selected area has to be moderately cold (20°C), with adequate sunshine during vegetative growth. However, in the reproductive phase, the optimum temperature is much higher (30–38°C) and sunshine is more preferable. The cold, cloudy, windy weather in this phenophase may have an adverse effect on the crop, reducing the flow of latex after lancing.

The cultivation of poppy in India starts after the rainy season. However, the optimum time for sowing varies according to the region. Manures and fertilizers are added at the time of soil

Table 14.8   Countries Involved in Illicit Cultivation of Poppy for Opium in 2006

Country	Area of Cultivation (ha)	Yield of Opium (tonnes)	Opium Yield/ha (kg)
Southwest Asia			
Afghanistan	165,000	6,100	36.9
Pakistan	1,545	39	12.9
Southeast Asia			
Lao PDR	2,500	20	8.0
Myanmar	21,000	315	15.0
Thailand	Trace	Trace	~
Vietnam	Trace	Trace	~
Latin America			
Colombia	1,023	13	12.7
Mexico	5,000	108	21.6

Source: INCB. 2008. *World drug report.* New York (Vienna): UN Office on Drugs and Crime, 2007.

preparation. Farmers prefer to apply 10–20 tonnes/ha farmyard manure, which helps to maintain the soil in good physical condition and accelerates the initial and vegetative growth of the plant. The application of other nutrients varies depending on the soil conditions, cultivation area, and the preceding crops (Jain and Solanki 1993).

The seed is sown in a very well-pulverized seed bed. If it is sown by broadcasting, 8 kg/ha seed rate is suggested, but 5–6 kg/ha is enough using 30-cm rows (or 25 cm in light soils). When the sowing is done by broadcasting, the seed is mixed with ash to give a uniform spreading of seeds in the bed. The seeds are covered afterward with a thin soil layer for ensuring conditions for good germination. Germination normally takes 5–10 days, depending upon the moisture content of the soil (Gupta 1974).

The plant grows very slowly in the first phase of its development. The right time for weeding and hoeing is about 3–4 weeks after sowing. It can be followed by irrigation. A second weeding should be carried out 50–60 days after sowing and can be used to combine with thinning to get 10 cm distance between plants in the rows. The best time for thinning is when the plants are 5–6 cm high with three

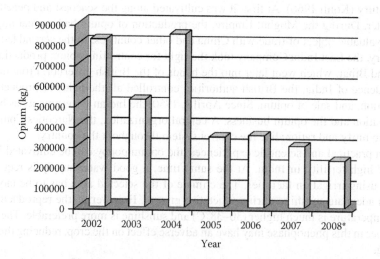

Figure 14.7   Changes in licit world production of opium between 2002 and 2008 according to the data of INCB. (INCB. 2008. *World drug report.* New York (Vienna): UN Office on Drugs and Crime, 2007.) (*Estimated data.)

to four leaves. The optimum spacing for opium poppy is reported to be 30 cm (row) × 10 cm (between plants); this provides a plant population of about 333,000 plants per hectare.

Usually, the crop needs as much as 7–12 irrigations during the cultivation period. The shortage of water in the rosette stage till the appearance of capsules causes enormous losses in latex and seed yield (Turkhede et al. 1981).

To protect plants against termites and shoot-cutting caterpillars, some insecticide is applied to the soil at the time of the land preparation. A presowing treatment of the seed by fungicides may be effective against fungi. Under Indian conditions, the crop can be damaged by many diseases and insects. The most important ones are downy mildew (*Peronospora arborescens*), powdery mildew (*Erysiphe polygoni*), and root rot (*Rhizoctina bataticola*). Curly leaves (resulting from a virus) are frequently observed in cultivation. The disease is spread by jassid and aphis species. Virus-affected plants should be burned. Soil nematodes together with action of fungi and bacteria have an adverse effect on latex production and seed yield, causing a special disease named plant leprosy (Ramanathan 1983). The poppy plants can be affected by root-lesion nematodes (*Pratylenchus* spp.) and root-knot nematodes (*Meloidogyne* spp.).

The optimum time for lancing is a cultivar-dependent phenomenon. At lancing, the capsule should reach its full size, but should be green and immature. For getting latex, a special knife (*naka*), which has three to four sharp-edged blades fitted at a distance of 1.5–2.0 mm from each other, is used. Making a longitudinal incision with this knife, a standard number of incisions of 1–2 mm depth are made on the immature capsule. The length of incision should be one-third but less than the full length of the capsule. Based on practical experiences, bright, sunny days are the most suitable for latex yield, especially in the early afternoon hours. The latex flows out from the cut end and is deposited on the capsule surface. Four lancings are usually made on each capsule at 2-day intervals (Turkhede and Singh 1981; Ramanathan 1982). The latex that exudes out through the lancing becomes thick during the cold night.

The collection of the latex is done early in the morning with a blunt-edged iron scoop called a *charpala*. The latex is usually collected in plastic containers. Using good varieties, 50–60 kg of opium yield is expected from 1 ha. The lanced crop is left for 20–25 days on the land while capsules reach the full maturity stage. Seeds are separated from the collected capsules by beating the capsules with a wooden rod. In India, the average seed yield is about 500–600 kg/ha.

14.5.2.2 Illicit Production of Opium

Around 7000–8000 tonnes of illicit opium are produced worldwide (INCB 2008). The area of illicit cultivation is increasing continuously (Figure 14.8). The opium is used by the inhabitants of the poppy growing areas for curing as well as opium addiction, religion ceremonies, etc. Based on the estimates of WHO and many countries where the opium poppy is grown, substantial segments of the rural populations are dependent upon opium (Smart and Archibald 1980). The rate of addiction among adults in these countries, including large areas of the Golden Triangle (Anonymous 1967) as well as areas of India, Iran and Pakistan, varies between 3 and 10%. However, poppy is not only cultivated in these countries for drug needs of the farmers but also is one of the few cash crops they have.

Southwest Asia (the Golden Crescent) became one of the regions experiencing the most illicit production (Table 14.8). The main producers in this region are Afghanistan and Pakistan. Little information is available from Iran and Lebanon. The production of opium is going on in 165,000 ha in Afghanistan and the amount of opium produced there is 6,100 tonnes (INCB 2008).

The poppy is cultivated in this region of the world mostly as a crop sown in autumn. Seeding takes place during October and December. The time period is dependent on the ecological conditions of the countryside. According to the data, the amount of seed used for plantation varies over a wide range between 5 and 10 kg. In the countryside, especially in Afghanistan and the dry

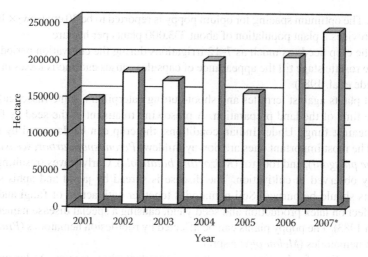

Figure 14.8 Changes in illicit world production of opium between 2001 and 2007 according to the data of INCB. (INCB. 2008. *World drug report.* New York (Vienna): UN Office on Drugs and Crime, 2007.) (*Estimated data.)

areas of Pakistan, the precondition of success of cultivation is proper irrigation. Irrigation is usually started only in spring and is repeated 8 to 14 times until the capsules ripen, according to the weather conditions.

The lancing of capsules begins after 15–20 days of flowering, when the maximum yield of latex and morphine is expected. Lancing is done in Afghanistan six times a year and repeated at intervals of 2–3 days. It is carried out during May and June in the majority of cases. However, data are also available about lancing of capsules as early as April in the extremely warm regions or postponing it till August, if the weather is cool. In Pakistan, plant development is accelerated due to warm conditions and the harvest can start earlier in March.

In this countryside, the high activity of clandestine laboratories is reported to be very active in processing opium to get commercial products of morphine and heroin. According to the government data of Lebanon (INCSR 1994), from 2.5 to 3.0 tonnes of morphine base are converted into heroin annually in that country.

A lot of information exists concerning the opium production of the ethnic groups in Southeast Asia (Golden Triangle). Analysis was carried out by several research groups supported by the World Health Organization, the UN Division of Narcotics, and other UN authorities (Suwanwela et al. 1976, 1978; Chindarsi 1976; Geddes 1971; Saihoo 1963; Crooker 1988; Crooker and Martin 1992). The main producer country is Myanmar (Table 14.9).

Poppy is cultivated in the surroundings of the settlements of the ethnic groups. However, the actual place of cultivation is dependent on many factors. By the estimation of UN experts (Anonymous 1967), the climatic conditions and altitude of the land have great importance. In some tropical regions, the cultivation is restricted to medium height (about 1200 m). The size of the field is dependent on the structure of farms, the spectra of other cultivated species, number of persons in the families, and availability of labor. Poppy fields are reported to be between 1.2 and 8.0 ha.

The poppy is usually propagated at the end of August or in September. In some districts, different cultivars (Taw, Phi, Glang) are used to get a longer harvesting season. Before sowing, the seeds are mixed with lettuce, parsley, or other small seeds. The mixture is sown by broadcasting and covered up lightly by soil and pressed with fingers. After germination, the poppy field needs constant weeding to keep the soil clean. Thinning is the next step of cultivation, connected with taking off the plant individuals of the vegetable that has been germinated from the mixture of seeds.

Table 14.9 Summary of Data about Inheritance of Plant Characteristics of *Papaver somniferum* L.

Characteristics	Feature of Inheritance	Original ref.
Plant height	Dominance, overdominance Heterosis Negative heterosis Dominance/additive gene actions Maternal effect	Levy and Milo 1998 Dános 1965; Sharma et al. 1997 Kálmán-Pál, Bernáth, and Tétényi 1987 Singh et al. 1999 Singh, Shukla, and Khanna 1995 Shukla et al. 1999 Khanna and Gupta 1989
Number of capsules	Maternal effect Heterosis Additive effects	Tétényi,Lőrincz, and Szabó 1961 Sharma, Lal, and Mishra 1988; Sharma et al. 1997 Shukla et al. 1999
Lacerate leaf	Double loci, recessive	Sharma et al. 1991
Size of flowers	Diploid genome	Levy and Milo 1998
Double petal	Monogenic, recessive Polyallelism	Levy and Milo 1998; Belyaeva 1988
Divided petal	Monogenic, dominant	Belyaeva 1988
Unusual stamina	Monogenic, partial dominance	Belyaeva 1988
Color of petals	Monogenic, polyallelism	Bhandari 1989
Ripening time	Nonadditive effects	Shukla et al. 1999
Capsule mass	Polygenic, nonadditive gene actions	Kandalkar et al. 1995
Capsule length	Heterosis	Singh et al. 1999
Capsule size (big capsule)	Monogenic, dominant	Patra et al. 1992
Seed yield	Polygenic Heterosis effect Heterosis Nonadditive effects	Levy and Milo 1998; Kálmán-Pál et al. 1987 Sharma et al. 1997 Singh et al. 1999 Shukla et al. 1999
Total alkaloid content	Monogenic, recessive Polygenic	Straka et al. 1993; Nyman and Hall 1974 Nothnagel, Straka, and Schultze 1996
Morphine content	Intermedier Heterosis effect Additive/dominance gene actions Negative heterosis Overdominance	Dános 1965; Morice and Louarn 1971; Kálmán-Pál et al. 1987; Singh et al. 1995 Tóthné-Lökös 1997 Srivastava and Sharma 1987 Lal and Sharma 1995 Kaicker 1985
Narcotine content	Heterosis effect Negative heterosis Its lack is dominant	Kálmán-Pál et al. 1987 Lal and Sharma 1995 Khanna 1985
Codeine content	Heterosis effect Negative heterosis Polygenic	Kálmán-Pál et al. 1987 Lal and Sharma 1995 Tookey et al. 1976
Thebaine content	Recessive Negative heterosis Heterosis effect	Böhm 1965 Lal and Sharma 1995 Khanna and Shukla 1989

(continued)

Table 14.9 Summary of Data about Inheritance of Plant Characteristics of *Papaver somniferum* L. (Continued)

Characteristics	Feature of Inheritance	Original ref.
Papaverine content	Dominance gene effects, heterosis	Shukla and Khanna 1987
	Additive × dominance gene actions	Shukla et al. 1999
Latex yield	Nonadditive effects	Lal and Sharma 1991; Shukla et al. 1999
	Heterosis effect	Lal and Sharma 1995
Oil content of the seeds	Heterosis effect	Singh and Khanna 1991
Peronospora resistance	Recessive, polygenic	Kandalkar et al. 1995

Source: Németh, É. 2002. In *Recent progress in medicinal plants, vol. 4. Biotechnology and genetic engineering*, ed. Govil, J. N., Kumar, A. P., and Singh, V. K., 129–141. Houston, TX: Sci. Tech. Publ

Four months after the sowing, the flowers appear, followed by the quick development of capsules. The optimum stage of green capsules for lancing starts in November and lasts until February or March in the case of late cultures. The plant height in this phase ranges between 40 and 100 cm, depending on soil and other ecological factors. The quality of the crop is characterized by the number of capsules formed in a single individual.

Young leaves of the plant are consumed in the early stages of the development similarly to the leaves of other edible vegetables. The main product is the opium taken from the capsule by tapping. Each capsule is tapped only once, but in very good areas it is possible to repeat this. The tapping has to be started after flowering and its effectiveness depends on many factors. The work usually begins in the early morning and continues till sunset. Bright and sunny days are considered to be the most effective ones for high resin yield. In contrast, rainfall, which is very rare in this period, can ruin the entire crop.

Special iron knives are used to take off the resin from the capsule. Its form and construction differ from one region to another. The tapper makes two or three incisions using this special knife on the capsule, from which the milky-white resin slowly flows out. The resin is left on the capsule to dry for 1 day and scraped off the next day. For scraping, broad, moon-shaped brass or iron plates are used. Large differences in yield are found in the data, which report yield from 10 up to 30 kg/ha. The semidry and sticky resin is collected in bamboo or iron containers. In the course of drying, the white color of the resin turns into dark brown.

Opium production in Latin America has less importance compared to the production of the Golden Triangle and Golden Crescent regions. The most important countries are Colombia and Mexico. As a result of increasing governmental and international control, the areas of cultivation and opium production are being reduced.

14.6 GERMPLASM ENHANCEMENT

14.6.1 Selection

Because of the large morphophenological and chemical diversity of poppy, selection is the most effective method of breeding today. A prerequisite for selection is the appropriate genetic variability of the starting population. Considering the wide distribution area and high degree of adaptability, several populations exist that are suitable for efficacious selection (Handrea 1996).

In the case of poppy, selection mostly means individual selection and is ensured by good self-pollination ability and seed production. However, because of the possibility of allogamous fertilization, in practice, isolation of the flowers proved to be necessary.

The efficacy of selection may be improved by selection pressure. Bernáth et al. (1988) describe an example. Under cool temperature and poor light conditions, narcotine and codeine cannot be detected in phenotype or have a low level. Selection under such conditions could help in identifying the genetic potential of the strains for accumulation of the alkaloids. Selection under provocative conditions is used also in resistance breeding (Hörömpöli 1998).

As a result of an intensive selection effort, the winter poppy cultivar Zeno Wintermohn has been developed from a landrace in Austria and seed yield is 30% more compared with traditional cultivars and may reach 1450 kg/ha (Dobos 1996). Since then, improved strains of this population have been developed.

The limit of selection has been assumed by Kopp, Csedő, and Mátyás (1961), who state that extra high alkaloid levels of parent individuals could never be reached by their progenies. A similar finding was described by Khanna (1985, cited in Németh 2002), who declared that the maintenance of high alkaloid level (morphine, papaverine) by continuous self-pollination is impossible. Also, our practical observations show that after reaching a certain level of alkaloid accumulation by simple selection, new methods are needed for increasing the effectiveness.

14.6.2 Combination Breeding

Hybridization started in poppy breeding in the 1940s. Both intra- and interspecific crossings of poppy may result in valuable new materials. The methodological basis of crossing is emasculation of flowers at the time when pollen grains are not ripened and self-pollination has not occurred. After removing the anthers, exposure of the stigma is necessary and at the same time or the following day pollination (crossing) can be fulfilled. Mórász (1979) declares a day after emasculation is optimal for crossing. In the hybrid generation, further selection is going on for the desired traits; this may last 3–4 years until stabilization of the genotypes is achieved. Some of the best known cultivars were developed this way (e.g., Mahndorfer, Strubes Blauer, and Eckendorfer Blausamiger in Germany; Kompolti M and BC-2 in Hungary; Reading in England). The widespread Hungarian cultivar Kék Duna has been developed by crossing *P. somniferum* and *P. orientale*, followed by selection of progenies for capsule size, seed color, and morphine content (Lőrincz and Tétényi 1970, cited in Németh 2002).

When crossing distant cultivars, evaluation of the early progenies may be difficult due to heterosis (Dobos and Vetter 1997). The results of the crossing cannot be predicted because of lack of appropriate information on inheritance for most features. Pedigree method was successful in combining and fixing capsule yield, opium and seed production, increased morphine content, and lodging resistance (Levy and Milo 1998).

Backcrossing is used in poppy mainly after interspecific crossings and for elimination of unwanted properties of the "wild" type. It happens also after mutation treatments for fixation of advantageous properties of the original cultivar. Shukla et al. (1999) suggested the possibility of developing genotypes of high papaverine content by backcrossing.

During crossings, the effects of the male partner may appear in the tissues of the mother plant. Bernáth et al. (2003) proved metaxenia for the alkaloid content of the capsules for the first time in poppy. If an alkaloid-free cultivar was pollinated by another one rich in morphine, the morphine content of maternal capsules increased by 0.9–7.5 mg/g. The appearance of narcotine in the capsules of narcotine-free cultivars could be explained as a consequence of pollination with a high narcotine containing partner.

Inheritance of several morphological characteristics of poppy has been studied and different gene actions described. Table 14.9 shows information on the inheritance of plant characteristics based on different references (Németh 2002).

14.6.3 Heterosis Breeding

Heterosis breeding can be applied in poppy with success because both self- and cross-pollinations are possible; it has high seed propagation rate and advantageous flower constitution and size

(Dános 1965). Heterosis breeding for improvement of different traits of poppy has been published, especially in the Indian literature. Heterotic effect can be observed even through two generations (Kaicker 1985). The author justified development of F_1 generation because he found mainly nonadditive genetic effects for almost every economically important characteristic (e.g., opium and seed yield). At the same time, he suggests further improvement by selection in later generations because additive effects may not be excluded in several cases. However, contradictory results have also been published. Singh, Tiwari, and Dubey (1999) found considerable inbreeding depression for six characters in F_2 generation.

Choosing appropriate genotypes was carried out after testing of the combination ability in diallel trials and improving it by recurrent or reciprocal recurrent selection (Heltmann and Silva 1978). After development of the parental lines, the technical method of heterosis breeding is similar to combination breeding (Mórász 1979). According to the author, the difficulty of heterosis poppy breeding lies less in choosing the right parental components than in producing enough seeds. Until now, no appropriate male-sterile source has been known and emasculation and pollination by hand require a lot of labor and expense. Formerly, experiments were carried out to produce male-sterile materials by the help of mutation and interspecific hybridization. However, there is no information about the results (Levy and Milo 1998). From time to time, some publications mention the existence of hybrid strains (Sharma et al. 1997). No hybrid cultivar has been produced at commercial scale yet because of huge costs.

Instead of classical hybrid breeding, development of synthetic varieties is often used and it is the most suitable method in poppy breeding (Khanna and Shukla 1989). These varieties may keep up with F_1 hybrids in production capacity; productivity does not decrease radically in the next generations and the cost requirement is much lower.

14.6.4 Polyploid Forms

Polyploid forms of poppy have been developed both for scientific and practical reasons. The applied methods include interspecific crosses and mutation treatments (colchicine, methane sulfonate, γ-irradiation). It was assumed that an increase in chromosome number would increase the content of active agents.

Polyploid materials exhibited an increased morphine level and more capsules (Andreev 1963, cited in Levy and Milo 1998; Chauhan and Patra 1993). Triploid strains, especially, showed considerable advantages. In Hungary, significant increase was expected by polyploid breeding (Kiskériné et al. 1977, cited in Németh 2002), but practical results were not achieved. The biggest problem was the decrease of seed production (from complete sterility to 7.0% fertility) in the tetraploid poppy forms compared to the diploid cytotype. This is a common phenomenon in other species.

14.6.5 Mutation

Several properties of poppy can be modified by mutation breeding (Khanna and Singh 1975, cited in Levy and Milo 1998; Belyaeva 1988; Nigam, Kandalkar, and Dhumale 1990). Male sterility, lack of opium production, increase in morphine accumulation, multiplication of capsule number, dwarf growth, and early flowering can be induced by chemical mutagenesis (ethylene-amide, nitroso-ethyl, ethyl-methane sulfonate) or by irradiation.

Shifting the biosynthetic pathway by mutations into a desired direction seems to be the most useful practice. In a screening study, the Swedish variety Indra proved to be almost morphine free, accumulating only 0.02–0.03% morphine in dry capsules and 1.0–2.0% in opium. It was supposed to be the result of a spontaneous mutation and proved that the morphine-free character is based on the lack of vesicles and latex production (Nyman and Hall 1974). Later, Indian researchers achieved the same result by using gamma rays (100–800 Gy) and ethyl-methane sulfonate (EMS 0.4%) and developed non-narcotic (alkaloid-free and opium-less) poppy with a high yield of seed and seed oil (Sharma et al. 1999).

The Swedish alkaloid-free cultivar Soma is the result of a spontaneous mutation. Today, it is not produced at commercial scale, but it has been used for development of further low-alkaloid varieties since the middle 1970s. Using it as a parent in Poland, the variety Przemko has been registered with a morphine content of 0.05% in the ripened capsules. Later, improvement of lines for even lower morphine content and characteristic marker traits such as laciniated and colored flowers was carried out (Liersch, Szymanowska, and Krzymanski 1996). As a result, varieties Michalko and Mieszko were developed for utilization in culinary activities. Their seeds are blue with a yield potential of 1.0–1.2 t/ha and contain 48–49% oil.

Recently, Australian researchers reported the discovery of a mutant that accumulates thebaine and oripavine and does not complete the biosynthesis into codeine and morphine (Millgate et al. 2004). Other mutants, like the one accumulating high levels of narcotine (noscapine) in addition to morphinanes, are also mentioned but are still without commercial importance (Ziegler and Kutchan 2005).

14.6.6 Enhancement by Biotechnology

Today, the quickly developing gene-manipulation technology has produced numerous results in the genetic development of poppy. Biotechnology of poppy in the last decades includes the following areas: search for molecular markers and identification of genotypes, enzymatic studies, gene sequence identification, and genetic transformation activities.

It is obvious that molecular markers may have a higher importance at an intraspecific level in variety testing and identification of cultivars. For this reason, several techniques have been applied in the last decades. In Hungary, isoenzyme studies were carried out examining 12 varieties and eight isoenzyme systems. Margl, Szatmáry, and Hajósné-Novák (2001) found that the GOT and 6-PGDH systems were identical in each examined cultivar, while the allelic constitutions of ACP, DIA, GDH, LAP, MDH, and SAD isoenzymes were different. The enzyme systems seem to be encoded by two to four loci and two to nine alleles. Except for GDH, all of the enzymes were active in each of the samples taken from flowers, green capsules, and seeds.

Straka and Nothnagel (2002), in Germany, examined the segregation of six morphological and 125 molecular markers in F_2 generation as a result of a low-morphine and high-morphine crossing. A total of 87 marker loci could be placed in 16 linkage groups (basic chromosome number of *P. somniferum* is $x = 11$). Among the morphological traits, only the number of stigmas could be linked. The aim of this study was to develop a linkage map of poppy, the first step of mapping genetic loci corresponding with the morphine content. However, there are still no data on the correlation with morphine contents with the mentioned molecular markers.

A series of studies was carried out in order to identify and isolate the enzymes involved in alkaloid formation of poppy. Enzymes of the general cellular metabolism were detected in isolated poppy latex in the 1970s and 1980s.

In Canada, research has been carried out for clearing up the genetic backgrounds of alkaloids with antibiotic effects localized in the roots of poppy (sanguinarine and berberine) and related species (e.g., *Eschscholzia californica*). More than 30 enzymes participating in biosynthesis of the benzylisoquinoline alkaloid have been isolated and described from cell cultures. Because of the high economic importance, the morphinane branch of the benzylisoquinoline pathway has been even more intensively studied and all enzymes of the pathway are now characterized and partially purified, with the exception of the demethylation step downstream of thebaine.

Data suggested the involvement of multiple cell types in alkaloid biosynthesis of poppy. While berberine bridge enzyme is localized to parenchyma cells of the root cortex, *O*-methyltransferases are found in the pericycle; typically, codeinone reductase is only found in organs where laticifers are present (Weid, Ziegler, and Kutchan 2004). The activity of the enzymes changes during ontogenesis.

Molecular cloning of pathway genes has been successful since the middle 1990s. One of the first genes identified from poppy was that encoding tyrosine-3,4-dihydroxyphenylalanine decarboxylase. The investigation proved that regulation is based on a family of 10–14 genes, from which several have already been isolated (Facchini et al. 1998). Their expression changes during ontogenesis and seems to be organ specific. According to Nessler (1998), the differential expression of the TyDc/DODC gene family is transcriptionally regulated and promoters from these genes may be useful for metabolic engineering of different alkaloid pathways in poppy.

In the past few years, the basic benzylisoquinoline pathway up to (S)-reticuline has been elucidated and for all enzymes cDNAs have been isolated from several plant species. Ounaroon, Frick, and Kutchan (2005) reported on the isolation of cDNAs for norcoclaurine 6-O-methyltransferase, (S)-N-methylcoclaurine 3'-hydroxylase, and berberine bridge enzyme. Although much less is known about the pathways downstream of (S)-reticuline, information on their genetic determination is accumulating. Recently, cDNAs for salutaridinol-7-O-acetyltransferase, salutaridine-reductase, and codeinone-reductase have been cloned (Ziegler et al. 2006). Based on the results, Facchini and De Luca (2008) concluded that an impressive collection of cDNAs encoding biosynthetic enzymes and regulatory proteins involved in the formation of benzylisoquinoline alkaloids is now available, and the rate of gene discovery has accelerated with the application of genomics.

Ziegler and Kutchan (2005) reported a trial for identification of special sequences responsible for morphinane biosynthesis. Stem sections 2–4 cm below the capsules have been used as a source for RNA and construction-specific cDNA library. More than 1,000 unique sequences could be created and used for gene expression analysis. Comparing the expression in *P. somniferum* strains containing either morphine or noscapine as well as in *P. bracteatum*, the authors found 39 sequences showing considerable differences in expression. Among them, 27 are highly expressed in morphine-containing plants.

In order to evaluate the regulatory consequences of genes in the morphinane pathway, silencing of the codeinone reductase (COR) has been carried out. Instead of accumulation of codeinone, surprisingly, it resulted in accumulation of (S)-reticuline, which is seven enzymatic steps upstream in biosynthesis. It suggests a feedback mechanism or the presence of a metabolism preventing intermediates from general benzylisoquinoline synthesis entering the morphinane pathway. The authors concluded a more complex regulation of the whole route than expected (Allen et al. 2004). For gene silencing, the virus-induced method seems to be applicable also in poppy. According to the report of Hileman et al. (2005), a vector based on tobacco rattle virus (TRV) sequences is effective at silencing the endogenous phytoene desaturase (PapsPDS) gene in poppy.

For genetic transformation of poppy, Nessler (1998) used a special strain of *Agrobacterium tumefaciens* in hypocotyl cultures. Well known antibiotic or herbicide resistance markers (e.g., hygromycine, cefotaxime) promoted the effective selection of transgenic cells. More recently, a reliable genetic transformation protocol via somatic embryogenesis has been developed for the production of fertile, herbicide-resistant poppy plants (Facchini, Loukanina, and Blanche 2008). Regeneration from callus by somatic embryogenesis has been proven to be relatively simple, and in 6–8 months the whole vegetation period of the plant can be regenerated (Nessler 1998).

First practical results of the genetic transformation of poppy have been known only from the last few years. Frick et al. (2004) transformed the berberine bridge enzyme cDNA into seedling explants of an industrial elite line. The selfed progenies of the regenerated plants showed an altered alkaloid profile, which was heritable.

In a further investigation, the activity of the codeinone reductase enzyme has been successfully increased as a result of genetic transformation of a high alkaloid yielding cultivar. Larkin et al. (2007) described a statistically significant increase (128% level in the best transformants) in morphinane alkaloids in transgenic whole plants in glasshouse and in open field as a consequence of overexpression of codeinone reductase.

Recently, molecular genetic manipulation of the gene regulating (S)-N-methycoclaurine-3'hydroxylase resulted in significant changes of alkaloid level in latex. Overexpression of this gene induced a 450% increase of the total alkaloid level, while silencing of it by antisence cDNA caused a reduction of up to 84% (Frick, Kramell, and Kutchan 2007). Further results of these experiments for the practice have not been reported until now. Similarly, using encoding genes isolated from *Arabidopsis*, soybean, and maize, Apuya et al. (2008) found that the overexpression of selected regulatory factors increased the levels of PsCor (codeinone reductase), Ps4'OMT (S-adenosyl-1-methionine:3'-hydroxy-N-methylcoclaurine-4'-O-methyltransferase), and Ps6OMT (R,S-norcoclaurine 6-O-methyltransferase) transcripts by 10- to more than 100-fold. This transcriptional activation translated into the enhancement of alkaloid production in poppy of up to at least 10-fold. Increase of morphine, codeine, and thebaine was achieved by Allen et al. (2008) in poppy plant in which the overexpression of the gene encoding the morphinane pathway enzyme salutaridinol 7-O-acetyltranferase (SalAT) was carried out.

14.7 UTILIZATION OF OPIUM AND OPIATE ALKALOIDS IN MEDICATION

14.7.1 Opium

The latex obtained by incision of the unripe capsule of the poppy *P. somniferum* and known as opium is the source of several pharmacologically important alkaloids. Its biological activity has been recognized even by ancient cultures. The Sumerians named opium *gil* ("happiness"). The Assyrians, like the Sumerians, also collected poppy juice and used it for similar purposes. The ancient Egyptians cultivated opium poppies; however, the use of opium was generally restricted to priests, magicians, and warriors and it was associated with religious cultism. The word "opium" has a Greek origin, derived from *opos* (juice) and *opion* (poppy juice). Its application was mentioned in the writings of Hippocrates (460–377 BC). The collection of crude opium was described in the first century AD by Dioscorides, who referred to both the latex (opos) and the total plant extract (*mekonion*) and to the use of oral and inhaled (pipe-smoked) opium to induce a state of euphoria and sedation. The Romans continued the use of opium for both medication and poisoning.

The real therapeutic application of opium started in medieval Europe as a part of various mixtures that contained numerous ingredients. Paracelsus (1493–1541) popularized the substance as an analgesic when he introduced various preparations utilizing the name of "laudanum" (Nemes 2008). However, the modern history of opium and scientific application of its alkaloids started at the beginning of the nineteenth century. In 1803, the German pharmacist Sertürner isolated morphine, which is one of the most important active ingredients of opium (Bernáth 1998).

Opium is the sun-dried latex of the unripe capsule taken by lancing and formulated afterward by a simple technological procedure. From the practical point of view, it contains five main alkaloids: morphine, codeine, thebaine, narcotine (noscapine), and papaverine. On a dry basis, opium contains 9–14% morphine, 0.7–2.5% codeine, 0.3–1.5% thebaine, 5.5–11.0% narcotine, and up to 1% papaverine (Madyastha and Bhatnagar 1982). The ratio of alkaloids may vary from country to country. For instance, Indian opium is richer in codeine compared to the codeine content of the opium produced under a temperate climate. However, there are large differences in composition of opium produced at the same location depending on the actual climate of the year. From the data of Kaicker, Saini, and Choudhury (1978), the morphine content of opium can reach even 18–20% in the Delhi region due to the higher ranges of temperature, both maximum and minimum, as well as the higher relative humidity.

Based on our investigations (Bernáth et al. 1988), the origin of the plant material can be identified by analysis of composition of opium. In spite of the fact that opium has proved to have a large therapeutic activity and is included in several pharmacopeias (Opii pulvis normatus, PhEur), its application seems to be declining. The opium is mostly used in preparations made by traditional practitioners and Galecic laboratories in Europe. In Eastern cultures, opium is more commonly used in the form of paregoric to treat diarrhea. The thinner solution of opium (tincture of opium) has been prescribed in Europe for, among other things, severe diarrhea.

Today, the manufacture, distribution, and use of tincture of opium are strictly regulated. It is available by prescription in the United States and the United Kingdom, although the drug's medicinal uses are generally confined to controlling diarrhea, alleviating pain, and easing withdrawal symptoms in infants born to mothers addicted to heroin or other opioids. Very rarely, the drug may be prescribed to alleviate pain although it has proved to be effective. Adverse effects of laudanum are generally the same as those of morphine and include euphoria, dysphoria, pruritis, constipation, reduced tidal volume, and respiratory depression, as well as psychological dependence, physical dependence, miosis, and xerostomia. Overdose can result in severe respiratory depression or collapse and death.

14.7.2 Pharmacological Activity of Opiate Alkaloids

14.7.2.1 Morphine

Morphine is the most important constituent of opium and has been employed as a pure alkaloid for over one and a half centuries. Morphine is a potent, reliable, and relatively inexpensive analgesic agent. Pharmacologically, morphine is active in all the standard bioassays for analgesia in animals or humans (Geller and Axelrod 1968; Kuhar and Pasternak 1984). It is able to change pain perception and reaction to pain. Morphine depresses the cerebral cortex and reduces the powers of concentration and fear. Morphine causes a sense of satisfaction and well-being (euphoria) and freedom from anxiety and distress. In addition, pain, particularly prolonged as opposed to acute pain, is reduced and these actions produce a feeling of contentment (Katzung 1995).

Morphine activates the brain stem chemoreceptor trigger zone to produce nausea and vomiting. Also, morphine has an action of vestibular apparatus. The vomiting center and associated center for salivation, sweat, and bronchial secretion are stimulated first, though they become depressed by large and subsequent doses. The sweating is associated with vasodilatation of the skin vessels, so morphine increases heat loss (Burgen and Mitchell 1985).

Respiratory depression is an undesirable effect of morphine. In humans, respiration is depressed by doses below the narcotic threshold. Large doses kill by stopping the respiration altogether (Rang, Dale, and Ritter 1995; Reisine and Pasternak 1996). Morphine therapy must therefore be used with particular care in obstetrics, where fetal respiration may be affected, and in respiratory ailments such as bronchial asthma.

Among the peripheral actions of morphine, constipation is one of the most important. The constipation produced is unaffected by enervation of the intestine or by atropine and is largely due to an increase of the tone of the gut and sphincters and an inhibitory action on Auerbach plexus. Other factors that probably increase this action of morphine are inhibition of the secretion of the intestinal glands and depression of the reflexes responsible for defecation (Burgen and Mitchell 1985; Katzung 1995; Rang et al. 1995).

Morphine also has antidiuretic action and inhibits urinary output. Urine retention is observed even with therapeutic doses. Morphine stimulates the release of the antidiuretic hormone, prolactine, and somatotropin but inhibits the release of the luteinizing hormone (Reisine and Pasternak 1996). It also causes retention of bile by closing the sphincters, raises the pressure in the common bile duct, and may cause biliary colic (Burgen and Mitchell 1985; Katzung 1995).

Morphine might produce hypotensive action in subjects whose cardiovascular system is stressed. This hypotensive effect is probably due to peripheral and arterial dilatation, which has been attributed to various factors (e.g., release of histamine and central depression of vasomotor stabilizing mechanisms).

Morphine inhibits the formation of rosettes by human lymphocytes. The administration of morphine to animals causes suppression of the cytotoxic activity of natural killer cells and enhances the growth of implanted tumors (Katzung 1995).

Humans or animals receiving morphine regularly are liable to become physically dependent on it. When this occurs, withdrawal of the drug produces symptoms within 15–20 hours. In addicts, morphine antagonists (e.g., naloxone) can produce withdrawal signs within 30 minutes. The withdrawal symptoms commence with yawning, sweating, running of the eyes and nose, restlessness, mydriasis, "goose flesh," cramps, nausea, insomnia, vomiting, and diarrhea. Tolerance to the drug is rapidly lost during this period, and the withdrawal symptoms may be terminated by a suitable dose of morphine (Martin 1984; Katzung 1995).

14.7.2.2 Codeine

Codeine is the 3-O-methyl ether of morphine. It is a less important alkaloid than morphine. Its analgesic activity in humans is about 10% that of morphine (Kosterlitz and Waterfield 1975). This is reflected in the human parenteral dose, where 60–120 mg of codeine is equivalent to 10 mg of morphine (Krueger 1955; Martin 1984). Codeine is used orally for the relief of mild to moderate pain and as an antitussive (Eddy, Halbach, and Braenden 1969). It is frequently combined with mild analgesics. Since its action is much weaker than that of morphine it appears less likely to elicit nausea, vomiting, constipation, or respiratory depression. It also has a lower potential than morphine for development of tolerance and physical dependence (Hoffmeister 1984).

14.7.2.3 Thebaine

Thebaine is an important opiate alkaloid, but it is rarely used in medication because of its high toxicity, which has been proved in several pharmacological investigations (Teraoka 1965). Investigating the pharmacological actions of thebaine, induction of a temporary decrease in blood pressure (Teraoka 1965), nervous system stimulation in the mouse, and modification of body temperature and respiration, as well as its convulsant action, were proved (Corrado and Longo 1961). It produced a moderate decrease of catecholamine levels in the heart and brain (Sloan et al. 1962). Detailed analgesic studies were performed in mice (Szegi et al. 1959), but it proved to be a much weaker analgesic than morphine. Thebaine is utilized by the pharmaceutical industry to be converted into other alkaloids and semisynthetic products of high importance.

14.7.2.4 Narcotine

Narcotine (noscapine), which belongs to the phthalideisoquinoline alkaloid group of opium, failed to produce antinociceptive activity, although it has central antitussive action better than that of codeine by inhibiting the cough reflex (La Barre and Plisnier 1959; Put et al. 1974). On the other hand, narcotine appears to be less toxic than codeine (Aurousseau and Navarro 1957; Winter and Flataker 1961). Like papaverine and many other isoquinoline alkaloids, narcotine exhibits mild local anesthetic properties. It has no significant actions on the central nervous system in doses within the therapeutic range.

14.7.2.5 *Papaverine*

Papaverine is an important member of the benzylisoquinoline group of opium. However, it is the only alkaloid of poppy that has production that is also economical by total synthesis. Papaverine is only slightly narcotic, and large doses tend to increase reflex excitability; it displays weak analgesic properties by parenteral or oral administration.

Papaverine decreases the tone of the smooth muscle and is a very effective agent against pathological spasms there. In patients with gastric and duodenal ulcers, papaverine decreased the bioelectrical potential of the stomach. In humans, pregnant and nonpregnant uterus papaverine has a strong spasmolytic effect. It has a vasodepressive effect on the vessels of perfused human placenta. Papaverine increases blood flow in the coronary arteries and causes their dilation, followed by an increase in the formation of creatine phosphate. In addition to the strong coronary vasodilating effect, papaverine diminishes the tendency to develop ventricular fibrillation. Papaverine has a marked vasodilating effect upon the vessels; the dilatation is more significant in atherosclerotic than in intact vessels or when the tension of the vessel walls is increased by epinephrine.

Intracavernous injection of papaverine to impotent men induces penile swelling, attributable to the smooth muscle relaxant action of this drug. Papaverine was found to be an effective histamine liberator (Feldberg and Paton 1951). Similarly to methylxanthines, papaverine relaxes the smooth muscle, presumably by inhibiting phosphorylation of myosin and by preventing breakdown of cAMP (Creed 1994).

14.8 FUTURE DIRECTIONS

Poppy seems to be one of the most important medicinal plants and has been utilized and cultivated since prehistoric times (Tétényi 1997). Both the nutritive property of the seeds and the biological activity of products made from different plant parts, especially from the capsules, have been highly appreciated by humans over the centuries. Poppy spread from its middle Asian gene center through the Roman Empire; later, its cultivation became very common in Europe and other parts of the world. At present, as a result of the large genetic, morphological, and physiological plasticity of poppy, the species is cultivated up to the Arctic Circle as well as under tropical conditions (Bernáth 1998).

The reason for the wide distribution of the poppy can be explained by the well-known narcotic, pharmacological, and nutritive value of its products. The utilization of opiate alkaloids (either taken from the opium or extracted from the dry capsule) is important, even today (Fürst and Hosztafi 1998). At the same time, poppy seed is utilized for human consumption, in industrial processing, and in the manufacture of animal feeds.

Since the isolation of morphine by Sertürner in 1805 (Bernáth 1998) more than 80 alkaloids have been detected in poppy. The increasing demand for poppy alkaloids is the consequence of the widening of the medical application of morphine and its related compounds. According to the INCB report (2008), the global demand for licit opiate raw materials rich in morphine and thebaine has increased over the last two decades. Since 1987, the global medical consumption of opioids increased about threefold. According to the present trends, the demand for opiates is also expected to increase steadily in the future.

However, the production of poppy for opium, dry capsule, or nutritional purposes needs some kind of narcotic control measure. As a consequence, all countries interested in the licit production of poppy must take into consideration the UN Convention on Narcotics, signed in 1988. This convention forced the countries not only to control their cultivation methods and to patrol cultivation areas, but also to build up a new strategy for developing appropriate cultivars (Németh et al. 2002).

In European countries interested in poppy cultivation and in Australia, utilizing the large biological diversity of the species, the selection work has been intensified into three main directions

(Bernáth and Németh 1999): to create cultivars with especially high alkaloid content (2.0–3.5%) for industrial utilization, making selection for low alkaloid content in capsules for culinary use (less than 0.01–0.02% morphine), and producing cultivars for special ornamental purposes. Especially, the importance of cultivars accumulating low alkaloid content increased because they are allowed to be cultivated without severe restriction and control. To gain these goals, parallel with traditional methods, up-to-date biotechnological tools, including gene manipulation, have been introduced. It is not a dream at all that in the near future GMO plants will appear producing rather high levels of alkaloids or free of them.

ACKNOWLEDGMENTS

The Hungarian poppy research has been supported since 2000 by the Hungarian Research Fund (OTKA), nos. T32393 and K62732.

REFERENCES

Allen, R. S., Miller, J. A. C., Chitty, J. A., Fist, A. J., Gerlach, W. L., and Larkin, P. J. 2008. Metabolic engineering of morphinan alkaloids by overexpression and RNAi suppression of salutaridinol 7-O-acetyltransferase in opium poppy. *Plant Biotechnology Journal* 6 (1): 22–30.

Allen, R., Millgate, A., Chitty, J., Thisleton, J., Miller, J., Fist, A., Gerlach, W., and Larkin, P. 2004. RNAi-mediated replacement of morphine with the non-narcotic alkaloid reticuline in opium poppy. *Nature Biotechnology* 22 (12):1559–1565.

Anonymous. 1967. *United Nations Survey Team Report on the economic and social needs of the opium producing areas in Thailand.* Vienna: UN.

Anonymous. 2007a. Geopolitics of illicit drugs in Asia. (www.geopium.org)

Anonymous. 2007b. European Union, Community Plant Variety Office. (www. cpvo.fr)

Anonymous 2007c European Union Community Plant Variety Office. (www.cpvo.fr, web resource accessed 1. July 2008).

Apuya, N. R., Park, J. H., Zhang, L., Ahyow, M., Davidow, P., Van Fleet, J., Rarang, J. C., Hippley, M., Johnson, T. W., Yoo, H. D., et al. 2008. Enhancement of alkaloid production in opium and California poppy by transactivation using heterologous regulatory factors. *Plant Biotechnology Journal* 6 (2): 160–175.

Aurousseau, M., and Navarro, J. 1957. Comparative acute toxicity of some alkaloids derived from opium. *Annales Pharmaceutiques Francaises* 15:640–653.

Basilevskaja, N. A. 1928. Osnovie botanikosistematitscheskoe gruppi olijovomaka (*Papaver somniferum* L.). *Trudi po prikladannoj botankie, geneike iselekcii* 5:185–196.

Belyaeva, R. G. 1988. Analysis of the inheritance of the mutations "anther in petal" and "dissected petal" in opium poppy. *Genetika* 24 (6): 1072–1080.

Bernáth, J. 1998. Poppy—Genus *Papaver.* Amsterdam: Harwood Academic Publisher.

Bernáth, J., Dános, B., Veres, T., Szántó, J., and Tétényi, P. 1988. Variation in alkaloid production in poppy ecotypes: Responses to different environments. *Biochemical Systematics and Ecology* 16:171–178.

Bernáth, J., and Németh, É. 1999. New trends in selection of poppy. *International Journal of Horticultural Science* 5 (3–4): 69–75.

Bernáth, J., Németh, É., and Petheő, F. 2003. Alkaloid accumulation in capsules of the selfed and cross-pollinated poppy. *Plant Breeding* 122: 263–267.

Boit, H. G., and Flentje, H. 1960. Nudaurin, Muramin und Amurensin, drei neue *Papaver*-Alkaloide. *Naturwissenschaften* 47:180.

Bhandari, M. M. (1989). Inheritance of petal colour in *Papaver somniferum* L. *Journal of Horticultural Sciences* 64 (3): 339–340.

Böhm H. V. (1965) Über *Papaver bracteatum* Lindl. II Mitteilung. Die Alkaloide des reifen bastards aus der reciproken kreuzeng dieser art mit *Papaver somniferum* L. *Plant Medica* 13(2): 234–240.

Burgen, A. S. V., and Mitchell, J. F. 1985. *Gaddum's pharmacology*, 9th ed. Oxford, England: Oxford University Press.

Chauhan, S. P., and Patra, N. K. 1993. Mutagenic effects of combined and single doses and EMS in opium poppy. *Plant Breeding* 110 (4): 342–345.

Chindarsi, N. 1976. *The religion of the Hmong Njua.* Bangkok: Siam Society. [299.5919 N942]

Corrado, A. P., and Longo, V. G. 1961. An electrophysiological analysis of the convulsant action of morphine, codeine, and thebaine. *Archives Internationales de Pharmacodynamie et de Therapie* 132:255–269.

Creed, K. E. 1994. Urinary tract. In *Pharmacology of smooth muscle*, ed. Szekeres, L., and Papp, G., 575–591. Berlin: Springer Verlag.

Crooker, R. A. 1988. Forces of change in the Thailand opium zone. *Geographical Review* 78 (3): 241–256.

Crooker, R. A., and Martin, R. N. 1992. Accessibility and illicit drug crop production: Lessons from Northern Thailand. *Journal of Rural Studies* 8 (4): 423–429.

Dachler, M. 1990. Varieties and nitrogen requirements of some medicinal and spice plants grown for seed. *Herba Hungarica* 29 (3): 39–44.

Danert, S. 1958. Zur Systematik von *Papaver somniferum* L. *Kulturpflanze* 6:61–88.

Dános, B. 1965. Wirkung der generativen Hybridisierung auf die Gestaltung des Alkaloidgehalts des Mohns. *Pharmazie* 20:727–730.

De Candolle, A. 1883. *Origine des plantes cultivées.* Paris: Bailliére.

Dittbrenner, A., Lohwasser, U., Mock, H. P., and Borner, A. 2008. Molecular and phytochemical studies of *Papaver somniferum* in the context of infraspecific classification. *Acta Horticulturae* 799:81–88.

Dobos, G. 1996. Winter poppy—A new genotype for seed production. *Beiträge zur Züchtungsforschung* 2 (1): 37–40.

Dobos, G., and Vetter, S. 1997. Variation des Morphingehaltes bei Wintermohn-Herkünften (*Papaver somniferum* L.). *Zeitschrift fur Arznei und Gewiizpflanzch. Pfl.* 2:87–89.

Dubey, M. K., Dhawan, O. P., and Khanuja, S. P. S. 2009. Downy mildew resistance in opium poppy: Resistance sources, inheritance pattern, genetic variability and strategies for crop improvement. *Euphytica* 165 (1): 177–188.

Eddy, N. B., Halbach, H., and Braenden, O. J. (1969) Synthetic substances with morphine-like effect. Relationship between analgesic action and addiction liability, with a discussion of the chemical structure of addiction-producing substances. *Bulletin of World Health Organization* 14:353–402.

Elkan, L. 1839. *Tentamen monographiae generis* Papaver. Königsberg: Regimentii Borussorum.

Espinasse, A. 1981. La production de la thébaine, codéine et morphine partir du genere *Papaver. Agronomie* 1:243–248.

Facchini, P. J., and De Luca, V. 2008. Opium poppy and Madagascar periwinkle: Model nonmodel system to investigate alkaloid biosynthesis in plants. *Plant Journal* 54 (4): 763–784.

Facchini, P. J., Loukanina, N., and Blanche, V. 2008. Genetic transformation via somatic embryogenesis to establish herbicide-resistant opium poppy. *Plant Cell Report* 27 (4): 719–727.

Facchini, P. J., Penzes-Yost, C., Samanani, N., and Kowalchuk, B. 1998. Expression pattern conferred by tyrosine/dihydrophenylalanine decarboxylase promoters from opium poppy are conserved in transgenic tobacco. *Plant Physiology* 118 (1): 69–81.

Fairbairn, J. W., and Williamson, E. M. 1978. Meconic acid as a chemotaxonomic marker in the Papaveraceae. *Phytochemistry* 17:2087–2089.

Fedde, F. 1909. Papaveraceae—Hypericoideae. In *Das Pflanzenreich*, ed. Engler, A., 1–430. Leipzig: Engelmannh.

———. 1936. Papaveraceae. In *Die natürlichen Pflanzenfamilien* 2, ed. Engler, A. and Pantl, K. Leipzig: Aufl.

Feldberg, W., and Paton, W. 1951. Release of histamine from skin and muscle in the cat by opium alkaloids and other histamine liberators. *Journal of Physiology* (London) 114:490–496.

Földesi, D. 1992. Poppy. In *Cultivation and processing of medicinal plants*, ed. Hornok, L., 119–128. Budapest: Akadémiai Kiadó.

———. 1994. *Papaver somniferum* In *Vadontermő és termesztett gyógynövények*, ed. Bernáth, J., 390–399. Budapest: Mezőgazda Kiadó.

———. 1995. Az őszi mák termesztése. *Új Kertgazdaság* 1 (1–2): 72–73.

Frick, S., Chitty, J., Kramell, R., Schmidt, J., Allen, R., Larkin, P., and Kutchan, T. 2004. Transformation of opium poppy (*Papaver somniferum* L.) with antisense berberine bridge enzyme gene (anti-bbe) via somatic embryogenesis results in an altered ratio of alkaloids in latex but not in roots. *Transgenic Research* 13 (6): 607–613.

Frick, S., Kramell, R., and Kutchan, T. 2007. Metabolic engineering with a morphine biosynthetic P450 in opium poppy surpasses breeding. *Metabolic Engineering* 9 (2): 167–169.

Fürst, Z., and Hosztafi, S. 1998. Pharmacology of poppy alkaloids. In *Poppy—The genus* Papaver, ed. Bernáth, L., 291–318. Amsterdam: Harwood Academic Press.

Gabra, C. S. 1956. *Papaver* species and opium through the ages. *Bulletin l'Institut d'Egypt* 37:40.

Geddes, W. R. 1971. Opium growing in northern Thailand. Seminar on Contemporary Thailand, Thailand.

Geller, I., and Axelrod, L. R. 1968. Methods for evaluating analgesics in laboratory animals. In *Pain*, ed. Soulairac, A., Cahn, J., and Charpenter, J., 153–170. London: Academic Press.

Goldblatt, P. 1974. Biosystematic studies in *Papaver* section *Oxytona. Annals of Missouri Botanical Garden* 61:264–296.

Gottlieb, O. R., Kaplan, M. A. C., and Zocher, D. H. T. 1993. A chemosystematic overview of Magnoliidae, Ranunculidae, Cariophyllidae and Hamamelididae. In *The families and genera of vascular plants*, vol. II, ed. Kubitzki, K., Rohwer, J. G., and Bittrich, V. Berlin: Springer Verlag.

Gumuscu, A., Arslan, N., and Sarihan, E. O. 2008. Evaluation of selected poppy (*Papaver somniferum* L.) lines by their morphine and other alkaloid contents. *European Food Research Technology* 226 (5): 1213–1220.

Günther, K. F. 1975. Beiträge zur Morphologie und Verbreitung der Papaveraceae 2.Teil. Die Wuchsformen der Papaverae, Eschcholzieae and Platystemonoideae. *Flora* 164:393–436.

Gupta, U. C. 1974. Effect of methods of sowing and nitrogen fertilization on yield and quality of selected varieties of opium poppy. MSc thesis. University of Udaipur, Udaipur.

Hagel, J. M., Weljie, A. M., Vogel, H. J., and Facchini, P. J. 2008. Quantitative H-1 nuclear magnetic resonance metabolite profiling as a functional genomics platform to investigate alkaloid biosynthesis in opium poppy. *Plant Physiology* 147 (4): 1805–1821.

Hammer, K. 1981. Problems of *Papaver somniferum*—Classification and some remarks on recently collected European poppy landraces. *Kulturpflanze* 29:287–296.

Hammer, K., and Fritsch, R. 1977. Zur Frage nach der Ursprungsart des Kulturmohns (*Papaver somniferum* L.). *Kulturpflanze* 25:113–121.

Handrea, D. 1996. Variability of quantitative characters in some landraces of *Papaver somniferum* L. *Beitrage zur Züchting Forschung* 2 (1): 127–130.

Hanelt, P., and Hammer, K. 1987. Einige infraspezifische Umkombinationen und Neubeschreibungen bei Kultursippen von *Brassica* L. und *Papaver* L. *Feddes Report* 98:553–555.

Heeger, E. F. 1956. *Handbuch des Arznei und Gewürzpflanzenbaues*. Erfurt, Germany: Deutscher Bauerverlag.

Hegnauer, R. 1990. *Chemotaxonomie der Pflanzen. Band 9*. Basel, Switzerland: Birkhäuser Verlag.

Heltmann, H., and Silva, F. 1978. Zur Züchtung leistungfahiger Inzuchtlinien für eine synthetische Mohnsorte. *Herba Hungary* 17 (2): 55–60.

Herbig, C. 2009. Recent archaeobotanical investigations into the range and abundance of Neolithic crop plants in settlements around Lake Constance and in Upper Swabia (south-west Germany) in relation to cultural influences. *Journal of Archaeological Science* 36 (6): 1277–1285.

Hileman, L. C., Drea, S., de Martino, G., Litt, A., and Irish, F. I. 2005. Virus induced gene silencing is an effective tool for assaying gene function in the basal eudicot species *Papaver somniferum. Plant Journal* 44 (2): 334–339.

Hoffmeister, F. 1984. Dependence potential of analgesic and psychotropic drugs. *Arzneimittel Forschung* 34:1096–107.

Hörömpöli, T. 1995. Amit a mák termesztéséről tudni kell. Regiokon Kft, Kompolt, Magyarország.

———. 1998. A mák tőkorhadása elleni rezisztencianemesítés hatása a 'Kompolti M' fajtára. Abstracts of J. Lippay and K. Vas. Scientific Session, Budapest, (1998 Sept. 16–17): 132–133.

Hosokawa, K., Shibata, T., Nakamura, I., and Hishida, A. 2004. Discrimination among species of *Papaver* based on the plasmid rp116 gene and the rp116-rp114 spacer sequence. *Forensic Science International* 139 (2–3): 195–199.

Hrishi, N. J. 1960. Cytogenetical studies on *Papaver somniferum* L. and *Papaver setigerum* DC and their hybrid. *Genetica, S-Gravenhage* 31 (1–2): 1–13.

INCB. 2006. World drug report. New York (Vienna): UN Office on Drugs and Crime, 2007.

INCB. 2008. *World drug report*. New York (Vienna): UN Office on Drugs and Crime, 2007.

———. 2009. *Report of International Narcotics Control Board for 2006*. UN Office on Drugs and Crime, 2008.

INCSR. 1994. International narcotics control strategy report. United States Department of State, Bureau of International Narcotics Matters, Washington, D.C.

Jain, P. M., and Solanki, N. S. 1993. Effect of preceding crops on fertilizer requirement of opium poppy (*Papaver somniferum*). *Indian Journal of Agronomy* 38 (1): 105–106.

Kadereit, J. W. 1986. *Papaver somniferum* L. (Papaveraceae): A triploid hybrid? *Botanische Jahrbücher für Systematik* 106:221–244.

———. 1987. A revision of *Papaver* sect. Carinatae (Papaveraceae). *Nordic Journal of Botany* 7:501–504.

———. 1988. Sectional affinities and geographical distribution in the genus *Papaver* L. (Papaveraceae). *Beitrage zur Biologie der Pflanzen* 63:139–156.

———. 1990. Some suggestions on the geographical origin of the central, west and north European synanthropic species of *Papaver* L. *Botanical Journal of Linnean Society* 103:221–231.

———. 1993. Papaveraceae. In *The families and genera of vascular plants*, vol. 2, ed. Kubitzki, K., Rohwer, J. G., and Bittrich, V., 20–33. Berlin: Springer Verlag.

Kaicker, U. S. 1985. Opium and breeding research at Indian Agricultural Research Institute. *Proceedings of 5th ISHS Symposium on Medicinal and Aromatic Plants*, Darjeeling, India, 75–81.

Kaicker, U. S., Saini, H. C., and Choudhury, B. 1978. Environmental effects on morphine content in opium poppy (*Papaver somniferum* L.). *Bulletin on Narcotics* 30:69–74.

Kalav, Y. N., and Sariyar, G. 1989. Alkaloids from Turkish *Papaver rhoeas*. *Planta Medica* 55:488.

Kálmán-Pál, Á., Bernáth, J., and Tétényi, P. 1987. Phenotypic variability in the production and alkaloid spectrum of the *Papaver somniferum* L. hybrid. *Herba Hungary* 26 (2–3): 75–82.

Kandalkar, V. S., Saxena, A. K., Khire, A. and Jain, Y. K. (1995). Genetical analysis of field resistance to downy mildew caused by *Peronospora arborescens* in opium poppy. *Journal of Hill Research* 8 (2): 252–255.

Katzung, B. G. 1995. *Basic and clinical pharmacology*, 6th ed., 1046. New York: Appleton Lange, Prentice Hall, Int.

Khanna, K. R. 1985. Breeding of some important medicinal plants with emphasis on the relevant genetics of biosynthetic pathways. *Proceedings of 5th ISHS Symposium on Medicinal and Aromatic Plants*, Darjeeling, India

Khanna, K. R., and Shukla, S. 1989. Gene action in opium poppy (*Papaver somniferum*). *Indian Journal of Agricultural Science* 59 (2): 124–126.

Khanna, K. R. and Gupta, R. K. (1989). Gene action in opium poppy (*Papaver somniferum*). *Indian Journal of Agricultural Science* 59, 124–126.

Köck, O., Bernáth, J., and Sárkány, S. 2001. Hazai mákfajták. In *A mák* (Papaver somniferum L.) *Magyarország Kultúrflórája*, ed. Sárkány, S., Bernáth, J., and Tétényi, P., 244–259. Budapest: Akadémiai Kiadó.

Kohler-Schneider, M., and Canapelle, A. 2009. Late Neolithic agriculture in eastern Austria: Archaeobotanical results from sites of the Baden and Jevidovice cultures (3600–2800 BC). *Vegetation History and Archaeobotany* 18 (1): 61–74.

Kohli, D. N. 1966. The story of narcotics control in India (opium). *Bulletin on Narcotics* 18 (3): 3–11.

Kopp, E., Csedő, K., and Mátyás, S. 1961. Weitere Versuche zur Züchtung einer alkaloidreichen Mohnsorte. *Pharmazie* 16 (4): 224–231.

Kosterlitz, H. W., and Waterfield, A. A. 1975. *In vitro* models in the study of structure–activity relationships of narcotic analgesics. *Annual Review of Pharmacology and Toxicology* 15:29–47.

Kritikos, P. G., and Papadaki, S. P. 1967. The history of the poppy and of opium and their expansion in antiquity in the eastern Mediterranean area. *Bulletin on Narcotics* 19 (3): 17–38.

Krueger, H. 1955. Narcotics and analgesics. In *The alkaloids. Chemistry and physiology*, vol. 5, ed. Manske, R. H. F., 1–77. New York: Academic Press Inc.

Kubitzky, K. 1993. *The families and genera of vascular plants*, vol. 2, 1–12. Berlin: Springer Verlag.

Kuhar, M. J., and Pasternak, G. W. 1984. *Analgesics: Behavioral and clinical perspectives*. New York: Raven Press.

La Barre, J., and Plisnier, H. 1959. Experimental study relating to the antitussive properties of narcotine hydrochloride. *Bulletin on Narcotics* 11 (3): 7–14.

Lal, R. K., and Sharma, J. R. 1991. Genetics of alkaloids in *Papaver somniferum*. *Planta Medica* 57:271–274.

———. 1995. Heterosis and its components for opium alkaloids in *Papaver somniferum*. *Current Research in Medicinal and Aromatic Plants* 17:165–170.

Larkin, P. J., Miller, J. A. C., Allen, R. S., Chitty, J. A., Gerlach, W. L., Frick, S., Kutchan, T. M., and Fist, A. J. 2007. Increasing morphinane alkaloid production by overexpressing codeinone reductase in transgenic *Papaver somniferum*. *Plant Biotechnology Journal* 5:26–37.

Laughlin, J. C., Chung, B., and Beattie, B. M. 1998. Poppy cultivation in Australia. In *Poppy—Genus* Papaver, ed. Bernáth, J., 249–278. Amsterdam: Harwood Academic Publishers.

Levy, A., and Milo, J. 1998. Genetics and breeding of *Papaver somniferum*. In *Poppy—The genus* Papaver, ed. Bernáth, J., 93–104. Amsterdam: Harwood Academic Publishers.

Liersch, J., Szymanowska, E., and Krzymanski, J. 1996. Valuation of stability and economical value of new strains of low morphine poppy. *Rosliny Oleiste* 17:391–396.

Linnaeus. 1753. Species *Plantarum*. Stockholm, 263.

Lörincz, G., Tétényi, P. (1970). Result of poppy breeding by the method of distant crossing. *Herba Hungarica* 9: 79–85.

Maas, H. 1986. Papaver. In *Verzeichnis landwirtschaftlicher und Gartnerischer Kulturpflanzer (ohne Zierpflanzen)* 2, ed. Mansfelds, R. Berlin: Neubearb. Aufl. Hrsg. J Schulze-Motel.

Madyastha, K. M., and Bhatnagar, S. P. 1982 Chemical and biochemical aspects of opium alkaloids. In *Cultivation and utilization of medicinal plants*, ed. Atal, C. K., and Kapur, B. M., 139–158. Jammu-Tawi: Regional Research Laboratory Council of Science and Industrial Research.

Margl, L., Szatmáry M., and Hajósné-Novák, M. 2001. Biokémiai-genetikai markerek alkalmazása a mák-nemesítésben és a fajtavédelemben: I. izoenzimek. VII. Növénynemesítési Tudományos Napok, Budapest, *Abstracts*, 110.

Markgraf, F. 1986. Familie Papaveraceae. In *Illustrierte Flora von Mitteleuropa. Band IV.* Teil 1. 3rd. ed., ed. Hegi, G. Parey Ed.

Martin, W. R. 1984. Pharmacology of opioids. *Pharmacology Reviews* 35:283–323.

Maturová, M., Pavlásková, D., and Šantavý, M. 1966. Isolierung der Alkaloide aus einiger arten der Gattung *Papaver. Planta Medica* 37:22–41.

Meshulam, H., and Lavie, D. 1980. The alkaloidal constituents of *Papaver bracteatum* 'Arya' II. *Phytochemistry* 19:2633–2635.

Millgate, A. G., Pogson, B. J., Wilson, I. A., Kutchan, T. M., Zenk, M. H., Gerlach, W. L., Fist, A. J., and Larkin, P. J. 2004. Analgesia: Morphine pathway block in top1 poppies. *Nature* 431:413–414.

Mórász, S. 1979. *A mák termesztése.* Budapest: Mezőgazdasági Kiadó.

Morice, J. and Louarn, J. (1971) Study of morphine content in the oil poppy (*P. somniferum* L.). *Am Amelior Plantes* 21 (4), 465–485.

Nemes, Cs. 2008. Orvostörténelem. Debreceni Egyetem Orvosi-és Egészségtudományi Centrum, Debrecen.

Németh, É. 2002. World tendencies, aims and results of poppy (*Papaver somniferum* L.) breeding. In *Recent progress in medicinal plants, vol. 4. Biotechnology and genetic engineering*, ed. Govil, J. N., Kumar, A. P. and Singh, V. K., 129–141. Houston, TX: Scientific Technological Publications.

Németh, É., Bernáth, J., Sztefanov, A., and Petheő F. 2002. New results of poppy (*Papaver somniferum* L.) breeding for low alkaloid content in Hungary. *Acta Horticulturae* 576:151–158.

Nessler, C. L. 1998. *In vitro* culture technologies. In *Poppy—The genus* Papaver, ed. Bernáth, J., 209–218. Amsterdam: Harwood Academic Publishers.

Nielsen, B., Röe, J., and Brochmann-Hanssen, E. 1983. Oripavine—A new opium alkaloid. *Planta Medica* 48:205–206.

Nigam, K. B., Kandalkar, V. S., and Dhumale, D. B. 1990. Induced mutants in opium poppy. *Indian Journal of Agricultural Science* 60 (4): 267–268.

Nothnagel, Th., Straka, P., and Schultze, W. 1996. Selection of a low morphine poppy *Papaver somniferum* L. *Beiträge zur Züchtungsforschung* 2 (1): 120–123.

Nyman, U., and Bruhn, J. G. 1979. Papaver bracteatum—A summary of current knowledge. *Planta Medica* 35:97–117.

Nyman, V., and Hall, O. 1974. Breeding oil poppy (*Papaver somniferum*) for low content of morphine. *Hereditas* 76:49–54.

Ounaroon, A., Frick, S., and Kutchan, T. 2005. Molecular genetic analysis of an *O*-methyltransferase of the opium poppy. *Acta Horticulturae* 675:167–171.

Öztekin, A., Baytop, A., Hutin, M., Foucher, J. P., Hocquemiller, R., and Cave, A. 1985. Comparison of chemical and botanical studies of Turkish *Papaver* belonging to the section *Pilosa. Planta Medica* 50:431–434.

Patra, N. K., Ram, R. S., Chauhan, S. P., and Singh, A. K. 1992. Quantitative studies on the mating system of opium poppy (*Papaver somniferum* L.). *Theoretical and Applied Genetics* 84 (3–4): 299–302.

Phillipson, J. D. 1983. Infraspecific variation and alkaloids of *Papaver* species. *Planta Medica* 48:87–192.

Phillipson, J. D., Scutt, A., Baytop, A., Özhatay, N., and Sariyar, G. 1981. Alkaloids from Turkish samples of *Papaver orientale* and *P. pseudo-orientale*. *Planta Medica* 43:261–271.

Phillipson, J. D., Thomas, O. O., Gray, A. I., and Sariyar, G. 1981. Alkaloids from *P. armeniacum, P. fugax* and *P. tauricola. Planta Medica* 41:105–118.

Popov, P., Dimitrov, J., Iliev, L., and Georgiev, S. 1971. Mak (*Papaver somniferum* L.) Izd. Sofia, Bulgaria: Bulgarskaja Akademija Naukite.

Preininger, V. 1986. Chemotaxonomy of Papaveraceae and Fumariaceae. *Alkaloids* 29:2–98.

Preininger, V., Novák, J., and Santavý, M. 1981. Isolierung und Chemie der Alkaloide aus Pflanzen der Papaveraceae, LXXXI. Glauca-eine neue Sektion der Gattung Papaver. *Planta Medica* 41:119–123.

Put, A., Wojcicki, J., Stanosz, S., and Gorecki, P. 1974. Antitussive properties of narcotine and its homologues. *Herba Polinica* 20:285–289.

Ramanathan, V. S. 1982. Utilization of opium and its alkaloids in medicine. In *Cultivation and utilization of medicinal plants*, ed. Atal, C. K. and Kapur, B. M., 159–172. Jammu-Tawi: Regional Research Laboratory Council of Science and Industrial Research.

———. 1983. Effect of nematicides on the yield of opium and morphine of opium poppy. *Indian Journal of Agricultural Science* 53 (3): 172–174.

Rändel, U. 1974. Beiträge zur Kenntnis der Sippenstruktur der Gattung *Papaver* L. sectio *Scapiflora Reihenb.* (Papaveraceae). *Feddes Report* 84:655–732.

Rang, H. P., Dale, M. M., and Ritter, J. M. 1995. *Pharmacology*, 3rd ed. New York: Churchill Livingstone.

Reckin, J. 1971. A contribution to the cytologie of *P. aculeatum, P. gracile* and proposals for the revision of the section Mecones. *Caryologie* 23:461–464.

Reisine, T., and Pasternak, G. 1996. Opioid analgesics and antagonists. In *The pharmacological basis of therapeutics*, 9th ed, ed. Hardman, J. G., Limbird, L. E., Molinoff, P. B., Ruddon, R. W., and Goodman-Gilman, A., 521–556. New York: McGraw–Hill.

Rosch, M. 2008. New aspects of agriculture and diet of the early medieval period in central Europe: Waterlogged plant material from sites in south-western Germany. *Vegetation History and Archaeobotany* 17 (Suppl.): 225–238.

Rothmaler, W. 1949. Notuale systematicae. 4. Papaveres. Index Sem. *Gaterslebense* 42–46.

Ruminska, A. 1973. Rosliny lecznicze. *Panstwowe Wydawnictwo Naukove* 96–100.

Saihoo, P. 1963. The hill tribes of northern Thailand and the opium problem. *Bulletin on Narcotics* 15 (2): 35–45.

Šantavý, M., Maturová, M., Němečková, A., Schröter, H-B., Potěšilová, P., and Preininger, V. 1960. Isolierung der Alkaloide aus einigen Mohnarten. *Planta Medica* 27:167–178.

Sariyar, G., and Baytop, T. 1980. Alkaloids from *Papaver pseudo-orientale* (*P. lasiothrix*) of Turkish origin. *Planta Medica* 46:378–380.

Sariyar, G., Baytop, T., and Phillipson, J. D. 1989. A new protopine alkaloid from Turkish *Papaver curviscapum* 1-methoxyallocryptopine. *Planta Medica* 55:89–90.

Sharma, J. R., Lal, R. K., and Mishra, H. O. 1988. Heterosis and gene action for important traits in opium poppy (*Papaver somniferum* L.). *Indian Journal of Genetics of Plant Breeding* 48:261–266.

Sharma, J. R., Lal, R. K., Mishra, H. O., Lohia, R. S., Pant, V., and Yadav, P. 1997. Economic heterosis for yield and feasibility of its exploitation in opium poppy (*Papaver somniferum* L.). *Journal of Medicinal and Aromatic Plant Science* 19:398–402.

Sharma, J. R., Lal, R. K., Mishra, H. O., Naqvi, A. A., and Patra, D. D. 1999. Combating opium linked global abuses and supplementing the production of edible seed and seed oil: A novel non-narcotic var. 'Sujata' of opium poppy (*Papaver somniferum* L.). *Current Science* 77 (12): 1584–1589.

Sharma, J. R., Lal, R. K., Singh, S. P., and Mishra, H. O. 1991. Duplicative gene control of leaf incision in opium poppy (*Papaver somniferum* L.). *Journal of Heredity* 82 (2): 174–175.

Shukla, S., and Khanna, K. R. 1987. Genetic association in opium poppy. *Indian Journal of Agricultural Science* 57 (3): 147–151.

Shukla, S., Singh, S. P., Yadav, H. K., and Chatterjee, A. 2006. Alkaloid spectrum of different germplasm lines in opium poppy (*Papaver somniferum* L.). *Genetic Research and Crop Evolution* 53 (3): 533–540.

Shulgjin, G. 1969. Cultivation of the opium poppy and the oil poppy in the Soviet Union. *Bulletin on Narcotics* 21:4.

Simmonds, Q. W. 1976. *Evolution of crop plants*. London: Longman.

Singh, H. G. 1982. Cultivation of opium poppy. In *Cultivation and utilization of medicinal plants*, ed. Atal, C. K., and Kapur, B. M., 120–138. Jammu-Tawi: Regional Research Laboratory Council of Science and Industrial Research.

Singh, S. P., and Khanna, K. R. 1991. Heterosis in opium poppy. *Indian Journal of Agricultural Science* 61:259–263.

Singh, S. P., Shukla, S., and Khanna, K. R. 1995. Diallel analysis for combining ability in opium poppy (*Papaver somniferum*). *Indian Journal of Agricultural Science* 65 (4): 271–275.

———. 1997. Characterization of Indian landraces and released varieties of opium poppy. *Journal of Medicinal and Aromatic Plant Science* 19:369–386.

Singh, S. P., Tiwari, R. K., and Dubey, T. 1999. Heterosis and inbreeding depression in opium poppy (*Papaver somniferum*). *Journal of Medicinal and Aromatic Plant Science* 21:23–25.

Shukla, S., Singh, S. P. and Shukla, S. (1999). Genetic systems involved in inheritance of papaverine in opium poppy. *Indian Journal of Agricultural* 69 (1), 44–47.

Sloan, J. W., Brooks, J. W., Eisenman, A. J., and Martin, W. R. 1962. Comparison of the effects of single doses of morphine and thebaine on body temperature, activity, and brain and heart levels of catecholamines and serotonin. *Psychopharmacologia* 3:291–301.

Smart, R. G., and Archibald, H. D. 1980. Intervention approaches for rural opium users. *Bulletin on Narcotics* 32 (4): 11–27.

Soó, R. 1968. *A magyar flóra és vegetáció kézikönyve*. 3. Budapest: Akadémiai Kiadó.

Srivastava, R. K., and Sharma, J. R. 1987. Estimation of genetic variance and parameters through biparental mating in opium poppy (*Papaver somniferum* L.). *Australian Journal of Agricultural Research* 38:1047–1052.

Straka, P., and Nothnagel, T. 2002. A genetic map of *Papaver somniferum* L. based on molecular and morphological markers. *Proceedings of the 2nd International Symposium on Breeding Research on Medicinal and Aromatic Plants* (July 11–16, Chania, Greece), Haworth Press, 235–241.

Straka, P., Schultze, W., and Nothnagel, Th. 1993. Stand der Arbeiten zur Entwicklung morphinarmer Mohnformen. *Vortr Pflanzenzüchtg* 26:36–41.

Suwanwela, Ch., Poshyachinda, V., Tasanapradit, P., and Dharmkrong, A. A. 1978. The hill tribes of Thailand, their opium use and addiction. *Bulletin on Narcotics* 30 (2): 1–19.

Suwanwela, Ch., Tasanapradit, P., Dharmkrong, A. A., Poshyachinda, V., and Artkampee, V. 1976. Opium use in a Meo village. Technical report of Institute of Health Research, Bangkok.

Szegi, J., Rausch, J., Magda, K., and Nagy, J. 1959. Relationship between the chemical structure and pharmacological activity of the opium alkaloids. *Acta Physica Hungarica* 15:325–335.

Takhtajan, A. 1959. *Die evolution der Angiospermen*. Jena: Fischer.

Teraoka, A. 1965. Supplementary studies on the pharmacological actions of thebaine. *Nippon Yakurigaku Zasshi* 61:396–406.

Tétényi, P. 1963. *Intraspecifikus kémiai taxonok és gyógynövénynemesítés*. Budapest: Akadémiai Doktori Értekezés.

———. 1985. Chemotaxonomic evaluation of the *Papaver* section *Macrantha*. *Acta Horticulturae* 188:35–47.

———. 1989. Morphinoides du genere *Papaver*. *Acte du Colloque Montréal, Chicoutimi*, 64–69.

———. 1993. Chemotaxonomy of the genus *Papaver*. *Acta Horticulturae* 344:154–165.

———. 1997. Opium poppy (*Papaver somniferum*): Botany and horticulture. *Horticultural Reviews* 19:373–405.

Tétényi, P., Lőrincz, C., and Szabó, E. 1961. Untersuchung der infraspecifischen chemischen Differenzen bei Mohn: Beitrage zur Characterisierung der Hybriden von *Papaver somniferum* L. × *Papaver orientale* L. *Pharmazie* 16:426–433.

Tookey, H. L. Spencer, G. F., Grove, M. D. and Kwolek, W. F. 1976. Codeine and morphine in *Papaver somniferum* grown in a controlled. *Planta Medica* 30 (4): 340–348.

Tóthné-Lökös, K. 1997. Növénynemesítési alapanyagok genetikai elemzése kvantitatív tulajdonságok alapján. Doktori értekezés, GATE, Gödöllő.

Turkhede, B. B., Rajat, D., and Singh, R. K. 1981. Consumptive water use by opium poppy. *Indian Journal of Agricultural Science* 51 (2): 102–107.

Turkhede, B. B., and Singh, R. 1981. Effect of number and methods of lancing on yield and quality of opium poppy. *Indian Journal of Agronomy* 26 (4): 461–462.

Vágújfalvi, D. 1970. Untersuchungen über die Lokalisation der Alkaloide in einigen Papaveraceae taxa. *Botanikai Közlemények* 57:113–120.

Vesselovskaya, M. A. 1975. The poppy, its variability, classification and evolution. *Trudi Prikladi Botaniki Genetiki Selekci* 55:175–223.

Weid, M., Ziegler, J., and Kutchan, T. M. 2004. The roles of latex and the vascular bundle in morphine biosynthesis in the opium poppy, *Papaver somniferum*. *Proceedings of the National Academy of Sciences* 101 (38): 13957–13962.

Winter, C. A., and Flataker, L. 1961. Toxicity studies on noscapine. *Toxicology and Applied Pharmacology* 3:96–106.

Yadav, H. K., Shukla, S., and Singh, S. P. 1999. Genetic systems involved in inheritance of papaverine in opium poppy. *Indian Journal of Agricultural Science* 69 (1): 44–47.

Ziegler, J., and Kutchan, T. M. 2005. Differential gene expression in *Papaver*–species in comparison with alkaloid profiles. *Acta Horticulturae* 675:173–177.

Ziegler, J., Voightlander, S., Schmidt, J., Kramell, R., Miersch, O., Ammer, C., Gesell, A., and Kutchan, T. 2006. Comparative transcript and alkaloid profiling in *Papaver* species identifies a short chain dehydrogenase/reductase involved in morphine biosynthesis. *Plant Journal* 48:177–192.

CHAPTER **15**

Ginger

K. Nirmal Babu, M. Sabu, K. N. Shiva, Minoo Divakaran, and P. N. Ravindran

CONTENTS

15.1 INTRODUCTION

Ginger (*Zingiber officinale* Roscoe) is the underground rhizome (Figure 15.1) of the herbaceous perennial, used as fresh and dry forms. It is an ancient and the third most important spice of the world and is prized for its flavor and medicinal properties. In addition to its use as spice and condiment, it is also used to treat liver complaints, flatulence, anemia, rheumatism, piles, and jaundice in Indian and Chinese systems of medicines. The genus *Zingiber*, consisting of about 150 species, is widely distributed in tropical and subtropical Asia. The other important taxa of this genus are turmeric, cardamom, and large cardamom, as well as several other species having economic and medicinal importance.

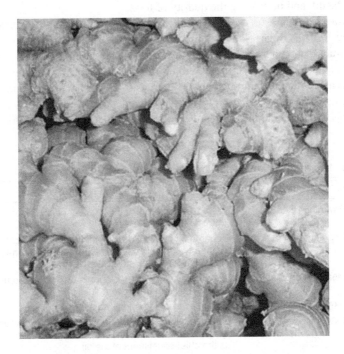

Figure 15.1 Fresh rhizomes of ginger.

15.2 ECONOMIC USES OF GINGER AND ITS RELATED SPECIES

Ginger was more valued for its medicinal properties by ancient peoples, and it played an important role in primary health care in ancient India and China. In European medicine, ginger was also among the most highly valued of all mild carminatives and entered into many pharmaceutical preparations.

Ginger is traded and consumed in many forms. The primary products of ginger are fresh ginger, preserved ginger, and dried ginger. The bulk of consumption is in the form of fresh immature and mature ginger used as a vegetable. Preserved ginger, prepared from the immature rhizomes, is used for culinary purposes and in processed foods such as jams, marmalades, cakes, and confectionary. Dried ginger is used directly as a spice in whole, split, or ground form and is used extensively in flavoring of processed foods. Ginger oils and oleoresins are value added products obtained from dried rhizomes by steam distillation/solvent extraction.

Ginger is used for culinary purposes in green and in dry form. Ginger is indispensable in the manufacture of a number of food products and certain soft drinks. It is used for spiced ginger vines, ginger beer, and several fresh ginger products (e.g., ginger preserves or Muraba, ginger candy, soft drinks like ginger cocktail, ginger squash, and products such as ginger pickles, salted ginger, salted in vinegar or vinegar mixed with lime, green chillies, etc.). The oil of ginger finds use in perfumery.

Ginger contains 2–3% protein, 0.9% fat, 1.2% minerals, 2.4% fiber, and 12.3% carbohydrate and is a good source of calcium, phosphorous, iron, and vitamins. The volatile oils (lisabolene, cinol, phellandrene, citrol, borneol, cibonellol, geranial, linaloal, limoline, zingiberol, zingiberene, and camphenes), oleoresin (gingenol and shogaol), phenol (gingeol and zingerone), proteolytic enzyme (zingibain), vitamin B6, vitamin C, calcium, magnesium, phosphorus, potassium, and linoleic acid are important constituents of ginger. The pungency of ginger is due to gingerol, while the aroma is due to volatile oils lisabolene, zingiberene, and zingiberol. Thus, ginger has all the constituents needed for good health and improving the quality of food.

Some important products of and constituents in ginger and their properties are given in Tables 15.1(a) and 15.1(b). Ginger is used in medicines especially as a carminative, stimulant of the gastrointestinal tract, rubifacient, diaphoretic, diuretic, anti-inflammatory, antiemetic, sialogogic emmangogus, abortifacient, and vermifuge. Ginger relieves flatulence, stimulates the gastrointestinal tract, and acts as a counterirritant. Alcoholic extracts of ginger have been found to stimulate the heart. Ginger increases fibrinolytic activity and thereby protects against coronary artery disease. An extract of ginger is used as an adjunct to many tonic and stimulating remedies. It is included among

Table 15.1(a) Important Products of Ginger and the Method of Preparation

SI no.	Product	Method of Preparation/Uses
1	Raw ginger	Fresh, cleaned product used for vegetable purpose immediately after harvest.
2	Dry ginger	Prepared by partially peeling off the outer skin of raw ginger by a sharp wood/split bamboo piece and then sun-drying for 10–12 days to a moisture level of 10%.
3	Bleached dry ginger	Prepared by dipping dry ginger in 2% fresh slaked-like solution followed by sun-drying repeatedly.
4	Ginger powder	Pulverized dry ginger to a mesh size of 50–60.
5	Ginger oil	Obtained by steam distillation of dry ginger powder or fresh ginger. It contains 1.25–2.5% oil. Zingiberine, a sesquiterpene, is the chief constituent of the oil.
6	Ginger oleoresin	Acetone, alcohol, or ethylene dichloride extract of dry/fresh ginger; the yield of oleoresin varies from 4–6%. Gingerol is the main constituent of ginger oleoresin.

Table 15.1(a)   Important Products of Ginger and the Method of Preparation (Continued)

SI no.	Product	Method of Preparation/Uses
7	Ginger ale	It is used as a soft drink or soft drink additive. Prepared from dry ginger syrup by fermenting and/or adding sugar.
8	Ginger candy	Raw ginger pieces syruped in sugar.
9	Ginger beer	Syrup of dry ginger and hops in hot water enriched with sugar, cooled, and fermented with yeast. Citric acid is added and kept for a few days. The supernatant is taken out, sealed in bottles, and marketed.
10	Brine ginger	Fresh ginger in salt solution.
11	Ginger wine	A combination of ginger, sugar, chillies, and water, fermented, and charred sugar (caramel) and citric acid added.
12	Ginger flakes	Flaked and dried fresh ginger.
13	Ginger squash	Squash prepared from equal volumes of ginger juice and lime juice, with sugar.
14	Ginger paste	A ready-to-use paste made out of fresh ginger.
15	Salted sugar	Half-mature ginger processed in citric acid and vinegar and then preserved in brine.

Table 15.1(b)   Important Constituents in Ginger and Their Medicinal Properties

SI no.	Constituents	Effect	References
1	Zingerone	Lipid peroxidation inhibited	Krishnakantha and Lokesh (1993)
2	Gingerols, shogaols	Antioxidant properties, antiulcer, decreasing blood pressure, cardiovascular effect, antispasmodic and neuromuscular effects	Wang (2001); Yamara et al. (1988); Suekawa et al. (1984)
3	Zerumbone	Chemoprevention	Duve (1980)
4	Curcumin, curcuminoids	Chemoprevention	Masuda et al. (1998)
5	6-Ginger sulfonic acid	Antitumor activity	Yoshikawa et al. (1994)
6	Proteolytic enzymes	Anti-inflammatory activity	Thompson, Wolf, and Allen (1974)
7	Ginger oil (+ eugenol)	Anti-inflammatory	Sharma et al. (1997)
8	Gingerols, diarylheptanoids	Antihepatotoxic activity	Hikino et al. (1985)
9	6-Shogaol	Cough depressant	Kiuchi, Shibuya, and Sankawa (1982)
10	Ginger oleoresins	Antimicrobial activity	Chen et al. (2001)
11	Ginger essential oils	Strong antibacterial effect	Singh et al. (2001)
12	Sesquiterpenes	Reduction of rhinoviral activity	Denyer et al. (1994)
13	A-zingiberene	Antifertility	Ni, Chen, and Yan (1988)
14	Ginger extracts	Ground regulators (effect of NAA and BAP)	Arimura et al. (2000)
15	Ginger oils + garlic oil	Insecticidal synergistic effect	Hus, Chang, and Jian (1999)
16	Ginger oil + insecticidal products	Insecticidal against the cockroach	Kawata (1998)
17	6-Dehydroshogaol, dehydrogingerone	Insecticidal, antifeedant, and antifungal against *Rhizoctonia solani*	Agarwal et al. (2001)

antidepressants and it forms an ingredient of some antinarcotic preparations. In veterinary practice, ginger is used as a stimulant and carminative, in tonics, for indigestion of horses and cattle.

Many species of *Zingiber* bear showy, long-lasting inflorescences with brightly colored bracts and shining leaves and hence are widely used as cut flowers in floral arrangements and ornamental foliage plants. Some, like *Z. zerumbet* and *Z. montanum*, are valuable medicinal species; *Zingiber mioga* is a popular vegetable.

15.3 ORIGIN AND DISTRIBUTION

Ginger is not found in the truly wild state. It is believed to have originated in Southeast Asia, but was under cultivation from ancient times in India (Bailey 1961; Purseglove 1972). There is no definite information on the primary center of origin or domestication. It was brought to the Mediterranean region from India by traders during the first century AD. During the thirteenth century AD, the Arabs took ginger to eastern Africa from India. Later, it was spread to West Africa by the Portuguese for commercial cultivation. Because of the ease with which ginger rhizomes can be transported long distances, it has spread throughout the tropical and subtropical regions in both hemispheres.

The main areas of ginger cultivation are India, China, Nigeria, Indonesia, Jamaica, Taiwan, Sierra Leone, Fiji, Mauritius, Brazil, Costa Rica, Ghana, Japan, Malaysia, Bangladesh, the Philippines, Sri Lanka, Solomon Islands, Thailand, Trinidad and Tobago, Uganda, Hawaii, Guatemala, and many Pacific Ocean islands (Table 15.2).

India is the largest producer of dry ginger in the world, contributing about 30% of the world's production. In 2009, India produced 382,600 metric tons of ginger. The other main production centers

Table 15.2　Countrywide Production of Ginger in the World during 2008

Rank	Country	Production (Int$1000)	Production (MT)
1	India	217856*	382600
2	China	187227*	328810 F
3	Indonesia	109520*	192341 F
4	Nepal	100558*	176602
5	Thailand	91962*	161505
6	Nigeria	79717*	140000 F
7	Bangladesh	43870*	77046
8	Japan	28356*	49800
9	Philippines	15680*	27538
10	Cameroon	6263*	11000 F
11	Malaysia	5887*	10340
12	Sri Lanka	5722*	10050
13	Bhutan	5620*	9870 F
14	Ethiopia	5124*	9000 F
15	Cote devoir	4669*	8200 F
16	Republic of Korea	2027*	3560
17	Fiji	1393*	2448
18	Costa Rica	543*	955
19	United States of America	464*	816
20	Mauritius	403*	709

*: Unofficial Figure; []: Official data; F: FAa estimate
Source: FAa 2011 Rome http://faostat.fao.org/site/339/default.aspx

of ginger are in South Asian and South East Asian countries. China is the largest producer (328,810 metric tons) after India followed by Indonesia (192,341 metric tons), Nepal (176602 metric tons), and Thailand (161,505 metric tons). Nigeria is the largest producer of ginger among African countries with a production of (140,000 metric tons) followed by Cameroon (11,000 metric tons) and Ethiopia (9,000 Metric tons). The production of ginger in various countries of the world according to FAO (Food and Agriculture Organization of the United Nations) estimates in 2008 are summarized in Table 15.2.

15.4 BOTANY AND SYSTEMATICS

15.4.1 Genus *Zingiber* Boehmer

The genus *Zingiber*, the type genus of the family Zingiberaceae, forms an important group of the order Zingiberales. The family Zingiberaceae consists of 47 genera and about 1,400 species. Among these, 22 genera and 178 species are endemic to India. The genus *Zingiber* consists of about 150 species. Among these *Z. officinale*, *Z. zerumbet*, and *Z. montanum* are medicinal species; *Zingiber mioga* is a vegetable; and *Z. officinale* is the cultivated ginger (Mohanty and Panda 1994).

The important floristic studies of *Zingiber* are those of Hooker (1890–92), Baker (1894), Gamble (1925), Holttum (1950), Mahanty (1970), Jain and Prakash (1995), Singh et al. (1998), Suryanarayana et al. (2001), and Sabu (2003, 2006). Sato (1960) believes ginger is probably a sterile hybrid between two distant species that survived because of the successful vegetative mode of propagation.

15.4.2 *Zingiber officinale* Roscoe

Ginger is an herbaceous perennial (Figures 15.2a and 15.2b) having an underground branched rhizome with small scales. The inner core of the rhizome is pale yellow to bluish tinge while the outer is light yellow. Adventitious roots and storage roots arise from among the nodes of these scales. The ancillary buds shoot up as a leafy stem known as a pseudostem, which dies out annually; however, the plant continues to live through its rhizome. Leaf sheathing is arranged alternatively; there is a linear lanceolate gradually becoming acuminate and glabrous. Flowers are borne on a spike produced in a peduncle, different from the aerial leafy stem, arising directly from the rhizome.

The spike is condensed, oblong, and cylindrical with numerous persistent imbricate bracts, each carrying two flowers. The flowers are trimerous bisexual, irregular, epigynous, and yellow in color with dark purplish spots. The outer perianth is cylindrical and three lobed; the inner perianth tube is cylindrical with lanceolate lobes. The androecium consists of outer three stamens which are reduced to staminoids and inner stamens are united and showy to form a deep-purple colored labellum. The posterior stamen of the inner whorl is the only fertile stamen and is enclosed by the labellum. The filament is flat and short with two prominent anther lobes. The style passes through the groove formed by the anther lobes and ends in a capitate stigma. Anther cells are contiguous and produced into a long beak. The ovary is inferior, three carpelled, and three celled. Many ovules are on axial placentation. The style is long, delicate, and lying in a groove in the stamen. The stigma is small and subglobose. The fruit, which is an oblong capsule, is very rarely produced. *Zingiber officinale* is not known to set seeds (Purseglove et al. 1981; Ravindran and Babu 2005; Ravindran, Nirmal Babu and Shiva. 2005).

15.4.3 Other Important Species of the Genus *Zingiber*

The genus *Zingiber* includes many species grown as ornamentals, but some are cultivated for valuable medicines. They bear showy, long-lasting inflorescences and often brightly colored bracts

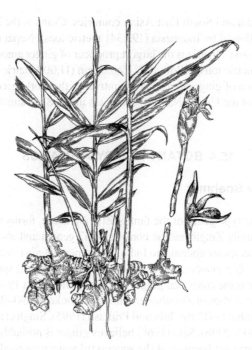

Figure 15.2(a)   Line diagram depicting ginger (*Zingiber officinale*) plant. (Source: Ravindran, P. N. et al. 2005. In *Ginger—The genus Zingiber*, ed. P. N. Ravindran and K. Nirmal Babu, 15–86. Boca Raton, FL: CRC Press.)

Figure 15.2(b)   Ginger (*Zingiber officinale*) plant showing inflorescence and rhizomes.

and floral parts; they are widely used as cut flowers in floral arrangements. Some of them are good foliage plants due to their arching form and shining leaves. Details of some important species of *Zingiber* and their distribution (Sabu and Skinner 2005) and uses are given in Table 15.3 and Figures 15.3(a) through 15.3(i).

15.4.3.1 *Zingiber americanus* Blume

Native to Malaysia, Z. *americanus* is an attractive, medicinally important garden plant widely grown in the United States. The rhizomes are used as an ingredient in various traditional medicines. The pounded rhizome is used as a poultice for women after delivery. In Java, young rhizomes are eaten as a vegetable (Prance and Sarket 1977).

15.4.3.2 *Zingiber aromaticum* Valeton

A native of tropical Asia, the rhizome of this species is strongly aromatic and fibrous, resembling Z. *americanus* in taste and aroma. It is widely cultivated in kitchen gardens and as an ornamental plant. Its fresh and tender shoots and flowers are eaten and used to flavor foods. The rhizomes are used as an ingredient in folk medicines as well as for poultices (Prance and Sarket 1977). The rhizomes contain zerumbone, which has HIV-inhibitory and cytotoxic activities (Dai et al. 1997).

15.4.3.3 *Zingiber argenteum* J. Mood & I. Theilade

This species is endemic to Malaysia and is related to Z. *lambii.* This very attractive plant is cultivated and highly valued for its floriferous habit.

15.4.3.4 *Zingiber bradleyanum* Craib

This plant is cultivated in the United States as an attractive foliage plant with a beautiful silvery stripe on the midrib of the leaves. The plant has a natural dormancy and is winter hardy.

15.4.3.5 *Zingiber chrysanthum* Roscoe

Zingiber chrysanthum produces inflorescence with long-lasting and colorful flowers with a spotted lip. The seed capsules are also ornamental, bright red, and remain for a long period of time.

15.4.3.6 *Zingiber citriodorum* J. Mood & I. Theilade

A native of Thailand, Z. *citriodorum* is a valuable ornamental plant used as a pot plant and for its cut flower. It is a beautiful plant, with sharply pointed bracts starting out green and maturing to bright red. The flowers are white. The pseudostems and foliage have a silvery gray color. Its horticultural variety, 'Chiang Mai Princess,' is cultivated in the United States.

15.4.3.7 *Zingiber clarkei* King ex Benth.

A native of Sikkim, India, Z. *clarkei* is a valuable ornamental plant suitable to subtropical and temperate regions. The plants are tall with a foot-long inflorescence appearing from the main stem. Among the *Zingiber* species, this is unique because the spike is produced laterally rather than radially as in other species.

Table 15.3 Important Species of *Zingiber*, Their Distribution, and Uses

SI no.	Species	Common Name	Origin and Distribution	Status	Uses	Important Constituents
1	Z. acuminatum Valeton		Malaysia	Wild, cultivated	Used as a poultice	
2	Z. albiflorum R.M. Sm.		Malaysia	Wild		
3	Z. amaricanus Blume		Southeast Asia	Wild	Ornamental, medicinal	
4	Z. argenteum J. Mood & I. Theilade		Malaysia	Wild, cultivated, endemic	Ornamental	
5	Z. aromaticum Valeton	Puyang	Tropical Asia	Cultivated	Ornamental, folk medicine	Zerumbone, 3",4"-o-diacetylafzelin
6	Z. aurantiacum I. Theilade		Malaysia	Wild		
7	Z. barbatum Wall.		Myanmar, Thailand	Wild		
8	Z. bradleyanum Craib		Thailand	Cultivated, wild	Foliage plant	
9	Z. capitatum Roxb.	Jangli ardrak	India, Nepal, Bangladesh	Wild, cultivated	Ornamental	
10	Z. cernum Dalz. = Z. nimmonii (J. Graham)		India			
11	Z. chlorobracteatum Mood & Theilade		Malaysia	Wild, endemic		
12	Z. chrysanthum Roscoe		India, Nepal	Wild	Long-lasting flowers	
13	Z. chrysostachys Ridley		Thailand, Malaysia	Wild, endemic		
14	Z. citriodorum J. Mood & I. Theilade	Chiang Mai Princess	Thailand	Wild, endemic	Ornamental	
15	Z. citrinum Ridley		Malaysia	Wild, endemic	Ornamental	
16	Z. clarkei King ex Benth.		India	Wild	Ornamental	
17	Z. collinsii I. Theilade & J. Mood		Vietnam	Wild, endemic	Ornamental	
18	Z. corallinum Hance	Karen pleklo	Thailand	Wild	Ornamental remedy after childbirth	
19	Z. cylindricum Moon		India	Wild, endemic		
20	Z. curtisii Holttum		Malaysia	Wild		
21	Z. eborium J. Mood & I. Theilade	White ginger, ivory ginger	Malaysia, Indonesia	Endemic	Ornamental	
22	Z. flagelliforme J. Mood & I. Theilade		Malaysia	Wild, endemic		

No.	Species	Common name	Distribution	Status	Uses	Compounds
23	Z. flammeum I. Theilade & J. Mood		Malaysia	Wild, endemic		
24	Z. flavovirens I. Theilade		Thailand	Wild		
25	Z. fraseri I. Theilade		Malaysia	Wild, endemic		
26	Z. georgei J. Mood & I. Theilade		Malaysia	Wild, endemic		
27	Z. gracile Jack	Mempoyang	Malaysia	Wild		
28	Z. gramineum Noronha ex Blume	Palm ginger	Thailand, Cambodia, Sumatra	Wild	Ornamental	
29	Z. griffithii Baker		Malaysia, Thailand, Singapore	Wild	Ornamental	
30	Z. incomptum Burtt & R.M. Sm.		Malaysia	Wild		
31	Z. intermedium Baker		India	Wild, endemic		
32	Z. junceum Gagnep.	Yellow delight	Cambodia, Thailand, Laos	Wild	Cut flowers	
33	Z. kerrii Craib		Thailand, Indochina	Wild		
34	Z. koshunense C.T. Moo		Taiwan	Wild, endemic		
35	Z. kuntstleri Ridley		Malaysia	Wild		
36	Z. lambii J. Mood & I. Theilade		Malaysia	Wild, endemic	Ornamental	
37	Z. larsenii I. Theilade		Thailand	Wild		
38	Z. latifolium J. Mood & I. Theilade		Malaysia	Endemic		
39	Z. leptostachyum Valeton		Malaysia	Wild		
40	Z. ligulatum Roxb.		India	Wild, endemic		
41	Z. longibracteatum I. Theilade		Thailand	Wild, endemic		
42	Z. longipedunculatum Ridley		Australia	Cultivated	Ornamental	
43	Z. macrostachyum Dalz.		India	Wild		
44	Z. malaysianum C.K. Lim	Midnight beauty	Malaysia	Wild, endemic	Ornamental	
45	Z. marginatum Roxb.		India			
46	Z. martini R.M. Sm.		Malaysia	Wild		
47	Z. mioga Roscoe	Japanese ginger	Japan	Endemic	Vegetables; used to treat fever and also as a vermifuge	Galanal A, B
48	Z. montanum (Koenig) Link ex Dietr. = Z. cassumunar Roxb.	Chocolate pinecone ginger; vanadraka; kadu shunti	India, Malay peninsula, Thailand, Sri Lanka	Abundant	Used to treat cough, asthma, diabetes, and as antidiarrheal	(E)-1-(3,4-dimethoxyphenyl)but-1-ene, zerumbone

(continued)

Table 15.3 Important Species of *Zingiber*, Their Distribution, and Uses (Continued)

Sl no.	Species	Common Name	Origin and Distribution	Status	Uses	Important Constituents
49	*Z. multibracteatum* Holttum		Malaysia			
50	*Z. neglectum* Valeton		United States	Cultivated	Ornamental	
51	*Z. newmanii* I. Theilade		Thailand	Wild, endemic		
52	*Z. neesanum* (J. Graham) Ramamurthy = *Z. macrostachyum* Dalz.		India	Wild, endemic		
53	*Z. niveum* J. Mood & I. Theilade	Milky way	United States	Cultivated	Ornamental	
54	*Z. nimmonii* (J. Graham) Dalz. = *Z. cernum* Dalz.		India	Wild, endemic		
55	*Z. odoriferum* Blume		India, Malaysia	Wild		
56	*Z. ottensii* Valeton	Lempoyang hitam or bonglai hitam	Southeast Asia	Cultivated, wild	Ornamental	Humulene, zerumbone
57	*Z. pachysiphon* B.L. Burtt & R.M. Sm.		Malaysia, Australia	Wild	Ornamental	
58	*Z. parishii* Hook. f.	Khin paa	India, Myanmar, Thailand	Wild		
59	*Z. pellitum* Gagnepo		Thailand, Laos, Malaysia	Wild		
60	*Z. pendulum* J. Mood & I. Theilade		Malaysia	Wild, endemic		
61	*Z. peninsulare* Theilade		Thailand	Wild		
62	*Z. petiolatum* (Holttum) I. Theilade		Thailand, Malaysia	Wild		
63	*Z. phillippsii* J. Mood & I. Theilade		Malaysia	Wild, endemic		
64	*Z. pseudopungens* R.M. Sm.		Malaysia	Wild		
65	*Z. puberulum* Ridley	Lempoyang ajing	Malaysia, Thailand, Singapore	Wild		
66	*Z. roseum* (Roxb.) Roscoe		India	Wild, endemic		
67	*Z. rubens* Roxb.	Sarg mang	India, Thailand, Vietnam, China	Wild, cultivated	Medicinal	
68	*Z. smilesianum* Craib		Thailand	Wild		
69	*Z. spectabile* Griffith	Beehive ginger	Malaysia, Thailand	Cultivated, wild	Cut flower, traditional Malay medicine	trans-d-Gergamontene,-β-elemene

No.	Species	Common name	Distribution	Status	Use	Compounds
70	*Z. squarrosum* Roxb.		India, Myanmar	Wild		
71	*Z. sulphureum* Burkill ex I. Theilade		Malaysia	Wild, endemic		
72	*Z. velutinum* J. Mood & I. Theilade		Malaysia	Wild, endemic		
73	*Z. villosum* I. Theilade		Thailand	Wild		
74	*Z. vinosum* J. Mood & I. Theilade		Malaysia	Wild, endemic	Ornamental	
75	*Z. viridiflorum* J. Mood & I. Theilade		Malaysia	Wild		
76	*Z. wightianum* Thwaites		India, Sri Lanka	Wild		
77	*Z. wrayii* Ridley		Malaysia, Thailand	Wild		
78	*Z. zerumbet* (L.) Smith	Shampoo ginger	Tropical Asia	Abundant	Cut flower	Zerumbone, 1,8-cineole

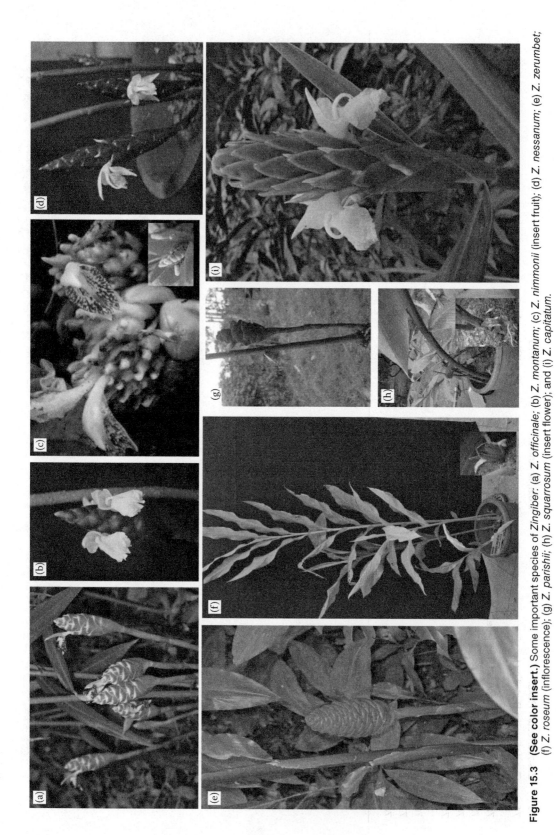

Figure 15.3 **(See color insert.)** Some important species of *Zingiber*: (a) *Z. officinale*; (b) *Z. montanum*; (c) *Z. nimmonii* (insert fruit); (d) *Z. nessanum*; (e) *Z. zerumbet*; (f) *Z. roseum* (inflorescence); (g) *Z. parishii*; (h) *Z. squarrosum* (insert flower); and (i) *Z. capitatum*.

15.4.3.8 *Zingiber collinsii* J. Mood & I. Theilade

A beautiful species, *Z. collinsii* was discovered and introduced by Mark Collins. It has silvery stripes across the leaves, somewhat similar to *Alpinia pumila*, and is very popular in the United States and Europe.

15.4.3.9 *Zingiber corallinum* Hance

Zin corallinum is valued as an ornamental and a medicinal plant of great promise. The rhizome is used in traditional Chinese medicine and is known to prevent skin infections (Shuxuan et al. 2001).

15.4.3.10 *Zingiber eborium* J. Mood & I. Theilade

Commonly known as white ginger, ivory ginger, or ivory spike ginger, this species is endemic to Borneo. The ivory white inflorescence bracts and orange flowers make this a very attractive garden plant in the West. Tolerant to freezing, the plant flowers profusely under cultivation, making it a valuable pot plant.

15.4.3.11 *Zingiber gramineum* Noronha

Cultivated in the United States under the common name, palm ginger, *Z. gramineum* is a thin-stemmed and narrow-leafed garden plant.

15.4.3.12 *Zingiber griffithii* Baker

This Malaysian species is a common ornamental plant, with a shortly peduncled cylindrical spike and bright red, obovate, obtuse bracts. The tip of the bract is yellowish white and three lobed.

15.4.3.13 *Zingiber junceum* Gagnep

This plant has symmetrical silvery stems and leaves and produces yellow and long-lasting basal or terminal inflorescences. The spikes are yellow and used as a cut flower also. This species is known in the United States under the name 'Yellow Delight.'

15.4.3.14 *Zingiber lambii* J. Mood & I. Theilade

Endemic to East Malaysia, *Z. lambii* is a small plant with beautiful silvery green leaves, orange spikes, and yellow flowers. The bracts are orange, greenish toward the apex, and; they turn pink with age. Related to *Z. argenteum*, this species has become popular as a valuable garden plant.

15.4.3.15 *Zingiber longipedunculatum* Ridley

Cultivated in Australia, this species is a valuable garden plant used for cut-flower purposes and often used in floral arrangements.

15.4.3.16 *Zingiber malaysianum* C.K. Lim

Widely sold in the United States under the name 'Midnight Beauty,' this much sought after ever-green, very attractive ornamental plant has become extremely popular with its shiny dark brown to almost black leaves and bright red inflorescences. It is used as a pot or bed plant and spikes are used as cut flowers and in floral arrangements.

15.4.3.17 *Zingiber mioga* Roscoe

Zingiber mioga, commonly known as myoga ginger or Japanese ginger, is a perennial species endemic to Japan. It is grown for its edible flowers and young shoots, both of which are used extensively as vegetables. This species is propagated using rhizome bits (Sterling et al. 2002). From Japan, myoga cultivation has spread to China, Vietnam, Taiwan, Thailand, Australia, and New Zealand. Many variants are available (Clark and Warner 2000). Two compounds, galanal A and galanal B, are known to contribute to the characteristic flavor of the myoga rhizomes. The pungent principle in myoga is myogadial. The pungency of myogadial depends on the presence of the α-β-unsaturated-1,4-dialdehyde group (Abe et al. 2002). In the Chinese pharmacopoeia, myoga ginger is used to treat fever and also as a vermifuge.

15.4.3.18 *Zingiber montanum* (Koenig) Link ex Dietr. (= *Z. cassumunar* Roxb.)

Native to India, *Z. montanum* is also present and cultivated throughout the Malay peninsula, Sri Lanka, and Java. Rhizome is used as an ingredient in many traditional medicines to relieve cough and asthma (Quisumbing 1978), as an antidiarrheal medicine (Saxena et al. 1981), as an antidote to snakebite, for cholera, and also as vermifuge (Barghava 1981). The oil has antibacterial and antifungal properties. The rhizome is also considered a good tonic and appetizer. Many studies have been made on the medicinal properties, especially of the anti-inflammatory effect of *Z. cassumunar* (Ozaki et al. 1991). Masuda, Jitoe, and Mabry (1995) isolated cassumunarins A, B, and C, three anti-inflammatory antioxidants. The rhizomes of *Z. cassumunar* exhibited insecticidal activity.

Bordoloi, Sperkova, and Leclercq (1999) investigated the essential oil composition of *Z. cassumunar* and it contained terpene-4-ol (50.5%), E-1-(3,4-dimethoxyphenyl) buta-1,3-diene (19.9%), E-1-(3,4-dimethoxyphenyl) but-1-ene (6.0%), and -sesquiphellandrene (5.9%) as major constituents out of the 21 compounds identified. In the leaf essential oil, 39 compounds were identified. The main components were 1(10),4-furanodien-6-one (27.3%), curzerenone (25.7%), and sesquiphellandrene (5.7%).

15.4.3.19 *Zingiber neglectum* Valeton

Zingiber neglectum is one of the most sought after and popular plants in cultivation in the United States. It has very long inflorescences with beautiful cup-shaped bracts similar to those of *Z. spectabile*, with purple flowers.

15.4.3.20 *Zingiber niveum* J. Mood & I. Theilade

Zingiber niveum is an attractive ornamental ginger with silvery grey leaves, very unusual looking milky-white, rounded spikes, and yellow flowers. It is sold in the United States under the name 'Milky Way'.

15.4.3.21 *Zingiber ottensii* Valeton

A native of Southeast Asia, *Z. ottensii* is widely cultivated as an ornamental both as a pot plant and for its cut flowers. It is close to *Z. zerumbet* and *Z. montanum*. The stem is reddish and attractive. The red inflorescence is more or less similar to that of *Z. zerumbet*. The rhizome is used as an appetizer and as a poultice in postnatal treatment. Three sesquiterpenes—humulene, humulene epoxide, and zerumbone—were isolated from dried rhizome.

15.4.3.22 *Zingiber pachysiphon* B.L. Burtt & R.M. Sm.

This is in cultivation in Australia. The plant has a beautiful purplish-colored inflorescence with white edges to the bracts. The species is rather rare and valued much by ginger lovers as a very attractive pot plant.

15.4.3.23 *Zingiber rubens* Roxb.

Native to the Indo-Malaysian region, this species was introduced to U.S. gardens as a pot plant. Its spike is globose with a small peduncle, bright red bracts, and red corolla with an oblong lip that is much spotted and streaked with red. The paste of the rhizome is used to treat giddiness (Saxena et al. 1981).

15.4.3.24 *Zingiber spectabile* Griff.

Known as beehive ginger due to the peculiar shape of the spike, *Z. spectabile* has a large and very attractive inflorescence. It is widely used as a cut flower because of its long shelf life. This species is widely used in Malay traditional medicine (Burkill and Haniff 1930). It is very popular in the United States as golden shampoo ginger and has two horticultural varieties; both are very much valued as pot plants and for cut-flower production.

15.4.3.25 *Zingiber vinosum* J. Mood & I. Theilade

Native to Sabah, East Malaysia, *Z. vinosum* is cultivated as an ornamental plant and is very popular in the United States. The attractive foliage and red inflorescence make it a very valuable garden plant. The rhizome is moderately aromatic.

15.4.3.26 *Zingiber zerumbet* (L.) Smith

Commonly known as shampoo ginger or pinecone ginger, *Z. zerumbet* is native to tropical Asia. The inflorescence resembles a tight pinecone and releases a thick juice when squeezed. This juice is used to make a shampoo. The inflorescence is widely used as a cut flower. Two of its popular horticultural varieties are named Darceyi and Twice as Nice.

The main components of the volatile oil of *Z. zerumbet* are zerumbone (56.48%), 1,8-cineole (1.07%), *o*-caryophyllene (2.07%), α-humulene (25.70%), caryophyllene oxide (1.41%), humulene epoxide (3.62%), and humulene epoxide 11 (2.45%) (Tewtraki et al. 1997). The sesquiterpene zerumbone is a neutraceutical compound and has anti-inflammatory, chemopreventive, and chemotherapeutic qualities (Murakami et al. 2002). Dai et al. (1997) reported that zerumbone has a potent HIV-inhibitory action.

15.4.4 Molecular Taxonomy

In recent times, molecular data have been increasingly used to understand the phylogeny and interrelationships among species in Zingiberaceae. Kress, Prince, and Williams (2002) studied the phylogeny of the gingers (Zingiberaceae) based on DNA sequences of the nuclear internal transcribed spacer (ITS) and plastid *matK* regions and suggested a new classification. Their studies suggest that at least some of the morphological traits based on which the gingers are classified are homoplasious and three of the tribes are paraphyletic. The former Alpinieae and Hedychieae for the most part are monophyletic taxa with the Globbeae and Zingiberaceae are included within the latter. They proposed a new classification of the Zingiberaceae that recognizes four subfamilies and four tribes: Siphonochiloideae (Siphonochileae), Tamijioideae (Tamijieae), Alpinioideae (Alpinieae, Riedelieae), and Zingiberoideae (Zingibereae, Globbeae).

A phylogenetic analysis was also performed by Ngamriabsakul, Newman, and Cronk (2003) using DNA sequence analysis data nuclear ribosomal DNA (ITS1, 5.8S, and ITS2) and chloroplast DNA (*trn*L [UAA] 5[prime prime or minute] exon to *trn*F [GAA]) to have a better understanding of the phylogeny and relationships in the tribe Zingibereae (Zingiberaceae). The study indicated that the tribe Zingibereae is monophyletic with the species separated into two major clades: the *Curcuma* and the *Hedychium*.

Gao et al. (2008) used SRAP (sequence-related amplified polymorphism) markers to analyze phylogenetic relationships of 22 species of Chinese *Hedychium*. Phylogenetic analysis showed that 22 species were grouped into three clusters.

15.5 CYTOLOGY

Except for Takahashi (1930), who claimed $2n = 24$ for the somatic chromosome number of ginger (*Z. officinale*), the number was reported as $2n = 22$ (Moringa et al. 1929; Sugiura 1936; Raghavan and Venkatasubban 1943; Darlington and Janaki Ammal 1945; Chakravorti 1948; Sharma and Bhattacharya 1959). The reports on chromosome numbers of various species are summarized in Table 15.4.

After a detailed study on cytology of *Z. officinale*, *Z. cassumunar*, and *Z. zerumbet*, Raghavan and Venkatasubban (1943) concluded that the chromosome morphology of *Z. officinale* was different from that of the other two species. Karyotype studies of 24 species belonging to 13 genera were conducted and it was concluded that the basic number of the genus *Zingiber* is $x = 11$ and that *Z. mioga*,

Table 15.4 Chromosome Numbers in the Genus *Zingiber*

| Species | Chromosome no. | | References |
	(*n*)	(2*n*)	
Z. clarkii King ex Benth.		22	Holttum (1950)
Z. cylindricum Moon		22	Mahanty (1970)
Z. mioga Roscoe		55	Moringa et al. (1929); Sato (1948)
Z. montanum Link = *Z. cassumunar* Roxb.		22	Chakravorti (1948); Raghavan and Venkatasubban (1943); Ratnambal (1979)
Z. neesanum (J. Graham) = *Z. macrostachyum* Dalz.		22	Ramachandran (1969)
Z. ottensii Valetor		22	Holttum (1950)
Z. officinale Roscoe		22	Raghavan and Venkatasubban (1943); Chakravorti (1948); Sharma and Bhattacharya (1959)
		22 + 2B	Darlington and Janaki Ammal (1945)
		24	Takahashi (1930)
	11	22	Ramachandran (1969)
	11	22	Ratnambal (1979)
Z. roseum (Roxb.) Roscoe.	11	22	Ramachandran (1969)
Z. spectabilis Griff.		22	Mahanty (1970)
Z. wightianum Thwaites	11	22	Ramachandran (1969)
Z. zerumbet (L.) Smith	11	22	Ratnambal (1979)

with $2n = 55$, is a pentaploid (Sato 1960; Chakravorti 1948). Ramachandran (1969) found a diploid number of $2n = 22$ in *Z. macrostachyum, Z. roseum, Z. wightianum, Z. zerumbet*, and *Z. officinale* and found evidence of structural hybridity involving interchanges and inversions in ginger. Mahanty (1970) reported $2n = 22$ for *Z. spectabile* and *Z. cylindricum* and concluded that the genus *Zingiber* appears to be much more correctly placed in the Hydychieae than in the Zingibereae.

Janaki Ammal (Darlington and Janaki Ammal 1945) reported two B chromosomes in certain types of ginger in addition to the normal complement of $2n = 22$. Beltram and Kam (1984) also observed B chromosomes in ginger.

Ratnambal (1979) investigated the karyotype of 32 cultivars of ginger (*Z. officinale*) and found that all of them possess a somatic chromosome number of $2n = 22$. An asymmetrical karyotype of 1B was found in all cultivars except in cultivars Bangkok and Jorhat, which have a karyotype asymmetry of 1A (Ratnambal 1979). The karyotypes of various cultivars exhibited only minor differences.

Ratnambal (1979) used the karyotype data in generalized distance D^2 statistics and found that geographical distances did not influence the clustering. Cultivars 'Tafingiwa', 'Jamaica', 'Rio de Janeiro', 'Thinladium', Thingpuri, 'Maran', and 'Himachal Pradesh' did not fall into any cluster, indicating their divergence from the rest of the cultivars. *Z. zerumbet* and *Z. cassumunar* did not fall into any group, but *Z. macrostachyum* fell into group B.

Ratnambal and Nair (1981) studied the process of meiosis in 25 cultivars of ginger. These cultivars exhibited much intercultivar variability in meiotic behavior. The presence of multivalent and chromatin bridges was found to be a common feature in most cultivars except Karakkal, where only bivalents were observed. The presence of univalents and multivalents in a diploid species indicates structural hybridity involving segmental interchanges, and four to six chromosomes are involved in the translocations as evidenced by quadrivalents and hexavalents. This structural hybridity leads to the production of gametes with deficiency and might be contributing to the sterility in ginger.

Structural chromosomal aberrations like laggards, bridges, and fragments; irregular chromosome separation and irregular cytokinesis at anaphase II; and micronuclei and supernumerary spores at the quartet stage occurred at all stages of microsporogenesis in ginger. Similar findings were found by P. Das, Rai, and A. Das (1999) in four ginger cultivars—namely, 'Bhaisey', 'Ernad Chernad', 'Gorubathany', and 'Thuria Local' and by Beltram and Kam (1984) in 33 species in Zingiberaceae, including 9 species of *Zingiber*. They observed various abnormalities such as aneuploidy, polyploidy, and B chromosomes.

Significant variation in nuclear DNA content at the cultivar level was observed. Structural alterations in the chromosomes without the changes in the numeric chromosome number ($2n = 22$), as well as loss or addition of highly repetitive sequences in the genome, caused variations in the DNA amount at cultivar level (Rai, A. Das, and P. Das 1997; A. Das, Rai, and P. Das 1998).

15.6 FLORAL BIOLOGY

Ginger flowers are produced in peduncled spikes arising directly from the rhizomes. The oval or conical spike consists of overlapping bracts from the axils of which flowers arise; each bract produces a single flower. The flowers are fragile, short lived, and surrounded by a scariose, glabrous bracteole. Each flower has a thin, tubular corolla that widens up at the top into three lobes. The colorful part of the flower is the labellum, the petalloid stamen. The labellum is tubular at the base, three lobed above, pale yellow outside, and dark purple inside the top and margins, and it is mixed with yellow spots. The single fertile anther is ellipsoid, two celled, and cream colored, and it dehisces by longitudinal slits. The inferior ovary is globose, the style is long and filiform, and the stigma is hairy. In general, ginger does not flower under subtropical or subtemperate climatic conditions. In India, most cultivars flower if sufficiently large rhizome pieces are used for planting. When rhizomes are left unharvested in pots, profuse flowering occurs in the next growing season.

Studies on the floral biology of ginger indicated that it takes about 20–25 days from the flower bud initiation to full boom, and blooming takes place in an acropetal succession (AICSCIP 1975; Jayachandran, Vijayagopal, and Sethumadhavan 1979; Das et al. 1999). The flowers open between 2:30 and 4:30 p.m. and anthesis takes place simultaneously between 1:30 and 3:30 p.m. Stigma is receptive at the time of anther dehiscence. The flower fades and falls the next morning. There is no fruit setting.

The nature of sterility in ginger is not clearly understood. Ramachandran (1969) suggested sterility is chromosomal. The pollen grains were heteromorphic and their size varied from 78 to 104 µm, with an average of 92 µm. Pollen sterility ranged up to an average of 76%. The stainability percentage of pollen grains ranges from 12.5 (cultivar 'Rio de Janeiro') to 28.5 (in cultivars 'Pottangi' and China) (Usha 1984). Pollen germination ranged from 8 (cultivar 'Sabarimala') to 24% (cultivar 'Maran'). The pollen tube growth under *in vitro* conditions was maximum in cultivar China (488 µm) and minimum in cultivar 'Nadia' (328 µm). The number of pollen tubes ranged from 6.5 (in cultivar 'Nadia') to 16.7 (in cultivar 'Varada') (Dhamayanthi, Sasikumar, and Remashree 2003).

Das et al. (1999) reported that flowers were hermaphroditic with "pin" and "thrum" type incompatibility and dehisced pollen grains did not reach the stigma. Selfing and cross-pollination did not produce any seed set. Dhamayanthi et al. (2003) reported that heterostyly with a gametophytically controlled self-incompatibility system exists in ginger. Flowers are distylous, there are long (pin) and short (thrum) styles. The pin type has a slender style that protrudes out of the floral parts, which are short, covering not even half the length of the style. The stigma is receptive before the anthesis, whereas the anthers dehisce after 15–20 hours. The anthers are situated far below and hence the pollen grains cannot reach the stigma. In case of the thrum style, the stigma is very short and the staminodes are long and facing inward. The occurrence of thrum styles is very rare among cultivated ginger. They also reported inhibition of pollen tube growth in the style, and this was interpreted to be due to incompatibility. This heterostyly coupled with incompatibility may be a contributing factor to the sterility in ginger.

The mechanism of pollination is different in other species of *Zingiber*. In *Z. zerumbet*, the flowers open from morning until evening in an inflorescence. The flowers are usually cross-pollinated. The pollination in the species of *Zingiber* is rather simple because of the specially modified anther structure and nature of staminodes. An insect visiting a flower first lands on the labellum and moves to the throat of the corolla tube. When the insect's front portion pushes the base of the anther, the anther bends forward and dusts the pollen grains on the backside of the insect. As it bends forward, the stigma protrudes and arches through the long anther crest and presses against the proboscis of the insect. Thus, pollen grains from other flowers deposited on the back of the insect stick to the stigma, and pollination is effected (Sabu and Skinner 2005).

15.7 PROPAGATION AND CULTIVATION

Though a perennial plant, ginger is usually grown as an annual crop (Figure 15.4). It requires a warm, humid climate and well-distributed rainfall.

15.7.1 Propagation

Ginger is propagated vegetatively using rhizomes pieces called seed sets. The seed rhizomes are to be selected after each harvest and stored properly in shade for planting in the next season. Rhizome pieces of 20–25 g size and 2.5–5.0 cm length with at least two growing buds are used as planting units (Figure 15.5). Since diseases and pests can be transmitted through seed rhizomes, pretreatment of seed rhizomes in fungicidal solution should be practiced.

In order to obtain good germination, bold and healthy rhizomes from disease-free plants are selected as seed rhizomes and collected immediately during the previous year's harvest. The seed

Figure 15.4 A view of ginger plantation in raised beds.

Figure 15.5 Rhizome bits of ginger used as planting material.

rhizomes are to be stored properly in pits under shade in layers, along with well-dried sand or saw-dust to provide adequate aeration.

15.7.2 Planting and Spacing

The field is loosened and prepared with the onset of monsoon. Ginger is planted on raised beds of about 1 m width and convenient length (Figure 15.4) or in ridges and furrows. Planting in flat lands is also practiced. The spacing is about 25 × 15 cm under a bed system or 40 × 20 cm under a ridges and furrows system. A seed rate of 1.5–2.5 t/ha is used. The seed rate varies from region to region and with the method of cultivation adopted. Ginger can also be transplanted. Transplanting can be adopted when the rhizomes are suspected of carrying pathogens. In such cases rhizomes can be initially planted in nursery beds under shade. When the healthy plantlets attain about 45 cm, they can be taken out, dipped in fungicidal solution, and transplanted to the main field. Proper drainage channels are to be provided when there is water stagnation.

15.7.3 Nutrition

Ginger is an exhaustive crop and it benefits greatly from organic manure application. Compost or farmyard manure (FYM) at the rate of 25–30 t/ha is applied at the time of preparing beds and application of NPK at the rate of 75:50:50 kg/ha in three split doses is recommended for better yields. Mulching with green leaves at the rate of 10–12 t/ha at the time of planting and 5 t/ha at the time of subsequent fertilizer application is recommended to prevent erosion of soil due to heavy rain. Studies also confirmed that application of green manure and FYM and groundnut cake/gingelly oil cake/cotton cake is good in registering high yields. Because ginger is a very exhaustive crop, it is not desirable to grow it in the same site year after year.

15.7.4 Diseases and Pests

The major diseases are soft rot (caused by *Pythium aphanidermatum*, *P. myriotylum*, and *P. vexans*), bacterial wilt (caused by *Ralstonia solanacearum*), leaf spot (caused by *Phyllosticta zingiberi*), and yellow disease (caused by *Fusarium oxysporum* and *F. solani*). Shoot borer (*Conogethes punctiferalis*) and rhizome scale (*Aspidiella hartii*) are the major pests affecting ginger. Treating seed rhizomes with 0.3% dithane M-45 and 200 ppm streptocycline before planting and drenching the affected sites with 1% Bordeaux mixture control rhizome rot and bacterial wilt, respectively. Spraying 0.1% malathion or 0.05% monocrotophos at monthly intervals during July through October controls shoot borer; treating the rhizomes with 0.1% quinalphos twice—once before storing and again during planting—controls rhizome scale.

15.7.5 Harvesting and Processing

The crop is ready for harvest 7–9 months after planting when the leaves start drying up. Harvesting is done using a spade or digging fork by carefully uprooting the underground rhizome without causing injury or damage. The rhizomes are separated from the dried up leaves. If the crop is for green ginger, it is to be harvested in 5–6 months. For curing, the rhizomes are soaked in water for 6–7 hours and washed to remove soil particles. The rhizomes are then rubbed well and washed. Average yield ranges from 15–30 t/ha.

The outer skin is removed with split bamboos having pointed ends. Only the outer skin is to be removed since the essential oil of ginger lies immediately under the skin. Rhizomes are dried in the sun for a week and stored in a cool, dry place. The yield of dry ginger is 20–25% of that of the green ginger, depending on the variety.

15.8 PHYSIOLOGY

15.8.1 Effect of Day Length on Flowering and Rhizome Growth

Ginger is grown under varying climatic conditions and in both hemispheres. Adaniya, Ashoda, and Fujieda (1989) report inhibition of vegetative growth of shoots and the underground rhizome, but with more rounded rhizome knobs, when light periods decreased from 16 to 10 hours, in three Japanese cultivars (Kintoki, Sanshu, and Oshoga). As the day length increased to 16 hours, the plants grew more vigorously and the rhizome knobs were slender and larger and active as new sprouts continued to appear. When the light period was further increased to 19 hours, there was reduction in all growth parameters.

The study indicates that the vegetative growth was promoted by a longer light period up to a certain limit, whereas rhizome swelling was accelerated under a relatively short day length. The results also suggested that a relatively short day length accelerated the progression of the reproductive growth, whereas relatively long day length decelerated it. Thus, ginger is described as a quantitative short-day plant for flowering and rhizome swelling. These researchers also observed intraspecific variations in photoperiodic response; cultivar Sanshu responded most sensitively, followed by Kintoki and Oshoga. They concluded that such an intraspecific response to the photoperiod could be related to traditional geographical distribution.

Sterling et al. (2002) studied the effect of photoperiod on flower bud initiation and development in *Zingiber mioga*. Plants grown under long-day conditions (16 hours) produced flower buds, whereas those under short-day conditions (8 hours) did not. This failure of flower bud production under short days was due to abortion of developing floral bud primordia rather than a failure to initiate inflorescences. Thus, flower development in myoga is dependent on long days, but flower initiation was day neutral. Short-day conditions also resulted in premature senescence of foliage and reduced foliage dry weight.

15.8.2 Chlorophyll Content and Photosynthetic Rate in Relation to Leaf Maturity

Xizhen, Zhenxian, Zhifeng, et al. (1998) investigated the chlorophyll content, photosynthetic rate (Pn), malondialdehyde (MDA) content, and the activities of the protective enzymes during leaf development. Both chlorophyll content and Pn increased with leaf expansion, reaching a peak on the 15th day, and then declined gradually. After 40 days, the MDA content increased markedly and SOD (Superoxide Dismutase) activity dropped substantially. Peroxidase (POD) and catalase activities exhibited a steady increase during a 60-day period.

Xizhen, Zhenxian, Shaohui, et al. (1998) reported that senescence of ginger leaf sets in when leaf age reaches about 40 days. They also studied the photosynthetic characteristic of different leaf positions and reported that the Pn of midposition leaves was the highest, followed by the lower leaves, and that Pn was lowest in upper leaves.

15.8.3 Photosynthesis and Photorespiration

Zhenxian et al. (2000) studied the diurnal variation of photosynthetic efficiency under shade and field conditions. There was marked photoinhibition under high-light stress at midday and this was severe in the seedling stage. The apparent quantum yield (AQY) and photochemical efficiency of PS II (Fv/Fm) decreased at midday, and there was a marked diurnal variation. In shade, the degree of photoinhibition declined markedly. However, under heavier shade, the photosynthetic rate declined because the carboxylation efficiency declined after shading.

Shi-jie et al. (1999) investigated the seasonal and the diurnal changes in photorespiration (Pr) and the xanthophyll cycle (L) in ginger leaves under field conditions in order to understand the role of L and Pr in protecting leaves against photoinhibitory damage. The results indicated that Pr and the xanthophyll cycle had positive roles in dissipating excessive light energy and in protecting the photosynthetic apparatus of ginger leaves from midday high-light stress. Xizhen et al. (2000) have also investigated the role of SOD in protecting ginger leaves from photoinhibition damage under high-light intensity. They concluded that midday high-light intensity imposed a stress on ginger plants and caused photoinhibition and lipid peroxidation. SOD and shading played important roles in protecting the photosynthetic apparatus of ginger leaves against high-light stress.

Xizhen, Zhenxian, and Shaohui (1998) have investigated the effect of temperature on photosynthesis of ginger leaf. They showed that the highest photosynthetic rate and apparent quantum efficiency were under 25°C. The light saturation point was also temperature dependent. The low-light saturation point was noted at temperatures below 25°C.

Xianchang et al. (1996) studied the relationship between canopy, canopy photosynthesis, and yield in ginger and found that canopy photosynthesis was closely related to yield. The canopies over 7,000 plants per 666.7 m² area, with the unit area of tillers and leaf area index, were over 150/m² and 6/m², respectively; had the criteria optimum yield; and there were no significant differences in height, tillers, leaf area index, and canopy photosynthesis.

15.8.4 Effect of Growth Regulators

Studies have been carried out to find out the effect of various growth regulators on ginger growth, rhizome development, flowering, and seed set. Furutani and Nagao (1986) investigated the effect of daminozide, gibberellic acid (GA3), and ethephon on flowering, shoot growth, and yield of ginger. Treating the field grown ginger plants with three weekly foliar sprays of GA3 (0, 1.44, and 2.88 mM), ethephon (0, 3.46, and 6.92 mM), or daminozide (0.3, 13, and 6.26 mM). GA3 inhibited flowering and shoot emergence, whereas ethepon and daminozide had no effect on flowering but promoted shoot emergence. Rhizome yields were increased with daminozide and decreased with GA3 and ethephon.

Exposure of ginger rhizome pieces to 35°C for 24 hours or to 250–500 ppm 2-chloroethyl phosphonic acid (ethrel or ethephon) for 15 minutes caused a substantial increase in shoot growth and number of roots (Islam et al. 1978). Ravindran et al. (1998) tested three growth regulators—triacontanol, paclobutrazole, and GA3 treatment on ginger and reported that paclobutrazole- and triacontanol-treated rhizomes resulted in thicker walled cortical cells and higher procambial activity compared to GA3 and control plants. Paclobutrazole-treated plants exhibited greater deposition of starch grains and higher frequency of oil cells. GA treatment also led to considerable increase in the number of fibrous roots and reduced fiber content in the rhizome.

15.8.5 Growth-Related Compositional Changes

Baranowski (1986) studied the growth-related changes in the rhizome cultivar Hawaii and found that the solid content of the rhizome increased throughout the season. The oleoresin content on a fresh weight basis was roughly constant in fresh and dried ginger. On a dry weight basis, gingerol content generally increased with maturity up to 24 weeks, followed by a steady decline through the rest of the period, indicating that it may be advantageous to harvest ginger early (i.e., by 24 weeks) for converting to various products. These results explain the reason for the gradual increase in pungency with maturity.

15.9 GENETIC RESOURCES

15.9.1 Conservation

Cultivar diversity is richest in China followed by India. Many of these cultivars have unique morphological markers for identification. The cultivar variability is much less in other ginger-growing countries. The important cultivars grown in different countries are given in Table 15.5.

Table 15.5   Cultivar Diversity of Ginger

Country	Important Cultivars
China	Gandzhou, Shandong Laiwu (sparse seedling type) Guangzhow, Zhejiang (dense seedling type) Fujian Red Bud, Hunan Yellow Heart, Chicken Claw Ginger, Xingguo Ginger (edible medicinal type) Guangzhou, Fuzhou, Tongling, Fujian Bamboo Ginger, Zunyi, Leifeng (edible processed type) Laishe Ginger, Flower Ginger, Tea Ginger, Strong Ginger, Hengchum Ginger, Hekou Ginger (ornamental ginger) Zaoyang, Zunji, Chenggu Yellow, Yulin, Bamboo Root Ginger, Mianyang, Xuanchang, Yuxi Yellow, Laiwu Slice Ginger, Yellow Claw, Taiwan Fleshy Zaoyang, Zunji Big White Ginger, Chenggu Yellow Ginger, Yulin Round Fleshy Ginger, Bamboo Root Ginger, Yang Ginger, Xuanchang Ginger, Yuxi Yellow Ginger, Taiwan Fleshy Ginger
India	Indigenous: Adoor, Ambala, Anamica, Assam, Bajpai, Bhaisae, Bhitarkatta, Burdwan-1, Cochin, Ernad Chernad, Edapalayam, Gurubathan, Himachal Pradesh, Jorhat, Jugiigan, Karakkal, Konni, Kuruppampadi, Kunduli Local, Lakhadong, Mananthodi, Maran, Moovattupuzha, Nadia, Narasapattam, Poona, Rajgarh Local, Sagar, Sargiguda, Sathara, Saw Thing laidum, Saw Thing pui, Shillong, Silcher, Singhihara, Thang Chang, Thinladium, Thingpui, Thodupuzha, Tura, Uttar Pradesh, Valluvanad, Vengara, Wynad Local, Wynad Kunnamangalam, Zahirabad Exotic: China, Jamaica, Rio de Janeiro, Sierra Leone, Taiwan, Taffingiva
Southeast Asia	Laiwu Slice Ginger, Laiwu Big Ginger, Sparse Ringed Big Fleshy Ginger, Dense Ringed Delicate Fleshy Ginger, Red-Claw Ginger, Yellow-Claw Ginger, Tongling White Ginger, Xingguo Ginger, Fuzhou Ginger, Laifeng Ginger
Malaysia	Halyia Betel, Halyia Bara, Halyia Udang, *Z. officinale* var. *Rubra* (pink ginger)
Japan	Kintoki (small-sized plants), Sanshu (medium-sized plants), Oshoga (large-sized plants), Sanshu (4×). *Z. mioga* (Japanese ginger)
Australia	Buderim local, Buderim Gold (4×), *Z. mioga* (Introduced)
Philippines	Native, Hawaiian
Nigeria	Taffingiva (High Yielder Bold, Yellow Ginger), Yasun Bari (Black Ginger)
Sierra Leone	Sierra Leone (slender rhizome, pungent)
Jamaica	St. Mary, Red Eye, Blue Turmeric, Bull Blue, China Blue Jamaica
China	China (extra bold rhizome)
Brazil	Rio de Janeiro (bold rhizome)

Table 15.6   Conservation of Ginger Germplasm in India

Center	No. of Accessions	
	Indigenous	Exotic
Indian Institute of Spices Research, Calicut, Kerala	708 Seven wild species	36
National Bureau of Plant Genetic Resources, Regional Station, Trichur, Kerala	173	—
Orissa University of Agriculture and Technology, Pottangi, Orissa	174	3
Y. S. Parmar University of Horticulture and Forestry, Solan, Himachal Pradesh	286	—
Rajendra Agricultural University, Dholi, Bihar	47	—
Uttar Banga Krishi Vishwa Vidyalaya, Pundibari, West Bengal	51	—
N. D. University of Agriculture and Technology, Kumarganj, Uttar Pradesh	61	—
Indira Gandhi Agricultural University, Regional Station, Raigarh, Chattisgarh	44	—
Central Agricultural Research Institute, Port Blair, Andamans	13	—
Department of Horticulture, Sikkim	58	—

Sources:
AICRPS. 2009. Annual report: 2008–09. All India Coordinated Research Project on Spices, Indian Institute of Spices Research, Calicut, Kerala, India.
IISR. 2009. Annual report 2008–09. Indian Institute of Spices Research, Calicut, Kerala, India.

In India excellent diversity of cultivated ginger is available. Most major growing tracts have cultivars that are specific to the area, and these cultivars are mostly known by place names. A good germplasm collection of ginger representing maximum diversity was maintained at the Indian Institute of Spices Research (IISR), which is the National Conservatory of Ginger germplasm, National Bureau of Plant Genetic Resources, and various centers of All India Coordinated Research Centers (Table 15.6).

In addition to local cultivars, a few high-yielding exotic cultivars were also conserved, such as Rio de Janeiro, with a bold rhizome with fair skin, pungent, flavored, and less fibrous; China, with a bold rhizome, yellowish-white skin, highly pungent, flavored, and fibrous; Jamaica, with a bold rhizome, white skin, moderately pungent, highly flavored, and fibrous; and Sierra Leone, with a slender rhizome, buff skin, pungent, and flavored. Other exotic collections from Nepal, Nigeria, the United States, New Zealand, and Fiji are also available in the germplasm conservatory of IISR, Calicut (Ravindran et al. 2005; Ravindran, Nirmal Babu and Shiva 2005).

The germplasm is usually maintained as clonal repositories in field gene banks, being replanted each year. This may lead to mixing up and loss of genetic purity. Hence, at IISR, a set of nucleus germplasm is maintained in culture tubs, with replanting only once in 5 years to maintain purity (Figure 15.6).

To augment the conservation efforts and also to minimize the loss, an *in vitro* gene bank of important genotypes is also operational at IISR and National Bureau of Plant Genetic Resources (NBPGR), New Delhi (Nirmal Babu et al. 2005). Slow-growth methods for short-term conservation of ginger and its related species are standardized. Technology for cryopreservation of ginger germplasm using encapsulation and vitrification was also developed (Yamuna et al. 2008).

15.9.2   Genetic Resources Management—GIS Technology and Biodiversity

Genetic resources management is a complex process. It includes a number of mutually dependent stages such as from the identification of a target gene pool for conservation to the use of genetic resources for economically useful traits. Analysis of geo-reference data of the flora and germplasm

Figure 15.6   Maintenance of nucleus germplasm of ginger to maintain genetic purity.

collections with geographical information system (GIS) technology can help to study the geographic distribution efficiently and effectively.

Ginger can be grown in diverse tropical conditions, from sea level to 1500 m elevation, at a temperature range of 20–30°C with an annual rainfall of 2500–3000 mm. Below 2000 mm, irrigation is necessary. Well-drained sandy loam or clayey soils rich in humus or laterite loam soil with good drainage is ideal; adequate supply of organic matter is essential. The geographic differentiation of plant populations reflects the dynamics of gene flow and natural selection. Sampling of geographically distinct populations is a practical approach to understand genetic diversity, and sampling in diverse ecogeographic areas will help in conservation of rare and wild species.

GIS analysis of the germplasm data helps to better understand and develop new strategies for exploiting geographic diversity and to predict where species naturally occur or may be successfully introduced. Habitat loss and fragmentation are among the most common threats facing endangered species, making GIS-based evaluations an essential component of population viability analysis. Utpala et al. (2007) used GIS to study the suitable areas for ginger cultivation and the effects of climate change on ginger. They prepared a land suitability map for ginger with the help of the Eco–Crop model of DIVA GIS. The study indicated that the northeastern and southwestern states of Orissa, West Bengal, Mizoram, and Kerala are highly suitable, while northwestern states like Gujarat, Rajasthan, Uttar Pradesh, and Madha Pradesh are marginally suitable for ginger cultivation. Future prediction of the Eco–Crop model shows rise of temperature by 1.5–2°C will drastically reduce the suitability of Orissa and West Bengal for ginger cultivation.

15.9.3 Genetic Resource Characterization and Cultivar Diversity

Ginger has been under cultivation since time immemorial. There is no natural seed set in ginger, which resulted in limited variability with regard to certain characters. However, reasonable variability exists for yield and quality attributes mainly in northeastern India and Kerala. Geographical spread accompanied by genetic differentiation into locally adapted populations, caused by mutation, could be the main factor responsible for the diversity in this clonally propagated crop. Several commercial cultivars, many landraces, and improved cultivars are available in India. They possess various quality attributes and yield potential and are prevalent in India, including a few of exotic origin. At present over 50 ginger cultivars are being cultivated in India. Most varieties are named after their place of origin, domestication, or collection (Mohanty and Sarma 1979; Ratnambal, Balakrishnan, and Nair 1982; Mohanty, Naik, and Panda 1990; Ravindran et al. 1994, 1999, 2005; Sasikumar et al. 1992, 1999; Suryanarayana et al. 2001; Ravindran, Nirmal Babu, and Shiva 2005). Some of the important cultivars of ginger are given in Table 15.6.

Ginger is location specific and the agroclimatic conditions have a great role to play in creating variability in the qualitative as well as the quantitative attributes. The same variety grown under different conditions produced marked differences in yield and quality attributes. The characterization and documentation of ginger germplasm are being carried out at various locations. Considerable variability exists for most of the traits. Highest variation within cultivated ginger occurs in northeastern India, which might be due to the geographical spread from its center of origin in Southeast Asia accompanied by genetic differentiation into locally adapted populations caused by mutations. Ravindran et al. (1994) characterized 100 accessions of ginger germplasm based on morphological, yield, and quality parameters. Moderate variability was observed for many yield and quality traits. Tiller number per plant had the highest variability, followed by rhizome yield per plant. Among the quality traits, the shogaol content recorded the highest variability, followed by crude fiber and oleoresin. None of the accessions possessed resistance to the causal organism of leaf spot disease, *Phyllosticta zingiberi*. Quality parameters such as dry recovery and oleoresin and fiber contents are known to vary with soil type, cultural conditions, and climate.

The high-yielding capacity of exotic variety Rio de Janeiro was proved in many locations. This variety has many quality attributes also (Khan 1959; Kannan and Nair 1965; Thomas 1966; Muralidharan and Kamalam 1973). The yields of Himachal Pradesh, Kuruppampadi, China, and Maran cultivars are comparable to Rio de Janeiro (Jogi et al. 1972; Nybe, Nair, and Kumaran 1982; Kumar et al. 1980; Mohanty, Das, and Sarma 1981; Thangaraj et al. 1983). Varieties like Nadia and Burdwan are high yielders in certain locations (AICSCIP 1978; Aiyadurai 1966; Khan 1959; Nybe et al. 1982; Arya and Rana 1990; Saikia and Shadeque 1992; Chandra and Govind 1999; Gowda and Melanta 2000). Varieties like Rio de Janeiro, Himachal Pradesh, Kuruppampadi, Maran Nadia, and Burdwan are still very popular with farmers. The Thingpuri cultivar produced the highest yield of 220 g/ha in Orissa (Panigrahi and Patro 1985). Under Nagaland conditions, Thinladium, Nadia, and Khasi Local varieties were the best (Singh et al. 1999).

Studies on genetic variability for yield indicated the existence of moderate variability and heritability in the germplasm of ginger. Little variability exists among the genotypes being grown in the same area; however, a good amount of genetic variability has been reported among the cultivars being grown in different states. Rhizome yield was positively and significantly correlated with number of stems, leaves, secondary rhizome fingers, tertiary rhizome fingers and total rhizomes; plant height; leaf breadth; girth of secondary rhizome fingers; and number and weight of adventitious roots. Straight selection was useful to improve almost all characters (Mohanty and Sarma 1979; Rattan, Korla, and Dohroo 1988; Sreekumar et al. 1982; Mohanty et al. 1981; Sasikumar et al. 1992; Pandey and Dobhal 1993; Chandra and Govind 1999; Singh 2001).

Yadav (1999) found that the genotypic coefficient of variation was high for length and weight of secondary rhizome, weight of primary rhizome, number of secondary and primary rhizomes, and rhizome yield per plant. The variability in the rhizome characters of a few ginger cultivars is shown in Figure 15.7.

Path analysis revealed that the strongest forces influencing yield are weight of fingers, width of fingers, and leaf width (Pandey and Dobhal 1993; Das et al. 1999). D^2 analysis indicated that the major forces for divergence were rhizome yield per plant and oleoresin and fiber content

(a)

(b)

(c)

Figure 15.7 Variability of rhizome characters of ginger cultivars and varieties: (a) Gurbathani, (b) Satara, and (c) Sagar.

Table 15.7 Promising Ginger Collections for Yield Attributes

Character/Trait	Variety/Cultivar/Accession	References
High yield	UP, Rio de Janeiro, Thingpuri, Karakkal, Suprabha, Anamika, Jugijan	Mohanty and Panda (1994)
	SG-646, SG-666	Rattan (1994)
	Rio de Janeiro, Suprabha, Suruchi, Suravi, Jugijan, Thingpuri, Wynad Local, Himachal, Karakkal, Varada, Maran, Acc. nos. 64, 117, 35	Sasikumar et al. (1994); Sarma et al. (2001)
	Rio de Janeiro, Maran, Nadia, Narasapattam	Paulose (1973)
	Rio de Janeiro, China, Ernad Chernad	Thomas (1966)
	Wynad, SG-700, SG-705, BDJR-1226, V2E4-5, PGS-43, SG-876, SG-882	AICRPS (1999, 2000, 2001)
Bold rhizome	China, Taffingiva, SG-35	Sasikumar et al. (1994)
	Varada, Gurubathan, Bhaise, China, Acc. nos. 117, 35, 15, 27, 142	Sasikumar et al. (1999); Sarma et al. (2001)
Slender rhizome	Suruchi, Kunduli Local	Mohanty and Panda (1994)
Short duration	Sierra Leone	Mohanty and Panda (1994)

Source: Ravindran, P. N. et al. 2006. In *Advances in spices research*, ed. P.N. Ravindran, K. Nirmal Babu, K. N. Shiva, and A. K. Johny, 365–432. Jodhpur: Agrobios.

(Singh et al. 2001). Multiple regression analysis using morphological characters indicated that the final yield could be predicted fairly accurately by taking into consideration plant height, number of leaves, and breadth of last fully opened leaf at day 90 and day 120 after planting (Ratnambal et al. 1982; Rattan 1994; Rai et al. 1999).

In addition, biochemical and molecular characterization are used to augment morphological characterization, yield, and quality attributes, and for documentation. For quality attributes, Rio de Janeiro and Maran had the highest oleoresins and Karakkal had the highest essential oil; crude fiber was least in China and Nadia and high in Kuruppampadi, Maran, Jugijan, Ernad Manjeri, Nadia, Poona, Himachal Pradesh, Tura, and Arippa (Jogi et al. 1972; Nybe et al. 1982; Kumar et al. 1980). The quality of ginger depends on the relative content of gingerol and shogaol. Zachariah, Sasikumar, and Ravindran (1993) classified 86 ginger accessions into high-, medium-, and low-quality types based on the relative contents of the quality components.

Based on yield trials at different locations, promising germplasm accessions were identified for further evaluation and selection of improved varieties with high yield, dry recovery, oleoresin, essential oils, crude fiber, and resistance to insect pests and diseases. Some of the promising cultivars and accessions identified for specific traits in India are listed in Tables 15.7, 15.8, and 15.9.

15.10 CROP IMPROVEMENT

The crop improvement in ginger is aimed at developing high-yielding varieties with wide adoption, high-quality parameters (oil, oleoresins), and low fiber, as well as resistance to major pests and diseases such as shoot borer and rhizome rot. In India, cultivars have been developed so far through introduction, selection, and mutation. Polyploidy breeding was also attempted in ginger. Hybridization in ginger is not feasible due to sterility.

15.10.1 Selection

Crop improvement work carried out so far has been confined to collection of cultivars from different localities and their comparative yield evaluation and selection. The cultivars differ

Table 15.8   Promising Selections of Ginger for Quality Attributes

Character/Trait	Variety/Cultivar/Accession	References
High dry recovery (over 20%)	Tura local, Thodupuzha, Nadia, Kuruppampadi, Maran, Thinladium, Jorhat, Vengara, Ernad (Chernad), Himachal Pradesh, Sierra Leone, Suruchi, Assam, China, Mowshom, Thingpui, Varada, Acc. nos. 27, 117, 204, 294, SG-685, Narasapattam, Rio de Janeiro	Mohanty (1984); Sreekumar et al. (1980); Nybe et al. (1982); Nair (1969); Thomas (1966); Muralidharan (1972); Sasikumar et al. (1994, 1999); Paulose (1973)
High oleoresin (%)	Assam, Mananthody, Wynad Local Kuruppampadi, Rio de Janeiro, Maran, Wynad Kunnamangalam, Ambalavayalan, Santhing Pui, Himachal, Varada, China, Santhing Pin (Manipur–I), Ernad Chernad, Erattupetta, Tamarassery Local, PGS-33 and PGS-1, Nadan Pulpally, Nadan, Acc. nos. 3, 6, 14, 57, 110, 118, 582, 236, 388, 414	Krishnamurthy et al. (1970); Natarajan et al. (1972); Muralidharan (1972); Nybe et al. (1982); Sreekumar et al. (1980); Sasikumar et al. (1994, 1999); Sarma and Sasikumar (2002); Zachariah et al. (1993); Zachariah, Sasikumar, and Nirmal Babu (1999)
High essential oil (%)	V1S1-8, BDJR-1226, Chanog-II, Mananthody, Karakkal, Rio de Janeiro, Vengara, Valluvanad, Elakallan, Sabarimala, Acc. nos. 118, 14, 64, 418, 399, 389, 205, 110, 236, 104, 296, Nadan Pulpally, Thodupuzha, BLP-6, SG-723, BDJR-1054, SG-55, Maran, Shilli, Bangi, Himgiri, Acc. nos. V1E4-4, PGS-23, SG-706	Krishnamurthy et al. (1970); Lewis et al. (1972); Nybe et al. (1982); Sarma et al. (2001); Sarma and Sasikumar (2002); Zachariah et al. (1999)
Low crude fiber (%)	China, UP, Himachal Pradesh, Nadia, Tura, Ernad Chernad, Zahirabad, Kuruppampadi, Mizo, PGS-16, Jamaica, Acc. nos.15, 27, 287, 288, 22, 18, 419, 386, 415, 200, 110, 336, Varada	Thomas (1966); Sreekumar et al. (1980); Nybe et al. (1982); Sasikumar et al. (1994, 1999); Sarma and Sasikumar (2002); Zachariah et al. (1999)
High gingerol and shogaol	Wynad Kunnamangalam, Ambalavayalan, Ernad Chernad, Sawthing Pui, Rio de Janeiro, Mizo, Nadia, Maran, Ernad, Chernad, Kada, Narianpara, Santhing Pin (Manipur-I), PGS-37, S-641, Maran, Erattupetta, Nadan Pulpally, Jorhat Local, PGS-16, Baharica, and Kunduli	Sasikumar et al. (1994, 1999); Sarma et al. (2001); Zachariah et al. (1993)
High zingiberene and (6) gingerol	Baharica, Amaravathy	Sasikumar et al. (1999)
Salted ginger	Rejatha, Varada	Sasikumar et al. (1999)

Source: Ravindran, P. N. et al. 2006. In *Advances in spices research*, ed. P.N. Ravindran, K. Nirmal Babu, K. N. Shiva, and A. K. Johny, 365–432. Jodhpur: Agrobios.

considerably in their rhizome characters and production potential. In ginger, yield and quality are influenced by various factors.

Most of the improved varieties of ginger are the result of direct selections from germplasm. Nine varieties of ginger have been released or are in the process of release so far. Rhizome characters of the released varieties from India are shown in Figure 15.8. The released varieties of ginger and their important characters are given in Table 15.10.

Varieties Suprabha, Suruchi, and Suravi have bold and plump rhizomes and are suitable for both rain-fed and irrigated conditions. Varada is the most promising, with bold rhizomes, wide

Table 15.9   Sources of Resistance in Ginger Germplasm for Pests and Diseases

Character	Cultivar/Accession	Reaction	References
Pests			
Shoot borer (*Conogethes punctiferalis*)	Rio de Janeiro	Tolerant	Nybe and Nair (1979) Nybe et al. (1982)
Rhizome scale (*Aspidiella hartii*)	Wild-2	Least infestation	Mohanty (1984)
	Anamika	Least infestation	Sasikumar et al. (1994)
Shoot fly (*Formosina flavipes*)	Suruchi	Low incidence (13.08%)	Chandramani and Chezhiyan (2002)
Storage pests	Varada, acc. nos. 215, 212	Resistant	Sarma et al. (2001)
Root knot nematode (*Meloidogyne incognita*)	Valluvanad, Tura, Himachal	Least infestation	Charles and Kuriyan (1980)
	Acc. nos. 36, 59, 221	Resistant	Sarma et al. (2001)
	IISR Mahima	Resistant	Sasikumar et al. (2003)
	Uttar Pradesh	Tolerant	Nehra and Trivedi (2005)
Diseases			
Rhizome rot (*Pythium aphanidermatum*)	Jorhat, Sierra Leone	Least incidence	AICSCIP (1975)
	Maran	Least infection	Nybe and Nair (1979)
	Narasapattam	Least susceptible	Mohanty (1984)
	Burdwan-1, Anamika,	Less susceptible	Sasikumar et al. (1994)
	Poona, Himachal Supraba, Himachal	Moderately resistant	Setty et al. (1995a)
	BDJR-1226, Jamaica, BLP-6	Less susceptible	AICRPS (1999)
Bacterial wilt (*Ralstonia solanacearum*)	V2E5-2, Rio de Janeiro	Least incidence	Pradeepkumar et al. (2000)
Leaf spot (*Phyllosticta zingiberi*)	Taiwan, Taffingiva,	Least susceptible	Nybe and Nair (1979)
	Maran, Bajpai, Nadia Maran , Kunduli Local	Less susceptible	Sasikumar et al. (1994)
	SG-554, V1S18, RGS-5	Field resistant	Singh et al. (2000)
	Narasapatam, Tura, Nadan, Tetraploid, Thingpoi	Moderately resistant	Setty et al. (1995b)

Source: Ravindran, P. N. et al. 2006. In *Advances in spices research*, ed. P.N. Ravindran, K. Nirmal Babu, K. N. Shiva, and A. K. Johny, 365–432. Jodhpur: Agrobios.

adaptability, and tolerance to rhizome rot. Mahima has low fiber content and is resistant to root knot nematode. Rejatha has round and bold rhizomes with low fiber content and high oil. Himagiri is best for green ginger and less susceptible to rhizome rot.

15.10.2 Mutation Breeding

Induction of variability through mutations, chemical mutagens, and ionizing radiation has been attempted by various workers (Gonzalez et al. 1969; Raju, Patel, and Shah 1980; Giridharan 1984; Jayachandran 1989; Mohanty and Panda 1991; Nwachukwu, Ene, and Mbanaso 1995). Use of chemical mutagen ethyl methane sulfonate (EMS) resulted in reduced growth and increased cytological irregularities (Rattan 1987). Use of Gamma rays also had similar effects (Rattan 1994). Almost all the induced changes appearing in the R_1 generation were in chimeric form, expressed a stunted or semidwarfing effect, and were inhibitory on production of rhizomes (Giridharan and Balakrishnan 1991, 1992; Jayachandran and Mohanachandran 1992). Mohanty and Panda (1991)

(a)

(b)

(c)

Figure 15.8 Rhizome characters of improved varieties of ginger: (a) Varada, (b) Mahima, and (c) Rejatha.

Table 15.10 Important Characters of Improved Varieties of Ginger from India

Cultivar Name	Pedigree	Mean Yield t/ha	Dry Recovery (%)	Oil Content (%)	Oleoresin (%)	Crude Fiber (%)	Mean Days to Maturity	Salient Features
Suprabha	Selection from Kunduli Local	16.6	20.5	1.9	8.9	4.4	230	Plump rhizome, less fiber, wider adaptability, suitable for both early and late sowing season
Suruchi	Selection from germplasm	11.6	23.5	2.0	10.0	3.8	218	Bold rhizome, suitable for both rain-fed and irrigated conditions
Suravi	Induced mutant of Rudrapur Local	17.5	23.0	2.1	10.2	4.0	225	Plump rhizome, suitable for both rain-fed and irrigated conditions
Varada	Selection from germplasm	22.6	19.5	1.7	6.7	3.3	200	Bold rhizome, tolerant to rhizome rot; wide adaptability
Mahima	Selection from germplasm	23.2	23.0	1.7	4.5	3.3	200	Plump, bold rhizome with low fiber content; resistant to root knot nematode
Rejatha	Selection from germplasm	22.4	19.0	2.4	6.2	4.0	200	Plump, round and bold rhizome with low fiber content and high oil type
Himagiri	Clonal selection from Himachal	14.0	20.6	1.6	4.3	6.0	230	Best for green ginger, less susceptible to rhizome rot, suitable for rain-fed condition
V3S1-8	Sodium azide mutant	29.0	22.2	1.8	10.8	3.2	—	Moderately tolerant to disease and pests, suitable for fresh and dry ginger, wide ecological adaptability
V1E8-2	Ethyl methyl sulfonaste mutant	32.9	21.4	1.8	10.8	3.5	—	A high-yielding mutant moderately tolerant to disease and pests; suitable for late planting under rain-fed conditions

Source: P. N. Ravindran. and Johny, A. K. 2006. *Spice India* 18:40–44.

developed and released a high-yielding yellow-fleshed mutant, Suravi, that is suitable for both rain-fed and irrigated conditions.

Two more mutant selections, V_3S_1-8 and V_1E_8-2, using mutagens sodium azide and ethyl methyl sulfonaste, respectively, are in the process of being released. Both have wide adaptability and are moderately tolerant to disease and pests.

15.10.3 Polyploidy Breeding

Ramachandran (1982) and Ramachandran and Nair (1992) reported successful induction of stable tetraploids having $2n = 44$ in ginger (cultivars Maran and Mananthody) by treating the sprouts with 0.25% aqueous colchicines. The polyploids were more vigorous than the diploids and flowered during the second year of induction. These autotetraploids had larger rhizomes and high yield (198.71 g/plant). However, oil content of these rhizomes was lower (2.3%) than that of the original diploid cultivar (2.8%).

The commercial ginger company in Queensland, Australia—Buderim Ginger Co.—has developed and released for cultivation a tetraploid line from the local cultivar. This line, named Buderim Gold, is much higher yielding and has plump rhizomes that are ideally suitable for processing (Buderim Ginger Co. 2003).

Nirmal Babu (1996) developed a promising tetraploid line of the cultivar Maran from somaclonal variants. This line is high yielding, with bolder rhizomes, larger and greener leaves, and taller plants.

15.11 BIOTECHNOLOGICAL APPROACHES

There is no seed set in ginger; this leads to limited variability and hampers conventional crop improvement programs. Rhizome rot caused by *Pythium aphanidermatum* and bacterial wilt caused by *Ralstonia solanacearum* are the major diseases affecting ginger. Thus, biotechnological tools can play an important role in ginger improvement.

15.11.1 Micropropagation

Clonal multiplication of ginger (Figure 15.9) from various explants has been reported by many workers (Table 15.11) (Hosoki and Sagawa 1977; Nadgauda et al. 1980; Nirmal Babu, Ravindran, and Peter 1997; Sharma and Singh 1997; Rout et al. 2001). Bhagyalakshmi and Singh (1988) and Rout et al. (2001) reported micropropagation of ginger through meristem culture. Cha-um et al. (2005) studied *ex vitro* survival and growth of ginger plantlets. The plantlets acclimatized under high RH with CO_2 enrichment conditions showed the highest adaptive abilities, resulting in 90–100% survival after planting out. Diseases of ginger are often spread through infected seed rhizomes. Micropropagation will help in multiplying disease- and virus-free planting materials of high-yielding varieties in ginger.

15.11.1.1 Micropropagation of Other Zingiberaceous Taxa

Protocols for micropropagation of a few economically and medicinally important *Zingiber* species are available. Chirangini and Sharma (2005) reported *in vitro* propagation of *Zingiber cassumunar* on MS (Murashige and Skoog) media fortified with 0.54–2.69 μM NAA (naphthaleneacetic acid) and 4.44 μM BAP (6-benzylaminopurine) with eight microshoots per explant. Prathanturarug et al. (2004)

Figure 15.9   Micropropagation of ginger.

reported *in vitro* propagation of *Zingiber petiolatum*—a rare plant from Thailand—using seedling explants on MS medium containing (17.8 µ*M* BAP) alone or in combination with (0.5 µM NAA). Rooting was spontaneously achieved in MS medium without plant growth regulators.

   Faria and Illg (1995) reported micropropagation of *Zingiber spectabile* on MS medium supplements with 10 µM BA and µM IAA (indole-3-acetic acid). Shoot formation occurred when buds were transferred to half-strength MS containing only 10 µM BA, at a rate of 15–20 new shoots every 30 days. Rooting was obtained when the shoots were placed in water or half-strength MS with 5 µM NAA or IAA.

15.11.2 Field Evaluation of Tissue-Cultured Plants

   Field evaluation of tissue-cultured (TC) plants indicated that micropropagated plants require at least two crop seasons to develop rhizomes of normal size that can be used as seed rhizomes for commercial cultivation (Smith and Hamill 1996; Nirmal Babu et al. 1998, 2005; Lincy, Jayarajan, and Sasikumar 2008). Molecular characterization of micropropagated plants indicated genetic uniformity (Rout et al. 1998) as well as certain polymorphisms among the micropropagated plants (Nirmal Babu, Ravindran, and Sasikumar 2003).

   Ma and Gang (2006) reported that metabolic profiling indicated that the biochemical mechanism used to produce the large array of compounds (gingerols and gingerol-related compounds, other diarylheptanoids, and methyl either derivatives of these compounds, as well as major mono- and sesquiterpenoids) found in ginger are not affected by *in vitro* propagation.

   Molecular characterization of micropropagated plants by Rout et al. (1998) indicated that RAPD (random amplified polymorphic DNA) profiles did not indicate any polymorphism among the micropropagated plants. However, Nirmal Babu et al. (2003) have reported RAPD profile differences among the micropropagated ginger.

Table 15.11 *In Vitro* Responses of Ginger

Crop and Explant Used	*In Vitro* Response	Media Composition	References
Vegetative buds and rhizome bits with axillary buds	MS + 1 mg/L NAA (liquid medium)	Multiple shoots and *in vitro* rooting	Nirmal Babu et al. (1996)
Vegetative buds	MS major elements + Ringe-Nitsch minor elements, vitamins, 1 ppm BA, 2% sucrose	Multiple shoots with roots	Hososki and Sagawa (1977)
Vegetative buds, ovary, rhizome, leaf sheath	MS + 0.5 ppm NAA	Callus	Choi (1991)
	MS + 2 mg/L 2,4-D	Callus	Nirmal Babu, et al. (1997)
Rhizome	MS + 0.5 mg/L 2,4-D + 0.5 mg/L BA	Plantlets	Ilahi and Jabeen (1987, 1992)
Flower, inflorescence	MS + 10 mg/L BA + 0.2 mg/L 2,4-D	Conversion of flowers to plants	Nirmal Babu Samsudeen, and Ravindran (1992)
Anther	MS + 1.5 mg/L 2,4-D + 200 mL/L coconut milk	Callus, roots	Ramachandran and Nair (1992)
	MS + 10 mg/L BA, 0.2 mg/L 2,4-D	Callus induction and plant regeneration	Nirmal Babu (1997)
Callus derived from bud, ovary, leaf	MS + 10 mg/L BA, 0.2 mg/L 2,4-D	Organogenesis and plantlet formation	Nirmal Babu, Samsudeen, and Ratnambal (1992); Nirmal Babu, Samsudeen, and Ravindran (1996), Nirmal Babu (1997).
Callus derived from vegetative buds	MS + 0.1 mg/L KT, 0.2 mg/L BA, 10% coconut milk	Shoot regeneration	Nadgauda et al. (1980)
In vitro plantlets	—	*In vitro* rhizomes	Bhat et al. (1994); Peter et al. (2002)
Other *Zingiber* species: *Zingiber cassumunar* (Roxb.); *Zingiber petiolatum*			
Z. cassumunar vegetative buds	MS + 0.54 – 2.69 μM NAA, 4.44 M BA	Multiple shoots	Chirangini and Sharma (2005)
Z. petiolatum seedling explants	MS + 17.8 μM BA + 0.5 μM NAA	Multiple shoots	Prathanturarug et al. (2004)

Note: MS = Murashige and Skoog medium. (Murashige, T., and F. Skoog. 1962. *Physiologia Plantarum* 15:473–497.)

15.11.3 Plant Regeneration from Callus Cultures and Somaclonal Variation

Regeneration of plantlets through callus phase via organogenesis as well as embryogenesis has been reported (Figure 15.10) from leaf, vegetative bud, ovary, and anther explants. The calli derived from various explants (namely, vegetative bud, young leaf, and ovary tissues) were cultured on MS medium 10 mg/L BAP and 0.2 mg/L 2,4-D (2,4-dichloroacetic acid) for morphogenesis and plant regeneration. In callus derived from ovary tissues, both organogenesis and embryogenesis were observed in the same culture.

The somatic embryos produced secondary embryoids by cyclic somatic embryogenesis resulting in a large number (100–300) of tiny embryoids. The development of these embryoids into complete plantlets was higher when NAA (1 mg/L) was added to the culture medium; this also enhanced rooting. This protocol with a production potential of a large number of tiny propagules is ideally suited for *in vitro* manipulations such as *in vitro* mutagenesis, *in vitro* polyploidization, and *in vitro* selection against biotic and abiotic stresses (Nirmal Babu, Samsudeen, and Ravindran 1996; Nirmal Babu 1997; Nirmal Babu et al. 2005).

Figure 15.10   Plant regeneration of ginger.

Successful plant regeneration in ginger was reported by various workers (Nadgauda et al. 1980; Ilahi and Jabeen 1987; Kulkarni, Khuspe, and Mascarenhas 1987; Malamug, Inden, and Asahira 1991; Kackar et al. 1993; Rout and Das 1997; Samsudeen 1996; Guo and Zhang 2005; Sumathi 2007; Jamil et al. 2007). Sultana et al. (2009) reported that plantlet regeneration from ginger calli showed that leaf explants gave best results over those of shoot tip and root explants. Direct organogenesis and indirect and direct somatic embryogenesis from aerial stem explants have been reported (Lincy 2007; Lincy, Remashree, and Sasikumar 2009).

These systems could be used for inducing somaclonal variability, which is very important in crops where conventional breeding is hampered by lack of seed set.

15.11.3.1 Somaclonal Variation

Nirmal Babu (1997) and Sumathi (2007) morphologically characterized plantlets obtained through micropropagation, callus regeneration, and microrhizome pathways and observed certain variations. The callus-regenerated plants show maximum variation in almost all the characters studied. The main characters that showed maximum variation were plant height, number of tiller per plant, number of leaves per plant, leaf length, yield, and size and shape of primary and secondary fingers (Figure 15.11). Biochemical characterization of ginger somaclones indicated significant variations in percentage of oil, oleoresin, starch, and fiber content. Variation observed between micropropagated, microrhizome, and conventional with regard to biochemical characters is minimal; the plants regenerated through callus pathway exhibited significant variation.

A few promising lines having important yield attributes, and other useful characters could be selected from callus-regenerated plants. One promising tetraploid with bold rhizomes was also

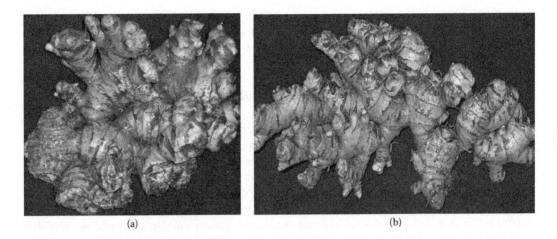

(a)

(b)

Figure 15.11 Variation of rhizome characters among somaclones of ginger: (a) somaclone with bold rhizomes, and (b) somaclone with lean and small rhizomes.

identified among the somaclones (Nirmal Babu, Samsudeen and Ravindran 1996; Nirmal Babu, 1997). RAPD characterization of these somaclones also indicated profile variations indicating genetic differences (Nirmal Babu et al. 2005; Sumathi 2007).

15.11.3.2 In Vitro *Selection*

Kulkarni et al. (1987) reported isolation of *Pythium*-tolerant ginger by using culture filtrate as the selecting agent. *In vitro* selection for types resistant to soft rot and bacterial wilt using culture filtrates of the pathogen or pathotoxin as the selecting agent was attempted by Dake et al. (1997) and Nirmal Babu, Samsudeen, and Ravindran (1996).

15.11.3.3 *Induction of Systemic Resistance*

Tilad, Sharma, and Singh (2002) treated ginger callus cultures with 104 µ*M* salicylic acid (SA) prior to selection with culture filtrate (CF) to induce insensitivity in the callus cultures of ginger against CF of *Fusarium oxysporum* f. sp. zingiberi. This increased the callus survival against the CF of the pathogen. *In vitro* antifungal activity of protein extract of calli treated with SA tested against the spores of *F. oxysporum* f. sp. zingiberi showed significant reduction in spore germination and germ tube elongation. They concluded that, in ginger, SA may result in the induction of resistance to *F. oxysporum* f. sp. zingiberi by inducing increased activity of peroxides, β-1,3-glucanase, and antifungal pathogenesis-related (PR) proteins.

15.11.3.4 In Vitro *Polyploidy*

Adaniya and Shirai (2001) reported *in vitro* induction of tetraploid ginger on MS medium containing 2.0 mg/L BA, 0.05 mg/L NAA, and 0.2% (w/v) colchicine for 8 days. Induced tetraploid strains, 4 × Sanshu and 4 × Philippine cebu 1, had higher pollen fertility and germinability than the diploid counterparts. Smith et al. (2004) reported development of autotetraploid ginger with improved processing quality through *in vitro* colchicine treatment. The autotetraploid lines had significantly wider, greener leaves than the diploids, and they had significantly fewer but thicker shoots. One superior line with an improved aroma/flavor profile and fiber content with consistently good rhizome yield was released as Buderim Gold. More importantly, it produced large rhizome sections, resulting in a higher recovery of premium grade confectionery ginger and a more attractive fresh-market product. Nirmal Babu, Samsudeen, and Ravindren (1996), Nirmal Babu et al. (2005) and Sumathi (2007) also reported selection of a tetraploid line with extra bold rhizomes from the somaclones of ginger.

15.11.4 Inflorescence Culture and *In Vitro* Pollination

Immature inflorescences form ideal explants for micropropagation of many crop species. In ginger, immature floral buds were converted into vegetative buds which later developed into complete plants. Vegetative shoots were produced in 70% of the explants when 1-week-old inflorescences were cultured on MS medium supplemented with 10 mg/L BAP and 0.2 mg/L 2,4-D. These shoots grew into complete plantlets in 7–8 weeks' time (Nirmal Babu, Samsudeen, and Ravindran 1992).

In nature, ginger fails to set fruit. However, by supplying required nutrients to young flowers, *in vitro* pollination could be effected and fruit developed. Subsequently, plants could be recovered from the fruit (Nirmal Babu, Samsudeen, and Ravindran 1992; Valsala, Nair, and Nazeem 1997). *In vitro* pollination was successfully attempted by Nazeem et al. (1996) to overcome prefertilization barriers like the spiny stigma, long style, coiling of pollen tube, etc. that interfered with natural seed set in ginger. Successful seed set was obtained. Development of seeds by *in vitro* pollination in ginger was also reported by Valsala, Nair, and Nazeem (1996, 1997). This will open up new possibilities of sexual reproduction and development of seed-derived progenies of ginger and new possibilities in ginger crop improvement programs.

15.11.5 Microrhizomes

In vitro induction of microrhizomes in ginger was reported by many workers (Sakamura et al. 1986; Sakamura and Suga 1989; Bhat, Chandel, and Kacker 1994; Sharma and Singh 1995; Nirmal Babu 1997; Nirmal Babu et al. 2003, Nirmal Babu, Minoo and Geetha et al., 2005; Nirmal Babu, Samsudeen and Minoo et al. 2005; Peter et al. 2002; Geetha 2002; Ravindran et al. 2004; Tyagi, Agarwal, and Yusuf 2006). Microrhizomes resembled the normal rhizomes in all respects, except for their small size (Figure 15.12).

The microrhizomes consisted of two to four nodes and one to six buds. They also had the aromatic flavor of ginger and resembled the normal rhizome in anatomical features in the presence of well-developed oil cells, fibers, and starch grains. These microrhizomes were directly planted in the field without any hardening with 85% success.

Most workers used 3–9% (w/v) sucrose after 2 months of culture. An increased photoperiod helped in the formation of more rhizomes. Zheng et al. (2008) studied the effect of kinetin (KT), gibberellic acid (GA), and naphthalene acetic acid (NAA) on increasing *in vitro* microrhizome production of ginger. The effect of GA on rhizome induction was larger than that of KT or NAA.

Quality analysis of *in vitro* developed rhizomes indicated that they contain the same constituents as the original rhizome but with quantitative differences. The composition of basal medium seems to affect the composition of oil (Sakamura et al. 1986; Sakamura and Suga 1989; Charlwood, Brown, and Charlwood 1988; Sumathi 2007).

The microrhizome-derived plants have more tillers but the plant height is smaller (Figure 15.3). These microrhizomes gave a fresh rhizome yield of 100–525 g per plant with an average yield per 3 m² bed of 10. 5 kg; the control gave yield per bed of 15 kg. For conventionally propagated plants, 20–30 g of seed rhizome was used; in microrhizomes, seed rhizome weight was 2–8 g. Thus, although microrhizomes gave lesser yield per bed, they gave very high recovery based on the weight of seed material used. *In vitro* formed rhizomes are genetically more stable compared to micropropagated plants. This, coupled with its disease-free nature, will make the microrhizome an

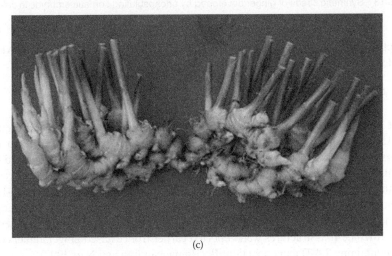

Figure 15.12 Induction of microrhizomes and microrhizomes in comparison with the rhizomes developed from them.

ideal source of planting material suitable for germplasm exchange, transportation, and conservation (Nirmal Babu et al. 2005; Sumathi 2007).

Chirangini and Sharma (2005) reported development of microrhizome induction weighing 0.81 g in *Zingiber cassumunar* for *in vitro* derived shoots cultured on MS media supplemented with 7–9% sucrose in about 8 weeks of incubation.

15.11.6 Synthetic Seeds

Sharma, Singh, and Chauhan (1994) successfully encapsulated ginger shoot buds in 4% sodium alginate gel for production of disease-free encapsulated buds. Encapsulated buds were germinated *in vitro* to form roots and shoots with high percentage of success.

Development of synthetic seeds in ginger by encapsulating the somatic embryos and *in vitro* regenerated shoot buds in 5% calcium alginate was reported by other workers (Figure 15.13). These synthetic seeds are viable up to 9 months at room temperature of 22 ± 2°C and germinated into

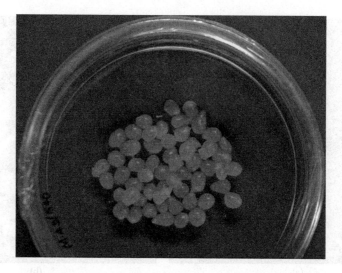

Figure 15.13 Synthetic seeds of ginger developed by encapsulating somatic embryos in calcium alginate.

normal plants with 80% success on MS medium supplemented with 1.0 mg/L BAP and 0.5 mg/L IBA (indole-3-butyric acid) (Sajina et al. 1997; Geetha 2002; Peter et al. 2002).

15.11.7 Anther Culture

Plant regeneration from anther callus was reported from diploid and tetraploid ginger (Figure 15.6) (Nirmal Babu 1997; Samsudeen et al. 2000). Ginger anthers collected at the uninucleate microspore stage were subjected to a cold treatment (0°C) for 7 days and induced to develop profuse callus on MS medium supplemented with 2–3 mg/L 2,4-D. Plantlets could be regenerated from these calli on MS medium supplemented with 5–10 mg/L BAP and 0.2 mg/L 2,4-D. The regenerated plantlets could be established in soil with 85% success. This protocol can be used for possible development of androgenic haploids and dihaploids in ginger. Callus formation and development of roots and rhizome-like structures were reported earlier from excised ginger anthers cultured on MS medium containing 2,4-D and coconut milk (Ramachandran and Nair 1992).

15.11.8 Protoplast Culture

Protoplasts could be isolated successfully from leaf tissues as well as from cell suspension culture ginger. A protoplast yield of 2.5×10^5/g of leaf tissue was obtained by digesting leaf tissue in an enzyme solution containing macerozyme R10 (0.5%), hemicellulase (3%), and cellulase Onozuka R10 (5%), when incubated for 10 hours at 15°C followed by 6 hours at 30°C. The protoplast size was 0.21 mm with a viability of 55%. Protoplast yield from cell suspension cultures was 1×10^5/g of callus when digested with an enzyme solution of 1% macerozyme R10, 3% hemicellulase, and 6% cellulase Onozuka R10 and incubated at 15°C for 10 hours and later at 30°C for 8 hours. Seventy-two percent of the protoplasts were viable with a size of 0.39 mm. These viable protoplasts could be successfully plated on culture media and made to develop up to microcalli stage (Nirmal Babu 1997; Geetha et al. 2000).

Guo, Bai, and Zhang (2007) described a procedure to regenerate plants from embryogenic suspension-derived protoplasts of ginger. Somatic embryogenic calli were induced from ginger shoot tips on solid MS medium with half the concentration of NH_4NO_3 and supplemented with 1.0 mg/L 2,4-D and 0.2 mg/L KT. Rapidly growing and well-dispersed suspension cultures were established by subculturing the embryogenic calli in the same liquid medium. Protoplasts were isolated from embryogenic suspensions with an enzyme solution composed of 4.0 mg/L cellulase, 1.0 mg/L macerozyme, 0.1 mg/L pectolyase, 11% mannitol, 0.5% CaCl2, and 0.1% 2-(N-morpholino) ethane sulfonic acid (MES) for 12–14 hours at 27°C with a yield of 6.27–106 protoplasts per gram of fresh weight. The protoplasts were cultured initially in liquid MS medium with 1.0 mg/L 2,4-D and 0.2 mg/L KT. Then the protoplast-derived calli (1.5 cm^2) were transferred to a basal MS medium containing 0.2 mg/L 2,4-D, 5.0 mg/L BA, 3% sucrose, and 0.7% agar. The white somatic embryos were transferred to MS medium lacking growth regulators for shoot development. Shoots developed into complete plantlets on a solid MS medium supplemented with 2.0 mg/L BA and 0.6 mg/L NAA.

15.11.9 Molecular Characterization and Diagnostics

15.11.9.1 Molecular Characterization

Very little information is available on molecular characterization of ginger. RAPD profiling was used to analyze 96 collections of ginger cultivars from India and interrelationships were studied. The polymorphism detected is moderate to low in ginger (Sasikumar and Zachariah 2003).

Lee et al. (2007) reported isolation and characterization of eight polymorphic microsatellite markers for *Zingiber officinale* Rosc. These were used to detect a total of 34 alleles across the 20 accessions with an average of 4.3 alleles per locus. The data generated indicate the existence of a moderate level of genetic diversity among the ginger accessions genotyped with eight markers.

Zingiber zerumbet is a potential source of resistance for soft rot disease in ginger caused by *Pythium aphanidermatum*. Kavitha and Thomas (2008b) studied the genetic diversity and *P. aphanidermatum* resistance of 74 *Z. zerumbet* accessions belonging to 15 populations. The disease index (DI) of the accessions varied from 0 to 72.24%, and the accessions could be separated into six frequency classes according to their DI values. Eight accessions were found to be immune to the infection. The relative frequency of resistant accessions was higher in the central and northern regions of Kerala.

AFLP (Amplified Fragment Length Polymorphism) analysis of *Z. zerumbet* accessions revealed a high genetic diversity in *Z. zerumbet*. In the UPGMA (Unweighted Pair Group Method with Arithmetic mean) dendrogram, accessions were clustered mostly according to their geographical origin. Though good variability for pathogen resistance among *Z. zerumbet* accessions was observed, no clear pattern was seen between the clustering of accessions and their responses to *Pythium aphanidermatum*. Though the sequence analysis of these bands confirmed the absence of target repeat motif, amplification of large numbers of polymorphic bands provided a basis for genetic diversity analysis.

15.11.9.2 Molecular Characterization of Traded Ginger for Checking Adulteration

A simple and rapid method for isolating good-quality DNA with fairly good yields from mature rhizome tissues of ginger has been reported by Syamkumar, Lowarence, and Sasikumar (2003). Genetic profiling of traded ginger from India and China using 20 RAPD primers and 15 ISSR (Interssimple Sequence Repeat) primers gave a consistent amplification pattern. Significant variation was observed between the products from the two countries (IISR 2008).

Jiang et al. (2006) used metabolic profiling and phylogenetic analysis for authentication of ginger. They used these tools to investigate the diversity of plant material within the ginger species and between ginger and closely related species in the genus *Zingiber*. Phylogenetic analysis demonstrated that all *Zingiber officinale* samples from different geographical origins were genetically indistinguishable. In contrast, other *Zingiber* species were significantly divergent, allowing all species to be distinguished clearly using this analysis. In the metabolic profiling analysis, the *Z. officinale* samples derived from different origins showed no qualitative differences in major volatile compounds, although they did show some significant quantitative differences in nonvolatile composition, particularly regarding the content of [6]-, [8]-, and [10]-gingerols, the most active anti-inflammatory components in this species. The metabolic profiles of other *Zingiber* species were very different, both qualitatively and quantitatively, when compared to *Z. officinale* and to each other.

Comparative DNA sequence and chemotaxonomic and phylogenetic trees showed that the chemical characters of the investigated species were able to generate essentially the same phylogenetic relationships as the DNA sequences. This supports the contention that chemical characters can be used effectively to identify relationships between plant species. Anti-inflammatory *in vitro* assays to evaluate the ability of all extracts from the *Zingiber* species examined to inhibit lipopolysaccharide (LPS)-induced prostaglandin E2 (PGE2) and tumor necrosis factor (TNF)-α production suggested that bioactivity may not be easily predicted by either phylogenetic analysis or gross metabolic profiling. These researchers suggested that identification and quantification of the actual bioactive compounds are required to guarantee the bioactivity of a particular *Zingiber* sample even after performing authentication by molecular and/or chemical markers.

15.11.9.3 Genetic Fidelity Testing of In Vitro Conserved Lines and Identification of Somaclonal Variants of Ginger

RAPD profiling was used to detect genetic variations within the replicates of *in vitro* conserved and cryopreserved lines of ginger. This did not detect any polymorphism between the conserved lines in any of the primers tested, indicating the genetic stability of conserved materials (Geetha 2002; Peter et al. 2002; Ravindran et al. 2004).

Rout et al. (1998) reported that RAPD profiling of micropropagated ginger plants did not detect any polymorphism, indicating genetic uniformity of micropropagated plants. Nirmal Babu et al. (2003) used RAPD profiles as an index for estimating genetic divergence of selected variants among micropropagated and callus-regenerated plants. They observed a certain amount of polymorphism in RAPD profiles in some of the micropropagated and callus-regenerated plants, indicating that these somaclonal variants were also genetic variants. Sumathi (2007) reported a higher number of variations in RAPD profiles in callus-regenerated plants.

15.11.9.4 Molecular Characterization and Detection of Pathogens

Kumar, Sarma, and Anandaraj (2004) characterized the genetic diversity of 33 strains of *Ralstonia solanacearum* causing bacterial wilt of ginger using repetitive sequence-based polymerase chain reaction (REP-PCR), Internal Transcribed spaces (ITS)-PCR, and PCR-RFLP (restriction fragment length polymorphism). Biovar characterization revealed the predominance of biovar 3 in India. Molecular analysis also revealed that ginger strains isolated from different locations during different years had 100% similarity according to Dice's coefficient. The analysis further revealed that the genetic diversity of *Ralstonia* is very low within ginger, confirming that the pathogen population is of clonal lineage and is distributed through rhizome transmission of the inoculum between location and also between seasons. Gosh and Purkauastha (2003) used polyclonal antibodies and antigens of host and pathogen for early diagnosis of rhizome rot disease of ginger caused by *Pythium aphanidermatum*. This pathogen was detected in ginger rhizome after 8 weeks of inoculation by agar gel double diffusion and immunoelectrophoretic tests, but only 1 week after inoculation by indirect ELISA (enzyme-linked immunosorbent assay).

15.11.10 Tagging and Isolation of Candidate Genes of Interest

15.11.10.1 Disease Resistance

Candidate genes responsible for pathogenesis can also be identified from sequence information available in databases. This approach using degenerate primers and functional genomics is more suitable for ginger improvement. In ginger, Aswati Nair and Thomas (2006) reported isolation, characterization, and expression of resistance gene candidates (RGCs) using degenerate primers based on conserved motifs from the NBS domains of plant resistance (R-) genes to isolate analogous sequences or RGCs from cultivated and wild *Zingiber* species.

Aswati Nair and Thomas (2007) reported evaluation of R-gene specific primer sets and characterization of R-gene candidates in ginger. Clones derived from three primers showed strong homology to cloned R-genes or RGCs from other plants and conserved motifs characteristic of the non-TIR subclass of NBS–LRR R-gene superfamily. Phylogenetic analysis separated ginger RGCs into two distinct subclasses corresponding to clades 3 and 4 of non-TIR NBS sequences described in plants. This study provides a base for future RGC mining in ginger and valuable insights into the characteristics and phylogenetic affinities of the non-TIR NBS–LRR R-gene subclass in the ginger genome.

Swetha Priya and Subramanian (2007) reported isolation and molecular analysis of the R-gene in resistant *Zingiber officinale* (ginger) varieties against *Fusarium oxysporum*. They observed that the R-gene is present only in resistant varieties. These cloned R-genes provide a new resource of molecular markers for marker-assisted selection (MAS) and rapid identification of ginger varieties resistant to *Fusarium* yellows.

Kavitha and Thomas (2006, 2008a) employed AFLP markers and mRNA differential display to identify genes whose expression was altered in a *Pythium*-resistant accession of *Zingiber zerumbet* before and after inoculating it with *P. aphanidermatum*. A few differentially expressed transcript-derived fragments (TDFs) were isolated, cloned, and sequenced. Homology searches and functional categorization of some of these clones revealed the presence of defense, stress, and signaling groups that are homologous to genes known to be involved actively in various pathogenesis-related functions in other plant species. They found *Z. zerumbet* shows adequate variability at DNA level and in response to *Pythium* (Kavitha and Thomas 2008a). Isolation of resistance genes from such related genera will help in ginger improvement via transgenic approaches.

15.11.10.2 Agronomically Important Traits

Chen et al. (2005) used primers designed from the conserved regions of monocot mannose-binding lectins and isolated, cloned, and characterized a mannose-binding lectin from cDNA derived from ginger rhizomes. The full-length cDNA (746 bp) of *Z. officinale* agglutinin (ZOA), a mannose-binding lectin, was cloned by rapid amplification of cDNA ends (RACE), and this contained a 510 bp open reading frame (ORF) encoding a lectin precursor of 169 amino acids with a signal peptide. Semiquantitative real time (RT)-PCR analysis revealed that ZOA expressed in leaf, root, and rhizome tissues of *Z. officinale*. ZOA protein was successfully expressed in *Escherichia coli* with the molecular weight as expected.

Yu et al. (2008) reported isolation and functional characterization of a β-eudesmol synthase from *Zingiber zerumbet*. They identified a new sesquiterpene synthase gene (ZSS2) from *Z. zerumbet* Smith. Functional expression of ZSS2 in *Escherichia coli* and *in vitro* enzyme assay showed that the encoded enzyme catalyzed the formation of β-eudesmol and five additional by-products. Quantitative RT-PCR analysis revealed that ZSS2 transcript accumulation in rhizomes has strong seasonal variations. They introduced a gene cluster encoding six enzymes of the mevalonate pathway into *E. coli* and coexpressed it with ZSS2 to further confirm the enzyme activity of ZSS2 and to assess the potential for metabolic engineering of β-eudesmol production. When supplemented with mevalonate, the engineered *E. coli* produced a similar sesquiterpene profile to that produced in the *in vitro* enzyme assay, and the yield of β-eudesmol reached 100 mg/L.

As the key enzyme of the xanthophyll cycle, violaxanthin de-epoxidase (VDE) plays an important role in protecting photosynthesis apparatus from the damage of excessive light. Huang, Cheng, and Zhang (2007) reported molecular cloning and characterization of VDE in ginger. A full-length (2000 bp) cDNA encoding violaxanthin deepoxidase (GVDE) (GenBank accession no. AY876286) was cloned from ginger using RT-PCR and 5′, 3′ rapid amplification of cDNA ends (RACE). Northern blot analysis showed that the GVDE was mainly expressed in leaves. GVDE mRNA level increased as the illumination time was prolonged under high light. For determining the GVDE function, its antisense sequence was inserted into tobacco plants via EHA105. PCR-Southern blot analysis confirmed the integration of antisense GVDE in the tobacco genome. Chlorophyll fluorescence measurements showed that transgenic plants had lower values of nonphotochemical quenching (NPQ) and the maximum efficiency of PSII photochemistry (Fv/Fm) compared with the untransformed controls under high light. The size and ratio of the xanthophyll cycle pigment pool were lower in tobacco (T)-VDE plants than in control, indicating that GVDE was suppressed in antisense (T)-VDE. These results showed that VDE plays a major role in alleviating photo inhibition.

15.11.11 Genetic Transformation

Transient expression of GUS (β-glucuronidase) was successfully induced in ginger embryogenic callus bombarded with plasmid vector pAHC 25 and promoter Ubi-1 (maize ubiquitin) callus tissue. (Nirmal Babu 1997; Nirmal Babu et al. 1998). Ginger embryogenic calli were bombarded with microprojectiles (1.6 μm gold particles) using the BioRad PDS-1000/He gene gun at 900 and 1100 psi helium pressure with the target distance of either 6 or 9 cm. The pAHC 25 vector used contained GUS and BAR (phosphinothricin-acetyl transferase) as reporter and selectable marker genes, respectively (Christensen and Quail 1996). The best GUS score was obtained when the target distance was 9 cm with 900 psi helium pressure. The GUS score of 133 blue spots per square centimeter indicated not only the optimization and efficiency of the biolistic process, but also the ability of the ubiquitin promoter to drive the expression of the reporter gene.

15.11.12 *In Vitro* Conservation

Minimal growth was induced and ginger plantlets could be successfully conserved for an extended period of over 12 months on half-strength MS basal medium supplemented with 10–15 g/L each of sucrose and mannitol. The cultures were sealed with aluminum foil and maintained at a temperature of 22 ± 2°C (Balachandran, Bhat, and Chandel 1990; Dekkers, Rao, and Goh 1991; Nirmal Babu et al. 1999; Geetha 2002; Peter et al. 2002). Use of *in vitro* rhizomes for germplasm conservation revealed that these rhizomes remain healthy and viable (capable of sprouting new shoots) for 22 months at 25°C (Tyagi et al. 2006). A total of 160 genotypes of cultivated and wild species of *Zingiber* could be conserved up to 16–20 months through *in vitro* rhizome formation under 16-hour light conditions. Sundararaj, Agrawal, and Tyagi (2010) reported *in vitro* short-term storage and exchange of ginger germplasm using encapsulated synthetic seeds. Synseeds dehydrated in 0.25 *m* sucrose liquid medium for 16 hours and stored in vials (without medium) at 25°C for 8 weeks and 12 weeks exhibited 53% and 13% conversion, respectively.

15.11.13 Cryopreservation

An efficient technique for cryopreservation *in vitro* grown shoots of ginger was developed based on encapsulation dehydration, encapsulation vitrification (Figure 15.14), and vitrification

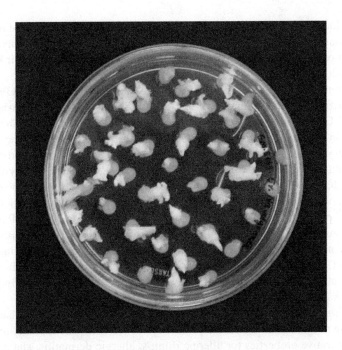

Figure 15.14 Germination of encapsulated and vitrified shoot tips of ginger after cryopreservation.

procedures. Plants could be successfully regenerated from cryopreserved shoots of ginger. The vitrification procedure resulted in higher regrowth (80%) when compared to encapsulation vitrification (66%) and encapsulation dehydration (41%). The genetic stability of cryopreserved ginger shoot buds was confirmed using ISSR and RAPD profiling (Ravindran et al. 2004; Yamuna 2007; Yamuna et al. 2007).

15.11.14 Production of Secondary Metabolites

Cultured plant cells produce a wide range of primary and secondary metabolites of economic value; hence, plant tissue culture can be used as an alternative to whole plants as a biological source of potentially useful metabolites and biologically active compounds. Successful establishment of cell suspension cultures and cell immobilization techniques in ginger for production of essential oils have been standardized in ginger (Ilahi and Jabeen 1992; Nirmal Babu 1997). Production of volatile constituents in ginger cell cultures was reported by Sakamura et al. (1986) and Sakamura and Suga (1989). Ilahi and Jabeen (1986) also reported preliminary studies on alkaloid biosynthesis in callus cultures of ginger, while Charlwood et al. (1988) have reported the accumulation of flavor compounds by cultures of Z. *officinale*.

15.12 PROTECTION OF PLANT VARIETIES AND IPR ISSUES

The International Treaty on Plant Genetic Resources for Food and Agriculture recognizes the importance of plant genetic resources and their conservation. The treaty was adopted by the FAO Conference on November 3, 2001, and entered into force on June 29, 2004. The Convention on Biological Diversity (CBD) (1992) is recognizing the sovereign rights of the states to use their own biological resources. The convention expects the parties to facilitate access to genetic resources by other parties subject to national legislation and on mutually agreed terms. The CBD also recognizes contributions of local and indigenous communities to the conservation and sustainable utilization of biological resources through traditional knowledge, practices, and innovations and provides for equitable sharing of benefits with such people arising from the utilization of their knowledge, practices, and innovations.

The International Union for the Protection of New Varieties of Plants (UPOV) is an intergovernmental organization with headquarters in Geneva (Switzerland). UPOV was established by the International Convention for the Protection of New Varieties of Plants. The convention was adopted in Paris in 1961 and revised in 1972, 1978, and 1991. The objective of the convention is the protection of new varieties of plants by an intellectual property right (IPR). Guidelines for the conduct of tests for distinctness, uniformity, and stability (DUS) were published by UPOV in 1996.

Similarly, India has adopted an effective sui generis system for protection of plant varieties, farmers', rights and those of the plant breeders, and development of new varieties. Registration and protection of plant varieties in India are covered under the Protection of Plant Varieties and Farmers' Rights Act, 2002. Conditions of registration are new (novel) and distinct, uniform, and stable (DUS) for new and extant varieties. The draft guidelines for DUS testing in ginger were prepared by PPV (Protection of Plant Varieties) and FRA (Farmers' Rights Authority), New Delhi, India) in collaboration with IISR (Indian Institute of Spices Research, Calicut, India) and are notified (PPV & FRA 2010).

Ginger is an important crop with many IPR issues. Over 44 patents were registered in India and many more are in progress. Most of these patents are concerned with formulations/herbal involving ginger with curative properties for allergic rhinitis, allergic dermatitis, and asthma; treatment of HIV and AIDS; anticough herbal formulations; treatment of cardiac ailments, gynecological disorders, diarrhea, stress; stomachache; dog bite; forming pharmaceuticals that boost immune suppression; an Ayurvedic preparation for curing urinary disorders, etc. Patents are also registered for synergistic compositions useful as fumigants, herbal feeds for milch animals, refreshing drinks, and bioenhancing composition of bioactive fraction. Many patents are available for processes for developing refreshing sugarcane and fruit juice formulations, increasing shelf life of sugarcane juice and coconut water, food and beverage preparations, improved processes for extraction of oleoresins, and improved snuff composition.

15.13 FUTURE OUTLOOK

Pollen sterility and lack of seed set make conventional breeding ineffective in ginger. Development of polyploidy lines to increase pollen fertility and use of *in vitro* technology for pollination and embryo rescue will open up new possibilities in ginger breeding. Until then, identification and selection of useful natural and induced mutants will play a major role in crop improvement in ginger. This can be supplemented by biotechnological approaches such as somaclonal variation, *in vitro* selection and mutation, and development of disease-resistant transgenics. Tissue culture techniques could also be used for *in vitro* pollination, embryo rescue, and possible seed production in ginger.

Molecular characterization of germplasm will lead to identification of duplicates in germplasm and short-list core collections. Development of diagnostic markers coupled with metabolic profiling will help in maintaining genetic purity and identifying adulterants in this medicinally important species. This also helps in maintaining genetic purity of planting material.

Use of marker technology must rely mostly on a genomics approach for identifying and tagging genes of interest. Application of genomic databases and information derived from conserved regions can be used as putative markers, as well as to validate and refine them as specific markers for specific traits in the target crop. This also helps in allele mining of useful genes, especially in ginger, which has many pharmaceutical attributes.

REFERENCES

Abe, M., Ozawa, Y. Uda, Y. Yamada, Y. Morimitsu, Y. Nakamura, and T. Osawa. 2002. Labdane-type diterpenoid dialdehyde, pungent principle of myoga, *Zingiber mioga* Roscoe. *Bioscience, Biotechnology, and Biochemistry* 66:2698–2700.

Adaniya, S., M. Ashoda, and K. Fujieda. 1989. Effect of day length on flowering and rhizome swelling in ginger (*Zingiber officinale* Rosc.). *Journal of Japan Society of Horticulture Science* 58:649–656.

Adaniya, S., and D. Shirai. 2001. *In vitro* induction of tetraploid ginger (*Zingiber officinalae* Roscoe) and pollen fertility and germinability. *Scientia Horticulturae* 83:277–287.

Agarwal, M., S. Walia, S. Dhingra, and B. P. S. Khambay. 2001. Insect growth inhibition, antifeedant and antifungal activity of compounds isolated/derived from *Zingiber officinale* Roscoe (ginger) rhizomes. *Pest Management Science* 57 (3): 289–300.

AICRPS. 1999. Annual report 1998–1999. All India Coordinated Research Project on Spices. Indian Institute of Spices Research, Calicut, Kerala, India.

———. 2000. Annual report 1999–2000. All India Coordinated Research Project on Spices. Indian Institute of Spices Research, Calicut, Kerala, India.

———. 2001. Annual report 2000–2001. All India Coordinated Research Project on Spices. Indian Institute of Spices Research, Calicut, Kerala, India.

———. 2009. Annual report 2008–2009. All India Coordinated Research Project on Spices. Indian Institute of Spices Research, Calicut, Kerala, India.

AICSCIP. 1975. Annual report 1974–75. All India Cashew and Spices Crops Improvement Project. Central Plantation Crops Research Institute, Kasaragod, Kerala, India.

———. 1978. Annual report 1976–77. All India Cashew and Spices Crops Improvement Project. Central Plantation Crops Research Institute, Kasaragod, Kerala, India.

Aiyadurai, S. G. 1966. A review of research on spices and cashew nut in India. Regional office (spices and cashewnut). Indian Council of Agricultural Research, Ernakulum, p. 228.

Arimura, C. T., F. L. Finger, Casali, and W. D. Vicente. 2000. Effect of NAA and BAP on ginger (*Zingiber officinale* Roscoe) sprouting in solid and liquid medium. *Revista Brasileira Plantas. Medicinais.* 2 (2): 23–26.

Arya, P. S., and K. S. Rana. 1990. Performance of ginger varieties in Himachal Pradesh. *Indian Cocoa, Arecanut Spices Journal* 14 (1): 16–19.

Aswati Nair, R., and G. Thomas. 2006. Isolation, characterization and expression studies of resistance gene candidates (RGCs) from *Zingiber* spp. *Theoretical and Applied Genetics* 116:123–134.

———. 2007. Evaluation of resistance gene (R-gene) specific primer sets and characterization of resistance gene candidates in ginger (*Zingiber officinale* Rosc.). *Current Science* 93 (1): 61–66.

Bailey, L. B. 1961. *Manual of cultivated plants.* Macmillan Co., New York, pp. 287–289.

Baker, J. G. 1894. Scitamineae. In *Flora of British India.* J. D. Hooker, London, pp. 168–264.

Balachandran, S. M., S. R. Bhat, and K. P. S. Chandel. 1990. *In vitro* clonal multiplication of turmeric (*Curcuma longa*) and ginger (*Zingiber officinale* Rosc.). *Plant Cell Reports* 3:521–524.

Baranowski, J. D. 1986. Changes in solids, oleoresin, and (6)-gingerol content of ginger during growth in Hawaii. *Horticultural Science* 21:145–146.

Barghava, N. 1981. Plants in folk life and folklore in Andaman and Nicobar Islands. In *Glimpses of ethnobotany*, ed. S. K. Jain, 329–344. Oxford & IBH, New Delhi.

Beltram, I. C., and Y. K. Kam. 1984. Cytotaxonomic studies in the Zingiberaceae. *Notes from the Royal Botanic Garden, Edinburgh* 41:541–557.

Bhagyalakshmi, B., and N. S. Singh. 1988. Meristem culture and micropropagation of a variety of ginger (*Zingiber officinale* Rosc.) with a high yield of oleoresin. *Journal of Horticultural Science* 63 (2): 321–327.

Bhat, S. R., K. P. S. Chandel, and A. Kacker. 1994. *In vitro* induction of rhizomes in ginger (*Zingiber officinale* Rosc.). *Indian Journal of Experimental Botany* 32:340–344.

Bordoloi, A. K., J. Sperkova, and P. A. Leclercq. 1999. Essential oils of *Zingiber cassumunar* Roxb. from northeast India. *Journal of Essential Oil Research* 11:441–445.

Buderim Ginger Co. (corporate author). 2002. Buderim gold. *Plant Varieties Journal* 15:85.

Burkill, T. H., and M. Haniff. 1930. The Malay village medicines. *Gardens Bulletin Singapore* 6:264–268.

Chakravorti, A. K. 1948. Multiplication of chromosome numbers in relation to speciation in Zingiberaceae. *Science & Culture* 14:137–140.

Chandra, R., and S. Govind. 1999. Genetic variability and performance of ginger genotypes under mid-hills of Meghalaya. *Indian Journal of Horticulture* 56:274–278.

Chandramani, P., and N. Chezhiyan. 2002. Evaluation of ginger varieties for resistance to shoot fly, *Fermosina flavipes* (Diptera: Chloropidal) *Pest Management in Horticultural Ecosystems* 8:131–132.

Charles, J. S., and K. J. Kuriyan. 1982. Relative susceptibility of ginger cultivars to the root knot nematode, *Meloidogyne incognita.* In *Ginger and turmeric,* ed. M. K. Nair, T. Premkumar, P. N. Ravindran, and Y. R. Sarma, 133–134. Proceedings of National Seminar, CPCRI, Kasaragod, India.

Charlwood, K. A., S. Brown, and B. V. Charlwood. 1988. The accumulation of flavor compounds by cultivars of *Zingiber officinale.* In *Manipulating secondary metabolites in culture,* ed. J. R. Richard, Michael and J. C. Rhodes, 195–200. AFRC Institute of Food Research, Norwich, UK.

Cha-um, S., N. M.Tuan, K. Phimmakong, and C. Kirdmanee. 2005. The *ex vitro* survival and growth of ginger (*Zingiber officinale* Rosc.) influence by *in vitro* acclimatization under high relative humidity and CO_2 enrichment conditions. *Asian Journal of Plant Sciences* 4 (2): 109–116.

Chen, Y., X. Zhou, S. Li, and T. Cai. 2001. Antimicrobial activity of ginger oleoresins. *Shipin Yu Fajiao Gongye* 27 (4): 30–34.

Chen, Z. H., G. Y. Kai, X. J. Liu, J. Lin, X. F. Sun and K. X. Tang. 2005. cDNA cloning and characterization of a mannose-binding lectin from *Zingiber officinale* Roscoe (ginger) rhizomes. *Journal of Biosciences* 30 (2): 213–220.

Chirangini. P., and G. J. Sharma. 2005. *In vitro* propagation and microrhizome induction in *Zingiber cassumunar* (Roxb.)—An antioxidant-rich medicinal plant. *Journal of Food, Agriculture & Environment* 3 (1): 139–142.

Choi, S. K. 1991. Studies on the rapid multiplication through *in vitro* culture of ginger (*Zingiber officinale* Rosc.). *Research Reports of Rural Development Administration of Biotechnology* 33 (1): 8–13.

Christensen, A. H., and P. H. Quail. 1996. Ubiquitin promoter-based vectors for high-level expression of selectable and/or screenable marker genes in monocotyledonous plants. *Transgenic Research* 5:213–218.

Clark, R. J., and R. A. Warner. 2000. Production and marketing of Japanese ginger (*Zingibermioga*) in Australia. RICRDC Pub. 00/117, Rural Ind. Res. & Dev. Corporation, Australia. Available from http://www.rirdc.gov.au/reports/AFO/00-117.sum.html (accessed on Oct. 5, 2003).

Dai, J. R., J. H. Cardellina, J. B. McMahan, and H. R. Boyd. 1997. Zerumbone, an HIV inhibitory and cytotoxic sesquiterpene of *Zingiber aromaticum* and *Z. zerumbet. National Product Letter* 10:115–118.

Dake G. N., K. Nirmal Babu, T. G. N. Rao, and N. K. Leela. 1997. *In vitro* selection for resistance to soft rot and bacterial wilt in ginger. In *Integrated plant disease management for sustainable agriculture.* International Conference, November 10–15, 1997, Indian Phytopathological Society—Golden Jubilee, New Delhi, abstract, p. 339.

Darlington, C. D., and E. K. Janaki Ammal. 1945. *Chromosome atlas of cultivated plants.* George Allen & Unwin, London, p. 397.

Das, A. B., S. Rai, and P. Das. 1998. Estimation of 4c DNA and karyotype analysis in ginger (*Zingier officinale* Rosc.). *Cytologia* 63:133–139.

Das, P., S. Rai, and A. B. Das. 1999. Cytomorphology and barriers in seed set of cultivated ginger (*Zingiber officinale* Rosc.). *Iranian Journal of Botany* 8:119–129.

Dekkers, A. J., A. Rao, and C. J. Goh. 1991. *In vitro* storage of multiple shoot cultures of gingers at ambient temperature of 24 to 29°C. *Scientia Horticulturae* 47:157–167.

Denyer, C., P. Jackson, D. M. Loakes, M. R. Ellis, and D. A. Young. 1994. Isolation of antirhinoviral sesquiterpenes from ginger (*Zingiber officinale* Rosc.). *Journal of Natural Products* 57:658–662.

Dhamayanthi, K. P. M., B. Sasikumar, and A. B. Remashree. 2003. Reproductive biology and incompatibility studies in ginger (*Zingiber officinale* Rosc.). *Phytomorphology* 53:123–131.

Duve, R. N. 1980. Highlights on the chemistry and pharmacology of wild ginger (*Zingiber zerumbet* Smith). *Fiji Agricultural Journal* 42 (1): 41–43. FAO 2011 Rome (http://faostat.fao.org/site/339/default.aspx)

Faria, R. T., and R. D. Illg. 1995. Micropropagation of *Zingiber spectabile* Griff. *Scientia Horticulturae* 62 (1–2): 135–137.

Furutani, S. C., and M. A. Nagao. 1986. Influence of daminozide, gibberellic acid and ethepon on flowering, shoot growth and yield of ginger. *Horticultural Science* 21:428–429.

Gamble, J. S. 1925. Flora of the presidency of Madras. II Botanical Survey of India, Calcutta.

Gao, L., N. Liu, B. Huang, and X. Hu. 2008. Phylogenetic analysis and genetic mapping of Chinese *Hedychium* using SRAP markers, *Scientia Horticulturae* 117:369–377.

Geetha, S. P. 2002. *In vitro* technology for genetic conservation of some genera of Zingiberaceae. Unpublished PhD thesis, University of Calicut, India.

Geetha, S. P., K. Nirmal Babu, J. Rema, P. N. Ravindran, and K. V. Peter. 2000. Isolation of protoplasts from cardamom (*Elettaria cardamomum* Maton.) and ginger (*Zingiber officinale* Rosc.). *Journal of Spices and Aromatic Crops* 9 (1): 23–30.

Giridharan, M. P. 1984. Effect of gamma irradiation in ginger (*Zingiber officinale* Rosc.). MSc (Hort.) thesis, Kerala Agricultural University, Vellanikkara, India.

Giridharan, M. P., and S. Balakrishnan. 1991. Effect of gamma irradiation on yield and quality of ginger. *Indian Cocoa, Arecanut and Spices Journal* 14 (3): 100–103.

————. 1992. Gamma ray induced variability in vegetative and floral characters of ginger. *Indian Cocoa, Arecanut and Spices Journal* 15:68–72.

Gonzalez, O. N., L. B. Dimaunahan, L. M. Pilac, and V. Q. Alabastro. 1969. Effect of gamma irradiation on peanuts, onions and ginger. *Philippine Journal of Science* 98:279–292.

Gosh, R., and R. P. Purkauastha. 2003. Molecular diagnosis and induced systemic protection against rhizome rot disease of ginger caused by *Pythium aphanidermatum. Current Science* 85:25.

Gowda, K. K., and K. R. Melanta. 2000. Varietal performance of ginger in Karnataka. In *Recent advances in plantation crops research*, ed. N. Muraleedharan and R. Rajkumar, 92–93. Allied Pub., New Delhi.

Guo, Y., and Z. Zhang. 2005. Establishment and plant regeneration of somatic embryogenic cell suspension cultures of the *Zingiber officinale* Rosc. *Ja Scientia Horticulture,* 107:90–96.

Guo, Y., J. Bai, and Z. Zhang. 2007. Plant regeneration from embryogenic suspension-derived protoplasts of ginger (*Zingiber officinale* Rosc.). *Plant Cell Tissue & Organ Cultures* 89:151–157.

Holttum, R. E. 1950. The Zingiberaceae of the Malay peninsula. *Gardens Bulletin (Singapore)* 13:1–50.

Hooker, J. D. 1890–92. *Flora of British India,* vol. 6, 198–264. Reeve, London (Rep.) Bishen Singh Mahendrapel Singh, Dehra Dun, India.

Hosoki, T., and Y. Sagawa. 1977. Clonal propagation of ginger (*Zingiber officinale* Roscoe.) through tissue culture. *HortScience* 12 (6): 451–452.

Huang, J. L., L. L. Cheng, and Z. X. Zhang. 2007. Molecular cloning and characterization of violaxanthin de-epoxidase (VDE) in *Zingiber officinale. Plant Science* 172:228–235.

Hus, H. J., H. L. Chang, and N. Jian. 1999. Synergistic natural pesticides containing garlic. Eur. Pat. Appl. Pat. 945,066, A1, 1999 09 29, Appl. 1999-302,286 (EP) 17 p. US 6231865,B1, 2001 05 15, Appl. 1999-273636.

Ilahi, I., and M. Jabeen. 1987. Micropropagation of Z. *officinale* Rosc. *Pakistan Journal of Botany* 19 (1): 61–65.

————. 1992. Tissue culture studies for micropropagation and extraction of essential oils from *Zingiber officinale* Rosc. *Pakistan Journal of Botany* 24 (1): 54–59.

IISR. 2008. Spices news. Indian Institute of Spices Research, Calicut, Kerala, India.

————. 2009. Annual report 2008–09. Indian Institute of Spices Research, Calicut, Kerala. India.

Islam, A. K. M. S., C. J. Asher, D. G. Edwards, and J. P. Evenson. 1978. Germination and early growth of ginger (*Zingiber officinale* Rosc.) 2. Effects of 2-chloroethyl phosphonic acid or elevated temperature pretreatments. *Tropical Agriculture* 55:127–134.

Jain, S. K., and V. Prakash. 1995. Zingiberacecae in India: Phytogeography and endemism. *Rheedea* 5 (2): 154–169.

Jamil, M., J. K. Kim, Z. Akram, S. U. Ajmal, and E. S. Rha. 2007. Regeneration of ginger plant from callus culture through organogenesis and effect of CO_2 enrichment on the differentiation of regenerated plant. *Biotechnology* 6 (1): 101–104.

Jayachandran, B. K. 1989. Induced mutations in ginger. Unpublished PhD thesis, Kerala Agricultural University, Kerala, India.

Jayachandran, B. K., and N. Mohanakumaran. 1992. Effect of gamma ray irradiation on ginger. *South Indian Horticulture* 40:283–288.

Jayachandran, B. K., P. Vijayagopal, and P. Sethumadhavan. 1979. Floral biology of ginger, *Zingiber officinale* R. *Agricultural Research Journal Kerala* 17:93–94.

Jiang, H., Z. Xie, H. J. Koo, S. P. McLaughlin, B. N. Timmermann, and D.R .Gang. 2006. Metabolic profiling and phylogenetic analysis of medicinal *Zingiber* species: Tools for authentication of ginger (*Zingiber officinale* Rosc.). *Phytochemistry* 67:1673–1685.

Jogi, B. S., I. P. Singh, N. S. Dua, and P. S. Sukhiya. 1978. Changes in crude fiber, fat and protein content in ginger (*Zingiber officinale* Rosc.) at different stages of ripening. *Indian Journal of Agricultural Science* 42:1011–1015.

Johny, A. K., and P. N. Ravindran. 2006. Over 225 high yielding spice varieties in India. Part 2. *Spice India* 18:40–44.

Kackar, A., S. R. Bhat, K. P. S. Chandel, and S. K. Malik. 1993. Plant regeneration via somatic embryogenesis in ginger. *Plant Cell Tissue and Organ Culture* 32 (3): 289–292.

Kannan, K., and K. P. V. Nair. 1965. Ginger (*Zingiber officinale* Rosc.) in Kerala. *Madras Agricultural Journal* 52:168–176.

Kavitha, P. G., and G. Thomas. 2006. *Zingiber zerumbet*, a potential donor for soft-rot resistance in ginger: Genetic structure and functional genomics. Extended abstract, XVIII Kerala Science Congress, pp.169–171.

———. 2008a. Defense transcriptome profiling of *Zingiber zerumbet* (L.) Smith by mRNA differential display. *Journal of Bioscience* 33 (1): 81–90.

———. 2008b. Population genetic structure of the clonal plant *Zingiber zerumbet* (L.) Smith (Zingiberaceae), a wild relative of cultivated ginger, and its response to *Pythium aphanidermatum*. *Euphytica* 160:89–100.

Kawata, H. 1998. Insecticidal baits containing ginger oil against cockroach. *Japan Kokai* Tokyo, 4 pp., no. 10,017,405, A2, 1998 01 20 (Date) 1996-172,020.

Khan, K. I. 1959. Ensure twofold ginger yields. *Indian Farming* 8 (2): 10–14.

Kiso, Y., Kato, N., Hamada, Y., Shioiri, T. and Ajyama, R. 1985. Anti-hepato toxic actions of Gingerols and Diarylhepanoids. *Ethnopharmacology* 14: 31–39.

Kiuchi, F., M. Shibuya, and U. Sankawa. 1982. Inhibitors of the biosynthesis of prostaglandins. *Chemical and Pharmaceutical Bulletin Tokyo* 30:754–757.

Kress, W. J., L. M. Prince, and K. J. Williams. 2002. The phylogeny and a new classification of the gingers (Zingiberaceae): Evidence from molecular data. *American Journal of Botany* 89:1682–1696.

Krishnakantha, T., and Lokesh, B. 1993. Scavenging for superoxide anions by spice principles. *Indian Journal of Biochemistry & Biophysics* 30:133–134.

Krishnamurthy, N., E. S. Nambudiri, A. G. Mathew, and Y. S. Lewis. 1970. Essential oil of ginger. *Indian Perfumer* 14:1–3.

Kulkarni, D. D., S. S. Khuspe, and A. F. Mascarenhas. 1987. Isolation of *Pythium* tolerant ginger by tissue culture. In *Proceedings VI Symposium on Plantation Crops*, ed. S. N. Potty, 3–13.

Kumar, A., Y. R. Sarma, and M. Anandaraj. 2004. Evaluation of genetic diversity of *Ralstonia solanacearum* causing bacterial wilt of ginger using REP-PCR and PCR-RELP. *Current Science* 87 (11): 10.

Kumar, V. S., P. P. Balasubramaniam, K. V. Kumar, and M. K. Mammen. 1980. A comparative performance of six varieties of ginger for second crop in Wynad. *Indian Spices* 17 (1): 10–11–114.

Lee, S.-Y., W. K. Fai, M. Zakaria, H. Ibrahim, R. Y. Othman, J.-G. Gwag, V. R. Rao, and Y.-J. Park. 2007. Characterization of polymorphic microsatellite markers, isolated from ginger (*Zingiber officinale* Rosc.). *Molecular Ecology Notes* 7:1009–1011.

Lewis, Y. S., A. G. Mathew, E. S. Nambudiri, and N. Krishnamurthy. 1972. Oleoresin ginger. *Flavor Industry* 3 (2): 78–81.

Lincy, A. K. 2007. Investigation on direct *in vitro* shoot regeneration from aerial stem explant of ginger (*Zingiber officinale* Rosc.) and its field evaluation. PhD thesis, Calicut University.

Lincy, A. K., K. Jayarajan, and B. Sasikumar. 2008. Relationship between vegetative and rhizome characters and final rhizome yield in micropropagated ginger plants (*Zingiber officinale* Rosc.) over two generations. *Scientia Horticulturae* doi:10.1016/j.scienta.2008.05.012.

Lincy, A. K., A. B. Remashree, and B. Sasikumar. 2009. Indirect and direct somatic embryogenesis from aerial stem explants of ginger (*Zingiber officinale* Rosc.). *Acta Botanica Croatica* 68 (1): 93–103.

Ma, X., and D. R. Gang. 2006. Metabolic profiling of *in vitro* micropropagated and conventionally greenhouse grown ginger (*Zingiber officinale*). *Journal of Phytochemistry* 67:2239–2255.

Madan, M. S. 2005. Production, marketing, and economics of ginger. In *Ginger—The genus Zingiber*, ed. P. N. Ravindran and K. Nirmal Babu, 181–210. CRC Press, Boca Raton, FL.

Mahanty, H. K. 1970. A cytological study of the Zingiberales with special reference to their taxonomy. *Cytologia* 35:13–49.

Malamug, J. J. F., H. Inden, and T. Asahira. 1991. Plant regeneration and propagation from ginger callus. *Scientia Horticulturae* 48 (1–2): 89–99.

Masuda, T., A. Jitoe, and M. J. Mabry. 1995. Isolation and structure determination of cassumunarins A, B and C: New anti-inflammatory antioxidants from a tropical ginger, *Zingiber cassumunar*. *Journal of the American Oil Chemists' Society* 72:1053–1057.

Masuda, T., H. Matsumura, Y. Oyama, and Y. Takeda. 1998. Synthesis of (+) cassumunins A and B, new curcuminoid antioxidants having protective activity on the living cell against oxidative damage. *Journal of Natural Products* 61: 609–613.

Mohanty, D. C. 1984. Germplasm evaluation and genetic improvement in ginger. Unpublished PhD thesis, Orissa Univ. Agri. Technology, Bhubaneswar.

Mohanty, D. C., R. C. Das, and Y. N. Sarma. 1981. Variability of agronomic characters in ginger (*Zingiber officinale* Rosc.). *Orissa Journal of Horticulture* 9:15–17.

Mohanty, D. C., B. S. Naik, and B. S. Panda. 1990. Ginger research in Orissa with reference to its varietal and cultural improvement. *Indian Cocoa, Arecanut and Spices Journal* 14 (2): 61–65.

Mohanty, D. C., and B. S. Panda. 1991. High yielding mutant V1K1-3 ginger. *Indian Cocoa, Arecanut & Spices Journal* 15:5–7.

———. 1994. Genetic resources in ginger. In *Advances in horticulture*, vol. 9: *Plantation crops and spices*, part 2, ed. K. L. Chadha and P. Rethinam, 151–168. Malhotra Pub., New Delhi.

Mohanty, D. C., and Y. N. Sarma. 1979. Genetic variability and correlation for yield and other variables in ginger germplasm. *Indian Journal of Agricultural Science* 49:250–253.

Moringa, T., E. Fukushina, T. Kanui, and Y. Tamasaki. 1929. Chromosome numbers of cultivated plants. *Botanical Magazine (Tokyo)* 43:589–594.

Murakami, A., D. Takahashi, T. Kiroshita, K. Koshimizu, H. W. Kim, A. Yoshihiro, Y. Nakamura, S. Jiwajinda, J. Terao, and H. Ohigashi. 2002. Zerumbone, a Southeast Asian ginger sesquiterpene, markedly suppresses free radical generation, pro-inflammatory protein production, and cancer cell proliferation accompanied by apoptosis: The α- and β-unsaturated carbonyl group is a prerequisite. *Carcinogenesis* 23:795–802.

Muralidharan, A. 1972. Varietal performance of ginger in Wynad, Kerala. *Journal of Plantation Crops* (Suppl.): 19–20.

Muralidharan, A., and N. Kamalam. 1973. Improved ginger means foreign exchange. *Indian Farming* 22:37–39.

Murashige, T., and F. Skoog. 1962. A revised medium for rapid growth and bioassays with tobacco tissue cultures. *Physiologia Plantarum* 15:473–497.

Nadgauda, R. S., D. B. Kulkarni, A. F. Mascarenhas, and V. Jaganathan. 1980. Development of plantlets from tissue cultures of ginger. In *Proceedings of Annual Symposium on Plantation Crops*, 143–147.

Nair, P. C. S. 1969. Ginger cultivation in Kerala. *Arecanut and Spices Bulletin* 1 (1): 22–24.

Natarajan, C. P., R. Padmabai, N. Krishnamurthy, B. Raghavan, N. B. Sankaracharya, S. Kuppuswamy, V. S. Govindarajan, and Y. S. Lewis. 1972. Chemical composition of ginger varieties and dehydration studies on ginger. *Journal of Food Science and Technology* 9:220.

Nazeem, P. A., L. Joseph, T. G. Rani, P. A. Valsala, S. Philip, and G. S. Nair. 1996. Tissue culture system for *in vitro* pollination and regeneration of plantlets from *in vitro* raised seeds of ginger—*Zingiber officinale* Rosc. *Acta Horticulturae* 426:467–472.

Ngamriabsakul, C., M. F. Newman, and Q. C. B. Cronk. 2003. The phylogeny of tribe *Zingibereae* (*Zingiberaceae*) based on its (nrDNA) and *trn*l–f (cpDNA) sequences. *Edinburgh Journal of Botany* 60:483–507.

Ni, M., Z. Chen, and B. Yan. 1988. Synthesis of optically active sesquiterpenes and exploration of their antifertility effect. *Huadong Huagong Xueyuan Xuebao* 14:675–679.

Nirmal Babu, K. 1997. *In vitro* studies in *Zingiber officinale* Rosc. PhD thesis. Calicut University, Kerala, India.

Nirmal Babu, K., S. P. Geetha, D. Minoo, P. N. Ravindran, and K. V. Peter. 1999. *In vitro* conservation of germplasm. In *Biotechnology and its application in horticulture*, ed. S. P .Ghosh, 106–129. Narosa Publishing House, New Delhi.

Nirmal Babu, K., D. Minoo, S. P. Geetha, P. N. Ravindran, and K. V. Peter. 2005. Advances in biotechnology of spices and herbs. *Indian Journal of Botanical Research* 1 (2): 155–214.

Nirmal Babu, K., D. Minoo, S. P. Geetha, K. Samsudeen, J. Rema, P. N. Ravindran, and K. V. Peter. 1998. Plant biotechnology—Its role in improvement of spices. *Indian Journal of Agricultural Sciences* 68 (special issue): 533–547.

Nirmal Babu, K., P. N. Ravindran, and K. V. Peter, eds. 1997. Protocols for micropropagation of spices and aromatic crops. Indian Institute of Spices Research, Calicut, Kerala, 35 pp.

Nirmal Babu, K., P. N. Ravindran, and B. Sasikumar. 2003. Field evaluation of tissue cultured plants of spices and assessment of their genetic stability using molecular markers. Final report, Department of Biotechnology, government of India. 94 pp.

Nirmal Babu, K., K. Samsudeen, D. Minoo, S. P. Geetha, and P. N. Ravindran. 2005. Tissue culture and biotechnology of ginger. In *Ginger—The genus Zingiber*, ed. P. N. Ravindran and K. Nirmal Babu, 181–210. CRC Press, Boca Raton, FL.

Nirmal Babu, K., K. Samsudeen, and M. J. Ratnambal. 1992. *In vitro* plant regeneration from leaf derived callus in ginger, *Zingiber officinale* Rosc. *Plant Cell Tissue and Organ Culture* 29:71–74.

Nirmal Babu, K., K. Samsudeen, M. J. Ratnambal, and P. N. Ravindran. 1996. Embryogenesis and plant regeneration from ovary derived callus cultures of *Zingiber officinale* Rosc. *Journal of Spices and Aromatic Crops* 5 (2): 134–138.

Nirmal Babu, K., K. Samsudeen, and P. N. Ravindran. 1992. Direct regeneration of plantlets from immature inflorescence of ginger (*Zingiber officinale* Rosc.) by tissue culture. *Journal of Spices and Aromatic Crops* 1:43–48.

———. 1996. Biotechnological approaches for crop improvement in ginger, *Zingiber officinale* Rosc. In *Recent advances in biotechnological applications on plant tissue and cell culture*, ed. G. A. Ravishanker and L. V. Venkataraman, 321–332. Oxford IBH Publishing Co., New Delhi.

Nwachukwu, E. C., L. S. O. Ene, and E. N. A. Mbanaso. 1995. Radiation sensitivity of two ginger varieties (*Zingiber officinale* Rosc.) to gamma irradiation. *Tropenlandwirt* 95:99–103.

Nybe, E. V., and P. C. S. Nair. 1979. Field tolerance of ginger types to important pests and diseases. *Indian Cocoa Arecanut Spices Journal* 2:190–111.

Nybe, E. V., P. C. S. Nair, and N. M. Kumaran. 1982. Assessment of yield and quality components in ginger. In *Proceedings of National Seminar on Ginger and Turmeric*. Calicut, April 8–9, 1980. CPCRI, Kasaragod, 24–29.

Ozaki, Y., N. Kawahara, and M. Harada. 1991. Anti-inflammatory effect of *Zingiber cassumunar* Roxb. and its active principles. *Chemical and Pharmaceutical Bulletin (Tokyo)* 39:2353–2356.

Pandey, G., and V. K. Dobhal. 1993. Genetic variability, character association and path analysis for yield components in ginger (*Zingiber officinale* Rosc). *Journal of Spices & Aromatic Crops* 2:16–20.

Panigrahi, U. C., and G. K. Patro. 1985. Ginger cultivation in Orissa. *Indian Farming* 33 (5): 3–4, 17.

Panthong, A., D. Kanhanapothi, W. Niwatananant, P. Tuntiwachurittikul, and V. Reutrakul. 1997. Anti-inflammatory activity of compound D ((E)-4-(3′,4′-dimethoxyphenyl)but-3-en-2-ol) isolated from *Zingiber cassumunar* Roxb. *Phytomedicine* 4:207–212.

Paulose, T. T. 1973. Ginger cultivation in India. In *Proceedings of Conference on Spices*. TPI, London, pp. 117–121.

Peter, K. V., P. N. Ravindran, K. Nirmal Babu, B. Sasikumar, D. Minoo, S. P. Geetha, and K. Rajalakshmi. 2002. Establishing *In Vitro* Conservatory of Spices Germplasm. ICAR project report. Indian Institute of Spices Research, Calicut, Kerala, India. 131 pp.

Pillai, P. K. T., G. V. Kumar, and M. C. Nambiar. 1978. Flowering behavior, cytology, pollen germination in ginger (*Zingiber officinale* Rosc.). *Journal of Plantation Crops* 6 (1): 12–13.

Poonsapaya, P., and K. Kraisintu. 2003. Micropropagation of *Zingiber cassumunar* Roxb. ISHS website. Available from www.Acta Hort.org/books/344/344-64.htm (accessed May 10, 2003).

PPV & FRA. 2010. Guidelines for the conduct of test for distinctiveness, uniformity and stability on ginger (*Zingiber officinale* Rosc.). Protection of Plant Varieties and Farmers' Rights Authority (PPV & FRA), government of India, p. 13.

Pradeepkumar, T., T. P. Manmohandas, M. Jayarajan, and K. C. Aipe. 2000. Evaluation of ginger varieties in Wayanad. *Spice India* 13 (1): 13.

Prance, M. S., and D. Sarket. 1997. Root and tuber crops. SDE-40, Bogor.

Prathanturarug, S., D. Angusumalee, N. Pongsiri, S. Suwacharangoon, and T. Jenjittikul. 2004. *In vitro* propagation of *Zingiber petiolatum* (Holttum) I. Theilade, a rare zingiberaceous plant from Thailand. *Journal of In Vitro Cell Development Biology—Plant* 40:317–320.

Purseglove, J. W. 1972. *Tropical Crops: Monocotyledons*. Halsted Pree Division, Wiley. 533–554.

Purseglove, J. W., E. G. Brown, C. L. Green, and S. R. J. Robbins. 1981. *Spices*, vol. 2. Longman Inc., New York, p. 813.

Quisumbing, E. 1978. *Medicinal plants of the Philippines*. JMC Press, Inc., Quezon City, Philippines, pp. 186–202.

Raghavan, T. S., and K. R. Venkatasubban. 1943. Cytological studies in the family Zingiberaceae with special reference to chromosome number and cytotaxonomy. *Proceedings of Indian Academy of Sciences* 17B:118–132.

Rai, S., A. B. Das, and P. Das. 1999. Variations in chlorophyll, carotenoids, protein and secondary metabolites amongst ginger (*Zingiber officinale*, Roscoe) cultivars and their association with rhizome yield. *New Zealand Journal of Crop and Horticultural Science* 27:79–82.

Raju, E. C., J. D. Patel, and J. J. Shah. 1980. Effect of gamma radiation in morphology of leaf and shoot apex of ginger, turmeric and mango ginger. *Proceedings of Indian Academy of Sciences* 89:173–178.

Ramachandran, K. 1969. Chromosome numbers in Zingiberaceae. *Cytologia* 34:213–221.

———. 1982. Polyploidy induced in ginger by colchicine treatment. *Current Science* 51:288–289.

Ramachandran, K., and P. N. C. Nair. 1992. Cytological studies on diploid and autotetraploid ginger (*Zingiber officinale* Rosc.). *Journal of Spices and Aromatic Crops* 1 (2): 125–130.

Ratnambal, M. J. 1979. Cytological studies in ginger (*Zingiber officinale* Rosc.). Unpublished PhD thesis, University of Bombay, India.

Ratnambal, M. J., R. Balakrishnan, and M. K. Nair. 1982. Multiple regression analysis in cultivars of *Zingiber officinale* Rosc. In *Ginger and turmeric*, ed. M. K. Nair, T. Premkumar, P. N. Ravindran, and Y. R. Sarma, 30–33. Central Plantation Crops Research Institute, Kasaragod, India.

Ratnambal, M. J., and M. K. Nair. 1981. Microsporogenesis in ginger (*Zingiber officinale* Rosc.). In *Proceedings of Plantation Crops Symposium* (PLACROSYM) VI, CPCRI, Kasaragod, India, pp. 44–57.

Rattan, R. S. 1988. Varietal performance of ginger. *Proceedings of Ginger Symposium*, Naban, Himachal Pradesh.

———. 1994. Improvement of ginger. In *Advances in horticulture*, vol. 9, *Plantation and spice crops*, part 1, ed. K. L. Chadha and P. Rethinam, 333–344. Malhotra Pub., New Delhi.

Rattan, R. S., B. N. Korla, and N. P. Dohroo. 1988. Performance of ginger varieties in Solan area of Himachal Pradesh. In *Proceedings of National Seminar on Chillies, Ginger and Turmeric*, ed. G. Satyanarayana, M. S. Reddy, M. R. Rao, K. M. Azam, and R. Naidu, 71–73. Spices Board, Cochin.

Ravindran, P. N., and A. K. Johny. 2000. High yielding varieties in spices. *Indian Spices* 37 (1): 17–19.

Ravindran, P. N., and K. Nirmal Babu. 2005. *Ginger—The genus Zingiber*. CRC Press, Boca Raton, FL, p. 310.

Ravindran, P. N., K. Nirmal Babu, K. V. Peter, Z. Abraham, and R. K. Tyagi. 2005. Spices. In *Plant genetic resources: Horticultural crops*, ed. B. S. Dhillon, R. K. Tyagi, S. Saxena, and G. J. Randhawa, 190–227. Narosa Publishing House, New Delhi.

Ravindran, P. N., K. Nirmal Babu, and K. N. Shiva. 2005. Botany and crop improvement of ginger. In *Ginger—The genus Zingiber*, ed. P. N. Ravindran and K. Nirmal Babu, 15–86. CRC Press, Boca Raton, FL.

Ravindran, P. N., B. Sasikumar, K. G. Johnson, M. J. Ratnambal, K. Nirmal Babu, J. T. Zachariah, and R. R. Nair. 1994. Genetic resources of ginger (*Zingiber officinale* Rosc.) and its conservation. *Plant Genetic Resources Newsletter* 98:1–4.

Ravindran, P. N., K. N. Shiva, K. Nirmal Babu, and B. N. Korla. 2006. Ginger. In *Advances in spices research*, ed. P. N. Ravindran, K. Nirmal Babu, K. N. Shiva, and A. K. Johny, 365–432, Agrobios, Jodhpur.

Rout, G. R., and P. Das. 1997. *In vitro* organogenesis in ginger. *Journal of Herbs, Spices and Medicinal Plants* 4 (4): 41–51.

Rout, G. R., P. Das, S. Goel, and S. N. Raina. 1998. Determination of genetic stability of micropropagated plants of ginger showing random amplified polymorphic DNA (RAPD) markers. *Botanic Bulletin of Academia Sinica* 39 (1): 23–29.

Rout, G. R., S. K. Palai, S. Samantaray, and P. Das. 2001. Effect of growth regulator and culture conditions on shoot multiplication and rhizome formation in ginger (*Zingiber officinale* Rosc.) *in vitro*. *In Vitro Cellular and Developmental Biology—Plant* 37 (6): 814–819.

Sabu, M. 2003. Revision of the genus *Zingiber* in South India. *Folia Malaysiana* 4 (1): 25–52.

———. 2006. Zingiberaceae and Costaceae of South India. Indian Association for Angiosperm Taxonomy, p. 281.

Sabu, M., and D. Skinner. 2005. Other economically important *Zingiber* species. In *Ginger—The genus Zingiber*, ed. P. N. Ravindran and K. Nirmal Babu, 533–545. CRC Press, Boca Raton, FL.

Saikia, L., and A. Shadeque. 1992. Yield and quality of ginger (*Zingiber officinale* Rosc.) varieties grown in Assam. *Journal of Spices and Aromatic Crops* 1:131–135.

Sajina, A., D. Minoo, S. P. Geetha, K. Samsudeen, J. Rema, K. Nirmal Babu, P. N. Ravindran, and K. V. Peter. 1997. Production of synthetic seeds in a few spice crops. In *Biotechnology of spices, medicinal and aromatic plants*, ed. S. Edison, K. V. Ramana, B. Sasikumar, K. Nirmal Babu, and S. J. Eapen, 65–69. Indian Society for Spices, Calicut, India.

Sakamura, F., K. Ogihara, T. Suga, K. Taniguchi, and R. Tanaka. 1986. Volatile constituents of *Zingiber officinale* rhizome produced by *in vitro* shoot tip culture. *Phytochemistry* 25 (6): 1333–1335.

Sakamura, F., and T. Suga. 1989. *Zingiber officinale* Roscoe (ginger): *In vitro* propagation and the production of volatile constituents. In *Biotechnology in agriculture and forestry*, vol. 7., ed. Y. P. S. Bajaj, 524–538. Medicinal and Aromatic Plants II. Springer–Verlag, Berlin.

Samsudeen, K. 1996. Studies on somaclonal variation produced by *in vitro* culture in *Zingiber officinale* Rosc. PhD thesis, University of Calicut, Kerala, India.

Samsudeen, K., K. Nirmal Babu, D. Minoo, and P. N. Ravindran. 2000. Plant regeneration from anther derived callus cultures of ginger (*Zingiber officinale* Rosc.). *Journal of Horticultural Science and Biotechnology* 75 (4): 447–450.

Sarma, Y. R., K. V. Ramana, S. Devasahayam, and J. Rema, eds. 2001. Ginger. In *The saga of spice research*. IISR, Calicut, pp. 75–89.

Sarma, Y. R., and B. Sasikumar. 2002. Indian spices: Accelerating production and export through proactive programs. In *National Consultative Meeting for Accelerated Production and Export of Spices*, May 29–30, 2002, Directorate of Arecanut and Spices Development, Calicut, 9–21.

Sasikumar, B. 2005. Genetic resources of *Curcuma*: Diversity, characterization and utilization. *Plant Genetic Resources: Conservation and Utilization* 3 (2): 230–251.

Sasikumar, B., B. Krishnamoorthy, K. V. Saji, K. G. Johnson, K. V. Peter, and P. N. Ravindran. 1999. Spice diversity and conservation of plants that yield of major spices in India. *Plant Genetic Resources Newsletter* 118:19–26.

Sasikumar, B., K. Nirmal Babu, J. Abraham, and P. N. Ravindran. 1992. Variability, correlation and path analysis in ginger germplasm. *Indian Journal of Genetics* 52:428–431.

Sasikumar, B., P. N. Ravindran, and K. J. George. 1994. Breeding ginger and turmeric. *Indian Cocoa, Arecanut and Spices Journal* 18:10–12.

Sasikumar, B., K. V. Saji, A. Antony, J. K. George, T. J. Zachariah, and S. J. Eapen. 2003. IISR Mahima and IISR Rejatha—Two high yielding and high quality ginger (*Zingiber officinale*) varieties. *Journal of Spices & Aromatic Crops* 12:34–37.

Sasikumar, B., and T. J. Zachariah. 2003. Organization of ginger and turmeric germplasm based on molecular characterization. Final report, ICAR ad hoc project, IISR, Calicut.

Sato, D. 1948. The karyotype and phylogeny of Zingiberaceae. *Japan Journal of Genetics* 23:44 (cited from Sharma, 1972).

———. 1960. The karyotype analysis in Zingiberales with special reference to the protokaryotype and stable karyotype. *Science Papers of the College of Education, University of Tokyo* 10 (2): 225–243.

Saxena, H. O., M. Brahamam, and P. K. Dutta. 1981. Ethnobotanical studies in Orissa. In *Glimpses of Indian ethnobotany*, ed. S. K. Jain, 232–244. Oxford and IBH Publishing Co., New Delhi.

Setty, T. A. S., T. R. Guruprasad, E. Mohan, and M. N. N. Reddy. 1995a. Susceptibility of ginger cultivars to rhizome rot at west coast conditions of Karnataka. *Environmental Ecology* 13:242–244.

———. 1995b. Suceptibility of ginger cultivars to phyllosticta leaf spot at west coast conditions of Karnataka. *Environmental Ecology* 13:443–444.

Sharma, A. K., and N. K. Bhattacharyya. 1959. Cytology of several members of Zingiberaceae and the study of the inconsistency of their chromosome complements. *La Cellule* 59:297–346.

Sharma, J. N., F. I. Ishak, A. P. M.Yusof, and K. C. Srivastava. 1997. Effects of eugenol and ginger oil on adjuvant arthritis and the kallikreins in rats. *Asia Pacific Journal of Pharmacology* 12 (1–2): 9–14.

Sharma, T. R., and B. M. Singh. 1995. *In vitro* microrhizome production in *Zingiber officinale* Rosc. *Plant Cell Reports* 15 (3/4): 274–277.

——— 1997. High frequency *in vitro* multiplication of disease free *Zingiber officinale* Rosc. *Plant Cell Reports* 17 (1): 68–73.

Sharma, T. R., B. M. Singh, and R. S. Chauhan. 1994. Production of disease-free encapsulated buds of *Zingiber officinale* Rosc. *Plant Cell Reports* 13:300–302.

Shi-jie, Z., A. Xizhen, W. Shaohui, Z. Zhenxian, and Z. Qi. 1999. Role of xanthophyll cycle and photorespiration in protecting the photosynthetic apparatus of ginger leaves from photoinhibitory damage. *Acta Agriculturae Boreali-Occidentalis Sinica* 8 (3): 81–85.

Shuxuan, J., H. Xiu Qin, X. Zheng, M. Li Mei, L. Ping, D. Juan, and M. Yicheng. 2001. Experimental study on *Zingiber corallium* Hance to prevent infection with *Schistosoma japonicum cercaria. Chinese Journal of Schistosomiasis Control* 13:170–172.

Singh, A. K. 2001. Correlation and path analysis for certain metric traits in ginger. *Annals of Agricultural Research* 22 (2): 285–286.

Singh, D. B., A. Subramanian, P. V. Sreekumar, and T. V. R. S. Sharma. 1998. Wild gingers of Andaman Nicobar Islands. *Indian Journal of Plant Genetic Resources* 11 (2): 249–250.

Singh, G., I. P. S. Kapoor, S. K. Pandey, O. P. Singh, U. K. Singh, and R. K. Singh. 2001. A note on antibacterial activity of volatile oils of some aromatic plants. *Indian Perfumer* 45 (4): 275–278.

Singh, P. P., V. B. Singh, A. Singh, and H. B. Singh. 1999. Evaluation of different ginger cultivars for growth, yield and quality character under Nagaland condition. *Journal of Medicinal and Aromatic Plant Sciences* 21 (3): 716–718.

Smith, M. K., and S. D. Hamill. 1996. Field evaluation of micropropagated and conventionally propagated ginger in subtropical Queensland. *Australian Journal of Experimental Agriculture* 36:347–354.

Smith, M. K., S. D. Hamill, B. J. Gogel, and A. A. Severn-Ellis. 2004. Ginger (*Zingiber officinale*) autotetraploid with improved processing quality produced by an *in vitro* colchicines treatment. *Australian Journal of Experimental Agriculture* 44:1065–1072.

Sreekumar, V., G. Indrasenan, and M. K. Mammen. 1980. Studies on the quantitative and qualitative attributes of ginger cultivars. In *Ginger and turmeric*, ed. M. K. Nair, T. Premkumar, P. N. Ravindran, and Y. R. Sarma, 46–49. Proceedings of the National Seminar, CPCRI, Kasaragod, India.

Sterling, K. J., R. J. Clark, P. H. Brown, and S. J. Wilson. 2002. Effect of photoperiod on flower bud initiation and development in myoga (*Zingiber mioga* Rosc.). *Scientia Horticulturae* 95:261–268.

Suekawa, M., A. Ishige, K. Yuasa, K. Sudo, M. Aburada, and E. Hosoya. 1984. I. Pharmacological actions of pungent constituents, (6)-gingerol and (6)-shogaol. *Journal of Pharmacobiodynamics* 7:836–848.

Sugiura, T. 1936. Studies on the chromosome numbers in higher plants. *Cytologia* 7:544–595.

Sultana, A., L. Hassan, S. D. Ahmad, A. H. Shah, F. Batool, M. A. Islam, R. Rahman, and S. Moonmoon. 2009. *In vitro* regeneration of ginger using leaf, shoot tip and root explants. *Pakistan Journal of Botany* 41 (4): 1667–1676.

Sumathi, V. 2007. Studies on somaclonal variation in zingiberaceous crops. PhD thesis, University of Calicut, Kerala, India.

Sundararaj, S. G., A. Agrawal, and R. K. Tyagi. 2010. Encapsulation for *in vitro* short-term storage and exchange of ginger (*Zingiber officinale* Rosc.) germplasm. *Scientia Horticulturae* 125:761–766.

Suryanarayana, M. A., K. N. Shiva, R. P. Medhi, T. Damodaran, and A. N. Sujatha. 2001. Genetic resources of plantation and spice crops in A&N Islands. *Journal of Andaman Science Association* 17 (1–2): 297.

Swetha Priya, R., and R. B. Subramanian. 2007. Isolation and molecular analysis of R-gene in resistant *Zingiber officinale* (ginger) varieties against *Fusarium oxysporum* f. sp. *zingiberi. Bioresource Technology* 99:4540–4543.

Syamkumar, S., B. Lowarence, and B. Sasikumar. 2003. Isolation and amplification of DNA from rhizomes of turmeric and ginger. *Plant Molecular Biology Reporter* 21 (2): 171–171.

Takahashi. 1930. Cited from Darlington and Janaki Ammal (1945).

Tewtraki, S., C. Sardsangjun, J. Itchaypruk, and P. Chaitongruk. 1997. Studies on volatile oil components in *Zingiber zerumbet* rhizomes by gas chromatography. *Songklanakarin Journal of Science Technology* 19:197–202.

Thangaraj, T., S. Muthuswamy, C. R. Muthukrishnan, and J. B. M. M. A. Khader. 1983. Performance of ginger (*Zingiber officinale* Rosc) varieties at Coimbatore. *South Indian Horticulture* 31:45–46.

Thomas, K. M. 1966. Rio de Janeiro will double your ginger yield. *Indian Farming* 15 (10): 15–18.

Thompson, E. H., I. D. Wolf, and C. E. Allen. 1974. Ginger rhizome a new source of proteolytic enzyme. *Journal of Food Science* 38:652–655.

Tilad, P., R. Sharma, and B. M. Singh. 2002. Salicylic acid induced insensitivity to culture filtrate of *Fusarium oxysporum* f. sp. *Zingiberi* in the calli of *Zingiber officinale* Roscoe. *European Journal of Plant Pathology* 108:31–39.

Tyagi, R. K., A. Agarwal, and A. Yusuf. 2006. Conservation of *Zingiber* germplasm through *in vitro* rhizome formation. *Scientia Horticulturae* 108:210–219.

Usha, K. 1984. Effect of growth regulators on flowering, pollination and seed set in ginger (*Zingiber officinale* Rosc.). Unpublished MSc (Ag.) thesis, Kerala Agricultural University, Vellanikkara, Trichur, India.

Utpala, P., K. Jayarajan, A. K. Johny, and V. A. Parthasarathy. 2007. Identification of suitable areas and effect of climate change on ginger—A GIS study. In *Threats and solutions to spices and aromatic crops industry*, ed. R. Dinesh, K. N. Shiva, D. Prasath, K. Nirmal Babu, and V. A. Parthasarathy, 360. Souvenir and abstracts, SYMSAC IV, Indian Society for Spices, Calicut, Kerala.

Valsala, P. A., G. S. Nair, and P. A. Nazeem. 1997. *In vitro* seed set and seed development in ginger, *Zingiber officinale* Rosc. In *Biotechnology of spices, medicinal and aromatic plants*, ed. S. Edison, K. V. Ramana, B. Sasikumar, K. Nirmal Babu, and S. J. Eapen, 106–108. Indian Society for Spices, Calicut, India.

———. 1996. Seed set in ginger (*Zingiber officinale* Rosc.) through *in vitro* pollination. *Journal of Tropical Agriculture* 34:81–84.

Wang, W. 2001. Antioxidant properties of four vegetables with sharp flavor. *Shipin Yu Fajiao Gongye* 27:28–31.

Xianchang, Y., X. Kun, A. Xizheng, C. Liping, and Z. Zhenxian. 1996. Study on the relationship between canopy, canopy photosynthesis and yield formation in ginger. *Journal of Shandong Agricultural University* 27 (1): 83–86.

Xizhen, A., Z. Zhenxian, and W. Shaohui. 1998. Effect of temperature on photosynthetic characters of ginger leaf. *China Vegetables* 3:1–3.

Xizhen, A., Z. Zhenxian, W. Shaohui, and C. Zhifeng. 1998. Study on photosynthetic characteristics of different leaf position in ginger. *Acta Agriculturae Boreali-Occidentalis Sinica* 7 (2): 101–103.

———. 2000. The role of SOD in protecting ginger leaves from photoinhibition damage under high light stress. *Acta Horticulturae Sinica* 27 (3): 198–201.

Xizhen, A., Z. Zhenxian, C. Zhifeng, and C. Liping. 1998. Changes of photosynthetic rate, MDA content and the activities of protective enzymes during development of ginger leaves. *Acta Horticulturae Sinica* 25:294–296.

Yadav, R. K. 1999. Genetic variability in ginger (*Zingiber officinale* Rosc.). *Journal of Spices and Aromatic Crops* 8:81–83.

Yamara, J., M. Mochizuki, H. Q. Rong, H. Matsuda, and H. Fujimara. 1988. The antiulcer effects in rats of ginger constituents. *Journal of Ethnopharmacology* 23:299–304.

Yamuna, G. 2007. Studies on cryopreservation of spices genetic resources. PhD thesis, University of Calicut, Kerala, India.

Yamuna, G., V. Sumathi, S. P. Geetha, K. Praveen, N. Swapna, and K. Nirmal Babu. 2007. Cryopreservation of *in vitro* grown shoot of ginger (*Zingiber Officinale* Rosc). *CryoLetters* 28 (4): 241–252.

Yoshikawa, M., S. Hatakeyama, N. Chatani, Y. Nishino, and J. Yamahara. 1994. Qualitative and quantitative analysis of bioactive principles in *Zingiberis rhizoma* by means of high performance liquid chromatography and gas liquid chromatography. On the evaluation of *Zingiberis rhizoma* and chemical change of constituents during *Zingiberis rhizoma* processing. *Yakugaky Zasshi* 114 (4): 307–310.

Yu, F., H. Haradab, K. Yamasakia, S. Okamotoa, S. Hirasec, Y. Tanakac, N. Misawab, and R. Utsumia. 2008. Isolation and functional characterization of a β-eudesmol synthase, a new sesquiterpene synthase from *Zingiber zerumbet* Smith. *FEBS Letters* 582:565–572.

Zachariah, T. J., B. Sasikumar, and K. Nirmal Babu. 1999. Variations for quality components in ginger and turmeric and their interaction with environment. In *Biodiversity conservation and utilization of spices, medicinal and aromatic plants*, ed. B. Sakummar, B. Kinshnamoorthy, J. Rema, P. N. Ravindran, and K. V. Peter, 116–120. ISS, IISR, Calicut, India.

Zachariah, T. J., B. Sasikumar, and P. N. Ravindran. 1993. Variability in gingerol and shogaol content of ginger accessions. *Indian Perfumer* 37 (1): 87–90.

Zheng, Y., Y. Liu, M. Ma, and K. Xu. 2008. Increasing *in vitro* microrhizome production of ginger (*Zingiber officinale* Roscoe). *Acta Physiologica Plant* DOI 10.1007/s11738-008-0149-3.

Zhenxian, Z., A. Xizhen, Z. Qi, and Z. Shi-jie. 2000. Studies on the diurnal changes of photosynthetic efficiency of ginger. *Acta Horticulturae Sinica* 27 (2): 107–111.

Color Figure 1.3 A cosmeceutical product VICCO produced from turmeric in India and marketed worldwide. This product was a gift from Shyamala Balgopal. (With permission from VICCO Laboratories; www.viccolabs.com; December 15, 2010.)

Color Figure 5.3(d) *Lycium barbarum* L. (Zhongning, Ningxia Province).

Color Figure 7.11 *Gentiana lutea.*

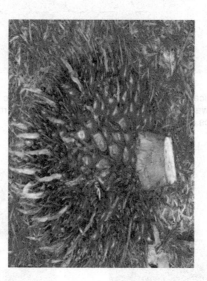

Color Figure 9.4 *Elaeis guineensis* (fruit).

Color Figure 12.2 Regeneration nursery, Pullman, Washington.

Color Figure 13.2 *Artemisia arborescens.*

Color Figure 14.2 Cultivar evaluation at the research station of the Hungarian Agricultural Quality Control Organization (Tordas) according to the new DUS evaluation method. (Köck et al. 2001. In *A mák (*Papaver somniferum *L.) Magyarország Kultúrflórája*, ed. Sárkány, S., Bernáth, J., and Tétényi, pp., 244–259. Budapest: Akadémiai Kiadó.)

Color Figure 15.3 Some important species of *Zingiber*: (a) *Z. officinale*; (b) *Z. montanum*; (c) *Z. nimmonii* (insert fruit); (d) *Z. nessanum*; (e) *Z. zerumbet*; (f) *Z. roseum* (insert inflorescence); (g) *Z. parishii*; (h) *Z. squarrosum* (insert flower); and (i) *Z. capitatum*.

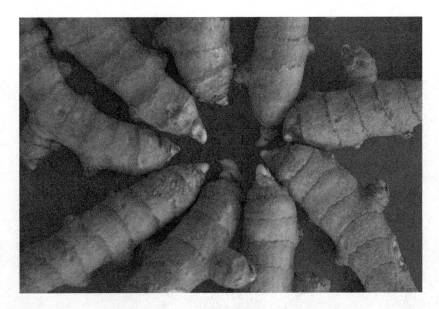

Color Figure 16.5 Rhizome bits of turmeric used as planting material.

(a)   (b)

Color Figure 17.2 (a) Ripe red berries of Korean ginseng. (b) Ripe yellow (white → yellow) berries of Korean ginseng.

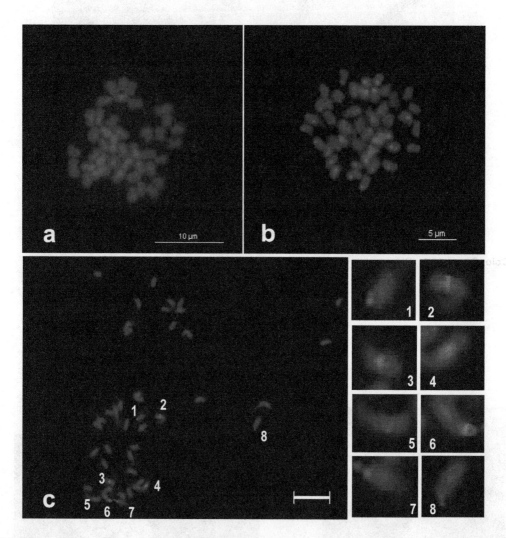

Color Figure 19.3 Coffee cytogenetics. (a), Typical metaphase spreads of *C. arabica* L.; (b) Somatic metaphase chromosomes of *C. arabica* subject to *in situ* hybridization with a repetitive BAC probe (red signals); (c) Example of GISH on metaphase chromosomes of an introgressed line of *C. arabica* L. Individual chromosomes carrying introgressed fragments from *C. liberica* species are indicated by numbers (1–8) and enlarged at the right. Total DNA from the *C. liberica* was used as probe (green signals). Chromosomes were counterstained with DAPI, in blue. Scale bar = 5 μm.

Color Figure 19.4 Example of phenotypic variation in coffee trees. (a) berry color and (b) seed shape.

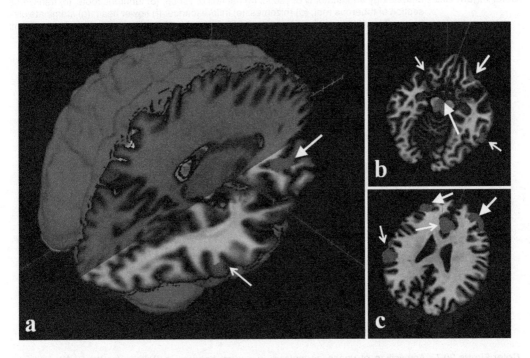

Color Figure 19.9 Brain stimulation by coffee aroma. Results of functional Magnetic Resonance Imaging (fMRI) analysis showing the effects of coffee aroma on specific neuronal centers in the brain (arrows) governing smell (a) pleasure (b) as well as attention and learning (c) See text for more explanation. Cerebral regions under activation are in yellow.

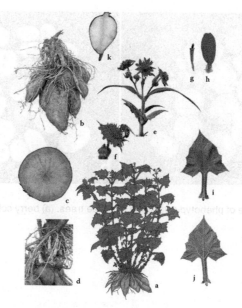

Color Figure 20.2 Morphology and anatomy of yacon: (a) habitus of yacon; (b) tuberous roots; (c) transverse section of tuberous root; (d) rhizomes; (e) inflorescence; (f) flower head; (g) staminate disk flower; (h) pistilate ray flower; (i) adaxial (upper) side of leaf; (j) abaxial (lower) side of leaf; (k) longitudinal section of tuberous root. (Photo courtesy of E. C. Fernández, 2005.)

Color Figure 20.7 Acquisition of yacon germplasm in natural habitats in Bolivia. (a) Yacon farmer in La Tranca, Chapare, Cochabamba, Bolivia; (b) European surveyor of genetic resources in La Tranca, Chapare, Cochabamba, Bolivia. (Composite photo courtesy of E. C. Fernández and I. Viehmannová.)

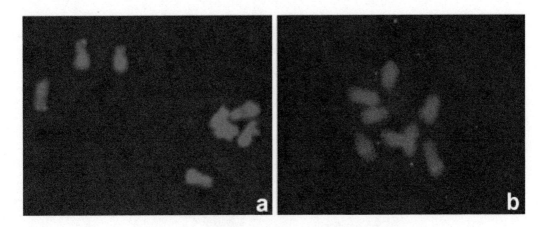

Color Figure 22.3 Localization of (a) 45S rDNA and (b) 5S rDNA on different chromosomes of *P. ovata* using FISH.

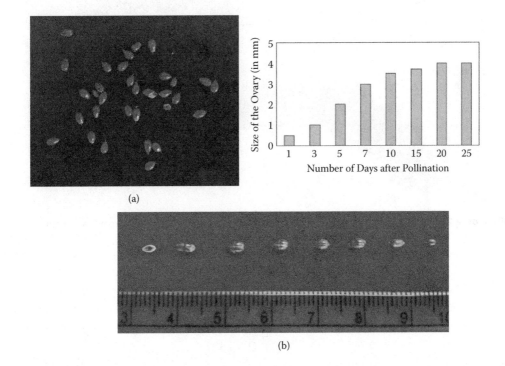

(a)

(b)

Color Figure 22.6 (a) Ovaries at different developmental stages. (b) Development of seeds from ovaries (see left to right).

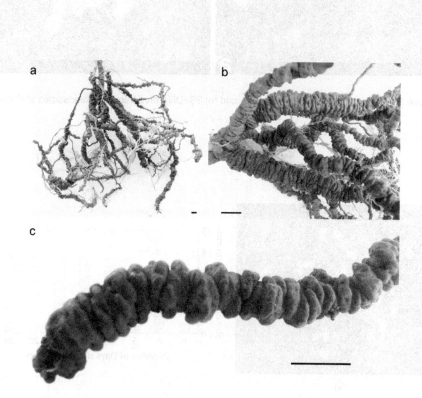

Color Figure 27.3 The root system and the characteristic rounded ridges of an ipecac root. Bar = 1 cm.

CHAPTER 16

Turmeric

K. Nirmal Babu, K. N. Shiva, M. Sabu, Minoo Divakaran, and P. N. Ravindran

CONTENTS

16.1 INTRODUCTION

Turmeric of commerce is the dried underground rhizome (Figure 16.1) of the perennial herb *Curcuma longa* L. (syn: *C. domestica* Val.) of the family Zingiberaceae. It is cultivated as an annual with the leafy shoot dying back during the dry period. Its generic name originated from the Arabic word *kurkum* meaning "yellow" and most likely refers to the deep yellow rhizome color of the true turmeric (*Curcuma longa* L.). Turmeric is traditionally used for medicinal and culinary purposes and also as a cosmetic and a natural dye. Turmeric is also credited with religious and magical rites in India and certain Southeast Asian countries. Some species of *Curcuma* are also recognized for their ornamental values by the floriculture industry.

Figure 16.1 Fresh (A) and dry (B) rhizomes of turmeric.

16.2 ECONOMIC USES OF TURMERIC AND RELATED SPECIES

Different species of turmeric are used in folk medicine, as a spice, as a vegetable in a variety of culinary preparations, pickles, and salads, in the production of arrowroot powder, and in toiletry articles. Many *Curcuma* species are highly valued as ornamentals. Turmeric oil is also now used in aromatherapy and the perfume industry. Many *Curcuma* species were recognized by local and tribal people all over Asia as valuable sources of medicine. In Ayurveda, turmeric is regarded as an aromatic stimulant, tonic, carminative, and antihelminthic. The essential oil of turmeric is antiseptic and it is used in treating gallstones and gall complaints.

Currently, all countries from India to the Philippines, China, and Japan to Indonesia use turmeric and other *Curcuma* species as a source of herbal remedies. In the recent past, modern medicine and the pharmacological industry have turned toward *Curcuma*. The traditional Indian Ayurvedic and Sidha systems of medicine have recognized the medicinal value of turmeric in its crude form since very ancient times. The last few decades have witnessed extensive research interests worldwide in the biological activity of turmeric and its compounds. Thus, *Curcuma* is now gaining importance all over the world as a cure to combat a variety of ailments because the genus carries molecules credited with anti-inflammatory, hypocholestremic, choleratic, anti-microbial, antirheumatic, antifibrotic, antivenomous, antiviral, antidiabetic, antihepatotoxic, antioxidant, and anticancerous properties as well as insect-repellent activity (Chempakam and Parthasarathy 2008).

Turmeric contains protein (6.3%), fat (5.1%), minerals (3.5%), carbohydrates (69.4%), and moisture (13.1%). The rhizomes contain curcuminoids (2.5–6%) and are responsible for the yellow color. Curcumin (diferuloylmethane) comprises curcumin I (curcumin), curcumin II (demethoxycurcumin), and curcumin III (bisdemethoxycurcumin), which are found to be natural antioxidants (Ruby et al. 1995 a,b). A new curcuminoid, cyclocurcumin, has been isolated from the nematocidally active fraction of turmeric. The fresh rhizomes also contain two new natural phenolics, which possess antioxidant and anti-inflammatory activities, and also two new pigments. The essential oil (5.8%) obtained by steam distillation of rhizomes has α-phellandrene (1%), sabinene (0.6%), cineol (1%), borneol (0.5%), zingiberene (25%), and sesquiterpines (53%) (Kapoor 1990). India is the global leader in value-added products of turmeric and exports. Value-added products from turmeric include curcuminoids, dehydrated turmeric powder, oils, and oleoresin.

Curcuminoids from turmeric and related species possess antioxidant and anti-inflammatory properties. Hence, studies on the chemical contents, structure, and composition of curcuminoids and essential oils of various species are gaining importance, and they are being extensively tested for their medicinal properties. Epidemiological observations suggest that turmeric and its constituent, curcumin, reduce certain forms of cancer and render other protective biological effects in humans (Radha et al. 2006). The curcuminoids, which are administered orally, enter the blood circulation and are present as glucuronides and glucuronide sulfate conjugate forms (Asai and Miyazawa 2000).

Curcuminoids counteract cancer, diabetes, cataractogenesis, liver diseases, and HIV. *Curcuma zanthorrhiza* has been reported for lipid reducing and sedative properties (Wichtl 1998) and to protect against hepatotoxins (Lin et al. 1995). *Curcuma zedoaria* has antiamoebic activity (Ansari and Ahmad 1991) and hepatoprotective constituents (Matsuda et al. 1998, 2001), *C. caesia* has anti-fungal and antibacterial properties (Garg and Jain 1998; Banerjee and Nigam 1976) and the Chinese species *C. wenyujin* exhibits antitumor active compounds (Shukui, Lin, and Jun 1995).

The main sources of East Indian arrowroot, an easily digestible starch used in much of Asia, including India, Bangladesh, Cambodia, and Indonesia, are *C. zedoaria*, *C. angustifolia*, and *C. aeruginosa*. This arrowroot powder is also used generally as a starch for culinary purposes, for making biscuits to use in feeding infants, and as a health drink to support convalescence in the elderly. It is also taken in the case of stomach disorders.

Some *Curcuma* species are beautiful and splendid garden plants. Their flowers are usually not very showy, but the colorful coma is quite attractive. Most species grow in clumps, serving also as good foliage plants for garden landscaping. Some species are widely cultivated for their foliage and very popular as cut flowers. Flowers of *Curcuma* species have a natural dormancy, which makes them convenient to package and ship safely (Skornickova, Rehse, and Sabu 2007).

16.3 ORIGIN AND DISTRIBUTION

The genus *Curcuma* consists of about 70–110 (true identity is unclear) species distributed chiefly in southern and southeastern Asia (Skornickova et al. 2007). In addition to *Curcuma longa*, the other economically important species of the genus are the following:

C. aromatica, which is used in medicine and in toiletry articles
C. kwangsiensis, C. ochrorhiza, C. pierreana, C. zedoaria, and C. caesia, which are used in folk medicines of the southern and southeastern Asian nations

C. alismatifolia, *C. elata*, and *C. roscoeana*, with floricultural importance

Curcuma amada, which is used as medicine and in a variety of culinary preparations, pickles, and salads

C. zedoaria, *C. pseudomontana*, *C. montana*, *C. angustifolia*, *C. rubescens*, *C. haritha*, and *C. caulina*, which are all used in manufacturing arrowroot powder

The other species of minor importance are *C. purpurescens*, *C. mangga*, *C. heyneana*, *C. zanthorrhiza*, *C. phaeocaulis*, and *C. petiolata* (Velayudhan, Asha, et al. 1999; Velayudhan, Muralidharan, et al. 1999; Shiva, Suryanarayana, and Medhi 2003; Sasikumar 2005; Skornickova et al. 2007).

The greatest diversity of the genus occurs in India, Myanmar, and Thailand and extends to Korea, China, Australia, and the South Pacific. This genus is also distributed in Cambodia, Indonesia, Malaysia, Laos, Madagascar, and the Philippines. Many species of *Curcuma* are economically valuable and different species are cultivated in China, India, Indonesia, and Thailand and throughout the tropics, including tropical regions of Africa, America, and Australia.

Genus *Curcuma* has about 42 species distributed in India, out of which *C. longa* is cultivated for turmeric, *C. aromatica* is grown for use in toiletry articles, and *C. amada* (mango ginger) is cultivated in limited areas for use as a vegetable. The country of origin of cultivated turmeric (*C. longa*) is presumed to be Southeast Asia. India is the single largest producer and exporter of turmeric in the world.

16.4 BOTANY AND SYSTEMATICS

16.4.1 Genus *Curcuma* L.

The genus *Curcuma* was established by Linnaeus in his *Species Plantarum* (1753). The generic epithet is derived from the Arabic word *karkum*, meaning yellow, and refers to the yellow color of the rhizome; *curcuma* is the Latinized version (Purseglove et al. 1981; Sirirugsa 1999). The earliest description of turmeric is found in Rheede's *Hortus Indicus Malabaricus*, under the local name *manjella kua*. He reinstated the name of *C. longa* L. and *C. domestica* Val. as its synonym. Recently, Skornickova, Sida, and Marhold (2010) studied the earlier literature and specimens and finally selected a specimen examined by Linnaeus as lectotype of *C. longa* (herb. Herman 3:5, no. 7 BM). Baker (1882) describes 27 species in The *Flora of British India*. He subdivided the genus into three sections: *Exantha*, *Mesantha*, and *Hitcheniopsis*. The section *Exantha* comprises 14 species, including *C. longa* L., *C. angustifolia* Roxb. (Indian arrow root), *C. aromatica* Salisb., and *C. zedoaria* Roscoe. Some of the important floristic studies on *Curcuma* are those of Roxburgh (1832), Hooker (1886), Valeton (1918), Gamble (1925), Holttum (1950), Mangaly and Sabu (1993), Velayudhan et al. (1999 a,b), and Skornickova et al. (2010). Some important and most common species of *Curcuma* occurring in India and neighboring countries are given in Table 16.1.

There are many species that are used by local and tribal people but full information is not available. However, the true identity of many *Curcuma* species is often unclear due to poor descriptions, lack of type specimens, and difficulties in preserving useful specimens. The appropriate scientific names rarely correspond even to some of the most commonly cultivated species. Most species are quite variable, but many of them look alike. The seed-setting species generally exhibit a range of natural variation in many characteristics that would be expected in populations. The non-seed-setting species are those for which seeds have rarely been reported that are generally consistent with clonal populations, with little variation in vegetative or reproductive characters. In addition, they may hybridize in the wild, and hybrids may get naturalized.

Some of the existing species are now being recognized as synonyms. It has now been established that the Chinese species *C. albicoma* and *C. chuanyujin* are synonyms of *C. sichuanensis* and *C. kwangsiensis*, respectively. The Chinese species *C. wenyujin* is now recognized as a

Table 16.1 Important Species of *Curcuma* and Their Distribution

Sl. No.	Name of Taxon	Distribution	Sl. No.	Name of Taxon	Distribution
1	*Curcuma aeruginosa* Roxb.	India, Thailand, Indochina, Malaysia, Indonesia, Sri Lanka, Myanmar	2	*Curcuma albiflora* Thwaites	Sri Lanka
3	*Curcuma alismatifolia* Gagnep.	Thailand, Laos, Cambodia	4	*Curcuma amada* Roxb.	India
5	*Curcuma amarissima* Roscoe	India, China	6	*Curcuma angustifolia* Roxb.	India
7	*Curcuma aromatica* Salisb.	India, China, Sri Lanka	8	*Curcuma aurantiaca* Zijp (= *Curcuma ecalcarata* Sivar. & Balach.)	India, Java, Thailand, Malaysia
9	*Curcuma australasica* Hook. f.	Australia	10	*Curcuma bhatii* (R.M. Sm.) Skornick. & M. Sabu	India
11	*Curcuma bicolor* Mood & Larsen	Thailand	12	*Curcuma burtii* K. Larsen & Smith	Thailand
13	*Curcuma caesia* Roxb.	India	14	*Curcuma cannanorensis* R. Ansari, V.J. Nair & N.C. Nair.	India
15	*Curcuma caulina* J. Graham.	India	16	*Curcuma codonantha* Skornick., M. Sabu, & Prasanthk.	India
17	*Curcuma comosa* Roxb.	Myanmar	18	*Curcuma cordata* Wall.	Myanmar
19	*Curcuma coriacea* Mangaly & M. Sabu	India	20	*Curcuma decipiens* Dalzell	India
21	*Curcuma ecomata* Craib.	Thailand	22	*Curcuma elata* Roxb.	India, Myanmar
23	*Curcuma exigua* N. Liu & S.J. Chen	China	24	*Curcuma ferruginea* Roxb.	India, Bangladesh
25	*Curcuma flaviflora* S. Q. Tong	China, Thailand	26	*Curcuma glans* K. Larsen & Mood	Thailand
27	*Curcuma gracillima* Gagnep.	Thailand	28	*Curcuma harita* Mangaly & M. Sabu	India
29	*Curcuma harmandii* Gagnep.	Thailand, Cambodia	30	*Curcuma inodora* Blatt.	India
31	*Curcuma karnatakensis* Amalraj., Velay. & Mural.	India	32	*Curcuma kudagensis* Velay, V.S. Pillai & Amalraj (*Curcuma thalakaveriensis* Velay., Amalraj & Mural.)	India
33	*Curcuma kwangsiensis* S.G. Lee & C.F. Liang	China	34	*Curcuma larsenii* C. Maknoi & T. Jenjittikul	Thailand, Laos, Vietnam

No.	Species	Distribution
35	Curcuma latifolia Roscoe	India
36	Curcuma leucorhiza Roxb.	India
37	Curcuma longa L.	Asia
38	C. mangga Valeton & Zijp	India, Malaysia, Indonesia
39	Curcuma montana Roxb.	India
40	Curcuma mutabilis Skornick., M. Sabu & Prasanthk. (= Curcuma nilamburensis Wight)	India
41	Curcuma neilgherrensis Wight	India
42	Curcuma oligantha Trimen	Sri Lanka
43	Curcuma parviflora Wall.	Thailand, Myanmar, Malaysia
44	Curcuma petiolata Roxb.	Myanmar
45	Curcuma picta Roxb. ex. Skornick.	India, Thailand, Sri Lanka, Peninsular Malaysia
46	Curcuma prakasha S. Tripathi	India
47	Curcuma pseudomontana J. Graham	India
48	Curcuma reclinata Roxb. (= Curcuma sulcata Roxb.)	India
49	Curcuma rhabdota Siriirugsa & M.F. Newman	Thailand, Laos
50	Curcuma rhomba Mood & K. larsen	Vietnam, Thailand
51	Curcuma roscoeana wall.	India, Bangladesh, Myanmar, Thailand
52	Curcuma rubescens Roxb.	India, Thailand, Bangladesh
53	Curcuma rubrobracteata Skornick., M. Sabu & Prasanthk.	India, Myanmar
54	Curcuma scaposa (Nimmo) Skornick., M. Sabu	India
55	Curcuma sessilis Gage	Myanmar
56	Curcuma sichuanensis X. X. Chen	China
57	Curcuma sparganifolia Gagnep.	Indochina, Cambodia, Thailand
58	Curcuma strobilifera Wall. ex. Baker	Myanmar
59	Curcuma thorelii Gagnep.	Thailand
60	Curcuma vamana M. Sabu & Mangaly	India
61	Curcuma viridiflora Roxb.	Indonesia, Malaysia, Thailand, China, Sumatra
62	Curcuma yunnanensis N. Liu & S.J. Chen	China
63	Curcuma zanthorrhiza Roxb.	India, Java, Peninsular Malaysia, Vietnam, China, Thailand, the Philippines
64	Curcuma zedoaria (Christm.) Roscoe (= Curcuma raktakanta Mangaly & M. Sabu; C. malabarica Velay. et al.)	India, Myanmar, Thailand, Malaysia

synonym of *C. aromatica*; *C. phaeocaulis* was misidentified in the past as *C. zedoaria*, *C. caesia*, and *C. aeruginosa*. *Curcuma kwangsiensis* var. *puberula* and var. *affinis* are not accepted and the identity of the Taiwan species *C. viridiflora* remains undecided. However, new species, such as *C. rhabdota* from Southeast Asia and *C. prakasha*, *C. codonantha*, *C. coriacea*, *C. haritha*, *C. karnatakensis*, *C. kudagensis*, *C. mutabilis*, *C. picta*, *C. rubrobracteata*, and *C. vamana* from India and *C. bicolor*, *C. glans*, and *C. rhomba* from Thailand are reported (Liu and Wu 1999; Sirirugsa and Newman 2000; Tripathi 2001; Mood and Larsen 2001; Pillai et al. 1990; Mangaly and Sabu 1989, 1993; Sabu and Mangaly 1988; Skornickova, Sabu, and Prasanth Kumar 2003a, 2003b, 2004; Skornickova et al. 2008). Several new combinations and typifications in *Curcuma* have been made: *C. bhatii*, *C. caulina*, and *C. scaposa* (Skornickova and Sabu 2005 a,b, 2007; Skornickova et al. 2010). Investigations on a taxonomic revision of the genus *Curcuma* is in progress to reinvestigate identities of individual species, to provide tools for their identification, and to ensure correct usage of the scientific names (Skornickova et al. 2007).

16.4.2 *Curcuma longa* L.

Turmeric (*C. longa*) is an erect perennial herb with an underground rhizome, but it is grown as an annual under cultivation. The leaves are distichous, obliquely erect, subsessile, oblong lanceolate, acuminate, and bright green to deep green above and usually paler green beneath, with long leaf sheaths forming a pseudostem (Figures 16.2a and 16.2b). Petioles of the outermost leaf are short or none, of inner leaves fairly long, channeled. The underground stem or rhizome is fleshy at the base of each aerial shoot, consisting of an erect, ovoid or ellipsoid structure (mother rhizome), ringed with the bases of old scale leaves bearing several horizontal or curved rhizomes (primary fingers), which are again (secondary fingers) and again (tertiary fingers) branched; the whole forms a dense clump. The rhizomes show a deep orange-colored inner core and yellowish-orange on the outer side and are rich in curcumin. Roots are fleshy; some of them bear ellipsoid tubers. Root tubers are

Figure 16.2(a) Line diagram depicting turmeric plant. (Source: Ravindran, P. N. et al. 2007.) In *Turmeric— The genus* Curcuma, ed. Ravindran, P. N., Nirmal Babu, K., and Sivaraman, K., 15–70. Boca Raton, FL: CRC Press.

Figure 16.2(b)   Turmeric (*C. longa*) plant showing inflorescence and rhizomes.

present in all *Curcuma* species, placed on the ends of roots, and their function is exclusively for the purpose of sustaining the plant during its dormant period when the leafy shoot dries up, storing energy for the next season.

The inflorescence is terminal and borne in between the leaf sheaths. Bracts are large and broad and each one is joined to those adjacent to it for about half of its length. The basal parts thus form enclosed pockets, the free ends more or less spreading, forming a cylindrical spike. The uppermost bracts, called coma, are usually larger than the rest, differently colored, and a few of them sterile. Coma bracts are white or white streaked with green, grading to light green bracts lower down (Figure 16.3a); bracts are adnate for less than half their length, elliptic, lanceolate. Flowers are pale yellow in color, in cincinni of two to seven in the axil of a bract. Bracteoles are thin, elliptic with the sides inflexed, each one at right angles to the last, quite enclosing the buds. The calyx is minutely pubescent, truncate, short, and tubular and divides above into three short teeth. The corolla tube is tubular at the base and cup shaped in the upper half, with corolla lobes at the distal end and the lip, and staminodes and stamen just above them. Corolla lobes are thin, translucent white, hooded, and ending in a hollow hairy point. The single epipetalous stamen arises from the throat of the corolla tube.

The flower of turmeric contains two whorls of three stamens; one becomes fertile and the rest become staminodes. The outer suppressed whorl is represented by two staminodes. Of the inner whorl, the posterior one is fertile, and the other two are sterile, petaloid, and fused together to form the labellum, the most conspicuous part of the flower. The labellum is obovate and creamy white, with a yellow median band on the lip. The filament of the stamen is short and broad, and constricted at the top. The anther is versatile and dorsifixed; the pollen sacs are parallel, with usually a curved spur at the base of each. A connective is sometimes produced at the apex into a small crest. The fertile stamen contains two anther lobes with broad connectives; spurs are very large, broad, and diverging, a little curved with the thin apex always recurved outward. The style is long and filiform and lies in a channel along the corolla tube and fertile stamen; the stigma is bilipped. There are two epigynous glands the inferior ovary is tricarpellary and syncarpous. There are many ovules on the

Figure 16.3 Flowering nature of some important species of *Curcuma:* (A) *C. longa*, (B) *C. aromatica*, (C) *C. haritha*, (D) *C. ecalcarata*, (E) *C. pseudomontana*, (F) *C. zedoaria*.

axile placenta. Fruiting is rare; fruit is ellipsoid, thin walled, dehiscing, and liberating the seeds in the mucilage of the bract pouch. Seeds are ellipsoid with a lacerate aril of few segments, which are free to the base (Holttum 1950; Nazeem et al. 1994).

16.4.3 Other Important *Curcuma* Species

There are many species that are used by local and tribal people but complete information is lacking. The seed-setting species generally exhibit a range of natural variation in many characters that would be expected in populations. The non-seed-setting species have rarely been reported and are generally consistent with clonal populations and with little variation in vegetative or reproductive characters. Skornickova et al. (2007) worked on a taxonomic revision of the genus *Curcuma* to reinvestigate identities of individual species, to provide tools for their identification, and to ensure correct usage of the scientific names.

16.4.3.1 Curcuma amada Roxb.

This species is known as mango ginger. The characteristic taste of the rhizome resembles green mango; this species resembles *C. longa* in its external morphology, but its rhizome is creamy yellow. It is similar to *C. mangga* of Indonesia by virtue of the mango smell of the rhizome. Mango ginger is native to eastern India and cultivated in many parts of India. Its leaves are plain green and glabrous on both sides. It is a late-flowering species with greenish-white inflorescence from the center of the leaves.

The rhizomes are considered cooling and are widely used as a vegetable, in pickles, and as a spice for adding a ginger-cum-mango flavor. They are used in traditional medicine as well as in local and tribal systems for various ailments such as prurigo nodularis and rheumatism (Watt 1889), as an expectorant, and as an astringent; it useful in treatment of diarrhea, as a carminative, and to promote digestion. The ether extract of *C. amada* lowered the cholesterol level in rabbits (Pachauri and Mukherjee 1970) and the crude extract from rhizomes of *C. amada* showed strong antifungal activity against *Trichophyton rubrum*, which causes skin infections in humans, and against *Aspergillus niger* (Gupta and Banerjee 1972). An improved cultivar, Amba, has been developed at the high-altitude research station at Pottangi, India.

16.4.3.2 *Curcuma aeruginosa* Roxb.

This species, distributed from India and the Andaman Islands to Cambodia, Malaysia, and Java, is one of the oldest named species of the genus. The specific epithet is derived from the striking greenish-blue color of the rhizome. It is quite a large and stately plant and can reach almost 2 m in height. The plant is native to Myanmar. The leaves are glabrous on both sides and have a feather-shaped dark maroon patch in the distal part of the lamina, along the midrib on the upper side. Inflorescences appear at the beginning of the rainy season, just after the leaves begin to sprout. In India, it is one of the most esteemed species for extracting arrowroot since it is believed that species with the blue-colored rhizome are of the best value. In Malaysia, it is used medicinally for asthma, cough, and as a paste with coconut oil for dandruff; in Indonesia, it is used as a purgative during childbirth. Several studies have investigated the chemical composition of *C. aeruginosa* from different parts of Asia—namely, India (Jirovetz et al. 2000), Malaysia (Sirat, Jamail, and Hussain 1998), and Vietnam (Tuyet, Dûng, and Leclercq 1995). The major chemical constituents are curzerenone and 1,8-cineole, while camphor, furanogermenone, curcumenol, isocurcumenol, and zedoarol compounds are present in lesser quantities.

16.4.3.3 *Curcuma alismatifolia* Gagnep.

This species is known as Siam tulip or tulip ginger. The main attraction of this beautiful plant is its inflorescence, where the coma bracts are larger and spreading, much like a tulip. It is one of the most utilized ornamental species of *Curcuma*. *Curcuma alismatifolia* is the mainstay of the Thai ginger horticultural industry, mass produced via tissue culture. It is exported to countries all over the world; it is used for mass plantings in landscapes and as a home pot plant, and has a long-lasting cut flower. There are countless cultivars, the coma varying anywhere between white and dark pink, with streaks or patches of coloration. Chiangmai Pink, Chiangmai Ruby, Tropic Snow, Thai Beauty, and Thai Supreme are some of the popular cultivars.

16.4.3.4 *Curcuma amarissima* Roscoe

This species is a stately species with a large, deep yellow colored rhizome with a circle of a bluish-green shade. Its specific epithet comes from its extremely bitter rhizome. The strikingly red-brown petioles and whitish coma with deep violet tips make this a species with great ornamental potential.

16.4.3.5 *Curcuma angustifolia* Roxb.

This species is one of the sources for East Indian arrowroot. It has narrow, green, glabrous leaves. Flowers appear first in the beginning of the rainy season before the leaves are fully developed. The inflorescences are small with yellow flowers and pink coma bracts near the ground. It is cultivated in small stretches for extraction of starch (Watt 1889) and used in the preparation of sweetmeats by the hill tribes.

16.4.3.6 *Curcuma aromatica* Salisb.

The specific epithet of this species is derived from the camphoraceous aroma of the rhizome. The mother rhizome is large, yellow within, and aromatic; there are many yellow to pale yellow sessile tubers. Inflorescence is lateral (Figure 16.3b) and produced early in the season. Coma bracts are large and pink with orbicular deep yellow labellum. The plant is cultivated widely in tropical Asia and used extensively as a cosmetic and as herbal medicine. It is known to be anti-inflammatory (Jangde, Phadnaik, and Bisen 1998), a stimulant, a tonic, and a carminative, and it is useful for curing leucoderma and blood diseases.

Analysis of the essential oils of *C. aromatica* has revealed variation among different collections (Kojima et al. 1998; Bordoloi, Sperkova, and Leclercq 1999; Skornickova et al. 2007). The Japanese samples contained curdione, germacrone, 1,8-cineole, (4S, 5S)-germacrone-4,5-epoxide, β-elemene, and linalool as the major components; those in the oil from Indian samples contained β-curcumene, arcurcumene, xanthorrhizol, germacrone, camphor, curzerenone, and, in some cases, α-turmerone, ar-turmerone, and 1,8-cineole. It is likely that the samples examined represent different species as there is still a lot of ambiguity in correct identification of this species. *Curcuma aromatica* is often confused with *C. zanthorrhiza*, which is a common species in the western part of South India (Skornickova and Sabu 2005b).

16.4.3.7 *Curcuma aurantiaca* Zijp

This species is native to Thailand and Java and is commonly known as the rainbow *Curcuma*. It is one of the most variable species, with high ornamental potential. *Curcuma aurantiaca* (syn: *Curcuma ecalcarata* Sivar. & Balach.) (Figure 16.3c) is a medium sized plant with pretty, sulcate leaves. The incredibly colorful inflorescence is borne from the center of the leaves. The color of inflorescence is from green to pink, orange with a red tinge to deep reddish brown. The coma is white to a deep pink. The anther is without spurs. Its specific epithet comes from the unusually striking orange color of its flowers.

16.4.3.8 *Curcuma australasica* Hook. f.

This species is native to Australia, contains attractive flowers, and is suitable for gardens as well as a cut flower. Aussie Plume is a popular horticultural variety.

16.4.3.9 *Curcuma bicolor* Mood and K. Larsen

This species has been recently described and is wild in Thailand. It is one of the most floriferous and showy curcumas in cultivation, with a small but strikingly colored basal red inflorescence. The flowering period can last for a couple of months; the flowers are attractively red-orange bicolored (Mood and Larsen 2001). One horticultural variety of this species is popularly known as Candy Corn in the United States.

16.4.3.10 *Curcuma caesia* Roxb.

This species is commonly called black turmeric because it produces a deeper blue rhizome and is known from Bengal, northeastern India, and Bangladesh. The leaves have a deep red-violet patch along the midrib. It is used extensively in tribal medicine against a variety of illnesses and especially to cure blood dysentery and rheumatic pain.

16.4.3.11 *Curcuma cordata* Wall.

This species is one of the most common plants utilized in the Thai horticultural trade. It is known as the pastel hidden ginger; terminal inflorescence is large and cylindrical, with a strongly pastel pink coma. The lower bracts are broadly rounded, waxy, and white to green, often rimmed with the pink color of the coma. It is prized for its long-lasting inflorescence as a cut flower. The flowering stem is seemingly resistant to many of the fungal pathogens of other horticultural *Curcuma* species.

16.4.3.12 *Curcuma elata* Roxb.

This species, native to Myanmar, is a tall plant that is ideal for landscaping. Its main attraction is the large spike with a deep rosy or even crimson coma, which comes out before the leaves appear. *Curcuma elata* is often called giant plume ginger in the horticultural trade.

16.4.3.13 *Curcuma glans* K. Larsen and Mood

This is a highly attractive species from Thailand. The plant is of medium size, with a clumping growth habit. It makes multiple pseudostems with about three shiny dark green leaves, which are silvery hirsute beneath. Unlike other *Curcuma* species, the flowers open in the late afternoon and will last the night until the next afternoon. The inflorescence is white with occasional red markings, and the flowers are a beautiful white and yellow with lavender glands, presumably used for nocturnal insect pollination (Mood and Larsen 2001).

16.4.3.14 *Curcuma harmandii* Gagnep.

The inflorescence of this species is very attractive, with green, long, linear bracts and delicate white flowers with a pink- and yellow-dotted labellum resembling that of an orchid. It has great potential in the cut-flower industry. One horticultural variety, Jade Pagoda, is available.

16.4.3.15 *Curcuma latifolia* Roscoe

This is one of the largest species in the genus, reaching over 3 m. The showy spikes are with a huge and bright pink coma. It differs from *C. elata* in having a red patch on the upper side of the leaf.

16.4.3.16 *Curcuma mangga* Valeton and Zijp

This species is found in Java. The rhizome has the smell of unripe mango, similar to *C. amada*. It is an early, laterally flowering species. It is used for curing fevers and for abdominal problems in the Malay Peninsula and Indonesia (Perry 1980).

16.4.3.17 *Curcuma montana* Roxb.

This species is native to the eastern ghat region of India. It superficially resembles *C. longa* and *C. amada* in producing only a central inflorescence. The spike is greenish white to white and the coma is usually also white or with a slight pink tinge. Its rhizome is, however, light yellow and the does not have any strong smell. It is used for extraction of East Indian arrowroot by the tribals.

16.4.3.18 Curcuma parviflora Wall.

This species occurs throughout Myanmar and Thailand. It is a small green plant with white and violet flowers and is popular as a pot plant. It is often confused with *Curcuma thorelii*, but this species has white lateral staminodes. The inflorescence is eaten as a vegetable and the pulp of the rhizome is applied to cuts (Sirirugsa 1999).

16.4.3.19 Curcuma petiolata Roxb.

This species is often grown for the foliage in landscaping. The leaves are narrowly elliptic and variegated with thick cream bands at the leaf margin in some cultivars. The terminal inflorescence is shortly pedunculate and has a white coma tinged with pink at the tips.

16.4.3.20 Curcuma pseudomontana J. Graham

This species is grown only in Maharashtra, India. It is a highly variable seed-setting species with a beautiful, well-developed coma and deep yellow flowers (Figure 16.3e). The leaves are broadly ovate and prominently sulcate, with a bright green color with ornamental value. The inflorescence is produced first laterally and later again from the center of the leaves. The tubers are big and sometimes eaten.

16.4.3.21 Curcuma rhabdota Sirirugsa and M.F. Newman

This species is a small plant native to Laos that is popular in Thailand for its striking inflorescence. The fertile bracts are white or greenish, beautifully striped with a deep red-brownish color. The coma bracts vary from white to pink with brown-red streaking. It is often known in the U.S. horticultural trade as Candy Cane. New hybrids have been produced between *C. rhabdota* and *C. alismatifolia* (Sirirugsa and Newman 2000).

16.4.3.22 Curcuma rhomba Mood and K. Larsen

This is a highly attractive species from Vietnam and Thailand. The diamond shape (rhomboid) name is derived from its lateral staminodes. The inflorescence is terminal and pushes the leaves of the pseudostem apart. The inflorescence consists of red fertile bracts with no coma and has deep orange flowers. The commercial name of this species is sri pak or sri pok (Mood and Larsen 2001).

16.4.3.23 Curcuma roscoeana Wall.

This species is known as the Jewel of Burma or Pride of Burma. This is one of the most beautiful and ornamental among *Curcuma* species. It occurs wild in Thailand and the Andaman Islands, India (Skornickova and Sabu 2005a). This species can be easily recognized by its bright orange to scarlet spike with no obvious coma, and especially by the peculiar arrangement of the bracts in three, four, or five serial rows. Flowers are usually a creamy color with a yellow center on the labellum. The pseudostem and petioles are often shaded with a dark reddish tinge; leaves are petiolate and cordate at the base. It is suitable for a garden as well as a cut flower because of the long-lasting nature of the inflorescence and is an important item in the horticultural trade of Thai, European, and U.S. markets.

16.4.3.24 Curcuma rubescens Roxb.

This species is found in Bengal, the northeastern region of India, and Bangladesh; it has beautiful and large foliage and is good for landscaping. The leaves are deep green while the petioles and midribs are of a deep red-wine color. The inflorescence is small and not attractive. *Curcuma rubescens* is one of the many species known to be used for extraction of East Indian arrowroot (Roxburgh 1832; Dymock, Warden, and Hooper 1893).

16.4.3.25 *Curcuma sparganifolia* Gagnep.

This species contains narrow leaves and occurs wild in Cambodia and Thailand. It is suitable for the garden as well as for planting in pots. A few horticultural varieties are available. The inflorescence is small, globose, and colored, with a few dark-tipped bright pink bracts on long stalks suitable for the cut-flower industry.

16.4.3.26 *Curcuma thorelii* Gagnep.

This species is a pretty, small to midsized plant with attractive inflorescence. The fertile bracts are green and arranged in definite rows; the coma is pure white, sometimes streaked by green. It often is confused with *C. roscoeana* and *C. gracillima*.

16.4.3.27 *Curcuma zanthorrhiza* Roxb.

This species is known as Temu lawak, Temu lawas, and Javanese turmeric and is also widely cultivated in Malaysia and Indonesia. *Curcuma zanthorrhiza* is one of the most extensively utilized species of the genus *Curcuma*. It has also been reported from Thailand, Vietnam, the Philippines, South India, and China. This species is often confused with *C. aromatica* and *C. zedoaria* (Skornickova and Sabu 2005b). Leaf blades have a dark purple stripe on either side of the midrib. It is an early flowering species and produces a beautiful lateral inflorescence. Coma bracts are purple, flowering bracts light green, and bracteoles up to 25 mm long. Flowers are as long as bracts. Corolla lobes are light red, staminodes whitish, lip yellowish with deeper median bands, and anthers short and broad. The primary rhizome is large; rhizomes are few and rather short, thick, with few branches, externally pale orange, internally deep orange or orange-red with large root tubers.

This species got its specific epithet from the deep yellow-orange color of its rhizome, which is medicinally valued, especially in Malaysia and Indonesia. The rhizomes are widely used in traditional and local medicines and in cosmetics to cure skin diseases, diseases of the digestive and urinary systems, and gallstones; as an antidiabetic; and for treatment of liver injuries (Prana 1977, 1978; Lin et al. 1995; Yasni, Imaizumi, and Sugano 1991). It is also used as a dye in textiles. The plant is rich in sesquiterpenes such as bisacurol, bisacurone, ar-curcumene, turmerones, xanthorrhizol, and zingiberene. Curcuminoids present to the extent of 1–2% consist mainly of curcumin and its derivatives (Bruneton 1999).

16.4.3.28 *Curcuma zedoaria* (Christm.) Roscoe

This species is native to and found in the wild as well as in cultivation in India and Bangladesh, especially in low-lying wastelands, and is reported from many other Asian countries. However, upon historical and taxonomic investigation based on flowering material from different parts of Asia, it is very clear that the name *C. zedoaria* is applied to several superficially similar species. All these share the presence of a nice pink coma and leaves with a red patch over the midrib on the upper side of the leaf (Figure 16.3f), which is conspicuous especially in the young leaves (Skornickova and Sabu 2005b; Skornickova et al. 2007). The lectotype of *C. zedoaria* as established by Burtt (1977) is a species described by Rheede (1692) as *kua*.

Curcuma zedoaria has a large, broadly ovoid primary rhizome with many secondary branches often curved upward. The rhizome is pale sulfur yellow to bright yellow inside, turning brownish when old; the taste is strong and bitter. There are many fleshy roots bearing tubers. Leaves have a long petiole and young leaves have purplish flush on both sides. Inflorescence is lateral, and coma bracts are purple or dark pink. A fertile bract is ovate, green with pink margin.

The rhizome is used in traditional and tribal medicines. It is used as a stomachic and is applied to bruises and sprains. It is used in alleviating colds and is mixed with lime and used in cases of scorpion bite. The fresh rhizome arrests inflammation of the intestines and purges the kidneys and blood, and cures gonorrhea; the juice of the leaves is a moderate laxative. Starch produced from this species is one of the most esteemed in India. Starch extracted from the rhizome as arrowroot is used for making cakes and as a substitute for barley starch given to children. In Bangladesh, a decoction is used to cure painful bowel movements and is known as a diuretic and a stimulant, and to have carminative and antihelmintic properties. It is also applied to bruises and sprains.

The medicinal uses of *C. zedoaria* have been summarized from various sources by Duke (2003) as an antipyretic, aromatic, carminative, demulcent, expectorant, stomachic, stimulant, and tonic. Fresh rhizomes have diuretic properties and are used in checking leucorrhea and gonorrheal discharge and for purifying blood. Rhizomes are chewed to alleviate coughs and clear the throat. Rhizomes are used in medicines given to women after childbirth. A paste of the rhizome mixed with alum is applied to sprains and bruises. It is also used in dermatitis, sprains, ulcer, and wounds in Asian countries.

Zedoary is used in the manufacture of liquors, various essences, bitters, cosmetics, and perfumes. Garg et al. (2005) reported the chemical composition of the leaf oil of *C. zedoaria* by GC-MS (gas chromatography-mass spectrometry) and identified 23 compounds. The major ones are α-terpenyl acetate (8.4%), iso-oborneol (7%), dihydrocurdione (9%), and selino-4(15),7(11)-dien-8-one (9.4%). Matsuda et al. (1998) analyzed the sesquiterpenes from the rhizome of *C. zedoaria* and their action on mice with induced acute liver injury. The principal sesquiterpenoids identified were furanodiene, germacrone, curdione, neocurdione, curcuminol, isocurcuminol, acrugideol, zedoarondiol, curcumenone, and curcumin.

Curcumol and curdione are regarded as anticancer agents, especially for cervical cancer and lymphosarcoma (Yoshioka et al. 1998; Bown 2001; Duke 2003). Hong et al. (2002) reported that the sesquiterpenoids β-turmerone and ar-turmerone inhibited lipopolysaccharide (LPS)-induced prostaglandin E2 production in cultured mouse macrophage cells in a dose-dependent manner. Kim, Shin, and Jun (2001) demonstrated strong dose-dependent lysosomal enzyme activity in the polysaccharides from the rhizomes. They hypothesized that *C. zedoaria* has macrophage-stimulating activity and that it can be used as a biological response modifier.

16.4.4 Molecular Taxonomy

With the advancement of molecular biology, molecular data are increasingly used to understand the phylogeny and interrelationships among species in Zingiberaceae. Kress, Prince, and Williams (2002) studied the phylogeny of the gingers (Zingiberaceae) based on DNA sequences of the nuclear internal transcribed spacer (ITS) and plastid *mat*K regions and suggested a new classification. Their studies suggest that at least some of the morphological traits based on which the gingers are classified are homoplasious and three of the tribes are paraphyletic. The African genus *Siphonochilus* and Bornean genus *Tamijia* are basal clades. The former Alpinieae and Hedychieae for the most part are monophyletic taxa with the Globbeae and Zingiberaceae.

Kress et al. (2002) proposed a new classification of the Zingiberaceae that recognizes four subfamilies and four tribes: Siphonochiloideae (Siphonochileae), Tamijioideae (Tamijieae), Alpinioideae (Alpinieae, Riedelieae), and Zingiberoideae (Zingibereae, Globbeae). A phylogenetic analysis was performed by Ngamriabsakul, Newman, and Cronk (2003) using DNA sequence analysis data nuclear ribosomal DNA (ITS1, 5.8S and ITS2) and chloroplast DNA (*trn*L [UAA] 5[prime prime or minute] exon to *trn*F [GAA]) to have a better understanding of the phylogeny and relationships in the tribe Zingibereae (Zingiberaceae). The study indicated that the tribe Zingibereae is monophyletic and the species is separated into two major clades: the *Curcuma* clade and the *Hedychium* clade.

16.5 CYTOLOGY

Cytological studies on *Curcuma* have been limited to determining chromosome numbers. The first report of diploid chromosome number of turmeric was $2n = 64$ (Sugiura 1931, 1936). The other reports of somatic chromosome numbers of *Curcuma* species are $2n = 42$ for *C. amada*, $2n = 42$ for *C. aromatica*, $2n = 62$ for *C. longa* (Raghavan and Venkatasubban 1943), and $2n = 64$ for *C. zedoaria* and *C. petiolata* (Venkatasubban 1946). An unusual number of $2n = 34$ was reported by Sato (1948). A detailed study on cytology of 48 taxa of Zingiberaceae was done by Chakravorti (1948), who reported somatic chromosome numbers of *C. amada* ($2n = 42$), *C. aromatica* ($2n = 42$), *C. angustifolia* ($2n = 42$), *C. longa* ($2n = 62, 63, 64$), and *C. zedoaria* ($2n = 63$, 64). He observed variation in chromosome numbers ranging from $2n = 62$ to 64 among seven turmeric cultivars.

Sharma and Bhattacharya (1959) reported a somatic chromosome number of $2n = 42$ for *C. amada* and $2n = 62$ for *C. longa*. Sato (1960) carried out karyotype analyses in Zingiberales and reported $2n = 32$ for *C. longa* and suggested that the species could be an allotetraploid with a basic number of $x = 8$. According to him, karyologically, one pair of A chromosomes has median constriction, five pairs of A chromosomes have submedian constriction, six pairs of B chromosomes have submedian and subterminal constrictions, and four pairs of C chromosomes have submedian and subterminal constrictions. One pair each of A and C has satellites. Table 16.2 shows the reported chromosome numbers of *Curcuma* species.

However, Ramachandran (1961, 1969) conducted a detailed study on the chromosome numbers in Zingiberaceae that included six species of *Curcuma* at both mitotic and meiotic phases. During meiosis, regular formation of bivalents in *C. decipiens* ($2n = 42$) and a high percentage of trivalent association in *C. longa* ($2n = 63$) were observed. This study confirmed that *C. longa* has a somatic chromosome number of $2n = 63$. He suggested that turmeric is a triploid and might have evolved as a hybrid between tetraploid *C. aromatica* ($2n = 84$) and an ancestral diploid *C. longa* ($2n = 42$) type or that one of these has evolved from the other by mutational steps, represented by the intermediate type that is known to occur. He also suggested that the basic chromosome number of 21 might have been derived by dibasic amphidiploidy by combination of lower basic numbers of 9 and 12 found in some genera in the family or by secondary polyploidy. The sterility in *C. longa* has been attributed to the triploidy. The herbaceous perennial habit of this species and the vegetative mode of propagation favored the perpetuation of polyploidy.

Nambiar (1979) studied the cytology of five *C. longa* types, six *C. aromatica* types, and one *C. amada* type. He found $2n = 63$ in all *C. longa* types, $2n = 84$ in *C. aromatica* types, and $2n = 42$ in *C. amada* type. However, he also reported that a certain percentage of cells exhibited varying numbers of chromosomes. For example, 7.0% of cells showed $2n = 62$ in *C. longa* 'Mydukur,' and 4.0% of cells showed $2n = 62$ in *C. longa* coll. no. 24. Among *C. aromatica*, 19.5% of the cells showed $2n = 82$ and 86 in *C. aromatica* 'Kasturi,' 14.0% of the cells showed $2n = 86$ in *C. aromatica* 'Kasturi Tanuku,' and 26.0% of the cells showed $2n = 86$ in *C. aromatica* 'Dahgi.'

R. Joseph, T. Joseph, and Jose (1999) studied the cytology of six species of *Curcuma* (*C. aeruginosa*, *C. caesia*, *C. comosa*, *C. haritha*, *C. malabarica*, and *C. raktakanta*). Three species (*C. aeruginosa*, *C. caesia*, and *C. raktacanta*) possessed $2n = 63$ chromosomes (triploids), while the

Table 16.2   Chromosome Numbers in the genus *Curcuma*

Name	Chromosome No.		References
	(*n*)	(2*n*)	
Curcuma amada Roxb.		42	Chakravorti (1948)
		42	Sharma and Bhattacharya (1959); Ramachandran (1961, 1969)
		42	
C. anguistifolia Roxb.		42	Chakravorti (1948)
		42	Sharma and Bhattacharya (1959)
C. aerugenosa Roxb.		63	Joseph et al. (1999)
C. caesia Roxb.		63	Joseph et al. (1999)
C. comosa Roxb.		42	Joseph et al. (1999)
C. decipiens Dalz.	21	42	Ramachandran (1961, 1969)
C. haritha Mangaly & M. Sabu		42	Joseph et al. (1999)
C. malabarica Velayudhan		42	Joseph et al. (1999)
C. neilgherrensis Wight		42	Ramachandran (1961, 1969)
C. parviflora Wall.		32	Weerapakdee and Krasaechai (1997)
C. petiolata Roxb.		64	Venkatasubban (1946)
C. raktacanta Mangaly & M. Sabu		63	Joseph et al. (1999)
C. rhabdota Sirirugsa and M.F. Newman		24	Eksomtramage et al. (2002)
C. aff. *oligantha* Trimen		42	Eksomtramage et al. (2002)
C. zeodaria Roscoe		64	Venkatasubban (1946); Chakravorti (1948)
		63, 64	Ramachandran (1961, 1969)
		63	
C. aromatica Salisb		42	Raghavan and Venkatasubban (1943)
		63, 86	Ramachandran (1961, 1969)
G.L. Puram type		63	Ramachandran (1961, 1969)
Polavaram		86	Ramachandran (1961, 1969)
Kasturi Amalapuram		86	Ramachandran (1961, 1969)
C. longa L.		64	Sugiura (1931, 1936)
		62	Raghavan and Venkatasubban (1943)
		32	Sato (1948)
		62, 63, 64	Chakravorti (1948)
		63	Ramachandran (1961, 1969)
Duggirala type		63	Ramachandran (1961)
Kovvur Desavali		63	Ramachandran (1961)
Tekurpeta		63	Ramachandran (1961)
G.L. Puram 2		63	Ramachandran (1961)
Kasturi Duggirala		63	Ramachandran (1961)
Nallakatla pasupu		63	Ramachandran (1961)
C. longa L.			Nair and Sasikumar (2009)
22 Collections		63	Nair and Sasikumar (2009)
1 Collection		61	Nair and Sasikumar (2009)
1 Collection		86	Nair and Sasikumar (2009)
C. longa L.		84, 63–86	Nair and Sasikumar (2009)
28 Open-pollinated seedling progenies			

others were diploids ($2n = 42$). The karyotypes were symmetrical in all the studied species. The chromosomes fall into three categories:

- Type A: comparatively short chromosomes with primary and secondary constrictions—one median and the other subterminal (size: 0.99–0.7 µm)
- Type B: short chromosomes with median and submedian primary constriction (size: 0.95–0.5 µm)
- Type C: very small chromosomes with median to submedian primary constriction (size: 0.49–0.24 µm)

Joseph et al. felt that *Curcuma* is karyologically more advanced than other members of the Zingiberaceae.

Meiosis exhibited varying degrees of chromosome abnormalities and chromosome associations. Quadrivalents, trivalents, bivalents, and univalents were recorded, but their relative frequency varied among cultivars. However, in *C. aromatica* the meiosis was relatively normal. Irrespective of multivalent formation the separation of chromosomes at anaphase-I was normal in 85.7% of cells in the cultivar Kasturi ($2n = 84$), 46.7% of cells in the cultivar Uadayagiri ($2n = 84$), and 66.7% of cells in the cultivar Kasturi Tanuku ($2n = 84$). These abnormalities in meiosis led to lower pollen fertility in *C. longa* types than in *C. aromatica* types. Pollen fertility varied among *C. longa* and *C. aromatica* types. Pollen fertility of *C. longa* types Kuchipudi, Nandyal, and collection no. 24 was 45.7, 46.4, and 48.5%, respectively. Pollen fertility of *C. aromatica* types Kasturi, Udayagiri, Katergia, Dahgi, and Amalapuram was much higher (74.5, 70.0, 73.6, 68.6, and 70.0%, respectively).

Based on this information, Nambiar (1979) concluded that turmeric is a natural hybrid between two species having $2n = 42$ and $2n = 84$ chromosomes. Though turmeric is relatively sterile, the chromosome pairing in meiosis is nearly normal, with few trivalents and univalents. In a triploid plant, meiosis is very irregular, with a high frequency of trivalents, univalents, bridge, and fragments. The presence of near-normal meiosis was explained by assuming that the putative parents with $2n = 42$ and $2n = 84$ were evolved from parents having secondary basic number of $x = 21$ and a primary basic number of $x = 9$ and $x = 12$; this supported the triploid origin of turmeric.

Sastrapradja and Aminali (1970) reported microsporogenesis and megasporogenesis in *C. aurantiaca* and *C. lorgengii* and they reported seed set in *C. aurantiaca*. The seed set is absent in *C. lorgengii*, possibly due to pollen abortion. Open-pollinated seed progenies were developed both in *C. longa* and *C. aromatica* as part of crop improvement programs in India. Nair and Sasikumar (2009) studied variation in chromosome number among 22 collections and 28 open-pollinated seedling progenies of turmeric and found 20 out of 22 collections showed $2n = 63$. One collection had $2n = 61$ and another $2n = 84$. The seedling progenies showed chromosome numbers ranging from $2n = 63$ to $2n = 86$, of which $2n = 84$ was the most frequent. They suggested that abnormalities during chromosomes segregation of triploid turmeric may be responsible for chromosome number variation among open-pollinated seedling progenies.

Eksomtramage et al. (2002) investigated the somatic chromosome numbers of 22 species belonging to 10 genera of Zingiberaceae distributed in Thailand. They reported the somatic number of *Curcuma* aff. *Oligantha* as $2n = 42$ and that of *C. rhabdota* as $2n = 24$.

Genome size and chromosome numbers are important cytological characters that significantly influence various organismal traits. Skornickova et al. (2007) investigated the chromosomal and genome size variation in 161 homogeneously cultivated plant samples classified into 51 taxonomic entities of *Curcuma* species from the Indian subcontinent using propidium iodide flow cytometry. The 2C values varied from 1.66 pg in *C. vamana* to 4.76 pg in *C. oligantha*, representing a 2.87-fold range. Three groups of taxa with significantly different homoploid genome sizes (Cx-values) and distinct geographical distribution were identified. Five species exhibited intraspecific variation in nuclear DNA content, reaching up to 15.1% in cultivated *C. longa*. They reviewed all the published chromosome counts and genome sizes in the genus *Curcuma* and concluded that the basic chromosome number in the majority of Indian *Curcuma* is $x = 7$, while the published counts correspond to $6x$, $9x$, $11x$, $12x$, and $15x$ ploidy levels.

16.6 FLORAL BIOLOGY AND SEED SET

Flowering in turmeric depends on the cultivars, climatic conditions, and cultivation practices. In a natural habitat, it is biennial, but it is cultivated as annual. It usually grows in raised beds or in ridges and furrows under rain-fed or irrigated conditions, respectively (Figure 16.4).

Flowering takes place between 109 and 155 days after planting, when mother rhizomes or bigger primary rhizomes are planted. This depends upon variety and environmental conditions. Turmeric inflorescence takes 7–11 days for blossoming after the emergence of the inflorescence. The mean number of flowers per inflorescence ranges from 26 to 35. Duration of flower opening within an inflorescence lasts for 7–12 days after the emergence of the inflorescence. The anthesis is from 7 to 9 a.m., with the maximum occurring around 8 a.m. Anther dehiscence takes place between 7:15 and 7:45 a.m. The pollen grains of turmeric are ovoid to spherical, light yellow, and slightly sticky. Pollen grains show heterogeneity in size among cultivars. Studies on pollen fertility among some turmeric cultivars revealed that the pollen stainability ranges from 71% (cultivar: Kodur) to 84.46% (cultivar: Kuchipudi). The mean pollen length ranges from 7.0 μm (cultivar: Amalapuram) to 7.2 μm (cultivar: Dindrigam), while the breadth varies between 4.2 μm (cultivar: Kodur) and 4.95 μm (cultivar: VK5). Modified ME3 medium at pH 6 gives the best result for *in vitro* pollen germination.

It has been found that pollen fertility and viability vary with the position of flowers in the inflorescence. Flowers in the lower portion show high fertility while those in the middle and upper portions have low fertility. The fruit of turmeric is a thick-walled trilocular capsule with many seeds. The arillate seeds have an outer thick and inner thin seed coat. Seeds are filled with endosperm and the embryo is seen toward the upper side of the ovule. Seed germination commences 17–26 days after sowing and continues up to 44 days. The seedlings produce mainly roots with root tubers and one very small (14–49 g) mother rhizome in the first year (Figure 16.4). The size of the mother rhizomes progressively increases over the years and full growth is observed during the third year. As

Figure 16.4 Rhizome development in turmeric seedling from 1 to 3 years.

the size of the rhizome increases, the number of root tubers declines, especially under cultivation (Pathak, Patra, and Mahapatra 1960; Nazeem et al. 1993; KAU 2000; Rao, Jagdeeshwar, and Sivaraman 2006)

Nambiar et al. (1982) studied flowering behavior, fruit set, and germination in *C. longa* and *C. aromatica* and reported seed propagation in turmeric, especially in *C. aromatica* types. The flowering period was July to September in *C. aromatica* and September to December in *C. longa*. In Kerala conditions, it was observed that the flowers open in the morning from 6.00 to 6.30 a.m. The number of days taken for flowering in *C. longa* was 118–143 days and in *C. aromatica* it was from 95 to 104 days. In *C. aromatica* types, the time taken from flowering to maturation of capsules and seeds ranged from 23 to 29 days.

Two distinct types of seeds with dark or light brown color with white aril, smooth surface and an apical micropylar ring with a wavy outline were obtained. The percentage of germination around 20 days after sowing varied from 70 to 90% depending on cultivar. The somatic chromosome number of the progenies in the parental clones was found to be $2n = 84$. Seedlings exhibited morphological similarities to the parental clones. Because *Curcuma aromatica* is a tetraploid semiwild species, it produces viable seeds; flowering and nonflowering types in the *C. longa* seed set were not observed. However, later seed-setting in cultivated turmeric (*C. longa*) types was reported (Nazeem et al. 1993; Lad 1993; Sasikumar et al. 1996; Nair and Sasikumar 2009).

16.7 SEED GERMINATION AND SEEDLING GROWTH

Turmeric seed germination commences 17–26 days after sowing, and its duration ranges from 10 to 44 days in various crosses. Crosses VK5 × Dindgam and VK5 × Kasthuri Tanuku gave 100% germination, while Amalapuram × Amruthapani Kothapetta gave only 17.22%. The seeds of the crosses, Amalapuram × Nandyal and Kodur × Kasthuri Tanuku, failed to germinate. The maximum mean height (70 cm) was in the progenies of VK5 × Kasthuri Tanuku and minimum mean height (44 cm) was in Kuchipudi × Amalapuram. The mean number of tillers varied between one (VK5 × Kasthuri Thanuku, Nadyal × Amalapuram) and two (Kuchipudi × Amalapuram, VK5 × Dindigam). The mean number of leaves was also highest in the preceding crosses (6.5), while it was lowest (2.18) in the progenies of Nadyal × Amalapuram. Flowering was observed in Amrithapani Kothapetta × Amalapuram and Nandyal open-pollinated progenies.

16.8 PROPAGATION AND CULTIVATION

16.8.1 Propagation

In turmeric, as is the case with most *Curcuma* species, vegetative propagation through rhizome bits with one or two buds is the most efficient. For commercial cultivation, both mother and finger rhizomes are used (Figure 16.5). The requirement of seed rhizome for turmeric is around 2–2.5 tonnes/hectare. The higher seed weight of the mother rhizomes promotes seedling vigor and better early growth of the plant and results in increased yield; however, it requires a higher seed rate. Due to their easy availability, planting of primary fingers is the most common practice. As planting material, primary fingers have been observed to remain better in storage, remain more tolerant to wet soil conditions at planting, and involve a 33% lower seed rate and lesser cost of seed material (Rao et al. 2006).

Rhizome rot caused by *Pythium* spp. is a major disease of turmeric that can spread through infected seed rhizomes; hence, it is important to ensure that planting materials are collected from disease-free gardens. Micropropagation and microtuber technology is also available for large-scale multiplication of disease-free planting materials (Nirmal Babu et al. 2007).

Figure 16.5   (See color insert.) Rhizome bits of turmeric used as planting material.

16.8.2 Planting and Spacing

Planting time varies with the regions and the cultivars. In most cases, planting is done with the onset of monsoon. The yield is influenced by the time of planting and delayed planting reduces the vegetative growth and yield and increases the incidence of leaf-spot diseases. The crop can be planted in flat beds or on ridges (Figure 16.6). The spacing of turmeric depends on the method of planting. Spacing ranges from 45 to 60 cm between rows and 20 to 25 cm between plants. Turmeric is cultivated either as a single crop or intercropped with other crops such as maize and castor, as well as among coconut plants for increased and diversified income.

16.8.3 Nutrition

Turmeric is a heavy feeder and requires higher levels of nutrients. Manure plays a very important role in the management of turmeric and there is good response to nitrogen for boosting vegetative growth. The requirement of nitrogen, phosphorus, and potassium (NPK) is dependent on the soil fertility; however, a fertilizer dose of 312.5 kg N, 125 kg P_2O_5, and 187.5 kg K_2O per hectare is recommended in Andhra Pradesh. Higher doses of NPK increase plant height, leaf number, leaf area, and rhizome production, but have an adverse effect on curing percentage and curcumin content (Rao et al. 2006). Turmeric responds well to organic matter and experimental evidence is available on the beneficial effects of organic matter either alone or in combinations with inorganic fertilizers. Usually, large quantities of organic manures in the form of FYM (farmyard manure), oil cakes, and green leaves (mulch) are applied in different turmeric growing areas (Rethinam, Sivaraman, and Sushma 1994).

16.8.4 Diseases and Pests

The crop suffers from foliar as well as rhizome diseases. Among the foliar diseases, leaf spot (*Colletotrichum capsici*) and leaf blotch (*Taphrina maculans*) are serious. Rhizome rot

Figure 16.6   A view of turmeric plantations in (A) raised beds and (B) ridges and furrows.

(*Pythium graminicolum*) is the most serious malady of the crop (Joshi and Sharma 1980; Sarma et al. 1994; Rao et al. 2006).

Leaf-spot disease is found in all turmeric growing areas. Disease initiation and development depend not only on initial inoculum and weather conditions but also on variety, location, cropping pattern, and soil fertility. The disease generally appears when the crop is about 4–5 months old and may result in up to 50% yield losses in cases of severe infection (Ramakrishna 1954).

The disease symptoms are elliptic to oblong spots of variable size, most evident on the upper surface, enlarging later to covering most of the leaf, which then dries up. Four sprays of 0.2% Captan or Dithane Z-78 at monthly intervals or six sprays of 0.25% Dithane M-45 fortnightly effectively control this disease.

Leaf blotch usually appears in lower leaves in October and November. The disease is characterized by the appearance of small, scattered, oily-looking translucent spots on both surfaces of the leaf (more on the upper surface). The infected leaves are distorted and have a reddish-brown appearance. In cases of severe attack, hundreds of spots appear on the leaves. Following crop rotation, selection of disease-free seed material and seed treatment with Mancozeb (3 g/L of water) or Carbendezim (1 g/L of water) reduces the disease. Spraying of the Dithane Z-78 (0.2%) followed by Dithane M-45, Blitox 50, Bavistin, and Cuman was also effective as a foliar spray (Srivastava and Gupta 1977).

Rhizome rot is the most serious disease affecting the turmeric crop and threatening its cultivation in some areas. The disease manifests initially in isolated patches in the field. The infected plants show progressive drying, beginning with older leaves. Leaves of the affected plants turn yellow and roll inward, giving an appearance of wilting symptoms (Rao and Rao 1988). All roots die, rhizome development is completely arrested, and fingers start rotting. Planting of healthy rhizomes, seed treatment with Mancozeb (0.25%) after harvest and before planting, crop rotation, and good drainage help in reducing the disease. The varieties Suguna and Sudarshana showed field tolerance to rhizome rot (Rao et al. 1994).

The caterpillar of turmeric shoot borer (*Dichocrocis punctiferalis*) bores into the shoots and causes dead hearts. Destruction of the affected shoot and spraying with Endosulphan or Carbaryl can control the pest. Philip and Nair (1981) reported that the variety Mannuthy Local showed lowest infestation of shoot borer.

Scale insect (*Aspediotus hartii*) infects rhizomes both in the field and in storage and causes considerable damage. In the field, in severe cases of infestation, the plants wither and dry. In storage, the pest infestation results in shrivelling of buds and rhizomes. If infestation is severe, it adversely affects sprouting of rhizomes. Dipping the seed rhizomes of turmeric in Quinalphos (0.1%) for 5 minutes after harvest and before planting was effective in controlling the pest infestation (Anonymous 1985).

Other pests, such as thrips (*Panchaetothrips indicus* Bagn), leaf feeders (*Creatonotus gangis* (L)), and rhizome maggots (*Mimegrella ceruleifrons*), have been reported in turmeric.

16.8.5 Harvesting and Processing

Turmeric is usually harvested 8–9 months after planting (Aiyadurai 1966). The harvesting time also depends on the variety. The short-duration varieties mature between 6 and 7 months, medium-duration varieties between 7 and 8 months, and long-duration varieties between 8 and 9 months. Processing of turmeric consists of boiling, drying, polishing, and coloring of raw turmeric rhizomes. Boiling is done to kill the biological activity of the cells as well as to get uniform color throughout the rhizome.

16.9 PHYSIOLOGY

Information is available on turmeric physiology and its relation to growth and yield attributes. Satheesan (1984) and Satheesan and Ramadasan (1987, 1988a,b) studied the physiology, growth, and productivity of turmeric. The turmeric plants showed three distinct stages in growth and development: The first phase (phase I) represents growth up to 8 weeks after planting; the second phase (phase II) represents the period from 8 to 10 weeks, characterized by initiation of finger and maximization of shoot growth; and the third phase (phase III) is dominated by rhizome growth. The maximum LAI (leaf area index) was gained by the 18th week in open conditions, while it took 22 weeks under shade. Considerable differences in genotype response to shade were also observed.

Turmeric genotype Cls. 24 (C1) showed considerable reduction in LAI under shade, while the cultivar Duggirala has significantly higher LAI under the shade. But the cultivar CLL 328 Sugadham did not show significant differences in LAI in open and in shade conditions. The LAI showed a significant positive association with crop growth rate (CGR) up to 3 months of growth. CGR and NAR (net assimilation rate) were found to be highly and positively correlated. A linear relationship also existed between cumulative leaf area development (LAD) and dry matter production. The increase in NAR is interpreted as a response of the photosynthetic apparatus to an increased demand for assimilates caused by the rapid bulking of the rhizomes. The cultivar differences in final rhizome yield can be well explained by the differences in CGR during phase III, which in turn is influenced by the NAR. The RGR (relative growth rate) of the cultivars was reduced significantly under shade, and the NAR rate was also reduced markedly. The cultivars showed higher solar energy conversion efficiency under shade than under open conditions. Significant variation in harvest index (HI) was also observed among cultivars. The HI was significantly higher in CLL-24, indicating a higher partitioning efficiency of this cultivar. The efficiency was higher in the open than under shade.

Li et al. (1997) also divided the growth of turmeric into three stages. Leaf formation takes place rapidly in the early germination stage. The increase of dry matter per day in roots and leaves is less in this stage. In the daughter rhizome formation stage, the leaf area and the net assimilation rate (NAR) reach the maximum. In the early germination stage and daughter rhizome formation stage, over 50–75% of dry matter is distributed in leaves. In the late growing period, the dry matter accumulation shifts to rhizomes.

Mehta, Raghava Rao, and Patel (1980) estimated curcumin content in leaves and rhizomes of *C. longa* and *C. amada* during various stages of growth and reported that it decreased while that of rhizome increased, with increasing maturity. This indicates that leaves are the site of curcumin biosynthesis and then it gets translocated to rhizome. In turmeric, the synthesis of curcumin starts as early as 120 days after planting and reaches an optimum at 180–190 days after flowering. Studies showed that if turmeric is planted in June and harvested in November, it yields 30% more curcumin per kilogram compared to regular harvest at full maturity. The relative reduction in curcumin content at full maturity of the plant can be attributed to accumulation of starch and fiber. Studies also showed that, though curcumin content is genotype specific, it is also location specific and highly dependent on the agroclimatic conditions in which it is grown (Zachariah, Sasikumar, and Nirmal Babu 1999).

Dixit and Srivastava (2000a) studied the physiological efficiency in relation to essential oil and curcumin accumulation. There were significant differences among the genotypes in yield components such as leaf area, leaf area ratio, and photosynthetic characters: namely, CO_2 exchange rate, initial transpiration rate, stomatal conductance, and chlorophyll content. Total $^{14}CO_2$ incorporation was the highest in the genotype Krishna and the lowest in CL-16. The highest percentage of incorporation of ^{14}C in oil was in genotype CL-315 and in curcumin in the cultivar Rashmi.

Dixit and Srivastava (2000a) investigated the distributions of photosynthetically fixed $^{14}CO_2$ into curcumin and essential oil in relation to primary metabolites in developing turmeric leaves. They determined the distribution of photosynthates and found that of the total $^{14}CO_2$ assimilated by plants, the first, second, third, and fourth leaves fixed 31, 23, 21, and 9%, respectively; roots 4%; rhizome 6%; oil 0.01%; and curcumin 4.6% of the fresh weight of rhizome. Leaf area, fresh and dry weight, and $^{14}CO_2$ exchange rate increased up to the third leaf. The incorporation of $^{14}CO_2$ into sugar was maximal, followed by incorporation into organic acid, amino acid, and essential oil at all stages of leaf development. The youngest leaf assimilated maximum $^{14}CO_2$. In the rhizome, curcumin constituted the major metabolite. The incorporation of $^{14}CO_2$ into metabolites and oil declined as the leaves matured and the major portion of $^{14}CO_2$ was translocated to roots and to rhizome for curcumin formation.

Dixit et al. (Dixit, Srivastava, and Sharma 1999; Dixit, Srivastava, and Kumar 2001; Dixit et al. 2002) studied boron-deficiency-induced changes in the essential oil and curcumin accumulation in turmeric and reported that boron deficiency resulted in decrease in leaf area, fresh and dry mass, chlorophyll content, and photosynthetic rate and in reduced accumulation of sugars, amino acids, and organic acids in leaves, decreasing translocation of metabolites toward rhizome and roots. Photoassimilate partitioning to essential oil in leaf and to curcumin in rhizome decreased. The overall yield of curcumin per unit area decreased.

Dixit and Srivastava (2000b, 2000c) studied the physiological effect of iron deficiency on six genotypes of turmeric and all the genotypes exhibited decrease in plant growth, fresh weight, rhizome size, photosynthetic rate, and chlorophyll content. But the curcumin content increased significantly in all the genotypes under iron deficiency. The oil content of rhizome increased in four out of the six genotypes studied.

Turmeric is cultivated for its underground rhizomes, and the photosynthates translocated to the rhizomes contribute to the economic yield. Increased productivity is associated with a higher HI in turmeric. Turmeric cultivars differed significantly in their HI, reflecting the physiological variations in their yield potentials. Duggirala (70.50%), followed by Tekurpet (68.70%), recorded higher values of HI, indicating higher physiological efficiency (Sreenivasulu and Rao 2002). Cultivars that are efficient in mobilizing dry matter content (i.e., with higher HI) are desired for obtaining higher economic yields.

Ravindran et al. (1998) studied effects of triacontanol, paclobutrazol, and GA3 application on growth of turmeric plants. Triacontanol application did not produce any significant change in growth of turmeric plants. Paclobutrazol produced a significant dwarfing effect, while GA3 increased plant height significantly. Application of Paclobutrazol (0.2 and 0.4%) resulted in 53% and 60% reduction in growth, respectively. However, the root length and number of primary, secondary, and tertiary rhizomes increased significantly. The leaves became darker and larger and rhizomes thicker with shorter internodes. In the treated plants, the girth of the mother rhizome, outer zone, inner zone, and intermediate layers were significantly higher. Paclobutrazol stimulated vascular tissue formation, resulting in increased (23–38%) rhizome growth. Due to higher procambial activity and the number and size of vascular elements, oil cells and curcumin cells increased significantly. Starch grains increased significantly, indicating increased photosynthate transport and higher carbohydrate accumulation.

Panja and De (2001) studied the stomatal frequency and its relationship with leaf biomass and rhizome yield in 12 genotypes of turmeric. Frequencies of stomata varied from 426 to 1773/cm^2 in the upper leaf surface and 7131 to 12,270/cm^2 in the lower surface. They observed that the stomatal frequency is positively correlated with rhizome yield.

16.10 GENETIC RESOURCES

16.10.1 Conservation

Curcuma is widespread in the tropics of Asia, Africa, and Australia from sea level to an altitude of 2000 m mean sea level. *Curcuma* consists of about 117 species, of which about 40 species are reported from around India (Purseglove et al. 1981; Velayudhan, Asha, et al. 1999; Velayudhan, Muralidharan, et al. 1999; Shiva et al. 2003; Sasikumar 2005; Ravindran, Nirmal Babu, and Shiva 2007) (Table 16.1). However, true identity is still unclear in many species and there may be around 70–110 species (Skornickova et al. 2007, 2010).

India is the world's largest producer, consumer, and exporter of turmeric. In India, turmeric is grown in a 161,230 ha area and produced 716,840 tonnes during 2004–2005, of which 43,000 tonnes was exported. Turmeric is grown in most of the states of India; the state of Andhra

Table 16.3 Conservation of Turmeric Germplasm at Various Centers in India

	Number of Accessions	
Center	**Cultivated Types**	**Wild and Related Species**
In Field Repository		
Indian Institute of Spices Research, Calicut, Kerala	1026	24
National Bureau of Plant Genetic Resources, Regional Station, Trichur, Kerala	954	
Regional Agriculture Research Station, Jagtial, Andhra Pradesh	273	
Tamil Nadu Agricultural University, Coimbatore	264	7
Orissa University of Agriculture and Technology, Pottangi, Orissa	199	
Y. S. Parmar University of Horticulture and Forestry, Solan, Himachal Pradesh	132	
Rajendra Agricultural University, Dholi, Bihar	90	2
Uttar Banga Krishi Vishwa Vidyalaya, Pundibari, West Bengal	152	18
N. D. University of Agriculture and Technology, Kumarganj Uttar-Pradesh	130	
Indira Gandhi Agricultural University, Regional Station, Raigarh, Chattisgarh	42	3
In-vitro Gene Bank		
Indian Institute of Spices Research, Calicut, Kerala	102	
National Bureau of Plant Genetic Resources, New Delhi and Trichur	100	

Sources: AICRPS, 2009. Annual report: 2008–09 of All India Coordinated Research Project on Spices, 89. Indian Institute of Spices Research, Calicut, Kerala, India.
IISR. 2009. Annual report: 2008–09. Indian Institute of Spices Research, Calicut, Kerala.

Pradesh is foremost in production, contributing to over 45% of turmeric production, followed by Tamil Nadu, Orissa, Karnataka, and West Bengal (Rao et al. 2006). India has good diversity in turmeric cultivars; almost every state has its own cultivars. Collection, evaluation, and conservation of genetic resources of turmeric have been given great importance in India, mainly at the Indian Institute of Spices Research (IISR), Calicut, which is the national conservatory of turmeric germplasm. Many regional collections are also maintained at various research centers (Table 16.3).

The germplasm is usually maintained as clonal repositories in field gene banks, being replanted each year. This may lead to mixing up and loss of genetic purity. Hence, at IISR, a set of nucleus germplasm is maintained in culture tubs with replanting only once in 5 years to maintain purity (Figure 16.7).

To augment the conservation efforts and also to minimize the loss, an *in vitro* gene bank of important genotypes is also operational at IISR and the National Bureau of Plant Genetic Resources (NBPGR), New Delhi (Ravindran et al. 2005, 2007).

16.10.2 Genetic Resources Management

Genetic resources management is a complex process. It includes a number of mutually dependent stages such as from the identification of a target gene pool for conservation to the use of genetic

Figure 16.7 Maintenance of nucleus germplasm of turmeric to maintain genetic purity.

resources for economically useful traits. Analysis of georeference data of the flora and germplasm collections with geographical information system (GIS) technology can be an efficient and effective way to study the geographic distribution.

16.10.3 GIS Technology and Biodiversity

Turmeric can be grown in diverse tropical conditions from sea level to 1500 m elevation at a temperature range of 20–30°C with an annual rainfall of 1500 mm or more under natural rainfall or irrigated conditions. Though it can be grown in different types of soils, it thrives best in well-drained sandy or clay loam soils. The geographic differentiation of plant populations reflects the dynamics of gene flow and natural selection. Sampling of geographically distinct populations is a practical approach to understand genetic diversity, and sampling in diverse ecogeographic areas will help in conservation of rare and wild species.

GIS analysis of the germplasm data helps in gaining better understanding, developing new strategies for exploiting geographic diversity, and predicting where species naturally occur or may be successfully introduced. Habitat loss and fragmentation are among the most common threats facing endangered species, making GIS-based evaluations an essential component of population viability analyses. Utpala et al. (2007) have prepared a land-suitability map for turmeric with the help of Eco–Crop Model of DIVA GIS. It shows that highly suitable places like Assam, parts of Andhra Pradesh, and Bihar Harbor have natural varieties with very high quality.

16.10.4 Genetic Resource Characterization and Cultivar Diversity

Cultivated turmeric is predominantly vegetatively propagated and seed set is rare due to its cultivation as an annual crop and absence in the wild. Thus, natural mutations preserved in the population by the clonal propagation are the main source of variation. Irrespective of this, over 70

Table 16.4 Popular Traditional Cultivars of Turmeric Grown in Different States of India

State	Cultivar
Andhra Pradesh	Armoor, Amalapuram, Amruthapani, Avanigadda, Chayapasupu, Cuddapah, Duggirala, G.L. Puram, Kothapett, Kasturi, Kodur, Kuchipudi, Mydukkur, Nandyal, Sugandham, Tekurpet, T. Sundar, Vontimitta
Karnataka	Amalapuram, Bangalore Local, Balaga, Cuddapah, Kasturi, Mundaga, Rajapuri, Shillong
Kerala	Alleppy, Armoor, Duggirala, Moovattupuzha, Tekurpetta, Wyanad Local
Tamil Nadu	Bhavanisagar, Erode, Katpadi, Salem
Maharashtra	Krishna, Rajapuri, Sugandham
Orissa	Dindigam, Duhgi, Jobedi, Katergia
Madhya Pradesh	Bilaspur, Jangir, Raigarh
Bihar	Bilaspur, Pusa
Uttar Pradesh	Gorakhpur
Northeast	Besar Along, Maran, Lakadong, Singhat Manipur

popular turmeric types belonging to both *C. longa* and *C. aromatica* are under cultivation in India, in addition to many lesser known local types. Most of these cultivars go by local names, derived mostly from the place of occurrence (Nair, Nambiar, and Ratnambal 1980). Table 16.4 shows traditionally grown cultivars of turmeric in India.

Curcuma collections and species differ in floral characters, aerial morphology, rhizome morphology, and chemical constituents (Valeton 1918; Velayudhan, Asha, et al. 1999; Velayudhan, Muralidharan, et al. 1999; Rao et al. 2005). Velayudhan, et al. (1999 a,b) listed distinguishing morphological features to characterize turmeric. Table 16.5 lists *Curcuma* species. Das et al. (2004)

Table 16.5 Important Distinguishing Features for Characterization of *Curcuma* Species

S. No.	Character	Character States
	I. Floral Characters (Qualitative)	
a	Spike position	Central, lateral
b	Presence of coma	Absent, present
c	Color of calyx	White, green, purple, light
d	Color of corolla	Light, white, orange, red, purple, pale
e	Color of staminode	White, yellow, red, pale yellow, orange yellow
f	Color of lip	White, yellow
g	Flower exertion	Exerted
h	Diversion of spur	Diverging/converging
	II. Floral Characters (Quantitative)	
a	Length of flower	Metric traits
b	Length of calyx	Metric traits
c	Length of corolla lobe	Metric traits
d	Width of corolla lobe	Metric traits
e	Length of lip	Metric traits
f	Width of lip	Metric traits
	III. Aerial Characters	
a	Plant type	Semierect, erect
b	Leaf habit	Semierect, erect, prostrate
c	Color of sheath	Purple, green, dark purple, brown, light, purple-brown

(continued)

Table 16.5   Important Distinguishing Features for Characterization of *Curcuma* Species (Continued)

S. No.	Character	Character States
	III. Aerial Characters	
d	Leaf margin waviness	Highly wavy, medium wavy, low wavy
e	Leaf veins arrangement	Close, distant
f	Presence of hair on dorsal side of leaf	Hairy, glabrous
g	Presence of hair on ventral side of leaf	Hairy, glabrous
h	Plant height	Metric traits
i	Sheath length	Metric traits
j	Petiole length	Metric traits
k	Leaf midrib color	Green, light purplish-brown, light purple, light purplish-green, green
l	Leaf midrib color fading	Absent, present
	IV. Underground Characters	
a	Shape of rootstock	Oblong, cylindrical
b	Color of rootstock	Yellow, orange-yellow, light orange-yellow, pale yellow, mustard yellow, pale blue, blue, verdis green, pale yellow-white, light orange-yellow
c	Length of rootstock	Metric traits
d	Thickness of rootstock	Metric traits
e	Presence of sessile tubers	Absent, present
f	Presence of stipitate tubers	Present, absent
g	Presence of stolons	Absent, present
h	Shape of stipitate tuber	Fusiform, long fusiform
i	Presence of beading in stipitate tuber	Present, absent
j	Length of stipitate tuber	Metric traits
k	Thickness of stipitate tuber	Metric traits
l	Color of stipitate tuber	Light orange-yellow, orange-yellow, white-yellow, green, white, pale yellow
m	Aroma of rhizome	No aroma, camphoraceous aroma, camphor and turmeric aroma, mango aroma
n	Taste of rhizome	No taste, slightly bitter, bitter

Source: Velayudhan, K. C., Muralidharan, et al. 1999. *Curcuma genetic resources*. Scientific monograph no. 4. National Bureau of Plant Genetic Resources, New Delhi, p. 149.

developed a simple key for identification of *Kaempferia galanga*, *C. amada*, *C. longa*, and *C. caesia* using morphological and anatomical characters. Velayudhan, et al. (1999 b) studied and analyzed distribution, habitat, flowering time, floral characters, quantitative characters of the floral parts, and features of above- and belowground characters of 31 *Curcuma* species from India using numerical taxonomy. These 31 species were clustered into nine groups.

Germplasm collections are evaluated, characterized, and utilized in crop improvement programs. A database on *Curcuma* species, which gives the details of the important species present in India and discriminative features of each species, is prepared at IISR. Useful information data can be retrieved by using the serial number and species name (IISR 2003).

Existence of wide variability among the turmeric cultivars with respect to growth parameters, yield attributes, resistance to biotic and abiotic stresses, and quality characters were reported (Govindarajan 1980; Jalgaonkar and Jamdagni 1989; Jana and Bhattacharya 2001; Jana, Dutta, and Chatterjee 2001; Kumar and Jain 1998; Kumar, Pandey, and Dwevedi 2004; Kumar 1999; Lynrah and Chakraborty 2000; Mohanty 1979; Natarajan 1975; Panja and De 2000; Panja, De, and Majumdar 2001; Nirmal Babu 1993; Palarpawar and Ghurde 1989; Pathania, Arya, and Singh 1988;

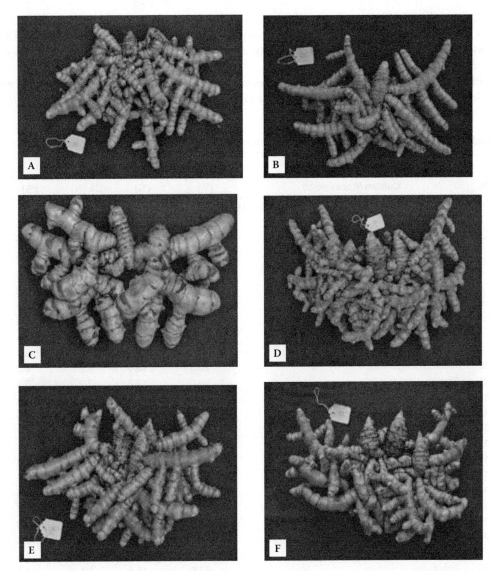

Figure 16.8 Variability in rhizome characters of turmeric cultivars: (A) Along, (B) Anogiri, (C) Jorhat, (D) Kuchipudi, (E) Grakhpur, (F) Waynadan.

Pathania, Singh, and Arya 1990; Philip and Nair 1981; Rakhunde, Munjal, and Patil 1998; Rao et al. 2006; Shahi et al. 1992, Shahi, Shahi, and Yadava 1994 a,b, Shahi, Yadava, and Sahi 1994; Shanmugasundaram and Thangaraj 2002; Shridhar, Prasad, and Gupta 2002; Singh and Tiwari 1995; Yadav and S. Singh 1989; Pandey et al. 2003, 2004 a, b, c). Figure 16.8 shows variability in rhizomes of important turmeric cultivars.

Rao et al. (1975) collected and evaluated 100 accessions from India and classified them into short-duration Kasthuri (short-duration type) type (maturing in 7 months) belonging to *C. aromatica*, long-duration types (maturing in 9 months) belonging to *C. longa*, and medium-duration Kesari types (maturing in 8 months). Clonal selections among the popular cultivars resulted in identification of eight promising short-duration types, six medium-duration types, and eight long-duration types (Rao et al. 1975; Subbarayudu, Reddy, and Rao 1976).

Ratnambal et al. (1986) studied quality variability in 100 accessions of turmeric collected from India belonging to *C. longa* and *C. aromatica* types and a few exotic and wild collections. They reported wide variability in the content of curcumin, oleoresin, essential oil, and dry recovery (Table 16.6).

Turmeric is affected by many foliar and rhizome diseases of fungal origin. Among them, leaf spot (*Colletotrichum capsici*), leaf blotch (*Taphrina maculans*), and rhizome rot (*Pythium graminicolum*)

Table 16.6   Diversity in Quality Parameters of Turmeric Accessions

Sl. No.	Cultivar/Accessions	Dry Recovery (%)	Oleoresin (%)	Oil (%)	Curcumin (%)
	Curcuma longa L.				
1	Maran	26.0	13.5	7.0	8.7
2	Jorhat	21.7	10.8	7.5	6.9
3	Dadra, Gauhati	23.2	16.6	7.0	7.7
4	Kaziranga, Jorhat	24.5	18.2	6.0	10.2
5	Anogiri, Garohills	26.9	13.6	6.0	5.2
6	Nowgong, Assam	20.0	10.0	5.0	4.0
7	Mekhozer	20.0	12.0	5.0	4.0
8	Hajo,Gauhati	21.0	13.0	7.5	5.5
9	Rajasagar	16.6	10.3	6.0	5.0
10	Teliamura, Agarthala	23.5	13.0	7.0	5.5
11	Barhola, Jorhat	25.0	13.3	7.0	5.3
12	Kahikuchi	21.2	12.1	9.5	3.1
13	Along	21.7	13.0	5.0	6.6
14	Besar, Along	20.6	11.6	5.0	6.1
15	Gaspani, Nagaland	24.4	11.0	8.0	4.5
16	Singhat, Manipur	19.7	15.0	7.0	7.9
17	Kongpopkri	21.4	13.2	7.5	5.6
18	Aigal	20.0	14.0	5.0	9.0
19	Amampuri, Jumpoi Hills	25.7	11.5	5.0	4.0
20	Amkara, Tripura	22.7	12.7	8.0	5.6
21	Torku	19.5	10.9	6.0	6.0
22	Barpather, Galoghat	23.3	12.0	3.0	4.3
23	Rorathong, E. Sikkim	22.8	16.8	4.0	4.7
24	C11 316, Gorakpur	18.7	15.0	6.0	6.0
25	Pusa	13.5	14.5	7.5	7.2
26	PTS 5	20.0	12.7	5.0	6.0
27	PTS 10	22.5	15.0	5.0	7.7
28	PTS 24	23.0	14.1	5.0	7.9
29	PTS 68	22.6	12.9	5.0	5.1
30	Amalapuram	20.2	16.0	8.0	6.0
31	Cls no. 34	23.8	14.0	5.0	5.0
32	Amalapuram II	20.0	13.3	4.0	4.3
33	Cls no. 15	21.8	16.0	5.5	4.8
34	Cls. no. 3	22.3	12.5	6.6	5.0
35	Amalapuram selection III	30.0	16.5	6.0	5.0
36	CII 390 Amalapuram	19.4	13.7	7.5	7.0
37	Amrithapani	23.2	19.0	7.0	7.0

Table 16.6   Diversity in Quality Parameters of Turmeric Accessions (Continued)

Sl. No.	Cultivar/Accessions	Dry Recovery (%)	Oleoresin (%)	Oil (%)	Curcumin (%)
	Curcuma longa L.				
38	Amrithapani, Kothapetta	32.4	15.0	4.0	7.0
39	Nandyal type	22.0	13.5	8.0	4.7
40	Cls no. 13	23.4	16.8	5.0	5.4
41	Vontimitta	18.0	10.5	6.5	5.4
42	Cls no. 11A	21.2	13.0	5.0	4.0
43	CII 322 Vontimitta	24.5	11.4	6.0	7.4
44	GL Puram II	19.6	13.1	6.0	6.2
45	Cls no. 5A	30.1	13.5	6.0	4.9
46	GL Puram III	21.8	12.9	6.0	5.1
47	CII 324 armoor	18.0	15.6	6.5	7.0
48	Cls no. 1	25.0	10.8	5.0	6.0
49	Cls no. 1A	24.0	15.8	5.0	6.4
50	Cls no. 1C	20.0	11.1	5.0	6.3
51	Ethamukula	22.5	15.0	5.0	5.5
52	Cls no. 26	24.6	12.0	5.0	5.7
53	CII 321 Ethamukula	26.2	11.3	6.5	6.0
54	Cls no. 27B	19.9	10.2	5.0	4.0
55	Duggirala	17.6	14.6	5.0	7.5
56	Cls no. 22	18.1	15.4	6.6	5.2
57	CII 325 Duggirala	21.0	11.7	8.0	5.0
58	Kuchipudi	18.6	14.0	7.0	7.5
59	Cls no. 8B	14.7	12.9	5.0	6.0
60	Cls no. 8C	19.4	14.3	6.0	7.9
61	T. Sundar	20.0	16.0	6.0	6.9
62	Sugandham	22.6	12.0	8.5	9.1
63	Cls no. 19	19.0	13.8	6.0	5.4
64	Dindigam	17.0	10.6	6.0	6.4
65	CII 327, Takkurpet	21.0	14.0	6.5	6.1
66	CII 326, Mydukkur	19.4	11.8	6.0	2.8
67	Karhadi Local	18.3	13.0	6.0	4.9
68	Cls no.7	26.9	12.7	7.0	5.0
69	CII 323 Avanigada	19.4	14.5	7.5	7.0
70	Cls no. 30	22.0	13.0	7.0	6.5
71	CIIs 328 Sugandam	20.0	12.8	6.0	9.0
72	Cls no. 9A	14.5	11.6	4.0	7.9
73	No. 24	23.0	16.0	5.0	5.0
74	Cls no. 24	22.0	14.3	6.0	4.5
75	Cls no. 6	21.0	13.9	4.0	6.7
76	Cls no. 6A	24.0	15.8	5.0	6.4
77	Palani	21.0	16.5	5.0	7.6
78	Kayyam, Gudalur	23.0	16.5	5.0	8.4
79	Pathavayal, Gudalur	25.0	14.0	4.0	3.6
80	Upper Dinamala	24.0	12.0	5.5	7.8
81	Rajpuri Local	16.3	13.8	7.0	7.8

(continued)

Table 16.6   Diversity in Quality Parameters of Turmeric Accessions (Continued)

Sl. No.	Cultivar/Accessions	Dry Recovery (%)	Oleoresin (%)	Oil (%)	Curcumin (%)
	Curcuma longa L.				
82	Cls no. 14B	18.5	13.5	5.0	6.5
83	CII 390 Rajpuri	18.3	12.2	8.0	6.0
84	Moovattupuzha	20.1	11.5	5.0	7.0
85	Varapetty, Kothamangalam	18.0	10.9	8.0	5.4
86	Pathanapuram	17.0	14.0	8.0	6.7
87	Karimala, Mannarghat	27.0	12.5	8.0	5.5
88	Ochira	21.8	12.0	5.0	5.6
89	Cls no. 29	23.5	11.7	4.0	4.0
90	Alleppey	17.2	13.0	8.0	6.0
91	Cls no. 21	25.5	12.1	8.0	6.2
92	Valra Falls, Adimali	20.0	14.0	6.5	6.0
93	Mundakkayam	23.4	10.5	8.5	3.2
94	Mananthody	22.5	16.5	8.5	9.1
95	Cls no. 16	25.0	18.3	9.0	7.0
96	Vandoor, Nilambur	22.6	10.5	4.0	5.4
97	Manjapally, Perumbavoor	16.4	10.6	6.5	5.6
98	Murangathapally, Meenachi	20.0	13.0	8.5	7.8
99	Puthuppadi, Meenangadi	24.4	11.3	5.9	5.4
100	Edapalayam	22.3	14.5	6.0	10.9
101	Erathupetta	20.0	11.2	5.0	6.0
102	Erathukunnam	21.0	12.0	6.0	10.3
103	Idukki no. 1	21.6	10.8	4.0	8.5
104	Idukki no. 2	28.5	13.7	4.0	9.0
105	Thodupuzha	21.2	14.8	6.9	9.5
106	Cls no. 28	24.0	13.0	5.0	5.7
107	Palapally, Trichur	21.0	15.3	6.0	10.7
108	Kolathuvayal	20.0	13.5	7.0	4.2
109	Elanji, Idukki	20.0	12.6	7.0	2.7
110	Karuvilangad	21.0	13.9	4.0	6.2
111	Ayur	21.5	12.4	4.6	5.1
112	Kothamangalam	23.5	10.7	5.0	4.2
113	Kakkayam Local	20.5	14.5	9.0	7.5
114	Chamakuchi	26.5	16.9	4.0	7.0
115	Anchal	28.0	12.4	5.0	5.4
116	Muringakalla	23.1	11.8	5.0	7.0
117	Mongam, Malappuram	26.2	12.0	7.0	5.7
118	Maramboor	18.0	13.0	4.0	5.6
119	Ernad	30.5	12.9	9.0	5.2
120	Wynad Local	20.0	15.3	7.0	9.4
	C. aromatica Salisb.				
1	Silapather, N. Lekhimpur	21.7	12.5	5.0	3.2
2	Burahazer, Dibrugarh	28.0	11.6	4.0	3.1
3	Tura, Garohills	25.0	13.9	4.0	4.3
4	Dibrugarh	20.3	12.5	6.5	8.0

(continued)

Table 16.6   Diversity in Quality Parameters of Turmeric Accessions (Continued)

Sl. No.	Cultivar/Accessions	Dry Recovery (%)	Oleoresin (%)	Oil (%)	Curcumin (%)
	C. aromatica Salisb.				
5	Hahim	27.2	13.0	6.0	2.3
6	Aseemgiri, Garohills	17.8	10.5	7.0	3.5
7	Bahumura, Agarthala	18.0	13.5	4.0	5.0
8	Nagsar, Titasar, Jorhat	18.5	10.5	5.0	2.5
9	Besar, Along	18.8	14.4	5.0	5.0
10	Kanchanpur, Tripura	24.1	12.1	5.5	4.1
11	Namachi	20.0	9.6	8.0	3.5
12	Pakyong	18.7	12.4	6.0	3.7
13	Nayabunglow, Meghalaya	22.4	12.5	5.5	3.9
14	Shillong	26.0	14.0	5.0	4.0
15	Phu, E. Sikkim	23.2	14.0	5.5	4.7
16	Pedong, Kalimpong	20.6	14.6	5.0	4.7
17	Ca 72 Udayagiri	21.0	13.0	5.5	4.0
18	Cas no. 57	22.5	12.9	9.0	7.4
19	G.L. Puram I	21.0	10.9	8.0	4.1
20	Ca 66 G.L. Puram	23.0	11.5	8.5	4.0
21	Armoor	17.8	12.8	9.0	3.5
22	Kodur	20.6	14.5	8.0	4.0
23	Cheyapasupu	20.3	14.2	8.0	2.5
24	Ca I Cheyapasupu	17.2	16.0	5.0	3.5
25	Ca 69 Dindigam	18.9	12.0	6.0	2.8
26	Ca 68 Dhagi	20.2	12.4	8.0	3.1
27	Katirgia	20.6	13.5	7.0	2.8
28	Jobedi	18.8	13.0	8.5	4.1
29	Kasturi	25.7	12.0	9.0	3.2
30	Kasturi Tanuka	18.8	10.5	6.5	2.9
31	Ca 73 Amalapuram	19.2	14.0	6.0	3.0
32	Cas no. 58	22.0	13.3	8.0	3.8
33	Cas no. 58B	14.0	11.7	8.0	6.0
34	Erode	20.9	12.0	7.0	3.1
35	Nadavayal	25.5	11.5	9.0	4.7
36	Keeranthode	19.8	10.3	5.0	5.0
37	Makkapuzha, Ranni	21.3	12.5	5.0	4.0
38	Konni	26.1	19.2	5.0	5.0
39	Thachanatukara, Mannarghat	24.2	10.7	8.5	6.6
40	Mampad, Nilambur	23.8	11.7	4.0	5.7
41	Chamakuchi	18.5	16.0	5.0	3.0
42	Adimali	18.0	12.0	5.0	2.3
43	Amnicad	20.0	10.3	8.5	4.0
44	Bigmathi, Meenachi	22.0	10.3	8.5	4.0
	Exotic types (Solomon Islands)				
1	Mamarei	20.0	13.2	6.0	3.0
2	Vatuloro	18.0	10.0	6.0	3.4

(*continued*)

Table 16.6 Diversity in Quality Parameters of Turmeric Accessions (Continued)

Sl. No.	Cultivar/Accessions	Dry Recovery (%)	Oleoresin (%)	Oil (%)	Curcumin (%)
	Exotic types (Solomon Islands)				
3	Vanagobulu	17.5	11.0	5.0	3.1
4	Cokuma	—	12.0	6.0	4.1
5	Tsavana	—	10.5	6.0	3.9
6	Tuva Vitalio	—	12.0	6.0	2.8
	***Curcuma* sp.**				
1	*C. angustifolia* Roxb.	30.0	7.6	3.0	0.2
2	*C. zanthorrhiza* Roxb.	25.8	10.0	2.0	1.5
3	*C. zedoaria* (Christn.) Roscoe	20.5	6.0	2.0	2.0
4	Wild unidentified (1)	22.7	6.4	2.0	0.3
5	Wild unidentified (2)	30.0	7.9	2.0	2.2
6	Wild unidentified (3) from Uttar Pradesh	23.6	7.2	5.0	1.3
7	Wild unidentified (4)	26.5	8.6	3.0	0.8
8	Wild unidentified (5)	30.8	4.5	4.0	0.2
9	Wild unidentified (6) from Nagsar, Titsar, Jorhat	24.0	6.2	1.0	0.5
10	Wild unidentified (7) (Dergroni, Jorhat)	21.4	5.8	4.0	1.6
11	Wild unidentified (8) (Kattapana, Idukki)	31.4	8.6	2.0	0.02
12	Wild unidentified (9) (Taranagar, Agarthala)	30.7	4.0	1.5	2.6
13	Wild unidentified (10) 9 (Sibsagar)	25.0	9.6	3.0	0.05
14	*C. amada* Roxb.	30.0	5.0	1.5	—

Source: Ratnambal, M. J. 1986. *Qualities Plantarum* 36:243–252.

are the most important. Turmeric collections exhibiting good variability for resistance to biotic stresses have been identified (Table 16.7).

High heritability with appreciable genetic advance was recorded for rhizome yield, crop duration, number of leaves, number of primary fingers, yield of secondary fingers, and height of pseudostem (Philip and Nair 1986; Subramanyam 1986; Reddy 1987; Yadav and Singh 1996). Singh, Mittal and Kotoch (2003) suggested that superior genotypes could be obtained through selection based on number and weight of mother, primary, and secondary rhizomes. The yield per plant was highly associated with length of primary fingers, mother rhizome diameter, and length and girth of secondary fingers.

The correlation coefficients of these yield components were positive and significant (George 1981; Jalgaonkar, Jamadagni, and Selvi 1990; Cholke 1993; Verma and Tiwari 2002). Number of leaves, number of primary fingers, and crop duration showed positive association with rhizome yield at both genotypic and phenotypic levels (Reddy 1987; Panja et al. 2002). The quality characters (curcumin, essential oils, and oleoresins) showed negative correlation. For crop improvement in turmeric, plant height and number of leaves determine the yield potential of the genotype (Narayanpur and Hanamashetti 2003). It is therefore concluded that plant height is the single most important morphological character on which selection for yield could be made.

Because yield is a complex and polygenically controlled character, direct selection for yield may not be a reliable approach because it is highly influenced by environmental factors. Jirali et al. (2003) reported that cultivars Amalapuram, Bidar-4, and Cuddapah are physiologically more efficient and can be used as genetic sources for yield improvement. A list of promising and superior genotypes is given in Table 16.8.

Table 16.7 Sources of Resistance to Pests and Diseases in Turmeric

Trait	Reaction	Promising line	Ref.
Nematode (*Meloidogyne incognita*)	Resistant	Acc. 31,43, 56, 57, 82, 84, 142, 178, 182, 198, 200	IISR (2005); Sarma (2001)
Rhizome rot (*Pythium graminicolum*)	Resistant	PCT-13, PCT-14, Shillong	Sarma and Anandaraj (2000); AICRPS (2003, 2005)
		GS, RH-5, NDH-18,	
	Resistant/tolerant	CLI-370, 320, 325, 330,	Rao et al. (1992)
		JTS-,12, 15, 303, 308, 314, 319, 320, 321, 325, 607 604, 612,	
		PCT-7, 10, 13, 14;	
		PTS-9, 10	
Leaf blotch (*Taphina maculans*)	Resistant/tolerant	CLL-324 Ethamukala,	Rao et al. (1992); Sarma and Anandaraj (2000)
		CLL-316 Gorakhpur,	
		CLL-326 Mydukur,	
		Alleppey, PCT-12 and PCT-13	
	Resistant	ACC. 360, 361, 585, 126, T4-11	AICRPS (2003)
	Resistant	PTS-11, 15, 52, 55, and 59, JTS-319	AICRPS (2003, 2005)
	Highly resistant	Kohinoor, G.L. Puram,	AICRPS (2003)
		Rajendra Sonia, RH-5	
		and RH-24	
	Immune	PTS-3, PTS-4, PTS-11,	Verma and Tiwari (2002a)
		Roma and Rashmi PTS-10 and PTS-36	
	Resistant		
Leaf spot and leaf blotch	Resistant	CL 32, CL 34, CL 54, CL 55	Subramanian et al. (2004); AICRPS (2003, 2005)
	Tolerant	Clone nos. 326, 327	Rao et al. (1992)
Leaf spot (*Colletotrichum capsici*)	Resistant	Krishna	Sarma and Anandaraj (2000)
	Resistant	JTS-10 to -15, 314–326, 606–612; Duggirala, PTS-11, CLI-317, PCT-13	AICRPS (2003, 2005)
	Resistant	Sudarshan and RRTS-1	AICRPS (2003)
	Resistant	PTS-39, 27	Verma and Tiwari (2002a)
	Resistant/tolerant	PCT cultures	Rao et al. (1992)

Source: Ravindran, P. N. et al. 2007. In *Turmeric — The genus* Curcuma, ed. P. N. Ravindran, K. Nirmal Babu, and K. Sivaraman, 15–70. Boca Raton, FL: CRC Press.

Studies on genetic divergence of turmeric based on D^2 statistics grouped cultivars of *C. longa* and *C. aromatica* into four clusters. The cultivars belonging to *C. aromatica* were grouped separately from those for *C. longa*. Estimation of oil and curcumin contents in different cultivars of *C. longa* and *C. aromatica* indicated that the variability was high in *C. longa* compared to that of *C. aromatica*. Genetic divergence studies (D^2 analysis) by Rao (2000) with 54 genotypes showed a wider diversity among genotypes and they were grouped in six clusters. The landraces of the northeast region were almost clustered together in low- to moderate-yield groups, while genotypes from the southern region were scattered among different complexes ranging from moderate to high yielders. Cultivars PCT-13 and Lakadong formed solitary groups and were genetically most distant (Chandra et al. 1997; Chandra, Govind, and Desai 1999).

Table 16.8   Promising Lines Identified at Various Locations for Various Traits in Turmeric

Trait	Promising Line/Cultivar/Variety	Location and State	Ref.
High yield	CA-9, C11-317, C11-326, C11-327	Kerala	Sasikumar et al. (2005)
	VK1 (Chayapasupu), D199, Karhadi Local, Kasthuri, Tekuripeta, Mannuthy Local, CO1, BSR-1, VK-5, VK-155, VK-31, VK-121, VK-116	Kerala	Philip et al. (1982); Ramakrishna et al. (1995); Kurian and Valsala (1996); Kurian and Nair (1996)
	Seed progeny of Amrithapani × Amalapuram, Nandyal—open pollinated	Kerala	Nazeem, Menon, and Valsala (1994)
	PTS-38, Duggirala, CLI-317, PCT-14, PCT-13, PTS-55, 11, 39	Andhra Pradesh	Reddy et al. (1989); Rao, et al. (2004); AICRPS (2003, 2005); Naidu et al. (2003)
	RH-5, JTS-12, 314		
	G.L. Puram II, CLL-328, Sugandham, CLL-320 Amalapuram	Andamans	Singh and Singh (1987)
	BD-7-105, AH6/2, DKH-26	Meghalaya	Pandey, Sharma, and Hore (1990)
	CI-2a, CI-15B	Goa	Dhandar and Varde (1980)
	Mydukar, Bangalore Local, Suvarna, Suroma, Suguna, BSR-1	Karnataka	Indiresh et al. (1990); Hegde (1992); Cholke (1993); Sheshagiri and Uthaiah (1994); Latha et al. (1995); Vadiraj et al. (1996); Jirali et al. (2003)
	Cuddapah, Bidar, Amalapuram, Bidar-4, CLI-315, CLI-327, VK-116, VK-31, VK-55	Maharashtra	Pujari et al. (1986); Jalgaonkar et al. (1988); Lawande et al. (1991)
	Cls-19, Cls-320, Cls no. 2, Ca-72, Ca91/2, Rajempeta, Kasturi, Rajempeta, Tekurpeta, Krishna, Waigaon, PCT-8		
	Jabedi, Sarangada, Udayagiri and Betul, IC 210265, PTS-11	Orissa	Aiyadurai (1966); Dikshit, Dabas, and Gautam (1999); AICRPS (2005)
	PTS-25, CLS-9	Punjab	Nandi (1990)
	IT-01-05	Chattisgarh	Tomar and Singh (2002)
	NDH-59, NDH-45, NDH-18, NDH-14, Rajendra Sonia	Utter Pradesh	Pandey et al. (2004a)
	Suguna, PTS-43, Sugandham TCP-9, TCP-56, 2, 11, 1	West Bengal	Dash and Jana (2003); Choudhuri and Hore (2004); AICRPS (2003)
	Chayapuspu, Meghalaya Local, Kasturi Tanuku	Mizoram	Pathak et al. (2003); AICRPS (2003, 2005)
	Acc. no. 53, 145, 74, Suguna, CL-101	Tamil Nadu	Chezhiyan et al. (1999); AICRPS (2003, 2005)
High yield and quality	VK1, 5, 47, 55, 56, 82, 88, 96, 107, 144, 145, 146, 159, 172, 230, PTS 10, 9 321 Ethamukulam, 15B, NBPGR 1, Jamaica, Wynad Local, Sugandham, Gorpan, and Amalapuram	Kerala	Nybe (2001)
High yield and high (>6.6%) curcumin	Acc. nos. 584 and 585, PCT-8, 2, 14, 13, 17	Kerala, Karnataka, Andhra Pradesh	Ratnambal and Nair (1986); Sarma (2001); Ratnambal et al. (1992)
	RH-10	Bihar	Maurya (1990)

Trait	Varieties/accessions	Location	Reference
High yield with high stability (hi > 1)	Lakadong, Megha turmeric	Mizoram	Pathak et al. (2003)
	Mannuthy Local, VK5 (Manuthy Local) and VK1 Chayapasupa	Trichur, Kerala	Philip et al. (1982)
	Kesar, Jugijan, Mydukur	Gujarat	Mehta and Patel (1982)
High curcumin, oleoresin, and oil	Acc. No. 220 (8% curcumin; 15% oleoresin; 6.8% oil), acc. no. 257 (7.1% curcumin; 15% oleoresin; 7.3% oil)	Kerala	Sarma (2001)
High yield and dry recovery	Clone no. 326, clone no. 327, clone no. 325, clone no. 324, clone no. 317, clone no. 69	Andhra Pradesh	Rao (1982)
	Suguna, PTS-43, Sugandham, Rajendra Sonia, Kalimpong Local	West Bengal	Choudhuri and Hore 2004
High curcumin (>6%)	Aieng, Wynad Local, Edapalayam, Thodupuzha, Manathody, Pulpally, Aizwal type, Sugantham, Lakadong	Kerala	Nybe (2001); Sarma (2001)
	PCT-9 PCT-13, Lakadong, Suvarna, Alleppey turmeric	Karnataka	Venkatesha et al. (1998); Vadiraj et al. (1996); Venkatesha, Jagadesh, and Umesha (2002); Lewis (1973)
	Alleppey, Ethamukkala, Mannuthy Local, VK 145 I, VK 188 Alleppey type, VK-96, VK-112	Kerala	Muralidharan and Kutty (1976); Pillai et al. (1980); Kurian and Valsala (1996)
	No. 24, CLL 323 (Avanigadda)	Northeastern hills of India	Ghosh and Govind (1982)
	Moovattipuzha, PCT-8, 13, 14, BSR-1, Rajapuri and Amalapuram	Karnataka	Hegde (1992); Choke (1993); Venkatesha (1994); Jirali et al. (2003)
	PTS-43	West Bengal	Choudhuri and Hore (2004)
	CL-67, CL18	Tamil Nadu	AICRPS (2003)
High curcumin and oil	Cls no. 24, Duggirala	Kerala	Satheesan and Ramadasan (1987)
High volatile oil	Kokraghar, Assam, Menangadi, Hajo, Gauhati, IC 29988	Himachal Pradesh	Pathania et al. (1990)
	PCT-14, PCT-13, Prabha, Prathibha	Kerala	Ratnambal et al. (1992); Sasikumar et al. (1996)
	PTS-20, PTS-27, PTS-49	Uttar Pradesh	Verma and Tiwari (2002a)
	SI-402 (7.75%)	Himachal Pradesh	AICRPS (2005)
High oleoresin (>15%)	Konni, PCT-14	Kerala	Ratnambal (1986); Ratnambal et al. (1992)
	IISR-Prabha, IISR-Prathibha	Kerala	Sasikumar et al. (1996)
	Acc. 773, 715, 772, 781, 727, 445, and 126	Kerala	IISR (2005)
	CL-Puram	Himachal Pradesh	AICRPS (2005)
High dry recovery (>25%)	D 311	Kerala	Nybe (2001)
	Sugandham, Mannanthody, CLL-328 Sugandham	Andamans	Singh and Singh (1987)
	BSR-1, Alleppey Finger turmeric, CO1, Suvarna	Karnataka	Vadiraj et al. (1996)

(continued)

Table 16.8 Promising Lines Identified at Various Locations for Various Traits in Turmeric (Continued)

Trait	Promising Line/Cultivar/Variety	Location and State	Ref.
	Suguna	West Bengal	Choudhuri and Hore (2004)
	JTS-2	Tamil Nadu	AICRPS (2003)
	Roma	Orissa	AICRPS (2003)
Early duration varieties	RH-5, Sugandham, ACC-360, Suguna, PTS-43	West Bengal	Choudhuri and Hore (2004)
Late duration varieties	Rajendra Sonia, Kalimpong Local	West Bengal	Choudhuri and Hore (2004)
Maximum flowering (%)	Dindrigam Ca-69 (95.28%), Amalapuram	Kerala	Nybe (2001)
Early flowering type	VK 70	Kerala	Nybe (2001)
Late flowering type	Karathi, Sobha	Kerala	Nybe (2001)
High pollen fertility (%) (>50%)	Kuchipudi, Nandyal, Kasthuri Tanuku, Amrithapani Kothapetta, Dindrigam	Kerala	Nambiar et al. (1982); Nazeem et al. (1993)
Maximum seed set through hybridization	Kasturi, Katergia, Amalapuram	Kerala	Nambiar et al. (1982)
	VK 5 × VK 59	Kerala	Nybe (2001)
Maximum seed set	Amalapuram x Amrithapani-Kothapetta	Kerala	Nazeem et al. (1993)
	Open-pollinated progeny of Nandyal	Kerala	Nazeem et al. (1993)
Seed set through stigmatic pollination	VK 70 × VK 76, VK 70 × VK 55, VK 70 × Suguna	Kerala	Nybe (2001)
Maximum germination (%)	54 Ca 68 Dahgi (60.3%) and 58 Ca 73 Amalapuram (62.5%)—C. aromatica VK5 × Dindrigam (100%), VK5 × Kasthuri-Thanuka (100%)	Kerala	Nambiar et al. (1982); Nazeem et al. (1993)
High harvest index	Duggirala (70.50%), Tekurpet (68.70%)	Andhra Pradesh	Sreenivasulu and Rao (2002)

16.11 CROP IMPROVEMENT

Crop improvement programs in turmeric were limited to identifying superior turmeric genotypes with high yield, high dry recovery, and high curcumin content through screening and selection. Sometimes mutation breeding was also applied.

16.11.1 Selection

The varietal improvement in turmeric is only through clonal selection because hybridization could not be practiced due to sterility, rare seed set, and, to a certain extent, incompatibility. Most of the selections have been from the landraces collected from various parts of the country. Improved selections were developed mainly in *C. longa* and, to a lesser extent, in the *C. aromatica* type and *C. amada*.

Pujari, Patil, and Sakpal (1986) released a high-yielding variety, Krishna, for cultivation in the Maharashtra and Konkan regions (Jalgaonkar, Patil, and Rajput 1988). PTS-10 and PTS-24 were identified for high rhizome yield and high curcumin content, essential oil, oleoresin contents, and high dry recovery by the high altitude research station in Pottangi. They are less susceptible to leaf spot, rhizome rot, and scale infection and were released as Roma and Suroma and recommended for large-scale cultivation to replace local varieties. At IISR, evaluation of a large number of accessions resulted in the development of five high-yield, high-quality lines (Suvarna, Suguna, Sudarshana, Kedaram, and Alleppey Supreme).

Suguna and Sudarshana are the short-duration varieties, are also tolerant to rhizome rot, and were found to be highly suitable for the northern Telangana zone of Andhra Pradesh (Ratnambal and Nair 1986; Reddy et al. 1989; Ratnambal et al. 1992; Ramakrishna, Reddy, and Padmanabham 1995; Sasikumar et al. 2005). Indiresh et al. (1992) found that the variety Suvarna gave high yields in Karnataka. Rajendra Sonia, a high-yield line with high curcumin, was developed and recommended for Bihar (Maurya 1990; Kumar et al. 1992). CO-1 and BSR-1 were developed for rain-fed conditions of Tamil Nadu and were found to perform well in Karnataka (Sheshagiri and Uthaiah 1994). Collections VK-5 and VK-155 were superior in rhizome yield. The yield of cured rhizome was highest in VK-116 and VK-121 and was superior under 25–30% shade; VK-31 and VK-55 were superior in open conditions (Latha, Giridharan, and Naik 1995).

So far about 24 improved varieties of turmeric and one in *C. amada* have been developed. They evolved through selection, clonal selection, mutant selection, and selection from open-pollinated progeny (Table 16.9; Figure 16.9; Johny and Ravindran 2006).

16.11.2 Mutation Breeding

Induced mutations using x-rays produced two mutants namely CO-1 and BSR-1 suitable for Tamil Nadu. They were identified from Erode Local and were released for large-scale cultivation (Balashanmugam, Chezhiyan, and Ahmad Shah 1986). A BSR-2 mutant was isolated from the same Erode Local using x-rays (Chezhiyan and Shanmuga Sundaram 2000). Suroma mutant was selected from Tsundur using x-rays. Preliminary trials using colchicine, ethyl methyl sulfonate (EMS) and *N*-methyl-*N'*-nitro-*N*-nitrosoguanidine (MNG) at 250, 500, and 1000 PPM on the cultivar Mydukar resulted in normal sprouting in all treatments. Plant height and yields were more in colchicine treatments; number of leaves was more in EMS at 1000 ppm, but yield was poor.

16.11.3 Hybridization

The pioneering studies (Nambiar et al. 1982) on blossom biology and viable seed set obtained through successful hybridization have opened up many ways for recombination breeding in turmeric. Nazeem et al. (1993) attempted various crosses involving both *C. aromatica* and *C. longa* types

Table 16.9 Improved Varieties of Turmeric and Their Important Characters

S. No.	Variety	Pedigree	Crop Duration (Days)	Mean Yield (Fresh t/ha)	Dry Recovery (%)	Curcumin (%)	Oleo Resin (%)	Essential Oil (%)	Important Characters
1	CO-1	Mutant (x-ray) selection from Erode Local	270	30.5	19.5	3.2	6.7	3.7	Bold, bright-orange rhizomes suitable for drought-prone areas
2	BSR-1	Mutant (x-ray) selection from Erode Local	285	30.7	20.5	4.2	4.0	3.7	Suitable for drought-prone areas
3	BSR-2	Mutant (x-ray) selection from Erode Local	245	32.7	—	—	—	—	Bold rhizomes resistant to scale insects
4	Krishna	Clonal selection from Tekurpeta	240	9.2	16.4	2.8	3.8	2.0	Moderately tolerant to pests and diseases
5	Sugandham	Germplasm selection	210	15.0	23.3	3.1	11.0	2.7	Moderately tolerant to pests and diseases
6	Roma	Clonal selection from Tsundur	250	20.7	31.0	6.1 areas	13.2	4.2	Suitable for hilly areas
7	Suroma	Mutant (x-ray) selection from Tsundur	253	20.0	26.0	6.1	13.	4.4	Field tolerant to leaf blotch, leaf spot, and rhizome scale
8	Ranga	Clonal selection from Rajpuri Local	250	29.0	24.8	6.3	13.5	4.4	Bold rhizomes, moderately resistant to leaf blotch and rhizome scales
9	Rasmi	Clonal selection from Rajpuri Local	240	31.3	23.0	6.4	13.4	4.4	Bold rhizomes
10	Rajendra Sonia	Local germplasm selection	225	23.0	18.0	8.4	—	5.0	Bold and plump rhizomes
11	Megha Turmeric	Selection from Lakadong types	300–315	20.0	16.37	6.8	—	—	Bold rhizomes
12	Pant Peetabh	Selection from germplasm	—	29.0 (potential)	18.5	7.5	—	1.0	Resistant to rhizome rot
13	Suranjana	Local germplasm selection	235	—	21.2	5.7	10.9	4.1	Tolerant to rhizome rot and leaf blotch resistant to rhizome scales and moderately resistant to shoot borer
14	Suvarana	Selection from germplasm	200	17.4	20.0	4.3	13.5	7.0	Bright-orange colored rhizome with slender fingers
15	Suguna	Germplasm selection	190	29.3	20.4	7.3	13.5	6.0	Short-duration type, field tolerant to rhizome rot
16	Sudarshana	Germplasm selection	190	28.8	20.6	5.3	15.0	7.0	Short-duration type, field tolerant to rhizome rot
17	IISR Prabha	Selection from open-pollinated seedlings	205	37.47	19.5	6.5	15.0	6.5	Rhizomes with close internodes

18	IISR Pratibha	Selection from open-pollinated seedlings	225	39.12	18.5	6.21	16.2	6.2	Rhizomes with close internodes
19	IISR Alleppey Supreme	Selection from Alleppey finger turmeric	210	35.4	19.0	5.55	16.0	—	Tolerant to leaf blotch
20	IISR Kedaram	Germplasm selection	210	35.5	18.9	5.9	13.6	—	Tolerant to leaf blotch
21	Kanthi	Clonal selection from Mydukur	240–270	37.65	20.15	7.18	8.25	5.15	Big mother rhizomes and bold fingers with short internodes
22	Sobha	Germplasm selection	240–270	35.88	19.38	7.39	9.65	4.24	Big mother rhizomes and bold fingers with short internodes
23	Sona	Germplasm selection	240–270	4.02 t dry	18.88	7.12	10.25	4.4	Field tolerant to leaf blotch
24	Varna	Germplasm selection	240–270	4.16 t dry	19.05	7.87	10.8	4.56	Bold rhizome with short internodes; field tolerant to leaf blotch

Source: Johny, A. K., and Ravindran, P. N. 2006. *Spice India* 18:40–44.

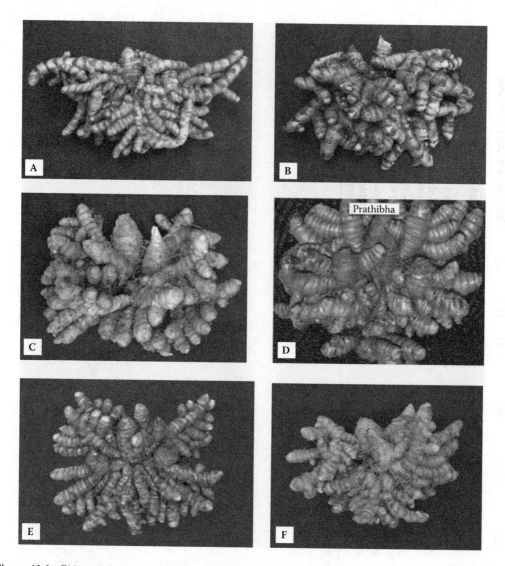

Figure 16.9 Rhizome characters of improved varieties of turmeric: (A) Suvarna, (B) Suguna, (C) Prabha, (D) Prathibha, (E) Kedaram, (F) Alleppey Supreme.

with varing degree of success. The seed germination ranged from 100% (VK5 × Dindgam, VK5 × Kasthuri Tanuku) to 17.22% (Amalapuram × Amruthapani Kothapetta). In crosses Amalapuram × Nandyal and Kodur × Kasthuri Tanuku, the seeds failed to germinate. The cross combinations also affected vigor of the seedlings.

The degree of success in cross combinations may be due to the differential cytotypes and pollen fertility among the parents. Failure of all the types to set seeds after selfing indicated that this may be due to self-incompatibility. Progenies generated through open pollination or through hybridization showed variability with respect to important agronomic characters (George 1981; Menon, Valsala, and Nair 1992; Renjith, Valsala, and Nybe 2001). Because turmeric is vegetatively propagated, any superior genotype obtained through hybridization is fixed immediately.

The seedling progenies obtained were evaluated and two of the most promising cultivars, Prabha and Prathibha, were released. These varieties were high yielding with high quality (Sasikumar et al. 1996; Table 16.8; Figure 16.9 C and D).

Hybridization and mutation breeding were vigorously employed by the horticulture industry to produce a few ornamental curcumas. For example, in *C. alismatifolia*, the flowers are highly valued as cut flowers and pot plants due to its range of vibrant colors and large, long-lasting inflorescences. *C. alismatifolia* 'Patumma,' which has strong stalks and large, symmetric inflorescences, is moderately resistant to fungal blight, has high yield, and has been used to produce numerous high-quality hybrids and mutants for commercial cultivation (Anuntalabhochai et al. 2007).

16.12 BIOTECHNOLOGICAL APPROACHES

16.12.1 Micropropagation

Protocols for micropropagation of turmeric (Figure 16.10) were published by Nadgauda et al. (1978); Nirmal Babu, Ravindran, and Peter (1997); and Sunitibala, Damayanti, and Sharma (2001). This technique could be used for production of disease-free planting material of elite plants. Salvi, George, and Eapen (2000) reported direct regeneration of shoots from immature inflorescence cultures of turmeric.

16.12.2 Field Evaluation of Tissue-Cultured Plants

Tissue-cultured plants of turmeric (kasturi turmeric, mango ginger, *Kaempferia galanga*) behave similarly to those of ginger and hence require at least two crop seasons to develop rhizomes of normal size that can be used as seed rhizomes for commercial cultivation. Salvi, George, and Eapen (2002) reported that the micropropagated plants showed a significant increase in shoot length, number of tillers, number and length of leaves, number of gingers, and total fresh rhizome weight per plant when compared with conventionally propagated plants.

16.12.3 Plant Regeneration from Callus Cultures and Somaclonal Variation

Organogenesis and plantlet formation were achieved (Figure 16.11) from the callus of turmeric (Shetty et al. 1982; Nirmal Babu et al. 1997; Sunitibala et al. 2001; Salvi, George, and Eapen 2001;

Figure 16.10 Micropropagation of turmeric.

Figure 16.11 Plant regeneration of turmeric through callus culture.

Salvi et al. 2002; Praveen 2005) and the micropropagated plants showed good morphological variation (Figure 16.12). Salvi et al. (2000, 2001) also reported plant regeneration from leaf callus of
turmeric, and random amplified polymorphic DNA (RAPD) analysis of regenerated plants showed
variation at the DNA level. Variants with high curcumin content were isolated from tissue-cultured
plantlets (Nadgauda, Khuspe, and Mascarenhas 1982).

Root rot disease-tolerant clones of the turmeric cultivar Suguna were isolated using continuous
in vitro selection techniques against pure culture filtrate in *Pythium graminicolum* (Gayatri, Roopa,
and Kavyashree 2005). Good variation in rhizome characters was obtained from somaclones of
turmeric (Praveen 2005; Sumathi 2007).

Successful plant regeneration and variations in plant and rhizome characters among regener-
ated plants were reported in *Kaempferia galanga* (Ajith Kumar and Seeni 1995); *C. domestica* var.
Koova; *C. aeruginosa* and *C. caesia* (Balachandran, Bhat, and Chandel 1990); *Alpinia conchigera*;
Alpinia galanga; *Curcuma domestica* [*C. longa*]; *C. zedoaria* and *Kaempferia galangal* (Chan and
Thong 2004); *Curcuma caesia* and *Curcuma zedoaria* (Raju et al. 2005); *C. zedoaria*; *C. domestica*
(*C. longa*) and *C. aromatica* (Yasuda et al. 1988); and *C. amada* (Prakash et al. 2004).

16.12.4 Inflorescence Culture and *In Vitro* Pollination

Salvi et al. (2000) reported direct shoot regeneration from cultured immature inflorescences of
C. longa 'Elite.' Renjith et al. (2001) reported *in vitro* pollination and hybridization between two
short duration types, VK-70 and VK-76, and observed seed set. This reduces the breeding time and
helps in recombination breeding, which had so far not been attempted in turmeric.

16.12.5 Microrhizomes

In vitro microrhizome formation (Figure 16.13) has been reported in turmeric (Raghu Rajan
1997; Nirmal Babu et al. 1997; Nirmal Babu, Ravindran, and Sasikumar 2003; Sanghamitra and

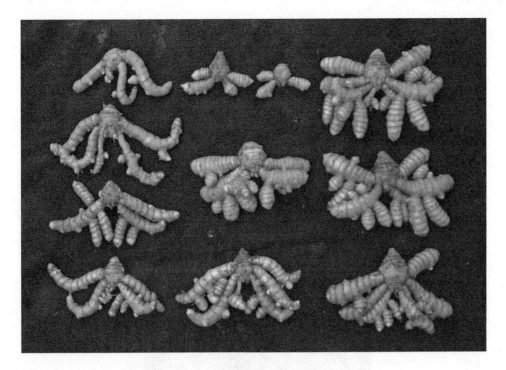

Figure 16.12 Variation in rhizome characters among somaclones of turmeric.

Nayak 2000 a,b; Sunitibala et al. 2001; Shirgurkar, John, and Nadgauda 2001; Peter et al. 2002). In turmeric, micropropagation by *in vitro* microrhizomes is an ideal method for the production of disease-free planting material and also for conservation and exchange of germplasm (Raghu Rajan 1997; Sunitibala et al. 2001). Since minimal levels of growth regulators are used and the number of subculture cycles is reduced in microrhizome production, the pathway is better suited for the production of genetically stable planting material. Microrhizomes can be produced *in vitro*, independently of seasonal fluctuations. Microrhizome production depends on size of the multiple shoots used. The microrhizomes produced varied in size (0.1–2.0 g) and could be planted directly in the field. Microrhizome-regenerated plantlets are smaller in size compared with the conventional method, but give reasonably big rhizomes (Nirmal Babu 2003).

16.12.6 Synthetic Seeds

Sajina et al. (1997) and Gayatri et al. (2005) reported development of synthetic seeds in turmeric by encapsulating the somatic embryos and shoot buds in calcium alginate.

16.12.7 Molecular Characterization and Diagnostics

Although a detailed morphological characterization of turmeric was done, its molecular characterization is still in a nascent stage except for some genetic fidelity studies of micropropagated plants and isozyme-based characterization (Sasikumar 2005). Isozyme markers were used to characterize turmeric germplasm (Shamina et al. 1998). Allozyme profiles, molecular markers, and sequence analysis were used to study genetic diversity in populations of *C. alismatifolia* (Paisooksantivatana, Kako, and Seko 2001).

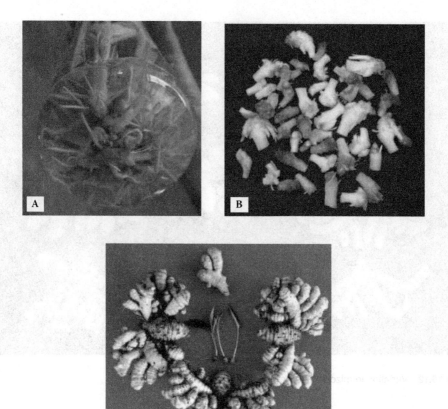

Figure 16.13 Induction of microrhizomes and microrhizomes in comparison with the rhizomes developed from them.

Interrelationships among 96 Indian cultivars and related species of turmeric were studied using the RAPD method. RAPD showed polymorphism among 96 accessions. The intraspecies polymorphism in curcuma was high as compared to the interspecies polymorphism (IISR 2003, 2004). Molecular fingerprints of 15 *Curcuma* species were developed using eight intersimple sequence repeats (ISSRs) and 39 RAPD markers to elucidate the genetic diversity/relatedness among the species. High polymorphism was observed among species. Dendrograms were constructed and cluster analysis of data using the unweighted pair group method with arithmetic mean (UPGMA) algorithm placed 15 species into seven groups, which is congruent with classification based on morphological characters. The maximum molecular similarity observed between two of the *Curcuma* species—namely, *C. raktakanta* and *C. montana*—suggests the need for reexamining the separate status given to these two species. Further, the status of *C. montana* and *C. pseudomontana*, the two species mainly discriminated based on the presence of sessile tubers, also needs to be reassessed (Syamkumar and Sasikumar 2007; Syamkumar 2008).

A polymerase chain reaction (PCR)-based detection of adulteration in market samples of turmeric powders was attempted (Remya, Syamkumar, and Sasikumar 2004; Sasikumar et al. 2004). The studies described an efficient method for the detection of extraneous *Curcuma* sp. contamination in the powdered market samples of turmeric using RAPD markers; this is not easily discriminated by other analytical techniques routinely used for the identification of such adulterants. Three market samples of turmeric powder studied revealed the presence of more *Curcuma zedoaria* (wild species) powder than *Curcuma longa* (the common culinary turmeric) powder, though the curcumin levels of the samples tallied with the quality standards prescribed for the commodity.

PCR-based techniques were also developed to differentiate the species of *Curcuma wenyujin*, *C. sichuanensis*, and *C. aromatica* to check adulteration (Chen et al. 1999). Molecular methods are being used for the authentication of Chinese and Japanese *Curcuma* drugs (Cao et al. 2001; Sasaki et al. 2002). Xia et al. (2005) used 5S-rRNA spacer domains and chemical fingerprints for quality control and authentication of Rhizoma and Curcumae, traditional Chinese medicine used in removing blood stasis and alleviating pain. Rhizomes of three species—*Curcuma wenyujin, C. phaeocaulis*, and *C. kwangsiensis*—were used.

Pimchai et al. (1999) developed molecular markers for identifying early flowering curcumas. Flowers of *Curcuma alismatifolia* are highly valued as a cut flower and pot plant due to its range of vibrant colors and long-lasting inflorescences. The variety Patumma of *C. alismatifolia* is most prefered because it has strong stalks and large, symmetric inflorescences, is moderately resistant to fungal blight, offers high yield, and has been used to produce numerous high-quality hybrids and mutants. In an effort to identify the hybrids of Patumma, a reproducible and robust sequence-characterized amplified region (SCAR) marker was developed. This SCAR amplified a 600 bp region in which were conserved all Patumma varieties and hybrids, but did not amplify in other 24 distinct *Curcuma* varieties (Anuntalabhochai et al. 2007).

16.13 PROTECTION OF PLANT VARIETIES AND IPR ISSUES

The International Treaty on Plant Genetic Resources for Food and Agriculture recognizes the importance of plant genetic resources and their conservation. The treaty was adopted by the FAO Conference on November 3, 2001, and entered into force on June 29, 2004. The Convention on Biological Diversity (CBD) (1992) is recognizing the sovereign rights of the states to use their own biological resources. The convention expects the parties to facilitate access to genetic resources by other parties subject to national legislation and on mutually agreed terms. The CBD also recognizes contributions of local and indigenous communities to the conservation and sustainable utilization of biological resources through traditional knowledge, practices, and innovations and provides for equitable sharing of benefits with such people arising from the utilization of their knowledge, practices, and innovations.

India has adopted an effective sui generis system for protection of plant varieties, farmers' rights and those of the plant breeders, and development of new varieties. Registration and protection of plant varieties in India are covered under the Protection of Plant Varieties and Farmers' Rights Act, 2002. Conditions of registration are new (novel) and distinct, uniform, and stable (DUS) for new and extant varieties. The draft guidelines for DUS testing in turmeric were prepared by the PPV (Protection of Plant Varieties) and FRA (Farmers' Rights Authority), New Delhi, India) in collaboration with IISR (Indian Institute of Spices Research, Calicut, India) and are notified (PPV & FRA 2010).

Turmeric is an important crop with many IPR (intellectual property right) issues. Over 61 patents were registered in India and many more are in progress. Most of these patents are concerned with processes for extraction and isolation of useful compounds, such as

a) Extraction of turmerone oil from turmeric oleoresin industry waste

b) Extraction of curcumins from turmeric rhizomes

c) Processes for the manufacture of pharmaceutical preparations for the treatment of disease involving turmeric

d) Process for making a synergistic composition for immunomodulatory activity with special references to rheumatic diseases, immunodeficiency diseases, and various forms of degenerative musculoskeletal diseases

e) Process for producing therapeutically active pure curcumin from *Curcuma longa*

f) Process for the preparation of curcuminoids mixture from spent turmeric oleoresin, which comprises a process for the isolation and purification of sesquiterpene compounds from a curcuma

g) Process for the extraction of an immunomodulating a fraction from *Curcuma longa* mainly containing artumerone and dihydrotumerone

Many other processes for preparing cough syrups, manufacture of herbal antimaggot composition, pest-repellent tablets, etc. using turmeric have been patented.

16.14 GEOGRAPHICAL INDICATIONS

Turmeric is a highly location-specific and agroclimatic crop and conditions in which it is grown affect the yield and quality. A geographical indication (GI) is a sign for goods that have a specific geographical origin and possess qualities or a reputation that is due to that place of origin. Indian turmeric varieties like Lakadang turmeric and Alleppey finger turmeric are known for their intrinsic quality characters and are protected by GI.

16.15 FUTURE OUTLOOK

Turmeric is the source of curcuminoids (curcumin) and has numerous pharmaceutical, nutraceutical, and phytoceutical properties in addition to being a nontoxic constituent of the diet and a natural colorant. The major priority is to develop varieties with high yields of curcumin. Organic turmeric is much desired. Identification of genotypes with resistance to biotic and abiotic stresses and responsive to organic cultivation is needed. Use of recombination breeding will open up hitherto nonavailable variability for developing new, improved varieties. Increasing the spectrum of variation through somaclonal variation and *in vitro* mutagenesis is an alternative approach. Value addition and identification of newer pharmaceutically and neutraceutically important compounds from turmeric using bioinformatic and biochemical tools is important. Many species of *Curcuma* are gaining importance as ornamentals.

REFERENCES

AICRPS. 2003. Annual report: 2002–03 of All India Coordinated Research Project on Spices, 9–12. Indian Institute of Spices research, Calicut, Kerala, India.

———. 2005. Annual Report: 2003–04 of All India Coordinated Research Project on Spices, 12–16. Indian Institute of Spices Research, Calicut, Kerala, India.

———. 2009. Annual report: 2008–09 of All India Coordinated Research Project on Spices, 89. Indian Institute of Spices Research, Calicut, Kerala, India.

Aiyadurai, S. G. 1966. *A review of research on spices and cashewnut in India.* Indian Council of Agricultural Research, New Delhi.

Ajith Kumar, P., and Seeni, S. 1995. Isolation of somaclonal variants through rhizome explant cultures of *Kaempferia galanga* L. In All India Symposium on Recent Advances in Biotechnological Applications of Plant Tissue and Cell Culture, 43. CFTRI Mysore, India.

Anonymous. 1985. Annual report 1983. Central Plantation Crops Research Institute, Kasargod.

Ansari, M. H., and Ahmad, S. 1991. *Curcuma zedoaria* root extract: *In vitro* demonstration of antiamoebic activity. *Biomedical Research (Aligarh)* 2:192–196.

Anuntalabhochai, S., Sitthiphrom, S., Thongtaksin, W., Sanguansermsri, M., and Cutler, R. W. 2007. Hybrid detection and characterization of *Curcuma* spp. using sequence characterized DNA markers. *Scientia Horticulturae* 111:389–393.

Asai, A., and Miyazawa, T. 2000. Occurrence of orally administered curcuminoid as lucuronide and glucuronide/ sulfate conjugates in rat plasma. *Life Sciences* 67:2785–2793.

Baker, J. G. 1882. Scitaminae. In *The flora of British India*, vol. VI, ed. Hooker, J. D., 198–264. Bishen Singh Mahendrapal Singh, Dehradun, Rep. 1978.

Balachandran, S. M., Bhat, S. R., and Chandel, K. P. S. 1990. *In vitro* clonal multiplication of turmeric (*Curcuma* spp.) and ginger (*Zingiber officinale* Rosc.). *Plant Cell Reports* 8:521–524.

Balashanmugam, P. V., Chezhiyan, N., and Ahmad Shah, H. 1986. BSR-1 Turmeric. *South Indian Horticulture* 34:60–61.

Banerjee, A., and Nigam, S. S. 1976. Antifungal activity of the essential oil of *Curcuma caesia* Roxb. *Indian Journal of Medical Research* 64:1318–1321.

Bordoloi, A. K., Sperkova, J., and Leclercq, P. A. 1999. Essential oil of *Curcuma aromatica* Salisb. from northeast India. *Journal of Essential Oil Research* 11:537–540.

Bown, D. 2001. *New encyclopedia of herbs and their uses.* D. K. Pub., New York.

Bruneton, J. 1999. *Pharmacognosy, phytochemistry, medicinal plants,* 2nd ed. Lavoisier Pub. Paris.

Burtt, B. L. 1977. *Curcuma zedoaria. Garden's Bulletin* (Singapore). 30:59–62.

Burtt, B. L., and Smith, R. M. 1983. Zingiberaceae. In *A revised handbook to the Flora of Ceylon*, vol. IV, ed. Dasanayake, M. D., 488–532. Amerind Publishers, New Delhi.

Cao, H., Komatsu, K., Yao Xue, and Xue, B. 2003. Molecular identification of six medicinal *Curcuma* plants produced in Sichuan: Evidence from plastid *trnK* gene sequences. *Biological and Pharmaceutical Bulletin* 22:871–875.

Cao, H., Sasaki, Y., Fushimi, H., Komatsu, K., and Cao, H. 2001. Molecular analysis of medicinally used Chinese and Japanese *Curcuma* based on 18S rRNA and *trnK* gene sequences. *Biological and Pharmaceutical Bulletin* 24:1389–1394.

Chakravorti, A. K. 1948. Multiplication of chromosome numbers in relation to speciation of Zingiberaceae. *Science and Culture* 14:137–140.

Chan, L. K., and Thong, W. H. 2004. *In vitro* propagation of Zingiberaceae species with medicinal properties. *Journal of Plant Biotechnology* 6:181–188.

Chandra, R., Desai, A. R., Govind, S., and Gupta, P. N. 1997. Metroglyph analysis in turmeric (*Curcuma longa* L.) germplasm in India. *Scientia Horticulturae* 70:211–222.

Chandra, R., Govind, S., and Desai, A. R. 1999. Growth, yield and quality performance of turmeric (*Curcuma longa* L.) genotypes in mid-altitudes of Meghalaya. *Journal of Applied Horticulture* (Lucknow) 1:142–144.

Chempakam, B., and Parthasarathy, V. A. 2008. Turmeric. In *Chemistry of spices*, ed. Parthasarathy, V. A., Chempakam, B., and John Zachariah, T., 97–123. CABI International, Cambridge, MA.

Chen, Y., Bai, S., Cheng, K., Zhang, S., and Nian, L. Z. 1999. Random amplified polymorphic DNA analysis on *Curcuma wenuujin* and *C. sichuanensis. Zhongguo Zhong Yao Za Zhi* (China Journal of Chinese Materia Medica) 24:131–133, 189.

Chezhiyan, N., and Shanmugasundaram, K. A. 2000. BSR-2—A promising turmeric variety from Tamil Nadu. *Indian Journal of Arecanut, Spices and Medicinal Plants* 2:24–26.

Chezhiyan, N., Thangaraj, T., Vijayakumar, M., Mohanalakshmi, M., and Ramar, A. 1999. Evaluation and selection for yield in turmeric. In *Biodiversity, conservation and utilization of spices, medicinal and aromatic plants*, ed. Sasikumar, B., Krishnamurthy, B., Rema, J., Ravindran, P. N., and Peter, K. V., 114–115. Indian Institute of Spices Research, Calicut, Kerala.

Cholke, S. M. 1993. Performance of turmeric (*Curcuma longa* L.) cultivars. MSc (Ag.) thesis, University of Agricultural Sciences, Dharwad, Karnataka.

Choudhuri, P., and Hore, J. K. 2004. Studies on growth, bulking rate and yield of some turmeric cultivars. *Journal of Plantation Crops* 32 (1): 47–50.

Dash, S. K., and Jana, J. C. 2004. Genetic variability in a collection of turmeric (*Curcuma longa* L.) genotypes. In *New perspectives in spices, medicinal and aromatic plants*, ed. Korikanthimath, V. S., John Zachariah, T., Nirmal Babu, K., Suseela Bhai, R., and Kandiannan, K., 56–61. ICAR Research Complex for Goa.

Dhandar, D. G., and Varde, N. P. S. 1980. Performance of selected clones of turmeric under Goa conditions. *Indian Cocoa, Arecanut and Spices Journal* 3:83–84.

Dikshit, N., Dabas, B. S., and Gautam, P. L. 1999. Genetic variation in turmeric germplasm of Kandhamal District of Orissa. In *Biodiversity, conservation and utilization of spices, medicinal and aromatic plants*, ed. Sasikumar, B., Krishnamurthy, B., Rema, J., Ravindran, P. N., and Peter, K. V., 110–113. Indian Institute of Spices Research, Calicut, Kerala.

Dixit, D., and Srivastava, N. K. 2000a. Distribution of photosynthetically fixed $^{14}CO_2$ into curcumin and essential oil in relation to primary metabolites in developing turmeric (*Curcuma longa*) leaves. *Plant Science* (Limerick) 152:165–171.

———. 2000b. Effect of iron deficiency stress on physiological and biochemical changes in turmeric (*Curcuma longa*) genotypes. *Journal of Medicinal and Aromatic Plant Science* 22:652–658.

———. 2000c. Partitioning of photosynthetically fixed $^{14}CO_2$ into oil and curcumin accumulation in *Curcuma longa* grown under iron deficiency. *Photosynthetica* 38:193–197.

Dixit, D., Srivastava, N. K., and Kumar, R. 2001. Intraspecific variation in yield capacity of turmeric, *Curcuma longa*, with respect to metabolic translocation and partitioning of $^{14}CO_2$ photoassimilate into essential oil and curcumin. *Journal of Medicinal and Aromatic Plant Science* 22–23:269–274.

Dixit, D., Srivatava, N. K., Kumar, R., and Sharma, S. 2002. Cultivar variation in yield, metabolite translocation and partitioning of $^{14}CO_2$ assimilated photosynthate into essential oil and curcumin of turmeric (*Curcuma longa* L.). *Journal of Plant Biology* 29:65–70.

Dixit, D., Srivastava, N. K., and Sharma, S. 1999. Effect of Fe deficiency on growth, physiology, yield and enzymatic activity in selected genotypes of turmeric (*Curcuma longa* L.). *Journal of Plant Biology* 26:237–241.

———. 2002. Boron deficiency induced changes in translocation of $^{14}CO_2$ photosynthate into primary metabolites in relation to essential oil and curcumin accumulation in turmeric (*Curcuma longa* L.). *Photosynthetica* 40:109–113.

Duke, J. A. 2003. *CRC handbook of medicinal spices.* CRC Press, Boca Raton, FL.

Dymock, W., Warden, C. J. H., and Hooper, D. 1893. *Pharmacographia indica.* London, Bombay, Calcutta.

Eksomtramage, L., Sirirugsa, P., Jivanit, P., and Maknoi, C. 2002. Chromosome counts of some Zingiberaceous species from Thailand. *Songklanakarin Journal of Science and Technology* 24 (2): 311–319.

Gamble, J. S. 1925. Flora of the presidency of Madras, II. Botanical Survey of India, Calcutta.

Garg, S. C., and Jain, R. K. 1998. Antimicrobial efficacy of essential oil from *Curcuma caesia*. *Indian Journal of Microbiology* 38:169–170.

Garg, S. N., Naquvi, A. A., Bansal, R. P., Bahl, J. R., and Kumar, S. 2005. Chemical composition of the essential oil from the leaves of *Curcuma zedoaria* Rosc. of Indian origin. *Journal of Essential Oil Research* 17:29–31.

Gayatri, M. C., Roopa, D. V., and Kavyashree, R. 2005. Selection of turmeric callus for tolerant to culture filtrate of *Pythium graminicolum* and regeneration of plants. *Plant Cell, Tissue and Organ Culture* 83:33–40.

George, H. 1981. Variability in the open pollinated progenies of turmeric (*Curcuma longa*). MSc (Horti.) thesis, Faculty of Agriculture, Department of Plantation Crops, Vellanikarra, Trichur.

Ghosh, S. P., and Govind, S. 1982. Yield and quality of turmeric in northeastern hills. *Indian Horticulture* 39:230–232.

Gopalam, A., and Ratnambal, M. J. 1987. Gas chromatographic evaluation of turmeric essential oils. *Indian Perfumer* 31:245–296.

Govindarajan, V. S. 1980. Turmeric—Chemistry, technology and quality. *CRC Critical Reviews in Food Science and Nutrition* 12:199–310.

Gupta, S. K., and Banerjee, A. B. 1972. Screening of West Bengal plants for antifungal activity. *Economic Botany* 26:255–259.

Hegde, G. S. 1992. Studies on the performance of turmeric (*Curcuma longa* L.) cultivars. MSc (Ag.) thesis, University of Agricultural Sciences, Dharwad.

Holttum, R. E. 1950. The Zingiberaceae of the Malay Peninsula. *Garden's Bulletin* (Singapore) 13:1–249.

Hong, C. H., Noh, M. S., Lee, W. Y., and Lee, S. K. 2002. Inhibitory effects of natural sesquiterpenoids isolated from the rhizomes of *Curcuma zedoaria* on prostglandin E-2 and nitric oxide production. *Planta Medica* 68:545–547.

Hooker, J. D. 1886. *The flora of British India*, vol. V. Reeve L and Co., London (Rep). pp. 78–95.

IISR. 2003. Annual report: 2002–03. Indian Institute of Spices Research, Calicut, Kerala.

———. 2004. Annual report: 2003–04. Indian Institute of Spices Research, Calicut, Kerala.

———. 2005. Annual report: 2004–05. Indian Institute of Spices Research, Calicut, Kerala.

———. 2009. Annual report: 2008–09. Indian Institute of Spices Research, Calicut, Kerala.

Indiresh, K. M., Uthaiah, B. C., Herle, P. S., and Rao, K. B. 1990. Morphological, rhizome and yield characters of different turmeric varieties in coastal Karnataka. *Mysore Journal of Agricultural Science* 24:484–490.

Indiresh, K. M., Uthaiah, B. C., Reddy, M. J., and Rao, K. B. 1992. Genetic variability and heritability studies in turmeric. *Indian Cocoa, Arecanut and Spices Journal* 16:52–54.

Jalgaonkar, R., and Jamdagni, B. M. 1989. Evaluation of turmeric genotypes for yield and yield determining characters. *Annals of Plant Physiology* 3:222–228.

Jalgaonkar, R., Jamadagni, B. M., and Selvi, M. J. 1990. Genetic variability and correlation studies in turmeric. *Indian Cocoa, Arecanut and Spices Journal* 14:20–22.

Jalgaonkar, J., Patil, M. M., and Rajput, J. C. 1988. Performance of different varieties of turmeric under Konkan conditions of Maharastra. In *Proceedings of the National Seminar on Chillies, Ginger and Turmeric*, Hyderabad, pp. 102–105.

Jana, J. C., and Bhattacharya, B. 2001. Performance of different promising cultivars of turmeric (*Curcuma domestica* Val.) under terai agro-climatic region of West Bengal. *Environment and Ecology* 19:463–465.

Jana, J. C., Dutta, S., and Chatterjee, R. 2001. Genetic variability, heritability and correlation studies in turmeric (*Curcuma longa* L.). *Research on Crops* 2:220–225.

Jangde, C. R., Phadnaik, B. S., and Bisen, V. V. 1998. Anti-inflammatory activity of extracts of *Curcuma aromatica* Salisb. *Indian Veteranary Journal* 75:76–77.

Jirali, D. I., Hiremath, S. M., Chetti, M. B., and Patil, B. C. 2003. Association of yield and yield components with growth, biochemical and quality parameters for enhancing the productivity in turmeric (*Curcuma longa* L.) genotypes. In *National Seminar on New Perspectives in Spices, Medicinal and Aromatic Plants*, ICAR Research Complex for Goa, Goa, abstract 5–6.

Jirovetz, L., Buchbauer, G., Puschmann, C., Shafi, M. P., and Nambiar, M. K. G. 2000. Essential oil analysis of *Curcuma aeruginosa* Roxb. leaves from South India. *Journal of Essential Oil Research* 12:47–49.

Johny, A. K., and Ravindran, P. N. 2006. Over 225 high yielding spice varieties in India. Part 2. *Spice India*, 18:40–44.

Joseph, R., Joseph, T., and Jose, J. 1999. Karyomorphological studies in the genus *Curcuma* Linn. *Cytologia* 64:313–317.

Joshi, L. K., and Sharma, N. P. 1980. Disease of ginger and turmeric. In *National Symposium on Spices Calicut*, 104–119.

Kapoor, L. D. 1990. *Handbook of Ayurvedic medicinal plants*. CRC Press, Boca Raton, FL, p. 185.

KAU. 2000. Three decades of spices research at KAU. Directorate of Extension, Kerala Agricultural University, Thrissur, Kerala.

Kim, K. I., Shin, K. S., and Jun, W. J. 2001. Effects of polysaccharides from rhizomes of *Curcuma zedoaria* on macrophage functions. *Bioscience, Biotechnology and Biochemistry* 65:2369–2377.

Kojima, H. T., Yanai, T., and Toyoya, A. 1998. Essential oil constituents from Japanese and indian curcuma aromatic, Rhizanes. *Plant Medica* 64: 380–381.

Kress, W. J., Prince, L. M., and Williams, K. J. 2002. The phylogeny and a new classification of the gingers (Zingiberaceae): Evidence from molecular data. *American Journal of Botany* 89:1682–1696.

Kumar, G. V., Reddy, K. S., Rao, M. S., and Ramavatharam, M. 1992. Soil and plant characters influencing curcumin content of turmeric. *Indian Cocoa, Arecanut and Spices Journal* 15:102–104.

Kumar, R., and Jain, B. P. 1998. Evaluation of growth and rhizome characters of some turmeric (*C. longa* L.) cultivars under plateau region of Bihar. *Souvenir, National Seminar on Recent Development in Spices Production Technology*, Bihar Agricultural College, Sabour, pp. 9–10.

Kumar, R., Pandey, V. P., and Dwevedi, A. 2004. Genetic variability in turmeric germplasm. In *Commercialization of spices, medicinal plants and aromatic crops*, Indian Institute of Spices Research, Calicut, Kerala, abstract 4.

Kumar, S. 1999. A note on conservation of economically important Zingiberaceae of Sikkim Himalaya. In *Biodiversity, conservation and utilization of spices, medicinal and aromatic plants*, ed. Sasikumar, B., Krishnamurthy, B., Rema, J., Ravindran, P. N., and Peter, K. V., 201–207. Indian Institute of Spices Research, Calicut, Kerala.

Kumar, T. V., Reddy, M. S., and Krishna, V. G. 1997. Nutrient status of turmeric growing soils in northern Telangana zone of Andhra Pradesh. *Journal of Plantation Crops* 25:93–97.

Kurian, A., and Nair, G. S. 1996. Evaluation of turmeric germplasm for yield and quality. *Indian Journal of Plant Genetic Resources* 9:327–329.

Kurian, A., and Valsala, P. A. 1996. Evaluation of turmeric types for yield and quality. *Journal of Tropical Agriculture* 33:75–76.

Lad, S. K. 1993. A case of seed-setting in cultivated turmeric types (*Curcuma longa* Linn.). *Journal of Soils and Crops* 3:78–79.

Latha, P., Giridharan, M. P., and Naik, B. J. 1995. Performance of turmeric (*Curcuma longa* L.) cultivars in open and partially shaded conditions. *Journal of Spices and Aromatic Crops* 4:139–144.

Lawande, K. E., Raijadhav, S. B., Yamgar, V. T., and Kale, P. N. 1991. Turmeric research in Maharashtra. *Journal of Plantation Crops* 18 (Suppl.): 404–408.

Lewis, Y. S. 1973. The importance of selecting the proper variety of a spice for oil and oleoresins extractions. In *Proceedings of Conference on Spices*, T.P.I., London, pp. 183–185.

Li, L., Zhang, Y., Oin, S., and Liao, G. 1997. Ontogeny of *Curcuma longa* L. *Zhongguo Zhong Yao Za Zhi* (China Journal of Chinese Materia Medica) 10:587–590.

Lin, S. C., Lin, C. C., Lin, Y. H., Supriyatna, S., and Teng, C. W. 1995. Protective and therapeutic effects of *Curcuma xanthorrhiza* on hepatotoxin-induced liver damage. *American Journal of Chinese Medicine* 23:243–254.

Linnaeus, C. 1753. *Specius* Plantarum. London.

Liu, N., and Wu T. L. 1999. Notes on *Curcuma* in China. *Journal of Tropical and Subtropical Botany* 7:146–150.

Lynrah, P. G., and Chakraborty, B. K. 2000. Performance of some turmeric and its close relatives/genotypes. *Journal of Agricultural Science. Society North East India* 13:32–37.

Mahadtanapk, S., Topoonyanont, N., Handa, T., Sanguansermsri, M., and Anuntalabhochai, S. 2006. Genetic transformation of *Curcuma alismatifolia* Gagnep. Using retarded shoots. *Plant Biotechnology* 23: 233–237.

Mangaly, J. K., and Sabu, M. 1993. A taxonomic revision of the South Indian species of *Curcuma* (Zingiberae). *Rheedea* 3:139–171.

Mathai, C. K. 1974. Quality studies in cashew and spices. Annual Report, CPCRI, Kasargod, pp. 166–167.

Matsuda, H., Morikawa, T., Ninomiya, K., and Yoshikawa, M. 2001. Hepatoprotective constituents from *Zedoariae rhizoma*: Absolute stereostructures of three new carabrane-type sesquiterpenes, curcumenolactones A, B, and C. *Bioorganc and Medicinal Chemistry Letters* 9:909–916.

Matsuda, H., Ninomiya, K., Morikawa, T., and Yoshikawa, M. 1998. Inhibitory effect and action mechanism of sesquiterpenes from *Zedoariae rhizoma* on d-galactosamine/lipopolysaccharide-induced liver injury. *Bioorganc and Medicinal Chemistry Letters* 8:339–344.

Maurya, K. R. 1990. RH 10, a promising variety of turmeric to boost farmers' economy. *Indian Cocoa, Arecanut and Spices Journal* 13:100–101.

Mehta, K. G., and Patel, R. H. 1980. Phenotypic stability for yield in turmeric. In *Proceedings of the National Seminar on Ginger and Turmeric*, ed. Nair, M. K., Premkumar, T., Ravindran, P. N., and Sarma, Y. R., 34–38. Central Plantation Crops Research Institute, Kasaragod.

Mehta, K. G., Raghava Rao, D. V., and Patel, S. H. 1980. Relative curcumin content during various growth stages in the leaves and rhizomes of three cultivars of *Curcuma longa* and *C. amada*. In *Proceedings of National Seminar on Ginger and Turmeric*, ed. Nair, M. K., Premkumar, T., Ravindran, P. N., and Sarma, Y. R., 76–78. Central Plantation Crops Research Institute, Kasaragod.

Menon, R., Valsala, P. A., and Nair, G. S. 1992. Evaluation of open pollinated progenies of turmeric. *South Indian Horticulture* 40:90–92.

Mohanty, D. C. 1979. Genetic variability and inter relationship among rhizome yield and yield components in turmeric. *Andhra Agricultural Journal* 26:77–80.

Mood, J., and Larsen, K. 2001. New *Curcuma* species from Southeast Asia. *New Plantsman* 8:207–217.

Muralidharan, A., and Kutty, R H. N. 1976. Performance of same selected elones of turmeric (*circuma longa*), *Agril. Res. J. Kerala* 10: 112–115.

Nadgauda, R. S., Khuspe, S. S., and Mascarenhas, A. F. 1982. Isolation of high curcumin varieties of turmeric from tissue culture. In *Proceedings of V Annual Symposium on Plantation Crops*, ed. Iyer, R. D., 143–144. Central Plantation Crops Research Institute, Kasargod.

Nadgauda, R. S., Mascarenhas, A. F., Hendre, R. R., and Jagannathan, V. 1978. Rapid clonal multiplication of turmeric *Curcuma longa* L. plants by tissue culture. *Indian Journal of Experimental Biology* 16:120–122.

Naidu, M. M., Murty, P. S. S., Kumari, P., and Murthy, G. N. 2003. Evaluation of turmeric selections for high altitude and tribal areas of Visakhapatnam (Dist.), Andhra Pradesh. In *National Seminar on New Perspectives in Spices, Medicinal and Aromatic Plants*, ICAR Research Complex for Goa, abstract 29.

Nair, M. K., Nambiar, M. C., and Ratnambal, M. J. 1980. Cytogenetics and crop improvement of ginger and turmeric. In *Proceedings of National Seminar on Ginger and Turmeric*, ed. Nair, M. K., Premkumar, T., Ravindran, P. N., and Sarma, Y. R., 15–23. Central Plantation Crops Research Institute, Kasaragod.

Nair, R. R., and Sasikumar, B. 2009. Chromosome number variation among germplasm collections and seedling progenies in turmeric, *Curcuma longa* L. *Cytologia* 74 (2): 153–155.

Nambiar, K. K. N., Sarma, Y. R., and Brahma, R. N. 1998. Field reaction of turmeric types to leaf blotch. *Journal of Plantation Crops* 5:124–125.

Nambiar, M. C. 1979. Morphological and cytological investigations in the genus *Curcuma* L. Unpublished PhD thesis, University of Bombay.

Nambiar, M. C., Thankamma Pillai, P. K., and Sarma, Y. N. 1982. Seedling propagation in turmeric (*Curcuma aromatica* Salisb). *Journal of Plantation Crops* 10:81–85.

Nandi, A. 1990. Evaluation of turmeric (*Curcuma longa*) varieties for northeastern plateau zone of Orissa under rain-fed conditions. *Indian Journal of Agricultural Science* 60:760–761.

Nandi, A., Lenka, D., and Singh, D. N. 1992. Path analysis in turmeric. *Indian Cocoa, Arecanut and Spices Journal* 27:54–55.

Narayanpur, V. B., and Hanamashetti, S. I. 2003. Genetic variability and correlation studies in turmeric (*Curcuma longa* L.). *Journal of Plantation Crops*, 31 (2): 48–51.

Natarajan, S. T. 1975. Studies on the yield components and gamma ray induced variability in turmeric (*Curcuma longa* L.). MSc (Ag.) thesis, Tamil Nadu Agricultural University, Coimbatore.

Nazeem, P. A., Menon, R., and Valsala, P. A. 1994. Blossom, biological and hybridization studies in turmeric (*Curcuma* spp.). *Indian Cocoa, Arecanut and Spices Journal* 16:106–109.

Ngamriabsakul, C., Newman, M. F., and Cronk, Q. C. B. 2003. The phylogeny of tribe *Zingibereae* (*Zingiberaceae*) based on its (nrDNA) and *trn*l-f (cpDNA) sequences. *Edinburgh Journal of Botany* 60:483–507.

Nirmal, S. V., and Yamgar, V. T. 1998. Variability in morphological and yield characters of turmeric (*Curcuma longa* L.) cultivars. *Advances in Plant Sciences* 11:161–164.

Nirmal Babu, K., Minoo, D., Geetha, S. P., Sumathi, V., and Praveen, K. 2007. Biotechnology of turmeric and related spcies. In *Turmeric—The genus* Curcuma, ed. Ravindran, P. N., Nirmal Babu, K., and Sivaraman, K., 107–125. CRC Press, Boca Raton, FL.

Nirmal Babu, K., Ravindran, P. N., and Peter, K. V. 1997. Protocols for micropropagation of spices and aromatic crops. Indian Institute of Spices Research, Calicut, Kerala, 35 pp.

Nirmal Babu, K., Ravindran, P. N., and Sasikumar, B. 2003. Field evaluation of tissue cultured plants of spices and assessment of their genetic stability using molecular markers. Final report submitted to Department of Biotechnology, government of India, 94 pp.

Nirmal Babu, K., Sasikumar, B., Ratnambal, M. J., George, J. K., and Ravindran, P. N. 1993. Genetic variability in turmeric (*Curcuma longa* L.). *Indian Journal of Genetics and Plant Breeding* 53:91–93.

Nybe, E. V. 2001. Turmeric. Three decades of spices research at KAU. Kerala Agricultural University, Thrissur, India, pp. 78–85.

Pachauri, S. P., and Mukherjee, S. K. 1970. Effect of *Curcuma longa* (Haridar) and *Curcuma amada* (Amragandhi) on the cholesterol level in experimental hypercholesterolemia of rabbits. *Journal of Research Indian Medicine* 5:27–31.

Paisooksantivatana, Y., Kako, S., and Seko, H. 2001. Isozyme polymorphism in *Curcuma alismatifolia* Gagnep. (Zingiberaceae) populations from Thailand. *Scientia Horticulturae* 88:299–307.

Palarpawar, M. Y., and Ghurde, V. R. 1989. Sources of resistance in turmeric against leaf spots incited by *Colletotrichum capsici* and *C. curcumae*. *Indian Phytopathology* 42:171–173.

Pandey, G., Sharma, B. D., and Hore, D. K. 1990. Metroglyph and index score analysis of turmeric germplasm in northeastern region of India. *Indian Journal of Plant Genetic Resources* 3:59–66.

Pandey, V. P., Dixit, J., Saxena, R. P., and Gupta, R. K. 2004a. Comparative performance of turmeric genotypes in eastern Uttar Pradesh. In *Commercialization of spices, medicinal plants and aromatic crops*, Indian Institute of Spices Research, Calicut, Kerala, abstract 4.

Pandey, V. P., Singh, T., and Srivastava, A. K. 2003. Genetic variability and correlation studies in turmeric. In *Proceedings of National Seminar on Strategies for Increasing Production and Export of Spices*, 195–200. IISR, Calicut, Kerala.

Pandey, V. S., Pandey, V. P., Pandey, S., and Dixit, J. 2004b. Divergence analysis of turmeric. In *New perspectives in spices, medicinal and aromatic plants*, ed. Korikanthimath, V. S., John Zachariah, T., Nirmal Babu, K., Suseela Bhai, R., and Kandiannan, K., 62–65. ICAR Research Complex for Goa.

Pandey, V. S., Pandey, V. P., and Singh, P. K. 2004c. Path analysis in turmeric. In *New perspectives in spices, medicinal and aromatic plants*, ed. Korikanthimath, V. S., John Zachariah, T., Nirmal Babu, K., Suseela Bhai, R., and Kandiannan, K., 32–36. ICAR Research Complex for Goa.

Pandey, V. S., Pandey, V. P., Singh, T., and Srivastava, A. K. 2003. Genetic variability and correlation studies in turmeric. In *Proceedings of National Seminar on Strategies for Increasing Production and Export of Spices*, 195–200. IISR, Calicut, Kerala.

Panja, B. N., and De, D. K. 2000. Characterization of blue turmeric: A new hill collection. *Journal of Interacademicia* 4:550–557.

———. 2001. Studies on stomatal frequency and its relationship with leaf biomass and rhizome yield of turmeric (*Curcuma longa* L.) genotypes. *Journal of Spices and Aromatic Crops* 10:127–134.

Panja, B. N., De, D. K., Basak, S., and Chattopadhyay, S. B. 2002. Correlation and path analysis in turmeric (*Curcuma longa* L.). *Journal of Spices and Aromatic Crops* 11:70–73.

Panja, B. N., De, D. K., and Majumdar, D. 2001. Evaluation of turmeric (*Curcuma longa* L.) genotypes for yield and leaf blotch disease (*Taphrina maculans* Butl.) for tarai region of West Bengal. *Environment and Ecology* 19:125–129.

Pathak, K. A., Kishore, K., Singh, A. K., and Bharali, R. 2003. Varietal evaluation of turmeric germplasm under Mizoram conditions. In *National Seminar on New Perspectives in Spices, Medicinal and Aromatic Plants*, ICAR Research Complex for Goa, Goa, abstract 50.

Pathak, S., Patra, B. C., and Mahapatra, K. C. 1960. Flowering behavior and anthesis of *Curcuma longa*. *Current Science* 29:402.

Pathania, N. K., Arya, P. S., and Singh, M. 1988. Variability studies in turmeric (*Curcuma longa* L.). *Indian Journal of Agricultural Research* 22:176–178.

Pathania, N. K., Singh, M., and Arya, P. S. 1990. Variation for volatile oil content in turmeric cultivars. *Indian Cocoa, Arecanut and Spices Journal* 14(1), 23–24.

Perry, L. M. 1980. Zingiberaceae. In *Medicinal plants of east and Southeast Asia: Attributed properties and uses*, 436–444. MIT Press, Cambridge, MA.

Philip, J., and Nair, P. C. S. 1981. Field reaction of turmeric types to important pests and diseases. *Indian Cocoa, Arecanut and Spices Journal* 4 (4): 107–109.

———. 1986. Studies on variability, heritability and genetic advance in turmeric. *Indian Cocoa, Arecanut and Spices Journal* 10:29–30.

Philip, J., Sivaraman Nair, P. C. S., Nybe, E. V., and Mohan, N. K. 1982. Variation of yield and quality of turmeric. In *Proceedings of National Seminar on Ginger and Turmeric*, ed. Nair, M. K., Premkumar, T., Ravindran, P. N., and Sarma, Y. R., 42–46. Central Plantation Crops Research Institute, Kasaragod.

Pimchai, A., Somboon, A. I., Puangpen, S., and Chiara, A. 1999. Molecular markers in the identification of some early flowering *Curcuma* L. (Zingiberaceae) species. *Annals of Botany* 84:529–534.

PPV & FRA. 2010. Guidelines for the conduct of test for distinctiveness, uniformity and stability on turmeric (*Curcuma longa* L.). Protection of Plant Varieties and Farmers' Rights Authority (PPV & FRA), government of India, 19 pp.

Prakash, S., Elangomathavan, R., Seshadri, S., Kathiravan, K., and Ignacimuthu, S. 2004. Efficient regeneration of *Curcuma amada* Roxb. plantlets from rhizome and leaf sheath explants. *Plant Cell, Tissue, Organ Culture* 78:159–165.

Prana, M. S. 1977. Studies on some Indonesian *Curcuma* species. PhD thesis, University of Birmingham.

———. 1978. Temu Lawak (*Curcuma xanthorrhiza* Roxb.). *Buletin Kebun Raya* 3 (6): 191–194.

Praveen, K. 2005. Variability in somaclones of turmeric (*Curcuma longa* L.). PhD thesis, University of Calicut, Kerala, India.

Pujari, P. P., Patil, R. B., and Sakpal, B. T. 1986. Krishna, a high yielding variety of turmeric. *Indian Cocoa, Arecanut and Spices Journal* 9:65–66.

———. 1987. Studies on growth yield quality components in different varieties. *Indian Cocoa, Arecanut and Spices Journal* 11:15–17.

Purseglove, J. W., Brown, E. G., Green, C. L., and Robin, S. R. J. 1981. Turmeric. In *Spices*, vol. 2, 532–580. Longman, New York.

Radha, K. M., Anoop, K. S., Gaddipati, J. P., and Richkab, C. S. 2006. Multiple biological activities of curcumin: A short review. *Life Sciences* 78 (18): 2081–2087.

Raghavan, T. S., and Venkattasubban, K. R. 1943. Cytological studies in the family Zingiberaceae with special reference to chromosome number and cytotaxonomy. *Proceedings of Indian Academy of Sciences*, Ser. B, pp. 118–132.

Raghu Rajan, V. 1997. Micropropagation of turmeric (*Curcuma longa* L.) by *in vitro* microrhizomes. In *Biotechnology of spices, medicinal and aromatic crops*, ed. Edison, S., Ramana, K. V., Sasikumar, B., Nirmal Babu, K., and Santhosh, J. E., 25–28. Indian Society for Spices, Calicut.

Raju, B., Anita D. and Kalitha M. C. 2005. *In vitro* clonal propagation of curcuma caesia Roxb. and *Curcuma Zedoaria* Ross. from rhizome bud explants. *J. Biochem. Biotechnol.* 14: 61–63.

Rakhunde, S. D., Munjal, S. V., and Patil, S. R. 1998. Curcumin and essential oil contents of some commonly grown turmeric (*Curcuma longa* L.) cultivars in Maharashtra. *Journal of Food Science and Technology* (Mysore) 35:352–354.

Ramachandran, K. 1961. Chromosome numbers in the genus *Curcuma* Linn. *Current Science* 30:194–196.

———. 1969. Chromosome numbers in Zingiberaceae. *Cytologia* 34:213–221.

Ramakrishna, M., Reddy, R. S., and Padmanabham, V. 1995. Studies on the performance of short duration varieties/cultures of turmeric in southern zone of Andhra Pradesh. *Journal of Plantation Crops* 23:126–127.

Ramakrishna, T. S. 1954. Leaf spot diseases of turmeric (*Curcuma longa* L.) caused by *Colletotrichum capsici* (Syd.) Butler and Bisby. *Indian Phytopathology* 7:111–117.

Rao, A. M. 2000. Genetic variability, yield and quality studies in turmeric (*Curcuma longa* L.). Unpublished PhD thesis, ANGRAU, Rajendranagar, Hyderabad.

Rao, A. M., Jagdeeshwar, R., and Sivaraman, K. 2006. Turmeric. In *Advances in spices research*, ed. Ravindran, P. N., Nirmal Babu, K., Shiva, K. N., and Johny, A. K., 433–492. Agribios, Jodhpur.

Rao, A. M., Rao, P. V., and Reddy, Y. N. 2004. Evaluation of turmeric cultivars for growth, yield and quality characters. *Journal of Plantation Crops* 32:20–25.

Rao, M. R., Reddy, K. R. C., and Subbarayudu, M. 1975. Promising turmeric types of Andhra Pradesh. *Indian Spices* 12:2–5.

Rao, P. S., Ramakrishna, M., Srinivas, C., Meena Kumari, K., and Rao, A. M. 1994. Short duration, disease resistant turmerics for northern Telangana. *Indian Horticulture* 39:55.

Rao, P. S., and Rao, T. G. N. 1988. Diseases of turmeric in Andhra Pradesh. In *Proceedings of the National Seminar on Chillies, Ginger and Turmeric*, 162–167.

Rao, P. S., Reddy, M. L. N., Rao, T. G. N., Krishna, M. R., and Rao, A. M. 1992. Reaction of turmeric cultivars to *Colletotrichum* leaf spot, *Taphrina* leaf blotch and rhizome rot. *Journal of Plantation Crops* 20:131–134.

Ratnambal, M. J. 1986. Evaluation of turmeric accession for quality. *Qualities Plantarum* 36:243–252.

Ratnambal, M. J., and Nair, M. K. 1986. High yielding turmeric selection PCT-8. *Journal of Plantation Crops* 14:94–98.

Ratnambal, M. J., Nirmal Babu, K. N., Nair, M. K., and Edison, S. 1992. PCT-13 and PCT-14—Two high yielding varieties of turmeric. *Journal of Plantation Crops* 20:79–84.

Ravindran, P. N., Nirmal Babu, K., and Shiva, K. N. 2007. Botany and crop improvement of turmeric. In *Turmeric—The genus Curcuma*, ed. Ravindran, P. N., Nirmal Babu, K., and Sivaraman, K., 15–70. CRC Press, Boca Raton, FL.

Ravindran, P. N., Nirmal Babu, K., Peter, K. V., Abraham, Z., and Tyagi, R. K. 2005. Spices. In *Plant genetic resources: Horticultural crops*, ed. Dhillon, B. S., Tyagi, R. K., Saxena, S., and Randhawa, G. J., 190–227. Narosa Publishing House, New Delhi.

Reddy, G. V., Reddy, M. L., and Naidu, G. S. 1989. Estimation of leaf area in turmeric (*Curcuma longa* Linn.) by nondestructive method. *Journal of Research*, APAU 17:43–44.

Reddy, M. L. N. 1987. Genetic variability and association in turmeric (*Curcuma longa* L.). *Progressive Horticulture* 19:83–86.

Reddy, M. L. N., Rao, A. M., Rao, D. V. R., and Reddy, S. A. 1989. Screening of short duration turmeric varieties/cultures suitable for Andhra Pradesh. *Indian Cocoa, Arecanut and Spices Journal* 12:87–89.

Remya, R., Syamkumar, S., and Sasikumar, B. 2004. Isolation and amplification of DNA from turmeric powder. *British Food Journal* 106:673–678.

Renjith, D., Valsala, P. A., and Nybe, E. V. 2001. Response of turmeric (*Curcuma domestica* Val.) to *in vivo* and *in vitro* pollination. *Journal of Spices and Aromatic Crops* 10:135–139.

Rethinam, P., Sivaraman, K., and Sushma, P. K. 1994. Nutrition of turmeric. In *Advances in horticulture, vol. 9, Plantation and Spices Crops*, 477–488.

Rheede, H. 1685. *Hortus indicus malabaricus*. Amsterdomi, Holland.

Roxburgh, W. 1832. *Flora indica or description of Indian plants*, ed. Carey, W. Serampur.

Ruby, A. J., Kuttan, G., Dinesh Babu, K., Rajasekharan, K. N., and Kuttan, R. 1995a. Antitumor and antioxidant activity of natural curcuminoids (*Curcuma longa*) on iron-induced lipid peroxidation in the rat liver. *Food and Chemical Toxicology* 32:279–283.

———. 1995b. Antitumor and antioxidant activity of natural curcuminoids. *Cancer Letters* 94:79–83.

Sabu, M. and Mangaly, J.K. 1988. *Curcuma vamana* (Zinziberaceae)-a new species from South India *J. Econ. Tax. Bot.* 12: 307–309.

Sajina, A., Minoo, D., Geetha, S. P., Samsudeen, K., Rema, J., Babu, K. N., and Ravindran, P. N. 1997. Production of synthetic seeds in a few spice crops. In *Biotechnology of spices, medicinal and aromatic plants*, ed. Ramana, K. V., Sasikumar, B., Babu, K. N. and Eapen, S. J., 65–69. Indian Society for Spices, Calicut.

Salvi, N. D., George, L., and Eapen, S. 2000. Plant regeneration from leaf base callus of turmeric and random amplified polymorphic DNA analysis of regenerated plants. *Plant Cell, Tissue and Organ Culture* 66:113–119.

———. 2001. Plant regeneration from leaf base callus of turmeric and random amplified polymorphic DNA analysis of regenerated plants. *Plant Cell, Tissue and Organ Culture* 66:113–119.

———. 2002. Micropropagation and field evaluation of micropropagated plants of turmeric. *Plant Cell, Tissue and Organ Culture* 68:143–151.

Sanghamitra, N., and Nayak, S. 2000a. *In vitro* microrhizome production in four cultivars of turmeric (*Curcuma longa* L.) as regulated different factors. In *Spices and aromatic plants—Challenges and opportunities in the new century*. Indian Society for Spices, Calicut, Kerala, India.

———. 2000b. *In vitro* multiplication and microrhizome induction in *Curcuma aromatica* Salisb. *Plant Growth Regulation* 32:41–47.

Sarma, Y. R. 2001. Research on major spices at IISR-blossomed and to be blossomed. *Indian Spices* 38 (3): 2–10.

Sarma, Y. R., Anandaraj, M., and Venugopal, M. N. 1994. Diseases of spice crops, In. Advances in Horticulture. Vol. 10, Plantation and Spice Crops Part 2 (Eds) K. L. Chadha and P. Rethinam. Malhotra Publishing House, New Delhi. pp. 1015–1057.

Sarma, Y. R., and Anandaraj, M. 2000. Disease of spice crops and their management. *Indian Journal of Arecanut, Spices and Medicinal Plants* 2:8–20.

Sasaki, Y., Fushimi, H., Cao, H., Cai, S. Q., and Komatsu, K. 2002. Sequence analysis of Chinese and Japanese *Curcuma* drugs on the 18S rRNA gene and *trnK* gene and the application of amplification-refractory mutation system analysis for their authentication. *Biological and Pharmaceutical Bulletin* 25:1593–9.

Sasaki, Y., Fushimi, H., and Komatsu, K. 2004. Application of single-nucleotide polymorphism analysis of the *trnK* gene to the identification of *Curcuma* plants. *Biological and Pharmaceutical Bulletin* 27:144–146.

Sasikumar, B. 2005. Genetic resources of *Curcuma*: Diversity, characterization and utilization. *Plant Genetic Resources: Characterization and Utilization* 3:230–251.

Sasikumar, B., George, J. K., Saji, K. V., and Zacharaiah, T. J. 2005. Two new high yielding, high curkumin turmeric (*Curcuma longa* L.) varieties—IISR Kedaram and IISR Alleppey Supreme. *Journal of Spices and Aromatic Crops* 14:71–74.

Sasikumar, B., George, J. K., Zachariah, T. J., Ratnambal, M. J., Nirmal Babu, K., and Ravindran, P. N. 1996. IISR Prabha and IISR Prathibha-two new high yielding and high quality turmeric (*Curcuma longa* L.) varieties. *Journal of Spices and Aromatic Crops* 5 (1): 41–48.

Sasikumar, B., Ravindran, P. N., and George, J. K. 1994. Breeding ginger and turmeric, Indian Cocoa Arecanut and Spices J. 18(1):10–12.

Sasikumar, B., Zachariah, T. J., Syamkumar, S., and Remya, R. 2004. PCR-based detection of adulteration in the market samples of turmeric powder. *Food Biotechnology* 18 (3): 299–306.

Sastrapradja, S., and Aminali, S. H. 1970. Factors affecting fruit production in *Curcuma* species. *Annals of Bogorensis*. 5:99–107.

Satheesan, K. V. 1984. Physiology of growth and productivity of turmeric (*Curcuma domestica* Val.) in monoculture and as an intercrop in coconut garden. PhD thesis, University of Calicut.

Satheesan, K. V., and Ramadasan, A. 1987. Curcumin and essential oil contents of three turmeric (*Curcuma domestica* Val.) cultivars grown in monoculture and as intercrop in cocunut gardens. *Journal of Plantation Crops* 15:31–37.

Satheesan, K. V., and Ramadasan, A. 1988a. Effect of growth retardant CCC on growth and productivity of turmeric under monoculture and in association with coconut. *Journal of Plantation Crops* 16:140–143.

———. 1988b. Changes in carbohydrate levels and starch/sugar ratio in three turmeric (*Curcuma domestica* Val.) cultivars grown in monoculture and as an intercrop in coconut garden. *Journal of Plantation Crops* 16:45–51.

Sato, D. 1948. The karyotype and phylogeny of Zingiberaceae. *Japanese Journal of Genetics* 23:44.

———. 1960. The karyotype analysis in Zingiberales with special reference to the protokaryotype and stable karyotype. *Science Papers of the College of General Education* Univ. Tokyo, 10:225–243.

Shahi, R. P., Shahi, B. G., and Yadava, H. S. 1994. Stability analysis for quality characters in turmeric (*Curcuma longa* L.). *Crop Research* (Hisar) 8:112–116.

Shahi, R. P., Yadava, H. S., and Sahi, B. G. 1994. Stability analysis for rhizome yield and its determining characters in turmeric. *Crop Research* (Hisar) 7:72–78.

Shamina, A., Zachariah, T. J., Sasikumar, B., and George, J. K. 1998. Biochemical variation in turmeric based on isozyme polymorphism. *Journal of Horticulture Science and Biotechnology* 73:477–483.

Sharma, A. K., and Bhattacharya, N. K. 1959. Cytology of several members of Zingiberaceae and study of the inconsistency of their chromosome complement. *La Cellule* 59:279–349.

Sheshagiri, K. S., and Uthaiah, B. C. 1994. Performance of turmeric (*Curcuma longa* L.) varieties in the hill zone of Karnataka. *Indian Cocoa, Arecanut Spices Journal* 3:161–163.

Shirgurkar, M. V., John, C. K., and Nadgauda, R. S. 2001. Factors affecting *in vitro* microrhizome production in turmeric. *Plant Cell, Tissue and Organ Culture* 64:5–11.

Shiva, K. N., Suryanarayana, M. A., and Medhi, R. P. 2003. Genetic resources of spices and their conservation in Bay Islands. *Indian Journal of Plant Genetic Resources* 16:91–95.

Shridhar, S. V., Prasad, S., and Gupta, H. S. 2002. Stability analysis in turmeric. In *National Seminar on Strategies for Increasing Production & Export of Spices*, IISR, Calicut, Kerala, abstract 13.

Shukui, Q., Lin, W., and Jun, Q. 1995. Elemene emulsion for advanced lung cancer (meeting abstract). 12th Asia Pacific Cancer Conference, Singapore, p. 298.

Singh, B., Yadav, J. R., and Srivastava, J. P. 2003. Azad Haldi-1 A disease resistant variety turmeric (*Curcuma longa*). *Plant Archives* 3:151–152.

Singh, S., and Singh, S. 1987. Evaluation of some turmeric types for the Andamans. *Journal of Andaman Science Association* 3:38–39.

Singh, Y., Mittal, P., and Katoch, V. 2003. Genetic variability and heritability in turmeric (*Curcuma longa* L.). *Himachal Journal of Agricultural Research* 29:31–34.

Sirat, H. M., Jamail, S., and Hussain, J. 1998. Essential oil of *Curcuma aeruginosa* Roxb. from Malaysia. *Journal of Essential Oil Research* 10:453–458.

Sirirugsa, P. 1999. Thai Zingiberaceae: Species diversity and their uses. International Conference on Biodiversity and Bioresources: Conservation, Utilization. Phuket, Thailand.

Sirirugsa, P., and Newman, M. 2000. A new species of *Curcuma* L. (Zingiberaceae) from S.E. Asia. *New Plantsman* 7:196–199.

Skornickova, J., Rehse, T., and Sabu, M. 2007. Other economically important *Curcuma*. In *Turmeric—The genus* Curcuma, ed. Ravindran, P. N., Nirmal Babu, K., and Sivaraman, K., 451–468. CRC Press, Boca Raton, FL.

Skornickova, J., and Sabu, M. 2005a. *Curcuma roscoeana* Wall. in India. *Garden's Bulletin* (Singapore) 57:187–198.

———. 2005b. The identity and distribution of *Curcuma zanthorrhiza* Roxb. *Garden's Bulletin* (Singapore) 57:199–210.

Skornickova, J., Sabu, M., and Prasanth Kumar, M. G. 2003a. *Curcuma codonantha.* A new species from Andaman Islands, India. *Garden's Bulletin* (Singapore) 55:219–228.

———. 2003b. A new species of *Curcuma* from Mizoram. *Garden's Bulletin* (Singapore) 55:89–95.

———. 2004. *Curcuma mutabilis.* A new species from Kerala. *Garden's Bulletin* (Singapore) 56:43–54.

Skornickova, J. L., Sida, O., Jarolimova, V., Sabu, M., Fer, T., and Travnieek. P. 2007. Chromosome numbers and genome size variation in Indian species of *Curcuma. Annals of Botany* 100:505–526.

Skornickova, J. L., Sida, O., and Marhold, K. 2010. Back to types! Towards stability of names in Indian *Curcuma* L. (Zingiberaceae). *Taxon* 69 (1): 269–282.

Skornickova, J. L., Sida, O., Sabu, M., and Marhold, K. 2008. Taxonomical and nomenclatural puzzles in Indian *Curcuma:* The identity and nomenclatural history of *C. zedoaria* (Christm.) Roscoe. and *C. zerumbet* Roxb. (Zingiberaceae). *Taxon* 57 (3): 949–962.

Sreenivasulu, B., and Rao, D. V. R. 2002. Physiological studies in certain turmeric cultivars of Andhra Pradesh. In *Plantation crops research and development in the new millennium*, ed. Rethinam, P., Khan, H. H., Reddy, V. M., Mandal, P. K., and Suresh, K., 442–443. Coconut Development Board, Kochi.

Srivastava, V. P., and Gupta, J. H. 1977. Fungicidal control of turmeric leaf spot incited by *Taphrina maculans*. *Indian Journal of Mycology and Plant Pathology* 7:76–77.

Subbarayudu, M., Reddy, K. R. C., and Rao, M. R. 1976. Studies on varietal performance of turmeric. *Andhra Agricultural Journal* 23:195–198.

Subramanian, S., Rajeswari, E., Balakrishnamoorthy, G., and Shiva, K. N. 2004. Genetic diversity in turmeric germplasm. In *Commercialization of spices, medicinal plants and aromatic crops*, November 1–2, 2004, Indian Institute of Spices Research, Calicut, Kerala, abstract 3.

Subramanyam, S. 1986. Studies on growth and development turmeric (*Curcuma longa* L.). MSc (Hort.) thesis, Tamil Nadu Agricultural University, Coimbatore.

Sugiura, T. 1931. *Bot. Maj.* Tokyo, 45:353. (Cited from Sugiura, 1936.)

———. 1936. Studies on the chromosome number of higher plants. *Cytologia* 7:544–595.

Sumathi, V. 2007. Studies on somaclonal variation in Zingeberacious crops. PhD thesis, University of Calicut, Kerala, India.

Sunitibala, H., Damayanti, M., and Sharma, G. J. 2001. *In vitro* propagation and rhizome formation in *Curcuma longa* Linn. *Cytobios* 105 (409): 71–82.

Syamkumar, S. 2008. Molecular, biochemical and morphological characterization of selected *Curcuma* accessions. Unpublished PhD thesis, University of Calicut, Kerala, India.

Syamkumar, S., and Sasikumar, B. 2007. Molecular marker based genetic diversity analysis of *Curcuma* species from India. *Scientia Horticulturae* 112:235–241.

Tomar, N. S., and Singh, P. 2002. Collection, evaluation and characterization of turmeric (*Curcuma longa* L.) germplasm. In *National Seminar on Strategies for Increasing Production and Export of Spices*, IISR, Calicut, Kerala, abstract 14–15.

Tripathi, S. 2001. *Curcuma prakasha* sp. nov. (Zingiberaceae) from northeastern India. *Nordic Journal of Botany* 21:549–550.

Tuyet, N. T. B., Dûng, N. X., and Leclercq, P. A. 1995. Characterization of the leaf oil of *Curcuma aeruginosa* Roxb. from Vietnam. *Journal of Essential Oil Research* 7:657–659.

Utpala, P., Johny, A. K., Jayarajan, K., and Parthasarathy, V. A. 2007. Site suitability for turmeric production in India—A GIS interpretation. *Natural Product Radiance* 6 (2): 142–147.

Vadiraj, B. A., Siddagangaiah, Sudharshan, M. R., and Krishnakumar, V. 1996. Performance of turmeric (*Curcuma longa* L.) under Malnad condition of Karnataka. *Journal of Plantation Crops* 24 (Suppl.): 483–486.

Valeton, T. H. 1918. New notes on Zingiberaceae of Java and Malaya. *Bulletin du Jardin Botanique de Buitenzorg* II 27:1–8.

Velayudhan, K. C., Asha, K. I., Mithal, S. K., and Gautam, P. L. 1999. Genetic resources of turmeric and its relatives in India. In *Biodiversity, conservation and utilization of spices, medicinal and aromatic plants*, ed. Sasikumar, B., Krishnamurthy, B., Rema, J., Ravindran, P. N., and Peter, K. V., 101–109. Calicut Indian Institute of Spices Research, Calicut, Kerala, India.

Velayudhan, K. C., Asha, K. I., Mithal, S. K., and Gautam, P. L. 1999a. Genetic resources of turmeric and its relatives in India. In *Biodiversity, conservation and utilization of spices, medicinal and aromatic plants*, Shsikumar, B., Krishnamurthy, B., Rema J., Ravindran, P. N., and Peter, K. V. (eds). 101–109. Calicut Indian Institute of Spices Research, Calicut, Kerala, India.

Velayudhan, K. C., Muralidharan, V. K., Amalraj, V. A., Gautam, P. L., Mandal, S., and Kumar, D. 1999 b. *Curcuma genetic resources*. Scientific monograph no. 4. National Bureau of Plant Genetic Resources, New Delhi,149 pp.

Venkatasubban, K. R. 1946. A preliminary survey of chromosome numbers in Scitamineae of Bentham and Hooker. *Proceedings of Indian Academy of Sciences* 23B:281–300.

Venkatesha, J. 1994. Studies on the evaluation of promising cultivars and nutrient requirement of turmeric (*Curcuma domestica* Val.). PhD thesis, Division of Horticulture, UAS, Bangalore.

Venkatesha, J., Khan, M. M. and Chandrappa, H. 1998. Studies on character association in turmeric. In. Mathew, N. M., Kuruvilla Jacob, C., Licy, J., Joseph, T., Meena Hoor, J. R. and Thomas, K. K. (eds.) Development in Plantation Crops Research, Allied Publishers Ltd. New Delhi. pp. 54–57.

Venkatesha, J., Jagadesh, S. K., and Umesha, K. 2002. Evaluation of promising turmeric cultivars for rain-fed conditions under hill zone of Karnataka. In *Plantation crops research and development in the new millennium*, ed. Rethinam, P., Khan, H. H., Reddy, V. M., Mandal, P. K., and Suresh, K., 249–251. Coconut Development Board, Kochi.

Verma, A., and Tiwari, R. S. 2002a. Genetic variability and character association studies in turmeric (*Curcuma longa* L.). In *National Seminar on Strategies for Increasing Production and Export of Spices*, IISR, Calicut, Kerala, abstract 8.

———. 2002b. Path coefficient analysis in turmeric (*Curcuma longa* L.). In *National Seminar on Strategies for Increasing Production and Export of Spices*, IISR, Calicut, Kerala, abstract 14.

Watt, G. 1889. *A dictionary of the economic products of India*, vol. II. Reprint ed. 1972. Cosmo Publications, Delhi.

Weerapakdee, W., and Krasaechai, A. 1997. Collection and development studies of certain *Curcuma* spp. *Journal of Agriculture* 13 (2): 127–136.

Wichtl, M. 1998. *Curcuma* (Turmeric): Biological activity and active compounds. In *Phytomedicines of Europe: Chemical and biological activity. ACS Symp. Ser.*, vol. 691, ed. Lawson, L. D. and Bauer, R., 133–139. American Chemical Society, Washington, D.C.

Xia, Q., Zhao, K. J., Huang, Z. G., Zang, P., Dong, T. T., Li, S. P., and Tsim, K. W. 2005. Molecular genetic and chemical assessment of Rhizoma Curcumae in China. *Journal of Agriculture Food Chemistry* 27; 53:6019–26

Yadav, D. S., and Singh, R. 1996. Studies on genetic variability in turmeric (*Curcuma longa*). *Journal of Hill Research* 9:33–36.

Yadav, D. S., and Singh, S. P. 1989. Phenotypic stability and genotype X environment interaction in turmeric (*Curcuma longa* L.). *Indian Journal of Hill Farming* 2:35–37.

Yasni, S., Imaizumi, K., and Sugano, M. 1991. Effects of an Indonesian medicinal plant, *Curcuma xanthorrhiza* Roxb., on the levels of serum glucose and triglyceride, fatty-acid desaturation, and bile-acid excretion in streptozotocin-induced diabetic rats. *Agricultural and Biological Chemistry* 55:3005–3010.

Yoshioka, T., Fujii, E., Endo, M., Wada, K., Tokunaga, Y., Shiba, N., Hohsho, H., Shibuya, H., and Muraki, T. 1988. Anti-inflammatory potency of dehydrocurdione, a zedoary-derived sesquiterpene. *Inflammation Research* 47:476–481.

Zachariah, T. J., Sasikumar, B., and Nirmal Babu, K. 1999. Variation for quality components in ginger and turmeric and their interaction with environments. In *Biodiversity, conservation and utilization of spices, medicinal and aromatic plants*, ed. Sasikumar, B., Krishnamurthy, B., Rema, J., Ravindran, P. N., and Peter, K. V., 116–120. Indian Institute of Spices Research, Calicut, Kerala.

CHAPTER 17

Ginseng

Kwang-Tae Choi and Gyuhwa Chung

CONTENTS

17.1 INTRODUCTION

Ginseng (*Panax ginseng* C.A. Meyer; Figure 17.1) has traditionally been considered a medicinal plant of mysterious powers and has several thousand years of history among Asian people. Especially in Korea, China, and Japan, ginseng has been recognized as the most prized medicine plant among all the herbal medicines. Therefore, Asian people have traditionally used ginseng roots and extracts to revitalize the body and mind, increase physical strength, slow the aging process, and increase vigor. From the pharmacological studies of animals and human beings for more than 40 years, the efficacy of ginseng, especially tonic effects, has been gaining popularity even in the Western countries. The term "tonic" refers to a drug that is meant to maintain normal physical tone or to restore a diseased state to normal.

Many scientists, including Brekhman (1957), have introduced a new pharmacological concept to the meaning of the tonic effect of ginseng, resulting in interest and attention by explaining the basic pharmacology of ginseng with adaptogen effects. Brekhman asserted that ginseng increases nonspecific resistance to various pathological factors of the body while helping to lower them in order to return to normal or to raise in order to return to normal by acting in a positive or negative way. The effectiveness lasts long and works better on an abnormal condition than a healthy one (Bittles et al. 1979; Zhou 1982; Brekhman and Dardymov 1969). Oh (1976) and Hong et al. (1970) reported that saponin fractions potentiate hypnosis, retard the onset of cocaine-induced convulsions, reduce body temperature, and enhance the process of sexual behavior in mice. Recently, a number of scholars have paid attention to elucidating the efficacy of ginseng scientifically.

Originally, the pharmacological efficacy of Korean ginseng was known on the basis of the theory of Oriental medicine. Gradually, modern scientific theory and technology have enabled many scientists to reconstruct and integrate into the architecture of the 1970s. Some academic scholars have long made efforts to identify which ingredient or substance of Korean ginseng gives rise to this mysterious efficacy, but their research has made little progress and uncovered no satisfactory evidence on account of the mystery of ginseng.

Figure 17.1 A mature Korean ginseng root.

For instance, the burgeoning pollution industries brought many harmful effects, such as degenerating the quality of human life, deteriorating man's living environment, and threatening human health. People worldwide are looking for safe food and natural medicines to defend and protect them from these harms. The importance of natural medicine is drawing attention. Likewise, the demand for Korean ginseng is growing rapidly for the prevention and cure of diseases.

Natural ginseng (*Panax ginseng* C.A. Meyer) is distributed in eastern Asia at latitude 30–48° north, such as Korea, the northeast of China, and the Russian Far East, because it enjoys the cool climate conditions of the temperate zone. Ginseng grows all across the Korean peninsula at latitude 33–46° north, especially in the Taebaek mountain range and the northern district of Korea. It is distributed in a range of 100–800 m above sea level. However, the number of wild ginseng plants has decreased gradually; alternatively, ginseng was artificially cultivated. Originally, artificial cultivation meant sowing the seeds and planting the seedlings of wild ginseng in the forest; later, ginseng grew in mountainous regions to flat areas, and currently ginseng cultivation prevails throughout almost all of Korea.

The idealistic ranges of temperature for Korean ginseng cultivation are 0.9–13.8°C, and in summer 20–25°C. The rainfall annually amounts to 1200 mm. The snowfall is relatively low. Direct sunlight is harmful, so diffused light is preferred—1/8 to 1/13 of the total sunlight capacity. Sandy loam is good for surface soil, but for subsoil it is desirable to have somewhat more clay. Old soil, neither too fertile nor damaged by blight or insects, with a pH of 5.5–6.0, is very suitable. It is extremely difficult to adapt and raise ginseng where climate or soil is incompatible with its natural surroundings. Even if it is possible to cultivate, it is recognized that unsuitably grown ginseng differs in shape, quality, and pharmacological effect.

Recently, Korea has been one of the largest *Panax ginseng* producers and exporters in the world. In 2008, 24,298 households cultivated ginseng over a total area of 19,408 ha. The estimated amount of cultivated ginseng crops reached 24,613 tons; among those, 2,128 tons were exported to more than 83 countries. Therefore, the use of ginseng is expanding all around the world.

This chapter will discuss various aspects of Korean ginseng, including its cultivation, breeding, pharmacological properties, and physiologically active components, as well as a broad range of comparative studies on the efficacy and ingredients of ginseng.

17.2   TAXONOMY OF GINSENG

The botanical classification of Korean ginseng is as follows:

• Phylum	• Embryophyta
• Subphylum	• Angiospermae
• Class	• Dicotyledoneae
• Subclass	• Anchichylamyeae
• Order	• Umbellifloreae
• Family	• Araliaceae
• Genus	• *Panax*

The scientific name of the Korean ginseng is currently *Panax ginseng* C.A. Meyer, and it belongs to family Araliaceae. Meyer, a Russian scholar, named the plants in 1843. *Panax* is a compound Greek word: *pan* denotes "all" and *axos* means "cure." Accordingly, *Panax* denotes a cure-all. The word "ginseng" originated from the Chinese pronunciation of ginseng plants.

The first scientific name for the Korean ginseng plant was *Panax schinseng* Nees (KGGA 1980). In 1833, Nees von Esenbeck of Germany described Korean ginseng as *P. schinseng* var. *coraiensis* Nees in his book, *Icones Plantarum Medicinalium*. Therefore, this was the first genuinely scientific name for Korean ginseng. However, *P. ginseng*, identified 10 years later, became more widely known and thus became widely used as a scientific name for Korean ginseng.

17.3 RELATED SPECIES

The genus *Panax* was established by Linnaeus. Taxonomically, there are six to eight species of the genus *Panax*. The six species, probably with little debate, are as follows:

Panax ginseng **C.A. Meyer.** *Panax ginseng* is found in Korea, North China, Manchuria, Siberia, and Japan and usually referred to as Korean or Asian ginseng. *Panax ginseng* has a long peduncle three to six times longer than the petiole. The chromosome number is $2n = 48$.

Panax quinquefolium **L.** *Panax quinquefolium* is found in America and Canada and commonly called American ginseng, Canadian ginseng, or North American ginseng. This species is a short-peduncled plant. Three flowering peduncles are equal to, shorter than, or only slightly larger than the petioles of the same plant. The chromosome number is $2n = 48$.

Panax trifolium **L.** *Panax trifolium* is found in North America and commonly called dwarf ginseng or ground nut. *Panax trifolium* is sexually specialized (Hu 1976). The male plants bear white flowers with slender long pedicels that are five or six times longer than the obconic float tube. The female plants bear pink flowers with short, stout pedicels.

Panax japonicum **C.A. Meyer.** *Panax japonicum* is found only in Japan and is called Japanese ginseng or chikusetsuninjien. This species has a long peduncle.

Panax notoginseng **Burkill.** *Panax notoginseng* is found in China and commonly called san chi or tien chi ginseng.

Panax pseudoginseng **Wallich.** *Panax pseudoginseng* Wallich is found in India, Nepal, Bhutan, and Sikkim.

Of the six ginseng species, *P. ginseng* and *P. quinquefolium* are assumed to have exceptional curative properties and therefore have the greatest commercial value.

17.4 BOTANICAL CHARACTERISTICS OF KOREAN GINSENG

Korean ginseng is a perennial herb with flesh roots, an annual stem bearing a whorl of palmate (palm-shaped) compound leaves, and a terminal simple umbel. Ginseng plants are basically self-pollinated, although they are sometimes cross-pollinated. Flowers that are wrapped prior to blooming show 95% self-pollination and percentage of cross-pollination is low.

Flowers start to bloom after 3 years of planting and flowering initiates in the middle of May. It is generally known to collect the seeds once from 4-year-old plants, and the flower buds are nipped off for better root growth. The ripe red (Figure 17.2a) or white (Figure 17.2b) berries are gathered two or three times in the middle of July. The mature berries usually contain two pale yellow seeds. At the time of picking, seeds do not have a clear embryo shape, are immature, and need a dormant period for ripening to enable germination. To accelerate the growth of immature embryos, toward the latter half of July, ginseng growers put a mixture of seeds and sands into specially designed containers. They cultivate the seeds, including controlled watering for approximately 100 days (from end of July to beginning of November) in the shade, and then sow the seeds in seedbeds during the beginning half of November. Generally, plants that have been cultivated in seedbeds are transplanted around the end of March or beginning of April and harvested 3 to 5 years after transplanting the seedlings—that is, 4- or 6-year-old roots are harvested (Figure 17.3).

The roots are corpulent and consist of a taproot (or main root), two to five lateral roots, and fine roots (Figure 17.4). The roots are light yellowish white in color. The size and the shape of the roots

Figure 17.2a   (See color insert.) Ripe red berries of Korean ginseng.

Figure 17.2b   (See color insert.) Ripe yellow (white → yellow) berries of Korean ginseng.

Figure 17.3 Six-year-old Korean ginseng plants.

Figure 17.4 Mature Korean ginseng roots.

vary due to soil property, fertilizer, water content in soil, method of transplantation, climate, and germplasm. The growth of the root and its shape differ during the process of aging. The roots are harvested after 4–6 years, but younger roots do not exhibit complete maturity in the growth of the main and lateral roots. The 6-year-old ginseng plant only has the rhizome and main and lateral roots that are grown in a balance like a human body (Figure 17.1). At 6 years, the rhizome becomes fat and the length of the main root is about 7 to 10 cm. The diameter of the main root is about 3 cm; additionally, it has many fat lateral roots at a length of about 35 cm. This mature root weighs 70–100 g or 300–500 g per root.

Korean ginseng is a perennial plant with buds sprouting from its roots each spring; leaves and stems fall and are dried in the autumn. The trace of an insertion, each year, is left on the rhizome, the head of the root. Its shape varies from the mortar shape to the hollow shape. This rhizome is an important factor in the quality assortment process of ginseng's manufactured goods; in the absence of the rhizome, it is judged as an inferior quality or becomes a low-priced item. For this reason, handling should be under special care. In particular, rhizome is the special feature of Korean ginseng and it is used as a manufacturing index to distinguish Korean ginseng from ginsengs of other countries.

The sprout of a Korean ginseng seedling has two cotyledons (primary leaves); between these cotyledons, a small stem supports three tiny leaves. The leaves of the ginseng plants are a palmate compound with a little longer petiole whorled on the tip of the stem. The number of leaves varies with the age of plants and cultivation factors. It is usually related to the age of the plants (one for yearlings, two for 2-year-old plants, three for 3-year-old plants, and so on). Each compound leaf has three to five leaflets. Yearlings have three leaflets, and the rest of the plants have five leaflets in each compound leaf.

Ginseng is a semishade plant and the aerial part above ground is fairly weak. Therefore, if it is exposed to direct sunlight, chlorophyll in the leaves is destroyed, leaving that part dried up and dead. For this reason, it is good to set up a shade in the high temperatures of summer, creating cooler conditions. Moreover, ginseng's growth is very slow in comparison with common plants. Fertilizer tolerance is very weak, so the use of chemical fertilizers often hinders the growth of the plants. Therefore, nutrition management should focus on the soil property with decomposed leaves. Another consideration is the nonconsecutive planting of ginseng. It usually takes 10–15 years for another turn of the planting. This results in damage from blight and harmful insects. The lack of particular nutrients in the soil and the accumulation of toxic components inhibit growth of ginseng. This suggests that the end-product of ginseng is expensive.

17.5 CULTIVATION OF KOREAN GINSENG

17.5.1 History of Korean Ginseng Cultivation

During the Three Kingdoms period the medicinal efficacies of ginseng were widely known. It was recorded in *Myeonguibyeollok* (ancient medical literature) that Baekje and Goguryeo sent wild ginseng to China in AD 513 and AD 435–546, respectively. In addition, there is a record in *Chaebuwongu*, a Chinese historical literary work, that Silla sent ginseng to China in AD 627 (KGGA 1980).

Wild ginseng, developed by private doctors through experience and used for medicinal purposes, became gradually depleted. It was, however, replaced by the cultivated ginseng. According to *Jeongbomunheonbigo* (Kim 1989), a collection of rearranged records, and *Junggyeongji*, a collection of historical records, ginseng was produced across the Korean peninsula and there were two types (KGGA 1980). One was mountain ginseng, which grew naturally (Sanjeong); the other ginseng was planted on the mountain and was harvested after a few years (Sanyang). It was not easy to find both of them, until a woman in Dongbok-Hyeon in Jeolla Province (presently Dongbok-Myeon, Hwasun-Gun, Jeollanam-Do) dug up ginseng on a mountain and planted ginseng seeds in the field. A man whose family name was Choi disseminated them. This explains the origin of the name *gasam*.

As Korean ginseng was considered precious in other parts of the world, Choi secretly sold it to a person from China's Qing Dynasty who regarded Korean ginseng as something very precious. By using ginseng as a medicine, an opium addict can be cured. However, taking ginseng occasionally has some side effects. After Choi understood the cause, he sold steamed ginseng and earned much profit. Choi became very rich in the province that is the origin of red ginseng (KGGA 1980).

According to some historical records, Sanyang ginseng was artificially cultivated during the reign of King Gojong (1214–1260) of the Joseon Dynasty. Ginseng was cultivated 1,200 years ago during the reign of King Soseong of the Silla Dynasty in Gyeongju City, Gyeongsangbuk-Do (KGGA 1980). Presumably, an area surrounding Mt. Mohu in Dongbok-Myeon, Hwasun-Gun, Jeollanam-Do, was the birthplace of full-fledged ginseng cultivation. Furthermore, the Dongbok ginseng was introduced to the city of Kaesong by Kaesong merchants, and Kaesong then became the hub of ginseng cultivation. In the folklore of ginseng, people prayed to the mountain god to cure diseases, have a child, or become rich. Under such guidance, they obtained ginseng seeds and planted them or relocated natural ginseng to other places, thus triggering ginseng cultivation.

When ginseng cultivation was still in its infancy stage, it was mainly referred to as wild ginseng. The government ginseng diggers were called *chaesamgun* (ginseng digging army). There were a considerable number of wild-ginseng diggers among the general public, other than the state diggers. Therefore, natural ginseng became depleted due to the proliferation of wild-ginseng digging. The shortages of wild ginseng became severe during the reign of King Seonjo (1567–1608) of the Joseon Dynasty (KGGA 1980). During the reign of Kings Sukjong and Yeongjo, the forest area was reduced due to deforestation, so the amount of natural ginseng sharply decreased. The demand for ginseng further increased, prompting cultivation of *gasam* (farmed ginseng) and *jangnoisam* (cultivated wild ginseng). There are records on *gasam* in *Jeongjongsillok* and other historical documents such as records of the cultivation method of *gasam* in *Imwonsipyukji* by Seo Yu-Gu. As such, artificial methods for cultivating ginseng on flat land were gradually developed (KGGA 1980).

In the 1900s, the cultivation in forests was replaced by cultivation in fields. Large-scale artificial cultivation methods were developed, mainly in Kaesong. To cultivate ginseng, a shade plant, modified soil cultivation methods were developed (KGGA 1980): Facilities to shade from the sun were erected. The ridges were dug toward the direction of the sunset on the lower ground to block direct sun rays. To change seedling growing technology from direct seeding to transplanting, Yakto (unique compound organic fertilizer mixed with fallen leaves in deep mountains, rice bran, oil cake, ash from the flue of hypocost, and ashes from grass and trees) was developed and mixed with original field soil to make bed soil. Ginseng cultivation technology has been developed over a long period of time through strenuous trial and error efforts.

17.5.2 Cultivation and Management of Korean Ginseng

Ginseng seeds are harvested in late July. After the flesh is removed, the seeds remain. The seeds undergo pregermination treatment in which they are mixed with sand for 100 days at 20°C and 10–15% moisture. Seeds are sown in November when the length of the embryo bud grows to more than 5 mm and its testa is wide open. The dehisced ginseng seeds are sown on the bed soil that comprises a mixture of clean organic fertilizer (e.g., leaves of broad-leaved trees) and clean soil without pathogenic bacteria. Only high-grade seeds of more than 4 mm in length are sown. After the 18-month cultivation period, ginseng seedlings grow 16 cm long and weigh 0.7 g.

The soil must be properly prepared to plant ginseng. Excessive fertilizers in the soil are absorbed and eliminated by growing rye, Sudan grass, and maize for 1–2 years. The cultivated green crops are supplied again to the soil (4.5 tons per 10 ha) to increase organic content. The soil is plowed more than 15 times a year to improve its physical properties and destroy pathogenic bacteria. The soil for the cultivated ginseng should be maintained 10% lower in the nutritional standards than the soil for other crops. When levels are higher than this, physiological impairment appears.

Transplanted to the main field, the superior ginseng seedlings of 0.68 g (Eulsam) are selected and planted obliquely at a 45° angle to produce the human-shaped white ginseng with a long stem length and two legs. The most suitable rate of water in ginseng fields is 60% of water or 22.1% of absolute water, although there are some differences depending on soil properties. Due to the narrow range of water in the ginseng fields, it is important to control the water rate with much caution. A high level of salt concentration also can be hazardous in ginseng fields. Thus, traditional organic fertilizers are preferred to chemical fertilizers.

Additionally, new crop protective agents or pesticides, which have less residue and much effectiveness, are used to control damage by harmful insects. In practice, clean ginseng requires the adoption of pest control with natural substances. Ginseng root rot impairs the fields for repeated cultivation. Treatments of a combined mixture of the sterilized soil, microbial agents, and crop rotation cultivation are applicable in some regions to control root rot.

17.5.3 Shapes of Korean Ginseng

The shape of ginseng is associated with age, variety, and environment. In a 4-year-old ginseng plant, the aboveground part has a flower, fruit, peduncle, leaves, and stem, while the underground part (root) has the latent bud, rhizome, fibrous root, main root, rootlets, and fine root.

Each year ginseng has a different number of palmate leaves. Generally, it is believed that the number of palmate leaves exactly matches the age of the plant. As for leaflets, 1-year-old ginseng has three leaflets and 2-year-old or older ginseng has five leaflets. Ginseng that is 7 years old or older has many missing plants, with a bigger body shape and lignified skin. Processed, it has an internal cavity and the internal color turns white.

17.6 BREEDING OF KOREAN GINSENG

Korean ginseng is the most economical crop at the highest level of quality. The yield of high-quality ginseng is very low because of difficulties associated with crop management and the lack of varietal differentiation, which is mainly due to the long generation time and difficulties with cultivation management. In the last decade, the production and consumption of ginseng have increased all over the world. Therefore, new varieties of ginseng with high quality and yield per unit area have to be developed.

The primary purpose of ginseng plant breeding is to develop varieties that have a morphologically good shape of root, high yield, large amount of pharmacologically active compounds, resistance to diseases and insects, resistance to environmental stress, and tolerance to high light intensity. We have used methods of ginseng breeding such as the pure-line method (individual plant selection), hybridization between lines or species (*P. ginseng* and *P. quinquefolius*), and genetic engineering in order to produce high-yielding ginseng. Eventually, we have selected many individual ginseng plants with desirable traits, such as high quality, high yield, and distinctive traits, since the 1970s (Ahn et al. 1985, 1987; Choi et al., 1994, 1998, 2001; Chung et al. 1992, 1993; Chung, Chung, and Choi 1995; Kwon et al. 1991, 1994, 1998; Kwon, Lee, and Choi 2000). Among them, a group of characteristic or promising lines, that is, KG (Korean ginseng), have been developed through comparative cultivation of several lines selected by pure-line separation from local races in the KT&G Central Research Institute, formerly the Korea Ginseng & Tobacco Research Institute (Choi et al. 1998, 2001; Kwon et al. 1991, 1994, 1998, 2000). Preliminary, advanced, regional, and adaptation yield trials were performed for 18 years.

17.6.1 Korean Ginseng Breeding by Pure-Line Separation Method

A large number of individual ginseng plants have been selected in farmers' fields since 1972 to develop new ginseng varieties with good quality and high yield. Among them, 10 promising lines have been developed through comparative cultivation of several lines that were selected with pure-line separation of local races at the Korea Ginseng & Tobacco Research Institute. Preliminary and advanced yield trials were performed for 8 years. Korean ginseng (KG) was tested in the regional yield and adaptation trials for 10 years. A local race of KG, Jagyeongjong, has a violet stem with red berry.

On the other hand, the KG101 line has a green stem with light violet and orange-yellow fruit and flowers 3–7 days later than the local race, Jagyeongjong. The KG102 line shows high frequency (40–50%) of multiple stems at above 3 years of age (Kim et al. 1989). The KG103 line has a dark violet stem. The KG104 line makes an early appearance and the KG105 line a late appearance. Hwangsukjong (yellow-berry variant) has a green stem with yellow berries. Morphological characteristics of 6-year-old roots in superior lines are presented in Table 17.1. The taproot diameter of KG102 was 3.3 cm. It was larger than other lines and Jagyeongjong. The taproots of KG lines and Hwangsukjong were longer than that of Jagyeongjong. Particularly, the KG101 line had the longest taproot among KG lines.

The quality of red ginseng is affected by the inside cavity inside white, cracking, and root shape. Among them, the quality of root shape depends on taproot length. In general, the quality of red ginseng has been evaluated with the excellent quality taproot (7–10 cm in length). Therefore, it was found that the KG 101 line had a superior root shape. The root weight of the KG102 line was 112.4 g/plant, which was 15% higher than that of Jagyeongjong. In general, survival rate and root weight per plant affect the yielding ability of ginseng (Choi, Ahn, and Shin 1980; Ahn et al. 1987).

The survival rate of ginseng plants decreases annually under biotic or abiotic stress (e.g., disease, water, salt, light, temperature). The survival rates of KG 101, KG 102, KG 103, and KG 105 were higher than that of the local race, Jagyeongjong (Table 17.2). KG 101 and KG 105 lines showed 73.1 and 77.6% survival rates, respectively. In this research, the promising KG lines were selected and subjected to the yield trials. They were tested through the regional adaptation trials for 10 years. KG 102 showed the highest yielding ability and averaged 2.8 kg/3.3 m² in fresh root yield. It was 27% higher than that of Jagyeongjong (Table 17.2). KG 101 and KG 102 lines were finally selected and registered in 1999 as Cheonpung and Yeonpung, respectively.

The key pharmacological component of Korean ginseng is ginseng saponin, which is called "ginsenoside," a combination of the words ginseng and glycoside. It is known that Korean ginseng contains more than 38 kinds of ginsenosides. Table 17.3 shows total amount and pattern of major ginsenosides in 6-year-old taproots of ginseng superior lines. Total ginsenosides in KG101, KG102, KG103, KG104, KG105, and Hwangsukjong amounted to 9.21 mg/g DW, 9.30 mg/g DW, 18.36 mg/g

Table 17.1 Characteristics of 6-Year-Old Roots in Superior Lines of Korean Ginseng (*Panax ginseng* C.A. Meyer)

Superior Lines	Diameter of Taproot (cm)	Length of Taproot (cm)	Length of Root (cm)	Weight of Root/Plant (g)	No. of Lateral Roots
Jagyeongjong	3.1	5.3	27.1	98.0	3.1
KG 101	2.8	7.0	35.1	87.5	2.8
KG 102	3.3	5.5	28.8	112.4	3.3
KG 104	2.9	6.5	29.6	85.5	2.7
Hwangsukjong	2.6	6.7	30.7	104.3	2.9
L.S.D. (0.05)	0.35	0.71	NS	11.45	NS

Source: From Ahn, S. D. et al. 1987. *Korean Journal of Ginseng Science* 11 (1): 46–55. With permission.

Table 17.2 Survival Rate, Yield, and Characteristics in Superior Lines of 6-Year-Old Korean Ginseng

Lines	Survival Rate (%)	Yield (kg/3.3m²)	Characteristics
Jagyeongjong	61.7	2.2 (100)[a]	Local variety
KG 101	73.1	2.4 (109)	Long taproot and good shape of root
KG 102	63.9	2.8 (127)	Multistem and large size of root
KG 103	64.3	2.3 (105)	Large amounts of ginsenosides
KG 104	57.4	2.5 (114)	Large size of root
KG 105	77.6	2.6 (118)	Late appearance

Source: From Ahn, S. D. et al. 1987. *Korean Journal of Ginseng Science* 11 (1): 46–55. With permission.
[a] Yield index to Jagyeongjong.

DW, 8.8 mg/g DW, 8.04 mg/g DW, and 8.96 mg/g DW, respectively. The ginsenoside content of the KG103 line was two times higher than that of other lines. Ginsenosides Rb_1, Rf, and Rg_1 in KG 101, KG 102, KG 103, and KG 104 lines and ginsenoside Rb_2 in KG 103 and Hwangsukjong showed higher content than those in other lines and Jagyeongjong (Table 17.3). The Rb and Rg content in KG lines was more than that of Jagyeongjong and Hwangsukjong. It was clarified that the KG 103 line had the highest production of total ginsenosides. KG 103 was registered in 2001 under the name Gopung.

Table 17.3 Amounts of Ginsenosides in 6-Year-Old Taproots of Superior Lines

Lines	Ginsenoside (mg/g)								
	Rb_1	Rb_2	Rc	Rd	Re	Rf	Rg_1	Rg_2	Total
Jagyeongjong	1.62	0.71	1.01	0.20	1.21	0.40	1.72	0.36	7.23
KG 101	2.44	0.65	0.92	0.27	1.23	0.66	2.66	0.38	9.21
KG 102	2.81	0.55	0.56	0.26	1.61	0.80	2.42	0.29	9.30
KG 103	4.88	1.56	1.79	0.48	2.62	1.31	5.20	0.51	18.36
KG 104	2.57	0.81	0.91	0.25	1.17	0.61	2.36	0.21	8.88
KG 105	1.84	0.56	0.76	0.19	1.91	0.47	1.85	0.45	8.04
Hwangsukjong	2.10	1.30	0.86	0.17	1.62	0.69	1.96	0.28	8.96

Source: From Choi, K. T. et al. 1998. *Proceedings of the 7th International Symposium on Ginseng*, Seoul, Korea, 96–102. With permission.

17.6.2 Production of New Korean Ginseng Varieties

Until recently, local varieties of Korean ginseng have mainly been cultivated. However, since 1997, eight new varieties (Cheonpung, Yeonpung, Geumpung, Gopung, Seonpung, Seonwon, Seonun, and Cheongseon) with characteristics unique to Korean ginseng have been developed. They are currently under cultivation and under strict management. Concerning characteristics of the roots of new varieties of Korean ginseng, yield was greater in the sequence of Yeonpung > Geumpung > Gopung > Cheonpung = Seonpung > native Jagyeong variety. Thus, the yield of Yeonpung was the greatest and individual weight showed the same trend. The stem diameter was greater in the sequence of Yeonpung > Geumpung > Seonpung > Cheonpung > Gopung > native Jagyeong variety and thus Yeonpung is the thickest. Regarding stem length, Cheonpung was the longest (8 cm), followed by Geumpung (7.6 cm); Yeonpung was the shortest (6.4 cm). Other varieties are between Geumpung and Yeonpung (Lee, Lee, and Ahn 2005; Lee 2007).

The root yield of Cheonsam, a new red ginseng variety (grade 1 ginseng), increased in the order of Cheonpung (highest) > Geumpung > Gopung > Seonpung > Yeonpung > Cheongseon > native Jagyeong variety. Regarding total ginsenoside content of 6-year-old new varieties per unit weight, the content increased in the order of Gopung (highest) > Yeonpung > Geumpung (Lee et al. 2005; S. S. Lee 2007a).

17.6.3 Characteristics of Intra- and Interspecific Hybrids

Two variants, Jagyeongjong and Hwangsukjong, were crossed to clarify the characteristics of the F_1 hybrid and the inheritance of the stem color. The characters of stem, leaf, and root were measured in parents F_1, and F_2 (Cheon et al. 1985). The stem diameter of Jagyeongjong and F_1 plants was larger than that of Hwangsukjong. Stem length, leaf length, and leaf width of Jagyeongjong and F_1 plants were longer or wider than those of Hwangsukjong. In the case of these characters, F_1 plants were similar to Jagyeongjong. However, the number of leaflets of F_1 plants was more than that of the parents (Jagyeongjong and Hwangsukjong; Table 17.4). Fresh weight of F_1 root was heavier than that of Jagyeongjong and Hwangsukjong and 100-seed weight of F_2 generation was heavier than that of parents (Table 17.4). In this experiment, we found that these traits showed hybrid vigor. The F_1 hybrids may achieve exceptional vigor due to heterosis. In many cases, the heterosis contributes also to seedling vigor, so the productivity of mature plants is much higher.

Table 17.4 Number of Leaflets, Fresh Weight of Roots, and Weight of 100 Seeds in Parents, F_1 and F_2 Generation

	No. of Leaflets of 3-year-old Plants	Fresh Weight of 3-year-old Roots (g/root)	Weight of 100 Seeds of 4-year-old Ginseng Plants (g)
Jagyeongjong	19.5	8.5	3.73
F_1 hybrids	21.0	9.5	
F_2			3.84
Hwangsukjong	19.0	3.8	3.74

Source: From Cheon, S. R. et al. 1985. *Korean Journal of Ginseng Science* 9 (2): 264–269. With permission.

As shown in Table 17.5, the stem color of the F_1 plants was violet. Table 17.6 shows the number of plants with green or violet stems and chi-square test in F_2 generation. The chi-square test is a useful method for testing goodness of fit of Mendelian ratios. In the F_2 generation of reciprocal crosses between two variants of Korean ginseng, Hwangsukjong and Jagyeongjong, producing green stem with yellow berry and violet stem with red berry, respectively, the number of plants was obtained in the two phenotypic classes (Table 17.6) and testing for deviations for the single factor ratio was done. In the counts of violet- and green-stemmed plants, we found $\chi^2 = 0.082$ and 1.421. As a matter of practical convenience, probability levels of 5% (0.05) and 1% (0.01) are commonly used in deciding whether to reject the null hypothesis. As seen from Table 17.6, these correspond to χ^2 lower than 3.841 and χ^2 lower than 6.635, respectively. In F_2 seedlings, it was found that the difference in the number of violet- and green-stemmed plants was not significant at the 5% level. Namely, the χ^2 test

Table 17.5 Stem Color of 4-Year-Old Ginseng Plants in Parents and F_1 Generation

Cross Combination	Stem Color
Jagyeongjong (J)	Violet
Hwangsukjong (H)	Green
(J × H) F_1	Violet

Source: From Cheon, S. R. et al. 1985. *Korean Journal of Ginseng Science* 9 (2): 264–269. With permission.

Table 17.6   Chi-square Test for the Ginseng Stem Color in F_2 Seedlings

Cross Combination	No. of Plants Observed	Observed Frequency of Stem Color		Expected Frequency of Stem Color		χ^2	P
		Violet	Green	Violet	Green		
$(J \times H)$ F_2	1044	787	257	783	261	0.0817	>0.75
$(H \times J)$ F_2	885	693	192	663.75	221.25	1.4211	>0.10

Source: From Cheon, S. R. et al. 1985. *Korean Journal of Ginseng Science* 9 (2): 264–269. With permission.
Notes: J: Jagyeongjong (violet-stem variant); H: Hwangsukjong (yellow-berry variant).

was consistent with the hypothesis of a 3:1 ratio of violet- to green-stemmed ginseng plants. From this result, it was elucidated that the violet color was controlled by a single dominant gene.

Panax ginseng (Jagyeongjong) was crossed by *P. quinquefolius* to develop the new ginseng varieties, but the fertile F_1 seeds could not be collected (Lee et al. 1997). Therefore, the floral structure, pollen viability, and germination and growth of pollen grains were examined in the pistil of several cross combinations of genus *Panax* and F_1, clarifying the sterility of F_1 hybrid. Table 17.7 shows the floral characteristics of ginseng species and interspecific F_1. The redundant anther was 0.2 mm longer than the pistil in F_1 hybrid and the size and structure of redundant carpel in F_1 hybrid were similar to those of *P. ginseng* and *P. quinquefolius* (Table 17.8). Therefore, the floral structure of F_1 hybrid was equal to that of normal parents.

Pollen grains did not germinate in (1) self-pollinated F_1 plants, (2) F_1 crossed by *P. ginseng* ($F_1 \times$ *P. ginseng*), and (3) *P. quinquefolius* crossed by F_1 (*P. quinquefolius* $\times F_1$). On the other hand, pollen grains germinated when *P. ginseng* was crossed by F_1 (*P. ginseng* $\times F_1$). Pollen tubes also grew

Table 17.7   Floral Characters of Ginseng Species and F_1

Species	Length of Pistil	Length of Anther	Difference Between Pistil and Anther
P. ginseng	1.25 ± 0.19	1.40 ± 0.16	0.15
P. quinquefolius	1.26 ± 0.23	2.20 ± 0.19	0.94
F_1	1.40 ± 0.25	1.60 ± 0.15	0.20

Source: From Lee, S. S. et al. 1997. *Korean Journal of Ginseng Science* 21 (2): 85–90. With permission.
Notes:
Measurement unit is the millimeter.
All values represent mean ± S.E.
F_1 is a hybrid plant between Panax ginseng and Panax quinquefolius.

Table 17.8   Carpel Characters of Ginseng Species and F_1

Species	Length Between Stigma and Basal Part of Style	Length Between Basal Part of Style and Entrance of Ovule	Ovule Length	Ovule Width
P. ginseng	2.25 ± 0.5	2.40 ± 0.38	1.10	0.75
P. quinquefolius	2.30 ± 0.4	2.50 ± 0.40	1.15	0.82
F_1	2.50 ± 0.63	2.60 ± 0.45	1.18	0.85

Source: From Lee, S. S. et al. 1997. *Korean Journal of Ginseng Science* 21 (2): 85–90. With permission.
Notes:
Measurement unit is the millimeter.
All values represent mean ± S.E.
F_1 is a hybrid plant between Panax ginseng and Panax quinquefolius.

Table 17.9 Number of Pollen Tubes at Stigma, Basal Part of Style, and Entrance of Ovule, and Ratio of Seed Harvest after Interspecific Pollination in *Panax* Genus

Cross Combination	Number of Pollen Tubes at:			Seed Harvest (%)
	Stigma	Basal Part of Style	Entrance of Ovule	
P. ginseng selfed	40.5 ± 6.3	20.3 ± 4.8	15.4 ± 2.5	75.8
P. quinquefolius selfed	31.5 ± 4.2	12.4 ± 1.8	8.5 ± 0.9	55.6
F₁ selfed	0	0	0	0
F₁ × *P. ginseng*	0	0	0	0
F₁ × *P. quinquefolius*	2.2 ± 0.1	0.2 ± 0.0	0	0
P. ginseng × F₁	5.6 ± 0.2	3.2 ± 0.1	2.2 ± 0.1	16.8
P. quinquefolius × F₁	0	0	0	0

Source: From Lee, S. S. et al. 1997. *Korean Journal of Ginseng Science* 21 (2): 85–90. With permission.
Notes:
Data obtained from observation of 60 carpels in each cross 15 hours after pollination.
All values represent mean ± S.E.

continuously when *P. ginseng* was used as the female parent and a small amount of seeds (16.8%) could be harvested (Table 17.9). In addition, the growth of the pollen tube was halted in the style and seed was not set when *P. quinquefolius* was used as the male parent. These results suggest that F₁ hybrid sterility was caused by the inhibitor of pollen germination on stigma (Lee et al. 1997).

17.7 ACTIVE COMPONENTS

Korean ginseng contains various medicinal ingredients. Carbohydrate, including free sugar and starch, comprises 60–70%. Ginsenosides, saponins unique to ginseng, have a different structure from saponins in other plants. Additionally, there are polysaccharides with immunostimulating activity; polyacetylene with anticarcinogenic effects; phenols with antioxidant effects; gomisin, which protects the liver; and acidic peptide, which acts similarly to insulin (Park 1996).

17.7.1 Ginseng Saponin (Ginsenosides)

The etymological origin of saponin is *sapo* (meaning "soap" in Greek). When saponins are dissolved and shaken in water, they have foaming characteristics. It is reported that there are approximately 750 species of natural saponins that are widely distributed in the plant kingdom. In general, saponins in plants normally show hemolysis action (dissolving blood corpuscles). Ginsenosides, ginseng saponins, show almost no hemolysis even if taken in large quantities (Namba et al. 1972). It has been known since ancient times that the long-term consumption of ginseng has no harmful effect on the body.

Saponins are glucosides that consist of sugar and aglycones. Depending on the frame structure of aglycones, saponins are classified into triterpenoid and steroid saponins. Triterpenoid saponins are further classified into lanostane, dammarane, ursane, and oleanane saponins. Saponins in plants are mostly oleananes. In addition to very small quantities of oleananes, ginseng mainly contains the dammarane type of triterpenoid saponin. Ginseng saponins are called "ginsenosides," which mean glucosides contained in ginseng. If the sugar part is eliminated from ginseng saponins, three types of aglycones remain: protopanaxadiol (PPD) (Shibata et al. 1966), protopanaxatriol (PPT) (Shibata et al. 1965), and oleanolic acid (Horhammer, Wagner, and Lay 1961) (Figure 17.5). Therefore, there are three groups of ginsenosides: PPD type, PPT type, and oleanolic acid type. Types of ginsenosides in Korean ginseng currently amount to 38 types in total: 22 ginsenosides (PPD type), 14 ginsenosides (PPT type), and 2 ginsenosides (oleanolic acid type).

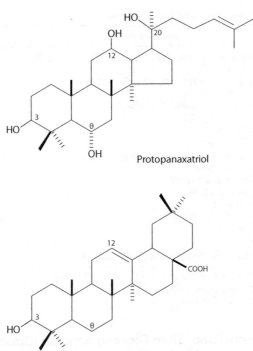

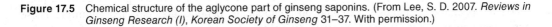

Protopanaxatriol

Oleanolic acid

Figure 17.5 Chemical structure of the aglycone part of ginseng saponins. (From Lee, S. D. 2007. *Reviews in Ginseng Research (I), Korean Society of Ginseng* 31–37. With permission.)

PPD type ginsenosides have glycosidic bonds at the hydroxyl group of C-3 and C-20 of protopanaxadiol (Figure 17.1). Major PPD type ginsenosides include Ra_1, Ra_2, Rb_1, Rb_2, Rc, Rd, and Rg_3 (Sanada et al. 1974). PPT type ginsenosides have glycosidic bonds in the hydroxyl group of C-6 and C-20 of protopanaxatriol in Figure 17.1; major PPT type ginsenosides include Re, Rf, 20 gluco Rf, Rg_1, Rg_2, and Rh_1 (Sanada et al. 1974). Oleanolic acid type ginsenosides include R_o and polyactyleneginsenoside R_o (Zhang et al. 2002). It has been found that 24 ginsenosides exist in white ginseng and 32 ginsenosides in red ginseng (Table 17.10) (Lee 2007).

Ginsenosides Rg_3, Rg_2, and Rh_2, which are common in red ginseng, do not exist in white ginseng. On the other hand, malonyl ginsenoside, Rb_1, Rb_2, Rc, and Rd exist only in white ginseng (Kitagawa et al. 1983). Ginsenoside Rh_2 in red ginseng had strong inhibitory action on various cancer cells (Tode et al. 1993). Other varieties of ginseng currently available in the global marketplace, along with Korean ginseng, are American ginseng (Hwagi-sam) and Chinese ginseng (Sanchi-sam). However, it has been determined that Korean ginseng has more kinds of saponins than American or Chinese ginseng (Table 17.11) (Lee 2007).

Korean ginseng, American ginseng, and Chinese ginseng (*P. notoginseng*) differ in the saponin content and content ratio of ginsenosides. Compared to American ginseng, Korean ginseng has less total saponin content, but it has a more balanced ginsenoside content. In other words, American ginseng's main ingredients are PPD type ginsenosides, such as Rb_1, and they have fewer PPT type ginsenosides. Compared to American ginseng, Korean ginseng has the balanced content of these ginsenosides (Han and Woo 1974). In particular, ginsenoside Rf, which alleviates pain, does not exist in American ginseng.

Table 17.10   Types of Ginsenosides in Korean Ginseng

Type	Ginsenosides	Total (38 types)
PPD type	Ginsenosides Ra_1, Ra_2, Ra_3, Rb_1, Rb_2, Rb_3, Rc, Rd, Rg_3 Quinquenoside R_1 Malonyl ginsenosides[a] Rb_1, Rb_2, Rc, Rd Koryoginsenoside[a] R_2 Ginsenosides Rh_2[b], Rs_1[b], Rs_2[b], Rs_3[b] 20S ginsenoside Rg_3[b] Notoginsenoside R_4[b] Ginsenoside Rg_5[b]	22 Types
PPT type	Ginsenosides Re, Rf, Rg_1, Rg_2, Rg_6[b], Rh_1 Notoginsenoside R_1 20 Gluco ginsenoside Rf Koryoginsenoside R_1 Ginsenoside Rf_2[b] 20R ginsenoside Rg_2[b], Rh_1[b] Ginsenoside Rh_4[b] 20E ginsenoside F_4[b]	14 Types
Oleanolic acid type	Ginsenoside R_o, polyactyleneginsenoside R_o[a]	2 Types

Source: From Choi, K. T. 2008. *Acta Pharmacologica Sinica* 29 (9): 1109–1118. With permission.
[a] Saponins unique to white ginseng (seven kinds).
[b] Saponins unique to red ginseng (13 kinds).

17.7.2   Active Components Other Than Ginseng Saponin (Ginsenosides)

In addition to saponins, ginseng contains a wide range of ingredients: polysaccharide, poly-acetylene, phenolic compounds, essential oil components, peptide, alkaloid, and vitamins. These nonsaponin components also have many medicinal benefits. Important nonsaponin components are antioxidant components such as maltol (Sanada et al. 1974b) and Arg-Fru-Glc (arginine fructose glucose) (Okuda et al. 1993), which are generated during the manufacturing process of red ginseng.

Polysaccharide in Korean ginseng improves immunofunction, thereby inhibiting tumor growth and lowering high blood sugar. In China, polysaccharide of ginseng is clinically used for treating cancers. Presumably, polysaccharide has an effect on stomach cancer and colon cancer to some degree. In the mid-1980s, about 20 kinds of polysaccharides—namely, panaxan A, B, C, D,...U—were isolated from Korean ginseng. Among them, there were neutral polysaccharides, such as glucose, galactose, rhamnose, arabinose, and mannose. Acidic polysaccharides included acidic sugars such as uronic acid and galacturonic acid (Konno et al. 1984).

Since 2000, acidic polysaccharide with strong immune activity has been isolated from Korean red ginseng and named red ginseng acidic polysaccharide (RGAP) (Park et al. 2001). The RGAP consists of 56.9% acidic sugar and 28.3% neutral sugar. Most of the acidic sugar is glucuronic acid with immunopotentiating activities. When RGAP extract is continuously purified, GFP, an active

Table 17.11   Comparison of Kinds of Ginsenosides of Korean Ginseng and Foreign Ginseng

Type	Korean Ginseng[a] *(Panax ginseng)*	American Ginseng (Hwagi-Sam) *(Panax quinquefolium)*	Chinese Ginseng (Sanchi-Sam) *(Panax notoginseng)*
PPD type	22	13	14
PPT type	14	5	15
Oleanolic acid type	2	1	—
Total	38	19	29

Source: From Choi, K. T. 2008. *Acta Pharmacologica Sinica* 29 (9): 1109–1118. With permission.
[a] Figures combine the numbers of ginsenosides in red ginseng and white ginseng.

substance with an immunopotentiating activity that is 10 times stronger, is produced. In a test with mice transplanted with cancer cells, GFP showed an effect of elongating life span more than RGAP. It has a strong ability to kill cancer cells through NK (natural killer) cells (Park et al. 2006).

Korean ginseng mainly contains polyacetylenes. Major polyacetylenes include panaxynol, panaxydol, and panaxytriol. In particular, panaxydol, which exists only in ginseng, has outstanding inhibitory effects on the proliferation of cancer cells (Hwang 1992). Panaxytriol, which is a polyacetylene unique to red ginseng, is produced when the epoxy ring located at C-9 of panaxydol is hydrolyzed in the red ginseng manufacturing process. Panaxytriol enhances the effect of anticancer drugs on stomach cancer cells (Matsunaga et al. 1994).

Maltol, a kind of antioxidant phenolic compound, was first discovered in Korean red ginseng (Han et al. 1979). This substance is produced by the pyrolysis of hexose when steamed at high temperatures in the red ginseng manufacturing process. Red ginseng contains antioxidant components, such as salicylic acid, ferullic acid (Han, Park, and Han 1981), gentisic acid (Wee et al. 1989), caffeic acid, and vanillic acid. Such phenolic compounds show antioxidant effects in tissues of animals with extreme oxidative injury. Compared with the phenolic compounds found in Korean ginseng and American ginseng, there are polyphenol components that exist only in Korean ginseng (Wee et al. 1998). In the manufacturing process of red ginseng, Arg-Fru-Glc (AFG) and Arg-Fru are produced in large quantities, in addition to maltol. Relatively high proportions of AFG and Arg-Fru exist in red ginseng: 5.37% and 0.2%, respectively. AFG has the medicinal benefits of inhibiting the action of maltase and dilating blood vessels (Okuda et al. 1993).

Korean ginseng contains nitrogen compounds: water-soluble protein, peptide, pyroglutamic acid, and adenosine. Ginseng protein defends against radiation hazards (Kim 1990) and the peptide has a hypoglycemic effect (Wang et al. 2003), while pyroglutamic acid inhibits lipolysis (Takaku et al. 1990).

Moreover, numerous studies have been conducted on essential-oil components as aromatic components unique to Korean ginseng, rather than their medicinal efficacies. Sesquiterpenes comprising 15 carbons take the majority of the essential-oil part. The sesquiterpene content ratio of American ginseng is concentrated in certain components, while different sesquiterpene components are balanced in Korean ginseng (Wee et al. 1997).

17.8 PHARMACOLOGICAL PROPERTIES

17.8.1 Preventing and Treating Alzheimer's Disease and Enhancing Cognitive Performance

According to "A Clinical Study on Effects of Korean Ginseng on the Enhancement of Cognitive Performance and Anti-Dementia," Lee et al. (2008) examined 30 Alzheimer's disease patients given Korean ginseng in quantities of 4.5 g or 9 g per day, starting in September 2006. After 3 months, evaluation was conducted to determine whether their cognitive performances were enhanced. The improvement of cognitive performance in MMSE (mini-mental-state examination) and CDR (clinical dementia rating) was more clearly observed in the 9 g/day group than in the 4 g/day group. This result may be used as clinical data on Korean ginseng's effects on improving symptoms of dementia (Lee et al. 2008).

Korean ginseng is effective in treating memory disorders by improving learning and memory. Ginseng saponin enhances learning and memory in the maze test and passive avoidance test (Yamaguchi et al. 1996; Jin et al. 1999). Recently, Korean red ginseng was effective in preventing and treating Alzheimer's disease. Korean red ginseng inhibits the generation of β-amyloid protein (Chen, Eckman, and Eckman 2006; Ji et al. 2006). It also enhances cognitive performance of dementia patients (Bao et al. 2005).

17.8.2 Improving Blood Circulation and Lowering Blood Pressure

Ginsenosides expand blood vessels, thus increasing blood flow and improving blood circulation and cerebral blood flow. By expanding the coronary arteries, they protect against myocardial infarction, myocardial ischemia, and angina pectoris. Ginsenosides also inhibit platelet aggregation, showing antithrombotic and antiarrhythmic effects (Kim et al. 1994; Ji et al. 2006; Tamura et al. 1993). Platelet aggregation of coronary artery disease patients has been inhibited; in particular, ginsenosides Rg_1 and Rg_3, the main saponin constituents, inhibit action and generation of thromboxane A2 (Lee and Park 1998; Kim et al. 1999). Nonsaponin constituents of red ginseng also have similar effects. Such effects work not only in coronary arteries, but also in the central nervous system, thus improving heart and cerebral ischemia (Park et al. 1995). Recently, the Korean Food and Drug Administration reported ginseng's functional effect of improving blood circulation by preventing blood clots.

Red ginseng had been traditionally known to raise blood pressure, so hypertension patients were advised not to consume it. However, there are now many reports that modern scientific evidence indicates that red ginseng actually lowers blood pressure. In fact, clinically speaking, red ginseng lowers both diastolic and systolic pressure of hypertension patients. In particular, ginsenosides, the major constituent of red ginseng, relax blood vessels by isolating NO (nitric oxide) originating from vascular endothelial cells (Jeon, Kim, and Park 2000; Sung et al. 2000; Kim et al. 2003).

17.8.3 Anticarcinogenic and Cancer-Preventive Effects

The epidemiologic study on anticarcinogenic and cancer-preventive effects of Korean ginseng shows that cancer rates were significantly low in the group with the intake of red ginseng. The 5-year study also showed that cancer rates were lower in the group with the intake of red ginseng and that frequencies of various types of cancers were also low. Korean ginseng is known to have cancer-preventive benefits through various ingredients, such as its antioxidant properties (Yun and Choi 1995, 1998; Keum et al. 2000). In particular, Korean ginseng boosts immune functions and strengthens functions of NK cells and T-cells, while inhibiting cancer metastasis. There is some controversy regarding whether Korean red ginseng directly attacks cancer cells and whether its actions bring about changes in cell cycle. There are also some controversies over its action on oncogene, its action to reduce resistance to multiple agents, the inhibition of angiogenesis, and inhibition of infiltration (Kenarova et al. 1990; Choi, Kim, and Jang 2002; Sonoda et al. 1998).

17.8.4 Preventing and Treating Diabetes

After one dose, glucose tolerance tests show that Korean red ginseng suppressed the increase in blood glucose. When Korean red ginseng is given to patients with diabetes, their blood sugar level after meals is reduced because Korean red ginseng improves secretion (Lai et al. 2006). Ginsenosides increase insulin action (Vuksan and Sievenpipper 2005). On the other hand, ginsenosides failed to lower blood sugar.

17.8.5 Alcohol Detoxification (Hangover Reduction) Effects

Korean red ginseng promotes alcohol metabolism and increases the activity of alcohol dehydrogenase, thereby lowering blood alcohol concentration (Kim 2007). Korean red ginseng reduces lactate after drinking alcohol and, when indexes related to liver toxicity that appear after taking alcohol are measured, reduces liver cell toxicity (Joo et al. 1985).

17.8.6 Improving Sexual Performance

When 300 mg of Korean red ginseng was given to patients with psychogenic erectile dysfunction for 2 months in the form of nine capsules daily, the result showed that their sexual performance improved. All categories, including sexual desire, frequency of sexual intercourse, penile erection, and sexual satisfaction, showed improvement (Choi and Seung 1995; Kim et al. 1998). In animal tests using mice and rats, Korean red ginseng not only reduced sexual behavior disorder caused by stress, but also enhanced sexual behavior. Moreover, it improved asthenospermia and oligospermia, causes of male infertility (Lee and Park 1998; Murphy et al. 1998). The action mechanism for improving erectile dysfunction is as follows: Ginseng expedites separation of NO from slices of penile corpus cavernosum, thereby inducing the relaxation of penile corpus cavernosum, which results in relaxation of blood vessels (Chen and Lee 1995).

17.8.7 Antifatigue and Antistress

Dr. Brekhman, a pharmacologist in Russia, explained the effect of ginseng for increasing nonspecific resistance among ginseng's medicinal properties as a concept that illustrates ginseng's tonic effect. He said that ginseng boosted immunity under harmful conditions of a living body, thereby promoting immunoability in a nonspecific manner. He dubbed this the "adaptogen effect" (Brekhman 1957).

In an animal test, restraint stress increased the concentration of polyamine in the brain, yet ginseng saponins inhibited polyamine (Kim et al. 2003; Lee et al. 2006). Restraint stress increases corticosteroid and IL-6 in blood and Korean red ginseng inhibited them. Ginseng's recovery function was also proved to be effective in diminished motor function in mice forced to walk a tightrope (Kim et al. 1998). Moreover, Korean red ginseng enhances cardiopulmonary function, thus speeding up recovery from fatigue. It also reduces anxiety symptoms (Han et al. 2005).

17.8.8 Antiaging Efficacy

Nonsaponin components of Korean red ginseng inhibit the damaging effects of free radicals that are related to the aging of cells. Moreover, phenolic compounds and maltol of ginseng are reported to play an important role in inhibiting aging (Han et al. 1979). Recently, a variety of red ginseng products designed to prevent skin aging have been under development (Chen and Lee 1995).

17.8.9 Efficacy of Enhancing Immune Function

Recently, research on Korean red ginseng's immunoeffects drew attention and more interest has been directed at the immunomodulation action of polysaccharide rather than saponins. Saponins and polysaccharide promote cell proliferation of T-lymphocyte and, in particular, it has been reported that polysaccharide fractions increased Tc lymphocytes' power to destroy cancer cells three- to fourfold (Mizuno et al. 1994; Rivera et al. 2005). It has been reported that polysaccharide separated from extracts of tissue-cultured ginseng induced proliferation of B-cells and T-cells. In addition, studies indicate that ginseng's saponins have an effect on recovery from inhibited immunity in cyclophosphamide-treated mice (Soloverva et al. 1989).

It has been reported that Korean red ginseng increases activity of NK cells, which inhibits metastasis of cancer cells. There are numerous studies that Korean red ginseng increases the activity of NL cells (Yun et al. 1987; Kim, Oh et al. 1990). In addition, numerous research results have been indicated that Korean red ginseng intervenes in the generation and inhibition of cytokines (Kim, Kang et al. 1990; Yun et al. 1993).

17.8.10 Anti-AIDS (Acquired Immune Deficiency Syndrome) or Antiviral Action

Korean red ginseng increased NL cell function for both AIDS patients and healthy persons. In particular, red ginseng was given to those infected with HIV (human immunodeficiency virus) in doses of 5.4 g daily and they survived for more than 20 years without taking other AIDS medications (Cho et al. 2001; Sung et al. 2005, 2007). The CD+4 cells after the intake of Korean red ginseng increased and the red ginseng inhibited manifestation of resistance against ZDV (zidovudine, an AIDS virus proliferation inhibitor). The combination of red ginseng and AIDS drugs was more effective (Han and Woo 1974; Konno et al. 1984; Okuda et al. 1993). Korean red ginseng is also effective for treating hepatitis C and inhibited inflammatory reactions caused by influenza virus infection (Cyong et al. 2000).

17.8.11 Obesity and Hyperlipidemia

The effect of Korean red ginseng on obesity, lipid metabolism, hyperlipidemia, and hypercholesterolemia has been extensively studied. In animal tests, ginseng saponins blocked the absorption of fat and cholesterol, thereby promoting metabolism (Yamamoto, Kumagai, and Yamamura 1983; Kim et al. 2009). Korean red ginseng lowers total cholesterol, triglyceride, and low-density lipoprotein (LDL) in blood and expedites breaking down of body fat and is thus effective in preventing and treating hyperlipidemia (Kim and Park, 2003).

17.8.12 Detoxifying Drug Addiction

Tests on lab animals addicted to various drugs such as morphine, cocaine, and methamphetamine demonstrated that Korean red ginseng had an effect in inhibiting the addictions, as well as mental and physical dependency (Kim et al. 1990; Bhargava and Ramarao 1991; Kim et al. 2005). Therefore, it was reported that Korean red ginseng may be effective in preventing and treating addiction symptoms caused by drug abuse (Oh 2008).

17.9 CONCLUSIONS

A large amount of research and findings dealing with the development of new varieties, cultivation, components, pharmacological characters, and effects of Korean ginseng (*Panax ginseng* C.A. Meyer) have been accumulated up to now. In particular, many scholars have investigated the pharmacological effects of ginseng through varied approaches, ranging from investigation of a single organ to that of the clinical effects on the whole body. Thanks to these efforts, the effects of ginseng have been confirmed, although not comprehensively. Korean scholars have asserted that the medical effects of ginseng reach all of the cells and smooth metabolism in the body through the stimulation of protein anabolism, which is the energy source of life.

Ginseng does not give rise to such problems as poisoning or drug dependency effects like those often caused by other foods; its effects are always preventive. It is true that the totality of the effects of ginseng has not been analyzed or understood yet in terms of modern pharmacological concepts. On the other hand, we already know that the whole function of the human body is surely returned to a normal level through long use of ginseng. We cannot undervalue the effects of ginseng, for they have already been proved by the long history of people's experiences.

Traditionally, ginseng has been used primarily in Southeast Asia. But it is now spreading to all countries in the world. It is produced in a variety of new forms to suit people's tastes. Ginseng will certainly be recognized as a safe health food by modern men who are troubled by various kinds of pollution. It is certain that ginseng will make significant contributions to the health and happiness of mankind.

REFERENCES

Ahn, S. D., Choi, K. T., Cheon, S. R., Chung, C. M., and Kwon, W. S. 1985. Coefficient of variability of agronomic characters in *Panax ginseng* C.A. Meyer. *Korean Journal of Ginseng Science* 9 (1): 9–14.

Ahn, S. D., Choi, K. T., Kwon, W. S., Chung, C. M., Chun, S. R., and Nam, K. Y. 1987. Estimation of yield in *Panax ginseng*. *Korean Journal of Ginseng Science* 11 (1): 46–55.

Bao, H. Y., Zhang, J., Yeo, S. J., Myung, C. S., Kim, H. M., Kim, J. M., Park, J. H., Cho, J., and Kang, J. S. 2005. Memory enhancing and neuroprotective effects of selected ginsenosides. *Archives of Pharmacology Research* 28:335–342.

Bhargava, H. N., and Ramarao, P. 1991. The effect of *Panax ginseng* on the development of tolerance to the pharmacological action of morphine in the rat. *General Pharmacology* 22:521–525.

Bittles, A. H., Fulder, S. J., Grant, E. C., and Nichills, M. R. 1979. The effect of ginseng on life span and stress response in mice. *Gerontology* 25:125.

Brekhman, I. I. 1957. *Panax ginseng*. Gosudaarst Isdat et Med. Lit., Leningrad: 1.

Brekhman, I. I., and Dardymov, I. V. 1969. New substances of plant origin which increase nonspecific resistance. *Annual Reviews in Pharmacology* 9:419–430.

Chen, F., Eckman, E. A., and Eckman, C. B. 2006. Reductions in levels of the Alzheimer's amyloid beta peptide after oral administration of ginsenosides. *FASEB* 20:1269–1271.

Chen, X., and Lee, T. J. 1995. Ginsenosides-induced nitric oxide-mediated relaxation of the rabbit corpus cavernosum. *British Journal of Pharmacology* 115:15–18.

Cheon, S. R., Ahn, S. D., Choi, K. T., and Kwon, W. S. 1985. Characters and inheritance of stem color in F_1 and F_2 of violet-stem variant × yellow-berry variant in *Panax ginseng* C.A. Meyer. *Korean Journal of Ginseng Science* 9 (2): 264–269.

Cho, Y. K., Sung, H., Lee, H. J., Joo, C. H., and Cho, G. J. 2001. Long-term intake of Korean red ginseng in HIV-1-infected patients: Development of resistance mutation to zidovudine is delayed. *International Immunopharmacology* 1:1295–1305.

Choi, H. K., and Seung, D. H. 1995. Effectiveness for erectile dysfunction after the administration of Korean red ginseng. *Korean Journal of Ginseng Science* 19:17–21.

Choi, K. T. 2008. Botanical characteristics, pharmacological effects and medicinal components of Korean *Panax ginseng* C.A. Meyer. *Acta Pharmacologica Sinica* 29 (9): 1109–1118.

Choi, K. T., Ahn, S. D., and Shin, H. S. 1980. Correlations among agronomic characters of ginseng plant. *Journal of Korean Society Crop Science* 25 (3): 63–67.

Choi, K. T., Kwon, W. S., Lee, S. S., Lee, J. H., Chung, Y. Y., Kang, J. Y., and Lee, M. G. 1998. Breeding of Korean ginseng (*Panax ginseng* C.A. Meyer). Proceedings of the 7th International Symposium on Ginseng, Seoul, Korea, 96–102.

Choi, K. T., Kwon, W. S., Lee, S. S., Lee, J. H., and Lee, M. G. 2001. Development of new varieties of Korean ginseng (*Panax ginseng* C.A. Meyer). Proceedings of the International Ginseng Workshop, Vancouver, Canada, 47–52.

Choi, K. T., Lee, M. G., Kwon, W. S., and Lee, J. H. 1994. Strategy for high quality ginseng breeding. *Korean Journal of Breeding* 26 (S): 125–138.

Choi, M. S., Kim, Y. H., and Jang, C. C. 2002. Effect of red ginseng extracts on the immune response in mouse. *Journal of Basic & Life Research* 2 (2): 47.

Chung, Y. Y., Chung, C. M., and Choi, K. T. 1995. The correlation of agronomic characters and path coefficient analysis in *Panax ginseng* C.A. Meyer. *Korean Journal of Ginseng Science* 19 (2): 165–170.

Chung, Y. Y., Chung, C. M., Kang, J. Y., Kim, Y. T., and Choi, K. T. 1992. The comparison of growth characteristics of *Panax ginseng* C.A. Meyer and *Panax quinquefolium* L. *Korean Journal of Breeding* 24 (1): 81–86.

———. 1993. Characteristics of early and late emergence groups in *Panax ginseng* C.A. Meyer. *Korean Journal of Breeding* 25 (3): 204–210.

Cyong, J. C., Ki, S. M., Iijima, K., Kobayashi, T., and Furuya, M. 2000. Clinical and pharmacological studies on liver diseases treated with kampo herbal medicine. *American Journal of Clinical Medicine* 28 (3–4): 351–360.

Han, B. H., Park, M. H., and Han, Y. N. 1981. Studies on the antioxidant components of Korean ginseng (III). Identification of phenolic acid. *Archives of Pharmaceutical Research* 4 (1): 53–58.

Han, B. H., Park, M. H., Woo, L. K., Woo, W. S., and Han, Y. N. 1979. Studies on the antioxidant components of Korean ginseng. *Korean Biochemical Journal* 12 (1): 33–40.

Han, B. H., and Woo, L. K. 1974. Dammarane glycosides of *Panax ginseng*. *Korean Journal of Pharmacognosy* 5 (1): 31–44.

Han, K., Shin, I. C., Choi, K. J., Yun, Y. P., Hong, J. T., and Oh, K. W. 2005. Korea red ginseng water extract increases nitric oxide concentrations in exhaled breath. *Nitric Oxide* 12 (3): 159–162.

Hong, S. A., Oh, J. S., Park, C. W., Chang, H. K., and Kim, E. C. 1970. Pharmacology of ginseng. *Korean Journal of Pharmacology* 6:1–9.

Horhammer, L., Wagner, H., and Lay, B. 1961. Constituents of the *Panax ginseng* root, preliminary report. *Pharm. Ztg. Ver. Apoth. Ztg.* 106:1307–1311.

Hu, S. Y. 1976. The genus *Panax* (ginseng) in Chinese medicine. *Economic Botany* 30:11–28.

Hwang, W. I. 1992. Anticancer effect of Korean ginseng. *Korean Journal of Ginseng Science* 16 (2): 170–171.

Jeon, B. H., Kim, C. S., and Park, K. S. 2000. Effects of Korea red ginseng on the blood pressure in conscious hypertensive rats. *General Pharmacology* 35:135–141.

Ji, Z. N., Dong, T. T., Ye, W. C., Choi, R. C., Lo, C. K., and Tsim, K. W. 2006. Ginsenoside Re attenuates beta-amyloid and serum-free induced neurotoxicity in PC12 cells. *Journal of Ethnopharmacology* 107:48–52.

Jin, S. H., Park, J. K., Nam, K. Y., Park, S. N., and Jung, N. P. 1999. Korean red ginseng saponins with low ratios of protopanaxadiol and protopanaxatriol saponin improve scopolamine-induced learning disability and spatial working memory in mice. *Journal of Ethnopharmacology* 66:123–129.

Joo, C. N., Tae, G. S., Joo, S. O., and Cho, K. S. 1985. The effect of saponin fraction of *Panax ginseng* C.A. Meyer on the liver of ethanol administered rat. *Korean Journal of Ginseng Science* 9 (1): 1–8.

Kenarova, B., Neychev, H., Hadjiivanova, C., and Petkov, V. D. 1990. Immunomodulating activity of ginsenoside Rg_1 from *Panax ginseng*. *Japan Journal of Pharmacology* 54 (4): 447–454.

Keum, Y. S., Park, K. K., Lee, J. M., Chun, K. S., Park, J. H., Lee, S. K., Kwon, H., and Surh, Y. J. 2000. Antioxidant and anti-tumor promoting activities of the methanol extract of heat-processed ginseng. *Cancer Letters* 150: 41–48.

Kim, C. 1990. Mechanisms of the radioprotective activity of ginseng protein fraction. *Korean Journal of Ginseng Science* 14:279–283.

Kim, D. H., Jung, J. S., Suh, H. W., Hur, S. O., Min, S. K., Son, B. K., Park, J. H., Kim, N. D., Kim, Y. H., and Song, D. K. 1998. Inhibition of stress-induced plasma corticosterone levels by ginsenosides in mice: Involvement of nitric oxide. *Neuroreport* 9:2261–2264.

Kim, D. H., Moon, Y. S., Lee, T. H., Jung, J. S., Suh, H. W., and Song, D. K. 2003. The inhibitory effect of ginseng saponins on the stress-induced plasma interleukin-6 level in mice. *Neuroscience Letters* 353:13–16.

Kim, H. C., Shin, E. J., Jang, C. G., Lee, M. K., Eun, J. S., Hong, J. T., and Oh, K. W. 2005. Pharmacological action of *Panax ginseng* on the behavioral toxicities induced by psychotropic agents. *Archives of Pharmaceutical Research* 28 (9): 995–1001.

Kim, H. J., Woo, D. S., Lee, G., and Kim, J. J. 1998. The relaxation effects of ginseng saponin in rabbit corporal smooth muscle: Is it a nitric oxide donor? *British Journal of Urology* 82:744–748.

Kim, H. S., Oh, K. W., Lee, M. K., Back, D. Y., Rheu, H. M., and Seong, Y. H. 1990. The role of ginseng total saponins in the inhibition of the development of analgesic tolerance to morphine. *Korean Journal of Ginseng Science* 14 (2): 178–188.

Kim, I. H. 1989. *Jeungbo Munheonbigo*, vol. 151 (translation). Memorial Association of the King Sejong, Compilation Committee of Classics Translation 151:174–233.

Kim, J. Y., Germolec, D. R., and Luster, M. I. 1990. *Panax ginseng* as a potential immunomodulator: Studies in mice. *Immunopharmacology and Immunotoxicology* 12 (2): 257–276.

Kim, N. D., Kang, S. Y., Kim, M. G., Park, J. H., and Schini, V. B. 1999. The ginsenoside Rg_3 evokes endothelium-independent relaxation in rat aortic rings: Role of K+ channels. *European Journal of Pharmacology* 367:51–57.

Kim, J. H., Kang, S. A., Han, S. M., and Shim, I. 2009. Comparison of the antiobesity effects of the protopanaxadiol-and protopanaxatriol-type saponins of red ginseng. *Phytotherapy Research* 23 (1): 78–85.

Kim, N. D., Kang, S. Y., and Schini, V. B. 1994. Ginsenosides evoke endothelium-dependent vascular relaxation in rat aorta. *General Pharmacology* 25:1070–1077.

Kim, N. D., Kim, E. M., Kang, K. W., Cho, M. K., Choi, S. Y., and Kim, S. G. 2003. Ginsenoside Rg$_3$ inhibits phenylephrine-induced vascular contraction through induction of nitric oxide synthase. *British Journal of Pharmacology* 140 (4): 661–670.

Kim, S. H., and Park, K. S. 2003. Effect of *Panax ginseng* extract on lipid metabolism in humans. *Pharmacology Research* 48 (45): 511–513.

Kim, S. S. 2007. Effects of Korean red ginseng on alcohol detoxification, lipid metabolism and recovery of fatigue function. *Reviews in Ginseng Research* (1), Korean Society of Ginseng: 149–159.

Kim, Y. S., Kang, K. S., and Kim, S. I. 1990. Study on antitumor and immunomodulating activities of polysaccharide fractions from *Panax ginseng*: Comparison of neutral and acidic polysaccharide fraction. *Archives of Pharmacology Research* 13:330–337.

Kim, Y. T., Chung, C. M., Kwon, W. S., Lee, J. H., and Chung, Y. Y. 1989. Annual report, Korea Ginseng &Tobacco Research Institute: 3–70.

Kitagawa, I., Taniyama, T., Hayashi, T., and Yoshikawa, M. 1983. Malonyl-ginsenosides Rb$_1$, Rb$_2$, Rc, and Rd, four new malonylated dammarane-type triterpene oligoglycosides from ginseng radix. *Chemical and Pharmacology Bulletin* 31:3353–3356.

Konno, C., Sugiyama, K., Kano, M., Takahashi, M., and Hikino, H. 1984. Isolation and hypoglycemic activity of panaxans A, B, C, D and E, glycans of *Panax ginseng* roots. *Planta Medica* 50:434–436.

KGGA (Korea Ginseng Growers Association). 1980. Korea ginseng history. Compilation Committee of the Korean Ginseng History: 59.

Kwon, W. S., Chung, C. M., Kim, Y. T., and Choi, K. T. 1991. Comparison of growth, crude saponin, ginsenosides, and anthocyanines in superior lines of *Panax ginseng* C.A. Meyer. *Korean Journal of Breeding* 23 (3): 219–228.

Kwon, W. S., Kang, J. Y., Lee, J. H., Lee, M. G., and Choi, K. T. 1998. Red ginseng quality and characteristics of KG101, a promising line of *Panax ginseng* C.A. Meyer. *Journal of Ginseng Research* 22 (4): 244–251.

Kwon, W. S., Lee, J. H., Kang, J. Y., Kim, Y. T., and Choi, K. T. 1994. Red ginseng quality of Jakyung-jong and Hwangsook-jong in *Panax ginseng* C. A. Meyer. *Korean Journal of Breeding* 26 (4): 400–404.

Kwon, W. S., Lee, M. G., and Choi, K. T. 2000. Breeding process and characteristics of Yunpoong, a new variety of *Panax ginseng* C.A. Meyer. *Journal of Ginseng Research* 24 (11): 1–7.

Lai, D. M., Tu, Y. K., Liu, I. M., Chen, P. F., and Yang, L. 2006. Mediation of beta-endorphin by ginsenoside Rh2 to lower plasma glucose in streptozotocin-induced diabetic rats. *Planta Medica* 72:9–13.

Lee, J. H., and Park, H. J. 1998. Effects of in-taking of red ginseng products on human platelet aggregation and blood lipids. *Journal of Ginseng Research* 22:173–180.

Lee, S. D. 2007. Reviews in ginseng research (I), Korean Society of Ginseng: 31–37.

Lee, S. H., Jung, B. H., Kim, S. Y., Lee, E. H., and Chung, B. C. 2006. The antistress effect of ginseng total saponin and ginsenoside Rg$_3$ and Rb$_1$ evaluated by brain polyamine level under immobilization stress. *Pharmacology Research* 54:46–49.

Lee, S. S. 2007a. Korean ginseng (ginseng cultivation). Annual report of Korea Ginseng & Tobacco Research Institute: 18–40.

———. 2007b. Review in ginseng research (I), research of ginseng cultivation, Korean Society of Ginseng: 3–27.

Lee, S. S., Chung, Y. Y., Lee, M. G., and Choi, K. T. 1997. Studies on incompatibility in interspecific hybrid between *Panax ginseng* C.A. Meyer and *Panax quinquefolium* L. *Korean Journal of Ginseng Science* 21 (2): 85–90.

Lee, S. S., Lee, J. H., and Ahn, I. O. 2005. The present status and prospect of new varieties of Korean ginseng. *Journal of Korean Society for Seed Science & Industry* 2 (1): 7–17.

Lee, S. R., Park, J. H., Choi, K. J., and Kim, N. D. 1997. Ginsenosides inhibit endothelium-dependent contraction in the spontaneously hypertensive rat aorta *in vitro*. *Korean Journal of Ginseng Science* 21 (2): 125–131.

Lee, S. T., Chu, K., Sim, J. Y., Heo, J. H., and Kim, M. 2008. *Panax ginseng* enhances cognitive performances in Alzheimer's disease. *Alzheimer Disease Association Disorders* 22 (3): 222–226.

Matsunaga, H., Katano, M., Saita, T., Yamamoto, H., and Mori, M. 1994. Potentiation of cytotoxicity of mitomycin C by polyacetylenic alcohol, panaxytriol. *Cancer Chemotherapy and Pharmacology* 33:291–297.

Mizuno, M., Yamada, J., Tarei, H., Kozukue, N., Lee, Y. S., and Tsuchida, H. 1994. Differences in immunomodulating effects between wild and cultured *Panax ginseng*. *Biochemical and Biophysics Research Communications*. 200:1672–1678.

Murphy, L. L., Cadena, R. S., Chavez, D., and Ferraro, J. S. 1998. Effect of American ginseng (*Panax quinque-folium*) on male copulatory behavior in the rat. *Physiology & Behavior* 64:445–450.

Namba, T., Yoshizaki, M., Tomori, T., Kobachi, K., and Hase, J. 1973. Hemolytic and its protective activity of ginseng saponins. *Chemical and Pharmacology Bulletin* (Tokyo) 21 (2): 459–461.

Oh, J. S. 1976. Pharmacology of ginseng. *Korean Journal of Ginseng Science* 1(1): 1–11.

Oh, S. 2008. Antinarcotic effect of ginseng. *Journal of Ginseng Research* 32 (1): 1–7.

Okuda, H., Zheng, Y., Matsuura, Y., Takaku, T., and Kameda, K. 1993. Studies on biologically active substances in nonsaponin fraction of Korean red ginseng. *Proceedings of 6th International Ginseng Symposium*, Seoul, Korea: 110.

Park, H. J., Lee, M. H., Park, K. M., Nam, K. Y., and Park, K. H. 1995. Effect of nonsaponin fraction from *Panax ginseng* on cGMP and thromboxane A2 in human platelet aggregation. *Journal of Ethnopharmacology* 49 (3): 157–162.

Park, J. D. 1996. Recent studies on the chemical constituents of Korean ginseng (*Panax ginseng* C.A. Meyer). *Korean Journal of Ginseng Science* 20:389–415.

Park, J. D., Shin, H. J., Kwak, Y. S., Wee, J. J., Song, Y. B., Kyung, J. S., Kiyohara, H., and Yamada, H. 2006. Partial chemical structure and immunomodulating antitumor activities of red ginseng acidic polysaccharide from Korean red ginseng. Proceedings of 9th International Symposium on Ginseng: 390–409.

Park, K. M., Kim, Y. S., Jeong, T. C., Joe, C. O., Shin, H. J., Lee, Y. H., Nam, K. Y., and Park, J. D. 2001. Nitric oxide is involved in the immunomodulating activities of acidic polysaccharide from *Panax ginseng*. *Planta Medica* 67:122–126.

Rivera, E., Ekholm, P. F., Inganils, M., Paulie, S., and Groenvik, K. O. 2005. The Rb$_1$ fraction of ginseng elicits a balanced Th1 and Th2 immune response. *Vaccine* 23:5411–5419.

Sanada, S., Kondo, N., Shoji, J., Tanaka, O., and Shibata, S. 1974a. Studies on the saponins of ginseng. I. Structures of ginsenoside-Ro, -Rb$_1$, -Rb$_2$, -Rc, and -Rd. *Chemical and Pharmacology Bulletin* 22 (2): 421–428.

———. 1974b. Studies on the saponins of ginseng. II. Structures of ginsenoside-Re, -Rf, and -Rg$_2$. *Chemical and Pharmacology Bulletin* 22 (10): 2407–2412.

Shibata, S., Tanaka, O., Ando, T. Sado, M., Thushima, S., and Ohsawa, T. 1966. Chemical studies on oriental plant drugs. XIV. Protopanaxadiol, a genuine sapogenin of ginseng saponins. *Chemical and Pharmacology Bulletin* 14:595–600.

Shibata, S., Tanaka, O., Soma, K., Iida, Y., Ando, T., and Nakamura, H. 1965. Studies on saponins and sapogenins of ginseng. The structure of panaxatriol. *Tetrahedron Letters* 3:207–213.

Soloverva, T. F., Besednova, N. N., Urarova, N. I., Faustov, V. S., Konstanyinova, N. A., Klyova, N. V., Tsybulski, A. V., Ovodov, I., and Eliakov, G. B. 1989. Phagocytosis-stimulating effect of polysaccharides isolated from ginseng tissue culture. *Antibiot Khimioter* 34 (10): 755–760.

Sonoda, Y., Kasahara, T., Mukaida, N., Shimizu, N., Tomota, M., and Takeda, T. 1998. Stimulation of interleukin-8 production by acidic polysaccharides from the root of *Panax ginseng*. *Immunopharmacology* 38:287–294.

Sung, H., Jung, Y., Kang, M. W., Bae, I. G., Chang, H. A., Woo, J. H., and Cho, Y. K. 2007. High frequency of drug resistant mutations in human immunodeficiency virus type 1-infected Korean patients treated with HAART. *AIDS Research in Human Tetroviruses* 23:1223–1229.

Sung, H., Kang, S. M., Lee, M. S., Kim, T. K., and Cho, Y. K. 2005. KRG slows depletion of CD4 T-cells in human immunodeficiency virus type 1-infected patients. *Clinical Diagnostic Laboratory Immunology* 12:497–501.

Sung, J., Han, K. H., Zo, J. H., Park, H. J., Kim, C. H., and Oh, B. H. 2000. Effects of red ginseng upon vascular endothelial function in patients with essential hypertension. *American Journal of Clinical Medicine* 28 (2): 205–216.

Takaku, T., Kameda, K., Matsuura, Y., Sekiya, K., and Okuda, H. 1990. Studies on insulin-like substances Korean red ginseng. *Planta Medica* 56:27–30.

Tamura, Y., Hirai, A., Terano, T., Thara, K., Saito, J., and Kondo, S. 1993. Effect of Korean red ginseng on eicosanoid biosynthesis in platelets and vascular smooth muscle cells. *Review of Medical Science* 16:55–60.

Tode, T., Kikuchi, Y., Kita, T., Hirata, J., Imaizumi, E., and Nagata, I. 1993. Inhibitory effects by oral administration of ginsenoside Rh$_2$ on the growth of human ovarian cancer cells in nude mice. *Journal of Cancer Research and Clinical Oncology* 120:24–26.

Vuksan, V., and Sievenpipper, J. L. 2005. Herbal remedies in the management of diabetes: Lessons learned from the study of ginseng. *Nutrition, Metabolism & Cardiovascular Diseases* 15:149–160.

Wang, B. X., Zhou, Q. L., Yang, M., Wang, Y., et al. 2003. Hypoglycemic mechanism of ginseng glycopeptide. *Acta Pharmacologica Sinica* 24:50–54.

Wee, J. J., Park, J. D., Kim, M. W., and Lee, H. J. 1989. Identification of phenolic antioxidant compound isolated from *Panax ginseng. Journal of Korean Agriculture Chemical Society* 32 (1): 50–56.

Wee, J. J., Shin, J. Y., Kim, S. K., and Kim, M. W. 1998. Comparison of phenolic components between Korean and American ginseng by thin-layer chromatography. *Journal of Ginseng Research* 22 (2): 91–95.

Wee, J. J., Shin, J. Y., Sohn, H. J., Heo, J. N., Kim, S. K., and Kim, M. W. 1997. Comparison of sesquiterpenes in Korean and American ginsengs. *Korean Journal of Ginseng Science* 21 (3): 209–213.

Yamaguchi, Y., Higashi, M., and Kobayashi, H., 1996. Effects of ginsenosides on impaired performance caused by scopolamine in rats. *European Journal of Pharmacology* 312:149–151.

Yamamoto, M., Kumagai, A., and Yamamura, Y. 1983. Metabolic actions of ginseng principles in bone marrow and testis. *American Journal of Chinese Medicine* 11 (1–4): 96–101.

Yun, T. K., and Choi, S. Y. 1995. Preventive effect of ginseng intake against various human cancers: A case-control study on 1987 pairs. *Cancer Epidemiology Biomarker Reviews* 4:401–408.

———. 1998. Non-organ-specific cancer prevention of ginseng: A prospective study in Korea. *International Journal of Epidemiology* 27:359–364.

Yun, Y. S., Lee, Y. S., Jo, S. K., and Jung, I. S. 1993. Inhibition of autochthonous tumor by ethanol insoluble fraction from *Panax ginseng* as an immunomodulator. *Planta Medica* 59:521–524.

Yun, Y. S., Moon, H. S., Oh, Y. R., Jo, S. K., Kim, Y. J., and Yun, T. K. 1987. Effect of red ginseng on natural killer cell activity in mice with lung adenoma induced by urethane and benzo[a]pyrene. *Cancer Detection and Prevention* 1:301–309.

Zhang, H., Lu, Z., Tan, G. T., Qiu, S., Farnsworth, N. R., Pezzuto, J. M., and Fong, H. S. 2002. Polyacetyleneginsenoside-R_o, a novel triterpene saponin from *Panax ginseng. Tetrahedron Letters* 43:973–977.

Zhou, D. H. 1982. Preventive geriatrics: An overview from traditional Chinese medicine. *American Journal of Chinese Medicine* 10 (1–4): 32.

CHAPTER 18

Tea

A. K. A. Mandal, S. Babu, and R. S. Senthil Kumar

CONTENTS

18.1 INTRODUCTION

Tea, the most widely consumed, nonalcoholic drink, is manufactured from the leaves of tea plant *Camellia sinensis* (L.) Kuntze. *Camellia sinensis* belongs to the family Theaceae. It originated in central China, somewhere in the Mongolian plateau (Weatherstone 1992) and is now cultivated in more than 30 countries worldwide. Tea cultivation plays an important role in the economies of China, India, Sri Lanka, Vietnam, Indonesia, Bangladesh, Japan, Kenya, Turkey, and Argentina.

It was first cultivated in China and consumed as drink. Today, it is consumed as a health drink because it contains a number of secondary metabolites like polyphenols, catechins, caffeine, theanine, saponin, etc. Tea consumption is beneficial in checking various pathological conditions, like cardiovascular disease, diabetes, obesity, and cancer.

The present review has been divided into three major divisions: botany and breeding, biochemistry, and biotechnology and molecular biology of tea. An attempt has been made to collate the up-to-date research efforts made in these areas, with a special emphasis on medicinal properties of tea and stress on the lacuna and future perspectives.

18.2 BOTANY AND BREEDING

18.2.1 Origin and Distribution

Although cultivation and usage of tea was known for the past two centuries, its origin and dispersal are still under speculation. In China, tea was consumed as a beverage about 3,000 years ago. The presence of of wild type plant, from time to time, in southern China suggests the possibility of cultivation in the Chinese provinces (Eden 1976). However, it has been suggested that tea originated in central China, somewhere in the Mongolian plateau (Weatherstone 1992). From this primary center, it is believed that tea dispersed toward the south of China and established a secondary center of origin near the Irrawadi River basin (Kingdon-Ward 1950). From the secondary center of origin, tea dispersed in three different directions along the sources of the three great rivers, Yang-Tse, Mekong, and Brahmaputra.

The distribution of tea from its center of origin to other parts is by natural or human transfer. It is now grown from tropical to subtropical regions, including tropical rain forest, savannah, and summer rain areas. It is cultivated at 27°S (Argentina) to 43°N (Georgia) and from 500 to 3000 m. Tea is now being, cultivated in at least 30 countries as a commercial beverage crop in 2,995,504 ha (FAO 2009; http://www.fao.org/).

18.2.2 Taxonomy and Botany

The botanical classification of tea was made differently by different authors. After a long debate, the great taxonomist Linnaeus listed the tea as *Camellia sinensis* in the *Index Kewensis*. He originally called tea *Thea sinensis* and later included two more species: *Thea bohas* and *Thea viridis*. Sometimes it was called *Camellia thea* Dyer (Wight and Barua 1939). Now it is commonly called *Camellia sinensis*.

Figure 18.1   Morphology of tea: (A) freely grown plant; (B) upper surface of leaf; (C) lower surface of leaf; (D) unopened bud; (E) fully opened flower; (F) upper view of mature fruit; (G) lower view of mature fruit.

Currently, the genus *Camellia* belongs to the family Theaceae along with eight other genera. Sealy (1958) revised the genus *Camellia* with 82 species. Recent literature available from China describes more than 200 species (Singh 1999). Most of the genera of Theaceae and *Camellia* species are found in Yunnan Province in southwest China and northeast India bordering Myanmar with the maximum diversity (Weatherstone 1992). Cytotaxonomical study revealed that most of the cultivated *Camellia* have comparable chromosomal structure and the numbers reveal that they have arisen from same basic genome (Janaki Ammal 1952). Most of the cultivated varieties of *Camellia sinensis* and other wild relatives have diploid chromosome number ($2n = 2x = 30$); some of the cultivars in this group registered as triploid ($2x = 45$; Purseglove 1968). The only hexaploid ($2x = 90$) species is *Camellia sasanqua* (Bezbaruah 1971; Bezbaruah and Gogoi 1972).

18.2.3 Vegetative Growth Phase

Tea is a unique crop grown for its leaf; the two or three leaves with a bud are the commercial part of the plant. A freely grown (unpruned) plant can grow up to 30 m (Figure 18.1a). This crop is always maintained in a vegetative phase as a small bush by pruning to produce leaves (Figure 18.1b, c) continuously.

The terminology of plant parts used in tea cultivation is not commonly matched with the standard botanical description. The vegetative growth is commonly termed "flush," which describes the apical growth of a tea shoot that occurs by a succession of flushes separated by short periods of dormancy (Bond 1942). The phase of rest or dormancy is normally termed the *banjhi* phase and the bud developed in this stage is called *banjhi* bud. The banjhi bud is usually small, without well-serrated margins, and is closed by two small bud scales called *janams* with shorter internodes. The position of *janam* scars or fish leaf plays a major role in considering the standard of plucking.

Janam, though, strictly refers to those cataphylls that separate one flush from other; the rest of the scale buds in the shoot are normal cataphylls (Barua 1970).

The next leaf part in the shoot is called fish leaf or go-pat, which normally occurs in between the *janam* at the base and the first fully unfurled, well-serrated, normal flush leaf. There are several intermediate stages between *janam* and normal leaf. If the affinity is more toward the latter, the flush is with rich green leaves, larger in size, with well-defined serration and venation. A gol-pat that is not developed prominently means lack of chlorophyll and more like a *janam* cataphyll.

Dormancy breaks by the fall of bud scales leaving a scar in the stem. As the bud continues developing, normal leaves will be produced with good size that will become a pluckable shoot with elongated internodes. After a certain period of normal flushing, the shoot again goes to *banjhi* and thus completes one production cycle. The cycle of *banjhi* and growth phase is inherently controlled and is independent of plucking. A typical growth of a shoot should have two scale leaves, a fish leaf, and four flush leaves.

Normally, the leaf flush in a year may be around five if the shoots are allowed to grow without forcing any external cultural practices. The first bud that initiates growth is called the beginning of the flushing cycle. The flushing sequence of almost all cultivars is common until the cultivars have vast genetic diversity. The flushing behavior of the crop plays a major role in organizing the harvesting schedule to maximum utilization of crop growth period (Portsmouth 1957), ensuring that the shoots are forced into a slow developmental phase.

Crop productivity depends on the growth behavior of the buds and shoots. To obtain higher productivity, the bush has to be maintained in continuous vegetative phase by pruning and training the plant. Flush shoots growing 10–20 cm above the plucking level of the bush should be broken to keep the bush level constant. The shoots developing from the axil of the broken shoots should be again harvested at 7- to 10-day intervals to stimulate continuous bud and shoot growth. The yield potential of the crop is generally related to the number and angle of the lateral branches (Barua 1970); therefore, in breeding programs, plants possessing many laterals with narrow angle and more flushing points would be ideal for higher productivity.

18.2.3.1 Dormancy

Tea enters into a phase of dormancy in northeastern India during winter (December to January), whereas in countries nearer the equator this will not happen. This is mainly because of short day length (Barua 1969). To get a normal crop growth, it is necessary to have a combination of short day length and low atmospheric temperature (Nakayama and Harada 1962).

18.2.3.2 Longevity

The life span of tea varies between 150 and 200 years under wild conditions; however, 350-year-old tea plants have been identified in Japan. Some of the gardens in Darjeeling and southern India have tea bushes more than 100 years old under cultivation. The economic life span of the tea depends on growing environment; hence, it is essential to select the cultivars that suit the local growing environment so that the economic life span of plants is relatively longer.

18.2.4 Reproductive Phase

All shoots on a tree make leafy growth when the season begins in the spring, but terminal buds apparently become dormant as the season advances. The axillary buds shed minute bud scales, leaving a number of scars on the stem, representing leafless cataphyllary flushes. The tea flower buds appear normally in the leaf axils of newly developed shoots of the current season. Flowers that appear to develop in leaf axils are, in fact, inserted in the axils of bud scales of the axillary buds. A distal floriferous cataphyllary flush appears as a terminal cluster of flowers. Thus, there is an acropetal succession of flowers, which is determined by the phasic activity of the apical bud. Each cataphyll axil bears one or two flowers. Initially, the flower buds are covered with one or two bracteoles. As the development of flower progresses, the bracteoles drop off and leave a scar on the flower stalk or pedicel. The floral parts, sepals, petals, stamens, and carpels remain attached to the receptacle. An unopened bud and a fully opened flower are shown in Figure 18.1(d) and 18.1(e), respectively; upper and lower views of fruit are shown in Figure 18.1(f) and 18.1(g), respectively.

The main crop of flowers exposes anthers from the end of the third flush (late September to early October) until the end of the winter period (late January to early February). In some parts it produces flowers between the end of the first and beginning of the second flushes. A slight variation in flowering behavior is noticed due to climatic changes. In spite of a considerable time lag between antheses, the fruits produced later also mature and dehisce at the same time during October and November.

Tea is a cross-pollinated crop and the pollen grains are not wind borne; normally the insects, syrphids, bees, and small flies take care of the pollination. To effect successful seed set and to avoid random pollination in the seed orchards, the plant needs to be kept away from the natural pollinators.

18.2.5 Grouping of Cultivars Based on Vegetative and Reproductive Morphology

The outcrossing behavior of the tea warranted scientists' grouping the cultivar in the early phase of tea growing, which created extreme variability in the population. The net result is that commercially grown tea today is heterogeneous (Hadfield 1975; Barua and Sarma 1982; Banerjee 1986, 1987). Sealy (1958) broadly classified species of *Camellia* based on leaf and floral traits and growth habits. Because of plasticity, the vegetative characters have often been used to differentiate the taxa (Wight 1959). Based on the foliar characteristics, Hadfield (1975) observed that the leaf pose played a major role in interception of sunlight into the bush canopy and its influence on photosynthetic efficiency and yield potential of the crop. He summarized and grouped tea plants into five categories (A, B, C, D, and E) based on distinct and measurable foliar characteristics (mean leaf area, length/ breadth ratio, and mean leaf angle from vertical, laminar angle, patina, and leaf area index).

Grouping of tea plants, based on morphological characters alone, was not agreeable to many scientists due to the presence of continuous variation in the population apart from the small-leaved China variety and broad-leaved Assam variety. Wight (1962) realized that, unlike foliar characters, flower morphology would give some clearer information in classifying the species. Apart from the reported China and Assam types (Sealy 1958), he noticed a third form with distinct semierect, leaved (oligophiles) plants. This was confirmed by various workers in different names like the southern form of tea (Roberts, Wight, and Wood 1958), Cambod race (Kingdon-Ward 1950), *Thea lassiocalyx* (Barua 1963a), and *Camellia assamica* ssp. *lasiocalyx* (Planchon ex Watt).

The classification of Wight (1962) mostly relied on stylar characteristics. He found that the style remains free for a greater part of its length in *C. sinensis*, whereas in *C. assamica* he observed the union of style for most of the length. In Cambod (*C. assamica* ssp. *lasiocalyx*), the style remains united half of the length and free for rest of the part.

The current commercial tea plantation consists of mostly the China, Assam, and Cambod types, from which the final product is manufactured and is generally traded as tea (Wight and Barua 1957). The morphological variation of these three variants of tea has commercial value; hence, it is necessary to focus on these species for further selection and improvement. Table 18.1 summarizes morphological and growth characteristics of three different taxa of tea and provides guidelines to scientists for selection and screening the germplasm.

18.2.6 Germplasm

The collection of germplasm of tea started in the eighteenth century after the discovery of wild Assam tea. The search for germplasm was extended to Vietnam, Cambodia, Japan, China, and

Table 18.1 Grouping of Cultivated Tea Taxa Based on Some Common Morphological, Floral, and Reproductive Characteristics

| Characteristics | Taxa | | |
	China	Cambod	Assam
Color (young leaves)	Dark green	Light green, pigmented with anthocyanin	Light green to dark green
Vestitute on lower surface	Lamina sparsely pubescent, midrib densely appressed with pubescent	Lamina glabrous, young leaves midrib sparsely pubescent	Lamina nearly glabrous to sparsely villous, young leaves pubescent to densely villous
Margin	Obtusely serrulate to biserrate, flat to wavy	Well serrulated, flat to slightly recurved	Obtuse to bluntly serrate or denticulate
Apex	Acute to shortly acuminate, acumen obtuse or reduse, downturned	Shortly acuminate, acumen obtuse, rarely retuse, mucronate, downturned	Shortly to gradually acuminate, acumen obtuse of retuse, mostly straight
Base posture	Cuneate to rounded	Acute to rounded	Obtuse to rounded
Leaf posture	Well upturned	Slightly upturned to horizontal	Horizontal to erect
Flowers			
Sepals	Five, glabrous to silvery villous	Five, glabrous to sparsely pubescent, ciliate along margin	Five, glabrous to silvery villous
Petals	Six to eight, obovate-oblong to suborbicular	Six to seven, obovate to suborbicular	Six to eight, obovate to suborbicular and glabrous
Ovary	Globular, sparsely to densely silvery hairs	Globular, glabrous to densely or sparsely pubescent	Globular to obovoid, sparsely villous to silvery villous
Style	Free up to half of its length, arms three to four, spreading horizontally, genticulate	Free up to one-third its length, shortly split in the upper portion, arms three, ascending	United to a greater part, arms three, spreading horizontally or ascendingly spreading

Sources: Mohanan, M., and Sharma V. S. 1981. In *Proceedings of Fourth Symposium of Plantation Crops* 4:391–400; Banerjee, B. 1987. *Ecology* 68:839–843.

Nepal northeastward to Taiwan and the Liu-Kiu Islands (Sealy 1958). It is believed that more than 60% of world tea areas are covered by the germplasm collected from Assam (Ukers 1935; Richards 1965; Singh 1979, 1999; Badrul et al. 1997).

Because it is an important operation in tea plant improvement, the collection of germplasm should continue. The existing variation in the old seedling population is a good source to diversify genetic materials. The germplasm not only provides useful genes accumulated from naturally mutated genes but is also essential to using the modern techniques to tag and study the exact role of the genes.

18.2.7 Breeding

Breeding of tea started when the crop was identified for commercial planting. Tea varietal improvement is problematic because of the perennial nature, self-incompatibility, and heterogeneity of the population (Watt 1907). Tea plants set seeds mostly through cross-pollination, with a limited percentage through selfing (Wight and Barua 1939). This results in poor or no seed set and small seeds with reduced germination (Mamedov 1961; Sebastian Pillai 1963). Based on the experience of obtaining inferior plants by selfing, the initial breeding strategies were to adopt artificial pollination between plants of different morphological features (Wight 1956).

The prime objective of tea breeding in the beginning was the production of more planting materials rather than high yield and quality. The first attempt of breeding was to start mass selection. Mass selection, however, often failed or was restricted in use (Visser and Kehl 1958) because this method produced tea of low quality and nonuniform morphological traits (Barua 1963b; Hasselo 1964; Green 1971).

Conventional and nonconventional methods have been used to develop high-yielding tea varieties. The conventional methods are clonal selection and hybridization (Bezbaruah 1974; Satyanarayana and Sharma 1986a). The nonconventional methods used in tea varietal improvement are mutation, polyploidy, nursery grafting, and tissue culture (Satyanarayana and Sharma 1993).

18.2.7.1 Clonal Selection

Seeds were the only propagating material until the beginning of the nineteenth century. However, due to outcrossing behavior, seedlings showed a wide variation in yield and quality. This forced breeders to develop a method of selection and multiplication of selected tea plants by vegetative means. The first attempt of vegetative propagation of tea, by budding and grafting, was reported in Indonesia. The method of vegetative propagation using single leaf and internode cuttings was initially tried by Tunstall (1931) and subsequently improved by various scientists (Tubbs 1932; Wellensick 1933; Wight 1955; Visser 1961; Venkatramani 1963; Sharma 1976) and is used commercially.

It is known that some bushes in seedling populations have higher yielding potential. The main objective of the clonal selection program is to develop cultivars for high-yielding potential irrespective of climate, soil, cultural practices, and incidence of biotic and abiotic stresses along with quality characters. It is more a matter of rejection rather than selection of plants. Without any compromise, the plants showing unsatisfactory agronomical features are straightaway rejected.

18.2.7.2 Selection for Yield

The selected bushes should have more vegetative growth and be less floriferous. Clonal selection is used for selecting the vigorous bush from the segregating populations. Bush vigor is a subjective character. It can be judged by several quantifiable traits such as bush size (spread), pruning weight, and height of plants and overall flushing behavior of the plant. All the leaf characters and plucking point density are grouped under flushing behavior. Large size (length and breadth of the leaf) is a major trait considered for the higher tea production (Barua 1963b). Leaf size indirectly relates to the leaf area index (LAI) (Visser 1969; Sharma and Ranganathan 1986). Generally, semierect leaves with an LAI of about 4.0 are ideal for producing a good-yielding cultivar (Barua and Sarma 1982). Leaf color (medium green color leaves are preferable) and leaf pose (medium-sized semierect (50–70°) have the greater productivity potential and are considered important criteria for selection (Banerjee 1991). The medium sized shoots with high pubescence are more preferable. Dry matter production of green leaf is the most important selection criterion (Banerjee 1991).

18.2.7.3 Selection for Drought Tolerance

Drought is a major abiotic stress that suppresses the growth and yield of tea (Araus et al. 2002). Drought directly affects the biochemical, molecular, and cellular mechanisms of plants (Neil and Burnett 1999; Suprunova et al. 2004). It is important to know the mechanism of drought stress before breeding for drought-tolerant plants (Griffiths and Pary 2002; Francois 2003). Physiological and biochemical (Mukherjee and Choudhury 1981; Handique 1992; Li 2000; Chakraborty, Dutta, and Chakraborty 2002; Martinez et al. 2003; Srivalli, Sharma, and Chopra 2003; Upadhyaya and

Panda 2004) and molecular (Hendry 1993; Thambussi et al. 2000; Sharma and Kumar 2005) changes due to water stress have been studied. From these studies, it is concluded that a high level of relative water content, proline, peroxidase, superoxide dismutase, Na^+: K^+ ratio, narrow leaf and branch angle (<45°), and medium-sized leaves with high wax content and less stomatal index are related to drought stress and can be used as selection criteria for screening and selection of tolerant bushes.

Drought parameter is highly influenced by the environmental conditions and the bushes behave differently in changing climate; one should screen for at least three consecutive years before concluding the identified and selected bushes as tolerant to drought. In southern India, clones UPASI-1, UPASI-2, UPASI-6, UPASI-12, UPASI-16, UPASI-20, UPASI-22, UPASI-24, UPASI-26, UPASI-27, UPASI-28, and ATK-1 are drought-tolerant clones (Muraleedharan, Hudson, and Durairaj 2007).

18.2.7.4 Selection for Pest and Disease Resistance

Tea is generally affected by various pests and diseases and crop loss may range from 5 to 20%. However, breeding clones for resistance to pests and diseases of tea has been unsuccessful. Banerjee (1987) reported that the leaf pose plays a role in the resistance level to pests and pathogens.

18.2.7.5 Selection for Quality

In the early breeding program, yield was the main objective; however, development of clones with quality without yield loss has gained importance. Earlier, scientists looked into morphological characters like pubescence (Wight and Barua 1954; Wu 1964), leaf color (Wight, Gilchrist, and Wight 1963), and anthocyanin pigmentation (Visser 1969) of the leaves before selecting bushes for quality. High correlation of pubescence with quality was observed by Wight and Gilchrist (1959). When considering the leaf greenness, medium green color is always considered good for liquor color, strength, and overall aroma compared to the pale or dark green leaf. By contrast, the mechanism of biosynthesis of complex volatile compounds related to quality does not have direct association with morphological characters (Wickremasinghe 1978; Takeo 1984). Presence of oxalate crystals in the parenchymatous cells and the phloem index (Wight 1958; Visser 1969) is also an important parameter for quality. However, because the characters are not visually seen and highly influenced by climate, they do not play a major role in visual selection of bushes for quality in the early stages of clonal selection.

The most expensive part of the selection process is the organoleptic tea quality determined by expert tea tasters. For organoleptic test, a minimum 200 g of prepared tea is required; this is difficult to get from a small number of bushes in the early generation. Hence, it is essential to develop methods of selection only by semiquantitative analysis (Apostolides, Nyirenda, and Mphangwe 2006).

Hilton and Ellis (1972) first reported that theaflavins decide the quality of tea. The theaflavins can be analyzed in the prepared tea, but it is difficult to screen the germplasm or accession in the early stage or nursery for theaflavins in made tea. Thus, biochemical markers in the green leaf that would correlate with the value of the final product (made tea) are being evaluated. The green leaf parameters, such as total chlorophyll content, carotenoids, polyphenols, total catechin and its fractions, and amino acid, play a major role in selecting quality clones in the early part of the selection process. The high chlorophyll content and chlorophyll inflorescence are related to theaflavins (Apostolides et al. 2006).

The association of chlorophyll and carotenoids with flavor in black tea and correlation of polyphenols and catechins with liquor characteristics were considered as selection criteria for quality clones (Babu 2007). Catechin fraction estimation (Swain and Hillis 1959; Robertson 1992; Gerats and Martin 1992; Saravanan et al. 2005) and studies of the volatile flavor compounds belonging to group I and group II (Ullah, Gogoi, and Baruah 1984; Owuor et al. 1988; Ranganath and Raju 1991) have been found to influence the quality and flavor profile of the teas precisely. But estimation of these compounds using HPLC and GC is costly.

18.2.7.6 Hybridization

In the earlier days, seed varieties were used for planting. This technique is less successful because of the wide heterogeneity of the plants due to natural uncontrolled cross-pollination. The situation gets complicated further due to use of seed varieties obtained from different orchards having diverse cultivar varieties (Wight 1961). Hence, seed population has a large number of genetically distinct plants. One of the advantages of this technique is that they perform well in a wide range of environments and cultural practices. A cross between seeds of diverse origin provides an opportunity for developing intraspecific hybrids with higher yield and quality tea (Barua 1963a), suggesting a biclonal hybridization program.

Although clonal selection improves the yield potential, selection in them deteriorates the genetic diversity and it does not allow any recombination (Bezbaruah and Saikia 1977), resulting in homogenous populations. Crosses between genetically diverse best clones for yield, quality, and tolerance to biotic and abiotic stresses based on their earlier performance are selected and hybridized by controlled hand pollination. The progenies developed are tested for agronomically important characters, quality attributes, and resistance to pests and pathogens (Bezbaruah 1981). The cross combination that performs better is selected; the parents of these crosses are planted side by side in isolated orchards and are allowed to set seeds naturally. The F_1 seeds are collected; seedlings are raised and planted in different locations to study their performance. If the performance of the F_1 population outyields any of the parent or standard variety they are released as hybrid seeds or biclonal seeds. However, they have only restrictive use because of a long reproductive cycle and does not lead to the recombination so essential for increasing the genetic base of the cultivars.

In the late 1970s and early 1980s, a hybridization program was initiated in southern India. About 146 crosses were carried out between 19 different United Planters, Association of South India (UPASI) and estate released clones. Hybrid progenies obtained from 63 successful combinations were field tested; five biclonal progenies among them were found to be good and released as UPASI: BSS-1 to UPASI: BSS-5 based on the multilocation trial for commercial cultivation (Satyanarayana and Sharma 1991).

18.2.7.7 Mutation Breeding

Mutation breeding is a method of artificial induction of genetic variability by mutagen (physical or chemical). The individual treatment of γ-ray (Singh 1980; Satyanarayana and Sharma 1986b) and chemical mutagens like ethyl methane sulfonate (EMS) and methyl methane sulfonate (MMS) were used in tea, and it was found that 2–6 kr of γ-irradiation was as good for creation of variability in tea seeds. The combined effect of γ-rays and chemical mutagens on biological damage was estimated by Yang and Lin (1992). Preliminary investigations were restricted to irradiating cuttings, seeds, and pollen grains. However, these treatments failed to produce superior mutants, and plants obtained showed reduced vigor, stunted growth, and fewer numbers of foliage and branches (Singh and Sharma 1982).

According to Sharma and Ranganathan (1985), use of irradiated pollens resulted in less fruit set. The reason for the failure may be the condition of plant material and doses. Liang, Zhou, and Yang (2007) elucidated the relationship between moisture content and radiation damage. Recent attempts of γ-ray irradiation using vegetative cuttings and seeds showed wide variation in morphological traits such as enlarged leaf size, quick branching, and increased plant height (Babu 2008). The same study revealed the importance of the correct dosage for creating variability. Hence, it is essential to fix the LD_{50} value for each and every material undergoing γ-irradiation. All this indicates that mutation breeding is one of the practical breeding approaches for tea plant improvement.

18.2.7.8 Polyploidy Breeding

Polyploidy breeding is a useful breeding tool to create variability and to induce vigor and resistance to pests and diseases. Tea is a diploid ($2n = 30$) and polyploidy production is difficult because of tea's heterogeneous nature. Being a self-incompatible, cross-pollinated crop, the open pollinated offspring often produce natural triploids, tetraploids, and aneuploids (Bezbaruah 1971, 1975; Karasawa 1935; Wachira 1991). Except for triploids, other polyploids are not attractive to plant breeders because of their poor rooting, leaf size, and dry matter content (Singh 1980). Triploid tea is more vigorous with good quality and is hardy and tolerant to cold compared to diploid tea (Simura 1956; Bezbaruah 1971; Sharma and Ranganathan 1986). Hence, breeding of polyploids in tea should be restricted to triploid production through crossing of tetraploid with diploid or by chemical (colchicine-colchiploidy) induction (Osone 1958; Choudhary 1979; Rashid, Choudhury, and Badrul Alam 1985; Singh 1999). Two triploid cultivars, UPASI-3 and TV-29, are under commercial cultivation in India.

18.2.7.9 Nursery Grafting

High-yielding and quality clones released in southern India were susceptible to drought because of their morphology. On the other hand, the drought-tolerant clones are always moderate yielders. This condition forces the breeder to develop a method suitable to ingrain some tolerance and high-yielding behaviors into a single clone or unit of plant material. Through hybridization, it is very difficult to transfer the characters and it takes more time. An alternative method, a nursery grafting technique, was adopted to develop drought-tolerant, high-yielding composite plants by grafting scions of high-yielding clones on fresh, unrooted, single-node cuttings of the drought-tolerant clones in the nursery (Anonymous 1972).

In southern India, UPASI is the pioneer in developing grafting techniques and releases successful graft partners for commercial planting. Raising composite plants involves making fresh scions of one leaf and one internode cuttings of the clone of the desired high yielder or quality and unrooted fresh one leaf and one internode cuttings of the same size of clone that is normally drought tolerant and best rooting clones as root stock. The cleft grafting method is adopted where graft union is made and the graft region is wrapped with strips of polythene. The grafted cuttings are allowed to root in the nursery. Shoots arising from the scion are only allowed to grow to get the plant.

Data on percent success of grafts and growth performance of composite plants with shoot systems of high-yielding, drought-susceptible clone and the root system of a drought-tolerant clone were evaluated in comparison with their straight cuttings in the nursery and field. Compatibility between clonal combinations could be established on the basis of data generated in the nursery and field, and promising combinations were suggested for commercial trials. Of 26 combinations fixed and evaluated, nine combinations were superior to ungrafted plants in all respects (Table 18.2) and were released for commercial cultivation with a yield improvement of 20–30% (Satyanarayana and Sharma 1991).

18.2.7.10 Micropropagation

Micropropagation could be adopted to develop a large number of uniform plants in a short period from a limited number of mother bushes. A detailed account is given later in this chapter.

Table 18.2 Root Stock and Scion Combinations for Grafting

Root Stock	Scion
UPASI-2, UPASI-6, and UPASI-9	UPASI-3
UPASI-2 and UPASI-9	UPASI-8
UPASI-2 and UPASI-9	UPASI-17
UPASI-9 and C-1	CR-6017

18.2.7.11 Application of Molecular Techniques

In tea, the early stage of selection is based on morphology, physiology, and biochemical basis. However, the use of molecular techniques in plant improvement is very rare. Molecular markers could be used for marker-aided selection (MAS) and construction of high-density genetic linkage maps to locate the quantitative trait loci (QTL) of agronomically important traits, quality, and biotic and abiotic resistance-related characters. The usefulness of the MAS technique should be explored to identify the clones of high value.

Functional genomics in tea should be strengthened in the wide germplasm collection, particularly for stress- and defense-related genes, including some important secondary metabolites. It is essential to deepen the research on functional genomics for better understanding the mechanism of genetic variation on the whole genome for breeding. Transfer of genes from one species or plant to other species or plants and development of new varieties with new characteristics need to be created. Attention should be given to using this technique and making transgenics an effective routine approach for varietal improvement and germplasm innovation. Furthermore, RNA-mediated posttranscriptional gene silencing techniques might be explored to make low- or zero-caffeine tea clones.

18.3 BIOCHEMISTRY

18.3.1 Chemical Composition of Tea Flush

The tea flush generally refers to the apical shoots consisting of the terminal bud and two to three adjacent leaves. In fresh tea flush, there exist a wide variety of nonvolatile compounds: polyphenols, flavonols and flavonol glycosides, flavones, phenolic acids and depsides, amino acids, chlorophyll and other pigments, carbohydrates, organic acids, caffeine and other alkaloids, minerals, vitamins, and enzymes. The chemical composition of the tea leaves depends upon leaf age, the clone being examined, soil and climatic conditions, and agronomic practices. Overall composition of tea flush has been studied to a greater detail, and much importance is given to those chemicals in the fresh flush that are important in determining the quality and hence the value of the made tea (Chen et al. 2002).

The total polyphenols in tea flush range between 20 and 35% on a dry weight basis. Catechins are the main components in this group of phenolic compounds. They are the main constituents in determining the color and taste of tea infusion and the key material basis in forming the tea quality. The polyphenols in tea mainly include the following six groups of compounds: flavanols, hydroxy-4-flavonols, anthocyanins, flavones, flavanol glycosides, and phenolic acids. Among these, the flavanols (mainly the catechins) are most important and occupy 60–80% of the total amount of polyphenols in tea. About 90–95% of the flavanols undergo enzymatic oxidation to products, which are closely responsible for the characteristic color of tea infusion and taste. They are generally water-soluble, colorless compounds.

Tea shoot contains a full complement of enzymes, biochemical intermediates, carbohydrates, proteins, and lipids. In addition, tea shoot is distinguished by its remarkable content of polyphenols and methyl xanthenes (caffeine and other purines, such as theobromine and theophyline). The great popularity of tea as a beverage may be due to the presence of these two groups of compounds, which are mainly responsible for the unique taste of tea, in addition to various compounds associated with tea aroma. The chemical composition of tea shoot varies with agroclimatic condition, season, cultural practice, and the type of material.

Flavanols, flavonols, flavonol glycosides, polyphenolic acids, and depsides are put together and referred to as total polyphenols; they make up about 30% of the dry weight in a tea shoot. Flavanols or catechins are the major compounds that are oxidizable in the tea leaf. (–)-Epigallo catechin (EGC) and (–)-epigallocatechin gallate (EGCG) are the predominant catechins present in tea leaf. The catechins are located in the cytoplasmic vacuoles and play a significant role during fermentation. The total content of polyphenols decreases to some degree during green tea processing. In comparison with fresh tea leaves, polyphenol contents of manufactured green tea generally decrease by around 15%.

Catechins, the major constituents of total polyphenols, are the substances responsible for the bitterness and astringency of green tea as well as the precursors of theaflavins in black tea. Tsujimura (1929) first isolated three catechins from tea: (–)-EC, (–)-ECG, and (–)-EGC. Bradfield et al. (1948) further isolated the (–)-EGCG. These four catechins constitute around 90% of the total catechins, and the (+)-C and (+)-GC occupy around 6% of the total catechins. The largest part of the catechins present in tea flushes is esterified with gallic acid in the 3-position. In addition, some minor catechins (less than 2% of the total catechins) have been reported (Coxon, Holmes, and Ollis 1972; Saijo 1982; Nonaka, Kawahara, and Nishioka 1983). The ether type catechins [(–)-ECG and (–)-EGCG] are stronger in bitterness and more astringent than (–)-EC and (–)-EGC. The relative amounts of various catechins and their gallates are genetically controlled and therefore a clonal characteristic. They also depend on the various seasons and other environmental factors. The catechin contents decrease with the increase in fiber in the shoot components (Chen et al. 2002). The biosynthesis of catechins was investigated tentatively in the former USSR and results were published as a monograph (Zaprometov and Bukhlaeva 1971).

The flavonol and flavonol glycosides occur in small quantities. There are three major flavonol aglycones in fresh tea flush: kaempherol, quercitin, and myricetin. The glucosidic group may be glucose, rhamnose, galactose, arabinose, or rutinose. These compounds are considered to contribute to bitterness and astringency in green tea (McDowell and Taylor 1993). The major phenolic acids present in tea flush are gallic acid, chlorogenic acid, and coumaryl quinic acid. The most important depside is 3-galloyl quinic acid (theogallin). It aroused attention due to its relatively high level in the tea flush and its statistic correlation to black tea quality (Cartwright and Roberts 1955; Roberts and Myers 1958).

Around 2–4% of amino acids are present in tea flush. They were considered to be important in the taste of green tea (Nakagawa 1970, 1975) and aroma of black tea (Ekborg-Ott, Taylor, and Armstrong 1997). The most abundant amino acid is theanine (5-N-ethyl glutamine), which is unique to tea and found at levels of 50% of the free amino acid fraction. The precursors in the biosynthesis of theanine in tea plant were identified as glutamic acid and ethlylamine (Sasaoka and Kito 1964; Takeo 1974). The site of biosynthesis of theanine is the root from where translocation occurred to younger leaves; thus, the roots have the highest concentration in the tea plant. It has been shown that theanine plays a role in protecting enzymes from inactivation by polyphenolic products (Wickremasinghe and Perera 1973). An additional 26 amino acids usually associated with proteins have been reported in tea flush.

The main pigments present in fresh tea flush are chlorophylls and carotenoids. The amount of chlorophyll in tea leaves is reported to be about 0.2–0.6% on dry basis. The proportion of chlorophyll A to chlorophyll B is around 2:1. The contents of chlorophyll are decreased during the black tea manufacturing process and some degradative products of chlorophyll, such as pheophytin A, pheophytin B, and pheophorbide, are produced. These products cause the blackness or brownness of black tea infusion due to the brown color of pheophorbides and black color of pheophytins. The total contents of carotenoids in tea flush were reported as 0.03–0.06% dry weight.

Sanderson and Perera (1965) investigated the carbohydrates in tea flush. The free sugar contents in tea flush were reported as 3–5% on a dry weight basis. They consisted of glucose, fructose, sucrose, raffinose, and stachyose. The monosaccharides and disaccharides present in tea flush are one of the components in tea infusion that give it a sweet taste. The polysaccharides were separated into hemicellulose, cellulose (6–8% dry weight basis), and other extractable polysaccharide fraction (1–3%) composed of different sugar residues (i.e., glucose, galactose, manose, arabinose, xylose, ribose, and rhamnose). The cellulose and hemicellulose contents were negatively correlated with the tenderness of the tea shoots. Starch was found mostly in the root system of the plant, with only a small amount in tea flush. Mori, Morita, and Kegaya (1988) in China and Wang, Xie, and Wang (1996) in Japan showed that the polysaccharides extracted from made tea could decrease blood glucose level and thus potentially be useful in the treatment of diabetes.

Tea contains 2–5% of caffeine in fresh flush. Caffeine is a trimethyl derivative of purine 2,6-diol. It was first isolated from tea by Runge and named *theine* in 1820. Later, it was found that tea theine was the same as the caffeine from coffee, so the theine name was abolished. It has been reported that during tea processing, caffeine reacts with theaflavins to form a compound that imparts "briskness" to the tea infusion. Caffeine is rapidly and completely absorbed after oral intake.

Caffeine is pharmacologically classified as a central nervous system stimulant and a diuretic. Caffeine possesses the ability to improve the elasticity of blood vessels, promoting blood circulation, increasing the efficient diameter of vessels, and stimulating urination and auto-oxidative activity (Dews 1982; Elias 1986; Shi and Dalal 1991). The quantity of caffeine ingested by average or moderate consumer of tea—three to six cups per day—is highly unlikely to induce undesirable effects unless significant additional amounts are obtained from other sources. Caffeine is the major purine alkaloid present in tea. Theobromine and theophylline are found in very small quantities. Traces of other alkaloids (e.g., xanthine, hypoxanthine, and tetramethyluric acid) have also been reported. Fatty acid composition and variation in lipid content with respect to different shoot components of clones UPASI-3, -15, and -17 have been studied by Ranganath, Raju, and Marimuthu (1993).

Aroma is one of the critical aspects of tea quality and can determine acceptance or rejection of a tea before it is tasted. Many volatile compounds, collectively known as the aroma complex, have been detected in tea. The aroma in tea can be broadly classified into primary or secondary products. The primary products are biosynthesized by the tea plant and are present in the fresh green leaf; the secondary products are produced during tea manufacture (Sanderson and Graham 1973). Some of the aroma compounds identified in fresh tea leaves are mostly alcohols, including Z-2-penten-1-ol, *n*-hexanol, Z-3-hexen-1-ol, E-2-hexen-1-ol, and linalool plus its oxides, nerol, geraniol, benzylalcohol, 2-phenylethanol, and nerolidol (Saijo and Takeo 1973). The aroma complex of tea varies with the country of origin. Slight changes in climate factors can result in noticeable changes in the composition of the aroma complex. Notably, teas grown at higher altitudes tend to have higher concentrations of aroma compounds and superior flavor, as measured by a flavor index (Owuor and Obanda 1991). Growing tea in a shaded environment may change the aroma composition and improves the flavor index. The aroma complex also varies with season, and these variations appear to be larger under temperate or subtropical climates (Gianturco, Biggers, and Ridley 1974).

Early research on tea aroma can be traced back 160 years (Mulder 1838). Many attempts have been made to look for the key compounds for the aroma of tea, but no single compound or group of compounds has been identified as responsible for the full tea aroma (Takei, Ishiwata, and Yamanishi 1976; Yamanishi 1978; Yamaguchi and Shibamoto 1981). It is generally believed that the characteristics of various kinds of tea consist of a balance of very complicated mixtures of aroma compounds in tea. An assessment of all data known shows that more than 630 compounds are responsible for tea aroma. Research on tea aroma has been well reviewed in a series of papers (Schreier 1988; Yamanishi 1995, 1996; Takeo 1996; Kawakami 1997).

18.3.2 Types of Tea

There are three basic types of tea: green, semifermented, and black. They differ mainly in the degree of fermentation. Green tea undergoes little or no fermentation and black tea is produced by full fermentation. Semifermented tea (oolong tea) is the product of partial fermentation.

18.3.2.1 Black Tea

Tea that is produced by a complete fermentation process and that appears black/brown in color is referred to as black tea. The majority of teas produced are of this kind.

18.3.2.2 Green Tea

Unfermented tea is known as green tea. The oxidizing enzymes in the freshly plucked leaves are inactivated or denatured by steam-blasting in perforated trays or by roasting in a hot iron pan. The steaming is carried out for less than a minute. The leaves are then subjected to further heating and rolling until they turn dark green. The leaves are finally dried to a moisture content of 3–4%. As the fermentation is arrested by the inactivation of polyphenol oxidase, the polyphenols are not oxidized and the leaves remain green. The beverage gives a weaker flavor than black tea due to the absence of theaflavins and thearubigins.

18.3.2.3 Semifermented Tea

Apart from black and green teas, partly fermented teas are also manufactured. When the fermentation is carried out in half the time as in the case for manufacturing black tea, the resultant teas are referred to as *oolong* teas, while for *pouchong* teas the fermentation is carried out for one-quarter time of the normal fermentation. Other processes are similar to that of the black tea manufacture. Oolong and pouchong teas are mainly consumed for their medical significance, and these types of teas are predominant in parts of Japan and China.

There are yet other types of teas, such as dark green, white, and yellow teas. These teas are not fermented as polyphenols in fresh leaves are less oxidized. Green tea is an absolutely nonfermented tea; however, polyphenols in yellow and dark green tea are nonenzymatically oxidized during the processing period. Each kind of tea has its characteristic flavor and appearance. Besides the preceding six teas, flower scented tea, compressed tea, instant tea, and herbal teas are classified as reprocessed teas.

Green tea is preferred in China, Japan, and the Middle East. Oolong tea is mainly consumed in the eastern part of China, China–Taiwan, and Japan, while 80% of the rest of the consumers prefer black tea. The manufacturing process in each type of tea has a pronounced impact on the formative and degradative reaction pathways, thus influencing the color, flavor, and aroma of the end product, which are dependent on different components in tea. The tea aroma is mainly dependent on the volatile compounds it contains, and the color and the taste of tea are mainly dependent on the nonvolatile compounds (Chen 2002).

18.3.3 Tea Leaf Processing

The various stages of processing consist of a series of operations:

- Withering refers to partial removal of moisture (partial desiccation).
- Rolling/cutting: In this stage, the tea leaves are macerated into small pieces facilitating mixing up of the cellular constituents and thereby initiating the subsequent step of fermentation.

- Fermentation/oxidation: During this stage, the rolled dhool is exposed to atmospheric oxygen, which converts the substrates (polyphenols) into the black tea quality constituents such as theaflavins and thearubigins with the help of the enzyme polyphenol oxidase.
- Drying/firing: This stage is carried out to arrest the enzyme action, thereby increasing the quality and shelf life of the product. The removal of moisture is complete at this stage.
- Sorting: This involves separation of tea particles into various size fractions with fiber removal and with the use of standard meshes.

While moisture removal and particle size determination are physical in nature, the biochemical changes occur during withering, fermentation, and drying. Based on the principle of operation of the machines employed in the factory, there are three types of processes:

- Orthodox process: The rolling machines used for this operation are of long-standing use and hence the name.
- CTC process: The crush, tear, and curl (CTC) process consists of a pair of contrarotating rollers that pulverize the leaves fed into them.
- LTP process: The Lawrie tea processor (LTP) machine has a rotating swing hammer-mill type disintegrator that macerates the leaves. This system of processing has become obsolete due to the operational difficulties arising out of its usage concomitant with drop in quality of the product due to high heat generation during the processes (Hampton 1992).

Quality of tea is determined by the presence or absence of chemical compounds that impart color, briskness, brightness, strength, and flavor in the infusion. The majority of the chemicals imparting quality are produced during processing of the tea leaves. Biogenesis of such precursors is influenced, on one hand, by the genetic and environmental factors, which cannot be controlled, and, on the other, by the cultural practices adopted in the field as well as by the conditions of processing, which can be controlled. Whether it is a green tea (nonfermented) or black tea (fully fermented), crop shoots (two/three leaves and a bud) are harvested for commercial tea production. A series of changes occur in the process of manufacturing. The major steps involved in the manufacture of black tea include withering, rolling, fermentation, and drying. Though biochemical degradation starts immediately after plucking (the crop shoots), the precise changes required for quality start from withering onward.

18.3.3.1 Withering

The changes that occur in the green leaf from the time of plucking to the time of rolling are collectively referred to as withering. Two aspects of withering are known to take place: physical and chemical. Physical withering refers to the loss of moisture from the shoots, which leads to changes in the cell membrane permeability. The increased permeability of the membrane has a great effect on the mixing of the substrates and enzymes during fermentation. The biochemical changes that occur during withering are collectively referred to as chemical withering. Air temperature, atmospheric vapor pressure deficits, and air velocity and direction are important in ensuring a good wither. The evenness of wither is also critical for tea manufacture. The loss of moisture ensures that the leaf is sufficiently flaccid for the rollers to produce the desired twist in the leaf, especially in the orthodox type of manufacture. Unwithered leaf produces unacceptable flaky leaf material and excessive moisture results in uneven fermentation.

Withering is also accompanied by the activation of oxidative and hydrolytic enzymes, causing significant changes in the chemical composition of green leaf. Proteins are acted upon by proteases, resulting in the production of more soluble forms and amino acids. Amino acids in turn are incorporated into the sugars, thereby forming volatile flavor compounds (VFCs). Amino acids form aldehydes and brown-red pigments in the course of their oxidative interaction with catechin during fermentation. This is very important for the formation of aroma and color of the infusion. Complex carbohydrates are broken down to simpler monosaccharides into which amino acids are incorporated to form VFCs. Lipids are acted upon by lipases and lipoxygenases, thereby converting them into C_6 aldehydes and alcohols, which are highly undesirable in aroma, which in turn will be lost during firing. Carotenoids are degraded into terpenoid flavor compounds, which are in turn converted to linalool and geraniol, which are highly desirable. Chlorophylls are broken down to phaeophytin, which increases the blackness of made tea. Caffeine is a degradation product of purine metabolism and is found to increase during withering. Peak activity of polyphenol oxidase (PPO) is attained after the normal withering period of about 18 hours. An adequate and even wither of the harvest and a uniform wither throughout the year pose fewer problems during fermentation.

18.3.3.2 Fermentation

Immediately upon rolling, the mixing up of cell constituents—that is, enzymes and substrates—occurs, thereby initiating fermentation. Enzymic oxidation of simple substrates into complex ones that impart characteristic features of tea is the main step in the process of fermentation, though a number of chemical reactions occur. During this stage, the polyphenols are acted upon by the enzymes PPO and peroxidase (PO), resulting in the production of theaflavins (TFs), thearubigins (TRs), and highly polymerized substances (HPSs), the major liquor parameters of black tea. During this stage, the leaf changes color and turns into a dark coppery tone. Typical aroma develops at this stage.

The enzyme polyphenol oxidase plays a key role in tea fermentation; it is present in chloroplasts. It has remarkable specificity for the *ortho*-dihydroxy functional group of the tea catechins. In intact plants, the enzyme is not in contact with the substrates, the flavanols. The basis of fermentation is to bring the enzyme and substrate together in the presence of oxygen by rupturing the membrane so that polyphenols can diffuse into the cytoplasm.

As a first step during fermentation, PPO oxidizes the catechins to highly reactive, transient *ortho*-quinones. The quinones thus derived from a simple catechin and gallo catechin dimerize to produce theaflavins, which are orange-red substances that contribute significantly to the astringency, briskness, brightness, and color of the tea beverage. Theaflavins comprise 0.3–2.0% of the dry weight of black tea as well as a number of fractions: namely, theaflavin, theaflavin monogallate and digallate, epitheaflavic acid, and isotheaflavin. As the gallation increases, the astringency also increases and the proportions of theaflavin fractions present in black tea depend upon the method and conditions of manufacture.

Further transformations of dicatechins and theaflavins yield the thearubigins. Thearubigins comprise about 9–19% of black tea, are red-brown in color, and contribute to color, strength, and mouth-feel of tea liquor. While the theaflavin content of tea increases during fermentation and starts declining after reaching a peak, the thearubigin content continuously increases throughout the fermentation.

18.3.3.2.1 Effect of Temperature on Tea Fermentation

Temperature is one of the most important factors influencing the complex series of enzymic and chemical reactions that take place during fermentation. Temperature affects not only the rate of fermentation but also the ultimate level of theaflavin in tea. Lower temperatures favor increase in the levels of accumulated theaflavins. The rate of thearubigin formation during fermentation is also markedly influenced by temperature—the higher the temperature is, the greater the rate and extent of thearubigin production are. Total color also increases throughout the fermentation period and the rate of color development depends upon temperature. PPO activity is found to decline over a period of fermentation. The rate of loss of PPO activity is markedly temperature dependent, being much more rapid at a high temperature. The flavor development also decreases as the temperature rises. The overall optimum working temperature for fermentation appears to be around 27°C.

Ramamoorthy and Venkateswaran (2003) studied the effect of temperature on tea quality in CTC tea manufacture. High temperature during manufacturing is known to affect the quality of black tea adversely. The optimum temperatures for green leaf handling, preconditioning, CTC rolling, humidification, drum fermentation, and drying have been worked out under Nilgiris conditions. The maximum retention of quality could be observed when the temperature was maintained at 25–30°C until the drying stage.

18.3.3.2.2 Effect of pH on Tea Fermentation

The pH of the fermenting dhool was found to vary from 5.1 to 5.8. A steep increase in theaflavin levels was observed when the fermentation was carried out at pH 4.7.

18.3.3.2.3 Effect of Humidity on Tea Fermentation

Humidification supplies oxygen for enzymic reactions and reduces the dhool temperature. Proper humidification will also prevent the surface drying of the fermenting dhool and ensures even fermentation. The ideal humidity should be more than 90–95% in the fermenting room.

18.3.3.2.4 Assessment of Fermentation

In the early days, assessment of fermentation was monitored by the color and aroma of fermenting dhool, followed by the course of temperature changes during fermentation, by observing the color intensity of liquor, by cell damage assessment using potassium dichromate solution, and by tannin estimation in the fermenting dhool. However, the best way of determining optimum fermentation is continuous monitoring of theaflavin formation for every 10 minutes with the help of a colorimeter. The time corresponding to maximum theaflavin level could be taken as the optimum fermentation time.

18.3.4 Role of Enzymes in the Quality of CTC Black Tea

Though the role of enzymes in tea processing was recognized several years ago, their exogenous addition to improve the quality of made tea has remained an important avenue for value addition in tea manufacture. The use of pectinase in improving the liquor characteristics has been well documented. Protease at a particular concentration influences the quality and aroma of made tea by activation of the PPO enzyme and by releasing free amino acids. Protease addition resulted in significant improvement in the theaflavin levels, briskness, and color indices. There also was a higher value of digallate equivalent of theaflavin (DGETF), which has a direct influence on the astringency of the liquor. An increase in the volatile flavor compounds, particularly aromatics and terpenoid compounds, was also noted due to protease addition. Trials conducted in pilot as well as factory scales confirmed the advantage of using a combination of pectinase and protease in improving the quality and aroma of black tea.

Tea leaves contain a large range of enzymes, but many have not been studied closely. Most enzymes are of importance because they influence and help in building tea quality during processing. Enzymes present in the harvested tea shoots play a vital role in the conversion of tea flush to consumable made tea. The role of enzymes in tea processing was recognized more than three decades ago (Jain and Takeo 1984). Isolation, purification, and characterization of various enzymes in tea leaves have been reported (Sanderson and Coggon 1977; Jain and Takeo 1984; Mahanta 1993; Finger 1994; Ravichandran and Parthiban 1998a, 1998b).

18.3.4.1 Hydrolytic Enzymes

Because only a few reports are available on the biochemical constituents of cell walls of the tea shoots and commercial black tea and their specific catalytic enzymes, a study on the cell wall polysaccharides of tea leaves and black tea was carried out by Senthil Kumar et al. (2002). Total carbohydrates and structural polysaccharides such as cellulose, hemicellulose, and pectic substances were estimated in green leaves as well as in made tea samples of six elite tea cultivars representing Assam, Cambod, and China types. Selvendran and Perera (1972) examined the chemical composition of the tea leaf cell wall. They reported that each stage of the extraction procedure removed a complex mixture of polysaccharides and proteins from cell-wall material. The structural constituents, because of their ability to absorb moisture, may play an important role in determining the storage characteristics of the black tea. The extraction procedure for the removal of a complex mixture of polysaccharides from the cell wall was reported (Selvendran and Perera 1971).

Hemicellulose is the maximum contributor toward the dry matter content of green leaves, followed by cellulose. Black tea contains lesser amounts of structural polysaccharides and carbohydrates compared to green leaves. Addition of hydrolytic enzymes during processing enhanced the level of carbohydrates (simple sugars) and decreased the structural polysaccharides in black tea samples of all the cultivars studied. An increase in the level of carbohydrates and a decrease in cellulose, hemicellulose, and pectic substances brought about by the enzyme treatments corroborates the enhanced quality of black tea. It has been demonstrated that external application of crude enzymes extracted from fungal sources, during processing, degrades the raw materials present in green tea shoots, thereby enhancing the quality of black tea.

Marimuthu, Manivel, and Abdul Kareem (1997) studied the effect of addition of cell wall degrading enzymes (to cut dhool) on quality parameters of made tea. The effect of adding cell wall degrading enzymes such as biocellulase ZK and biopectinase OK to the fermenting dhool on quality constituents of made tea has been studied. Addition of enzymes to the cut dhool increased the total soluble solids without affecting liquor characteristics of made tea. Pectinase proved superior to cellulase. Tea made from pectinase-treated dhool had higher TF level compared to the untreated control. Application of pectinase improved the cuppage, color, and value realization.

Murugesan, Angayarkanni, and Swaminathan (2002) studied the use of cellulolytic enzymes, cellulases, pectinase, and xylanases isolated from the tea fungus and laccase from *Trametes versicolor* for the improvement of black tea quality. The effects of these enzymes on black tea quality revealed that the purified cellulase improved the tea quality most effectively. The TF content was increased by 52.4% over that of the control due to the addition of purified cellulase. It was concluded that the exogenous application of purified cellulase of tea fungus and laccase of *Trametes versicolor* mixed in the ratio of 3:2 (v/v) could increase the quality of black tea and its market value. The tea processed by this treatment increased the TF by 57.1%, HPS by 25.7%, and TLC by 22.8% over those of the control.

18.3.4.2 Pectinases

Pectinases constitute a unique group of enzymes that catalyze the degradation of pectic polymers present in the plant cell walls (Fogarty and Kelly 1983). Pectin methyl esterase reduces oxygen uptake during processing and adversely affects the quality of tea (Lamb and Ramaswamy 1958). Pectinases are produced by organisms such as bacteria (Horikoshi 1972). In the industrial sector, acidic pectinases are used in the extraction and clarification of fruit juices (Rombouts and Pilnik 1986). Recently, the use of fungal pectinases for the improvement of tea quality has been reported (Angayarkanni et al. 2002; Murugesan et al. 2002). A soil isolate of *Bacillus* species DT7 has been found to produce significant amounts of an extracellular pectinase characterized as pectin lyase

(EC 4.2.2.10). The presence of 100 mM concentrations of $CaCl_2$ and mercaptoethanol significantly enhanced pectinase activity of the purified enzyme (Kashyap et al. 2000).

A follow-up study on the role of pectinase on the native level of peroxidase (PO) and polyphenol oxidase (PPO) during CTC black tea manufacturing and its influence on other biochemical quality constituents of made tea was conducted (Marimuthu et al. 2000). Observations revealed that the addition of biopectinase during processing did not affect the native level of peroxidase and polyphenol oxidase activities. The enzyme-treated tea samples contained significantly higher amounts of theaflavins, total liquor color, total soluble solids, simple sugars, and free amino acids; however, unoxidized polyphenols, catechins, and chlorophyll contents were very low compared to the untreated samples. Digallate equivalent of theaflavins (DGETF), a major contributor for the astringency, was found to be high in enzyme-treated tea samples. Organoleptic evaluation substantiated the analytical results.

Pectinase enzymes isolated from *Aspergillus* spp., *A. indicus*, *A. flavus*, and *A. niveus* were purified, characterized, and used for the improvement of tea quality (Angayarkanni et al. 2002). Crude as well as purified enzyme preparations were used in the study. The crude enzyme preparations obtained from ethanol precipitation were found to be more effective in improving tea quality than the purified pectinase enzymes. The use of crude enzyme preparation of *A. indicus*, *A. flavus*, and *A. niveus* resulted in a maximum increase in TF content by 43.81, 62.86, and 59.05%, respectively, over that of the control.

Polygalacturonases (PGs) are part of the group of enzymes involved in pectin degradation. Their substrate is the pectic substances that occur as structural polysaccharides in the primary cell walls and in the middle lamella of higher plants. Pectin is a complex heteropolysaccharide composed mainly of d-galacturonic acid residues joined by α-1,4 linkages that form homogalacturonan chains. This backbone structure ("smooth regions") alternates with branched regions ("hairy regions"), which contain rhamnose, arabinoses, and arabinogalactane as side chains. In addition to PG, pectinases include pectin methyl esterase (PME), which removes methoxyl groups, and pectin lyase (PL), which cuts the internal glycosidic bond of highly esterified pectic polymers by a β-elimination reaction. PG (EC 3.2.1.15) catalyzes the random hydrolysis of nonmethylated pectin or polygalacturonic acid. Lang and Dorenburg (2000) have reported the technological application of polygalactronases. Rhamnogalacturonase, which specifically cuts at the rhamnose side chains, has been characterized (Suykerbuyk et al. 1995). *Endo*-PG catalyzes the hydrolysis of 1,4-α-galacturosiduronic linkages between two nonmethylated galacturonic acid residues, while *exo*-PG removes terminal residues.

The commercial enzyme preparations used for food processing are almost exclusively derived from *Aspergillus* spp. and are traditionally mixtures of PG, PL, and PME. The yeast *Pichia pinus* was found to be capable of growing on mango wastes and produced pectinase (pectin lyase, EC-4.2.2.10) and lactase (β-galactosidase, EC-3.2.1.23) enzymes. Both the enzymes have been purified and characterized. On the basis of the evaluation tests, the enzymes were considered to have a potential technological use in treating mango pastes (residues left after mango juice preparation); enzyme treatment resulted in an increase in the color intensity, total carbohydrate content, and juice yield (Moharib, El-sayed, and Jwanny 2000).

18.3.4.3 Proteases

Peptidase enzyme breaks down protein into amino acids during the withering of black tea (Sanderson and Roberts 1964). The activity of the enzyme varies in plant parts and among the clones. Addition of *Agaricus bisporus* protease at a 0.01% level on the cut dhool increased the levels of TF and TR and the briskness index considerably. A higher value of DGETF was also noted. An increase in amino acids that eventually led to the production of volatile flavor constituents was also reported. Treatment with PPO invariably improves the quality of tea in terms of TF, TR, TLC, and the briskness index. No undesirable odor or taint was noticed either in the made tea or in tea brew due to the treatments. Addition of lipase resulted in an increase in the total soluble solid, theaflavin 3,3'-digallate (TFDG), and DGETF. It was also reported that addition of PPO during tea processing enhanced the quality of black tea in terms of cuppage and creaming properties. Trials conducted in pilot and in factory confirmed the advantage of using a combination of pectinase, protease, and PPO in improving the quality and aroma of black tea (Senthil Kumar et al. 2001).

Joseph Lopez et al. (2002) studied the reconstitution of tea polyphenol oxidase by limited proteolysis. Partially purified tea PPO was treated with commercial protease enzyme at different concentrations. Protease enzyme at 1 g/L concentration increased the PPO enzyme activity in six tea clones under *in vitro* conditions. Membrane-bound pro-PPO in tea is in an inactive precursor form and this was converted to active PPO enzyme by protease treatment. These observations were verified using SDS-PAGE.

External application of protease at various concentrations was also studied and maximum PPO activity was observed in the dhool treated with protease. Phenylalanine ammonia lyase (PAL) plays a vital role in the biosynthesis of flavanols, the prime substrate for PPO and is intimately involved in tea quality (Roberts and Fernando 1981). Variations in PPO and PAL activities with respect to different cultural and manufacturing processes and their effects on black tea quality were studied (Ravichandran and Parthiban 1998a). A wide variation between the enzyme activities of different clones, as well as variation due to seasonal changes and shoot maturity, was reported. The enzyme activities positively correlated with tasters' scores. Supplementation of enzymes enhanced the black tea quality markedly in terms of cuppage and creaming properties. The addition of PPO markedly improved the TF, TR, HPS, color index, and total soluble solids. Addition of pectinase increased the liquor color and water-soluble solids (water extract). It is also suggested that a combination of PPO and pectinase might further enhance the quality and price realization because their mode of action is totally different.

18.3.4.4 Lipases and Lipoxygenases

Lipid acyl hydrolases increased aldehydes, alcohols, and their derivatives during tea processing (Ramarethinam and Latha 2000). A thorough study was carried out on some of the properties of tea lipoxygenases, their variation due to cultural and manufacturing practices, and their contribution to tea aroma (Ravichandran and Parthiban 1998b). An increase in lipoxygenase activity was observed with leaf maturity. The changes of lipoxygenase activity during CTC manufacture revealed that it increased with the degree of withering and upon rolling. A gradual decline was recorded during the fermentation and drying processes. The residual lipoxygenase activity in black tea is expected to affect its quality upon storage. High lipoxygenase activity leads to the production of greater amounts of group I VFCs and hence lack of flavor (aroma) of tea and vice versa.

18.3.4.5 Phenol Oxidases

Tea leaf polyphenol oxidase (*o*-diphenol:O_2 oxidoreductase 1.10.3.1) is of crucial importance during the manufacture of black tea. Its activity increases with the increase in copper concentrations and it is believed to be a copper-containing protein (Sreerangachar 1943a, 1943b) consisting of at least four isoenzymes (Bendall and Gregory 1963; Gregory and Bendall 1966; Takeo and Uritani 1966); the major component has a molecular weight of $144,000 \pm 16,000$ and contains 0.32% (w/w) copper. The polyphenol oxidase activity varies between mature and young tea leaves and seasonally as well as during different stages of tea processing (Takeo 1966; Takeo and Baker 1973).

Thanaraj and Seshadri (1990) studied the variations in PPO activity and levels of total polyphenols and catechins with respect to different clones and shoot components and their effect on quality of black tea. A good nonlinear relationship was found between PPO activity of fresh tea shoots of different clones and TF content of corresponding black teas. Among different shoot components, buds and first leaves had higher levels of ployphenols and catechins than internodes. However, PPO activity showed a reverse trend: The internodes exhibited a higher activity compared with other components. HPLC analysis of TF fraction in tea brew of black tea made from different components of tea shoot showed that buds resulted in black tea with the highest amount of theaflavin gallates, whereas tea produced from internodes had the lowest amount of TF gallates. Based on this, a new factor—theaflavin digallate equivalent (DGETF)—was developed and the significance of this factor for chemical evaluation of black tea quality has been reported.

18.3.4.6 Peroxidase

Peroxidase is an oxidative enzyme of significance in tea manufacture. An insoluble form of peroxidase in tea leaf homogenates (Roberts 1952) was later isolated in a soluble form (Takeo and Kato 1971). This enzyme also exists in the form of several isoenzymes. Ethylene and iodoacetate cause activation of the different isoenzymes (Saijo and Takeo 1974).

18.3.4.7 Tannase

Tannase, an acyl hydrolase enzyme preparation from *Aspergillus oryzae*, has been used as a processing aid for the manufacture of cold water-soluble tea beverages. Toxicity and gene mutation study using *Salmonella typhimurium* have been performed by Lane et al. (1997) to establish the safety of the enzyme preparation for the consumer. General toxicity was low with no adverse effects at the highest dose of 1% of the diet. There was no evidence of mutagenic potential with any strain at any concentration with or without metabolic activation. These results, together with knowledge of the production organism and the chemical and microbiological characterization of the enzyme preparation, indicate that tannase can be regarded as safe for processing tea.

Wright (2005) studied the effect of a commercial tannase extract on the TF profile of black tea produced from tea clones developed by the Tea Research Foundation of Central Africa (TRF CA). The effect of the added tannase on the characteristics of the manufactured black tea, including quality and the effect of the treatment on creaming down of tea liquor, was investigated. It was reported that the apparent increase in the gallated flavan-3-ols is most possible due to some reactions taking place during the fermentation process. The addition of 0.1g/kg MT tannase resulted in a significant increase in the amount of TF-f (free theaflavins) in the black tea.

18.3.4.8 Other Enzymes

Leaf chlorophyllase activity is inversely related to the chlorophyll contents and is partly responsible in determining the proportions of pheophytin and pheophorbide during the processing of tea. 5-Dehydro shikimate reductase, acid phosphatases, alcohol dehydrogenase, and leaf ribonuclease have also been studied (Sanderson 1966; Tirimanna 1967; Yamanishi, Wickremasinghe, and Perera 1968; Baker and Takeo 1974; Gianturco et al. 1974; Tsushida and Takeo 1976; Hazarika, Mahanta, and Takeo 1984).

18.3.4.9 Enzymatic Degradation of Caffeine

Caffeine is a purine alkaloid that is a major constituent of tea and other beverages. Caffeine has negative withdrawal effects and hence decaffeinated beverages are used to overcome these negative effects. To solve the problem of chemical extraction of caffeine in food products as well as treating the caffeine containing waste products, development of a process involving an enzymatic degradation of caffeine to nontoxic compound is necessary. Different microbial and enzymatic methods of caffeine removal have been developed (Gokulakrishnan et al. 2005). The oxidative degradation of caffeine to trimethyluric acid appears to be efficient for development of enzymatic degradation of caffeine.

Development of a process involving an enzymatic (specific) degradation of caffeine to nontoxic compound is necessary to solve the problems of chemical extraction of caffeine in food products as well as treating the caffeine containing waste products. The literature revealed that major caffeine-degrading strains belong to *Pseudomonas* and *Aspergillus* genera. Though the enzymes involved in degradation of caffeine by microorganisms are known, *in vitro* enzymatic studies for caffeine degradation have not yet been reported.

Microbial degradation of caffeine using a strain of *Pseudomonas alcaligenes* CFR 1708 isolated from coffee plantation soil has been reported (Sarath Babu et al. 2005). The enzymes responsible for caffeine degradation were found to be inducible. Preinduction of the microbial cells in a medium containing caffeine as the sole source of carbon and nitrogen was carried out for 48 hours. The induced bacterial cells were found to be capable of completely degrading caffeine (1 g/L) from solutions containing caffeine within 4–6 hours at $30 \pm 2°C$ in the pH range of 7.0–8.0. To make the process of decaffeination application oriented, immobilization of cell debris was done and used in a designed bioreactor for a continuous decaffeination process. An environment-friendly biodecaffeination process has been reported.

18.3.5 Chemical Composition of Made Tea

The major catechin oxidative products are theaflavins and thearubigins. Roberts (1950) and his co-workers (Roberts, Cartright, and Oldschool 1957) were the first to use paper chromatography to isolate the ethyl acetate fraction and named chemical components as theaflavin and theaflavin gallate. Theaflavin (TF) has been shown to be a bis-flavan substituted 1′,2′-dihydroxy-3,4-benzo-tropolone, orange-red in color, constituting about 0.3–1.8% of black tea dry weight and 1–6% of the solids in tea infusion. It contributes significantly to the bright color and brisk taste of tea infusions. The theaflavins are formed by the enzymatic oxidation and condensation of catechins with di- and trihydroxylated B rings, referred to from now on as simple catechins [(+)-catechin, (–)-epicatechin, (–)-epicatechin-3-gallate)] and gallocatechins [(–)-epigallocatechin-3-gallate, (–)-epigallocatechin, and (+)-gallocatechin] (Takino et al. 1964).

Bryce, Collier, and Fowlis (1970) and Bryce, Collier, and Mallows (1972) discovered the epitheaflavic acid, epitheaflavic acid-3-gallate, theaflavin-3-3′-digallate (TFDG). A gallated version and a nongallated version of theaflavate-A have been isolated from black tea (Wan and Nursten 1997). Until now, there have been 11 theaflavins reported. Approximate relative proportions of the theaflavins in black tea were theaflavins (18%), theaflavin-3-gallate (18%), theaflavin-3′-gallate (20%), theaflavin-3,3′-digallate (40%), and isotheaflavin + theaflavic acids (4%).

Roberts (1950), Takino et al. (1964), Collier, Bryce, and Mallows (1973), and Robertson (1983) investigated the mechanism and pathways of theaflavin formation. It was considered that the oxidation of the catechins to their respective *o*-quinones was catalyzed by polyphenol oxidase and a net uptake of molecular oxygen was observed. Low-oxygen conditions may inhibit this reaction, thus resulting in poor recovery of theaflavins during fermentation (Robertson 1983). The quinones rapidly reacted with each other and other compounds to form theaflavin. The formation of theaflavin during fermentation reached a maximum and then declined. For CTC tea, this maximum usually occurred between 90 and 120 minutes. It was generally recognized that the theaflavins play a premium role in determining the characteristic cup quality of black tea infusion described by tea tasters as "brightness" and "briskness" (Roberts 1962; Hilton and Ellis 1972). The digallate is believed to contribute the most, while theaflavin itself is considered to contribute the least.

Thearubigin (TR) is the name originally assigned to a heterogeneous group of orange-brown, weakly acidic pigments formed by enzymatic oxidative transformation of flavanols during the fermentation process of black tea manufacture (Roberts and Myers 1958). However, the withering process also showed an obvious impact on the formation of theaflavins and thearubigins (Xiao 1987). Thearubigins comprise between 10 and 20% of the dry weight of black tea and between 30 and 60% of the solids of the black tea infusion, which is 10–20 times higher than the dry weight of theaflavins. Unlike the theaflavins, thearubigins have still not been characterized. The chemical structure of the thearubigins remains a mystery. The major difficulty is that they are diverse in their chemistry and possibly in their molecular size.

Hazarika, Chakravarty, and Mahanta (1984) separated the TRs from 60% acetone CTC black tea extracts by using the Sephadex LH-20 column chromatography and divided them into three components: TR1 (with high molecular weight), TR2 (with moderate molecular weight), and TR3 (with low molecular weight). The contents of low molecular weight TRs were decreased and the contents of high molecular weight TRs were increased with the time of fermentation. Apart from the theaflavins, which impart the briskness and brightness to the black tea infusion, the TRs also make an important contribution to the color, strength, and mouth feel (Roberts and Smith 1963; Millin, Swaine, and Dix 1969). They are responsible for body, depth of color, richness, and fullness of tea infusion. The amounts of TRs were increased with the decrease of TFs during the fermentation process in black tea manufacture, indicating that the TFs are probably the intermediate of TRs. Dix, Fairley, and Millin (1981) proved the transformation of TFs to TRs with the action of peroxidase.

Polyphenols undergo marked changes during black tea processing. As a result of enzymatic oxidation of the catechins by polyphenol oxidase, two groups of polyphenol compounds—theaflavins and thearubigins—are formed; these are thought to be unique to black tea. The enzyme oxidizes the catechins to their respective o-quinones, which rapidly react with each other and other compounds to form theaflavins. Theaflavins account for about 0.3–1.8% of the dry weight of black tea and 1–6% of the solids of tea liquor. Various theaflavins and their precursors are

theaflavin (EC + ECG)
theflavin-3-gallate (EC + EGCG)
theaflavin-3'-gallate (ECG + EGC)
theaflavin-3,3'-digallate (ECG + EGCG)
isotheaflavin (EC + GC)

The approximate relative proportions of the theaflavins in black tea were theaflavin (18%), theaflavin-3-gallate (18%), theaflavin-3'-gallate (20%), theaflavin-3.3'-digallate (40%), and isotheaflavin together with the theaflavic acids (approximately 4%). The exact levels of different theaflavins vary with fermentation conditions applied (Robertson 1983). Theaflavins are bright red pigments giving the tea liquor the characteristics determined by tasters as "brightness" and "briskness." The contribution made by these compounds to tea quality differs with individual theaflavins. It is believed that the digallate contributes the most, while theaflavin itself contributes least.

The thearubigins constitute between 10 and 20% of the dry weight of black tea and represent approximately 30–60% of the solids in tea liquor (Robertson 1992). Unlike the theaflavins, the thearubigins have still not been characterized. They are diverse in their chemistry and their molecular size may range from 700 to 40,000 Da. The level of free amino acids may increase during tea processing. Because of an increase in the activity of proteolytic enzymes, proteins are hydrolyzed with a concurrent increase of free amino acids. Accordingly, there is a change in the qualitative composition of free amino acids in the manufactured tea as compared with the fresh leaves. Reportedly, 25 amino acids are found in made tea; among those, theanine is the highest in quantity. The contents of theanine in made tea averaged 2.37%.

Theanine has two enantiomers: l- and d-theanine. The average level of d-theanine is around 1.85% of the total theanine. The relative amounts of d-theanine display inverse correlation to tea quality. The ratio of d-theanine to l-theanine increases under high temperature in storage. Therefore, the ratio of theanine enantiomers might be used as an indicator for long-term storage or as a tool in the grading of tea. Theanine is considered to be important in the taste of green tea. Recently, theanine has become a substance of interest for its bioactivity. Theanine showed a lowering effect on blood pressure in hypertensive rats (Yokogoshi 1995). Theanine may also act as a biochemical modulator to enhance the antitumor effect of doxorubicin (Sadzuka et al. 1996).

The level of caffeine may increase during withering, an early step of tea processing, and due to the losses of other components (nucleic acids). However, caffeine content may decrease during the firing process of tea manufacturing. Caffeine accounts for 5–10% of the solid material extracted from tea when mixed with boiling water. Theobromine accounts for about a further 0.3% of the total extracted solids; however, contrary to common belief, there is no theophylline in tea as drunk ordinarily (Scott, Chakraborty, and Marks 1989). The pharmacological properties of caffeine have been recognized for many years. Caffeine has marked stimulatory effects upon the central nervous system and has been employed therapeutically for this purpose (Aranda et al. 1977). Caffeine relaxes smooth muscle of the bronchi, making it of value for the treatment of asthma and the bronchospasm of chronic bronchitis. In humans, caffeine can increase the capacity of muscular work. This may, at least in part, be due to its stimulatory effect upon the nervous system.

There has been some concern about the untoward effects of caffeine. A number of investigations on caffeine intake have been reported. According to data cited by Scott et al. (1989), the estimated average daily caffeine intake per capita was 50 mg worldwide, 186–325 mg in the United States, and 359 mg in the UK. The no-effect dosage of caffeine was recommended as 40 mg/kg/day (Elias 1986). As reported, the average caffeine content of tea (mg/cup) was 55 mg/cup for bagged tea or leaf tea (Scott et al. 1989). Notably, the caffeine content goes up with the infusion time. The contents of caffeine per cup were 48 mg (2 minutes) and 80 mg (5 minutes) for bagged tea as well as 38 mg (2 minutes) and 60 mg (5 minutes) for leaf tea. Apparently, the quantity of caffeine ingested by average or moderate tea consumers, taking three to six cups a day, is below the previously mentioned dosage level unless significant additional amounts of caffeine come from other sources. Clonal variation of some biochemical parameters in CTC teas has been studied by Ranganath and Marimuthu (1992). The clones were classified into three groups: high, medium, and average quality according to the taster's score.

The contents of tea aroma in manufactured tea differ from those of the fresh tea leaves. During the process of tea manufacturing, the amounts of various aroma compounds are changing differently; some increase, while some others decrease. While some of the aroma compounds of black tea are present in the fresh leaf, most of them are formed during tea manufacture via enzymatic, redox, or pyrolytic reactions. So far, over 600 aroma compounds have been found. These include hydrocarbons, alcohols, aldehydes, ketones, acids, esters, lactones, phenols, nitrogenous compounds, sulfur compounds, and miscellaneous oxygen compounds probably derived from carotenes, amino acids, lipids, and terpene glycosides.

The compounds produced from carotenes have a major effect on the aroma of tea. High-flavor teas are normally produced from green leaf with high carotene contents. It is of interest to investigate the bioactivity of the aroma compounds. Some of them have been reported to display a variety of biological effects. Geraniol, for example, is an inhibitor of mevalonate biosynthesis, causing reduction of cholesterol. Geraniol inhibits the proliferation of cultured tumor cells and exerts an inhibitory effect on the growth of transplanted hepatoma in rats and melanoma in mice (Yu, Hildebrandt, and Elson 1995). Geraniol also shows antifungal activity. Since tea aroma is composed of a complicated group of compounds, the bioactivity of the majority of tea aroma compounds remains to be further explored.

As is well known, there are a great variety of teas that differ in their origin, the plant cultivars, and the manufacturing practices. Different kinds of tea may vary in their compositions, so the biological effects may vary to some degree. As estimated by Gill (1992), there were over 250 different specialty tea products included in a grocer price list. At the present time in China, more than 200 varieties exist. These include 138 green teas, 10 black teas, 13 semifermented teas, and many miscellaneous teas (Cheng 1992).

18.3.6 Tea and Health

The medicinal use of tea was known long before it was used as a beverage. It is therefore no wonder that now tea is one of the three major nonalcoholic drinks and is popular due to its stimulating property. The effect of tea on immune functions, the process of aging, detoxification, and chemoprevention of some human diseases has been studied and documented well through several scientific reports and publications. A number of reports concerning tea and human health have appeared in the past 10 years. More than 500 chemical constituents are present in tea. The compounds closely related to human health are flavonoids, amino acids, vitamins, caffeine, and polysaccharides. Tea also contains many essential micronutrients. About 90–100% of the vitamins of the B group are extracted into the infusion during brewing. The presence of vitamins C, E, and K is also reported.

Tea is a fluorine bioconcentrating plant and ingestion of fluorine from tea drinking has been reported to be beneficial for preventing tooth caries. Tea is also known to be a modulator of immune function. On an overall basis, tea is not only a nutritional, stimulating, and flavorful beverage, but also a physiological function-modulating drink. Flavonoids found in tea show 20 times more powerful antioxidant activity than vitamin C (Vinson et al. 1995). The promotion of skin cancer is inhibited by black tea and its polyphenols (Katiyar and Mukhtar 1997; Lu et al. 1997).

In addition to the probable application as medicine, tea used as a daily beverage has made great contributions to human health in at least two major aspects (Zhu 1992). First, tea drinking changes the habits of people who consume water. In ancient times, when people felt thirsty they would simply drink natural, unprocessed water that might contain pathogenic microbes. Since the adoption of tea drinking, people have used boiling water to make tea infusion. In fact, this practice helped people avoid a variety of infectious diseases. Second, tea appears to be a good substitute for alcoholic beverages. Those people who very much enjoy tea drinking might avoid alcohol overconsumption that causes severe damage to the human body.

18.3.6.1 Antioxidants

Both green and black teas are rich sources of flavanols, which are a subclass of antioxidant flavonoids. They have potentially beneficial effects on human health in preventing or repairing the damage in cells caused by free radicals. Green tea is one of the most powerful antioxidants because it contains 30% polyphenols including catechins, the building blocks of antioxidants. In addition to polyphenols, β-carotene, a provitamin (which, when derived, gets converted into vitamin A in the body), another potent antioxidant, is also present. Vitamins A, C, and E (commonly known as ACE vitamins), which act as scavengers of free radicals, are also reported to be present in tea. As a part of phytochemicals or flavonoids found in tea, these antioxidants reduce the risk of developing cardiovascular disease, cancers, cataracts, macular degeneration, cognitive impairment, osteoporosis, and Alzheimer's and liver diseases. Several polyphenolic substances in green and black teas have been shown to have antioxidant activities (Han and Xu 1990; Xu and Han 1990). Epigallocatechingallate is a potent antioxidant component in tea. Addition of low concentrations of EGCG to the extracts of green and black teas enhances the capacity to scavenge H_2O_2 and inhibition of UV-induced 8-hydroxy-2'-deoxyguanosine (8-OHdG; Wei et al. 1999).

18.3.6.2 Cardiovascular Diseases

Several studies have provided evidence that polyphenols and flavonoids have a beneficial effect on two long established heart disease risk factors: high blood cholesterol and high blood pressure (hypertension). Chemical studies revealed that the average blood cholesterol level and systolic blood pressure decreased with increased amounts of tea consumption. The risk of myocardial infarction was reduced to half in people who drank about three or more cups of tea per day. This benefit was attributed to the high concentration of flavonoids, which reduced blood clotting and the deposition of cholesterol in the blood vessels. It was also found that regular consumption of tea produces an increase of antioxidants in the blood plasma level. Recent scientific investigations have found that both short-term and long-term tea consumption improves the endothelial function and reduces cholesterol levels, thus reducing the risk of heart disease (Amarakoon 2004).

18.3.6.3 Cancer

Epidemiological studies have indicated that a high concentration of phytochemicals may help in stabilizing free radicals, thereby decreasing the risk of cancer. Studies in the laboratory and in humans have shown that theaflavin and polyphenols inhibited the growth of pancreatic and prostate tumor cells; this has been confirmed by the National Center for Toxicological Research in the United States. It is also reported that tea could play an important role in changing the genes involved in the process of causing cancer, which has also been supported by chemical studies. The cancer and tumor causing activity takes place because of cellular interaction of the pancreas, liver, and prostate, which is inhibited by tannin and its constituents like flavonoids and phenolic acid, enothin, tannic acid, chlorogenic acid, ellagic acid, etc. found in green tea. A constituent of green tea is EGCG, which destroys the enzyme urokinase, which is one of the causal factors for cancer. There are consistent findings that tea or tea polyphenol administration prevented carcinogen-induced increases in the oxidized DNA base (8-OHdG) in animal models of skin, lung, colon, liver, and pancreatic cancers (Frei and Higdon 2003).

18.3.6.4 Tooth Decay and Plaque Formation

Since tea contains fluoride, it is considered to be teeth friendly. It protects against tooth decay and inhibits plaque formation (Ooshima et al. 1994). It also improves clinical manifestation of oral lesions. Tea is useful in preventing kidney stone formation and also increases the resistance of the human body against bacterial infection. It is also reported that tea may be considered a "functional food for oral health" by controlling, through prevention, the infectious disease of caries (Wu and Wei 2002).

18.3.6.5 Antibacterial Activity

Tannin and its constituents can efficiently destroy bacteria, viruses, and fungi such as tobacco mosaic virus (TMV), herpes simplex virus (HSV), poliomyelitis virus (PMV), and influenza virus. It has been observed that extracts of green tea are somehow effective against HIV by activating V-lymphocytes in human blood. Tea was also found to control human amoebiasis. Polyphenols present in tea are susceptible to oxidation to quinones and in principle can reduce the proliferation and growth of amoeba in humans (Ghoshal et al. 1993).

18.3.6.6 Osteoarthritis Treatment

Tea is helpful in treating rheumatoid arthritis and osteoporosis, especially in postmenopausal women. Osteoporosis and liver disease may be reduced by green tea. Tea also reduces the injury induced by radiation.

18.3.6.7 Diabetic Treatment

Constituents of oolong tea increase the secretion level of the estrogen hormone, which stimulates the secretion of the insulin hormone and protects the human being from diabetes. The enzyme α-amylase catalyzes the conversion of starch in food to glucose in the digestive process. Polyphenols in tea inhibit α-amylase activity and could contribute to reducing blood glucose levels (Amarakoon 2004).

18.3.6.8 Bleeding

Bleeding is checked if a tea bag is applied at the scar. Tea tannin activates the thrombocytes for rapid clotting.

18.3.6.9 Diarrheal Treatment

Tea tannin can bind the mucosal membrane of intestines and controls diarrhea in children.

18.3.6.10 Hair Tonic, Perfume, and Food Flavoring

Tea oil from *Camellia sasanqua* or *Camellia sinensis* is an attractive perfume. It is also a food flavor in spices. It may be used as a good hair tonic also. This oil contains 97% eugenol (isoeugenol), which is also known as karyophilic acid ($C_{10}H_{22}O_2$).

18.3.6.11 Preservative

Tea tannin contains tannic acid, which can be used as a preservative for storage of some fruits such as apples, plums, and oranges.

18.4 BIOTECHNOLOGY AND MOLECULAR BIOLOGY

18.4.1 Micropropagation

Conventionally, tea is propagated through either seeds or cuttings. Seeds are collected from orchards and germinated in the nursery. Because tea is a cross-pollinated species, seedling populations show a high degree of variability. Alternative to this, clonal propagation of an elite variety from single node has been in practice recently. Clonal population gives uniformity of characters. Grafting, where a scion is grafted to root stock, has recently become a popular choice. Although all these vegetative propagation methods are successful and commercially viable, they are limited by several factors, like slower rate of propagation, poor survival rate in the nursery due to poor rooting in certain clones, and seasonal dependence for rooting (Mondal et al. 2004). Therefore, tissue culture could be an alternative choice to overcome these problems.

Several reports available on micropropagation of tea and related species were reviewed previously (Kato 1989; Vieitez et al. 1992; Dood 1994; Das 2001; Mondal et al. 2004; Mondal 2007). Literature shows that in tea, tissue culture was first initiated by Forrest (1969) and that Kato (1985) conducted a systematic study on micropropagation. Different aspects of micropropagation, like explants, media, growth regulators, rate of multiplication, hardening, and field performance, were studied by Mondal et al. (2004).

18.4.1.1 Explants

Choice of explants as starting material is crucial in any micropropagation program. Factors involved at this stage are type, origin, and availability of explants throughout the year. Different types of explants were tried in tea—for example:

stem node (Phukan and Mitra 1984; Agarwal, Singh, and Banerjee 1992; Jha and Sen 1992; Bag, Palni, and Nandi 1997; Mondal et al. 1998; Prakash et al. 1999)
shoot tip (Phukan and Mitra 1984; Nakamura 1987a; Kuranuki and Shibata 1992)
cotyledon node (Jha and Sen 1992)
stem internode (Nakamura 1989)
stem segment without epidermis (Kato 1985)
epidermal layer (Kato 1985)
flower stalk (Sarwar 1985)
immature cotyledon (Sarwar 1985)
petiole (Sarwar 1985)
zygotic embryo (Iddagoda et al. 1988)
axillary bud (Nakamura 1987b; Tahardi and Shu 1992)

Most commonly used and successful among these are shoot tip and stem node with the axillary bud.

18.4.1.2 Basal Media

Standardization of media is critical for micropropagation. Different formulations of basal media were attempted for micropropagation of tea; full- or half-strength MS media (Murashige and Skoog 1962) was most common. Other media, like Heller's medium (Heller 1953), B5 medium (Gamborg, Miller, and Ojima 1968), and the Nitsch and Nitsch medium (Nitsch and Nitsch 1969), were also tried. Among them, MS medium is most suitable for micropropagation of tea. A comparative study among different basal media (MS, B5, and Nitsch and Nitsch media) was made by Nakamura (1987a); MS medium was found to be the best for tea micropropagation. Although full-strength MS salts were widely used, half-strength MS salts were also attempted (Phukan and Mitra 1984; Banerjee and Agarwal 1990; Agarwal et al. 1992). Pandidurai et al. (1997) reported uniform and better growth of shoots in half-strength MS medium compared to full-strength MS medium.

18.4.1.3 Growth Regulator

Growth regulator is another factor for success in micropropagation. Combinations of cytokinin and auxin were used successfully for micropropagation of tea. Different combinations of 6-benzyladenine (BA, 1–6 mg/L) with either indole-3-butaric acid (IBA, 0.001–2.0 mg/L) or indole-3-acetic acid (IAA, 0.001–0.5 mg/L) were most successful for shoot initiation and further multiplication in tea (Kato 1985; Agarwal et al. 1992; Jha and Sen 1992; Bag et al. 1997; Prakash et al. 1999). In some cases, gibberellic acid (GA$_3$, 0.1–10 mg/L) in combination with other auxins/cytokinins was used (Nakamura 1987a, 1987b, 1989; Kuranuki and Shibata 1993). Shoot proliferation was also reported using thidiazuron (TDZ, in MS medium) in combination with naphthaleneacetic acid (NAA) (Mondal et al. 1998) or IBA and BA (in WPM medium; Tahardi and Shu 1992).

Mondal et al. (1998) compared the effectiveness of TDZ and BA and found a higher number of shoot initiations in media containing TDZ compared to media containing BA, but the multiplication rate was more or less the same during subsequent subcultures. Callus formation and subsequent differentiation of shoots from callus were reported in media containing NAA + BA (Phukan and Mitra 1984; Bag et al. 1997); only auxin (NAA or 2,4-dichlorophenoxyacetic acid) was found ineffective (Nakamura 1988a). Auxins like picloram and 2,4,5-trichlorophenoxyacetic acid (2,4,5-T) were found helpful for elongation of shoots (Nakamura 1987a, 1987b, 1989; Jain, Das, and Barman 1991; Kuranuki and Shibata 1993). Apart from semisolid media, Sandal, Bhattacharya, and Ahuja (2001) reported an efficient liquid culture system for shoot proliferation. A significant role of growth adjuvants like coconut milk, yeast extract, casein acid hydrolysate, serine, and glutamine was found in tea micropropagation (Phukan and Mitra 1984, 1990; Agarwal et al. 1992; Jha and Sen 1992).

Tea is a woody perennial plant that contains high levels of polyphenol. Therefore, explants collected from field-grown plants turn brown due to exudation of phenolic compounds from the cut ends to media. Different strategies have been adopted to overcome this problem—for example, use of ascorbic acid, polyvinyl pyrolidone-10, catechol, 1-cystein, phloroglucinol, phenyl-thiourea, sodium diethyl dithio-carbonate, sodium fluoride, thiourea, different strengths of MS salts, etc. (Creze and Beauchesne 1980; Agarwal et al. 1992). Different degrees of success have been achieved.

18.4.1.4 Rooting

Efficient rooting of *in vitro* shoots is important for establishment of plants in soil. Successful rooting of tea microshoots was achieved in liquid or semisolid, full-strength or half-strength MS medium containing IBA (0.5–8 mg/L; Kato 1985; Nakamura, 1987a, 1987b; Phukan and Mitra 1990; Agarwal et al. 1992; Sandal et al. 2001). Induction and elongation of roots are better in half-strength MS salts. Half-strength liquid MS medium with NAA was also successful in rooting (Bag et al. 1997). Efficiency of rooting is also influenced by genotype (Murali et al. 1996). Root initiation was noticed within 4 weeks in clones UPASI-26 and UPASI-27 followed by UPASI-3, while a minimum of 6–8 weeks was required for BSB-1. Highest rooting percentage was recorded in clone UPASI-26 and lowest in BSB-1. Banerjee and Agarwal (1990) observed a favorable effect of low light and low pH (4.5–4.6) in rooting.

18.4.1.5 Hardening and Field Transfer

The most critical phase of micropropagation is the establishment of *in vitro* plants in *ex vitro* conditions. A major limitation of large-scale micropropagation of tea is the high rate of mortality at this stage. Generally, tea microshoots are hardened on various soil mixtures like sand, cow dung, and soilrite. Rooted shoots are transferred to potting mixture under high humidity created by misting or fogging units. Gradually, the moisture level is reduced to a natural level. In addition to different abiotic factors, sudden exposure of aseptically grown plants to a soil microbial community is one important cause of low survival of micropropagated plants during hardening. An attempt was made to overcome this "transient transplant shock" using a biological hardening technique (Pandey, Palni, and Bag 2000). A few antagonistic bacterial isolates (*Bacillus subtilis*, *Bacillus* sp., *Pseudomonas corrugate*, and *P. corrugate*) were isolated from tea rhizosphere and obtained 100, 96, and 88% survival against 50, 52, and 36% in control during rainy, winter, and summer seasons, respectively.

Thomas et al. (2010) studied in detail the effects of native isolates of *Pseudomonas fluorescens*, *Azospirillum brasilense*, and *Trichoderma harzianum* on acclimatization of micropropagated tea shoots. These plant growth-promoting rhizobacteria (PGPR) were found to influence the survival rate, plant growth, biometric parameters, key enzymes, nutrient use efficiency (NUE), and systemic resistance. Another technique of hardening (i.e., micrografting) was also attempted in tea and was successful. *In vitro* raised microshoots were grafted on *in vitro* or *in vivo* raised root stocks (Prakash et al. 1999; Mondal et al. 2004). A savings of 1 year was possible by grafting *in vitro* raised shoots on 3-month-old seedlings (Mondal et al. 2004).

18.4.2 Somatic Embryogenesis

Several reports are available on plant regeneration via somatic embryogenesis in tea. Induction of somatic embryos depends upon several factors, such as explant type, physiological stage of explant, and media formulation, including growth regulators.

18.4.2.1 Explants

Somatic embryogenesis in tea was reported from different types of explants; the most popular among them were zygotic embryos or mature cotyledons. Several other types of explants were also used for somatic embryogenesis in tea, such as immature cotyledons (Abraham and Raman 1986; Haridas et al. 2000), decotylenated embryos (Paratasilpin 1990; Mondal et al. 2000), deembryonated cotyledons (Ponsamuel et al. 1996), nodal cuttings (Akula and Dodd 1998; Akula and Akula 1999), juvenile leaves (Sarathchandra, Upali, and Wijeweardena 1988), and leaf stalk (Hua, Dai, and Hui 1999).

Direct and indirect somatic embryogenesis was reported for many *Camellia* species, like *C. reticulata* (Zhuang and Liang 1985a; Plata and Vieitez 1990), *C. japonica* (Kato 1986; Vieitez and Barciela 1990), *C. sasanqua* (Zhuang, Duan, and Zhou 1988), and *C. chrysantha* (Zhuang and Liang 1985b).

Physiological maturity of seeds played an important role in somatic embryogenesis of tea. Seeds of different cultivars mature at different time and therefore optimum time for initiation of embryogenic culture from seed cotyledon explants varies. Late September to mid-October was found to be an optimum time for cotyledon culture in *C. sinensis* in Shizuoka, Japan, when the seeds are physiologically mature (Nakamura 1988b). Mondal et al. (2000) reported that best time for cultivars TG 270/2/B and UPASI-9 is July to August, for Tuckdah-78 is September to October, and for Kangra Jat is November to December. Effect of genotype on somatic embryogenesis was also reported in tea (Nakamura 1988b; Paratasilpin 1990; Kato 1996).

18.4.2.2 Basal Media, Growth Regulators, and Adjuvants

MS medium was most commonly used for somatic embryogenesis in tea, but reports are available on use of Nitsch and Nitsch (1969) medium. High concentration of cytokinin with low concentration of auxin or low concentration of cytokinin alone was most effective for somatic embryo induction. BA is most commonly used for somatic embryogenesis, and reports on use of kinetin are also available (Bano, Rajaratnam, and Mohanty 1991; Wachira and Ogado 1995). Among different auxins, IBA was most common; however, other auxins like NAA (Paratasilpin 1990; Bag et al. 1997; Balasubramanian et al. 2000a), 2,4-D (Bano et al. 1991), and IAA (Sood et al. 1993) were also used. Ponsamuel et al. (1996) induced somatic embryos with tetraphenylboron (TPB) and phenylboronic acid (PBOA). Mondal et al. (2000) reported that half-strength MS salts with NAA and BA were best for somatic embryogenesis in tea. Inclusion of ABA in the induction medium for rapid induction of somatic embryos from mature seeds was also reported (Akula, Akula, and Bateson 2000).

Other than growth regulators, a number of adjuvants, like yeast extracts, coconut milk, and betaine, were tried (Yamaguchi, Kunitake, and Hisatomi 1987; Arulpragasam, Latiff, and Seneviratne 1988; Sarathchandra et al. 1988; Akula et al. 2000). Apart from primary embryogenesis, secondary embryogenesis is reported in tea (Abraham and Raman 1986; Kato 1986; Jha, Jha, and Sen 1992; Mondal, Bhattachary, and Ahuja 2001b; Mondal, Bhattachary, Ahuja, et al. 2001).

18.4.2.3 Maturation and Germination

Maturation and germination of somatic embryos at different levels of sucrose were tried for different time periods. Sucrose at 3% level yielded significantly high germination percentages after 5 weeks of treatment. High percentages of somatic embryo germination were reported when 4% maltose was included along with 3 mg/L *trans*-cinamic acid (*t*-CA; Mondal et al. 2002). Desiccation of somatic embryos was found ineffective in improving germination of somatic embryos (Mondal et al. 2004). A number of growth regulators like ABA, GA$_3$, and BA with IBA or IAA were tried to improve maturation and germination of somatic embryos (Mondal et al. 2002; Kato 1986; Jha et al. 1992). BA with IBA or IAA was found effective for germination of tea somatic embryos.

18.4.3 Somaclonal Variation

Although somaclonal and gametoclonal variation are important tools for crop improvement, particularly in perennial plants like tea, no serious effort has been made to develop somaclonal variants in tea. Few reports are available on identification of somaclonal variants from *in vitro* plants. Raj Kumar et al. (2001; Raj Kumar, Balasarvanan, and Pius 2002) reported identification of somaclonal plants regenerated through somatic embryogenesis (from cotyledonary explants). These somaclonal plants varied in morphological, physiological, and biochemical characters (Thomas, Raj Kumar, et al. 2006). Presence of genetic variation in these plants was also proved using ISSR marker (Thomas, Vijayan, et al. 2006).

18.4.4 Anther Culture

Because tea is highly heterozygous and heterogeneous, production of homozygous diploid plants is of great importance in tea improvement. Although several attempts have been made to regenerate haploid plants through anther culture, success remained up to microcalli development (Okano and Fuchinone 1970; Saha and Bhattacharya 1992), except for one where plant regeneration was successful from anther derived calli (Chen and Liao 1982, 1983). Among nine cultivars tested, plants were developed only from the cultivar Fuyun No-7. Attempts were also made to regenerate haploid plants through pollen culture (Shimokado, Murata, and Miyaji 1986; Raina and Iyer 1992) but they failed to develop haploid plants.

18.4.5 Protoplast Culture

Protoplast culture in tea is important for intergenetic transfer of valuable genes because many wild relatives of tea have biotic and abiotic resistance characters. These characters can be transferred to cultivated clones through somatic hybrids or cybrids. Protoplast culture was attempted by Nakamura (1983), Purakayastha and Das (1994), and Balasubramanian et al. (2000b); success was achieved up to microcalli development.

18.4.6 Genetic Transformation

Transfer of important characters through genetic engineering is successful in many plants. The technique has many advantages over conventional methods of gene transfer. Using this technique, desirable exogenous genes from other organisms can be introduced into tea and new varieties with new characters can be developed.

18.4.6.1 Agrobacterium tumefaciens-Mediated Transformation

Agrobacterium tumefaciens-mediated transformation was attempted in tea by using leaves and somatic embryos (Matsumoto and Fukui 1998; Mondal et al. 1999; Mondal, Bhattachary, Ahuja, et al. 2001; Luo and Liang 2000; Joseph Lopez et al. 2004). Although stable transformed callus was reported, a transgenic plant was first developed by Mondal et al. (2001b). Stable transformation and integration of transgene was confirmed by molecular analysis (e.g., PCR amplification and Southern hybridization) (Matsumoto and Fukui 1998; Mondal et al. 2001b; Joseph Lopez et al. 2004). Kanamycin at 75 mg/L was found optimum for selection of transformed internode explant (Tosca, Pondofi, and Vasconi 1996) and somatic embryo (Mondal et al. 2001b). Among different antibiotics tried as bactericidal to check the growth of *A. tumefaciens*, sporidex (400 mg/L) was found to be best, followed by carbenicillin and cefotaxime (Mondal et al. 2001a).

18.4.6.2 Biolistic Gun-Mediated Transformation

An attempt was made with biolistic gun-mediated genetic transformation, but regeneration of transformed plants has not yet been achieved (Akula and Akula 1999).

18.4.6.3 Agrobacterium rhizogenes-Mediated Transformation

Large-scale production of secondary metabolites through hairy root culture is successful in many plants. In tea, a preliminary attempt was made to induce hairy roots through *Agrobacterium rhizogenes* (Zehra et al. 1996; Konwar et al. 1998), but commercial exploitation of the technique has not yet been achieved. Recently John et al. (2009) induced hairy root in leaf explants and recorded higher catechin levels in hairy root in comparison to leaf.

18.4.7 Molecular Markers

During the last two decades, a number of molecular (DNA-based) markers (e.g., restriction fragment length polymorphism [RFLP], random amplified polymorphic DNA [RAPD], amplified fragment length polymorphism [AFLP], simple sequence repeat [SSR], intersimple sequence repeat [ISSR], and cleaved amplified polymorphic sequence [CAPS]) have been developed and used for marker-assisted selection (MAS), study of diversity, molecular phylogeny, genetic fidelity testing, and genome mapping.

18.4.7.1 RAPD

A number of reports are available on the RAPD marker, primarily to characterize tea cultivars and to determine their genetic relationship. Wachira et al. (1995) estimated the genetic diversity and taxonomic relationship of 38 Kenyan tea cultivars. Based on average linkage cluster analysis, they separated the cultivars into Assam, Cambod, and China types. From a preliminary study among Korean, Japanese, Chinese, Indian, and Vietnamese teas, Tanaka, Sawai, and Yamaguchi (1995) concluded that Korean tea has undergone very little diversification after being introduced from China and Japanese tea showed a closer relationship with both Chinese and Indian tea, indicating their possible sources of introduction. Mondal (2000) studied 25 Indian tea cultivars and two ornamental species using the RAPD marker and separated them into three clusters (China, Assam, and ornamental types). Principal component analysis (PCA) revealed more diversity among the Chinese clones.

Kaundun, Zhyvoloup, and Park (2000) studied 27 accessions from Korea, Japan, and Taiwan and reported the highest diversity among the Korean group, followed by the Taiwanese and Japanese groups. They indicated that the origin of Taiwanese tea might be different from that of Korean and Japanese tea. Lai, Yang, and Hsiao (2001) studied the genetic relationship of 37 tea samples composed of 21 clones of China, 3 clones of Assam, 7 hybrid clones between China and Assam tea, and 6 individual samples of native Taiwanese wild tea using RAPD and ISSR markers. They revealed three major groups: cultivars of China tea and the cultivars developed in Taiwan from hybridization and selection, Assam tea, and native Taiwanese wild tea. Ramakrishnan et al. (2009) studied the genetic relationship of 12 commercial tea accessions. Based on Ward's method of cluster analysis, they separated these accessions into Assam, China, and Cambod types; this confirmed the present taxonomic framework of *Camellia* species.

An assessment of genetic diversity of cultivated tea and its wild relatives using RAPD and AFLP markers revealed that population diversity (H) decreased in the order of wild camellia > India > China > Kenya > Srilanka > Vietnam > Japan > Taiwan, and analysis of molecular variance (AMOVA) revealed that most variation resided among individuals within populations (Wachira, Tanaka, and Takeda 2001). Other studies showed more genetic diversity present in China than any other country (Chen et al. 1998, 2005a; Chen and Yamaguchi 2002; Luo et al. 2002; Shao et al. 2003; Duan et al. 2004; Huang et al. 2004).

RAPD markers were also used for identification of true-crossing progenies, cultivars, species, pollen parent, commercial teas (Wachira et al. 1995, 2001; Wachira, Powell, and Waugh 1997; Wright, Apostolides, and Louw 1996; Singh, Bandana, and Ahuja 1999; Kaundun et al. 2000; Liang, Tanaka, and Takeda 2000; Lai et al. 2001; Tanaka, Yamaguchi, and Nakamura 2001; Chen and Yamaguchi 2002, 2005; Kaundun and Matsumoto 2003a; Chen, Gao, et al. 2005), genetic fidelity of tea plants raised *in vitro* (Mondal and Chand 2002; Devarumath et al. 2002), and intraclonal genetic variability (Singh, Saroop, and Dhiman 2004).

18.4.7.2 ISSR

Zietkiewicz, Rafalski, and Labuda (1994) used the ISSR marker for genetic fingerprinting of 25 diverse tea cultivars and separated them into China, Assam, and Cambod types, supporting the known taxonomical classification of tea. Mondal (2002) and Yao et al. (2005) also used the ISSR marker for genetic fingerprinting of tea. Thomas, Vijayan, et al. (2006) used this marker for assessment of somaclonal variation.

18.4.7.3 RFLP

Matsumoto et al. (1994) employed the RFLP marker to assess the genetic variation of Japanese green tea using phenylalanine ammonia-lyase (PAL) as a probe and could differentiate Japanese green tea cultivars from Assam hybrids. The study also indicated the origin of present-day green tea cultivars of Japan by crossing and selection from a narrow genetic background.

18.4.7.4 AFLP

Diversity of 32 tea clones from India and Kenya was studied by Paul et al. (1997) using the AFLP marker. The study showed that diversity is found more in the China type compared to the Assam or Cambod types. In principal coordinate (PCO) analysis, close clustering of Assam clones from both India and Kenya indicates that Korean teas originated from India. Mishra and Sen-Mandi (2001, 2004) employed the AFLP marker for characterization of 29 poplar Darjeeling tea cultivars and divided them into Assam, China, and Cambod groups. Their results showed high similarity (68–92%) among the cultivars grown in the Darjeeling area and that they belong to a narrow genetic pool that originated through intra- and interspecific hybridization. Balasarvanan et al. (2003) analyzed 49 tea cultivars from southern India and distinguished those into Assam, China, Cambod, and an intermediate group; clones under the China group were more diverse. Sharma, Negi, and Sharma (2010) assessed the genetic diversity of 123 commercially important tea accessions representing major populations in India. They used seven AFLP primer combinations and recorded 51% genetic similarity among the accessions. They did not notice any significant differences in average genetic similarity among populations cultivated in different geographic regions of India.

18.4.7.5 SSR

Kaundun and Matsumoto (2002) used heterologous chloroplast and nuclear microsatellite (SSR) markers to study polymorphism in 24 tea clones. Using factorial correspondence analysis of SSR data, they separated the *assamica* and *sinensis* genotypes into two groups and demonstrated the value of these markers in establishing the genetic relationship between tea clones. Freeman et al. (2004) of the UK were pioneers in isolating and developing 15 tea-specific microsatellite loci from tea genomic libraries.

18.4.7.6 CAPS

An attempt was made to develop a CAPS marker based on three key genes—phenylalanine ammonialyase (PAL), chalcone synthase (CHS), and dihydroflavonol 4-reductase (DFR)—involved in the phenylpropanoid pathway in tea (Kaundun and Matsumoto 2003b). The study revealed that variety *sinensis* is more variable than variety *assamica* and that a higher proportion of overall diversity resides within varieties as compared to between varieties. Joshi (2007) converted RAPD markers into CAPS markers associated with the blister blight tolerance character of tea.

18.4.7.7 Organelle DNA

Few attempts have been made to study the organelle DNA. Wachira et al. (1997) used five different organelle-specific primers in 19 taxa and nine tea cultivars and analyzed the species' introgression into the cultivated gene pool of tea. Thakor (1997) used five cpDNA sequences from 25 *Camellia* species covering four subgenera of tea to reexamine their taxonomic relationship. Prince and Parks (1997) used two cpDNA regions (*rbc*L and *mat*K sequences) for four species of the subfamily Ternstroemioideae and 24 species from Theoideae to analyze the evolutionary relationship. Nucleotide sequences of the *mat*K region of cpDNA were also used for diversity study in 118 cultivated teas from India, Bangladesh, Myanmar, Thailand, Laos, Vietnam, China, and Japan by Katoh et al. (2003). The study indicates that the cultivated teas of India, Bangladesh, eastern China, and Japan belong to the group of *C. sinensis* and that the native cultivars in Myanmar and southern China are genetically more similar to *C. taliensis* and *C. irrawadiensis.*

18.4.8 Genomics

The isolation and cloning of potentially important functional genes and study of their functions will play an important role in genetic improvement of tea. Although substantial progress has been made in genomics of many plants, in tea it is still in infancy. Research in tea genomics was initiated during the early 1990s and some progress has been made. Up to now, the genomic information and the gene expression patterns of the tea plant have not been studied in detail.

At least 25 full-length genes from tea have been cloned and deposited in the GenBank (accession nos.: DQ194358, DQ194356, AY641729, AY641731, AB115184, AB115183, AB117640, AB114913, AY830416, AY169404, AY169402, AY641733, AY665295, AY741464, AB031281, AF537127, AY659975, AY656677, AY641730, AY648027, AB088027, AF462269, AF380077, AX138777, AB041534, AF315492, AB015047, and D26596). Most of them are important enzymes related to secondary metabolism (enzymes involved in polyphenols [catechins] and caffeine biosynthesis), quality (floral aroma formation), and stress tolerance (disease and defense) of tea (Chunlei and Liang 2007).

Takeuchi, Matsumoto, and Hayatsu (1994) sequenced three full-length complementary DNAs (cDNAs)—CHS1, CHS2, and CHS3—encoding the chalcone synthase (CHS) gene from the cultivar Yabukita and studied organ-specific and sugar-responsive expression. These cDNAs (389 amino acid residues each) showed high similarity with the CHS gene of many other species. The transcripts were expressed more in leaves and stems and were moderate in flowers while low in roots. Matsumoto et al. (1994) sequenced the complete cDNA fragment (2,344 nucleotides including poly-A tail) of phenylalanine ammonia-lyase (PAL) from tea leaves, which have high homology with several dicotyledonous plants and low homology with many monocotyledonous plants and gymnosperms. The caffeine synthase gene from tea leaf was cloned by Kato et al. (2000) using the RACE (rapid amplification of complementary DNA ends) technique with degenerate gene-specific primer. The isolated cDNA termed as TCS1 consists of 1,438 base pairs and encodes a protein of 369 amino acids.

Li, Jiang, and Wan (2004) and Yu et al. (2004) studied the expression pattern of TCS gene mRNA. Another important gene of caffeine biosynthesis, the S-adenosylmethionine synthetase (SAM) gene, was cloned using the real-time polymerase chain reaction (RT-PCR) technique; it had one open reading frame (ORF) encoding 394 amino acids (Feng and Liang 2001). Mizutani et al. (2002) isolated and cloned the cDNA of the β-primeverosidase enzyme and expressed in *Escherichia coli*; complete cDNA of β-glucosidase enzyme was cloned by Li, Jiang, Yang, et al. (2004). Coding cDNA for the enzyme flavonol synthase (FLS) was cloned from tea and expressed in *E. coli* (Lin et al. 2007). Zhao, Liu, and Xi (2001) first cloned the polyphenol oxidase (PPO) gene in tea. A full-length cDNA (697 bp) of nucleoside diphosphate kinase (NDPK), designated as *CsNDPK1*, was cloned from tea leaves (Prabu et al. 2011; GenBank accession no. GU332636, unpublished) that consisted of 448 bp ORF encoding a 147-amino acid polypeptide. We have also deposited 149 complete and partial CDS in the GenBank (GenBank accession nos.: HM003230 to HM003378, unpublished).

According to dbEST release 052810 (May 28, 2010), there are only ~7,318 ESTs available in tea in the NCBI database (http://www.ncbi.nlm.nih.gov/dbEST/dbEST_summary.html). Chen, Zhaoa, and Gao (2005) sequenced 1,684 ESTs from tender shoots of tea and further classified them into 12 putative cellular roles based on the functional categories established for *Arabidopsis*. EST analysis of the genes involved in secondary metabolism in tea was reported by Park et al. (2004). They sequenced 588 cDNA clones from a substractive cDNA library; 8.7% were involved in secondary metabolism with a particularly high abundance of flavonoid-metabolism proteins (5.1%). They isolated and characterized a few genes (e.g., chalcon synthase [CHS], flavanol 3-hydroxilase [F3H], flavonoid 3'5'-hydroxylase [F3'5'H], flavonol synthase [FLS], dihydroflavonol 4-reductase [DFR], and leucoanthocyanidin reductase [LCR]) involved in the falvonoid pathway in tea. Among them, three genes—namely, F3H, DFR and LCR—expressed more in young leaves than mature leaves, which indicates that flavonoid biosynthesis genes are differentially regulated in the developmental stages in tea. Sharma and Kumar (2005) studied drought-modulated ESTs in tea and identified three ESTs; this could be an initial step to cloning these full-length genes and their promoters. We have deposited 809 ESTs in the GenBank (GenBank accession nos.: FD800891 to FD 800906, GT054202 to GT054263, GW690681 to GW691037, GW696671 to GW696870, GW787565 to GW787738, unpublished) from tea.

18.4.9 Genome Mapping

Tanaka (1996) first constructed an RAPD-based linkage map in tea and detected markers related with theanin content, date of bud sprouting, resistance to anthracnose, and tolerance to cold (Tanaka 1996). Hackett et al. (2000) developed a linkage map based on RAPD and AFLP markers covering 1349.7 cM, with an average distance of 11.7 cM between loci on the map. Another AFLP-based linkage map was developed by Huang et al. (2005). It is apparent that linkage mapping in tea needs more attention for effective use of MAS in tea breeding programs.

18.5 CONCLUSION

Tea is a wonderful plant with plentiful secondary metabolites. Many of these secondary metabolites are potential antioxidants. Tea was consumed as a beverage in earlier days. But, in recent years, people have been considering tea as a potential health drink. Much attention has been given to its potential health benefit, particularly its anticancerous activity. Considerable effort has been made by breeders to develop new cultivars with improved characters without losing yield through suitable breeding, molecular, or integrated approaches. Yet, further effort is required to develop new cultivars that suit the changing climatic conditions and can withstand biotic and abiotic stresses. Vast germplasms available in countries like India, China, and Japan are required to be exploited systematically by using classical and molecular techniques to meet these challenges. Research efforts also should focus on improving traits like important secondary metabolites.

Although a great advance has been made on the aspect of tea biotechnology, it is still far behind that of cereal and other crops. More attention is required to integrate molecular techniques with breeding, genomics, and proteomic approaches in identifying important genes and development of transgenic tea. To achieve greater height in tea crop, it is essential that all approaches have a thorough understanding of the molecular biology and biochemistry of the tea plant, which at present are very poorly understood.

ACKNOWLEDGMENTS

All the authors are thankful to their respective present employers for encouragement and support.

REFERENCES

Abraham, G. C., and Raman, K. 1986. Somatic embryogenesis in tissue culture of immature cotyledons of tea (*Camellia sinensis*). In *6th International., Congress on Plant Tissue and Cell Culture*, ed. D. A. Somersm, B. G. Gengenbach, D. D. Biesboor, et al., p. 294. Univ. Minnesota, Minneapolis.

Agarwal, B., Singh, U. and Banerjee, M. 1992. *In vitro* clonal propagation of tea (*Camellia sinensis* (L.) O. Kuntze). *Plant Cell Tissue and Organ Culture* 30:1–5.

Akula, A., and Akula, C. 1999. Somatic embryogenesis in tea (*Camellia sinensis* (L) O. Kuntze). In *Somatic embryogenesis in woody plants*, vol. 5, ed. S. M. Jain, P. K. Gupta, and R. J. Newton, 239–259. Kluwer Academic Publishers, the Netherlands.

Akula, A., Akula, C., and Bateson, M. 2000. Betaine, a novel candidate for rapid induction of somatic embryogenesis in tea (*Camellia sinensis* (L.) O. Kuntze). *Plant Growth Regulation* 30:241–246.

Akula, A., and Dodd, W. A. 1998. Direct somatic embryogenesis in a selected tea clone, 'TRI-2025' (*Camellia sinensis* (L) O. Kuntze) from nodal segment. *Plant Cell Reports* 17:804–809.

Amarakoon, T. 2004. *Tea for health.* The Tea Research Institute of Sri Lanka, ISBN 955-9023-07-1, pp. 23, 30.

Angayarkanni, J., Palaniswamy, M., Murugesan, S., and Swaminathan, K. 2002. Improvement of tea leaves fermentation with *Aspergillus* spp. pectinase. *Journal of Bioscience and Bioengineering* 94 (4): 299–303.

Anonymous.1972. Cleft grafting of tea cuttings in the nursery—Annual report for 1972. UPASI Scientific Department, p. 63. Chinchona, India.

Apostolides, Z., Nyirenda, H. E., and Mphangwe, N. I. K. 2006. Review of tea (*Camellia sinensis*) breeding and selection in southern Africa. *International Journal of Tea Science* 5 (1–2): 13–19.

Aranda, J. V., Gorman, W., Bergsteinsson, H., and Gunn, T. 1977. Efficacy of caffeine in treatment of apnea in the low-birth-weight infant. *Journal of Pediatrics* 90:467–472.

Araus, J. L., Slafer, G. A., Reynolds, M. P., and Royo, C. 2002. Plant breeding and drought in C3 cereals: What should we breed for? *Annals of Botany* 89:925–940.

Arulpragasam, P. V., Latiff, R., and Seneviratne, P. 1988. Studies on tissue culture of tea (*Camellia sinensis* (L.) O. Kuntze). 3. Regeneration of plants from cotyledon callus. *Sri Lanka Journal of Tea Science* 57:20–23.

Babu, S. 2007. A new high-quality clone TRF-2. *Planters' Chronicle* 103 (12): 6–7.

————. 2008. Studies on induced mutation and fixing of LD_{50} value in tea (*Camellia* sp. (L)). *Journal of Plantation Crops* 36 (3): 200–203.

Badrul Alam, A. F. M., Dutta, M. J., and Harque. 1997. Tea germplasm in Bangladesh and their conservation. *Tea Journal of Bangladesh* 33 (1–2): 29–35.

Bag, N., Palni, L. M. S., and Nandi, S. K. 1997. Mass propagation of tea using tissue culture methods. *Physiology and Molecular Biology of Plants* 3:99–103.

Baker, J. E., and Takeo, T. 1974. Acid phosphatases in plant tissues; changes in activity and multiple forms in tea leaves and tomato fruit during maturation and senescence. *Study Tea* 46:63–75.

Balasarvanan, T., Pius, P. K., Raj Kumar, R., Muraleedharan, M., and Shasany, A. K. 2003. Genetic diversity among south Indian tea germplasm (*Camellia sinensis*, *C. assamica* and *C. assamica* ssp. *Lasiocalyx*) using AFLP markers. *Plant Science* 165:365–372.

Balasubramanian, S., Marimuthu, S., Raj Kumar, R., and Vinod Haridas. 2000a. Somatic embryogenesis and multiple shoot induction in *Camellia sinensis* (L). O. Kuntze. *Journal of Plantation Crops* 28:44–49.

Balasubramanian, S., Marimuthu, S., Raj Kumar, R., and Balasaravanan, T. 2000b. Isolation, culture and fusion of protoplast in tea. In *Recent advance in plants crops research*, ed. N. Muraleedharan and R. Rajkumar, 3–9. Allied Publishers Ltd., India.

Banerjee, B. 1986. Biotechnology—A panacea for tea? In *Plantation crops*, vol. II, ed. H. C. Srivastava et al., 57–61. Oxford and IBH Publishing Co. Pvt. Ltd., New Delhi.

————. 1987. Can leaf aspect alter herbivory? A case study with tea. *Ecology* 68:839–843.

————. 1991. *Tea: Production and processing dynamics*. Oxford IBH Publishing Co., New Delhi.

Banerjee, M., and Agarwal, B. 1990. *In vitro* rooting of tea (*Camellia sinensis* (L) O Kuntze). *Indian Journal of Experimental Biology* 28:936–939.

Bano, Z., Rajaratnam, S., and Mohanty, B. D. 1991. Somatic embryogenesis in cotyledon culture of tea (*Thea sinensis* L.). *Journal of Horticultural Science* 66:465–470.

Barau, D. N. 1963a. Classification of the tea plant. *Two and a Bud* 10:3–11.

————. 1963b. Selection of vegetative clone. *Tea Encyclopedia* 163:32–88.

————. 1969. Seasonal dormancy in tea (*Camellia sinensis* L.). *Nature* 224:514.

Barua, D. N., and Sarma, P. C. 1982. Effect of leaf-pose and shade on yield of cultivated tea. *Indian Journal of Agricultural Science* 52:653–656.

Barau, P. K. 1970. Flowering habit and vegetative behavior in tea (*Camellia sinensis* L.) seed trees in North India. *Annals of Botany* 34:721–735.

Bendall, D. S., and Gregory, R. P. F. 1963. Purification of phenol oxidases. In *Enzyme chemistry of phenolic compounds*, ed. J. B. Pridham, 7–24. Pergamon, Oxford, England.

Bezbaruah, H. P. 1971. Cytogenetical investigations in the family Theaceae. I. Chromosome number in some *Camellia* species and allied genera. *Caryologia* 24:421–426.

————. 1974. Tea breeding—A review. *Indian Journal of Genetics and Plant Breeding (Suppl.)* 34A:90–100.

————. 1975. Development of flower, pollination and seed set in tea in northeast India. *Two and a Bud* 22:25–30.

————. 1981. Origin and history of development of tea. In *Global advances in tea science*, ed. I. N. K. Jain, 383–392. Aravali Books International (P) Ltd., New Delhi, India.

Bezbaruah, H. P., and Gogoi S. C. 1972. An interspecific hybrid between tea (*Camellia sinensis* L.) and *C. japonica* L. *Proceedings of the Indian Academy of Sciences* B:219–220.

Bezbaruah, H. P., and Saikia, L. R. 1977. Variation in self and cross compatibility in tea (*C. sinensis* L.), summary of forty years of population results. *Two and a Bud* 24 (1): 21–26.

Bond, T. E. T. 1942. Studies in the vegetative growth and anatomy of the tea plant (*Camellia thea* Link) with a special reference to the phloem. I. The flush shoot. *Annals of Botany* 10:153–158.

Bradfield, A. E., Penney, M., and Wright, W. B. 1948. The catechins of green tea, part. I. *Journal of Chemical Society* 32:22–49.

Bryce, T., Collier, P. D., and Fowlis, I. 1970. The structures of the theaflavins of black tea. *Tetrahedron Letters* 26:2789–2792.

Bryce, T., Collier, P. D., and Mallows, R. 1972. Three new theaflavins from black tea. *Tetrahedron Letters* 28:463–466.

Cartwright, R. A., and Roberts, E. A. H. 1955. Theogallin as a galloyl ester of quinic acid. *Journal of the Society of Chemical Industry (London)* 7:230–231.

Chakraborty, V., Dutta, S., and Chakraborty, B. N. 2002. Responses of tea plant to water stress. *Biologia Plantarum* 45 (4): 557–562.

Chen, L., Gao, Q. K., Chen, D. M., and Xu, C. J. 2005. The use of RAPD markers for detecting genetic diversity, relationship and molecular identification of Chinese elite tea genetic resources [*Camellia sinensis* (L.) O. Kuntze] preserved in a tea germplasm repository. *Biodiversity and Conservation* 14:1433–1444.

Chen, L., and Yamaguchi, S. 2002. Genetic diversity and phylogeny of tea plant (*Camellia sinensis*) and its related species and varieties in the section *Thea* genus *Camellia* determined by randomly amplified polymorphic DNA analysis. *Journal of Horticultural Science and Biotechnology* 77:729–732.

———. 2005. RAPD markers for discriminating tea germplasms at the inter-specific level in China. *Plant Breeding* 124:404–409.

Chen, L., Yang, Y. J., Yu, F. L., Gao, Q. K., and Chen, D. M. 1998. Genetic diversity of 15 tea (*Camellia sinensis* (L.) O. Kuntze) cultivars using RAPD markers. *Journal of Tea Science* 18:21–27.

Chen, L., Zhaoa, L. P., and Gao, Q. K. 2005. Generation and analysis of expressed sequence tags from the tender shoots cDNA library of tea plant (*Camellia sinensis*). *Plant Science* 168:359–363.

Chen, M. L. 2002. Tea and health—An overview. In *Tea bioactivity and therapeutic potential*, ed. Y. Zhen, 1–13. Taylor & Francis, London.

Chen, Z., and Liao, H. 1982. Obtaining plantlet through anther culture of tea plants. *Zhongguo Chaye* 4:6–7.

———. 1983. A success in bringing out tea plants from the anthers. *China Tea* 5:6–7.

Chen, Z. M., Wang, H. F., You, X. Q., and Xu, N. 2002. The chemistry of tea nonvolatiles. In *TEA bioactivity and therapeutic potential*, ed. Yong-su Zhen, 57–88. Taylor & Francis, London.

Cheng, Q. K. 1992. Contemporary famous tea. In *China tea book*, ed. Z. M. Chen, 128–258. Shanghai Culture Publishers, Shanghai.

Choudhary, T. C. 1979. Studies on the morphology and cytology of the progenies of triploid tea. PhD thesis, Assam Agriculture University, Jorhat, Assam, India.

Chunlei, M. A., and Liang, C. 2007. Research progress on isolation and cloning of functional genes in tea plants. *Frontiers of Agriculture in China* 1:449–455.

Collier, P. D., Bryce, T., and Mallows, R. 1973. The theaflavins of black tea. *Tetrahedron Letters* 29:125–142.

Coxon, D. T., Holmes, A., and Ollis, W. D. 1972. Flavanol digallates in green tea leaf. *Tetrahedron Letters* 28:2819–2826.

Creze, J., and Beauchesne, M. G. 1980. *Camellia* cultivation *in vitro*. *International Camellia Journal* 12:31–34.

Das, S. C. 2001. Tea. In *Biotechnology of horticultural crops*, vol. 1, ed. V. A. Parthasarathy, T. K. Bose, P. C. Deka, P. Das, S. K. Mitra, and S. Mohandas, 526–546. Naya Prokash, India.

Devarumath, R. M., Nandy, S., Rani, V., Marimuthu, S., Muraleedharan, N., and Raina, S. N. 2002. RAPD, ISSR and RFLP fingerprints as useful markers to evaluate genetic integrity of micropropagated plants of three diploid and triploid elite tea clones representing *Camellia sinensis* (China type) and *C. assamica* ssp. *assamica* (Assam-India type). *Plant Cell Reports* 21:166–173.

Dews, P. B. 1982. Caffeine. *Annual Review of Nutrition* 2:323–341.

Dix, M. A., Fairley, C. J., and Millin, D. J. 1981. Fermentation of tea in aqueous suspension. Influence of tea peroxidase. *Journal of the Science of Food and Agriculture* 32:920–932.

Dood, A. W. 1994. Tissue culture of tea (*Camellia sinensis* (L.) O. Kuntze)—A review. *International Journal of Tropical Agriculture* 12 (3–4): 212–247.

Duan, H. X., Shao, W. F., Wang, P. S., Xu, M., Pang, R. H., Zhang, Y. P., and Cui, W. R. 2004. Study on the genetic diversity of peculiar tea germplasm resource in Yunnan by RAPD. *Journal of Yunnan Agricultural University* 19:246–254.

Eden, T. 1976. *Tea*. Longman Group Limited, London, 236 pp.

Ekborg-Ott, K. H., Taylor, A., and Armstrong, D. W. 1997. Varietal differences in the total and enantiomeric composition of theanine in tea. *Journal of Food and Agricultural Chemistry* 45:353–363.

Elias, P. S. 1986. Current biological problems with coffee and caffeine. *Café Cacao The* 30:121–138.

FAO 2009. http://www.fao.org/

Feng, Y. F., and Liang, Y. R. 2001. Cloning and sequencing of S- adenosylmethionine synthase gene in tea plant. *Journal of Tea Science* 21:21–25.

Finger, A. 1994. *In vitro* studies on the effect of polyphenol oxidase and peroxidase on the formation of black tea constituents. *Journal of the Science of Food and Agriculture* 66:293–305.

Fogarty, W. M., and Kelly, C. T. 1983. In *Microbial enzymes and biotechnology*, ed. W. M. Fogarty and C. T. Kelly, 131–182. London and New York, Elsevier Applied Science Publishers.

Forrest, G. I. 1969. Studies on the polyphenol metabolism of tissue culture derived from the tea plant (*C. sinensis* L.). *Biochemical Journal* 113:765–772.

Francois, T. 2003. Virtual plants: Modeling as a tool for the genomics of tolerance of water deficit. *Trends in Plant Science* 8 (1): 9–14.

Freeman, S., West, J., James, C., Lea, V., and Mayes, S. 2004. Isolation and characterization of highly polymorphic microsatellites in tea (*Camellia sinensis*). *Molecular Ecology Notes* 4:324–326.

Frei, B., and Higdon, J. V. 2003. Antioxidant activity of tea polyphenols *in vivo*: Evidence from animal studies. *Journal of Nutrition* 133:3275S–3284S.

Gamborg, O., Miller, R., and Ojima, K. 1968. Nutrient requirements of suspension cultures of soyabean root cells. *Experimental Cell Research* 50:157–158.

Gerats, A. M., and Martin, C. 1992. Flavonoid synthesis in *Petunia* hybrid: Genetics and molecular biology of flower color. In *Phenolic metabolism in plants*, ed. H. A. Stafford and R. K. Ibrahim, 167–175. Plenum Press, New York.

Ghoshal, S., Krishna Prasad, B. N., Raj, K., and Bhaduri, A. P. 1993. Antiamoebic and antibacterial properties of black tea. *Proceedings of the International Symposium on Tea Science and Human Health*, Tea Research Association, Calcutta, pp. 119–124.

Gianturco, M. A., Biggers, R. E., and Ridley, B. H. 1974. Seasonal variations in the composition of the volatile constituents of black tea. A commercial approach to the correlation between composition and the quality of tea aroma. *Journal of Agricultural and Food Chemistry* 22:758–764.

Gill, M. 1992. Specialty and herbal teas. In *Tea: Cultivation to consumption*, ed. K. C. Wilson and M. N. Clifford, 517–534. Chapman and Hall, London.

Gokulakrishnan, S., Chandraraj, K., Sathyanarayana, N., and Gummadi, S. 2005. Microbial and enzymatic methods for the removal of caffeine. *Enzyme and Microbial Technology* 37:225–232.

Green, M. J. 1971. An evaluation of some criteria used in selecting large–yielding tea clones. *Journal of Agricultural Science* (Cambridge) 76:143–156.

Gregory, R. P. F., and Bendall, D. S. 1966. The purification and some properties of the polyphenol oxidase from tea (*Camellia sinensis* L.). *Biochemistry Journal* 101:569–581.

Griffiths, H., and Parry, M. A. J. 2002. Plant responses to water stress. *Annals of Botany* 89:801–802.

Hackett, C. A., Wachira, F. N., Paul, S., Powell, W., and Waugh, R. 2000. Construction of a genetic linkage map for *Camellia sinensis* (tea). *Heredity* 85:346–355.

Hadfield, W. 1975. Shade in northeast Indian tea plantation. II. Foliar illumination and canopy characteristics. *Journal of Applied Ecology* 11:179–199.

Hampton, M. G. 1992. Production of black tea. In *Tea cultivation to consumption*, ed. K. C. Willson and M. N. Clifford, 459–511. Chapman & Hall, London.

Han, C., and Xu, Y. 1990. The effect of Chinese tea on the occurrence of esophageal tumors induced by *N*-nitrosomethylbenzylamine in rats. *Biomedical and Environmental Sciences* 3:35–42.

Handique, A. C. 1992. Some salient features in the study of drought resistance in tea. *Two and a Bud* 39 (2): 16–18.

Haridas, V., Balasaravanan, T., Rajkumar, R., and Marimuthu, S. 2000. Factor influencing somatic embryogenesis in *Camellia sinensis* (L.) O. Kuntze. In *Recent advances in plantation crops research*, ed. N. Muraleedharan and R. Raj Kumar, 31–35. Allied Publishers Ltd., India.

Hasselo, H. N. 1964. Productivity gradient in sloping tea land in Ceylon. *Tea Quarterly* 35:307–317.

Hazarika, M., Chakravarty, S. K., and Mahanta, P. K. 1984. Studies on the thearubigin pigments in black tea manufacturing systems. *Journal of the Science of Food and Agriculture* 35:1208–1218.

Hazarika, M., Mahanta, P. K., and Takeo, T. 1984. Studies on some volatile flavor constituents in orthodox black teas of various clones and flushes in northeast India. *Journal of the Science of Food and Agriculture* 35:1201–1207.

Heller, R. 1953. Recherches sur la nutrition minerale des tissus vegetaux cultives *in vitro*. *Annales des Sciences Naturelles-Botanique et Biologie Vegetale* 14:1–223.

Hendry, G. A. F. 1993. Oxygen free radical process and seed longevity. *Seed Science Research* 3:141–153.

Hilton, P. J., and Ellis, R. T. 1972. Estimation on market value of Central Africa tea by theaflavin analysis. *Journal of the Science of Food and Agriculture* 23:227–232.

Horikoshi, K. 1972. Production of alkaline enzymes by alkalophilic micro organisms. Part III. Alkaline pectinase of "Bacillus" no. p-4-N. *Agricultural and Biological Chemistry* 36:285–293.

Hua, L. D., Dai, Z. D., and Hui, X. 1999. Studies on somatic embryo and adventitious bud differentiation rate among different tissues of *Camellia sinensis* L. *Acta Agronomica Sinica* 25:291–295.

Huang, F. P., Liang, Y. R., Lu, J. L., Chen, R. B., and Mamati, G. 2004. Evaluation of genetic diversity in oolong tea germplasms by AFLP fingerprinting. *Journal of Tea Science* 24:183–189.

Huang, J. A., Li, J. X., Huang, Y. H., Luo, J. W., Gong, Z. H., and Liu, Z. H. 2005. Construction of AFLP molecular markers linkage map in tea plant. *Journal of Tea Science* 25:7–15.

Iddagoda, N., Kataeva, N. N., and Butenko, R. G. 1988. *In vitro* clonal micropropagation of tea (*Camellia sinensis* L.) 1. Defining the optimum condition for culturing by means of a mathematical design technique. *Indian Journal of Plant Physiology* 31:1–10.

Jain, C., and Takeo, T. 1984. A review. Enzymes of tea and their role in tea making. *Journal of Food Biochemistry* 8:243–279.

Jain, S. M., Das, S. C., and Barman, T. S. 1991. Induction of roots from regenerated shoots of tea (*Camellia sinensis* L.). *Acta Horticulturae* 289:339–340.

Janaki Ammal, E. K. 1952. Chromosome relationship in cultivated species of *Camellia. Camellia Year Book* 106–114.

Jha, T. B., and Sen, S. K. 1992. Micropropagation of an elite Darjeeling tea clone. *Plant Cell Reports* 11:101–104.

Jha, T. B., Jha, S., and Sen, S. K. 1992. Somatic embryogenesis from immature cotyledon of an elite Darjeeling tea clone. *Plant Science* 84:209–213.

John, K. M., Joshi, S. D., Mandal, A. K. A., Ram Kumar, S., and Raj Kumar, R. 2009. *Agrobacterium rhizogenes*—Mediated hairy root production in tea leaves [(*Camellia sinensis* (L) O. Kuntze]. *Indian Journal of Biotechnology* 8:430–434.

Joseph Lopez, S., Marimuthu, S., Ramakrishnan, M., and Rajkumar, R. 2002. Reconstitution of tea polyphenol oxidase by limited proteolysis. In *Plantation crops research and development in the new millennium*, ed. P. Rethinam, H. H. Khan, V. M. Reddy, P. M. Mandal, and K. Suresh, Coconut Development 121–125.

Joseph Lopez, S., Raj Kumar, R., Pius, P. K., and Muraleedharan, N. 2004. *Agrobacterium tumefaciens*–mediated genetic transformation in tea (*Camellia sinensis* (L.) O. Kuntze). *Plant Molecular Biology Reporter* 22:201a–201j.

Joshi, S. D. 2008. Biotechnological studies on tea blister blight pathogen. PhD thesis, Bharathiar University, Coimbatore, India.

Karasawa, K. 1935. On the somatic chromosome number of triploid tea (Japanese). *Japanese Journal of Genetics* 11:101–104.

Kashyap, D. R., Chandra, S., Kaul, A., and Tewari, R. 2000. Production, purification and characterization of pectinase from a "Bacillus sp." DT 7. *World Journal of Microbiology and Biotechnology* 16:277–282.

Katiyar, S. K., and Mukhtar, H. 1997. Inhibition of phorbol ester tumor promoter 12-otetradecanoylphorbol-13-acetate-caused inflammatory responses in SENCAR mouse skin by black tea polyphenols. *Carcinogenesis* 18:1911–1916.

Kato, M. 1985. Regeneration of plantlets from tea stem callus. *Journal of Breeding* 35:317–322.

———. 1986. Micropropagation through cotyledon culture in *Camellia japonica* L. and *Camellia sinensis* L. *Journal of Breeding* 36:31–38.

———. 1989. *Camellia sinensis* L. (tea): *In vitro* regeneration. In *Biotechnology in agriculture and forestry, Vol. 7, Medicinal and aromatic plants II*, ed. Y. S. P. Bajaj, Springer–Verlag, Berlin, Heidelberg.

———. 1996. Somatic embryogenesis from immature leaves of *in vitro* grown tea shoots. *Plant Cell Reports* 15:920–923.

Kato, M., Mizuno, K., Crozier, A., Fujimura, T., and Ashihara, A. 2000. Caffeine synthase gene from tea leaves. *Nature* 406:956–957.

Katoh, Y., Katoh, M., Takeda, Y., and Omori, M. 2003. Genetic diversity within cultivated teas based on nucleotide sequence comparison of ribosomal RNA maturase in chloroplast DNA. *Euphytica* 134:287–295.

Kaundun, S. S., and Matsumoto, S. 2002. Heterologous nuclear and chloroplast microsatellite amplification and variation in tea, *Camellia sinensis. Genome* 45:1041–1048.

———. 2003a. Identification of processed Japanese green tea based on polymorphisms generated by STS-RFLP analysis. *Journal of Agricultural and Food Chemistry* 51:1765–1770.

———. 2003b. Development of CAPS markers based on three key genes of the phenylpropanoid pathway in tea, *Camellia sinensis*(L.) O. Kuntze, and differentiation between assamica and sinensis varieties. *Theoretical and Applied Genetics* 106:375–383.

Kaundun, S. S., Zhyvoloup, A., and Park, Y. G. 2000. Evaluation of the genetic diversity among elite tea (*Camellia sinensis* var. *sinensis*) accessions using RAPD markers. *Euphytica* 115:7–16.

Kawakami, M. 1997. Comparison of extraction techniques for characterizing tea aroma and analysis of tea by GC-FTIR-MS. In *Plant volatile analysis*, ed. H. F. Linkskens and I. F. Jackson, 211–229. Springer, Saladruck, Berlin.

Kingdon-Ward, F. 1950. Does wild tea exist? *Nature* 165:297–299.

Konwar, B. K., Das, S. C., Bordoloi, B. J., and Dutta, R. K. 1998. Hairy root development in tea through *Agrobacterium rhizogenes*–mediated genetic transformation. *Two and a Bud* 45:19–20.

Kuranuki, Y., and Shibata, M. 1992. Effect of concentration of plant growth regulators on the shoot apex culture of tea plant. *Bulletin Shizuoka Tea Experimental Station* 16:1–6.

———. 1993. Improvement of medium components for *in vitro* cuttings of tea plant. 2. Optimum concentration of plant growth regulators. *Journal of Tea Science* 77:39–45.

Lai, J. A., Yang, W. C., and Hsiao, J. Y. 2001. An assessment of genetic relationships in cultivated tea clones and native wild tea in Taiwan using RAPD and ISSR markers. *Botanical Bulletin of Academia Sinica* 42:93–100.

Lamb, J., and Ramaswamy, M. S. 1958. Fermentation of Ceylon tea XI. Relations between polyphenol oxidase activity and pectin methyl esterase activity. *Journal of the Science of Food and Agriculture* 9:51–56.

Lane, R. W., Yamakoshi, J., Kikuchi, M., Mizusawa, K., Henderson, L., and Smith, M. 1997. Safety evaluation of tannase enzyme preparation derived from *Aspergillus oryzae*. *Food and Chemical Toxicology* 35:207–212.

Lang, C., and Dornenburg, H. 2000. Perspectives in the biological function and the technological application of polygalacturonases. *Applied Microbiology and Biotechnology* 53:366–375.

Li, C. 2000. Population differences in water-use efficiency of *Eucalyptus microtheca* seedling under different watering regimes. *Physiologia Plantarum* 108:134–139.

Li, Y. H., Jiang, C. J., and Wan, X. C. 2004a. Study on the expression of caffeine synthase gene mRNA in tea plant. *Journal of Tea Science* 24:23–28.

Li, Y. H., Jiang, C. J., Yang, S. L., and Yu, Y. B. 2004b. β-Glucosidase cDNA cloning in the tea (*Camellia sinensis*) and its prokaryotic expression. *Journal of Agricultural Biotechnology* 12:625–629.

Liang, C., Zhou, Z-X., and Yang, Y-J. 2007. Genetic improvement and breeding of tea plant (*Camellia sinensis*) in China: From individual selection to hybridization and molecular breeding. *Euphytica* 154:239–248.

Liang, Y-R., Tanaka, J. C., and Takeda, Y. Y. 2000. Study on diversity of tea germplasm by RAPD method. *JI Zejiang Forestry College* 17 (2): 215–218.

Lin, G. Z., Lian, Y. L., Ryu, J. H., et al. 2007. Expression and purification of His-tagged flavonol synthase of *Camellia sinensis* from *Escherichia coli*. *Protein Expression and Purification* 55:287–292.

Lu, Y. P., Leu, Y. R., Xie, J. G., Yen, P., Huang, M. T., and Conney, A. H. 1997. Inhibitory effect of black tea on the growth of established skin tumors in mice: Effects on tumor size, apoptosis, mitosis and bromodeoxyuridine incorporation into DNA. *Carcinogenesis* 18:2163.

Luo, J. W., Shi, Z. P., Shen, C. W., Liu, C. L., Gong, Z. H., and Huang, Y. H. 2002. Studies on genetic relationships of tea cultivars (*Camellia sinensis* (L.) O. Kuntze) by RAPD analysis. *Journal of Tea Science* 22:140–146.

Luo, Y.-Y., and Liang, Y.-R. 2000. Studies on the construction of Bt gene expression vector and its transformation in tea plant. *Journal of Tea Science* 20 (2): 141–147.

Mahanta, P. K., 1993. Quality control during black tea manufacturing. Paper presented at International Symposium on Tea and Human Health, TRA, Calcutta, pp. 44–55.

Mamedov, M. A. 1961. Tea selection in Azarbajdzan. *Agrobiologia* 1:62–67.

Marimuthu, S., Manivel, L., and Abdul Kareem, A. 1997. Hydrolytic enzymes on the quality of made tea. *Journal of Plantation Crops* 25 (1): 88–92.

Marimuthu, S., Senthil Kumar, R. S., Balasubramanian, S., Raikumar, R., and Aneetha Christie, S. 2000. Effect of addition of biopectinase on biochemical composition CTC black tea. In *Recent advances in plantation* ed. crops research, ed. N. Muralidharan and R. Raj kumar, 265–269, Allied Publishers Limited, Mumbai India.

Martinez, J. P., Lendent, J. F., Bajji, M., Kinet, J. M. and Lutls, S. 2003. Effect of water stress on growth Na$^+$ and K$^+$ accumulation and water use efficiency in relation to osmotic adjustment in two populations of *Artiplex haliums* L. *Plant Growth Regulation* 41: 63–73.

Matsumoto, S., and Fukui, M. 1998. *Agrobacterium tumefaciens* mediated gene transfer in tea plant (*Camellia sinensis*) cells. *Japan Agricultural Research Quarterly* 32:287–291.

Matsumoto, S., Takeuchi, A., Hayastsu, M., and Kondo, S. 1994. Molecular cloning of phenylalanine ammo-nia-lyase cDNA and classification of varieties and cultivars of tea plants (*Camellia sinensis*) using the tea PAL cDNA probes. *Theoretical and Applied Genetics* 89:671–675.

McDowell, I., and Taylor, S. 1993. Tea: Types, production and trade. In *Encyclopedia of food science, food technology and nutrition*, vol. 7, ed. R. Macrae, R. K. Robinson, and M. J. Sadler, 4521–4533. Academic Press, London.

Millin, D. J., Swaine, D., and Dix, P. L. 1969. Separation and classification of the brown pigments of aqueous infusions of black tea. *Journal of the Science of Food and Agriculture* 20:296–307.

Mishra, R. K., and Sen-Mandi, S. 2001. DNA fingerprinting and genetic relationship study of tea plants using amplified fragment length polymorphism (AFLP) technique. *Indian Journal of Plant Genetic Resources* 14 (2): 148–149.

———. 2004. Genetic diversity estimates for Darjeeling tea clones based on amplified fragment length poly-morphism markers. *Journal of Tea Science* 24:86–92.

Mizutani, M., Nakanishi, H., Ema, J., et al. 2002. Cloning of β-primeverosidase from tea leaves, a key enzyme in tea aroma formation. *Plant Physiology* 130:2164–2176.

Mohanan, M., and Sharma V. S. 1981. Morphology and systematics of some tea (*Camellia* spp.) cultivars. In *Proceedings of Fourth Symposium of Plantation Crops* 4:391–400.

Moharib, S. A., El-sayed, S. T., and Jwanny, E. W. 2000. Evaluation of enzymes produced from yeast. *Nahrung* 44:47–51.

Mondal, T. K. 2000. Studies on RAPD marker for detection of genetic diversity, *in vitro* regeneration and *Agrobacterium*-mediated genetic transformation of tea (*Camellia sinensis*). PhD thesis, Utkal University, India.

———. 2002. Assessment of genetic diversity of tea (*Camellia sinensis* (L.) O. Kuntze) by intersimple sequence repeat polymerase chain reaction. *Euphytica* 128:307–315.

———. 2007. Tea. In *Biotechnology in agriculture and forestry, Vol. 60 Transgenic Crops V*, ed. E. C. Pua and M. R. Davey. Springer–Verlag, Berlin, Heidelberg.

Mondal, T. K., Bhattachary, A., and Ahuja, P. S. 2001a. Development of a selection system for *Agrobacterium*-mediated genetic transformation of tea (*Camellia sinensis*). *Journal of Plantation Crops* 29:45–48.

———. 2001b. Induction of synchronous secondary embryogenesis of tea (*Camellia sinensis*). *Journal of Plant Physiology* 158:945–951.

Mondal, T. K., Bhattacharya, A., Ahuja, P. S., and Chand, P. K. 2001. Factors affecting *Agrobacterium tumefaciens* mediated transformation of tea (*Camellia sinensis* (L). O. Kuntze). *Plant Cell Reports* 20:712–720.

Mondal, T. K., Bhattacharya, A., Laxmikumaran, M., and Ahuja, P. S. 2004. Recent advances of tea (*Camellia sinensis*) biotechnology. *Plant Cell Tissue and Organ Culture* 76:195–254.

Mondal, T. K., Bhattacharya, A., Sood, A., and Ahuja, P. S. 1998. Micropropagation of tea using thidiazuran. *Plant Growth Regulation* 26:57–61.

———. 1999. An efficient protocol for somatic embryogenesis and its use in developing transgenic tea (*Camellia sinensis* (L) O. Kuntze) for field transfer. In *Plant biotechnology and in vitro biology in 21st century*, ed. A. Altman, M. Ziv, and S. Izhar, 101–104. Kluwer Academic Publishers, Dordrecht, the Netherlands.

———. 2000. Factors affecting induction and storage of encapsulated tea (*Camellia sinensis* L. O. Kuntze) somatic embryos. *Tea* 21:92–100.

———. 2002. Factors affecting germination and conversion frequency of somatic embryos of tea. *Journal of Plant Physiology* 159:1317–1321.

Mondal, T. K., and Chand, P. K. 2002. Detection of genetic instability among the miocropropagated tea (*Camellia sinensis*) plants. *In Vitro Cellular and Developmental Biology Plant* 37:1–5.

Mori, M., Morita, N., and Kegaya, K. 1988. Polysaccharides from tea for manufacture of hypoglycemics, antidiabetics and health foods. Japan Patent 63308001.

Mukherjee, S. P., and Choudhuri, M. A. 1981. Effect of water stress on some oxidative enzymes and senescence in *Vigna* seedlings. *Physiologia Plantarum* 52:37–42.

Mulder, G. J. 1838. Chemische untersuchung des chinesischen and javanis chen Thees. *Annalen Physik Chemic Leipzig* XIII (161): 161–180.

Muraleedharan, N., Hudson, J. B., and Durairaj, J. 2007. Clonal propagation. In *Guidelines on tea culture in South India*, 20–21. United Planters' Association of Southern India, Coonoor, Tamilnadu, India.

Murali, K. S., Pandidurai, V., Manivel, L., and Raj Kumar, R. 1996. Clonal variation of tea through tissue culture. *Journal of Plantation Crops* 24:517–52.

Murashige, T., and Skoog, F. 1962. A revised medium for rapid growth and bioassays with tobacco tissue cultures. *Physiologia Plantarum* 15:473–493.

Murugesan, G. S., Angayarkanni, J., and Swaminathan, K. 2002. Effect of tea fungal enzymes on the quality of black tea. *Food Chemistry* 79:411–417.

Nakagawa, M. 1970. Constituents in tea leaf and their contribution to the taste of green tea liquors. *Japan Agricultural Research Quarterly* 5:43–47.

———. 1975. Chemical components and tastes of green tea. *Japan Agricultural Research Quarterly* 9:156–160.

Nakamura, Y. 1983. Isolation of protoplasts from tea plant. *Tea Research Journal* 58:36–37.

———. 1987a. Shoot tip culture of tea cultivar Yabukita. *Tea Research Journal* 65:1–7.

———. 1987b. *In vitro* rapid plantlet culture from axillary buds of tea plant (*C. sinensis* (L.) O. Kuntze). *Bulletin Shizuoka Tea Experimental Station* 13:23–27.

———. 1988a. Effects of the kinds of auxins on callus induction and root differentiation from stem segment culture of *Camellia sinensis* (L.) O. Kuntze. *Tea Research Journal* 68:1–7.

———. 1988b. Efficient differentiation of adventitious embryos from cotyledon culture of *Camellia sinensis* and other *Camellia* species. *Tea Research Journal* 67:1–12.

———. 1989. Differentiation of adventitious buds and its varietal difference in stem segment culture of *Camellia sinensis* (L.) O. Kuntze. *Tea Research Journal* 70:41–49

Nakayama, A., and Harada, S. 1962. Studies on the effect temperature upon the growth the tea plant. IV. *Bulletin of the Tea Research Station, Japan* 1:28–39.

Neil, S. J., and Burnett, E. C. 1999. Regulation of gene expression during water deficit stress. *Plant Growth Regulation* 29:23–28.

Nitsch, J. P., and Nitsch, C. 1969. Haploid plants from pollen grains. *Science* 163:185.

Nonaka, G., Kawahara, O., and Nishioka, I. 1983. Tannins and related compounds. XV. A new class of dimeric flavan-3-ol-gallates, theasinensis A and B, and proanthocyanidin gallate from green tea leaf. (I). *Chemical and Pharmaceutical Bulletin* 31:3906.

Okano, N., and Fuchinone, Y. 1970. Production of haploid plants by anther culture of tea *in vitro*. *Japanese Journal of Breeding* 20:63–64.

Ooshima, T., Minami, T., Aono, W., et al. 1994 Reduction of dental plaque deposition in humans by oolong tea extract. *Caries Research* 28:146.

Osone, K. 1958. Studies on the breeding of triploid plants by diploidizing gametic cells. *Japanese Journal of Breeding* 8:171–177.

Owuor, P. O., Tushida, T., Hortia, H., and Murai, T. 1988. Effects of geographical area of production on the composition of volatile flavor compounds in Kenyan clonal black CTC teas. *Experimental Agriculture* 24 (2):227–235.

Owuor, P. O., and Obanda, A. M. 1991. Effect of altitude of manufacture on black tea quality. Tea Research Foundation of Kenya Annual Report 1991, pp 135–137.

Pandey, A., Palni, L. M. S., and Bag, N. 2000. Biological hardening of tissue culture raised tea plants through rhizosphere bacteria. *Biotechnology Letter* 22:1087–1091.

Pandidurai, V., Murali, K. S., Manivel, L., and Raj Kumar, R. 1997. Factors influencing *in vitro* morphogenesis of tetraploid tea cultivar. *Indian Journal of Plant Physiology* 2:158–159.

Paratasilpin, T. 1990. Comparative studies on somatic embryogenesis in *Camellia sinensis* var. *sinensis* and *C. sinensis* var. *assamica* (Mast.) Pierre. *Journal of the Science Society of Thailand* 16:23–41.

Park, J. S., Kim, J. B., Hahn, B. S., Kim, K. Y., Ha, S. H., Kim, J. B., and Kim, Y. H. 2004. EST analysis of genes involved in secondary metabolism in *Camellia sinensis* (tea), using suppression substractive hybridization. *Plant Science* 166:953–961.

Paul, S., Wachira, F. N., Powell, W., and Waugh, R. 1997. Diversity and genetic differentiation among population of Indian and Kenyan tea (*Camellia sinensis* (L.) O. Kuntze) revealed by AFLP markers. *Theoretical and Applied Genetics* 94:255–263.

Phukan, M. K., and Mitra, G. C. 1984. Regeneration of tea shoots from nodal explants in tissue culture. *Current Science* 53:874–876.

———. 1990. Nutrient requirement for growth and multiplication of tea plants *in vitro*. *Bangladesh Journal of Botany* 19:65–71.

Plata, E., and Vieitez, A. M. 1990. *In vitro* regeneration of *Camellia reticulata* by somatic embryogenesis. *Journal of Horticultural Science* 65 (6): 707–714.

Ponsamuel, J., Samson, N. P., Ganeshan, P. S., Satyaprakash, V., and Abrahan, G. C. 1996. Somatic embryogenesis and plant regeneration from the immature cotyledonary tissues of cultivated tea (*Camellia sinensis* (L.) O. Kuntze). *Plant Cell Report* 16:210–214.

Portsmouth, G. B. 1957. Factors affecting shoot production in tea (*Camellia sinensis*) when grown as a plantation crops. IV. Interclonal variation in the effects of apical dominance. *Tea Quarterly* 36:183–186.

Prabu, G. R., Thirugnanasambantham, K., Mandal, A. K. A., and Saravanan, A. 2011. Molecular cloning and characterization of nucleoside diphosphate kinase 1 (NDPK1) cDNA in tea [*Camellia sinensis* (L.) O. Kuntze]. *Biologia Plantarum*. [in press].

Prakash, O., Sood, A., Sharma, M., and Ahuja, P. S. 1999. Grafting micropropagated tea (*Camellia sinensis* (L.) O. Kuntze) shoots on tea seedling—A new approach to tea propagation. *Plant Cell Reports* 18:137–142.

Prince, L. M., and Parks, C. R. 1997. Evolutionary relationships in the tea subfamily Theoideae based on DNA sequence data. *International Camellia Journal* 29:135–144.

Purakayastha, A., and Das, S. C. 1994. Isolation of tea protoplast and the culture. *Proceedings of 32nd Tocklai Conference*, Tea Research Association, Tocklai Experimental Station, Jorhat, India.

Purseglove, J. W. 1968. *Tropical crops. Dicotyledons* 2, 599–611. Taylor & Francis, London.

Raina, S. K., and Iyer, R. D. 1992. Multicell pollen proembryoid and callus formation in tea. *Journal of Plantation Crops* 9:100–104.

Raj Kumar, R., Balasarvanan, T., and Pius, P. K. 2002. Morphological variation and diversity among tea (*Camellia sinensis* (L). O. Kuntze) somaclones. *Proceedings of Placrosym* XV:24–31.

Raj Kumar, R., Balasusaravanam, S., Jayakumar, D., Haridas, V., and Marimuthu, S. 2001. Physiological and biochemical feathers of field grown somaclonal variants of tea. *United Planters' Association of Southern India, Tea Research Foundation Bulletin* 54:73–81.

Ramakrishnan, M., Rajanna, L., Narayanaswamy, P., and Simon, L. 2009. Assessment of genetic relationship and hybrid evaluation studies in tea (*Camellia* spp.) by RAPD. *International Journal of Plant Breeding* 3:144–148.

Ramamoorthy, G., and Venkateswaran, G. 2003. Effect of temperature on tea quality in CTC tea manufacture. *Planter's Chronicle* 99 (2): 57–60.

Ramarethinam, S., and Latha, K., 2000. Lipid acyl hydrolases during tea processing. *Planter's Chronicle* 96: 389–392.

Ranganath, S. R., and Marimuthu, S. 1992. Clonal variation of some biochemical parameters in CTC teas. *Journal of Plantation Crops B* 20 (2): 85–93.

Ranganath, S. R., and Raju, K. 1991. Biochemical evaluation of Nilgiri teas. *United Planters' Association of Southern India, Scientific Department Bulletin* 44:88–96.

Ranganath, S. R., Raju, K., and Marimuthu, S. 1993. Fatty acid composition in the crop shoots of *Camellia* spp. L. *Journal of Plantation Crops (Supplement)* 21:333–335.

Rashid, A., Choudhury, M., and Badrul Alam, A. F. M. 1985. Studies on the progenies of a cross between diploid and tetraploids tea. *Sri Lankan Journal of Tea Science* 54:54–61.

Ravichandran, R., and Parthiban, R. 1998a. Changes in enzyme activities (polyphenol oxidase and phenylalanine ammonia lyase) with type of tea leaf and during black tea manufacture and the effect of enzyme supplementation of dhool on black tea quality. *Food Chemistry* 62 (3): 277–281.

———. 1998b. Occurrence and distribution of lipoxygenase in *Camellia sinensis* (L) O. Kuntze and their changes during CTC black tea manufacture under southern Indian conditions. *Journal of the Science of Food and Agriculture* 78:67–72.

Richards, A. V. 1965. Origin of popular TRI clones. *Tea Quarterly* 36:183–188.

Roberts, E. A. H. 1950. The phenolic substances of manufactured tea. II. Their origin as enzymatic oxide products in fermentation. *Journal of the Science of Food and Agriculture* 9:212–216.

———. 1952. The chemistry of tea fermentation. *Journal of the Science of Food and Agriculture* 3:193–198.

———. 1962. Economic importance of food substances: Tea fermentation. In *The chemistry of flavonoid compounds*, ed. T. A. Geissmann, 468–512. Pergamon, Oxford.

Roberts, E. A. H., Cartright, R. A., and Oldschool, M. 1957. Fractionation and paper chromatography of water soluble substances from manufactured tea. *Journal of the Science of Food and Agriculture* 8:72–80.

Roberts, E. A. H., and Myers, M. 1958. Theogallins a polyphenol occurring in tea. II. Identification as a galloyl quinic acid. *Journal of the Science of Food and Agriculture* 9:701–705.

Roberts, E. A. H., and Smith, R. F. 1963. The phenolics of manufactured teas—Spectrophotometric evaluation of tea liquors. *Journal of the Science of Food and Agriculture* 14:687–700.

Roberts, E. A. H., Wight, W., and Wood, D. J. 1958. Paper chromatography as an aid to the identification of Thea camellias. *New Physiologist* 57:211–225.

Roberts, G. R., and Fernando, R. S. S. 1981. Some observations on the correlation of polyphenol content to the quality of tea clones. *Tea Quarterly* 50 (1): 30–34.

Robertson, A. 1983. Effect of physical and chemical conditions on the *in vitro* oxidation of tea leaf catechins. *Phytochemistry* 22:889–896.

———. 1992. The chemistry and biochemistry of black tea production—The nonvolatiles. In *Tea: Cultivation to consumption*, ed. K. C. Willson and M. N. Clifford, 555–601. Chapman and Hall, London.

Rombouts, F. M., and Pilnik, W. 1986. Pectinases and other cell wall degrading enzymes of industrial importance. *Symbiosis* 2:79–89.

Sadzuka, T., Sugiyama, T., Miyagishima, A., Mazawa, Y., and Hirota, S. 1996. The effects of theanine, as a novel biochemical modulator, on the anti tumor activity of adriamycin. *Cancer Letters* 105:203–209.

Saha, S. K., and Bhattacharya, N. M. 1992. Stimulating effect of elevated temperatures on callus production of meristemoids from pollen culture of tea (*Camellia sinensis* (L.) O. Kuntze). *Indian Journal of Experimental Biology* 30:83–86.

Saijo, R. 1982. Isolation and chemical structures of two new catechins from fresh tea leaves. *Agricultural and Biological Chemistry* 46:1969–1970.

Saijo, R., and Takeo, T. 1973. Volatile and nonvolatile forms of aroma compounds in tea leaves and their changes due to injury. *Agricultural and Biological Chemistry* 37:1367–1373.

———. 1974. Induction of peroxidase activity by ethylene and indole-3 acitic acid in tea shoots. *Agricultural and Biological Chemistry* 38:2283–2284.

Sandal, I., Bhattacharya, A., and Ahuja, P. S. 2001. An efficient liquid culture system for tea shoot proliferation. *Plant Cell Tissue and Organ Culture* 65:75–80.

Sanderson, G. W. 1966. 5-Dehydroshikimate reductase in the tea plant (*Camellia sinensis* L.). *Biochemistry Journal* 98:248–252.

Sanderson, G. W., and Coggon, P. 1977. Use of enzymes in the manufacture of black tea and instant tea. In *Enzymes in food and beverage processing*, ed. R. L. Ory and A. J. St. Angelo, 12–26. ACS Symposium Series.

Sanderson, G. W., and Graham, H. N. 1973. On the formation of black tea aroma. *Journal of Agricultural and Food Chemistry* 21:576–585.

Sanderson, G. W., and Perera, B. P. H. 1965. Carbohydrates in tea plants. I. The carbohydrate on tea shoot tips. *Tea Quarterly* 36:6–13.

Sanderson, G. W., and Roberts, G. R. 1964. Peptidase activity in shoot tips of the tea plant (*Camellia sinensis* L.). *Biochemistry Journal* 93:419–423.

Sarath Babu, V. R., Patra, S., Thakur, M. S., Karanth, N. G., and Varadaraj, M. C, 2005. Degradation of caffeine by *Pseudomonas alcaligens* CFR 1708. *Enzyme and Microbial Technology* 37:617–624.

Sarathchandra, T. M., Upali, P. D., and Wijeweardena, R. G. A. 1988. Studies on the tissue culture of tea (*Camellia sinensis* (L.) O. Kuntze) 4. Somatic embryogenesis in stem and leaf callus cultures. *Sri Lankan Journal of Tea Science* 52:50–54.

Saravanan, M., Mariya John, K. M., Raj Kumar, R., Pius, P. K., and Sasikumar, R. 2005. Genetic diversity of UPASI tea clones (*Camellia sinensis* (L) O Kuntze) on the basis of total catechins and their fractions. *Phytochemistry* 66 (5): 561–565.

Sarwar, M. 1985. Callus formation from explanted organs of tea (*Camellia sinensis* L.). *Journal of Tea Science* 54:18–22.

Sasaoka, K., and Kito, M. 1964. Synthesis of theanine by tea seedling homogenate. *Agricultural and Biological Chemistry* 28:313–317.

Satyanarayana, N., and Sharma, V. S. 1986a. Tea (*Camellia* L. spp.) germplasm in South India. In *Plantation crops*, vol. II, ed. H. C. Srivastava, 173–179. Oxford and IBH Publishing Co. Pvt. Ltd., New Delhi, India.

———. 1986b. Triploid breeding in tea. *Two and a Bud* 31:55–59.

———. 1991. Tea plant improvement in South India. *United Planters' Association of Southern India Scientific Department Bulletin* 44:69–70.

———. 1993. An overview of tea plant improvement in South India. In *Tea culture, processing and marketing*, ed. M. J. Mulky and V. S. Sharma, 36–44. Oxford IBH Publishing, New Delhi, India.

Schreier, P. 1988. Analysis of black tea volatiles. In *Analysis of nonalcoholic beverages*, ed. H. F. Linskens and I. F. Jackson, 296–320. Springer, Saladruck, Berlin.

Scott, N. R., Chakraborty, J., and Marks, V. 1989. Caffeine consumption in the United Kingdom: A retrospective survey. *Food Science and Nutrition* 42:183–191.

Sealy, J. 1958. *A revision of the genus* Camellia. Royal Horticultural Society: Taylor & Francis, London.

Sebastian Pillai, A. R. 1963. Report on plant breeding. *Annual Report. Tea Research Institute of Ceylon* 87–89.

Selvendran, R. R., and Perera, B. P. M. 1971. The chemical composition of the cell wall material of tea waste. *Tea Quarterly* 42:16–18.

———. 1972. Chemical composition of tea leaf cell-wall. *Tea Quarterly* 43 (1–2): 18–20.

Senthil Kumar, R. S., Marinuthu, S., Rajkumar, R., Joseph Lopez, S., and Ramakrishnan, M. 2001. Role of certain enzymes on the quality of CTC black tea. In *Bulletin of UPASI Tea Research Foundation* 54:109–118.

Senthil Kumar, R. S., Swaminathhan, K., Marimuthu, S., and Rajkumar, R. 2000. Microbial enzymes for tea processing. In *Recent Advances in Plantation Crops Research*, ed. N. Muraleedharan, and R. Raj Kumar, 273–276, Allied Publishers Limited, Mumbai, India.

Shao, W. F., Pang, R. H., Wang, P. S., Xu, M., Duan, H. X., Zhang, Y. P., and Li, J. H. 2003. RAPD analysis of tea relationship in Yunnan. *Scientia Agricultura Sinica* 36:1582–1587.

Shao, W. F., Sharma, P., and Kumar, S. 2005. Differential display-mediated identification of three drought-reponsive expressed sequence tags in tea (*Camellia sinensis* (L) O. Kuntze). *Journal of Bioscience* 30:231–235.

Sharma, R. K., Negi, M. S., and Sharma, S. 2010. AFLP-based genetic diversity assessment of commercially important tea germplasm in India. Biochemical Genetics DOI 10.1007/s10528-010-9338-z.

Sharma, V. S. 1976. Tea nursery. *Planters' Chronicle* 71:211–213.

Sharma, V. S., and Ranganathan, V. 1985. The world of tea today. *Outlook on Agriculture* 14:35–40.

———. 1986. Present status and future needs of tea research. In *Plantation crops*, vol. II, ed. H. C. Srivastava, 37–50. Oxford and IBH Publishing Co. Pvt. Ltd., New Delhi, India.

Shi, X., and Dalal, N. S. 1991. Antioxidant behavior of caffeine: Efficient scavenging of hydroxyl radicals. *Food and Chemical Toxicology* 29:1–6.

Shimokado, T. T., Murata, T., and Miyaji, Y. 1986. Formation of embryoid by anther culture of tea. *Japanese Journal of Breeding* 36:282–283.

Simura, T. 1956. Breeding of polyploid varieties of tea with special reference to their cold tolerance. *Proceedings of International Genetic Symposium* Tokyo, Japan, pp. 321–324.

Singh, I. D. 1979. Indian tea germplasm and its contribution to the world's tea industry. *Two and a Bud* 26:23–26.

———. 1980. Nonconventional approaches to the breeding of tea in northeast India. *Two and a Bud* 27:3–6.

———. 1999. Plant improvement. In *Global advances in tea science*, ed. I. N. K. Jain, 427–448. Aravali Books International (P) Ltd., New Delhi, India.

Singh, I. D., and Sharma, P. C. 1982. Studies in radiation breeding in tea plants. *Proceedings of 4th Annual Symposium on Plantation Crops*, pp. 295–301.

Singh, M., Bandana, and Ahuja, P. S. 1999. Isolation and PCR amplification of genomic DNA from market samples of dry tea. *Plant Molecular Biology Reporter* 17:171–178.

Singh, M., Saroop, J., and Dhiman, B. 2004. Detection of intraclonal genetic variability in vegetatively propagated tea using RAPD markers. *Biologia Plantarum* 48:113–115.

Sood, A., Palni, L. M. S., Sharma, M., Rao, D. V., Chand, G., and Jain, N. K. 1993. Micropropagation of tea using cotyledon culture and encapsulated somatic embryos. *Journal of Plantation Crops* 21: 295–300.

Sreerangachar, H. B. 1943a. Studies on the fermentation of Ceylon tea. 6. The nature of the tea oxidase system. *Biochemistry Journal* 37:661–667.

———. 1943b. Studies on the fermentation of Ceylon tea. 7. The prosthetic group of tea oxidase. *Biochemistry Journal* 37:667–674.

Srivalli, B., Sharma, G., and Chopra, R. K. 2003. Antioxidative defense system in an upland rice cultivar subjected to increasing intensity of water stress followed by recovery. *Physiologia Plantarum* 199:503–512.

Suprunova, T., Krugman, T., Fahima, T., Cheng, G., Shams, I., Korol, A., and Nevo, E. 2004. Differential expression of dehydrin genes in wild barley, *Hordeum spontaneum*, associated with resistance to water deficit. *Plant Cell Environment* 27: 1297–1308.

Suykerbuyk, M. E. G., Schaap, P. J., Stam, H., Musters, W., and Visser, J. 1995. Cloning, sequence and expression of the gene coding for rhamnogalacturonase of *Aspergillus aculeatus*, a novel pectinolytic enzyme. *Applied Microbiology and Biotechnology* 43:861–870.

Swain, T., and Hillis, W. E. 1959. The phenolic constituents of *Prunus domestica*. 1. The quantitative analysis of phenolic constituents. *Journal of Science of Food and Agriculture* 10:63–68.

Tahardi, J. S., and Shu, W. 1992. Commercialization of clonal micropropagation of superior tea genotypes using tissue culture technology. USAID/CDR network meeting on tea crop biotechnology, Costa Rica.

Takei, Y., Ishiwata, K., and Yamanishi, T. 1976. Aroma components characteristic of spring green tea. *Agricultural and Biological Chemistry* 40:2151–2157.

Takeo, T. 1966. Tea leaf polyphenol oxidase part III. Studies of the changes of polyphenol oxidase during black tea manufacture. *Agricultural and Biological Chemistry* 30:529–535.

———. 1974. l-Alanine as a precursor of ethylamine in *Camellia sinensis*. *Phytochemistry* 13:1401–1406.

———. 1984. Effect of withering process on volatile compound formation during black tea manufacture. *Journal of the Science of Food and Agriculture* 35:84–87.

———. 1996. The relation between clonal characteristic and tea aroma. *Foods and Food Ingredients Journal of Japan* 168:35–45.

Takeo, T., and Baker, J. E. 1973. Changes in multiple forms of polyphenol oxidase during maturation of tea leaves. *Phytochemistry* 12:21–24.

Takeo, T., and Kato, Y. 1971. Tea leaf peroxidase isolation and partial purification. *Plant Cell Physiology* 12:217–223.

Takeo, T., and Uritani, 1966. Tea leaf polyphenol oxidase. II. Purification and properties of solubilized polyphenol oxidase in tea leaves. *Agricultural and Biological Chemistry* 30:155–163.

Takeuchi, A., Matsumoto, S., and Hayatsu, M. 1994. Chalcone synthase from *Camellia sinensis*, isolation of the cDNAs and the organ specific and sugar responsive expression of the genes. *Plant Cell Physiology* 35:1011–1018.

Takino, Y., Imagawa, H., Hikikawa, H., and Tanaka, A. 1964. Studies on the mechanism of the oxidation of tea leaf catechins. Part III. Formation of a reddish orange pigment and its spectral relationship to some benzotropolone derivatives. *Agricultural and Biological Chemistry* 28:64–71.

Tanaka, J. 1996. RAPD linkage map of tea plant and the possibility of application in tea genetics and breeding. *Tea Research Journal* 84S:44–45.

Tanaka, J., Yamaguchi, N., and Nakamura, Y. 2001. Pollen parent of tea cultivar 'Sayamakaori' with insect and cold resistance may not exist. *Breeding Research* 3:43–48.

Tanaka, J. I., Sawai, Y., and Yamaguchi, S. 1995. Genetic analysis of RAPD markers in tea. *Journal Japanese Breeding* 45 (2): 198–199.

Thakor, B. H. 1997. A reexamination of the phylogenetic relationships within the genus *Camellia*. *International Camellia Journal* 29:130–134.

Thambussi, E. A., Bartoli, C. G., Beltrano, J., Guiamet, J. J., and Arans, J. L. 2000. Oxidative damage to thylakoid proteins in water stresses leaves of water (*Triticum aestivum*). *Physiologia Plan* 108: 398–404.

Thanaraj, S. N. S, and Seshadri, R. 1990. Influence of polyphenol oxidase activity and polyphenol content of tea shoot on quality of black tea. *Journal of the Science of Food and Agriculture* 51:57–69.

Thomas, J., Ajay, D., Raj Kumar, R., and Mandal, A. K. A. 2010. Influence of beneficial microorganisms during *in vivo* acclimatization of *in vitro* derived tea (*Camellia sinensis*) plants. *Plant Cell Tissue and Organ Culture* 101:365–370.

Thomas, J., Raj Kumar, R., and Mandal, A. K. A. 2006. Metabolic profiling and characterization of somaclonal variants in tea (*Camellia* spp.) for identifying productive and quality accession. *Phytochemistry* 67:1136–1142.

Thomas, J., Vijayan, D., Joshi, S. D., Joseph Lopez, S., and Raj Kumar, R. 2006. Genetic integrity of somaclonal variants in tea (*Camellia sinensis* (L). O. Kuntze) as revealed by inter simple sequence repeats. *Journal of Biotechnology* 123:149–154.

Tirimanna, A. S. L. 1967. Acid phosphatase of the tea leaf. *Tea Quarterly* 38:331–334.

Tosca, A., Pondofi, R., and Vasconi, S. 1996. Organogenesis in *Camellia × williamsii*: Cytokinin requirement and susceptibility to antibiotics. *Plant Cell Reports* 15:541–544.

Tsujimura, M. 1929. On tea catechin isolated from green tea. Scientific Papers of the Institute of Physical and Chemical Research. 10: 252.

Tsushida, T., and Takeo, T. 1976. Partial purification and properties of tea leaf ribonuclease. *Agricultural and Biological Chemistry* 40:1279–1285.

Tubbs, F. R. 1932. A note on vegetative propagation of tea. *Tea Quarterly* 5:154–156.

Tunstall, A. C. 1931. A note on the propagation of tea by green shoot cuttings. *Quarterly Journal of Indian Tea Association* Part I 49–51.

Ukers, W. H. 1935. *Tea in other countries. All about tea*, vol. I. The Tea and Coffee Trade Journal, New York, pp. 449–463.

Ullah, M. R., Gogoi, N., and Baruah, D. 1984. The effect of withering on fermentation of tea leaf and development of liquor characters of black teas. *Journal of the Science of Food and Agriculture* 35:1142–1147.

Upadhyaya, H., and Panda, S. K. 2004. Responses of *Camellia sinensis* to drought and rehydration. *Biologia Plantarum* 48 (4): 115–120.

Venkataramani, K. S. 1963. The principles of tea clonal selection and propagation and some practical considerations in clonal planting. *United Planters' Association of South India, Tea Scientific Department, Bulletin* 22: 2–14.

Vieitez, A. M., and Barciela, J. 1990. Somatic embryogenesis and plant regeneration from embryonic tissues of *Camellia japonica* L. *Plant Cell Tissue and Organ Culture* 21:267–274.

Vieitez, A. M., Vieitez, M. L., Ballester, A., and Vieitez, E. 1992 Micropropagation of *Camellia* spp. In *Biotechnology in agriculture and forestry, Vol. 19. High tech and micropropagation III*, ed. Y. P. S. Bajaj, 361–387. Springer–Verlag, Berlin, Heidelberg.

Vinson, J. A., Dabbagh, Y., Serry, M., and Jang, J. 1995. Plant flavonoids, especially tea flavonols, *are* powerful antioxidants using an *in vitro* oxidation model for heart disease. *Journal of Agricultural Food Chemistry* 43:2800–2802.

Visser, T. 1961. Some aspects of the propagation of tea cuttings. *Proceedings of 15th International Horticultural Congress*, Nice, 1958, International Society of Horticultural Science, III:185.

———. 1969. Tea *Camellia sinensis* (L.) O. Kuntze. In *Outline of perennial crop breeding in the tropics*, ed. F. P. Frewards and F. Wit, 459–493. H. Veenman and Zonen, N.V., Wageningen, the Netherlands.

Visser, T., and Kehi, F. H. 1958. Selection and vegetative propagation of tea. *Tea Quarterly* 29:76–86.

Wachira, F., and Ogado, J. 1995. *In vitro* regeneration of *Camellia sinensis* (L.) O. Kuntze by somatic embryo. *Plant Cell Reports* 14:463–466.

Wachira, F., Tanaka, J., and Takeda, Y. 2001. Genetic variation and differentiation in tea (*Camellia sinensis*) germplasm revealed by RAPD and AFLP variation. *Journal of Horticultural Science and Biotechnology* 76:557–563.

Wachira, F. N. 1991. Newly identified Kenyan polyploid tea strains. *Tea* 12:103.

Wachira, F. N., Powell, W., and Waugh, R. 1997. An assessment of genetic diversity among *Camellia sinensis* L. (cultivated tea) and its wild relatives based on randomly amplified polymorphic DNA and organelle specific STS. *Heredity* 78 (6): 603–611.

Wachira, F. N., Waugh, R., Hackett, C. A., and Powell, W. 1995. Detection of genetic diversity in tea (*Camellia sinensis*) using RAPD markers. *Genome* 38:201–210.

Wan, X. C., and Nursten, H. E. 1997. A new type of tea pigment—From the chemical oxidation of epicatechin gallate and isolated from tea. *Journal of the Science of Food and Agriculture* 74:401–408.

Wang, D. F., Xie, X. F., and Wang, S. L. 1996. Composition and the physical and chemical characteristics of tea polysaccharide (Chinese). *Journal of Tea Science* 16:1–8.

Watt, G. 1907. Tea and tea plant. *Journal of Royal Horticulture Society* 32:64–96.

Weatherstone, J. 1992. Historical introduction. In *Tea: Cultivation to consumption*, ed. K. C. Wilson and M. N. Clifford, 1–23. Chapman and Hall, London.

Wei, H., Zhang, X., Zhao, J. F., et al. 1999. Scavenging of hydrogen peroxide and inhibition of ultraviolet light-induced oxidative DNA damage by aqueous extracts from green and black teas. *Free Radical Biology and Medicine* 26:1427–1435.

Wellensick, S. J. 1933. Rooting of cuttings. *Archives of Tea Culture* 7:1–47.

Wickremasinghe, R. L. 1978. Tea. In *Advances in food research*, vol. 24, ed. E. M. Mark and G. F Steward, 229–286. Academic Press, New York.

Wickremasinghe, R. L., and Perera, K. P. W. C. 1973. Factors affecting quality, strength and color of black tea liquors. *Journal of the National Science Council of Sri Lanka* 1:111–112.

Wight, W. 1955. Commercial propagation of *Camellia sinensis* in India. *American Camellia Year Book* 88–110.

———. 1956. Commercial selection and breeding of tea in India. *World Crops* 8:263–268.

———. 1958. The agrotype concept in tea taxonomy. *Nature* 181:893–895.

———. 1959. Nomenclature and classification of the tea plant. *Nature* 183:1726–1728.

————. 1961. Tea seed resources in northeast India. *Commercial Times (Calcutta)* 11:21–24.

————. 1962. Tea classification revised. *Current Science* 31:298–299.

Wight, W., and Barua, D. N. 1939. The tea plant industry: Some general principles. *Tropical Agriculture* (Ceylon) 93:4–13.

————. 1954. Morphological basis of quality in tea. *Nature* 173:630–631.

————. 1957. What is tea? *Nature* 179:506–507.

Wight, W., and Gilchrist, R. C. H. H. 1959. Concerning the quality and morphology of tea. *Annual Report, Tocklai Experimental Station* pp. 69–86.

Wight, W., Gilchrist, R. C. H. H., and Wight, J. 1963. Note on the color and quantity of tea leaf. *Empire Journal of Experimental Agriculture* 31:124–126.

Wright, L. P. 2005. Biochemical analysis for identification of quality in black tea (*Camellia sinensis*). PhD thesis, University of Pretoria, South Africa.

Wright, L. P., Apostolides, Z., and Louw, A. I. 1996. DNA fingerprinting of tea clones. *Proceedings of the 1st Regional Tea Research Seminar*, Blantyre, Malawi.

Wu, C. T. 1964. Studies on the heredity variation and morphology of pubescence on young tea shoots. *Journal of the Agricultural Association of China* 47 (1): 22.

Wu, D. C., and Wei, G. X. 2002. Tea as a food for oral health. *Nutrition* 18:443–444.

Xiao, W. X. 1987. Progress in the investigation of thearubigin in black tea (Chinese). *Foreign Agriculture (Tea)* 1:1–6.

Xu, Y., and Han, C. 1990. The effect of Chinese tea on the occurrence of esophageal tumors induced by *N*-nitrosomethylbenzylamine formed *in vivo*. *Biomedical and Environmental Sciences* 3:406–412.

Yamaguchi, K., and Shibamoto, T. 1981. Volatile constituents of green tea Gyokuro (*Camellia sinensis* L. var. Yabukita). *Journal of Agricultural and Food Chemistry* 29:366–370.

Yamaguchi, S., Kunitake, T., and Hisatomi, S. 1987. Interspecific hybrid between *Camellia japonica* cv. choclidori and *C. chrysantha* produced by embryo culture. *Japanese Journal of Breeding* 37:203–206.

Yamanishi, T. 1978. Flavor of green tea. *Japan Agricultural Research Quarterly* 12:05–210.

————. 1995. Flavor of tea. *Food Research International* 11:477–525.

————. 1996. Tea aroma. *Foods and Food Ingredients Journal of Japan* 168:23–34.

Yang, Y. H., and Lin, S. Q. 1992. The study on artificial mutation techniques of the tea plant. In *Tea Science Research Proceedings*, ed. Tea Research Institute CAAS, 45–54. Shanghai Scientific and Technical Publishers.

Yao, M. Z., Huang, H. T., Yu, J. Z., and Chen, L. 2005. Analysis on applicability of ISSR in molecular identification and relationship investigation of tea cultivars. *Japanese Tea Science* 25:153–157.

Yokogoshi, H. 1995. Reduction effect of theanine on blood pressure and brain 5-hydroxyindoles in spontaneous hypertensive rats. *Bioscience, Biotechnology and Biochemistry* 59:615–618.

Yu, S. G., Hildebrandt, L. A., and Elson, C. E. 1995. Geraniol, an inhibitor of mevalonate biosynthesis, suppresses the growth of hepatomas and melanomas transplanted to rats and mice. *Journal of Nutrition* 125:2763–2767.

Yu, Y. B., Jiang, C. J., Wang, C. X., Li, Y. Y., and Wan, X. C. 2004. Expression of tea caffeine synthase in *Escherichia coli*. *Journal of Nanjing Agricultural University* 27:105–109.

Zaprometov, M. N., and Bukhlaeva, V. 1971. The effectiveness of the use of various ^{14}C-precursors of the biosynthesis of flavanoids in the tea plants (Russian). *Biochemistry Journal* 36:272–276.

Zehra, M., Banerjee, S., Mathur, A. K., and Kukreja, A. K. 1996. Induction of hairy roots in tea (*Camellia sinensis* (L.) using *Agrobacterium rhizogenes*. *Current Science* 70:84–86.

Zhao, D., Liu, Z. S., and Xi, B. 2001. Cloning and alignment of polyphenol oxidase cDNA of tea plant. *Journal of Tea Science* 21:94–98.

Zhu, Z. Z 1992. Tea affairs in ancient times. In *China tea book*, ed. Z. M. Chen, 1–9. Shangai Culture Publishers, Shanghai.

Zhuang, C., Duan, J., and Zhou, J. 1988. Somatic embryogenesis and plantlets regeneration of *Camellia sasanqua*. *Acta Biologie Experimentalis Sinica* 10:241–244.

Zhuang, C., and Liang, H. 1985a. *In vitro* embryoid formation of *Camellia reticulata* L. *Acta Biologie Experimentalis Sinica* 18:275–281.

————. 1985b. Somatic embryogenesis and plantlet formation in cotyledon culture of *Camellia chrysantha*. *Acta Botanica Yunnan* 7:446–450.

Zietkiewicz, E., Rafalski, A., and Labuda, D. 1994. Genome fingerprinting by simple-sequence repeat (SSR) anchored polymerase chain reaction amplification. *Genomics* 20:176–183.

CHAPTER 19

Coffee (*Coffea* spp.)

Juan Carlos Herrera, Hernando A. Cortina, François Anthony,
Nayani Surya Prakash, Philippe Lashermes, Alvaro L. Gaitán,
Marco A. Cristancho, Ricardo Acuña, and Darci R. Lima

CONTENTS

19.1 INTRODUCTION

The history of the coffee tree and of coffee as a beverage is closely linked to the growth of great empires and trade—first under the influence of the Arabs, then the Turks in the fifteenth century, and, finally, the European colonizers since the eighteenth century.

Coffee is one of the world's most valuable export commodities, ranking second on the world market after petroleum products. The total retail sales value exceeded US$70 billion per year and about 125 million people depend on coffee for their livelihood in Latin America, Africa, and Asia (www.ico.org). Commercial production relies on two species: *Coffea arabica* L. (also called "arabica" coffee) and *Coffea canephora* Pierre ex A. Froehner (also called "robusta" coffee). The cup quality (low caffeine content and fine aroma) of *C. arabica* makes it by far the most important species, representing 70% of world production. Arabica production is constrained by numerous diseases and pests like coffee leaf rust (CLR; *Hemileia vastatrix* Berk & Br.), coffee berry disease (CBD; *Colletotrichum kahawae*), coffee berry borer (CBB; *Hypothenemus hampei*), stem borer (*Xylotrechus quadripes* Chevr.), and nematodes (*Meloidogyne* spp. and *Pratylenchus* spp.). In contrast, robusta is more tolerant to these diseases and pests. Hence, transfer of desirable genes, in particular for disease resistance, from coffee species into arabica cultivars without affecting quality traits has been the main objective of arabica breeding.

The scope of this chapter is to give a brief review on the taxonomy, origin, and distribution of coffee species, as well as to provide a current perspective about coffee germplasm resources, breeding strategies, and use of biotechnological methods, including recent significant advances in structural and functional genomics. Finally, the unsettled controversy between coffee consumption and human health will be discussed in the framework of the available scientific information.

19.2 TAXONOMY, ORIGIN, AND DISTRIBUTION OF COFFEE SPECIES

The genus *Coffea*, which belongs to the Ixoroideae subfamily, is economically the most important genus of the Rubiaceae family that includes around 500 genera and more than 6,000 species, most of them being trees and bushes. Their origin is mainly tropical and they have a wide distribution. Garden bushes like *Ixora* and medicinal plants such as ipecacuana (*Psichoria ipecacuanha*) or cinchona, from which quinine is obtained, belong to this family (Bridson 1988).

The first description of a coffee plant was made in 1592 by Prospero Alpini; a century later Antoine de Jussieu considered it to be a jasmine and called it *Jasminium arabicum laurifolia*. In 1737 Linnaeus classified the coffee plant into a new genus, the *Coffea*, with just one known species: *C. arabica* (Charrier and Berthaud 1985).

The *Coffea* genus, subgenus *Coffea*, contains 103 species; 59 are found in the wild in Madagascar, 41 in tropical Africa, and 3 in the Mascarenes (Mauritius and Reunion). The phenotype of coffee plants can vary from small perennial bushes to thick, hard, wooden trees, characterized by plagiotropic branches, paired inflorescences, hermaphrodite flowers, white or slightly pink corollas, and a long style that sticks out the corolla (Figure 19.1). The fruit is an indehiscent drupe with two seeds. Each seed exhibits a characteristic deep groove (invagination) in the ventral part (Davis et al. 2006).

Figure 19.1 Some differences in fruit and flower structurs in the genus *Coffea*. (a) *Coffea* arabica; (b) *C. pseudozanguebariae*; (c) *C. canephora;* and (d) a wild *Coffea* spp.

19.2.1 Taxonomic Classification

At the end of the nineteenth century, numerous coffee species were discovered in the tropical forests of Africa. One of the first classifications of the *Coffea* genus was made by Chevalier in 1947 based on morphological characteristics (i.e., leaves texture, plant size, fruit color) and on geographical distribution. This author divided the genus into four sections: *Argocoffea, Paracoffea, Mascarocoffea,* and *Eucoffea*. The *Eucoffea* section, which includes the true coffee trees, was in turn divided by Chevalier into five subsections: *Pachycoffea, Nanocoffea, Melanocoffea, Mozambicoffea,* and *Erythrocoffea*. This classification has been reconsidered by diverse authors, especially by Leroy (1980) and Bridson (1987), as a consequence of the advances in taxonomy and the discovery of new coffee species in the last century (Davis et al. 2006). Nowadays, most of the species of the *Argocoffea* section (bushes or climber plants from West and Central Africa) are considered to belong to the *Argocoffeopsis* Lebrun genus, whereas most of the species of the *Paracoffea* section (bushes found in India, Burma, and Southeast Asia) belong to the *Psilanthus* Hook genus (Leroy 1980). The *Lachnastoma* Korth genus (*Nostolachma* Durand) was also excluded (Bridson 1988; Davis et al. 2006). Of the four sections, only *Mascarocoffea* and *Eucoffea* are consistent with the modern concept of coffee trees.

Leroy (1980) considers that coffee plants belong to the tribe Coffeeae, which is characterized by ovaries with two carpels, each one with just one ovule; an axillary placentation; a hard endocarp and seeds covered with a thin parenchyma; and exhibiting a deep invagination on its ventral side, called the *coffeanum* suture (characteristic of "coffee bean" morphology). This tribe includes two closely related genera: *Psilanthus* and *Coffea*. However, based on molecular and morphological data, Davis et al. (2006) suggested that the tribe Coffeeae consists of nine additional genera, although none of them exhibited the *coffeanum* suture.

Differences between *Psilanthus* and *Coffea* lie in the floral and pollen morphology. Thus, in *Coffea*, the anther and the style generally emerge, the tube of the corolla has the same length as the lobes, the pollen is predominantly three colporate, and the flowers are in axillary inflorescences (Bridson 1988; Stoffelen 1998). In *Psilanthus*, the anthers do not emerge, the style is short, the tube of the corolla is longer than the lobes, the pollen is four to five colporate, and the flowers are rather terminal and axillary. Such differences are not absolute and their taxonomic fusion or separation is a matter that needs further research (Lashermes et al. 1997; Davis et al. 2006).

The genus *Coffea* is divided into two subgenera: *Coffea* and *Baracoffea*. Similarly, *Psilanthus* comprises the subgenera *Psilanthus* and the *Afrocoffea* (Bridson 1987). Table 19.1 summarizes main differences among groups. Such differences are related with the location of the flowers (i.e., axillary

Table 19.1   Morphological and Physiological Criteria to Differentiate Genera and Subgenera of the Tribe Coffeeae

	Genus *Psilanthus*[a] (22 species)	Genus *Coffea*[b] (103 species)
	Short Style (anthers do not emerge)	Long style (anthers emerge)
Axillary flowers; monopodic growth	Subgenus *Psilanthus*; West Africa, Zaire	Subgenus *Coffea*; tropical Africa, Madagascar, Mascarens (95 species)
Terminal flowers; simpodic	Subgenus *Afrocoffea* (Moens)[c]; África, south and south east, north of Australia	Subgenus *Baracoffea* (Leroy); West Madagascar (8 species)

Note: The geographical distribution is indicated for each subgenera.
[a] Geographical distribution according to Bridson, D. M. 1987. *Kew Bulletin* 42:453–460.
[b] Distribution and number of species according to Davis, A. P. et al. 2006. *Botanical Journal of the Linnean Society* 152:465–512.
[c] Paracoffea (Miquel) in Leroy, J. F. 1980. In *IX Colloque Scientifique International sur le Café*, 473–477. ASIC, London, UK.

or terminal), the type of growth (i.e., monopodic or simpodic), the location of the reproductive organs, and the type of reproductive strategy (i.e., allogamy or autogamy). Modern molecular analyses suggest the existence of four phylogenetic groups inside the subgenus *Coffea* that, in general, are consistent with the geographical distribution of species (Lashermes et al. 1997). According to several pieces of evidence, the origin of the genus *Coffea* is African and relatively recent, long after the separation of Gondwana (Anthony 1992). This has also been confirmed by the phylogenetic analysis based on cpDNA polymorphism (Lashermes, Cros, et al. 1996).

19.2.2 Origin and Distribution of *Coffea* Species

The species of the subgenus *Coffea* are distributed in four geographical regions of the intertropical African forests: the west and central Africa regions that hold not only the two most important species, *Coffea arabica* and *C. canephora*, but also *C. liberica;* Central Africa with *C. eugenioides;* the east African region that includes *C. salvatrix*, *C. racemosa*, and *C. costatifructa*, among others; and, finally, the Madagascar region (Lashermes et al. 1997). While some species, such as *C. liberica* or *C. canephora*, show a wide distribution, others, like *C. arabica* or *C. stenophylla*, are only found in a rather reduced area (Charrier and Berthaud 1985).

Coffea arabica L.: Cultivation of *C. arabica*, the most important species in the genus, started in southwestern Ethiopia about 1,500 years ago (Wellman 1961), and its origin is found in the southeast of Ethiopia, south of Sudan, and north of Kenya. The primary center of genetic diversity of *C. arabica* is situated in the highlands of southwest Ethiopia and the Boma Plateau of Sudan. Most of the wild populations of this species have also been reported in Mount Imatong (Sudan) and Mount Marsabit (Kenya) (Anthony et al. 1987).

Coffea canephora Pierre ex Froehner: This species exhibits a wide geographical distribution and is found in the wild in West Africa, Zaire, Sudan, Uganda, northwest Tanzania, and Angola (Berthaud 1986b; Davis et al. 2006). Coffee that used to be commercialized under the name of robusta and that represents the 30% of world commerce belongs to this species. Unlike *C. arabica*, *C. canephora* grows well in altitudes under 1000 m. Phenotypes exhibiting small organs (leaves, flowers, fruits, and grains) are known as conilon (or Koulliou) and quillou types (Porteres 1959). From a botanic point of view, two major groups are recognized, Guineen and Congolais, which in general correspond to the respective biogeographical areas, but not with the classification robusta-conilon (Berthaud 1986b).

Coffea congensis Froehn: This species comes from Central Africa and is found in the deltas and flooding zones of the Congo River and its affluents. It is adapted to conditions of high humidity (Berthaud 1986b), and it is easily crossed with *C. canephora*, allowing production of fertile hybrids.

Coffea eugenioides S. Moore: This species can be found at an altitude between 1000 and 2000 m (Davis et al. 2006) and is located from Sudan to the south of Zaire, but mainly in the south of Uganda. It is a conic bush with small lance-shaped and ovate leaves, small fruits, and grains with low caffeine content (Wellman 1961; Charrier and Berthaud 1985).

Coffea liberica Bull ex Hiern: This species is adapted to low altitudes. Trees with large coriacea leaves and big fruits with outsized navels can be found in this species. Its distribution area is similar to that of *C. canephora*. Two botanic groups are recognized: the libericas (liberians and abeokutae), from the west and central west of Africa, and the dewevrei (excelsa) from Central Africa (Berthaud 1986b; Davis et al. 2006).

Coffea racemosa Lour: This bush can be from 1 to 2 meters tall and comes from Tanzania and Mozambique (Guerreiro-Filho 1992), as well as from Zimbabwe and southern Africa (Davis et al. 2006). This species is adapted to long drought periods and is characterized by its low caffeine content (Guerreiro-Filho 1992).

Coffea stenophylla G. Don: This species, a small tree, is found in the forest massif of the African western region. It is recognized not only for its drought tolerance (Berthaud 1986b), but also because the ripe fruits are black (*Melanocoffea*, in the ancient classification made by Chevalier) and leaves are small and narrow. Two main forms are distinguished: one from the east and other from the west of the Ivory Coast.

19.3 GENOME STRUCTURE AND CYTOGENETICS

Cytogenetic studies in coffee species have been limited because chromosomes are small and morphologically similar. The basic chromosome number in the *Coffea* genus is $x = 11$ (Krug 1934), one of the most common numbers in the family Rubiaceae. All known coffee species are diploid ($2n = 2x = 22$ chromosomes) and generally self-incompatible, except *C. arabica*, which is the only tetraploid ($2n = 4x = 44$ chromosomes) and self-compatible species in the genus. At present, it is accepted that the allotetraploid *C. arabica* species has evolved from a cross between two ancestral ecotypes close to the actual diploid species *C. eugenioides* and *C. canephora* (Figure 19.2). Indeed, both molecular as well as cytogenetic studies have provided complementary evidence of the existence of two low-differentiated subgenomes (named E*a* and C*a*) as a result of a recent speciation event (Lashermes et al. 1999; Herrera, D'Hont, and Lashermes 2007; Tesfaye et al. 2007; Clarindo and Carvalho 2008). The allopolyploid condition of *C. arabica* has also been supported by molecular studies showing its diploid-like meiotic behavior, which allows preferential pairing between homologous chromosomes (Lashermes, Paczek, at al. 2000).

The genome structure among all the *Coffea* species is quite similar. Early studies conducted on several species revealed that somatic chromosomes are rather small (1–3 μm) and exhibit a pronounced morphological similarity (Krug 1937; Bouharmont 1963). As mentioned before, this similarity in chromosome size and shape represents one of the most important obstacles that have made it difficult to identify and distinguish individual species through chromosome length or arm ratios (Figure 19.3a). Furthermore, experiments using several chromosome banding methods have showed that coffee chromosomes exhibit uniform pericentromeric patterns of heterochromatin, suggesting that during evolution, most of the coffee species retained not only the same number but also the same (or a very similar) pattern of repetitive sequences (Pierozzi et al. 2001). Recent improvements in cytological methods allowing high-quality chromosome preparations in *C. canephora* and *C. arabica* have revealed slight but evident morphological differences between chromosomes for

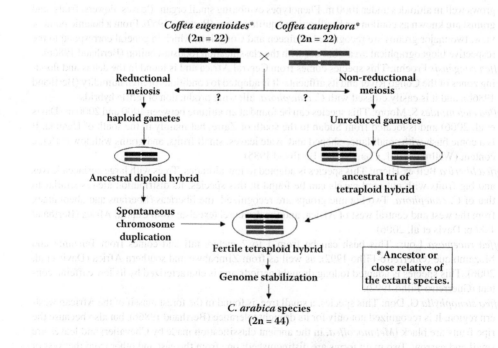

Figure 19.2 A possible scheme for the origin of the allotetraploid species *C. arabica* L.

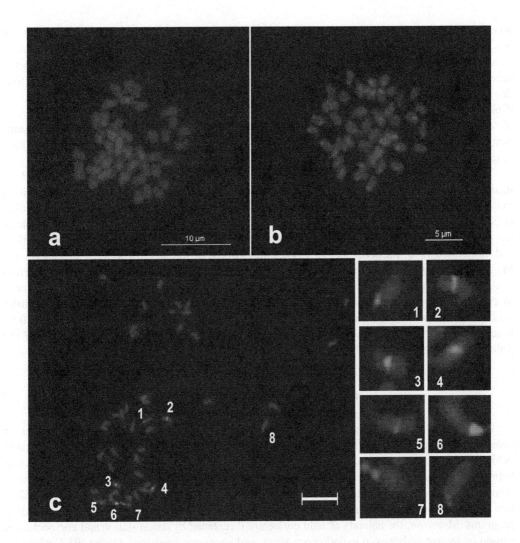

Figure 19.3 **(See color insert.)** Coffee cytogenetics. (a) Typical metaphase spreads of *C. arabica* L.; (b) Somatic metaphase chromosomes of *C. arabica* subject to *in situ* hybridization with a repetitive BAC probe (red signals); (c) Example of GISH on metaphase chromosomes of an introgressed line of *C. arabica* L. Individual chromosomes carrying introgressed fragments from *C. liberica* species are indicated by numbers (1–8) and enlarged at the right. Total DNA from the *C. liberica* was used as probe (green signals). Chromosomes were counterstained with DAPI, in blue. Scale bar = 5 μm.

these two species, opening the possibility to assemble future karyograms in other coffee species (Clarindo and Carvalho 2006, 2008).

Using flow cytometry analysis, important differences in DNA content have been observed among different species of *Coffea* (Cros et al. 1995). Indeed, 2C values of DNA amount vary from 0.95 ± 0.13 pg to 1.78 ± 0.33 pg for diploid species, and 2.6 ± 0.23 pg for *C. arabica*. Based on geographic distribution and phenology and excluding differences due to ploidy level, such variation in DNA content among coffee species is probably due to changes in the copy number of repeated DNA sequences along its evolutionary history.

Different cytogenetic studies have been carried out on triploid and tetraploid interspecific hybrids involving the cultivated *C. arabica* and several diploid relatives (Medina 1972; Boaventura and Cruz 1987). Such studies demonstrated that in spite of its differences on ploidy level, almost all crosses between *C. arabica* and the diploid species allowed successful regeneration of more or less fertile interspecific hybrids (Carvalho and Monaco 1968; Berthaud and Charrier 1988; Herrera, Combes, Cortina, et al. 2002; Herrera et al. 2004). Additionally, cytological analysis on microsporogenesis or meoitic behavior of chromosomes in F_1 plants also revealed important genomic affinities between species, suggesting that, overall, coffee species shared a basic genetic background (Charrier 1978a; Louarn 1992). More recently, molecular analysis carried out in arabusta hybrids (*C. arabica* × *C. canephora* 4×) revealed that recombination between chromosomes is not significantly restricted by genetic differentiation between chromosomes belonging to either the *C. arabica* or the *C. canephora* genomes (Herrera, Combes, Anthony, et al. 2002).

The advent of molecular cytogenetics protocols in plants opened the possibility to address chromatin regions of individual chromosomes on the basis of DNA sequence information in addition to mere morphological features. Cytomolecular techniques such as fluorescent *in situ* hybridization (FISH) have progressively been implemented in coffee (Figure 19.3b and c). The FISH technology has proved to be useful for studies focusing on genome organization of *C. arabica* (Raina, Mukai, and Yamamoto 1998; Lashermes et al. 1999), chromosome differentiation between species (Pinto-Maglio et al. 2001), detection of alien chromatin in interspecific hybrids (Barre et al. 1998; Herrera et al. 2007), and, more recently, the localization of introgressed fragments on specific chromosomes by combination of repeated and genomic probes (Herrera et al. 2007). Current evidence of similar genome as well as chromosome structures between coffee and solanaceous species (Lin et al. 2005; Mueller et al. 2005) will provide additional genomic information useful for development of the future physical and cytogenetic maps for coffee.

19.4 GERMPLASM RESOURCES

Diversification in the genus *Coffea* has produced more than 100 taxa and the inventory is not yet complete since new species are still being discovered (Stoffelen et al. 2008, 2009). As mentioned before, all species share a common genome, which makes possible interspecific hybridizations and hybrid production; thus, wild coffee trees constitute a valuable reservoir of genes for breeding. About 50% of known species are conserved worldwide in field gene banks, but their genetic evaluation is weakly documented. This explains why so few genetic resources have been used in breeding programs. Moreover, the habitats of wild coffee are threatened by deforestation and encroachment caused by human activities, population pressures, and economic hardships. Of the 103 *Coffea* species now recognized, 73 (70.9%) are considered critically endangered, 36 are endangered, and 23 (including *C. arabica*) are vulnerable (Davis et al. 2006). Regarding the socioeconomic importance of coffee, the conservation of genetic resources has not received much attention until now.

19.4.1 *Ex Situ* Conservation

Coffee growers have known for a long time that seeds remain viable during a short period (only a few months). Storage behavior of coffee seeds does not conform to orthodox or recalcitrant types, but it is considered as being intermediate (Ellis, Hong, and Roberts 1990; Hong and Ellis 1995). These seeds can be dried to some extent, but their long-term conservation remains problematic. That is why field gene banks were established in most coffee-producing countries; the oldest collections are in Indonesia (1900), Brazil (1924), and India (1925).

Conservation was initially focused on the cultivated species *C. arabica*, *C. canephora*, and *C. liberica*. Efforts to collect and preserve wild species were initiated in the 1960s and 1970s by the Institute de Récherche pour le Développement (IRD, formerly ORSTOM), the Food and Agriculture Organization, (FAO), and the International Board for Plant Genetic Resources (IBPGR) (Anthony, Dussert, and Dulloo 2007). Two field gene banks were set up in Madagascar for the

Madagascan species and in Ivory Coast for the African species. However, genetic resources conserved in field gene banks are subject to strong selection pressure, which generates genetic erosion (Anthony, Astorga, et al. 2007).

As an alternative to field conservation, research on *in vitro* techniques has been performed for medium- and long-term preservation. Attempts for *in vitro* conservation under slow growth conditions have been reported by Kartha et al. (1981) and Bertrand-Desbrunais, Noirot, and Charrier (1991, 1992). However, adaptation to *in vitro* conditions can induce a genotypic selection, which causes a genetic drift (Dussert, Chabrillange, et al. 1997). Cryopreservation (i.e., storage at ultralow temperature, −196°C in liquid nitrogen) protocols have been developed using seeds, zygotic embryos, apices, or somatic embryos (Dussert, Engelmann, et al. 1997). A simple and efficient protocol of seed cryopreservation was developed for the cultivated species *C. arabica* (Dussert et al. 2007) and then applied successfully to a subset of a core collection (Vasquez et al. 2005). Cryopreservation appears to be more cost efficient and effective than field conservation, but its impact as an *ex situ* conservation method remains limited.

19.4.2 Genetic Resources in Cultivated Coffee Species

19.4.2.1 *Coffea arabica*

According to their origin, the genetic resources of *C. arabica* comprise wild plants collected in the center of diversity (i.e., Ethiopia, Kenya, and Sudan), landraces conserved on farms in Ethiopia (Labouisse et al. 2008), plants grown in the primary center of dispersion (i.e., Yemen) (Eskes 1989), varieties and mutants selected worldwide in the genetic populations Typica and Bourbon spread in the eighteenth century (Krug, Mendes, and Carvalho 1939), and hybrids. In Ethiopia, the largest and most documented collections were carried out by the FAO in 1964 and 1965 (Meyer et al. 1968) and IRD in 1966 (Guillaumet and Hallé 1978). Other surveys were carried out by individual workers such as Lejeune (1958), Sylvain (1958), and Meyer (1965). The wild and semiwild accessions collected were spread worldwide and are now conserved in several countries (Anthony, Astorga, et al. 2007). The main field gene banks of *C. arabica* are established in Ethiopia (>4,000 accessions), Brazil (>4,000), Costa Rica (about 1,800), and Ivory Coast (about 1,700). Other collections of importance are located in Cameroon, Kenya, Madagascar, and Colombia.

19.4.2.2 *Coffea canephora*

Compared to *C. arabica*, the history of *C. canephora* cultivation is much more recent and fewer accessions have been accumulated in gene banks. The genetic resources of *C. canephora* are composed of wild plants collected in the center of diversity (i.e., West and Central Africa), genotypes with particular traits, mutants selected in plantations, and hybrids. Wild plants were collected in Ivory Coast (Berthaud 1986), Guinea (Le Pierrès et al. 1989), Cameroon (Anthony, Couturon, and Namur-de 1985), Central African Republic (Berthaud and Guillaumet 1978), and the Republic of Congo (de Namur et al. 1987). The main field gene banks of *C. canephora* are located in Indonesia (1,300 accessions), Brazil (>1,000), and Ivory Coast (800); minor centers exist in Cameroon and India.

19.4.2.3 *Coffea liberica*

The genetic resources of *C. liberica* comprise wild plants collected in the center of diversity (i.e., West and Central Africa), genotypes selected worldwide in plantations, and hybrids. The plants cultivated worldwide belong mainly to *C. liberica* var. *dewevrei*, but most plants cultivated in West Africa originated from local forests and were identified as *C. liberica* var. *liberica* (Bridson 1985). Wild plants were collected in Ivory Coast (Berthaud 1986) and Guinea (Le Pierrès et al. 1989) for *C. liberica* var. *liberica* and in the Central African Republic (Berthaud and Guillaumet 1978), Cameroon (Anthony et al. 1985), and the Republic of Congo (de Namur et al. 1987) for *C. liberica* var. *dewevrei*. The richest gene bank of *C. liberica* is located in Ivory Coast (1,000 accessions), but smaller gene banks also exist in India, Costa Rica, Colombia, and Brazil.

19.4.3 Wild Coffee Species

Conservation of noncultivated species received little attention until the 1960s, when explorations in Madagascan forests were initiated jointly by the French Museum of Natural History, the Centre de Coopération Internationale en Recherche Agronomique pour le Développement (CIRAD), and IRD (Charrier 1978a). Accessions of about 50 species are now conserved in a large field gene bank at Kianjavato.

In Africa, survey missions were organized in Guinea (Le Pierrès et al. 1989), Ivory Coast (Berthaud 1986), Cameroon (Anthony et al. 1985), Central African Republic (Berthaud and Guillaumet 1978), Republic of Congo (de Namur et al. 1987), and Tanzania and Kenya (Anthony et al. 1987). The collected material was introduced in field gene banks in Ivory Coast and in each visited country. The accessions belong to 12 well-identified species and to new species that could not be identified using available flora and botanical classifications (Table 19.2). Other wild species are also conserved in Ivory Coast, but with a low number of plants (Anthony 1992):

Species	Genotypes
C. heterocalyx	1
C. humblotiana	169
C. kapakata	2
C. millotii	1
C. perrieri	1
C. racemosa	65
C. salvatrix	42
C. sakarahae	6

Species such as *C. racemosa* have been distributed in Brazil, Colombia, Costa Rica, Ethiopia, India, and Tanzania.

From the field gene bank in Ivory Coast, a core collection comprising 32 genetic groups (19 species) was introduced *in vitro* (Dussert, Chabrillange, et al. 1997) and in greenhouse at Montpellier. Recently, the accessions conserved in greenhouse were duplicated by cutting and introduced in a field gene bank in La Réunion (E. Couturon, pers. comm.).

19.4.4 Germplasm Evaluation

19.4.4.1 Phenotypic Studies

Phenotypic evaluation based on observations and measurements has been carried out in field gene banks with a double objective: to define groups of similar accessions and to identify genotypes with features that could be of interest for coffee-breeding programs (Figure 19.4). The use of agromorphological characters failed to structure phenotypic variations in Ethiopian accessions of *C. arabica* (Berthou et al. 1978; Berthaud and Pernès 1978; Montagnon and Bouharmont 1996), but

Table 19.2   Wild Coffee Species Collected in Africa and Conserved in the Field Gene
Bank in The Ivory Coast

Species	Country	Number of Populations	Number of Accessions
C. anthonyi	Cameroon, Congo	5	50
C. brevipes	Cameroon	4	120
C. charrieriana	Cameroon	1	10
C. congensis	Cameroon, Central African Republic, Congo	15	1560
C. costatifructa	Tanzania	5	40
C. eugenioides	Kenya	6	990
C. fadenii	Kenya	1	4
C. humilis	Côte d'Ivoire	39	360
C. pocsii	Tanzania	1	40
C. pseudozanguebariae	Kenya, Tanzania	6	330
C. sessiliflora	Kenya	1	10
C. stenophylla	Côte d'Ivoire	9	140
Coffea sp. (several taxa)	Cameroon, Congo	18	450

Source: Updated from Anthony, F. 1992. Les ressources génétiques des caféiers: Collecte, gestion
d'un conservatoire et évaluation de la diversité génétique. Collection Travaux & Documents
Microfichés no. 81, ORSTOM, Paris.

they were used successfully to characterize the genetic Guinean and Congolese groups in *C. canephora* (Leroy et al. 1993).

A major trait of interest has been the biochemical composition of coffee beans since the presence or absence of some components could greatly modify cup quality. Variations in caffeine, trigonelline, sucrose, and seven isomers of chlorogenic acids were estimated for *C. arabica* and *C. canephora* (Ky et al. 2001). The results showed that genotypes with extreme contents (e.g., low caffeine content) can be selected in the cultivated species and used as progenitors in breeding programs. In the field gene bank managed by the Instituto Agronômico de Campinas (IAC), the caffeine content of 724 *C. arabica* plants from Ethiopia varied from 0.42 to 2.90% dry matter basis (dmb) (Silvarolla, Mazzafera, and de Lima 2000). In addition, three plants almost completely free of caffeine were discovered in this gene bank (Silvarolla, Mazzafera, and Fazuoli 2004). Regarding *C. canephora*, the caffeine content ranged from 1.16 to 3.33% dmb (Charrier and Berthaud 1975; Ky et al. 2001).

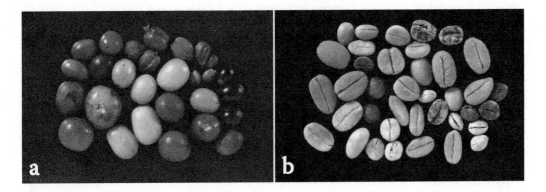

Figure 19.4   (See color insert.) Example of phenotypic variation in coffee trees. (a), berry color and (b), seed shape.

Evaluation for resistance to the main coffee pests and diseases has been an intensive and continuous activity in field gene banks for decades. For *C. arabica*, resistance to pests and diseases is a breeding objective of the highest priority (van der Vossen 2001a), but resistance to few pathogens has been found in the species. Ethiopian accessions were found resistant to the root-knot nematodes *M. arabicida* (Bertrand et al. 2002) and *M. paranaensis* (Anzueto et al. 2001; Boisseau et al. 2009). Incomplete resistance to the coffee leaf rust (*Hemileia vastatrix*) was observed in Ethiopian accessions and could have an action complementary to that of specific resistance (Eskes 1983; Gill, Berry, and Bieysse 1990). By contrast, resistance to several coffee pathogens was found in *C. canephora*—mainly against the coffee leaf rust (Kushalappa and Eskes 1989; Montagnon and Leroy 1993), coffee berry disease (*Colletotrichum kahawae*) (Carvalho 1988; Rodrigues, Varzea, and Medeiros 1992), and root-knot nematodes (for a synthesis, see Bertrand and Anthony 2008). Other species of interest are *C. liberica*, for its resistance to the coffee leaf rust (Bettencourt and Rodrigues 1988; Prakash et al. 2004), and *C. racemosa*, for its resistance to the leaf miner (*Leucoptera coffeella*) (Medina-Filho, Carvalho, and Medina 1977; Medina-Filho, Carvalho, and Monaco 1977).

Another important trait for breeders is male sterility, which has been observed in many field gene banks—initially as a curiosity, but recently with interest. For autogamous species such as *C. arabica*, the use of male sterile progenitors constitutes a cheap alternative to somatic embryogenesis for F$_1$ hybrid seed propagation. Male sterile genotypes were identified in *C. arabica* (Dufour et al. 1997), *C. canephora* (Mazzafera et al. 1989), and *C. liberica* (Santaram and Ramachandran 1981).

19.4.4.2 Genotypic Studies

Genotypic evaluation based on neutral markers has been carried out using enzyme and/or DNA markers (Table 19.3). Genetic groups of related accessions could be defined after cluster and discriminate analyses. Species diversity has thus been revealed for the cultivated species and some

Table 19.3 Studies of Genetic Diversity within Coffee Species Involving Accessions from Different Origins[a]

Species	Enzyme Alleles	DNA Polymorphisms
C. anthonyi	Stoffelen et al. (2009)	
C. arabica	Berthou and Trouslot (1977)	Anthony et al. (2001)
		Aga et al. (2003)
		Chaparro et al. (2004)
		Aga and Bryngelsson (2006)
		Silvestrini et al. (2007)
C. canephora	Berthou et al. (1980)	Dussert et al. (2003)
	Berthaud (1986)	Prakash et al. (2005)
	Leroy et al. (1993)	Cubry et al. (2008)
C. congensis	Berthou et al. (1980)	
C. costatifructa	Anthony (1992)	
C. eugenioides	Berthou et al. (1980)	
C. humilis	Berthou et al. (1980)	
C. liberica	Berthou et al. (1980)	N'Diaye et al. (2005)
	Berthaud (1986)	
C. pseudozanguebariae	Hamon et al. (1984)	
	Anthony (1992)	
C. stenophylla	Berthou et al. (1980)	
C. sessiliflora	Anthony (1992)	

[a] Based on enzyme and DNA markers.

wild species. In *C. arabica*, genetic diversity (as expressed by the number of molecular markers) and polymorphism were low (Anthony et al. 2002). This is doubtless related to the recent origin of the species (Lashermes et al. 1999) and a consequence of the predominant autogamy, which favors homozygous genetic structure (Carvalho et al. 1991). Low polymorphism was also detected in another autogamous species, *C. anthonyi* (Stoffelen et al. 2009).

By contrast, allogamous species displayed a higher level of diversity (Moncada and McCouch 2004; Poncet et al. 2004; Cubry et al. 2008). Some polymorphic species, like *C. canephora* or *C. stenophylla*, exhibit discontinuities in morphological types; however, these are not submitted to genetic barriers (Charrier and Eskes 2009). Other species, like *C. costatifructa* and *C. pseudo-zanguebariae*, are monomorphic; each population contains a large part of the species diversity without discontinuities between populations (Hamon, Anthony, and Le Pierrès 1984; Anthony 1992).

In *C. arabica*, genetic diversity appeared to be much weaker in the cultivars and mutants derived from the Typica and Bourbon populations than in wild accessions collected in the center of diversity (Anthony et al. 2001, 2002; Chaparro et al. 2004; Silvestrini et al. 2007). This has been related to the history of coffee dissemination, which is characterized by the successive reductions of diversity within the two populations of wild coffee introduced from Ethiopia to Yemen and then spread worldwide (Anthony et al. 2002). The genetic diversity of wild accessions appears to be structured into two groups separated by the tectonic rift that cuts Ethiopia from the northeast to the southwest (Anthony et al. 2001). Regarding Ethiopian landraces, their genetic diversity was similar to that of wild accessions (Anthony et al. 2001). By contrast, the accessions from the primary center of dispersion (i.e., Yemen) exhibited a very low genetic diversity (Silvestrini et al. 2007) and grouped unambiguously with the Typica-derived cultivars (Anthony et al. 2002). Mutants were classified according to their genetic origin (i.e., Typica or Bourbon) (Anthony et al. 2002).

In *C. canephora*, the species diversity has been structured by enzyme markers into two groups: the Guinean and Congolese groups, which are distributed in West and Central Africa, respectively (Berthaud 1986; Leroy et al. 1993). Using molecular markers, the Congolese group was then subdivided into four smaller groups (Dussert et al. 2003). In addition, one group population (i.e., Pelezi) was recently revealed within the Guinean group, increasing known diversity.

Molecular evaluation of *C. liberica* clearly indicated that *C. liberica* var. *liberica* and *C. liberica* var. *dewevrei* constitute distinct genetic groups more differentiated than botanical varieties (N'Diaye et al. 2005). However, taxonomists have adopted the simplification of Lebrun (1941) in two varieties (Bridson 1985; Davis et al. 2006). Numerous varieties, forms, and races have been described in the past, so it has become difficult to relate possible taxonomic characters to geographical areas in many instances (Bridson and Verdcourt 1988). Regarding the wild species, intraspecific diversity was evaluated in the 1980s by using enzyme markers (Table 19.3). New analyses should be carried out involving DNA markers and accessions collected in the future.

19.5 CONVENTIONAL BREEDING STRATEGIES

19.5.1 Commercial Coffee Varieties: Initial Efforts in Coffee Breeding

Apart from their diverse ecological adaptation to the locations situated at high and low altitudes, *C. arabica* and *C. canephora*, the most important commercial species in the genus *Coffea*, differ significantly in terms of their morphology, vegetative vigor, ploidy level, breeding behavior, genetic diversity, yielding potential, bean quality traits, and in genes' conditioning resistance for major diseases and pests.

The *C. arabica* species is characterized by a low genetic diversity that has been attributed to its allotetraploid origin, reproductive biology, and evolutionary process (Lashermes, Cros, et al. 1996).

Nevertheless, most of the arabica cultivars are derived from the few trees that survived various efforts to spread the growing of arabica worldwide that contributed further to a narrow genetic base. Therefore, *C. arabica* manifests susceptibility to major diseases and pests like coffee leaf rust, coffee berry disease, stem borer, and nematodes. On the other hand, the diploid species of *Coffea* represent considerable variability and some of these species form valuable gene reservoirs for different breeding purposes (Berthaud and Charrier 1988). *Coffea canephora*, for example, is more tolerant to most of coffee diseases and pests and has potential to give consistently high yields. But, vulnerability to drought conditions, small bean size, and poor cup quality are the negative features of this species.

It is well believed that original arabica coffee from southern Arabia (now Yemen) gave rise to two distinctive types of coffee—Typica and Bourbon—that were spread worldwide in the eighteenth century (Krug et al. 1939) (Figure 19.5). Historically, Typica was cultivated in Latin America and Asia while Bourbon was initially introduced into South America from the French colony of Bourbon (now Reunion Island) (Carvalho et al. 1969) and later to eastern Africa. The Bourbon type has a more compact and upright growth habit than the typical coffee and is generally higher yielding, with good-quality coffee (van der Vossen 1985). Because of the self-pollinating nature of *C. arabica*, both these varieties have remained genetically stable. It is also pertinent to believe that the agromorphological variation observed, which formed the source for so many named varieties during the twentieth century, was the outcome of the spontaneous mutations in a few major genes conditioning plant, fruit, and seed characters rather than of residual heterozygosity (Carvalho et al. 1969; Carvalho 1988).

In arabica coffee, initial breeding efforts were limited to simple selection or crossing within genetically homogeneous base populations and the breeding objectives were yield, quality, and adaptability to local conditions. In general, breeding for disease resistance was given low priority except in India, where rust resistance breeding has been given prime importance since the 1920s. These early programs, carried out from 1920 to 1940, resulted in considerable success in developing vigorous and productive cultivars that suggested a somewhat larger degree of genetic variability of

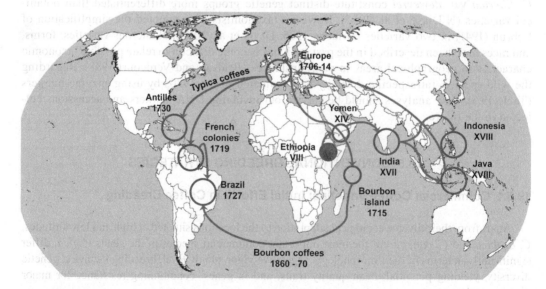

Figure 19.5 Origin and route of dispersal of the Arabica coffee around the world.

the base populations (van der Vossen 1985). In fact, several of these varieties are still under commercial cultivation. Varieties like Mundo Novo, Caturra, and Catuai from Brazil; Kents and S.288 from India; and Blue Mountain from Jamaica are some examples of known commercial cultivars.

The epidemics of CLR (*Hemileia vastatrix* Berk & Br) in Southeast Asia between 1870 and 1900 had a devastating effect on arabica coffee cultivation in several coffee-growing countries on this continent. This paved the way for the introduction of other tolerant species, especially *C. canephora*, in countries like Indonesia and India; in Sri Lanka, there was a shift toward tea cultivation. In the backdrop of these developments, robusta breeding programs aimed at yield and bean-quality improvement were initiated even earlier than in arabica. In fact, the pioneering work on coffee biology and selection carried out in east Java in the early years of the twentieth century (reviewed by Cramer 1957) formed the basis for subsequent breeding programs of robusta coffee not only in India but also in Africa (van der Vossen 1985).

19.5.2 Threat of Coffee Diseases—Shift in Breeding Focus

The gradual spread of CLR to other countries across the continents—Brazil in 1970, Central American countries in 1976, Colombia in 1983, Papua New Guinea and Jamaica in 1986—necessitated shifting the focus of arabica breeding, worldwide, toward rust resistance. Anticipating the spread of the disease, in countries like Colombia, preemptive and precocious breeding programs for rust resistance were initiated in 1970 (Castillo, Moreno, and Lopez 1976; Castillo and Moreno 1988), years before the appearance of the disease. Furthermore, appearance of the coffee berry disease (CBD) caused by *Colletotrichum kahawae* in highlands of eastern and central Africa, as was first detected in Kenya in 1922 (McDonald 1926), and its further spread to other countries in Africa like Angola, Cameroon (Müller 1964), Uganda (Butt and Butters 1966), Tanzania (Fernie 1970), and finally Ethiopia (Mulinge 1975) prompted a number of new breeding programs (van der Vossen 2001a), especially in Kenya (van der Vossen and Walyaro 1980, 1981) and Ethiopia (van der Graaf 1981).

Thus, the main focus of arabica breeding was shifted to disease resistance coupled with productivity and quality. This phase of coffee improvement from the 1950s to 1980s was critical and very productive, as considerable developments were achieved in a relatively short period. This success could be largely attributed to the frantic efforts of different coffee research groups across the continents to enrich valuable genetic resources (Meyer 1965; Guillaumet and Hallé 1978) and exploitation of new genetic diversity by application of advanced selection and breeding methods (van der Vossen 1985). The availability of basic information on coffee genetics, as well as the classical work of the Coffee Rust Research Center (CIFC) in Oeiras (Portugal) on various aspects of the coffee rust pathogen *Hemileia vastatrix* (Rodrigues, Bettencourt, and Rijo 1975), formed the basis for rust resistance breeding programs in many countries.

In addition, key developments included identification of several mutants of arabica coffee, like Caturra, San Ramon, and Villasarchi, with agronomic significance and also the discovery of spontaneous natural hybrids between robusta and arabica, like Hibrido de Timor (HdT, discovered in the Timor Islands; Bettencourt 1973) and Devamachy (discovered in India; Sreenivasan, Ram, and Prakash 1993), with high levels of disease resistance and their successful utilization in coffee breeding. These efforts have resulted in the evolution of several derived cultivars (namely, Catimor and Sarchimor) with high potential of production and disease tolerance that are suitable for high-density planting.

As regards robusta, the history indicates that the nucleus robusta stock introduced into Java in 1901 came from trees already under cultivation in Zaire in 1895, originating in the Lomani River region. This Java nucleus formed the major source for early selection of superior plant types

(Cramer 1957; van der Vossen 1985) and was later enriched with material from Gabon (Kouillou) and Uganda (Charrier and Berthaud 1988). The material selected in Java was reintroduced in the Belgian Congo around 1916 at INEAC (Institut National pour l'Etude Agronomique du Congo Belge), which had become the major selection center of *C. canephora* from 1930 to 1960 (Montagnon, Leroy, and Eskes 1998). It has also been believed that the improved seed from Java was used to establish robusta plantations in India, Uganda, Ivory Coast, and Zaire, where robusta coffee originated (van der Vossen 1985). Likewise, several African countries in the area of origin of *C. canephora* began cultivation with similar types of introductions from Java and Zaire. As robusta was hardy with high levels of field tolerance to major diseases and pests, yield, plant vigor, and bean quality have been the main characteristics for selection in breeding programs.

19.5.3 Traditional Breeding Strategies

We have tried to summarize, briefly, the most important methods adopted in traditional coffee breeding of arabica and robusta, the salient achievements in variety development, and their implications in devising strategies with respect to ongoing and future breeding programs.

19.5.3.1 Coffea arabica Breeding

Breeding methods common to self-pollinated diploid crops, including simple pure-line selection to complex intra- and interspecific hybridization, have been successfully employed in *C. arabica* improvement. The resultant hybrids were sufficiently homogeneous to permit propagation by seed, and the majority of the arabica cultivars grown in the world today have been established from seedlings (van der Vossen 1985).

19.5.3.1.1 Pure-Line Selection

This strategy was applied in initial years of coffee improvement wherein the elite plants with agronomically useful traits were identified in the populations and these individual plants were advanced by selfing to derive commercial lines. Varieties like Caturra from Brazil, Kent and S.288 from India, SL.28 from Kenya, and Java from Cameroon were the outcome of this strategy (Figure 19.6).

Figure 19.6 Coffee plants of *C. arabica* growing under the shade (left) and unshade (right) conditions.

19.5.3.1.2 Intraspecific Hybridization Followed by Pedigree Selection

This strategy was mainly applied to integrate and improve the agronomically desirable characters like yield, tolerance to diseases, and improved quality. Crosses were carried out between the proven genotype and a donor parent and the hybrid progenies were subjected to selection by pedigree method. In certain cases, the hybrids were backcrossed to one of the parents followed by pedigree selection. The well known varieties like Catuai, Tupi, Obata from Brazil, and S.795 from India; various types of Catimor lines derived in different countries from the cross of Caturra × HdT; the varieties Colombia, Tabi, and Castillo from Colombia, and Costa Rica 95 from Costa Rica are some successful examples. In a majority of these cases, it was the introgressive breeding because the genes conferring disease or pest resistance were transferred into selected arabica strains from the introgressed genotypes like HdT and S.288 (spontaneous tetraploid interspecific hybrids involving *C. arabica* and other diploid species).

19.5.3.1.3 Interspecific Hybridization

The potential of diploid species of *Coffea* as important sources of variability was well recognized and some of these species form valuable gene reservoirs for different breeding purposes (Berthaud and Charrier 1988). Among several diploid species, *C. canephora* provides the main source for resistance to CLR, CBD, and root-knot nematode (*Meloidogyne* spp.). Similarly, other diploid species like *C. liberica* have been successfully used as sources for rust resistance in India (Srinivasan and Narasimhaswamy 1975), while *C. racemosa* constitutes a potential source of resistance to coffee leaf minor caused by *Perileucoptera coffeella* (Guerreiro-Filho, Silvarolla, and Eskes 1999). Hence, interspecific hybridization with an objective to transfer the desirable genes (in particular, for disease resistance to species like *C. canephora* and *C. liberica*) into arabica cultivars without affecting quality traits has been the main objective of arabica coffee breeding (Carvalho 1988; van der Vossen 2001b) for a long time.

However, the ploidy difference between *C. arabica* (tetraploid) and other species (diploids) is the main limitation for developing fertile hybrids from direct crossings. The occurrence of spontaneous hybrids between tetraploid arabica and diploid species is common when these species are growing in close proximity. Exploitation of such natural tetraploid interspecific hybrids (e.g., HdT, Devamachy) has gained priority in coffee breeding. Both are natural hybrids between *C. arabica* and *C. canephora* and very much resemble arabica coffee, with high levels of resistance to major coffee diseases and pests. Consequently, the HdT has been extensively used worldwide in arabica coffee breeding as the source of resistance to diseases and pests (Lashermes, Andrzejewski, et al. 2000; van der Vossen 2001a) and still assumes greater significance for introgressive breeding.

Regarding synthetic hybrids, two breeding strategies—tetraploid breeding and triploid breeding methods—were followed and fertile tetraploid interspecific hybrid derivatives were obtained. In the tetraploid breeding method, in order to maintain gametic balance, the chromosome number of the diploid species (*C. canephora*) was doubled and tetraploid robusta types were obtained that were used for crossing with normal arabicas. The resultant hybrids were backcrossed to arabicas and selection was exercised in the progeny for desirable types. The descendants of the tetraploid breeding method tend to be very vigorous with characters more like those of robusta (i.e., yield and resistance) but with bean quality traits of arabica. The famous Icatu variety from Brazil was developed by using this method (Monaco, Carvalho, and Fazuoli 1974; Carvalho et al. 1989).

Under the triploid breeding strategy that was followed in India (Vishveshwara 1974) and Colombia (Orozco 1989), direct crosses were made between tetraploid *C. arabica* and diploid *C. canephora*. The resultant triploids were subjected to recurrent backcrossings with arabica types for fertility restoration and selection was exercised for desirable types. Commercial lines were developed in India (Santaram 2005) and Colombia (Alvarado and Cortina 1997) by using this trip-loid breeding strategy. These lines resemble the arabica phenotype, with the vigor and fruit cluster characters and resistance of robusta and the bean and cup quality of arabica.

19.5.3.1.4 Heterosis Breeding—F₁ Hybrids

Production of hybrids between wild populations of Sudan–Ethiopia was suggested as a strat-egy for increasing diversity (Charrier 1978b). Creation of hybrids between genetically diverse sub-populations, such as crosses between common cultivars and Ethiopian accessions, was suggested to enhance the chances of exploiting transgressive hybrid vigor (Lashermes, Trouslot, et al. 1996; van der Vossen 2001b). Further, development of F_1 hybrids between diverse genetic groups like wild Sudan–Ethiopian origins in order to exploit the heterosis has been envisaged as one of the promis-ing strategies in arabica breeding (Lashermes et al. 2008). This strategy was successfully applied in Kenya, Ethiopia, and Central America. The varieties Ababuna, developed in Ethiopia, and Ruiru II, a composite F_1 hybrid variety developed in Kenya, are the best examples.

In Central America, a program for development of F_1 hybrids between high-yielding and culti-vated varieties of arabica (Caturra, Catuai, Catimor, Sarchimor) and semiwild trees originating in Ethiopia or Sudan was initiated in 1992 (Etienne et al. 1997). Field trials of these F_1 hybrids recorded 30% higher productivity, along with resistance to CLR and nematodes, in addition to maintaining an excellent cup quality (Bertrand et al. 2005). Based on field performance, the selected F_1 hybrids are being multiplied in large numbers using a somatic embryogenesis technique for commercial cultivation (Lashermes et al. 2008).

19.5.3.2 Coffea canephora Breeding

In *C. canephora*, a strictly allogamous species, breeding methods common to cross-pollinated crops, such as mass selection and intra- as well as interspecific hybridization, have been followed. In mass selection strategy, the outstanding plants in terms of vigor, yield, and bean quality characters marked in base populations were selected and advanced through open pollinated seed. Varieties like Apoata of Brazil, S.274 of India, and Nemaya of Central America were derived accordingly. Further, superior plants, which yielded higher than the family mean yields in single plant prog-enies, were selected and used for establishing bi- or polyclonal gardens. Either the seed mixture of these clones or the mixture of clones was released as a clonal variety for commercial cultivation in different countries. The Balehonnur robusta clones (BR series) of India, SA and BP selections of Indonesia, and IF clones of Ivory Coast are some of the examples.

In intraspecific hybridization strategy, emphasis has been on exploiting the available diversity within the species. Two major diversity groups—the Guinean group (from West Africa) and the Congolese group (from central African countries)—have been identified in robusta. The Congolese coffee types generally show better agronomic value than Guinean types, and the majority of the cul-tivated *C. canephora* populations in the world constitute Congolese genotypes. The Guinean types are limited to Ivory Coast and Guinea in either wild or cultivated forms. More extensive studies of the *C. canephora* populations in field collections in Ivory Coast, based on morphological characters and molecular markers, revealed two genetic groups (SG 1 and SG 2) and four subgroups within the Congolese types (Dussert et al. 1999). The value of Guinean genotypes for *C. canephora* breeding was emphasized by Berthaud (1986) as the most productive clones selected in Ivory Coast during 1960s were the hybrids between Congolese and Guinean types. Taking into account these clues, a reciprocal recurrent selection program based on hybridization between Congolese and Guinean genotypes was initiated in Ivory Coast in 1984 (Montagnon et al. 2001). A high amount of heterosis for vigor and yield in intergroup hybrids compared to intragroup hybrids was achieved, as reported by Leroy et al. (1993).

Another important breeding strategy, interspecific hybridization, has also been tried for robusta improvement and was mainly aimed at improving the bean size and liquor quality. Exploitation of this strategy was started in Java at the beginning of twentieth century with the discovery of a spontaneous diploid interspecific hybrid between *C. canephora* var. *ugandae* and *C. congensis*, called Congusta or Conuga, which proved to be of considerable commercial value (Cramer 1957). Some of the interesting agronomic characters of these fertile hybrids, like vigor, productivity, adaptation to sandy soils, tolerance to temporary water logging, and good bean size and cup quality, generated interest in coffee breeders in different countries outside Indonesia for commercial cultivation in place of robusta.

The efforts of Indian breeders were successful in developing a highly fertile C × R hybrid variety through controlled crosses between *C. congensis* and robusta followed by backcrossing to robusta and sib-mating of the selected plants from backcross progeny. With good vigor and compact bush stature, this variety is suitable for planting in closer spacing and possesses bean and liquor characteristics superior to those of robusta. A selection program undertaken in Madagascar also resulted in clonal hybrid varieties HA, HB, H865, etc. Similarly, in Ivory Coast a program of controlled crosses between *C. canephora* and *C. congensis* has been initiated recently.

In Brazil and Ivory Coast, the tetraploid breeding strategy was adopted to develop tetraploid interspecific hybrids between arabica and duplicated robusta coffee (colchicines-induced autotetraploids) with the twin objectives of improving quality of robusta and transferring the vigor and disease resistance into arabica (Capot 1972; Monaco et al. 1974). Popularly called "arabustas," these hybrids' performance with respect to vigor, adaptation to tropical lowlands, and cup quality was initially encouraging. Based on this, the breeders in Ivory Coast have resorted to clonal selection among highly variable F_1 populations, targeting for good fertility and productivity (Capot 1975; Duceau 1980). However, the high expectations of this program have not been fulfilled to the desired level because of persistent problems of genetic instability and low fertility (van der Vossen 2001a).

19.5.4 Future Outlook

As described in earlier sections, there has been remarkable success with respect to the development of new varieties that rise up to the expectations of coffee farmers worldwide in terms of realizing higher yields and improved product quality in both robusta and arabica. Further, availability of varieties suitable for new cropping systems, such as high-density planting, and technical advances in coffee agronomy enabled remarkable increases in production per unit area. Hence, development of new hybrid varieties, especially by selecting parents from genetically diverse populations, is the best strategy for further increases in productivity. Development of new F_1 hybrids of *C. arabica* and mass propagation of the same by *in vitro* methods for generation of genetically uniform planting material, as is practiced in Central America and Tanzania (Bertrand et al. 2005), appears to be one promising option for the future.

Selection for and improvement of quality traits such as bean size and cup characteristics were given top priority in both arabica and robusta breeding programs. However, when introgressive breeding became the focus of arabica coffee breeding for transfer of genes imparting disease resistance, there was an apprehension that the introgressed diploid genome might have a negative impact on quality traits. Hence, much rigorous selection was made for keeping up the quality traits of introgressed and resistant strains and hybrid varieties to be on par with the traditional cultivars, and good success was achieved as observed in disease-resistant varieties like Colombia, Tabi, and Castillo® (Colombia); Ruiru II (Kenya); Sln.9 (India); and IAPAR 59 (Brazil).

Recently, van der Vossen (2008) presented a detailed account on cup quality of disease-resistant coffee genotypes based on analysis of accumulated scientific evidence and established that, with all nongenetic factors at optimum level, the CLR- and CBD-resistant cultivars can be equal in quality to the best of traditional varieties in professional cup testing trials. The inference was that the prejudices of the coffee trade appear vastly exaggerated. However, exercising rigorous selection for high quality before the release of disease-resistant cultivars, as well as use of wild arabica genotypes from Ethiopia and Sudan as donor parents for quality traits, is a prospective option in traditional breeding for arabica. In the case of robusta, use of elite genotypes from the Congolese group as donor parents in breeding programs and exploitation of interspecific Congusta hybrids appear to be promising options for the future. The compact C × R hybrid variety of India has been gaining popularity in growers' fields owing to its superior bean and cup qualities and the convenience for preparation of washed coffees.

The genes imparting resistance to the two most destructive diseases—the CLR and the CBD present in *C. canephora* and *C. liberica*—have been successfully introgressed into *C. arabica* after several decades of breeding and selection efforts. However, the adaptive capacity of the rust patho-gen *H. vastatrix* with the ability to overcome host resistances has resulted in the gradual loss of resistance in some cultivars under field conditions. As a result, all the nine major S_H genes (S_H1 to S_H9) identified so far have been defeated by progressive appearances of new virulent physiological races of rust pathogen (45 races to date). Though the Catimor derivatives performed well in terms of productivity, the quick breakdown of resistance to CLR due to the appearance of new rust races (initially in India and subsequently in other countries) was a great disappointment. The resistance sourced from HdT (CIFC 832/1) thus proved to be less durable.

However, the performance of Colombia, a multiline variety derived from the crosses involving Caturra and HdT (CIFC 1343), exhibited much better behavior in durability of resistance as it took nearly 20 years to see rust susceptibility on various populations of this variety; the disease progres-sion was also observed to be slow under Colombian conditions (Alvarado 2005). The Sarchimor derivatives developed from another HdT source (CIFC 832/2) have remained rust free so far. The variable performance of different sources of resistance (different HdT introductions) is a prospective aspect for breeding for durable rust resistance. In the backdrop of this variable performance of dif-ferent resistance sources, the search for new sources of resistance is a constant breeding priority. In this context the natural interspecific hybrids between *C. arabica* and *C. canephora* recently reported from New Caledonia (Mahé et al. 2007) offer great scope for exploitation in breeding programs.

With regard to coffee pests, there appear to be no naturally occurring resistance genes for cer-tain important pests like coffee berry borer and stem borer. However, in cases of certain pests like leaf minor, a severe pest in Brazil, host resistance was found in *C. stenophylla* and *C. racemosa*. Breeding efforts to introgress the resistance into arabica were successful only with respect to *C. rac-emosa* (Carvalho 1988). Similarly, encouraging results were obtained in identifying some resistance sources for root-knot nematodes (*Meloidogyne* spp.) and using the genetic resistance in breeding programs (Bertrand and Anthony 2008). Apart from these, no other cases of useful host resistance to important insect pests have been detected in coffee.

The latest advancements in plant biotechnology and molecular biology have opened up new possibilities for genetic analysis and provide new tools for plant breeding with the potential of increasing selection efficiency. In the perspective of coffee breeding, molecular markers and genetic transformation technology are the two main promising applications. In this context, good success has been achieved in use of the whole range of different DNA molecular marker techniques for genetic diversity analysis, detecting genetically divergent breeding populations, and establishing alien genome introgression.

Furthermore, the recent development of molecular markers closely linked to resistance genes against *M. exigua* (Noir et al. 2003), CLR (Prakash et al. 2004; Mahé et al. 2008; Herrera et al. 2009), and CBD (Gichuru et al. 2008) offered new possibilities for molecular marker-assisted selection in coffee. Similarly, the advancements in coffee genetic engineering during the last decade—especially for elucidating the structure and function of agronomically interesting genes through functional genomics approach and as a tool to introduce desirable genes into commercial genotypes through a transgenic approach—have potential implications for future genetic improvement of coffee.

19.6 MOLECULAR MARKERS AND GENETIC MAPPING

19.6.1 Molecular Markers

In coffee, a whole range of techniques has been used to detect polymorphism at the DNA level, including randomly amplified polymorphic DNA (RAPD), cleaved amplified polymorphism (CAP), restriction fragment length polymorphism (RFLP), amplified fragment length polymorphism (AFLP), inverse sequence-tagged repeat (ISR), and simple sequence repeat (SSR) or microsatellites. Further, during the last few years, the number of codominant markers has been considerably increased by SSR mining in coffee expressed sequence tag (EST) databases, offering new possibilities not only for marker cross-amplification within species but also for genetic analysis (Table 19.4).

19.6.2 Physical and Genetic Mapping Approaches

Several genetic maps have been constructed. The low polymorphism has been a major drawback for developing genetic maps of the *C. arabica* genome. Hence, the work reported so far is often restricted to alien DNA introgressed fragments into *C. arabica* (Prakash et al. 2004). Nevertheless, Pearl et al. (2004) recently obtained a genetic map from a cross between Catimor and Mokka cultivars of *C. arabica*. Furthermore, to overcome the limitation of low polymorphism, efforts were directed to the development of genetic maps in *C. canephora* or interspecific crosses. The earliest attempt to develop a linkage map was based on *C. canephora* doubled haploid (DH) segregating populations (Paillard, Lashermes, and Pétiard 1996; Lashermes et al. 2001). *Coffea canephora* is a strictly allogamous species consisting of polymorphic populations and of strongly heterozygous individuals. Conventional segregating populations are therefore somewhat difficult to generate and analyze. However, the ability to produce DH populations in *C. canephora* offers an attractive alternative approach. The method of DH production is based on the rescue of haploid embryos of maternal origin occurring spontaneously in association with polyembryony (Couturon 1982).

Table 19.4 Use of Molecular Marker Techniques in Coffee Genome Analysis

Technique	Focused Polymorphism	Ref.
RAPDs	Genetic diversity between cultivated and wild accessions; introgressive breeding	Orozco-Castillo et al. (1994); Lashermes, Trouslot, et al. (1996); Anthony et al. (2001); Aga et al. (2003); Sera et al. (2003); Chaparro et al. (2004)
CAPs	Taxonomic studies in *Coffea*	Lashermes, Cros, et al. (1996); Orozco-Castillo et al. (1996)
RFLPs	Genetic diversity in cultivated accessions; linkage mapping studies; origin of *C. arabica* species	Paillard et al. (1996); Lashermes et al. (1999); Dussert et al. (2003)
AFLPs	Introgressive breeding; origin of cultivated varieties	Lashermes, Andrzejewski, et al. (2000); Anthony et al. (2002); Steiger et al., (2002); Prakash et al. (2005)
ISRs	Genetic diversity in wild species	Aga and Bryngelsson (2006)
SSRs	Genetic diversity; introgressive breeding; marker cross-amplification within species	Mettulio et al. (1999); Combes et al. (2000); Anthony et al. (2002); Baruah et al. (2003); Moncada and McCouch (2004); Poncet et al. (2004)
ESTs	Identification of gene-derived markers; marker cross-amplification within species	Bhat et al. (2005); Poncet et al. (2006); Aggarwal et al. (2007)

Two complementary segregating plant populations of *C. canephora* were produced from the same genotype. One population comprised 92 doubled haploids derived from female gametes, while the other was a test cross consisting of 44 individuals derived from male gametes. A genetic linkage map of *C. canephora* was constructed spanning 1041 cM of the genome (Lashermes et al. 2001). This genetic linkage map comprised more than 40 specific STS markers, either single-copy RFLP probes or SSRs that are distributed on the 11 linkage groups. These markers constituted an initial set of standard landmarks of the coffee genome that have been used as anchor points for map comparison (Herrera, Combes, Anthony, et al. 2002) and coverage analysis of bacterial artificial chromosome (BAC) libraries (Leroy et al. 2005; Noir et al. 2004).

Furthermore, the recombination frequencies in both populations were found to be almost indistinguishable. These results offer evidence in favor of the lack of significant sex differences in recombination of *C. canephora*. More recently, Crouzillat et al. (2004) reported the development of a canephora consensus genetic map based on a segregating population of 93 individuals from the cross of two highly heterozygous parents. Backcross genetic maps were established for each parent (i.e., elite clones BP409 and Q121) and then a consensus map was elaborated. More than 453 molecular markers, such as RFLP and SSRs, were mapped covering a genome of 1258 cM. Recently, this map was used to map COS (conserved orthologous sequence) markers and perform comparative mapping between coffee and tomato (Wu et al. 2006). In parallel, several diploid interspecific maps were constructed. Those maps are based on an F_1 hybrid population resulting from a cross between coffee diploid species (López and Moncada 2006) or on progenies obtained by backcrossing of hybrid plants to one of the parental species (Ky et al. 2000; Coulibaly et al. 2003).

Finally, the construction of an integrated genetic and physical map of the coffee genome has been undertaken. A BAC library of the allotetraploid species *C. arabica* was constructed (Noir et al. 2004). This large insert DNA library derived from a multidisease-resistant line contains 88,813 clones, with an average insert size of 130 kb, and represents approximately eight times the *C. arabica* haploid genome equivalents. The mapping approach combines hybridization with mapped markers and BAC fingerprinting. For instance, hybridizations with both low-copy RFLP markers distributed on the 11 chromosomes and probes corresponding to disease resistance gene analogs (Noir et al. 2001) were completed. Positive BAC clones from subgenomes E*a* and C*a* were assembled into separate contigs, showing the efficiency of the combined approach for mapping purposes.

19.6.3 Future Developments

The recent development of high-capacity methods for analyzing the structure and function of genes, which may be collectively termed "genomics," represents a new paradigm with broad implications. Although currently available for only a few model plants, it seems likely that such information will rapidly become available for most widely studied plant species such as *C. arabica*. The advent of large-scale molecular genomics will provide access to previously inaccessible sources of genetic variation that could be exploited in breeding programs. Anticipated outcomes in coffee breeding include (1) rapid characterization and managing of germplasm resources, (2) enhanced understanding of the genetic control of priority traits, (3) identification of candidate genes or tightly linked genomic regions underlying important traits, (4) identification of accessions in genetic collections with variants of genomic regions or alleles of candidate genes having a favorable impact on priority traits. In this way, the recent efforts to set up an international commitment (ICGN, http://www.coffeegenome.org) to work jointly for the development of common sets of genomic tools, plant populations, and concepts would be extremely useful.

19.7 FUNCTIONAL GENOMICS AND TRANSCRIPTOME ANALYSIS

Current methods to study functional genomics and transcriptome analysis involve the use of large amounts of EST data derived from Sanger sequencing of cDNA libraries. For coffee, there are so far three large efforts to obtain this information, and they correspond to the Nestle initiative with *C. canephora*, the Brazilian coffee EST project with *C. arabica*; *C. canephora*, and *C. racemosa*, and the Colombian coffee genome project, centered on *C. arabica* but also with *C. liberica* and *C. kapakata* sequences. Considering additional libraries constructed in Europe and India that also include differential display techniques (Lashermes et al. 2008), the total number of ESTs could be estimated to be over 300,000. The significant number of ESTs requires techniques for arranging the clones and massively screening gene activity in order to find those few candidates responsible for phenotypic traits of interest. Research groups have relied on macro- and microarrays, complemented with quantitative real-time polymerase chain reaction (qRT-PCR) analysis and, more recently, on proteomics to offer for the first time a transcriptome analysis of coffee focused on disease and pest resistance (leaf rust, leaf miner, and berry borer) as well as on seed development and drought tolerance.

19.7.1 Pathogen Resistance

While genetics and structural genomics search for markers and genes that once transmitted from parents to offspring to ensure a resistance response, functional genomics aims at the identification of mechanisms of disease resistance—either structural or chemical or constitutive or induced—that correlate to signal transduction pathways, transcription activation and silencing, and metabolic changes for defense response that enable the detection of pathogens and conclude in a resistance reaction. The understanding of these mechanisms can result in new markers for improving varieties through coffee breeding and in the design of new strategies for disease and pest management.

To understand the early reactions in the interactions between the biotrophic parasite *Hemileia vastatrix* and coffee leaves, Fernandez et al. (2004) constructed suppression subtractive hybridization (SSH) libraries from *C. arabica* var. S4 Agaro inoculated with compatible and incompatible races (XIV and II); in a macroarray of 960 clones, they identified 28 unique cDNAs expressed mostly 24 and 48 hours after rust challenge. For 13 of these sequences, RT-PCR experiments on *C. arabica* var. *caturra* confirmed enhanced transcript accumulation during the interactions, with some of the genes clearly associated to known defense pathways such as NDR and WRKY homologs. Using qRT-PCR, Ganesh et al. (2006) observed that, in greenhouse plants of *C. arabica* var. *caturra*, the transcription patterns of two of these genes—the homolog CAWRKY and CAR111—could be induced 1 hour after wounding or 3 hours after infiltration with 0.5 mM SA, as well as after inoculations with *H. vastatrix* race VI (incompatible).

Differential expression has also been reported among intact leaves and disks from detached leaves of *C. arabica* var. *obata* (resistant) and *ouro verde* (susceptible) when inoculated with *H. vastatrix* (Ramiro et al. 2006), which warns on the extrapolation of results obtained in disk systems toward the explanation of gene expression under field conditions. Microarray technology has been applied in the resistance source HdT, with a cDNA array of 36,480 probes hybridized with RNA samples from compatible (Caturra) and incompatible (TH) interactions with the coffee leaf rust (Cardenas et al. 2008). Out of 1,644 differentially expressed clones, 715 could be associated with previously annotated homologs, resulting in gene clusters related to metabolic changes required to the adjustment of plant metabolism to facilitate or block fungal nutrition and development.

Besides coffee leaf rust responses, expression patterns in systemic acquired resistance (SAR) have also been evaluated through microarray hybridization experiments. On a 1,554 clone array (886 from embryonic root, 670 from leaf, and 48 representing NBS-containing resistance genes), de Nardi et al.

(2006) monitored the effect of benzo (1,2,3) thiadiazole-7-carbothionic acid *s*-methyl ester (BTH) in roots and leaves of *C. arabica* var. *bourbon* 16 hours after inoculation. In leaves, 55 genes were overexpressed and 16 underexpressed, while in roots 37 were overexpressed and 42 underexpressed; very few genes (12) were common to both tissues. Although defense-related genes are activated, the effectiveness of these responses against leaf and root pathogens remains to be determined.

The biochemical pathways characterized in these transcriptome analyses are turning into important candidates to provide race-independent resistance to coffee leaf rust and, hopefully, other pathogens. They are the basis for the development and preservation of durable resistance to limiting biotic stresses under field conditions.

19.7.2 Pest Resistance

Two of the most limiting pests worldwide are the coffee leaf miner (*Leucoptera coffeella*) and the coffee berry borer (*Hypothenemus hampei*), to which no resistance sources have been identified among commercial coffee varieties. In order to analyze gene expression after leaf miner infestations, a macroarray with 768 clones was spotted on nylon membranes, after construction of substractive cDNA libraries using a resistant fifth backcross progeny of (*C. racemosa* × *C. arabica*) × *C. arabica* (Mondego et al. 2005). With this approach, and using RNA blot to validate candidate genes, 157 clones were found to be differentially expressed, many of them associated to the susceptible genotype (probably biased by the origin of the clones from subtractive libraries). In resistant reactions, up-regulated genes were identified after oviposition but not induced after eclosion. Defense genes were also overexpressed after oviposition in susceptible varieties, although at insufficient amounts to impact insect development.

For the coffee berry borer, patterns of gene expression have been studied in *C. liberica*, where discrete levels of antibiosis have been detected. Gongora et al. (2008) used a cDNA microarray constructed from normalized libraries, with 19,074 clones from a five-tissue mixture of Caturra and 14,189 clones from leaves and berries of *C. liberica* hybridized with RNA samples from both genotypes, taken 24 and 72 hours after infestation. A total of 2,585 genes were found to be differentially expressed, especially some involved in the Jasmonic acid pathway, which is known to be part of the defense mechanisms in other insect–plant interactions. More detailed studies are required to determine the importance of inducible defense responses against the constitutive expression of genes that affect nutrition quality for the insect in order to use this information in future assisted selection in a breeding program.

19.7.3 Seed Development

Very few genes related to cup quality have been identified in green coffee beans besides those encoding for enzymes involved in the biosynthesis of caffeine, chlorogenic acid, sucrose, oleosins, and albumins. These products act during roasting as aroma and flavor precursors. In order to obtain a wider picture of gene expression during seed development, Salmona et al. (2008) constructed a cDNA array with 266 selected clones from *C. canephora*, amplified by PCR and spotted on nylon membranes. The macroarrays were hybridized with RNA samples from *C. arabica* var. *laurina* comprising six stages of seed development that included perisperm and endosperm development and pericarp maturation.

For half of the genes analyzed, the hybridization signals were above background, with 33% (88 genes) displaying differential expression along the seed development time line and confirmed through qRT-PCR. The expression patterns support the respiratory burst and ripening process produced by an autocatalyzed ethylene synthesis cascade and reflect differences in the endosperm and perisperm transcriptional programs, probably due to a sporophytic-to-gametophytic transition.

In addition, the gene clusters identified provided phenological markers for each of the developmental stages, as well as put in evidence the sequential pattern of precursor accumulation inside the seed.

19.7.4 Drought Tolerance

Exposure to long drought periods is becoming more frequent to coffee plantations around the world, with a negative consequence on yield. Therefore, breeders are interested in finding sources of drought tolerance, particularly among *C. canephora* varieties, and transferring these traits into commercial varieties. Using digital expression analysis and qRT-PCR, Alves-Ferreira et al. (2008) identified two homeobox gene homologs to *Arabidopsis* AtHB1 and AtHB12 genes whose RNAs hybridized *in situ* either in the phloem of lateral roots and in the upper part of the main root (for CAHB12), or along the root (for CAHB1). Paiva et al. (2008) also reported the differential expression in leaves of SnRK2 (an SNF1 related kinase) involved in abscisic acid (ABA) signaling under conditions of water deficit.

Introducing differential proteomic analyses in coffee, the study of leaf tissue in *C. canephora* by Almeida, Guimaraes, et al. (2008) has indicated that drought acclimation mechanisms involve a reduction in the activity of photosystem II (PSII) through lower expression of the PSII PsbP protein, which in turn decreases the formation of reactive oxygen species caused by drought stress. Similarly, proteomics of stressed roots identified 81 proteins with differential expression under drought stress conditions; some were involved in cytoskeleton and root architecture, as well as in multiple changes in metabolism and stress response (Almeida, Soares, et al. 2008). The identification of functional markers for drought tolerance can assist the selection process and explain mechanisms for successfully overcoming this abiotic stress.

19.7.5 Future Prospects

Coffee is still far behind model species and many crops in both ESTs and genetic map markers, as well as in phenotypic characterization of the diversity present in germplasm collections. Very few limiting diseases have been studied so far, and problems such as the coffee berry disease and the white stem borer (*Xylotrechus quadripes*) still remain poorly understood and continuously pose serious threats for many coffee-producing countries. Efforts should be made toward the application of functional genomics on the preservation and improvement of cup quality, nutrient assimilation, and adaptation to climatic change. For efficient transcriptomic analysis, EST information is currently very fragmented, and a unified database could be very helpful in the design of more comprehensive research tools such as oligo- and protein arrays in order to implement technologies to be used in specific studies related to the particular needs of each country or market. Such a database will help to coordinate further sequencing initiatives, for tissues and genotypes, that can complement the existing data set and avoid duplicated efforts.

Role characterization of candidate genes will require more sophisticated techniques, primarily based on the identification of orthologous relationships with genes deeply studied in other genomes, but also on the development of powerful annotation pipelines linked to known biochemical pathways. More solid evidence of gene function will be obtained through modifications of gene expression— either stable or, preferably, transient—for which technology tools such as gene transformation vectors and cell introduction methods for these constructs still need to be developed.

19.8 BIOINFORMATICS RESOURCES

The most important goal of bioinformatics is the management, analysis, interpretation, and visualization of data from genomics research. Most of bioinformatics work includes database development and implementation, genomics data analysis and data mining, and biological interpretation

(Moore 2007). Genomics research is a field that continuously faces the problem of storing, indexing, and retrieving these large amounts of data being produced; fortunately for bioinformaticians, there is a trend in the field to rely to a greater extent on standard methods for the analysis of these data. It is possible nowadays to share bioinformatics resources between different research groups so that the integrity of the data is not jeopardized in any way (Teufel et al. 2006).

As an example of the genomics data being generated, ESTs are being produced for a number of plants as a rapid method for gene discovery. For instance, rice has more than 1 million EST sequences and there are 11 plant species, most of them grasses, with at least 300,000 EST sequences in dbEST (release 101,708; October 2008). The ultimate aim in most projects is to catalogue all the expressed genes in a particular genome.

Despite coffee's global importance, very little information has been gathered from this genus at the genetic level. As of October 2008, roughly 1,500 EST sequences from the species *C. arabica* had been deposited in the GenBank database. Only in recent years has a large expressed sequenced tag data set from the species *C. canephora* developed jointly by Nestle and Cornell University scientists been deposited in public databases (Wei et al. 2005).

19.8.1 Data Analysis

EST assembly and the viewing of assemblies, as well as consensus sequences, are the prime goals in bioinformatics for most coffee genome projects that are underway in several laboratories around the world. In addition to the databases, most systems include tools for the analysis of data including local implementations of BLAST, wEMBOSS, Primer3, and InterProScan. Web-based bioinformatics systems that function as a genomics information resource for coffee have been implemented by several projects that study the coffee genome. Information produced by coffee genome projects can be accessed from several web-based sites (Table 19.5). The advantage of web-based delivery is that anyone with an Internet connection can have access, with systems developed in a user-friendly environment.

The bioinformatics platforms usually include laboratory integrated management systems (LIMS), tools for data analysis, and databases of annotated sequences. The main backbone of the systems

Table 19.5　Main Coffee Bioinformatics Resources

Site	URL	Description
SGN-USA	http://sgn.cornell.edu/	*C. canephora* ESTs assemblies
IRD plant bioinformatics platform—France	http://www.mpl.ird.fr/bioinfo/	*Coffea* spp. ESTs
CoffeeDNA-Trieste, Italy	http://www.coffeedna.net/	*Coffea* spp. specific SSRs and ESTs (~13,686), retrotransposons (43), and SNPs
CCMB-India	http://www.ccmb.res.in/	*Coffea* spp. SSRs profiles (~500), EST-SSRs, and AFLP profiles
CBP&D-Brazil	http://www.lge.ibi.unicamp.br/cafe/	*Coffea* spp. SSRs, SNPs, and transposons; an important number of ESTs (~215,000) from *C. arabica, C. canephora,* and *C. racemosa*
Embrapa (mirror)	http://www.cenargen.embrapa.br/biotec/genomacafe/	
Cenicafe—Colombia	http://bioinformatics.cenicafe.org/	*C. arabica, C. liberica, C. kapakata* ESTs; *C. arabica* SSRs and BAC end sequences; *C. arabica* chloroplast GBrowse display
NCBI—USA	http://www.ncbi.nlm.nih.gov/dbEST/	*Coffea* spp. ESTs and genes

includes analysis of ESTs, molecular markers databases, and BAC end-sequences databases. Most systems are based on relational databases and work in Linux environments.

As an example of such systems, the National Coffee Research Center (CENICAFE) database includes 81,378 ESTs representing the coffee transcriptome with 30,646 unigenes. The majority of the coffee ESTs were derived from cDNA tissue-specific and normalized libraries of *C. arabica* and some from the diploid species *C. liberica* and *C. kapakata*. In addition, CENICAFE has generated 4,186 *Beauveria bassiana* ESTs, representing 2,401 unigenes, and 4,870 *Hypothenemus hampei* (coffee berry borer) ESTs representing 1,766 unigenes. The coffee database also includes over 80,000 BAC end sequences of *C. arabica* (Restrepo et al. 2009). As a recent development, coffee gene families have been annotated after analysis with the algorithm OrthoMCL. A GBrowse graphics visualization tool for the display of large sequences such as whole BAC sequences and chloroplast DNA has also been incorporated (Cristancho et al. 2006).

Clustering and assembly of Brazilian ESTs resulted in 14,886 clusters and 24,426 singletons from *C. arabica*, 2,147 clusters and 4,622 singletons for *C. canephora*, and 949 clusters and 3,107 singletons for *C. racemosa* (Vieira et al. 2006). The Solanaceae Genomics Network (SGN) (http://sgn.cornell.edu/) serves as another example of these online resources. SGN is dedicated to the biology of Solanaceae species, including tomato, potato, tobacco, eggplant, pepper, petunia, and other related plants such as coffee.

Bioinformatics groups have to be aware that they need to operate in two major areas: service and production (development). The service routines include the analysis of sequence and other types of data produced by genomics scientists. The development activities include the setting up and administration of bioinformatics servers, construction of structured databases, development of web-based interfaces for the display of data, and writing of scripts in perl and other languages for the manipulation of data.

Coffee ESTs have been analyzed and assembled in a very similar way (Cristancho et al. 2006; Vieira et al. 2006). In synthesis, chromatograms are called with phred, assemblies are performed with CAP3, amino acid prediction is accomplished by ESTScan, functional annotation of sequences is performed with several databases (among them GenBank), and additional functional annotation of gene ontology terms is performed with InterProScan. Other bioinformatics analyses include the discovery of SSR markers, development of specific PCR primers, prediction of SNPs, and homology comparisons between large sets of sequences. An additional goal of the groups is the construction of genetic linkage maps and physical maps.

19.8.2 Storage of Information

A final aspect that needs special consideration by bioinformatics groups is the backup storage of information. There are several ways that this task can be accomplished, and among them we can give two examples:

- *External hard drives/tape backups:* Tapes and external hard drives can store large amounts of information in a safe and reliable way. There are tapes in the market that already can store over 40 Tb of information. It is advisable to keep backup information in a secure environment, at least 80 km (50 miles) away from the central repository of data.
- *Mirror sites:* This is another way of keeping backups, with the advantage of having the up-to-date information in two different geographic places. These sites represent an update duplicate of a central repository site of genomics data. The Brazilian Coffee Genomics Consortium has a central repository site (http://www.lge.ibi.unicamp.br/) at the Laboratório de Genômica e Expressão (LGE) of the State University at Campinas and a mirror site at the Embrapa Recursos Genéticos e Biotecnologia's bioinformatics group (http://www.cenargen.embrapa.br/biotec/genomacafe/).

Another advantage of mirroring and having the possibility of handling the data by two different bioinformatics groups is the development of derivatives of the core data (e.g., gene-specific oligonucleotides, protein sequences, promoters, gene expression data, etc.) to meet the most frequent needs of users through databases that may be customized to take into consideration the preferences of coffee scientists.

19.8.3 Future Prospects

Current genomics and bioinformatics research is facing new challenges. Recent advances in technology and equipment are allowing scientists to produce large amounts of sequence, gene expression, protein, metabolite, genotype, and phenotype data to develop resourceful and integrated databases for a better understanding of biological processes. In the short term, the new sequencing technologies, including 454 and Solexa, will generate thousands of DNA sequences from several coffee species that will revolutionize genomics research in this plant.

The genomics data produced will be fundamental to study and find solutions to the major constraints faced by the coffee agribusiness, such as control of pre- and postharvest physiological factors involved in quality, disease, and pest control and management of plant responses to environmental changes (e.g., limited water availability and adverse temperatures) (Lashermes et al. 2008). Biologists and bioinformaticians will have to find clever ways of dealing with and analyzing this large amount of data and making biological sense of them.

19.9 TISSUE CULTURE AND GENETIC TRANSFORMATION

19.9.1 Tissue Culture in Coffee

Tissue culture techniques have already been applied for the improvement of many plant species, both annuals and perennials, including *Coffea*. Plant tissue culture is a required procedure in projects for the genetic manipulation of coffee plants by recombinant DNA technology. However, other applications could be included in programs for *Coffea* improvement: Problems with the multiplication and testing of cultivars that are difficult to propagate sexually can be circumvented by *in vitro* propagation—for example, sterility of hybrids obtained from the crossing of different *Coffea* species. Two systems for coffee *in vitro* propagation are employed: "micropropagation," which refers to propagation from buds or from tissue with the capacity for rapid cell division, and "somatic embryogenesis," which refers to the production of organized embryo-like tissue in response to taking plant segments through a disorganized callus phase.

19.9.1.1 Micropropagation

The coffee plant presents one apical meristem and each axil leaf has four to five dormant orthotropic buds and two plagiotropic buds. The plagiotropic buds only start development from the 10th and 11th nodes. For apical meristem culture and the culture of dormant buds, both orthotropic and plagiotropic buds give rise to plantlets, which can be used as initial explants for coffee micropropagation. Microcuttings or nodal culture comprise a tissue culture approach that entails culturing nodal stem segments carrying dormant auxiliary buds and stimulating them to develop. Since this method involves the exploitation of buds already present on the parent stock plant, it provides means of clonal multiplication. Each single segment can produce seven to nine microcuttings every 80 days. Most of this work was carried out during the 1980s. These aspects have been covered extensively in an earlier review (Carneiro 1999). Culture of microcuttings in temporary immersion systems has resulted in a sixfold increase in the multiplication rate in comparison with microcuttings multiplied on solid medium (Berthouly et al. 1994; Teisson et al. 1995).

19.9.1.2 Somatic Embryogenesis

Efficient methods for the regeneration of plants from cultured cells of most of the important species of crop plants were developed only in the 1980s. This was made possible by the induction of embryogenic cultures, which formed somatic embryos that germinated to give rise to normal and fertile plants. It is this technology, combined with the ability to transform plant cells genetically, that forms the basis of today's commercial plant biotechnology.

Somatic embryogenesis has been well documented in coffee (Staritsky 1970; Sharp et al. 1973; Pierson et al. 1983; Dublin 1984; Yasuda, Fujii, and Yamaguchi 1985; Berthouly and Michaux-Ferrière 1996; Berthouly and Etienne 1999). The callus induction was more efficient in the absence of light. During the last 35 years, a number of protocols for somatic embryogenesis have been developed for various genotypes of coffee (reviewed by Carneiro 1999). Two types of processes have generally been described using leaf sections as explants (Figure 19.7):

- Low-frequency somatic embryogenesis (LFSE; also called "direct somatic embryogenesis"): somatic embryos are obtained quickly (approximately 2 months) on only one medium without producing callus. Generally, a small number of somatic embryos are obtained.
- High-frequency somatic embryogenesis (HFSE; also called "indirect somatic embryogenesis"): Induction of somatic embryos by this way requires two media—an induction medium for primary callogenesis and a secondary regeneration medium to produce embryogenic friable callus regenerating several hundred thousand somatic embryos per gram of callus. The high-frequency procedure (from leaf explants to somatic embryos) takes about 7–8 months for *C. canephora* and arabustas (arabica × robusta derived crosses) and 9–10 months for *C. arabica*. This process enables the use of a liquid medium for both embryogenic tissue proliferation and the regeneration phase, and it was consequently preferred for scale-up and development of mass propagation procedures (Söndahl and Sharp 1977).

The LFSE and HFSE were established from different types of explants of *C. arabica* and *C. canephora* (Söndahl and Sharp 1977; Dublin 1980a, 1980b; Molina et al. 2002). Even from coffee seed integument (perisperm) tissues, embryogenesis was successful (Sreenath et al. 1995). The time required for embryogenesis in coffee reported by various groups ranged from 8 months to more than a year (reviewed by Carneiro 1999). A number of attempts have been made to reduce this time. Triacontanol, silver nitrate (AgNO3), salicylic acid, thidiazuron, and 2iP are the widely used growth regulators in coffee embryogenesis. Triacontanol incorporated at 4.55 and 11.38 μM in half-strength MS basal medium containing 1.1 μM 6-benzylaminopurine (BA) and 2.28 μM indole-3-acetic acid (IAA) induced direct somatic embryos in both species of *C. arabica* and *C. canephora* (Giridhar, Indu, Ravishankar, et al. 2004). Direct somatic embryogenesis was achieved from hypocotyl explants of *in vitro* regenerated plantlets of *C. arabica* and *C. canephora* on modified MS medium containing 10–70 μM AgNO3 supplemented with 1.1 μM N6-benzyladenine and 2.85 μM indole-3-acetic acid (Giridhar, Indu, Vinod, et al. 2004). Somatic embryogenesis in just 2–4 months was the major breakthrough in these reports.

Commercialization of large-scale somatic embryogenesis is still in its infancy, mainly because of the difficulty in maintaining aseptic conditions for the long duration of *in vitro* culture. Apart from that, maintaining synchrony in induction of embryogenesis and maturation stages is difficult in coffee. Other problems include highly variable conversion rates from embryos to plants among different genotypes and explants. However, direct somatic embryogenesis has several advantages over propagation by microcuttings and *in vitro* nodal culture. It does not have the problems of phenolic accumulation in the medium, and the contamination rates are generally very low. Furthermore, the

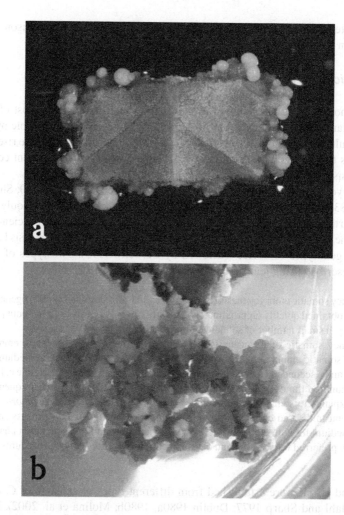

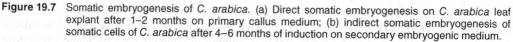

Figure 19.7 Somatic embryogenesis of *C. arabica*. (a) Direct somatic embryogenesis on *C. arabica* leaf explant after 1–2 months on primary callus medium; (b) indirect somatic embryogenesis of somatic cells of *C. arabica* after 4–6 months of induction on secondary embryogenic medium.

genetic variability is relatively low and the method permits the development of plantlets in larger numbers than those possible with the microcuttings method. For example, one leaf segment can produce hundreds of plantlets in a single culture cycle, whereas the microcutting method requires several subculturing cycles in order to get a comparable number of plantlets (Muniswamy and Sreenath 1996).

The field performance of embryo-regenerated plants has been reported and found to show normal response in terms of physiology and yield. In order to validate the propagation technology of *C. canephora* var. *robusta* via somatic embryogenesis in liquid medium, the clonal fidelity of regenerated trees was assessed for the first time in large-scale field trials. For the observed morphological traits and the yield characteristics, no significant differences were seen between the somatic seedlings and the microcutting-derived trees (Ducos et al. 2003).

A somatic embryogenesis procedure using a temporary immersion bioreactor was developed for *C. arabica* F_1 hybrids, enabling mass and virtually synchronous production of mature somatic embryos, without the need for selection before acclimatization (Etienne et al. 1997; Etienne-Barry et al. 1999). Depending on the genotypes, yields ranging from 15,000 to 50,000 somatic embryos per gram of embryogenic suspension cell mass were recorded. The proportion of normal torpedo type embryos produced by temporary immersion is usually ≥90%. Moreover, using short immersion times (1 minute) overcame hyperhydricity (Etienne and Berthouly 2001; Albarrán et al. 2005).

Conditions for direct sowing of *C. arabica* somatic embryos produced in a temporary immersion bioreactor on horticultural soil have also been developed (Etienne-Barry et al. 1999). Plant conversion rates for mature somatic embryos (i.e., possessing a pair of open cotyledons), along with a well-developed chlorophyllous embryonic axis, frequently reach 70% in the nursery. It was shown that 86% of embryos in the same 1l-RITA® bioreactor reached the "mature" stage and could be directly sown (Barry-Etienne et al. 2002). However, they revealed morphological heterogeneity in terms of cotyledon area, which resulted in heterogeneity in the nursery, mainly related to retarded growth in plantlets derived from somatic embryos with small cotyledons.

Direct sowing reduces handling time to 13% and shelving area requirements to 6.3% of the values obtained with conventional acclimatization of plants developed on semisolid media (Etienne-Barry et al. 1999). Moreover, a physiological approach has recently shown that even though the plantlets obtained with direct sowing were still small compared with seedlings (ratio 1:3), they were readily acclimatized to nursery conditions and had vigorous aerial and root systems and active growth (Barry-Etienne et al. 2002). The economic viability of direct sowing was then proved along with the quality of the regenerated plantlets.

19.9.2 Genetic Transformation of *Coffea* Species

Transgenic research has opened up new avenues for genetic studies and improvements in *Coffea* (Figure 19.8). A detailed review on the development of transgenic coffee can be obtained from Etienne et al. (2008). The early reports on coffee genetic transformation referred to cocultivation of protoplasts with different strains of *Agrobacterium tumefaciens* and electroporation-mediated gene delivery of coffee protoplasts. Plantlets were recovered from electroporated embryos using kanamycin as a selective marker (Barton, Adams, and Zorowitz 1991).

But now it is well known that kanamycin selection is not very efficient as hygromycin for the selection of transgenic coffee. Transgenic plants of *C. canephora* and *C. arabica* were obtained using genetic transformation mediated by *A. rhizogenes* (Spiral et al. 1993; Sugiyama, Matsuoka, and Takagi 1995). Transgenic work on coffee was continued with different approaches. Some of them include *A. tumefaciens*-mediated genetic transformation (Grezes, Thomasset, and Thomas 1993; Madhava Naidu et al. 1998), biolistic gene delivery (van Boxtel et al. 1995), PEG-mediated direct DNA uptake (De Peña 1995), electroporation-mediated gene delivery (Fernandez and Menendez 2003), and *Agrobacterium rhizogenes*-mediated genetic transformation of *C. canephora* (Kumar et al. 2003, 2004, 2006).

In a recent report by Canche-Moo et al. (2006), the *Agrobacterium*-mediated plant transformation protocol was evaluated as a fast method to obtain genetically modified *C. canephora* plantlets. Leaf explants were used as source material for *A. tumefaciens*-mediated transformation involving a vacuum infiltration protocol, followed by a step of somatic embryogenesis induction and a final selection of the transformed plants. For the first time, transformation efficiency of 33% was achieved in coffee. The authors claim to have generated transgenic embryos in just 2 months (Canche-Moo et al. 2006). Stable transformation of *C. canephora* was obtained by particle bombardment of embryogenic tissue (Ribas, Galvao, et al. 2005). The somatic embryos and embryogenic tissue were bombarded with tungsten particles (M-25) carrying the plasmid pCAMBIA3301 (containing the *bar* and *uid A* genes) using a high-pressure helium microprojectile device. The transgenic plants were successfully transferred to the greenhouse (Ribas, Galvao, et al. 2005).

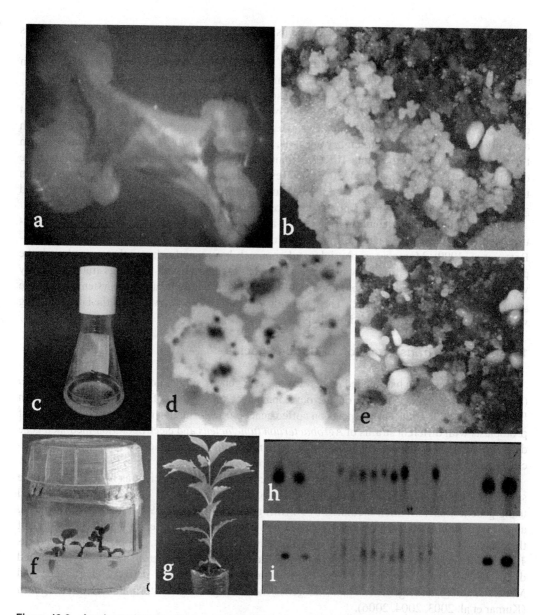

Figure 19.8 *Agrobacterium tumefaciens*-mediated transformation of *C. arabica*. (a) Callus on leaf explant after 30 days on primary callus induction medium. (b) embryogenic callus after 4 months on secondary callus medium. (c) Embryogenic cell suspension co-cultivated for 3–4 days with the LBA4404 strain of *A. tumefaciens* harboring the binary vector pCambia 1301. (d) Transgenic embryogenic calli showing blue color indicating strong *gus* gene expression. (e) Selective proliferation of cream colored transgenic embryogenic calli in the presence of 50 mgl −1 hygromycin. The nontransformed calli have turned brown. Transgenic calli were obtained from embryogenic calli co-cultivated with *A. tumefaciens* carrying a binary plasmid with *hpt* gene driven by CaMV35S promoter. (f) Germination of transgenic somatic embryos differentiated from calli selected in presence of hygromycin. (g) *C. arabica* transgenic plant from *A. tumefaciens*-mediated transformation, selected in presence of hygromycin. (h and i) Southern blot analysis of leaves from transformed *C. arabica* plants. Genomic DNA was digested with *Nco*I and *Hind*III and the blot was hybridized with the P[32]-labeled *gus* (h) and *hpt* (i) genes as probes.

19.9.2.1 Engineering Coffee Physiological Traits

Reduced caffeine levels have been obtained in transgenic coffee by the simultaneous down-regulation of caffeine biosynthesis of three distinct methylation steps of the caffeine biosynthetic pathway using RNAi. The caffeine content of the transgenic plants was reduced by up to 70%, indicating that it is possible to produce coffee beans that are intrinsically deficient in caffeine (Ogita et al. 2003, 2004). In a study reported by Kumar et al. (2004), *A. tumefaciens*-mediated genetic transformation was achieved in *C. canephora* for silencing of *N*-methyl transferase involved in caffeine biosynthesis.

Ribas, Kobayashi, et al. (2005) achieved inhibition of the ethylene burst by introducing the 1-aminocyclopropane-1-carboxylic acid (ACC) oxidase gene in antisense orientation; this technique would permit the understanding of genes involved in fruit maturation and ethylene production.

With regard to molecular breeding for agronomic qualities, transgenic coffee has been developed for resistance to the Lepidopteran coffee leaf miner, *Leucoptera coffeella*. The possible use of *Bacillus thuringiensis* in coffee biotechnology and the production of genetically modified coffee plants expressing the *cry1Ac* gene for resistance to leaf miner were reported (Oliveiro et al. 1998; Dufour et al. 2000; Leroy et al. 2000). Perthuis et al. (2005) reported that a field trial of transgenic clones of *C. canephora* transformed for resistance to the leaf miner was installed in French Guiana. Fifty-eight clones produced by transformation of the *C. canephora* were planted. They were harboring the pEF1*a* constitutive promoter of *Arabidopsis thaliana* controlling either the *B. thuringiensis* native gene for the *cry1Ac* insecticidal protein (eight clones) or a synthetic *cry1Ac* gene (53 clones). Over a 4-year period after plantation, a majority of the independent transformed clones harboring the synthetic gene displayed resistance against *L. coffeella* when compared to controls.

Root-knot nematodes significantly reduce yields on coffee plantations throughout the world. The addition of nematode resistance genes, a modified rice cystatin with or without a cowpea trypsin inhibitor, reduced the *Meloidogyne konaensis* population by over 70% to 26% in a biossay with transgenic *C. arabica* plants (Cabos et al. 2007).

On the other hand, production of transformed coffee plants with resistance to CBB was conducted by Cruz et al. (2004) with the α-*AI1* gene from common bean. The authors achieved transformation of *C. canephora* plants with this gene, and bioassays with the insect are underway to confirm functional validation of its proteins in coffee (Cruz et al. 2004).

Transgenic plants of *C. canephora* resistant to the herbicide ammonium glufosinate were regenerated from leaf explants after coculture with *A. tumefaciens* strain EHA105 harboring pCAMBIA3301, a plasmid that contains the bar, and the *uidA* genes, both under control of 35S promoter or bombarded with the pCAMBIA3301 plasmid. Presence and integration of the bar gene were confirmed by PCR and Southern blot analysis. Selected transgenic coffee plants sprayed with up to 1600 mg/L of Finale™, a herbicide containing glufosinate as the active ingredient, did not show any symptom of toxicity, retained their pigmentation, and continued to grow normally during *ex vitro* acclimation (Ribas et al. 2008).

19.9.2.2 Gene Discovery

19.9.2.2.1 Heterologous Genes

Collections of *B. thuringiensis* strains have been screened for toxicity against the coffee berry borer (CBB). Only those strains that showed a comparable protein content in their parasporal crystals to the *israelensis* type strain showed high levels of toxicity toward CBB. An accurate LC_{50} was estimated, using purified parasporal crystals from *B. thuringiensis* serovar *israelensis* type strain, at 219.5 ng cm^{-2} of diet. All the statistical requirements for a reliable estimator were fulfilled. This is the first report of *B. thuringiensis* serovar *israelensis* being active against a Coleopteran species (Mendez-Lopez, Basurto-Ríos, and Ibarra 2003).

Plant α-amylase inhibitors are proteins found in several plants, and they play a key role in natural defenses. Genes encoding α-amylase inhibitors were isolated from *Phaseolus* species and cloned into plant transformation vectors under the control of constitutive and seed-specific promoters. Transgenic plants of *Nicotiana* spp. were regenerated and its expression in transformed generations T_0 and T_1 was monitored by PCR amplification, enzyme-linked immunosorbent assay (ELISA), and immunoblot analysis, respectively. Seed protein extracts from selected transformants reacted positively with a polyclonal antibody raised against αAI-1, while no reaction was observed with untransformed tobacco plants. Immunological assays showed that the αAI-Pc1 gene product represented up to 0.05% of total soluble proteins in T_0 plants' seeds. Furthermore, recombinant αAI-Pc1 expressed in tobacco plants was able to inhibit 650% of digestive *H. hampei* α-amylases. The data suggest that the protein encoded by the α-amylase gene has potential to be introduced into coffee plants to increase their resistance to the CBB (Pereira et al. 2006). Coffee transformation experiments with α-amylase inhibitors are being developed (J. R. Acuña, 2008, pers. comm.).

19.9.2.2.2 Homologous Genes

Recent advances in coffee genomics consist of a huge collection of ESTs and BAC libraries. In the past few years, several projects of sequencing coffee ESTs have been initiated in the United States (Mueller et al. 2004), Brazil (http:// www.cenargen.embrapa.br/biotec/genomacafe/), Europe (Fernandez et al. 2004; Pallavicini et al. 2004), and Colombia (Cristancho et al. 2006). Libraries were constructed from various organs and tissues sampled at different developmental stages. Functional analysis has been initiated (Fernandez et al. 2004; Pallavicini et al. 2004). With these efforts in progress on EST sequencing, the number of coffee ESTs in the public domain will continue to increase in the future. In fact, the reports compiled to date, from different research groups working worldwide, indicate that around 250,000 good-quality ESTs from at least four different coffee species have already been produced (see Table 19.5).

19.10 COFFEE COMPOSITION AND HEALTH

The first coffee of the day is a make-or-break moment. A robust, flavorful cup can clear the mind, cheer the soul, and boost self-confidence. More than one billion people start their day by drinking a cup of coffee, making it the most popular drink worldwide, after water, and the coffee industry second in the worldwide economy, after oil. But coffee is not just a flavored drink or a commodity with unstable value at the stock exchange. Coffee is truly a functional plant that gives a beverage made from the seeds of its fruits that can have healthy effects if it is consumed in moderation (Santos and Lima 2007).

Eating a diet rich in fruits and vegetables may reduce risk for stroke and perhaps other cardiovascular diseases. Additionally, a diet rich in fruits and vegetables may reduce risk for type 2 diabetes, and may protect against certain cancers, such as stomach, liver, and colorectal cancer. Increasing evidence suggests that certain plant foods, such as cruciferous vegetables (e.g., cabbage, broccoli, Brussels sprouts, and cauliflower), carrots, and green leafy vegetables, among others, can protect us against many diseases. A plant may consist of several components, including leaves, roots, fruits, flowers, bark, stems, or seeds. Any of these parts may contain the active ingredients that give the plant its medicinal properties. Plant products constitute almost 95% of the human diet, with only 30 crop species providing most of the world's calories and proteins; thus, the identification of old species and the search for new ones with either nutritional or medicinal values, or both, is imperative. But compared with all other fruits, coffee is unique. So far growers have had to throw away the fruit because it is too perishable to process; they harvest only the bean, which has far more bioactive compounds than any other fruit or seed.

19.10.1 Chemical Constituents of the Coffee Bean

Concerns about potential health risks of coffee and caffeine consumption raised by epidemiological research in the past were likely exacerbated by associations between high intakes of coffee and unhealthy behaviors such as cigarette smoking and physical inactivity (Lima 1989; Lima et al. 1989, 1990a, 1990b). But coffee is not only caffeine. It has many compounds, such as the chlorogenic acids and their roast-dependent derivates (quinides, caffeic acid), as well as trigonelline and its derivate vitamin PP (niacin), minerals (K, Fe, Zn), melanoidins, diterpenes (cafestol, kahweol), and hundreds of volatiles (Trugo, Macrae, and Dick 1984), which cause one to experience pleasure from its aroma (Table 19.6).

19.10.1.1 Caffeine

The 1,3,7-trimethylxanthine is a purine alkaloid that occurs naturally in coffee beans. At intake levels associated with coffee consumption, caffeine appears to exert most of its biological effects through the antagonism of the A1 and A2A subtypes of the adenosine receptor. Adenosine is an endogenous neuromodulator with mostly inhibitory effects, and adenosine antagonism by caffeine results in effects that are generally stimulatory (Ascherio, Zhang, and Hernan 2001). Some physiological effects associated with caffeine administration include central nervous system stimulation, acute but transient elevation of blood pressure, increased metabolic rate, and a light diuresis. Caffeine is rapidly and almost completely absorbed in the stomach and small intestine and distributed to all tissues, including the brain. Caffeine metabolism occurs primarily in the liver, where the activity of the cytochrome P450 isoform CYP1A2 accounts for almost 95% of the primary metabolism of caffeine (Corrao et al. 2001). Caffeine concentrations in coffee beverages can be quite variable depending on whether arabica coffee (0.8–1.8%) or robusta coffee (1.5–3.3%) (Anthony et al. 1993; Bertrand et al. 2003; Campa et al. 2005) is used.

Table 19.6   Average Content of Substances in Arabica and Robusta Roasted Coffee Beans

Compound	Roasting Stability[a]	Arabica Coffee (%)	Robusta Coffee (%)
Caffeine	Thermostable	1.2	2.2
Trigonelline	Depends on roasting	1	0.75
Chlorogenic acids	Depends on roasting	6.5	10
Proteins	Depends on roasting	11	11
Free amino acids[b]	Depends on roasting	0.5	0.8
Mineral salts	Depends on roasting	4.2	4.4
Sugars[b]	Depends on roasting	52	52
Lipids[b]	Depends on roasting	16	10
Others[c]	Depends on roasting	7	8

[a] Coffee roasted too much (very dark roasting) can have mainly caffeine, ashes, and traces of the many other compounds.

[b] Proteins, amino acids, sugars, and lipids are volatilized during the roasting process, giving birth to almost 1,000 compounds (aroma and taste) yet to be identified.

[c] Cafestol, oils, pigments, ashes, water, etc.

Coffee blends made with arabica and robusta may explain the wide range of caffeine content in brewed coffee. The caffeine content of the same type of coffee purchased from stores can vary from 130 to 282 mg per 8-ounce serving. Nowadays, caffeine citrate in a loading dose of 20 mg/kg is the drug of choice for the treatment of apnea in premature infants; caffeine is used to relieve headaches after spinal puncture and is part of the medicines used to treat migraine (Lima 1990a).

19.10.1.2 Chlorogenic Acids

Chlorogenic acids are a family of esters formed between quinic and *trans*-cinnamic acids, which are an important group of dietary phenols. Phenolic acids occur widely in nature as mixtures of esters, ethers, or free acids. The caffeic, ferulic, and coumaric acids are phenolic compounds derived from cinnamic acid and occur naturally in the form of mono- or diesters with the aliphatic alcohols of quinic acid, under the common name of chlorogenic acid (CGA). Coffee is made up of 10% of these compounds. Chlorogenic acids were first isolated from green coffee in 1908, almost 100 years after caffeine (Clifford 1999).

Although these compounds have no chloride in their formula, they received this name because early studies about the chemistry of coffee showed that green coffee extracts produced a green color after the addition of a solution of ferric chloride. This reaction was used as evidence for the presence of tannic acid in coffee beans. The most common individual chlorogenic acid is 5-*O*-caffeoylquinic acid, which is still often called chlorogenic acid. For those who drink it, coffee represents the richest dietary source of chlorogenic acids and cinnamic acids (caffeic acid). The CGA content of a 200 mL (7 oz) cup of coffee has been reported to range from 70 to 350 mg, which would provide about 35–175 mg of caffeic acid. Studies in human volunteers show that the ingested chlorogenic acid and the caffeic acid are absorbed intestinally.

Thus, about two-thirds of ingested chlorogenic acid reaches the colon, where it may be metabolized by the colonic microflora. In the colon, chlorogenic acid is likely hydrolyzed to caffeic acid and quinic acid. The presence of bacterial metabolites of chlorogenic acid in the urine suggests that they are absorbed in the colon. Chlorogenic acid and caffeic acid have antioxidant activity *in vitro* and this may explain many healthy effects of regular coffee consumption (Takeda et al. 2002, 2003). A general nomenclature system classifies chlorogenic acids into five major groups:

1. Caffeoylquinic acids (CQAs) are esters of caffeic acid and quinic acid.
2. Dicaffeoylquinic acids (diCQAs) are esters involving two residues of caffeic acid attached to the same residue of quinic acid.
3. Feruloylquinic acids (FQAs) are represented by esters of ferulic acid and quinic acid and monomethyl ethers of the CQA.
4. *p*-Coumaroylquinic acids (CoQAs) are esters of *p*-coumaric acid (4-hydroxycinnamic acid) with quinic acid.
5. Caffeoyferuloylquinic acids (CFQAs) are a poorly studied group of esters consisting of one residue of caffeic acid and one residue of ferulic acid attached to the same residue of quinic acid corresponding to monomethyl ethers of the diCQA.

The total CGA contents of coffee can vary from 6.6 to 14.4% in robusta coffee (Anthony et al. 1993; Campa et al. 2005) and from 5.2 to 8.3% in *C. arabica* (Anthony et al. 1993; Bertrand et al. 2003). Roasting has a marked effect on the final content of all compounds found in coffee except caffeine (Trugo, Macrae, and Dick 1984). During the roasting process, the chlorogenic acids lose a molecule of water and thus give birth to the chlorogenic acid lactones (CGLs) or quino-lactones.

These isomers (lactones or quinides) constitute a new group of compounds that seems to have healthy properties that aid the human body in many ways, such as having an antidiabetes effect (increased glucose uptake by cells), fighting alcoholism and cirrhosis by blocking μ-opioid receptors into the brain (thus blocking the craving for drugs), having antiadenosine effects (thus protecting cells against caffeine), and improving microcirculation (Wynne et al. 1987; Boublik et al. 1993; De Paulis et al. 2004). These compounds may well explain the epidemiological data showing an inverse correlation between regular coffee intake and adult diabetes type 2, depression and suicide (Kawachi et al. 1996), alcoholism and cirrhosis (O'Brien 1996; O'Malley 1996; Lima et al. 1991), and Parkinson's and Alzheimer's diseases, as well as some forms of cancer (La Vecchia 2005).

19.10.1.3 Trigonelline

Trigonelline is a substance identified in several species of fruits and seeds (Trugo, Macrae, and Dick 1983). Depending on the coffee species, the amount varies from 1.0 to 1.3% in *C. arabica* (Campa et al. 2004) and around 1.0% in robusta (Bertrand et al. 2003; Campa et al. 2004). It rapidly degrades during roasting. Trigonelline is important for coffee flavor and aroma and, during the roasting process, leads to the formation of different by-products, such as nicotinic acid or niacin (vitamin PP) and a volatile fraction in which pyridine and methylpyrrole have been identified. Methylpyrrole is a basic structure of pheromones secreted in minute amounts by insects and vertebrates that may be shed freely into the environment or deposited in carefully chosen locations.

Pheromones are widely used to promote aggregation. Among termites and ants, different pheromones with methylpyrrole transmit messages needed to coordinate the complex activities of a colony. Pheromones play a role in sexual attraction and copulatory behavior, and they have been shown to influence the sexual development of many mammals as well as of insects. Pheromones may be involved in human sexual response. The human female's sensitivity to musk-like odors is greatest around the time of ovulation, which some researchers interpret as proof of the ancestral presence of a musky pheromone in the male (Keller et al. 2009).

Recent and pioneer studies using functional magnetic resonance imaging (fMRI) in the evaluation of neural correlates with coffee aroma have shown that areas of pleasure such as the ventral tegmental area (VTA), the nucleus accumbens, and amygdala are stimulated by coffee aroma (Figure 19.9). The mesolimbic pathway is one of the dopaminergic pathways in the brain. The pathway begins in the VTA of the midbrain and connects to the limbic system via the nucleus accumbens, the amygdala, and the hippocampus, as well as to the medial prefrontal cortex. It is known to be involved in modulating behavioral responses to stimuli that activate feelings of reward, motivation, and reinforcement.

This suggests that coffee volatiles such as methylpyrrole (among many others yet to be identified) seem to cause not only olfactory impact but also far more sophisticated effects in humans. Coffee aroma seems to recruit at least two more types of reward systems: the VTA–striatum mesolimbic network, which is involved in basic animal rewards such as nutrition and sex for reproduction as well as areas that play key roles in social attachment and affiliative reward mechanisms. This may explain why coffee is a unique socializing beverage and the alluring aroma of steaming hot coffee can drag pedestrians to meet in coffeehouses they never forget (Moll et al. 2006; Lima and Moll 2009).

Coffee is the only plant capable of forming a vitamin—niacin—at high temperatures because in all other plants there is destruction. Niacin was shown to be the pellagra-preventing or "PP" vitamin, a word derived from the Italian "pelle agra." Pellagra is characterized by skin, gastrointestinal, and central nervous system abnormalities—a triad frequently referred to as the "three Ds" (dermatitis, diarrhea, and dementia). Pellagra is now considered a rare disease, but it still can occur. Niacin is widely distributed in plant and animal foods, but in relatively small amounts, except in meat (especially organs), fish, and whole-meal cereals. Cooking causes little destruction of nicotinic acid but considerable amounts may be lost in cooking water and "drippings" from cooked meat if these are discarded. Commercial processing and storage of foodstuffs cause little loss. Roasted coffee beans and instant coffee powder may contain 10–40 mg niacin per 100 g of coffee. Pellagra is common among alcoholics, and coffee can help fight both alcoholism and pellagra.

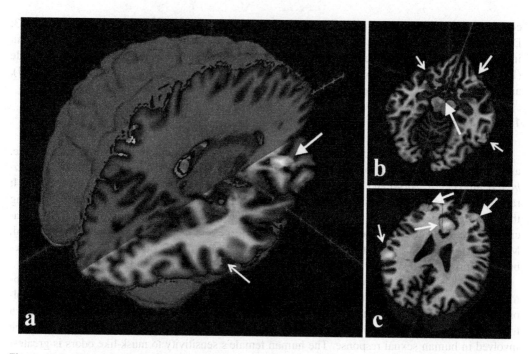

Figure 19.9 **(See color insert.)** Brain stimulation by coffee aroma. Results of functional Magnetic Resonance Imaging (fMRI) analysis showing the effects of coffee aroma on specific neuronal centers in the brain (arrows) governing smell (a), pleasure (b), as well as attention and learning (c). See text for more explanation. Cerebral regions under activation are in yellow.

19.10.1.4 Proteins

The final composition of proteins and free amino acids of green, roasted, and instant coffee varies. Total protein content is in the range of 11–13% for green, 13–15% for roasted, and 16–21% for instant coffee. The protein composition of coffee appears to vary very little with the variety of the bean, but a variation exists in the total protein content between green and roasted coffee as protein content decreases and the physicochemical properties of the proteins are altered as a result of denaturation. Green coffee also has small quantities of amino acids, such as alanine, arginine, asparagine, cysteine, glutamic acid, glycine, histidine, isoleucine, leucine, lysine, methionine, phenylalanin, proline, serine, threonine, tyrosine, and valine. Also, some amino acids, such as arginine, cysteine, lysine, serine, and threonine, decrease markedly during the roasting process. Others, like glutamic acid, leucine, phenylalanine, and praline, have their amount increased, although significant levels are not present in the final beverage. The new products formed are volatiles and aromatic. The enzymes of green coffee represent only a minor part of the proteins and they are destroyed during roasting.

19.10.1.5 Carbohydrates

Carbohydrates are also present in coffee (Trugo and Macrae 1982; Bertrand et al. 2003). Sucrose is the major free sugar (ranging from 6 to 7%) with low levels of glucose (0.5–1%) and fructose (0.3%). The free sugar composition of green coffee changes considerably during roasting. Sucrose can be destroyed while small amounts of glucose, fructose, and arabinose and traces of galactose can be detected in roasted coffee. Free sugars like mannose, ribose, and xylose can also be found. Green coffee beans have a very high content of polysaccharides (50–60%), mainly represented by mannose, arabinose, galactose, and glucose. Cellulose is also present in the green coffee bean (approximately 5%). The polysaccharides found in roasted coffee have basically the same structure as those found in green coffee, containing mannose, galactose, glucose, and arabinose. Arabica and robusta varieties have almost the same polysaccharide composition. Different polysaccharide fractions have been isolated in green coffee consisting of galactan, araban, and mannan together with cellulose. Roasting can increase the amount of mannan but decrease araban and galactan, as well as decrease the total amount of carbohydrates from a light to heavy roast. An enormous amount of research has yet to be done concerning the sugars found in coffee and their relationship with human pleasure and health.

19.10.1.6 Lipids

Lipids are present in green coffee at relatively high levels. Significant differences are found in their total content between species, with arabica coffee containing 14–15% (Bertrand et al. 2003) and robusta containing 7–10%. They are important vehicles for coffee aroma and are relatively stable during the roasting process. Apparently there is even an increase in their levels in roasted coffee. Triglycerides account for 70–80% of the content of coffee oils. Free fatty acids are also present at very low levels (range of 1%). However, significant amounts are liberated during the roasting process, reaching levels up to 3%. Linoleic acid is the main component followed by palmitic, oleic, stearic, arachidic, and linolenic acids. Small amounts of myristic and traces of palmitoleic, gadoleic, lignoceric, and margaric acids are also present. After roasting, the proteins, sugars, and lipids take part in complex processes such as Maillard (amino acids with reducing sugars) and Strecker (amino acids with diketones) reactions and part is destroyed.

These reactions create almost 1,000 volatile compounds responsible for the unique aroma and taste of coffee. To most animals, the sense of smell is a matter of life and death. Smell is vital to mankind. The sense of a specific scent from our childhood can bring us back long forgotten memories such as the tar of the fishing boats from long ago or the scent of honeysuckle roses. The aroma of coffee and the physiology of olfaction are related to pleasure as well as to love and passion. The larger the number of different kinds of smells one learns from childhood to adulthood, the more one's brain will be able to translate into pleasure and emotions. For this, among other reasons, regular intake of coffee is recommended to all youth (Lima et al. 2000; Santos and Lima 2007, 2009).

The diterpenes cafestol and kahweol have been associated with higher serum total and LDL cholesterol concentrations in some observational studies but not in others. The observation that the positive association between coffee consumption and serum cholesterol was more consistent in Scandinavia, where boiled coffee was popular at the time, than in other European countries and the United States, where filtered coffee was more popular, led to the hypothesis that the brewing method was critical to the cholesterol-raising effect of coffee. The cholesterol-raising factors, first isolated in coffee oil, were later found to be cafestol and kahweol. These diterpenes are extracted from ground coffee during brewing, but are mostly removed from coffee by paper filters. Although diterpene concentrations are relatively high in espresso coffee, the small serving size makes it an intermediate source of cafestol and kahweol. The mechanisms for the effects of these diterpenes on lipoprotein metabolism are not yet clear, but *in vitro* studies have shown that the diterpenes also have an anticancer effect (La Vecchia 2005).

19.10.1.7 Other Compounds

Coffee is also a rich source of minerals, mainly potassium (K, 1.5 wt%), magnesium (Mg, 0.15 wt%), and calcium (Ca, 0.12 wt%), together with a great number of others minerals in smaller amounts (range of 0.0005–0.002% w/w), such as sodium (Na), iron (Fe), manganese (Mn), rubidium

(Rb), zinc (Zn), copper (Cu), and trace elements (parts per million) like chromium (Cr), vanadium (V), nickel (Ni), cobalt (Co), lead (Pb), cadmium (Cd), and molybdenum (Mo). It is important to observe that while coffee is rich in potassium, it is low in sodium, contrary to many energy beverages not recommended for people with cardiovascular diseases such as hypertension (Willett et al. 1996).

19.10.2 Coffee Quality and Health

Quality of coffee is important as is the proper roasting of the beans. As to health, coffee can be classified into three major groups: (1) dominance of antioxidant chlorogenic acids (those roasted within colors Agtron SCAA #95 to #75, (2) harmonic compounds (those roasted within colors Agtron SCAA #65 to #45), and (3) caffeine dominance (those roasted within colors Agtron SCAA # 35 and #25), according to the classification of the Specialty Coffee Association of America (http://scaa.org)

Ongoing studies with students at all levels in Brazil are showing higher rates of participation in school breakfast programs that offer coffee with (5–10 years of age) or without (10–20 years of age) milk in the short term and long term. There is an improvement in students' performance in a broad range of psychosocial and academic measures as well as lower incidence of obesity, apathy, depression, and alcohol consumption. Considering the relationship of junk food, which includes soft drinks, and the great incidence of obesity among youth worldwide, coffee with or without milk might be the healthiest option as a beverage for school breakfast programs. Coffee can even be of help fighting many other important health problems, such as later adult diabetes, depression/suicide, alcoholism/cirrhosis, and Parkinson's and Alzheimer's diseases, as well as cancer (Lima 1990a, 1990b; Klatsky et al. 1993; Higdon et al. 2006).

19.11 USEFUL COFFEE WEBSITES

http://www.coffeegenome.org
http://www.thecoffeeguide.org
http://www.coffeeresearch.org
http://www.coffeeinstitute.org
http://www.coffeekids.org
http://www.sustainable-coffee.net
http://www.cosic.org
http://www.coffeescience.org
http://www.cenicafe.org
http://www.ico.org

Note: All authors contributed equally to this chapter. This review was completed in the middle of 2009.

REFERENCES

Aga, E., and Bryngelsson, T. 2006. Inverse sequence-tagged repeat (ISTR) analysis of genetic variability in forest coffee (*Coffea arabica* L.) from Ethiopia. *Genetic Resources and Crop Evolution* 53:721–728.
Aga, E., Bryngelsson, T., Bekele, E., and Salomon, B. 2003. Genetic diversity of forest arabica coffee (*Coffea arabica* L.) in Ethiopia revealed by random amplified polymorphic DNA (RAPD) analysis. *Hereditas* 138:36–46.
Aggarwal, R. K., Hendre, P. S., Varshney, et al. 2007. Identification, characterization and utilization of EST-derived genetic microsatellite markers for genome analyses of coffee and related species. *Theoretical and Applied Genetics* 114:359–372.

Albarrán, J., Bertrand, B., Lartaud, M., and Etienne, H. 2005. Cycle characteristics in a temporary immersion bioreactor affect regeneration, morphology, water and mineral status of coffee (*Coffea arabica*) somatic embryos. *Plant Cell Tissue and Organ Culture* 81:27–36.

Almeida, A., Guimarães, B., Soares, C., et al. 2008. Differential proteomic analysis indicates that modulation in the expression of photosystem ii proteins may be involved in differential drought response in *Coffea canephora* genotypes. In *Proceedings of XXII International Conference on Coffee Science*, PB624. ASIC, Campinas, Brazil.

Almeida, A., Soares, C., Guimarães, B., et al. 2008. Differential proteomic analysis indicates several proteins involved in drought stress response in root of *Coffea canephora* genotypes. In *Proceedings of XXII International Conference on Coffee Science*, PB625. ASIC, Campinas, Brazil.

Alvarado, A. G. 2005. Evolution of *Hemileia vastatrix* in Colombia. In *Durable resistance to leaf rust*, ed. Zambolin, L., Zambolin, E. M., and Varzea, V. M. P., 99–116. Universidade Federal de Vicosa, Vicosa, Brazil.

Alvarado, A. G., and Cortina, G. H. A. 1997. Comportamiento agronómico de progenies de híbridos triploides de *Coffea arabica* var. Caturra × (Caturra × *Coffea canephora*). *Cenicafé* 48:73–91.

Alves-Ferreira, M., Cruz, F., Kalaoum, S., et al. 2008. Functional genomics studies in "genoma café" identified two coffee homeobox genes involved in drought stress responses. In *Proceedings of XXII International Conference on Coffee Science*, PB632. ASIC, Campinas, Brazil.

Anthony, F. 1992. Les ressources génétiques des caféiers: Collecte, gestion d'un conservatoire et évaluation de la diversité génétique. Collection Travaux & Documents Microfichés no. 81, ORSTOM, Paris.

Anthony, F., Astorga, C., Avendaño, J., and Dulloo, E. 2007. Conservation of coffee (*Coffea* spp.) genetic resources in the CATIE field gene bank. In *Complementary strategies for ex situ conservation of coffee (Coffea arabica L.) genetic resources. A case study in CATIE, Costa Rica*, ed. Engelmann, F., Dulloo, E., Astorga, C., Dussert, S., and Anthony, F., 23–34. Topical Reviews in Agricultural Biodiversity, Bioversity International, Rome.

Anthony, F., Berthaud, J., Guillaumet, J-L., and Lourd, M. 1987. Collecting wild *Coffea* species in Kenya and Tanzania. *Plant Genetic Resources Newsletters* 69:23–29.

Anthony, F., Bertrand, B., Quiros, O., et al. 2001. Genetic diversity of wild coffee (*Coffea arabica* L.) using molecular markers. *Euphytica* 118:53–65.

Anthony, F., Clifford, M. N., and Noirot, M. 1993. Biochemical diversity in the genus *Coffea* L.: Chlorogenic acids, caffeine and mozambioside contents. *Genetic Resources and Crop Evolution* 40:61–70.

Anthony, F., Combes, M-C., Astorga, C., Bertrand, B., Graziosi, G., and Lashermes, P. 2002. The origin of cultivated *Coffea arabica* L. varieties revealed by AFLP and SSR markers. *Theoretical and Applied Genetics* 104:894–900.

Anthony, F., Couturon, E., and de-Namur, C. 1985. Les caféiers sauvages du Cameroun. Résultats d'une mission de prospection effectuée par l'ORSTOM en 1983. In *Proceedings of XI Colloque Scientifique International sur le Café*, 495–505. ASIC, Lausane.

Anthony, F., Dussert, S., and Dulloo, E. 2007. The coffee genetic resources. In *Complementary strategies for ex situ conservation of coffee (Coffea arabica L.) genetic resources. A case study in CATIE, Costa Rica*, ed. Engelmann, F., Dulloo, E., Astorga, C., Dussert, S., and Anthony, F., 12–22. Topical Reviews in Agricultural Biodiversity, Bioversity International, Rome.

Anzueto, F., Bertrand, B., Sarah, J-L., Eskes, A. B., and Decazy, B. 2001. Resistance to *Meloidogyne incognita* in Ethiopian *Coffea arabica* accessions. *Euphytica* 118:1–8.

Ascherio, A., Zhang, S. M., and Hernan, M. A. 2001. Prospective study of caffeine consumption and risk of Parkinson's disease in men and women. *Annals of Neurology* 50:56–63.

Barre, P., Laissac, M., D'Hont, A., et al. 1998. Relationship between parental chromosomic contribution and nuclear DNA content in the *Coffea* interspecific hybrid: *C. pseudozanguebariae* × *C. liberica* var. dewevrei. *Theoretical and Applied Genetics* 96:301–305.

Barry-Etienne, D., Bertrand, B., Vasquez, N., and Etienne, H. 2002. Comparison of somatic embryogenesis-derived coffee (*Coffea arabica* L.) plantlets regenerated *in vitro* or *ex vitro*: Morphological, mineral and water characteristics. *Annals of Botany* 90:77–85.

Barton, C. R., Adams, T. L., and Zorowitz, M. A. 1991. Stable transformation of foreign DNA into *Coffea arabica* plants. In *Proceedings of XIV International Conference on Coffee Science*, 460–464. ASIC, Lausane.

Baruah, A., Naik, V., Hendre, P. S., et al. 2003. Isolation and characterization of nine microsatellite markers from *Coffea arabica* L., showing wide cross-species amplifications. *Molecular Ecology Notes* 3:647–650.

Berthaud, J. 1986a. Les ressources génétiques pour l'amélioration des caféiers Africains diploïdes. Collection Travaux et Documents, no.188, ORSTOM, Paris.

————. 1986b. Les ressources génétiques pour l'amélioration des caféiers Africains diploides. Evaluation de la richesse génétique des populations sylvestres et de ses mécanismes organisateurs: Conséquences pour l'application. Travaux et Documents de l'ORSTOM, no. 188, France, 383.

Berthaud, J., and Charrier, A. 1988. Genetics resources of *Coffea.* In *Coffee,* ed. Clarke, R. J., and Macrae, R., vol. 4, *Agronomy,* 1–42. Elsevier Applied Science, London.

Berthaud, J., and Guillaumet, J-L. 1978. Les caféiers sauvages en Centrafrique: Résultats d'une mission de prospection (Janvier–Février 1975). *Café Cacao Thé* 3:171–186.

Berthaud, J., and Pernès, J. 1978. Variabilité lue sur les variables qualitatives. *Bulletin IFCC* 14:63–65.

Berthou, F., and Trouslot, P. 1977. L'analyse du polymorphisme enzymatique dans le genre *Coffea:* Adaptation d'une méthode d'électrophorèse en série, premiers résultats. In *Proceedings of VIII Colloque Scientifique International sur le Café,* 373–383. ASIC, Abidjan, Ethiopia.

Berthou, F. Chaure, R., and Pernès, J. 1978. Variabilité lue sur les variables quantitatives Bull. *IFCC* 14: 57–62.

Berthou, F., Trouslot, P., Hamon, S., Vedel, F., and Quetier, F. 1980. Analyse en électrophorèse du polymorphisme biochimique des caféiers: Variation enzymatique dans dix-huit populations suavages. Variation de l'ADN mitochondrial dans les espèces *C. canephora, C. eugenioides* et *C. arabica. Café Cacao Thé* 16:313–326.

Berthouly, M., Alvarad, D., Carrasco, C., and Teisson, C. 1994. *In vitro* micropropagation of coffee sp. by temporary immersion. In *Abstracts 8th International Congress of Plant Tissue and Cell Culture,* 162. Florence, Italy.

Berthouly, M., and Etienne, H. 1999. Somatic embryogenesis of coffee. In *Somatic embryogenesis in woody plants,* ed. Jain, S. M., Gupta, P. K., and Newton, R. J., 259–288. Kluwer Academic Publishers, London.

Berthouly, M., and Michaux-Ferrière, N. 1996. High frequency somatic embryogenesis in *Coffea canephora:* Induction conditions and histological evolution. *Plant Cell Tissue and Organ Culture* 44:169–176.

Bertrand, B., and Anthony, F. 2008. Genetics of resistance to root-knot nematodes (*Meloidogyne* spp.) and breeding. In *Plant-parasitic nematodes of coffee,* ed. Souza, R. M., 165–190. APS Press & Springer, New York.

Bertrand, B., Etienne, H., Cilas, C., Charrier, A., and Baradat, P. 2005. *Coffea arabica* hybrid performance for yield, fertility and bean weight. *Euphytica* 141:255–262.

Bertrand, B., Guyot, B., Anthony, F., and Lashermes, P. 2003. Impact of the *Coffea canephora* gene introgression on beverage quality of *C. arabica. Theoretical and Applied Genetics* 107:387–394.

Bertrand, B., Ramirez, G., Topart, P., and Anthony, F. 2002. Resistance of cultivated coffee (*Coffea arabica* and *C. canephora*) to the corky-root caused by *Meloidogyne arabicida* and *Fusarium oxysporum,* under controlled and field conditions. *Crop Protection* 21:713–719.

Bertrand-Desbrunais, A., Noirot, M., and Charrier, A. 1991. Minimal growth *in vitro* conservation of coffee (*Coffea* spp.). 1. Influence of low concentrations of 6-benzyladenine. *Plant Cell Tissue and Organ Culture* 27:333–339.

————. A. 1992. Slow growth *in vitro* conservation of coffee (*Coffea* spp.). 2. Influences of reduced concentrations of sucrose and low temperature. *Plant Cell Tissue and Organ Culture* 31:105–110.

Bettencourt, A. J. 1973. Consideracoes gerais sobre o Híbrido de Timor. Instituto Agronómico de Campinas, 20. Sao Paulo, Brazil, circular no. 23.

Bettencourt, A. J., and Rodrigues, C. J., Jr. 1988. Principles and practice of coffee breeding for resistance to rust and other disease. In *Coffee. Agronomy,* vol. 4, ed. Clarke, R. J., and Macrae, R., 199–234. Elsevier, London.

Bhat, P. R., Krishnakumar, V., Hendre, P. S., et al. 2005. Identification and characterization of gene-derived EST–SSR markers from robusta coffee variety 'CxR' (an interspecific hybrid of *Coffea canephora* and *Coffea congensis*). *Molecular Ecology Notes* 5:80–83.

Boaventura, Y. M. S., and Cruz, N. D. da. 1987. Citogenética do híbrido interespecífico (*Coffea arabica* L. var Bourbo x *C. canephora* Pierre ex Froehner var. Robusta – Linden-Chev) que originou o café Icatu. *Turrialba,* 37: 171–178.

Boisseau, M., Aribi, J., Carneiro, R. M. D. G., and Anthony, F. 2009. Resistance to *Meloidogyne paranaensis* in wild *Coffea arabica* L. *Tropical Plant Pathology* 34:38–41.

Boublik, J. H. et al. 1983. Coffee contains potent opiate receptor binding activity. *Nature* 301:246–248.

Bouharmont, J. 1963. Somatic chromosomes of *Coffea* species. *Euphytica* 12:254–257.

Bridson, D., and Verdcourt, B. 1988. *Coffea.* In *Flora of tropical east Africa—Rubiaceae (part 2),* ed. Polhill, R. M., 703–727. A. A. Balkema, Rotterdam, the Netherlands.

Bridson, D. M. 1985. The lectotypification of *Coffea liberica* (Rubiaceae). *Kew Bulletin* 40:805–807.

————. 1987. Nomenclatural notes on *Psilanthus,* including *Coffea* sect. *Paracoffea* (Rubiaceae tribe *Coffeeae*). *Kew Bulletin* 42:453–460.

————. 1988. Classification. In *Coffee*, ed. Wrigley, G., 61–75. Longman Scientific and Technical, Harlow, England.

Butt, D. J., and Butters, B. 1966. The control of coffee berry disease in Uganda. Specialist meeting on coffee research in east Africa (Nairobi), 11, 8–11.

Cabos, R. Y., Sipes, B. S., Schmitt, D. P., Atkinson, H. J., and Nagai, C. 2007. Plant proteinase inhibitors as a natural and introduced defense mechanism for root-knot nematodes in *Coffea arabica*. *Journal of Nematology* 39:100.

Campa, C., Ballester, J. F., Doulbeau, S., Dussert, S., Hamon, S., and Noirot, M. 2004. Trigonelline and sucrose diversity in wild *Coffea* species. *Food Chemistry* 88:39–43.

Campa, C., Doulbeau, S., Dussert, S., Hamon, S., and Noirot, M. 2005. Diversity in bean caffeine content among wild *Coffea* species: Evidence of a discontinuous distribution. *Food Chemistry* 91:633–637.

Canche-Moo, R. L. R., Ku-Gonzalez, A., Burgeff, C., Loyola-Vargas, V. M., Rodríguez-Zapata, L. C., and Castaño, E. 2006. Genetic transformation of *Coffea canephora* by vacuum infiltration. *Plant Cell Tissue and Organ Culture* 84:373–377.

Capot, J. 1972. L'amélioration du caféier en Cote d'Ivoire. Les hybrides Arabusta. *Café Cacao Thé* 16:3–18.

————. 1975. Obtention et perspectives d'un nouvel hybride de cafeier en Cote d'Ivoire: l'Arabusta. In *Proceedings of VII International Conference on Coffee Science*, 449–457. ASIC, Lausane.

Cardenas, F., Galeano, N., O'Brien, K., Yepes, M., Geoffroy, M., Buell, R., and Gaitan, A. 2008. Transcript profiles in compatible and incompatible host-coffee leaf rust interactions. In *Proceedings of XXII International Conference on Coffee Science*, PB623. ASIC, Campinas, Brazil.

Carneiro, M. F. 1999. Advances in coffee biotechnology—Review. Ag Biotechnet, vol. 1, ABN 006, 1–14. http://www.agbiotecnet.com/reviews/Jan99/HTML/Carneiro.htm

Carvalho, A. 1988. Principles and practice of coffee plant breeding for productivity and quality factors: *Coffea arabica*. In *Coffee. Agronomy*, vol. 4, ed. Clarke, R. J., and Macrae, R., 129–165. Elsevier, London.

Carvalho, A., Eskes, A. B., and Fazuoli, L. C. 1989. Breeding for rust resistance in Brazil. In *Coffee rust epidemiology, resistance and management*, ed. Kushalappa, A. C., and Eskes, A. B., 295–307. CRC Press Inc., Boca Raton, FL.

Carvalho, A., Ferwerda, F. P., Frahm-Leliveld, J. A., Medina, D. M., Mendes, A. J. T., and Monaco, L. C. 1969. Coffee: *Coffea arabica* and *Coffea canephora* Pierre ex Froehner. In *Outlines of perennial crop breeding in the tropics*, 189–241. Wageningen Agricultural University, the Netherlands.

Carvalho, A., Medina-Filho, H. P., Fazuoli, L. C., Guerreiro-Filho, O., and Lima, M. M. A. 1991. Aspectos genéticos do cafeeiro. *Revista Brasileira de Genetica* 14:135–183.

Carvalho, A., and Monaco, L. C. 1968. Relaciones genéticas de especies seleccionadas de *Coffea*. *Café (Lima) IICA* 9:1–19.

Chaparro, A. P., Cristancho, M. A., Cortina, H. A., Gaitán, A. L. 2004. Genetic variability of *Coffea arabica* L. accessions from Ethiopia evaluated with RAPDs. *Genetic Resources and Crop Evolution* 51:291–297.

Castillo, Z. J., and Moreno, R. L. 1988. La variedad Colombia: Selección de un cultivar compuesto resistente a la roya del cafeto, 171. Cenicafé. Chinchiná, Colombia.

Castillo, Z. J., Moreno, R. G., and Lopez, D. S. 1976. Uso de resistencia genetica a *Hemileia vastatrix* Berk & Br. existente en germoplasma de café en Colombia. *Cenicafé* 27:3–25.

Charrier, A. 1978a. La structure génétique des caféiers spontanés de la région Malgache (*Mascarocoffea*). Leurs relations avec les caféiers d'origine Africaine (*Eucoffea*). Mémoires ORSTOM 87. Paris.

————. 1978b. Synthèse de neuf années d'observation e d' expérimentation sur les *Coffea arabica* collectés en Ethiopie par une mission ORSTOM. In Etude de la structure et de la variabilité génétique des caféiers, 4–10, Bulletin no. 14, ORSTOM–IRCC, París.

Charrier, A., and Berthaud, J. 1975. Variation de la teneur en caféine dans le genre *Coffea*. *Café Cacao Thé* 19:251–263.

————. 1985. Botanical classification of coffee. In *Coffee: Botany, biochemistry and production of beans and beverage*, ed. Cliffort, M. N., and Wilson K. C., 13–47. Croom Helm, London.

————. 1988. Principles and methods in coffee plant breeding: *Coffea canephora* Pierre. In *Coffee. Agronomy*, 167–197. Elsevier Applied Science, London.

Charrier, A., and Eskes, A. B. 2009. Botany and genetics of coffee. In *Coffee: Growing, processing, sustainable production. A guidebook for growers, processors, traders, and researchers*, 2nd ed., ed. Wintgens, J. N. Wiley–VCH, Weinheim.

Chevalier, A. 1947. Les caféiers du globe systématique des caféiers et faux-caféiers et faux-caféirs maladies et insect nuisibles. *Encyclopédie Biologique XXVIII.* Paris, 356 pp.

Clarindo, W. R., and Carvalho, C. R. 2006. A high quality chromosome preparation from cell suspension aggregates culture of *Coffea canephora. Cytologia* 71:243–249.

———. 2008. First *Coffea arabica* karyogram showing that this species is a true allotetraploid. *Plant Systematics and Evolution* 274:237–341.

Clifford, M. N. 1999. Chlorogenic acids and other cinnamates: Nature, occurrence and dietary burden. *Journal of the Science of Food and Agriculture* 79:362–372.

Combes, M-C., Andrzejewski, S., Anthony, F., et al. 2000. Characterization of microsatellite loci in *Coffea arabica* and related coffee species. *Molecular Ecology* 9:1178–1180.

Corrao, G. et al. 2001. Coffee, caffeine and the risk of liver cirrhosis. *Annals of Epidemiology* 11:458–465.

Coulibaly, I., Revol, B., Noirot, M., et al. 2003. AFLP and SSR polymorphism in a *Coffea* interspecific backcross progeny [(*C. heterocalyx* × *C. canephora*) × *C. canephora*]. *Theoretical and Applied Genetics* 107:1148–1155.

Couturon, E. 1982. Obtention d'haploïde spontanés de *Coffea canephora* Pierre par l'utilisation du greffage d'embryons. *Café Cacao Thé* 26:155–160.

Cramer, P. J. S. 1957. A review of literature of coffee research in Indonesia. Turrialba, Costa Rica, 262, IICA.

Cristancho, M. A., Rivera, C., Orozco, C., Chalarca, A., and Mueller, L. 2006. Development of a bioinformatics platform at the Colombia National Coffee Research Center. In *Proceedings of XXI International Conference on Coffee Science*, 638–643. ASIC, Montpellier, France.

Cros, J., Combes, M. C., Chabrillange, N., Duperray, C., Monnot des Angles, A., and Hamon, S. 1995. Nuclear DNA content in the subgenus *Coffea* (Rubiaceae): Inter- and intraspecific variation in African species. *Canadian Journal of Botany* 73:14–20.

Crouzillat, D., Rigoreau, M., Bellanger, L., et al. 2004. A robusta consensus map using RFLP and microsatellite markers for the detection of QTL. In *Proceedings of XX International Conference on Coffee Science*, 546–553. ASIC, Bangalore, India.

Cruz, A. R. R., Paixao, A. L. D., Machado, F. R., et al. 2004. Metodologia para obtencão de plantas transformadas de *Coffea canephora* por co-cultivo e calos embriogenicos com A. *tumefaciens.* 15, Boletin de Pesquisa e Desenvolvimento #58. Embrapa, Brasilia, Brazil.

Cubry, P., Musoli, P., Legnaté, H., et al. 2008. Diversity in coffee assessed with SSR markers: Structure of the *Coffea* genus and perspectives for breeding. *Genome* 51:50–63.

Davis, A. P., Govaerts, R., Bridson, D. M., and Stoffelen, P. 2006. An annotated taxonomic conspectus of the genus *Coffea* (Rubiaceae). *Botanical Journal of the Linnean Society* 152:465–512.

De Nardi, B., Dreos, R., Del Terra, L., et al. 2006. Differential responses of *Coffea arabica* L. leaves and roots to chemically induced systemic acquired resistance. *Genome* 49:1594–1605.

De Paulis, T., Commers, P., Farah, A., et al. 2004. 4-Caffeoyl-1,5-quinide in roasted coffee inhibits [3H]naloxone binding and reverses antinociceptive effects of morphine in mice. *Psychopharmacology* 176:146–153.

De Peña, M. 1995. Development of stable transformation procedures for the protoplasts of *Coffea arabica* cv. Colombia. 75, PhD diss. University of Purdue.

Dublin, P. 1980a. Multiplication végétative *in vitro* de l'Arabusta. *Café Cacao Thé* 24:281–290.

———. 1980b. Induction de bourgeons néoformés et embryogenése somatique. Deux voies de multiplication végétative *in vitro* des caféiers cultivées. *Café Cacao Thé* 24:121–130.

———. 1984. Techniques de reproduction végétative *in vitro* et amelioration génétique chez les caféiers cultivés. *Café Cacao Thé* 25:237–244.

Duceau, P. 1980. Criteres de sélection pour l'amélioration des hybrides Arabusta en Côte d'Ivoire. In *Proceedings of IX International Conference on Coffee Science*, 603–608. ASIC, London.

Ducos, J. P., Alenton, R., Reano, J. F., Kanchanomai, C., Deshayes, A., and Pétiard, V. 2003. Agronomic performance of *Coffea canephora* P. trees derived from large-scale somatic embryo production in liquid medium. *Euphytica* 131:215–223.

Dufour, M., Anthony, F., Bertrand, B., and Eskes, A. B. 1997. Identification de caféiers mâle-stériles de *Coffea arabica* au CATIE, Costa Rica. *Plantations, Recherche, Développement* 4:401–407.

Dufour, M., Leroy, T., Carasco-Lacombe, C., et al. 2000. Coffee (*Coffea* sp.) genetic transformation for insect resistance. In *Coffee biotechnology and quality*, ed. Sera, T., Soccol, C. C. R., Pandey, A., and Roussos, S., 209–217. Kluwer, Dordrecht, the Netherlands.

Dussert, S., Chabrillange, N., Anthony, F., Engelmann, F., Recalt, C., and Hamon, S. 1997. Variability in storage response within a coffee (*Coffea* spp.) core collection under slow growth conditions. *Plant Cell Reports* 16:344–348.

Dussert, S., Engelmann, F., Chabrillange, N., Anthony, F., Noirot, M., and Hamon, S. 1997. *In vitro* conservation of coffee (*Coffea* spp.) germplasm. In *Conservation of genetic resources in vitro*, vol. 1, ed. Razdan, M. K., and Cocking, E. C., 287–305. Science Publishers, New York.

Dussert, S., Lashermes, P., Anthony, F., et al. 1999. Le caféier, *Coffea canephora*. In *Diversité génétique des plantes cultivées*, ed. Hamon, P., Seguin, M., Perrier, X., and Glaszmann, C., 175–194. Repères, CIRAD, France.

Dussert, S., Lashermes, P., Anthony, F., et al. 2003. Coffee (*Coffea canephora*). In *Genetic diversity of cultivated tropical plants*, ed. Hamon, P., Seguin, M., Perrier, X., and Glaszmann, J. C., 239–258. Science Publishers Inc., New York.

Dussert, S., Vasquez, N., Salazar, K., Anthony, F., and Engelmann, F. 2007. Cryopreservation. In *Complementary strategies for ex situ conservation of coffee* (Coffea arabica *L.*) *genetic resources. A case study in CATIE, Costa Rica*, vol 1., ed. Razdan, M. K., and Cocking, E. C., 49–58. Science Publishers, New York.

Ellis, R. H., Hong, T. D., and Roberts, E. H. 1990. An intermediate category of seed storage behavior? 1. Coffee. *Journal of Experimental Botany* 41:1167–1174.

Eskes, A. B. 1983. Incomplete resistance to coffee leaf rust (*Hemileia vastatrix*). PhD diss., University of Wageningen.

———. 1989. Identification, description and collection of coffee types in P.D.R. Yemen. Bioversity International, Rome.

Etienne, H., and Berthouly, M. 2001. Temporary immersion systems in plant micropropagation—Review. *Plant Cell Tissue and Organ Culture* 69:215–231.

Etienne, H., Bertrand, B., Anthony, F., Côte, F., and Berthouly, M. 1997. Somatic embryogenesis: A tool for coffee breeding. In *Proceedings of XVII International Conference on Coffee Science*, 457–465. ASIC, Nairobi, Kenya.

Etienne, H., Lashermes, P., Menendez-Yuffa, A., De Guglielmo-Croquer, Z., Alpizar, E., and Sreenath, H. 2008. Coffee. In *Compendium of transgenic crop plants. Transgenic plantation crops, ornamentals and turf grasses*, vol. 8, ed. Kole, C., and Hall, T. C., 219. Blackwell Publishing Ltd., London.

Etienne, H., Solano, W., Pereira, et al. 1997. Utilización de la embriogénesis somática en medio líquido para la propagación masal de los híbridos F1 de *Coffea arabica*. In *Simposio latinoamericano sobre Caficultura*, 18. San José, Costa Rica.

Etienne-Barry, D., Bertrand, B., Vásquez, N., and Etienne, H. 1999. Direct sowing of *Coffea arabica* somatic embryos mass-produced in a bioreactor and the regeneration of plants. *Plant Cell Reports* 19:111–117.

Fernandez, D., Santos, P., Agostini, C., et al. 2004. Coffee (*Coffea arabica* L.) genes early expressed during infection by the rust fungus (*Hemileia vastatrix*). *Molecular Plant Pathology* 5:527–536.

Fernandez, R., and Menendez, A. 2003. Transient gene expression in secondary somatic embryos from coffee tissues electroporated with the genes GUS and BAR. *Electronic Journal of Biotechnology* 6:29–38.

Fernie, L. M. 1970. The improvement of arabica coffee in east Africa. In *Crop improvement in east Africa*, ed. Leakey, C. L. A., 231–249. Technical communication of the Commonwealth Bureau of Plant Breeding and Genetics, no. 19 CAB, Farnham Royal.

Ganesh, D., Petitot, A-S., Silva, M. C., Alary, R., Lecouls, A., and Fernandez, D. 2006. Monitoring of the early molecular resistance responses of coffee (*Coffea arabica* L.) to the rust fungus (*Hemileia vastatrix*) using real-time quantitative RT-PCR. *Plant Science* 170:1045–1051.

Gichuru, E., Agwnada, C. O., Combes, M. C., et al., 2008. Identification of molecular markers linked to a gene confering resistance to coffee berry disease (*Colletotrichum kahawae*) in *Coffea arabica*. *Plant Pathology* 57:1117–1124.

Gill, S. L., Berry, D., and Bieysse, D. 1990. Recherche sur la résistance incomplète à *Hemileia vastatrix* Berk et Br. dans un groupe de génotypes de *Coffea arabica* L. d'origine éthiopienne. *Café Cacao Thé* 34:105–133.

Giridhar, P., Indu, E. P., Ravishankar, G. A., et al. 2004. Influence of Triacontanol on somatic embryogenesis in *Coffea arabica* L. and *Coffea canephora* P. ex. Fr. *In Vitro Cellular and Development Biology—Plant* 40:200–203.

Giridhar, P., Indu, E. P., Vinod, K., et al. 2004. Direct somatic embryogenesis from *Coffea arabica* L. and *Coffea canephora* P. Ex. Fr. under the influence of ethylene action inhibitor silver nitrate. *Acta Physiolgiae Plantarum* 26:299–305.

Gongora, C., Idarraga, S., Sanchez, M., et al. 2008. Differential gene expression response of *Coffea arabica* and *C. liberica* to coffee berry borer attack. In *Proceedings of XXII International Conference on Coffee Science*, B207. ASIC, Campinas, Brazil.

Grezes, J., Thomasset, B., and Thomas, D. 1993. *Coffea arabica* protoplast culture: Transformation assays. In *Proceedings of XV International Conference on Coffee Science*, 745–747. ASIC, Montpellier, France.

Guerreiro-Filho, O. 1992. *Coffea racemosa* lour: Une revue. *Café Cacao Thé* 36:171–186.

Guerreiro-Filho, O., Silvarolla, M. B., and Eskes, A. B. 1999. Expression and mode of inheritance of resistance in coffee to leaf miner *Perileucoptera coffeella*. *Euphytica* 105:7–15.

Guillaumet, J. L., and Hallé, F. 1978. Echantillonnage du materiel récolté en Ethiopie. *Bulletin IFCC* 14:13–18.

Hamon, S., Anthony, F., and Le Pierrès, D. 1984. La variabilité génétique des caféiers spontanés de la section *Mozambicoffea* A. Chev. 1. Précisions sur deux espèces affines: *Coffea pseudozanguebariae* Bridson et *C.* sp. A Bridson. *Adansonia* 2:207–223.

Herrera, J. C., Alvarado, G., Cortina, H., Combes, M.C., Romero, G., and Lashermes, P. 2009. Genetic analysis of partial resistance of coffee leaf rust (*Hemileia vastatrix* Berk & Br.) introgressed into the cultivated *Coffea arabica* L. from the diploid *C. canephora* species. *Euphytica* 167:57–67.

Herrera, J. C., Combes, M.C., Anthony, F., Charrier, A., and Lashermes, P. 2002. Introgression into the allotetraploid coffee *Coffea arabica* L.: Segregation and recombination of the *C. canephora* genome in the tetraploid interspecific hybrid *C. arabica* × *C. canephora*. *Theoretical and Applied Genetics* 104:661–668.

Herrera, J. C., Combes, M. C., Cortina, H., Alvarado, G., and Lashermes, P. 2002. Gene introgression into *Coffea arabica* by way of triploid hybrids (*C. arabica* × *C. canephora*). *Heredity* 89:488–494.

Herrera, J. C., Combes, M. C., Cortina, H., and Lashermes, P. 2004. Factors influencing gene introgression into the allotetraploid *Coffea arabica* L. from its diploid relatives. *Genome* 47:1053–1060.

Herrera, J. C., D'Hont, A., and Lashermes, P. 2007. Use of fluorescence *in situ* hybridization as a tool for introgression analysis and chromosome identification in coffee (*Coffea arabica* L.). *Genome* 50:619–626.

Higdon, J. V. et al. 2006. Coffee and health: A review of recent human research. *Critical Reviews in Food Science and Nutrition* 46:101–123.

Hong, T. S., and Ellis, R. H. 1995. Interspecific variation in seed storage behavior within two genera—*Coffea* and *Citrus*. *Seed Science and Technology* 23:165–181.

Kartha, K. K., Mroginski, L. A., Pahl, K., and Leung, N. L. 1981. Germplasm presrvation of coffee (*Coffea arabica* L.) by *in vitro* culture of shoot apical meristems. *Plant Science Letters* 22:301–307.

Kawachi, I. et al. 1996. A prospective study of coffee drinking and suicide in women. *Archives of Internal Medicine* 11:521–525.

Keller, M., Baum, M. J., Brock, O., Brennan, P. A., and Bakker, J. 2009. The main and the accessory olfactory systems interact in the control of mate recognition and sexual behavior. *Behavior and Brain Research* 200:268–76.

Klatsky, A. L. et al. 1993. Coffee, tea, and mortality. *Annals of Epidemiology* 3:375–381.

Krug, C. A. 1934. Contribuçao para o estudo da citologia do gênero *Coffea*. Campinas, Instituto Agronomico, Boletim Técnico, 11:1–10.

———. 1937. Observaçoes citologicas em *Coffea*. III. Campinas, Instituto Agronomico, Boletim, 37:1–19.

Krug, C. A., Mendes, J. E. T., and Carvalho, A. 1939. Taxonomia de *Coffea arabica* L. Bolétim Técnico no. 62. Instituto Agronômico do Estado, Campinas.

Kumar, V., Sathyanarayana, K. V., Indu, E. P., et al. 2003. Stable transformation and direct regeneration in *Coffea canephora* by *Agrobacterium rhizogenes* mediated transformation. In *Proceedings of X Congress of Federation of Asian and Oceanian Biochemists and Molecular Biologists*, 10.

Kumar, V., Sathyanarayana, K. V., Indu, E. P., et al. 2004. Posttranscriptional gene silencing for down regulating caffeine biosynthesis in *Coffea canephora* P ex Fr. In *Proceedings of XX International Conference on Coffee Science*, 769–774. ASIC, Bangalore, India.

Kumar, V., Satyanarayana, K. V., Indu, E. P., et al. 2006. Stable transformation and direct regeneration in *Coffea canephora* by *Agrobacterium rhizogenes* mediated transformation without hairy root phenotype. *Plant Cell Reports* 25:214–222.

Kushalappa, C. A., and Eskes, A. B. 1989. Advances in coffee rust research. *Annual Review of Phytopathology* 27:503–531.

Ky, C. L., Barre, P., Lorieux, et al. 2000. Interspecific genetic linkage map, segregation distortion and genetic conversion in coffee *Coffea* sp. *Theoretical and Applied Genetics* 101:669–676.

Ky, C. L., Louarn, J., Dussert, S., Guyot, B., Hamon, S., and Noirot, M. 2001. Caffeine, trigonelline, chlorogenic acids and sucrose diversity in wild *Coffea arabica* L. and *C. canephora* P. accessions. *Food Chemistry* 75:223–230.

Labouisse, J. P., Bellachew, B., Kotecha, S., and Bertrand, B. 2008. Current status of coffee (*Coffea arabica* L.) genetic resources in Ethiopia: Implications for conservation. *Genetic Resources and Crop Evolution* 55:1079–1093.

Lashermes, P., Andrade, A. C., and Etienne, H. 2008. Genomics of coffee, one of the world's largest traded commodities. In *Genomics of tropical crop plants*, ed. Moore, P. H., and Ming, R., 203–225. Springer, New York.

Lashermes, P., Andrzejewski, S., Bertrand, B., et al. 2000. Molecular analysis of introgressive breeding in coffee (*Coffea arabica* L.). *Theoretical and Applied Genetics* 100:139–146.

Lashermes, P., Combes, M. C., Prakash, N. S., et al. 2001. Genetic linkage map of *Coffea canephora*: Effect of segregation distortion and analysis of recombination rate in male and female meioses. *Genome* 44:589–595.

Lashermes, P., Combes, M. C., Robert, J., et al. 1999. Molecular characterization and origin of the *Coffea arabica* L. genome. *Molecular Genomics and Genetics* 261:259–266.

Lashermes, P., Combes, M. C., Trouslot, P., and Charrier, A. 1997. Phylogenetic relationships of coffee-tree species *Coffea* L. as inferred from ITS sequences of nuclear ribosomal DNA. *Theoretical and Applied Genetics* 94:947–955.

Lashermes, P., Cros, J., Combes, M. C., et al. 1996. Inheritance and restriction fragment length polymorphism of chloroplast DNA in the genus *Coffea* L. *Theoretical and Applied Genetics* 93:626–632.

Lashermes, P., Paczek, V., Trouslot, P., Combes, M. C., Couturon, E., and Charrier, A. 2000. Single-locus inheritance in the allotetraploid *Coffea arabica* L. and interspecific hybrid *C. arabica* × *C. canephora*. *Journal of Heredity* 91:81–85.

Lashermes, P., Trouslot, P., Anthony, F., Combes, M-C., and Charrier, A. 1996. Genetic diversity for RAPD markers between cultivated and wild accessions of *Coffea arabica*. *Euphytica* 87:59–64.

La Vecchia, C. 2005. Coffee, liver enzymes, cirrhosis and liver cancer. *Journal of Hepatology* 42:444–46.

Lebrun, J. 1941. Recherches morphologiques et systématiques sur les caféiers du Congo. *Publications de l'Institut National pour l'Etude Agronomique du Congo Belge* 11:1–186.

Lejeune, J. B. H. 1958. Rapport au gouvernement impérial d'Ethiopie sur la production caféière. FAO, Rome.

Le Pierrès, D., Charmetant, P., Yapo, A., Leroy, T., Couturon, E., Bontemps, S., and Tehe, H. 1989. Les caféiers sauvages de Côte d'Ivoire et de Guinée: Bilan des missions de prospection effectuées de 1984 à 1987. In *Proceedings of XIII Colloque Scientifique International sur le Café*, 420–428. ASIC, Paipa, Colombia.

Leroy, J. F. 1980. Les grandes lignées de caféiers. In *IX Colloque Scientifique International sur le Café*, 473–477. ASIC, London, UK.

Leroy, T., Henry, A. M., Royer, M., et al. 2000. Genetically modified coffee plants expressing the *Bacillus thuringiensis cry1Ac* gene for resistance to leaf miner. *Plant Cell Reports* 19:382–389.

Leroy, T., Marraccini, P., Dufour, M., et al. 2005. Construction and characterization of a *Coffea canephora* BAC library to study the organization of sucrose biosynthesis genes. *Theoretical and Applied Genetics* 111:1032–1041.

Leroy, T., Montagnon, C., Charrier, A., and Eskes, A. B. 1993. Reciprocal recurrent selection applied to *Coffea canephora* Pierre. I. Characterization and evaluation of breeding populations and value of intergroup hybrids. *Euphytica* 67:113–125.

Lima, D. R. 1989. Coffee as a medicinal plant and vitamin source for smokers. *Italian Journal of Chest Diseases* 43:56–58.

———. 1990a. *Coffee, a medicinal plant.* Vantage Press, New York.

———. 1990b. Is coffee good for drug addiction? Maybe. *African Coffee* 46–48.

Lima, D. R. et al. 1989. Cigarettes and caffeine. *Chest* 95:255–256.

Lima, D. R. et al. 1990a. How to give up smoking by drinking coffee. *Chest* 97:254.

Lima, D. R. et al. 1990b. Smoking, drug addiction, opioid peptides and coffee intake. *Yonago Acta Medica, Japan* 33:79–82.

Lima, D. R. et al. 1991. Effects of coffee in alcoholics. *Annals of Internal Medicine* 115:499.

Lima, D. R. et al. 2000. Can coffee help fighting the drug problem: Preliminary results of the Brazilian Youth Drug Study (BYDS). *Acta Pharmacologica Sinica* 21:1059–1070.

Lima, D. R., and Moll, J. 2009. Aroma of coffee and the physiology of passion. *Revista do Café, Centro do Comércio do Café do Rio de Janeiro* 88:26–27.

Lin, C., Mueller, L. A., McCarthy, J., Crouzillat, D., Pétiard, V., and Tanksley, S. D. 2005. Coffee and tomato share common gene repertoires as revealed by deep sequencing of seed and cherry transcripts. *Theoretical and Applied Genetics* 112:114–130.

López, G., and Moncada, M. D. P. 2006. Construction of an interspecific genetic linkage map from a *Coffea liberica* x *C. eugenioides* F1 Population. In *Proceedings of XXI International Conference on Coffee Science*, 644–652. ASIC, Montpellier, France.

Louarn, J. 1992. La fertilité des hybrides interspécifiques et les relations génomiques entre caféiers diploïdes d'origine Africaine (Genre *Coffea* L. sous-genre *Coffea*). PhD diss., Université d'Orsay, Paris XI. France.

Madhava-Naidu, M., Veluthambi, K., Srinivasan, C. S., et al. 1998. Agrobacterium mediated transformation in *Coffea canephora*. *Development Plant Crops Research* 46–50.

Mahé, L., Combes, M. C., Varzea, V. M. P., Guilhaumon, C., and Lashermes, P. 2008. Development of sequence characterized DNA markers linked to leaf rust (*Hemileia vastatrix*) resistance in coffee (*Coffea arabica* L.). *Molecular Breeding* 21:105–113.

Mahé, L., Varzea, V. M. P., Le Pierres, D., Combes, M. C., and Lashermes, P. 2007. A new source of resistance against coffee leaf rust from New Caledonian natural interspecific hybrids between *Coffea arabica* and *Coffea canephora*. *Plant Breeding* 126:638–641.

Mazzafera, P., Eskes, A. B., Parvais, J. P., and Carvalho, A. 1989. Stérilité mâle détectée chez *Coffea arabica* et *C. canephora* au Brésil. In *Proceedings of XIII Colloque Scientifique International sur le Café*, 466–473. ASIC, Paipa, Colombia.

McDonald, J. 1926. A preliminary account of a disease of green coffee berries in Kenya. *Transactions of the British Mycological Society* 11:145–154.

Medina, D. M. 1972. Caracterização de híbridos interespecíficos de *Coffea*. Piracicaba, Escola Superior de Agricultura "Luiz de Queiroz" da USP, PhD diss.

Medina-Filho, H. P., Carvalho, A., and Medina, D. M. 1977. Germoplasma de *C. racemosa* e seu potential no melhoramento do cafeeiro. *Bragantia* 36:43–46.

Medina-Filho, H. P., Carvalho, A., and Monaco, L. C. 1977. Observações sobre a resistência docafeeiro ao bicho mineiro. *Bragantia* 36:131–137.

Mendes, A. J. T. 1938. Morfologia dos cromossomos de *Coffea excelsa* Chev. Campinas, Instituto Agronômico. *Boletin Técnico* 56:1–8.

Mendez-Lopez, I., Basurto-Ríos, R., and Ibarra, J. E. 2003. *Bacillus thuringiensis* serovar israelensis is highly toxic to the cofee berry borer, *Hypothenemus hampei* Ferr. (Coleoptera: Scolytidae). *FEMS Microbiology Letters* 226:73–77.

Mettulio, R., Rovelli, P., Anthony, F., et al. 1999. Polymorphic microsatellites in *Coffea arabica*. In *Proceedings of XVIII International Conference on Coffee Science*, 344–347. ASIC, Lausane.

Meyer, F. G. 1965. Notes on wild *Coffea arabica* from southwestern Ethiopia, with some historical considerations. *Economic Botany* 19:136–151.

Meyer, F. G., Fernie, L. M., Narasimhaswamy, R. L., Monaco, L. C., and Greathead, D. J. 1968. FAO coffee mission to Ethiopia 1964–1965. FAO, Rome.

Molina, D. M., María, E., Aponte, M. E., et al. 2002. The effect of genotype and explant age on somatic embryogenesis of coffee. *Plant Cell Tissue and Organ Culture* 71:117–123.

Moll, J., Krueger, F., Zahn, R., Pardini, M., Oliveira-Souza, R., and Grafman, J. 2006. Human fronto–mesolimbic networks guide decisions about charitable donations. *Proceedings of the National Academy of Sciences, USA* 103:15623–15628.

Monaco, L. C., Carvalho, A., and Fazuoli, L. C. 1974. Melhoramento do cafeeiro. Germoplasma do café Icatu e seu potencial no melhoramento. In *Resumos II Cogresso Brasileiro de pesquisas cafeeiras*. IBC/MIC, Rio de Janeiro, Brazil.

Moncada, M. D. P., and McCouch, S. 2004. Simple sequence repeat diversity in diploid and tetraploid *Coffea* species. *Genome* 47:501–509.

Mondego, J. M. C., Guerreiro-Filho, O., Bengtson, M., et al. 2005. Isolation and characterization of *coffea* genes induced during coffee leaf miner *Leucoptera coffeella* infestation. *Plant Science* 169:351–360.

Montagnon, C., and Bouharmont, P. 1996. Multivariate analysis of phenotypic diversity of *Coffea arabica*. *Genetic Resources and Crop Evolution* 43:221–227.

Montagnon, C., and Leroy, T. 1993. Résultats récents sur la résistance de *Coffea canephora* à la sécheresse, à la rouille orangée et au scolyte des branchettes en Côte d'Ivoire. In *Proceedings of XV Colloque Scientifique International sur le Café*, 309–317. ASIC, Montpellier, France.

Montagnon, C., Leroy, T., Charmetant, P., et al. 2001. Outcome of two decades of reciprocal recurrent selection applied to *C. canephora* in Côte d'Ivoire: New outstanding hybrids available for growers. In *Proceedings of XIX International Conference on Coffee Science*. ASIC, Trieste, Italy.

Montagnon, C., Leroy, T., and Eskes, A. B. 1998. Amélioration variétale de *Coffea canephora*. II. Les programmes de sélection et leurs résultats. *Plantations, Recherche, Développement* 5:89–98.

Moore, J. H. 2007. Bioinformatics. *Journal of Cell Physiology* 9999:1–5.

Mueller, L., Lin, C., Mc Carthy, J., et al. 2004. Generation and analysis of a coffee EST database: Deductions about genome content and comparison with tomato/potato. In *Proceedings of XX International Conference on Coffee Science*, B222. ASIC, Bangalore, India.

Mueller, L., Solow, T. H., Taylor, N., et al. 2005. The SOL genomics network. A comparative resource for Solanaceae biology and beyond. *Plant Physiology* 138:1310–1317.

Mulinge, S. K. 1975. Plant pathology. In *Annual report 74. Coffee Research Foundation*, 60–64, Ruiru, Kenya.

Müller, R. A. 1964. L'anthracnose des baies du caféier d'arabie (*Coffea arabica*) due a *Colletotrichum coffeanum* Noack. au Cameroun. *IFCC Bulletin* 6:38, Paris.

Muniswamy, B., and Sreenath, H. L. 1996. Effect of kanamycin on callus induction and somatic embryogenesis in cultured leaf tissues on *Coffea canephora* Pierre (robusta). *Journal of Coffee Research* 26:44–51.

Namur-de, C., Couturon, E., Sita, P., and Anthony, F. 1987. Résultats d'une mission de prospection des caféiers sauvages du Congo. In *Proceedings of XII Colloque Scientifique International sur le Café*, 397–404. ASIC, Lausane.

N'Diaye, A., Poncet, V., Louarn, J., Hamon, S., and Noirot, M. 2005. Genetic differentiation between *Coffea liberica* var. *liberica* and *C. liberica* var. *Dewevrei* and comparison with *C. canephora*. *Plant Systematics and Evolution* 253:95–104.

Noir, S., Anthony, F., Bertrand, B., Combes, M. C., and Lashermes, P. 2003. Identification of a major gene (*Mex-1*) from *Coffea canephora* conferring resistance to *Meloidogyne exigua* in *Coffea arabica*. *Plant Pathology* 52:97–103.

Noir, S., Combes, M. C., Anthony, F., and Lashermes, P. 2001. Origin, diversity and evolution of NBS-type disease-resistant gene homologues in coffee trees (*Coffea* L.). *Molecular Genetics and Genomics* 265:654–662.

Noir, S., Patheyron, S., Combes, M. C., Lashermes, P., and Chalhoub, B. 2004. Construction and characterization of a BAC library for genome analysis of the allotetraploid coffee species (*Coffea arabica* L.). *Theoretical and Applied Genetics* 109:225–230.

O'Brien, C. P. 1996. Endogenous opioids in the treatment of alcohol dependence. Meeting report. *Alcohol* 13:1–39.

Ogita, S., Uefuji, H., Morimoto, M., et al. 2004. Application of RNAi to confirm theobromine as the major intermediate for caffeine biosynthesis in coffee plants with potential for construction of decaffeinated varieties. *Plant Molecular Biology* 54:931–941.

Ogita, S., Uefuji, H., Yamaguchi, Y., et al. 2003. RNA interference: Producing decaffeinated coffee plants. *Nature* 423:823.

Oliveiro, G. F., Peter, D., Marnix, P., et al. 1998. Susceptibility of the coffee leaf miner (*Perileucoptera* spp.) to *Bacillus thuringiensis* O-endotoxins: A model for transgenic perennial crops resistant to endocarpic insects. *Current Microbiology* 36:175–179.

O'Malley, S. S. 1996. Opioid antagonists in the treatment of alcohol dependence: Clinical efficacy and prevention of relapse. *Alcohol and Alcoholism* 31:77–81.

Orozco, F. J. 1989. Utilización de los híbridos triploides en el mejoramiento genético del café. In *Proceedings of XIII International Conference on Coffee Science*, 485–494. ASIC, Paipa, Colombia.

Orozco-Castillo, C., Chalmers, K. J., Powell, W., and Waugh, R. 1996. RAPD and organelle specific PCR reaffirms taxonomic relationships within the genus *Coffea*. *Plant Cell Reports* 15:337–341.

Orozco-Castillo, C., Chalmers, K. J., Waugh, R., and Powell, W. 1994. Detection of genetic diversity and selective gene introgression in coffee using RAPD markers. *Theoretical and Applied Genetics* 87:934–940.

Paillard, M., Lashermes, P., and Pétiard, V. 1996. Construction of a molecular linkage map in coffee. *Theoretical and Applied Genetics* 93:41–47.

Paiva, R., Freitas, R., Lopes, F., and Loureiro, M. 2008. Identification and characterization of gene expression of different members of SnRK kinase family in *Coffea canephora* and *Coffea arabica*. In *Proceedings of XXII International Conference on Coffee Science*, PB630. ASIC, Campinas, Brazil.

Pallavicini, A., de Nardi, B., Dreos, R., et al. 2004. Transcriptomics of resistance response in coffee (*C. arabica* L.). In *Proceedings of XX International Conference on Coffee Science*, B206. ASIC, Bangalore, India.

Pearl, H. M., Nagai, C., Moore, P. H., et al. 2004. Construction of a genetic map for arabica coffee. *Theoretical and Applied Genetics* 108:829–835.

Pereira, R. A., Batista, J. A., Da-Silva, M. C., et al. 2006. An α-amylase inhibitor gene from *Phaseolus coccineus* encodes a protein with potential for control of coffee berry borer (*Hypothenemus hampei*). *Phytochemistry* 67:2009–2016.

Perthuis, B., Pradon, J. L., Montagnon, C., et al. 2005. Stable resistance against the leaf miner *Leucoptera coffeella* expressed by genetically transformed *Coffea canephora* in a pluriannual field experiment in French Guiana. *Euphytica* 144:321–329.

Pierozzi, N. I., Pinto-Maglio, C. A. F., Silvarola, B., and Barbosa, R. L. 2001. Karyotype characterization in some diploid coffee species by acetic orcein, C-band and AgNO3 techniques. *Chromosome Research* 9:99.

Pierson, E. S., Van Lammeren, A. A. M., Schel, J. H. N., and Starisky, G. 1983. *In vitro* development of embryoids from punched leaf discs of *Coffea canephora*. *Protoplasma* 15:208–216.

Pinto-Maglio, C. A. F., Barbosa, R. L., Cuellar, T., et al. 2001. Chromosome characterization in *Coffea arabica* L. using cytomolocular techniques. *Chromosome Research* 9:100.

Poncet, V., Hamon, P., Minier, J., Carasco, C., Hamon, S., and Noirot, M. 2004. SSR cross-amplification and variation within coffee trees (*Coffea* spp.). *Genome* 47:1071–1081.

Poncet, V., Rondeau, M., Tranchant, C., et al. 2006. SSR mining in coffee tree EST databases: Potential use of EST-SSRs as markers for the *Coffea* genus. *Molecular Genetics Genomics* 276:436–449.

Porteres, R. 1959. Valeur agronomique des cafeiers des types Kouillou et robusta cultives en Cote d'Ivoire. *Café Cacao Thé* 3:3–13.

Prakash, N. S., Combes, M. C., Dussert, S., Naveen, S., and Lashermes, P. 2005. Analysis of genetic diversity in Indian robusta coffee genepool (*Coffea canephora*) in comparison with a representative core collection using SSRs and AFLPs. *Genetic Resources and Crop Evolution* 52:333–343.

Prakash, N. S., Marques, D. V., Varzea, V. M. P., Silva, M. C., Combes, M-C., and Lashermes, P. 2004. Introgression molecular analysis of a leaf rust resistance gene from *Coffea liberica* into *C. arabica* L. *Theoretical and Applied Genetics* 109:1311–1317.

Raina, S. N., Mukai, Y., and Yamamoto, M. 1998. *In situ* hybridization identifies the diploid progenitor species of *Coffea arabica* (Rubiaceae). *Theoretical and Applied Genetics* 97:1204–1209.

Ramiro, D., Toma-Braghini, M., Petitot, A., Maluf, M., and Fernandez, D. 2006. Use of leaf-disk technique for gene expression analysis of the coffee responses to *Hemileia vastatrix* infection. In *Proceedings of XXI International Conference on Coffee Science*, PB263. ASIC, Montpellier, France.

Restrepo, S., Pinzón, A., Rodríguez, R. L. M., Sierra, R., Grajales, A., Bernal, A., Barreto, E., Moreno, P., Zambrano, M., Cristancho, M., et al. 2009. Computational biology in Colombia. PLoS *Computational Biology* 5 (10): e1000535. doi:10.1371/journal.pcbi.1000535.

Ribas, A. F., Galvao, R. M., Pereira, L. F. P., and Vieira, L. G. E. 2005. Transformacao de *Coffea arabica* com o gene da ACC-oxidase em orientacao antinsenso. In *50th Congreso Brasileiro de Genetica*. September 7–10, Sao Paulo, Brazil, p. 492.

Ribas, A. F., Kobayashi, A. K., Pereira, L. F. P., and Vieira, L. G. E. 2005. Genetic transformation of *Coffea canephora* by particle bombardment. *Biologia Plantarum* 49:493–497.

———. 2008. Produção de plantas transgênicas de café resistentes ao herbicida glufosinato de amônio, 413–419. II Simposio de pesquisas dos cafes do Brasil.

Rodrigues, C. J., Jr., Bettencourt, A. J., and Rijo, L. 1975. Races of the pathogen and resistance to coffee rust. *Annual Review of Phytopathology* 13:49–70.

Rodrigues, C. J., Jr., Varzea, V. M., and Medeiros, E. F. 1992. Evidence for the existence of physiological races of *Colletotrichum coffeanum* Noack sensu Hindorf. *Kenya Coffee* 57:1417–1420.

Salmona, J., Dussert, S., Descroix, F., De Kochko, A., Bertrand, B., and Joët, T. 2008. Deciphering transcriptional networks that govern *Coffea arabica* seed development using combined cDNA array and real-time RT-PCR approaches. *Plant Molecular Biology* 66:105–124.

Santaram, A. 2005. Breeding coffee for leaf rust resistance—The Indian experience. *Indian Coffee* LXIX: 10–14.

Santaram, A., and Ramachandran, M. 1981. Male sterility in *Coffea liberica*. *Journal of Coffee Research* 11:142–145.

Santos, R. M., and Lima, D. R. 2007. *Coffee, the revolutionary drink for pleasure and health.* Xlibris/Random House, New York.

———. 2009. *101 Reasons to drink coffee without guilt.* Xlibris/Random House, New York.

Sera, T., Ruas, P. M., Ruas, C. F., et al. 2003. Genetic polymorphism among 14 elite *Coffea arabica* L. cultivars using RAPD markers associated with restriction digestion. *Genetics and Molecular Biology* 26:59–64.

Sharp, W. R., Caldas, L. S., Crocomo, O. J., et al. 1973. Production of *Coffea arabica* callus of three ploidy levels and subsequent morphogenesis. *Phyton* 31:67–74.

Silvarolla, M. B., Mazzafera, P., and de Lima, M. M. A. 2000. Caffeine content of Ethiopian *Coffea arabica* beans. *Genetics and Molecular Biology* 23:213–215.

Silvarolla, M. B., Mazzafera, P., and Fazuoli, L. C. 2004. A naturally decaffeinated arabica coffee. *Nature* 429:826.

Silvestrini, M., Junqueria, M. G., Favarin, A. C., et al. 2007. Genetic diversity and structure of Ethiopian, Yemen and Brazilian *Coffea arabica* L. accesions using microsatellites markers. *Genomic Resources and Crop Evolution* 54:1367–1379.

Söndahl, M. R., and Sharp, W. 1977. High frequency induction of somatic embryos in cultured leaf explants of *Coffea arabica* L. *Zeitschrift fur Pflanzen* 81:395–408.

Spiral, J., Thierry, C., Paillard, M., et al. 1993. Obtention de plantules de *Coffea canephora* Pierre (robusta) transformés par *Agrobacterium rhizogenes*. *Comptes Rendus de l'Académie des Sciences* 3 (316): 1–6.

Sreenath, H. L., Shantha, H. M., Harinath-Babu, K., et al. 1995. Somatic embryogenesis from integument (perisperm) cultures of coffee. *Plant Cell Reports* 14:670–673.

Sreenivasan, M. S., Ram, A. S., and Prakash, N. S. 1993. Tetraploid interspecific hybrids in coffee breeding in India. In *Proceedings of XV International Conference on Coffee Science*, 226–233. ASIC, Montpellier, France.

Srinivasan, K. H., and Narasimhaswamy, R. L. 1975. A review of coffee breeding work done at the Government Coffee Experiments Station, Balehonnur. *Indian Coffee* 34:311–321.

Staritsky, G. 1970. Embryoid formation in callus cultures of coffee. *Acta Botanica Neerlandica* 19:509–514.

Steiger, D. L., Nagai, C., Moore, P. H., Morden, C. W., Osgood, R. V., and Ming, R. 2002. AFLP analysis of genetic diversity within and among *Coffea arabica* cultivars. *Theoretical and Applied Genetics* 105:209–215.

Stoffelen, P. 1998. *Coffea* and *Psilanthus* in tropical Africa: A systematic and palynological study, including a revision of the west and central African species. PhD diss., Katholieke Universiteit Leuven, 187–209.

Stoffelen, P., Noirot, M., Couturon, E., and Anthony, F. 2008. A new caffeine-free coffee species from Cameroon. *Botanical Journal of the Linnean Society* 158:67–72.

Stoffelen, P., Noirot, M., Couturon, E., Bontems, S., De Block, P., and Anthony, F. 2009. *Coffea anthonyi* Stoff. & F. Anthony, a new self-compatible Central African coffee species, closely related to an ancestor of *C. arabica* L. *Taxon* 58:133–140.

Sugiyama, M., Matsuoka, C., and Takagi, T. 1995. Transformation of coffee with *Agrobacterium rhizogenes*. In *Proceedings of XVI International Conference on Coffee Science*, 853–859. ASIC, Kyoto, Japan.

Sylvain, P. G. 1958. Ethiopian coffee. Its significance to world coffee problems. *Economic Botany* 12:111–139.

Takeda, H., Tsuji, M., et al. 2002. Caffeic acid produces antidepressive like effect in the forced swimming test in mice. *European Journal of Pharmacology* 449:261–267.

Takeda, H. et al. 2003. Caffeic acid produces antidepressive and/or anxiolitic effects through indirect modulation of the alpha A-1 adrenoceptor system in mice. *Neuroreport* 14:1067–1070.

Teisson, C., Alvarad, D., Berthouly, M., et al. 1995. Culture *in vitro* par immersion temporaire: Un nouveau récipient. *Plantations Recherche Dévelop* 2:29–31.

Tesfaye, K., Borsch, T., Govers, K., and Bekele, E. 2007. Characterization of *Coffea* chloroplast microsatellites and evidence for the recent divergence of *C. arabica* and *C. eugenioides* chloroplast genomes. *Genome* 50:1112–1129.

Teufel, A., Krupp, M., Weinmann, A., and Galle, P. 2006. Current bioinformatics tools in genomic biomedical research. *International Journal of Molecular Medicine* 17:967–973.

Trugo, L., and Macrae, R. 1982. The determination of carbohydrates in coffee products using high performance liquid chromatography. In *Proceedings of X International Conference on Coffee Science*. ASIC, Lausane.

Trugo, L., Macrae, R., and Dick, J. 1983. Determination of purine alkaloids and trigonelline in instant coffee and other beverages using high performance liquid chromatography. *Journal of the Science of Food and Agriculture* 34:300–306.

————. 1984. Chlorogenic acid composition of instant coffee. *Analyst* 109:263–266.

Van Boxtel, J., Berthouly, M., Carasco, C., et al. 1995. Transient expression of beta-glucuronidase following biolistic delivery of foreign DNA into coffee. *Plant Cell Reports* 14:748–752.

Van der Graaf, N. A. 1981. The principles of scaling and the inheritance of resistance to coffee berry disease in *Coffea arabica*. *Euphytica* 31:735–740.

Van der Vossen, H. A. M. 1985. Coffee selection and breeding. In *Coffee: Botany, biochemistry and production of beans and beverage*, 48–96. Croom Helm, London.

————. 2001a. Coffee breeding and selection: Review of archievements and challenges. In *Proceedings of XIX International Conference on Coffee Science*, A001. ASIC, Trieste, Italy.

————. 2001b. Coffee breeding practices. In *Coffee—Recent developments*, ed. Clarke, R. J., and Macrae, R., 184–201. Blackwell Science, Oxford, England.

————. 2008. Disease resistance and cup quality in arabica coffee: The persistent myths in the coffee trade versus scientific evidence. In *Proceedings of XXII International Conference on Coffee Science*, A114. ASIC, Campinas, Brazil.

Van der Vossen, H. A. M., and Walyaro, D. J. 1980. Breeding for resistance to coffee berry disease in *Coffea arabica* L. II. Inheritance of the resistance. *Euphytica* 29:777–791.

————. 1981. The coffee breeding program in Kenya: A review of progress made since 1971 and plan of action for the coming years. *Kenya Coffee* 46:113–130.

Vasquez, N., Salazar, K., Anthony, F., Chabrillange, N., Engelmann, F., and Dussert, D. 2005. Variability in response of seeds to cryopreservation within a coffee (*Coffea arabica* L.) core collection. *Seed Science Technology* 33:293–301.

Vieira, L. G., Andrade, A., Colombo, C., et al. 2006. Brazilian Coffee Genome Project: An EST-based genomic resource. *Brazilian Journal of Plant Physiology* 18:95–108.

Vishveshwara, S. 1974. Periodicity of *Hemileia* in arabica selection—S.795. *Indian Coffee* 38:49–51.

Wei, C., Mueller, L. A., McCarthy, J., Crouzillat, D., Pétiard, V., and Tanksley, S. D. 2005. Coffee and tomato share common gene repertoires as revealed by deep sequencing of seed and cherry transcripts. *Theoretical and Applied Genetics* 112:114–130.

Wellman, F. L. 1961. *Coffee; botany, cultivation and utilization*, 41–66. Leonard Hill, London.

Willett, W. C., Stampfer, M. J., Manson, J. E., et al. 1996. Coffee consumption and coronary heart disease in women. A ten-year follow-up. *Journal of the American Medical Association* 275:458–462.

Wu, F., Mueller, L. A., Crouzillat, D., Pétiard, V., and Tanksley, S. D. 2006. Combining bioinformatics and phylogenetics to identify large sets of single-copy orthologous genes (COSII) for comparative, evolutionary and systematic studies: A test case in the euasterid plant clade. *Genetics* 174:1407–1420.

Wynne, K. N. et al. 1987. Isolation of opiate receptor ligands in coffee. *Clinical Experimental Pharmacology and Physiology* 14:785–790.

Yasuda, T., Fujii, Y., and Yamaguchi, T. 1985. Embryogenic callus induction from *Coffea arabica* leaf explants by benzyladenine. *Plant Cell Physiology* 26:595–597.

CHAPTER 20

Yacon (Asteraceae; *Smallanthus sonchifolius*)

A. Lebeda, I. Doležalová, E. Fernández, and I. Viehmannová

CONTENTS

20.1 INTRODUCTION

The Andean region is considered one of the most important centers of crop origin and diversity in the world (Vavilov 1950, 1987, 1995). Many of the important food crops grown recently worldwide (e.g., potato) were domesticated in this region (Veilleux and De Jong 2007). At the time of the Spanish conquest, the Incas cultivated almost as many species of plants as the farmers of all of Asia and Europe. Andean Indians domesticated as many as 70 crop species (Popenoe et al. 1989). Long-lasting human selection of different plant species in various agroecological conditions has led to very broad intra- and interspecific diversity, microclimatic adaptability, and utilization of landraces.

The Andean region is unique taxonomically for a very diverse group of root and tuber crops. About 17 species of root and tuber crops were domesticated in the Andes, and they represent the largest known geographical concentration of underground crops (Flores et al. 2003). These crops substantially contributed to the development of the Incan Empire (AD 1450–1535) population that reached about 12 million people (King 1987). Tuber crops are considered a common food for about 25 million people in the Andean area and an occasional food for more than 100 million people from various parts of South America (Flores et al. 2003). Unfortunately, the majority of these valuable plants are little known worldwide (Popenoe et al. 1989).

Andean tuber and root crops belong to at least nine plant families: Asteraceae, Basellaceae, Brassicaceae, Fabaceae (Leguminosae), Nyctaginaceae, Oxalidaceae, Solanaceae, Tropaeolaceae, and Umbelliferae. The genetic diversity and potential of these species are remarkable. These crops represent a great and mostly untapped pool of variation in type and content of starch, amino acids, other nutritional factors, and natural pesticides. These crops have great adaptability and potential for cultivation outside the area of origin (Flores et al. 2003). The most important tuber crops are achira (*Canna edulis*), ahipa (*Pachyrhizus ahipa*), mashua (*Tropaeolum tuberosum*), oca (*Oxalis tuberosa*), potato (*Solanum tuberosum*), ulluco (*Ullucus tuberosus*), and yacon (*Smallanthus sonchifolius*); the most important root crops are maca (*Lepidium meyenii*) and mauka (*Mirabilis expansa*) (Popenoe et al. 1989).

Yacon, *Smallanthus sonchifolius* (Poepp. & Endl.) H. Robinson (syn. apple of the earth, jacon, jiquama), of the family Asteraceae, is a traditional crop of the Andean Indians and was little known on the European continent until it was introduced as a plant with antidiabetic, nutritive, and fertility-enhancing properties in the 1990s. About 20 wild species of *Smallanthus* have been described (Robinson 1978). Wild forms of yacon occur in the Andean regions of Colombia, Ecuador, and Peru, where crops are cultivated at high elevations, but not much above 3000 m (Rubatzky and Yamaguchi 1997).

Lebeda et al. (2008) suggested that yacon is used as food crop, medicinal plant, and industrial crop. In Andean countries, yacon has also been used as a symbolic and religious plant since the time of the old Inca civilization (Seminario, Valderrama, and Manrique 2003).

Outside the Andean region, yacon is considered a tuber vegetable crop (Rubatzky and Yamaguchi 1997). Surprisingly, it has been sold as a fruit crop together with apples, avocados, and pineapples in the countries of origin for a long time (Grau and Rea 1997). The tuberous roots are juicy, consumed as a fresh food, frequently eaten raw in snacks and salads, and also boiled and baked in the Andean area. Preservation by sun-drying is used. The skin must be peeled before consumption because it is resinous (Rubatzky and Yamaguchi 1997) and has a strong laxative property.

Yacon is commonly consumed by people with diabetes and suffering from digestive problems. Its antidiabetic property has been attributed to yacon leaves, which are used to brew medicinal tea (Kakihara et al. 1996). The antioxidative property of yacon is believed to prevent some forms of cancer (Lebeda, Doležalová, Valentová, et al. 2003).

Industrial utilization of yacon is relatively broad (Fernández 2005). The tuberous roots are used as a source of inulin and fructose (Valentová et al. 2006), serve in the production of fresh drinks and alcohol (Grau and Rea 1997), and are used for the production of dry chips, jam and bottled stewed fruit (Fernández 2005), and some milk products (yogurts, muesli bars, and sweet flavoring) (Lebeda et al. 2008).

Yacon possesses an attractive set of features advantageous to producers, processors, consumers, and the environment. The challenge for other regions outside the Andes resides in a growing system that is environmentally friendly and emphasizes low input and high yield (Lebeda et al. 2008).

In this chapter, we summarize taxonomy, centers of origin, geographic distribution and ecology, germplasm management, chemical composition, diseases, and pests of yacon. We discuss the varietal improvement by conventional breeding and biotechnological approaches, including processing and utilization of the crop.

20.2 TAXONOMY

20.2.1 Evolutionary and Taxonomic Relationships at the Familial Level

The family Asteraceae (Compositae) contains the largest number of described species (24,000–30,000) of any plant family and is distributed in all the continents except Antarctica. Assuming 200,000–300,000 species of the flowering plants, 1 out of every 8–12 species is in the Asteraceae (about 10%) (Funk et al. 2005). Although the family is monophyletic, there is a great diversity in habitus, habitats, life cycles, pollination mechanisms, and chromosome numbers. The Asteraceae includes many important crops as well as common and rare taxa. The origin of extant members of the Asteraceae took place in southern South America; subsequent radiation in Africa gave rise to most extant tribes. The African radiation was followed by the movement of individual clades into Asia, Eurasia, and Australia. The origin and diversification of the Heliantheae *sensu lato* (*s.l.*) is likely to have occurred in North America, subsequent to separation from Gondwana (Funk et al. 2005).

The species within the Asteraceae have often been arranged into three subfamilies: Asteroideae, Barnadesioideae, and Cichorioiodeae, which represent 17 tribes and 1,535 genera (Bremer et al. 1994). The subfamily Asteroideae (with approximately 16,200 species in 1,135 genera) is monophyletic and represents the largest clade within the family, whereas most of the subfamily's genera are easily classified into 10 tribes. According to Bremer and colleagues (1994), the generic classification of the North American Heliantheae is well understood. They put approximately 2,500 species into 189 genera. Recent studies confirmed the North American origin and diversification of the Heliantheae *s.l.* that involved repeated incursion into Mexico and South America (Funk et al. 2005).

20.2.2 General Characterization of the Tribe Heliantheae

Commercially important plants in the Heliantheae include yacon, sunflower, and Jerusalem artichoke. Many garden flowers—*Coreopsis, Cosmos, Echinacea, Rudbeckia,* and *Zinnia*—are also in this group. The traditional circumscription of the Heliantheae arises from Cassini's 19th century classification of the Asteraceae (Cassini 1818). Bentham (1873), Cronquist (1955), and Turner (1977) assumed that the Heliantheae was the most primitive tribe of the family and the ancestor was a perennial herb with opposite leaves and a yellow flower, radiate, and capitula. Robinson (1981), in his revision, divided the group into 35 subtribes. Baldwin, Wessa, and Patero (2002) proposed a new classification system for the large paraphyletic tribe Helianteae *s.l.* within the Asteroideae. Baldwin et al. (2002) and Panero and Funk (2002) recognized new and previously described subfamilies (11 in total) and tribes so that now there are 10 and 35, respectively; however, the new classification is still open for acceptance (Funk et al. 2005).

This broad Heliantheae has been divided into 12 smaller tribes: Bahieae, Chaenactideae, Coreopsideae, Eupatorieae, Helenieae, Heliantheae *sensu stricto,* Madieae, Millereae, Neurolaeneae, Perityleae, Polymnieae, and Tageteae (Baldwin et al. 2002; Panero and Funk 2002). Nearly all of the species in 7 of the 10 subtribes illustrated here may be readily recognized due to the association of a receptacular bract or chaff scale with each disk floret in the head (Chaenactinidae, Gaillardiinae, and Pectidinae; mostly lacking true chaff scales). The heads usually include bisexual, actinomorphic disk florets with tubular corollas that have four or five distal lobes and peripheral zygomorphic female or sometimes sterile florets with strap-shaped corollas that have three or fewer distal teeth. However, the ray flowers are sometimes absent and the heads are then discoid, containing only bisexual florets with tubular corollas. The pappus is absent or more commonly ranges from scales to stiff bristles (Flora of North America 2006).

20.2.3 Delimitation and Characterization of the Genus *Smallanthus*

Originally, yacon and its relatives were placed in *Polymnia* (tribe Heliantheae, subtribe Melampodiinae), a genus founded by Linnaeus in 1751. In his revision of the genus, Wells (1965) maintained *Polymnia* with other coarse herbs or shrubs in the subtribe Melampodiinae, based upon the characters of opposite leaves and ray achenes not enclosed in a bract. Robinson (1978) reevaluated the closely related genus *Smallanthus,* originally classified by Mackenzie (1933). Robinson separated the species into two genera—*Smallanthus* and *Polymnia*—and kept both within the subtribe Melampodiinae.

One North American species, most Central American species, and all the South American species were placed in the genus *Smallanthus,* while a few North American species remained in *Polymnia.* The primary difference recognized between the two genera was that, in *Polymnia,* the achene walls are smooth without striations while in *Smallanthus* they have shallow grooves on the surface. Other differences cited by Robinson (1978) are that *Polymnia* lacks the distinct whorl of outer involucral bracts prominent in *Smallanthus* and that *Polymnia* has glands on thin anther appendages and *Smallanthus* lacks these glands. Additionally, the shape of the achenes differs between the two genera. Some of those characters place *Polymnia* as the most isolated genus within the subtribe, while *Smallanthus* is closer to other genera in the group, such as *Melampodium* and *Espeletia*, than to true *Polymnia*. Robinson's (1978) classification has been widely accepted (Brako and Zarrucchi 1993; Grau and Rea 1997; Jørgensen and León 1997). Robinson's concept of the genus *Polymnia* and its phylogenetic relationships was supported by molecular studies of members of the Heliantheae *s.l.* (Panero and Funk 2002) and the *Espeletia* complex (Rauscher 2002).

According to phylogenetic classification proposed by Cronquist (1955) and Meza (2001), the taxonomy of yacon is

- Kingdom: Plantae
- Subkingdom: Embryobionta
- Division: Magnoliophyta
- Class: Magnoliopsida
- Subclass: Asteridae
- Order: Asterales
- Family: Asteraceae
- Subfamily: Asteroidae
- Tribe: Helianthae
- Subtribe: Melampodiinae
- Genus: *Smallanthus* Mackenzie
- Species: *Smallanthus sonchifolius* (Poepp. & Endl.) H. Robins

Yacon, *Smallanthus sonchifolius* (Poepp. & Endl.) H. Robinson, has previously been named as a *Polymnia sonchifolia* Poepp. & Endl., nov. gen. sp. pl. 3:47:1845 and *Polymnia edulis* Wedd., Ann. Sci. Nat. Bot. IV. 7:114:1857. Common names include *arboloco* (Colombia), *aricoma* (Peru and Bolivia; Aymara name), *jíquima* and *jiquimilla* (Venezuela and Colombia), *yacón*, *llacon*, *llacjon*, and *llag'on* (Peru, Bolivia, and Argentina—from the Quechua *yakku* = insipid and *unu* = water evoking to watery roots, which are tasteless when just harvested) (Cárdenas 1969; Hanelt 2001). Yacon is also called *poirre de terre* (France), *polimnia* (Italy), *erbine* (Germany), and yacon strawberry (United States).

Yacon has been grown in the Yacones of Salta Province in Argentina. The locality in south Colombia near the Caqueta River is called Yaco. Unfortunately, it is not clear if there is a relationship between the localities' names and the yacon plant (Bredemann 1948).

The genus *sensu* Robinson includes at least 21 species, all of American origin, ranging mostly through southern Mexico and Central America and the Andes (Robinson 1978). Some of them are shrubs, small trees, or perennial herbs and only rarely annuals (Table 20.1).

20.2.4 Taxonomy and Relationships among *Smallanthus* Species

Except for a taxonomic study of Wells (1965) and rearrangement of Robinson (1978), there is not a comprehensive taxonomic overview of the species within the genus *Smallanthus*. The North American–Mexican–Central American species are herbs and included in the best studied group. Wells (1965) described several varieties in some species. *S. maculatus* is related to *S. uvedalius*; *S. macvaughii* is close to *S. oaxacanus*, and *S. quichensis* to *S. latisquamus* and *S. lundellii*. The only Central American species that extends to the Andean region is *S. riparius*. Regarding the species geography, habitus of plants, and shoot morphology, only six species, forming a yacon group (*S. connatus*, *S. macroscyphus*, *S. riparius*, *S. meridensis*, *S. suffruticosus*, and *S. siegesbeckius*), are closely related to *S. sonchifolius* (Grau and Rea 1997). *Smallanthus riparius* resembles two species: *S. siegesbeckius* and *S. macroscyphus*. Four species—*S. glabratus*, *S. microcephalus*, *S. parviceps*, and *S. fruticosus*—are treated into a *glabrata* complex and include shrubs or small trees reaching up to 10 m or more sometimes. *Smallanthus jelskii* and the related *S. pyramidalis*

Table 20.1 Species in the Genus *Smallanthus*

Species	Distribution	Habitus	Chromosome No.
Smallanthus apus (Blake)	Mexico		32
S. connatus (Spreng.)	Brazil, Paraguay, Uruguay, eastern Argentina	Annual herb	32
S. fruticosus (Benth.)	Southern Colombia, Ecuador, northern Peru	Shrub or tree	>50
S. glabratus (D.C.)	Peru, Ecuador, Chile	Shrub or tree	
S. jelskii (Hieronymus)	Peru	Shrub or tree	58
S. latisquamus (Blake)	Costa Rica	Herb	
S. lundelii H. Robinson	Guatemala	Herb	
S. macroscyphus (Baker ex Martinus) A. Grau, comb. nov.	Bolivia and northwestern Argentina	Perennial herb	32
S. maculatus (Cav.)	Mexico, Guatemala, Honduras, El Salvador, Nicaragua, Costa Rica	Coarse herb	32, 68
S. macvaughii (Wells)	Mexico	Herb	
S. meridensis (Steyerm.)	Venezuela, Colombia	Herb	
S. microcephalus (Hieron.)	Ecuador	Shrub or small tree	54, 60
S. oaxacanus (Sch. Bip. ex Klatt)	Mexico, Guatemala, Honduras	Herb	32
S. parviceps (Blake)	Southern Peru, northern Bolivia	Shrub or tree	58
S. pyramidalis (Triana)	Venezuela, Colombia, Ecuador	Tree	60, 58
S. quichensis (Coult.)	Costa Rica, Quatemala	Herb	
S. riparius H(Kunth)H.Rob	Southern Mexico to northern Bolivia	Herb or shrub	30, 32
S. siegesbeckius (D.C.)	Peru, Bolivia, Brazil, Paraguay	Perennial herb	
S. sonchifolius (Poepp. & Endl.) H. Robinson	Peru, Bolivia, Argentina, Ecuador, Venezuela, Colombia	Perennial herb	58
S. suffruticosus (Baker)	Venezuelan Amazonia	Herb or shrub	
S. uvedalius (L.) Mackenzie	Eastern United States from New York to Florida and Texas	Perennial herb	32

Source: Modified according to Robinson, H. 1978. *Phytologia* 39:47–53.

also reach up to tree size, but both are close to the yacon group more than *glabrata* complex (Grau and Rea 1997).

Molecular phylogenetics study of the *Espeletia* complex (Asteraceae) (Rauscher 2002) supports Robinson's (1978) rearrangement of the genus *Smallanthus*. In Rauscher (2002) study the three species of *Polymnia*, form a distinct clade which is phylogenetically distant from *Smallanthus* and from all taxa traditionally included in or considered close to Melampodiinae. Rauscher (2002) reported at least four distinct lineages (based on the ITS tree) within *Smallanthus*. The first includes five taxa: four South American species—*S. jelskii* (Hieron.) H. Rob., *S. pyramidalis* (Triana) H. Rob., *S. sonchifolius* (Poepp. & Endl.) H. Rob., and *S. microcephalus* (Hieron.) H. Rob.—and one Mexican–Central American species, *S. oaxacanus* (Sch. Bip. ex Klatt) H. Rob., the sister taxon to the South American clade.

In two other major lineages, a split between South American and Central American species was also found. In the first, the South American species *S. siegesbeckius* (DC.) H. Rob. group with the Central American species *S. fruticosus* (Benth.) H. Rob. and *S. connatus* (Spreng.) H. Rob.; in the second, *S. meridensis* (Steyerm.) H. Rob. and *S. riparius* (Kunth) H. Rob. (northwestern South America) group with *S. maculatus* (Cav.) H. Rob. and *S. quichensis* (Coult.) H. Rob. (Central America). The fourth major clade contains a single species, *S. uvedalius* (L.) Mackenzie—the only species distributed as far north as the United States (Rauscher 2002).

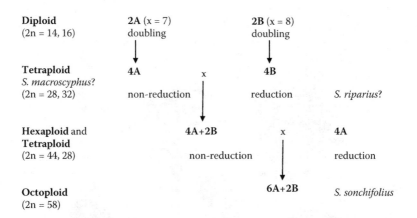

| Diploid (2n = 14, 16) | **2A** (x = 7) doubling | | **2B** (x = 8) doubling | |

| **Tetraploid** *S. macroscyphus?* (2n = 28, 32) | **4A** non-reduction | x | **4B** reduction | *S. riparius?* |

| **Hexaploid** and **Tetraploid** (2n = 44, 28) | **4A+2B** non-reduction | x | **4A** reduction | |

| **Octoploid** (2n = 58) | **6A+2B** | | *S. sonchifolius* | |

Figure 20.1 Hypothetical evolution of the yacon genome. (Grau, A., and J. Rea. 1997. In *Andean roots and tubers: Ahipa, arracacha, maca and yacon: Promoting the conservation and use of underutilized and neglected crops*, ed. M. Hermann and J. Heller, 199–242. Gatersleben, Germany: Institute of Plant Genetic and Crop Plant Research, Rome, Italy: IPGRI; Ishiki, K. et al. 1997. *Revision of chromosome number and karyotype of yacon (Polymnia sonchifolia)*. Resúmenes del Primer Taller Internacional sobre Recursos Fitogenéticos del Noroeste Argentino. Salta, Argentina: INTA.)

Heiser (1963), who worked with yacon materials of Ecuadorian origin, published chromosome number (2*n* = 60). Leon (1964) studied yacon grown at La Molina University in Peru and reported, in this material, 2*n* = 32. Wells (1967) published the first review of *Smallanthus* (*Polymnia*) chromosome numbers, and the most common number was 2*n* = 32. Talledo and Escobar (1996), working with the Peruvian material, and Grau and Rea (1997) described yacon as a tetraploid. Grau and Slanis (1996) and Ishiki, Salgado, and Arellano (1997), in their detailed study, considered yacon as an allopolyploid derived from hybridization between *S. macroscyphus* and/or similar related wild species (2*n* = 28, A = 7) and *S. riparius* (2*n* = 32, B = 8; A and B = two different genomes) (see Figure 20.1). The octoploid (2*n* = 6A + 2B = 58) and dodecaploid (2*n* = 9A + 3B = 87) organization of yacon genome could be explained by its hybrid origin (Figure 20.1). Ishiki et al. (1997) proposed an octoploid structure dominant in most yacon clones of 2*n* = 8*x* = 58, while a dodecaploid structure would explain 2*n* = 87.

Yacon, as a clonally propagated crop, could exhibit significant diversity in chromosome numbers. The second phenomenon, which may affect number of chromosomes, is the presence of B chromosomes, reported in other *Smallanthus* taxa (Wells 1971; Ishiki et al. 1997). However, chromosome number in some species remains still in question (Carr et al. 1999).

20.2.5 Description of Yacon (*S. sonchifolius*)

Yacon is a perennial herb, 1.5–3 m tall (Figure 20.2). The most important edible underground part of the plant is formed by roots of an adventitious nature, growing from a developed and ramified stem system formed by short, thick sympodial rhizomes or rootstock (corona, crown) (Figure 20.2). From this part of a plant grow shoots, gentle absorbent roots, and plenty of tuberous roots. If a plant is cultivated from the seed, it possesses one main stem. Sometimes stems are branched from the base or from the upper part of the plant. If the yacon plant is derived from vegetative propagation, it possesses a number of stems (Seminario et al. 2003).

20.2.5.1 Morphology and Characterization of Root System

The root system of yacon is composed of 4–20 fleshy tuberous roots that reach a length of 25 cm by 10 cm in diameter and an extensive system of thin fibrous roots. The absorbent (fibrous) roots,

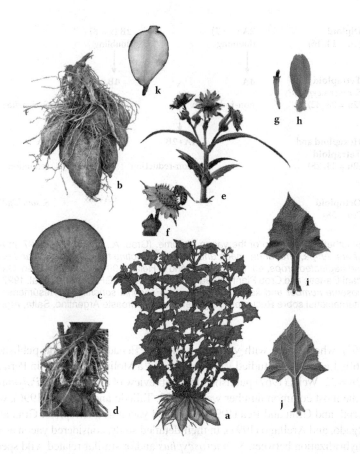

Figure 20.2   (See color insert.) Morphology and anatomy of yacon: (a) habitus of yacon; (b) tuberous roots; (c) transverse section of tuberous root; (d) rhizomes; (e) inflorescence; (f) flower head; (g) staminate disk flower; (h) pistilate ray flower; (i) adaxial (upper) side of leaf; (j) abaxial (lower) side of leaf; (k) longitudinal section of tuberous root. (Photo courtesy of E. C. Fernández, 2005.)

growing up from the base of the plant stem, suck up the water and nutrients from the soil. Thick storage tubers are formed at the base of rhizomes within 90–120 days after planting. In the storage, tuberous roots synthesize storage substances, which constitute an economically valuable part of yacon (Meza 2001).

Tuberous root growth is caused by the proliferation of parenchymatous tissue in the root cortex and particularly in the vascular cylinder. At first, tuberous roots are mainly fusiform; later, their shape becomes elliptical or even circular. However, tuberous roots often acquire irregular shapes due to contact with soil stones or the pressure of neighboring roots. Bredemann (1948) and Calvino (1940) reported that different shapes of tuberous roots are due to tuber development, not due to characters of cultivars. It is also possible to find several shapes of tuberous roots within one ecotype. The parenchyma accumulates sugars and, in some cases, pigments typical of certain clone groups. Flesh color is related with the pigments (white, cream, white with purple striations, purple, pink, and yellow) and is typical for each genotype. The tuberous root bark is brown, pink, purple, cream, or ivory white and thin (1–2 mm); it contains resin tubes filled with yellow crystals (Grau and Rea 1997). Vascular bundles are bounded by an endodermis and pericycle is well developed, mainly in young tubers.

At the cross-tissue section, xylem and phloem tissues are observed arranged from four to six bundles in young tubers. At first, they are separated by thin parenchyma rays and later on continue the differentiation of xylem and phloem cells, subsequently coming up to the reduction of xylem and phloem segments, which are separated by broad intercalated parenchyma rays. The growth of bark tissues is not well developed and development of tuberous roots is limited by intercalated parenchyma tissue (León 1964; Meza 2001) (Figure 20.2).

The stem tubers of yacon are also named rhizomes and they serve for vegetative propagation of the crop. They are branched and have an irregular shape, and they contain plenty of buds in the upper part. Yacon is reproduced through buds flushing in main stems in young plants. Thin fibrous roots are found in the bottom stem tubers. The pulpy stem root later becomes bulky and lignified. The rhizomes are often white, creamy, and purple colors. Shape is irregular, influenced by the soil type and genotype. The weight of rhizomes ranges from 0.5 to 4.5 kg and depends on the size, thickness, ecotype, and environmental conditions.

20.2.5.2 Morphology and Characterization of Stem

The number of shoots ranges from 11 to 53 (Meza 2001). The aerial stems are cylindrical or subangular, hollow at maturity with few branches in most clones or ramified in others, densely pubescent, and green to purplish. Diameter in the middle part of the stem varies between 10 and 28 mm. The number of internodes depends on earliness of ecotypes, size of plant, and size of stem tuber used for planting. Yacon ecotypes contain from one to six main (primary) stems in one plant. There are mostly three main stems in one plant, and sometimes the number of main stems reaches up to five. Except for the main stem, there are between 4 and 12 secondary stems growing from the primary stems (Figure 20.2). The branching of the stems starts 4–5 months after planting. Three types of branching are recognized: (1) basal—occurring on the main stem, (2) overall—branches form along the whole stem, and (3) top branching in the upper third of the plant. The number of secondary stems ranges from 2 to 10. Height of yacon is an ecotype characteristic and depends on natural conditions.

20.2.5.3 Morphology and Characterization of Leaves

Opposite leaves are narrow and grow out of aerial nodes. Lower leaves are broadly ovate and hastate or subhastate, connate and auriculate at the base; upper leaves are ovate-lanceolate, without lobes and hastate base, and upper and lower surfaces are densely pubescent (Figure 20.2). Lower and upper epidermis is covered by trichomes (0.8–1.5 mm long, 0.05 mm diameter) and glands that contain terpenoid compounds (Figure 20.3). At the flowering stage, the yacon plant possesses from 13 to 16 pairs of leaves. After the full-flowering stage, plants only produce small leaves (Seminario et al. 2003).

20.2.5.4 Morphology and Characterization of Inflorescence

Inflorescences are terminal and composed of one to five axes, each one with three capitula; peduncles are densely pilose. There are five uniseriate and ovate phyllaries. The head is composed of two types of flowers. Ligulae (ray flowers) form the apparent part of inflorescence, are placed marginally, and number about 16. They are yellow to bright orange. Ray flowers are two- or three-toothed, depending on the clone, to 12 mm long and 7 mm broad, and pistillate (Figure 20.2). Disc flowers numbering from 80 to 90 are about 7 mm long and staminate. They contain five stamens with free filaments. Anthers are black with fine yellow strips. The diameter of the heads reaches up to 30 mm. Pollen grains are globular, echinate, sometimes with three poles, bright yellow, glairy at the surface, and about 27 μm in diameter. The female flowers are zygomorphic, open earlier than male flowers, and mostly cease blossoming later than male flowers (León 1964; Meza 2001; Seminario et al. 2003).

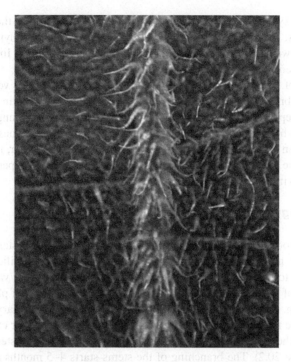

Figure 20.3   Trichomes on the adaxial part of yacon leaves. (Photo courtesy of E. C. Fernández.)

20.2.5.5 *Morphology and Characterization of Fruit and Seed*

Fruit of yacon is indehiscent cypsela (achene) of trapezoid shape. Immature cypselas (achenes) are purple and turn dark brown or black at maturity. Pericarp is dry and thin at maturity; the outer side is marked by a longitudinal serration forming parallel striae. Achenes are approximately 3.7 mm long and 2.2 mm wide. The endosperm is not present and storage components are located in cotyledons (Meza 2001).

Flower production in yacon is much more reduced than in other wild *Smallanthus* species. Limited flowering and fruit set are features commonly present in other clonally propagated tuber crops. During yacon evolution, continued vegetative propagation and selection for root yield may have affected flowering and fruit set. Yacon is a typical protogyny in which female gametes mature and shed before maturation of male gametes. Female flowers open before the male flowers release pollen. This means that yacon is heterogamous and needs pollinators (Grau and Rea 1997). Moreover, the presence of echinate and viscid pollen grains, conspicuousness of female flowers, and secretion of sugar substances indicate that inflorescences are visited by many insect species (Table 20.2) (Seminario et al. 2003).

Flowering strongly depends on environment of the growing area. In some regions, such as northwestern Argentina, flowering occurs very late in the growing cycle or not at all. By contrast, flowering is intense in most clones in northern Bolivia, the growing areas around Cusco, southern and northern Peru, and Cajamarca. In the Cajamarca region, flowering begins 6–7 months and culminates 8–9 months after planting. Generative reproduction in yacon is problematic due to low pollen fertility. The low pollen fertility was observed in yacon clones originating from Argentina and Ecuador (Grau 1993; Grau and Slanis 1996; Grau and Rea 1997). But, even in the areas where flowering is abundant, seed set is frequently poor or nonexistent and a high proportion of the seeds are nonviable. Yacon produces a limited number of seeds and germination rate is 15–30%. Seed viability, determined by tetrazolium, in yacon clones of different origin indicated that low seed

Table 20.2 Pollinators Observed in Yacon Inflorescence

Order	Family	Genus/Species	Frequency
Coleoptera	Dasytidae	*Astyllus* sp.	High
	Chrysomelidae	*Diabrotica* sp.	Low
Diptera	Syrphidae	*Syrpus* sp.	High
		Allograta sp.	Medium
Hemiptera	Miridae	Not identified	Medium
	Lygaeidae	Not identified	Low
Hymenoptera	Apidae	*Apis mellifera*	High
	Andrenidae	*Bombus* sp.	Low
Lepidoptera	Nymphalidae	Not identified	Low
	Pieridae	*Leptophobia* sp.	Low
		Not identified	Low

Source: Seminario, J. M. et al. 2003. http://www.cipotato.org/market/
PDFdocs/Yacon_Fundamentos_password.pdf

viability was associated with pollination, fertilization, seed development, and seed quality (Chiata 1998; Soto 1998; Fernández 2005).

20.3 CENTER OF ORIGIN, GEOGRAPHICAL DISTRIBUTION, ECOLOGY, VARIATION, AND GENETIC DIVERSITY

20.3.1 Center of Origin

It has been hypothesized that a species ancestral to yacon arose through hybridization of two or more *Smallanthus* species colonized in disturbed areas. This presumption is based on the fact that several wild *Smallanthus* species (*S. glabratus*, *S. riparius*, *S. siegesbeckius*, *S. macroscyphus*, and *S. connatus*) grow in the areas with vegetation gaps such river banks, landslides, and roadsides. Yacon has been associated with humans since prehistoric times, when the ancient Andean people practiced slash-and-burn agriculture and thus provided ideal conditions for colonizing the new areas. Grau and Rea (1997) reported that a similar situation from the present time can be observed in the Vilcanota River basin, Peru, where *S. siegesbeckius* is a common invader of abandoned fields and is a weed in coffee plantations. The same strategy is used by *S. macroscyphus* in northwestern Argentina. This species invades abandoned sugarcane fields and shrubby vegetation patches between cultivated fields.

The probable area of domestication of yacon with the largest clone diversity is the eastern humid slopes of the Andes, the region extending from northern Bolivia to central Peru (Figure 20.4). The original occurrence of yacon is situated in humid mountain forests from the Apurímac River basin of Peru (12°S) to the La Paz River basin of Bolivia (17°S) extended from north to south along Andean slopes and mountain valleys in elevations from 1000 to 3770 m. The diversity of yacon clones is more reduced in Ecuador. It seems that from the humid mountain area of Bolivia and Peru, yacon may have expanded to the north and south, into the inter-Andean valleys and to the Peruvian coast.

The native names indicate that yacon was cultivated a long time ago and was probably one of the most often cultivated crops in the Inca's time and a postal/peru stamp was issued in (Figure 20.5). The rest of the drains (e.g., La Cotaña, Nevado de Illimani [La Paz]) demonstrate agriculture at that time and small yacon is still found in the fields (Rea 2000a). Yacon depicted on textiles and ceramic material was found in the coastal archeological places of Nazca (AD 500–1200)

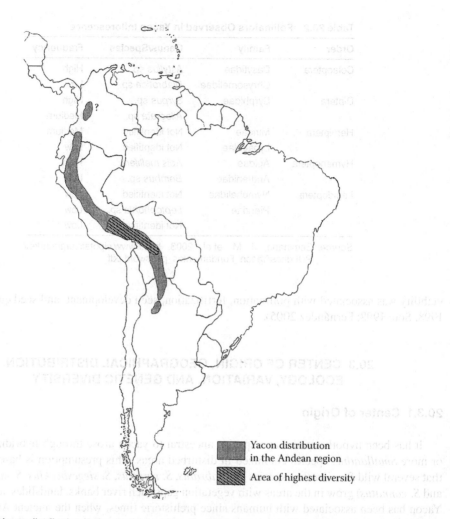

Figure 20.4   Yacon distribution in the Andean region. Doubtful presence in Colombia at present is indicated
by a question mark. (According to Grau, A., and J. Rea. 1997. In *Andean roots and tubers:
Ahipa, arracacha, maca and yacon. Promoting the conservation and use of underutilized and
neglected crops*, ed. M. Hermann and J. Heller, 199–242. Gatersleben, Germany: Institute of
Plant Genetic and Crop Plant Research, Rome, Italy: IPGRI.)

(Safford 1917; Yacovleff 1933; O'Neal and Whitaker 1947). The ceramic is kept in the collection of
National Archaeological Museum in Lima (Perú) (Calvino 1940). Yacon was identified in prehistoric embroideries of the Nazca culture (Towle 1961). Presumable relics of yacon root tubers were
found in the area of the Calendaria culture (AD 1–1000) developed in northwestern Argentina at
Pampa Grande, Serranía Las Pirgulas, Guachipas, and Salta Province (Zardini 1991).

The first written record on yacon (*llacum*) is by Felipe Guaman Poma de Ayala in 1615; he
listed 55 native crops cultivated by Andean inhabitants, including eight crops introduced from Spain
(Grau and Rea 1997). The chronicler Parde Bernabé Cobo (1653) recommended eating yacon as a
fruit because of its sweetness, sappiness, and postharvest life; he stressed its suitability to withstand several days' longer durability of transport by sea. The indigenous people used to appraise
and consume yacon during the relic rituals in the pre-Hispanic time. Currently, the Andeans eat
yacon tuberous roots at Corpus Christi (Meza 2001), which replaced the original Cojapac-raimi
days (Herrera 1943).

Figure 20.5   Yacon tuberous roots depicted on postal stamp of Peru. (Photo I: Viehmannová, SERPOST, 2007.)

Yacon was introduced for the first time in Europe at the Paris exhibition under the name *jiqui-milla* (Perez Arbeláez 1956). Grau and Rea (1997) stated that European interest was not very significant. In Italy, there was a serious attempt to cultivate yacon in the late 1930s, which faded during World War II (Calvino 1940). Yacon was little known on the European continent until it was introduced as a plant with antidiabetic, nutritive, and fertility-enhancing properties in the 1990s (Valentová, Frček, and Ulrichová 2001).

20.3.2 Geographic Distribution

Yacon is being grown in many localities scattered throughout the Andes, from mountain zones of Merida in Venezuela to northwestern Argentina (Grau and Rea 1997; Fernández 2005). In these areas, the crop is cultivated at high elevations but not above 3000 m (Rubatzky and Yamaguchi 1997).

The greatest diversity of yacon occurs in Peru, where it is cultivated from the valleys of the Peruvian coast (500 m above sea level [a.s.l.]) to the villages of Totorani and Sicuani, Cusco (3700 m a.s.l.). The maximum concentration is in the north (Cajamarca) and in the southeastern mountains (Cusco) between 2000 and 3000 m a.s.l. (Meza 2001). In the Cusco region, yacon is cultivated in the Mainlycin valleys of Marcapata, Lares, and La Convención and in the basin of the Mapato River (Paucartambo). According to Seminario et al. (2003), the cultivation of yacon is centralized in 18 of 22 regions and the total area under cultivation is 600 ha. Yacon is mainly sold at local markets.

Within Ecuador, yacon is predominantly grown between the mountain ridges of Andes, from 2400 to 3000 m a.s.l., in the southern provinces of Loja, Azuay, and Cañar. The crop is also present in the central highland provinces, such as in Bolivar and Chimborazo, and in the north of the country—namely, in Pichincha, Imbabura, and Carchi (Rea 1992; Tapia et al. 1996; Meza 2001).

Fernández and Pérez (2003) reported the greatest concentration of yacon in altitudes of about 2500 m with maximum to 3770 m in Bolivia (department Charcas, Potosí). It is grown in department

La Paz (provinces Camacho, Inquisivi, Larecaja, Loayza, Muñecas, Murillo, Nor Yungas, Saavedra, and Sud Yugas), Cochamamba, Potosí (provinces Bilbao Rioja, Charcas, Tand, and Linares), Chuquisaca, Santa Cruz, and Taria (Rea 1992; Fernández and Pérez 2003).

Yacon is a rare crop in northwestern Argentina and is cultivated only in a few localities in the Salta and Jujuy Provinces, where it was reported close to extinction (Zardini 1991).

In Columbia, yacon is grown at Cundinamarca, Boyacá, Nariño, and Huila plateaus at altitudes of 2600–3000 m in the Páramo area (Cárdenas 1969; Rea 1992). Cultivation of yacon in Venezuela and Colombia has been reported in the literature (Popenoe et al. 1989; Zardini 1991; Rea 1992).

Today, yacon is grown outside the Andes and its cultivation has extended to the other continents (Figure 20.6). Yacon is an adaptable crop and has been cultivated in regions between latitude 55°N and 46°S around the world. Bredemann (1943) reported that yacon was first introduced and cultivated in France in 1861. The first successful introduction to Europe was realized in 1927 in San Remo, Italy. After 13 years of acclimatization, yacon was recommended for cultivation, primarily for nutritious purposes, as a fodder crop and for sugar (Calvino 1940). Yacon was cultivated in Germany (Hamburg and Wulsdorf) in 1941 (Bredemann 1948).

Grau (1993) reported that cultivars grown in the North Island of New Zealand originated from Ecuador and their tuberous roots were sold in the supermarkets as a specialty vegetable. From New Zealand, yacon was introduced into Japan and the area of cultivation reached approximately 100 ha (Frček 1996). The first introduction to Japan came in the 1970s via Korea, where it was cultivated as a medicinal plant for stimulants and longevity. The second period of introduction passed through New Zealand in 1985 (Tsukihashi, Myiamoto, and Suzuki 1991). From Japan, it was introduced to Brazil. In Brazil, the crop has been cultivated commercially as a medicinal crop in São Paulo state since 1991 (natural nourishment). Fresh tuberous roots, dry tuberous root chips, and dry leaves are sold for preparation of medicinal tea against diabetes (Grau et al. 2001).

Yacon is cultivated in several U.S. states (California, Oregon, New Mexico, Florida, Alabama and north Virginia) (Popenoe et al. 1989). In 1994, yacon was introduced from Japan to Russia and also from Argentina in 1994–1995 (Tjukavin 2001). Today, yacon is cultivated in the Krasnodarsky and Stavropolsky regions for industrial usage. The crop has also been cultivated in Estonia, Iran, England, Paraguay, and Taiwan. However, it has apparently failed in the Czech Republic (Matejka 1994) and it is likely to fail in most of Central Europe because of the long winter period (Grau and Rea 1997).

Nevertheless, in 1993, the suitability of the species for cultivation under moderate continental European conditions was reported (Valentová and Ulrichová 2003). The possibility of cultivation in the Czech Republic was first demonstrated in 1995 (Frček 2001; Frček et al. 1995). It has been successfully cultivated in three localities of the Czech Republic, including the experimental plots of the Czech University of Life Sciences in Prague, the Potato Research Institute in Havlíčkův Brod, Ltd., and Palacký University in Olomouc-Holice (Lebeda, Doležalová, Dziechciarková, et al. 2003; Lebeda, Doležalová, Valentová, et al. 2003; Lebeda, Doležalová, and Doležal 2004; Lebeda et al. 2008; Fernández 2005; Valentová et al. 2006). The popularity of yacon has increased as a vegetable crop having health-optimizing effects among European populations.

20.3.3 Variation and Genetic Diversity of *S. sonchifolius*

Yacon has the greatest diversity in Peru, although it is cultivated from Ecuador to northwestern Argentina. In most cases, only a few yacon plants were cultivated for family consumption in kitchen gardens and, less frequently, as a cash crop to be marketed at the local farmers' market. Even in this situation, farmers rarely cultivated yacon as a main crop and seldom dedicated a high proportion of their arable land (Grau and Rea 1997). Recently, increased market demand has sustained cultivation

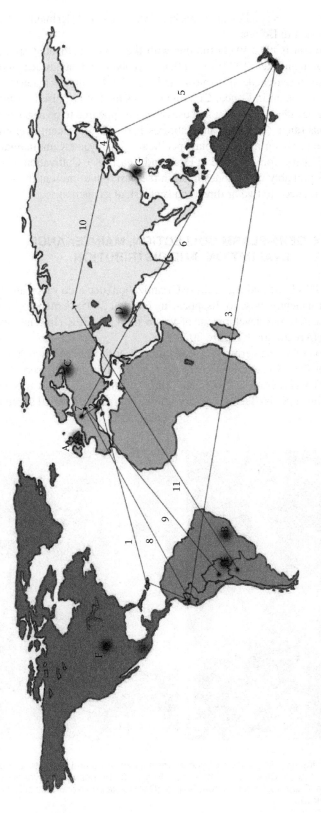

Figure 20.6 Geographic distribution of yacon outside the Andes. (Fernández, C. E. 2005. Habilitation thesis, 154 pp. Prague: Czech Agricultural University, Institute of Tropics and Subtropics.) Line (dissemination from → to, year): (1) Dominican Republic → Italy, 1927; (2) Italy → Germany, 1941; (3) Ecuador → New Zealand, 1960s; (4) Korea → Japan, 1970s; (5) New Zealand → Japan, 1985; (6) New Zealand → Czech Republic, 1993; (7) Germany → Czech Republic, 1994; (8) Ecuador → Czech Republic, 1994; (9) Bolivia → Czech Republic, 1995; (10) Japan → Russia, 1994; (11) Argentina → Russia, 1994–1995. A–G: Other countries where the yacon was disseminated (the UK, Brazil, Estonia, Iran, Paraguay, the United States, and Tchai-Wan).

of yacon for sale in open markets and/or supermarkets (Manrique and Hermann 2003). This may be the cause of genetic erosion in Bolivia.

The La Paz department is most likely the one with the largest cultivated area and the largest germplasm diversity. Grau and Rea (1997) stated that Cochabamba, Chuquisaca, and Tarija departments have been poorly researched, up to now, and may hold valuable materials. Wells (1965) indicated the presence of yacon in Cauca, Colombia—presumably because he studied specimens from that region. Three decades ago, traditional use of yacon appeared to be restricted to the eastern Colombian mountainous range (Debouck and Libreros Ferla 1995). Recent explorations have not confirmed the presence of yacon in Colombia (specifically in Boyaca, Cundinamarca, Huila, and Nariño) (Hermann 1997 pers. comm., cited in Grau and Rea 1997). Cultivation of yacon in northwestern Argentina was probably more widespread in the past, but at present the number of clones available in the area is reduced to two or three and is at risk of extinction.

20.4 GERMPLASM COLLECTION, MAINTENANCE, EVALUATION, AND DISTRIBUTION

In 1981, FAO (the UN's Food and Agriculture Organization) declared yacon an endangered species because it was permanently close to disappearing (Arbizu 2005; Fernández et al. 2005). FAO decided to support research and development of yacon through ICFR (International Council for Fitogenetic Resources) (Arbizu 2009; Figure 20.7).

Systematic collection of yacon germplasm was initiated in the 1980s, sponsored by the IBPGR (International Board for Plant Genetic Resources), and a second exploration period was initiated in 1993, guided by the RTA (Program of Roots and Tubers in the Andes) and funded by COTESU (Cooperación Técnica Suiza [Swiss Technical Cooperation]) for a 5-year period (Grau and Rea 1997).

Figure 20.7 (See color insert.) Acquisition of yacon germplasm in natural habitats in Bolivia. (a) Yacon farmer in La Tranca, Chapare, Cochabamba, Bolivia; (b) European surveyor of genetic resources in La Tranca, Chapare, Cochabamba, Bolivia. (Composite photo courtesy of E. C. Fernández and I. Viehmannová.)

20.4.1 Gene Bank and Other Significant Germplasm Collections

Recently, a network of institutions in Latin America began to collect and to document systematically yacon germplasms in gene banks (Holle and Talledo 2001), based on a collaborative program of the RTA initiated in 1992 (Polreich 2003). However, there is no central gene bank to maintain the entire germplasm of yacon. A list of all institutions conserving yacon is unavailable in the Andean region and beyond. Nevertheless, the most significant collections are located in Peru, Ecuador, and Bolivia.

Peru. Arbizu and Robles (1986; in Grau and Rea 1997) initiated a collection of yacon efforts in Peru and in different places of the country; a total of more than 200 accessions was found. The National University of Cajamarca (UNC) collected 120 accessions of yacon between 1993 and 2002. CIP (Centro Internacional de la Papa/International Potato Center, Lima, Peru) maintains 47 accessions (Seminario, Valderrama, and Romeo 2004), the National Research Institute of Agriculture (INIA) 119 accessions, and the National University San Cristóbal de Huamanga (UNSCH) 7 accessions (CIP 1997). The University of Ayacucho re-collected 10 accessions and the Agricultural Research Institute at the Canaan Center 6 accessions of yacon (De la Cruz 1995, in Grau and Rea 1997). The Breeding Center of Andean Crops (CICA, in Cusco) maintains 46 accessions (Seminario et al. 2004) and the Regional Center of Andean Biodiversity (CRIBA) at the Universidad Nacional de San Antonio Abad del Cusco (UNSAAC) maintains 157 accessions from southern Peru in Ahuabamba, Cusco (Polreich 2003). The other institutions that conserve the germplasm of yacon are the Universities of Huánuco, Huancayo, Cerro de Pasco, and La Molina (Mansilla et al. 2006; in Seminario et al. 2003).

Ecuador. A collection of 32 yacon accessions is maintained by INIAP (Autonomous National Institute of Agricultural Research) in Pichincha (Tapia et al. 1996). By July 1994, INIA (experimental Estacion de Santa Catalina) was maintaining 36 accessions *ex situ* (Rea 1997).

Bolivia. The Bolivian Institute of Agricultural Technology (IBTA) mantains 30 accessions of yacon (CIP 1997). The PROINPA Foundation (Promotion and Research on Andean Crops), in collaboration with the Faculty of Agronomy, Universidad Mayor de San Simón (UMSS), maintains around 40 accessions of yacon in the locality of Locotal (Cochabamba) in field conditions. The gene bank of Andean crops in Toralapa, Cochabamba, administered by the PROINPA Foundation, conserves five accessions of yacon (Seminario et al. 2004).

Argentina. The Cerrillos Research Station in Salta (INTA—National Agricultural Technology Institute) maintains only one yacon accession (Grau and Rea 1997). Although there is enough information on yacon cultivation outside the Andean region, details about the amount of genetic material and its preservation are unavaible.

Czech Republic. The Institute of Tropics and Subtropics (ITS), Czech University of Life Sciences, in Prague, maintains 25 accessions obtained from Peru (16 accessions), Bolivia (5 accessions), Ecuador (1 accession), New Zealand (2 accessions), and Germany (1 accession). Accessions from New Zealand originally are from Ecuador (Fernández, Robles, and Riessová 2004; Fernández et al. 2006; Zámečníková 2007; Viehmannová 2009). At the ITS, a related species of yacon *S. connatus* (syn. *Polymnia connata*) is also being maintained. It was gained via Index Seminum from the Institute of Plant Genetics and Crop Plant Research, Gatersleben, Germany (Viehmannová 2009).

20.4.2 Acquisition and Exploration

Yacon grows in different ecological habitats along the entire length of the Andes, from mountain ranges of Merida in Venezuela to northwestern Argentina (Meza 2001). The largest diversity of yacon was found in Peru (Rea 1994; Meza 2001), where the cultivars are grown from 500 m (in valleys of the Peruvian coast) to 3700 m a.s.l. (village of Totorani, Sicuani, Cusco); the biggest concentration is in the north (Cajamarca) and the southeastern mountains (Cusco) (Meza 2001; Seminario et al. 2003, 2004). In Bolivia, yacon is mostly grown at altitudes of around 2500 m and is also found up to 3770 m (Charcas Province, Potosí) (Fernández et al. 2005). The concentration is dense mainly in the La Paz region (provinces: Camacho, Inquisivi, Larecaja, Loayza, Muñecas, Murillo, Nor Yungas, Saavedra, and Sud Yungas), Cochabamba (from Pocono to the south), Potosí (provinces: Bilbao Rioja, Charcas, Linares, and Saavedra), Chuquisaca, Tarija, and Santa Cruz (Rea 1992; Fernández et al. 2005) (Figure 20.7).

In Ecuador, yacon is cultivated in the valleys between the mountain massifs of the Andes at altitudes from 2400 to 3000 m, mostly in the southern provinces of Loja, Azuay, and Cañar. Yacon is cultivated also in the Andean provinces in central Ecuador—Bolívar, Chimborazo, and Tungurahua—and in the north of the country, especially in the provinces of Pichincha, Imbabura, and Carchi (Rea 1992; Tapia et al. 1996; Meza 2001). In Colombia, yacon is grown on the plateaus of Cundinamarca, Boyaca, Nariño, and Cauca at altitudes of 2600–3000 m in the zone of Páramo (Cárdenas 1969; Rea 1992) and in Argentina northwest of Buenos Aires and Salta (Rea 1992).

It is assumed that the most valuable yacon collection is maintained *in situ* in the countries of the Andean region. Local farmers conserve the biodiversity of yacon, a new source of clones with favorable characteristics of economic importance (Rea 2000a). The color of the edible part of tuberous roots (flesh) is a qualitative characteristic, which describes and distinguishes each cultivar. Landraces are distinguished by the color of the flesh (Figure 20.8), such as yellow (Amarillo), orange (Anaranjado), white (Blanco), and purple (Morado).

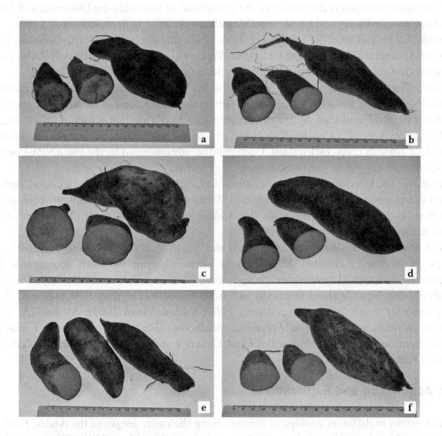

Figure 20.8 Morphological variability of tuberous roots. (a) Locotal-morado (origin Bolivia): epidermis [gray-orange with light purple pigmentation]; flesh [yellow-white with secondary purple pigmentation]; (b) Cajamarca (origin Peru): epidermis [gray-purple]; flesh [white]; (c) Yanayo Grande (origin Bolivia): epidermis [gray-orange]; flesh [yellow-orange]; (d) BOL (origin Bolivia): epidermis [gray-purple]; flesh [white]; (e) PER 75 (origin Peru): epidermis [red-purple]; flesh [white]; (f) PER 55 (origin Peru): epidermis [gray-orange]; flesh [yellow-orange.] (Photo courtesy of F. Pudil.)

Peru. Meza (1995, 2001) described the six original cultivars in the yacon collection in the south of Peru (Cusco, Puno, and Apurímac). They were distinguished by color of skin (epidermis) and flesh of tuberous roots. These characteristics are genotypically dependent:

- Q'ello llakjum: epidermis of tuberous roots is creamy and the flesh is yellow.
- Ch'ecche llakjum: epidermis of tuberous roots is creamy and the flesh is yellow with violet spots. It is very sweet and, for consumption, does not need to be exposed to the sun.
- Yurac ch'ecche: epidermis of tuberous roots is dark and creamy and the flesh is white.
- Yurac llakjum: epidermis of tuberous roots is pink and the flesh is white.
- Culli llakjum: epidermis of tuberous roots is purple red and the flesh is crystal white. It has a slightly sweet taste and, for consumption, requires exposure to the sun.
- Laranja Llakjum: epidermis of tuberous roots is cream and the flesh is orange.

Seminario et al. (2004) distinguished six grown morphotypes in yacon collection in the north of Peru (Cajamarca). These morphotypes are maintained at the local gene bank under the names Hualqui (local name of Amarillo), San Ignasio (Morado), Aachen, Taquia (Hueso Negro), Moteado (Blanco con Morado), and Ancash.

Bolivia. In La Paz, 8 provinces of the total 19 were examined, covering the 30 cantons where around 100 families cultivating yacon lived. The biggest concentration of yacon was described in the altitude from 3200 to 3400 m. Morphotypes described were yellow (Amarillo), white (Blanco), and purple (Kulli) (Rea 2000a, 2000b).

Fernández et al. (2005) screened yacon at the north of Potosí department in the villages of Bilbao Rioja and Charcas provinces (Table 20.3). Geographically, they are located between 65°45′31″ and 66°03′29″ of the east longitude and 18°01′01″ and 18°08′47″ of the north latitude. Local people recognized two local cultivars: K'ellu (yellow) and Yurak (white), grown for their own consumption on the small fields. The studies conducted in the provinces involved 51 families of 16 villages at the altitudes of 2680–3776 m. Five landraces were examined for flesh color such as yellow, orange, white, crystal white, and clearly violet (Fernández et al. 2005; Table 20.3).

Ecuador. Most of the accessions of the yacon collection in Santa Catalina come from the provinces of Cañar, Azuay, and Loja. In Ecuador, yacon accessions have been identified as four morphotypes: purple, light green, dark green with white flesh, and dark green with yellow flesh (Morillo et al. 2002).

20.4.3 Collection, Regeneration, and Maintenance

Yacon is propagated by rhizomes used for commercial purposes (see Section 20.8). Elaborated protocol for the storage of the rhizomes has not been established in the Andean region—place of the origin of yacon—nor anywhere else. Temperature and humidity are supposed to be the most important parameters for storage of tuberous roots and rhizomes of yacon (Fernández et al. 2006). After the harvest of tuberous roots (May to July), small farms in the Andean region (Bolivia, Peru) store rhizomes in dark and dry rooms until planting (August to October). When harvesting of the tuberous roots is finished, farmers in some regions leave the rhizomes in the soil (Meza 2007, personal communication).

The farmers in Jujuy Province, Argentina, store the pieces of rhizomes underground covered by straw from late May to mid-August (Zardini 1991). In the Czech Republic, the rhizomes and tuberous roots are generally stored in boxes covered with peat in a cool (around 10°C) and humid (60–70%) room (Fernández et al. 2006). There are also modern technologies for maintenance of germplasm of yacon such as *in vitro* cultures (see Section 20.7.2). Research on long-term conservation by cryopreservation of *S. sonchifolius* started in cooperation with the Crop Research Institute, Prague, in 2008. The research is focused on the use of vitrification methods for cryopreservation (Zámečníková 2009, personal communication).

Table 20.3 Evaluation of Yacon

Province	Village	No. of Families[a]	No. of Plants[b]	Morphotypes[c]	No. of Roots[d]	Bx[e]
Bilbao Rioja	Carapacayma	3	108	Yellow	7	11.50
				White	9	10.50
	Collpa Pampa	2	100	Yellow	7	11.25
				White	8	10.75
	Challa Villque	6	60	White	8	12.50
	Lujani	1	50	White	10	13.00
	Niño Kollo	1	70	White	7	11.25
	Quirusillani	3	160	White	6	10.25
	Tuquiza	7	248	Yellow	12	11.00
				White	9	10.50
				Crystal white	9	10.25
				Clearly violet	11	12.00
	Villa Paraíso	3	250	Yellow	9	12.00
				White	10	11.50
	Yanayo Chico	5	120	Yellow	12	13.00
				Orange	14	14.00
	Yanayo Grande	6	300	Yellow	9	12.00
				White	10	11.75
Charcas	Ararian	1	370	Yellow	15	11.50
				White	10	10.25
	Estrillani	3	150	White	9	11.75
	Hacienda Loma	3	170	Yellow	8	12.25
				White	9	11.50
	Vaquería	2	90	Crystal white	12	10.50
	Villa K'asa	3	69	Yellow	8	12.25
				Crystal white	10	10.50
	Yunguma	2	120	Yellow	12	11.75
				White	11	11.00
Total	16 Villages	51	2435	Yellow (10x)	7–15	11.00–13.00
				Orange (1x)	14	14.00
				White (13x)	6–11	10.25–13.00
				Crystal white (3x)	9–12	10.25–10.50
				Clearly violet (1x)	11	12.00

Source: Fernández, C. E. et al. 2005. Agricultura Tropica et Subtropica 38:6–11.
[a] Number of families in a village growing yacon.
[b] Total number of plants grown in a village.
[c] Morphotypes of yacon, distinguished according to flesh color.
[d] Average number of tuberous roots on a plant.
[e] Bx (Brix grade) determines a concentration of soluble matter (saccharides) in yacon's tuberous roots.

20.4.4 Germplasm Characterization

Morphological descriptors of yacon have been developed for agricultural practices and varietal improvement (Table 20.4). Methods have been established to determine variability and the relationship between genotypes using polymorphism of isoenzymes and molecular markers. The caryological and chemical analysis of different genotypes has also been conducted.

Table 20.4   Descriptors for Yacon

1. Predominant color of the stems	**11. Leaf margin**
1. Yellow-green	1. Crenate
2. Light gray-purple	2. Dentate
3. Gray-purple	3. Double dentate
2. Distribution of the secondary stem pigmentation	**12. Habit of the inflorescence**
0. Absent	0. Absent
1. Purple pigmentation in the terminal nodes and internodes	1. Sparse
2. Purple pigmentation distributed along the whole stem	2. Moderate
3. Ramification of the stems	3. Frequent
1. Predominant apical	**13. Color of the ray flowers**
2. Within the complete stem	1. Light yellow-orange
4. Foliage color	2. Dark yellow-orange
1. Yellow-green	**14. Shape of the ray flowers**
2. Dark yellow-green	1. Ovate
5. Purple pigmentation in the apical leaves	2. Elongated
0. Absent	3. Elliptic
1. Present	**15. Teeth number of the ray flowers**
6. Color of the leaf blade (adaxial surface)	0. Absent
1. Yellow-green	1. Two
2. Dark yellow-green	2. Three
3. Green with purple pigmentation	**16. Seed production**
7. Color of the petiole (adaxial surface)	0. Absent
1. Yellow-green	1. Present
2. Gray-red	**17. Color of the tuberous roots (peel)**
3. Gray-purple	1. White
8. Leaf blade shape	2. Gray-purple
1. Triangular	3. Red-purple
2. Deltoid	4. Gray-orange
3. Cordate	**18. Color of the tuberous roots (flesh)**
9. Shape of the leaf blade base	1. White
1. Truncate	2. Yellow-white
2. Cordate	3. Yellow-white with secondary purple pigmentation
3. Hastate	4. Yellow
4. Subhastate	5. Orange
5. Sagittate	6. Yellow-orange
10. Shape of the leaf tip	**19. Hendidures in the tuberous roots**
1. Narrowly acute	0. Absent
2. Broadly acute	1. Present
	20. Color of the rhizomes
	1. White
	2. Red-purple with white
	3. Gray-purple
	4. Red-purple

Source: Seminario, J. et al. 2004. *Arnoldia* 11:139–160.

Although limited information about biological, chemical, and molecular variability is available, comprehensive research aimed at improving the cultivation of yacon and acquiring knowledge of morphology, isoenzyme polymorphism, yield parameters, saccharide content in the tuberous roots, disease and pest resistance, DNA content, and the content of phenolics in the leaves has been undertaken (Lebeda, Doležalová, Dziechciarková, et al. 2003; Lebeda, Doležalová, Valentová, et al. 2003; Lebeda et al. 2004; Valentová et al. 2001, 2006). Three years' running field experiments have shown large variations in yield parameters and morphological variability of yacon's underground parts (Lebeda et al. 2004; Valentová et al. 2006). Isoenzyme polymorphism and variation in nuclear DNA content in 25 yacon accessions have been evaluated (Doležalová, Lebeda, and Gasmanová 2004; Valentová et al. 2006). From 2001 to 2003, the aerial parts of the plants were observed (every fortnight) during the vegetative period (Lebeda et al. 2004; Valentová et al. 2006).

For the underground portion of yacon, the tuberous roots were weighed. Tuber shape was characterized following the morphological descriptors of yacon (Frček 2001; the most frequent shapes are in Figure 20.9), and the following traits were subsequently evaluated: color, tuberous root skin character, color of flesh, shape of tuberous roots in transverse section, and tuberous root yield. Nine shapes of genotypes were found of the total 15 morphotypes described (Frček 2001). Four of 25 genotypes studied were found to be the most stable because they retained a napaceous shape within the 3 years of running experiment. The same was recorded for one genotype that retained the shape of an irregularly transverse oblong. Napaceous shape was observed in six genotypes over 2 years of cultivation. The most frequent was the irregularly napaceous shape, broaden in the middle, recorded in 11 genotypes, and the shape of an irregularly transverse oblong recorded in six genotypes. An oblong tapered shape was described in three genotypes.

These types of root tubers were the most often noted, although the resulting shape depended on soil texture and structure; in addition, several different root shapes concur in one genotype (Fernández 2005). On the other hand, once in 3 years, globular, oblong, and carrot shapes were recorded. As for yacon tuberous root production, the most productive was one genotype with an average weight of tuberous roots 2.8 kg per plant and two genotypes with 2.7 and 2.6 kg per plant, respectively. In contrast, the lowest average weights—1 and 1.7 kg per plant—were recorded (Figure 20.11). Meza (2001) reported that the maximal yield of tuberous roots obtained from a hectare was 25–111 t in the Andean region (Cusco, Peru); outside the Andes (Brazil), the maximum was 120 t per hectare (Anonymous 2003).

The 38–66 t of tuberous roots were the maximum yield obtained from a hectare under conditions in the Czech Republic (Frček et al. 1995). The data already reported (Lebeda, Doležalová, Dziechciarková, et al. 2003; Valentová et al. 2006) are consistent with the results summarized by Fernández (2005), who reported that high yield of yacon is influenced by a number of factors (e.g., ecotype, ecosystem, region of cultivation, land type, and environmental factors). From the agricultural point of view, possible factors include the method of cultivation, time of growth establishment, and the size of plantation.

Twenty-five genotypes of *S. sonchifolius* were screened using 16 enzymes:

ACP: acid phosphatase (EC 3.2.1.30)
ADH: alcohol dehydrogenase (EC 1.1.1.1)
DIA: diaphorase (EC 1.6.99.1)
EST: esterase (EC 3.1.1.1)
GDH: glutamate dehydrogenase (EC 1.4.1.2)
GOT: glutamate-oxaloacetate transaminase (EC 2.6.1.1)
GPI: glucose-6-phosphate isomerase (EC 5.3.1.9)
IDH: isocitrate dehydrogenase (EC 1.1.1.42)
LAP: leucine aminopeptidase (EC 3.4.11.1)
MDH: malate dehydrogenase (EC 1.1.1.37)
ME: malic enzyme (EC 1.1.1.40)

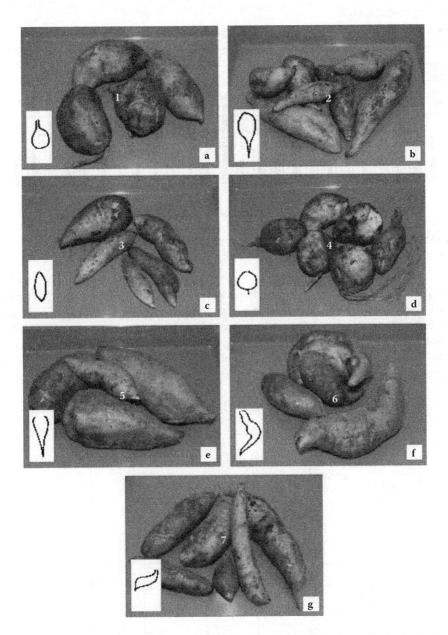

Figure 20.9 Yacon germplasm. Basic shapes of tuberous roots (numerical codes according to Frček, J. 2001. El cultivo del yacón y la maca en la República Checa. Resumenes workshop sobre la agricultura sostenible del 17 al 25 de Septiembre 2001, 54–58. Praga: Instituto de Agricultura Tropical y Subtropical, Universidad Checa de Agricultura Praga; descriptor state according to Doležalová and Lebeda, unpubl.) recorded in yacon. (a) pyriform; (b) turnip-like; (c) oblong; (d) globose; (e) carrot-like; (f) incurvate; (g) cylindrical. (Composite photo courtesy of I. Doležalová and J. Doležal.)

NADHDH: NADH dehydrogenase (EC 1.6.5.3)
PGM: phosphoglucomutase (EC 5.4.2.2)
SHDH: shikimate dehydrogenase (EC 1.1.1.25)
PGD: 6-phosphogluconate dehydrogenase (EC 1.1.1.44)
SOD: superoxide dismutase (EC 1.15.1.1)

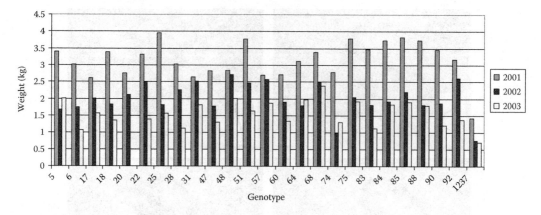

Figure 20.10   Variation in weight of tuberous roots in different yacon genotypes. Plants were cultivated in the Czech Republic in 2001–2003.

Only ACP and EST showed a relatively large degree of polymorphism (Figures 20.11 and 20.12). The remaining 14 systems were monomorphic. Among all the staining systems, 55 bands (isoforms) were observed. The variation in morphotypes recorded does not correspond unambiguously to EST polymorphisms, which showed six separate groups. The largest group of EST isoforms represented about 70% of all studied genotypes. The majority of observed morphotypes (seven) were included in this group (Lebeda et al. 2008). The remaining isoforms were only represented by one to three

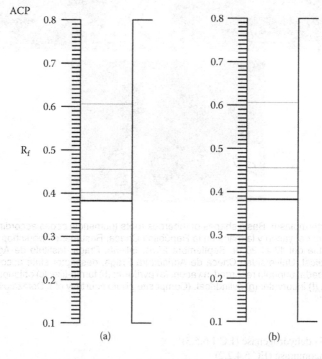

Figure 20.11   Zymograms of acid phosphatase (ACP) isozyme spectra. (a) Genotypes 17, 18, 22, 28, 31, 47, 48, 51, 57, 60, 64, 68, 74, 75, 83, 84, 85, 88, 92, and 1237; (b) genotypes 5, 6, 20, 25, and 90. (Adopted from Lebeda, A., Doležalová, Dziechciarková, et al. 2003. *Czech Journal of Genetics and Plant Breeding* 39:1–8.)

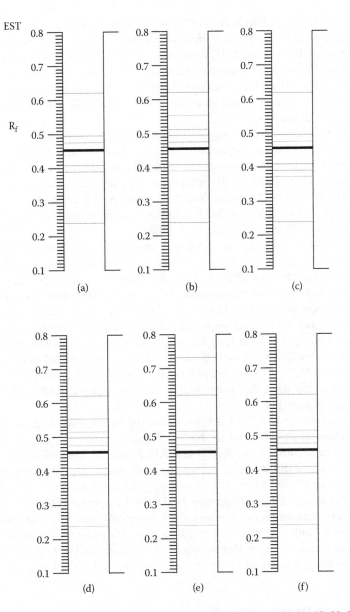

Figure 20.12   Zymograms of esterase (EST) isozyme spectra. (a) Genotypes 17, 18, 22, 28, 31, 47, 48, 51, 57, 60, 64, 68, 74, 75, 83, 85, 88, and 92; (b) genotypes 5 and 84; (c) genotype 1237; (d) genotypes 25, 20, and 90; (e) genotype 6; (f) genotype 22. (Adapted from Lebeda, A., I. Doležalová, M. Dziechciarková, et al. 2003. *Czech Journal of Genetics and Plant Breeding* 39:1–8.)

genotypes. Very slight differences in zymogram spectra of individual groups can give only relatively weak predictions about variability. The same can be said for ACP polymorphism. When results of isoenzyme polymorphism were compared with the results of morphological studies (shape of tuberous roots), it was concluded that the isoenzymes' variability is relatively very small compared to the wide variability of morphological characters (Figure 20.11 and 20.12).

Variability in yacon was also examined by karyological analysis. Chromosome number in somatic cells varies depending on the plant materials and the authors. In the material of Ecuador, chromosome number was $2n = 60$ (Heiser 1963). Materials from Peru also contained $2n = 60$

(Talledo and Escobar 1996). Ishiki et al. (1997) cytologically examined clones from Ecuador, Peru, Bolivia, and Argentina and reported $2n = 58$ chromosomes, while clones from Argentina contained $2n = 87$ chromosomes. Frías et al. (1997) reported $2n = 58$ chromosomes in one clone from northwestern Argentina. The lowest chromosome number, $2n = 32$, in yacon was reported for material grown at the University of La Molina in Peru (León 1964). Yacon plants maintained in the Czech Republic (from Bolivia, Germany, and New Zealand) contained $2n = 58$ (Fernández and Kučera 1997) and $2n = 87$ (Viehmannová 2009).

Relative nuclear 2C DNA content in yacon ranged from 5.82 to 6.12. Statistical analysis (LSD test), based on relative DNA data, separated five significantly distinct groups of yacon genotypes ($p \leq 0.5$). The majority of yacon genotypes (15 of the 25 analyzed) belonged to the group with low relative DNA content (from 5.82 to 5.99) within the studied set. The second group, with three genotypes, showed 6.01 relative DNA content. The third group, with four genotypes, contained relative DNA content ranging from 6.02 to 6.04. The largest amount of nuclear DNA was observed for genotypes 83 (6.06) and 92 (6.12) (Lebeda et al. 2008).

Intraspecific genetic variability in yacon was investigated by molecular markers. Mansilla et al. (2006) determined genetic diversity of 30 Peruvian genotypes of yacon by the random amplification of polymorphic DNA (RAPD) method. By using 34 decamer primers, 166 bands were observed and more than 30% were polymorphic. Milella et al. (2004) determined genetic variability of five landraces grown at the Czech University of Life Sciences in Prague—Institute of Tropics and Subtropics by using RAPD and amplified fragment length polymorphism (AFLP) markers. Of 61 RAPD primers, 28 showed at least one polymorphic band and the remaining primers were monomorphic. A total of 282 bands were detected with the mean number of 4.9 accountable fragments per primer, and 28.7% of them were polymorphic. For the AFLP analysis, 16 combinations of *EcoRI/MseI* primers were used; they generated 983 reliable markers, of which 7.4% were polymorphic among the yacon genotypes.

The primer combination M-CTC/E-ACA revealed the highest percentage of polymorphic fragments (38.5%). Dendograms obtained for both molecular markers showed almost identical clustering. This study demonstrated that RAPD and AFLP markers are useful tools for determining genetic variability. These results are in accordance with morphological variability (Milella et al. 2004) and the clustering based on contents of total polyphenols in leaves and rhizomes in the previously mentioned landrace of yacon (Lachman et al. 2007). Lachman et al. (2004) also determined variability in saccharide content in tuberous roots over several years and found that the saccharide content is affected not only by genotype but also by climatic conditions during the year.

The previously mentioned methods can help in determining variability in yacon and can furnish information about the genetic material that could be used in the breeding of the crop.

20.4.5 Germplasm Descriptors

Descriptors were created on the basis of morphological differences of particular clones (Arbizu et al. 2001; Polreich 2003; Seminario et al. 2004; Table 20.4). Holle and Talledo (2001) emphasized the need for international standardization of morphological descriptors for effective management of the yacon collection to avoid duplication of samples in the collection. Moreover, institutions could better share data.

Seminario et al. (2004) proposed seven qualitative descriptors and grouped nine morphotypes on the basis of visual assessment of 108 accessions of yacon from Peru. These descriptors are the color of the main stem, presence of purple pigmentation on the top leaves, margin of leaf blades, leaf blade shape, color of tuberous roots (peel), color of the flesh, and rhizome color. Seminario et al.

(2004) also evaluated plants using 20 qualitative descriptors (Table 20.4) as defined by PRTA-UNC, one of which concerned the whole plants, three the stem, seven the foliage, four the inflorescence and flower, one the seed, three the tuberous roots, and one the rhizomes. Cluster analysis was performed in NTSYS 2.1. The resulting dendrogram split the file into nine clusters corresponding to the original allocation by seven visual qualitative descriptors.

20.5 CHEMICAL COMPOSITION

There are large differences in the content of various compounds in aerial and underground parts of the yacon plant. These values were summarized by Lachman, Fernández and Orsák (2003) and Valentová et al. (2001). Chemical composition of yacon varies greatly (Valentová et al. 2001). Table 20.5 shows values of minimum versus maximum variation of each compound.

The major portion of yacon plant biomass is composed of water (fresh tuberous roots: 69.5–92.7%, leaves: 83.2%, stems: 86.7%). Due to high water content (Lachman, Fernández, et al. 2003), the root energy value is very low (619–937 kJ/kg of fresh matter) (Quemener, Thibault, and Cousement 2003). In tuberous roots, the most substantial compound is saccharides (19.67%). Ash is represented by 0.26–3.5%, proteins by 0.3–3.7%, lipids by 0.1–1.5%, and fiber by 0.2–3.4%. In dry material, saccharides represent 67.5%, ash 3.6–23%, proteins 2.4–24.4%, lipids 0.4–9.9%, and fiber 3.7–22.4%. Fresh leaves contain 1.4% saccharides, 2.7% ash, 2.7% proteins, 1.2% lipids, and 1.7% fiber. The main part of dry leaves is proteins (17.1–21.2%), followed by ash (12.5–16%), fiber (10–11.6%), saccharides (8.6%), and lipids (4.2–7.4%). Composition of the stem is similar to that of leaves, with the exception of fiber material, which is 23.8–26.8% in dry stems (Table 20.5). Table 20.6 shows microelements (Ca, P, Cu, Fe, Mn, and Zn in milligrams per 100 g) and vitamins (retinol, thiamine, ascorbate, carotene, riboflavin, and niacin) in yacon plants (Valentová et al. 2001). Chemical composition of individual parts of yacon is summarized next.

20.5.1 Tuberous Roots

Surveys of the most important compounds in yacon tuberous roots are summarized in Tables 20.5 and 20.6. A substantial part of tuberous roots is *saccharides* (approximately 70%, Table 20.5)—mainly fructose (350 ± 42 mg/g dry matter [dm]), glucose (158 ± 29 mg/g dm) and saccharose (75 ± 19 mg/g dm) (Ohyama et al. 1990; Valentová et al. 2001). As storage compounds, the main part of tuberous roots are fructo-oligosaccharides (FOS) (β-(2-1)-fructo-oligosaccharide/inulin type

Table 20.5 Chemical Composition of Yacon Tuberous Roots, Leaves, and Stems

Compound (%)	Tuberous roots		Leaves		Stems	
	Fresh	Dry	Fresh	Dry	Fresh	Dry
Water	69.5–92.7	—	83.2	—	86.7	—
Ash	0.3–3.5	3.6–23.0	2.7	12.5–16.0	1.4	9.6–10.2
Proteins	0.3–3.7	2.2–24.3	2.9	17.1–21.2	1.5	9.7–11.4
Lipids	0.1–1.5	0.4–9.9	1.2	4.2–7.4	6.3	2.0–2.3
Fiber	0.3–3.4	3.7–22.4	1.7	10.0–11.7	3.6	23.8–26.9
Saccharides	19.8	67.5	1.4	8.6	1.6	11.7

Sources: Lachman, J., J. Dudjak, et al. 2003. In I. Mezinárodní seminář "Andské plodiny" v České republice, 12.5.2003, Praha, *Proceedings* (47–54), ed. E. C. Fernández. Praha, Czech Republic: Czech University of Agriculture Prague (in Czech); Lachman, J., E. C. Fernández, et al. 2003. *Plant Soil Evironment* 49:283–290.

Table 20.6 Content of Compounds in Yacon

Compound	Tuberous Roots	Leaves	Stems
	Percent		
Water	93–70	10.47	
Proteins	0.4–2.0	21.48[a]	9.73[a]
Saccharides	12.58		
Lipids	0.1–0.3	4.2	1.98
Ashes	0.3–0.2	12.52	9.60
Fiber	0.3–1.7	11.63	23.82
	Milligrams/100 g		
Calcium	23	1805	967
Phosphorus	21	543	415
Iron	0.3	10.82	7.29
Copper	0.963	<0.5	<0.5
Manganese	0.541	3.067	<0.5
Zinc	0.674	6.20	2.93
Retinol	10		
Thiamine	0.01		
Ascorbate	13.10		
Carotene	0.02		
Riboflavin	0.11		
Niacin	0.34		

Source: Modified by Valentová, K. et al. 2001. *Chemické Listy* 95:594–601.
[a] Content of N compounds.

oligofructans) (Figure 20.13), representing mainly oligosaccharides from trisaccharide to decasaccharide with terminal sucrose (inulin type FOS) (Goto et al., 1995). Underground storage organs accumulate over 60% (in dry matter) of inulin type β (2→1) fructans, mainly oligomers (GF$_2$–GF$_6$) (Ohyama et al. 1990; Itaya, De Carvalho, and Figueiredo-Ribeiro 2002; Valentová 2004). Yacon fructans are of low molecular mass (Hermann, Freire, and Pazos 1998). Cisneros-Zevallos et al. (2002) reported high variability in FOS content (2.1–70.8 g/100 g dm) and the highest fructan content was recorded in dodecaploid lines.

Developmental studies (Itaya et al. 2002) showed that the activities of enzymes involved in the metabolism of saccharides were the highest at the beginning of tuberization (e.g., 3-month-old plants) and at the flowering of plants (e.g., 7-month-old plants). There are known substantial changes in the content of FOS during the storage of tuberous roots that depend on length and temperature (Asami et al. 1991; Fukai et al. 1997; Cisneros-Zevallos et al. 2002). The quality of tuberous roots and metabolic processes during storage is significantly influenced by activity of FOS-metabolizing enzymes in tuberous roots (Narai-Kanayama, Tokita, and Aso 2007). Experiments focused on short-term storage (on-farm experiments) of yacon tuberous roots showed that the effect of shading and the effect of traditional sunlight exposure ("sunning") have a substantial effect on carbohydrate composition in tuberous roots (Graefe et al. 2004). This indicates that partial hydrolysis of FOS starts shortly after harvesting. Sunning of tuberous roots effectively reduces water content (approximately 40%) and thus allows energy to be saved if yacon is processed into dehydrated products (Graefe et al. 2004).

The difference in the content of FOS also depends substantially on yacon ecotypes or varieties and growing season. Experiments in the Czech Republic with four yacon ecotypes originating from

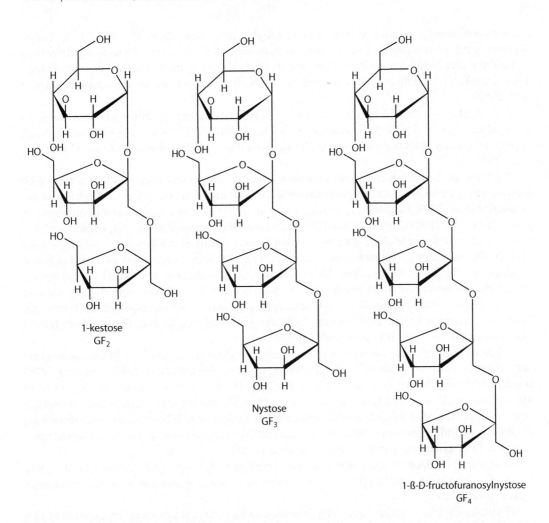

1-kestose
GF$_2$

Nystose
GF$_3$

1-ß-D-fructofuranosylnystose
GF$_4$

Figure 20.13 Examples of three main FOS (GF$_2$–GF$_4$) from tubers of yacon. GF$_2$: trisaccharide consisting of one molecule of glucose and two molecules of fructose; GF$_3$: tetrasaccharide consisting of one molecule of glucose and three molecules of fructose; GF$_4$: pentasaccharide consisting of one molecule of glucose and four molecules of fructose. (Lachman, J., Fernández, et al. 2003. *Plant Soil Environment* 49:283–290.)

Bolivia, Ecuador, Germany, and New Zealand, in 1995, 1996, 2000, and 2001, demonstrated these differences (Lachman et al. 2004). Considerable differences among ecotypes were observed in their content of inulin (141–289 mg/kg dry matter) and lesser for fructose (195–217 mg/kg dm). The year of cultivation has a statistically significant effect on the content of all saccharides (Lachman et al. 2004).

Beta (2→1)-fructans have activities similar to those of β-glucans which act as unspecific stimulators of immune system. These compounds specifically bind on macrophages and thus initiate the cascade of the immune system. They are recommended for medical treatment for deficiency of the immune system, infections, and allergies and also as a supplementary treatment during therapy for cancer (Valentová 2004). Beta-fructans are low energetic, are part of dietary fiber, and are considered prebiotics (Suzuki et al. 1998). It was shown that consumption of prebiotics has a positive effect on intestinal microflora. This effect was demonstrated by extracts from tuberous roots fermented by *Lactobacillus plantarum*, *L. acidophilus* (completely utilized the GF$_2$ molecule), and

Bifidobacterium bifidum (utilized molecules with higher DP) (Pedreschi et al. 2002, 2003). Fructo-oligosaccharide consumption by these microbes apparently depends on the degrees of polymerization and the initial FOS composition (Pedreschi et al. 2003). Components from tuberous roots have a positive effect on the digestive system and due to its effect against cancer (Lachman, Fernández et al. 2003).

The sweetness is caused by fructose, which does not stimulate insulin production and does not produce a glycemic effect (Cisneros-Zevallos et al. 2002). This is the reason why yacon has been consumed by diabetics and those suffering from digestive disorders (Lachman, Fernández, et al. 2003).

Polyphenols. Yacon tuberous roots contain substantial amounts of phenols (2030 mg/kg) with predominating chlorogenic acid (approximately 50 mg/kg) (Lachman, Dudjak et al. 2003; Lachman, Fernández et al. 2003; Lachman, Hejtmánková et al. 2003) as a major antioxidant compound of yacon (Yan et al. 1999) (Figure 20.14). However, some other phenolic acids (e.g., caffeic acid and ferulic acid) and other radical scavenging compounds were identified in tuberous roots (Simonovska et al. 2003); for example, five new water-soluble polyphenols were isolated and identified in yacon tuberous roots by Takenaka and Ono (2003). Detailed study of the presence of caffeic acid derivatives in the tuberous roots confirmed five different derivatives. Two of these were chlorogenic acid (3-caffeoylquinic acid) and 3,5-dicaffeoylquinic acid, common phenolic compounds in Asteraceae. Three were esters of caffeic acid with the hydroxyl groups of aldaric acid, derived from hexose (Takenaka and Ono 2003; Takenaka et al. 2003).

Tuberous roots are also the source of polyphenol oxidase (E.C. 1.14.18.1; PPO), the enzyme that catalyzes oxidation of phenolic compounds on quinone (Yoshida et al. 2002). Currently, PPO is extracted from tuberous roots and partially purified. The enzyme had a molecular weight of 45 490 ± 3500 Da (Dalton) and Km values of 0.23, 1.14, 1.34, and 5.0 mM for the substrates caffeic acid, chlorogenic acid, 4-methylcatechol, and catechol, respectively. When assayed with resorcinol; dl-DOPA; pyrogallol; protocatechuic, p-coumaric, ferulic, and cinnamic acids; catechin; and quercetin, the PPO showed no activity (Neves and da Silva 2007).

Amino acids. Yacon tuberous roots contain tryptophan (15 mg/kg) (Valentová et al. 2001). Except for chlorogenic acid, tryptophan was identified as a major antioxidant in yacon tuberous roots (Yan et al. 1999).

Phytoalexins. Phytoalexins have been identified in tuberous root extracts. Inoculation of this organ with *Pseudomonas cichorii* resulted in the formation of antifungal compounds that were identified as derivatives from 4-hydroxyacetophenone: (1) 4'-hydroxy-3'-(3-methylbutanoyl)aceto-phenone, (2) 4'-hydroxy-3'-(3-methyl-2-butenyl)acetophenone, and (3) 5-acetyl-2-(1-hydroxy-1-methylethyl)benzofuran (Takasugi and Masuda 1996).

Various other compounds were identified in the extract of tuberous roots (e.g., *flavonoid quercetin* and an *unidentified flavonoid*) (Simonovska et al. 2003) and other aglycones of glycoflavonoids. However, their chemical structure is not yet known (Valentová 2004).

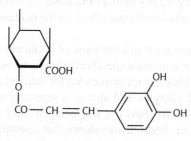

Figure 20.14   Chlorogenic acid—a major antioxidant compound of yacon. (Lachman, J., Fernández et al. 2003. *Plant Soil Environment* 49:283–290.)

Recently, *essentials oils* were identified in yacon roots; 61 components, representing 97.7% of the total oil, were identified. The main constituents of the oil were (1) α-pinene (33.5%), (2) neo-abienol (15.7%), (3) *cis*-abienol (15.4%), (4) *trans*-abienol (7.0%), and (5) 8,12-epoxy-14-labden-13-ol (2.8%) (Miyazawa and Tamura 2008).

Toxicology and safe use. Yacon tuberous roots have not been studied in detail for toxicology. Genta et al. (2005) focused on analysis of the effect of subchronic (4-month) oral consumption of dried yacon root flour as a diet supplement using normal Wistar rats. Flour administered as a diet supplement was well tolerated and did not produce any negative effect, toxicity, or adverse nutritional effect at both intake levels used (340 and 6800 mg FOS/body weight), suggesting lack of toxicity and certain beneficial metabolic activity (Genta et al. 2005).

20.5.2 Leaves and Stems

Various compounds present in yacon leaves and stems differ substantially from the tuberous roots (Tables 20.5 and 20.6). The most important compounds are *phenolics, catechol, terpenes, phytoalexins*, and *flavonoids* (Valentová et al. 2001; Lachman, Fernández et al. 2003). However, *essentials oils* have also been isolated from dried leaves of yacon (Adam et al. 2005).

The content of *phenolic compounds* in the yacon leaves is about twice as high as that in the tuberous roots (Lachman, Dudjak et al. 2003; Valentová 2004), and the leaves are considered a promising source of these compounds (Valentova et al. 2003). Phytochemical analysis of yacon leaves showed four important acids: chlorogenic, dicaffeoylquinic, caffeic, and ferulic (Lachman, Dudjak et al. 2003; Lachman, Hejtmánková et al. 2003; Simonovska et al. 2003). Chlorogenic (Figure 20.14) and caffeic acids are known as important antioxidants and scavengers of ROS (reactive oxygen species) (e.g., Kono et al. 1998; Chen and Ho 1997). Hot water extracts of the aerial part of yacon showed potent free radical-scavenging activity and inhibitory effects on lipid peroxidation in rat brain homogenate (Terada et al. 2006) and rat hepatocytes (Valentová et al. 2003). The structure of the major component in the fraction was identified as 2,3,5-tricaffeoylaltraric acid (TCAA). The antioxidative activity of TCAA was superior to that of natural antioxidants such as catechin, α-tocopherol, and ellagic acid (Terada et al. 2006).

Hypoglycemic activity of yacon extract has been described previously (Aybar et al. 2001; Ulrichová et al. 2003). Studies showed that the aerial part of yacon has strong antioxidative activity; this may encourage its potential use as a food supplement to prevent type II diabetes mellitus (Terada et al. 2006). It is supposed that extracts from yacon leaves could also have a significant effect in regulation of glucose metabolism in humans and could be used as a compound with antidiabetic effects (Valentová 2004); that is, they may have effects similar to those of yacon tuberous roots.

The relationship between phenolic content and antioxidant activity has been evaluated. Recent results indicate that a certain correlation between phenolic content and antioxidant activity exists. Nevertheless, in many cases, there was also antioxidant activity in the herbs that may be attributable to other unidentified substances or to synergistic interactions (Valentová et al. 2003).

Various secondary metabolites have been identified in yacon leaves, such as *terpenes* (e.g., *ent*-kaurenic acid) and their related diterpene compounds (Figure 20.15). Kaurene-type diterpenoids (15-α-angeloyoxy *ent*-kauren-19-oic acid 16 epoxide) were isolated from the glandular trichome exudate and leaf extracts. These diterpenes have an important role as phytoalexins and they are involved in defense mechanisms of the yacon glandular trichomes (Kakuta et al. 1992; Goto et al. 1995). *Ent*-kaurenic acid belongs to the intermediates in biosynthesis of gibberellins in *Gibberella fujikuroi* (Barrero et al. 2001). Inoue et al. (1995) speculated that the aerial part of yacon should contain some antifungal and pesticidal compounds because it is

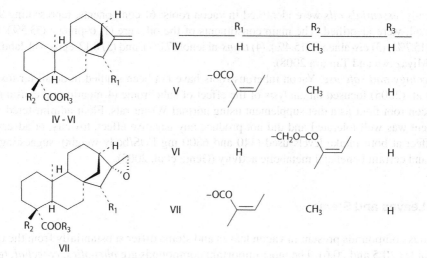

Figure 20.15 *Ent*-caurenic acid and its derivatives from yacon leaves. (Lachman, J., Fernández et al. 2003. *Plant Soil Environment* 49:283–290.)

not necessary to use pesticides in the cultivation of yacon. These substances are called melampolides (Kakuta et al. 1992).

Previously, three known melampolides (polymatin B, uvedalin, and enhydrin) and one new melampolide (sonchifolin) were reported by Inoue et al. (1995). Fungicidal activities of these four compounds were studied against *Pyricularia oryzae*. Among these four melampolides, the sonchifolin exhibited the highest fungicidal activity (Inoue et al. 1995). Four known (sonchifolin, fluctuanin, uvedalin, and enhydrin) and two new antibacterial melampolide-type sesquiterpene lactones (8β-tigloyloxymelampolid-14-oic acid methyl ester and 8β-methacryloyloxymelampolid-14-oic acid methyl ester) were isolated and identified by Lin, Hasegawa, and Kodawa (2003) from yacon leaves. The newly identified compound, 8β-methacryloyloxy-melampolid-14-oic acid methyl ester, exhibited antimicrobial activity against *Bacillus subtilis* and *P. oryzae* (Lin et al. 2003).

Among these six compounds, fluctuanin exhibited the strongest antibacterial activity (against *B. subtilis*); the second and third were uvedalin and enhydrin. All of these three compounds have an acetoxy group at the C-9 position, and therefore it is expected that the acetoxy group is necessary for strong antibacterial activity (Lin et al. 2003). A new dimeric melampolide was isolated from the leaf rinse extract of yacon (Schorr, Merfort, and Da Costa 2007). In this study, the anti-inflammatory properties of uvedalin were assessed and confirmed. Current studies showed (Hong et al. 2008) that 13 different melampolides are in yacon leaves (Figure 20.16). Except for melampolides, antifungal activities of yacon leaf extracts are also related to the presence of essential oil content (Inoue et al. 1995; Adam et al. 2005).

Melampolides and some other compounds were also isolated from the leaves of *Smallanthus connatus* and the constituents profile was related to taxonomical position (Bach et al. 2007). Chemical evidence from this study concluded that (1) *S. connatus* and *S. macroscyphus* are distinct taxa, (2) *S. connatus* is more closely related to *S. sonchifolius* than *S. macroscyphus*, and (3) *S. connatus* is more closely related to *S. sonchifolius* than was suggested by Rauscher (2002) in his phylogenetic study.

The other acids found in the leaves are gallic, gentisic, protocatechuic, and rosmarinic acids, as well as quercetin and the isomers of dicaffeoylquinic and chlorogenic acid and nonidentified flavonoids (Valentová 2004; Jandera et al. 2005).

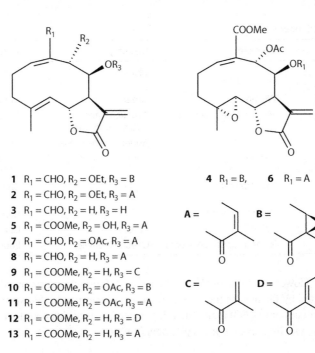

1 R_1 = CHO, R_2 = OEt, R_3 = B
2 R_1 = CHO, R_2 = OEt, R_3 = A
3 R_1 = CHO, R_2 = H, R_3 = H
5 R_1 = COOMe, R_2 = OH, R_3 = A
7 R_1 = CHO, R_2 = OAc, R_3 = A
8 R_1 = CHO, R_2 = H, R_3 = A
9 R_1 = COOMe, R_2 = H, R_3 = C
10 R_1 = COOMe, R_2 = OAc, R_3 = B
11 R_1 = COOMe, R_2 = OAc, R_3 = A
12 R_1 = COOMe, R_2 = H, R_3 = D
13 R_1 = COOMe, R_2 = H, R_3 = A

4 R_1 = B,   6 R_1 = A

A =   B =

C =   D =

Figure 20.16 Survey of known melampolides from yacon leaves. (Hong, S. S. et al. 2008. *Chemical & Pharmaceutical Bulletin* 56:199–202.)

Metabolic derivatives of *cinnamic acid*, which are produced by the activity of epiphytic bacteria *Klebsiella oxytoca* and *Erwina uredovora*, were identified in leaves (Hashidoko et al. 1993), as well as derivatives of *benzoic acid* (Jandera et al. 2005).

Three predominant essentials oils in dry yacon leaves are β-pinene, caryophylene, and γ-cadinene. The relative content was important for specification of yacon varieties (Adam et al. 2005).

20.6 DISEASES AND PESTS

20.6.1 Diseases

Information on diseases and pests of yacon is still limited. Several fungal and bacterial diseases in aboveground and underground parts of yacon have been observed (Table 20.7). Root rot is caused by fungi of genera *Fusarium* and *Rhizoctonia*, which has been associated with excessive soil moisture (Seminario et al. 2003). Black rot (charcoal rot) caused by *Macrophomina phaseolina* (Tassi) Goidanich was reported in Japan (Sato et al. 1999). Wilting in Peru has been by fungus *Fusarium* (Lizárraga et al. 1997) and in Japan by bacterium *Erwinia chrysanthemi* (Mizumo, Nakanishi, and Nishiyama 1993). In Peru, the genus *Sclerotinia* causes soft rot for the tuberous roots (Lizárraga et al. 1997). Fungi of the genus *Alternaria* cause marginal necrosis in the leaves in Ayacucho and Cajamarca, Peru (Barrantes 1998a). In Peru, the fungus *Bipolaris* sp. attacks yacon in hot areas of the Andean valley (Barrantes 1998b). The pycnidial fungus (*Sphaeropsidales*, the species until now not exactly determined) infection was recorded at a low level, with the exception of 2 of 25 yacon genotypes, which were free of infection (Lebeda et al. 2008).

Only a few studies are available on viruses affecting yacon that is free of several common tuber viruses (Grau and Rea 1997), including those affecting potato (potato leaf roll X, Y, S, M;

Table 20.7   Yacon Diseases

Pathogen	Disease	Country	References
Cucumovirus		Japan	Kuroda and Ishihara (1993)
Erwinia chrysanthemi	Wilting	Japan	Mizumo et al. (1993)
Alternaria sp.	Marginal leaf necrosis	Ayacucho and Cajamarca, Peru	Barrantes (1998a)
Bipolaris sp.		Andean valleys, Peru	Barrantes (1998b)
Fusarium sp.	Wilting	Peru	Lizárraga et al. (1997), Seminario et al. (2003)
Macrophomina phaseolina	Charcoal rot	Japan	Sato et al. (1999)
Rhizoctonia sp.	Common rot infection		Seminario et al. (2003)
Sclerotinia sp.	Soft rot	Peru	Lizárraga et al. (1997)
Sphaeropsidales sp.	Necrotic spots on leaves	Czech Republic	Lebeda et al. (2008)

Popenoe et al. 1989). However, clonal decline (*cansancio* = fatigue) and the need to rejuvenate the seed have been reported by farmers in the La Paz region, a phenomenon that strongly suggests viral infection. Kuroda and Ishihara (1993) mentioned that cucumber mosaic cucumovirus infected yacon under field conditions, yielding less vigorous plantlets *in vitro*.

20.6.2 Pests

Pest pressure in yacon is much lower in the dry valleys of the Andes than in humid areas (Grau and Rea 1997). Yacon has two pest protective system trichomes and increased density of glands that excrete sesqui- and diterpenes to repel insects on its leaves and stems (Figure 20.3). These are responsible for little insect infestation (Lizárraga et al. 1997). Their active ingredient is *ent*-kaurenic acid and its esters and derivatives (Valentová et al. 2001). Thus, cultivation of yacon needs almost no pesticides; this is currently called "organic cultivation" and is beneficial for medicinal values and dietetic foods.

Grau and Rea (1997) reported that agoutis and rodents of the genus *Dasiprocta* attack tuberous roots in the La Paz region in Bolivia. Table 20.8 lists selected pests of yacon. Fernández (2005) reported 10 pests attacking yacon in Peru and Brazil but did not mention the occurrence of pests and diseases in Europe. Lebeda et al. (2008) were probably the first to report the natural attack by pests and diseases in yacon under Central European conditions in the Haná region, Olomouc, Czech Republic. In 3 years of field experiments, the pest *Trialeurodes vaporariorum* was recorded every year. The degree of insect attack was moderate (according to a scale of zero to four) in 2001 and 2002; in 2003, it was low in all studied genotypes. *Citatella atropunctata* occurred sporadically on yacon genotypes during the 3 years.

20.6.3 Disease and Pest Control

A specific method to control diseases and pests has not been developed for yacon as it has been in some other crops. Nevertheless, three main approaches—mechanical, chemical, and biological—have been used to control diseases and pests (Fernández 2005). For chemical control, contact or systemic insecticides (pyrethroids or neonicotinoids) have been used against some insects (*Acordulecera* sp., *Epitrix* sp., *Diabrotica* sp.). Biological protection is represented by bacteria (*Klebsiella oxytoca* and *Erwinia uredovora*) that are able to control leaf microflora by conversion of hydroxycinnamic acids on dicarboxylates (Hashidoko et al. 1993). Presence of the insect predator *Cycloneda* spp. (Coccinellidae) on yacon in the fields has been recorded (Fernández 2005).

Table 20.8   Yacon Pests

Pest	Damage/Activity	Region, Country	References
Liriomyza sp. (Agromyzidae, Diptera)	Leaf mining	Cusco, Peru	Lizárraga et al. (1997)
Stink bugs (Pentatomidae, Coridae, Hemiptera)	Leaf sucking	Cusco, Peru	Lizárraga et al. (1997)
Diabrotica sp. (Chrysomelidae, Coleoptera)	Flower chewing	Cusco, Peru	Lizárraga et al. (1997)
Unknown larvae (Scarabeidae, Coleoptera)	Tuberous root boring	Cusco, Peru	Lizárraga et al. (1997)
Papilio sp. (Lepidoptera)	Leaf chewing	São Paulo, Brazil	Kakihara et al.(1996)
Nematodes	Root damage	São Paulo, Brazil	Kakihara et al. (1996)
Tetranychus sp.	Leaf sucking	Cajamarca, Peru	Seminario et al. (2003)
Empoasca sp.	Leaf sucking	Cajamarca, Peru	Seminario et al. (2003)
Aphis sp.	Leaf sucking	Cajamarca, Peru	Seminario et al. (2003)
Myzus persicae	Leaf sucking	Cajamarca, Peru	Seminario et al. (2003)
Trialeurodes vaporariorum	Leaf sucking	Olomouc, Czech Republic	Lebeda et al. (2008)
Citatella atropunctata	Not clarified	Olomouc, Czech Republic	Lebeda et al. (2008)

Sources: Modified according to Grau, A., and J. Rea. 1997. In *Andean roots and tubers: Ahipa, arracacha, maca and yacon. Promoting the conservation and use of underutilized and neglected crops,* ed. M. Hermann and J. Heller, 199–242. Gatersleben, Germany: Institute of Plant Genetic and Crop Plant Research, Rome, Italy: IPGRI; Fernández, C. E. 2005. Habilitation thesis, 154 pp. Prague: Czech Agricultural University in Prague, Institute of Tropics and Subtropics; Lebeda, A. et al. 2008. *Acta Horticulturae* 765:127–136.

20.7 BREEDING AND APPLICATION OF BIOTECHNOLOGICAL APPROACHES

20.7.1 Breeding Strategy

Yacon is mainly propagated vegetatively. This method maintains genetic stability of the clones. However, monoculture produces susceptibility to pathogens and pests, thus limiting productivity. Yacon has a high degree of ploidy ($8x = 6A + 2B = 58$; $12x = 9A + 3B = 87$; see Figure 20.1). This reduces fertility significantly and restricts the use of conventional breeding techniques. Only positive selection has helped to increase yield or the saccharide content.

A new cultivar of yacon, Sarada otome (yellowish-white flesh), was developed in Japan. It was obtained through selection after crossing between two lines (Sugiura et al. 2007). In 2003, two new cultivars (Andes no Yuki—white flesh—and Salad Okame—orange flesh), registered in Japan, were also obtained by hybridization and subsequent selection (Fujino et al. 2008).

Biotechnology will take a main role in the varietal improvement of yacon, where some related wild species could be used to enrich its genome. Based on geographical distribution, growth habit, and morphology of aboveground parts, it can be deduced that the following wild species are closely related to yacon: *S. connatus, S. macroscyphus, S. riparius, S. meridensis, S. suffruticosus,* and *S. siegesbeckius* (Grau and Rea 1997). These species could not be utilized in conventional breeding techniques because of chromosome number incompatibility; most of them have a lower number of chromosomes than yacon. Sexual hybridization between allopolyploid yacon and synthesized polyploid wild species holds a possibility for breaking the chromosomal barrier, but this has not been attempted.

Polyploidization of yacon genotypes has produced clones with higher FOS content. Hermann et al. (1998) found that dodecaploid ($2n = 12x = 87$) accession had a higher fructan yield than octoploid ($2n = 8x = 58$) plants. The positive effect of higher ploidy level on dm content and yield of tuberous roots was also investigated.

The future breeding program of yacon depends upon the purposes of utilization; production of fresh tuberous roots, for industrial processing (e.g., production of purified oligofructans, syrups, and chips); and market location (Andean or international market). For example, throughout the Andes, markets prefer cultivars with yellow flesh, while other cultivars are grown only for domestic consumption. The international market may accept a wider range of flesh color. Another important feature for trade and processing of tuberous roots is their size and storability.

The main breeding goals in yacon are

- Higher yields
- Higher content of inulin-type FOS in tuberous roots
- Higher content of antioxidants in leaves
- Shortening of growing period and increased frost tolerance (for conditions of temperate climate and high altitude)
- Proper shape (oval) of tuberous roots
- Thicker skin of tuberous roots

The Andean region is the logical center for breeding because an extremely wide diversity of yacon and wild species of the genus *Smallanthus* has been found. In this area, germplasm of yacon is also preserved *in situ* and can be effectively used for breeding and research (Meza 1995; Rea 2000b; Seminario et al. 2003, 2004; Mansilla et al. 2006). Evaluation and *ex situ* preservation of yacon genotypes is also being conducted in the Czech Republic (Lebeda, Doležalová, Dziechciarková et al. 2003; Lebeda, Doležalová, Valentová et al. 2003; Milella et al. 2005; Fernández 2005).

20.7.2 Biotechnological Approaches

A prerequisite of biotechnological manipulation in yacon is to regenerate plants through *in vitro* conditions. By using *in vitro* methods, new genotypes can be produced by polyploidization, dihaploids, somatic hybridization, transgenosis, induced mutations, and somaclones. Biotechnology in yacon is at the infancy stage; in the future, this technology will be certainly used not only in the varietal improvement and multiplication of the crop but also for *in vitro* sanitation of plant materials.

Induction of somaclonal variability is helpful for developing variable genotypes from which individuals with improved characteristics can be selected. This method is mainly used for those plants where the genetic variability is difficult to ensure through conventional breeding methods (Jha and Ghosh 2005). This procedure for yacon has been described (Kuroda 1992; Kuroda and Ishihara 1993). After regeneration of shoots from calluses, somaclones with higher sugar content have been selected. Protocols for indirect somatic embryogenesis and organogenesis have been established (Matsubara et al. 1990; Kuroda and Ishihara 1993; Matsumoto, Matsubara, and Murakami 1996; Derid et al. 1997; Santarém 2007). However, somaclones have not been evaluated.

Induced polyploidization in yacon was described by Viehmannová (2009) and Viehmannová et al. (2009). *In vitro* chromosome doubling was induced by treating nodal segments of yacon ($2n = 8x = 58$) by oryzalin and colchicine (mitotic spindle inhibitors). Out of 240 nodal segments treated, 3.33% hexadecaploid ($2n = 116$) plants were regenerated. The greatest proportions of hexadecaploid plants (1.6%) were obtained after 48 h of 25 μM oryzalin treatment. Colchicine treatment (3 mM/24 h) produced only 0.42% hexadecaploid plants. In hexadecaploid yacon, significantly higher levels of saccharides were detected (FOS 13.9 g/100 g f.m., fructose 4.6 g/100 g f.m., and glucose 2.1 g/100 g f.m.) compared to the octoploid control (FOS 5.3 g/100 g f.m., fructose 2.9 g/100 g f.m., and glucose 1.0 g/100 g f.m. (fresh matter)). This suggests that polyploidy breeding can help to increase the FOS content in the tuberous roots.

However, *in vitro* is a supportive method for yacon breeding, particularly for micropropagation (Hamada, Hosoki, and Kusabirki 1990; Matsubara et al. 1990; Estrella and Tapia 1994; Fernández

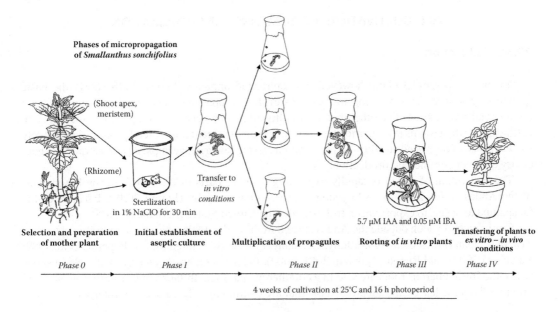

Figure 20.17   Phases of yacon micropropagation. (Fernández, C. E. 1997. PhD thesis, 131 pp. (in Czech), Czech University of Agriculture, Prague, Czech Republic.)

and Opatrný 1996; Niwa et al. 2002). These studies are mainly focused on optimization of nodal and shoot cultures. Fernández (1997) optimized the phases (I–IV) of micropropagation in yacon (Figure 20.17): phase I: establishment of aseptic culture, phase II: multiplication of plant material by nodal cultures, phase III: root induction, and phase IV: transfer of *in vitro* plants to *ex vitro* conditions. It was concluded that the most important factors for each phase of micropropagation were the composition of cultivation medium, the time of cultivation, and genotype.

Yacon can be propagated from both apical and axillary buds even on basic MS (Murashige and Skoog 1962) medium, without growth hormones. However, the addition of growth regulators can influence positive growth such as compactness of calluses and rooting of regenerated plants, thus increasing the successful transfer of regenerants into nonsterile conditions. After 30 days of cultivation, the explants formed 2.4–2.9 nodes (micropropagation coefficient) per plant. The optimum temperature for rapid propagation was 25°C with 16 h photoperiod. The saccharose in a 3% (w/v) concentration was better as a carbon and energy source than glucose and fructose in the same concentration. For rooting, 5.7 μM indole-3 acetic acid and 0.05 μM indole-3-butyric acid were the optimal concentrations. Fernández and Opatrný (1996) and Fernández (1997) developed a protocol for establishing callus cultures and de novo regeneration. The formation of callus in stem internodes was very intensive (80–90%) on MS medium supplemented with 0.3 μM α-naphthaleneacetic acid and 9.3 μM kinetin or 8.9 μM 6-benzylaminopurine (BAP), after 4–5 weeks of induction at 21°C and 16 h photoperiod. Plants were regenerated from calluses on the medium of Binding et al. (1981). The plant regenerability trait was genotypically dependent.

Recently, plant regeneration through leaf culture of yacon was developed (Niwa et al. 2002). Leaf pieces were cultured on the gelrite-solidified MS medium, supplemented with various combinations of auxins and cytokinins. The most effective means of inducing calluses was achieved on an MS medium that was supplemented with 0.1 mg/L 2,4-D and 0.1 mg/L BAP. Shoot regeneration occurred when the calluses were subcultured on the HaR medium (Paterson and Everett 1985). The regenerated shoots rooted and the plantlets were acclimatized and transplanted in the field. AFLP analysis of regenerants indicated that the somaclonal mutation was induced during callusing and plant regeneration (Niwa et al. 2002).

20.8 CULTIVATION, PROCESSING, AND UTILIZATION

20.8.1 Cultivation

Yacon is a perennial plant; however, it is cultivated annually. Figure 20.18 shows the basic growth stages of yacon (sprouting to senescence). It is propagated vegetatively by rhizomes for commercial propagation. The other means of cultivation is by vegetative propagation such as stem cuttings, the whole stems (Seminario et al. 2003), and *in vitro* techniques. Seed propagation is rarely used because of high seed sterility. To preserve virus-free material, germplasm conservation is necessary even for commercial production.

The cultivation of yacon is rapidly expanding to other parts of the world. In the lower areas of the Andes (1800–2500 m a.s.l.) yacon is grown commercially (Figure 20.19). Farmers of the Oxapampa area (Peru) export yacon to Japan and the United States (Fernández 2005).

Cultivation of yacon outside the Andes started in the early 1980s when it was introduced to New Zealand, Japan (Frček et al. 1995), Italy, Germany, France, the United States (Popenoe et al. 1989; Valentová 2004), and Russia (Tjukavin 2001, 2002). In some countries, yacon is available in shops where the fresh tuberous roots are sold in a commercial package. In Japan, it is usually sold as dried products from the leaves and tuberous roots that are commonly offered over the Internet.

20.8.1.1 Plantation and Seeding

The time of plantation or seeding depends on the area of production, climate, system of irrigation, market demand, and the altitude. In frost-free areas, yacon may be planted throughout the year; planting is limited to one season in the areas where frost occurs. Planting or seeding in dry areas starts at the beginning of the rainy season, usually from September to October. In northwest Argentina, it is planted from May to August (Zardini 1991). In Central Europe, planting is done in May (Figure 20.20), after the end of spring frosts (Fernández et al. 2006). Preferably, yacon should be planted in deep, loamy, light soil, with a good supply of organic matter and good drainage.

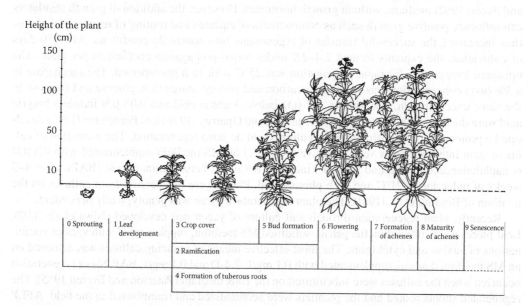

Figure 20.18 Basic growth stages of yacon. (Fernández, C. E. et al. 2007. *Agricultura Tropica et Subtropica* 40:71–77.)

Figure 20.19 Cultivation and natural occurrence of yacon plants in the native Andean region. (Composite photo courtesy of E. C. Fernández and T. Klásková.) (a) Yacon growing in Andean mountains (2900 m a.s.l.), Potosí, Bolivia; (b) yacon growing in Cajamarca, Peru; (c) yacon plant growing in the ruins of Machu Picchu, Peru; (d) field of yacon in Cusco, Peru; (e) yacon growing in Cajamarca, Peru.

These conditions facilitate development of tuberous roots without distortion and the risk of decay (Seminario et al. 2003). Planting density depends on the production area, terrain slope, fertility, type of soil, temperature, cultivar characteristics, and length of growing period. Table 20.9 lists selected density plantations.

20.8.1.2 Fertilization

Application of N (nitrogen) and K (potassium) fertilizers depends on the purpose of cultivation. For high production of tuberous roots, the optimal dose of nitrogen is 160 kg/ha and of potassium is 100 kg/ha; for high yields of rhizomes, 100 kg/ha potassium (without nitrogen) is needed (Araújo and Amaya 2003). Yacon responds well to organic fertilizers (cattle manure or compost)—an excellent trait for the expanding market of organic food.

20.8.1.3 Irrigation

Supplementary irrigation is required for development and growth of plants in the first months of vegetation and, if possible, moist soil should be maintained during the growth period. It is recommended to water within 3–4 weeks after planting; irrigation is not required in the rainy season (Zardini 1991). Optimal annual rainfall of 800 mm is needed for the development of yacon plants

Figure 20.20 Cultivation of yacon plants in Czech Republic. (a) Overwintered rhizomes; (b, c) young plants in the glasshouse; (d) yacon growing in the field. (Composite photo courtesy of I. Doležalová and J. Doležal.)

(Grau and Rea 1997). The main factor that influences yield of yacon is the total precipitation during the vegetative growth period (Fernández et al. 2006).

20.8.1.4 Weed Control

First weeding is done 30 days after planting and the second weeding after 30 more days. Further weeding is not required 60 days after planting. After first weeding, the ridges are formed like

Table 20.9 Density of Yacon Plantation

Spacing (m)	Density of Plants (Plants/ha)	Region, Country	References
0.90 × 0.30	37,000	San Remo, Italy	Calvino (1940)
0.70 × 0.60	23,800	Peru	Lizárraga et al. (1997)
0.90 × 0.40	27,800	Cusco, Peru	Meza (2001)
1.00 × 0.80	12,500	Cusco, Peru	Meza (2001)
0.75 × 0.70	19,000	Havlíčkův Brod, Czech Republic	Frček et al. (1995)
0.70 × 0.70	20,400	Prague, Czech Republic	Fernández, Riessová, and Čespiva (2003)
1.40 × 0.90	8,000	Brazil	Anonymous (2003)

Source: Fernández, C. E. 2005. Habilitation thesis, 154 pp. Prague: Czech Agricultural University in Prague, Institute of Tropics and Subtropics.

potatoes. Further weed control is carried out, as required, depending on the state of vegetation (Meza 2001; Fernández et al. 2006). In Japan, weeds are controlled by mulching (straw, black PE film). However, it has been demonstrated experimentally that yield is lower in mulching conditions than by conventional cultivation (Tsukihashi et al. 1991).

20.8.1.5 Harvest

Yacon is traditionally cultivated for tuberous roots, but is now also grown for the leaves. In the lower areas of the Andes (0–2000 m a.s.l.), an early harvest (6–10 months; Meza 2001; Manrique and Hermann, 2003) and, in higher altitudes (3000–3750 m a.s.l.), a late harvest (10–12 months; Fernández et al. 2005) are carried out. Harvesting is usually done in the Andes in May to July and in Europe in October (Figure 20.21). The authors are not aware of any mechanized harvesting system in the Andes. Potato harvesting machines have been successfully used in Brazil (Kakihara et al. 1996). Leaves may be harvested 2 or 2.5 months after plantation or when plants have four or five pairs of leaves. It is important to harvest fully developed leaves to increase their total dry weight. The best time to harvest leaves is when stems and petioles form a right angle. Overripe leaves showing chlorotic color and drooping petioles are not suitable for harvesting (Seminario et al. 2003).

20.8.1.6 Yield of Tuberous Roots

Yield of yacon varies according to ecotypes, area of cultivation, soil type, environmental conditions, method of cultivation, cultivation period, length of growing period, and size of planting

Figure 20.21 Harvest of yacon tuberous roots under conditions in the Czech Republic. (a) Dry plants before harvest; (b) harvest of tuberous roots; (c) freshly harvested tuberous roots; (d) tuberous roots prepared for morphological characterization. (Composite photo courtesy of I. Doležalová and J. Doležal.)

Table 20.10 Maximum Yacon Yields Obtained in Different Climatic Conditions

Region	Tuberous Root (t/ha)	Region, Country	References
Andean region	15–100	Bolivia	Rea (1995)
	74	Santa Catalina, Quito, Ecuador	Castillo, Nieto, and Peralta (1988)
	70	Santa Catalina, Quito, Ecuador	Nieto (1991)
	41 (roots with peel) 34 (roots without peel)	Ecuador	Rea (1992)
	16[a]	Santa Catalina, Quito, Ecuador	Ramos et al. (1999)
	31[a]	Los Eucaliptos, Cajamarca, Peru	León (1983); not listed
	7–55	Baños del Inca, Cajamarca, Peru	Huamán (1991)
	95	Cajamarca, Peru	Seminario (1995)
	52[a]	Baños del Inca, Cajamarca, Peru	Franco and Rodríguez (1997)
	14–28	Ahuabamba, Cusco, Peru	Lizárraga et al. (1997)
	10–107	Oxapampa, Pasco, Peru	Melgarejo (1999)
	25–111	Cusco, Peru	Meza (2001)
	51[a]	Hualqui, Cajamarca, Peru	Seminario, Valderrama, and Honorio (2001)
	27[a]	Hualqui, Cajamarca, Peru	Seminario, Valderrama, and Seminario (2002)
Outside the Andes	38	San Remo, Italy	Calvino (1940)
	54	Japan	Ohyama et al. (1990)
	35–45	Japan	Ogiso, Naito, and Kurasima (1990)
	28–47	Ibaraki, Japan	Tsukihashi et al. (1989, 1991)
	100	Capão Bonito, Brazil	Kakihara et al. (1996)
	80	Capão Bonito, Brazil	Vilhena et al. (2000)
	15–56	Botucatu, Sao Paulo, Brazil	Amaya (2000)
	44–66	Botucatu, Sao Paulo, Brazil	Amaya (2002)
	120	Brazil	Anonymous (2003)
	25–31	Chonju, Korea	Doo et al. (2001)
	57–86	Russia	Tjukavin (2001)
	38–66	Havlíčkuv Brod, Czech Republic	Frček et al. (1995)
	8–49[b]	Prague, Czech Republic	Fernández et al. (2003)

Source: Fernández, C. E. 2005. Habilitation thesis, 154 pp. Prague: Czech Agricultural University in Prague, Institute of Tropics and Subtropics.
[a] Average yield.
[b] Organic cultivation system.

material. Table 20.10 shows available data on yacon yield in a fresh mass. Dry matter ranges from 15 to 30% of fresh material.

20.8.1.7 Yield of Leaves

Little information on yield of leaves is available in the literature. In climatic conditions of the Andean region, while the density of planting is 18,500 plants per hectare, the yield of dry leaves ranges between 3 and 4 t/ha (Seminario et al. 2003). In the Czech Republic, the yield of dry leaves ranges between 1.4 and 1.9 t/ha (Fernández et al. 2004; Fernández 2005).

20.8.1.8 Phenological Scale

Based on knowledge of morphology and dynamics of plant development, a macrophenological scale (according to BBCH scale; Strauss 1994) has been developed (Table 20.11, Figure 20.18). There are 10 principal growth stages: 0, sprouting; 1, leaf development; 2, ramification; 3, crop cover; 4, formation of tuberous roots; 5, bud formation; 6, flowering; 7, formation of achenes; 8, maturity of achenes; and 9, senescence (Fernández et al. 2007).

20.8.2 Postharvest Treatment and Processing

20.8.2.1 Selection and Classification of Tuberous Roots

Bad, deformed, or damaged tuberous roots are removed after harvesting. Currently, no official standard exists for classification of tuberous roots. In order to assess the types of tuberous roots and their relative proportions during harvesting in Cajamarca (Peru), tuberous roots are classified into three categories (Seminario et al. 2003):

First grade: the tuber length is over 200 mm, and the range of diameter is 70–100 mm. The weight of the tuber is below 300 g/tuber.
Second grade: length ranges from 120 to 200 mm and diameter ranges from 50 to 60 mm; average weight ranges from 120 to 130 g/tuber.
Third grade: the length is less than 120 mm, diameter is less than 50 mm, and weight is less than 120 g. This grade is considered for marketing.

For the Czech Republic, a classification according to the quality and size of tuberous roots, described next, has been established (Frček 2001):

- Quality classes
  - Class I: Yacon tuberous roots for commercial grade should
    - Be a spheroidal or an oblong shape with a slight curvature
    - Not be cracked
    - Be free of side roots
    - Be smooth
    - Not be bruised
    - Have a fresh appearance
    - Have flesh that is pinkish, orange, white, or cream with livid dots around the perimeter
  - Class II: This class includes yacon tuberous roots excluded from class I that satisfy the minimum commercial grade:
    - Slight variations in shape (more curved)
    - Slight bruising
- Small healed cracks

  Taking into account the special provisions for each class and the deviations allowed, tuberous roots of both classes must

- Be healthy
  - Be sanitary
  - Have a fresh appearance
  - Be clean
  - Be free of pests and diseases
  - Be free of abnormal external moisture

Table 20.11   Macrophenological Scale of Yacon

Code (Two Digits)	Description
	Principal growth stage 0: sprouting
	Principal growth stage 1: leaf development
11	Formation of the first couple of leaves
13	Formation of the second couple of leaves
15	Formation of the third couple of leaves
17	Formation of the fourth couple of leaves
19	Formation of the fifth couple of leaves
	Principal growth stage 2: ramification
21	Beginning of ramification
27	Ramified half of stem
29	All stem ramified
	Principal growth stage 3: crop cover
31	Beginning of crop cover: 10% of the plants meet between rows
32	20% of plants meet between rows
33	30% of plants meet between rows
34	40% of plants meet between rows
35	50% of plants meet between rows
36	60% of plants meet between rows
37	70% of plants meet between rows
38	80% of plants meet between rows
39	Crop cover complete: about 90% of plants meet between rows
	Principal growth stage 4: formation of tuberous roots
40	Beginning of tuberous roots formation
41–48	Formation of tuberous roots
49	Maturity of tuberous roots
	Principal growth stage 5: bud formation
	Principal growth stage 6: flowering
61	Beginning of flowering
65	Full flowering (50% of flowers in the first inflorescence)
67	70% of flowers in the first inflorescence
69	End of flowering in the first inflorescence
	Principal growth stage 7: formation of achenes
	Principal growth stage 8: maturity of achenes
	Principal growth stage 9: senescence

Source: Fernández, C. E. et al. 2007. *Agricultura Tropica et Subtropica* 40:71–77.

- Be free of any foreign smell and strange taste
- Have tuberous roots in a proper stage of development that can be delivered to destinations in satisfactory condition

- Size classes (measured by the diameter in the upper third of the tuberous root)
  - Class I: root diameter 20 mm
  - Class II: root diameter at least 15 mm

Permitted deviation:
- Class I: 10% by weight of tuberous roots not satisfying the requirements of the class, but meeting the requirements for class II; exceptionally, the tolerances are allowed
- Class II: 10% by weight of the tuberous roots not satisfying the requirements of the class or the minimum requirements, with the exception of tuberous roots affected by rotting or other damage, rendering them not suitable for human consumption

Size tolerances for both classes: 10% of the weight of the roots not set the size sort

20.8.2.2 Storage of Tuberous Roots

Storage methods for yacon have not been extensively developed. However, yacon deteriorates due to high temperature and high relative humidity. High air humidity helps to minimize weight loss, especially in combination with cold, but it can speed up the process of decaying at high temperatures. It was found that weight loss and perishability of tuberous roots in storage significantly reduces at around 5–6°C. In addition, if paper or plastic bags (perforated with tiny holes) are used, the storability is much better. By using this system, tuberous roots almost do not lose water and perishability of tuberous roots after 90 days of storage is less than 15% (Graefe 2002; Fernández 2005).

Important factors that affect the FOS content during storage are the temperature and storage time. Conversion of fructo-oligosaccharides to monosaccharides is slower when the tuberous roots are stored at low temperatures (4–5°C) than at higher temperatures (Asami et al. 1991; Vilhena, Camara, and Kakihara 2000). Even at low temperatures, however, a certain proportion of the FOS is decomposed to monosaccharides (Table 20.12).

20.8.2.3 Processing

Many processed yacon products are available in Brazil, Japan, and Peru: dried products from the tuberous roots (obtained by drying in the air) (Grau and Rea 1997; Kakihara, Câmara, and

Table 20.12 Effects of Temperature and Storage Period on Saccharide Content and the Degree of Polymerization (DP) of FOS as Percentage of Dry Matter

Duration of Storage	FOS (% dry)	Fructose (% dry)	Glucose (% dry)	Saccharose (% dry)	DP
Harvest, day 0	67	1.2	1.3	4.4	4.3
Storage at 5°C					
28 Days	45	21	8.3	5.9	4.0
194 Days	13	36	16.6	6.4	3.2
Storage at 25°C					
31 Days	39	30	10.2	8.2	3.7
74 Days	23	28	14.5	11.8	3.4

Source: Asami, T. et al. 1991. *Japanese Journal of Soil Science and Plant Nutrition* 62:621–627.

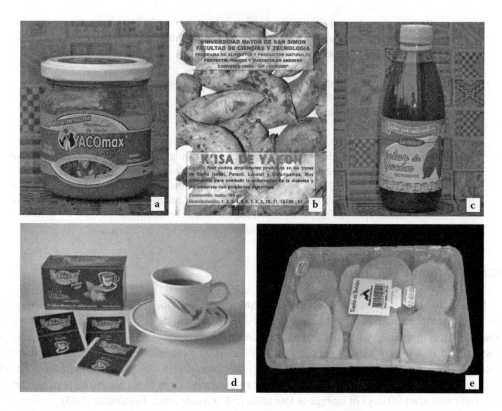

Figure 20.22   Yacon products. (a) Yacon jam; (b) chips; (c) juice; (d) tea from leaves; (e) slices from tuberous roots. (Composite photo courtesy of E. C. Fernández and I. Viehmannová.)

Vilhena 1997), unrefined yacon syrup (which has the consistency of honey and can be sold as a dietary sweetener) (Hermann et al. 1998), and juice (Figure 20.22) (without adding sweeteners, synthetic dyes and preservatives, only containing a small amount of vitamin C). Tuberous roots serve as a source of raw material for confectionery products, can be pickled as a vegetable, and serve as a source for acquiring ethanol (biofuel). They can also be used for the production of dehydrated chips (Figure 20.22) and jam (Figure 20.22). Fermentation of juice by bacteria *Acetobacter pasteurianus* produces high-quality yacon vinegar with the natural FOS (Hondo, Okumura, and Yamaki 2000).

Yacon leaves were first used as tea (Figure 20.22) in Japan, where leaves and stems are now mixed with tea leaflets for marketing (Anonymous 2002). Presently, the largest producers of yacon tea are Brazil and Japan. Leaves can be dried either in open air or in ovens (Grau et al. 2001). The suitable temperature for efficient drying of yacon leaves is 60°C. Water content in dry leaves should be around 5%. After drying, followed by grinding and sifting, tea is packed into the filter bags (Espinoza 2003).

20.8.3 Utilization and Commercialization

Yacon is suitable for various purposes, such as utilization in human nutrition, commerce, and industry. It possesses an attractive set of features advantageous to producers, processors, consumers, and the environment (protection against soil erosion). Fernández (2005) summarized its use as a food crop (tuberous roots eaten raw, sun dried, cooked, and baked), medicinal plant (remedy against diabetes, prevention against some types of cancers and circulatory disorders), and for industrial purposes (source of inulin, fructose, alcohol, and raffinate). Additionally, the plant may be useful in agroforestry because it grows well beneath a canopy of trees.

In Andean countries, yacon has been used as a symbolic and religious plant since the time of the old Inca civilization (Seminario et al. 2003). In the Moche era, it may have been food for a special occasion. Effigies of edible food may have been placed at Moche burials for the nourishment of the dead, as offerings to lords of the otherworld, or in commemoration of a certain occasion. The Moche depicted these yacon in their ceramics (Benson and Berrin 1997). Cárdenas (1969) reported usage of yacon as a food during the Corpus Christi fest in the area spread out from Peru to northwestern Argentina. In Ecuador, it has been eaten during All Soul's Day (Popenoe et al. 1989). In some areas of northern Peru, it is used as an oblation and decoration on the crucifix celebration and celebration of St. San Isidro Labrador, patron of the harvest (Seminario et al. 2003).

In general, yacon is considered a tuber vegetable crop outside the Andean region (Rubatzky and Yamaguchi 1997). Surprisingly, it has been sold as a fruit crop together with apples, avocados, and pineapples in the countries of origin for a long time (Grau and Rea 1997).

Small agricultural industries focused on the dietetic and medicinal properties of yacon soon after its introduction to Japan (Grau et al. 2001). In Brazil (state of San Pablo), it is called food "in natura" with possible medicinal properties; raw tuberous roots, chips, and dry leaves are sold for preparing diabetic tea. The medicinal effect and pleasant taste of yacon entice the attention of modern society, which looks up to natural products as medicinal alternative remedies (Araújo and Amaya 2003). Seminario et al. (2003) mentioned that farmers in Jaén (Cajamarca, Peru) burn yacon leaves (fumigation) to control mosquitoes (reservoir host of *Plasmodium* sp., causers of malaria). Ashes from stems together with stems of species *S. macroscyphus* are used for preparation of "yista" ("llicta" or "llipta") applied at coca chewing in Salta Province of Argentina.

The potential properties of yacon have been utilized only on a small scale in the Andean areas of its origin (3000–3700 m a.s.l.), where it is grown for growers' own needs on small fields (family gardens). It is in danger of possible extinction because the cultivated area has been gradually decreasing and, for local farmers, it does not represent a priority crop—mainly because of poor knowledge of its nutrition, medicinal, and feeding significance. Yacon has been sold in small amounts at provincial markets (Figure 20.23) from Columbia to northern Argentina. Because it is grown mainly for growers' own consumption, its commerce potential is minimal; only occasionally does it serve as a product for exchanging with other products (e.g., potatoes, wheat, vegetables, etc.). Of 51 families growing yacon, only one family of farmers grows it commercially in Ararian Village (Charcas Province, Bolivia) (Fernández et al. 2005). This family maintains the greatest number of plants.

In recent years, in the Andean region as well as outside the Andes, the yacon field has been widely distributed. In the lower elevation of the Andes, in the altitudes from 1800 to 2500 m, yacon has been cultivated commercially. The farmers from the province of Oxapampa (Peru) have been exporting yacon and its products to the United States and Japan. Outside the Andes area, yacon is grown in New Zealand, where it is commonly sold in supermarkets (Endt 1983). In Japan, dry products from leaves and tuberous roots are sold and currently supplied through the Internet (Fernández 2005).

20.8.3.1 Pharmacology and Medicine

Utilization of yacon in pharmacology and medicine is summarized in Table 20.13. It is commonly consumed by diabetics and people suffering from digestive problems (Meza 2001). In some

Figure 20.23 Tuberous roots of yacon sold at the local market in Colomi, Cochabamba, Bolivia. (Photo courtesy of E. C. Fernández.)

Table 20.13 Utilization of Yacon in Pharmacology and Medicine

Utilized Parts of Yacon	Therapeutic Effects	References
Tuberous roots		
	Diabetes	Frček et al. (1999), Grau et al. (2001), Meza (2001), Fernández (2005)
	Dietetic effect	Rea (1992, 1994, 1997), Grau et al. (2001), Cisneros-Zevallos et al. (2002), http://en.wikipedia.org/wiki/Smallanthus_sonchifolius
	Long-lived effect	Rea (1992, 1994, 1997), Grau et al. (2001)
	Obesity	Rea (1992, 1994, 1997), Grau et al. (2001)
	Prebiotics	Grau et al. (2001), Pedreschi et al. (2002)
Leaf extractions		
	Antioxidative properties (cancer prevention)	Zardini (1991), Grau et al. (2001), Lebeda et al. (2003b), Fernández (2005)
	Cardiovascular disease prevention	Grau et al. (2001), Fernández (2005)
	Diuretic and healing effects	Grau et al. (2001), Valentová et al. (2003, 2004)
	Diabetes	Kakihara et al. (1996), Volpato et al. (1997), Aybar et al. (2001), Grau et al. (2001)
	Dietetic effects	Aybar et al. (2001), Grau et al. (2001)
	Protective effects	Grau et al. (2001), Valentová (2004)

productive areas of Bolivia, people have been using yacon as a medication against diabetes and obesity and for its dietetic and long-lived effects (Rea 1992, 1994, 1997). The tuberous roots contain inulin, an indigestible sugar; thus, although it has a sweet flavor, it contains fewer calories. This is an attractive marketing feature to dieters and diabetics. The low-sugar characteristic is due to FOS, a special type of fructose that the human body cannot absorb (http://en.wikipedia.org/wiki/Smallanthus_sonchifolius). Therefore, tuberous roots also may have potential as a diet food. The sweetness is caused by fructose, which is 70% sweeter than table sugar, does not stimulate insulin production, and does not bring a glycemic reaction (Cisneros-Zevallos et al. 2002).

Yacon provides two nutritional products: the syrup and tea. The syrup has been applied also as a prebiotic (due to FOS content): It feeds the friendly bacteria in the colon that boost the immune system and help digestion. Fructo-oligosaccharides are used as food ingredients and prebiotics (Pedreschi et al. 2002). Pedreschi et al. (2002) found that *Lactobacillus plantarum* NRRL B-4496 and *L. acidophilus* NRRL B-1910 completely utilized the GF2 molecule, while *Bifidobacterium bifidum* was apparently able to utilize molecules with higher DP. Fructo-oligosaccharides (two to nine molecules of fructose) have received much attention as prebiotics due to their small utilization by the body and their ability to enhance the growth of probiotics.

The suitability of tuberous roots for preparation of diabetic nourishment of patients suffering from chronic hepatic diseases was shown in Olomouc (Czech Republic) (Frček, Psotová, and Šimánek 1999; Valentová 2004). This antidiabetic property is attributed to yacon leaves used to brew medicinal tea (Kakihara et al. 1996). Volpato et al. (1997) demonstrated the hypoglycemic activity of water extract of dried leaves in feeding experiments with rats with induced diabetes. Aybar et al. (2001) tested the hypoglycemic effect of water extract of leaves in normal, transiently hyperglycemic, and streptozocin-induced diabetic rats. A significant decrease in plasma glucose levels in normal rats was produced by 10% yacon decoction administered intraperitoneally.

The antioxidative properties of yacon are believed to be prevention against some types of cancer (Zardini 1991; Lebeda, Doležalová, Valentová et al. 2003). Yacon water extracts induced an increase in plasma insulin concentration (Aybar et al. 2001). Valentová et al. (2004) demonstrated a strong protective effect of tea prepared from yacon leaves. Diuretic and healing effects on the skin were also mentioned (Valentová et al. 2001). Valentová et al. (2003, 2004) observed that a yacon extract reduced the proportion of glucose in cultures of hepatocytes, acting in a manner similar to insulin. The possibility to prevent cardiovascular diseases by using yacon has been mentioned by Fernández (2005). He showed identical polyphenol content in yacon and red wine. Grau et al. (2001) summarized potential positive effects of yacon as follows:

- An effective antidiabeticum, yacon reduces the level of sugar in blood due to active hypoglycemic ability.
- It reduces the level of cholesterol and triglycerides in blood (against arteriosclerosis).
- It supports progression of bifidobacteria and *Bacillus subtilis* in the colon.
- It inhibits propagation of putrescent microorganisms that cause diarrhea.
- It improves calcium assimilation.
- It is a prop for synthesis B vitamins.
- It contributes to health by low caloric content.
- Presented sugars do not induce caries.
- It modifies bowel movement.
- It prevents cancer diseases.
- Dry leaves contain flavonoids, kaurenoids, and sesqui- and diterpenes with antioxidative properties.

20.8.3.2 Nutrition and Food

Yacon tuberous roots have a sweet flavor and are crunchy to eat, like traditional fruit. The texture and flavor have been described as a cross between a fresh apple and watermelon—sometimes referred to as the apple of the earth. Owing to their crispy texture, the tuberous roots are juicy and are consumed as a fresh food; they are frequently eaten raw in snacks and salads and also boiled and baked in the Andean area. Sun-drying preservation is used; exposure to sunlight is believed to improve the flavor and sugar concentration. The skin must be peeled because it is resinous (Rubatzky and Yamaguchi 1997) and has strong laxative properties. The tuberous roots can also be stewed or can be grated and squeezed through a cloth to produce a drink. Sometimes, this is concentrated to form dark-brown blocks of candy called *chancaca* (http://www.cipotato.org/artc/cip_crops/making_yacon_candy.pdf). Consumption of yacon in some areas is linked to particular cultural or religious festivals. Undamaged tuberous roots keep well, and in Spanish colonial times, yacon was used as a food for sailors (http://www.nap.edu/catalog.php?record_id=1398). The main stem and leaves of the plant are used as a cooked vegetable.

Sometimes, yacon is used for cattle and sheep feed (Grau and Rea 1997) and it may have potential as a forage crop. The foliage is luxuriant, and the leaves have 11–17% (on a dry-weight basis) protein content (Popenoe et al. 1989). When cut, the foliage sprouts again from the underground stems. The tuberous roots may also be a good cattle feed, for inulin is rapidly metabolized by ruminants (http://www.nap.edu/catalog.php?record_id=1398).

20.8.3.3 Industry

Industrial exploitation of yacon is relatively broad (Fernández 2005). Utilization of yacon FOS is focused on substitution of lipids (in salad dressings and low-calorie cheeses), on decreasing of caloric contents (imitation of chocolate), and on increasing retention of water (confectionery, production of gingerbread, and pork). It is also used for prevention of crystal set formation in ice cream production (Fernández-Jeri 2003). Fernández (2005) cited the main firms of Orafti and Cosucra in the Netherlands and Belgium, which deal with yacon FOS. Peru has initiated a project of yacon syrup commercialization for production of sweetener to yogurts sold in the United States and European Union. Brazilian and Japanese farmers produce many products from yacon: dehydrated products from tuberous roots, which are prepared by simple air-drying (Grau and Rea 1997; Kakihara et al. 1997); nonrefined syrup of honey consistency sold as dietetic sweetener (Hermann et al. 1998); and juice without addition of any sweeteners, synthetic colorants, and conservatives—only with small content of vitamin C. The juice also can be concentrated at low pressure, with the addition of sodium bisulphate to inhibit enzymatic darkening. The final product is dense syrup similar to sugarcane syrup but with significantly lower energy value for humans (Chaquilla 1997).

Hondo, Nakano et al. (2000) and Hondo, Okumura et al. (2000) demonstrated preparation of yacon juice treated by active carbon for clarification and elimination of coloring and odor and proposed fermentation of juice by bacterium *Acetobacter pasteurianus* to obtain high-quality vinegar-harboring natural FOS. Another promising processing technique is the production of dry chips. In this case tuberous roots are peeled and cut in thin slices; the slices are first dried in a plastic tunnel and then in an oven at 60°C (Kakihara et al. 1996). Dried chips can be stored indefinitely. The technique of *pasas* (*k'isa*) *de yacón* consists of slice preparation of dehydrated tuberous roots eaten as sweets or used in confectionery and bakery (Vera and Alfaro 2000). The tuberous roots are used as a source of inulin and fructose (Valentová et al. 2006). They also serve in the production of alcohol (Grau and Rea 1997), jam and bottled stewed fruit (Fernández 2005), some milk products (e.g., yogurts), muesli bars, and sweet flavoring (Lebeda et al. 2008). The tea from leaves is used in Japan and stems and leaves mixed with tea leaflets (Anonymous 2002). Recently, Brazil and Japan have been the largest producers of yacon tea. A procedure for drying and yacon tea processing has been described (Grau et al. 2001; Espinoza 2003).

20.9 SUMMARY AND FUTURE PROSPECTS

Yacon is a vigorous, herbaceous perennial plant originating in the Andean highlands of South America. Some limitations, prospects, and research needs regarding the yacon were very clearly summarized and specified in a review article by Grau and Rea (1997). Here we will focus our attention on the most recent achievements and developments in yacon research, including the most crucial aspects of future research and breeding programs.

Taxonomy of the genus *Smallanthus* is relatively well known. Recent molecular studies on phylogenetics of the *Espeletia* complex (Asteraceae) (Rauscher 2002) well supported Robinson's (1978) rearrangement of the genus. It is evident that the three species of *Polymnia* belong to a distinct clade and form a phylogenetically distant group from *Smallanthus* (Rauscher 2002). Further research must be focused on taxonomical and phylogenetical studies in the genus *Smallanthus* of South America and based on conventional taxonomy (e.g., morphology, anatomy, cytology, species distribution, etc.), systematics, chemical analysis, and molecular approaches (Bach et al. 2007). More research can contribute to elucidating relationships among different *Smallanthus* taxa and their relatives, including their adaptive radiation (Rauscher 2002).

Information on geographic distribution, ecology, variation, and genetic diversity in the center of origin of yacon is rather rare. This is also true for other *Smallanthus* species occurring in South and Central America (Robinson 1978; Grau and Rea 1997). Detailed research involving all known species is required. We have only the basic knowledge of ecology, biology, and variation of autochthonous *Smallanthus* species.

Collection, conservation, and evaluation of yacon and wild *Smallanthus* relatives are limited. The most material of *ex situ* and *in situ* collections of this species was found in Andean countries (Grau and Rea 1997; Fernández 2005). However, a number of accessions in all these collections is relatively low and clones of *S. sonchifolius* and related species are presented in a minority. In existing yacon collections, the wild accessions are most likely wild *Smallanthus* species other than *S. sonchifolius*. The accessions of *S. sonchifolius* material would play an irreplaceable role in further yacon breeding, while wild relatives may represent an interesting source of useful genes. Probably, the most valuable collections of local landraces are still found in the fields of Andean farmers. This is the main reason for field collection missions, conservation, and evaluation of local clones. Similarly, field collection and observation of wild *Smallanthus* species should be continued. Only this approach could contribute to exploitation of the enormous genetic potential of the exciting Andean crop. For detailed characterization of yacon germplasm, it is also very important to develop internationally unified and accepted descriptor lists.

Roots and tubers of Andean crops perform specialized roles as chemical factories that can synthesize, store, and secrete a diverse array of biologically active compounds (Flores, Vivanco, and Loyola-Vargas 1999; Flores et al. 2003). Extracts and chemical constituents of the tuberous roots and dried leaves have antimicrobial, antifungal, antihyperglycemic, and antioxidative activity. The tuberous roots contain FOS and phenolic compounds, and the leaves have several kaurene diterpenoids, acetophenone-type phytoalexins, and melampolide-type sesquiterpene lactones. Enhydrin, a melampolide-type sesquiterpene lactone, has been shown to possess anti-inflammatory activity and inhibitory effects of the transcriptional factor NF-κB (Hong et al. 2008). This subject belongs to the most exciting and innovative areas of yacon research, as well as to practical applications in pharmacology and nutritional industry.

According to recent knowledge, diseases and pests are not an extremely limiting factor of yacon production. Information about diseases and pests is rather limited. Only a limited number of viral, bacterial, fungal diseases and insect pests are known. Probably one of the reasons for this phenomenon is that yacon cultivated for commercial production is not frequently and seriously attacked by diseases and pests. One explanation of this may be attributed to chemical compounds in yacon (e.g., melampolides) (Hong et al. 2008). Another explanation regarding the cultivation of yacon in regions outside South America is that there are no natural enemies of this plant species (Lebeda et al. 2008). Nevertheless, more detailed phytopathological and entomological research of yacon (Grau and Rea 1997), including wild growing relatives, must be seriously conducted.

Yacon has been suggested as an industrial crop, particularly as a source of sugar and syrup (Popenoe et al. 1989). It is tempting to speculate that yacon is being transformed into a modern industrial crop with the application of modern technology of agronomy, fertilization, and genetic engineering (Grau and Rea 1997). From this viewpoint, the yacon breeding and *Smallanthus* germplasm exploitation is considered very important. Until now, the breeding of yacon was not very well developed. No specialized breeding company is focusing on this crop and only different ecotypes (selected clones) and/or local landraces are being cultivated. Sophisticated breeding strategies for this crop, like sunflower, have not been developed (Sarrafi and Gentzbittel 2005; Jan and Seiler 2006). Recent breeding strategies must be more oriented on creating genetic variation by artificial crossing between different clones and different species of *Smallanthus* (Grau and Rea 1997). The reduced fertility in interspecific crosses is a major difficulty in yacon breeding. However, the presence of pistillate and staminate flowers helps control hybridization more easily.

Plant callus, cell suspension, and root cultures are amenable for the large-scale production of pharmaceutical enzymes (Flores et al. 1999) and offer a viable method for the production of antimicrobial proteins (Flores et al. 2003). These methods were reported for some Andean root and tuber crops, such as mashua, oca, and mauka (Flores et al. 2003), and they may generally be applied also to yacon. Current biotechnological manipulations showed that yacon is especially suitable for *in vitro* cultures for rapid propagation of plant materials (Niwa et al. 2002). For this purpose, axillar and nodal explants are suitable on MS medium supplemented with or without phytohormones. Selection of lines with higher sugar content by tissue culture was reported by Kuroda and Ishihara (1993).

The advantage of vegetative propagation is the preservation of stability in individual clones; the disadvantage is the risk of development of a monoculture susceptible to mass infection that may limit productivity. Techniques of *in vitro* cultivation offer the possibility of induction and selection of somaclones with various properties in yacon (Viehmannová 2009) like those shown for many crops (Larkin and Scowcroft 1981). However, future breeding and biotechnological approaches depend on the wide scale of objectives: type of farmers, production scale, fresh produce, industrial processing, and targeting Andean and international markets.

Yacon has a great potential to be a medicinal crop because of its nutritional (exceptional qualities for low-calorie diets) and pharmacological properties. It has adaptability to a wide range of climates and soils, fits in agroforestry systems, controls erosion, can be potentially used as a forage crop, can process alternations, and has good postharvest life (Grau and Rea 1997). To conclude, the presented review suggests that the yacon (*S. sonchifolius*), a traditional crop of Andean Indians, is a challenging crop for the future (Lebeda et al. 2008).

ACKNOWLEDGMENTS

The work of A. Lebeda and I. Doležalová in this chapter was supported by the project MSM 6198959215 (Ministry of Education, Youth and Sports of the Czech Republic). C. E. Fernández and I. Viehmannová were supported by the project MSM 419056017. The authors thank Dr. O. Blahoušek for his kind help with preparation of structure formulas of chemical compounds.

REFERENCES

Adam, M., M. Juklová, T. Bajer, A. Eisner, and K. Ventura. 2005. Comparison of three different solid-phase microextraction fibers for anylysis of essentials oils in yacon (*Smallanthus sonchifolius*) leaves. *Journal of Chromatography A* 1084:2–6.

Amaya, J. 2000. Efeitos de doses crescentes de nitrogênio e potássio na produtividade de yacón (*Polymnia sonchifolia* Poepp. & Endl.). Tese de título de Mestre em Agronomia—Area de concentração em horticultura. Universidade Estadual Paulista Julio de mesquita Filho, Brasil, 58 pp.

———. 2002. Desenvolvimiento de yacón (*Polymnia sonchifolia* Poepp. & Endl.) a partir de rizóforos e de gemas axilares, em diferentes espaçamentos. Tese do título de Doutor em Agronomia—Area de concentração em horticultura. Universidade Estadual Paulista Julio de mesquita Filho, Brasil, 89 pp.

———. 2002. Report on the use of yacon in Japan. http://www.yaconcha.com (accessed May 2004).

———. 2003. Yacón (*Polymnia sonchifolia*). http://peshp.vilabol.uol.com.br/yacon.htm (accessed August 2004).

Araújo, C. F. L., and R. J. E. Amaya. 2003. Efecto de dosis crecientes de nitrógeno y potasio en la calidad y productividat de yacón (*Polymnia sonchifolia*) (Effect of dose of nitrogen and potassium on quality and yacon yield) (in Portuguese). UNEP, Botucatu, Brasil, 2003, http://www.horticom.com (accessed April 2003).

Arbizu, C. 2005. El yacón (in Spanish). http:www.galeon.com/florindaguerrero/yacon.htm (accessed February 2005).

———. 2009. Ficha técnica del yacón (in Spanish). http:www.lindavida.com (accessed January 2009).

Arbizu, C., and E. Robles. 1986. La colección de raíces y tubérculos andinos de la Universidad de Huamanga. V Congreso International de Sistemas Agropecuarios Andinos. Anales—INIPA—UNA—Puno.

Arbizu, C., J. Seminario, M. Valderrama, F. F. Santos, M. L. Ugarte, L. Lizárraga, and M. Aguirre. 2001. Descriptores para yacón. Lista 3. In *Conservación y uso de la biodiversidad de raíces y túberculos andinos*, ed., M. Holle, and D. Talledo. Lima, Peru.

Asami, T., K. Minamisawa, T. Tsuchiya, K. Kano, I. Hori, T. Ohyama, M. Kubota, and T. Tsukihashi. 1991. Fluctuations of oligofructan contents in tubers of yacon (*Polymnia sonchifolia*) during growth and storage. *Japanese Journal of Soil Science and Plant Nutrition* 62:621–627.

Aybar, M. J., A. N. S. Riera, A. Grau, and S. S. Sanchez. 2001. Hypoglycemic effect of the water of *Smallanthus sonchifolius* (yacon) leaves in normal and diabetic rats. *Journal of Ethnopharmacology* 74:125–132.

Bach, S. M., C. Schuff, A. Grau, and C. A. N. Catalán. 2007. Melampolides and other constituents from *Smallanthus connatus*. *Biochemical Systematics and Ecology* 35:785–789.

Baldwin, B. G., B. L. Wessa, and J. L. Patero. 2002. Nuclear rDNA evidence for major lineages of Helenioid Heliantheae (Compositae). *Systematic Botany* 27:161–198.

Barrantes del Águila, F. 1998a. Diseases of Andean crops in Ayacucho, Peru. 4th International Congress of Andean Crops (in Spanish). Quito, Ecuador: INIAP.

———. 1998b. Pathology of Andean roots and corms, vol. 17. In *Yacon: Foundations of the advantage of a promissory resource*, ed. J. Seminario, M. Valderrama, and I. Manrique. Lima, Peru: CIP, UNC, COSUDE.

Barrero, A. F., J. E. Oltra, E. Cerdá-Olmedo, J. Ávalos, and J. Justicia. 2001. Microbial transformation of *ent*-kaurenoic acid and its 15-hydroxy derivatives by the SG138 mutant of *Gibberella fujikuroi*. *Journal of Natural Products* 64:222–225.

Benson, T., and K. Berrin, eds. 1997. *The spirit of ancient Peru: Treasures from the Museo Arqueológico Rafael Larco Herrer.* Larco Museum, New York: Thames and Hudson.

Bentham, G. 1873. Notes on the classification, history and geographical distribution of Compositae. *Journal of the Linnean Society of Botany* 13:335–577.

Binding, H., R. Nehls, R., Koch, J. Finger, and G. Mordhorst. 1981. Comparative studies on protoplast regeneration in herbaceous species of the Dicotyledoneae class. *Zeitschrift für Pflanzenphysiologie* 101:119–130.

Brako, L., and J. L. Zarucchi. 1993. *Catalogue of the flowering plants and gymnosperms of Peru. Monographs in systematic botany*, vol. 45, 1286 pp. St. Louis, MO: Missouri Botanical Garden.

Bredemann, G. 1948. Über *Polymnia sonchifolia* Poepp. & Endl. (*Polymnia edulis* Wedd.), die Yacon—Erdbirne. *Botanica Oeconomica (Hamburg)* 1 (2): 65–85.

Bremer, K., A. A. Anderberg, P. O. Karis, B. Nordenstam, J. Lundberg, and O. Ryding. 1994. *Asteraceae: Cladistic and classification*. Portland, OR: Timber Press.

Calvino, M. 1940. A new plant, *Polymnia edulis*, for forage or alcohol. *Industria Saccharifera Italiana (Genova)* 33:95–98.

Cárdenas, M. 1969. *Manual de plantas económicas de Bolivia (Manual of economic plants from Bolivia)*. Cochabamba, Bolivia: Editorial Los Amigos del Libro.

Carr, G. D., R. M. King, A. M. Powell, and H. Robinson. 1999. Chromosome numbers in Compositae. XVIII. *American Journal of Botany* 86:1003–1013.

Cassini, F. 1818. De la classification naturelle des synanthérées. In *Dictionnaire des sciences naturelles*, vol. 10, ed. G. F. Cuviér, 152–57. Paris: Le Normant.

Castillo, R., C. Nieto, and E. Peralta. 1988. El germoplasma de cultivos andinos en Ecuador. VI Congreso Latino-americano de Cultivos Andinos. Memorias INIAP, Quito, Ecuador.

Chaquilla, G. 1997. Obtención de azucar a partir de yacon (*Polymnia sonchifolia*) y su potencial. IX Congreso Internacional de Cultivos Andinos, Cusco, Perú.

Chen, J. H., and C.-T. Ho. 1997. Antioxidant activities of caffeic acid and related hydroxycinnamic acid compounds. *Journal of Agricultural and Food Chemistry* 45:2374–2378.

Chiata, N. 1998. Variabilidad de la semilla botánica y compración de progenies y clones provenientes del germoplasma de yacón (*Polymnia sonchifolia*). Tesis de Grado, Universidad Nacional de San Antonio Abad del Cusco, Perú.

CIP (International Potato Center). 1997. Andean root and tuber crops: A report on collaborative research in biodiversity, 1993–1997, 27 pp.

Cisneros-Zevallos, L. A., R. Nunez, D. Campos, G. Noratto, R. Chirinos, and C. Arvizu. 2002. Characterization and evaluation of fructo-oligosaccharides on yacon roots (*Smallanthus sonchifolia* Poepp. & End.) during storage. Abstract session 15E-27, nutraceuticals and functional foods. Annual Meeting Food Expo, Anaheim, CA.

Cobo, B. (1653) 1890–1895. *Historia del nuevo mundo*, vol. 4, ed. M. J. de la Espada. Sevilla: Sociedad de Bibliófilos Andaluces.

Cronquist, A. 1955. Phylogeny and taxonomy of the Compositae. *American Midland Naturalist* 53:478–511.

Debouck, D. G., and D. Libreros Ferla. 1995. Neotropical montane forests: A fragile home of genetic resources of wild relatives of New World crops. In *Biodiversity and conservation of neotropical montane forests. Proceedings of a symposium, New York Botanical Garden*, eds S. P. Churchill, H. Balslev, E. Forero, and J. L. Luteyn, June 21–26, 1993, 561–577. New York: New York Botanical Garden.

De la Cruz, G. 1995. Informes sobre yacón conservado en invernadero. In *Memorias RTA-1993–94*.

Derid, E., A. Ceborari, M. Bodrug, and M. Soroni. 1997. *In vitro* propagation of yacon (*Polymnia sonchifolia*). In *Biological and technical development in horticulture*, International Horticultural Scientific Conference, Lednice na Morave, Czech Republic, 9.-12.9.1997, 42–47. Lednice na Moravě: Faculty of Horticulture, Mendel Agriculture and Forestry University in Brno.

Doležalová, I., A. Lebeda, and N. Gasmanová. 2004. Variation in relative DNA content in maca and yacon germplasm. In *Genetic variation for plant breeding. Proceedings of the 17th EUCARPIA General Congress*, September 8–11, 2004, Tulln, Austria, ed. J. Vollmann, H. Grausgruber, and P. Ruckenbauer, 127–130. Vienna, Austria: BOKU-University of Natural Resources and Applied Life Sciences.

Doo, H. S., J. H. Ryu, K. S. Lee, and S. Y. Choi. 2001. Effect of plant density on growth responses and yield in yacon. *Korean Journal of Crop Science* 46:407–410.

Endt, A. 1983. Two new vegetable crops from the tobacco enthusiasts. Landsedt New Crop Development, Oratia.

Espinoza, C. R. 2003. Elaboración de filtrantes de la hoja de yacón (Processing of yacon leaves for tea). 2nd Latino-American Symposium of Tubers and Roots, November, 28–30, 2001. In *Yacon: Foundations of the advantage of a promissory ressource* (in Spanish), ed. J. Seminario, M. Valderrama, and I. Manrique. Lima, Peru: CIP, UNC, COSUDE.

Estrella, J., and C. Tapia. 1994. Micropropagación y conservación del germoplasma de jícama (*Polymnia sonchifolia*) en Ecuador (Micropropagation and germplasm conservation of jícama in Ecuador) (In Spanish). *8th International Congress of Andean Agropecuarial Systems* 22:25, Valdivia, Chile.

Fernández, C. E. 1997. Využití tkáňových kultur u jakonu (*Polymnia sonchifolia* Poeppig & Endlicher) (Use of plant tissue culture on yacon). PhD thesis, 131 pp. (in Czech), Czech University of Agriculture, Prague, Czech Republic.

Fernández, C. E. 2005. Jakon (*Smallanthus sonchifolius* (Poeppig & Endlicher) H. Robinson), Pěstování v klimatických podmínkách České republiky (Yacon (*Smallanthus sonchifolius* (Poeppig & Endlicher) H. Robinson)) (Growing in the climatic conditions of the Czech Republic). Habilitation thesis, 154 pp. Prague: Czech Agricultural University in Prague, Institute of Tropics and Subtropics.

Fernández, C. E., and L. Kučera. 1997. Determination of number of chromosomes in select ecotypes of yacon (*Polymnia sonchifolia* Poeppig & Endlicher) cultivated *in vitro*. *Agricultura Tropica et Subtropica* 30:89–93.

Fernández, C. E., and Z. Opatrný. 1996. Organ specifity of regeneration in explant cultures of yacon, *Polymnia sonchifolia* Poeppig & Endlicher. *Agricultura Tropica et Subtropica* 29:99–108.

Fernández, C. E., and W. Pérez. 2003. Mapování jakonu (*Smalllanthus sonchifolius* Poepping & Endlicher) H. Robinson, v provinciích Bilbao Rioja a Charas kraje Potosí—Bolívie. Sborník referátů I. Mezinárodní seminář "Andské plodiny" v České republice, 87–95. Prague: Czech Agricultural University in Prague, Institute of Tropics and Subtropics.

Fernández, C. E., W. Pérez, C. H. D. Robles, and I. Viehmannová. 2005. Screening of yacon (*Smallanthus sonchifolius*) in the Bilbao Rioja and Charcas provinces of Departament Potosí in Bolivia. *Agricultura Tropica et Subtropica* 38:6–11.

Fernández, C. E., M. Riessová, and J. Čespiva. 2003. Yacón [*Smallanthus sonchifolius* (Poeppig & Endlicher)] H. Robinson: Výnos biomasy v klimatických podmínkách České republiky. Sborník referátů I. Mezinárodní seminář "Andské plodiny" v České republice, 75–85. Prague: Czech Agricultural University in Prague, Institute of Tropics and Subtropics.

Fernández, C. E., C. H. Robles, and M. Riessová. 2004. Yacon [*Smallanthus sonchifolius* (Poeppig & Endlicher)] H. Robinson: Yield of biomass under climatic conditions in the Czech Republic. *Agricultura Tropica et Subtropica* 37:5–9.

Fernández, C. E., I. Viehmannová, M. Bechyne, J. Lachman, L. Milella, and G. Martelli. 2007. Cultivation and phenological growth stages of yacon (*Smallanthus sonchifolius*). *Agricultura Tropica et Subtropica* 40:71–77.

Fernández, C. E., I. Viehmannová, J. Lachmann, and L. Milella. 2006. Yacon [*Smallanthus sonchifolius* (Poepp. et Endl.) H. Robinson]: A new crop in Central Europe. *Plant Soil Environment* 52:564–570.

Fernández-Jeri, A. 2003. Yacón: Imprortancia prebiótica y tecnológica, AGROENFOQUE, ed. No. 139. http://barrioperu.Terra.com.pe/agroenfoque

Flora of North America North of Mexico. 2006. Volume 19: Magnoliophyta: Asteridae, part 6: Asteraceae, part 1, ed. by Flora of North America Editorial Committee. Oxford, CT: Oxford University Press.

Flores, H. E., J. M. Vivanco, and V. M. Loyola-Vargas. 1999. Radicle biochemistry: The biology of root specific metabolism. *Trends in Plant Sciences* 4:220–226.

Flores, H. E., T. S. Walker, R. L. Guimarães, H. P. Bais, and J. M. Vivanco. 2003. Andean root and tuber crops: Underground rainbows. *HortScience* 38:161–167.

Franco, P. S., and C. J. Rodríguez. 1997. Caracterización y evaluación del germoplasma de llacón (*Polymnia sonchifolia* Poepp. & Endl.) del INIA en el valle de Cajamarca. IX Congreso Internacional de Cultivos Andinos. Libro de resúmenes, 22–25 de abril 1997. Universidad Nacional de San Antonio Abad del Cusco (UNSAAC), Centro de Investigaciones en Cultivos Andinos (CICA), Asociación ARARIWA, Cusco, Perú, 55 pp.

Frček, J. 1996. Jakon (*Polymnia sonchifolia*)—Nová kořenová zelenina v dietě diabetiků. Seminář Alternativní a maloobjemové plodiny pro zdravou lidskou výživu, Listopad 1996. Praha–Ruzyně, Czech Republic: Výzkumný ústav rostlinné výroby.

———. 2001. El cultivo del yacón y la maca en la República Checa. Resumenes workshop sobre la agricultura sostenible del 17 al 25 de Septiembre 2001, 54–58. Praga: Instituto de Agricultura Tropical y Subtropical, Universidad Checa de Agricultura Praga.

Frček, J., J. Michl, J. Pavlas, and J. Šupichová. 1995. Yacon (*Polymnia sonchifolia* Poepp. and Nedli.)—New prospective root and fodder crop (in Czech). Plant Genetic Resources (annual report 1995), 73–77. Nitra, Slovakia: Univerzity of Nitra.

Frček, J., J. Psotová, and V. Šimánek. 1999. Jakon (*Smallanthus sonchifolius*)—Nová kořenová zelenina v dietě (Yacon (*Smallanthus sonchifolius*)—A new root vegetable in diet). Sborník referátů z konference s mezinárodní účastí "Výživa rostlin, kvalita produkce a zpracovatelské využití." Mendelova zemědělská a lesnická univerzita v Brně, 29.-30.6.1999, 300–302. Brno, Czech Republic: Mendel Agricultural and Forestry University.

Frías, A. M., A. Grau, E. Lozzia, and S. Caro. 1997. Estudio citológico del yacón (*Polymnia sonchifolia*) y el yacón del campo (*Polymnia macroscypha*) (Cytological study of yacon and yacon del campo) (in Spanish). 28 Congreso Argentino de Genética, Tucumán, Argentina.

Fujino, M., T. Nakanishi, J. Ishihara, S. Ono, Y. Doi, M. Sugiura, and K. Tomioka. 2008. New yacon cultivars, 'Andes no Yuki' and 'Salad Okame.' *Bulletin of National Agriculture Research Center for Western Region, Japan* 7:131–143.

Fukai, K., S. Ohno, K. Goto, F. Nanjo, and Y. Hara. 1997. Seasonal fluctuations in fructans content and related activities in yacon (*Polymnia sonchifolia*). *Soil Science and Plant Nutrition* 43:171–177.

Funk, V. A., R. J. Bayer, S. Keeley, R. Chan, L. Watson, B. Gemeinholzer, E. Schilling, J. L. Panero, B. G. Baldwin, N. Garcia-Jacas, et al. 2005. Everywhere but Antarctica: Using a supertree to understand the diversity and distribution of the Compositae. *Biologiske Skrifter Kongelige Danske Videnskabernes Selskat* 55:343–374.

Genta, S. B., W. M. Cabrera, A. Grau, and S. S. Sanchez. 2005. Subchronic 4-month oral toxicity study of dried *Smallanthus sonchifolius* (yacon) roots as diet supplement in rats. *Food and Chemical Toxicology* 43:1657–1665.

Goto, K., K. Fukai, J. Hikida, F. Nanjo, and Y. Hara. 1995. Isolation and structural analysis of oligosaccharides from yacon (*Polymnia sonchifolia*). *Bioscience, Biotechnology and Biochemistry* 59:2346–2347.

Graefe, S. 2002. Post-harvest compositional changes of yacon roots (*Smallanthus sonchifolius* Poepp. & Endl.) as affected by storage conditions and kultivar. MSc thesis, 63 pp., Kassel, University of Kassel, Germany.

Graefe, S., M. Hermann, I. Manrique, S. Golombek, and A. Buerkert. 2004. Effects of post-harvest treatments on the carbohydrate composition of yacon roots in the Peruvian Andes. *Field Crops Research* 86:157–165.

Grau, A. 1993. *Yacon: A highly productive root crop*. New plants dossier no. 9, Crop and Food Research Institute Internal Report, Biodiversity Program. Invermay, New Zealand: Crop and Food Research Institute.

Grau, A., A. M. Kortsarz, J. M. Abyar, R. A. Sánchez, and S. S. Sánchez. 2001. El retorno del yacón. *Ciencia Hoy, Revista de Divulgación Científica y Tecnológica de le Asociacón Ciencia Hoy*, vol. 11: no. 63 Junio/Julio. http://www.ciencia-hoy.retina.ar/hoy63/yacon.htm

Grau, A., and J. Rea. 1997. Yacon—*Smallanthus sonchifolius* (Poepp. & Endl.) H. Robinson. In *Andean roots and tubers: Ahipa, arracacha, maca and yacon. Promoting the conservation and use of underutilized and neglected crops*, ed. M. Hermann and J. Heller, 199–242. Gatersleben, Germany: Institute of Plant Genetic and Crop Plant Research, Rome, Italy: IPGRI.

Grau, A., and A. Slanis. 1996. Is *Polymnia sylphioides* var. *perennis* a wild ancestor of yacon? *Resumos I Congresso Latino Americano de Raízes Tropicais*. São Pedro, Brasil: CERAT-UNESP.

Hamada, M., T. Hosoki, and Y. Kusabirki. 1990. Mass propagation of yacon (*Polymnia sonchifolia*) by repeated node culture. *Plant Tissue Culture Letters (Japan)* 7:35–37.

Hanelt, P., ed. 2001. *Smallanthus*. In *Mansfeld's encyclopedia of agricultural and horticultural crops*, vol. 4, 2135. Berlin, Germany: Springer.

Hashidoko, Y., T. Urashima, T. Yoshida, and J. Nizutani. 1993. Decarboxylative conversion of hydroxycinnamic acids by *Klebsiella oxytoca* and *Erwinia uredovora*, epihytic bacteria of *Polymnia sonchifolia* leaf, possibly associated with formation of microflora on the damaged leaves. *Bioscience, Biotechnology and Biochemistry* 57:215–219.

Heiser, C. 1963. Numeratión cromosómica de plantas ecuatorianas. *Ciencia y Naturaleza* 6:2–6.

Hermann, M., I. Freire, and C. Pazos. 1998. Compositional diversity of the yacon storage root. CIP Program Report 1997–1998, Lima, Peru, 425–432.

Herrera, F. L. 1943. Plantas alimenticias y condimenticias indígenas del departamento del Cuzco. *Boletín Dirección Genetica Agricultura Perú* 14:48–51.

Holle, M., and D. Talledo. 2001. Conservación y uso de la biodiversidad de raíces y túberculos andinos (Conservation and utilization of the biodiversity of Andean root and tuber crops) (in Spanish). Lima, Peru.

Hondo, M., A. Nakano, Y. Okumura, and T. Yamaki. 2000. Effects of activated carbon powder treatment on clarification, decolorization deodorization and fructo-oligosacharide content of yacon juice. *Journal of the Japanese Society for Food Science and Technology* 47:148–154.

Hondo, M., Y. Okumura, and T. Yamaki. 2000. A preparation of yacon vinegar containing natural fructo-oligosaccharides. *Journal of the Japanese Society for Food Science and Technology* 47:803–807.

Hong, S. S., S. A. Lee, Y. H. Han, M. H. Lee, J. S. Hwang, J. S. Park, K.-W. Oh, K. Han, M. K. Lee, H. Lee, et al. 2008. Melampolides from the leaves of *Smallanthus sonchifolius* and the inhibitory activity of LPS-induced nitric oxide production. *Chemical and Pharmaceutical Bulletin* 56:199–202.

Huamán, W. 1991. Caracterización y evaluación de 45 entradas de germoplasma de llacón (*Smallanthus sonchifolius*, H. Robinson) en Cajamarca. Tesis de Grado, Universidad nacional de Cajamarca, Perú, 70 pp.

http://en.wikipedia.org/wiki/Smallanthus_sonchifolius

http://www.cipotato.org/artc/cip_crops/making_yacon_candy.pdf

http://www.nap.edu/catalog.php?record_id=1398

Inoue, A., S. Tamogami, H. Kato, Y. Nakazato, M. Akiyama, O. Kodama, T. Akatsuka, and Y. Hashidoko. 1995. Antifungal melampolides from leaf extracts of *Smallanthus sonchifolius*. *Phytochemistry* 39:845–848.

Ishiki, K., M. V. X. Salgado, and J. Arellano. 1997. *Revision of chromosome number and karyotype of yacon (Polymnia sonchifolia)*. Resúmenes del Primer Taller Internacional sobre Recursos Fitogenéticos del Noroeste Argentino. Salta, Argentina: INTA.

Itaya, N. M., M. A. M. De Carvalho, and R. D. I. Figueiredo-Ribeiro. 2002. Fructosyl transferase and hydrolase activities in rhizophores and tuberous roots upon growth of *Polymnia sonchifolia* (*Asteraceae*). *Physiologia Plantarum* 116:451–459.

Jan, Ch.-Ch., and G. J. Seiler. 2006. Sunflower. In *Genetic resources, chromosome engineering and crop improvement: Oilseed crops*, vol. 4, ed. R. J. Singh, 103–165. Boca Raton, FL: CRC Press.

Jandera, P., V. Škeříková, L. Řehová, T. Hájek, L. Baldriánová, G. Škopová, V. Kellner, and A. Horna. 2005. RP-HPLC analysis of phenolic compounds and flavonoids in beverages and plant extracts using a CoulArray detector. *Journal of Separation Science* 28:1005–1022.

Jha, T. B., and B. Ghosh. 2005. *Plant tissue culture (basic and applied)*. India: Universities Press.

Jórgensen, P. M., and C. Leon. 1997. Catalogue of vascular plants of Ecuador. *Monographs in Systematic Botany*. St. Louis, MO: Missouri Botanical Garden.

Kakihara, T. S., F. L. A. Câmara, and S. M. C. Vilhena. 1997. Cultivation and processing of yacon: A Brazilian experience. 1st Yacon Workshop, Botucatú (SP), Brazil (in Portuguese).

Kakihara, T. S., F. L. A. Câmara, S. M. C. Vilhena, and L. Riera. 1996. Cultivation and industrialization of yacon: A Brazilian experience (Cultivo e industrilização de yacon: Uma experiência brasileira) (in Portuguese). 1st Latino-American Congress of Tropical Roots, CERAT-UNESP, São Pedro, Brasil.

Kakuta, H., T. Seki, Y. Hashidoko, and J. Mizutani. 1992. *Ent*-kaurenic acid and its related-compounds from glandular trichome exudate and leaf extracts of *Polymnia sonchifolia*. *Bioscience, Biotechnology and Biochemistry* 56:1562–1564.

King, S. R. 1987. Four endemic Andean tuber crops: Promising food resources for agricultural diversification. *Mountain Research and Development* 7:43–52.

Kono, Y., S. Kaseine, T. Yoneyama, and Y. Sakamoto. 1998. Iron chelation by chlorogenic acid as a natural antioxidant. *Bioscience, Biotechnology and Biochemistry* 62:22–27.

Kuroda, S. 1992. Improvement of sugar content in tuberous roots of yacon, *Polymnia sonchifolia*, using tissue culture. Tsukuba, Japan: National Institue of Agrobiological Resources (NIAR), 185–190.

Kuroda, S., and J. Ishihara. 1993. Field growth characteristics of plantlets propagated *in vitro* and line selection for increased percentage of sugar in tuberous roots of taxon, *Polymnia sonchifolia*. *Bulletin of the Shikoku National Agricultural Experiment Station* 57:111–121.

Lachman, J., J. Dudjak, E. C. Fernández, and V. Pivec. 2003. Obsah polyfenolických antioxidantů ve vybraných orgánech jakonu [*Smallanthus sonchifolius* (Poepp. et Endl.) H. Robinson] (The contents of polyphenolic antioxidants in different organs of yacon [*Smallanthus sonchifolius* (Poepp. et Endl.) H. Robinson]). In I. Mezinárodní seminář "Andské plodiny" v České republice, 12.5.2003, Praha, *Proceedings* (47–54), ed. E. C. Fernández. Praha, Czech Republic: Czech University of Agriculture Prague (in Czech).

Lachman, J., E. C. Fernández, and M. Orsák. 2003. Yacon [*Smallanthus sonchifolius* (Poepp. et Endl.) H. Robinson] chemical composition and use—A review. *Plant Soil Environment* 49:283–290.

Lachman, J., E. C. Fernández, I. Viehmannová, M. Šulc, and P. Cepková. 2007. Total phenolic content of yacon (*Smallanthus sonchifolius*) rhizomes, leaves and roots affected by genotype. *New Zealand Journal of Crop and Horticultural Science* 35:117–123.

Lachman, J., B. Havrland, E. Fernández, and J. Dudjak. 2004. Saccharides of yacon [*Smallanthus sonchifolius* (Poepp. et Endl.) H. Robinson] tubers and rhizomes and factors affecting their content. *Plant Soil Environment* 50:383–390.

Lachman, J., Hejtmánková, J. Dudjak, E. C. Fernández, and V. Pivec. 2003. Zastoupení dominantních fenolick-ých kyselin v různých částech rostliny jakonu [*Smallanthus sonchifolius* (Poepp. et Endl.) H. Robinson] (Presence of dominant phenolic acids in different parts of yacon [*Smallanthus sonchifolius* (Poepp. et Endl.) H. Robinson]). In I. Mezinárodní seminář "Andské plodiny" v České republice, 12.5.2003, Praha, *Proceedings* (67–73), ed. E. C. Fernández. Praha, Czech Republic: Czech University of Agriculture Prague (in Czech).

Larkin, P. J., and W. R. Scowcroft. 1981. Somaclonal variation a novel source of variability from cell cultures for plant improvement. *Theoretical and Applied Genetics* 60:197–214.

Lebeda, A., I. Doležalová, and K. Doležal. 2004. Variation in morphological and biochemical characters in genotypes of maca and yacon. In *Proceedings XXVI IHC—IVth International Symposium Future for Medicinal and Aromatic Plants*, ed. L. E. Craker, J. E. Simon, A. Jatisatienr, and E. Lewinsohn. *Acta Horticulturae* 629:483–490.

Lebeda, A., I. Doležalová, M. Dziechciarková, K. Doležal, and J. Frček. 2003. Morphological variability and isozyme polymorphisms in maca and taxon. *Czech Journal of Genetics and Plant Breeding* 39:1–8.

Lebeda, A., I. Doležalová, K. Valentová, M. Dziechciarková, M. Greplová, H. Opatová, and J. Ulrichová. 2003. Biologická a chemická variabilita maky a jakonu (Biological and chemical variability of maca and yacon). *Chemické Listy* 97:548–556.

Lebeda, A., I. Doležalová, K. Valentová, N. Gasmanová, M. Dziechciarková, and J. Ulrichová. 2008. Yacon (*Smallanthus sonchifolius*)—A traditional crop of Andean Indians as a challenge for the future—The news about biological variation and chemical substances' contents. In *Proceedings XXVII IHC—Plants as Food and Medicine*, ed. G. Gardner and L. E. Craker. *Acta Horticulturae* 765:127–136.

León, J. 1964. Plantas alimenticias andinas (Andean food crops). Instituto Interamericano de Ciencias Agrícolas Zona Andina, Lima, Perú. *Boletín Técnico* 6:57–62.

———. 1983. Caracteres agronómicos de cinco cultivares de llacón (*Polymnia sonchifolia*) bajo condiciones de la campiña de Cajamarca. Tesis de Grado, Universidad Nacional de Cajamarca, Perú, 82 p. http://www.cipotato.org/market/PDFdocs/Yacon_Fundamentos_password.pdf

Lin, F., M. Hasegawa, and O. Kodawa. 2003. Purification and identification of antimicrobial sesquiterpene lactones from yacon (*Smallanthus sonchifolius*) leaves. *Bioscience, Biotechnology and Biochemistry* 67:2154–2159.

Linnaeus, C. 1751. *Species Plantarum*, vol. 1. Impensis Laurentii Salvii: Stockolm.

Lizárraga, L., R. Ortega, W. Vargas, and A. Vidal. 1997. The cultivation of yacon (*Polymnia sonchifolia*). 9th International Congress of Andean crops (in Spanish), 65–67, Cusco, Peru.

Mackenzie, K. K. 1933. Carduaceae 56. *Smallanthus*. In *Manual of the southeastern Flora*, ed. J. K. Small, 1307–1406. Chapel Hill, NC: University of North Carolina Press.

Mansilla, S. R. C., B. C. López, S. R. Blas, W. J. Chia, and J. Baudoin. 2006. Análisis de la variabilidad molecular de una colección peruana de *Smallanthus sonchifolius* (Poepp & Endl) H. Robinson 'Yacón' (Molecular variability analysis of a peruavian *Smallanthus sonchifolius* (Poepp & Endl) H. Robinson 'Yacon' collection) (in Spanish). *Ecología Aplicada* 5:75–80.

Manrique, I., and M. Hermann. 2003. El potencial del yacón en la salud y la nutrición (The potential of yacon on health and nutrition) (in Spanish). 11th International Congress of Andean Crops, Cochabamba, Bolivia, October 15–19, 2003. http://www.cipotato.org/artc/docs/Manrique_Hermann_2004_yacon_en_la_salud.pdf (accessed April 2005).

Matejka, V. 1994. On the possibilities of growing yacon (*Polymnia sonchifolia* Poepp. & Endl.) in the Czech Republic. *Agricultura Tropica et Subtropica* 27:22–30.

Matsubara, S., Y. Ohmori, Y. Takada, T. Komasadomi, and H. Fukazawa. 1990. Vegetative propagation of yacon by shoot apex, node and callus culture. *Scientific Reports of the Faculty of Agriculture-Okayama University (Japan)* 76:1–6.

Matsumoto, Y., S. Matsubara, and K. Murakami. 1996. Regeneration of adventitious embryo from mesophyll callus in *Polymnia sonchifolia*. *Journal of the Japanese Society for Horticultural Science* 65 (Suppl 1): 182–183.

Melgarejo, D. 1999. Potencial productivo de la colección nacional de yacón (*Smallanthus sonchifolius* Poeppig & Endlicher), bajo condiciones de Oxapampa. Tesis de Grado, Universidad Nacional Daniel Alcides car-rión, Perú, 96 pp. http://www.cipotato.org/market/PDFdocs/Yacon_Fundamentos_password.pdf

Meza, Z. G. 1995. Variedades nativas de llacón (*Polymnia sonchifolia*) en Cusco (Landraces of yacon, *Polymnia sonchifolia*, in Cusco) (in Spanish). Peru: UNSAAC–CICA, CIP, COTESU.

————. 2001. Cultivo de llacon (*Smallanthus sonchifolius* H. Robinson) en Cusco (Cultivation of yacon in Cusco) (p. 31). UNSAAC, CICA, Cusco, Peru: Facultad de Agronomía y Zootecnia.

Milella, L., J. Salava, G. Martelli, I. Greco, E. Fernandez, and J. Ovesna. 2004. Comparison of two PCR-based marker systems for the analysis of genetic relationships in yacon, In *Proceedings of International Symposium: Advances in Molecular Biology: Methods for Genotype Identification, Plant Breeding and Product Control*. November 3, 2004, Prague, Czech Republic, 58–60. Prague, Czech Republic: Research Institute for Crop Production.

Milella, L., J. Salava, G. Martelli, I. Greco, C. E. Fernández, and I. Viehmannová. 2005. Genetic diversity between yacon landraces from different countries based on random amplified polymorphic DNAs. *Czech Journal of Genetics and Plant Breeding* 41:73–78.

Miyazawa, M., and N. Tamura. 2008. Characteristic odor components in the essentials oil from yacon tubers (*Polymnia sonchifolia* Poepp. et Endl.). *Journal of Essential Oil Research* 20:12–14.

Mizumo, A., T. Nakanishi, and K. Nishiyama. 1993. Bacterial wilt of yacon strawberry caused by *Erwinia chrysanthemi*. *Annals of the Phytopathological Society of Japan* 59:702–708.

Morillo, E., C. Tapia, J. Estrella, and R. Castillo. 2002. Caracterización agromorfológica de la colección de jícama (*Smallanthus sonchifolius* P. & E.) del banco de germoplasma del INIA (Agromorphological characterization of yacon collection in gene bank of INIA) (in Spanish), Ecuador.

Murashige, T., and F. Skoog. 1962. A revised medium for rapid groth and bioassays with tabacco tissue cultures. *Physiologia Plantarum* 15:473–497.

Narai-Kanayama, A., N. Tokita, and K. Aso. 2007. Dependence of fructo-oligosaccharides content on activity of fructo-oligosaccharide-metabolizing enzymes in yacon (*Smallanthus sonchifolius*) tuberous roots during storage. *Journal of Food Science* 72:S381–S387.

Neves, V. A., and M. A. da Silva. 2007. Polyphenol oxidase from yacon roots (*Smallanthus sonchifolius*). *Journal of Agricultural and Food Chemistry* 55:2424–2430.

Nieto, C. 1991. Estudios agronómicos y bromatológicos en jícama (*Polymnia sonchifolia* Poeppig & Endlicher). Instituto Nacional de Investigaciones Agropecuarias, Quito, Ecuador: *Archivos Latinoamericanos de Nutrición*, XLI (2): 213–221.

Niwa, M., T. Arai, K. Fujita, W. Marubachi, E. Inoue, and T. Tsukihashi. 2002. Plant regeneration through leaf culture of yacon. *Journal of the Japanese Society for Horticultural Science* 71:561–567.

Ogiso, M., H. Naito, and H. Kurasima. 1990. Planting density, harvesting time and storage temperature of yercum. *Research Bulletin of the Aichiken Agricultural Research Center* 22:161–164.

Ohyama, T., O. Ito, S. Yasuyoshi, T. Ikarashi, K. Minamisawa, M. Kubota, T. Tsukinashi, and T. Asami. 1990. Composition of storage carbohydrate in tuber roots of yacon (*Polymnia sonchifolia*). *Soil Science and Plant Nutrition* 36:167–171.

O'Neal, L. M., and T. W. Whitaker. 1947. Embroideries of the early Nazca period and the crops depicted on them. *Southwestern Journal of Anthropology* 3:294–321.

Panero, J., and V. A. Funk. 2002. Toward a phylogenetic subfamilial classification for the Compositae (Asteraceae). *Proceedings of the Biological Society of Washington* 115:909–922.

Paterson, K. E., and N. P. Everett. 1985. Regeneration of *Helianthus annuus* inbred plants from callus. *Plant Science* 42:125–132.

Pedreschi, R., D. Campos, G. Noratto, R. Chirinos, and L. A. Cisneros-Zevallos. 2002. Fermentation of fructo-oligosaccharides from yacon (*Smallanthus sonchifolia* Poepp. End.) by *L. acidophilus* NRRL B-1910, *L. brevis* NRRL B-4527, *L. gasseri* NRRL B-14168, *L. plantarum* NRRL B-4496 and *B. bifidum*. Abstract session 84, nutraceuticals and functional foods, 2002 Annual meeting, Food Expo-Anaheim, CA, 84–89.

————. 2003. Andean yacon root (*Smallanthus sonchifolius* Poepp. Endl.) fructo-oligosaccharides as a potential novel source of prebiotics. *Journal of Agricultural and Food Chemistry* 51:5278–5284.

Perez Arbeláez, E. 1956. *Plantas útiles de Colombia*. Librería Colombiana, Camacho Roldán, Bogota.

Polreich, S. 2003. Establishment of a classification scheme to structure the post-harvest diversity of yacon storage roots (*Smallanthus sonchifolius* (Poepp. & Endl.) H. Robinson). Thesis (58 pp.), University of Kassel, Faculty of Agriculture, International Rural Development and Environmental Protection, Germany.

Popenoe, H., S. R. King, J. León, and K. L. Sumar. 1989. *Lost crops of the Incas: Little known plants of the Andes with promise for worldwide cultivation*. National Research Council. National Academy Press: Washington D.C.

Quemener, B., J. F. Thibault, and P. Cousement. 2003. Determination of inulin and oligofructose in food products and integration in the AOAC method for measurement of total dietary fiber. *Lebensmittel-Wissenschaft Technology* 27:125–132.

Ramos, R., J. Galarza, R. Castillo, and C. Nieto. 1999. Respuesta de tres raíces andinas: Zanahoria blanca (*Arracacia xanthorrhiza*), miso (*Mirabilis expansa*) y jícama (*Polymnia sonchifolia*); dos pastos y una mezcla forrajera al efecto de tres sistemas agroforestales, en Santa Catalina, Quito. http://www.cipotato.org/market/PDFdocs/Yacon_Fundamentos_password.pdf

Rauscher, J. T. 2002. Molecular phylogenetics of the Espeletia complex (Asteraceae): Evidence from nrDNA ITS sequences on the closests relatives of an Andean adaptive radiation. *American Journal of Botany* 89:1074–1084.

Rea, J. 1992. Raíces Andina: Yacón (*Polymnia sonchifolia*) In *Cultivos marginados: Otra perspectiva de 1942*, ed. J. Leon, and J. E. Hernández, 163–179. Rome: FAO, pp. 174–177.

———. 1994. Manejo *in situ* de germoplasma de tubérculos y raíces andinas en Bolivia. Primera Reunión Internacional de Recursos Genéticos de Papa, Raíces y Tubérculos Andinos, 7–10 de Febrero 1994, Cochabamba, Bolivia, 81–89.

———. 1995. Conservación y Manejo *in situ* de Recursos Fitogenéticos Agrícolas en Bolivia. http://condesan.org/Biodiver/InSitu/rea.htm (accessed April 1996).

———. 1997. Recursos genéticos del yacón (*Smallanthus sonchifolius*). La Paz, Bolivia: Instituto Internacional de Recursos Fitogenéticos.

———. 2000a. Recursos genéticos del yacón. Raíces Andinas, Fascículo 29. Manual de capacitación. Lima, Peru: Centro International de la Papa (CIP).

———. 2000b. Conservación y manejo *in situ* de recursos fitogenéticos agrícolas en Bolivia conservation and maintenance of plant genetic resources in Bolivia) (in Spanish). Raíces Andinas, Fascículo 3. Manual de capacitación. Lima, Peru: Centro International de la Papa (CIP).

Robinson, H. 1978. Studies in the Heliantheae (Asteraceae). XII. Reestablishment of the genus *Smallanthus*. *Phytologia* 39:47–53.

———. 1981. A revision of the tribal and subtribal limits of the Heliantheae (Asteraceae). *Smithsonian Contributions to Botany* 51:1–102.

Rubatzky, V. E., and M. Yamaguchi. 1997. *World vegetables. Principles, production, and nutritive values*, 2nd ed. New York: Chapman and Hall.

Safford, N. E. 1917. Food plants and textiles of ancient America. *Proceedings of International Congress of America* 19:12–30.

Santarém, E. R. 2007. Embriogênese somática em yacón (*Smallanthus sonchifolius*) (Somatic embryogenesis in yacon) (in Portuguese). http:www.unicruz.tche.br/biotecnologia (accessed June 2007).

Sarrafi, E. A., and L. Gentzbittel. 2005. Genomic as efficient tools: Example sunflower breeding. In *Biotechnology in agriculture and forestry*, vol. 55: *Molecular marker systems in plant breeding and crop improvement*, ed. H. Lörz and G. Wenzel, 107–120. Berlin-Heidelberg, Germany: Springer-Verlag.

Sato, T., K. Tomioka, T. Nakanishi, and H. Koganizawa. 1999. Charol root of yacon [*Smallanthus sonchifolius* (Poepp. et Endl.) H. Robinson], oca (*Oxalis tuberosa* Molina) and pearl lupin (Tarwi, *Lupinus mutabilis* Swet), caused by *Macrophomina plaseolina* (Tassi), *Goid. Bulletin of the Shikoku National Agricultural Experiment Station* 64:1–8.

Schorr, K., I. Merfort, and F. B. Da Costa. 2007. A novel dimeric melampolide and further terpenoids from *Smallanthus sonchifolius* (Asteraceae) and the inhibition of the transcription factor NF-kappa B. *Natural Products Communications* 2:367–374.

Seminario, J. 1995. Descriptores básicos para la caracterización de chago (*Mirabilis expansa*), llacón (*Smallanthus sonchifolius*) y achira (*Canna edulis*). I. Congreso Peruano de Cultivos Andinos, Uamanga, Ayacucho.

Seminario, J., M. Valderrama, and H. Honorio. 2001. Propagación por esquejes de tres morfotipos de yacón [*Smallanthus sonchifolius* (Poepp. & Endl.)] H. Robinson. *Agronomía* XLVII:12–20.

Seminario, J., M. Valderrama, and I. Manrique. 2003. El yacón: Fundamentos para el aprovechaminento de un recurso promisorio (Yacon: Foundations of the advantage of a promissory resource). Centro International de la Papa (CIP), Universidad Nacional de Cajamarka (UNC), Agencia Suiza para el Desarrollo y la Cooperación (COSUDE), Lima, Perú. http://www.cipotato.org/market/PDFdocs/Yacon_Fundamentos_password.pdf

Seminario, J., M. Valderrama, and J. Romeo. 2004. Varibilidad morfológica y distribución geográfica del yacón, *Smallanthus sonchifolius* (Poepp. & Endl.) H. Robinson, en el norte peruano (Morphological variability and geographical distribution of yacon in northern Peru) (In Spanish). *Arnoldia* 11:139–160.

Seminario, J., M. Valderrama, and A. Seminario. 2002. Prueba de rendimiento de dos morfotipos de yacón [*Smallanthus sonchifolius* (Poepp. & Endl.)] H. Robinson. Propagados por esquejes y cepas. *Cajamarca* 10:99–108.

Simonovska, B., I. Vovk, S. Andrenšek, K. Valentová, and J. Ulrichová. 2003. Investigation of phenolic acids in yacon (*Smallanthus sonchifolius*) leaves and tubers. *Journal of Chromatography A* 1016:89–98.

Soto, F. R. 1998. Estudio de la biología floral del germoplasma regional. Tesis de Grando. Universidad Nacional de Cajamarka, Perú, p. 51.

Strauss, R. 1994. *Compendium of growth stage identification keys for mono- and dicotyledonous plants (extended BBCH scale).* Basel: Ciba-Geigy AG.

Sugiura, M., T. Nakanishi, T. Kameno, Y. Doi, and M. Fujino. 2007. A new yacon cultivar, Sarada Otome. *Bulletin of the National Agricultural Research Center for Western Region, Japan* 6:1–13.

Suzuki, T., H. Hara, T. Kasai, and F. Tomita. 1998. Effects of difructose anhydrite III on calcium absorption in small and large intestines of rats. *Bioscience, Biotechnology and Biochemistry* 62:837–841.

Takasugi, M., and T. Masuda. 1996. Studies on stress metabolites. 22. Three 4-hydroxyacetophenone-related phytoalexins from *Polymnia sonchifolia*. *Phytochemistry* 43:1019–1021.

Takenaka, M., and H. Ono. 2003. Novel octulosonic acid derivatives in the composite *Smallanthus sonchifolius*. *Tetrahedron Letters* 44:999–1002.

Takenaka, M., X. J. Yan, H. Ono, M. Yoshida, T. Nagata, and T. Nakanishi. 2003. Caffeic acid derivatives in the roots of yacon (*Smallanthus sonchifolius*). *Journal of Agricultural and Food Chemistry* 51:793–796.

Talledo, D., and M. Escobar. 1996. *Caracterización cariotípica de germoplasma RTA.* Laboratorio de Biología Celular y Genetica. Memorias, Quito, Ecuador: INIAP.

Tapia, B. C., R. Castillo T., and N. O. Mazón. 1996. *Catálogo de recursos genéticos de raíces y tubérculos Andinos en Ecuador (Catalogue of genetic resources of Andean roots and tubers in Ecuador).* Quito, Ecuador: INIAP-DENAREF.

Terada, S., K. Ito, A. Yoshimura, N. Noguchi, and T. Ishida. 2006. The constituents relate to antioxidative and alpha-glucosidase inhibitory activities in yacon aerial part extract. *Journal of the Pharmaceutical Society of Japan* 8:665–669.

Tjukavin, G. B. 2001. *Introdukcija jakona v Rossii* (Monografija). Moskva, Russia: Vympel.

———. 2002. Productivity and morphological determination of yacon under the influence of growing conditions and in connection with date of harvesting. *Selskochozjajstvennaja Biologia* 3:81–87.

Towle, M. A. 1961. *The ethnobotany of pre-Colombian Perú.* Chicago, IL: Aldine Publishers.

Tsukihashi, T., M. Myiamoto, and N. Suzuki. 1991. Studies on the cultivation of yacon, III. Effect of the planting methods on the growth and yield of yacon. *Japanese Journal of Farm Work Research* 26:185–189.

Tsukihashi, T., T. Yohida, M. Miyamoto, and N. Suzuki. 1989. Studies on the cultivation of yacon. I. Influence of different planting densities on tuber yield. *Japanese Journal of Farm Work Research* 24:32–38.

Turner, B. L. 1977. Fossil history and geography. In *The biology and chemistry of the Compositae*, vol. 1, ed. V. H. Heywood, J. B. Harborne, and B. L. Turner, 19–39. London: Academic Press.

Ulrichová, J., A. Lebeda, N. Škottová, K. Valentová, V. Krečman, and V. Šimánek. 2003. Obsahové látky jakonu (*Smallanthus sonchifolius*) a jeho chemoprotektivní účinky (Substances of yacon (*Smallanthus sonchifolius*) and their chemoprotective effect). In I. Mezinárodní seminář Andské plodiny v České republice, 12-5-2003, Praha, *Proceedings*, ed. E. C. Fernández, 41–45. Praha, Czech Republic: Czech University of Agriculture Prague (in Czech).

Valentová, K. 2004. Biologická aktivita andských plodin *Smallanthus sonchifolius* a *Lepidium meyenii* (Biological activity of Andean crops *Smallanthus sonchifolius* and *Lepidium meyenii*). PhD thesis (in Czech). Olomouc, Czech Republic: Faculty of Medicine, Palacký University in Olomouc.

Valentová, K., L. Cvak, A. Muck, J. Ulrichová, and V. Šimánek. 2003. Antioxidant activity of extracts from the leaves of *Smallanthus sonchifolius*. *European Journal of Nutrition* 42:61–66.

Valentová, K., J. Frček, and J. Ulrichová. 2001. Jakon (*Smallanthus sonchifolius*) a maka (*Lepidium meyenii*), tradiční andské plodiny jako nové funkční potraviny na evropském trhu (Yacon (*Smallanthus sonchifolius*) and maca (*Lepidium meyenii*), (traditional Andean crops as new functional foods on the European market). *Chemické Listy* 95:594–601.

Valentová, K., A. Lebeda, I. Doležalová, D. Jirovský, B. Simonovska, I. Vovk, P. Kosina, N. Gasmanová, M. Dziechciarková, and J. Ulrichová. 2006. The biological and chemical variability of yacon. *Journal of Agricultural and Food Chemistry* 54:1347–1352.

Valentová, K., A. Moncion, I. de Waziers, and J. Ulrichová. 2004. The effect of *Smallanthus sonchifolius* leaf extracts on rat hepatic metabolism. *Cell Biology and Toxicology* 20:109–120.

Valentová, K., and J. Ulrichová. 2003. *Smallanthus sonchifolius* and *Lepidium meyenii*—Prospective Andean crops for the prevention of chronic diseases. *Biomedical Papers of Medical Faculty of University Palacky in Olomouc, Czech Republic* 147:119–130.

Vavilov, N. I. 1950. *The origin, variation, immunity and breeding of cultivated crops.* Waltham, MA: Chronica Botanica, 13.

———. 1987. *Origin and geography of cultivated plants.* Leningrad, USSR: Nauka

———. 1995. *Five continents.* Rome, Italy: International Plant Genetic Resources Institute.

Veilleux, R. E., and H. De Jong. 2007. Potato. In *Genetic resources, chromosome engineering, and crop improvement series,* vol. 3: *Vegetable crops,* ed. R. Singh, 17–58. CRC Press, Boca Raton, FL.

Vera, B., and G. Alfaro. 2000. Yacón deshidratado (k'isa de yacón). Memorias de la 2da Reunión Boliviana sobre Recursos Fitogenéticos de Cultivos Nativos, 33–34. Cochabamba, Bolivia.

Viehmannová, I. 2009. *In vitro* polypoidization and protoplast cultures in yacon [*Smallanthus sonchifolius* (Poeppig & Endlicher) H. Robinson]. PhD thesis, 111 pp. (in Czech), Czech University of Life Sciences, Prague, Institute of Tropics and Subtropics.

Viehmannová, I., E. F. Cusimamani, M. Bechyne, M. Vyvadilová, and M. Greplová. 2009. *In vitro* induction of polyploidy in yacon (*Smallanthus sonchifolius*). *Plant Cell, Tissue and Organ Culture* 97:21–25.

Vilhena, S. M. C., F. L. A. Camara, and S. T. Kakihara. 2000. Yacon cultivation in Brazil. *Horticulture Brasileira* 18:5–8.

Volpato, G. T., F. C. G. Vieira, F. Almeida, F. Câmara, and I. P. Lemonica. 1997. Study of the hypoglycemic effects of *Polymnia sanchifolia* leaf extracts in rats, II. World Congress on Medicinal and Aromatic Plants for Human Welfare, Mendoza, Argentina.

Wells, J. R. 1965. A taxonomic study of *Polymnia* (Compositae). *Brittonia* 17:144–159.

———. 1967. A new species of *Polymnia* (Compositae: Heliantheae) from Mexico. *Brittonia* 19:391–394.

———. 1971. Variation in *Polymnia* pollen. *American Journal of Botany* 38:124–130.

Yacovleff, E. 1933. La jiquima, raíz comestible extinguida en el Perú. *Revue del Museo Nacional (Lima)* 3:376–406.

Yan, X. J., M. Suzuki, M. Ohnishi-Kameyama, Y. Sada, T. Nakanishi, and T. Nagata. 1999. Extraction and identification of antioxidants in the roots of yacon (*Smallanthus sonchifolius*). *Journal of Agricultural and Food Chemistry* 47:4711–4713.

Yoshida, M., H. Ono, Y. Mori, Y. Chuda, and M. Mori. 2002. Oxygenation of bisphenol A to quinones by polyphenol oxidase in vegetables. *Journal of Agricultural and Food Chemistry* 50:4377–4381.

Zámečníková, J. 2007. Study of landraces and utilization of *Smallanthus sonchifolius* (Poepp. et Endl.) H. Robinson. Diploma thesis, Institut of Tropics and Subtropics, Czech University of Life Science Prague, Czech Republic.

Zardini, E. 1991. Ethnobotanical notes on yacon. *Economic Botany* 45:72–85.

CHAPTER 21

Mint

Arthur O. Tucker and Sherry Kitto

CONTENTS

21.1 INTRODUCTION

The genus *Mentha* includes 18 species. Five species in section *Mentha* have produced 11 named hybrids (Tucker and Naczi 2007):

- Species
  - *M. aquatica* L.
  - *M. arvensis* L.

- *M. longiflolia* (L.) L.
- *M. spicata* L.
- *M. suaveolens* Ehrh.
- Hybrids
  - *M. ×carinthiaca* Host = *M. arvensis* × *M. suaveolens*
  - *M. ×dalmatica* Tausch = *M. arvensis* × *M. longifolia*
  - *M. ×dumetorum* Schultes = *M. aquatica* × *M. longifolia*
  - *M. ×gracilis* Sole = *M. arvensis* × *M. spicata*
  - *M. ×maximilinea* F.W. Schultz = *M. aquatica* × *M. suaveolens*
  - *M. ×piperita* L. = *M. aquatica* × *M. spicata*
  - *M. ×rotundifolia* (L.) Huds. = *M. longifolia* × *M. suaveolens*
  - *M. ×smithiana* = *M. aquatica* × *M. arvensis* × *M. spicata*
  - *M. ×verticillata* L. = *M. R. A. Graham aquatica* × *M. arvensis*
  - *M. ×villosa* Huds. = *M. spicata* × *M. suaveolens*
  - *M. ×villoso-nervata* Opiz = *M. longifolia* × *M. spicata*

In addition, evidence points to *M. canadensis* as a hybrid of *M. arvensis* × *M. longifolia* from the Lower Tertiary, but it is nominally accepted as a species because of its ancient origin and high fertility (Tucker and Chambers 2002). From these six species and 11 hybrids, five taxa have achieved economic importance for their essential oils: peppermint (*M. ×piperita*), spearmint (*M. spicata*), Scotch spearmint (*M. ×gracilis*), American spearmint (*M. ×villoso-nervata*), and Japanese cornmint or Japanese peppermint (*M. canadensis*, often misidentified as *M. arvensis*) (Lawrence 2007). Pennyroyal (*M. pulegium* L.) has been cultivated on a very minor scale in the past. Efforts directed to breeding in three of the hybrids are hampered by their intense sterility, and thus efforts have been directed to finding their origins. Murray, Lincoln, and Marble (1972) resynthesized *M. ×piperita* from *M. aquatica* × *M. spicata*, while Tucker and Fairbrothers (1990) and Tucker et al. (1991) resynthesized *M. ×gracilis* from *M. arvensis* × *M. spicata*.

Since 1753, more than 3,000 names have been published for the genus *Mentha*, reflecting, in part, the extreme variation encountered in nature and in cultivation. Heterozygosity and hybridization are promoted by genes for gynodioecy. Additionally, not only are plants phenotypically plastic, but the genus displays transgressive segregation in which the hybrids exceed the parents. Complement fractionation, whereby the chromosomes segregate in multiples of the monoploid number ($x = 12$) during meiosis (via cytomixis), produces dosage effects of the genes, in turn fostering transgressive segregation (Tucker and Naczi 2007; Tucker 2009).

21.2 MENDELIAN GENETICS AND CORRELATION OF GENES WITH BIOCHEMICAL PATHWAYS IN *MENTHA*

21.2.1 Breeding for Monoterpenes

One pioneer in the inheritance of essential oil patterns in *Mentha* stands out: Merritt J. Murray. With his associates, he was the very first, beginning in 1954, to apply Mendelian concepts of inheritance and genome analysis comprehensively to any plant with essential oils. Murray was employed at the A. M. Todd Co., Kalamazoo, Michigan. Most of the herbarium vouchers of the crosses by Murray are currently housed in the Claude E. Phillips Herbarium (DOV) at Delaware State University, Dover. Murray attempted to correlate his genes in *Mentha* with proposed biogenetic schemes, particularly the exemplary work originating from the laboratory of Rodney Croteau at Washington State University, Pullman.

The postulated inheritance of the monoterpenes of *Mentha* is intimately connected to their postulated biosynthesis (each providing evidence for the other); quite literally, the story of genetic inheritance cannot be told without the story of biosynthesis and chirality of the constituents of the essential oils. Thus, we have summarized the voluminous literature from over almost a century in Table 21.1. This table tries to separate the evidence from the hypothesis, which is not always easily teased out of the literature, and attempts to undermine assumptions made on flimsy evidence. The work of Murray and his associates allows a genotypic characterization of his clones, which is summarized in Table 21.2. Finally, integrated putative biogenetic pathways, with principal genes and characterized enzymes, are given in Figures 21.1–21.3.

Murray (1972) outlined his genetic methods to determine the inheritance of monoterpenes in *Mentha*:

> Find a strain that allows the conversion to occur and another that prevents or nearly prevents the conversion.
> The gene or genes that inhibit the conversion must be found in a fertile species or extracted into one by backcrossing to the fertile species.
> Colchicine-induced polyploid strains must be made for all sterile species.
> To avoid the great amount of emasculation, male-sterile strains of all principal species and tester strains were developed by substituting one or several male-sterile genes discovered in natural strains.
> A dominant [or recessive or male-sterile] gene from one species may be substituted into another species by four to six backcrosses to determine the effects of the gene on oil biogenesis when the gene is present in a different genotype.

Years of breeding by Murray resulted in the release of the cultivar "Todd 664", sometimes called American lavender (U.S. Plant Patent 2907), a clone of *M. aquatica* var. *citrata* with the proper levels of linalool and linalyl acetate to mimic lavender (Todd and Murray 1968). However, while the A. M. Todd Co. contracted for experimental fields of this in the western United States in the late 1970s, it never achieved the forecasted market potential (Lawrence 2007). The A. M. Todd Co. also released two similar clones: "661" (U.S. Plant Patent 2811) and "66-13" (U.S. Plant Patent 2908); the former was a hybrid of *M. aquatica* var. *citrata* with *M. spicata* var. *crispata* and *M. longifolia*; the latter was derived from *M. aquatica* var. *citrata* and resistant to wilt. Neither attained any economic significance. George Sturtz in Albany, Oregon, also selected an S_1 seedling of *M. spicata* var. *crispata* from Murray (U.S. Plant Patent 8645): 'Erospicata,' a spearmint with low menthol.

Interspecific hybridization has resulted in at least one named clone with potential for commercial plantations. Patra et al. (2001) crossed *M. canadensis* 'Kalka' × *M. spicata* 'Neera' to produce 'Neerkalka'. This hybrid has 64–67% carvone and 0.66–2.5% menthol, disease resistance, and high oil and herb yield.

Kieft Seeds Holland has introduced two clones of *Mentha*—'Snowcones Purple' (U.S. Plant Patent 12971) and 'Snowcones White' (U.S. Plant Patent 12975)—for the U.S. herb plant trade. The claim was made in the patent that these were the result of a breeding program with *M. pulegium* by Pieter den Haan. However, these clones are indistinguishable from naturally occurring color variants of *M. cervina*, rather than *M. pulegium*.

Another avenue to create clones with new complements of monoterpenes has been irradiation by x-rays and gamma rays, primarily led by S. N. Kak and B. L. Kaul at the Regional Research Laboratory, Jammu, India (Kak and Kaul, 1978, 1979, 1980, 1981, 1982a, 1982b, 1984, 1988). Patra, Chauhan, and Mandal (1988) at the Central Institute of Medicinal and Aromatic Plants (CIMAP) at Lucknow, India, also isolated seven promising strains of *M. ×piperita* by gamma irradiation. However, while this method has intense promise, it has not yet resulted in any significant new commercial strains.

21.2.2 Breeding for Disease Resistance

The most devastating disease for commercial mint plantations is caused by the wilt fungus *Verticillium dahliae* Kleb. and secondarily by the rust fungus *Puccinia menthae* Pers., so naturally attention has been focused upon finding resistant strains. The pioneer was Ray Nelson, who began in the early 1940s. Nelson (1947, 1950) surveyed wilt- and rust-resistant strains and collected seeds

Table 21.1. Observations, Hypotheses, and References Relating to the Genetic Inheritance of Monoterpenes in *Mentha*

Observation	Hypothesis	Ref.
Menthone and menthyl esters increase while menthone decreases as *M.* ×*piperita* flowers	Menthone→menthol and esters	Charabot (1900)
Esters and menthol increase as *M.* ×*piperita* flowers	Menthone→menthol and esters	Rabak (1916)
Menthol, menthone, pulegone, and pulegol associated in some plants of *Mentha*; carvone, dihydrocarveol, and carveol associated in some plants of *Mentha*	Theoretical biogenetic scheme proposed for oils of *M. spicata* versus *M.* ×*piperita*	Kremers (1922)
Carvone and menthone never in same plants of *Mentha*; carvone and dihydrocarvone always associated; pulegone and menthone always associated; gene (*C*) for carvone dominant over gene (*c*) for menthone in *M. spicata* var. *crispata*	*C* acts on key intermediate for carvone (versus menthone)	Murray and Reitsema (1954)
Total menthol and esters increase with plant development of *M.* ×*piperita*	Menthone→menthol and esters	Watson and St. John (1955)
[14]C-labeled β,β-dimethyl-acrylic acid inc. into pulegone and phytol in *M. pulegium*; no incorporation ^{14}C-labeled geranic acid	β,β-Dimethyl-acrylic acid→pulegone and phytol	Sanderman and Stockman (1956, 1958)
Pulegone high in young leaves, menthone in older leaves, and menthol in oldest leaves of *M.* ×*piperita*; carvone high in young leaves, dihydrocarvone in older leaves of *M. spicata*	Pulegone→menthone→menthol; carvone→dihydrocarvone	Reitsema et al. (1957)
Distribution of predominant monoterpenes in different species of *Mentha*	Linear intermediate→cyclic intermediate and acylic types; cyclic intermediate→spearmint and peppermint types; carvone→dihydrocarvone; piperitenone→piperitone, pulegone, and piperitenone oxide; piperitone→menthone and piperitone oxide; menthone→menthol; pulegone→menthofuran	Reitsema (1958)
[14]CO₂ to leaf of *M.* ×*piperita* shows pari passu development of pulegone to menthone and menthol; labeled terpenes to tissue slices show pulegone to menthone and piperitenone to piperitone	Geranyl-PP→Blimonene→2-oxygenated monoterpenes; geranyl-PP→terpinolene→3-oxygenated monoterpenes; pulegone→menthone→menthol; piperitenone→piperitone	Battaile (1960)
Menthol and menthyl esters increase as menthone decreases in *M.* ×*piperita*; oils of *M. pulegium*, *M. canadensis*, *M. suaveolens*, *M. spicata*, and *M. longifolia*	Linalool→α-terpineol→limonene→isopiperitenone→piperitenone→ piperitone→piperitone oxide→disophenol; piperitenone→piperitenone oxide→disophenolene; disophenol→disophenol; α-terpineol→1,8-cineole; carvone→dihydrocarvone or carveol→dihydrocarveol	Fujita (1960a)
Oils of *M. aquatica* var. *citrata*, *M. arvensis*, *M.* ×*verticillata*, and *M. spicata*	Theoretical biogenetic pathway proposed for oils of *Mentha*	Fujita (1960b)
Hybrids of *M. suaveolens*, *M. spicata*, *M. arvensis*, and *M. canadensis*	*K*[c] [C] acts on cyclic C₁₀ intermediate→carvone; *K*[d] [c] acts on cyclic C₁₀ intermediate→piperitenone→piperitone; *K*[l] [*P*s] acts on piperitenone→piperitone *K*[t] [o] acts on piperitone→rotundifolone [piperitenone oxide]; *K*[g] [*P*s] acts on pulegone→menthone; *K*[g] [*A*l] acts on piperitone→menthone; *K*[h] [*R*l] acts on menthone→menthol	Ikeda and Shimizu (1960)

Gene (R) for menthol dominant over gene (r) for menthone in M. canadensis; M. canadensis (ccRr) × M. spicata var. crispata (Ccrr)	R acts on reduction menthone→menthol; R cannot convert carvone (C) to menthol; suggested mode of action is that R activates a reductase to reduce product rather than basal substance	Murray (1960a)
C dominant over R, rr, A, and aa; R dominant over cc; cc dominant over A and aa in M. spicata var. crispata, M. spicata, M. longifolia, and M. suaveolens	A acts on piperitenone→pulegone; aa acts on piperitenone→piperitone	Murray (1960b)
$^{14}CO_2$ to leaves of M. ×piperita and M. pulegium shows pari passu menthone to menthol to menthol acetate; labeled terpenes to leaf slices show piperitenone to piperitone, pulegone to menthone, and pulegone to menthofuran	Menthone→menthol→menthyl acetate; piperitenone→piperitone; pulegone→menthone; pulegone→menthofuran	Battaile and Loomis (1961)
Oils of M. suaveolens, M. ×verticillata, and M. ×piperita	Limonene→isopiperitenone→piperitenone→piperitone; isopiperitenone→neoisopulegol; piperitenone→piperitenone oxide→piperitone oxide; piperitone→piperitone oxide; isopiperitenone→isopulegone or isopiperitenol→neoisopulegol; carvone→dihydrovarvone or carveol; linalool→(−)-α-terpineol; (−)-limonene→(−)-carvone; (−)-limonene→isopiperitenone; piperitenone→pulegone	Fujita (1961a)
Oil of M. canadensis	Caprylic acid→n-heptane and β-oxycarpulic acid→3-octanone	Fujita (1961b)
^{14}C-labeled terpenes show menthone to menthol and pinene and limonene to menthol, pulegone, and menthone in M. ×piperita	Menthone→menthol	Reitsema et al. (1961)
^{14}C-labeled acetate shows slow conversion of menthone to menthol in M. ×piperita	Menthone→menthol	Campbell (1962)
Oils of M. ×piperita and M. canadensis	Piperitone→2-isopropylcyclopentanone; pulegone→1-methyl-3-cyclohexanone	Fujita (1962)
Older leaf oil of M. aquatica has more hydrogenated compounds	→Hydrogenated compounds with age of leaf	Hefendehl (1962)
Oil of M. canadensis	(+)-Piperitone or (+)-pulegone→(+)-isomenthone	Shimizu and Ikeda (1962)
Oils of M. ×piperita and M. spicata	Extension of Fujita (1961a)	Fujita (1963a)
Oil of M. pulegium	Piperitenone→pulespenone [p-mentha-4(8),6-dien-3-one]	Fujita (1963b)
Oil of M. aquatica var. citrata	Linalool→citral or linalyl acetate	Fujita (1963c)
Oils of M. spicata, M. suaveolens, and M. pulegium	Extension of Fujita (1961a) (limonene→isopiperitenone→piperitenone or isopulegone, etc.); (linalool→sabinene hydrate, α- and β-pinene; nerolidol→sesquiterpenes)	Fujita (1965a, 1965b)
^{14}C-labeled acetate and mevalonate fed to leaves of M. ×piperita show pari passu piperitenone to piperitone and pulegone, piperitone to menthone to menthol to menthyl actate, and pulegone to menthofuran	Piperitenone→piperitone and pulegone; piperitone→menthone→menthol→menthyl acetate; pulegone→menthofuran	Battu and Youngken (1966)

(continued)

Table 21.1 Observations, Hypotheses, and References Relating to the Genetic Inheritance of Monoterpenes in *Mentha* (Continued)

Observation	Hypothesis	Ref.
Oils of many species of *Mentha*	Theoretical biogenetic scheme proposed for oils of *Mentha* based on stereochemical considerations	Katsuhara (1966)
Short nights or cool nights enhance menthofuran and depress accumulation of menthofuran or pulegone in *M. ×piperita*	Pulegone→menthone	Burbott and Loomis (1967)
$^{14}CO_2$ fed to *M. ×piperita* shows pari passu limonene to 3-oxygenated monoterpenes	Limonene→3-oxygenated monoterpenes	Hefendehl (1967a)
^{14}C-labeled acetate, mevalonate, and CO_2 fed to shoots of *M. canadensis* and *M. ×piperita* show pari passu piperitenone to piperitone to menthone to menthol and piperitone to menthone to menthofuran; also evidence for unknown precursor to isomenthone to menthone to menthol	Piperitenone→piperitone→menthone→menthol; piperitenone→pulegone→menthone and menthofuran; isomenthone→menthone→menthol	Hefendehl, Underhill, and von Rudloff (1967)
Review of previous work	Gene difference of carvone and pulegone is one determining double bond; intercoversions are nonspecific enzymes that act readily on terpinolene and carvone but cannot handle double bonds of limonene and carvone; α-terpineol or carbonium ion key step in biosynthesis of cyclic monoterpenes	Loomis (1967)
Oil of *M. ×piperita*, *M. aquatica*, and *M. spicata*	Theoretical biogenetic scheme proposed	Shimizu (1967)
^{14}C-labeled pulegone fed to cell-free extracts of *M. ×piperita*, *M. canadensis*, and *M. longifolia* shows conversion of pulegone to menthone and isomenthone; $NADPH_2$ essential cofactor	Pulegone→menthone and isomenthone	Battaile et al. (1968)
Hybrids of *M. aquatica*, *M. arvensis*, and *M. japonica*	Extension of Fujita (1961a); geranyl-PP→linalyl-PP→limonene equivalent; K^c [C] acts on cyclic C_{10} intermediate→carvone; K^t [P^s] acts on piperitenone→(+)-pulegone or (+)-piperitenone; K^g [P^s and A^r] acts on (+)-pulegone→(−)-menthone or (+)-piperitone→(−)-menthone; K^h [R^r] acts on (−)-menthone→(−)-menthol	Fujita (1968)
Review of literature	Theoretical biogenetic scheme proposed for oils of *Mentha*	Rothbächer (1968)
Menthol increases pari passu with decrease of menthone, while menthyl acetate increases pari passu with decrease of menthol in *M. ×piperita*	Ion→menthone→menthol→menthyl acetate	Schantz and Norri (1968)
Geraniol in cell-free extracts of *M. ×piperita* shows conversion to geranyl-P and geranyl-PP	Geraniol→geranyl-P and geranyl-PP	Madyastha and Loomis (1969)
M. pulegium shows (+)-pulegone in young leaves; (−)-menthone and (+)-isomenthone later, and later (−)-menthol	(+)-Pulegone→menthone→(+)-isomenthone→(−)-menthol	Fujita and Fujita (1970)
Oil of *M. pulegium*	(−)-Piperitinone and (−)-pulegone→(+)-isomenthone and (−)-menthone	Hefendehl (1970)
1,2-Epoxymenthyl acetate increases in age of leaves in *M. suaveolens*	Menthone and isomenthone→menthol and neomenthol	Hendriks (1970)

Menthol and neomenthol increase while menthone and isomenthone decrease as leaves of M. ×piperita age	Menthone and isomenthone→menthol and neomenthol	Manning (1970)
Gene (I) for linalool dominant over gene (i) for cyclic ketones and alcohols in M. aquatica var. citrata	I acts on precursor→linalool; ii acts on precursor→cyclic ketones and alcohols	Murray and Lincoln (1970)
Glucose-U-^{14}C efficient precursor for monoterpenes but not sesquiterpenes; mevalonate-2-^{14}C poor precursor for monoterpenes but readily incorporated into sesquiterpenes; rapid turnover for both labels	Separate sites mono- and sesquiterpene biosynthesis; endogenous DMAPP pool	Croteau, Burbott, and Loomis (1971)
Gene (Lm) for limonene dominant over gene (lm) for 2- and 3-oxygenated monoterpenes in M. aquatica var. citrata; l dominant over Lm in M. spicata, M.longifolia, M. spicata var. crispata, and M. aquatica var. citrata; C and cc interact with Lm and lm	Lm acts on precursor→limonene; Lm prevents formation of 2-oxygenated and 3-oxygenated compounds; l intercepts oil biogenesis at earlier stage than Lm; lmc acts on precursor→3-oxygenated terpenes; lmC acts on precursor→2-oxygenated terpenes	Lincoln et al. (1971)
Gene (F) for pulegone dominant over gene (f) for menthofuran; gene (P) for (−)-menthone dominant over gene (p) for pulegone in M. gattefossei and M. pulegium	ff acts on pulegone→menthofuran; P acts on pulegone→(−)-menthone; pp prevents reduction pulegone→menthone; F limits oxidation pulegone→menthofuran	Murray, Marble, and Lincoln (1971)
Extensive decarboxylation of MVA; 0.05% MVA-2-^{14}C maximum incorporation into M. pulegium and no inc. MVA-1-^{14}C; ^{14}C-geraniol and ^{14}C-labeled C-5 precursors derived from MVA-2-^{14}C little degraded to CO_2	MVA→geraniol	Banthorpe and Charlwood (1972)
^{14}C-MVA fed to shoots of M. pulegium shows labeling in IPP portion of (+)-pulegone	MVA→terpinolene→piperitenone→(+)-pulegone	Banthorpe, Charlwood, and Young (1972)
^{14}C-label in IPP portion of pulegone essentially unlabeled in M. ×piperita derived from glucose-^{14}C and $^{14}CO_2$; metabolic turnover of monoterpenes	Mono- and sesquiterpenes produced separate sites; endogenous DMAPP pool used in monoterpene biosynthesis	Croteau, Burbott, and Loomis (1972a)
MVA-2-^{14}C incorporated mono- and sesquiterpenes of M. ×piperita when fed to cuttings; enhanced with additional sucrose or 5% CO_2 + light or Na acetate but proportion of label in mono- and sesquiterpenes shifted	Mono- and sesquiterpenes produced separate sites; MVA "spared" rather than satisfying energy requirement when extra energy supplied	Croteau et al. (1972b)
MVA-2-^{14}C incorporated into caryophyllene and other sesquiterpenes more extensively than into monoterpenes in M. ×piperita	Separate sites biosynthesis mono- and sesquiterpenes	Croteau and Loomis (1972)
M. gattefossei shows piperitenone in young leaves, (+)-pulegone and piperitone in older leaves, and (−)-menthone later	Piperitenone→(+)-pulegone→(−)-menthone; piperitenone→piperitone	Fujita et al. (1972)
Gene (R) for menthone, carvone, or dihydrocarvone dominant over gene (r) for menthol, carveol, or dihydrocarveol in M. aquatica; interaction of R with C, A, F	R acts on menthone→menthol; R acts on carvone→carveol; R acts on dihydrocarvone→dihydrocarveol	Hefendehl and Murray (1972)
Gene (C) for carvone dominant over gene (c) for 3-oxygenated monoterpenes in M. suaveolens and M. longifolia	Supports Murray and Reitsema (1954)	Hendriks and van Os (1972)

(continued)

Table 21.1   Observations, Hypotheses, and References Relating to the Genetic Inheritance of Monoterpenes in *Mentha* (Continued)

Observation	Hypothesis	Ref.
Review of previous work	Summarizes genes and biogenesis of monoterpenes	Murray (1972)
Senescent and overmature foliage more dihydrocarvone than mature foliage in *M. spicata*, *M. ×gracilis*, and *M. spicata* var. *crispata*; gene *E* allows rapid conversion of all spearmint alcohols to esters	Carvone→dihydrocarvone; *E* allows carveol→carvyl esters and dihydrocarveol→dihydrocarvyl esters	Murray, Faas, and Marble (1972a)
Mature herbage more menthol and menthyl acetate than juvenile herbage in *M. canadensis*	Menthone→menthol→menthyl acetate	Murray et al. (1972b)
Genotype *FF* allows less than 0.1% menthofuran, *Ff* allows 0.4–25% menthofuran, and *ff* allows 60–80% menthofuran in *M. canadensis* and *M. aquatica*	*ff* acts on pulegone→menthofuran; *F* is a single incompletely dominant gene	Murray and Hefendehl (1972)
Gene (*O*) for piperitone dominant over gene (*o*) for piperitone oxide in *M. longifolia*	Genes *CAFR* further confirmed in *M. aquatica* and *M. longifolia*; *oo* acts in piperitone→piperitone oxide	Murray and Lincoln (1972)
Hybrids of *M. aquatica* and *M. spicata*	Confirms genes *ACFPR*	Murray et al. (1972)
Hybrids of chemical selections of *M. spicata*	Confirms genes *A* and *C*	Shimizu, Karasawa, and Ikeda (1972)
Cell-free systems of *M. ×piperita* and *M. spicata* convert neryl-PP to (−)-(4S)-α-terpineol and nerol	Neryl-PP→(−)-(4S)-α-terpineol key reaction in synthesis of cyclic monoterpenes	Croteau, Burbott, and Loomis (1973)
High limonene strain of *M. ×piperita*	Confirms genes *ACFILmPR*	Hefendehl and Murray (1973a)
Review of previous work	Summarizes genes and biogenesis of monoterpenes	Hefendehl and Murray (1973b)
Sesquiterpenes inc. more exogenous MVA-2-^{14}C, while monoterpenes inc. more glucose-^{14}C, sucrose-^{14}C and phs-fixed $^{14}CO_2$; metabolic turnover in few hours; MVA-2-^{14}C inc. into IPP-half of monoterpenes	Separate sites monoterpene and sesquiterpene biosynthesis; endogenous pool of DMAPP	Loomis and Croteau (1973)
Gene (*C*) for carvone dominant over gene (*c*) for limonene; gene (*lm*) for acyclic and 2-oxygenated monoterpenes dominant over gene (*lm*) for 3-oxygenated monoterpenes in *M. aquatica*	Supports previous work on genes *c* and *lm* and indicates a close coupling phase in the linkage of these two genes	Murray and Hefendehl (1973)
Selfs of *M. spicata*	Confirms gene *C*	Bugaenko and Reznickova (1974)
Enzyme system in *M. ×piperita* converts geraniol and geranyl-PP to nerol and neryl-PP with FAD and reducing atmosphere required; cell-free preparation of *M. ×piperita* and *M. spicata* converts nerol and neryl-PP to α-terpineol and terpinen-4-ol, α-terpineol to terpinolene in *M. ×piperita* and to limonene in *M. spicata*; cell-free	Geraniol and geranyl-PP→nerol and neryl-PP; nerol and neryl-PP→α-terpineol and terpinene-4-ol; α-terpineol→terpinolene or limonene; pulegone→isomenthone or menthone; menthone→menthol; piperitenone→piperitone	Burbott et al. (1974)

preparation of M. ×piperita converts pulegone to isomenthone or menthone to menthol and piperitenone to piperitone		
Crosses of M. suaveolens and M. longifolia	Confirms genes C and R	Hendriks (1974)
Selfs of M. spicata and hybrids of M. ×piperita	Confirms genes C and R	Reznickova et al. (1974)
Oils of many species of Mentha	Theoretical biogenetic scheme proposed for oils of Mentha	Shimizu (1974)
Cell-free system catalyzed trans-cis isomerization of geraniol and geranyl-PP to nerol and neryl-PP in presence of FAD or FMN, a thiol, or sulfide and light; reversible but equilibrium favors geraniol and geranyl-PP in M. ×piperita	Neryl-PP→geranyl-PP	Shine and Loomis (1974)
$^{14}CO_2$ uptake and degradation (+)-pulegone shows C-2 units labeled widely different extents in M. pulegium	Metabolic pools of acetyl CoA and/or acetoacetyl-CoA	Banthorpe et al. (1975)
Review of previous literature and work on Mentha and other genera in Lamiaceae	IPP + DMAP→neryl-PP or geranyl-PP; geranyl-PP→neryl-PP Neryl-PP or linalyl-PP→carbonium ion→terpinolene, α-terpineol, or limonene Carbonium ion→bornane skeleton, pinane skeleton, or thujane skeleton; pinane skeleton→fenchane skeleton Neryl-PP→α-terpineol→limonene→carvone→dihydrocarvone; α-terpineol→terpinolene→piperitenone→piperitone→(+)-isomenthone→(+)-isomenthol or (+)-neoisomenthol; piperitenone→(+)-pulegone→(−)-menthone→(+)-neomenthol or (−)-menthyl esters; (+)-pulegone→(+)-menthofuran or (+)-isomenthone	Croteau and Loomis (1975)
Gene for piperitenone (O) dominant over gene (o) for piperitenone oxide in crosses of M. suaveolens and M. longifolia	Confirms gene O Mp [A] and mp [a] act on piperitenone→piperitone	Hefendehl and Nagell (1975)
Gene for piperitenone (Mp) dominant over gene (mp) for piperitone in crosses of M. suaveolens and M. longifolia		
Selfs of M. spicata, M. spicata var. crispate, and hybrids with M. arvensis	Confirms genes ACILmOPR nn acts on precursor→1,8-cineole (with ii)	Tucker (1975); Tucker et al. (1991)
Gene for 1,8-cineole (n) recessive		
Partial degradation of (+)-pulegone after uptake of 3,3-dimethylacrylate-[Me-14C] in M. pulegium not a direct source of monoterpene skeleton; l-valine-[U-]14C and l-leucine-[U-14C] results in negligible incorporation tracer into (+)-pulegone	Two metabolic pools of intermediates for monoterpene biosynthesis, one-protein bound	Allen et al. (1976)
Review of previous work	Summarizes previously reported genes	Hefendehl and Murray (1976)
Crosses of chemotypes of M. suaveolens	Gene C epistatic over A	Hendriks and van Os (1976)
Increase neo- and dihydrocarveol and acetates with maturation of leaves in one chemotype; 1,2-epoxymenthyl acetate increases with maturation of leaves in other chemotype		

(continued)

Table 21.1 Observations, Hypotheses, and References Relating to the Genetic Inheritance of Monoterpenes in *Mentha* (Continued)

Observation	Hypothesis	Ref.
Seasonal variation of oils of *M. spicata* var. *crispata*	Carvone→dihydrocarvone→neodihydrocarvyl acetate; carvone→dihydrocarvone→dihydrocarvyl acetate; carvone→carveol→*cis*-carvyl acetate	Nagasawa et al. (1976)
Carvone polyploidy strain of *M. longifolia*	*oo* converts piperitone to piperitone oxide in *M. longifolia*; cannot cause epoxidation of carvone	Hefendehl (1977)
Review of previous work	Summarizes previous research	Hendriks (1977)
	Diosphenolene→carvone or piperitenone or piperitenone oxide; carvone→dihydrocarvone; piperitenone→isopiperitenone→piperitone→menthone; piperitenone→pulegone→menthone; pulegone→isopulegone; piperitenone oxide→piperitone oxide	Nickolaev et al. (1977); Nickolaev (1978, 1979, 1982); Nickolaev and Peljah (1981)
Hybrids of *M. canadensis*, *M. longifolia*, *M. spicata*, and *M. aquatica*	*C* epistatic over *A* No chi-square analysis but supports previous reports of genes	
Mentha callus cultures convert pulegone to isomenthone	Pulegone→isomenthone	Aviv and Galun (1978)
[14]C- and ^{3}H-geraniol and nerol fed to foliage of *M. spicata* and *M. ×piperita*	Questionable results, as 1,8-cineole reported as principal constituent of *M. ×piperita*	Banthorpe et al. (1978)
Review of previous literature on *Mentha* and other genera of Lamiaceae	Neryl-PP→1,8-cineole; neryl-PP→γ-terpinene; neryl-PP→bornyl-PP→borneol→camphor; neryl-PP→(+)-bornyl-PP	Croteau (1978)
[3]H-menthol converted to menthyl acetate in leaf discs and leaf protein n.w. 37,000 in *M. ×piperita*; neryl Co-A-dependent acetylation	Menthol→menthyl acetate	Croteau and Hooper (1978)
Analysis of *M. dahurica* with 30% α-terpineol and *M. ×verticillata* with 75% α-terpinyl acetate; 1,8-cineole and α-terpineol inversely related but co-occur with limonene, ocimene, and acyclic terpenes; isopiperitenone and terpinolene only in minor proportions in *Mentha* oils; piperitenone, piperitone, and pulegone can coexist; high proportions of (+)-(1*R*)-pulegone and (−)-(1*R*:4*S*)-menthone coexist but (+)-(4*S*)-piperitone and (−)-(1*R*:4*S*)-menthone do not co-occur; *cis*- and *trans*-dihydrocarvone can co-occur; (+)-(4*S*)-piperitone exists in *Mentha* but confused with (−)-(4*R*)-piperitone, which does not occur, so piperitone cannot be a precursor of isomenthone; neither diosphenolene nor diosphenol native constituents of *Mentha* oils; (+)-(1*R*)-pulegone and (+)-(1*R*:4*S*)-isomenthone can coexist; *cis*-isopulegone, *trans*-isopulegone, and pulegone can coexist;	*T*[z] (cyclase) acts on neryl-PP→(−)-(4*S*)-α-terpineol; nery-PP→1,8-cineole; *Lm* acts on neryl-PP-limonene; *L*[i] [*lm*?] acts on neryl-PP→terpinolene; disophenolene and disophenol not in biosynthetic pathway; *V* (oxidase) acts on carbonium ion→terpinolene→piperitenone or carbonium ion→piperitenone; *C* (oxidase) acts on carbonium ion→(−)-carvone or limonene→(−)-carvone; *A*[i] (reductase) acts on piperitenone→(+)-pulegone and (+)-piperitone→(−)-menthone and (−)-(4*S*)-carvone→(+)-*trans*-(1*R*:4*S*)-dihydrocarvone; *P*[s] (reductase) acts on piperitenone→(+)-piperitone and (+)-pulegone→(−)-menthone; *A*[s] (reductase) acts on (−)-(4*S*)-carvone→(+)-*cis*-(1*S*4*S*)-dihydrocarvone; *P*[r] (reductase) acts on (+)-(1*R*)-pulegone→(+)-(1*R*:4*R*)-isomenthone; *H*[r] (isomerase) acts on (+)-(1*R*)-pulegone→(+)-(1*R*:4*R*)-isopulegone; *H*[s] (isomerase) acts on (+)-(1*R*)-pulegone→(−)-*trans*-(1*R*:4*S*)-isopulegone; *f* (oxidase) acts on (+)-*cis*-(1*R*)-pulegone→(+)-(1*R*)-menthofuran; *R*[r](reductase) acts on (+)-(1*R*:4*R*)-isomenthone→(+)-(1*R*:3*R*:4*R*)-	Lawrence (1978, 1979, 1981)

menthofuran, pulegone, and cis- and trans-isopulegone can coexist but no examples of the co-occurrence of menthofuran and isopiperitenone are known; crosses of M. canadensis, M. spicata, and M. aquatica; piperitenone oxide a major component in some oils; cis- and trans-piperitone oxide can coexist; 1,2-epoxymenthyl acetate, 1,2-epoxyneomenthol, and 1,2-epoxymenthyl acetate coexist with piperitenone oxide; diosphenol and diosphenolene are artifacts of acid hydrolysis during distillation	neoisomenthol or (−)-(1R:4S)-menthone→(−)-(1R:3R:4S)-menthol or (+)-cis-(1R:4R)-isopulegone→(+)-(1R:3R:4R)-neoiso(iso)pulegol or (−)-trans-(1R:4S)-isopulegone→(−)-(1R:3R:4S)-isopulegol or (−)-(4S)-carvone→(−)-(2S:4S)-carveol or (+)-cis-(1S:4S)-dihydrocarveol or (1S:2S:4S)-neoisodihydrocarveol or (+)-trans-(1R:4S)-dihydrocarvone→(+)-(1R:2S:4S)-dihydrocarveol or (−)-trans-(1S:2S:4S)-piperitone oxide→(1S:2S:3S:4S)-1,2-epoxymenthol; Rˢ (reductase) acts on (+)-(1R:4R)-isomenthone→(+)-(1R:3S:4R)-isomenthol or (−)-(1R:4S)-menthone→(+)-(1R:3S:4S)-neomenthol or isomenthol or (−)-(1R:4R)-isopulegone→(1R:3S:4R)-iso(iso)pulegol or (−)-trans-(1R:4S)-isopulegone→(1R:3S:4S)-neoisopulegol or (−)-(4S)-carvone→(+)-trans-(2R:4S)-carveol or (+)-cis-(1S:4S)-dihydrocarvone→(+)-(1S:2R:4S)-isodihydrocarveol or (+)-trans-(1R:4S)-dihydrocarvone→(+)-(1R:2R:4S)-neodihydrocarveol or (−)-trans-(1S:2S:4S)-piperitone oxide→(+)-1,2-epoxyneomenthol; E (esterase) acts on alcohols→esters; oo acts on (+)-(4S)-piperitone→(−)-trans-(1S:2S:4S)-piperitone oxide or piperitenone→(+)-(1S:2S)-piperitenone oxide; Pⁱ acts on (+)-(1S:2S)-piperitenone oxide→(−)-cis-(1S:2S:4R)-piperitone oxide; Bˢ (synthetase) acts on neryl-PP→bornyl-PP; l (synthetase) acts on neryl-PP→linalyl-PP	Lincoln and Murray (1978)
Crosses of M. canadensis with other species	P dominant or epistatic; acts (+)-pulegone→(−)-menthone; accumulation of pulegone occurs when lacking P for pulegone→menthone and ff for pulegone→menthofuran	Croteau and Martinkus (1979)
(−)-[G-³H]menthone converted to (−)-[³H]menthol in leaf discs of M. ×piperita but major portion of tracer in (+)-neomenthyl-β-d-glucoside	(−)-Menthone→(−)-menthol and (+)-neomenthyl-β-d-glucoside	Reznickova and Bugaenko (1979)
Selected selfs of M. spicata and hybrids of M. spicata × M. ×piperita	Confirms genes Aa, f, R, and P; Pⁱ [Aⁱ] acts on piperitone→menthone; Im acts on piperitone→isomenthone	Umemoto and Nagasawa (1979)
Seasonal variation of oil of M. ×gracilis	cis-Pulegol→8-hydroxy-3-p-menthone	
Degradation of (−)-carvone in M. spicata from ³H- and ¹⁴C-labeled geraniol and MVA	Oxidation of limonene or equivalent→carvone involves shift of endocyclic double bond; limonene and carvone biogenetically related but probably on separate pathways; exocyclic double bond of carvone is not formed regiospecifically	Akhila, Banthorpe, and Rowan (1980)
Review of previous literature and work in Mentha and other genera of Lamiaceae	DMAPP + IPP→geranyl-PP; geranyl-PP + IPP→farnesyl-PP→(+)-bornyl-PP→(+)-borneol→(+)-camphor; geranyl-PP→terpinolene→piperitenone→(+)-pulegone; (−)-menthone→(−)-menthol; dehydrogenase acts on (−)-menthone→(−)-menthol or (+)-neomenthol; acetyl transferase acts on (−)-menthol→(−)-menthyl acetate; glucosyl transferase acts on (+)-neomenthol→(+)-neomenthyl glucoside; sesquiterpene biosynthetic routes postulated	Croteau (1980a, 1980b)

(continued)

Table 21.1 Observations, Hypotheses, and References Relating to the Genetic Inheritance of Monoterpenes in *Mentha* (Continued)

Observation	Hypothesis	Ref.
Pulegone high in *M. ×piperita* harvested early in season, with menthone and menthol increasing later	Pulegone→menthone and menthol	Farley and Howland (1980)
Ontogenetic variation of oil of *M. spicata*	(–)-Carvone→(+)-dihydrocarvone afer blooming; (+)-dihydrocarvone→(+)-neodihydrocarveol and (–)-dihydrocarveol (3:1)→acetates	Fujita and Nezu (1980)
Crosses of *M. longifolia × M. spicata* var. *crispata*	P^r and P^s incompletely dominant alleles at single locus; P^r acts on (+)-pulegone→(+)-isomenthone or piperitenone→(–)-piperitone; P^s acts on (+)-pulegone→(–)-menthone or piperitenone→(+)-piperitone	Murray, Lincoln, and Hefendehl (1980)
Seasonal variation of oil of *M. canadensis*	Menthyl glucoside highest at preflowering; menthone highest at first half of preflowering; menthol increases with age of leaf until onset of flowering	Sakata and Koshimizu (1980)
Seasonal variation of oil of *M. spicata* var. *crispata*	(–)-Limonene→(–)-carvone; (–)-carvone→carveol→carvyl acetate; dihydrocarvone→(–)-dihydrocarveol→dihydrocarvyl acetate; dihydrocarvone→neodihydrocarveol	Umemoto and Nagasawa (1980)
(–)-Menthone fed to cell lines of *M. aquatica, M. aquatica* var. *citrata, M. ×piperita, M. ×villosa,* and *M. canadensis* transformed to (+)-neomenthol; none converted (+)-isomenthone to corresponding alcohol	(–)-Menthone→(+)-neomenthol	Aviv et al. (1981)
(–)-[3-³H]menthol and (+)-[3-³H]neomenthol fed to leaf discs of *M. ×piperita* form menthyl and neomenthyl acetates and glucosides with equal facility; same substrates with acetyl Co-A:monoterpenol acetyltransferase produced similar results	Compartmentalization of enzyme activities	Martinkus and Croteau (1981)
Seasonal variation of oil of *M. canadensis*	Confirms Sakata and Koshimizu (1980); theoretical biosynthetic schemes proposed	Sakata et al. (1981)
Compartmentalization (–)-menthone metabolism reveals neomenthol dehydrogenase and glucosyl transferase in mesophyll layer and menthol dehydrogenase and acetyl transferase in epidermis; compartmentation intercellular, not intracellular	(–)-Menthone→(–)-menthol and (–)-menthyl acetate in epidermis; (–)-menthone→(+)-neomenthol and neomenthyl glucoside in mesophyll	Croteau and Winters (1982)
(–)-[G-³H]menthone incorporated with purified enzymes from leaves of *M. ×piperita* converted to (–)-menthol and (–)-menthyl acetate and (+)-neomenthol and (+)-neomenthyl-β-glucoside; NAPDH-dependent dehydrogenases	Menthol dehydrogenase compartmentalization with acetyl transferase, neomenthol dehydrogenase with glucosyl transferase; (–)-menthone→(–)-menthol and (–)-menthyl acetate or (+)-neomenthol and (+)-neomenthyl-β-d-glucoside	Kjonaas, Martinkus-Taylor, and Croteau (1982)
Hybrids of *M. canadensis* and *M. ×piperita* with *M. spicata*	Confirms genes *A, P, P^t, Im,* and *R*	Reznickova, Bugaenko, and Melnickov (1982)
Menthol increases as menthone decreases in age of development in *M. canadensis*	Menthone→menthol	Abad Farooqi, Misra, and Naqvi (1983)

(+)-Pulegone fed to cell lines of *M. aquatica, M. aquatica var. citrata, M. ×piperita, M. ×villosa,* and *M. canadensis* transformed to (+)-isomenthone in only some cell lines	(+)-Pulegone→(+)-isomenthone	Aviv et al. (1983)
Piperitone from *M. ×piperita* is 76–85% (+)-(4S)-, remaining (−)-(4R)-	Piperitenone→(+)-piperitone preferentially	Burbott et al. (1983)
Limonene in immature leaves of *M. ×piperita* is ~80% (−)-(4S)- and ~20% (+)-(4R); incorporation [U-^{14}C]sucrose into shoot tips show limonene converted to pulegone to menthone; (±)-[9-^{3}H]limonene readily incorporated into pulegone, menthone, etc. In *M. ×piperita* and carvone in *M. Spicata*; [9,10-^{3}H]terpinolene was not; soluble enzymes from epidermis of immature leaves of *M. ×piperita* convert [1-^{3}H]geranyl-PP to limonene and no free intermediates detected	Geranyl-PP→limonene→carvone or geranyl-PP→limonene→pulegone→menthone; no terpinolene involved	Kjonaas and Croteau (1983)
^{14}C- and ^{3}H-labeled linalyl-PP, geranyl-PP, and neryl-PP in cell-free extract of *M. spicata* show most label in monoterpenoids from linalyl-PP	Linalyl-PP most pivotal and immediate precursor for cyclic monoterpenoids	Suga et al. (1983)
Selfs of *M. spicata, M. longifolia,* and *M. aquatica*	Confirms genes *C* and *I*	Bugaenko and Reznickova (1984a)
Hybrids of high menthol species with wild-growing species of *Mentha*	Confirms gene *R*	Bugaenko and Reznickova (1984b)
(−)-[G-^{3}H]menthone applied to midstem of *M. ×piperita* shows transport to rhizomes; (+)-[G-^{3}H]neomenthyl-β-d-glucoside on excised rhizomes shows hydrolysis; (+)-[G-^{3}H]neomenthol on excised rhizomes shows lactonization; confirmed *in vivo*	(+)-Neomenthyl glucoside→(+)-neomenthol→(−)-menthone→(−)-3,4-menthone lactone	Croteau et al. (1984)
Selfs of *M. spicata* and hybrids of *M. aquatica × M. spicata*	Confirms genes *C, R*	Reznickova, Bugaenko, and Makarov (1984)
(+)-[G-^{3}H]neomenthol converted to acyl and isoprenoid lipids and water-soluble products in rhizomes of *M. ×piperita*	(+)-Neomenthol degraded to acetyl CoA (via (−)-3,4-menthone lactone) and carbohydrates in rhizomes	Croteau and Sood (1985)
Mentha cells arrested in cell division by irradiation convert (−)-menthone to (+)-neomenthol	(−)-Menthone→(+)-neomenthol	Galun et al. (1985)
Soluble enzyme preparations from leaves of *M. ×piperita* oxidize isopiperitenol to isopiperitenone and isomerize isopiperitenone to piperitenone; dehydrogenase and isomerase m.w. 66,000 and 54,000, respectively; NAD-dependent dehydrogenase	Geranyl-PP→(−)-limonene→(−)-*trans*-isopiperitenol→(−)-isopiperitenone→piperitenone→(+)-pulegone→(−)-menthone→(−)-menthol; (+)-pulegone→(+)-neomenthol; (+)-pulegone→(+)-isomenthone→(+)-isomenthol or (+)-neoisomenthol	Kjonaas, Venkatachalam, and Croteau (1985)
Hybrids of *M. canadensis × M. spicata*	Confirms gene *R*	Reznickova, Bugaenko, and Rodov (1985)
Review of previous literature in *Mentha* and other genera of Lamiaceae; discusses catabolism	Reviews and extends previous work	Croteau (1986a, 1986b)

(continued)

Table 21.1 Observations, Hypotheses, and References Relating to the Genetic Inheritance of Monoterpenes in *Mentha* (Continued)

Observation	Hypothesis	Ref.
[3H]piperitenone fed to leaf discs of *M. ×piperita* converted to (+)-piperitone; (−)-[3H]isopiperitenone converted to (+)-*cis*-isopulegone, (+)-pulegone, (−)-menthone, and (+)-isomenthone with poor labeling in piperitenone; (+)-*cis*-[3H]isopulegone converted to (+)-pulegone, (−)-menthone, and (+)-isomenthone; (+)-[3H]pulegone converted to (−)-menthone and (+)-isomenthone	(−)-Isopiperitenone→piperitenone and (+)-*cis*-isopulegone→(+)-pulegone→(−)-menthone and (+)-isomenthone	Croteau and Venkatachalam (1986)
[3H] almost entirely prevents development of monoterpenes with *p*-menthane carbon skeleton substituted at 2- and/or 3-position; *I*s and *I*s linked	*I*s acts on β-pinene→isopinocamphone; *A*/*a*s controlling α-terpineol may be in linkage group	Lincoln, Murray, and Lawrence (1986)
[2-14C,5-3H2]MVA and [1-14C,1-3H2]geraniol fed to *M. spicata* and *M. ×piperita* reveal interconversions of acyclic-PP	Geranyl-PP, neryl-PP, and linalyl-PP interconvertible	Suga et al. (1986)
Chiral analysis of oils of *M. ×piperita*	100% (−)-Menthol, 99–100% (+)-neomenthol, 99% (+)-neoisomenthol, 99% (+)-isomenthol, 99–100% (−)-menthone, 100% (+)-isomenthone	Werkhoff and Hopp (1986)
Menthol, menthyl acetate, and menthofuran incorporated as *M. ×piperita* matures while menthone decreases	Menthone→menthol, menthyl acetate, and menthofuran	White, Iskandar, and Barnes (1987)
Suspension cultures of *M. ×piperita* show 36.5% glycosylation of (−)-menthol and 44.8% glycosylation of (+)-menthol	Glycosylation may facilitate the transport to catabolic enzymic sites	Berger and Drawert (1988)
Review of catabolism of monoterpenes	Reviews and extends previous work	Croteau (1988)
Glandular trichomes of leaves of *M. spicata* have almost all cyclase and hydrolase activities; 30% of carveol dehydrogenase activity throughout leaf	Carvone biosynthesis takes place exclusively in glandular trichomes	Gershenzon, Maffei, and Croteau (1988, 1989); Maffei, Gershenzon, and Croteau (1988)
Essential oil and age of *M. canadensis*	Menthone→menthol→menthyl acetate	Singh (1988)
Oil of *M. canadensis*	Piperitenone→(+)-piperitenone oxide→(−)-piperitone oxide	Umemoto and Tsuneya (1988)
Young shoots of *M. spicata* fed [14C-1-3H2]-labeled geraniol, nerol, (1R)-geraniol, or (1S)-geraniol showed maximum uptake of geraniol and nerol into carvone and limonene with day/night temperature of 6°/18°C	Redox process controlled by NADH and NAD+ or other cofactors and activities of appropriate enzymes	Akhila and Thakur (1989)
Microsomal preparation from epidermal oil glands of *M. ×piperita* catalyzes NADPH- and O2-dependent allylic hydroxylation of (−)-limonene to (−)-*trans*-isopiperitenol from *M. spicata* to (−)-*trans*-carveol	(−)-Limonene→(−)-*trans*-isopiperitenol or (−)-*trans*-carveol	Karp et al. (1990)
Hybrids of *M. ×verticillata* × *M. spicata*	Confirms genes *A*r, *C*, *E*, *o*, *P*r, *P*s, and *R*r	Maffei (1990)
Crosses of *M. canadensis*, *M. longifolia*, *M. spicata*, and *M. aquatica* var. *citrata*	No chi-square analysis; proposed theoretical scheme of monoterpene biosynthesis	Nickolaev, Pysova, and Lollo (1990)

Glucoside of *M. canadensis*	(−)-Menthyl 6′-O-acetyl-β-d-glucoside	Shimizu, Shibata, and Maejima (1990)
Glucosides of *M. ×gracilis*	Dihydrocarvyl-, neodihydrocarvyl-, *trans*-carvyl-, *cis*-carvyl-, and 3-octyl-β-d-glucosides	Shimizu et al. (1990)
Isotope ratios of menthone, menthol, (−)-mintlactone, and (−)-isomintlactone biosynthesized in *M. ×piperita* from [2-^{14}C,5-^{3}H$_2$]-MVA, [1-^{14}C]geraniol and nerol; [4,10-^{14}C$_2$]geraniol and nerol; and 1*R*,1*S*[1-^{3}H$_1$]geraniol and nerol	Terpinyl-carbonium ion or equivalent→(−)-mintlactone and (+)-isomintlactone via (−)-piperitenol→(−)-*cis*-pulegol and (+)-piperitenol→(+)-*trans*-pulegol; double labeled precursors MVA geraniol and nerol incorporated into terpinolene such that C-5 of MVA and C-1 of geraniol become oxygenated C of cyclohexane ring of *p*-menthane skeleton; *gem*-methyls of DMAPP retain their identity until final steps of oxidation and lactonization, implying that C-2 of MVA incorporates at that methyl of DMAPP, which was oxidized	Akhila et al. (1991)
In ontogenetic stages of leaves of *M. ×piperita*, first cyclic terpene is limonene; menthone appears later, then menthol, then menthyl acetate	Limonene→menthone→menthol→menthyl acetate	Brun et al. (1991)
Review of metabolism of monoterpenes in *Mentha*	Reviews and extends previous work	Croteau (1991)
Enzyme activity of mutant of *M. ×gracilis* high in piperitenone oxide/pulegone; C6-hydroxylase produced (−)-*trans*-carveol replaced by C3-hydroxylase produced (−)-*trans*-isopiperitenol; mutant could carry out P-450-dependent epoxidation of α,β-unsaturated bond of ketones formed via C3-hydroxylation	Reviews and extends previous work; (−)-isopiperitenone→(−)-*cis*-isopiperitenone oxide→(+)-piperitenone oxide→(+)-*trans*-piperitone oxide + (−)-*cis*-piperitone oxide; piperitenone→(+)-piperitone + (−)-piperitone→(−)-*cis*-piperitone oxide	Croteau et al. (1991)
Metabolic turnover of monoterpenes labeled from ^{14}CO$_2$ in whole plants of *M. ×piperita*	Turnover of monoterpenes, previously observed in detached stems, not observed in intact plants	Mihaliak, Gershenzon, and Croteau (1991)
Oil of *M. canadensis*	Theoretical biogenetic scheme proposed	Shimizu (1991)
Crosses of *M. arvensis* × *M. spicata*	Gene (*N*) hypothesized for inheritance of 1,8-cineole	Tucker et al. (1991)
Monoterpene cyclase isolated from *M. ×piperita* and *M. spicata* is 4S-limonene synthase and identical from both sources; [1-^{3}H]geranyl-PP converted to [3-^{3}H]limonene with this enzyme; P-450-dependent limonene-6-hydroxylase from *M. spicata* converts (−)-(4S)-limonene into (−)-*trans*-(4R,6S)-carveol or (+)-(4R)-limonene to (+)-*cis*-(4S,6S)-carveol	Geranyl-PP→(−)-4S-limonene via 4S-limonene synthase; (−)-(4S)-limonene→(−)-*trans*-(4R,6S)-carveol; (+)-(4S)-limonene→(+)-*cis*-(4S,6S)-carveol via limonene-6-hydroxylase	Alonso et al. (1992)
Characterized (−)-limonene synthase from *M. spicata*	Geranyl-PP→linalyl-PP→via carbonium ions	Alonso and Croteau (1992)
Geranyl-PP synthase isolated from *M. spicata* incorporates [4-^{14}C]IPP into geraniol; farnesyl-PP synthase incorporates [4-^{14}C]IPP into farnesol	Geranyl-PP synthase incorporates IPP into geraniol	Endo and Suga (1992)
(4S)-limonene synthase from *M. ×piperita* characterized; histidine at or near active site	Geranyl-PP→(+)-(3S)linalyl-PP→(4S)-limonene via carbonium ions	Rajaonarivony et al. (1992a, 1992b)

(continued)

Table 21.1 Observations, Hypotheses, and References Relating to the Genetic Inheritance of Monoterpenes in *Mentha* (Continued)

Observation	Hypothesis	Ref.
Chiral analysis of carvone in *M. spicata* and *M. longifolia*	(*R*)-(−)-carvone	Ravid et al. (1992)
Antibodies to 4S-limonene synthase from *M. spicata*	Similar enzyme in all *Mentha* species surveyed; geranyl-PP→(−)-(3S)-linalyl-PP→(−)-(4S)-limonene	Alonso, Crock, and Croteau (1993)
4S-limonene synthase from *M. spicata* converts geranyl-PP to (−)-(4S)-limonene with smaller amounts of α- and β-pinene and myrcene; several nucleotide differences in 5'-untranslated region suggest presence of several genes and/or alleles	Geranyl-PP→(−)-(4S)-limonene	Colby et al. (1993)
Review biosynthesis of limonene in *Mentha* species	Reviews and extends previous work	Croteau (1993)
Analogue of geranyl-PP inhibits (−)-(4S)-limonene synthase from *M. spicata*	Functional enzyme necessary to effect complex formation; analogue serves as a mechanism-based inactivator that must undergo both ionization-dependent isomerization and cyclization steps to reveal an allylic cation that alkylates the protein	Croteau et al. (1993)
Review of genetic literature	Reviews and extends previous work	Franz (1993)
Chirality at C-4 of limonene in *M. spicata*	(4S)-(−)-limonene	Hiraga et al. (1993)
New procedures for the isolation and quantification of the intermediates of the MVA pathway	DMAPP and IPP incorporated in the cytoplasm for the pathway to sesquiterpenes, leucoplasts for the pathway to monoterpenes	McCaskill and Croteau (1993, 1994, 1995a)
Menthol fed to suspension cells of *M. ×piperita*	(−)-Isopiperitenone→(−)-7-hydroxyisopiperitenone and (6*R*)- and (6*S*)-hydroxyisopiperitenone	Park et al. (1993)
Glycosidic-bound volatiles in *M. ×piperita* include (−)-menthol, 1-octen-3-ol, and 3-octanol; (+)-neomenthol maximum 4% of that fraction; leaves fed menthone show increase in menthol and neomenthol in aglycone fraction	Menthone→menthyl glucoside	Stengele and Stahl-Biskup (1993, 1994)
$[2\text{-}^{14}C,5\text{-}^3H_2]$MVA, $[1\text{-}^3H_2, ^{14}C]$geraniol and nerol fed to *M. canadensis* converted to menthone, menthol, and menthyl acetate; $[2\text{-}^{14}C,5\text{-}^3H_2]$ MVA, $[1\text{-}^{14}C]$geraniol and nerol, $[4,10\text{-}^{14}C_2]$geraniol, $[1\text{-}^3H2]$geraniol and nerol fed to *M. ×piperita* converted to (−)-mintlactone and (+)-isomintlactone; degradation of terpenes in *M. ×gracilis* fed ^3H- and ^{14}C-labeled MVA and geraniol	Geranyl-PP→menthone→menthol→menthyl acetate; geranyl-PP→terpinolene→(−)-piperitenol→(−)-cis-pulegol→mintlactone or terpinolene→(−)-piperitenol→(+)-trans-pulegol→(+)-isomintlactone; (−)-limonene→(−)-carvone; (+)-limonene→(+)-carvone; (−)-limonene→(−)-trans-carveol→(−)-dihydrocarvone→(+)-dihydrocarvone→(−)-dihydrocarveol→(−)-dihydrocarvyl acetate; (+)-dihydrocarvone→(+)-carvomenthone	Thakur and Akhila (1993)
Suspension cultures of *M. canadensis* and *M. ×piperita* convert acetate ions and leucine into cyclic monoterpenes; limonene degraded; piperitone converted to piperityl glucoside and (+)-isomenthone; (+)-pulegone converted to piperitone and (+)-isomenthone; (−)-menthone converted to (+)-isomenthone and (+)-neomenthol; (−)-menthol and (+)-neomenthol partially glucosylated	Piperitone→(+)-isomenthone; (+)-pulegone→piperitone and (+)-isomenthone; (−)-menthone→(+)-isomenthone and (+)-neomenthol	Werrmann and Knorr (1993)

Subject	Observation	Reference
Chiral analysis of commercial oils of M. ×piperita	(−)-(1R:3R:4S)-menthol, (−)-(1R:4S)-menthone, (+)-(1R:3S:4S)-neomenthol, (+)-(1R:4R)-isomenthol, (+)-(1R:3S:4R)-isomenthol, (+)-(1R:3R:4R)-neoisomenthol, menthofuran, and (−)-(1R:3R:4S)-menthyl acetate; (+)-menthol and (+)-menthyl acetate interpreted as adulterants	Faber et al. (1994)
(−)-Isopiperitenone fed suspension cells of M. ×piperita	(−)-Isopiperitenone→(−)-7-hydroxyisopiperitenone	Park et al. (1994)
Chiral analysis of piperitone in M. longifolia, M. ×piperita, and M. canadensis	79–99% (4S)-(+)-piperitone but one population of M. longifolia with 94% (4R)-(−)-piperitone	Ravid, Putievsky and Katzir (1994a)
Chiral analysis of menthone and isomenthone in Mentha species	99–100% (1R:4S)-(−)-menthone and (1R,4R)-(+)-isomenthone	Ravid et al. (1994b)
Chiral analysis of pulegone in M. ×piperita, M. longifolia, and M. pulegium	95–100% (1R)-(+)-pulegone	Ravid et al. (1994c)
Chiral analysis of linalyl acetate in M. aquatica var. citrata	(R)-(−)-linalyl acetate	Ravid et al. (1994d)
Limonene-6-hydroxylase and cytochrome P450 reductase purified from glands of M. spicata; amino acid sequences from purified hydroxylase utilized to design primers	Deduced amino acid sequence shows significant homology to other cytochrome P450 sequences of plant origin	Lupien et al. (1995)
Secretory cells isolated from glandular cells of M. ×piperita incorporate [14C]glucose-6-phosphate and [14C]pyruvic acid into mono- and sesquiterpenes; exogenous [14C]MVA and [14C]citrate or [14C]acetyl-CoA results in the accumulation of HMG-CoA	Cytoplasmic MVA pathway is blocked at HMG-CoA and IPP utilized for both mono- and sesquiterpene biosynthesis	McCaskill and Croteau (1995b)
(−)-4S-limonene synthase from M. spicata compared to (+)-pinene and (+)-bornyl-PP synthases from Salvia officinalis	Highly homologous cyclases	McGeady and Croteau (1995)
NADPH-cytochrome c (P450) reductase from M. spicata purified and characterized	NADPH-cytochrome c reductase reconstitutes NADPH-dependent (−)-4S-limonene-6-hydroxylase activity in presence of cytochrome P450	Ponnamperuma and Croteau (1996)
Stereochemistry of geranyl-PP→(4S)-limonene by (4S)-limonene synthase in M. spicata studied	re-Facial, anti-proton elimination at the cis methyl group of geranyl-PP(4S)-limonene	Coates et al. (1997)
cDNA from M. ×piperita encoding (E)-β-farnesene synthase cloned and expressed in E. coli	Farnesyl diphosphate→(E)-β-farnesene	Crock, Wildung, and Croteau (1997)
[1-13C]glucose and [U-13C]glucose fed to shoots of M. ×piperita	Monoterpenes not of MVA origin	Eisenreich et al. (1997)
(−)-(4R)- and (+)-(4S)-isopiperitenone fed suspension cultures of M. ×piperita	(−)-(4R)- and (+)-(4S)-isopiperitenone→(−)-(4R)-7-hydroxyisopiperitenone	Park et al. (1997)
Sequence relatedness and phylogenetic reconstruction, based on 33 members of the Tps gene family (plant terpenoid synthases), delineated and compared	(−)-Limonene synthase of M. spicata and M. candicans closely related to each other and that of (−)-limonene synthase from Perilla frutescens; (E)-β-farnesene synthase from M. ×piperita in family of sesquiterpene synthases from other genera and families	Bohlmann, Meyer-Gauen, and Croteau (1998)
1-Deoxy-d-xylulose-5-phosphate synthase from M. ×piperita cloned and expressed in E. coli catalyzes the first reaction of this pyruvate/glyceraldehyde 3-phosphate pathway	Synthesis of isopentyl diphosphate is via a highly conserved transketolase	Lange et al. (1998)

(continued)

Table 21.1　Observations, Hypotheses, and References Relating to the Genetic Inheritance of Monoterpenes in *Mentha* (Continued)

Observation	Hypothesis	Ref.
Geranyl diphosphate synthase purified from isolated oil glands of *M. spicata* and compared with *M. ×piperita*	Geranyl diphosphate synthase is a heterodimer and similar existing prenyltransferases	Burke, Wildung, and Croteau (1999)
[²H₃]-pulegone enantiomers fed shoot tips and first leaf pairs of *M. ×piperita*	(*R*)-pulegone→2 (–)-menthone: 3 (+)-isomenthone; (*S*)-pulegone (which does not occur in *Mentha*) →2 (+)-menthone:3 (–)-isomenthone	Fuchs, Beck, Burkardt, et al. (1999); Fuchs, Beck, Sandvoss, et al. (1999)
[²H₂]- and [²H₂]/[¹⁸O]-pulegone enantiomers fed shoots and first leaf pairs of *M. ×piperita*	(*R*)-pulegone→(*R*)-menthofuran; (*S*)-pulegone→(*S*)-menthofuran	Fuchs, Zinn, et al. (1999)
Reductoisomerase for conversion of 1-deoxy-d-xylulose-5-phosphate to 2-*C*-erythritol-4-phosphate from *M. ×piperita* and cloned *E. coli*	Reductoisomerase encodes a preprotein that directs the enzyme to plastids, where mevalonate-independent pathway operates	Lange and Croteau (1999a)
Kinase that catalyzes phosphorylation of isopentyl monophosphate to isopentyl diphosphate from *M. ×piperita* cloned in *E. coli*; secretory cells incubated with isopentyl monophase	Kinase phosphorylates isopentenol to isopentyl monophosphate, which is terminal intermediate of deoxyxylulose 5-phosphate pathway	Lange and Croteau (1999b)
Review of genetic engineering or essential oil production in *Mentha*	Updated biogenetic pathways provided	Lange and Croteau (1999c)
Review of 30 years of work, primarily involving *Mentha*	Updated biogenetic pathways provided	Little and Croteau (1999)
Microsomal limonene-6-hydroxylase from oil glands of *M. spicata* compared with cDNA of *M. ×piperita*	Enzymes similar but identified as limonene-3-hydroxylase to catalyze (–)-limonene→(–)-*trans*-isopiperitenol in *M. ×piperita* versus limonene-6-hydroxylase to catalyze (–)-limonene→(–)-*trans*-carveol in *M. spicata*	Lupien et al. (1999)
Immunogold labeling using polyclonal antibodies raised to (4*S*)-limonene synthase localized in leucoplasts in *M. ×piperita* oil gland secretory cells during period of essential oil production; labeling absent from all other plastid types examined	Leucoplastidome of the oil gland secretory cells is exclusive location of limonene synthase, but succeeding steps of monoterpene metabolism appear to occur outside the leucoplasts	Turner et al. (1999)
[²H₃]-piperitone and [²H₃]-piperitenone fed shoots and first leaf pairs of *M. ×piperita*	(±)-Piperitone→(–)-menthone/(+)-isomenthone; piperitenone→(–)-menthone/(+)-isomenthone	Fuchs et al. (2000)
Monoterpene accumulation in *M. ×piperita* restricted leaves of 12–20 days of age, rate of monoterpene biosynthesis by ¹⁴CO₂ inc. closely correlated with monoterpene accumulation; RNA blot analyses indicate genes encoding enzymes of early pathway steps are transcriptionally activated in a coordinated fashion	Efforts to improve production should focus on the genes, enzymes, and cell differentiation processes that regulate monoterpene biosynthesis	Gershenzon, McConkey, and Croteau (2000); McConkey, Gershenzon, and Croteau (2000)
Sequence information from1,316 randomly selected cDHA clones, or expressed sequence tags (ESTs), from *M. ×piperita* oil gland secretory cell cDNA library	ESTs involved in essential oil metabolism represent about 25% of the described sequences; 7% recognized genes code for proteins involved in transport process, and a subset of these are likely involved in secretion of essential oil terpenes from site of synthesis to storage cavity of oil glands; summary of biogenetic sequence with enzymes provided; updated biogenetic pathways provided	Lange et al. (2000)

Limonene-3-hydroxylase from M. ×piperita and limonene-6-hydroxylase from M. spicata 70% identical at amino acid level	Exchange of a single residue (F363I) in the limonene-6-hydroxylase led to the complete conversion to the regiospecificity and catalytic efficacy of the limonene-3-hydroxylase	Schalk and Croteau (2000)
(−)-4S-limonene synthase expressed in E. coli and examined with noncyclizable substrate analogs 6,7-dihydrogeranyl diphosphate and 2,3-cyclopropylgeranyl diphosphate	Normal cyclization of geranyl-PP by (−)-4S-limonene synthase proceeds via preliminary isomerization to the bound tertiary intermediate 3S-linalyl diphosphate	Schwab, Williams, and Croteau (2000)
Period of 20–30 hours required for filling of foliar 10-celled peltate glandular trichomes in M. ×piperita; SER associated with leucoplasts	SER involved in transport of monoterpenes from secretory cells to storage space	Turner, Gershenzon, and Croteau (2000a, 2000b)
Microsomal preparations from oil gland secretory cells of M. pulegium in presence of NADPH and O_2 transform (+)-pulegone to (+)-menthofuran; abundant cytochrome P450 clone from oil gland of M. ×piperita cell cDNA library functionally expressed in S. cerevisiae and E. coli and shown to encode (+)-menthofuran synthase [(+)-pulegone-9-hydroxylase] with 35% identity to (+)-limonene-3-hydroxylase	(+)-Pulegone→(+)-menthofuran via (+)-menthofuran synthase [(+)-pulegone-9-hydroxylase]	Bertea et al. (2001)
[^{2}H]-pulegone enantiomers fed shoots and first leaf pairs of M. spicata	(R)-pulegone→3(−)-(1R,4S)-menthone:1(+)-(1R,4S)-isomenthone, opposite of M. ×piperita	Fuchs, Beck, and Mosandl (2001)
Deoxyxylulose phosphate reductoisomerase and menthofuran synthase in M. ×piperita expression altered	Updated biogenetic pathways provided	Mahmoud and Croteau (2001)
cDNA encoding (−)-3R-linalool synthase from M. aquatica cloned and sequenced	Geranyl diphosphate→(−)-3R-linalool by (−)-3R-linalool synthase	Crowell et al. (2002)
Suspension cell culture of M. ×piperita fed (+)- or (−)-isopiperitenone and (−)- or (+)-(4S)-carvone	(+)-Isopiperitenone or (−)-isopiperitenone→(4S,6R)-6-hydroxy- and (4S,8R)-8,9-epoxyisopiperitenone and (+)-hydroxyisopiperitenone; (−)-(4R)-carvone→(1R,2S,4R)-neodihydrocarveol and (1R,2R,4R)-dihydrocarveol; (+)-(4S)-carvone→→(1R,2R,4S)-neodihydrocarveol and (1S,4S)-dihydrocarvone	Kim et al. (2002)
(−)-(4S)-limonene hydroxylated by (−)-(4S)-limonene-3-hydroxylase and (−)-(4S)-limonene-6-hydroxylase at designated C3- and C6-allylic positions to produce (−)-trans-isopiperitenol and (−)-trans-carveol, respectively; (+)-(4R)-limonene hydroxylated by (−)-(4S)-limonene-3-hydroxylase to (+)-trans-isopiperitenol but (−)-(4S)-limonene-6-hydroxylase yields (+)-cis-carveol and (±)-trans-carveol	(−)-(4S)-limonene→(−)-trans-isopiperitenol by (−)-(4S)-limonene-3-hydroxylase; (−)-(4S)-limonene→(−)-trans-carveol by (−)-(4S)-limonene-6-hydroxylase; (+)-(4R)-limonene→(+)-trans-isopiperitenol by (−)-(4S)-limonene-3-hydroxylase; (+)-(4R)-limonene→(+)-cis-carveol and (±)-trans-carveol by (−)-(4S)-limonene-6-hydroxylase	Wüst et al. (2001); Wüst and Croteau (2002)
Short review of work on (−)-limonene synthase	Geranyl diphosphate→(−)-limonene by (−)-limonene synthase; (−)-limonene→(−)-menthone→(−)-menthol	Gershenzon, McConkey, and Croteau (2002)

(continued)

Table 21.1 Observations, Hypotheses, and References Relating to the Genetic Inheritance of Monoterpenes in *Mentha* (Continued)

Observation	Hypothesis	Ref.
Gamma-irradiation of *M. ×gracilis* created mutant line with oil of peppermint with mutant enzyme more closely related to peppermint limonene-3-hydroxylases than to the spearmint limonene-6-hydroxylases	Mutation occurred at a regulatory site within genome that controls expression of one or other regiospecific variants	Bertea et al. (2003)
Modified expression of menthofuran synthase in transformed cells of *M. ×piperita*; overexpression or cosuppression results in respective increase or decrease of menthofuran; menthofuran does not inhibit pulegone reductase activity	Metabolic fate of (+)-pulegone controlled through transcriptional regulation of menthofuran synthase, and menthofuran influences this process by down-regulating transcription from gene for pulegone reductase and/or decreasing message stability	Mahmoud and Croteau (2003)
Random sequencing of essential oil gland secretory cDNA library of *M. ×piperita* reveals redox-type enzymes	(–)-Isopiperitenone reductase member of short-chain dehydrogenase/reductase superfamily, while (+)-pulegone reductase member of medium-chain dehydrogenase/reductase superfamily, implying very different evolutionary origins	Ringer et al. (2003)
DNA and protein of geranyl diphosphate synthase examined from *M. ×piperita* and *M. spicata*; inhibited by farnesyl diphosphate synthase and geranylgeranyl diphosphate synthase	Dimethylallyl diphosphate + isopentyl diphosphate→geranyl diphosphate via geranyl diphosphate synthase, which employs different mechanism for chain-length determination than do other short-chain prenyltransferases	Burke, Klettke, and Croteau (2004)
37 Clones from open-pollinated *M. ×piperita* 'Kukrail' fall into two categories: high menthol or high menthofuran	Kurkail genotype *Ff*	Kumar, Khanuja, and Patra (2004)
Cosuppression of limonene-3-hydroxylase in *M. ×piperita* results in accumulation of limonene	Limonene does not impose negative feedback on the synthase or apparently influence other enzymes of monoterpene biosynthesis	Mahmoud, Williams, and Croteau (2004)
Immunochemical localization of three enzymes from *M. ×piperita* [geranyl diphosphate synthase, (–)-trans-isopiperitenol dehydrogenase, (+)-pulegone reductase] and enzyme from *M. spicata* [(–)-(4S)-limonene-6-hydroxylase] were localized in secretory cells of the peltate glandular trichomes with abundant labeling corresponding to the secretory phase of gland development; labeling of geranyl diphosphate synthase in secretory cell leucoplasts, (–)-4S-limonene-6-hydroxylase labeling associated with ER, (–)-trans-isopiperitenol dehydrogenase labeling restricted to secretory cell mitochondria, while (+)-pulegone reductase labeling occurs only in secretory cell cytoplasm	Pathway compartmentalization and intracellular movement of monoterpene metabolites	Turner and Croteau (2004)
Review of (–)-menthol biosynthesis in *M. ×piperita*	IPP + DMAPP→geranyl diphosphate via geranyl pyrophosphate synthase in plastids→(–)-limonene via (–)-limonene synthase in plastids→(–)-trans-isopiperitenol via (–)-limonene-3-hydroxylase in ER→(–)-isopiperitenone via (–)-trans-isopiperitenol dehydrogenase in mitochondria→(+)-cis-isopulegone via (–)-isopiperitenone reductase in cytosol→(+)-pulegone via (+)-cis-isopulegone isomerase→(–)-menthone	Croteau et al. (2005)

	via (+)-pulegone reductase in cytosol→(+)-menthofuran via menthofuran synthase in ER or (−)-menthol via (−)-menthone reductase in cytosol	
Menthone:(+)-(3S)-neomenthol reductase characterized from M. ×piperita	(−)-Menthone→94% (−)-(3S)-neomenthol + 6% (−)-(3R)-menthol or (+)-isomenthone→86% (+)-(3S)-isomenthol + 14% (+)-(3R)-neoisomenthol via menthone:(+)-(3S)-neomenthol reductase	Davis et al. (2005)
(−)-trans-Carveol dehydrogenase characterized from M. spicata and homologous to (−)-trans-isopiperitenol dehydrogenase from M. ×piperita	Genes arose from different ancestors and not by simple duplication and differentiation of a common progenitor; trans-carveol→(−)-carvone via (−)-trans-carveol dehydrogenase; (−)-trans-isopiperitenol→(−)-isopiperitone via (−)-trans-isopiperitenol dehydrogenase	Ringer, Davis, and Croteau (2005)
cDNA of (E)-β-farnesene synthase from M. ×piperita cloned and sequenced	Farnesyl diphosphate→γ-cadinene + α-cadinene or cis-muurola-3,5-diene + cis-muurola-4(14),5-diene via intermediates	Prosser et al. (2006)
(4S)-limonene synthase from M. spicata metal ion-dependent cyclase	Geranyl-PP→(4S)-limonene	Hyatt et al. (2007)
Racemic linalool fed to soluble, cell-free extracts of M. aquatica var. citrata 99% converted to (R)-linalyl acetate	(RS)-linalool→(R)-linalyl acetate by alcohol acetyl transferase	Larkov et al. (2008)
(+)-Menthofuran weak competitive inhibitor of pulegone reductase in M. ×piperita but (+)-menthofuran selectively retained in secretory cells in low light and accumulates to high levels but very low levels under normal conditions	(+)-Pulegone→(+)-menthofuran via menthofuran synthase and (+)-menthofuran inhibits (+)-pulegone reductase	Rios-Estepa et al. (2008)

Notes: Where alternate designations of genes have been proposed, those of Murray are inserted in brackets. Abbreviations: CoA = coenzyme A, DMAPP = dimethylallyl diphosphate, ER = endoplasmic reticulum, FAD = flavin-adenine dinucleotide, FMN = flavin mononucleotide, HMG-CoA = 3-hydroxy-3-methyl-glutaryl-CoA, IPP = isopentenyl diphosphate, MVA = mevalonic acid, NADPH = reduced nicotinamide adenine dinucleotide phosphate, P = phosphate, phs = photosynthesis, PP = diphosphate, U = uniformly labeled, SER = smooth endoplasmic reticulum.

Table 21.2   Genotypes of Clones of M. J. Murray et al.

Taxon	Abbreviation	Origin	2n	Genotype and Predominant Monoterpene(s)	Ref.
M. aquatica	Du Aq	Leiden, the Netherlands	96	$AAc_1c_2c_2ffi_1i_1i_2i_2lm_1lm_1lm_2lm_2PPRR$ Menthofuran	Murray and Lincoln (1970); Lincoln et al. (1971); Murray et al. (1972); Murray and Hefendehl (1972, 1973)
M. aquatica	Du Aq ms	Leiden, the Netherlands	96	$AAc_1c_2c_2ffi_1i_1i_2i_2lm_1lm_1lm_2lm_2PPRR$ Menthofuran	Murray and Lincoln (1970); Lincoln et al. (1971); Murray et al. (1972); Murray and Hefendehl (1972, 1973)
M. aquatica	2n Aq	Kew Gardens, England	96	$AAccffi_1i_1i_2l_2PPRR$ Menthofuran	Reitsema (1958); Handa et al. (1964); Murray and Lincoln (1970, 1972); Murray et al. (1972); Lawrence (1978)
M. aquatica	Baq Aq #1	Schleswig-Holstein, Germany	96	$AAc_1c_2c_2ffi_1i_1i_2i_2lm_1lm_1lm_2lm_2PPRR$ Menthofuran	Hefendehl (1967b); Murray and Lincoln (1970); Lincoln et al. (1971); Murray et al. (1972); Murray and Hefendehl (1972, 1973)
M. aquatica var. *citrata*	2n Cit	Europe	96	$AAccl_1i_1i_2l_2lm_1lm_1lm_2lm_2$ Linalool/linalyl acetate	Murray and Lincoln (1970); Lincoln et al. (1971); Lawrence (1978)
M. aquatica var. *citrata*	Loomis Cit	Europe	96	$AAccl_1i_1i_2l_2lm_1lm_1lm_2lm_2$ Linalool	Murray and Lincoln (1970); Lincoln et al. (1971)
M. aquatica var. *citrata*	Derk Cit	Michigan	96	$l_1i_1i_2$ Linalool/linalyl acetate	Murray and Lincoln (1970)
M. aquatica var. *citrata*	Herb's Cit	Michigan	96	$l_1i_1i_2$ Linalool/linalyl acetate	Murray and Lincoln (1970)
M. aquatica var. *citrata*	Cal Norb Cit	Oregon	96	$AAccl_1i_1i_2l_2Lm_1Lm_1lm_1lm_2lm_2$ Linalool	Murray and Lincoln (1970); Lincoln et al. (1971)
M. aquatica var. *citrata*	Thomas Cit	Adelaide, S. Australia	96	$l_1l_1l_2$ Linalool	Murray and Lincoln (1970)
M. aquatica var. *citrata*	10199	Michigan	96	$i_1i_1i_2l_2isisIm_1lm_1Lm_2lm_2$ Limonene/1,8-cineole	Lincoln et al. (1971)
M. arvensis	Arv Gent Kemp ms	Kempton, Pennsylvania	72	$Aacclinnpp$ Linalool	Tucker et al. (1991)
M. canadensis	2n Arv	Michigan	96	$Aacci_1i_1i_2l_2lm_1lm_1lm_2lm_2$ Pulegone	Ikeda and Udo (1966); Murray and Lincoln (1970); Lincoln et al. (1971)
M. canadensis	2n Jap	Japan	96	$AAccFFiiPPRr$ Menthol/menthyl acetate	Murray (1960a); Murray and Lincoln (1970); Murray et al. (1971); Murray et al. (1972b); Lawrence (1978)
M. ×*gracilis*	2n Sc	Scotland	84	$A-CcE-iirr$ Carvone/limonene	Ikeda, Shimizu, and Udo (1963); Murray and Lincoln (1970); Murray et al. (1972a); Hefendehl and Murray (1973); Lawrence (1978)

M. longifolia	2n Syl	Europe	24	*aaccFFi$_1$i$_1$i$_2$i$_2$lm$_1$lm$_1$lm$_2$lm$_2$rr* trans-Piperitone oxide/beta-caryophyllene	Murray (1960b); Murray and Lincoln (1970, 1972); Lincoln et al. (1971); Lawrence (1978)
M. spicata	2n Line 1	Kew, England	48	*AaCcFFi$_1$i$_1$i$_2$i$_2$lm$_1$lm$_1$lm$_2$lm$_2$ooPPrr* Carvone	Murray (1960b); Murray and Lincoln (1970); Lincoln et al. (1971); Murray et al. (1972); Tucker et al. (1991)
M. spicata	Missiones	Argentina	48	*AaccFFi$_1$i$_1$i$_2$i$_2$PPrr* Menthone	Murray and Lincoln (1970); Murray, Lincoln, et al. (1972)
M. spicata	62-199 or Strain 199	Michigan	48	*AaccFFi$_1$i$_1$i$_2$i$_2$lm$_1$lm$_1$lm$_2$lm$_2$PPrr* Menthone/piperitone	Murray and Lincoln (1970); Lincoln et al. (1971); Murray et al. (1972)
M. spicata var. *crispata*	2n Cr	U.S.A.	48	*AaCci$_1$i$_1$i$_2$i$_2$lm$_1$lm$_1$lm$_2$lm$_2$ooPPrr* Carvone	Murray (1960a, 1960b); Murray and Lincoln (1970); Lincoln et al. (1971); Murray et al. (1972, 1972a, 1972b); Hefendehl and Murray (1972, 1973); Tucker et al. (1991)
M. suaveolens	2n Rot	Maryland	24	*aaccii* 1-Octen-3-yl acetate/germacrene D/piperitenone oxide	Murray (1960b); Murray and Lincoln (1970); Lawrence (1978)
M. ×villoso-nervata	2n Amer Sp or Nat Sp	Michigan	36	*aaCceerr* Carvone	Murray and Lincoln (1970); Murray et al. (1971); Murray et al. (1972, 1972a)

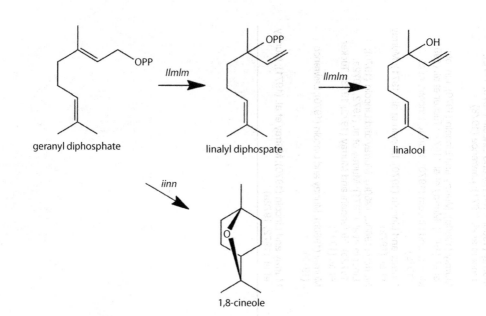

Figure 21.1 Putative biosynthetic pathway of acyclic and bicyclic monoterpenes in *Mentha* with principal genes. Additional details are provided in Table 21.1.

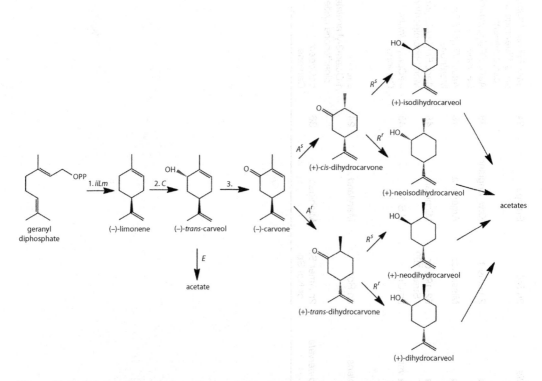

Figure 21.2 Putative biosynthetic pathway of 2-oxygenated monoterpenes in *Mentha* with characterized enzymes and principal genes. Table 21.1 provides additional details. 1. (−)-limonene synthase (in leucoplasts); 2. (−)-limonene-6-hydroxylase (in endoplasmic reticulum); 3. (−)-*trans*-carveol dehydrogenase (in mitochondria).

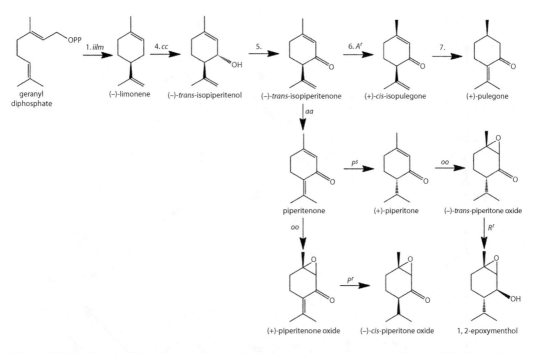

Figure 21.3(a)   Putative biosynthetic pathway of 3-oxygenated monoterpenes in *Mentha* with characterized enzymes and principal genes. Table 21.1 provides additional details. 1. (–)-limonene synthase (in leucoplasts); 4. (–)-limonene 3-hydroxylase (in endoplasmic reticulum); 5. (–)-*trans*-isopiperitenol dehydrogenase (in mitochondria); 6. (–)-*trans*-isopiperitenone reductase (in cytosol); 7. (+)-*cis*-isopulegone isomerase (in cytosol).

from mass insect-pollinated plants of spearmint, but nothing further was published, partly because of inadequate oil quality (Murray 1969).

This work was then taken up by M. J. Murray (1961), who found that resistance to rust was due to a dominant gene (*S*) in *M. aquatica, M. aquatica* var. *citrata, M.* ×*piperita, M. spicata*, and *M. spicata* var. *crispata*. This genetic work was confirmed in Japan with *M. canadensis* by Ikeda, Aoki, and Udo (1961). Selfs from *M. canadensis* 'Vainhos' and 'Samby' selected by Donalísio, D'Andréa Pinto, and de Souza (1985) in Brazil produced clones resistant to rust with high menthol and low pulegone. Russian researchers (Sidorenko, Zav'yalova, and Vorob'eva 1985; Shilo et al. 1986) selected three clones of *M.* ×*piperita* resistant to rust with high menthol: "Prilukskaya 6," "Prilukskaya 14," and 'Zarya'.

The A. M. Todd Co. patented six hybrids of *M.* ×*piperita, M. canadensis, M. spicata* var. *crispata*, and/or *M. longifolia* (U.S. Plant Patents 1612, 1613, 1614, 1926, 1927, 1928), primarily for disease resistance; however, none attained any economic significance. Donald Roberts, who used to be in charge of the USDA National Clonal Germplasm Repository for *Mentha* in Corvallis, Oregon, selected a seedling from a cross of a polyploid strain of *M.* ×*piperita* from Murray (U.S. Plant Patent 11788)—'Cascade Mitcham'—that is resistant to both wilt and rust.

The most successful approach to finding *Verticillium*-resistant mint has been x-ray and gamma irradiation (Murray 1969, 1971). During 1955–1959, over 100,000 plants were set into wilt-infested soil after irradiation at Brookhaven National Laboratory. Eventually, seven highly wilt-resistant strains of peppermint were selected, finally resulting in the registration of 'Todd's' Mitcham' and 'Murray Mitcham' peppermints with satisfactory oil (Murray and Todd 1972; Todd, Green, and Horner 1977). Japanese cornmint (Ono and Ikeda 1970), Scotch spearmint (Horner and Melouk 1977), and native spearmint (Johnson and Cummings 2000) were also irradiated, and wilt-resistant selections were identified.

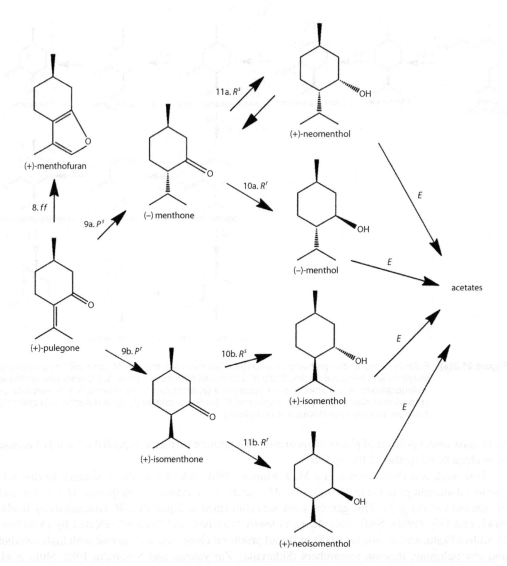

Figure 21.3(b) Putative biosynthetic pathway of 3-oxygenated monoterpenes in *Mentha* with characterized enzymes and principal genes. Table 21.1 provides additional details. 8. (+)-menthofuran synthase (in endoplasmic reticulum); 9a. (+)-pulegone reductase[(−)-menthone-forming activity] (in cytosol); 9b. (+)-pulegone reductase[(+)-isomenthone-forming activity] (in cytosol); 10a. (−)-menthone:(−)-menthol reductase[(+)-neoisomenthol-forming activity] (in cytosol); 10b. (−)-menthone:(−)-menthol reductase[(+)-neoisomenthol-forming activity] (in cytosol); 11a. (−)-menthone:(+)-neomenthol reductase[(+)-neomenthol-formingactivity](incytosol);11b.(−)-menthone:(+)-neomentholreductase [(+)-isomenthol-forming activity] (in cytosol).

21.2.3 Breeding for Yield

Breeding for yield has been primarily focused on heritability and coheritability in Japanese cornmint (*M. canadensis*). In research pioneered by Srikant Sharma and Bali Tyagi at the Central Institute of Medicinal and Aromatic Plants (CIMAP) in Lucknow, India, the oil yield showed the highest co-inheritance with leaf/stem ratio, followed by menthone and menthol contents (Sharma and Tyagi 1990, 1991), and several selections were identified (Sharma et al. 1992, 1996; Tyagi, Ahmad, and Bahl 1992; Tyagi and Naqvi 1987). This work has been continued by Birendra Kumar,

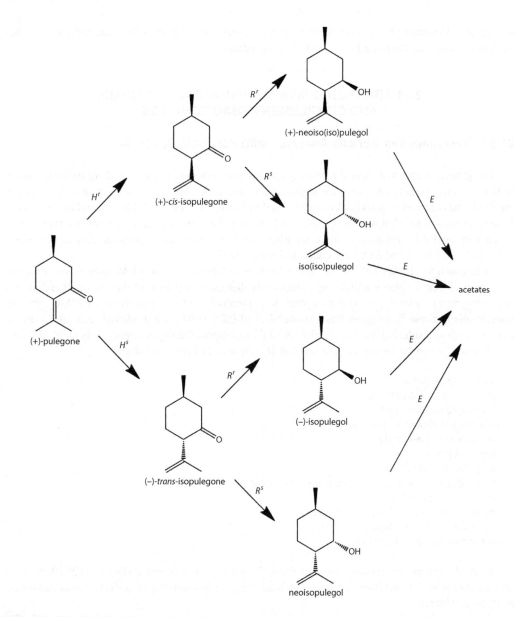

Figure 21.3(c)   Putative biosynthetic pathway of the pulegones in *Mentha* with principal genes. Additional details are provided in Table 21.1.

H. P. Singh, and N. K. Patra at CIMAP (Kumar et al. 2005; Kumar, Singh, and Patra 2007; Patra et al. 2001a, 2001b).

Thus, Sharma et al. (1996) looked at genetic divergence of 38 genotypes of *M. ×piperita*. They identified 12 different genotype clusters based upon herb and oil yield and chemical constituents. A similar approach was applied by Singh, Sharma, and Tewari (1998) to 38 half-sib progeny from *M. canadensis* 'Shivalik.' They identified eight genotype clusters based upon herb and oil yield and chemical constituents.

Another avenue that has been explored for increased yield is polyploidy effected with colchicine (e.g., Ammal and Sobti 1962; Ikeda and Konishi 1954; Ikeda and Udo 1956; Murray 1952). For example, Lavania et al. (2006) treated stolons of *M. canadensis* 'Kosi' with colchicine and observed

43 variants. However, this has not resulted in any significant releases with increased yield but has been more useful in effecting fertile pollen for breeding.

21.3 TRANSGRESSIVE SEGREGATION, CYTOMIXIS, AND COMPLEMENT FRACTIONATION

21.3.1 Oversimplified Genetic Analyses with Polyploid Species

The genetic analysis of Merritt J. Murray and his associates involved trained organoleptic panels rather than analytical techniques such as gas chromatography. This was necessary in order to examine thousands of crosses quickly. As summarized in Tables 21.1 and 21.2, Murray and his associates found that selfs of his clones of *M. spicata* (2n Line 1) and *M. spicata* var. *crispata* (2n Cr) produced 12 carvone and dihydrocarvone: three menthone and pulegone: one piperitone and piperitenone. Thus, both 2n Line 1 and 2n Cr are characterized as *AaCc*.

In research for Tucker (1975), the selfed progeny of Murray's clone of *M. spicata* var. *crispata* (2n Cr) produced 50 plants with these compounds that exceeded 10% of the oil, analyzed by gas chromatography: 40 carvone, five pulegone, four menthol, and one piperitone. A 40:10 carvone: noncarvone in these 50 progeny has a probability of 0.25–0.50 for a predicted ratio of 3:1 and is close enough (considering the low number of selfs) to support the organoleptic analysis of Murray.

However, a similar analysis of the selfs of *M. spicata* (2n Line 1) produced:

twenty-one carvones
one carvone, dihydrocarvone
three carvones, limonene
six carvones, limonene, 1,8-cineole
four carvones, 1,8-cineole
three pulegones
two pulegones, menthone
three pulegones, menthone, isomenthone, 1,8-cineole
one pulegone, piperitone
one menthone, isomenthone, 1,8-cineole
one menthone, isomenthone, piperitone
four menthones, piperitone oxide

A 35:15 carvone/noncarvone ratio in these 50 progeny has a probability of 0.25–0.50 for a predicted ratio of 3:1 and is close enough (considering the low number of selfs) to the organoleptic analysis of Murray.

However, selfs of 2n Line 1 have a greater array of essential oil constituents than selfs of 2n Cr, and characterization of both 2n Cr and 2n Line 1 simply as identical (*AaCciilmlmPPrr*) is probably more practical than realistic.

21.3.2 Polyploidy and Genes

Murray and his associates also disregarded the implication of polyploidy in their genetic work. Thus, while their clone of *M. longifolia* (2n Syl) with $2n = 24$ was characterized as $aaccFFi_1i_1i_2$-$i_2lm_1lm_1lm_2lm_2rr$, their principal clone of *M. spicata* (2n Line 1) with $2n = 48$ was characterized as $AaCcFFi_1i_1i_2i_2lm_1lm_1lm_2lm_2PPrr$. However, if these two lines were crossed, then the genes for *M. spicata* should be doubled to indicate the doubled chromosome number accurately (e.g., *AAaaCCcc*... for 2n Line 1).

21.3.2.1 The Reitsema Rule

In 1958, Reitsema wrote that "no evidence has been obtained in the present program for the coexistence of both 2- and 3-oxygenated-p-menthanes in the oil from a single mint plant." This later became known as the "Reitsema rule" that peppermint and spearmint oils cannot coexist in the same plant because they are on two alternative biogenetic pathways; that is, "doublemints" are impossible. In spite of the Reitsema rule, Tucker et al. (1991) found a doublemint clone of *M. ×gracilis* (Card Hilltop, $2n = 96$, now sold as Madalene Hill in the U.S. herb plant trade) with 13.9% limonene, 21.4% menthol, and 39.5% carvone + dihydrocarveol.

21.3.3 Cytomixis and Complement Fractionation

In 1981, Tucker and Fairbrothers crossed *M. arvensis* $2n = 72 \times$ *M. spicata* $2n = 48$ and observed in the F$_1$ progeny somatic chromosomes of 48, 60, 72, 84, and 96. Kundu and Sharma (1985, 1988) confirmed cytomixis in both *M. ×piperita* and *M. canadensis*; in the former, cytomixis was observed in almost every stage of meiosis and, in the latter, it was observed only in meiosis I. Tyagi and Ahmad (1989) also confirmed this phenomenon in a cross of *M. spicata* $2n = 72 \times$ *M. ×piperita* $2n = 120$ with F$_1$ progeny of somatic chromsomes of 36, 96, 110, and 115. In 2003, Tyagi observed microsporogenesis in *M. spicata* $2n = 48$ and found cytomixis in leptotene to pachytene stages of meiosis I with migration of nuclear content involving all or part of the chromosomes.

Cytomixis was defined in 1911 by Gates as "an extrusion of chromatin from the nucleus of one mother-cell through cytoplasmic connections, into the cytoplasm of an adjacent mother-cell." Contrary to Maheshwari (1950), who wrote, "It is believed, however, that it is a pathological phenomenon, or that such appearances are caused by faulty fixation," cytomixis has since been observed in animals, mosses, ferns, conifers, and many flowering plant families, and in both mitotic and meiotic cells (Guo and Zheng 2004; Table 21.3). However, since this type of cytomixis results in chromosomes in multiples of the monoploid number ($x = 12$) in *Mentha*, we prefer to use the term "complement fractionation" of Thompson (1962) to better describe the entire phenomenon, regardless of the cause.

If complement fractionation is a common phenomenon in *Mentha*, this would upset Mendelian ratios by providing dosage effects of genes. Thus, the unusual clone of *M. ×gracilis* (Card Hilltop)

Table 21.3 Brief Survey of Cytomixis in Plant Families

Family	Ref.
Agavaceae	Lattoo et al. (2006)
Brassicaceae	Malallah and Attia (2003)
Fabaceae	Bellucci, Roscini, and Mariani (2003)
Hemerocallidaceae	Narain (1979)
Isoetaceae	Wang, Li, and He (2007)
Lamiaceae	Carlson and Stuart (1936); Datta, Mukherjee, and Iqbal (2005); Kundu and Sharma (1985, 1988); Tyagi (2003); Tyagi and Ahmad (1989)
Liliaceae	Zheng et al. (1985)
Onagraceae	Gates (1908, 1911)
Papaveraceae	Singhal and Kumar (2008a)
Poaceae	Boldrini, Pagliarini, and do Valle (2006); Ghaffari (2006); Sheidai and Fadaei (2005); Wang and Cheng (1983)
Pinaceae	Guzicka and Wozny (2005)
Solanaceae	Cheng, Yang, and Zheng (1982); Datta et al. (2005); Siddiqui, Khan, and Rao (1979); Singhal and Kumar (2008b)

Table 21.4   Fertility of 18 Natural Clones of *M. ×gracilis*

2n	60 (Six Clones)	72 (Five Clones)	84 (Five Clones)	96 (Two Clones)
Average percent fertile pollen	0%	0–4%	0–14%	0%
Average percent fertile seeds	0–0.1%	0–0.2%	0.–0.2%	0%

Source: Tucker, A. O., and D. E. Fairbrothers. 1990. *Economic Botany* 44:183–213.

might be explained by multiple copies of genes, resulting in biogenetic pathways of both 2- and 3-oxygenated monoterpenes in the same plant. The degree of complement fractionation may also influence the phenotypic expression of the S_1 progeny of 2n Line 1 and 2n Cr, as stated earlier.

21.3.4 Complement Fractionation Fuels Transgressive Segregation

Rieseberg, Archer, and Wayne (1999) defined transgressive segregation as "the presence of phenotypes that are extreme relative to those of the parental line…a major mechanism by which extreme or novel adaptations observed in new hybrid ecotypes or species are thought to arise." Thus, complement fractionation provides the fuel for the phenomenon of transgressive segregation in *Mentha*.

The attempt to resynthesize *M. ×piperita* 'Mitcham' by Murray et al. (1972) is an example of transgressive segregation. Murray crossed a high-menthone strain of *M. spicata* with high-mentho-furan strains of *M. aquatica* to produce 200,000 seedlings; however, only a few approached the commercial "Mitcham" clone high in menthol. As an additional example of transgressive segregation, the attempt to resynthesize *M. ×gracilis* produced hundreds of F_1 hybrids, but only one clone completely matched a selected clone from cultivated and wild-collected material in morphology, essential oil chemistry, and chromosome number (Tucker and Fairbrothers 1990; Tucker et al. 1991).

In another example, in the attempt to resynthesize *M. canadensis*, Tucker and Chambers (2002) crossed *M. arvensis* high in pulegone or linalool with *M. longifolia* high in *trans*-piperitone oxide. While most F_1 hybrids had the monoterpenes of the parents, individual plants could be isolated high in *trans*-carveol, carvone, β-caryophyllene, citronellyl acetate, geranyl acetate, isomenthone, *trans*-isopulegone, limonene, linalool, menthol, menthone, neomenthol, (Z)-β-ocimene, 3-octanol, 3-octanone, *cis*-piperitenone oxide, *trans*-piperitenone oxide, pulegone, *trans*-sabinene hydrate, and/or terpinen-4-ol.

21.3.5 Complement Fractionation Restores Fertility to Hybrids

Complement fractionation may also restore fertility to interspecific hybrids. Table 21.4 (Tucker and Fairbrothers 1990) gives the percentage of fertile pollen and seeds in 18 natural clones of *M. ×gracilis*. As can be seen, the expected chromosome number of *M. arvensis* $2n = 72 \times M.$ *spicata* $2n = 48$ is $2n = 60$, and this has 0% pollen fertility and 0–0.1% fertile seeds. However, the clones with $2n = 72$ and 84 have up to 14% pollen fertility and 0.2% fertile seeds.

21.4 BIOTECHNOLOGY

21.4.1 Introduction

Crop improvement of *Mentha* species by way of tissue culture requires the development of micropropagation and regeneration protocols. Micropropagation involves the production of clonally identical plants through the proliferation of microshoots, resulting from axillary bud break, which are subsequently rooted and reestablished. Regeneration in *Mentha* species involves the culture of plant parts, explants, on defined medium with the result that individual or groups of cells divide and organize to generate organs (organogenesis; e.g., shoots). The generation of new variation in *Mentha*, whether through somaclonal variation, protoplast fusion, or transformation, requires the development of repeatable, reliable regeneration protocols. Regenerated shoots can then be mass proliferated using the previously developed micropropagation protocols, thus facilitating the release and study of new mint genotypes. Crop improvement within *Mentha* also has relied on callus and cell suspension cultures to study essential oil biosynthesis.

21.4.2 Micropropagation

Generally, *Mentha* species have been easy to work with *in vitro*. Proliferating shoot cultures have been initiated by culturing seedling parts (Kitto 2008; Tisserat 1996), nodal segments (Bhat et al. 2001; Chishti et al. 2006; Mucciarelli et al. 1995; Rech and Pires 1986), shoot tips (Hirata et al. 1990; Repcáková et al. 1986), or stolons (Poovaiah, Weller, and Jenks 2006). They may derive from growth chamber-, greenhouse- (Berry, Van Eck, and Kitto 1997; Holm et al. 1989; Repcáková et al. 1986), or field-maintained (Chishti et al. 2006; Rech and Pires 1986) stock plants. The more sanitary the *ex vitro* environment is, the easier it has been to establish clean *in vitro* cultures (Berry et al. 1997).

Simple surface disinfestation regimes coupled with commonly used medium formulations (Table 21.5A) have been used to micropropagate mints. Once shoot cultures have been established, verification that they are free of contaminants, especially if they are to serve as a germplasm repository or are to be shared with other laboratories, is critical. Buckley, DeWilde, and Reed (1995) and Reed, Buckley, and DeWilde (1995) have developed procedures for identifying and eliminating endophytic bacteria contaminating *in vitro* maintained *Mentha* accessions.

Typically, medium devoid of growth regulators results in cultures of unbranched, elongated shoots, while medium supplemented with growth regulators results in axillary bud proliferation (Towill 2002). Mints have proliferated on basal medium supplemented with cytokinin alone, auxin alone, or a combination of cytokinin and auxin (Table 21.5A); however, butyric acid alone or in conjunction with an auxin most commonly has been used (Table 21.5A). Hirata et al. (1990) found that B5 medium (Gamborg, Miller, and Ojima 1968) supplemented with NAA (naphthaleneacetic acid) and thiamine hydrochloride influenced monoterpene concentration in *M. spicata* microshoots compared to mother stocks.

In *Mentha* species, electric field (Matsuo and Uchino 1993), light color (Bhat et al. 2001), and physical environment (culture chamber capacity, medium volume, and culture density; Tisserat and Silman 2000) have influenced shoot proliferation. Tisserat and Vaughn (2008) found that shoot proliferation and overall biomass were greatest when *M. spicata* shoots were cultured in liquid medium. Frequency of medium replenishment and of shoot immersions was directly correlated with increased growth (Tisserat and Vaughn 2008). Growth of *M. spicata* shoots in an automated plant culture system using liquid medium was the most cost-effective protocol studied (Tisserat 1996).

21.4.3 Rooting, Acclimation, and Reestablishment

Microshoots of *Mentha* have been quick to root, *in vitro* or *ex vitro*, and easy to transplant and reestablish under greenhouse and field conditions. Microshoots have rooted *in vitro* in growth regulator-free medium (Table 21.5A) or in basal medium supplemented with low concentrations of auxin (Table 21.5A). Microshoots have rooted *ex vitro* under mist or high humidity in a greenhouse (Kitto 2008; Kukreja et al. 1991), in techniculture plugs (McCown 1986), or any number of soil-less greenhouse mixes (Caissard et al. 1996; Ghanti et al. 2004; Holm et al. 1989; Phatak and Heble 2002;

Table 21.5A   Historical Perspective of Micropropagation of *Mentha* spp.

| Genotype | Basal Medium[1] | Growth regulator type and concentration[2] | | Rooting environment[3] | Citation |
		Proliferation	Rooting		
M. canadensis, M. ×piperita, M. pulegium, M. spicata	MS	4.4 or 8.8 µM BA or 9.3 µM K	4.4 or 8.8 µM BA or 9.3 µM K	EV laboratory	Rech & Pires, 1986
M. ×piperita 'Bulharská,' *M. ×piperita* 'Krasnodarská'	LS	NR	0	IV	Repcáková *et al.*, 1986
M. ×piperita 'Mitcham,' *M. ×piperita, M. aquatica* var. *citrata, M. spicata, M. canadensis, M. aquatica × M. spicata* (M312)	MS	2.2 µM BA + 0.5 µM IBA	NR		Towill, 1988
M. ×piperita	MS	9.3 µM K + 0.05 µM NAA	9.3 µM K + 0.05 µM NAA	IV	Holm *et al.*, 1989
M. ×piperita 'Perpeta'	LS	2.3 µM K	0	IV	Nádaská, *et al.*, 1990
M. ×piperita	MS	4.4 µM BA	0	EV laboratory	Van Eck & Kitto, 1990
M. aquatica var. *citrata*	MS	0	NA		Spencer *et al.*, 1990
M. aquatica × M. spicata	MS	2.2 µM BA + 0.5 µM IBA	NA		Towill, 1990
M. canadensis "CIMAP/Hybrid-77"	MS	8.8 – 13.3 µM BA + 5.7 µM IAA	5.4 µM NAA	IV	Kukreja *et al.*, 1991
M. ×piperita	LS	8.8 µM BA	0	IV	Čellárová, 1992
M. canadensis "CIMAP/Hybrid-77"		see Kukreja *et al.*, 1991			Kukreja *et al.*, 1992
M. aquatica var. *citrata, M. ×piperita* 'Mitcham'	MS	4.4 µM BA	0	GH	Van Eck & Kitto, 1992
M. ×piperita 'Blackmint'	B5	0	0	IV	Sato *et al.*, 1993
M. aquatica var. *citrata, M. ×piperita* nothovar. *piperita*	MS	0	NA		Spencer *et al.*, 1993
M. ×piperita	B5	2.2 µM BA + 0.5 µM NAA	0	IV	Sato *et al.*, 1994
M. spicata, M. suaveolens, M. spicata var. *crispata, M. canadensis, M. longifolia, M. aquatica* var. *citrata, M. ×smithiana, M. ×villosa, M. suaveolens × M. longifolia, M. aquatica* var. *citrata × M. aquatica, M. aquatica × M. spicata, M. canadensis × M. spicata, M. spicata × M. suaveolens*	MS	2.2 µM BA	NA		Reed *et al.*, 1995

M. spicata, M. suaveolens, M. spicata var. crispata, M. canadensis, M. longifolia, M. aquatica var. citrata, M. ×smithiana, M. ×villosa, M. suaveolens × M. longifolia, M. aquatica var. citrata × M. aquatica, M. aquatica × M. spicata, M. canadensis × M. spicata, M. spicata × M. suaveolens	MS	2.2 µM BA	NA		Buckley et al., 1995
M. ×piperita 'Maine et Loire'	MS	0	0	IV	Mucciarelli et al., 1995
M. ×piperita 'Mitcham' "19"	MS modified	0	NR	IV	Caissard et al., 1996
M. ×piperita 'Mitcham-Milly', M. ×piperita 'Hongrie'	MS	0	0	IV	Chaput et al., 1996
M. aquatica var. citrata 'Bergamote-Latour'	MS	0	0	IV	
M.×piperita 'Mitcham', M. ×gracilis	MS	0	NA	NA	Berry et al., 1996
M. spicata	MS	0		NA	Tisserat, 1996
M. ×piperita, M. arvensis 'Variegata'	B5	2.2 µM BA + 0.54 µM NAA	0	IV	Sato et al., 1996
M. ×piperita "MP-1"	MS	9.3-13.9 µM K & 5.7 µM IAA or 8.8 - 13.2 µM BA & 1.4 µM IAA	1.4 µM IAA	IV	Kukreja, 1996
M. ×gracilis, M. spicata	MS	44.4 µM BA	0	GH	Berry et al., 1997
M. ×piperita 'Mitcham'	MS	44.4 µ BA	0	GH	
M. spicata, M. ×piperita	MS	0	0	IV	Faure et al., 1998
M. ×piperita 'Mitcham', M. ×villoso-nervata	MS	0	0	IV	Krasnyanski et al., 1998
M. ×piperita 'Mitcham'	MS	0	0	IV	Krasnyanski et al., 1999
M. ×piperita 'Mitcham' "Digne 38," M. ×piperita 'Mitcham' "Ribecourt 19," M. ×piperita 'Todd Mitcham'	B5	0	0	IV	Jullien et al., 1998
M. ×piperita	MS	0.4 µM BA	0.05 µM NAA	IV	Niu et al., 1998
M. ×piperita 'Mitcham' "38"	MS modified	0	0	IV	Diemer et al., 1998
M. canadensis 'Kalka'	MS	22.2 µM BA + 5.4 µM NAA	0	IV	Shasany et al., 1998
M. canadensis 'Himalaya'	MS	44.4 µM BA + 10.7 µM NAA	0	IV	
M. spicata	MS	0.05 µM NAA	0.05 µM NAA	IV	Li et al., 1999
M. spicata	MS	0	0	IV	Hirai, & Sakai, 1999

(continued)

Table 21.5A Historical Perspective of Micropropagation of *Mentha* spp. (Continued)

Genotype	Basal medium[1]	Growth regulator type and concentration[2]		Rooting environment[3]	Citation
		Proliferation	Rooting		
M. spicata, M. canadensis		see Caissard *et al.*, 1996			Diemer *et al.*, 1999
M. canadensis, M. spicata, M. suaveolens	MS	2.2 µM BA + 0.5 µM IBA	NA		Reed, 1999
M. canadensis "CIMAP/Hybrid-77"		see Kukreja *et al.*, 1991			Kukreja *et al.*, 2000
M. spicata	MS	0	0	IV	Tisserat & Silman, 2000
M. xpiperita 'Mitcham'		see Niu *et al.*, 1998			Niu *et al.*, 2000
M. canadensis 'Gomti,' *M. canadensis* 'Shivalik,' *M. canadensis* "MAS-1," *M. canadensis* 'Kalka,' *M. canadensis* 'Himalaya,' *M. canadensis* "MAH-3"	MS	10 µM TDZ	0	IV	Bhat *et al.*, 2001
M. xpiperita 'Mitcham' "38," *M. canadensis* "101"		see Diemer *et al.*, 1998, 1999			Diemer *et al.*, 2001
M. xpiperita 'Mitcham'	MS	0.05 µM NAA	0.05 µM NAA	IV	Li *et al.*, 2001
M. xpiperita 'Mitcham'		see Niu *et al.*, 1998,2000			Mahmoud & Croteau, 2001
Mentha sp.	MS	0	0	IV	Tisserat & Vaughn, 2001
M. canadensis	MS: proliferation; 1/2 MS rooting	0	11.4 µM IAA	IV	Phatak & Heble, 2002
M. xpiperita 'Mitcham'		see Niu *et al.*, 1998,2000			Mahmoud & Croteau, 2003
M. xvillosa accessions MEN 204 & MEN 148, *M. xpiperita* accessions MEN 186 & MEN 166	MS	0	0	IV	Islam *et al.*, 2003
M. xpiperita	MS: proliferation; B5: rooting	0	0	IV	Inoue *et al.*, 2003a
M. xpiperita	MS	0	0	IV	Inoue *et al.*, 2003b

M. xpiperita 'Mitcham'	see Niu *et al.*, 1998				Mahmoud *et al.*, 2004
M. xpiperita	MS	4.4 µM BA	5.4 µM NAA	IV	Ghanti *et al.*, 2004
M. spicata	MS	0	0	IV	Tisserat & Vaughn, 2004
M. spicata	MS	0	0.05 µM NAA	IV	Poovaiah *et al.*, 2006
M. canadensis	MS	8.9 µM BA	4.4 µM IAA	IV	Chishti *et al.*, 2006
M. spicata	MS	0	0	IV	Tisserat & Vaughn, 2008
Mentha spp.	MS	5 µM BA	0	GH	Kitto, 2008
M. xpiperita 'Mitcham'	see Niu *et al.*, 1998				Wang *et al.*, 2009
M. canadensis	MS: proliferation; 1/2 MS: rooting	0.4 µM BA + 2.7 µM NAA	1.1 µM NAA	IV	Kumar *et al.*, 2009

[1]MS = Murashige and Skoog (1962); LS = Linsmaier and Skoog (1965); B5 = Gamborg *et al.* (1968)

[2]BA = N[6]-benzyladenine [also known as N-(phenyl-methyl)-1*H*-purin-6-amine; benzyladenine; or 6-benzylaminopurine]; 4-CPPU=N-(2-chloro-4-pyridyl)-N'-phenylurea; IAA = indole-3-acetic acid; IBA = indole-3-butyric acid; 2iP = N[6]-(2-isopentyl) adenine; K = kinetin [also known as 6-furfurylaminopurine]; NAA = α-naphthaleneacetic acid; 4-PU = N-phenyl-N'-4-pyridyl urea; TIBA = triiodobenzoic acid; TDZ = thidiazuron (also known as N-phenyl-N'-1,2,3-thidiazol-5'-ylurea; Z = zeatin. NR = not reported; NA = not applicable; 0 = no growth regulators used.

[3]GH = greenhouse; IV = *in vitro*, EV = *ex vitro*.

Table 21.5B Historical Perspective of Regeneration of *Mentha* spp.

Genotype	Explant	Direct/ indirect	Basal medium[1]	Growth regulator type and concentration[2]	Environ.	Addenda	Citation
M. xpiperita 'Bulharská,' *M. xpiperita* 'Krasnodarská'	leaf	direct	LS	0.4 - 22 µM BA or 0.46 - 4.6 µM K	2500 lux		Repčáková et al., 1986
M. xpiperita 'Perpeta'	leaf	direct	LS	2.3 µM K			Nádaská, et al., 1990
M. xpiperita	immature embryo	indirect	MS	4.4 µM BA + 2.0 µM TIBA	dark		Van Eck & Kitto, 1990
	mature embryo	indirect	MS	2.2 µM BA + 2.7 µM NAA	dark		
M. aquatica var. citrata	stem	indirect	MS	0	2000 lux		Spencer et al., 1990
M. canadensis "CIMAP/ Hybrid-77"	node	indirect	MS	13.3 µM BA + 5.7 µM IAA	58.5 µM·m^{-2}·s^{-1}		Kukreja et al., 1991
M. canadensis "CIMAP/ Hybrid-77"				see Kukreja et al., 1991			Kukreja et al., 1992
M. aquatica var. citrata, M. xpiperita 'Mitcham'	leaf disk	direct	MS	44.4 µM BA	dark	25% CW[3]	Van Eck & Kitto, 1992
M. xpiperita 'Blackmint'	leaf protoplast	indirect	B5	22.2 µM BA + 0.54 µM NAA	3000 lux	0.2 M mannitol	Sato et al., 1993
M. aquatica var. citrata, M. xpiperita nothovar. piperita	stem	indirect	MS	0	2000 lux		Spencer et al., 1993
M. xpiperita	protoplast	indirect	B5	2.3 µM 4-PU + 0.5 µM NAA	3000 lux	0.2 M mannitol	Sato et al., 1994
M. xpiperita 'Mitcham' "19"	internode	direct & indirect	MS	0, 8.8, 22 µMBA or 8.8 µM BA + 10.7 µM NAA	dark		Caissard et al., 1996
M. xpiperita 'Mitcham-Milly,' *M. xpiperita* 'Hongrie'	leaf protoplast	indirect	MS	13.2 µMBA or 3.6 µMBA → 4.4 µMBA	dark → 60 µM·m^{-2}·s^{-1}		Chaput et al., 1996
M. aquatica var. citrata 'Bergamote-Latour'	leaf protoplast	indirect		1.8 µM TDZ → 1.8 µM or 3.6 µM BA → 4.4 µM BA	dark → 60 µM·m^{-2}·s^{-1}		
M. xpiperita, M. arvensis 'Variegata'	fused leaf protoplasts	indirect	B5	2.3 µM 4-PU + 0.54 µM NAA	50 µM·m^{-2}·s^{-1}	0.2 M mannitol	Sato et al., 1996
M. xpiperita "MP-1"	node		MS	13.9 µM K + 5.7 µM IAA or 8.8 - 13.2 µM BA+ 5.7 µM IAA or 1.3 µM NAA	40 µM·m^{-2}·s^{-1}		Kukreja, 1996

Species/cultivar	Explant	Type	Medium	PGR	Light	Osmoticum	Reference
M. ×gracilis, M. spicata	petiole	direct	MS	44.4 µM BA	dark	25% CW	Berry *et al.*, 1997
M. ×piperita 'Mitcham'	petiole	direct	MS	44.4 µM BA or 24.6, 49.2, 123 µM 2iP	dark	25% CW	
M. spicata, M. ×piperita	leaf disk	indirect	MS	2 µM BA + 2 µM IBA → 0.5 µM NAA + 9.0 µM BA + 0.5 µM TDZ	dark	300 mM mannitol	Faure *et al.*, 1998
M. ×piperita 'Mitcham,' *M. ×villoso-nervata*	fused stem + leaf protoplasts	indirect	MS	8.8 µM BA + 13.6 µM TDZ → 4.4 µM BA + 9.1 µM TDZ	dark → 25 µM·m⁻²·s⁻¹	25% CW	Krasnyanski *et al.*, 1998
M. ×piperita 'Mitcham' "Digne 38," *M. ×piperita* 'Mitcham' "Ribecourt 19," *M. ×piperita* 'Todd Mitcham'	protoplast	indirect	1/2 B5	2.3 µM TDZ + 0.5 µM NAA	10 µM·m⁻²·s⁻¹	300 mM mannitol	Jullien *et al.*, 1998
M. ×piperita	leaf	direct & indirect	MS	8.4 µM TDZ	dark	25% CW	Niu *et al.*, 1998
M. ×piperita 'Mitcham' "38"	leaf disk	indirect	MS	2 µM BA + 2 µM IBA	dark	300 mM mannitol	Diemer *et al.*, 1998
M. canadensis 'Kalka'	internode	direct	MS	22.2 µM BA + 5.4 µM NAA	400 - 600 lux		Shasany *et al.*, 1998
M. canadensis 'Himalaya'	internode	direct	MS	44.4 µM BA + 10.7 µM NAA	400 - 600 lux		
M. spicata	leaf	direct & indirect	MS	16.8 µM TDZ	dark	25% CW	Li *et al.*, 1999
	internode	direct & indirect	MS	4.5, 11.4, 22.7 or 34.0 µM TDZ	dark	25% CW	
M. spicata, M. canadensis				see Faure *et al.*, 1998			Diemer *et al.*, 1999
M. ×piperita 'Mitcham'	internode	indirect	MS	8.8 µM BA + 16.8 µM TDZ	dark	25% CW	Krasnyanski *et al.*, 1999
	internode protoplast	indirect	MS	8.8 µ M BA + 12.6 µM TDZ	dark	25% CW	
M. canadensis "CIMAP/Hybrid-77"				see Kukreja *et al.*, 1991			Kukreja *et al.*, 2000
M. ×piperita 'Mitcham'				see Niu *et al.*, 1998			Niu *et al.*, 2000
M. canadensis 'Gomti,' *M. canadensis* 'Shivalik,' *M. canadensis* "MAS-1," *M. canadensis* 'Kalka,' *M. canadensis* 'Himalaya,' *M. canadensis* "MAH-3"	leaf	indirect	MS	60 µM BA + 0.5 µM NAA	red light		Bhat *et al.*, 2001
	internode	direct	MS	60 µM BA + 0.05 µM NAA	red light		

(continued)

Table 21.5B   Historical Perspective of Regeneration of *Mentha* spp. (Continued)

Genotype	Explant	Direct/ indirect	Basal medium[1]	Growth regulator type and concentration[2]	Environ.	Addenda	Citation
M. ×piperita 'Mitcham' "38," *M. canadensis* "101"	internode	direct		40 μM BA + 0.5 μM NAA / see Diemer et al., 1998, 1999			Diemer et al., 2001
M. ×piperita 'Mitcham'	leaf	direct & indirect	MS	8.4 μM TDZ	dark	25% CW	Li et al., 2001
M. ×piperita 'Mitcham'				see Niu et al., 1998,2000			Mahmoud & Croteau, 2001
M. canadensis	leaf	NR	MS	22.2 μM BA + 2.7 μM NAA	light		Phatak & Heble, 2002
M. ×piperita 'Mitcham'				see Niu et al., 1998,2000			Mahmoud & Croteau, 2003
M. ×piperita	internode	indirect	1/2 MS	10 μM 4-CPPU + 1 μM NAA	2500 lux		Inoue et al., 2003a
M. ×piperita	leaf & internode	indirect	1/2 B5	10 μM 4-CPPU + 1 μM NAA	2500 lux		Inoue et al., 2003b
M. ×piperita 'Mitcham'	see Niu et al., 1998						Mahmoud et al., 2004
M. spicata	internode	indirect	MS	4.4 μM TDZ + 4.54 μM Z	dark	10% CW	Poovaiah et al., 2006
M. ×piperita 'Mitcham'	internode	direct & indirect	MS	11.35 μM TDZ + 4.54 μM Z	dark	10% CW	Wang et al., 2009
M. canadensis	leaf	direct	MS	5 μM BA + 1.1 μM NAA	dark	25% CW	Kumar et al., 2009

[1] MS = Murashige and Skoog (1962); LS = Linsmaier and Skoog (1965); B5 = Gamborg et al. (1968)

[2] BA = N^6-benzyladenine [also known as N-(phenyl-methyl)-1H-purin-6-amine; benzyladenine; or 6-benzylaminopurine]; 4-CPPU=N-(2-chloro-4-pyridyl)-N'-phenylurea; IAA = indole-3-acetic acid; IBA = indole-3-butyric acid; 2iP = N^6-(2-isopentyl) adenine; K = kinetin [also known as 6-furfurylaminopurine]; NAA = α-naphthaleneacetic acid; 4-PU = N-phenyl-N'-4-pyridyl urea; TIBA = triiodobenzoic acid; TDZ = thidiazuron (also known as N-phenyl-N'-1,2,3-thidiazol-5'-ylurea; Z = zeatin. NR = not reported; NA = not applicable; 0 = no growth regulators used.

[3] CW = coconut water

Table 21.5C Historical Perspective of Transformation of *Mentha* spp.

Genotype	Method[1]	Agrobacterium strain	Plasmid	Gene	Product	Explant	Citation
M. aquatica var. citrata	At	T37, C58	nopaline pTi		shooty teratomas		Spencer *et al.*, 1990
M. aquatica var. citrata, M. xpiperita nothovar. *piperita*		see Spencer et al., 1990					Spencer *et al.*, 1993
M. xpiperita 'Mitcham' "19"	At	GV2260/GI, C58MP90/GI	p35S GUS INT	*gus*	transient	leaf	Caissard *et al.*, 1996
M. xpiperita 'Mitcham'; *M. xgracilis*	At	A281	pTiBo542		callus	petiole	Berry *et al.*, 1996
M. aquatica var. citrata	At	A281	pTiBo542		callus	greenhouse leaf disk	
M. xpiperita	At	EHA105	pBISNI	*gus*	stable plants	leaf	Niu *et al.*, 1998
M. xpiperita 'Mitcham' "38"	At	C58pMP90/GI, GV2260/GI, AGL1 BGI, EHA105/MOG	p35S GUS INT, pB+GIN	*gus, nptII*	stable plants	leaf disk	Diemer *et al.*, 1998
M. spicata, M. canadensis	At	GV2260/GI, EHA105/MOG	p35S GUS INT	*gus, nptII*	stable plants	leaf disk	Diemer *et al.*, 1999
M. xpiperita 'Mitcham'	D, At	GV3101 (pMP90)	pGA643	*gus, nptII, 4S-ls*	stable plants	internode protoplast	Krasnyanski *et al.*, 1999
M. xpiperita 'Mitcham'	At	EHA105	pBISN1, pOC	*gus, nptII*	stable plants	leaf	Niu *et al.*, 2000
M. xpiperita 'Mitcham' "38," *M. canadensis* "101"	At	C58pMP90	pBILS	*nptII, 4S-ls*	stable plants	leaf disk	Diemer *et al.*, 2001
M. xpiperita 'Mitcham'	At	EHA105, GV3103	pGPTV, pCASI	*bar*	stable plants	leaf	Li *et al.*, 2001
M. xpiperita 'Mitcham'	At	EHA105	pGAdekG/Nib.L	*dxr*, antisense *mfs*	stable plants	leaf disk	Mahmoud & Croteau, 2001
M. xpiperita 'Mitcham'	At	EHA105	pGAMFSS	*mfs*		leaf disk	Mahmoud & Croteau, 2003
M. xpiperita	Ar	MAFF0301724			stable plants	internode	Inoue *et al.*, 2003a
M. xpiperita	Ar	IFO14554	pBI 121	*gus, rol*	stable plants	leaves	Inoue *et al.*, 2003b
M. xpiperita 'Mitcham'	At	EHA105	pGADXRS	*smls, pmlh*	stable plants	leaf disk	Mahmoud *et al.*, 2004
M. canadensis	At	LBA4404	pCAMBIA-CpGS	*gus, nptII*	stable plants	leaf	Kumar *et al.*, 2009

[1] Method: D = direct; At = *Agrobacterium tumefaciens*; Ar = *Agrobacterium rhizogenes*

Van Eck and Kitto 1990). Regardless of medium or environment, microshoots have rooted within 2–4 weeks. Reported survival of transplanted *in vitro* generated plants has been 90% or greater (Ghanti et al. 2004; Jullien et al. 1998; Kukreja 1996; Li et al. 1999).

21.4.4 Conservation of Germplasm

Two approaches to germplasm preservation with mints are long-term storage and cryopreservation. Both approaches are concerned with reducing the costs, space, and time associated with conventional maintenance regimes whether field or tissue culture based. Reed (1999) found that *M. canadensis* and *M. spicata* can be stored for 2–3 years (4°C, 12 h photoperiod) on MS medium (Murashige and Skoog 1962) containing 50% nitrogen; *M. suaveolens* can be stored for 1.5 years (25°C, dark) prior to being subcultured. Islam et al. (2003) studied storage at 2°C for 6 months and found that cultivars as well as species differ in their response.

Towill (2002) recently provided an overview of cryopreservation and mints. Early work demonstrated that there were survival differences among several genotypes cryopreserved (shoot tips ≈ 0.3–0.5 mm in size). However, each genotype after exposure to liquid nitrogen did survive and grow; within a month, shoots were transplantable (Towill 1988, 1990). Cryopreservation of *M. spicata* and *M. canadensis* was improved when alginate-encapsulated apical domes (0.8 mm in size) were vitrified prior to cryopreservation (Hirai and Sakai 1999). Plants grown from cryopreserved shoot tips have been true to type (Hirai and Sakai 1999; Towill 2002).

21.4.5 Regeneration

Although ease of regenerability in *Mentha* is genotype dependent (Bhat et al. 2001; Jullien et al. 1998; Shasany et al. 1998; Van Eck and Kitto 1990), many genotypes have been regenerated and most of the work has been done on economically important genotypes (Table 21.5B). Embryos (Van Eck and Kitto 1990), leaves (Faure et al. 1998; Li et al. 1999; Nádaská, Erdelsky, and Cupka 1994), internodes (Poovaiah et al. 2006; Shasany et al. 1998; Wang et al. 2009), nodes (Kukreja 1996), and roots (Inoue et al. 2003a) have been used as primary explant sources (Table 21.5B). Stock plants maintained in a greenhouse (Repcáková et al. 1986; Van Eck and Kitto 1992) and *in vitro* (Kukreja 1996; Niu et al. 1998, 2000) have served as explant sources. Leaves and internodes commonly have been collected near shoot apices and there appears to be a gradient within leaves, with the basal portion more regenerative (Repcáková et al. 1986; Kukreja 1996; Niu et al. 1998; Van Eck and Kitto 1992). Caissard et al. (1996) found leaves obtained from greenhouse-maintained stock plants to be more regenerative compared to leaves obtained from *in vitro* maintained cultures of *M.* × *piperita*.

Mentha species regenerate through a shoot organogenic pathway that may be direct or indirect (Table 21.5B). In direct regeneration, progenitors of regeneration derive from palisade parenchyma cells associated with the vascular tissue (Van Eck and Kitto 1992) as well as densely cytoplasmic cells at the petiole–leaf boundary (Caissard et al. 1996). Repcáková et al. (1986) found leaf mesophyll cells to be mitotically active. Bhat et al. (2001) found that *M. canadensis* shoots regenerated from within nodal explants where vascularized regions arose directly from the vasculature, and from within internodal explants where vascularized regions arose directly in the cortical region. Indirect regeneration involves the generation of callus from which shoots differentiate. Early work demonstrated that *M.* ×*piperita* could be regenerated from embryo-generated callus (Van Eck and Kitto 1990) and from shoot-generated teratomas (Spencer et al. 1993). Within *M. canadensis* leaves, regeneration was from callus generated from palisade cells (Bhat et al. 2001).

An MS-based medium supplemented with at least a cytokinin most commonly has been used to regenerate mints, although the cytokinin type and concentration have varied based on laboratory and it has been customary to add an auxin (Table 21.5). Other features of successful regeneration protocols have included the addition of supplements such as coconut water or mannitol and placing the cultures under low light or dark conditions (Table 21.5B).

21.4.6 Variation

Regeneration for the purpose of creating variation in *Mentha* species has taken several approaches. Researchers have taken advantage of the variation that may occur in somatic cell regenerants, somaclonal variation, or have overtly tried to create variation via gamma irradiation, protoplast fusion, or transformation. The success of all of these approaches depends on regeneration frequency—the greater the number of shoots to regenerate from an explant the better.

Typically, regenerants have not been phenotypically variable (Niu et al. 1998; Wang et al. 2009); however, genotypic variation in plants regenerated from protoplasts (Jullien et al. 1998) and phenotypic variation in one regenerant from leaf callus (Diemer 2001) have been reported. Kukreja et al. (1991, 1992, 2000) evaluated the regenerants from cultured *M. canadensis* var. *piperascens* nodes and found variation in the agronomic traits (e.g., plant height, herb yield) of the plants. Phatak and Heble (2002) found essential oil profiles of *M. canadensis* regenerants did not match that of the control stock plants while *in vitro*; however, when field grown, the terpenoid profile was similar to that of the control stock plants. The plants regenerated from gamma-irradiated petioles of peppermint, native spearmint, and Scotch spearmint were not visibly variant (Berry et al. 1997).

21.4.7 Protoplasts

In an effort to improve mints, regeneration protocols using individual protoplasts and protoplast-fusion products have been developed. Leaf protoplasts of *M. aquatica* 'Bergamote-Latour' (Chaput et al. 1996) and *M. ×piperita* (Chaput et al. 1996; Jullien et al. 1998; Sato et al. 1993, 1994) have regenerated shoots. After one growing season, all protoplast-derived *M. ×piperita* 'Mitcham-Milly' ssp. *vulgaris* plants compared to the control had decreased menthone and menthol and increased carbone (Chaput et al. 1996). Leaf protoplasts of peppermint (*M. ×piperita* 'Blackmint') were electrofused with leaf protoplasts of gingermint (*M. arvensis* 'Variegata') (Sato et al. 1996). While plating efficiency was only 5%, analysis of regenerants showed that one plant was an interspecific somatic hybrid. Krasnyanski, Ball, and Sink et al. (1998) created somatic hybrids by fusing stem protoplasts of *M. ×piperita* 'Mitcham' with leaf mesophyll protoplasts of *M. ×villoso-nervata*. To date, plants resulting from fusion products have not been improved compared to parent genotypes.

21.4.8 Transformation

Concurrent with the development of regeneration protocols has been the development of transformation protocols. Three transformation approaches have been used on mints: microprojectile bombardment, direct DNA cocultivation, and *Agrobacterium* cocultivation. *M. ×piperita* 'Mitcham' was not transformed using bombardment (Caissard et al. 1996; Niu et al. 1998), and while stem-derived protoplasts were directly transformed after cocultivation with the limonene synthase gene (Krasnyanski et al. 1999), the majority of *Mentha* transformation studies have used *Agrobacterium* cocultivation (Table 21.5C). Early work with *Agrobacterium tumefaciens* transformation of *Mentha* spp. focused solely on the development of protocols and, as such, the plasmid constructs contained no economically important genes (Berry et al. 1996; Diemer et al. 1998, 1999; Niu et al. 1998, 2000; Spencer et al. 1993). Transformed plants of *Mentha ×piperita* generated from *A. rhizogenes*-inoculated internodes had altered essential oil profiles (Inoue et al. 2003a, 2003b).

Recent articles provide very clear, extensive explanations of what needs to be done and has been done to improve mints with regard to essential oil yield and quality and agronomic traits such as weed and disease resistance (Jullien 2007; Lange and Croteau 1999c; Veronese et al., 2001; Wildung and Croteau 2005). To better understand monoterpene metabolism, the limonene synthase cDNA has been transformed into plants of *M.* ×*piperita* 'Mitcham' (Krasnyanski et al. 1999), *M.* ×*piperita* "Mitcham 38" (Diemer 2001), and *M. canadensis* "101" (Diemer 2001). Mahmoud et al. (2004) used cDNA clones of limonene synthase and limonene-3-hydroxylase and Mahmoud and Croteau (2001, 2003) used sense and antisense versions of (+)-menthofuran synthase cDNA to transform *M.* ×*piperita* 'Mitcham.' The resulting plants, while phenotypically normal, have had altered essential oil yield or quality and have helped to clarify the critical metabolic pathways necessary for improving oil composition. *M.* ×*piperita* 'Mitcham' has been transformed with the *bar* gene to confer resistance to bialaphos herbicide (Li et al. 2001), and *M. canadensis* has been transformed with the glutathione synthetase gene to confer resistance to the heavy metal cadmium sulfate (Kumar et al. 2009).

21.4.9 Essential Oil Biosynthesis/Biotransformation via Callus and Cell Suspension

In an effort to better understand essential oil biosynthesis/biotransformation, callus and cell suspension cultures have been studied. The use of callus and cell suspension cultures to elucidate essential oil biosynthetic pathways has been reviewed by Banthorpe (1996). Early work developed protocols for initiating and maintaining callus and cell suspension cultures (Lin and Staba 1961; Wang and Staba 1963) to study and understand their capability for essential oil biosynthesis (Staba, Laursen, and Büchner 1965) and biotransformation (Park et al. 1994). Callus cultures, for the most part, have had oil spectra dissimilar to those of intact plants (Becker 1970; Suga, Hirata, and Yamamoto 1980; Brown and Banthorpe 1992); however, essential oil synthesis has depended on species and cell line within a species (Charlwood and Charlwood 1983). Also, Kireyeva et al. (1978) found peppermint callus synthesized monoterpenes similar to those within intact plants.

Many have worked on refining peppermint culture protocols to improve essential oil synthesis (Chae and Park 1994; Chakraborty and Chattopadhyay 2008; Galun et al. 1985; Jin and Chae 1991; Kim et al. 1996; Park and Chae 1990). Intermediate compounds within the menthol pathway have been fed to suspension cultures with mixed biotransformation success (Staba et al. 1965; Park et al. 1994). Aviv and Galun (1978) found that biotransformation was specific; *Mentha* cell lines could convert pulegone while tomato cell lines could not.

21.5 SUMMARY

Mendelian inheritance of monoterpenes and disease resistance has been demonstrated; however, with the knowledge that polyploidy, complement fractionation, and transgressive segregation occur in *Mentha*, perhaps the best method for conventional breeding is simply to raise large numbers of progeny from open-pollinated superior clones. For example, utilizing alternate rows of *M. canadensis* 'Kalka' and 'Gomti' in the field, open-pollinated seeds were collected, and the subsequent progeny was evaluated for yield and disease resistance, resulting in the release of two superior clones: Himalaya and Kosi (Kumar et al. 1997, 1999). Himalaya was also patented in the United States (U.S. Plant Patent 10935). Open-pollinated seed progenies from *M. canadensis* 'Shivalik' resulted in 'Damroo', a cultivar with a high homogeneity of seed progeny (Patra et al. 2001). Likewise, open-pollinated progeny resulted in a hybrid of *M. aquatica* called 'Aquamint', which can be blended with peppermint oil to increase the menthofuran oil and thus match Yakima oil (Roberts and Plotto 2002).

Gamma irradiation of large numbers of vegetative clones is another possibility, as demonstrated with *M.* ×*piperita* 'Todd's Mitcham' and 'Murray Mitcham' (Murray and Todd 1972; Todd et al. 1977). In both cases, "large numbers" are the key words, implying the generation and analysis of thousands of progeny. This should not imply a daunting task because the plants could be screened by trained organoleptic panels for evaluation of the oils before chemical analysis or selection for disease resistance, thereby reducing the numbers significantly.

Utilizing far fewer progeny, Jim Westerfield, who operates Westerfield House, a bed-and-breakfast inn in Freeburg, Illinois, selected open-pollinated seedlings of *Mentha* for the U.S. herb plant trade. This first resulted in the patented 'Hillary's Sweet Lemon' (U.S. Plant Patent 9197). Lately, using trademarks rather than patents, Westerfield has released Berries & Cream™, Candied Fruit™, Candy Lime™, Pink Candypops™, Citrus Kitchen™, Cotton Candy™, Jim's Fruit™, Fruit Sensations™, Fruitasia™, Italian Spice™, Julia's Sweet Citrus™, Margarita™, Marilyn's Salad™, Oregon-Thyme™, Sweet Pear™, and Wintergreen™. Westerfield's mints are currently sold by Richters in Goodwood, Ontario, Canada.

In regard to biotechnology, *Mentha* species have been shown to be very malleable in their response *in vitro*, whether the goal has been micropropagation, long-term storage/cryopreservation, regeneration, or transformation. In a relatively short time frame, this malleability *in vitro*, when dovetailed with gene discovery work, has resulted in the generation of transformants having commercially interesting traits. Continued improvement of mints will be limited by the discovery and generation of commercially important genetic constructs.

REFERENCES

Abad Farooqi, A. H., A. Misra, and A. A. Naqvi. 1983. Effect of plant age on quality and quantity of oil in Japanese mint. *Indian Perfumer* 27:80–82.

Akhila, A., D. V. Banthorpe, and M. G. Rowan. 1980. Biosynthesis of carvone in *Mentha spicata. Phytochemistry* 19:1433–1437.

Akhila, A., R. Srivastava, K. Rani, et al. 1991. Biosynthesis of (–)-mintlactone and (+)-isomintlactone in *Mentha piperita. Phytochemistry* 30:485–489.

Akhila, A., and R. S. Thakur. 1989. Effect of seasonal variation on the *E-Z* isomerization of acyclic allylic alcohols in the biosynthesis of cyclic monoterpenes in higher plants. *Fitoterapia* 60:429–437.

Allen, K. G., D. V. Banthorpe, B. V. Charlwood, et al. 1976. Metabolic pools associated with monoterpene biosynthesis in higher plants. *Phytochemistry* 15:101–107.

Alonso, W. R., J. E. Crock, and R. Croteau. 1993. Production and characterization of polyclonal antibodies in rabbits to 4*S*-limonene synthase from spearmint (*Mentha spicata*). *Archives of Biochemistry and Biophysics* 301:58–63.

Alonso, W. R., and R. Croteau. 1992. Comparison of two monoterpene cyclases isolated from higher plants; γ-terpinene synthase from *Thymus vulgaris*, and limonene synthase from *Mentha × piperita*. In *Secondary-metabolite biosynthesis and metabolism*, ed. R. J. Petroski and S. P. McCormick, 239–251. New York: Plenum Press.

Alonso, W. R., J. I. M. Rajaonarivony, J. Gershezon, et al. 1992. Purification of 4*S*-limonene synthase, a monoterpene cyclase from the glandular trichomes of peppermint (*Mentha × piperita*) and spearmint (*Mentha spicata*). *Journal of Biological Chemistry* 267:7582–7587.

Ammal, E. K. J., and S. N. Sobti. 1962. The origin of the Jammu mint. *Current Science* 31:387–388.

Aviv, D., A. Danes, E. Krochmal, et al. 1983. Biotransformation of monoterpenes by *Mentha* cell lines: Conversion of pulegone-substituents and related unsatured α-β ketones. *Planta Medica* 47:7–10.

Aviv, D., and E. Galun. 1978. Conversion of pulegone to isomenthone by cell suspension lines of *Mentha* chemotypes. In *Production of natural compounds by cell culture methods: Proceedings of an international symposium on plant cell culture*, ed. A. W. Alfermann and E. Reinhard, 60–67. München: Gesellschaft für Strahlen und Umweltforschung mbH.

Aviv, D., E. Krochmal, A. Dantes, et al. 1981. Biotransformation of monoterpenes by *Mentha* cell lines: Conversion of menthone to neomenthol. *Planta Medica* 42:236–243.

Banthorpe, D. V. 1996. *Mentha* species (mints): *In vitro* culture and production of lower terpenoids and pigments. In *Biotechnology in agriculture and forestry 37. Medicinal and aromatic plants IX*, ed. Y. P. S. Bajaj, 202–225. Berlin: Springer–Verlag.

Banthorpe, D. V., and B. V. Charlwood. 1972. The catabolism of biosynthetic precursors by higher plants. *Planta Medica* 22:428–433.

Banthorpe, D. V., B. V. Charlwood, and M. R. Young. 1972. Terpene biosynthesis: IV. Biosynthesis of (+) -pulegone in *Mentha pulegium* L. *Journal of Chemical Society Perkin Transactions* 12:1532–1534.

Banthorpe, D. V., O. Ekundayo, J. Mann, et al. 1975. Biosynthesis of monoterpenes in plants from ^{14}C-labeled acetate and CO_2. *Phytochemistry* 14:707–715.

Banthorpe, D. V., B. M. Modawi, I. Poots, and M. G. Rowan. 1978. Redox interconversions of geraniol and nerol in higher plants. *Phytochemistry* 17:1115–1118.

Battaile, J. 1960. Biosynthesis of terpenes in mint. *Dissertation Abstracts* 21:442.

Battaile, J. A., J. Burbott, and W. D. Loomis. 1963. Monoterpene interconversions: Metabolism on pulegone by a cell-free system from Mentha piperita. *Phytochemistry* 7:1159–1163.

Battaile, J., and W. D. Loomis. 1961. Biosynthesis of terpenes. II. The site and sequence of terpene formation in peppermint. *Biochimica et Biophysica Acta* 51:545–552.

Battu, R. G., and H. W. Youngken. 1968. Biogenesis of terpenoids in *Mentha piperita*. Part II. Di-, tri- and tetraterpenoids. *Lloydia* 31:30–37.

Becker, H. 1970. Untersuchungen zur Frage der Bildung flüchtiger Stoffwechselprodukte in Calluskulturen. *Biochemie und Physiologie der Pflanzen* 161:425–441.

Bellucci, M., C. Roscini, and A. Mariani. 2003. Cytomixis in pollen mother cells of *Medicago sativa* L. *Journal of Heredity* 94:512–516.

Berger, R. G., and F. Drawert. 1988. Glycosilation of terpenols and aromatic alcohols by cell suspension cultures of peppermint (*Mentha piperita* L.). *Zeitschrift fur Naturforschung C* 43:485–490.

Berry, C., J. M. Van Eck, and S. L. Kitto. 1997. *In vitro* irradiation and regeneration in mints. *Journal of Herbs Spices & Medicinal Plants* 4 (4): 17–26.

Berry, C., J. M. Van Eck, S. L. Kitto, et al. 1996. *Agrobacterium*-mediated transformation of commercial mints. *Plant Cell Tissue Organ Culture* 44:177–181.

Bertea, C. M., M. Schalk, F. Karp, et al. 2001. Demonstration that menthofuran synthease of mint (*Mentha*) is a cytochrome P450 monooxygenase: Cloning, functional expression, and characterization of the responsible gene. *Archives of Biochemistry and Biophysics* 390:279–286.

Bertea, C. M., M. Schalk, C. J. D. Mau, et al. 2003. Molecular evaluation of a spearmint mutant altered in the expression of limonene hydroxylases that direct essential oil monoterpene biosynthesis. *Phytochemistry* 64:1203–1211.

Bhat, S., S. K. Gupta, R. Tuli, et al. 2001. Photoregulation of adventitious and axillary shoot proliferation in menthol mint, *Mentha arvensis*. *Current Science* 80:878–881.

Bohlmann, J., G. Meyer-Gauen, and R. Croteau. 1998. Plant terpenoid synthases: Molecular biology and phylogenetic analysis. *Proceedings of National Academies of Science* 95:4126–4133.

Boldrini, K. R., M. S. Pagliarini, and C. B. do Valle. 2006. Cell fusion and cytomixis during microsporogenesis in *Brachiaria humidicola* (Poaceae). *South African Journal of Botany* 72:478–481.

Brown, G. D., and D. V. Banthorpe. 1992. Characteristic secondary metabolism in tissue cultures of the Labiatae: Two new chemotaxonomic markers. In *Advances in labiate science*, ed. R. M. Harley and T. Reynolds, 367–373. Kew, UK: Royal Botanic Gardens.

Brun, N., M. Colson, A. Perrin, et al. 1991. Chemical and morphological studies of the effects of ageing on monoterpene composition in *Mentha × piperita* leaves. *Canadian Journal of Botany* 69:2271–2278.

Buckley, P. M., T. N. DeWilde, and B. M. Reed. 1995. Characterization and identification of bacteria isolated from micropropagated mint plants. *In Vitro Cellular and Developmental Biology—Plants* 31:58–64.

Bugaenko, L. A., and S. A. Reznikova. 1974. Variation in chemical characters following self-pollination of mint [in Russian]. Trudy Vsesoyuznogo Nauchno-Issledovatel'skogo Insituta Efirnomaslichnykh Kul'tur 7:25–31.

———. 1984a. Genetic control of terpene biosynthesis in mint. I. The study of genotypes in certain *Mentha* species and intraspecific varieties for genes controlling biosynthesis of main essential oil components. *Genetika* 20:1857–1863.

———. 1984b. Genetic control of terpene biosynthesis in mint. II. Variability and inheritance of terpene composition at interspecific crossings. *Genetika* 20:2018–2024.

Burbott, A. J., R. Croteau, W. E. Shine, et al. 1974. Biosynthesis of cyclic monoterpenes by cell-free extracts of *Mentha piperita* L. In *VI International Congress of Essential Oils*, San Francisco, California. September 8-12, 1974, paper no. 17.

Burbott, A. J., J. P. Hennessey, W. C. Johnson, et al. 1983. Configuration of piperitone from oil of *Mentha piperita*. *Phytochemistry* 22:2227–2230.

Burbott, A. J., and W. D. Loomis. 1967. Effects of light and temperature on the monoterpenes of peppermint. *Plant Physiology (Lancaster)* 42:20–28.

Burke, C. C., K. Klettke, and R. Croteau. 2004. Heteromeric geranyl diphosphate synthease from mint: Construction of a functional fusion protein and inhibition by biphosphate substrate analogs. *Archives of Biochemistry and Biophysics* 422:52–60.

Burke, C. C., M. R. Wildung, and R. Croteau. 1999. Geranyl diphosphate synthase: Cloning, expression, and characterization of this prenyltransferase as a heterodimer. *Proceedings of National Academies of Science* 96:13062–13067.

Caissard, J.-C., O. Faure, F. Jullien, et al. 1996. Direct regeneration *in vitro* and transient GUS expression in *Mentha × piperita*. *Plant Cell Reports* 16:67–70.

Campbell, A. N. 1962. Biosynthesis of terpenes: Carbon-14 incorporation in *Mentha piperita* and *Pelargonium graveolens*. *Dissertation Abstracts* 22:4170–4171.

Carlson, E. M., and B. C. Stuart. 1936. Development of spores and gametophytes in certain New World species of *Salvia*. *New Phytologist* 35:68–91.

Čellárová, E. 1992. Micropropagation of *Mentha* L. In *Biotechnology in agriculture and forestry 19. High-tech and micropropagation III*, ed. Y. P. S. Bajaj, 262–276. Berlin: Springer–Verlag.

Chae, Y. A., and S. U. Park. 1994. Effects of subculture and cold temperature on the oil content and menthol synthetic ability in the callus cells of *Mentha piperita* L. In *Toward enhanced and sustainable agricultural productivity in the 2000s: Breeding research and biotechnology*, 515–520. Proceedings of the 7th International Congress of the Society for Advancement of Breeding Researchers in Asia and Oceania (SABRAO) and International Symposium of World Sustainable Agriculture Association (WSAA). Taichung District Agric. Improvement Sta. & SABRAO.

Chakraborty, A., and S. Chattopadhyay. 2008. Stimulation of menthol production in *Mentha piperita* cell culture. *In Vitro Cellular and Developmental Biology—Plants* 44:518–524.

Chaput, M.-H., H. San, L. de Hys, et al. 1996. How plant regeneration from *Mentha × piperita* L. and *Mentha × citrata* Ehrh. leaf protoplasts affects their monoterpene composition in field conditions. *Journal of Plant Physiology* 149:481–488.

Charabot, E. 1900. Récherchés sur la genèse des composés de la série du menthol dans les plantes. *Comptes-Rendus Hebdomadaires des Seances de l'Académie des Sciences* 130:518–519.

Charlwood, B. V., and K. A. Charlwood. 1983. The biosynthesis of mono- and sesquiterpenes in tissue culture. *Biochemical Society Transactions* 11:592–593.

Cheng, K-C., Q-L. Yang, and Y.-R. Zheng. 1982. The relation between the patterns of cytomixis and variation of chromosome numbers in pollen mother cells of jimsonweed (*Datura stramonium* L.). *Acta Botanica Sinica* 24:103–110.

Chishti, N., A. S. Shawl, Z. A. Kaloo, et al. 2006. Clonal propagation of *Mentha arvensis* L. through nodal explant. *Pakistan Journal of Biological Science* 9:1416–1419.

Coates, R. M., C. S. Elmore, R. B. Croteau, et al. 1997. Stereochemistry of the methyl-methylene elimination in the enzyme-catalyzed cyclization of geranyl diphosphate to (4S)-limonene. *Chemistry Communications* 1997:2079–2080.

Colby, S. M., W. R. Alonso, E. J. Katahira, et al. 1993. 4S-limonene synthase from the oil glands of spearmint (*Mentha spicata*). *Journal of Biological Chemistry* 268:23016–23024.

Crock, J., M. Wildung, and R. Croteau. 1997. Isolation and bacterial expression of a sesquiterpene synthase cDNA clone from peppermint (*Mentha × piperita*, L.) that produces the aphid alarm pheromone (E)-β-farnesene. *Proceedings of National Academies of Science* 94:12833–12838.

Croteau, R. 1978. Biogenesis of flavor components: Volatile carbonyl compounds and monoterpenoids. In *Postharvest biology and biotechnology*, ed. H. Hultin and M. Milan, 400–432. Westport CT: Food & Nutrition Press.

———. 1980a. The biosynthesis of terpene compounds. In *Fragrance and flavor substances*, ed. R. Croteau, 13–36. Pattensen, Germany: D&PS Verlag.

———. 1980b. The biosynthesis of terpene compounds. *Perfumer Flavorist* (5):35–38, 45–48, 50–52, 54–69.

———. 1986a. Biosynthesis of cyclic monoterpenes. In *Biogeneration of aromas*, ed. T. H. Parliment and R. Croteau, 134–156. Washington, D.C.: American Chemical Society

————. 1986b. Biochemistry of monoterpenes and sesquiterpenes of the essential oils. In *Herbs, spices, and medicinal plants: Recent advances in botany, horticulture, and pharmacology*, vol. 1, ed. L. E. Craker and J. E. Simon, 81–133. Phoenix, AZ: Oryx Press.

————. 1988. Catabolism of monoterpenes in essential oil plants. In *Flavors and fragrances: A world perspective*, ed. B. M. Lawrence, B. D. Mookherjee, and B. J. Willis, 65–84. Amsterdam: Elsevier.

————. 1991. Metabolism of monoterpenes in mint (*Mentha*) species. *Planta Medica* 57:10–14.

————. 1993. The biosynthesis of limonene in *Mentha* species. In *Progress in flavor precursor studies*, ed. P. Schreier and P. Winterhalter, 113–122. Carol Stream, IL: Allured Publ. Corp.

Croteau, R., W. R. Alonso, A. E. Koepp, et al. 1993. Irreversible inactivation of monoterpene cyclases by a mechanism-based inhibitor. *Archives of Biochemistry and Biophysics* 307:397–404.

Croteau, R., A. J. Burbott, and W. D. Loomis. 1971. Compartmentation of lower terpenoid biosynthetic sites in peppermint. *Plant Physiology (Lancaster)* 47 (Suppl.): 21 (abstract).

————. 1972a. Biosynthesis of mono- and sesquiterpenes in peppermint from glucose-^{14}C and ^{14}CO$_2$. *Phytochemistry* 11:2459–2467.

————. 1972b. Apparent energy deficiency in mono- and sesquiterpene biosynthesis in peppermint. *Phytochemistry* 11:2937–2948.

————. 1973. Enzymatic cyclization of neryl pyrophosphate to α-terpineol by cell-free extracts from peppermint. *Biochemical and Biophysical Research Communications* 50:1006–1012.

Croteau, R., E. M. Davis, K. L. Ringer, et al. 2005. (–)-Menthol biosynthesis and molecular genetics. *Naturwissenschaften* 92:562–577.

Croteau, R., and C. L. Hooper. 1978. Metabolism of monoterpenes. Acetylation of (–)-menthol by a soluble enzyme preparation from peppermint (*Mentha piperita*) leaves. *Plant Physiology (Lancaster)* 61:737–742.

Croteau, R., F. Karp, K. C. Wagschal, et al. 1991. Biochemical characterization of a spearmint mutant that resembles peppermint in monoterpene content. *Plant Physiology (Lancaster)* 96:744–752.

Croteau, R., and W. D. Loomis. 1972. Biosynthesis of mono- and sesquiterpenes in peppermint from mevalonate-2-^{14}C. *Phytochemistry* 11:1055–1066.

————. 1975. Biosynthesis and metabolism of monoterpenes. *International Flavors and Food Additives* 6:292–296.

Croteau, R., and C. Martinkus. 1979. Metabolism of monoterpenes. Demonstration of (+)-neomenthyl-β-d-glucoside as a major metabolite of (–)-menthone in peppermint (*Mentha piperita*). *Plant Physiology (Lancaster)* 64:169–175.

Croteau, R., and V. K. Sood. 1985. Metabolism of monoterpenes. Evidence for the function of monoterpene catabolism in peppermint (*Mentha piperita*) rhizomes. *Plant Physiology (Lancaster)* 77:801–806.

Croteau, R., V. K. Sood, B. Renstrøm, et al. 1984. Metabolism of monoterpenes. Early steps in the metabolism of d-neomenthyl-β-d-glucoside in peppermint (*Mentha piperita*) rhizomes. *Plant Physiology (Lancaster)* 76:647–653.

Croteau, R., and K. V. Venkatachalam. 1986. Metabolism of monoterpenes: Demonstration that (+)-*cis*-isopulegone, not piperitenone, is the key intermediate in the conversion of (+)-isopiperitenone to (+)-pulegone in peppermint (*Mentha piperita*). *Archives of Biochemistry and Biophysics* 249:306–315.

Croteau, R., and J. N. Winters. 1982. Demonstration of the intercellular compartmentation of l-menthone metabolism in peppermint (*Mentha piperita*) leaves. *Plant Physiology (Lancaster)* 69:975–977.

Crowell, A. L., D. C. Williams, E. M. Davis, et al. 2002. Molecular cloning and characterization of a new linalool synthase. *Archives of Biochemistry and Biophysics* 405:112–121.

Datta, A. K., M. Mukherjee, and M. Iqbal. 2005. Persistent cytomixis in *Ocimum basilicum* L. (Lamiaceae) and *Withania somnifera* (L.) Dun. (Solanaceae). *Cytologia* 70:309–313.

Davis, E. M., K. L. Ringer, M. E. McConkey, et al. 2005. Monoterpene metabolism. Cloning, expression, and characterization of menthone reductases from peppermint. *Plant Physiology (Lancaster)* 137:873–881.

Diemer, F. 2001. Altered monoterpene composition in transgenic mint following the introduction of 4S-limonene synthase. *Plant Physiology and Biochemistry* 39:603–614.

Diemer, F., J. C. Caissard, S. Moja, et al. 1999. *Agrobacterium tumefaciens*-mediated transformation of *Mentha spicata* and *Mentha arvensis*. *Plant Cell Tissue Organ Culture* 57:75–78.

Diemer, F., F. Jullien, O. Faure, et al. 1998. High efficiency transformation of peppermint (*Mentha ×piperita* L.) with *Agrobacterium tumefaciens*. *Plant Science* 136:101–108.

Donalísio, M. G. R., A. J. D'Andréa Pinto, and C. J. de Souza. 1985. Variacão na resistência à ferrugem e na composição do óleo essencial de dois clones de menta. *Bragantia* 44:541–547.

Eisenreich, W., S. Sagner, M. H. Zenk, et al. 1997. Monoterpenoid essential oils are not of mevalonoid origin. *Tetrahedron Letters* 38:3889–3892.

Endo, T., and T. Suga. 1992. Demonstration of geranyl diphosphate synthase in several higher plants. *Phytochemistry* 31:2273–2275.

Faber, B., A. Dietrich, and A. Mosandl. 1994. Chiral compounds of essential oils. XV. Stereodifferentiation of characteristic compounds of *Mentha* species by multidimensional gas chromatography. *Journal of Chromatography* 666:161–165.

Farley, D. R., and V. Howland. 1980. The natural variation of the pulegone content in various oils of peppermint. *Journal of the Science of Food and Agriculture* 31:1143–1151.

Faure, O., F. Diemer, S. Moja, et al. 1998. Mannitol and thidiazuron improve *in vitro* shoot regeneration from spearmint and peppermint leaf disks. *Plant Cell Tissue Organ Culture* 52:209–212.

Franz, Ch. 1993. Genetics. In *Volatile oil crops: Their biology, biochemistry and production*, ed. R. K. M. Hay and P. G. Waterman, 63–96. Essex, England: Longman Scientific & Technical.

Fuchs, S., T. Beck, S. Burkardt, et al. 1999. Biogenetic studies in *Mentha × piperita*. 1. Deuterium-labeled monoterpene ketones: Synthesis and stereoselective analysis. *Journal of Agricultural and Food Chemistry* 47:3053–3057.

Fuchs, S., T. Beck, and A. Mosandl. 2001. Different stereoselectivity in the reduction of pulegone by *Mentha* species. *Planta Medica* 67:260–262.

Fuchs, S., T. Beck, M. Sandvoss, et al. 1999. Biogenetic studies in *Mentha × piperita*. 2. Stereoselectivity in the bioconversion of pulegone into menthone and isomenthone. *Journal of Agricultural and Food Chemistry* 47:3058–3062.

Fuchs, S., A. Gross, T. Beck, et al. 2000. Monoterpene biosynthesis in *Mentha × piperita* L.: bioconversion of piperitone and piperitenone. *Flavor Fragrance Journal* 15:84–90.

Fuchs, S., S. Zinn, T. Beck, et al. 1999. Biosynthesis of menthofuran in *Mentha × piperita*: Stereoselective and mechanistic studies. *Journal of Agricultural and Food Chemistry* 47:4100–4105.

Fujita, S.-I., and Y. Fujita. 1970. Studies on the essential oils of the genus *Mentha*. Part V. A biochemical study of the essential oils of *Mentha pulegium* Linn. *Journal of the Agricultural Chemical Society of Japan* 44:293–298.

Fujita, S.-I., T. Nakano, and Y. Fujita. 1972. A biochemical study of the essential oils of *Mentha Gattefossei* Maire (studies on the essential oils of the genus *Mentha* part VIII). *Journal of the Agricultural Chemical Society of Japan* 46:393–397.

Fujita, S.-I., and K. Nezu. 1980. On the components of the essential oils of *Mentha spicata* Linn. (a pilose form, longifolia type) (studies on the essential oils of the genus *Mentha* part XII). *Journal of the Agricultural Chemical Society of Japan* 54:341–344.

Fujita, Y. 1960a. Problems in the genus *Mentha* (II) [in Japanese]. *Koryo* 58:15–20.

———. 1960b. Problems in the genus *Mentha* (3) [in Japanese]. *Koryo* 59:41–42.

———. 1961a. Problems in the genus *Mentha* (4) [in Japanese]. *Koryo* 61:43–46.

———. 1961b. Problems in the genus *Mentha* (5) [in Japanese]. *Koryo* 64:61–62.

———. 1962. Problems in the genus *Mentha* (6) [in Japanese]. *Koryo* 66:29–32.

———. 1963a. Problems in the genus *Mentha* (7) [in Japanese]. *Koryo* 70:33–38.

———. 1963b. Problems in the genus *Mentha* (9) [in Japanese]. *Koryo* 73:51–53.

———. 1963c. On *Mentha citrata* Ehrh. *Journal of Japanese Botany* 38:171–174.

———. 1965a. Problems in the genus *Mentha* 9 [in Japanese]. *Koryo* 77:53–57.

———. 1965b. Problems in the genus *Mentha* (10) [in Japanese]. *Koryo* 80:49–53.

———. 1968. Problems in the genus *Mentha* (part 11). Some considerations to the biogenesis of the essential oil in the *Mentha* species. *Koryo* 88:23–27.

Galun, E., D. Aviv, A. Dantes, et al. 1985. Biotransformation by division-arrested and immobilized plant cells: Bioconversion of monoterpenes by gamma-irradiated suspended and entrapped cells of *Mentha* and *Nicotiana*. *Planta Medica* 1985:511–514.

Gamborg, O. L., R. A. Miller, and K. Ojima. 1968. Nutrient requirements of suspension cultures of soybean root cells. *Experiments in Cell Research* 50:151–158.

Gates, R. R. 1908. A study of reduction in *Oenothera rubrinervis*. *Botanical Gazette* 46:1–34.

———. 1911. Pollen formation in *Oenothera gigas*. *Annals of Botany* 25:909–940.

Gershenzon, J., M. Maffei, and R. Croteau. 1988. Biochemical localization of monoterpene synthesis in *Mentha spicata* L. (spearmint). *American Journal of Botany* 75 (6–2): 129 (abstract).

———. 1989. Biochemical and histochemical localization of monoterpene biosynthesis in the glandular trichomes of spearmint (*Mentha spicata*). *Plant Physiology (Lancaster)* 89:1351–1357.

Gershenzon, J., M. E. McConkey, and R. B. Croteau. 2000. Regulation of monoterpene accumulation in leaves of peppermint. *Plant Physiology (Lancaster)* 122:205–213.

———. 2002. Biochemical and molecular regulation of monoterpene accumulation in peppermint (*Mentha* × *piperita*). *Journal of Herbs, Spices & Medicinal Plants* 9 (2/3): 153–156.

Ghaffari, S. M. 2006. Occurrence of diploid and polyploidy microspores in *Sorghum bicolor* (Poaceae) is the result of cytomixis. *African Journal of Biotechnology* 5:1450–1453.

Ghanti, K., C. P. Kaviraj, R. B. Venugopal, et al. 2004. Rapid regeneration of *Mentha piperita* L. from shoot tip and nodal explants. *Indian Journal of Biotechnology* 3:594–598.

Guo, G.-Q., and G.-C. Zheng. 2004. Hypotheses for the functions of intercellular bridges in male germ cell development and its cellular mechanisms. *Journal of Theoretical Biology* 229:139–146.

Guzicka, M., and A. Wozny. 2005. Cytomixis in shoot apex of Norway spruce [*Picea abies* (L.) Karst.]. *Trees* 18:722–724.

Handa, K. L., D. M. Smith, I. C. Nigam, et al. 1964. Essential oils and their constituents XXIII. Chemotaxonomy of the genus *Mentha*. *Journal of Pharmaceutical Science* 53:1407–1409.

Hefendehl, F. W. 1962. Zusammensetzung des ätherischen Öls von *Mentha piperita* im Verlauf der Ontogenese und Versuche zur Beeinflussung der Ölkomposition. *Planta Medica* 10:241–266.

———. 1967a. Beiträge zur Biogenese ätherischer Öle. *Planta Medica* 15:121–131.

———. 1967b. Zusammensetzung des ätherischen Öls von *Mentha aquatica* L. Beiträge zur Terpenbiosenese. *Archiv der Pharmazie (Berlin)* 300:438–448.

———. 1970. Beiträge zur Biogenese ätherischer Öle Zusammensetzung zweier ätherischer Öle von *Mentha pulegium* L. *Phytochemistry* 9:1985–1995.

———. 1977. Monoterpene composition of a carvone containing polyploid strain of *Mentha longifolia* (L.) Huds. *Herba Hungarica* 16:39–43.

Hefendehl, F. W., and M. J. Murray. 1972. Changes in monoterpene composition in *Mentha aquatica* produced by gene substitution. *Phytochemistry* 11:189–195.

———. 1973a. Monoterpene composition of a chemotype of *Mentha piperita* having high limonene. *Planta Medica* 23:101–109.

———. 1973b. Relations between biogenesis and genetic data of essential oils in genus *Mentha*. *Rivista Italiana Essenze, Profumi, Piante Officinali, Aromi, Saponi, Cosmetici, Aerosol* 55:791–796.

———. 1976. Genetic aspects of the biosynthesis of natural odors. *Lloydia* 39:39–52.

Hefendehl, F. W., and A. Nagell. 1975. Unterschiede in der Zusammensetzung der ätherischen Öle von *Mentha rotundifolia, Mentha longifolia* und des F₁-Hybriden beider Arten. *Parfümerie & Kosmetik* 56:189–193.

Hefendehl, F. W., E. W. Underhill, and E. von Rudloff. 1967. The biosynthesis of the oxygenated monoterpenes in mint. *Phytochemistry* 6:823–835.

Hendriks, H. 1970. Een epoxymonoterpeenester uit de vluchtige olie van *Mentha rotundifolia* (L.) Hudson. *Pharmaceutisch Weekblad* 105:733–736.

———. 1974. De vluchtige olie van enkele chemotypen van *Mentha suaveolens* Ehrh. en van hybriden met *Mentha longifolia* (L.) Hudson. Dissertation thesis, Rijksuniversiteit te Groningen.

———. 1977. A new relation scheme between the species and hybrids of the genus *Mentha* subgenus *Menthastrum*. *Pharmaceutisch Weekblad* 112:48–54.

Hendriks, H., and F. H. L. van Os. 1972. The heredity of the essential oil composition in artificial hybrids between *Mentha rotundifolia* and *Mentha longifolia*. *Planta Medica* 21:421–425.

———. 1976. Essential oil of two chemotypes of *Mentha suaveolens* during ontogenesis. *Phytochemistry* 15:1127–1130.

Hendriks, H., F. H. L. van Os, and W. J. Feenstra. 1976. Crossing experiments between some chemotypes of *Mentha longifolia* and *Mentha. Medica* 30:154–162.

Hiraga, Y., W. Shi, D. I. Ito, et al. 1993. Biosynthetic generation of the species-specific chirality of limonene in *Mentha spicata* and *Citrus unshiu*. *Journal of Chemical Society—Chemical Communications* 1993:1370–1371.

Hirai, D., and A. Sakai. 1999. Cryopreservation of *in vitro* grown axillary shoot-tip meristems of mint (*Mentha spicata* L.) by encapsulation vitrification. *Plant Cell Reports* 19:150–155.

Hirata, T., S. Murakami, K. Ogihara, et al. 1990. Volatile monoterpenoid constituents of the plantlets of *Mentha spicata* produced by shoot tip culture. *Phytochemistry* 29:493–495.

Holm, Y., R. Hiltunen, K. Jokinen, and T. Törmälä. 1989. On the quality of the volatile oil in micropropagated peppermint. *Flavor Fragrance Journal* 4:81–84.

Horner, E., and H. A. Melouk. 1977. Screening, selection and evaluation of irradiation-induced mutants of spearmint for resistance to *Verticillium* wilt. In *Induced mutations against plant diseases*, International Atomic Energy Agency, 253–262. Vienna.

Hyatt, D. C., B. Youn, Y. Zhao, et al. 2007. Structure of limonene synthase, a simple model for terpenoid cyclase catalysis. *Proceedings of National Academies of Science* 104:5360–5385.

Ikeda, N., Y. Aoki, and S. Udo. 1961. Studies on mint breeding. XI. Resistance to rust. *Japanese Journal of Breeding* 11:269–276.

Ikeda, N., and T. Konishi. 1954. Studies of mint breeding III. Induced polyploidy of Japanese mint (*Mentha arvensis* L. var. *piperascens* Mal.) by the colchicine method. *Scientific Reports of the Faculty of Agriculture, Okayama University* 5:1–9.

Ikeda, N., and S. Shimizu. 1960. Study on components of essential oil by cross-breeding species of peppermint, part 1 [in Japanese]. *Koryo* 58:25–31.

Ikeda, N., S. Shimizu, and S. Udo. 1963. Studies on *Mentha gentilis* L. *Japanese Journal of Breeding* 13:31–41.

Ikeda, N., and S. Udo. 1956. Studies of mint breeding. VI. Utility of tetraploid Japanese mint, *Mentha arvensis* L. var. *piperascens* Mal. *Bulletin of Okayama Prefecture Agricultural Experiment Station* 54:51–66.

———. 1966. Studies on *Mentha arvensis* L. *Japanese Journal of Breeding* 16:251–259.

Inoue, F., H. Sugiura, A. Tabuchi, et al. 2003a. Plant regeneration of peppermint, *Mentha piperita*, from the hairy roots generated from microsegment infected with *Agrobacterium rhizogenes*. *Plant Biotechnology* 20:169–172.

———. 2003b. Alternation of essential oil composition in transgenic peppermint *(Mentha piperita)* carrying T-DNA from *Agrobacterium rhizogenes*. *Breeding Science* 53:163–167.

Islam, M. T., S. Lenufna, D. P. Dembele, et al. 2003. *In vitro* conservation of four mint (*Mentha* spp.) accessions. *Plant Tissue Culture* 13:37–46.

Jin, S.-T., and Y.-A. Chae. 1991. Effect of plant growth regulator on essential oil and its composition in *Mentha piperita* L. *in vitro* culture. *Korean Journal of Breeding* 23:173–179.

Johnson, D. A., and T. F. Cummings. 2000. Evaluation of mint mutants, hybrids, and fertile clones for resistance to *Verticillium dahliae*. *Plant Disease* 84:235–238.

Jullien, F. 2007. Mint. In *Biotechnology in agriculture and forestry 59. Transgenic crops IV*, ed. E. C. Pua and M. R. Davey, 435–466. Berlin: Springer–Verlag.

Jullien, F., F. Diemer, M. Colson, et al. 1998. An optimizing protocol for protoplast regeneration of three peppermint cultivars (*Mentha* × *piperita*). *Plant Cell Tissue Organ Culture* 54:153–159.

Kak, S. N., and R. L. Kaul. 1978. Mutation breeding of some novel chemotypes in Japanese mint, *Mentha arvensis* L. *Indian Perfumer* 22:249–251.

———. 1979. Mutation studies in *Mentha citrata* Ehrh. In *VII International Congress of Essential Oils*, October 7–11, 1977, 131–133. Kyoto, Japan.

———. 1980. Radiation induced useful mutants of Japanese mint (*Mentha arvensis* L.). *Zeitschrift fur Pflanzenzuchtung* 85:170–174.

———. 1981. Mutation studies in *Mentha spicata* L. *Proceedings of Indian Academy of Science (Plant Science)* 90:211–215.

———. 1982a. Commercial capability of new mutant strain of Japanese mint. In *Cultivation and utilization of aromatic plants*, ed. C. K. Atal and B. M. Kapur, 283–286. Jammu-Tawi, India: Regional Res. Lab.

———. 1982b. Radiation induced mutants of *Mentha arvensis*. In *VIII International Congress of Essential Oils*, October 12–17, 1980, Cannes-Grasse, France, paper no. 51.

———. 1984. Induction of genetic variability in *Mentha* species through radiations. In *Recent trends in botanical researches*, ed. R. P. Sinha, 141–145. Patra, India: R. P. Roy Commemoration Fund.

———. 1988. Improvement of essential oil and aroma chemical bearing plants through induced mutations—Accomplishments and prospects. *Parfümerie Kosmetik* 69:102, 104–105.

Karp, F., C. A. Mihaliak, J. L. Harris, et al. 1990. Monoterpene biosynthesis: Specificity of the hydroxylations of (–)-limonene by enzyme preparations from peppermint (*Mentha piperita*), spearmint (*Mentha spicata*), and perilla (*Perilla frutescens*) leaves. *Archives of Biochemistry and Biophysics* 276:219–226.

Katsuhara, J. 1966. An aspect on biogenesis of terpenoids in *Mentha* species. *Koryo* 83:51–63.

Kim, G.-S., S.-H. Park, Y.-J. Chang, et al. 2002. Transformation of menthane monoterpenes by *Mentha piperita* cell culture. *Biotechnology Letters* 24:1553–1556.

Kim, T., T. Y. Kim, G. W. Bae, et al. 1996. Improved production of essential oils by two-phase culture of *Mentha piperita* cells. *Plant Tissue Culture Letters* 13:189–192.

Kireyeva, S. A., V. N. Melnikov, S. A. Reznikova, et al. 1978. Accumulation of volatile oil in the callus tissue of peppermint. *Fiziologiya Rastenii* 25:564–566.

Kitto, S. L. 2008. Micropropagation of mint (*Mentha* spp.). In *Plant propagation—Concepts and laboratory exercises*, ed. C. A. Beyl and R. N. Trigiano, 347–354. Boca Raton, FL: CRC Press.

Kjonaas, R., and R. Croteau. 1983. Demonstration that limonene is the first cyclic intermediate in the biosynthesis of oxygenated *p*-menthane monoterpenes in *Mentha piperita* and other *Mentha* species. *Archives of Biochemistry and Biophysics* 220:79–89.

Kjonaas, R., C. Martinkus-Taylor, and R. Croteau. 1982. Metabolism of monoterpenes: Conversion of l-menthone to l-menthol and d-neomenthol by stereospecific dehydrogenases from peppermint (*Mentha piperita*) leaves. *Plant Physiology (Lancaster)* 69:1013–1017.

Kjonaas, R., K. V. Venkatachalam, and R. Croteau. 1985. Metabolism of monoterpenes: Oxidation of isopiperitenol to isopiperitenone, and subsequent isomerization to piperitenone by soluble enzyme preparations from peppermint (*Mentha piperita*) leaves. *Archives of Biochemistry and Biophysics* 238:49–60.

Krasnyanski, S., T. M. Ball, and K. C. Sink. 1998. Somatic hybridization in mint: Identification and characterization of *Mentha piperita* (+) *M. spicata* hybrid plants. *Theoretical and Applied Genetics* 96:683–687.

Krasnyanski, S., R. A. May, A. Loskutov, T. M. Ball, et al. 1999. Transformation of the limonene synthase gene into peppermint (*Mentha piperita* L.) and preliminary studies on the essential oil profiles of single transgenic plants. *Theoretical and Applied Genetics* 99:676–682.

Kremers, R. E. 1922. The biogenesis of oil of peppermint. *Journal of Biological Chemistry* 50:31–34.

Kukreja, A. K. 1996. Micropropagation and shoot regeneration from leaf and nodal explants of peppermint (*Mentha piperita* L.). *Journal of Spices & Aromatic Crops* 5:111–119.

Kukreja, A. K., O. P. Dhawan, P. S. Ahuja, et al. 2000. Yield potential and stability behavior of *in vitro* derived somaclones of Japanese mint (*Mentha arvensis* L.) under different environments. *Journal of Genetics and Breeding* 54:109–115.

———. 1992. Genetic improvement of mints: On the qualitative traits of essential oil of *in vitro* derived clones of Japanese mint (*Mentha arvensis* var. *piperascens* Holmes). *Journal of Essential Oil Research* 4:623–629.

Kukreja, A. K., O. P. Dhawan, A. K. Mathur, et al. 1991. Screening and evaluation of agronomically useful somaclonal variations in Japanese mint (*Mentha arvensis* L.). *Euphytica* 53:183–191.

Kumar, A., A. Chakraborty, S. Ghanta, et al. 2009. *Agrobacterium*-mediated genetic transformation of mint with *E. coli* glutathione synthetase gene. *Plant Cell Tissue Organ Culture* 96:117–126.

Kumar, B., S. P. S. Khanuja, and N. K. Patra. 2004. Study of genetic variations in open-pollinated seed progenies (OPSPs) of the *Mentha piperita* cv. Kukrail. *Journal of Medicinal and Aromatic Plant Science* 26:84–88.

Kumar, B., H. P. Singh, Y. Kumar, et al. 2005. Analysis of characters associated with high yield and menthol content in menthol mint (*Mentha arvensis*) genotypes. *Journal of Medicinal and Aromatic Plant Science* 27:435–438.

Kumar, B., H. P. Singh, and N. K. Patra. 2007. Phenotypic stability in Japanese mint (*Mentha arvensis* L.). *Crop Improvement* 34:95–99.

Kumar, S., N. K. Patra, H. P. Singh, et al. 1999. Kosi—An early maturing variety requiring late planting of *Mentha arvensis*. *Journal of Medicinal and Aromatic Plant Science* 21:56–58.

Kumar, S., B. R. Tyagi, J. R. Bahl, et al. 1997. Himalaya—A high menthol yielding hybrid clone of *Mentha arvensis*. *Journal of Medicinal and Aromatic Plant Science* 19:729–731.

Kundu, A. K., and A. K. Sharma. 1985. Chromosome characteristics and DNA content in *Mentha* Linn. *Nucleus* 28:89–96.

———. 1988. Cytomixis in Lamiaceae. *Cytologia* 53:469–474.

Lange, B. M., and R. Croteau. 1999a. Isoprenoid biosynthesis via a mevalonate-independent pathway in plants: Cloning and heterologous expression of 1-deoxy-d-xylulose-5-phosphate reductoisomerase from peppermint. *Archives of Biochemistry and Biophysics* 365:170–174.

———. 1999b. Isopentenyl diphosphate biosynthesis via a mevalonate-independent pathway: Isopentenyl monophosphate kinase catalyzes the terminal enzymatic step. *Proceedings of National Academies of Science* 96:13714–13719.

———. 1999c. Genetic engineering of essential oil production in mint. *Current Opinion in Plant Biology* 2:139–144.

Lange, B. M., M. R. Wildung, D. McCaskill, et al. 1998. A family of transketolases that directs isoprenoid biosynthesis via a mevalonate-independent pathway. *Proceedings of National Academies of Science* 95:2100–2104.

Lange, B. M., M. R. Wildung, E. J. Stauber, et al. 2000. Probing essential oil biosynthesis and secretion by functional evaluation of expressed sequence tags from mint glandular trichomes. *Proceedings of National Academies of Science* 97:2934–2939.

Larkov, O., A. Zaks, E. Bar, et al. 2008. Enantioselective monoterpene alcohol acetylation in *Origanum, Mentha* and *Salvia* species. *Phytochemistry* 69:2565–2571.

Lattoo, S. K., S. Khan, S. Bamotra, et al. 2006. Cytomixis impairs meiosis and influences reproductive success in *Chlorophytum comosum* (Thunb) Jacq.—an additional strategy and possible implications. *Journal of Biosciences* 31:629–637.

Lavania, U. C., N. K. Misra, S. Lavania, et al. 2006. Mining de novo diversity in palaeopolyploids. *Current Science* 90:938–941.

Lawrence, B. M. 1978. A study of the monoterpene interrelationships in the genus *Mentha* with special reference to the origin of pulegone and menthofuran. Dissertation thesis. Rijksuniversiteit te Groningen.

———. 1979. A fresh look at the biosynthetic pathways for most compounds found in *Mentha* oils. In *VII International Congress of Essential Oils*, October 7–11, 1977, Kyoto, Japan, 121–126.

———. 1981. Monoterpene interrelationships in the *Mentha* genus: A biosynthetic discussion. In *Essential oils*, ed. B. D. Mookherjee and C. J. Mussinan, 1–81. Wheaton, IL: Allured Publ. Corp.

———. 2007. *Mint: The genus* Mentha. Boca Raton, FL: CRC Press.

Li, X., Z. Gong, H. Koiwa, et al. 2001. *Bar*-expressing peppermint (*Mentha × Piperita* L. var. Black Mitcham) plants are highly resistant to the glufosinate herbicide Liberty. *Molecular Breeding* 8:109–118.

Li, X., N. Xiamou, R. A. Bressan, et al. 1999. Efficient plant regeneration of native spearmint (*Mentha spicata* L.). *In Vitro Cellular and Developmental Biology—Plants* 35:333–338.

Lin, M.-L., and E. J. Staba. 1961. Peppermint and spearmint tissue cultures. I. Callus formation and submerged culture. *Lloydia* 24:139–145.

Lincoln, D. E., P. M. Marble, F. J. Cramer, et al. 1971. Genetic basis for high limonene-cineole content of exceptional *Mentha citrata* hybrids. *Theoretical and Applied Genetics* 41:365–370.

Lincoln, D. E., and M. J. Murray. 1978. Monogenic basis for reduction of (+)-pulegone to (–)-menthone in *Mentha* oil biogenesis. *Phytochemistry* 17:1727–1730.

Lincoln, D. E., M. J. Murray, and B. M. Lawrence. 1986. Chemical composition and genetic basis for the isopinocamphone chemotype of *Mentha citrata* hybrids. *Phytochemistry* 25:1857–1863.

Linsmaier, E. M. and F. Skoog. 1965. Organic growth factor requirements of tobacco tissue cultures. *Physiologia Plantarum* 18:100–127.

Little, D. B., and R. B. Croteau. 1999. Biochemistry of essential oil terpenes. In *Flavor chemistry: 30 Years of progress*, ed. R. Teranishi, E. L. Wick, and I. Hornstein, 239–253. New York: Kluwer Acad.

Loomis, W. D. 1967. Biosynthesis and metabolism of monoterpenes. In *Terpenoids in plants*, ed. J. B. Pridham, 59–82. New York: Academic Press.

Loomis, W. D., and R. Croteau. 1973. Biochemistry and physiology of lower terpenoids. *Recent Advances in Phytochemistry* 6:147–185.

Lupien, S., F. Karp, K. Ponnamperuma, et al. 1995. Cytochrome P450 limonene hydroxylases of *Mentha* species. *Drug Metabolism Drug Interactions* 12:245–260.

Lupien, S., F. Karp, M. Wildung, et al. 1999. Regiospecific cytochrome P450 limonene hydroxylases from mint (*Mentha*) species: cDNA isolation, characterization, and functional expression of (–)-4S-limonene-3-hydroxylase and (–)-4S-limonene-6-hydroxylase. *Archives of Biochemistry and Biophysics* 368:181–192.

Madyasrtha, K. M., and W. D. Loomis. 1969. Phosphorylation of geraniol by cell-free enzymes from *Mentha piperita*. *FASEB Journal* 28:665 (abstract).

Maffei, M. 1990. F_1 and F_2 hybrids from *Mentha* × *verticillata* clone 7303 × *Mentha spicata* L. A chemogenetic study. *Flavor Fragrance Journal* 5:211–217.

Maffei, M., J. Gershenzon, and R. Croteau. 1988. Histochemical localization of a monoterpene dehydrogenase in *Mentha spicata* L. (spearmint). *American Journal of Botany* 75 (6–2): 131 (abstract).

Maheshwari, P. 1950. *An introduction to the embryology of angiosperms.* New York: McGraw–Hill Book Co.

Mahmoud, S. S., and R. B. Croteau. 2001. Metabolic engineering of essential oil yield and composition in mint by alternating expression of deoxyxylulose phosphate reductisomerase and menthofuran synthase. *Proceedings of National Academies of Science* 98:8915–8920.

———. 2003. Menthofuran regulates essential oil biosynthesis in peppermint by controlling a downstream monoterpene reductase. *Proceedings of National Academies of Science* 100:14481–14486.

Mahmoud, S. S., M. Williams, and R. Croteau. 2004. Cosuppression of limonene-3-hydroxylase in peppermint promotes accumulation of limonene in the essential oil. *Phytochemistry* 65:547–554.

Malallah, G. A., and T. A. Attia. 2003. Cytomixis and its possible evolutionary role in a Kuwaiti population of *Diplotaxis harra* (Brassicaceae). *Botanical Journal of Linnean Society* 143:169–175.

Manning, T. D. R. 1970. The composition of oil from peppermint grown in New Zealand. *New Zealand Journal of Science (Wellington)* 13:18–26.

Martinkus, C., and R. Croteau. 1981. Metabolism of monoterpenes. Evidence for compartmentation of l-menthone metabolism in peppermint (*Mentha piperita*) leaves. *Plant Physiology (Lancaster)* 68:99–106.

Matsuo, M., and T. Uchino. 1993. Effect of electric field on culture multiplication of tissue-cultured plant. *Journal of Japanese Society of Agricultural Machinery* 55:93–100.

McCaskill, D., and R. Croteau. 1993. Procedures for the isolation and quantification of the intermediates of the mevalonic acid pathway. *Analytical Biochemistry* 215:142–149.

———. 1994. Recent advances in terpenoid biosynthesis: Implications for essential oil production. In *Proceedings of the 4th International Conference on Aromatic and Medicinal Plants*, ed. N. Verlet, 92–101. Nyons, France: CERDEPPAM.

———. 1995a. Isoprenoid synthesis in peppermint (*Mentha* × *piperita*): Development of a model system for measuring flux of intermediates through the mevalonic acid pathway in plants. *Biochemical Society Transactions* 23:290S.

———. 1995b. Monoterpene and sesquiterpene biosynthesis in glandular trichomes of peppermint (*Mentha* × *piperita*) rely exclusively on plastid-derived isopentenyl diphosphate. *Planta* 187:49–56.

McConkey, M. E., J. Gershenzon, and R. E. Croteau. 2000. Development regulation of monoterpene biosynthesis in the glandular trichomes of peppermint. *Plant Physiology (Lancaster)* 122:215–223.

McCown, D. D. 1986. Plug systems for micropropagules. In *Tissue culture as a plant production system for horticultural crops*, ed. R. H. Zimmerman, R. J. Griesbach, F. Hammerschlag, and R. H. Lawson, 53–60. Dordrecht: Martinus Nijhoff Publ.

McGeady, P., and R. Croteau. 1995. Isolation and characterization of an active-site peptide from a monoterpene cyclase labeled with a mechanism-based inhibitor. *Archives of Biochemistry and Biophysics* 317:149–155.

Mihaliak, C. A., J. Gershenzon, and R. Croteau. 1991. Lack of rapid monoterpene turnover in rooted plants: Implications for theories of plant chemical defense. *Oecologia* 87:373–376.

Mucciarelli, M., T. Sacco, M. Brincarello, et al. 1995. Oli essenziali di *Mentha* × *piperita* micropropagata *in vitro*. *Rivista Italiana EPPOS* 6 (17): 19–30.

Murashige, T., and F. Skoog. 1962. A revised medium for rapid growth and bioassays with tobacco tissue cultures. *Physiologia Plantarum* 15:473–497.

Murray, M. J. 1952. Colchicine-induced tetraploids of fifteen species of *Mentha*. *Genetics* 37:609.

———. 1960a. The genetic basis for the conversion of menthone to menthol in Japanese mint. *Genetics* 45:925–929.

———. 1960b. The genetic basis for a third ketone group in *Mentha spicata* L. *Genetics* 45:931–937.

———. 1961. Spearmint rust resistance and immunity in the genus *Mentha*. *Crop Science (Madison)* 1:175–179.

———. 1969. Successful use of irradiation breeding to obtain *Verticillium*-resistant strains of peppermint, *Mentha piperita* L. In *International Atomic Energy Agency, induced mutations in plants*, 345–371. Vienna: International Atomic Energy Agency.

————. 1971. Additional observations on mutation breeding to obtain *Verticillium*-resistant strains of peppermint. In *International Atomic Energy Agency, mutation breeding for disease resistance*, 171–195. Vienna: International Atomic Energy Agency.

————. 1972. Genetic observations on *Mentha* oil biogenesis. *Anais Academia Brasileira de Ciências* 44 (Suppl.): 24–30.

Murray, M. J., W. Faas, and P. Marble. 1972a. Effects of plant maturity on oil composition of several spearmint species grown in Indiana and Michigan. *Crop Science (Madison)* 12:723–728.

————. 1972b. Chemical composition of *Mentha arvensis* var. *piperascens* and four hybrids with *Mentha crispa* harvested at different times in Indiana and Michigan. *Crop Science (Madison)* 12:742–745.

Murray, M. J., and F. W. Hefendehl. 1972. Changes in monoterpene composition of *Mentha aquatica* produced by gene substitution from *M. arvensis*. *Phytochemistry* 11:2469–2474.

————. 1973. Changes in monoterpene composition of *Mentha aquatica* produced by gene substitution from a high limonene strain of *M. citrata*. *Phytochemistry* 12:1875–1880.

Murray, M. J., and D. E. Lincoln. 1970. The genetic basis of acyclic oil constituents in *Mentha citrata* Ehrh. *Genetics* 65:457–471.

————. 1972. Oil composition of *Mentha aquatica–M. longifolia* F_1 hybrids and *M. dumetorum*. *Euphytica* 21:337–343.

Murray, M. J., D. E. Lincoln, and F. W. Hefendehl. 1980. Chemogenetic evidence supporting multiple allele control of the biosynthesis of (–)-menthone and (+)-isomenthone stereoisomer in *Mentha* species. *Phytochemistry* 19:2103–2110.

Murray, M. J., D. E. Lincoln, and P. M. Marble. 1972. Oil composition of *Mentha aquatica* × *M. spicata* F_1 hybrids in relation to the origin of *M.* × *piperita*. *Canadian Journal of Genetics and Cytology* 14:13–29.

Murray, M. J., P. M. Marble, and D. E. Lincoln. 1971. Intersubgeneric hybrids in the genus *Mentha*. *Journal of Heredity* 62:363–366.

Murray, M. J., and R. H. Reitsema. 1954. The genetic basis of the ketones, carvone, and menthone in *Mentha crispa* L. *Journal of American Pharmaceutical Association, Science Education* 43:612–613.

Murray, M. J., and W. A. Todd. 1972. Registration of Todd's Mitcham peppermint (reg. no. 1). *Crop Science (Madison)* 12:128.

Nádaská, M., K. Erdelsky, and P. Cupka. 1994. Improvement of *Mentha piperita* L. cs. cv. Perpeta by means of *in vitro* micropropagation and stabilization of contained substances. *Biologia (Bratislava)* 45:955–959.

Nagasawa, T., K. Umemoto, T. Tsuneya, et al. 1976. Essential oil of *Mentha spicata* L. var. *crispa* Benth. grown in winter season (studies on essential oil *Mentha spicata* L. var. *crispa* Benth. Part IV). *Journal of the Agricultural Chemical Society of Japan* 50:287–289.

Narain, P. 1979. Cytomixis in the pollen mother cells of *Hemerocallis* Linn. *Current Science* 48:996–998.

Nelson, R. 1947. Production of mint species hybrids resistant to *Verticillium* wilt. *Phytopathology* 37:16–17 (abstract).

————. 1950. Development of varieties of spearmint resistant to *Verticillium* wilt and to rust. *Phytopathology* 40:20 (abstract).

Nickolaev, A. G. 1978. Variability of the composition of monoterpenoids in some hybrid posterities of mint. In *IUPAC 11th International Symposium on Chemistry of Natural Products. Volume 1. Bioorganic chemistry*, 205–208. Sofia: Bulgarian Academy of Science.

————. 1979. Inheritance of terpenoids in hybrids of mint. In *VII International Congress of Essential Oils*, October 7–11, 1977, Kyoto, Japan, 121.

————. 1982. Inheritance of monoterpenoids by sterile hybrids of mint no. 401. In *VIII International Congress of Essential Oils*, October 12–17, 1982, Cannes-Grasse, France, paper no. 16.

Nickolaev, A. G., A. V. Dizdar, E. M. Pelyakh, et al. 1977. Inheritance of monoterpenoid composition in mint and carrot intraspecific crosses. Izvestija Akademiya Nauk Moldaveskoi SSR, Seriya Biologicheskikh i Khimicheskikh Nauk 6:16–21.

Nickolaev, A. G., and E. M. Peljah. 1981. Character inheritance in interspecific mint hybrids from *Mentha sachalinensis* (Briq.) Kudo × *M. caucasica* (Briq.) Gandg. [in Russian]. *Herba Hungarica* 20:27–41.

Nickolaev, A. G., M. T. Pysova, and L. V. Lollo. 1990. Monoterpenoids transformation in *Mentha* leaves by means of distant hybridization. In *Proceedings of the 11th International Congress of Essential Oils, Fragrances and Flavors*, New Delhi, India, November 12–16, 1989, vol. 3, ed. S. C. Bhattacharyya, N. Sen, and K. L. Sethi, 43–48. London: Aspect Publ.

Niu, X., X. Li, R. A. Veronese, et al. 2000. Factors affecting *Agrobacterium tumefaciens*-mediated transformation of peppermint. *Plant Cell Reports* 19:304–310.

Niu, X., K. Lin, P. M. Hasegawa, et al. 1998. Transgenic peppermint (*Mentha × piperita* L.) plants obtained by cocultivation with *Agrobacterium tumefaciens. Plant Cell Reports* 17:165–171.

Ono, S., and N. Ikeda. 1970. Studies on the radiation breeding in the genus *Mentha*. VII. The sensibility to rust in radiation-induced variation of Japanese mint. *Scientific Reports of the Faculty of Agriculture, Okayama University* 35:7–10.

Park, S.-H., and Y.-A. Chae. 1990. Factors affecting on callus induction and growth in peppermint. *Korean Journal of Breeding* 22:53–57.

Park, S.-H., Y.-A. Chae, H. J. Lee, et al. 1993. Menthol biosynthesis pathway in *Mentha piperita* suspension cells. *Journal of Korean Agriculture & Chemistry Society* 36:358–363.

———. 1994. Production of (–)-7-hydroxyisopiperitenone from (–)-isopiperitenone by a suspension culture of *Mentha piperita. Planta Medica* 60:374–375.

Park, S.-H., K.-S. Kim, Y. Suzuki, et al. 1997. Metabolism of isopiperitenones in cell suspension culture of *Mentha piperita. Phytochemistry* 44:623–626.

Patra, N. K., S. P. Chauhan, and S. Mandal. 1988. Stability analysis in peppermint. *Crop Improvement* 15:187–191.

Patra, N. K., S. Kumar, S. P. S. Khanuja, et al. 2001a. Damroo—A seed producing menthol mint variety (*Mentha arvensis*). *Journal of Medicinal and Aromatic Plant Science* 23:141–145.

———. 2001b. Neerkalka—An interspecific hybrid variety of *Mentha arvensis* and *M. spicata* for commercial cultivation of spearmint. *Journal of Medicinal and Aromatic Plant Science* 23:129–132.

Patra, N. K., H. Tanveer, S. P. S. Khanuja, et al. 2001b. A unique interspecific hybrid spearmint clone with growth properties of *Mentha arvensis* L. and oil qualities of *Mentha spicata* L. *Theoretical and Applied Genetics* 102:471–476.

Phatak, S. V., and M. R. Heble. 2002. Organogenesis and terpenoid synthesis in *Mentha arvensis. Fitoterapia* 73:32–39.

Ponnamperuma, K., and R. Croteau. 1996. Purification and characterization of an NADPH-cytochrome P450 (cytochrome c) reductase from spearmint (*Mentha spicata*) glandular trichomes. *Archives of Biochemistry and Biophysics* 329:9–16.

Poovaiah, C. R., S. C. Weller, and M. A. Jenks. 2006. *In vitro* adventitious shoot regeneration of native spearmint using internodal explants. *Hortscience* 41:414–417.

Prosser, I. M., R. J. Adams, M. H. Beale, et al. 2006. Cloning and functional characterization of a *cis*-muuroladiene synthase from black peppermint (*Mentha × piperita*) and direct evidence for a chemotype unable to synthesize farnesene. *Phytochemistry* 67:1564–1571.

Rabak, F. 1916. The effect of cultural and climatic conditions on the yield and quality of peppermint oil. *USDA Bulletin* (1915–1923): 454.

Rajaonarivony, J. I. M., J. Gershenzon, and R. Croteau. 1992a. Characterization and mechanism of (4S)-limonene synthase, a monoterpene cyclase from the glandular trichomes of peppermint (*Mentha × piperita*). *Archives of Biochemistry and Biophysics* 296:49–57.

Rajaonarivony, J. I. M., J. Gershenzon, J. Miyazaki, et al. 1992b. Evidence for an essential histidine residue in 4S-limonene synthase and other terpene cyclases. *Archives of Biochemistry and Biophysics* 299:77–82.

Ravid, U., E. Putievsky, and I. Katzir. 1994a. Enantiomeric distribution of piperitone in essential oils of some *Mentha* spp. *Calamintha incána* (Sm.) Hetdr. and *Artemisia judaica* L. *Flavor Fragrance Journal* 9:85–87.

———. 1994b. Chiral GC analysis of menthone and isomenthone with high enantiomeric purities in laboratory-made and commercial essential oils. *Flavor Fragrance Journal* 9:139–142.

———. 1994c. Chiral GC analysis of (1R)(+)- pulegone with high enantiomeric purity in essential oils of some Lamiaceae aromatic plants. *Flavor Fragrance Journal* 9:205–207.

———. 1994d. Chiral GC analysis of enantiomerically pure (R) (–)-linalyl acetate in some Lamiaceae, myrtle and petitgrain essential oils. *Flavor Fragrance Journal* 9:275–276.

Ravid, U., E. Putievsky, I. Katzir, et al. 1992. Chiral GC analysis of (S) (+)- and (R) (–)-carvone with high enantiomeric purity in caraway, dill and spearmint oils. *Flavor Fragrance Journal* 7:289–292.

Rech, E. L., and M. J. P. Pires. 1986. Tissue culture propagation of *Mentha* spp. by the use of axillary buds. *Plant Cell Reports* 5:17–18.

Reed, B. M. 1999. *In vitro* storage conditions for mint germplasm. *HortScience* 34:350–352.

Reed, B. M., P. M. Buckley, and T. N. DeWilde. 1995. Detection and eradication of endophytic bacteria from micropropagated mint plants. *In Vitro Cellular and Developmental Biology—Plants* 31:53–57.

Reitsema, R. H. 1958. A biogenetic arrangement of mint species. *Journal of American Pharmaceutical Association, Science Edition* 47:267–269.

Reitsema, R. H., F. J. Cramer, and W. E. Fass. 1957. Chromatographic measurement of variations in essential oils within a single plant. *Journal of Agricultural and Food Chemistry* 5:779–780.

Reitsema, R. H., F. J. Cramer, N. J. Scully, et al. 1961. Essential oil synthesis in mint. *Journal of Pharmaceutical Sciences* 50:18–21.

Repcáková, K., M. Rychlová, E. Cellárová, et al. 1986. Micropropagation of *Mentha piperita* L. through tissue cultures. *Herba Hungarica* 25:77–88.

Reznickova, S. A., and L. A. Bugaenko. 1979. Genetic control of terpene biosynthesis in mint breeding. In *VII International Congress of Essential Oils*, October 7–11, 1977, Kyoto, Japan, 117–120.

Reznickova, S. A., L. A. Bugaenko, L. N. Lishtvanova, et al. 1974. The influence of genetic factors on the biosynthesis of terpenoids in mint. In *VI International Congress of Essential Oils*, San Francisco, California. September 8–12, 1974, paper no. 180.

Reznickova, S. A., L. A. Bugaenko, and V. V. Makarov. 1984. Characterization of menthol oil formation in *Mentha × piperita* L. in connection with its resynthesis [in Russian]. *Rastitel'n. Resursy* 20:544–552.

Reznickova, S. A., L. A. Bugaenko, and V. N. Melnickov. 1982. Genetical determination of high menthol content in the interspecific *Mentha* hybrids. In *VIII International Congress of Essential Oils*, October 12–17, 1982, Cannes-Grasse, France, paper no. 18.

Reznickova, S. A., L. A. Bugaenko, and V. S. Rodov. 1985. Genetic control of terpene biosynthesis in mint. III. Special feature of genetic regulation of menthol biosynthesis in *Mentha canadensis* L. *Genetika* 21:95–102.

Rieseberg, L. H., M. A. Archer, and R. K. Wayne. 1999. Transgressive segregation, adaptation and speciation. *Heredity* 83:363–372.

Ringer, K. L., E. M. Davis, and R. Croteau. 2005. Monoterpene metabolism. Cloning, expression, and characterization of (–)-isopiperitenol/(–)-carveol dehydrogenase of peppermint and spearmint. *Plant Physiology (Lancaster)* 137:863–872.

Ringer, K. L., M. E. McConkey, E. M. Davis, et al. 2003. Monoterpene double-bond reductases of the (–)-menthol biosynthetic pathway; isolation and characterization of cDNAs encoding (–)-isoperipenone reductase and (+)-pulegone reductase of peppermint. *Archives of Biochemistry and Biophysics* 418:80–92.

Rios-Estepa, R., G. W. Turner, J. M. Lee, et al. 2008. A systems biology approach identifies the biochemical mechanisms regulating monoterpenoid essential oil composition in peppermint. *Proceedings of National Academies of Science* 105:2818–2823.

Roberts, D., and A. Plotto. 2002. A unique *Mentha aquatica* mint for flavor. *Perfumer Flavorist* 27 (6): 24, 26–29.

Rothbächer, H. 1968. Zur Biosynthese einiger Terpene des Öls von *Mentha piperita* L. *Pharmazie Beih Ergänsungsband* 23:389–391.

Sakata, I., T. Higashiyama, H. Iwamura, et al. 1981. Studies on menthyl glycosides. *Koryo* 130:79–91.

Sakata, I., and K. Koshimizu. 1980. Seasonal variations in menthyl glucoside, menthol, menthone and related monoterpenes in developing Japanese peppermint. *Journal of the Agricultural Chemical Society of Japan* 54:1037–1043.

Sanderman, W., and H. Stockmann. 1956. Untersuchungen über die Biogenese von Terpenen. *Naturwissenschaften* 43:580–581.

———. 1958. Über die Biogenese von Pulegon. *Chemische Berichte* 91:930–933.

Sato, H., S. Enomoto, S. Oka, et al. 1993. Plant regeneration from protoplasts of peppermint (*Mentha piperita* L.). *Plant Cell Reports* 12:546–550.

———. 1994. The effect of 4-PU on protoplast culture of peppermint (*Mentha piperita* L.) [in Japanese]. *Plant Tissue Culture Letters* 11:134–138.

Sato, H., K. Yamada, M. Mii, et al. 1996. Production of an interspecific somatic hybrid between peppermint and gingermint. *Plant Science* 115:101–107.

Schalk, M., and R. Croteau. 2000. A single amino acid substitution (F3631) converts the regiochemistry of the spearmint (–)-limonene hydroxylase from a C6- to a C3-hydroxylase. *Proceedings of the National Academy of Sciences USA* 97:11948–11953.

Schantz, M. V., and R. Norri. 1968. Über die Verïnderungen der Ölzusammensetzung der Blattinsertionen der ukrainischen Pfefferminze während der Entwicklung. *Scientia Pharmaceutica* 36:187–199.

Schwab, W., D. C. Williams, and R. Croteau. 2000. On the mechanism of monoterpene synthases: Stereochemical aspects of (–)-4S-limonene synthase, (+)-bornyl diphosphate synthase and (–)-pinene synthase. In *Frontiers of flavor science*, ed. P. Schieberle and K.-H. Engel, 445–451. Garching: Deutsche Forschungsanstalt Lebensmittelchemie.

Sharma, S., and B. R. Tyagi. 1990. Heritability and coheritable variation in Japanese mint. *Journal of Genetic Breeding* 44:81–84.

———. 1991. Character correlation, path coefficient and heritability analyses of essential oil and quality components in Japanese mint. *Journal of Genetic Breeding* 45:257–262.

Sharma, S., B. R. Tyagi, S. Mandal, et al. 1996. Cluster analysis of 38 genotypes of peppermint (*Mentha piperita*) based on essential oil yield and quality traits. *Journal of Medicinal and Aromatic Plant Science* 18:280–286.

Sharma, S., B. R. Tyagi, A. Naqvi, et al. 1992. Stability of essential oil yield and quality characters in Japanese mint (*Mentha arvensis* L.) under varied environmental conditions. *Journal of Essential Oil Research* 4:411–416.

Shasany, A. K., S. P. S. Khanuja, S. Dhawan, et al. 1998. High regenerative nature of *Mentha arvensis* internodes. *Journal of Bioscience* 23:641–646.

Sheidai, M., and F. Fadaei. 2005. Cytogenetic studies in some species of *Bromus* L., section *Genea* Dum. *Journal of Genetics* 84:189–194.

Shilo, N. P., O. M. Sidorenko, L. E. Zav'yalova, et al. 1986. New mint hybrids [in Russian]. *Maslichnye Kul'tury* 1986 (2): 35.

Shimizu, S. 1967. Chemistry of *Mentha* [in Japanese]. *Chemistry (Japan)* 22:67–74.

———. 1974. Chemistry and biology of *Mentha* plants [in Japanese]. *Chemistry & Biology (Japan)* 12:659–666.

Shimizu, S., and N. Ikeda. 1962. (+)-Isomenthone in the essential oils of *Mentha* hybrids which involve *Mentha arvensis* L. var. *piperascens* Mal. (Japanese mint). (Studies on the essential oils of interspecific hybrids in the genus *Mentha*. Part III). *Journal of the Agricultural Chemical Society of Japan* 36:907–912.

Shimizu, S., D. Karasawa, and N. Ikeda. 1972. Studies on the essential oils of interspecific hybrids in the genus *Mentha*, part VI. On a new chemical strain in the hybrids involving *Mentha spicata* var. *crispa* Benth. *Journal of the Faculty of Agriculture Shinshu University* 9:72–81.

Shimizu, S., H. Shibata, D. Karasawa, et al. 1990. Carvyl- and dihydrocarvyl-β-d-glucosides in spearmint (studies on terpene glycosides in *Mentha* plants, part II). *Journal of Essential Oil Research* 2:81–86.

Shimizu, S., H. Shibata, and S. Maejima. 1990. A new monoterpene glucoside l-menthyl 6'-O-acetyl-β-d-glucoside in *Mentha arvensis* var. *piperascens* Mal. (studies on terpene glucosides in *Mentha* plants, part 1). *Journal of Essential Oil Research* 2:21–24.

Shine, W. E., and W. D. Loomis. 1974. Isomerization of geraniol and geranyl pyrophosphate by enzymes from carrot and peppermint. *Phytochemistry* 13:2095–2101.

Siddiqui, N. H., R. Khan, and G. R. Rao. 1979. A case of cytomixis in *Solanum nigrum* L. complex. *Current Science* 48:118–119.

Sidorenko, O. M., L. E. Zav'yalova, and G. V. Vorob'eva. 1985. A promising mint variety [in Russian]. *Maslichnye Kul'tury* 1985 (5): 34.

Singh, A. K. 1988. Transformations of menthol, menthone and menthyl acetate in *M. arvensis* L. with relation to the age of the plant. *Current Science* 57:480–481.

Singh, S. P., S. Sharma, and R. K. Tewari. 1998. Genetic improvement of *Mentha arvensis* based on essential oil yield and quality traits. *Journal of Herbs, Spices & Medicinal Plants* 6 (2): 79–86.

Singhal, V. K., and P. Kumar. 2008a. Impact of cytomixis on meiosis, pollen viability and pollen size in wild populations of Himalayan poppy (*Meconopsis aculeate* Royle). *Journal of Bioscience* 33:371–380.

———. 2008b. Cytomixis during microsporogenesis in the diploid and tetraploid cytotypes of *Withania somnifera* (L.) Dunal, 1852 (Solanaceae). *Comparative Cytogenetics* 2:85–92.

Spencer, A., J. D. Hamill, and M. J. C. Rhodes. 1990. Production of terpenes by differentiated shoot cultures of *Mentha citrata* transformed with *Agrobacterium tumefaciens* T37. *Plant Cell Reports* 8:601–604.

———— 1993. *In vitro* biosynthesis of monoterpene by *Agrobacterium* transformed shoot cultures of two *Mentha* species. *Phytochemistry* 32:911–919.

Staba, E. J., P. Laursen, and S. Büchner. 1965. Medicinal plant tissue culture. In *Proceedings of an International Conference on Plant Tissue Culture*, ed. P. R. White and A. R. Grove, 191–210. Berkeley: McCutchan Publ. Corp.

Stengele, M., and E. Stahl-Biskup. 1993. Glycosidically bound volatiles in peppermint (*Mentha piperita* L.). *Journal of Essential Oil Research* 5:13–19.

————. 1994. Influencing the level of glycosidically bound volatiles by feeding experiments with a *Mentha ×* *piperita* L, cultivar. *Flavor Fragrance Journal* 9:261–263.

Suga, T., T. Hirata, T. Aoki, et al. 1986. Interconversion and cyclization of acyclic allylic pyrophosphates in the biosynthesis of cyclic monoterpenoids in higher plants. *Phytochemistry* 25:2769–2775.

Suga, T., T. Hirata, H. Okita, et al. 1983. Biosynthesis of cyclic monoterpenoids. The pivotal acyclic precursor for the cyclization leading to the formation of α-terpineol and limonene in *Mentha spicata* and *Citrus natsudaidai*. *Chemistry Letters* 1983:1491–1494.

Suga, T., T. Hirata, and Y. Yamamoto. 1980. Lipid constituents of callus tissues of *Mentha spicata*. *Agricultural Biology and Chemistry* 44:1817–1820.

Thakur, R. S., and A. Akhila. 1993. Biosynthetic studies on some isoprenoid compounds of perfumer and flavor values from the plants cultivated CIMAP. In *Newer trends in essential oils and flavors*, ed. K. L. Dhar, R. K. Thappa, and S. G. Agarwal, 51–67. New Delhi: Tata McGraw–Hill Publ. Co.

Thompson, M. M. 1962. Cytogenetics of *Rubus*. II. Meiotic instability in some higher polyploids. *American Journal of Botany* 49:575–582.

Tisserat, B. 1996. Growth responses and construction costs of various tissue culture systems. *HortTechnology* 6:62–68.

Tisserat, B., and R. Silman. 2000. Interactions of culture vessels, media volume, culture density, and carbon dioxide levels on lettuce and spearmint shoot growth *in vitro*. *Plant Cell Reports* 19:464–471.

Tisserat, B., and S. F. Vaughn. 2001. Essential oils enhanced by ultra-high carbon dioxide levels from Lamiaceae species grown *in vitro* and *in vivo*. *Plant Cell Reports* 20:361–368.

————. 2004. Techniques to improve growth, morphogenesis and secondary metabolism responses from Lamiaceae species *in vitro*. *Acta Horticulturae* 629:333–339.

————. 2008. Growth, morphogenesis, and essential oil production in *Mentha spicata* L. plantlets *in vitro*. *In Vitro Cellular and Developmental Biology—Plants* 44:40–50.

Todd, W. A., R. J. Green, and C. E. Horner. 1977. Registration of Murray Mitcham peppermint (reg. no. 2). *Crop Science (Madison)* 17:188.

Todd, W. A., and M. J. Murray. 1968. New essential oils from hybridization of *Mentha citrata* Ehrh. *Perfumery & Essential Oil Research* 59:97–102.

Towill, L. E. 1988. Survival of shoot tips from mint species after short-term exposure to cryogenic conditions. *Hortscience* 23:839–841.

————. 1990. Cryopreservation of isolated mint shoot tips by vitrification. *Plant Cell Reports* 9:178–180.

————. 2002. Cryopreservation of *Mentha* (mint). In *Biotechnology in agriculture and forestry 50. Cryopreservation of plant germplasm II*, ed. L. E. Toweill and Y. P. S. Bajaj, 151–163. Berlin: Springer–Verlag.

Tucker, A. O. 1975. Morphological, cytological, and chemical evaluation of the *Mentha × gracilis* L. *s.l.* hybrid complex. PhD thesis, Rutgers Univ., New Brunswick, NJ.

————. 2009. Transgressive segregation and cytomixis: Important genetic and cytological phenomena in the phenotypic expression of essential oil constituents and patterns in *Mentha* (mint). *HortScience* 44:550 (abstract).

Tucker, A. O., and H. L. Chambers. 2002. *Mentha canadensis* L. (Lamiaceae): A relict amphidiploid from the Lower Tertiary. *Taxon* 51:703–718.

Tucker, A. O., and D. E. Fairbrothers. 1981. A euploid series in an F_1 interspecific hybrid progeny of *Mentha* (Lamiaceae). *Bulletin of Torrey Botanical Club* 108:51–53.

————. 1990. The origin of *Mentha ×gracilis* (Lamiaceae). I. Chromosome numbers, fertility, and three morphological characters. *Economic Botany* 44:183–213.

Tucker, A. O., H. Hendriks, R. Bos, et al. 1991. The origin of *Mentha* × *gracilis* (Lamiaceae). II. Essential oils. *Economic Botany* 45:200–215.

Tucker, A. O., and R. F. C. Naczi. 2007. *Mentha*: An overview of its classification and relationships. In *Mint: The genus* Mentha, ed. B. M. Lawrence, 1–39. Boca Raton, FL: CRC Press.

Turner, G. W., and R. Croteau. 2004. Organization of monoterpene biosynthesis in *Mentha*. Immunocytochemical localizations of geranyl diphosphate synthase limonene-6-hydroxylase, isopiperitenol dehydrogenase, and pulegone reductase. *Plant Physiology (Lancaster)* 136:4215–4227.

Turner, G. W., J. Gershenzon, and R. B. Croteau. 2000a. Distribution of peltate glandular trichomes on developing leaves of peppermint. *Plant Physiology (Lancaster)* 124:655–663.

———. 2000b. Development of peltate glandular trichomes of peppermint. *Plant Physiology (Lancaster)* 124:665–679.

Turner, G. W., J. Gershenzon, E. E. Nielson, et al. 1999. Limonene synthease, the enzyme responsible for monoterpene biosynthesis in peppermint, is localized to leucoplasts of oil gland secretory cells. *Plant Physiology (Lancaster)* 120:879–886.

Tyagi, B. R. 2003. Cytomixis in pollen mother cells of spearmint (*Mentha spicata* L.). *Cytologia* 68:67–73.

Tyagi, B. R., and T. Ahmad. 1989. Chromosome number variation in an F₁ interspecific hybrid progeny of *Mentha* (Lamiaceae). *Cytologia* 54:355–358.

Tyagi, B. R., T. Ahmad, and J. R. Bahl. 1992. Cytology, genetics and breeding of commercially important *Mentha* species. *Current Research in Medicinal and Aromatic Plants* 14:51–66.

Tyagi, B. R., and A. A. Naqvi. 1987. Relevance of chromosome number variation to yield and quality of essential oil in *Mentha arvensis*. *Cytologia* 52:377–385.

Umemoto, K., and T. Nagasawa. 1979. Essential oil of *Mentha gentilis* L. containing pulegone-8-hydroxy-3-*p*-menthene as a major components (studies on chemical constituents of wild mints, part XIII). *Journal of the Agricultural Chemical Society of Japan* 53:269–271.

———. 1980. Constituents in the essential oil of *Mentha spicata* L. var. *crispa* Benth. (Studies on the wild mints of Tokai districts. Part X). *Nagoya Gakuin Daigaku Ronshu Jinbun, Shizen Kagakuhen* 17 (1): 139–153.

Umemoto, K., and T. Tsuneya. 1988. *Mentha arvensis* containing piperitenone oxide and piperitone oxide as major components (studies on chemical constituents of wild mints, part XIV). *Journal of the Agricultural Chemical Society of Japan* 62:1073–1076.

Van Eck, J. M., and S. L. Kitto. 1990. Callus initiation and regeneration in *Mentha*. *Hortscience* 25:804–806.

———. 1992. Regeneration of peppermint and orange mint from leaf disks. *Plant Cell Tissue Organ Culture* 30:41–49.

Veronese, P., X. Li, X. Niu, et al. 2001. Bioengineering mint crop improvement. *Plant Cell Tissue Organ Culture* 64:133–144.

Wang, C.-J., and E. J. Staba. 1963. Peppermint and spearmint tissue culture II. Dual-carboy culture of spearmint tissues. *Journal of Pharmaceutical Sciences* 52:1058–1062.

Wang, H.-C., J.-Q. Li, and Z.-C. He. 2007. Irregular meiotic behavior in *Isoetes sinensis* (Isoetaceae), a rare and endangered fern in China. *Caryologia* 60:358–363.

Wang, Y.-Z., and K-C. Cheng. 1983. Observations on the rate of early embryogenesis and cytomixis in spring wheat. *Acta Botanica Sinica* 25:115–121.

Wang, X., Z. Gao, Y. Wang, et al. 2009. Highly efficient *in vitro* adventitious shoot regeneration of peppermint (*Mentha* × *piperita* L.) using intermodal explants. *In Vitro Cellular and Development Biology—Plants* 45:435–440.

Watson, V. K., and J. L. St. John. 1955. Relation of maturity and curing of peppermint hay to yield and composition of oil. *Journal of Agricultural and Food Chemistry* 3:1033–1038.

Werkhoff, P., and R. Hopp. 1986. Isolation and gas chromatographic separation of menthol and menthone enantiomers from natural peppermint oils. In *Progress in essential oils*, ed. E.-J. Brunke, 529–549. Berlin: Walter de Gruyter.

Werrmann, U., and D. Knorr. 1993. Plant cell cultures—Model systems for biosynthetic pathways? Conversion of monoterpenes in *Mentha* cell suspension cultures. In *Progress in flavor precursor studies*, ed. P. Schreier and P. Winterhalter, 195–199. Carol Stream, IL: Allured Publ. Corp.

White, J. G. H., S. H. Iskandar, and M. F. Barnes. 1987. Peppermint: Effect of time of harvest on yield and quality of oil. *New Zealand Journal of Experimental Agriculture* 15:73–79.

Wildung, M. R., and R. B. Croteau. 2005. Genetic engineering of peppermint for improved essential oil composition and yield. *Transgenic Research* 14:365–372.

Wüst, M., and R. B. Croteau. 2002. Hydroxylation of specifically deuterated limonene enantiomers by cytochrome 1450 limionene-6-hydroxylase reveals the mechanism of multiple product formation. *Biochemistry* 41:1820–1827.

Wüst, M., D. B. Little, M. Schalk, et al. 2001. Hydroxylation of limonene enantiomers and analogs by recombinant (–)-limonene 3- and 6-hydroxylases from mint (*Mentha*) species: Evidence for catalysis with sterically constrained active sites. *Archives of Biochemistry and Biophysics* 387:125–136.

Zheng, G.-C., X.-W. Nie, Y.-X. Wang, et al. 1985. Cytochemical localization of adenosine triphosphatase activity during cytomixis in pollen mother cells of David lily and its relation to the intercellular migrating chromatin substance. *Acta Botanica Sinica* 27:26–32.

CHAPTER 22

Plantago ovata
Cultivation, Genomics, Chemistry, and Therapeutic Applications

Manoj K. Dhar, Sanjana Kaul, Pooja Sharma, and Mehak Gupta

CONTENTS

22.1 INTRODUCTION

Plantago is the only genus on which family Plantaginaceae is based (Rahn 1996). Commonly known as plantains, these plants are mostly annual or perennial herbs or subshrubs. Most of the species belonging to the genus exist as weeds, like *P. lanceolata* (*L.*) and *P. major* (*L.*), which are common lawn weeds and are distributed throughout the world. Out of more than 200 species of the genus, only two—namely, *P. ovata* (Figure 22.1) and *P. psyllium* (*L.*)—have been extensively used for the production of seed husk. The husk of *P. ovata* is colorless and is known as isabgol in Hindi and blonde psyllium in English (Figure 22.1), while that of *P. psyllium* (*L.*) is dark brown and is called French psyllium. Since blonde psyllium (hereafter referred to as psyllium) is colorless and has higher mucilage content, it has gained preference and popularity over French Psyllium in the world market.

The word *Plantago* is from Latin, meaning "sole of the foot"; this refers to the typical shape of the leaf. Isabgol is derived from two Persian words: *isap* and *ghol*, meaning "horse ear," referring to its boat-shaped seeds. Psyllium is a Greek word meaning "flea" (referring to the size, color, and shape of the seeds) and *ovata* refers to the ovate shape of the leaves. *Plantago ovata* is known by several names, such as isabgul, isabgul gola, issufgul, jiru, aswagolam, aspaghol, psyllium, bazarqutuna, blonde psyllium, ch'-ch'ientzu, ghoda, grappicol, Indian plantago, indische psylli-samen, obeko, spogel seeds, and plantain.

In India, *P. ovata* is cultivated on a large scale in north Gujarat on about 144,000 ha of land, yielding 720 kg/ha. The major areas of cultivation include districts of Banaskantha and Mehsana. Of late, cultivation has also been initiated in a few more areas in states adjoining Gujarat. Outside India, it is cultivated on a limited scale in some western parts of Pakistan. Efforts have also been made to cultivate this species in the United States. India holds monopoly in the world trade of psyllium. More than 90% of the total Indian produce is exported to various countries of the world; the largest buyer of husk from India is the United States, accounting for nearly 75%. Most of the seed is exported to Germany. During 2002–2003, India exported 25,583 tonnes of husk worth 2.5 billion Indian rupees (INR) (US$1 = 44 INR) and 404 tonnes of seed worth INR 2 million.

Figure 22.1 (A) A plant of *Plantago ovata*. (B) Spikes showing initiation and completion of flowering. (C) Plants grown in experimental plots. (D, E) Seeds and seed husk (psyllium).

Psyllium has been extensively studied in view of its several health benefits and its applications in the pharmaceutical, food, and cosmetic industries. The laxative property of psyllium has been recognized for ages. Recently, its other potential benefits—namely, lowering of blood cholesterol and hyperglycemia, reducing the risk of colon cancer and treatment for irritable bowel syndrome—have caught the attention of many workers throughout the world (Dhar et al. 2005; Singh 2007). In addition to the health benefits, psyllium has found applications in food and other industries. It has been used as a deflocculant in paper and textile manufacturing, an emulsifying agent, a binder or lubricant in meat products, and a replacement of fat in low-calorie foods. Therefore, it is not surprising that psyllium has been incorporated into breakfast cereals, ice cream, instant beverages, and bakery and other dietary products (Chan and Wypyszyk 1988; Dhar, Kaul, and Jamwal 2000). Jain and Babbar (2005) used psyllium as a gelling agent in tissue culture media and found it to be a cheaper replacement for agar.

22.2 THE PLANT

Plantago ovata, commonly called "desert Indian wheat," is a small annual herb of temperate sandy regions (between 26 and 36°N latitudes) (Figure 22.1). It is distributed in the Canary Islands, across southern Spain, North Africa, the Middle East, Pakistan, the contiguous areas of western India, and the southern part of the former Soviet Union and central Asia (Stebbins and Day 1967). The species is indigenous to the Mediterranean region and west Asia, extending up to Sutlej and Sind in west Pakistan (Singh and Virmani 1982). The plant is believed to be of west Asian origin and was introduced in India during the Muslim rule (Dastur 1952; Husain 1977). The latter used it as a remedy for chronic dysentery and other intestinal problems. Initially, the seeds of this plant were collected from the wild, but due to difficulties in collection, it was brought under cultivation. Cultivation was first started in Lahore and Multan (Pakistan) and later in Bengal (India and Bangladesh), Mysore, and Coromandel Coast (India) (Dhar et al. 2005).

22.2.1 Botany

Plantago ovata is a short-stemmed plant, about 10–45 cm tall (Jamwal 2000) (Figure 22.1A). Leaves are borne alternately on the stem or in rosettes appressed to the soil surface. The number of leaves per plant varies between 20 and 60. The leaves are strap shaped, recurved, linear, 6.0–25.0 cm long, and 0.3–1.9 cm broad. The leaf surface is glabrous or slightly pubescent. Scapes are axillary, glabrous, or slightly pubescent; there are 5–60 per plant, usually exceeding the length of the leaves. Spikes are cylindrical or ovoid and measure 0.6–5.6 cm. Flowers are arranged on the spike in four spiral rows. The flowers are tetramerous, actinomorphic, hermaphrodite, and hypogynous. These are subtended by ovate, acuminate, sparsely hairy, or highly glabrous bracts (Sharma 1990).

Sepals are light to dark green and are persistent. The four sepals are free, concave, glabrous, and elliptic; the four petals are glabrous, reflexed, and white. The four stamens are exserted and epipetalous; the gynoecium is bicarpellary syncarpous; style is filiform, shorter than the filaments; the ovary is bilocular, with one ovule per locule; and placentation is axile. The capsule is ovate or ellipsoid, dehiscing along the ring of abscission tissue that develops around the capsule; the pyxidium splits into an upper lid and lower base. Seeds are cymbiform, translucent, and concavo-convex. The concave side is boat shaped and the convex side is covered with a thin, white membrane. Seeds are pinkish-gray, brown, or pinkish-white with a brown streak. The outer papery covering of the seed is termed husk; it is odorless and tasteless and constitutes the isabgol or psyllium.

22.2.2 Reproductive Biology

In *P. ovata*, flowering starts during the first week of February (Figure 22.1B). On the basis of type of flower, Sharma, P. Koul, and A. Koul (1992) categorized the plants into morphoforms A and B. In both morphoforms, the anther dehiscence synchronizes with the opening of the flower. In morphoform A, flowers are protogynous; the stigma remains clothed with prominent papillae and protrudes through gaps formed between corolla lobes. On the second day of protrusion, the stigma elongates and turns receptive. In morphoform B, stigma receptivity and anther dehiscence overlap. The stigma remains receptive until it is compatibly pollinated. The spike and scape continue to elongate even after the initiation of anthesis of the first few flowers. Anthesis completes in each spike within 10 days and it takes a total of 70 days for completion in each plant. As the flower opens, petals expand, leaving gaps between where the anthers protrude. Dehiscence of the anthers takes place through lateral longitudinal slits.

22.2.3 Pollination Mechanism

The breeding systems of different species of *Plantago* are highly diverse; there is a complete range from endogamy to allogamy (Wolff, Friso, and Van Damme 1988; Sharma et al. 1992). *Plantago ovata* combines the two processes; this is largely due to its floral structure. The species has chasmogamous flowers and practices a blend of self- and cross-pollination. Morphoform A is pollinated by wind and insects, while the morphoform B is autogamous. The features that favor wind pollination are plumose stigma, versatile anthers, and smooth-walled, dry pollen grains. Insect pollination has been reported to be less common; only *Apis dorsata* and a few dipteran flies visit the plant when most of the spikes are in bloom.

Except for the first maturing flowers on the spike, which may indulge in cross-pollination due to weak protogyny, in general, *P. ovata* can be considered to be an inbreeder. Loss of protogyny in morphoform B has been considered to represent a major step toward the evolution of autogamy in this species. Various researchers have explored the breeding system of *P. ovata* (Mital and Bhagat 1979; Patel, Sriram, and Dalal 1980; Sharma 1990; Jamwal 2000). Male sterile plants of the species are not visited by the pollinators (Sharma 1990), indicating that pollen is the major attraction for the insects visiting psyllium flowers.

22.2.4 Male Sterility

In *P. ovata*, male sterile individuals have been reported in natural populations by several workers (Atal 1958; Jamwal, Dhar, and Kaul 1998). Atal (1958) reported that male sterility is cytoplasmic. Subsequently, Mital and Issar (1970) opined that a dominant sterility gene (Ms) induces male sterility but its action is suppressed by another dominant gene (Sp). Since the sterility gene is expressed in S-cytoplasm only, sterility was considered to be gene cytoplasmic. Jamwal (2000) undertook an extensive breeding program for 5 years and progenies of as many as 70 male sterile individuals were raised and studied. The results of these crosses conclusively proved that inheritance of male sterility is complex gene cytoplasmic. Jamwal et al. (1998) reported the occurrence of partially male sterile individuals in this species that exhibit labile sex expression. The studies indicated that there are several mechanisms operative in this species. This has been explained on the basis of involvement of several nuclear genes, either dominant or recessive, in the restoration of male fertility and partially male sterile individuals are most likely heterozygous at these loci.

22.3 *PLANTAGO* GENOME

22.3.1 Genome Size

Recently, Dhar, Fuchs, and Houben (2009) used flow cytometry to determine the genome size of *P. ovata*. Earlier efforts by several workers using Feulgen densitometry (Badr, Labani, and Elkington 1987) resulted in contradictory results on the genome size in diploid and tetraploid *P. ovata*. Using Feulgen densitometry, Badr et al. (1987) reported the mean total 1C DNA content of 13 independent measurements of *P. ovata*. The estimated value was 0.635 pg (621 Mbp), 30% larger than the value reported for diploid *P. ovata*. The AT frequency of *P. ovata* was calculated to be 59.7%, which falls in the range of values determined for other plant species.

22.3.2 Genome Organization

The diploid chromosome number of the species is $2n = 2x = 8$. The karyotype consists of two metacentric, two submetacentric, and four subtelocentric chromosomes. The latter carry nucleolus organizer regions (NORs) in the short arms. The chromosome size ranges from 2.5 to 2.9 μm in Feulgen-stained preparations. The size of the NOR varies in the two chromosome pairs. This difference is reflected in the size of the nucleoli that they organize (Sareen, Koul, and Langer 1988). Dhar et al. (2002) used C-banding and fluorescence *in situ* hybridization (FISH) for identification of different chromosomes in *P. ovata*. C-banding resulted in distinct eu- and heterochromatin patterns of each chromosome. One of the arms of chromosome 1 was found to be euchromatic. Besides the prominent centromeric band, it carries a dark band toward the terminal end of the arm. Chromosome 2 is distinct and shows a prominent centromeric band. Chromosome 3 is nucleolar and has a centromeric band. Chromosome 4 is also nucleolar and has a dark band at the end of the short arm, in addition to the centromeric band.

Dhar et al. (2009) investigated the distribution of eu- and heterochromatin in chromosomes of *P. ovata* using a combination of propidium iodide (PI) staining, Giemsa banding, and immunostaining techniques (Figure 22.2). The PI staining of prometaphase chromosomes clearly differentiated

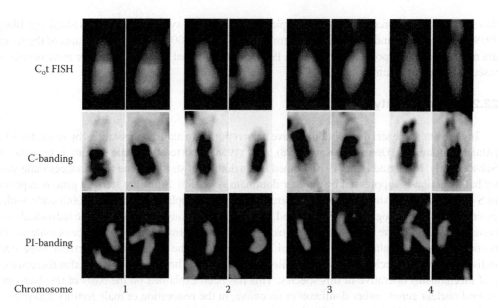

C_0t FISH

C-banding

PI-banding

Chromosome   1   2   3   4

Figure 22.2 Comparative mapping of repetitive DNA elements on *P. ovata* chromosomes using C_0t FISH, C-banding, and propidium iodide banding.

the dark, highly condensed heterochromatic from light, less condensed euchromatic regions on all chromosomes. Giemsa C-banding revealed dark bands in the pericentromeric regions of all chromosomes and at the NORs. In both cases, euchromatic portions were only lightly stained. In interphase nuclei, the heterochromatin was organized in highly condensed and DAPI stainable chromocenters.

Out of various fractions isolated through HAP chromatography, C_0t-1 DNA fraction was selected and used for *in situ* hybridization experiments (Dhar et al. 2009). Since the constitutive heterochromatin is known to be composed mainly of rapidly reannealing, highly repetitive DNA, it was assumed that the heterochromatic regions of *P. ovata* would be composed of DNA within the C_0t-1 fraction. After FISH with labeled C_0t-1 DNA, fluorescent signals were detected specifically at the C-banding positive chromosome regions. The remaining parts of the chromosomes were less densely covered by the probe. On the whole, one of the arms of chromosome 1, terminal regions of the non-nucleolar chromosomes (in both the arms), and the long arms of nucleolar chromosomes were poorly labeled with the C_0t-1 fraction.

Dhar et al. (2009) determined the distribution of histone H3 methylated at lysine 4, lysine 9, and lysine 27 on flow-sorted leaf nuclei. The specific antibodies exclusively labeled the euchromatin, while the heterochromatic chromocenters were free of signals. On the other hand, antibodies against H3K9me1, H3K9me2, and H3K27me1 labeled the entire chromatin. H3K9me3, however, showed a homogeneous distribution over the entire chromatin. In somatic chromosomes, H3K4me2 signals were especially pronounced at euchromatic terminal portions, thereby suggesting the presence of genes in these regions. In addition, one arm of the metacentric chromosome 1 was entirely labeled, suggesting a high proportion of euchromatin. This was consistent with data obtained after FISH with C_0t-1 DNA. The chromosomal distribution of the heterochromatin-associated H3K9me2 showed an opposite pattern; signals were enriched around the pericentromeric regions. According to Houben et al. (2003), this enrichment is typical for heterochromatin-specific marks in plants with relatively small genomes.

Interestingly, in *P. ovata* H3K9me3 showed a uniform labeling with no clear preference for either the euchromatin or the heterochromatin. Such distribution has been observed only for heterochromatin-associated marks in plants with large genomes (Houben et al. 2003; Fuchs et al. 2006) and is considered to be a modification required for the silencing of transposable elements dispersed throughout the genome. Therefore, the *P. ovata* genome is unique in having qualities of both the types of genomes.

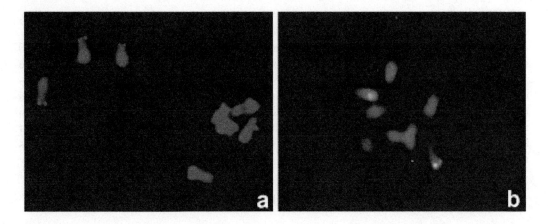

Figure 22.3   (See color insert.) Localization of (a) 45S rDNA and (b) 5S rDNA on different chromosomes of *P. ovata* using FISH.

The distribution of the cell cycle dependent histone modifications H3S10ph and H3T11ph was interesting in *P. ovata* because it was restricted to the pericentromeric regions. While the distribution pattern of H3S10ph was similar to those of other plants, the distribution pattern of H3T11ph resembled that of mammals (Houben et al. 2005). In plants, H3T11ph along the chromosomal arms is linked to chromosome condensation, while H3S10ph is connected to centromere cohesion (Houben et al. 2007), which is the reverse in mammals. Therefore, existence of H3T11ph in centromeric regions in *P. ovata* is unique and may have some role in centromere cohesion.

FISH analysis was carried out by Dhar et al. (2002, 2006) using several DNA probes. While 45S ribosomal RNA genes were mapped on chromosomes 3 and 4 (Figure 22.3a), 5S rRNA genes were located in one of the arms of chromosome 1 (metacentric) (Figure 22.3b). FISH analysis revealed that *P. ovata* chromosomes are capped by the telomeres composed of TTTAGGG repeat sequences, as in *Arabidopsis*. One of the interesting cytological features in *P. ovata* is the somatic association of NOR-bearing chromosomes observed by Sharma, Langer, and Koul (1989) in an artificially induced aneutetraploid individual. Dhar and Kaul (2004) used FISH to demonstrate satellite association in normal diploid plants. Their study provided conclusive evidence in favor of the involvement in association of rDNA rather than proteins, as proposed by earlier workers in animals.

During pollen mother cell meiosis, four perfect bivalents—two rod shaped and two ring shaped—are formed. The chiasmata frequency is low (1.4 per bivalent and 5.6 per cell). Rarely, one or two bivalents undergo delayed disjunction, which upsets the otherwise orderly disjunction of chromosomes at anaphase-I (Koul and Sharma 1986).

22.4 GENETIC DIVERSITY

According to Koul and Sharma (1986), the narrow genetic base of *P. ovata* is due to a few small, heterochromatic chromosomes having low chiasmata frequency and recombination index. It has now become clear that a major portion of the genome is composed of repetitive DNA, which accounts for a low recombination index. Consequently, the species lacks inherent variability. Earlier efforts to induce variations using physical and chemical mutagens have produced limited results (Koul, Sareen, and Dhar 1994). It seems that this recalcitrant nature of the crop emanates from its peculiar genome organization. Dhar et al. (unpublished data) analyzed genetic diversity in *P. ovata* and several of its wild allies using randomly amplified polymorphic DNA (RAPD) and amplified fragment length polymorphism (AFLP) markers. RAPD analysis did not reveal much polymorphism among 10 different accessions, while the level of polymorphism was very high among wild species. For AFLP analysis, seven primer combinations were used. A total of 800 bands were analyzed. Interestingly, AFLP analysis was successful in bringing out hidden differences among various accessions.

Earlier, Singh, Lal, and Shasany (2009) analyzed phenotypic and molecular diversity in 80 accessions of *Plantago*. The material consisted of 69 accessions from India and 11 exotics. Three species—*P. ovata*, *P. lanceolata*, and *P. major*—were used for the study. RAPD analysis was conducted on 36 accessions selected on a geographical and morphological basis. Biometric analysis revealed a higher degree of divergence among accessions than within accessions. The dendrogram revealed seven clusters. Based on the RAPD data, these workers have identified highly divergent lines that can be used for further breeding. They concluded that the genetic diversity in *P. ovata* is not related to the geographical origin.

22.5 GENETIC IMPROVEMENT

Plantago ovata has a narrow genetic base because it is known only under cultivation and no wild plants have been reported so far. Several efforts have been made in the past for the genetic improvement of this economically important plant. Keeping in view the limited success of these experiments, fresh breeding efforts for improvement in several quantitative and qualitative characteristics need to be undertaken. Some of the breeding objectives that need to be achieved are increase in the seed size and yield, development of compact and nonshattering spikes, and production of seeds with higher swelling factor and genotypes resistant to biotic and abiotic stresses. Some of the achievements already made in the improvement of psyllium are given next.

22.5.1 Selection

The prospects of evolving better varieties through selection alone are limited since the species has a narrow genetic base and lacks variability. Nevertheless, efforts have been made to exploit whatever little variability exists to select better genotypes. As a result, several varieties have been developed: RI-87, RI-89, AMB-2, GI-1, GI-2, MI-4, MIB-121, HI-34, HI-2, HI-1, HI-5, JI-4, and NIHARIKA. The Gujarat Isabgol-1 variety yields 800–900 kg of seeds per hectare. The new variety, Gujarat Isabgol-2, has a potential to yield 1000 kg of seeds per hectare. Punia, Sharma, and Verma (1985) observed genotypic variation in seed yield, husk percentage, and downy mildew index. Some workers found cultivars from Palanpur (Banaskantha) superior to those from Mehsana, Visnagar, and Sidhpur. An improved cultivar, Gujarat Isabgol-1, has been developed at Pilwai Research Station of Gujarat Agricultural University, Anand. It yields about 10–12% more seed than the local cultivars. HI-5 is another high-yielding strain that has been developed at Hissar (Punia et al. 1985).

Several exotic strains of psyllium were introduced in India. Mital and Singh (1986) isolated three new genotypes, EC42706-1, EC42706-2, and EC42706-B, from the Italian strain EC42706. The new genotypes were found to be superior to other local genotypes in terms of compact spikes, synchronous maturity, and dwarf plant size. Recently, Singh et al. (2009) evaluated 60 accessions of *P. ovata*, of which four were exotics from Pakistan, Balochistan, and the United States.

22.5.2 Induced Polyploidy

Polyploidy has been reported to increase the seed size in several plants. Since seed constitutes the commercially and medicinally important part in *P. ovata*, one of the approaches to increase the seed size was thought to be through induction of polyploidy. With this aim in mind, many workers have attempted to utilize this approach for breeding superior psyllium (Dhar et al. 2005). Polyploids are characterized by the formation of multivalents during meiosis that very often leads to irregular anaphasic segregation and low seed output. Therefore, the advantages of polyploids get neutralized by the low production and in many cases get reverted into diploids after several generations. However, in psyllium this disadvantage gets buffered due to small, heterochromatic chromosomes and low chiasmata frequency.

Despite multivalent formation, chromosome disjunction has generally been found to be regular, as is the seed set (Zadoo and Farooqi 1977; Sharma 1984). Zadoo and Farooqi (1977) compared the time taken for germination by tetraploid and diploid seeds and found no difference; however, the rate of germination of tetraploid seeds was considerably reduced. The induced tetraploids were superior to diploids in terms of number of tillers and inflorescences, seed weight, and volume; yet, the seed output was less by 29.2%. Koul and Sharma (1986) reported that tetraploid seed was superior to diploid seed in mucilage content and characteristics such as overall pentosan content, absorbency, and viscosity. These characters improve the medicinal properties of the mucilage and confer a distinct advantage to the autotetraploid. Unfortunately, tetraploid superiority gets eroded beyond third or fourth generations, thereby neutralizing the advantages of polyploidy.

22.5.3 Mutation Breeding

The major bottleneck in the genetic improvement of *P. ovata* is the lack of variability. Therefore, one of the primary concerns of plant breeders has been to induce mutations. Sareen (1991) and Sareen and Koul (1999) conducted detailed investigations on induction of mutations and characterization of mutants. The seeds were exposed to γ-rays or treated with ethyl methane sulfonate (EMS). While EMS did not prove effective, γ-rays induced significant variation in mean and variance of yield-related characters—namely, tiller and spike count, seed size, seed count, and seed weight per individual. The M2 plants had fewer spikes per plant, although their seed output was significantly higher.

In addition to gross morphological changes, γ-rays also induced structural and numerical chromosome changes. One of the mutants in M2 progeny of the 50 kRad treated seeds was desynaptic. The mutant was characterized by the presence of two to six univalents in more than 75% of pollen mother cells. Sareen and Koul (1991) reported a triploid plant in *P. ovata*. As expected, meiosis of this individual was characterized by anaphase irregularities, which resulted in the isolation of a trisomic in its progeny. Sareen (1991) also reported three other trisomics, two isolated from progeny of a complex translocation heterozygote and a third from the progeny of a chromosome mosaic. The additional chromosome was involved in trivalent association in all three plants.

Sareen (1991) raised a few translocation heterozygotes from seeds exposed to γ-rays. In two translocation heterozygotes, four chromosomes (two nucleolar and two non-nucleolar chromosomes) were involved in interchange (Koul et al. 1994). In continuation of this study, Padha (1996) isolated trisomics in the progeny of a cross between translocation heterozygote (female) and normal disomic (male). The trisomics were of two types: tertiary trisomics and interchange telotrisomics. Lal, Sharma, and Misra (1998) developed Niharika, a new variety of *P. ovata*, using γ-irradiation. It was found to be superior to other varieties in some traits.

22.5.4 Somaclonal Variability

One of the methods for widening the genetic base of the crop is to induce somaclonal variability through *in vitro* culture. Barna and Wakhlu (1988) for the first time standardized a protocol for micropropagation in *P. ovata*. Subsequently, Wakhlu and Barna (1989) developed a protocol for callus initiation from hypocotyl explants. Barna and Wakhlu (1989) regenerated plants from callus derived from root cultures using MS medium supplemented with IAA and kinetin.

Figure 22.4 *In vitro regeneration of P. ovata.*

The regenerated plants were successfully transferred to the field. Cytological studies of the callus revealed variability in chromosome number, ranging from 8 to 48. Yet, all plants raised *in vitro* had a diploid chromosome number—probably because the diploid cells out compete the polyploid cells (Wakhlu and Barna 1988). Pramanik, Chakraborty, and Raychaudhuri (1995) raised *in vitro* cultures of *P. ovata* and characterized the regenerants using isozyme analysis; no significant differences were observed. Recently, Dhar et al. (unpublished) developed a rapid protocol for efficient regeneration of *P. ovata* (Figure 22.4). The genetic fidelity of the regenerated plants was assessed by RAPD markers. Obviously, somaclonal variability is of little help in widening the genetic base of this species.

Somatic embryogenesis is an important method for multiplication and genetic improvement of plants. This is especially so in those cases where true-to-type plants are needed. Das and Raychaudhuri (2001) used casein hydrolysate and coconut water as additives in the medium for developing somatic embryos. They reported a significant increase in the embryogenesis due to these additives.

22.6 DISEASES

Plantago ovata suffers from three major fungal diseases: wilt, damping off, and downy mildew. Among these, downy mildew caused by *Peronospora plantaginis* is the most serious, striking at the time of spike emergence (Mandal and Geetha 2001). The symptoms include development of chlorotic areas on the upper leaf surface and ashy-white frost-like mycelial growth on the undersurface. Powdery mildew attacks psyllium occasionally. The disease is controlled by Karathane W.D. (0.2%), Dithane M-45, or Dithane Z-78 at 2.0–2.5 g/L or Bordeaux mixture 6:3:100. Rathore and Rathore (1996) first reported that the diseased plant produces abnormal spikes. Recently, Mandal et al. (2010) reported that powdery mildew infection leads to male and female sterility in *P. ovata*. The infected plants produced long spikes with sterile florets. Androecium was affected more than the gynoecium; however, overall seed yield was reduced by about 73%.

Wilt of psyllium is caused by *Alternaria* sp., *Fusarium oxysporum*, and *F. solani*. Wilting starts from outer leaves and then spreads to the whole plant. The leaves change their color to silver. Wilt disease can be controlled by seed treatment with Bavistin or Benlate at 2.5 g/kg of seed. *Pythium ultimum* causes damping off of psyllium (Chastagner, Ogawa, and Sammeta 1978). Fenaminosulf and CGA-48 988 treatments have been recommended for protection against *P. ultimum*. Generally, aphids attack the crop during maturity, leading to severe losses. Spraying with Endosulfan (0.5%) or Dimethodafe (0.2%) fortnightly can control the aphids.

22.7 CULTIVATION

22.7.1 Seed Sowing and Germination

In India, *P. ovata* is grown on a commercial scale in the western state of Gujarat because its dry climate is best suited for the plant. A large number of small-scale industries located in Gujarat and Rajasthan are involved in processing the seed husk. *Plantago ovata* is grown as a 120- to 130-day winter crop. The seeds are sown between the end of October and middle of December, when the day temperature ranges between 18 and 26°C. Late sowing (in December) cuts short the period required for full growth and endangers the plant with spring rains that cause seed shattering. Sowing dates were also found to be important under Iranian conditions (Karimzadeh and Omidbaigi 2004), where early sowing exposed plants to chilling conditions. To ensure dense population and high yield, the seed is mixed with fine sand or sieved farmyard manure before it is broadcast. The seed rate is 4–13 kg/ha. Irrigation of the field prior to seeding helps in weed control because weed seeds germinate quickly; this is followed by shallow tillage. The first de-weeding is undertaken within 20–25 days of sowing. Weed control is normally achieved by one or more approaches, including

high seed rate of about 12 kg/ha
preplant application of 1.5 kg Diuron (N'-3,4-dichlorophenyl-N,N-dimethyl urea) per hectare
application of 0.1 kg paraquat (N,N'-dimethyl-4,4'-bipyridinium ion) per hectare, 28 days after sowing
   (McNeil 1989)

Patel and Mehta (1990) suggest the use of Isoproturon at 0.5 kg/ha for controlling weeds.

22.7.2 Soil

Although *P. ovata* can be grown on various soil types, well-drained, loamy soil is best suited for its cultivation. The other requirements include 7.2–7.9 soil pH, organic carbon, potash, and low phosphate level in the soil. The field is generally irrigated prior to seed sowing to ensure quicker germination. It is divided into plots of convenient size depending upon texture of soil, gradient of the field, and irrigation facilities available. Although attempts have been made to cultivate psyllium in saline-alkaline soils and in hydroponics, heavy soils with poor drainage (Kalyansundram et al. 1984) or salinity (Pal, Jadaun, and Parmar 1988) do not suit the crop.

22.7.3 Irrigation

Immediately after sowing, light irrigation is necessary. The seeds germinate within 4–5 days. Once the seeds germinate, the beds are irrigated only when it is necessary. Modi, Mehta, and Gupta (1974) rate the irrigation requirements of psyllium as medium; however, according to Randhawa et al. (1985), these are very low. Depending upon soil type and climate, three to six irrigations are sufficient. However, frequent watering after transplantation establishes the seedlings better.

22.7.4 Fertilizer Application

According to Kalyansundram, Patel, and Dalal (1982), *P. ovata* has a low nitrogen requirement, which can be easily met by planting a leguminous crop prior to its cultivation. Increase in the quantity of nitrogen from 0 to 50 kg/ha was reported to lead to reduction in the swelling factor of seeds. Randhawa et al. (1985) reported increase in seed yield following increased nitrogen application. Singh and Nand (1988) recorded the highest seed yield in the Tarai area after application of 40–80 kg/ha nitrogen. Therefore, generally, 25 kg/ha nitrogen and 25 kg/ha phosphorus are applied as a basal dose. Karimzadeh and Omidbaigi (2004) found 100 kg/ha nitrogen (of which half is to be applied presowing and other half at the time of flowering) most suitable for good plant growth and higher swelling factor of seeds.

22.7.5 Harvesting and Yield

Flowering starts 2 months after sowing and the crop is ready for harvesting in March or April. Plants are cut, with the help of hand sickles, 15 cm above ground level. These are heaped, dried for 2 days, and then threshed. As a result, the seeds separate and fall apart. The seeds are collected and winnowed repeatedly to separate out the undesirable parts. Thereafter, the seed is marketed and the straw is fed to cattle. The average seed yield is 1000 kg/ha.

22.7.6 Processing

The harvested plants are threshed and the seed is collected and transported to factories or mills for processing. The seeds are cleaned by passing them through large sieving pans to separate out dust particles and shriveled and deformed seed. The remaining seed is passed through grinding mills. The ground seed material is again sieved and winnowed to separate out the husk from coarse remains of the kernel. The best grade of isabgol is 70 mesh. Husk-free full or broken seeds are used as cattle feed.

22.8 MUCILAGE

Psyllium is well renowned for its mucilaginous properties. When the seeds are soaked in water, they increase to 8–14 times their original size (Figure 22.5). The swelling factor of 1 g of dry psyllium seed has been reported to be 15.25 mL and the mucilage content as 198 mg (Sharma and Koul 1986).

22.8.1 Mucilage-Producing Cells

Isabgol or psyllium has been used since ancient times for regulation of bowel functions (Cummings 1993). The seed coat of the psyllium seed consists of mucilage-producing cells filled with mucilage (Hyde 1970). These cells undergo a complex differentiation process, beginning with an undifferentiated parenchyma leading to thin-walled containers of almost pure mucilage and increase in the number and size of Golgi vesicles. Both microscopic and histochemical evidence has shown that mucilage is present within Golgi vesicles while they are still attached to the Golgi

Figure 22.5 Development of mucilage in *P. ovata* seeds on exposure to water.

apparatus. Mucilage deposition is another complex process that involves cell expansion, separation of the protoplast from the cell wall, fusion of vacuoles and extra protoplasmic space, and the disappearance of starch. Seed wetting leads to almost instantaneous hydration of the hydrophilic mucilage, followed by rupture of the primary cell wall and the release of mucilage to form a gel capsule around the seed. Hyde (1970) has divided the developmental series of mucilage producing cells into two parts:

1. The first series includes the processes involved in cell enlargement resulting in the increase in volume of the nucleus, particularly the nucleolus; the appearance of rough endoplasmic reticulum and the polyribosomal arrays; and the synthesis of large starch grains.
2. The second series includes the processes involved in the digestion of starch, synthesis and extrusion of mucilage, and degeneration of the protoplast.

In our laboratory, systematic studies were initiated in order to understand the developmental process of formation of mucilage and to correlate it with the functional organization of the genes involved in the process (Figure 22.6; Table 22.1).

22.8.2 Composition of Mucilage

The seeds contain mucilage, fatty oil, albuminous matter, a glucoside (aucubin) (Chopra 1930; Khorana, Prabhu, and Rao 1958), planteose sugar (French et al. 1953), and proteins (Patel et al. 1979). Laidlaw and Percival (1949) prepared both cold- and hot-water extracts of mucilage from seeds. The cold-water extract contained polysaccharides composed of 20% uronic anhydride, 52% pentosan, and 18% methyl pentosan. Hot-water extract of seeds contained polysaccharides PI and PII. On hydrolysis,

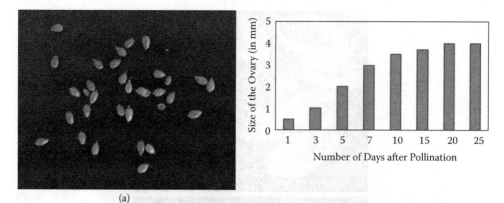

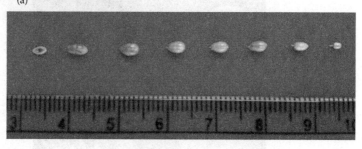

(b)

Figure 22.6 **(See color insert.)** (a) Ovaries at different developmental stages. (b) Development of seeds from ovaries (see left to right).

Table 22.1   Time Line of Development of the Mucilage in *Plantago ovata*

Days after Pollination (DAP) (Present Work)	Size of Ovule (mm) (Present Work)	Morphological Observations (Present Work)	Description[a]
0	0.6	Soft, fragile, green	At the time of pollination
1	0.7	Soft, fragile, green	Rapid growth occurs
2	0.9	Soft, green	Size of the ovule ranges between 0.8–1 mm, during rapidly growing stages
3	1.3	Soft, green	Protoplast begins to pull away from cells
4	1.7	Soft, green	Continued stretching of cell walls and rapid accumulation of mucilage, primarily in distal end of the cells
5	2.0	Soft, green	In the size range of 1.5–1.9 mm, vacuoles largely disappear from protoplast
6	2.4	Soft, green	Protoplast largely composed of remnants of starch; some evidence of layers of mucilage
7	3.0	Soft, green	Nearly mature mucilage containing cells
10	3.5	Soft, unripe, green	Thin-walled mucilage containers developed
15	3.7	Soft, unripe, green	Mucilage fully developed
20	4.0	Hard, half mature, purplish ovules	Mucilage fully developed
25	4.0	Hard, purplish, immature seeds	Mucilage fully developed

[a] Modified from Hyde, B. B. 1970. *American Journal of Botany* 57:1197–1206.

Table 22.2   Monosacharide Composition of the Gel-Forming
Polysaccharide in *Plantago ovata*

Sugar	After Marlett and Fischer (2002)	After Fischer et al. (2004)	After Saghir et al. (2008)
Arabinose	22.5[a]	22.6	23.3
Xylose	73.3	74.6	75.1
Galactose	1.77	1.5	1.2
Mannose	tr[b]	n.d.[c]	n.d.
Rhamnose	0.99	0.4	0.8
Uronic acid	2.22	0.7	n.d
Glucose	0.11	0.3	n.d

Sources: Marlett, J. A., and M. H. Fischer. 2002. *Journal of Nutrition* 132:2638–2643; Fischer, H. M. et al. 2004. *Carbohydrate Research* 339:2009–2017; Saghir, S. et al. 2008. *Carbohydrate Polymers* 74:309–317.

[a] Percent content.
[b] tr = trace.
[c] n.d. = not determined.

the former yields d-xylose (46%), l-arabinose (7%), an aldobiuronic acid (40%) called 2d-galacturono-side-l-rhamnose, and an insoluble residue. PII yields l-arabinose (14%), d-xylose (80%), aldobiuronic acid (0.3%), trace of d-galactose, and an insoluble residue (Anonymous 1969). The fatty acids in the seed oil are linolenic, oleic, palmitic, stearic, and lignoceric (Anonymous 1969). Jamal et al. (1987) identified two oxygenated fatty acids—9-hydroxyactadec-*cis*-12-enoic acid and 9-oxoactadec-*cis*-12-enoic acid—in the seed oil. According to Kennedy, Sandhu, and Southgate (1979), psyllium contains 15% nonpolysaccharide material and the remaining 85% appears to be a single polysaccharide comprising d-xylose (62%), l-arabinose (20%), l-rhamnose (9%), and d-galactose (9%) (Table 22.2).

Romero-Baranzini et al. (2006) analyzed *P. ovata* seeds for composition and *in vitro* digestibility. The seeds contained 17.4% protein, 6.7% fat, 24.6% total dietary fiber, 19.6% insoluble fiber, 5.0% soluble fiber, and a combustion heat of 4.75 kcal/g. Fractionation of the material yielded albumin (35.8%), globulin (23.9%), and prolamin (11.7%). The oil had high amounts of linoleic (40.6%) and oleic (39.1%) acids and a minor quantity of linolenic acid (6.9%). *In vitro* protein digestibility of the psyllium seed was observed as 77.5%, thereby suggesting it to be a highly digestible protein. Lysine content was 6.82 g/100 g of protein, higher than wheat and oats (2.46 and 4.20 g/100 g of protein, respectively). Rat bioassays conducted by these workers revealed 89.6% digestibility of dry matter, 86.0% apparent digestibility, 88.1% true digestibility, and 4.40 net protein ratios corrected (NPRc). The importance of these findings is that *Plantago* whole grain shows favorable nutritional quality when compared to cereals and legumes.

In 1979, Kennedy and coworkers reported that the polysaccharide has a linear backbone of β-d-xylose residues in the pyranose ring form and disaccharide side chains with a terminal d-galacturonic acid linked to C(O)-2 of an l-rhamnose. All the three side chains are attached to either C(O)-2 or C(O)-3 of xylose in the polymer backbone. The backbone has (1→3) and (1→4) β-linkages but their sequence and the distribution of side chains have not yet been determined.

Sandhu, Hudson, and Kennedy (1981) obtained the mucilage from *Plantago* seed husk by an alkali extraction method and concluded that the preparation, although polydisperse, represented a single species of polysaccharide—a highly branched, acidic arabinoxylan. Based on the gel-forming fraction of the alkali extractable polysaccharides of *Plantago ovata* seed husk, Fischer et al. (2004) concluded that the main chain of arabinoxylan consists of β-(1→4) linked d-xylopyranose residues, some of which carry a single Xylp moiety at position 2 with hydroxyl groups at positions 2, 3, and 4, which are available for chemical modifications. Other Xylp residues bear trisaccharide branches at position 3 with the sequence l-Araf-α-(1→3)-d-Xylp-β-(1→3)-l-Araf.

The carboxymethyl derivative of psyllium husk is a fibrous mass, light in odor and mucilaginous in taste. It swells readily in water, forming viscous mucilage of pH 6.7 and swelling factor 90. It is superior to the intact husk in viscosity, spreadability, film-forming characteristics, and smooth texture (Jain and Mithal 1976). The grayish aqueous extract of the husk is an excellent thickening agent. Saghir et al. (2008) carboxymethylated seed husk of *P. ovata* with sodium monochloroacetate and characterized it by means of Fourier transform infrared spectroscopy (FTIR) spectra. They obtained a white gel, arabinoxylan, that showed pH-dependent water solubility and high swelling ability in water by simple alkali extraction and freeze drying. This carboxymethylation represents a useful way to get water-soluble arabinoxylan derivatives with anionic functions. By simple extraction, the seed husk of *P. ovata* can become a source for this biopolymer, arabinoxylan.

The husk mucilage is a white fibrous material that, although insoluble in water, swells to give a thick solution. This solution is neutral to litmus. It does not reduce Fehling's solution and gives no test for starch. Qualitative tests proved the absence of methoxyl groups, methyl pentose sugars, and hexose sugars. The husk obtained by milling the seeds of *P. ovata* contains a high proportion of hemicelluloses composed of a xylan backbone linked with arabinose, rhamnose, and galacturonic acid units (Sandhu et al. 1981). Marlett and Fischer (2002) demonstrated that the gel-forming fraction composing 55–60% of the husk acts as a laxative and also lowers the blood cholesterol. Seeds of psyllium also contain 17–19% protein (Patel et al. 1979). The amino acids separated from psyllium seed by paper chromatography include valine, alanine, glycine, glutamic acid, cystine, lysine, leucine, tyrosine, and serine. The mucilage contained in psyllium is an arabinosyl (galactosyluronic acid) rhamnosyxylan (Sandhu et al. 1981). The sugar and mucilage content of seed bear an inverse relation at different stages of seed development (Chinoy, Mehta, and Mehta 1978); to begin with, the sugar content of seed is high and the mucilage content low and the reverse is true during later stages.

The husk displays negative thixotrophy when it is dispersed in water, but when it is heated to 60°C it gets transformed into thixotropic gel with bulges and spurs. The mucilage has emulsifying properties and is a good suspension agent. The 2% (w/v) mucilage of seed husk powder in cold water compares well, whereas 1.5% (w/v) mucilage in hot water is superior to 10% (w/v) starch mucilage in binding properties. It is comparable to methyl cellulose and is superior to sodium alginate and sodium carboxymethyl cellulose in suspending properties (Mithal and Gupta 1965).

22.9 PSYLLIUM: A THERAPEUTIC AGENT

Most of the species of the genus *Plantago* are highly useful as herbal medicines. Phytochemical investigations of these plants have revealed their high potential to produce a wide array of bioactive secondary metabolites. *Plantago ovata* is the source of psyllium or isabgol, which contains a high proportion of hemicellulose, rhamnose, and glacturonic acid residues. It has been classified as a mucilaginous fiber due to its powerful ability to form gel in water. This ability is due to the water retention property of its endosperm, which prevents the seeds from drying. Psyllium is known to be a very important therapeutic agent. Singh (2007) has presented a very exhaustive review on its therapeutic applications. Psyllium has been extensively used for the treatment of constipation, diarrhea, irritable bowel syndrome, ulcerative colitis, type 2 diabetes hypercholesterolemia, and, recently, colon cancer. A wide range of biological activities have been found in the plant extract or isolated compounds, including wound healing and antimicrobial, anti-inflammatory, and antileukemic activity. Moreover, allergic pollens of the plant are of particular interest to clinicians.

22.9.1 Constipation

Psyllium is known to be a very effective laxative. It has been used for a long time to provide relief during constipation, which is the result of absorption of excessive amounts of water so that the stools become hard and difficult to expel. Similarly, constipation can also be caused by improper contraction of the colon, leading to its failure to push the stools to the rectum. Regular use of psyllium under such conditions provides much needed relief since it acts as a dietary fiber. It increases the frequency and weight of stools, softens hard stools, and reduces pain at defecation. Psyllium was found to be more effective than sodium docusate (McRorie et al. 1998).

Interestingly, psyllium has properties of relieving constipation by increasing stool weight as well as treating chronic diarrhea. Scientific explanation of the laxative property of psyllium comes from several studies conducted on humans. It has been shown that the fiber shortens gastrointestinal transit time and increases the number of bowel movements per day, the amount of stool passed, and stool weight. In an experiment to study the effect of psyllium on chronic constipation, 20 patients were selected and 10 were given psyllium and 10 a placebo. After 1 month of treatment, the patients receiving psyllium showed good results against the placebo group (i.e., the frequency of stools increased from 2.5 ± 1 to 8 ± 2.2 stools per week). A decrease in consistency of stools was also observed. In the placebo group, no significant change was observed.

After swelling in the stomach, the mucilaginous mass promotes peristalsis; hydrates the feces, thereby increasing its volume and viscosity; helps in the bowel movement; and, finally, relieves chronic constipation. Similarly, it is also useful for treating other gastrointestinal problems. The mechanism of action of psyllium is mechanistic rather than physiological. Therefore, for efficient action of mucilage, it is necessary to drink enough water so that psyllium is able to absorb the water and swell. The mechanism of action of psyllium has been found to be different from those of the other bowel-regulating dietary fiber sources. The others act by increasing the bulk in the colon, which stimulates propulsion of the stools; psyllium increases the water concentration in the stool, leading to soft or slick stool that passes easily (Singh 2007).

Marlett, Kajs, and Fischer (2000) conducted an experiment whereby the stools of the patients were collected after every dose of psyllium. They observed that the stools were gelatinous and the gel fraction isolated from stool contained 75% carbohydrate. Most of this carbohydrate was xylose (64%) and arabinose (27%), the same two sugars that account for the majority (79%) of the carbohydrate in psyllium.

One of the properties of psyllium is that, unlike other viscous fibers, psyllium is not completely fermented in the colon. The unfermented or incompletely fermented fiber holds moisture and together these contribute toward the increased stool mass (Cummings 1993) and provide substrate for microbial growth. The greater bacterial mass and accompanying water further increase stool weight (Stephen and Cummings 1980; Chen et al. 1998). In most of the studies, the additional stool mass produced by consumption of more dietary fiber contains the same proportion of moisture as do low-fiber stools (Prynne and Southgate 1979; Eastwood, Brydon, and Tadesse 1980). It has also been reported by some workers that psyllium increases stool frequency and weight and decreases stool consistency in constipated patients. These effects are not associated with significant changes in colorectal motility. The clinical parameters were not significantly affected by treatment with psyllium, although there was a significant decrease in transit time.

22.9.2 Diarrhea

Several studies have shown that psyllium is useful in controlling diarrhea by increasing the number of normal stools and decreasing the number of liquid stools (Belknap, Davidson, and Smith 1997). According to Wenzl et al. (1995), under normal circumstances the intestine maintains a percentage of fecal water in the stools, although the quantity varies a great deal. Diarrhea is caused due to imbalance in the ratio of fecal water to water-holding capacity of insoluble solids leading to looseness of stools. Patients with normal stool weight have loose stools due to low output of insoluble solids, although there is no reduction in water output. Psyllium intake plays a very important role because it delays gastric emptying, possibly by increasing the meal viscosity, and delays the colon transit, possibly by delaying the production of gaseous fermentation products. A combination of psyllium and calcium seems to be a cheap and effective alternative to conventional treatment of chronic diarrhea (Qvitzau, Matzen, and Madsen 1988).

In patients with painful irritable bowel syndrome, supplementation with psyllium or gum arabic was associated with improvement of stool consistency. Improvements in stool consistency did not appear to be related to unfermented dietary fiber (Bliss et al. 2001).

One of the common causes of diarrhea is the medication, especially protease inhibitors (PIs). Psyllium is one of the major agents used for the treatment of PI-associated diarrhea (Rachlis et al. 2005). The etiology and treatment of chronic nonspecific diarrhea of children is still poorly understood. In those children where other etiologies of chronic diarrhea had been ruled out, treatment with psyllium and normalization of the diet seem to be an effective therapy. Various studies on use of psyllium for the treatment of diarrhea in animals have been reported (Neiffer 2001) as well.

22.9.3 Obesity

Increased use of junk food and changes in lifestyle have led to increased incidence of obesity. The prevalence of obesity is increasing worldwide. Obesity is associated with other major health problems like type 2 diabetes, heart diseases, stroke, etc. It is very important to treat obese individuals by every means. Unfortunately, despite being effective in the short term, drug treatment of obesity is often associated with rebound weight gain. Psyllium-containing products, which have already been used to control bowel functions, are useful supplements in weight control diets, leading to the feeling of fullness in the stomach and resulting in intake of less food. Also, dietary fiber, particularly psyllium, does not stimulate dietary enzymes like pepsin, trypsin, α-amylase, etc. Because these enzymes are decisive for food digestion, psyllium has a direct role in the management of obesity. In a study carried out on obese Zucker rats fed on *P. ovata* husk-supplemented diet for 10 weeks, it was observed that tumor necrosis factor (TNF)-α and reduced adiponectin secretion by adipose tissue found in obese Zucker rats was significantly improved (Galisteo et al. 2010).

In children and adolescents from developed countries, obesity prevalence has strongly increased in the last decades and insulin resistance and impaired glucose tolerance are frequently observed. Some dietary components, such as low glycemic index foods and dietary fiber, could be used to improve glucose homeostasis in these children (Frati Munari et al. 1998).

22.9.4 Diabetes

In humans, blood glucose level is maintained by the insulin produced by the pancreas. In recent years, type 2 diabetes has become highly prevalent, especially in developed countries. Diabetes is a lifestyle disorder that has been attributed to the increase in the population, urbanization, and stress of modern life. It is highly essential for diabetics to maintain good glycemic control because this can delay the onset of other complications or diseases. Type 2 diabetes is known to increase the risk of cardiovascular disease dramatically; therefore, reduction in atherogenic lipids could greatly reduce mortality and morbidity from cardiovascular disease in individuals with type 2 diabetes (Laakso 1996; Savage 1996).

Regular use of psyllium has been suggested to be a potential treatment for lowering blood sugar levels and thereby controlling diabetes. Several studies have demonstrated reduction in blood sugar levels with a single dose of psyllium; however, the long-term effects are not clear (Dhar et al. 2005). Psyllium has been shown to decrease postprandial glucose concentrations and decrease serum cholesterol concentrations in men with type 2 diabetes (Florholmen et al. 1982; Anderson et al. 1999). In developed countries, a significant increase in insulin resistance and impaired glucose tolerance has been observed, especially in young adults and children. Dietary fiber has been shown to improve glucose homeostasis in such individuals. Psyllium improves glucose homeostasis and lipid and lipoprotein profiles.

Gupta et al. (1994) conducted an uncontrolled study that suggested that psyllium improved gly-cemic and lipid control in individuals with type 2 diabetes. Earlier, Pastors et al. (1991) carried out a detailed study on the effect of psyllium on blood sugar levels in type 2 diabetes mellitus patients. In one set of patients, psyllium was given before breakfast and dinner; in the other set, a cellulose pla-cebo was supplemented. The results were highly encouraging: Postprandial serum glucose values were 14, 31, and 20% lower after breakfast, lunch, and dinner, respectively, in the psyllium-treated group. According to a study, after psyllium supplementation, the percentage change in postprandial glucose in type 2 diabetes patients ranged from –12.2 to –20.2% (Moreno et al. 2003). That psyllium increases peripheral insulin sensitivity is exemplified by the reduction of postprandial glucose levels even in nondiabetic individuals (Jarjis et al. 1984).

22.9.5 Cholesterol Lowering

Several studies have revealed a positive association of plasma LDL cholesterol levels and risk of coronary heart disease (Anderson, Castelli, and Leby 1987; Anderson, Jones, and Riddell-Mason 1994). It has been shown by many workers that regular intake of dietary fibers can lower blood cholesterol levels, especially LDL, and consequently reduce the risk of coronary heart disease. Psyllium is the best possible mucilaginous soluble fiber to treat such conditions because it is not habit forming and is highly effective with almost no side effects. There are many studies to sub-stantiate this claim. Psyllium intake has been shown to result in 10–24% reductions in cholesterol (LDL) levels, which is quite significant (Fernandez et al. 1997). Sprecher et al. (1993) demonstrated a 3.5% reduction in total cholesterol and a 5.1% reduction in LDL levels when 5.1 g of psyllium husk were consumed twice daily for 8 weeks. These changes are associated with a significant increase in serum levels of cholesterol precursors. In hypercholesterolemic children, the effect of psyllium in LDL cholesterol serum concentration ranged from 2.78 to –22.8%, the effect in HDL cholesterol ranged from –4.16 to 3.05%, and the effect of triglycerides from 8.49 to –19.54%. Thus, it was evi-dent that psyllium improves the lipid and lipoprotein profiles (Moreno et al. 2003).

The mechanism of action of psyllium in reducing blood cholesterol levels has not been fully elucidated. However, many studies on hypercholesterolemic men have suggested that it binds to bile acids in the intestinal lumen (Van Rosendaal et al. 2004). A well-established mechanism of reduc-tion of serum cholesterol levels is the conversion of hepatic cholesterol toward bile acid produc-tion. This process is regulated by the bile acids returning to the liver via enterohepatic circulation (Buhman et al. 1998) by inhibiting the enzyme 7α-hydroxylase. Psyllium was shown to stimulate bile acid synthesis by increasing the 7α-hydroxylase activity in animal and human models (Horton, Cuthbert, and Spady 1994; Matheson, Colon, and Story 1995). Hepatic 7α-hydroxylase is the ini-tial and rate-limiting enzyme in the bile biosynthetic pathway that is the major regulated pathway whereby cholesterol is eliminated from the body.

Induction of hepatic 7α-hyrdoxylase activity by dietary psyllium may account, in large part, for the hypocholesterolemic effect of this soluble dietary fiber. When rats were fed with cholestyramine (a bile acid sequestrant), an increase in transcription, translation, and activity of hepatic 7α-hydroxylase was observed, while these parameters were low in rats fed bile acids (Singh 2007). Similarly, it has been demonstrated that rats fed with 5% psyllium had greater activity of 7α-hydroxylase and higher amounts of bile acid as compared to controls. In another study, psyllium has been shown to increase 7α-hydroxylase activity and mRNA levels coordinately in hamsters (Horton et al. 1994).

Molecular studies have identified a bile acid response element in the promoter of the 7α-hydroxylase gene, thereby suggesting a molecular mechanism involved in transcriptional regu-lation of 7α-hydroxylase (Hoekman et al. 1993). The hypocholesterolemic effect of psyllium has been confirmed in several animal models, including hamsters, guinea pigs, and rats, and has been attributed to greater excretion of bile acids and total steroids leading to an up-regulation of bile acid biosynthesis. Intake of psyllium has been suggested to increase viscosity in the intestine, thus preventing absorption of bile acids and neutral steroids, a phenomenon reported in other viscous sources of dietary fiber.

Daily fecal bile acid excretion was 400% greater in hamsters fed with 6% psyllium, whereas CHY caused an 11-fold increase compared to controls (Bergman and van der Linden 1967). The mechanism of this protection is not fully elucidated but seems to be related to changes in the bile acid profile affecting the hydrophobicity of the bile acid pool. Psyllium and CHY caused distinct alterations in the bile acid profile. Psyllium caused a selective reduction of taurine-conjugated bile acids, especially of taurochenodeoxycholate. As a result, the glycine:taurine conjugation and the cholate:chenodeoxycholate ratios were significantly higher in psyllium-fed hamsters. There is consistent evidence from a number of studies that psyllium, like CHY, leads to an increase in cholesterol 7α-hydroxylase activity in parallel with 7α-hydroxylase mRNA (Fernandez 1995; Matheson et al. 1995).

In a pilot study in post- and premenopausal hypercholesterolemic women, it was observed that the former women benefited more from the addition of psyllium to their diets in reducing the risk for heart disease. It was observed that 7.5% dietary psyllium produces effects on hepatic 7α-hydroxylase and LDL metabolism that are similar to those produced by 1% cholestyramine.

22.9.6 Ulcerative Colitis

Higher levels of dietary fiber have been shown to decrease the relative risk of ulcerative colitis. In a trinitrobenzenesulfonic acid model of colitis, dietary fiber was shown to decrease the production of TNF-α and nitric oxide (known as mediators of inflammation). Butyrate and other short chain fatty acids such as acetate, propionate, and valerate are created when soluble fibers are fermented in the intestine. The high concentration of these short chain fatty acids plays a role in decreasing the production of certain inflammatory mediates and also contributes to the repair and regeneration of the damaged colonic cells. A recent comparative study of *P. ovata* seeds to oral mesalamine demonstrated equal effectiveness in the maintenance of remission in patients with ulcerative colitis.

22.9.7 Irritable Bowel Syndrome

Psyllium has been used for ages as a dietary fiber supplement to promote the regulation of bowel function. In addition to constipation, irritable bowel syndrome is another problem associated with the functioning of intestines. To a great extent, this affects the large intestine (colon); however, other parts of the gastrointestinal track can also get involved. In medical terminology, the disease is referred to by several names, including colitis, spastic colon, spastic, mucous colitis, etc. In this syndrome, the contraction of the colon is abnormal—so much so that sometimes the patient suffers a great deal. In a healthy individual, the colon dehydrates and stores the stool and subsequently, as a result of rhythmic contractions, propels the stool from the right side over to the rectum, storing it there until it can be evacuated. In irritable bowel syndrome, the colon does not contract normally; instead, the contraction is disorganized. The contractions may be violent and last for longer periods, resulting in abdominal pain or discomfort. The net result of such abnormal contractions is a change in the bowel patterns that leads to other problems, such as constipation, diarrhea, and bloating. The benefit of psyllium is twofold: It forms a matrix that resists hydrolysis and also resists colonic bacterial degradation (Marteau et al. 1994).

Intake of psyllium may be effective in treatment of chronic constipation in those patients where transit through the colon is slow or who suffer from disordered constipation. Similarly, fiber with lactulose may improve stool consistency in patients with irritable bowel syndrome with constipation (Fernandez-Banares 2006). According to Misra et al. (1989), treatment with a combination of psyllium and propantheline was effective in relieving symptoms and in the maintenance of remission. There are several reports on the assessment of utility of psyllium for treating irritable bowel syndrome (Greenbaum and Stein 1981; Singh 2007). The well-recognized benefit of psyllium in irritable bowel syndrome is partly due to its treatment of constipation, but psyllium also benefits those with diarrhea and pain (Kumar et al. 1987).

The major reason for constipation in irritable bowel syndrome patients may be the delayed intestinal transit. Therefore, fibers like psyllium that accelerate intestinal transit have proved beneficial in these patients. Psyllium seeds were shown to be superior to wheat bran with regard to stool frequency and abdominal distension; therefore, it is preferred in the treatment of irritable bowel syndrome and constipation. Prior and Whorwell (1987) found the optimum dose of psyllium in irritable bowel syndrome to be 20 g/day. The easing of bowel dissatisfaction appears to be a major reason for the therapeutic success of psyllium in irritable bowel syndrome. It was suggested that psyllium might modify the response to rapidly fermentable, poorly absorbed dietary carbohydrates such as lactose, fructose, and sorbitol, which have been implicated in some studies of irritable bowel syndrome (Symons, Jones, and Kellow 1992).

22.9.8 Colon Cancer

Colon cancer is one of the leading forms of cancer, after lung, breast, or stomach cancer. However, it is perhaps the only type of cancer that is curable if detected early. The genetic basis of this disease has been established (Rustgi 2007). It begins in the cells that line the colon. Some people are genetically predisposed to develop polyps, which can show uncontrolled growth and develop into cancer. However, there are other reasons for development of this type of cancer, such as ulcerative colitis—a chronic inflammation in the colon. Several studies have revealed an inverse correlation between high-fiber consumption and a lower incidence of colon cancer. Several possible mechanisms have been postulated to explain the inhibition of cancer by dietary fiber. These include (1) dilution and adsorption of any carcinogens, bile acids, and other potential toxins contained within the intestinal lumen; (2) modulation of metabolic activity of the microbial flora of the colon; (3) binding the essential nutrients, which are anticarcinogenic; and (4) biological modification of intestinal epithelial cells (Jacobs 1986).

One of the pronounced effects of fermentation of dietary fibers within the large intestine is the production of short chain fatty acids (SCFA), which stimulates cell proliferation *in vivo* and *n*-butyrate, which is antineoplastic *in vitro*. The role of dietary fiber during colorectal carcinogenesis might therefore be related to its fermentation to *n*-butyrate. The risk of colon cancer has been observed to be related to the ratio of *n*-butyrate production to total SCFA production from fiber. Generally, the ratio was low in cancer patients compared to the healthy controls (Clausen, Bonnen, and Mortensen 1991). Interestingly in a majority of cases in both humans and rodent cancer models, tumor formation occurs in the distal colon. With the oral administration of psyllium, the colonic flora gets adapted to increasing the production of *n*-butyrate (and acetate).

Butyrate has antineoplastic properties against colorectal cancer cells and is the preferred oxidative substrate for colonocytes. The butyrate is produced by colonic fermentation of dietary fibers. It has long been shown that fiber is beneficial in human inflammatory bowel disease, which is associated with an increased production of SCFA in the distal colon. Based on that, the intestinal anti-inflammatory effect of psyllium was associated with lower TNF-α levels and lower NO synthase activity in the inflamed colon. Intake of psyllium delayed the fermentation rate of high-amylose cornstarch in the cecum and shifted the fermentation site of starch toward the distal colon, leading to the higher *n*-butyrate concentration in the distal colon and feces (Morita et al. 1999). The presence of *n*-butyrate in the distal colon may be important in the prevention of colon cancer.

SCFA have a range of effects that may be relevant to colonic health (Cummings 1981; Pouillart 1998). Of these, *n*-butyrate is of particular interest because it exerts a concentration-dependent slowing of the rate of cancer cell proliferation and promotes expression of differentiation markers *in vitro*, leading to reversion of cells from a neoplastic to a non-neoplastic phenotype. Fermentation is normally more active in the cecum and proximal colon than in the distal colon (Cummings and Englyst 1987). For these reasons, highly fermentable dietary fibers such as pectin, guar gum, and oat bran are fully fermented in the cecum and proximal colon and do not contribute n-butyrate to the distal colon.

Regular physical exercise and the use of psyllium and aspirin reduced the risk of colon cancer. Psyllium strongly reduced the tumorigenicity of 1, 2-dimethylhydrazine and psyllium-fed rats had the highest fecal aerobic counts, lowest β-glucuronidase, and highest 7-α-dehydroxylase activities (Roberts-Andersen, Mehta, and Wilson 1987). Psyllium fiber provided colonocytes some protection from deoxycholic acid-induced lysis. Propionic acid, a product of fiber breakdown, was a potent colonocyte mitogen, suggesting that fiber could indirectly protect the colon by providing colonocyte nutrients (Friedman, Lightdale, and Winawer 1988).

22.9.9 Drug Interaction

Levodopa combined with cabidopa constitutes one of the most frequent medications in the treatment of Parkinson's disease. *P. ovata* husk improves levodopa absorption conditions, but when this drug is administered with cabidopa, fiber could reduce its effectiveness.

22.9.10 Wound Healing

Mucopolysaccharides derived from the husk of psyllium have properties beneficial for wound cleansing and wound healing. Recent studies indicate that they also limit scar formation.

22.9.11 Effect on Skin Keratinocytes and Fibroblasts

Endogenous carbohydrates, especially oligo- and polysaccharides, participate in the regulation of a broad range of biological activities (e.g., signal transduction) and act as *in vivo* markers for the determination of cell types. The ethanolic extract of husk of psyllium seed restores gap junctional intercellular communication in V-Ha-ras transfected rat liver epithelial WB-F344 cell lines. The active compound of the ethanolic extract was identified as β-sitosterol based on gas chromatography and electron mass spectroscopy (Nakamura et al. 2005).

As pointed out earlier, isabgol seeds consist of water-soluble and gel-forming polysaccharides. It was observed that only water-soluble polysaccharides exhibited strong and significant effects on cell physiology of keratinocytes and fibroblasts. Proliferation of the cells of the spontaneously immortalized keratinocytes' cell lines HaCaT was significantly up-regulated in a dose-independent manner. RNA analysis of activated signal pathways proved an effect of the acidic arabinoxylan on the expression of keratinocyte growth of normal humans. Differentiation behavior by involucrin was slightly influenced due to the enhanced cell proliferation leading to cell–cell-mediated induction of early differentiation (Deters et al. 2005).

22.9.12 Allergy

Psyllium is frequently used in powdered form. When it is mixed or poured, fine dust particles are readily dispersed into the air; they can be inhaled and cause sensitization. Adverse effects caused by psyllium have been relatively infrequent; however, the drug is known to cause IgE-mediated hypersensitivity reactions in susceptible individuals. Allergic symptoms can be severe and can include rhino-conjunctivitis, skin reactions, asthma, gastrointestinal symptoms, or anaphylaxis. Health care workers involved with dispensing the product and workers in pharmaceutical firms that manufacture it are generally exposed to psyllium dust. Orally ingested psyllium seems to be less likely to induce hypersensitivity, but if earlier sensitization has occurred, subsequent oral ingestion can cause severe allergic reactions (Lantner et al. 1990). The first case of allergic reaction to *P. ovata* seed was described by Ascher (1941). The prevalence of occupational asthma in health care professionals was described as 4%. In pharmaceutical manufacturing workers, the prevalence of occupational asthma was 3.6% and sensitization to *P. ovata* seed was 27.9% (Bardy et al. 1987). Freeman conducted a study and suggested that the allergens in psyllium appear to be protein in nature and are derived from the inner seed endosperm and embryo (Freeman 1994).

Aleman, Quirce, and Bombin (2001) reported that psyllium may act as a potent inhalant allergen capable of eliciting asthma symptoms, not only in an occupational context but also in a domestic environment, thus affecting consumers of this laxative or relatives who handle it. The psyllium granules when taken with insufficient liquid may cause esophageal obstruction; but intake of a sufficient amount of water may prevent it. To study the effect of *P. ovata* on humoral immune response, a study was conducted on rabbits. The rabbits were administered the aqueous extract of *P. ovata* consisting of a mixture of polysaccharides and glycosides. The study indicated that the extract of *P. ovata* can suppress the humoral immune response (Rezaeipoor, Saeidnia, and Kamalinejad 2000).

22.9.13 Bleeding Hemorrhoids

In addition to being useful in treating constipation and loose stools, psyllium is also beneficial for hemorrhoids. In a study based on patients with internal bleeding hemorrhoids, the effect of psyllium was determined. Individuals treated with psyllium showed significant improvement in reduced bleeding and a dramatic reduction of congested hemorrhoidal cushions. Also, bleeding on contact stopped after treatment with psyllium (Perez-Miranda et al. 1996).

22.9.14 Antimicrobial Activity

Plantago ovata has been used as an antimicrobial agent in traditional medicine. Reports indicate that ethanolic and methanolic extracts of the plant have antimicrobial activities against Gram-positive and Gram-negative bacteria. According to Motamedi et al. (2010), *Plantago* is more effective against Gram-positive than Gram-negative bacteria. They inferred that the diameter of the inhibitory zone around the more active extracts was comparable with standard antibiotics. The extracts were found to be most active against *Staphylococcus aureus* and *S. epidermidis*. The plant extract is also reported to have broad-spectrum activity against pathogens causing urinary tract infections, particularly against *Providencia pseudomallei* (Sharma, Verma, and Ramteke 2009).

22.9.15 Toxicity

The use of psyllium as a laxative does not induce any side effects. However, the plants cause allergies in the work environment. Occasional asthma has been reported among people who work with psyllium (Bernton 1970). Unsoaked seeds may cause gastrointestinal irritation, inflammation, mechanical obstruction, and constipation; powdered and chewed seeds release a pigment that is injurious to the kidneys (Singh and Virmani 1982).

22.10 USAGES

Psyllium seed has been used in medicine for ages. The Persian physician Alhervi prescribed its use as early as the tenth century for the treatment of chronic dysentery and intestinal fluxes. The use of two or three heaped dessert spoonfuls of seed twice a day for a few days to a couple of months resolves all symptoms of chronic constipation, dysentery, and other complaints. The husk is consumed alone or mixed with different chemicals, such as powdered anhydrous dextrose, sodium bicarbonate, citric acid, etc. (Virmani, Singh, and Husain 1980). Powdered or granulated husk with added fruit essence or other additives to improve its palatability is sold under different brand names.

Isabgol is diuretic; it alleviates kidney and bladder complaints, gonorrhea, urethritis, and hemorrhoids. It removes burning sensations in feet, relieves polyuria and difficult micturition, and tones up the bladder. It is also effective in checking spermatorea. It is recommended for use for termination of pregnancy (Khanna et al. 1980). According to Siddiqui, Kapur, and Atal (1964), isabgol oil is more potent than safflower oil for reducing serum cholesterol levels. The efficacy is increased when the husk is consumed with metrondazole. Ingestion of 10 g of isabgol a day for a month reduces serum cholesterol level by 9.6% and triglyceride by 8.6% (Dhar et al. 2005).

A decoction of *P. ovata* mixed with honey is good for the treatment of sore throats and bronchitis. The liquid obtained after boiling psyllium seed is chilled and used as eye drops. It is also used to get rid of pimples and check hair loss. Regular use of isabgol removes all kinds of blemishes of the skin. Psyllium is useful in treating frequent griping in the belly caused by stomach ulcers. The seeds are one of the most effective remedies for piles due to rich fiber and tannin content. The oil from the embryo of the seeds prevents artherosclerosis. The oil contains 50% linoleic acid. Psyllium has also been found useful for treatment of gonorrhea because of its diuretic and soothing properties. The poultice of psyllium is effective in the treatment of whitlow. A poultice made of psyllium and vinegar with oil is useful for rheumatism and gout.

The psyllium is used as a substitute for sodium alginate in the making of ice cream. In addition, the mucilage is used as an ingredient in making chocolate, sizing textiles, manufacturing cosmetics, and setting and dressing hair. The seed and husk are used in dyeing. Psyllium is also a source of commercial gum. Isabgol has ethnobotanical value. Several tribal communities use it for various purposes. Members of the Santhals tribe use it to relieve pain and treat bronchitis (Jain and Trafder 1970). The tribal inhabitants of north Gujarat consume seed decoctions of psyllium as a cooling demulcent to cure diarrhea and dysentery. Isabgol-Gola, a by-product of psyllium, is used as cattle feed. Its consumption has no adverse effect on production and composition of milk and the body weight of milch cows.

22.11 FUTURE PERSPECTIVE

Plantago ovata is one of the most important medicinal and commercial plants in India. Most of the psyllium produced in the country is exported to various parts of the world. Although the global demand is increasing, production is less. In order to bridge this gap, efforts are being made to increase production. Development of high-yielding varieties is the need of the hour. However, the major bottleneck in achieving this objective is the lack of genuine variability. Therefore, intensive efforts are being made in the authors' laboratory for microidentification of variability at the molecular level. This may facilitate identification of novel genotypes having value in the breeding of improved varieties. Genomics and functional genomics approaches are underway to identify new genes in *P. ovata* and its wild relatives. There are 200 wild allies, which could be the source of many useful genes. The potential of these wild species needs to be exploited for genetic improvement of *Plantago ovata*.

ACKNOWLEDGMENTS

The authors are grateful to the Department of Biotechnology (DBT), the Government of India, for financial assistance in the form of a research project to the senior author. The authors are thankful to Prof. A. K. Koul for his encouragement and advice.

REFERENCES

Aleman, A. M., S. Quirce, and C. Bombin. 2001. Asthma related to inhalation of *Plantago ovata. Medicina Clinica (Barc)* 116:20–22.

Anderson, J. W., L. D. Allgood, J. Turner, et al. 1999. Effects of psyllium on glucose and serum lipid responses in men with type 2 diabetes and hypercholesterolemia. *American Journal of Clinical Nutrition* 70:466–447.

Anderson, J. W., A. E. Jones, and S. Riddell-Mason. 1994. Ten different dietary fibers have significantly different effects on serum and liver lipids of cholesterol-fed rats. *Journal of Nutrition* 124:78–83.

Anderson, K. M., W. P. Castelli, and D. Leby. 1987. Cholesterol and mortality: 30 years of follow-up from the Framingham study. *JAMA* 257:2176-2180.

Anonymous. 1969. *The wealth of India. Raw materials*, vol. VIII. New Delhi: Publications and Information Directorate, CSIR, pp. 146–154.

Ascher, M. S. 1941. Psyllium seed sensitivity. *Journal of Allergy* 12:607–609.

Atal, C. K. 1958. Cytoplasmic male sterility in psyllium (*P. ovata* Forsk.). *Current Science* 27:268.

Badr, A., R. Labani, and T. T. Elkington. 1987. Nuclear DNA variation in relation to cytological features of some species in the genus *Plantago* L. *Cytologia* 52:733–737.

Bardy, J. D., J. L. Malo, P. Seguin, et al. 1987. Occupational asthma and IgE sensitization in a pharmaceutical company processing psyllium. *American Review of Respiratory Disease* 135 (5): 1033–1038.

Barna, K. S., and A. K. Wakhlu. 1988. Axillary shoot induction and plant regeneration in *Plantago ovata* Forsk. *Plant Cell, Tissue and Organ Culture* 15:169–173.

———. 1989. Shoot regeneration from callus derived root cultures of *Plantago ovata* Forsk. *Phytomorphology* 39:353–355.

Belknap, D., L. J. Davidson, and C. R. Smith. 1997. The effects of psyllium hydrophilic mucilloid on diarrhea in enterally fed patients. *Heart Lung* 26:229–237.

Bergman, F., and W. van der Linden. 1967. Diet-induced cholesterol gallstones in hamsters. Prevention and dissolution by cholestyramine. *Gastroenterology* 53:418–421.

Bernton, S. H. 1970. The allergenicity of psyllium seed. *Medical Annals of District of Columbia* 39:313–317.

Bliss, D. Z., H. J. Jung, K. Savik, et al. 2001. Supplementation with dietary fiber improves fecal incontinence. *Nursing Research* 50:203–213.

Buhman, K. K., E. J. Furumoto, S. S. Donkin, et al. 1998. Dietary psyllium increases fecal bile acid excretion, total steroid excretion and bile acid biosynthesis in rats. *Journal of Nutrition* 128:1199–1203.

Chan, J. K. C., and V. Wypyszyk. 1988. A forgotten natural dietary fiber: Psyllium mucilloid. *Cereal Foods World* 33:919–922.

Chastagner, G. A., J. M. Ogawa, and K. P. V. Sammeta. 1978. Cause and control on damping off of *Plantago ovata. Plant Disease Reporter* 62:929–932.

Chen, H. L., V. S. Haack, C. W. Janecky, et al. 1998. Mechanisms by which wheat bran and oat bran increase stool weight in humans. *American Journal of Clinical Nutrition* 68:711–719.

Chinoy, J. J., K. G. Mehta, and D. H. Mehta. 1978. Some biochemical changes associated with the biosynthesis of mucilage seed development in *Plantago* (*P. ovata* Forsk.). In *Physiology of sexual reproduction of flowering plants*, ed. C. P. Malik, A. K. Srivastava, and N. C. Bhattacharya, 248–257. India: Kalyani Publishers.

Chopra, R. N. 1930. *Plantago ovata* (isbaghul) in chronic diarrheas and dysenteries. *Indian Medical Gazette* 65, 428–433.

Clausen, M. R., H. Bonnen, and P. B. Mortensen. 1991. Colonic fermentation of dietary fiber to short chain fatty acids in patients with adenomatous polyps and colonic cancer. *Gastroenterology* 32:923–928.

Cummings, J. H. 1981. Short chain fatty acids in the human colon. *Gastroenterology* 22:763–779.

———. 1993. The effect of dietary fiber on fecal weight and composition. In *Dietary fiber in human nutrition*, ed. G. A. Spiller, 263–349. Boca Raton, FL: CRC Press.

Cummings, J. H., and H. N. Englyst. 1987. Fermentation in the human large intestine and the available substrates. *American Journal of Clinical Nutrition* 45:1243–1255.

Das, M., and S. S. Raychaudhuri. 2001. Enhanced development of somatic embryos of *Plantago ovata* Forsk. by additives. *In Vitro Cellular and Developmental Biology* 37:568–571.

Dastur, J. F. 1952. *Medicinal plants of India and Pakistan.* Bombay: D. B. Taraporewal Sons & Co.

Deters, A. M., K. R. Schroder, T. Smiatek, et al. 2005. Ispaghula (*Plantago ovata*) seed husk polysaccharides promote proliferation of human epithelial cells (skin keratinocytes and fibroblasts) via enhanced growth factor receptors and energy production. *Planta Medica* 71 (1): 33–39.

Dhar, M. K., B. Friebe, S. Kaul, et al. 2006. Characterization and physical mapping of 18S-5.8S-25S and 5S ribosomal RNA gene families in *Plantago* species. *Annals of Botany* 97:541–548.

Dhar, M. K., J. Fuchs, and A. Houben. 2009. Distribution of eu- and heterochromatin in *Plantago ovata*. *Cytogenetics and Genome Research* 125:235–240.

Dhar, M. K., and S. Kaul. 2004. FISH reveals somatic association of nucleolus organizing regions in *Plantago ovata*. *Current Science* 87:1336–1337.

Dhar, M. K., S. Kaul, B. Friebe, et al. 2002. Chromosome identification in *Plantago ovata* Forsk. through C-banding and FISH. *Current Science* 83:150–152.

Dhar, M. K., S. Kaul, and S. Jamwal. 2000. *Plantago ovata* Forsk. In *Plant breeding: Theory and techniques*, ed. S. K. Gupta, 56–67. Jodhpur: India Agrobios.

Dhar, M. K., S. Kaul, S. Sareen, et al. 2005. *Plantago ovata:* Genetic diversity, cultivation, utilization and chemistry. *Plant Genetic Resources: Characterization and Utilization* 3:252–263.

Eastwood, M. A., W. G. Brydon, and K. Tadesse. 1980. Effect of fiber on colon function. In *Medical aspects of dietary fiber*, ed. G. A. Spiller and R. M. Kay, 1–26. New York: Plenum Medical.

Fernandez, M. L. 1995. Distinct mechanisms of plasma LDL lowering by dietary fiber in the guinea pig: Specific effects of pectin, guar gum and psyllium. *Journal of Lipid Research* 36:2394–2404.

Fernandez, M. L., M. Vergara-Jimenez, K. Conde, et al. 1997. Regulation of apolipoprotein B-containing lipo-proteins by dietary soluble fiber in guinea pigs. *American Journal of Clinical Nutrition* 65:814–822.

Fernandez-Banares, F. 2006. Nutritional care of the patient with constipation. *Best Practice and Research Clinical Gastroenterology* 20:575–587.

Fischer, H. M., Y. Nanxiong, R. G. J. Ralph, et al. 2004. The gel-forming polysaccharide of psyllium husk (*Plantago ovata* Forsk). *Carbohydrate Research* 339:2009–2017.

Florholmen, J., R. Arvidsson-Lenner, R. Jorde, et al. 1982. The effect of Metamucil on postprandial blood glucose and plasma gastric inhibitory peptide in insulin-dependent diabetics. *Acta Medica Scandinavica* 212:237–239.

Frati Munari, A. C., W. Benitez Pinto, C. Raul Ariza Andraca, et al. 1998. Lowering glycemic index of food by acarbose and *Plantago psyllium* mucilage. *Archives of Medical Research* 29:137–141.

Freeman, G. L. 1994. Psyllium hypersensitivity. *Annals of Allergy* 73:490–492.

French, D., G. M. Wild, B. Young, et al. 1953. Constitution of planteose. *Journal of American Chemical Society* 75:709–712.

Friedman, E., C. Lightdale, and S. Winawer. 1988. Effects of psyllium fiber and short-chain organic acids derived from fiber breakdown on colonic epithelial cells from high-risk patients. *Cancer Letters* 43:121–124.

Fuchs, J., D. A. Demidov, D. A. Houben, et al. 2006. Chromosomal histone modification patterns—From con-servation to diversity. *Trends in Plant Science* 11:199–208.

Galisteo, M., R. Moron, L. Rivera, et al. 2010. *Plantago ovata* husks—Supplemented diet ameliorates meta-bolic alterations in obese Zucker rats through activation of AMP-activated protein kinase. Comparative study with other dietary fibers. *Clinical Nutrition* 29 (2): 261–267.

Greenbaum, D. S., and G. E. Stein. 1981. Psyllium and the irritable bowel syndrome. *Annals of Internal Medicine* 95:660.

Gupta, R. R., C. G. Agarwal, G. P. Singh, et al. 1994. Lipid-lowering efficacy of psyllium hydrophilic mucil-loid in non-insulin-dependent diabetes mellitus with hypercholesterolemia. *Indian Journal of Medical Research* 100:237–241.

Hoekman, M. F. M., J. M. J. Rientjes, J. Twisk, et al. 1993. Transcriptional regulation of the gene encoding cholesterol 7α-hydroxylase in the rat. *Gene* 130:217–223.

Horton, J. D., J. A. Cuthbert, and D. K. Spady. 1994. Regulation of hepatic 7α-hydroxylase expression by dietary psyllium in the hamster. *Journal of Clinical Investigation* 93:2084–2092.

Houben, A., D. Demidov, A. D. Caperta, et al. 2007. Phosphorylation of histone H3 in plants—A dynamic affair. *Biochimica et Biophysica Acta* 1769:308–315.

Houben, A., D. Demidov, D. Gernand, et al. 2003. Methylation of histone H3 in euchromatin of plant chromo-somes depends on basic nuclear DNA content. *Plant Journal* 33:967–973.

Houben, A., D. Demidov, T. Rutten, et al. 2005. Novel phosphorylation of histone H3 at threonine 11 that temporally correlates with condensation of mitotic and meiotic chromosomes in plant cells. *Cytogenetics and Genome Research* 109:148–155.

Husain, A. 1977. Achievements in the research on medicinal plants, their present and future value in India. In *Proceedings of the Fourth Symposium of Pharmacognosy and Chemistry of Natural Products.* Leiden: Gorlaeus Laboratories.

Hyde, B. B. 1970. Mucilage producing cells in the seed coat of *Plantago ovata*: Development fine structure. *American Journal of Botany* 57:1197–1206.

Jacobs, L. R. 1986. Relationship between dietary fiber and cancer: Metabolic, physiologic, and cellular mechanisms. *Proceedings of the Society for Experimental Biology and Medicine* 183:299–310.

Jain, A. K., and B. M. Mithal. 1976. Derivatives of *Plantago ovata* seed husk gum. Part II. Methoxy derivative. *Indian Journal of Pharmacology* 38:15–17.

Jain, R., and S. Babbar. 2005. Guar gum and isabgol as cost effective alternative gelling agents for *in vitro* multiplication of an orchid *Dendrobium chrysotoxum. Current Science* 88:292–295.

Jain, S. K., and C. R. Trafder. 1970. Medicinal plant lore of the sandals (a revival of P. O. Bodding's work). *Economic Botany* 24:241–278.

Jamal, S., I. Ahmad, R. Agarwal, et al. 1987. A novel oxo fatty acid in *Plantago ovata* seed oil. *Phytochemistry* 26:3067–3069.

Jamwal, S. 2000. Studies on some aspects of male sterility in *Plantago ovata* Forsk. PhD thesis, University of Jammu, Jammu, India.

Jamwal, S., M. K. Dhar, and S. Kaul. 1998. Male sterility in *Plantago ovata* Forsk. *Current Science* 74:504–505.

Jarjis, H. A., N. A. Blackburn, J. S. Redfern, et al. 1984. The effect of ispaghula (Fybogel and Metamucil) and guar gum on glucose tolerance in man. *British Journal of Nutrition* 51:371–378.

Kalyansundram, N. K., P. B. Patel, and K. C. Dalal. 1982. Nitrogen need of *Plantago ovata* Forsk. in relation to the available nitrogen in the soil. *Indian Journal of Agricultural Sciences* 52:240–242.

Kalyansundram, N. K., S. Sriram, B. R. Patel, et al. 1984. Psyllium: A monopoly of Gujarat. *Indian Horticulture* 28:35–37.

Karimzadeh, G., and R. Omidbaigi. 2004. Growth and seed characterization of isabgol (*Plantago ovata* Forsk.) as influenced by some environmental factors. *Journal Agricultural Science and Technology* 6:103–110.

Kennedy, J. F., J. S. Sandhu, and D. A. T. Southgate. 1979. Structural data for the carbohydrate of ispaghula husk *ex Plantago ovata* Forsk. *Carbohydrate Research* 75:265–274.

Khanna, N. M., J. P. S. Sarin, R. C. Nandi, et al. 1980. Isaptent—A new cervical dilator. *Contraception* 21:29–40.

Khorana, M. L., V. G. Prabhu, and M. R. R. Rao. 1958. Pharmacology of an alcoholic extract of *Plantago ovata. Indian Journal of Pharmacology* 20:3–6.

Koul, A. K., S. Sareen, and M. K. Dhar. 1994. Cytogenetic studies on blond psyllium. In *Plant cytogenetics in India*, ed. S. Ghosh, 77–88. Calcutta: *Cell and Chromosome Research.*

Koul, A. K., and P. K. Sharma. 1986. Cytogenetic studies of *Plantago ovata* Forsk. and its wild allies. In *Genetics and crop improvement*, ed. P. K. Gupta and J. R. Bahl, 359–366. Meerut, India: Rastogi and Co.

Kumar, A., N. Kumar, J. C. Vij, et al.1987. Optimum dosage of ispaghula husk in patients with irritable bowel syndrome: Correlation of symptom relief with whole gut transit time and stool weight. *Gut* 28:150–155.

Laakso, M. 1996. Glycemic control and the risk for coronary heart disease in patients with non-insulin-dependent diabetes mellitus. *Annals of Internal Medicine* 124:127–130.

Laidlaw, R. A., and E. G. V. Percival. 1949. Studies on polysaccharide extracted from the seeds of *Plantago ovata* Forsk. *Journal of the Chemical Society (London)* 6:1600–1607.

Lal, R. K., J. R. Sharma, and H. O. Misra. 1998. Development of a new variety Niharika of isabgol (*Plantago ovata*). *Journal of Medicinal and Aromatic Plant Sciences* 20:1067–1070.

Lantner, R. R., B. R. Baltazar, P. Zumerchik, et al. 1990. Anaphylaxis following ingestion of a psyllium-containing cereal. *JAMA* 264:2534–2536.

Mandal, K., and K. A. Geetha. 2001. Floral infection of downy mildew of isabgol. *Journal of Mycology and Plant Pathology* 31:355–357.

Mandal, K., P. R. Patel, S. Maiti, et al. 2010. Induction of male and female sterility in isabgol (*Plantago ovata*) due to floral infection of downy mildew (*Pernospora plantaginis*). *Biologia* 65:17–22.

Marlett, J. A., and M. H. Fischer. 2002. A poorly fermented gel from psyllium seed husk increases excreta moisture and bile acid excretion in rats. *Journal of Nutrition* 132:2638–2643.

Marlett, J. A., T. M. Kajs, and M. H. Fischer. 2000. An unfermented gel component of psyllium seed husk promotes laxation as a lubricant in humans. *American Journal of Clinical Nutrition* 72:784–789.

Marteau, P., B. Flourie, C. Cherbut, et al. 1994. Digestibility and bulking effect of ispaghula husks in healthy humans. *Gut* 35:1747–1752.

Matheson, H. B., I. S. Colon, and J. A. Story. 1995. Cholesterol 7α-hydroxylase activity is increased by dietary modification with psyllium hydrocolloid, pectin, cholesterol and cholestyramine in rats. *Journal of Nutrition* 125:454–458.

McNeil, D. L. 1989. Factors affecting the field establishment of *Plantago ovata* Forsk. in northern Australia. *Tropical Agriculture* 66:61–64.

McRorie, J. W., B. P. Daggy, J. G. Morel, et al. 1998. Psyllium is superior to docusate sodium for treatment of chronic constipation. *Alimentary Pharmacology and Therapeutics* 12:491–497.

Misra, S. P., V. K. Thorat, G. K. Sachdev, et al. 1989. Long-term treatment of irritable bowel syndrome: Results of a randomized controlled trial. *Quarterly Journal of Medicine* 73:931–939.

Mital, S. P., and N. R. Bhagat. 1979. Studies on the floral biology in *Plantago ovata* Forsk. and other species. *Current Science* 48:261–263.

Mital, S. P., and S. C. Issar. 1970. Fertility restorer for cytoplasmic male sterility in isaphgul (*Plantago ovata* Forsk.). *Science and Culture* 36:550–551.

Mital, S. P., and B. Singh. 1986. Introduction of genetic resources of some important medicinal and aromatic plants in India. *Indian Journal of Genetics* 46:209–216.

Mithal, B. M., and V. D. Gupta. 1965. Suspending properties of *Plantago ovata* seed husk (isapghula) mucilage. *Indian Journal Pharmacology* 27:331–334.

Modi, J. M., K. G. Mehta, and R. Gupta. 1974. Isabgol—A dollar earner of North Gujarat. *Indian Farming* 23:17–19.

Moreno, L. A., B. Tresaco, G. Bueno, et al. 2003. Psyllium fiber and the metabolic control of obese children and adolescents. *Journal of Physiology and Biochemistry* 59:235–242.

Morita, T., S. Kasaoka, K. Hase, et al. 1999. Psyllium shifts the fermentation site of high amylose corn-starch towards the distal colon and increases fecal butyrate concentration in rats. *Journal of Nutrition* 129:2081–2087.

Motamedi, H., E. Darabpour, M. Gholipour, et al. 2010. Antibacterial effect of ethanolic and metanolic extracts of *Plantago ovata* and *Oliveria decumbens* endemic in Iran against some pathogenic bacteria. *International Journal of Pharmacology* 6 (2): 117–122.

Nakamura, Y., N. Yoshikawa, I. Hiroki, et al. 2005. Beta-sitosterol from psyllium seed husk (*Plantago ovata* Forsk.) restores gap junctional intercellular communication in Ha-ras transfected rat liver cells. *Nutrition and Cancer* 51 (2): 218–225.

Neiffer, D. L. 2001. *Clostridium perfringens* enterotoxicosis in two Amur leopards (*Panthera pardus orientalis*). *Journal of Zoo and Wildlife Medicine* 32:134–135.

Padha, L. 1996. Cytogenetic studies on *Plantago ovata* Forsk. and some of its allies. PhD thesis, University of Jammu, Jammu, India.

Pal, B., S. P. S. Jadaun, and A. S. Parmar. 1988. Note on the effect of water salinity on yield and nutrient composition of isabgol. *New Botanist* 15:277–278.

Pastors, J. G., P. W. Blaisdell, T. K. Balm, et al. 1991. Psyllium fiber reduces rise in postprandial glucose and insulin concentrations in patients with non-insulin-dependent diabetes. *American Journal of Clinical Nutrition* 53:1431–1435.

Patel, N. H., S. Sriram, and K. C. Dalal. 1980. Floral biology and stigma pollen maturation schedule in isabgol *Plantago ovata*. *Current Science* 49:688–691.

Patel, P. H., and H. M. Mehta. 1990. Effect of irrigation timing and isoproturon application on weeds and yield of isabgol (*Plantago ovata* Forsk.). *Gujarat Agricultural University Research Journal* 15:46–48.

Patel, R. B., N. G. Rana, M. R. Patel, et al. 1979. Chromatographic screening of proteins of *Plantago ovata* Forsk. *Indian Journal of Pharmaceutical Sciences* 41:249.

Perez-Miranda, M., A. Gomez-Cedenilla, T. Leon-Colombo, et al. 1996. Effect of fiber supplements on internal bleeding hemorrhoids. *Hepatogastroenterology* 43 (12): 1504–1507.

Pouillart, P. R. 1998. Role of butyric acid and its derivatives in the treatment of colorectal cancer and hemoglobinopathies. *Life Science* 63:1739–1760.

Pramanik, S., S. Chakraborty, and S. Raychaudhuri. 1995. *In vitro* clonal propagation and characterization of clonal regeneration of *Plantago ovata* Forsk. by isozyme analysis. *Cytobios* 82:123–130.

Prior, A., and P. J. Whorwell. 1987. Double blind study of ispaghula in irritable bowel syndrome. *Gut* 28:1510–1513.

Prynne, C. J., and D. A. T. Southgate. 1979. The effects of a supplement of dietary fiber on fecal excretion by human subjects. *British Journal of Nutrition* 41:495–503.

Punia, M. S., G. S. Sharma, and P. K. Verma. 1985. Genetics and breeding of *Plantago ovata* Forsk.—A review. *International Journal of Tropical Agriculture* 3:255–264.

Qvitzau, S., P. Matzen, and P. Madsen. 1988. Treatment of chronic diarrhea: Loperamide versus ispaghula husk and calcium. *Scandinavian Journal of Gastroenterology* 23:1237–1240.

Rachlis, A., J. Gill, J. G. Baril, et al. 2005. Effectiveness of step-wise intervention plan for managing nelfinavir-associated diarrhea: A pilot study. *HIV Clinical Trials* 6:203–212.

Rahn, K. 1996. A phylogenetic study of the Plantaginaceae. *Botanical Journal of the Linnean Society* 120:145–198.

Randhawa, G. S., R. K. Mahey, S. S. Saini, et al. 1985. Studies on irrigation requirements of psyllium (*Plantago ovata* Forsk.). *Indian Journal of Agronomy* 30:187–191.

Rathore, B. S., and R. S. Rathore. 1996. Downy mildew of isabgol in Rajasthan. *PKV Research Journal* 20:107.

Rezaeipoor, R., S. Saeidnia, and M. Kamalinejad. 2000. The effect of *Plantago ovata* on humoral immune response in experimental animals. *Journal of Ethnopharmacology* 72 (1–2): 283–286.

Roberts-Andersen, J., T. Mehta, and R. B. Wilson. 1987. Reduction of DMH induced colon tumors in rats fed psyllium husk or cellulose. *Nutrition and Cancer* 10:29–36.

Romero-Baranzini, A. L., O. G. Rodriguez, G. A. Yanez-Farias, et al. 2006. Chemical, physicochemical, and nutritional evaluation of *Plantago* (*Plantago ovata* Forsk). *Cereal Chemistry* 83:358–362.

Rustgi, A. K. 2007. The genetics of hereditary colon cancer. *Genes and Development* 21:2525–2538.

Saghir, S., M. S. Iqbal, M. A. Hussain, et al. 2008. Structure characterization and carboxymethylation of arabinoxylan isolated from ispaghula (*Plantago ovata*) seed husk. *Carbohydrate Polymers* 74:309–317.

Sandhu, J. S., G. K. Hudson, and J. F. Kennedy. 1981. The gel nature and structure of the carbohydrate of ispaghula husk *ex Plantago ovata* Forsk. *Carbohydrate Research* 93:247–259.

Sareen, S. 1991. Mutation studies in *Plantago ovata* Forsk. PhD thesis, University of Jammu, Jammu, India.

Sareen, S., and A. K. Koul. 1991. Gamma ray induced variation in *Plantago ovata* Forsk. *Crop Improvement* 18:144–147.

———. 1999. Mutation breeding in improvement of *Plantago ovata* Forsk. *Indian Journal Genetics and Plant Breeding* 59:337–344.

Sareen, S., A. K. Koul, and A. Langer. 1988. NOR size in relation to nucleogenesis. *Nucleus* 31:21–23.

Savage, P. J. 1996. Cardiovascular complications of diabetes mellitus: What we know and what we need to know about their prevention. *Annals of Internal Medicine* 124:123–126.

Sharma, N. 1990. Genetic systems in *Plantago ovata* Forsk. and some of its allies. PhD thesis, University of Jammu, Jammu, India.

Sharma, N., P. Koul, and A. K. Koul. 1992. Reproductive biology of *Plantago*: Shift from cross to self pollination. *Annals of Botany* 69:7–11.

Sharma, P. K. 1984. Cytogenetic studies on some Himalayan species of genus *Plantago* L. PhD thesis, University of Jammu, Jammu, India.

Sharma, P. K., and A. K. Koul. 1986. Mucilage in seeds of *P. ovata* and its wild allies. *Journal of Ethnopharmacology* 17:289–295.

Sharma, P. K., A. Langer, and A. K. Koul. 1989. Satellite association in *Plantago ovata*. *Current Science* 58:321–323.

Sharma, A., R. Verma, and P. Ramteke. 2009. Antibacterial activity of some medicinal plants used by tribals against uti-causing pathogens. *World Applied Sciences Journal* 7:332–339.

Siddiqui, H. H., K. K. Kapur, and C. K. Atal. 1964. Studies on Indian seed oils. Part II. Effect of *Plantago ovata* embryo oil on serum cholesterol levels in rabbits. *Indian Journal of Pharmacology* 26:266–268.

Singh, A. K., and O. P. Virmani. 1982. Cultivation and utilization of isabgol (*Plantago ovata* Forsk.)—A review. *Current Research in Medicinal and Aromatic Plants* 4:109–120.

Singh, B. 2007. Psyllium as therapeutic and drug delivery agent. *International Journal of Pharmaceutics* 334:1–14.

Singh, J. N., and K. Nand. 1988. Effect of nitrogen levels and row spacings on seed yield of psyllium. *Indian Drugs* 25:459–461.

Singh, N., R. K. Lal, and A. K. Shasany. 2009. Phenotypic and RAPD diversity among 80 germplasm accessions of the medical plant isabgol (*Plantago ovata*, Plantaginaceae). *Genetics and Molecular Research* 8:1273–1284.

Sprecher, D. L., B. V. Harris, A. C. Goldberg, et al. 1993. Efficacy of psyllium in reducing serum cholesterol levels in hypercholesterolemic patients on high- or low-fat diets. *Annals of Internal Medicine* 119:545–554.

Stebbins, G. L., and A. Day. 1967. Cytogenetic evolution for long continued stability in genus *Plantago*. *Evolution* 21:409–428.

Stephen, A. M., and J. H. Cummings. 1980. Mechanism of action of dietary fiber in the human colon. *Nature* 284:283–284.

Symons, P., M. P. Jones, and J. E. Kellow. 1992. Symptom provocation in irritable bowel syndrome. Effects of differing doses of fructose–sorbitol. *Scandinavian Journal Gastroenterology* 27:940–944.

Van Rosendaal, G. M., E. A. Shaffer, A. L. Edwards, et al. 2004. Effect of time of administration on cholesterol-lowering by psyllium: A randomized cross-over study in normocholesterolemic or slightly hypercholesterolemic subjects. *Nutrition Journal* 28:17.

Virmani, O. P., P. Singh, and A. Husain. 1980. Current status of medicinal plant industry in India. *Indian Drugs* 17:318–340.

Wakhlu, A. K., and K. S. Barna. 1988. Chromosome studies in hypocotyls callus cultures and regenerated plants of *Plantago ovata* Forsk. *Nucleus* 31:14–17.

———. 1989. Callus initiation, growth and plant regeneration in *Plantago ovata* Forsk. cv. GI.2. *Plant Cell, Tissue and Organ Culture* 17:235–241.

Wenzl, H. H., K. D. Fine, L. R. Schiller, et al. 1995. Determinants of decreased fecal consistency in patients with diarrhea. *Gastroenterology* 108:1729–1738.

Wolff, K., B. Friso, and M. M. J. Van Damme. 1988. Outcrossing rates and male sterility in natural populations of *Plantago coronopus*. *Theoretical and Applied Genetics* 76:190–196.

Zadoo, S. N., and M. I. H. Farooqi. 1977. Performance of autotetraploid blond psyllium. *Indian Journal of Horticulture* 34:294–300.

CHAPTER 23

Christmas Candle Senna
An Ornamental and a Pharmaceutical Plant

J. Bradley Morris

CONTENTS

23.1 INTRODUCTION

Christmas candle (*Senna alata* L.) is a shrub or tree in the Fabaceae family that is native to French Guiana, Guyana, Suriname, Venezuela, Brazil, and Colombia (USDA 2009). The plant has become naturalized in tropical Africa, tropical Asia, Australia, Mexico, the West Indies, Melanesia, Polynesia, Florida, and Hawaii. Christmas candle is only one of several common names, including bajagua, bois darter, café-beirao, candlebush, candlestick senna, darters, empress-candleplant, fedegoso, fedegosao, fedegoso-gigante, fedegoso-grande, mangerioba-do-para, mangerioba-grande, mata-pasto, mocote, ringworm bush, ringworm shrub, seven-golden-candlesticks (USDA 2009), craw-craw plant, king of the forest, dartrier, casse ailee, plante des cros-cros, buisson de la gale, quatre epingles, Dartial, cortalinde, upupu wa mwitu (Bosch 2007), and candelabra plant. Christmas candle is cultivated in India for export to Japan; however, trade statistics are unavailable (Bosch 2007). Christmas candle was previously known as *Cassia alata*. During the early 1980s, species within the genus *Cassia* were separated into three new genera: *Cassia*, *Chamaecrista*, and *Senna*. *Senna* spp., including Christmas candle, consist of three short and straight adaxial stamens and pedicels without bracteoles, which are distinguishing characteristics separating it from *Cassia* or *Chamaecrista* spp. Christmas candle is primarily adapted to subtropical and tropical climates worldwide.

23.2 ECOLOGY AND CULTIVATION

Christmas candle plants prefer growing in disturbed habitats, along roadsides, on river banks, at rain forest edges, on lake shores and pond and ditch margins, in open forest, in orchards, and around villages. It is mostly found at lower elevations but can grow in areas exceeding 2000 m and is tolerant of annual rainfall ranging from 600–4300 mm. Christmas candle also tolerates average temperatures of 15–30°C and grows well in acid to moderately alkaline, well drained soils (Bosch 2007). Christmas candle seed requires scarification to enhance germination prior to planting. Scarification is accomplished by placing seeds in boiling water for approximately 3 seconds, followed by quickly removing the seeds from the boiling water, drying the damp seeds on a paper towel, and planting them into potting soil.

While Christmas candle plants will grow in Griffin, Georgia, field soil (Figure 23.1), optimum plant growth producing multiple branches and prolific flower numbers occurs when they are grown in pots, gardens, or fields amended with potting soil consisting of pine bark, perlite, vermiculite, and sphagnum moss (Figure 23.2). Pot-to-pot cultivation is the best approach to use when growing Christmas candle plants in subtropical environments such as that in the U.S. state of Georgia (Figure 23.3). This can best be managed by transplanting seedlings to starter size jiffy pots containing sphagnum moss and perlite. After several weeks of growth, Christmas candle seedlings can then be transplanted to larger pots (1 gal and greater) containing pine bark, perlite, vermiculite, and sphagnum moss potting soil. They can also be propagated by placing approximately 2–4 cm long stem cuttings (cut below the node) in potting soil consisting of sphagnum moss and perlite (Figure 23.4) and can be produced in the same pot-to-pot method described previously.

Greater plant size, multiple branching, and flower production occur when Christmas candle plants are grown from cuttings. These plants will produce greater seed numbers than those grown from seeds due to its inherent perennial characteristic. Clonal stock provides a way to bypass juvenile flowering, which enhances Christmas candle's ability to increase in size, produce more flowers,

Figure 23.1   Christmas candle growing in a field in Griffin, Georgia.

Figure 23.2   Christmas candle (PI 322311) growing in a garden containing potting soil.

Figure 23.3   Christmas candle (PI 279691) growing in large pots containing potting soil.

Figure 23.4 Christmas candle (PI 279691) rooted cuttings in a greenhouse.

and produce greater seed numbers. Adequate fertilization using osmocote at a rate of about 1 tsp. per potted plant and gradually increased to 10 tsp. per plant growing in the larger containers is recommended, especially when the plants begin to develop yellow leaves. Mature seeds are generally produced during October in Georgia. The seeds should be placed in air tight aluminum bags and stored at −18°C after threshing and cleaning.

23.3 ORNAMENTAL CHARACTERISTICS

Christmas candle plants produce very ornate leaves and flowers (Figure 23.5). However, limited flower production without seed development may occur due to apparent photoperiod requirements by some types of Christmas candle plants such as PI 279691 (Figure 23.3). Others, such as PI 322311, are not photoperiod sensitive and will produce multitudes of leaves, flowers, and seeds (Figure 23.2). Christmas candle plants are grown at resorts in Disney World, Orlando, Florida. However, these plants exhibited low vigor and were very low seed producers based on personal observation. The accession PI 322311 is a prime candidate for use as an ornamental plant because of its beautiful yellow flowers, ornate compound leaves, and quality growth in subtropical areas such as Georgia or tropical areas including central and southern Florida, southern Alabama, Mississippi, Louisiana, and Texas. Both PI 279691 and 322311 produce very beautiful compound leaves (Figures 23.2 and 23.3). Drawings of Christmas candle leaves and flowers have been used to decorate stamps from various countries (Figures 23.6 and 23.7).

23.4 ETHNOBOTANICAL USES

Crushed leaves of Christmas candle are topically used for skin rashes, and a decoction is used for constipation by traditional medicine users in Martinique, French West Indies (Longuefosse and

Figure 23.5 Christmas candle (PI 322311) with ornate flowers and leaves growing in a garden containing potting soil.

Nossin 1996). In Akwa Ibom, Nigeria, Christmas candle leaves are used locally (in powder form) and orally (as a decoction) against skin diseases (Ajibesin et al. 2008). Leaves are used for dermatologic purposes as well as for anti-infection uses primarily aimed at *Herpeszoster*, eczema, and mycosis in Akwapim, Ghana (Pesewu, Cutler, and Humber, 2008). Christmas candle fruit is used as a decoction against infectious diseases in Guinean traditional medicine (Magassouba et al. 2007).

Figure 23.6 Postage stamp from the République Gabonaise showing the Christmas candle senna drawing.

Figure 23.7   Postage stamp from Singapore showing the Christmas candle senna drawing.

Leaf macerations of Christmas candle are also used to treat people with diabetes in southwestern Nigeria (Abo, Fred-Jaiyesimi, and Jaiyesimi 2008). Christmas candle leaves are used against fungal infections (McClatchey 1996) and pounded together with sulfur and applied to skin infected with ringworms or sores (Ong and Norzalina 1999).

23.5 POTENTIAL PHARMACEUTICAL USES

An aqueous ethanol extract of Christmas candle plants showed significant antitumor and antioxidant effects on carcinomatous cells in rats without toxic effects (Pieme et al. 2008). For horses infected with bovine dermatophytosis, an application of a formulation consisting of leaves from both Christmas candle and turmeric plus calcium oxide resulted in complete recovery of 70% of the animals (Vijayan et al. 2009). Christmas candle extracts showed anti-MCV (mouse corona virus), the surrogate for human SARS virus, and anti-herpes-simplex virus activities at a concentration as low as 0.4 µg/mL (Vimalanathan, Ignacimuthu, and Hudson 2009). Extracts of water, methanol, chloroform, and petroleum ether of Christmas candle flowers have produced *in vitro* antimicrobial activities in assays against isolates of *Staphylococcus aureus*, *Candida albicans*, *Escherichia coli*, *Proteus vulgaris*, *Pseudomonas aureginosa*, and *Bacillus subtilis* (Idu, Omonigho, and Igeleke 2007). Recently, a leaf extract of Christmas candle plants protected against cutaneous photodamage after repetitive UV irradiation in healthy females (Danoux et al. 2007).

23.6 CONCLUSIONS

Even though Christmas candle plant is known primarily for its ornamental and phytochemical characteristics, additional studies are needed to develop its full potential as a source of natural pharmaceuticals. Perhaps the most interesting and useful challenge of all will be the production of high-quality flowers and leaves with an attractive appearance that will contain a reasonably high content of antioxidants, anticancer properties, antimicrobial properties, and skin protectants. Some functional and clinically active compounds have been identified in Christmas candle plants.

For example, the flavonoid kaempferol found in Christmas candle leaves has been shown to inhibit pancreatic cancer (Nothlings et al. 2007). The steroid β-sitosterol found in Christmas candle leaves has been found to diminish benign prostatic hyperplasia significantly when combined with cernitin, saw palmetto, and vitamin E (Preuss et al. 2001). It has also been shown to inhibit acne-inducing bacteria (Chomnawang et al. 2005). Chrysoeriol found in Christmas candle seeds has potential for antioxidation and anti-inflammatory activities for humans (Choi et al. 2005). Rhein found in Christmas candle leaves and pods has been shown to inhibit nasopharyngeal carcinoma cells (Lin et al. 2009) as well as additional human cancer cells (Shi, Huang, and Chen 2008).

However, to my knowledge none of these phytochemicals are extracted from Christmas candle and used as pharmaceutical agents. A Christmas candle accession in the U.S. germplasm repository has been successfully characterized for morphological and seed reproductive descriptors (Morris 2009), providing further evidence of its potential as a new ornamental. Its use as an ornamental as well as a potential pharmaceutical species should be encouraged as well.

REFERENCES

Abo, K. A., Fred-Jaiyesimi, A. A., and Jaiyesimi, A. E. A. 2008. Ethnobotanical studies of medicinal plants used in the management of diabetes mellitus in Southwestern Nigeria. *Journal of Ethnopharmacology* 115:67–71.

Ajibesin, K. K., Ekpo, B. A., Bala, D. N., Essien, E. E., and Adesanya, S. A. 2008. Ethnobotanical survey of Akwa Ibom state of Nigeria. *Journal of Ethnopharmacology* 115:387–408.

Bosch, C. H. 2007. *Senna alata* (L.) Roxb. [Internet]. Record from protabase, ed. G. H. Schmelzer and A. Gurib-Fakim. PROTA (Plant Resources of Tropical Africa/Resources vegetales de l'Afrique tropicale), Wageningen, the Netherlands. http://database.prota.org/search.htm

Choi, D. Y., Lee, J. Y., Kim, M. R., Woo, E. R., Kim, Y. G., and Kang, K. W. 2005. Chrysoeriol potently inhibits the induction of nitric oxide synthase by blocking AP-1 activation. *Journal of Biomedical Science* 12:949–959.

Chomnawang, M. T., Surassmo, S., Nukoolkarn, V. S., and Gritsanapan, W. 2005. Antimicrobial effects of Thai medicinal plants against acne-inducing bacteria. *Journal of Ethnopharmacology* 101:330–333.

Danoux, L., Jeanmaire, C., Bardey, V., Perie, G., Vazquez-Duchene, M. D., Gillon, V., Henry, F., Moser, P., and Pauly, G. 2007. *Protecting the genome of skin cells from oxidative stress and photoaging. Naturals and organics in cosmetics from R & D to the marketplace.* Carol Stream, IL: Allured Publishing Corp.

Idu, M., Omonigho, S. E., and Igeleke, C. L. 2007. Preliminary investigation on the phytochemistry and antimicrobial activity of *Senna alata* L. flower. *Pakistan Journal of Biological Sciences* 10:806–809.

Lin, M. L., Chung, J. G., Lu, Y. C., Yang, C. Y., and Chen, S. S. 2009. Rhein inhibits invasion and migration of human nasopharyngeal carcinoma cells *in vitro* by down-regulation of matrix metalloproteinases-9 and vascular endothelial growth factor. *Oral Oncology* 45:531–537.

Longuefosse, J. L., and Nossin, E. 1996. Medical ethnobotany survey in Martinique. *Journal of Ethnopharmacology* 53:117–142.

Magassouba, F. B. et al. 2007. Ethnobotanical survey and antibacterial activity of some plants used in Guinean traditional medicine. *Journal of Ethnopharmacology* 114:44–53.

McClatchey, W. 1996. The ethnopharmacopoeia of Rotuma. *Journal of Ethnopharmacology* 50:147–156.

Morris, J. B. 2009. Characterization of medicinal *Senna* genetic resources. *Plant Genetic Resources: Characterization and Utilization* 7:257–259.

Nothlings, U., Murphy, S. P., Wilkens, L. R., Henderson, B. E., and Kolonel, L. N. 2007. Flavonols and pancreatic cancer risk: The multiethnic cohort study. *American Journal of Epidemiology* 166:924–931.

Ong, H. C., and Norzalina, J. 1999. Malay herbal medicine in Gemencheh, Negri Sembilan, Malaysia. *Fitoterapia* 70:10–14.

Pesewu, G. A., Cutler, R. R., and Humber, D. P. 2008. Antibacterial activity of plants used in traditional medicines of Ghana with particular reference to MRSA. *Journal of Ethnopharmacology* 116:102–111.

Pieme, C. A., Penlap, V. N., Nkegoum, B., and Ngogang, J. 2008. *In vivo* antioxidant and potential antitumor activity of aqueous ethanol extract of leaves of *Senna alata* (L.) Roxb (Ceasalpiniaceae) on bearing carcinomatous cells. *International Journal of Pharmacology* 4:245–251.

Preuss, H. G., Marcusen, C., Regan, J., Klimberg, I. W., Welebir, T. A., and Jones, W. A. 2001. Randomized trial of a combination of natural products (cernitin, saw palmetto, β-sitosterol, vitamin E) on symptoms of benign prostatic hyperplasia (BPH). *International Urology and Nephrology* 33:217–225.

Shi, P., Huang, Z., and Chen, G. 2008. Rhein induces apoptosis and cell cycle arrest in human hepatocellular carcinoma BEL-7402 cells. *American Journal of Chinese Medicine* 36:805–813.

USDA, ARS, National Genetic Resources Program. 2009. *Germplasm resources information network—(GRIN)* [online database]. National Germplasm Resources Laboratory, Beltsville, MD. http://www.ars-grin.gov/cgi-bin/npgs/html/taxon.pl?100063 (accessed December 9, 2009).

Vijayan, R., Saritha, A. L., Anoop, S., Ramakrishnan, A., and Nisha, M. 2009. Therapeutic efficacy of an ethnoveterinary formulation in the treatment of clinical cases of bovine dermatophytosis. *Animal Science Reporter* 3:33–36.

Vimalanathan, S., Ignacimuthu, S., and Hudson, J. B. 2009. Medicinal plants of Tamil Nadu (southern India) are a rich source of antiviral activities. *Pharmaceutical Biology* 47:422–429.

CHAPTER 24

Fenugreek (*Trigonella foenum-graecum* L.)

S. K. Malhotra

CONTENTS

24.1 INTRODUCTION

The fenugreek (*Trigonella foenum-graecum* L.) belongs to the family Fabaceae and is a multiuse and commercially important spice crop grown for its seeds, tender shoots, and fresh leaves. It is an annual plant, extensively cultivated as a food crop in India, the Mediterranean region, North Africa, and Yemen. Fenugreek seeds and herbs are well known for their distinct aroma and slightly bitter taste. The cultivation of this crop is confined to areas with moderate or low rainfall and a cool growing season without extreme temperatures. It can tolerate 10–15°C of frost (Duke 1986). It has been reported to be grown under a wide range of soil and climatic conditions in many countries of Europe, Asia, Africa, Australia, and America (Table 24.1). Fenugreek has economic value as food, fodder, medicine, and in cosmetics. According to the natural medicines comprehensive database, 523 commercial products containing fenugreek have been reported, including 50 Canadian licensed products (Natural Database 2010).

Fenugreek is a cultivated crop in India, Turkey, Egypt, Portugal, Spain, Ethiopia, Kenya, Tanzania, Israel, Lebanon, Morocco, Tunisia, Pakistan, China, Japan, Russia, Argentina, Australia, the UK, Canada, and the United States (Rosengarten 1969; Smith 1982; Petropoulos 2002; Malhotra and Rana 2008). Large numbers of countries have been reported to cultivate fenugreek, but India, Ethiopia, Egypt, and Turkey are the major countries for seed production. India is the leading producing country in the world and currently is responsible for more than 68% of world production. The area and production statistics of the world are very limited; because fenugreek is cultivated in a relatively small area, perhaps its pertinent data are not recorded or highlighted in the agricultural statistics of different countries.

However, according to the current available information, fenugreek was grown in a 68,290 ha area with a production of 76,580 metric tons and productivity of 1121 kg/ha during 2008–2009 in India (NHB 2009). The yield of fenugreek in India increased to 38% during 2008–2009 compared to 2005–2006. Considering India's production, the world production can be estimated to be between 100,000 and 120,000 metric tons. In Turkey, 670 tonnes of fenugreek seed are produced annually from a 700 ha area with productivity of 957 kg/ha of seed (Altuntas, Ozgoz, and Faruk Taser 2005). As of 2006, fenugreek was grown in western Canada from 140–500 ha, mainly for seed production (Slinkard et al. 2006; Thomas, Basu, and Acharya 2006). During 2008–2009, a quantity of 20,750 metric tons, worth 15.81 million US$, was exported from India to more than 20 countries such as Sri Lanka, UAE, Japan, South Africa, Nepal, Saudi Arabia, the UK, Morocco,

Table 24.1　Countries on Different Continents That Grow Fenugreek (*Trigonella foenum-graecum* L.)

Continent	Countries
Africa	Egypt, Ethiopia, Kenya, Morocco, Sudan, Tanzania, and Tunisia
Asia	China, India, Iran, Israel, Japan, Lebanon, and Pakistan
Australia	Parts of Australia
Europe	Austria, France, Germany, Greece, Portugal, Russia, Spain, Switzerland, Turkey, and the UK
North America	Canada and the United States
South America	Argentina

the United States, Malaysia, Bangladesh, France, Egypt (A.R.E.), the Netherlands, Israel, and a few others (Spices Board 2009).

On analysis of the list of fenugreek-producing and -importing countries, Japan, the UK, Morocco, the United States, Egypt, and Israel fell into both categories. The reason may be that these countries take up production occasionally or that there is insufficient production in relation to increasing demand. The world market for fenugreek seeds is estimated to range from 30,000 to 50,000 tonnes (McCormick, Norton, and Eagles 2006). Considering these figures, India contributes the lion's share of more than 50% of world demand.

Fenugreek is becoming popular around the world; its extract is used to flavor cheese in Switzerland, artificial maple syrup and bitter-rum in Germany, roasted seeds as coffee substitute in Africa, seed powder mixed with flour as fortification to make flat-bread in Egypt, and as an antidiabetic herb in Israel. The whole seed and dried plant are used as insect- and pest repellents in grain storage, and the oil is used in perfumery in France. Its leaves are used as greens and leafy vegetables and its seeds as spice and as powder in flour in India (Rajagopalan 2001); it is used as forage in Australia, Canada, and India (Moyer et al. 2003; Acharya, Thomas, and Basu 2008). In the recent past, use of fenugreek has been well recognized and demand is increasing due to its multifarious uses as spice, forage, medicinal plant, and a source for natural diosgenin and galactomannan in the pharmaceutical and steroid industries.

Till the 1980s, the production of fenugreek was restricted to a few countries; however, with time, demand for new crops like fenugreek with potential nutraceutical value has secured it a place as an alternative crop to extend the diversity of the farming system. Recent examples of increase in area sown have been in Australia, due to high grain prices (>$400/t) and adaptation to gray vertisol soil zones (McCormick et al. 2000), and in the dry growing conditions of prairies of western Canada (Moyer et al. 2003). Moreover, it is capable of producing consistent yield under a short growing season and is adaptable to semiarid conditions.

Fenugreek has been a neglected crop and has been underused; the fact that it is a locally important crop has restricted large-scale production. The information related to genetic diversity, intra- and interspecific variability, and genetic relationships among varieties is meager. The presence of few chromosomes ($2n = 16$) and the large and varying sizes of chromosomes (longest = 26.28 μm and shortest = 13.52 μm) (Reasat, Karapetyam, and Nasirzadeh 2002) offer good scope for cytogenetic and molecular studies.

This chapter primarily covers information on the origin and domestication of fenugreek, management of plant genetic resources, available genetic and cytogenetic tools for exploiting germplasm for crop improvement, and medicinal properties of the crop.

24.2 DOMESTICATION AND DISSEMINATION OF FENUGREEK

24.2.1 History and Domestication

Fenugreek is one of the oldest cultivated spice crops in the world; it has been grown for its medicinal value and forage in India, western Africa, and the Nile Valley since remote antiquity and is used for human and animal consumption. The species name, *foenum-graecum*, means Greek hay (Sinskaja 1961), indicating its historical use as forage crop in the ancient world. In Egypt, it has been cultivated since 1000 BC, and it has been part of the Indian diet for over 3,000 years. It is found growing wild in parts of north India and cultivated all over the subcontinent for its green leaves and seeds. Fenugreek is a native of southeastern Europe and western Asia. India has also been reported to be the native home of fenugreek (Shanmugavelu, Kumar, and Peter 2002), which exists in wild form in Kashmir, Punjab, and the upper Gangetics Plains. The carbonized fenugreek seed from a Rohira village in the Sangrur district of Punjab, India, suggests its use in trade by people of the Harappan civilization as far back as 2000–1700 BC (Saraswat 1984).

Fenugreek seeds were found in the tomb of Tutankhamun (Manniche 1989). Portius Cato, a Roman authority on animal husbandry in the second century BC, ordered *foenum-graecum*, which was today's fenugreek, to be sown as fodder for oxen (Fazli and Hardman 1968). In North Africa, it has been cultivated around the Saharan oasis since very early times (Duke 1986). References to the utilization of fenugreek were found as far back as 1578 in the famous *Kolozsvar Herbarium* compiled by Melius in 1578 and cited by Hidvegi et al. (1984). Fenugreek was introduced into Chinese medicine in the Sung dynasty, AD 1057 (Jones 1989). According to a citation by Miller (1969), in his examination of the definition and function of spices in his materia medica, Dioscorides, a Greek physician of Anazarbus in Cilicia and considered "father of pharmacology" (AD 65), writes that fenugreek is an active compound of ointments and mentions fenugreek as a spice crop in his texts. Dioscorides also mentions that the Egyptians called fenugreek *itasin* (Manniche 1989). Leaves of fenugreek were one of the components of the celebrated Egyptian incense *kuphi*, a holy smoke used in fumigation and embalming rites (Rosengarten 1969).

According to Fazli and Hardman (1968), Charlemagne encouraged cultivation of fenugreek. Fenugreek is probably one of the forages cultivated before the era of recorded history. As a fodder plant, it is said to be the *Hedysarum* of Theophrastus and Dioscorides (Leyel 1987). In the Middle Ages, it is recorded that fenugreek was added to inferior hay because of its peculiarly pleasant smell (Howard 1987). Fenugreek was known and used for different purposes in ancient times, especially in Greece and Egypt (Rouk and Mangesha 1963). Rosengarten (1969) reported that the Romans obtained the plant from the Greeks and that it became a commercial commodity of the Roman Empire (Miller 1969); Stuart (1986) and Howard (1987) support the contention that Benedictine monks introduced the plant into medieval Europe. Fenugreek was introduced into central Europe at the beginning of the ninth century (Schauenberg and Paris 1990). However, it is not mentioned in any herbal literature until the sixteenth century, when it was recorded as grown in England.

24.2.2 Origin and Dissemination

According to Vavilov (1926), the Near East region, extending from Israel through Syria and southern Turkey into Iran and Iraq, and the Mediterranean center including Spain, Morocco, and Turkey are the centers of origin of *Trigonella*, *Trifolium*, and *Medicago* species. In a study by Dangi et al. (2004), *Trigonella foenum-graecum* and *T. caerulea* accessions from Turkey exhibited more diversity and support Vavilov's hypothesis. It was also revealed that *T. foenum-graecum*, the large seeded cultivar, is abundant around the Mediterranean region, whereas the small seeded cultivars are predominant eastward. Species of the genus exist wild in the countries of Europe, Macronesia (Canary Islands), North and South Africa, central Asia, and Australia.

Indigenous species of this genus have also been reported (Anonymous 1994): six for Asia (*T. caelesyriaca*, *T. calliceras*, *T. emodi*, *T. geminiflora*, *T. glabra*, *T. kotschyi*), five for Europe (*T. graeca*, *T. striata*, *T. polycerata*, *T. monspeliaca*, *T. procumbens*), one for Africa (*T. laciniata*), and one for Australia (*T. suavissima*), where it has adapted well to the wet swampy habitat (Allen and Allen 1981). The rest of the species exist on more than one continent. Of this genus, 23 species have been reported for Europe (Ivimey-Cook 1968); 15 occur in the Balkan area (Polunin 1988), including 14 in Greece (Kavadas 1956). Of the latter, four species are found on the famous island of Kefallinia (Phitos and Damboldt 1985). However, the most interesting species of the genus is the widely cultivated *T. foenum-graecum* (fenugreek). *Trigonella balansae*, an annual legume of Eurasian origin, is vegetative productive and able to regenerate on alkaline soils receiving <400 mm of annual rainfall in temperate Australia (Loi et al. 2000). The distribution of various species of *Trigonella* is reported to occur scattered in various parts near the Mediterranean region, Europe, and Asia (summarized in Table 24.2, which has been developed from the text of Seidemann 2005).

Table 24.2   Major *Trigonella* Species and Their Distribution

Trigonella Species	Distribution
T. arabica Delile (syn. *T. pectin* Schenk)	N. Africa, especially in Arabia; Syria to NE Egypt
T. caerulea (L.) Ser. (syn. *Melilotus caeruleus* Desr., *Trifolium caeruleum* Moench., *Trigonella melilotus caoerulea* L. Aschers. et Graebn)	E. Mediterranean region; SE Europe, origin in Mediterranean region
T. caerulea (L.) Ser. ssp. *Caerulea*	C., W., and S. Europe; N. Africa, widely cultivated in gardens
T. corniculata (L.) L. (syn. *Medicago corniculata* L. Trautv., *Trifolium corniculata* L., *Trigonella esculenta* L.)	Mediterranean region; Near East countries
T. stellata Forsk.	N. Africa, Arabia, Egypt, Tunisia, Algeria, Morocco, Canary Islands, W. Asia, Iran, Iraq, Middle East, Israel, Lebanon, Kuwait
T. foenum-graecum L. (syn. *Foenum graecum officinale* Moench, *T. graeca* St. Lag.)	Caucasus, ex Soviet Union, C. Asia, E. Europe

Source: Table developed from text of Seidemann, J. 2005. *World spice plants, economic usages, botany and taxonomy*, 372–374. Berlin: Springer Verlag.

Broadly, two origins of fenugreek are reported to be the Indian subcontinent and the eastern Mediterranean area (Sinskaja 1961; Acharya et al. 2008). The genus *Trigonella* consists of 50 species, most of which have an oriental origin in the Iranian/Indian region; 11 spices occur in India, out of which *Trigonella foenum-graecum* L. (fenugreek) and *Trigonella corniculata* L. are commercially cultivated in India (Singhania et al. 2006). Fenugreek is native to an area extending from the Mediterranean basin to South Asia. It is now widely cultivated in North and East Africa, China, Ukraine, and Greece. In the natural flora of Turkey, 49 species of fenugreek are found (Davis 1978).

De Candolle (1964) and Fazli and Hardman (1968) observed that fenugreek grows wild in Punjab and Kashmir in India, in the deserts of Mesopotamia and Persia, in Asia Minor, and in some countries in southern Europe such as Greece, Italy, and Spain. De Candolle (1964) believed that the origin of fenugreek should be Asia rather that southern Europe because if a plant of fenugreek nature were indigenous in southern Europe, it would be far more common and not be missing in insular floras of Sicily, Ischia, and the Balearic Isles. Sinskaja (1961) reported that the direct wild ancestor of cultivated fenugreek belonging to the species *T. foenum-graecum* has not been exactly determined and the existence of these wild forms (that have not escaped from cultivation) is problematic. Many authors maintain that the direct ancestor of cultivated fenugreek is the wild *T. gladiata* Stc., which differs from *T. foenum-graecum* in respect of the entire aggregate of characters, of which seed tuberculation and the small size of pods are only the most striking. It is possible that the species *T. foenum-graecum* evolved from *T. gladiata*, which has possibly given rise to some newly extinct forms of *T. foenum-graecum*.

In contrast to the idea of the Mediterranean region as the origin, deviating notions about the nativity have been expressed by different authors, such as origin in Asia (De Candolle 1964; Fazli and Hardman 1968), Turkey (Dangi et al. 2004), or India (Shanmugavelu et al. 2002) and the issue is still debatable. Fenugreek is also grown very satisfactorily in central Europe, the UK, and the United States. This wide distribution and cultivation in the world are characteristic of its adaptation to variable climatic conditions and growing environments. Sinskaja (1961) reported that, in Transcaucasia, fenugreek reaches mountain altitudes of up to 1300–1400 m above mean sea level (a.m.s.l.) and in Ethopia 3000 m (a.m.s.l), although the main zone of distribution in that country is between 2150 and 2400 m (a.m.s.l.). Fenugreek is found growing in varied climates ranging from cool-temperature steppe to wet tropical and very dry forest; it is reported to tolerate an annual precipitation of 3.8–15.3 dm and an annual mean temperature of 7.8–27.5°C (Duke 1986).

24.3 BOTANY

24.3.1 Taxonomy

Fenugreek belongs to the family Fabaceae (Leguminosae), subfamily Papilionaceae, and genus *Trigonella*. Different species of the genus have different chromosome numbers. The taxonomy of the fenugreek is as follows:

Order	Fables (Leguminales)
Family	Fabaceae (Leguminosae)
Subfamily	Faboideae (Papilionaceae)
Tribe	Trifolieae
Subtribe	Trigonellinae
Genus	*Trigonella*
Species	*T. foenum-graecum* Linn.

According to Hutchinson (1964), the genus *Trigonella*, along with five other genera—*Parachetus, Melilotus, Factorovekya, Medicago,* and *Trifolium*—of the tribe Trifolieae of the family Fabaceae is within the order Fables (Leguminales). The genera *Trigonella, Medicago, Trifolium,* and *Melilotus* belong to subtribe Trigonellinae, tribe Trifolieae, in the Leguminosae family (Small, Lassen, and Brookes 1987). The genus *Trigonella* has been described as mostly annual or perennial plants. More than 100 species have been reported (Vasil'chenko 1953). *Trigonella* is one the largest genera of the tribe Trifolieae; *Trigonella foenum-graecum* is an extensively grown species (Balodi and Rao 1991). There is a controversy about the number of species that comprise the genus *Trigonella*. Although 260 species (182 from Linnaeus to 1885 and 78 from 1886 to 1965) are listed under this genus, a close scrutiny reveals about 97 distinct species (Fazli 1967). The 60 species of genus *Trigonella* as reported by Petropoulos and Kouloumbis (2002) are given in Table 24.3.

On the basis of form and shape of the calyx and pod, Tutin and Heywood (1964) divided the genus *Trigonella* into three subgenera: *Trigonella* (*T. graeca, T. cretica, T. maritime,* and *T. corniculata*), *Trifoliastrum* (*T. caerulea* and *T. procumbens*), and *Foenum-graecum* (*T. foenum-graecum* and *T. coerulescens*). According to color of corolla in fenugreek, Furry (1950) classified three groups as corolla blue type (*T. caerulea*), corolla whitish (*T. foenum-graecum*), and corolla yellow (*T. polycerata, T. monspeliaca,* and *T. sitavissima*). The *T. polycerata* and *T. monspeliaca* are annual, whereas *T. sitavissima* is perennial. Hocker and Jackson (1955) also reported three species of *Trigonella* (*T. gladiate, T. cariensis,* and *T. monspeliaca*) as having been described as *T. foenum-graecum*.

Serpukhova (1934) showed the polymorphic character of fenugreek and divided *T. foenum-graecum* into two types: from Yemen (*T. foenum-graecum* L. ssp. *iemensis* with short vegetation period) and from Albyssinia (*T. foenum-graecum* L. ssp. *culta* with long vegetation period). Moschini (1958) divided the cultivated fenugreek into three ecotypes in Italy: Sicilian (high precocity and high yield), Toscanian (late in maturity, tolerant to cold, and high yielding), and Moroccon (high precocity, tolerant to cold, and low yielding). According to shape, size, and color, Serpukhova (1934) classified *T. foenum-graecum* into three groups: Indicae (*nano-fulva*), Anatolicae (*magno-fulva*), and Aethiopicae (*fulva, punctato-fulva, olivacea, punctato-olivacea, lecosperma,* and *griseo-coerulescence*); Sinskaja (1961) confirmed this classification.

Furry (1950) divided fenugreek into six types (Yemenese, Transcaucasian, African, Afghan, Chinese-Persian, and Indian). Based on critical examination of a rich collection of fenugreek from different countries, Petropoulos (1973, 2002) classified fenugreek into four groups: fluorescent type, Ethiopian type, Indian type, and Mediterranean type, which are summarized in Table 24.4 with information of seed fluorescence behavior under UV light, pigment in seed coat, 1,000 seed weight, and origin and name of representative cultivars.

Table 24.3   List of Species of the Genus *Trigonella*[a]

T. anguina Del.	*T. laciniata* (L.) Desf. (Boiss. & Noe.) (Lassen)
T. arabica Del.	*T. lilacina* Boiss.
T. arcuata C.A. Mey	*T. marginata* Hochst. & Steud.
T. aristata Vass.	*T. maritima* Poiret or Delile ex Poiret in Lam.
T. auradiaca Boiss. (*T. aurantiaca* Boiss.)	*T. melilotus caeruleus* (L.) Ascherson & Graebner[c]
T. balansae Boiss. and Reut. in Boiss. (*T. corniculata* L.)	*T. monantha* C.A. Mey
T. berythaea Boiss. and Blanche	*T. monspeliaca* L. (*T. monspeliana* L.)[d]
T. brachycarpa (Fisch) Moris	*T. nosana* Boiss.
T. caelesyriaca Boiss.	*T. occulta* Ser. Del.
T. caerulea (L.) Ser. (*T. coerulea* L.)	*T. ornithopoides* (L.) DC.[e]
T. calliceras Fisch ex Bieb.	*T. orthoceras* Kar. & Kir.
T. cancellata Dest.	*T. pamirica* Gross. in Kom.
T. cariensis Boiss.	*T. platycarpos* L.
T. coerulescens (Bieb.) Halacsy Hal.	*T. polycerata* L.
T. corniculata (L.) L. (*T. balansae T. rechingeri* Sirj. Boiss. & Reut.)	*T. popovii* Kor.
T. cretica (L.) Boiss.[b]	*T. procumbens* (Besser) Reichenb.
T. cylindracea Desv. (*T. culindracea* Desv.)	*T. radiata* Boiss.
T. emodi Benth.	*T. rigida* Boiss. & Bal.
T. erata	*T. ruthenica* L.
T. fischeriana Ser.	*T. schlumbergeri* Buser (Boiss.)
T. foenum-graecum L.	*T. sibthorpii* Boiss.
T. geminiflora Bunge	*T. smyrnaea* Boiss.
T. gladiata Stev. or Stev. ex Bieb. (*T. tortulosa* Gris.)	*T. spicata* Sibth. and Sm. (*T. homosa* Bess.)
T. graeca (Boiss. and Spruner) Boiss.	*T. spinosa* L.
T. grandiflora Bunge	*T. sprunerana* Boiss. (*T. spruneriana* Boiss.)
T. hamosa L.	*T. stellata* Forssk.
T. hybrida Pourr.	*T. striata* L.
T. incisa Benth.	*T. suavissima* Lindl.
T. kotschyi Fenzl. ex Boiss.	*T. tenuis* Fisch ex Bieb.
	T. tortulosa Gris. (*T. sprunerana* or *spruneriana* Boiss.)
	T. uncata Boiss. & Noe. (*T. glabra* subsp. *Uncata*)

Source: Petropoulos, G. A., and P. Kouloumbis. 2002. In *Fenugreek: The genus* Trigonella, *medicinal and aromatic plants—Industrial profiles*, ed. G. A. Petropoulos, 9–7, London: Taylor & Francis Inc

[a] The botanical names have been completed according to the Hocker, J. B., and D. Jackson. 1955. *Index Kewensis*, Tomus II, 1116–1117 (1895) Suppl. XII, 146 (1951–1955). Oxford, England: Clarendon Press.

[b] Transformed to the genus *Melilotus* under the name *M. creticus.*

[c] Fused with the species *T. caerulea* under the name *T. caerulea.*

[d] Transformed to the genus *Medicago* under the name *M. mospeliaca* or *monspeliana.*

[e] Transformed to the genus *Trifolium* under the name *T. ornithopoides.*

24.3.2 Plant Structure and Growth Habit

Fenugreek is an annual herb. The plants are spreading, moderately branched, and weak. The plant has taproot growth initially, followed by a large number of secondary roots. The roots form root nodules and establish a symbiotic relation with the *Rhizobium* bacterium. The stem is erect, branched, green, smooth, and herbaceous. The leaves are petiolate, alternate, compound, trifoliolate, fragrant, and stipulate (Subramanian 1996; Malhotra and Rana 2008). The genus *Trigonella* contains mostly annual and perennial plants, often strongly scented. For detailed descriptions of taxonomic characters, readers may consult Hutchinson (1964), Sinskaja (1961), and Heywood (1967).

Table 24.4   Classification of Fenugreek Types Based on Seed Fluorescence, Pigment in Seed Coat, 1,000 Seed Weight, and Origin/Place of Collection

Category of Fenugreek	Seed Fluorescence Under UV Light	Pigment in Seed Coat	Seed Size	1,000 Seed Weight	Origin	Representative Cultivars
Fluorescent type	Fluorescent	Absence of any pigment in seed coat	Large seeds (5–6 × 3–4 mm), rounded in outline	High 1,000 seed weight (27–32 g) and germ./husk	Spontaneous mutation from Ethiopian populations because most of characters are controlled by recessive genes.	Fluorescent and Barbara
Ethiopian type	Nonfluorescent	With at least four different pigments in seed coat	Moderate in seed size (4.0–4.5 × 3.0–3.5 mm)	1,000 Seed weight is 22–25 g	Natural mixture of Serpukhova's olivacea and punctato-olivacea; most of the samples are from Ethiopia and neighboring countries.	Ethiopian breeding cultivars
Indian type	Nonfluorescent	With at least four pigments in seed coat	Very small seeds (2.5–3.5 × 2.0–2.5 mm), rectangular in outline	1,000 Seed weight is 15–20 g	Nanofulva according to Serpukhova's classification; most of the samples are from India, Pakistan, China, and Kenya, the latter being bigger than the rest.	Kenyan breeding cultivars
Mediterranean type	Nonfluorescent	—	Large seeds, (4.5–6.0 × 3.5–5.0 mm), rectangular in outline	1,000 Seed weight is 25–31 g	A natural mixture of magnofulva, fulva, and punctatofulva (Serpukhova's classification). In this type, samples are from Israel (magnofulva is dominant), Morocco, Portugal, Spain, France (punctatofulva is dominant), and Greece and Turkey (the fulva is dominant).	Moroccan breeding cultivars

Sources: Table developed from the text of Petropoulos, G. A. 1973. PhD diss. Bath University, England; Petropoulos, G. A. 2002. *Fenugreek—The genus* Trigonella, 1–255. London: Taylor & Francis Inc.

The leaves are pinnately trifolioliate, stipules adulate to the petiole, leaflets are usually toothed and nerves often run out into teeth. The taxonomic characters of *Trigonella foenum-graecum*, the most commonly cultivated species worldwide, have been described. Plant height is 20–130 cm with straight growth, rarely ascending and branching. The stem is rarely simple, sparsely pubescent, usually hollow, and rarely completely green; anthocyanin is present at base or covered on whole plant (Petropoulos 2002). Fenugreek plants in North America typically grow 40–60 cm high (Slinkard et al. 2006) and are trifoliolate with branched stems.

The first leaf of the fenugreek plant is simple, sometimes weakly trifoliolate, and oval or orbicular, with the entire margin and a long petiole. The stipules are fairly large and covered with soft hair. Leaf petioles are thickened at the top and attenuate beyond point of attachment of lateral leaflets. Petioles and petiolules are vested on the underside with simple, soft, sparse hair and are a little cartilaginous. The shape of leaflets varies from ovate–orbicular to oblong–lanceolate, 1–4 cm long, almost equal, finely haired, dentate, and near the apex; dentation is more strongly developed in upper than lower leaves. The petioles and blades of the leaflets are anthocyanin tinged to a varying degree of green (Sinskaja 1961; Hutchinson 1964). Figure 24.1 shows a fenugreek plant exhibiting different morphological traits of rooting, branching, leaves, pods, etc.

24.3.3 Flowering and Floral Biology

Fenugreek plants enter into the reproductive phase following a vegetative growth phase. Two types of flowering shoots, indeterminate and determinate, have been observed. The axillary flowers showing indeterminate growth habit are the most common; near-determinate or determinate types (blind shoots bearing axillary and terminal flowers) are not common (Basu 2006).

Figure 24.1   Fenugreek (*T. foenum-graecum* L.) plant.

Figure 24.2 Fenugreek (*T. foenum-graecum* L.) plant at flowering stage.

Two types of flowers—cleistogamous (closed type) and aneictgamous (open type)—have been reported for fenugreek (Petropoulos 2002). The germplasm being maintained at the National Research Center on Seed Spices in Ajmer, India, revealed that 98.4% of the accessions have cleistogamous flowers and 1.6% accessions have aneictgamous flowers (Malhotra 2010). Cleistogamous flowers ensure a high level of self-pollination. However, the rate of cross-pollination in *T. foenum-graecum* has ranged from 2.1 to 7.0% (Choudhary 2003).

The inflorescence is racemose and flowers are bracteates, pedicellate (Figure 24.2), complete, zygomorphic, bisexual, and hypogynous. The five sepals are gamosepalous, green, and valvate. The five petals are polypetalous, papilionaceous, and standard, with two wings; anterior petals form a boat-shaped keel, with descending imbricate aestivation. There are 10 diadelphous stamens; 9 are united to form a tube around the ovary and the 10th is free. Anthers are basified, introse, and enclosed in the keel. The gynoecium is monocarpellary, ovary superior, and unilocular; there are numerous ovules, placentation is marginal, style is long, and stigma is terminal.

The large-seeded fenugreek (*T. foenum-graecum*) bears white flowers with straight pods, whereas Kasuri Methi (*T. corniculata*), the small-seeded type, bears yellow to orange flowers with sickle-shaped pods (Malhotra 2003). Petropoulos (2002) mentioned that flowers appear in leaf axils, mostly twin, more rarely solitary. The calyx is 6–8 mm, soft and hairy with teeth as long as the tube and half as long as the corolla. The corolla is 13–19 mm long, pale yellow (white at the end of a flowering period, sometimes lilac colored at the base). The standard leans backward and is oblong emarginated at the apex with bluish spots (these spots are absent from some genotypes); wings are half as long as the standard and the keel is obtuse and split at the base.

The genus *Trigonella* is a typical self-pollinating crop in which double fertilization occurs with the unopened flower buds. Pollen fertility ranges from 95 to 99% in the unopened flower buds and 67 to 80% in the opened flower buds. The flowers open between 9 a.m. and 6 p.m., with peak at 11.30 a.m. The peak of anthesis is between 11.30 a.m. and 12.30 p.m. The stigma remains receptive for about 10 hours after flowering. The pollen grains are oval (70–90%) to circular or orbicular, ellipsoidal grain (10–30%), hyaline, and stain pink or red when treated with 0.5% acetocarmine (Malhotra 2003; Basu 2006).

Petropoulos (2002) recommended that cross-pollination be done in closed flowers of fenugreek at initiation of the second stage of floral development, when stamens are lower in position than the stigma and anthers are closed but stigma is receptive. The fenugreek has been described as a rarely cross-pollinated plant because its stigma becomes receptive before the anthers mature. Based on floral structure, Nair et al. (2004) described the inability of *Trigonella balansae* to self-pollinate and the requirement for an external vector to transfer pollen effectively from the anther onto stigmas of this species. The self rate was also estimated to be 0.13, but the level of inbreeding depression recorded was 0.48 and self-compatibility index was 0.51, which indicate occurrence of cross-pollination in this species of *Trigonella*.

24.3.4 Pods and Seeds

Pods are sickle shaped and the pod color turns from green or light purple to brown or yellow-brown at maturity. Pods are with beak, 9.5–18.6 cm long, 0.2–10.4 cm broad, curved, and rarely straight, with transient hairs. Each pod contains about 10–20 seeds. The seeds vary in shape from rectangular to round with a deep groove between the radical and cotyledons with dimensions of 3.5–6 mm in length and 2.5–4 mm in width. The seeds are light grayish, brown, olive green, or cinnamon color, depending upon the type and variety, with a pronounced radical that is half the length of the cotyledons (Figures 24.3 and 24.4). The minute hilum lies partly obscured with a deep notch (Petropoulos 1973; Basu 2006).

Seeds are surrounded by the seed coat, which is separated from the embryo by a well developed endosperm, the principal storage organ. The majority of the endospermic cells are nonliving in nature seeds and the cytoplasmic contents are occluded by the store reserves, called galactomannan (Bewley et al. 1993), which support seedling growth. This tissue is surrounded by a one-cell layer of living tissue—the aleuronic layer in which the cells are small and thick walled and contain aleurone grains that disappear during the course of seed germination (Reid and Meier 1972).

Figure 24.3   Seeds of *Trigonella foenum-graecum* (variety NRCSS AM 1).

Figure 24.4   Seeds of *Trigonella corniculata* (variety Pusa Kasuri).

24.3.5 Seed Development

After fertilization, a series of changes takes place in the ovule and, as a result, the seed is formed. The seed development of fenugreek to full-size, well-filled grains takes place about 120 days after anthesis. The role of galactomannan in endosperm starts 30 days after anthesis and ends at about 55 days after anthesis; at this stage the fresh weight of seed starts decreasing. The galactomannan supports the seedling growth during germination and also regulates the water balance of the embryo (Campbell and Reid 1982).

The galactomannan has its use in food, pharmaceuticals, cosmetics, and paper products. Thus, study of biochemistry of synthesis and mobilization has attracted the attention of many researchers (Rosser 1985; Scherbukhin and Anulov 1990); Petropolous (2002) has reviewed the results. Among the eight *Trigonella* species studied, all have a mannose-to-galactose ratio of approximately 1:1; only *T. erata* has a ratio of 1.6:1 (Reid and Meier 1970). Galactomannan biosynthesis has been studied using cell-free extracts and whole endosperm tissue (Edwards et al. 1989, 1992). Through genetic engineering, it is possible to transform fenugreek with galactomannan having a required ratio of mannose to galactose (i.e., 4:1), which is suitable for use in industry.

24.4 CYTOGENETICS

24.4.1 Chromosomes: Description and Number

Trigonella foenum-graecum, the most common species of fenugreek cultivated largely the world over, is a diploid with $2n = 16$. The basic chromosome number of the genus *Trigonella* is $x = 8, 9,$ 11, and 14 (Darlington and Wylie 1945). The majority of species reported are diploid ($2n = 2x = 16$).

Table 24.5   Chromosome Numbers (2n) for Different Species of Trigonella

Trigonella Species	Chromosome Numbers (2n)	Ref.
T. foenum-graecum	16	Srivastava and Raghuvanshi (1987)
T. balansae	16	Dundas et al. (2006)
T. corniculata	16	Singh and Roy (1970)
T. sprunerana Boiss., T. monspeliaca L., T. uncata, T. anguina, T. stellata, and T. astroites	16	Reasat et al. (2002)
T. gladiata, T. cariensis, T. berythea, T. macrorrhyncha, T. cassia, and T. foenum-graecum	16	Ladizinsky and Vosa (1986)
T. homosa (Egypt)	16, 44	Darlington and Wylie (1945)
T. ornithpodioides (Europe)	18	
T. polycerata (Mediterranean and Southeast Asia)	28, 30, or 32	Petropoulos (2002)
T. geminiflora (Persia)	44	
T. grandiflora (Turkestan)	44	

However, *T. hamosa* from Egypt contains $2n = 16$ and 44 chromosomes, *T. ornithoides* from Europe contains $2n = 18$, and *T. polycerata* from the Mediterranean and southwest Asia contains $2n = 28$, 30, and 32. *Trigonella geminiflora* from Persia in Asia Minor and *T. grandiflora* from Turkestan contain $2n = 44$ chromosomes (Petropoulos 2002).

Cytogenetical studies conducted by different workers revealed variation in chromosome numbers in different species of *Trigonella* (Table 24.5). The karyotype study involving 13 species of *Trigonella* suggested that all, except *T. neoana* ($2n = 30$), have $2n = 16$ chromosomes (Singh A. and Singh D. 1976). The cytological studies suggest that some of the *Trigonella* species have undergone several rounds of chromosome duplication that occurred followed by diploidization with gene and chromosome elimination. Marc and Capararu (2008) studied cytogenetic effects induced by a sodium phosphate food additive in meristematic cells of fenugreek root tips and found that an increase of food additive decreases the mitotic index, while the frequency and type of chromosome aberrations are much higher in treated variants.

Variation in chromosome number has been reported in fenugreek. An extra B chromosome in some fenugreek lines has been reported (Raghuvanshi and Joshi 1964; Joshi and Raghuvanshi 1968). Presence of the B chromosome is known to modify the growth of plants (Petropoulos 2002). Lakshmi, Rao, and Rao Venkateswara (1984) studied the occurrence and cytological behavior of B chromosomes in *T. corniculata* and reported two types of pollen mother cells (PMCs): one with $2n = 16$ and the other with $2n = 16+$ B chromosomes. On the other hand, Singh (1973) and A. Singh and D. Singh (1976) examined *T. corniculata* and did not observe B chromosomes. It is possible that the B chromosome may be absent in *T. corniculata* growing in certain regions of India. Srivastava and Raghuvanshi (1987) described that the B chromosome acts as a buffering agent, neutralizing the variability affect of different soil types. Das et al. (2000) observed significant variation in chromosome length, volume, and total form percentage at the cultivar level. Variations in genomic structure and the nuclear DNA content of fenugreek cultivars, despite the same somatic chromosome number, suggest a genetic drift among the cultivars.

24.4.2 Interspecific Relationships

The genus *Trigonella* is considered to have the primary basic number of $x = 8$. Karyotypes of *T. procumbens* and *T. gladiata* are considered to be primitive and those of *T. foenum-graecum* and *T. neoana* are considered to be highly specialized; other species fall between these two extreme groups. Karyotype specialization has been achieved by the shift of centromere from the median to submedian or subterminal positions in *T. radiate*, *T. rigida*, and *T. corniculata*. This increases in the differences between the shortest and the longest chromosomes in the complement of *T. arabica* and *T. coerulescens* and by both ways in *T. foenum-graecum*, *T. neoana*, and *T. colliceras* (A. Singh and D. Singh 1976). During speciation of *Trigonella*, polyploidy has played a major role that has also conferred wider adaptability to *T. polycerata* and *T. neoana*.

Ladizinsky and Vosa (1986) studied six species of the section *foenum-graecum* of *Trigonella* and found 2*n* = 16. *Trigonella gladiata* and *T. cariensis* have fairly symmetrical karyotypes, while *T. foenum-graecum*, *T. ferythea*, *T. macrorrhyncha*, and *T. cassia* have asymmetrical karyotypes. C-bands were present in all six species but number of bands and position varied considerably among species. Karyotype evidence suggests that none of the available species can be considered as the wild progenitor of fenugreek. Dundas, Nair, and Verlin (2006) reported that *T. balansae* that contains eight chromosomes (the haploid complement). The satellite chromosome and possibly a subtelocentric chromosome were identifiable. The remaining six chromosomes were not identified by Karyotype analysis (Figure 24.5A and B).

The mitotic chromosome number (2*n* = 16) of an accession from Nisyros, on the Aegean Sea (*T. balansae* Boiss. & Reuter), an annual pasture legume of Eurasian origin, was first reported by Kamari and Papatson (1973). Contrary to some studies reporting one pair of satellite chromosomes in fenugreek (Wanjari 1976; Agarwal and Gupta 1983), two pairs of satellite chromosomes at metaphase (Figure 24.6a) and prometaphase (Figure 24.6b) have been reported. This has been confirmed in 23 accessions of diverse origin by Ahmad et al. (1999). Singh and Roy (1970) and Singh A. and Singh D. (1976) reported that *T. corniculata* contains 2*n* = 16. The chromosome length ranged from 34.8 to 29.74 µm. Reasat et al. (2002) reported that eight species (*T. elliptica* [perennial type], *T. foenum-graecum*, *T. spruneriana*, *T. monspeliaca*, *T. uncata*, *T. anguina*, *T. stellata*, and *T. astroites* [annuals]) were diploid with 2*n* = 16. They carried metacentric and submetacentric chromosomes. *Trigonella foenum-graecum* and *T. stellata* had the longest and shortest chromosomes of 26.28 and 13.52 µm, respectively.

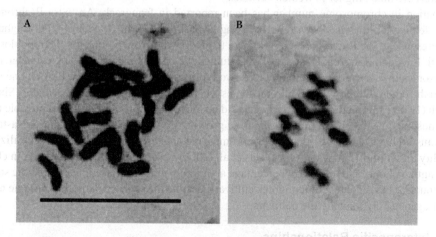

Figure 24.5 *Trigonella balansae* chromosome spreads showing (A) mitotic cell with 16 chromosomes; (B) meiotic metaphase I cell with eight bivalent chromosomes. Both photographs are of the same magnification. Bar represents 10 µm. (Source: Dundas, I. S. et al. 2006. *New Zealand Journal of Agricultural Research* 49:55–58.)

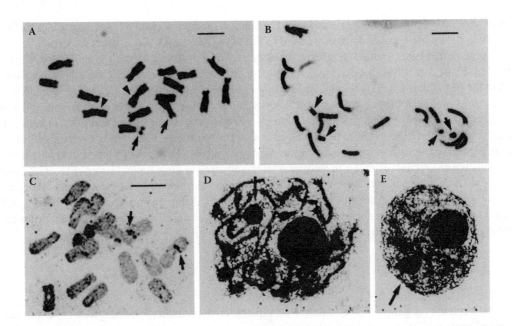

Figure 24.6   (A, B) Feulgen-stained and (C, D, E) silver-stained mitotic cells of fenugreek. (A) Metaphase cell showing morphology of the 16 chromosomes. Note the intercalary location of secondary constrictions on the metacentric chromosome pair (arrowheads) and the near terminal location on the acrocentric chromosome pair (arrows). (B) Prometaphase cell clearly showing four satellite (arrows) chromosomes. (C) Silver-stained metaphase cell showing dark transcriptionally active NOR bands on two pairs of satellite chromosomes. The NOR site on the metacentric chromosome (arrows) has less activity than the one on the acrocentric chromosome. (D) Prometaphase and (E) interphase nucleus showing two different sizes of nucleoli. The smaller one is marked by an arrow bar: 5 u. (Source: Ahmad, F. et al. 1999. *Theoretical and Applied Genetics* 98:179–185.)

24.4.3 Polyploidy

A few euploid and aneuploid cytogenetic stocks, primarily in the form of tetraploids, triploids, and simple primary trisomics, have been reported in fenugreek. Natural and chemically induced autoploidy is tolerated within the species. D. Singh and A. Singh (1976) isolated five double trisomics ($2n + 1 + 1 = 18$) and primary trisomics ($2n + 1$) from the progeny of autotetraploids of *Trigonella*. Roy and Singh (1968) produced autotetraploids by treating fenugreek shoot apices with colchicine. Increased ploidy levels often result in larger seeds and may contain higher levels of chemical constituents and other valuable morphometric traits not present in genotypes with smaller seeds. Gopinath (1974) isolated autotriploidy in excised root cultures of fenugreek. Autopolyploidy has been isolated from natural populations and induced by ethyl methane sulfonate (EMS) (Petropoulos 2002).

Lakshmi et al. (1984) have suggested the evolutionary trend within this genus from large to small chromosome and from symmetrical to asymmetrical karyotype; polyploidy played no part. They reported that fenugreek strain CS960, which was unique in having subterminal and submedian primary constriction (Kinetochore) on all chromosomes, was the most asymmetrical of the karyotypes studied.

Improved seed set and an increase in the frequency of bivalents in the C_7 and C_8 generation of six autotetraploid strains compared to corresponding diploid was observed by Gopal, Singh, and Gopal (1979). Genotype was found to be more important than ploidy for determining pollen and seed fertility. Singh, Singh, and Raghuvanshi (1986) induced autotetraploids by colchicine and selected over 12 generations (C_1–C_{12}). Seed sterility proved to be a problem. Number of pods per plant and seeds per pod improved sharply in C_4; however, seeds per plant did not improve in further generations despite continuous improvement in vegetative traits and regularity of meiosis.

Arya et al. (1988) studied synthesized autotetraploid fenugreek for four generations (C_0, C_1, C_2, C_3). Plant growth was reduced in C_0 plants, but improved in subsequent generations. Bivalents were the most common chromosome association in all generations and the frequency of quadrivalents was low. The decrease in chiasma frequency and the unequal distribution of chromosomes at anaphase I in C_0 came down substantially ameliorated in succeeding generations. Kesarwani and Raghuwanshi (1988) reported a significant decrease in chiasmata in tetraploid over diploid. Ravi (1987) observed significant increase of chloroplasts per guard cell, guard cell length and width, and nuclear and chloroplast diameter in autotetraploids compared to three diploid strains (T8, IC 74, and IC 7496). Singh, Singh, and Raghuvanshi (1992) reported that fertility in autotetraploid fenugreek is controlled by certain genetic-physiological factors. Basu, Acharya, and Thomas (2007) selected tetraploid genotypes from survivors and for increased vigor and forage and seed yield from the population of colchicines-treated fenugreek plants (Tristar cultivar).

24.5 GENETICS

Information on genetics of qualitative and quantitative characters in fenugreek is scanty and limited to variability, heritability, and genetic advance.

24.5.1 Genetics of Qualitative Characters

Bakshi and Hameed (1971) evaluated fenugreek accessions from Algeria, Morocco, Ethiopia, and India for diosgenin content and reported 0.35, 0.25, 0.20, and 0.10%, respectively. P. Makai, S. Makai, and Kismanyoky (2004) evaluated 12 cultivars from Syria, Spain, France, Libya, India, and Hungary and recorded significant difference in diosgenin content among cultivars; high-yielding cultivars were Qvari Gold from Hungary, Ghahkamon from Libya, and Metha from India. Similarly, Taylor et al. (2002) observed 0.98% variability for diosgenin content.

Ten accessions of fenugreek were assessed to determine the influence of genetic and environmental factors on the level of diosgenin. Analysis of variance on combined diosgenin levels from the three sites after 2 years revealed that accession, accession × year, and site × year effects were significant for diosgenin content, whereas site, year, and site × year effects were not observed (Taylor et al. 2002).

24.5.2 Association Analysis

Singh, Lodhi, and Arora (1993) carried out association analysis in 32 genotypes of exotic and Indian origin fenugreek under two dates of sowing. Correlation and path analysis indicated that seed yield could be improved by selecting for greater plant height, more tertiary branches, longer pods, more seeds per pod, and greater 100-seed weight. Seed yield per plant is positively correlated with plant height (Mahey, Raje, and Singhania 2003), number of primary branches (Banerjee and Kole 2004), number of tertiary branches, number of pods per plant (Raje, Singhania, and Singh 2003; Ayanoglu, Arslan and Mert 2004), and pod length (number of seeds per pod [test weight]).

Path coefficient analysis indicated that seed yield per plant has positive association with plant height (Mahey et al. 2003), number of branches per plant, number of pods per plant (Datta, Chatterjee, and Mukherjee 2005; Ayanoglu, Arslan, and Mert 2004), pod length (Mahey et al. 2003), number of seeds per pod, and seed test weight. Avtar, Jatasra, and Jhorar (2003) studied the association of biochemical parameters with powdery mildew severity in fenugreek and reported that polyphenol oxidase and peroxidase were positively and significantly associated with each other traits. A positive and significant relationship was also observed between catalase activity, total chlorophyll, chlorophyll a, chlorophyll b, carotenoids, total phenols, and orthodihydric phenols. According to Kumar and Choudhary (2003), yield was positively and significantly correlated with plant height, length, number of seeds per pod, and yield per plant.

Path analysis indicated that pod length, seeds per pod, and protein content exerted positive direct effect on as well as positive correlation with seed yield (Kaushik 2002). Results of genotypic path analysis revealed high positive direct effects of branches per plant, straw yield, pod length, and pods per plant on grain yield (Saha and Kole 2001). Sastry, Kumhar, and Singh (2000) recorded a significant positive association of seed yield with days to maturity, number of branches per plant, number of pods per plant, and biological yield per plant. According to Chandra, Sastry, and Singh (2000), based on correlations and path coefficient analysis, the number of pods per plant, test weight, and plant height were the most important traits that directly or indirectly influence seed yield. Path coefficient analysis indicated direct positive effects of straw yield, test weight, grains per pod, and pods per plant on grain yield per plant (Dash and Kole 2000). Path analysis revealed that plant height, seeds per pod, and 1,000 seeds per weight were the most important characters contributing to seed yield (Berwal et al. 1996).

24.6 GERMPLASM RESOURCES

The germplasm pool has a great significance in breeding because it will serve as a gene complex reservoir for all characters, particularly for yield, biotic and abiotic stresses, and quality. Good collections of fenugreek are maintained in Turkey (Oltraco and Sabanci 1996), Australia (McCormick et al. 2000), Canada (Basu et al. 2004; Acharya et al. 2006), India (Malhotra 2010), and Russia (Provorov et al. 1996). One of the important factors restricting large-scale production of fenugreek and development of better varieties is that little information is available about genetic diversity, intra- and interspecific variability, and genetic relationship among the species. This is essential for crop improvement.

The exotic material is expected to possess several promising traits such as yield, quality, and resistance/tolerance to biotic and abiotic stresses, which could be used for crop improvement programs. Crop resources are reservoirs of irreplaceable genes and gene complexes that are essential for the progressive improvement of crop plants. With a view to assembling the variability, collections made by the researchers in various countries such as India, Canada, Australia, and Russia are available. Other countries maintaining fenugreek germplasm are Iran, Pakistan, Bangladesh, Syria, Spain, France, Libya, Hungary, Turkey, Egypt, and Poland. The germplasm collections of fenugreek are being maintained at Plant Genetic Resources Collection Center, Saskatoon, Canada; USDA–ARS Plant Introduction Station, Washington, D.C.; National Bureau of Plant Genetic Resources (NBPGR), New Delhi, India; National Research Center on Seed Spices, Ajmer, Rajasthan, India; All-Union Research Institute for Plant Breeding, Leningrad, Russia; Department of Field Crops, University of Ankara, Turkey; Division of Genetic Resources, Institute of Biotechnology, Monastir, Tunisia; and Center for Crop Improvement, Longerenong College, University of Melbourne, Horsham, Victoria, Australia.

The germplasm of fenugreek currently has not been subjected to intensive selection through modern breeding programs and still expresses a high level of variability among genotypes present in world collections (Acharya, Blade, and Moyer 2007; Dangi et al. 2004; Moyer et al. 2003; Huang and Liang 2000; Taylor et al. 1997). World accessions of fenugreek from 20 countries have been evaluated under rain-fed and irrigated field conditions to select for early maturing and high-yielding lines with potential for use in western Canada (Basu et al. 2007). A wide variability in forage and seed yield was observed for the 73 world accessions grown under irrigation and rain-fed conditions in western Canada. The accessions X 92-23-3 (under irrigation) and L3308 (under rain-fed

Figure 24.7   An early flowering line of fenugreek (*Trigonella foenum-graecum* L.).

conditions) were among the top five high yielder genotypes, whereas PI 143504 and L3312 were consistent with their high seed yield under both irrigation and rain-fed conditions. The standard check cultivar Tristar was not among the top five producers under rain-fed or irrigated conditions (Basu 2006).

The germplasm of fenugreek including varieties grown in western Canada does not exhibit great variations in flower color or morphology and does not show variability in seed and forage yield across years and differing environmental conditions (Basu et al. 2004; Acharya et al. 2008). The variability among the collections has been observed for earliness (Figure 24.7) and also shape, size, and color of seeds (Malhotra 2011a) (Figure 24.8).

Evaluation of 132 accessions of fenugreek germplasm revealed significant variability for days to flowering, days to maturity, plant height, branches per plant, pods per plant, seed and straw yield per plant (Sharma and Sastry 2001). McCormick et al. (2000) evaluated 200 accessions of fenugreek collected from 20 countries and observed significant variation in growth habit, biomass, maturity, and seed color, size, and yield. Seventy-five accessions outyielded 150% or more of check line yield and also for relative nitrogen fixation efficacy (McCormick, Norton, and Eagles 1998; McCormick et al. 2000). McCormick et al. (1998) evaluated 207 accessions of fenugreek from 13 countries, and the range values of minimum and maximum for days to first flowering, flower duration, growth habit, and seed yield are given in Table 24.6.

Acharya, Thomas, and Basu (2007) observed considerable variability among genotypes collected from different countries, which differed in morphology, growth habit, biomass, and seed production. The genotypes differed in chemical constituents of the seed (e.g., saponins, fiber, protein, amino acids, and fatty acids). International varieties varied considerably in the chemical composition of seed—notably, the diosgenin content (Taylor et al. 2002). These observations suggest that evolution in fenugreek has generated variation in phenotype of the plants, which can be used in selective breeding (Acharya et al. 2006). The content and types of sapogenins vary across the species of fenugreek

Figure 24.8   Variability in seed size, shape, and color in fenugreek seeds.

Table 24.6   Characteristics of Higher Yielding Lines Compared to the Check Line

Origin	Number of Accessions	Days from Sowing to First Flower		Flowering Duration (days)		Growth Habit Score		Seed Yield (kg/ha)	
		Range	Mean	Range	Mean	Range	Mean	Range	Mean
All	207	98–147	111	14–52	40	1-5	3	0–3487	1613
Check line	1	—	110	—	39	—	4	—	1278
Afghanistan	11	113–129	122	27–40	36	1–3	1	721–2068	1524
Algeria	3	112–114	113	35–52	39	1–3	2	236–1288	621
Egypt	22	102–110	106	38–51	44	2–5	4	377–2688	1274
Ethiopia	16	103–125	109	37–47	42	1–4	3	200–2228	1300
India	30	103–112	108	33–44	39	1–4	3	467–3003	1873
Iran	33	98–123	114	33–50	41	1–4	3	814–3483	1788
Morocco	5	105–110	108	36–44	41	3–3	3	1353–3279	2484
Oman	13	106–113	110	34–45	40	3–5	4	475–1683	1025
Pakistan	23	103–125	110	31–45	38	1–4	2	575–2619	1796
Spain	4	112–147	122	14–40	32	1–2	2	2–2369	1469
Syria	3	103–110	106	38–43	41	2–3	3	1357–3122	1967
Turkey	19	109–120	111	35–47	40	1–3	3	233–2903	2075
USSR	9	113–133	118	20–44	41	1–4	2	221–1424	697

Source: McCormick, K. M. et al. 1998. In *Proceedings of 9th Australian Agro Conference*, 1998. Wagga Wagga, Australia.

Note: Check line was supplied by Revell Seeds Dimboola and is the dominant line grown commercially in Victoria.

as well the accessions of different types of cultivated forms found in different parts of the world (Skaltsa 2002).

Toshniwal (1984) observed that, in fenugreek, the effects of the environment and its interactions with genotypes were significant with respect to all the morphophysiological characters as well as seed yield. He suggested that the evaluation of the germplasm must be carried out over different environments to ascertain the real potential of germplasm entries.

In India, a large collection of fenugreek accessions has been made under the All India Coordinated Spices Improvement Project. The accessions are evaluated and utilized by the different coordinated centers spread throughout the country in different agroclimatic regions. One important weakness of fenugreek collection is that most germplasm expeditions have been made at random and by the breeders, whose prime objective was to develop improved varieties (Malhotra 2010). Bias for superior types might have occurred in such collections. Germplasm evaluation and utilization in crop improvement have developed many new varieties of fenugreek in different countries (Table 24.7).

Marzougui et al. (2007) evaluated the 38 Tunisian cultivars of fenugreek based on vegetative and reproductive morphological and chemical markers. The results were found significant to all morphological characters except for the flowers' standard color. Hierarchical classification based on

Table 24.7 Fenugreek Varieties Reported from Different Countries

Fenugreek Varieties	Country	Text citation ref.
Tristar, Amber, Quatro, Canagreen, Canafen	Western Canada	Acharya et al. (2006), Mir et al. (1997) Slinkard et al. (2006)
Giza 3	Egypt	Ahmed et al. (2000)
Co1, Co2, Rajendra Kanti, Lam sel. 1, RMt 1, RMt 303, RMt 143, Hisar Sonali, Guj. Methi-1, Rajendra Khushba, Hisar Suvarna, Hisar Mukta, Hisar Madhvi, RMt 305, GM2, GM351, RMt 361, NRCSS AM1, NRCSS AM2, LFC 84, Pusa Early Bunching, Pusa Kasuri	India (varieties are for different agroclimatic regions)	Malhotra (2010)
Dilba	Yemen	Ahmed et al. (2000)
19 X, H 26, D 19	Syria	
Blidot, Ciadoncha, Obanos	Spain	
Gers	France	
Ghahkamon	Libya	Makai and Makai (2004)
Ovari 4, Ovari Gold	Hungary	
Hebar, Metha	India	
Margaret, Ionia, Barbara, Paul	Slovenia	Zupancic et al. (2001)
Ionia	Greece	Vaitsis (1985)
Barbara, Margaret, Paul	UK	Hardman et al. (1980)
Gouta	France	Haefele, Bonfils, and Sauvaire (1997)
Sidi Khiar II, Kssar, Douar Lehmailia I, Tel Elgozlan II, Douar Lehmailia II, Sidi Khiar I, Borj Berrzig, Tel Gozlan II, Ben Bechir, Sidi Hamed, Beja I, Beja II, Beja III, Beja IV, Nefza I, Nefza II, Nefza III, Mater I, Mater II, Mater III, Mater IV, Mater V, Mhamdia I, Mhamdia II, Mhamdia III, Bizerte, Manzel Tmim I, Manzel Tmim II, Manzel Tmim III, Manzel Tmim IV, Manzel Tim V, Sidi Maaouia, Zaouia, Merguaz, Dakhla I, Dakhla II, Dakhla III, Manzel Hbib I, Manzel Hbib II	Tunisia (varieties are for different regions and altitudes within the country)	Marzougui et al. (2007)
Nakhichevanskaya Shambala	Russia	Provorov et al. (1996)
150,000	Australia	McCormick et al. (2000)
Tfg-15-PN, Tfg-18-PN, De Muntenia	Romania	Floria and Ichim (2006)
Sel. no. 3 and Sel. no. 18	Turkey	Khawar, Gulbitti Onarici and Cocu et al. (2002)

morphological characters and on the set of leaf content of K, Na, P, Ca, Mg, and Fe did not succeed to the same grouping for 53% of the studied cultivars, whereas 47% of these cultivars were grouped in the same way as well as in the grouping based on the morphological and chemical data.

Diversity analysis based on randomly amplified polymorphic DNA (RAPD) markers of 17 accessions of *T. foenum-graecum* from Afghanistan, Canada, Egypt, Ethiopia, India, Nepal, Pakistan, Turkey, and Yemen and 9 accessions of *T. caerulea* from Turkey, Spain, Australia, and some unknown sources was conducted (Dangi et al. 2004). The *T. foenum-graecum* accessions from Pakistan and Afghanistan were grouped in one cluster and accessions from India and Nepal were grouped in another. Accessions from Turkey did not group and fell into different clusters. Provorov and Tikhonovich (2004) evaluated germplasm and found that wild-growing populations and local varieties exceed the agronomically advanced cultivars in the activity of N_2 fixation that occurs in symbiosis with nodule bacteria.

24.7 GERMPLASM ENHANCEMENT

24.7.1 Breeding Objectives

The objectives vary with specific situations and with production areas but generally are for high seed yield, saponins, diosgenin, adaptation to different climates, and resistance in fenugreek. The four basic objectives identified are production breeding, breeding for resistance to biotic and abiotic stresses, breeding for agronomic and physiological characters, and high nutraceutical value (Table 24.8).

24.7.1.1 Production Breeding

So far, the ultimate objective of the fenugreek breeding program has been is to develop varieties that produce higher seed yield or forage yield. The selection, based on determinants such as number of branches, number of pods per plant, number of seeds per pod, and test weight, may result in identification of promising genotypes. The high-yielding genotypes could be developed by combining suitable morphophysiological characters. A yield potential of 2.5–3.5 tonnes per hectare has been reported for fenugreek and only 60% of the genetic potential at present is harvested. The genetic potential of the cultivated varieties is limited due to a lack of one or more of the attributes. In addition to seed yield in fenugreek, there is demand for varieties exclusively for forage or green production. Because it is a single-cut fodder crop, the dual-purpose objectives cannot be served because it is difficult to harvest for economic yield of both seed and forage.

Table 24.8 Breeding Objectives and Selection Criteria

Breeding Objectives	Selection Criteria
Production breeding	Seed yield, forage yield, maturity (early, mid, and late maturity), large seeded or small seeded, green seeded or yellow seeded
Resistance breeding	Resistance to root rot, powdery mildew, downy mildew, fusarium wilt, aphids, mites, leaf minor
	Tolerance to drought, frost, salinity, or any other microlocation specificity or for suitability to climate change
Breeding for agronomic/ physiological characters	Determinant type, indeterminate type, high photosynthetic efficiency, shattering resistance, efficient nutrient and water use, suitability for intercropping, high biological N fixation
Breeding for special use (pharmaceutical purposes, etc.)	High bitter type, low bitter type, high steroidal sapogenins (diosgenin, tigogenin), amino acids (hydroxyl-isoleusine), polysaccahrides (galactamannan), trigonelline, high protein and high essential oil

Source: Malhotra, S. K. 2003. In *Breeding field crops*, ed. A. K. Sharma. Bikaner: Academic Publishers.

The fenugreek varietal development program involves suitability of maturity period for a certain environment or zone for specific adaptation. Late, early, and medium varieties are required for full seasons, as well as late planting, and to fit into multiple cropping systems depending upon the situation. Desirable variability does not exist for this trait in the germplasm to the elite background for transferring it. Breeding for greater stability and wider adaptability coupled with high yield is of major consideration in the development of new cultivars. One of the breeding objectives based on production breeding is large size and color of seed. On the basis of the author's discussion with exporters, it is revealed that fenugreek seed markets also have demand for small seed size. The colors of seed also have varied demand in the market; most of the demand is for yellow seeds, although demand for green seeds is increasing in the pharmaceutical industry.

24.7.1.2 Resistance Breeding

The fenugreek crop suffers from several diseases and pests. The disease spectra vary in different agroclimatic zones in different countries. In the early years of fenugreek cultivation, diseases were less a problem, but with the introduction of crops in new areas, several diseases and pests are now problems (Table 24.9) and resistance breeding is required. Root rot, collar rot, fusarium wilt, and their complex of pathogens are serious diseases and cause 20–70% yield loss in India. Precise screening techniques are needed to isolate sources of resistance. The resistance has not been so far successfully transferred to commercial cultivars. The other two important diseases are powdery mildew and downy mildew, which are known to cause losses to yield and quality of fenugreek. Screening for resistance should normally be carried out in the field, and laboratory or glasshouse procedures are followed only as a supplement to field screening. The screening method should be simple, practical, and efficient. There is an urgent need to develop varieties resistant to such diseases.

Among insects, the aphid is a serious pest, causing severe losses and deteriorating the quality of seeds. The progress in breeding for insect resistance in comparison to disease resistance is negligible due to lack of effective insect-rearing methods, germplasm screening technologies, and resistant sources. Concerted efforts toward incorporating disease- and pest-resistant genes in adapted varieties are required.

Table 24.9   Major Diseases and Pests of Fenugreek

Name of Disease	Causal Organism
Leaf spot	Cercospora traversiana, Ascochyta sp.
Powdery mildew	Erisiphe polygoni, Leveillula taurica, Oidiopsis
Downy mildew	Peronospora trigonellae
Fusarium wilt	Fusarium oxysporum
Collar rot	Rhizoctonia solani
Root rot	Alternaria alternate
Bacterial leaf spot	Pseudomonas syringae pv. syringae
Xanthomonas	Xanthomonas alfalfa
Viral diseases	Bean yellow mosaic virus, alfa mosaic virus, cowpea mosaic virus, soy bean mosaic virus, pea mosaic virus, potato virus A and Y, clover vein mosaic virus
Root knot nematode	Meloidogyne incognita
Aphids	Aphis cracivora, Myzus persicae
Leaf minor	Liriomyza trifolii

Source: Malhotra, S. K. 2010. Souvenir, National Conference on Horticulture Biodiversity, Livelihood, Economic Development and Health Care, University of Horticultural Sciences Campus, Bangalore, May 28–31, 2010, 83–97.

Adverse weather conditions in the form of drought or heat ultimately reduce the seed yield. The stresses have a major influence on morphological, physiological, and biochemical processes. Higher root density, turgidity, frequent stomata closure, stability in nitrate reductase activity, protein synthesis, proline accumulation, and decrease in abscisic acid are the desirable traits. Genetic variability for various characters contributing to resistance for these stresses has been observed and exploited. In the growing region, occurrence of frost during winter has been reported to cause damage to crops. Therefore, to obtain consistently good yields in these conditions, the development of a variety that is hardy in low temperatures is essential. Fenugreek is sensitive to salinity problems in the area of cultivation, but suitably tolerant cultivars can be developed through effective breeding procedures.

24.7.1.3 Breeding for Agronomic and Physiological Characters

The most efficient plant type is that which can utilize soil nutrients and solar energy to the maximum extent and can divert most of the photosynthates to the economic part—that is, the grains. Most of the varieties under cultivation are indeterminate types and there is no uniformity in maturity of seeds in pods. But indeterminate types are good for high forage yield. For seed types, it is advisable to breed cultivars with determinate growth habit which assimilates to plant parts for more economic yield (seeds). For higher production setting and better grain filling, there is a need to develop varieties that remain photosynthetically active up to flowering and fruiting stages.

Biological efficiency of plants with favorable balance between the vegetables and reproductive development should also be given due consideration for selection programs in fenugreek breeding. Moreover, varieties with higher nutrient and water efficiency and biological nitrogen fixation will contribute to high yield of biomass and seeds. Fenugreek has a tendency to dehisce; this results in loss of grain before and during harvest. The genes of shattering resistance may be introduced into adapted varieties. The early maturity type will help the crop to fit into multiple cropping systems. The uniform maturity of the crop will help in harvesting at one time.

24.7.1.4 Breeding for Nutraceuticals/Special Use

Fenugreek seed is valued for a number of pharmaceutical properties, mainly hypoglycemic and hypocholesterolemic action. These properties are high due to the presence of diosgenin, trigonelline, galactomannans, fiber, protein, and bitter traits. Breeding of fenugreek thus involves developing cultivars with such contents for special use in health care. Just like other legumes, such as groundnut (25–34% protein), soybean (40% protein), and pea (23–40% protein), protein content in fenugreek is an important nutritional quality and has been reported by different workers to be 27% (Khawar, Cocu, et al. 2002), 36% (Slinkard et al. 2006), 31% (Mir et al. 1997), 26% (Sharma 1986), 38.6% (Pruthi 2001), and 28.5% (Gupta et al. 1989).

In a study at the Central Food Technology Research Institute (CCFTRI), in India, an analysis of 25 samples of fenugreek seeds drawn from different regions showed variations of 27.7–38.6% crude protein (Pruthi 2001). Acharya et al. (2006) reported variations in crude protein in five varieties from international collections and recorded high protein (31.6%) in varieties Amber and L 3314, which was higher in comparison to 24% observed in varieties from Egypt (Gerhartz 1987). Primarily, protein is the major nutritional content due to which fenugreek seed is a very good food and feed and is thus used for human and animal consumption. The previous research has indicated success of fenugreek in dairy (Shah and Mir 2004) and beef (Mir et al. 1997). The studies conducted by various workers in different locations indicate the possibility of improving this food character through breeding programs.

Fenugreek seeds are rich in steroidal sapogenins well known for medicinal uses. The prominent saponin compound isolated from fenugreek seeds that is used in the pharmaceutical industry as protective medicine for treatment of diabetes and cholesterolemia. Diosgenin is often used as a raw precursor for the production of steroidal drugs and hormones such as testosterone, glucocorticoids, and progesterone (Fazli and Hardman 1968; Raghuraman, Sharma, and Sivakumar 1994). The diosgenin content in fenugreek seed varies from 0.78 to 1.9% (Bochannon et al. 1974; Bhavsar, Kapadia, and Patel 1980; Sharma and Kamal 1982) depending upon genotypes as well as on cultural practices.

Provorov et al. (1996) succeeded in identification of promising genotypes with diosgenin content ranging from 1.45 to 1.64% in fenugreek seeds. Anis and Amimuddin (1985) found improvement in diosgenin content in seeds of induced autoploid *T. foenum-graecum*. Acharya et al. (2006) found variations in sapogenin content in five varieties tested: 0.4% in Amber, a Canadian variety and 0.5% in an Indian variety were significantly higher than other lines, such as F 70, F 86, and L 3314, which had about 0.3% sapogenin content in seeds. In this study, diosgenin was found to be the predominant steroidal sapogenin in all lines, but they were not significantly different from each other in the five lines tested. The same set of genotypes grown in three locations in western Canada by Taylor et al. (2002) in two consecutive years revealed genotype–environment interaction for diosgenin content in fenugreek varieties.

Acharya et al. (2006) reported that mutants showing signs of high proportion of double or twin pods are indicative of high diosgenin content in the seeds. They concluded with evidence of genotype–environment interaction for variation in nutraceutical compositions. Such interactions indicated that seed produced in all environments will not be of similar quality and all genotypes will not produce high-quality seed every year. To maintain high quality in nutraceutical products, it is necessary to produce stable cultivars showing the least amount of genotype–environment interaction and to continue maintaining quality every year.

An amino acid, 4-hydroxyisoleucine (4-HIL), and a polysaccharide in fenugreek are known to increase insulin secretion when a carbohydrate-rich diet is ingested (Sauvaire et al. 1996; Broca et al. 2000). However, 4-hydroxyisoleucine has been found to be the major free amino acid in the seeds of fenugreek (Sauvaire et al. 1984). The scientific reports for genetic control of amino acids and galactomannans are not available, but it is speculated that genotype as well as environment interaction greatly influences the quantity. Cultural practices and methodology of extraction play major roles in recovery of high content of isoleucine and amino acid or galactomannan, a polysaccharide.

Fiber-rich fraction of fenugreek seeds and its use as pharmaceutical excipient have been disclosed. The multifunctional fiber-rich fraction (FRF) and highly purified FRF are useful as excipients for pharmaceutical dosages forms for various routes of administration. These excipients can be used as binder, disintegrant, filler, dispersing agent, coating agent, film forming agent, thickener, and the like for preparation of a variety of dosage forms in food and cosmetic formulations (Pilgaonkar et al. 2005).

24.7.2 Breeding Methods

Fenugreek breeding has been mainly directed toward enhancing the yield of seeds and greens for vegetable and fodder purpose. The seed yield is directly related to the number of pods per plant and number of seeds per pod, and 1,000 grain weights are increased. For fodder purposes, the plant green yield is directly related to plant height and number of branches, and leaf sizes are improved. Most of the fenugreek varieties available are the result of selection. Breeders have developed the high-yielding and desirable genotypes of fenugreek through selection from the indigenous or exotic materials. Fenugreek is a highly self-pollinated crop, and generally two methods—pure line selection or single plant selection—and mass selection have been reported to be practiced largely for improvement in this crop.

Single plant selection is recognized as one of the best methods for breeding new fenugreek lines since the crop is predominantly self-pollinated and there is no inbreeding depression reported so far. This method has proven useful for selection of highly heritable traits such as seed size, color, growth habit, and number of seeds per pod (Del'Gaudio 1953; Green et al. 1981; Petropoulos 1973, 2002; Saleh 1996). Hisar Methi 350, a powdery-mildew-resistant and downy-mildew-tolerant variety in India, was developed through single plant selection from PLME 46-1 (Pratap, Rana, and Mehra 2002). The mass selection method of breeding is practiced for improvement of fenugreek varieties (Petropoulos 1973; Fehr 1993, Saleh 1996). It is expected that more than 100 varieties of fenugreek are available for cultivation in varied agroclimatic conditions of different countries such a Tunisia (40), India (23), western Canada (5), Slovenia (4), Syria (3), and a few others (Table 24.6). Of the 23 varieties developed in India, 20 are through selection and three varieties—RMt 303, RMt 305, and RMt 351—were through mutation breeding (Malhotra 2010). About 40 varieties in Tunisia have been developed for different regions and altitudes (Marzougui et al. 2007) and most of them are through selection.

Emasculation and manual pollination has been used effectively for crossing different lines of fenugreek, but the success rate is low because of its adaptation for self-pollination. Few of the fenugreek varieties have been developed by this method (Petropoulos 1973, 2002; Cornish, Hardman, and Sadler 1983; Provorov et al. 1996; Saleh 1996). Singh, Raghuvanshi, and Singh (1991) carried out hybridization followed by pedigree selection between two autotetraploid strains of fenugreek and described that this method is helpful in regularizing the meiotic behavior; in addition, vigor and seed fertility improved. There is enough scope of breeding for resistance to powdery mildew, Fusarium wilt, root rot, downy mildew, and aphids and high nutraceuticals through interspecific hybridization in fenugreek by using related species.

Mutation breeding has become increasingly popular in recent times as an effective tool for crop improvement. Some legume crops that have been improved through mutation breeding are soybeans, string beans, french beans (Sigurbjornsson and Micke 1974), navy peas, haricot beans (Sigurbjornsson 1983), peas, fenugreek (Petropoulos 1973, 2002; Acharya Thomas, and Basu 2007), and lupines (Gaul 1961). The majority of induced mutations in these plants are recessive and segregate in a 3:1 ratio (Gaul 1961; Petropoulos 1973; Singh and Singh 1974). An inheritance study by Raje, Singh, and Singhania (2003) concluded that normal plant type was controlled by a single dominant gene and giant mutant plant type by a recessive allele. Treatment of seed with chemical mutagens such as ethyl methane sulfonate (EMS) has the potential to cause chromosome damage, resulting in individual mutations as well as addition or loss of chromosomes from cells (Petropoulos 2002).

The higher concentrations of sodium azide (0.006 and 0.009%) in 2 and 4 h treatment resulted in low mitotic index value (Mendhulkar, Gupta, and Raut 2005). Koli and Ramakrishna (2002) reported that combined treatments of 40 krad gamma radiation and 0.045% EMS were effective in inducing chlorophyll mutation; six types of spectra of macromutations observed were tendrils, bunchy, creeper, giant leaf, sterile flower, and thin and elongated leaves. Choudhary and Singh (2001) isolated a determinate type mutant from a population of 400 plants of fenugreek in M_2 generation of EMS-treated (0.1%) seeds that have potential of improvement in seed yield. Pratap, Rana, and Mehra (2002) isolated an induced mutant line, IL-355-1, with resistance to downy mildew (*Peronospora trigonella*). Saini et al. (1998) isolated induced mutants 36, 38, and 42 that had significantly better symbiotic nitrogen fixation than Rmt-1.

Janardhan and Nizam (1995) found that a low dose of 5 krad for irradiation treatment of seeds was most effective in inducing chlorophyll mutations. Two- to fourfold increases in steroidal sapogenins (diosgenin and tigogenin) were observed in M_2 generation derived from seeds treated with 0.1 and 0.2 EMS, MMS, and sodium azide; however, higher concentration reduced sapogenin content (0.3 and 0.4 M) (Jain and Agarwal 1991).

Increasing radiation doses decreased seed germination and seedling survival and height and increased chromosome aberrations. In mature plants, increasing radiation doses delayed flowering, and maturity and reduced plant height and yield components. In another study, Choubey and Trar

Figure 24.9 A mutant with high pigmentation in *Trigonella foenum-graecum*.

(1990) used gamma rays at 10 and 20 krad and observed no effect on amino acid contents in 7-day-old seedlings, while irradiation at 60 and 70 krad markedly decreased the amino acid contents. Lakshmi and Datta (1989) observed unifoliolate to hexafoliolate types and even occasionally octa-foliate types. The natural mutation from fenugreek germplasm, anthocyanin pigment, and albino types has been reported (Malhotra 2010) (Figures 24.9 and 24.10).

Basu et al. (2007) succeeded in selection of determinate growth height, early maturity, and high seed yield mutants from generations of fenugreek seed treated with EMS. Floria and Ichim (2006) identified 11 mutant lines from M_2 to M_4 generations from seeds of fenugreek treated

Figure 24.10 An albino type natural mutant of *Trigonella foenum-graecum*.

with 25–15 krad gamma rays, EMS (011–0.5%, 2 h), or ethylene imine (0.01–0.1%, 2 h) with better yield potential and higher diosgenin content. Al-Rumaih and Al-Rumaih (2008) studied the influence of ionizing radiation on antioxidant enzymes in three species (*T. stellata*, *T. hamosa*, *T. anguina*) and found that three species differed in their radio sensitivity. *Trigonella stellata* was more radio sensitive than *T. hamosa* and *T. anguina*: It showed higher stimulation in ascorbate peroxidase (APOX), superoxide dismutase (SOD), and glutathione reductase (GR) activities under irradiation stress.

24.8 BIOTECHNOLOGY

Anther culture, ovule culture, micropropagation, *in vitro* selection, somaclonal variations, and genetic transformation are biotechnological methods that can be used in fenugreek breeding programs. These techniques offer new tools for opening up new research avenues to develop novel genotypes with unlimited potential to resolve various problems and are being employed for strengthening and supplementing the conventional fenugreek improvement programs.

24.8.1 Callus Culture and Micropropagation

The tissue culture technique can be used to create variation. Micropropagation, callus regeneration, and somatic embryogenesis have been reported in fenugreek (Provorov et al. 1996; Aasim et al. 2009; Azam and Biswas 1989). *In vitro* shoot regeneration was successfully achieved using apical meristem on Murashige and Skoog (1962) (MS) medium containing thidiazuron (TDZ) without indole-3-butyric acid (IBA) (Aasim et al. 2009). The medium containing 0.1 mg/L each of benzylaminopurine (BAP) and zeatin and addition of glutamine and asparagines was found essential for rapid cell division, callus growth, and differentiation (Shekhawat and Galston 1983). Callus culturing on MS medium supplemented with naphthaleneacetic acid (NAA); 2,4-D; kinetin; and coconut water was successfully achieved in fenugreek (Azam and Biswas 1989). Best growth was on MS medium containing 3% sucrose and 2 mg 2,4-D (El Bahr 1989). Multani (1981) used tissue culture in maintenance of diploid and autotetraploid strains of fenugreek. Fenugreek possesses limited variability for earliness and resistance to diseases and pests.

The frequency of somaclonal variation can be enhanced if coupled with *in vitro* mutagenesis and screening of population at large scale for resistance to biotic and abiotic stresses. The regeneration is also a little difficult and reproducibility is poor. These studies indicate that germplasm enhancement is possible using tissue culture techniques. However, this needs stringent selection procedures to be followed for the useful regenerates. Studies conducted by various workers on *in vitro* multiplication and plant regeneration in seed spices revealed that regeneration can be successfully induced from callus cultures has been viewed by Malhotra (2011b). Moreover, it can also quicken the domestication of uncultivated/wild plants and their development into commercial crops. Differentiation in callus of leaves of two species of *Trigonella* was reported by Sen and Gupta (1979). Regeneration of shoots has also been achieved from fenugreek protoplasts (Xu, Davey, and Cocking 1982, 1985).

Callus formation and shoot regeneration were achieved from hypocotyl explants of *T. foenumgraecum* using MS medium and described the important role of kinetin or NAA for proliferation and development of shoots from calli of fenugreek (Khawar, Cocu, et al. 2002; Madhuri et al. 1988; and Oncina et al. 2000). Callus was initiated callus from root stem and explants on MS medium supplemented with 0.5 mg/L 2,4-D and 10% coconut milk or 2,4-D and kinetin. Ortuno (2000) obtained callus on 10.4 mM GA$_3$ and 20 ppm BAP and 50 ppm ethephon. Ahmed et al. (2000) obtained callus proliferation from leaf, stem, and root explants of fenugreek variety Giza 3. No callus production was observed in the absence of 2,4-D. Root explants were better for callus induction.

Owing to poor success in hand pollination, somatic hybrids can be achieved easily between distantly related species. This may lead to progeny showing greater variability than is possible by sexual means. The fenugreek crop has not been sufficiently studied for fundamental aspects for which somatic hybridization offers a great scope for future studies. Androgenic haploids have been produced in many horticultural plants and used in crop improvement. Moreover, coupling of this technique with conventional breeding methods has resulted in the release of improved varieties of a large number of crop plants and offers scope for application in fenugreek for addressing complex problems of breeding for resistance to biotic and abiotic stresses.

Traditional breeding programs are costly and take years to achieve results; therefore, under such situations, double haploid breeding holds promise and particularly offers better scope for improvement in productivity and diosgenin and galactomannan content in fenugreek. Reports are available on *in vitro* selection for salt tolerance in fenugreek on media containing 0.025–1.5% NaCl (Settu, Ranjitha Kumari, and Jeya Mary 1997). Such *in vitro* methods of screening could prove highly useful in screening large germplasm collections or cell lines for resistance to prevalent fungal diseases and tolerance to drought and salt stress.

24.8.2 Production of Secondary Metabolites and Biotransformation

The grain spices crops are known for their peculiar spicy aroma due to presence of essential oils and phytochemicals, particularly secondary metabolites of medicinal and industrial importance. There has always been a need of accurate and rapid analytical methods for isolation and characterization of steroids in fenugreek. In recent years, plant cell suspension cultures and immobilized cells have been used for production of phytochemicals. Production of secondary metabolites through cell culture has the advantage over the extraction from whole plant or seeds. The yield and quality of the product are more consistent in all cultures because they are not influenced by the environment and the production schedule can be predicted and controlled in the laboratory or industry.

Bioproduction of diosgenin, the steroidal sapogenin, has been achieved successfully through callus cultures in fenugreek (Oncina et al. 2000). Steroidal sapogenin content was stimulated with treatment of physical and chemical mutagens (Jain and Agarwal 1994). The development of fenugreek calluses has been achieved after shoot or root culture from 4-day-old seedlings upon culturing on Gamborg's B-5 modified medium supplemented with hormones. Content of trigonelline was appreciably higher in cell suspensions than that found in calluses (Radwan and Kokate 1980). In fenugreek, the highest total alkaloid content was recorded in nonhabituated callus in explants (Sharon and Bhaskare 1998) and biosynthesis of rotenoid in callus tissues obtained from seeds (Kamal, Yadav, and Mehra 1997). Addition of 40 mg chitosan/L in media elevated diosgenin content three times compared to that found in nonelicited hairy roots of fenugreek (Merkli, Christen, and Kapetanidis 1997). Morpholines (tridemorph and fenpropimorph) and diniconazole were also reported as sterol biosynthesis inhibitors in fenugreek *in vitro* secondary metabolite production (Cerdon et al. 1995, 1996).

Galactomannan biosynthesis has been studied in fenugreek using cell-free extracts and whole endosperm tissue (Edwards et al. 1989, 1992). Identification of the mechanisms of fenugreek galactomannan biosynthesis (during seed development) and hydrolysis (during germination) in order to produce transformed fenugreek plants, where the ratio of galactomannan to mannan is appropriate for industrial use (Reid and Meier 1970; Li et al. 1980), opened the ways for further research. The biological model to study the biosynthetic pathway of sotolone in fenugreek through hairy root cultures (Peraza et al. 2001) and for trigonelline content of fenugreek calluses has been reported (Reda et al. 2000). Ahmed et al. (2000) determined that trigonelline content of leaves, stems, and roots *in vivo* was 0.45, 0.21, and 0.29 mg/g dry weight, respectively, compared with 0.61, 0.30, and 0.40 mg/g dry weight in callus derived from their tissues.

24.9 NUTRACEUTICAL CONSTITUENTS

The fenugreek plant and its products are known to have nutraceutical properties due to the presence of natural, bioactive chemical compounds that have health-promoting, disease-preventing, or medicinal properties of wide variety. Different types of medicinal properties in fenugreek have been reported by various workers and such properties are expressed due to the presence of many bioactive compounds in seeds and leaves. Broadly, such important bioactive compounds can be divided into four groups: natural steroidal sapogenins, polysaccharide galactomannans, hydroxyl-isoleucine, and antioxidants and other compounds (Table 24.10).

Table 24.10 Nutraceutical Compounds of Fenugreek

Major Group of Compounds	Constituent Compound	Ref.
Steroidal sapogenins		
Monohydroxy sapogenins	Diosgenin	Acharya et al. (2006) Shah and Mir (2004)
	Yamogenin, tigogenin, gitogenin, neotigogenin, smilagenin, and sarsasapogenin	Sauvaire et al. (1991) Taylor et al. (1997)
Dihydroxy sapogenins	Yuccagenin, gitogenin, and neotigogenin	Taylor et al. (1997)
Nitrogen compound	Trigonelline	Radwan and Kokate (1980)
Saponenins	Choline, six trigoneosides (based on furostaniol aglycones)	Yoshikawa et al. (1997) Varshney and Sood (1969)
Sapogenin peptide easter (pirostanol saponins)	Fenugreekine Graecunin B, C, D, E, and G	Ghosal et al. (1974) Varshney and Jain (1979)
Polysaccharides	Galactomannans	Raghuraman et al. (1994) Vats et al. (2003) Ramesh et al. (2001)
	Kaempferol glycosides lilyn	Han et al. (2001), Petropoulos (2002)
Amino acids	4-Hydroxyl-isoleucine	Sauvaire et al. (1996) Petropoulos (2002)
	Argine, alanine, glycine	Gopal and Singh (1978), Sharma (1984)
Lipids	Neutral lipids Glycolipids Phospholipids	Hemavathy and Prabhakar (1989)
Aroma compounds Sesquiterpenes	Pyrazines	Leela (2009)
	Cadinene, cadinol, eudesmol, bisabool, murolene, liguloxide, cubenol, murolol and epi-α-gluoulol	Girardon et al. (1985) Ahmadiani et al. (2004)
Sotolone	3-Hydroxy-4-5-dimethyl-2 (5H)-Furanone	Peraza et al. (2001)
	3-Amino-4, 5, dimethyl-2 (5H)-Furanone	Shang et al. (1998)
	Triterpenoids	
Total antioxidants and phenols	Total phenols Flavonoids Orthodihydroxy phenols Coumarines	Varshney and Sharma (1996) Chawla et al. (2004) Joshi et al. (2009)
Phenolic compounds	Scopoletin, chlorogenic, caeffic acids, p-coumaric acids, hymercromone, coumarin, and trigocoumarin	Petropoulos (2002)
Flavonoids	Kaemfrol, afroside, quercetin, isoquercitrin, vitexin, isivitexin, orientin, and luteolin	Huang and Liang (2000), Petropoulos (2002)
Isoflavonoid phytoalexins	Medicarpin, maackiaian, vestitol, and sativan	Petropoulos (2002)

24.9.1 Natural Steroidal Sapogenins

Fenugreek is an identified source of natural steroid sapogenin compounds and thus valued in the pharmaceutical industry due to inherent nutraceutical qualities. The diosgenin is the major steroidal compound reported in production of steroidal drugs and hormones. Fenugreek is primarily recognized for hypocholesterolemia, a disorder often associated with high cholesterol and hyperglycemia (McAnuff et al. 2002).

The sapogenins derived from fenugreek have antimicrobial properties that suppress acetate-producing bacteria but leave propionate-producing bacteria unchanged (Devant, Anglada, and Bach 2007). The other sources for preparation of natural diosgenin are yam (*Dioscorea* spp.), cluster bean (*Cymaposis tetragonoloba* L.), fenugreek, and a few other legumes (Rosser 1985; Mathur and Mathur 2006). The diosgenin production from yam is costly and takes years to wait for tuber growth to accumulate diosgenin to desired levels. Therefore, fenugreek, which is an annual, can prove to be a very good alternative source for production of diosgenin, a source of commercial and pharmaceutical reagents. Diosgenin is found at 0.5 dm in fenugreek seeds (Shah and Mir 2004). Mathur and Mathur (2006) have reported fenugreek steroidal sapogenin and diosgenin, which are in good demand for making the sex hormone, cortisone, used in making oral contraceptives. Due to the presence of these sapogenins, the fenugreek whole seed powder has been used to regulate blood sugar in healthy, obese, and non-insulin-dependent or type-2 diabetic persons.

Fenugreek seeds are a rich source of saponins such as diosgenin, yamogenin, gitogenin, tigogenin, and neotigogens (Taylor et al. 1997). The other bioactive constituents of fenugreek, including mucilage, volatile oils, and alkaloids (such as choline and trigonelline production from tissue culture of fenugreek), have also been studied (Radwan and Kokate 1980; Brain and Williams 1983). The seeds of fenugreek have pharmaceutical interest due to their content of monohydroxysapogenines (diosgenin, yamogenin), which are precursors used in steroid hormone semisynthesis (Anis and Amimuddin 1985; Elujoba and Hardman 1985; Bruneton 1995). The range of diosgenin recorded by Floria and Ichim (2006) was from 0.26 to 0.38 mg/g dry weight. The level of sapogenins in fenugreek seed ranged from 0.3 to 0.5%, whereas diosgenin ranged from 41 to 47.8% (Acharya et al. 2006).

Through analytical techniques based on gas chromatography coupled with mass spectrometry (GC-MS), 10 different sapogenins were detected by Brenac and Sauvaire (1996). The presence of a sapogenin peptide ester, fenugreekine, was reported by Ghosal et al. (1974). Six trigoneosides that were novel saponins based on furostaniol aglycones were isolated by Yoshikawa et al. (1997). The diosgenin content in fenugreek may increase the growth of cattle through its natural steroidal properties (Mir et al. 1997; Acharya et al. 2008). Decreased reliance on synthetic steroids for efficient cattle production could decrease input costs for beef producers (Acharya et al. 2006, 2008).

24.9.2 Polysaccharide Galactomannans

Galactomannans are the major polysaccharide type found in fenugreek seeds and represent more than 50% of the seed weight (Raghuraman et al. 1994). The galactomannans extracted from fenugreek seed may control type 2 diabetes by hyperglycemic action in animals (Tayyaba, Nazrul Hasnain, and Hasnain 2001; Vats, Yadav, and Grover 2003) and in humans (Raghuraman et al. 1994; Sharma et al. 1996). The fenugreek galactomannans have a high water-binding capacity and form highly viscous solutions at relatively low concentrations; this appears to reduce glucose absorption in the digestive tract (Raghuraman et al. 1994). Fenugreek gum (galactomannan) is an underexploited compound in the pharmaceutical industry (Ramesh et al. 2001). Galactomannan, a soluble dietary fiber, plays an important role in reducing sugar level together with other hypoglycemic constituents (Ali et al. 1995).

Fenugreek is a viable source of galactomannan gum (McCormick et al. 2006), which can be used as a thickening agent in foods (Slinkard et al. 2006) or as a food emulsifier (Garti et al. 1997). Galactomannans often found in the endosperm cell wall (Meier and Reid 1977) are the main polysaccharide in fenugreek seeds.

Mathur and Mathur (2006) have given a detailed review on fenugreek galactomannan production. Galactomannan has unique application in food as well as nonfood industries. Galactomannan from fenugreek has acquired great commercial value because it is edible and is allowed as a food additive. Fenugreek galactomannan is variously referred to as fenugreek gum or fenugreek polysaccharide and is interchangeable for its endospermic powder. In the case of fenugreek, both dry and wet extraction of galactomannan is feasible, but the patented methods are generally based on wet extraction because it is separated by differential grinding, sifting, and sieving. Mathur and Mathur (2006) claimed to have developed a method to produce highly pure fenugreek endosperm split by a differential grinding in the process involving complete lack of odor and without a bitter taste; yield of gum is >25%, which is better and a higher quantity than in a wet process.

24.9.3 Hydroxyl-isoleucine (HIL)

The isoleucine, an amino acid, is the precursor of 4-hydroxyisoleucine (4-HIL). It has been reported as a by-product in the drying process of galactomannan extraction of the seed germ (protein) and husk, which are also useful for extraction of neutraceuticals (e.g., fenugreek oleoresins) and saponins. The latest addition to fenugreek products is 4-hydroxyisoleucine, which is known to increase insulin secretion (insulinotropic) when a carbohydrate-rich diet is ingested. It catabolizes by depositing sugars as muscle glycogen. Fenugreek whole seed powder has two constituents (gum + 4-HIL) that act in synergy to regulate sugar metabolism (Mathur and Mathur 2006). Most hypoglycemic and antiglycemic effects of fenugreek are attributed to the gastrointestinal effect of dietary fiber and systematic effects of amino acids like 4-HIL present in the seed (Broca et al. 2000; Sauvaire, Baccou, and Besancon 1976; Sauvaire et al. 1984; Petropoulos 2002; Madar 1984). Like other legumes, fenugreek is reported to be rich in arginine, alanine, and glycine, but poor in lysine content (Gopal and Singh 1978; Sharma 1984). However, 4-hydroxyisoleucine has been found to be the major free amino acid in the seeds of fenugreek (Sauvaire et al. 1984).

24.9.4 Antioxidants and Other Compounds

The seeds of fenugreek contain less starch but a higher proportion of minerals (Ca, P, Fe, Zn, and Mn) when compared with other legumes (Sankara and Deosthale 1981). The total lipid content (7.5%) of the seeds consists of neutral lipids, glycolipids, and phospholipids (Hemavathy and Prabhakar 1989). Leela (2009) reported that fenugreek seeds contain 0.02–0.05% volatile oil. Toasted fenugreek seeds owe their flavor to another type of heterocyclic compound, called pyrazines. Another aroma-imparting chemical found in fenugreek is sotolone. The aerial parts of fenugreek have 0.3% light yellow oil. The chief constituents are δ-cadinene (27.6%), α-cadinol (12.1%), γ-eudesmol (11.2%), α-bisabolol (10.5%), α-murolene (3.9%), liguloxide (7.6%), cubenol (5.7%), α-murolol (4.2%), and epi-α-globulol (5.7%). The aromatic constituents of the seeds have been elucidated (Girardon et al. 1985) and include n-alkanes, sesquiterpenes, and some oxygenated compounds such as hexanol and γ-nonalactone. Girardon et al. (1986) identified 39 components including n-alkanes, sesquiterpenes, and some oxygenated compounds in the volatile oils of fenugreek.

The elements muurolen and γ- and δ-lactone that are present in small quantities could be of great importance in the aroma of seeds because of their olfactory properties, and the characteristic compound with typical flavor from solvent extract was identified as 3-hydroxy-4,5-dimethyl-2(5H)-furanone. Fresh aerial parts of fenugreek plant yielded 0.3% light yellow oil. The main constituents of the oil were δ-cadinene, α-cadinol, γ-eudesmol, and α-bisabolol (Ahmadiani et al. 2001). Peraza et al. (2001) detected two compounds, 3-hydroxy-4,5-dimethyl-2(5H)-furanone (sotolone) and 3-amino-4,5-dimethyl-2(5H)-furanone, the postulated precursor of sotolone, from hairy root cultures of fenugreek.

The seeds are also known to contain flavonoids, carotenoids, coumarines, and other components with very low LD_{50} values (Varshney and Sharma 1996). Chawla, Kanwar, and Sharma (2004) have reported higher content of dry matter (18.90%), total phenols (0.62%), orthodihydroxy phenols (0.80%), and flavonoids (2.13%) in fenugreek leaves.

The medicinal values of fenugreek lie in its phytochemical components, which produce definite physiological actions on the human body. In a study by Joshi et al. (2009), the total phenolic contents in fenugreek seeds ranged from 38 to 41 mg/g GAE, whereas flavonoid contents ranged from 1.2 to 2.3 mg/g QE. The highest free radical scavenging activity in DPPH assay observed was 11.25%. As reviewed by Khawar, Cocu, et al. (2002), fenugreek contains 73% digestible dry matter, 27% protein, 7–10% oil, and 0.36% trigonellin, with 20–45% choline, phytine, flavonoids, lecithin, and mucilage. Slinkard et al. (2006) described the chemical composition of fenugreek seeds as 32% insoluble dietary fiber, 13% soluble dietary fiber, 36% protein, 3% ash, 1.6% starch, and 0.4% sugar. The seeds also contain calcium, iron, and β-carotene (Sauvaire et al. 1976). The seeds of fenugreek may contribute to the relatively high protein content of the plant (Mir et al. 1997).

24.9.5 Nutraceutical Properties and Industrial Uses

Fenugreek is one of the few spices used extensively for medicinal purposes. The fenugreek herb, seed, powders, and extracts are known to possess several pharmacological effects, like hypoglycemic, hypocholesterolemic, antinociceptive, antioxidative, laxative, and fungicidal effects, as well as appetite stimulation (Table 24.11). Owing to these properties, fenugreek is an essential factor in health care both in the modern system and through the ages in Indian Ayurvedic and traditional Chinese medicines (Tiran 2003). Fenugreek seeds contain substantial amounts of the steroid diosgenin, which is used as a starting material in the synthesis of natural sex hormones for simulative and contraceptive purposes (Banyai 1973). In addition, seeds of fenugreek are used in the Ayurvedic system as carminatives, antipyretics, anthelmintics, and galactogogues; its commercial formulations are popular in India for combating dandruff, curing badness, and as hair conditioners.

Alcoholic extract of fenugreek significantly reduced the sugar level with a favorable glucose deposition in alloxan-induced diabetes (Tayyaba, Siddiqui, and Nazrul Hasnain 2001; Vats et al. 2003). It is helpful in the normalization of altered levels of glycolytic, gluconeogenic, and lipogenic enzymes in diabetes (Raju et al. 2001) and also decreases levels of creatine kinase activity in heart, skeletal muscle, and liver (Genet, Kale, and Baquer 1999). Just like Indians, the natives of Oriental Morocco highly rely on fenugreek for controlling diabetes (Ziyyat et al. 1997). Constituents of the leaf also reduced the sugar level in diabetic patients hours after ingestion with a profound hypokalemic effect (Abdel and Al-Hakiem 2000). The reduction in plasma cholesterol level is mainly attributed to the action of steroidal saponins (Petit et al. 1995). Inhibition of tumor cell growth of Ehrlich ascites carcinoma occurred with the peritoneal administration of the extract (Sur et al. 2001). The extract also showed significant analgesic and anti-inflammatory activity (Javan et al. 1997).

In addition to the preceding activity, the leaves also exhibited antipyretic activity (Ahmadiani et al. 2001). The herb is found to be useful in calcium oxalate urolithiasis induced by glycolic acid (Ahsan et al. 1989). Inclusion of fenugreek in the diet will also supply a good amount of vitamin C (Saleh et al. 1977). Immunomodulatory activity has been recorded for fenugreek in mice. The plant extract elicited a significant increase in phagocytic index and phagocytic capacity of macrophages (Bin-Hafeez et al. 2003).

Table 24.11   Medicinal Properties of Fenugreek

Medicinal Properties	Ref.
Hyperglycemic and antidiabetic	Raghuraman et al. (1994), Broca et al. (2004), Devi, Kamalakkannan, and Prince (2003), Hannan et al. (2003), Suboh, Bilto, and Aburjai (2004)
Hypercholesterolemic	McAnuff et al. (2002), Suboh et al. (2004), Thomson and Ernst (2003)
Reducing hypertension	Balaraman et al. (2006)
Antihyperthyroidism	Tahiliani and Kar (2003)
Anti-inflammatory	McAnuff et al. (2002)
Anti-inflammatory and antipyretic	Ahmadiani et al. (2001)
Antimicrobial (suppresses acetate producing bacteria)	Devant et al. (2007) Alkofahi et al. (1996)
Anticarcinogenic	Devasena and Menon (2003)
Anthelmentic	Ghafghazi, Farid, and Pourafkari (1980)
Gastric antiulcer action	Al-Meshal et al. (1985)
Antifertility and antiandrogenic	Kamal et al. (1993)
Antinociceptive	Javan, Ahmadiani, and Semnanian (1997)
Nematicidal	Zia et al. (2001)
Molluscicidal	Singh, Singh, and Singh (1997)
Galactogogue	Tiran (2003)
Antitumor	Alkofahi et al. (1996)
Antioxidative	Al-Rumaih and Al-Rumaih (2008) Joshi et al. (2009), Mansour and Khalil (2000)
Hyperglycemic, hypercholesterolemic, antioxidative, laxative, fungicidal, and appetite stimulant	Marzougui et al. (2007)
Antiallergic	Thiel (1997)
Wound healing	Taranalli and Kuppast (1996)

The major constituents—trigonelline; protein; linoleic, oleic, linolenic, and palmitic acids; choline; coumarin; and nicotinic acid—contributed to reducing cholesterol and triglycerides, blood sugar level, and platelet aggregation in patients with coronary artery disease (Li 1980).

The medicinal properties of fenugreek were recorded by the Egyptians and Hippocrates (Lust 1974), making it one of the oldest recorded plants used in medicine (Sinskaja 1961; Acharya et al. 2008). It is referred to in Indian Ayurvedic (Sur et al. 2001), Greek, Chinese, and Arabian medicines (Evidente et al. 2007) and has also been used for veterinary purpose (Sinskaja 1961). The medicinal uses of fenugreek are various and include wound-healing, bust enhancement, aphrodisiac (Tiran 2003; Acharya et al. 2006), galactogogue (Tiran 2003), and expectorant. It is used in the treatment of bronchial ailments, sore throats, sciatica, wounds, sores, irritation of the skin (Lust 1974), tumors (Sur et al. 2001), head lice (El-Bashier and Fouad 2002), and sickness caused by air pollution and UVA/UVB radiation damage to skin cells (Singh, Shaswat, and Kapur 2004). Seeds are used in the treatment of hypertension (Balaraman, Dangwal, and Mohan 2006); hyperglycemia, which can help in the regulation of type 2 diabetes (Raghuraman et al. 1994); and hypocholesterolemia (McAnuff et al. 2002), as well as an anti-inflammatory.

In addition to these medicinal effects, the nutritional value and physiological properties of fenugreek seeds and leaves have been extensively studied by Billaud and Adrian (2001). Fenugreek is extensively used as a galactogogue for releasing milk in mothers feeding infants in humans and cattle. The suitability of the development of food products based on millets, legumes, and fenugreek seed for use in diabetic diets has been advocated (Pathak, Srivastava, and Grover 2000). The synergistic effect of the multiple plant extracts of *Trigonella foenum-graecum*, *Hibiscus cannabinus*,

Lawsonia alba, and *Artemisia cina* (El-Bashier and Fouad 2002) was extremely successful and removed head lice from infected patients within a week.

The seed constituents of fenugreek have important uses as functional foods. According to the USDA (2001), fenugreek seed contains 20% protein, 50% carbohydrate, 5% fat, and 25% dietary fibers. Lipids, cellulose starch, ash, calcium, iron, β-carotene (Sauvaire et al. 1976), and ascorbic acid (Riddoch, Mills, and Duthie 1998) have also been reported to be present in the seed. Fenugreek seed endosperm is characterized by soluble fibers, and approximately half the dry weight of fenu-greek seed has soluble dietary fibers (SDFs), and edible dietary fiber (Aspinall 1980). The fenugreek seeds are added to foods like ground meat and baked goods as a potential functional food besides their use in nutritional supplementation (Mansour and El-Adawy 1994).

Seed galactomannan is industrially used as a food emulsifier (Garti et al. 1997); mucilage is used as an ice cream stabilizer (Balyan et al. 2001) and also as a viscosity builder (Seghal, Chauhan, and Kumbhar 2002). In a study where the composition of raw, soaked, and germinated fenugreek seed was compared, Hooda and Jood (2003a) found that the nutritional quality of fenugreek seed could be improved through careful processing and subsequent reduction in bitterness. Improvement in oxidative stability of raw food products such as eggs could be achieved by adding fenugreek to a seed mix (Armitage, Hettiarachchy, and Monsoor 2002). Fenugreek leaf has been reported to have significant nutritional quality (Gupta et al. 1998).

With an increasingly health-conscious population depending upon herbal medicine for health care and due to several inherent medicinal properties, the demand for fenugreek as a protective food is increasing. Therefore, many value-added products involving fenugreek fibers and extracts are available in the market as dietary supplements in combination with grain and pulses. Fenugreek fibers, psyllium husk, and wheat bran could be used as dietary supplements to increase roughage in the human diet (Al-Khalidi, Martin, and Prakash 1999). In India, fenugreek flour-blended rice bran was found to improve the physical and sensory properties of breads and cookies while improving their quality (Sharma and Chauhan 2000). Corn bread mixed with a small amount of fenugreek (3%) or with wheat flour (30%) is used as staple food in Egypt (Galal 2001). Good-quality wheat breads with high nutritional characteristics and higher acceptability have been produced in Egypt by supplementing wheat flour with 4% fenugreek flour (Bakr 1997).

Bhatia and Khetrapaul (2002) recorded a significant ($P < 0.05$) reduction in phytic acid levels and a simultaneous *in vitro* increase in calcium and iron content in fenugreek-supplemented Indian bread with higher temperature and longer duration of the fermentation process. The supplementa-tion of wheat flour with ground debittered fenugreek improves the physicochemical, nutritional, and rheological properties of wheat dough (Sharma and Chauhan 2000). In another study, Hooda and Jood (2003b) found that the physiological, rheological, and organoleptic characteristics of wheat–fenugreek blends had an increased protein and fat content.

Every part of this multipurpose, crop is useful as food, fodder, medicine, and cosmetics. Its green fresh leaves and tender immature pods are used as green cooked vegetables and the role of its seed in preserving food, reducing spoilage cannot be overlooked. The seeds are also used for making dye and extraction of alkaloids or steroids. In certain areas, the green herbage as well as the seeds of fenugreek and other green or dry fodder are usually fed to animals. It is recognized well for its use in industry in syrups, pickles, baked foods, condiments, chewing gums, icings, cooked food seasonings, cosmetics, and hair conditioning.

24.10 CONCLUSIONS

Fenugreek, a multiuse and commercially important spice crop, is one of the few species exten-sively used for medicinal purposes. The fenugreek herb, seeds, powder, and extracts are known to possess several medicinal properties, such as antidiabetic, hypocholesterolemic, antinociceptive, antioxidative, laxative, and antimicrobial effects. It is established that sapogenins (diosgenin and tigogenin), isoleucine (hydroxyl-isoleucine), trigonelline, and galactomannans contribute profusely not only for regulation of hypoglycemia and hypocholesterolemia but also for synthesis of natural sex hormones as stimulants and as contraceptives, in addition to their use as galactogogues, carminatives, and hair tonic.

The commercial cultivation at large scale is done in India, Ethiopia, Egypt, Turkey, and Tunisia; occasional growing is taken up by a few other countries. India is the leading producer and exporter of fenugreek in the world. More than 100 *Trigonella* species have been reported but *Trigonella foenum-graecum* L. is the extensively grown species. Little information is available about genetic diversity, intra- and interspecific variability, and genetic relationships among species. There is a need to study genetic relationships among the different species of *Trigonella* for further exploring possibilities for genetic improvement of the crop, particularly for nutraceutical values such as the steroidal sapogenins, trigonelline, galactomannan, and isoleucine content for which there is demand in the pharmaceutical industry. There is every possibility to increase steroidal sapogenins (diosgenin and tigogenin), isoleucine (hydroxyl-isoleucine), and trigonelline content through chemical or physical mutagenic treatment of fenugreek seeds. But concerted studies are required to standardize such treatments along with cultural practices so as to realize more constituents of bioactive compounds from fenugreek.

Owing to little success in hybridization, the genes from related species have not been exploited in fenugreek, but here the double haploid breeding can offer better scope for improvement in productivity and medicinal quality. Bioproduction of sapogenins, polysaccharides, and amino acids has been successfully achieved through callus cultures in fenugreek. Scientists need to understand the ways in which several factors affect bioactive constituents in fenugreek seed. The major emphasis should be tolerance to major stress and to ensure that these factors are durable and the yield and quality potential are retained. Concerted efforts will be required to develop and validate effective screening techniques, develop enhanced resistant sources, and determine the nature of inheritance to biotic stresses. Besides root rot and wilt, powdery mildew, downy mildew, aphids, and leaf minor are the major problems; interventions are required for breeding resistant varieties. Large variability for physiological parameters offers selection opportunities for breeding for adaptation to several abiotic stress factors—mainly drought and salinity. Improving quality for nutraceuticals and accelerating research into alternative uses of fenugreek in food, feed, and health-protecting products is necessary in order to harness the potential.

REFERENCES

Aasim, M., M. K. Khawar, C. Sancak, and S. Ozcan. 2009. *In vitro* shoot regeneration of fenugreek. *American-Eurasia Journal of Sustainable Agriculture* 3 (2): 135–138.

Abdel-Barry, J. A., and M. H. H. Al-Hakiem. 2000. Acute intraperitoneal and oral toxicity of the leaf glycoside extract of *Trigonella foenum-graecum* L. in mice. *Journal of Ethnopharmacology* 70 (1): 65–68.

Acharya, S. N., S. Blade, Z. Mir, and J. R. Moyer. 2007. Tristar fenugreek. *Canadian Journal of Plant Science* 87:901–903.

Acharya, S. N., A. Srichamroen, S. Basu, B. Ooraikul, and T. Basu. 2006. Improvement in the nutraceutical properties of fenugreek. *Songhlanakarin Journal of Science & Technology* 28 (1): 1–9.

Acharya, S. N., J. E. Thomas, and S. K. Basu. 2007. Improvement in the medicinal and nutritional properties of fenugreek. In *Advances in medicinal plant research*, ed. S. N. Acharya and J. E. Thomas. Trivandrum, Kerala, India: Research Signpost.

———. 2008. Fenugreek, an alternative crop for semiarid regions of North America. *Crop Science* 48:841–853.

Agarwal, K., and P. K. Gupta. 1983. Cytological studies in the genus *Trigonella* Linn. *Cytologia* 48:771–779.

Ahmad, F., S. N. Acharya, Z. Mir, and S. Mir. 1999. Localization and activity of rRNA genes on fenugreek chromosomes by fluorescent *in situ* hybridization and silver staining. *Theoretical and Applied Genetics* 98:179–185.

Ahmadiani, A. M., S. Javan, E. Semnanian Barat, and M. Kamalinejad. 2001. Anti-inflammatory and anti-pyretic effects of *Trigonella foenum-graecum* L. leaves extract in the rat. *Journal of Ethnopharmacology* 75:283–286.

Ahmed, F. A., S. A. Ghanem, A. A. Reda, and M. Solaiman. 2000. Effect of some growth regulators and subcultures on callus proliferation and trigonelline content of fenugreek. *Bulletin of National Research Center Cairo* 25 (1): 35–46.

Ahsan, S. K., M. Tariq, et al. 1989. Effect of *Trigonella foenum-graecum* L. and *Ammi majus* on calcium oxalate urolithiasis in rats. *Journal of Ethnopharmacology* 26 (3): 249–254.

Ali, L., A. K. Khan, Z. Hasan, M. Mosihuzzaman, N. Nahar, T. Nasreen, M. Nure-e-Alam, and B. Rokeya. 1995. Characterization of the hypoglycemic effects of *Trigonella foenum-graecum* seed. *Planta Medica* 61:358–360.

Al-Khalidi, S. F., S. A. Martin, and L. Prakash. 1999. Fermentation of fenugreek fiber, psyllium husk and wheat bran by *Bacteroides ovatus* V975. *Current Microbiology* 39:231–232.

Alkofahi, A., R. Batshour, W. Owais, and N. Najib. 1996. Biological activity of some Jordanian medicinal plant extracts. *Fitoterapia* 67 (5): 435–437.

Allen, O. N., and E. K. Allen. 1981. *The Leguminosae.* London: Macmillan Co.

Al-Meshal, I. A., N. S. Parmar, M. Tariq, and A. M. Aqueel. 1985. Gastric antiulcer activity in rats of *Trigonella foenum-graecum* L. (hu-lu-pa). *Fitoterapia* 56:232–235.

Al-Rumaih, M. M., and M. M. Al-Rumaih. 2008. Influence of ionizing radiation on antioxidant enzymes in three spices of *Trigonella. American Journal of Environmental Science* 4 (2): 151–156.

Altuntas, E., E. Ozgoz, and O. Faruk Taser. 2005. Some physical properties of fenugreek seeds. *Journal of Food Engineering* 71:36–43.

Anis, M., and E. Amimuddin. 1985. Estimation of diosgenin in seeds of induced autoploid *Trigonella foenum-graecum. Fitoterapia* 56:51–52.

Anonymous. 1994. *Plants and their constituents, phytochemical dictionary of the Leguminosae*, vol. 1. London: Chapman & Hall.

Armitage, D. B., N. S. Hettiarachchy, and M. A. Monsoor. 2002. Natural antioxidants as a component of an egg-albumen film in the reduction of lipid oxidation in cooked and uncooked poultry. *Journal of Food Science* 67:631–634.

Arya, I. D., S. R. Rao, and S. N. Raina. 1988. Cytomorphological studies of *Trigonella foenum-graecum* L. autotetraploids in three (C1, C2, C3) generations. *Cytologia* 53 (3): 525–534.

Aspinall, G. O. 1980. *The biochemistry of plants*, 473–500. New York: Academic Publisher.

Avtar, R., D. S. Jatasra, and B. S. Jhorar. 2002. Analysis of gene effects for some important yield components in fenugreek (*Trigonella foenum-graecum* L.). *Forage Research* 28 (2): 59–62.

Ayanoglu, F., M. Arslan, and A. Mert. 2004. Correlation and path analysis of the relationship between yield and yield components in fenugreek (*Trigonella foenum-graecum* L.). *Turkish Journal of Field Crops* 9 (10): 11–15.

Azam, M., and A. K. Biswas. 1989. Callus culturing its maintenance and cytological variations in *Trigonella foenum-graecum. Current Science* 58 (15): 844–847.

Bakr, A. A. 1997. Production of iron-fortified bread employing some selected natural iron sources. *Nahrung* 41:293–298.

Bakshi, V. M., and Y. K. Hameed. 1971. Isolation of diosgenin from fenugreek seeds. *Indian Journal of Pharmacy* 33 (3): 55–56.

Balaraman, R., S. Dangwal, and M. Mohan. 2006. Antihypertensive effect of *Trigonella foenum-graecum* seeds in experimentally induced hypertension in rats. *Pharmaceutical Biology* 44:568–575.

Balodi, B., and R. R. Rao. 1991. The genus *Trigonella* L. (Fabaceae) in northwest Himalaya. *Journal of Economic and Taxonomic Botany* 5 (1): 11–16.

Balyan, D. K., S. M. Tyagi, D. Singh, and V. K. Tanwar. 2001. Effects of extraction parameters on the properties of fenugreek mucilage and its use in ice cream as stabilizer. *Journal of Food Science Technology* 38:171–174.

Banerjee, A., and P. C. Kole. 2004. Genetic variability, correlation and path analysis in fenugreek. *Journal of Spices and Aromatic Crops* 13 (1): 44–48.

Banyai, L. 1973. Botanical and qualitative studies on ecotypes of fenugreek (*Trigonella foenum-graecum* L.). *Agrobotanica* 15:175–187.

Basu, S. K. 2006. Seed production technology for fenugreek in the condition prairies. Thesis submitted for MSc, Department of Biological Science, University of Lethbridge. Lethbridge, Alberta, Canada.

Basu, S. K., S. N. Acharya, M. Bandara, and J. Thomas. 2004. Agronomic and genetic approaches for improving seed quality and yield of fenugreek (*Trigonella foenum-graecum* L.) in western Canada. In *Proceedings of Science of Changing Climates—Impact on Agriculture, Forestry*, 38. Wetlands, July 20–23, 2004, University of Alberta, Edmonton, Alberta, Canada.

Basu, S. K., S. N. Acharya, and J. E. Thomas. 2007. Genetic improvement of fenugreek through EMS-induced mutation breeding for higher seed yield under western Canada prairie condition. *Euphytica* 160 (2): 249–258.

Berwal, K. K., J. V. Singh, B. S. Jhorar, G. P. Lodhi, and C. Kishore. 1996. Character association studies in fenugreek. *Annals of Agriculture and Biology Research* 1 (1/2): 93–99.

Bewley, J. D., D. W. M. Leung, S. MacIsaak, J. S. G. Reid, and N. Xu. 1993. Transient starch accumulation in the cotyledons of fenugreek seeds during galactomannan mobilization from the endosperm. *Plant Physiology and Biochemistry* 31:483–490.

Bhatia, A., and N. Khetrapaul. 2002. Effect of fermentation on phytic acid and *in vitro* availability of calcium and iron of *doli-ki roti*—An indigenously fermented Indian bread. *Ecology and Food Nutrition* 41:243–253.

Bhavsar, G. C., N. S. Kapadia, and N. M. Patel. 1980. Studies on *Trigonella foenum-graecum* L. *Indian Journal of Pharmacology Science* 42 (2): 39–40.

Billaud, C., and J. Adrian. 2001. Fenugreek: Composition, nutritional value and physiological properties. *Sciences de Aliments* 21:3–26.

Bin-Hafeez, B., R. Haque, S. Parvez, S. Panda, I. Sayeed, and S. Raisuddin. 2003. Immunomodulatory effects of fenugreek extract in mice. *International Immunopharmacology* 3 (2): 257–265.

Bochannon, M. B., J. W. Hageman, F. R. Earle, and A. S. Barclay. 1974. Screening seed of *Trigonella* and three related genera for diosgenin. *Phytochemistry* 13:1513–1514.

Brain, K. R., and M. H. Williams. 1983. Evidence for an alternative route from sterol to sapogenin in suspension cultures from *Trigonella foenum-graecum*. *Plant Cell Reports* 2:7–10.

Brenac, P., and Y. Sauvaire. 1996. Chemotaxonomic value of sterols and steroidal sapogenins in the genus *Trigonella*. *Biochemical Systematics and Ecology* 24 (2): 157–164.

Broca, C., M. Manteghetti, R. Gross, et al. 2000. 4-Hydroxyisoleucine effects of synthetic and natural analogues on insulin secreation. *European Journal of Pharmacology* 390:339–345.

Bruneton, J. 1995. *Pharmacognosy, phytochemistry, medicinal plants*, 95. Paris: Lavoisier Publ. Ins.

Campbell, J. M. A., and J. S. G. Reid. 1982. Galactomannan for motion and guanosine 5'diphosphate-mannose: Galactomannan mannosytransferase in developing seeds of fenugreek. *Planta* 155:105–111.

Cerdon, C., A. Rahier, M. Taton, and Y. Sauvaire. 1995. Effect of diniconazole on sterol composition of roots and cell suspension cultures of fenugreek. *Phytochemistry* 39 (4): 883–893.

———. 1996. Effect of tridemorph and fenpropimorph on sterol composition in fenugreek. *Phytochemistry* 41 (2): 423–431.

Chandra, K., E. V. D. Sastry, and D. Singh. 2000. Genetic variation and character association of seed yield and its component characters in fenugreek. *Agricultural Science Digest* 20 (2): 93–95.

Chawla, N., J. S. Kanwar, and S. Sharma. 2004. A study on the phenolic compounds in *methi* and *metha* leaves. *Journal of Research PAU* 41 (4): 454–456.

Choubey, R., and J. L. Trar. 1990. Effect of gamma rays on amino acid contents of *Trigonella foenum-graecum* seedling. *Annals of Plant Physiology* 4 (2): 133–138.

Choudhary, A. K. 2003. Outcrossing behavior in fenugreek (*Trigonella foenum-graecum* L.). *Indian Journal of General Plant Breeding* 63 (2): 178.

Choudhary, A. K., and V. V. Singh. 2001. An induced determinate mutant in fenugreek (*Trigonella foenum-graecum* L.). *Journal of Spices and Aromatic Plants* 10 (1): 51–53.

Cornish, M. A., R. Hardman, and R. M. Sadler. 1983. Hybridization for genetic improvement in the yield of diosgenin from fenugreek seed. *Planta Medica* 48:149–52.

Dangi, R. S., M. D. Lagu, B. L. Choudhary, P. K. Ranjekar, and V. S. Gupta. 2004. Assessment of genetic diversity in *Trigonella foenum-graecum* and *Trigonella caerulea* using ISSR and RAPD markers. *BMC Plant Biology* 4:4–13.

Darlington, C. D., and A. P. Wylie. 1945. *Chromosome atlas of flowering plants.* London: George Allen & Unwin Ltd.

Das, A. B., S. Mohanty, T. Thangraj, and P. Das. 2000. Variations of 4c DNA content and karyotype in nine cultivars of fenugreek. *Journal of Herbs Spices and Medicinal Plants* 7 (1): 25–32.

Dash, S. R., and P. C. Kole. 2000. Association analysis of seed yield and its components in fenugreek. *Crop Research* 20 (3): 449–452.

Datta, S., R. Chatterjee, and S. Mukherjee. 2005. Variability, heritability and path analysis studies in fenugreek. *Indian Journal of Horticulture* 62 (1): 96–98.

Davis, P. H. 1978. *Flora of Turkey.* Edinburgh, Scotland: Edinburgh University Press.

De Candolle, A. 1964. *Origin of cultivated plants.* New York: Hafner.

Del'Gaudio, S. 1953. Ricerrche sui consume idrici e indugini sull outofertilite del fieno g. reco. Annals Sper *Agricultura* 7:1273–1287.

Devant, M., A. Anglada, and A. Bach. 2007. Effects of plant extract supplementation on rumen fermentation and metabolism on young Holstein bulls consuming high level of concentrate. *Animal Feed Science Technology* 137:46–57.

Devasena, T., and V. P. Menon. 2003. Fenugreek affects the activity of β-glucuroxidase and mucinase in the colon. *Phytotherapy Research* 17 (9): 1088–1091.

Devi, B. A., N. Kamalakkannan, and P. S. Prince. 2003. Supplementation of fenugreek leaves to diabetic rats—Effect on carbohydrate metabolic enzymes in diabetic liver and kidney. *Phytotherapy Research* 17 (10): 1231–1233.

Duke, A. J. 1986. *Handbook of legumes of world economic importance.* New York: Plenum Press.

Dundas, I. S., R. M. Nair, and D. C. Verlin. 2006. First report of meiotic chromosome number and karyotype analysis of an accession of *Trigonella balansae.* *New Zealand Journal of Agricultural Research* 49:55–58.

Edwards, M., I. C. M. Dea, P. V. Bulpin, and J. S. G. Reid. 1989. Biosynthesis of legume seed galactomannan *in vitro. Planta* 178:41–51.

Edwards, M., C. Scott, M. J. Gidley, and J. S. G. Reid. 1992. Control of mannose/galactose ratio during galactomannan formation in developing legume seed. *Planta* 187:67–74.

El-Bahr, M. K. 1989. Influence of sucrose and 2,4-D on *Trigonella foenum-graecum* tissue culture. *African Journal of Agricultural Science* 16 (1–2): 87–96.

El-Bashier, Z. M., and M. A. Fouad. 2002. A preliminary pilot study on head lice pediculosis in Shakaria Governorate and treatment of lice with natural plant products. *Journal of Egypt Society of Parasitology* 32:725–776.

Elujoba, A. A., and R. Hardman. 1985. Fermentation of powdered fenugreek seeds for increased sapogenin yield. *Fitoterapia* 56 (6): 368–370.

Evidente, A., A. M. Fernandez, A. Andolfi, D. Rubiales, and A. Molta. 2007. Trigoxazonanae, a momosubstituted trioxazonanae from *Trigonella foenum-graecum* L. root exudates, inhibits *Orobranche crenata* seed germination. *Phytochemistry* 68:2487–2492.

Fazli, F. R. Y. 1967. Studies in steroid-yielding plants of the genus *Trigonella.* PhD thesis, University of Nottingham, England.

Fazli, F. R. Y., and R. Hardman. 1968. The spice fenugreek (*Trigonella foenum-graecum* L). Its commercial varieties of seed as a source of diosgenin. *Tropical Science* 10:66–78.

Fehr, W. R. 1993. *Principles of cultivar development: Theory and technique,* vol.1. New York: MacMillan Publishing Company.

Floria, F., and M. C. Ichim. 2006. Valuable fenugreek mutants induced by gamma rays and alkylating agent. *Plant Mutation Reports* 1 (2): 30–31.

Furry, A. 1950. Les cahiers de la recherché. *Agronomique* 3:25–317.

Galal, O. M. 2001. The nutrition transition in Egypt: Obesity, undernutrition and the food consumption context. *Public Health Nutrition* 5:141–148.

Garti, N., Z. Madar, A. Aserin, and B. Sternheim. 1997. Fenugreek galactomannans as food emulsifiers. *Lebensmittel-Wissenschaft und-Technologie* 30 (3): 305–311.

Gaul, H. 1961. Mutation and plant breeding. *Proceedings of Symposium on Muation and Plant Breeding,* Cornell Univ., Nov.–Dec., 1960.

Genet, S., R. K. Kale, and N. Z. Baquer. 1999. Effect of vanadate, insulin and fenugreek on creative kinase levels in tissues of diabetic rat. *Indian Journal of Experimental Botany* 37(2): 200.

Gerhartz, W. 1987. *Ullmann's encyclopedia of industrial chemistry*, 5th ed., A8: 597, A13: 110, 117, 135–138. Weinheim, Germany: VCH.

Ghafghazi, T., H. Farid, and A. Pourafkari. 1980. *In vitro* study of the anthelmintic action of *Trigonella foenum-graecum* L. grown in Iran. *Iranian Journal of Public Health* 9:21–26.

Ghosal, S., R. S. Srivastava., D. C. Chatterjee, and S. K. Datta. 1974. Fenugreekine, a new steroidal saprogenin-peptide ester of *Trigonella foenum-graecum*. *Phytochemistry* 13:2247–2251.

Girardon, P., J. M. Bessiere., J. C. Baccou, and Y. Sauvaire. 1985. Volatile constituents of fenugreek seeds. *Planta Medica* 6:533–534.

Girardon, P., Y. Sauvaire., J. C. Baccou, and J. M. Bessiere. 1986. Identification of 3-hydroxy-4,5-diethyl-2(5H)-furanone in aroma of fenugreek seeds. *Lebensmittel Wissenschaft und Technologie* 19 (1): 44–46.

Gopal, J., and A. Singh. 1978. Barrier to crossability between *Trigonella foenum-graecum* and *T. corniculata*. *Crop Improvement* 5:81–83.

Gopal, J., A. Singh, and J. Gopal. 1979. Meiotic behavior and seed fertility in advanced generation autotetraploids of fenugreek. *Indian Journal of General Plant Breeding* 39 (2): 323–329.

Gopinath, P. M. 1974. A case of rare triploidy in excised root culture of *Trigonella foenum-graecum*. *Current Science* 43:524–525.

Green, J. M., D. Sharma, L. J. Reddy, K. B. Saxena, S. C. Gupta, K. C. Jain, B. V. S. Reddy, and M. R. Rao. 1981. Methodology and the progress in the ICRISAT in pigeon pea breeding program. In *Proceedings of Indian Workshop on Pigeon Peas*, 1981, Patancheru, India.

Gupta, K. G., G. K. Barat., D. S. Wagle, and H. K. L. Chawla. 1989. Nutrient contents and antinutritional factors in conventional and nonconventional leafy vegetables. *Food Chemistry* 31 (2): 105.

Gupta, U., P. Rudrama, E. R. Rati, and R. Joseph. 1998. The nutritional quality of lactic fermented bitter gourd and fenugreek leaves. *International Journal of Food Science and Nutrition* 49:101–108.

Haefele, C., C. Bonfils, and Y. Sauvaire. 1997. Characterization of a dioxygenase from *Trigonella foenum-graecum* L. involved in 4-hydroxy isoleucine biosynthesis. *Phytochemistry* 44 (4): 563–566.

Han, Y., S. Nishibe, Y. Noguchi, and Z. Jin. 2001. Flavonol glycosides from the stem of *Trigonella foenum-graecum* L. *Phytochemistry* 58:577–580.

Hannan, J. M., B. Rokeya, O. Faruque, N. Nahar, M. Moshihuzzaman, A. K. Azad Khan, and L. Ali. 2003. Effect of soluble dietary fiber fraction of *Trigonella foenum-graecum* L. on glycemic, insulenimic, lipidemic and platelet aggregation status of type 2 diabetic model rats. *Journal of Ethnopharmacology* 88:73–77.

Hardman, R., J. Kosugi, and R. T. Parfitt. 1980. Isolation and characterization of a furostanol glycoside from fenugreek. *Journal of Phytochemistry*: 19(4):698–700.

Hemavathy, J., and J. V. Prabhakar.1989. Lipid composition of fenugreek seeds. *Food Chemistry* 31 (1): 1–8.

Heywood, V. H. 1967. *Plant taxonomy—Studies in biology, No. 5*. London: Edward Arnold Ltd.

Hidvegi, M., A. El-Kady, R. Lásztity, F. Bekes, and L. Simon-Sarkadi. 1984. Contribution to the nutritional characterization of fenugreek (*Trigonella foenum-graecum* L.). *Acta Alimentaria* 13 (4): 315–324.

Hocker, J. B., and D. Jackson. 1955. *Index Kewensis*, Tomus II, 1116–1117 (1895) Suppl. XII, 146 (1951–1955). Oxford, England: Clarendon Press.

Hooda, S., and S. Jood. 2003a. Effect of soaking and germination on nutrient and antinutrient contents of fenugreek (*Trigonella foenum-graecum* L.). *Journal of Food Biochemistry* 27:165–176.

———. 2003b. Physicochemical, rheological, and organoleptic characteristics of wheat-fenugreek supplemented blends. *Nahrung* 47:265–268.

Howard, M. 1987. *Traditional folk remedies. A comprehensive herbal*. London: Century Hutchinson Ltd.

Huang, W. Z., and X. Liang. 2000. Determination of two flavone glycosides in the seeds of *Trigonella foenum-graecum* L. from various production locations. *Journal of Plant Research & Environment* 9 (4): 53–54.

Hutchinson, J. 1964. *The genera of flowering plants*, vol. 1. Oxford, England: Clarendon Press.

Ivimey-Cook, R. B. 1968. *Trigonella* L. In *Flora Europaea—Rosaceae to Umbelliferae*, vol. 2, ed. T. G. Tutin, V. H. Heywood, N. A. Burges, D. M. Moore, D. H. Valentine, S. M. Walters, and D. A. Webb, 150–152. Cambridge: University Press.

Jain, S. C., and M. Agarwal. 1991. Generation of variants for steroidal sapogenin production in *Trigonella* species by chemical mutagens. *Plant Physiology and Biochemistry* 18 (2): 109–111.

———.1994. Effect of mutagens on steroidal sapogenins in *Trigonella foenum-graecum* tissue cultures. *Fitoterapia* 65 (4): 367–375.

Janardhan, K., and J. Nizam. 1995. Effect of gamma rays on fenugreek. *Advances in Plant Science* 8 (1): 152–156.

Javan, M., A. Ahmadiani, and S. Semnanian. 1997. Antinociceptive effects of *Trigonella foenum-graecum* L. leaves extract. *Journal of Ethnopharmacology* 58 (2): 125–129.

Jones, C. P. 1989. *Extracts from nature.* London: Marks and Spencer, PIC, Tigerprint.

Joshi, R., C. Mansi, S. K. Malhotra, and M. M. Anwer. 2009. Antioxidant activity, phenol and flavonoid contents in fenugreek varieties under semi-arid conditions. *Proceedings of International Conference on Horticulture for Livelihood Security and Economic Growth*, PNASF, Bangalore, November 9–12, 2009.

Joshi, S., and S. S. Raghuvanshi. 1968. B-chromosome, pollen germination *in situ* and connected grains in *Trigonella foenum-graecum* L. *Beitrage zur Biologie der Pflanzen* 44 (1): 161–166.

Kamal, R., R. Yadav., and J. D. Sharma. 1993. Efficiency of the steroidal fraction of the fenugreek seed extract on the fertility of male albino rats. *Phytotherapy Research* 7: 134–138.

Kamal, R., R. Yadav, and P. Mehra. 1997. Rotenoid biosynthesis potentiality of *Trigonella* species *in vivo* and *in vitro*. *Journal of Medicinal and Aromatic Plant Science* 19 (4): 988–993.

Kamari, G., and S. Papatson. 1973. Chromosome studies in some Mediterranean angiosperms. *Botaniska Notiser* 126:266–268.

Kaushik, S. K. 2002. Correlation and path analysis in M-7 lines of fenugreek (*Trigonella foenum-graecum* L.). *Annals of Agriculture and Biology Research* 7 (2): 165–170.

Kavadas, D. S. 1956. *Illustrated botanical–phytological dictionary*, vol. XII, Athens: Pegasus, 3929–3933.

Kesarwani, R., and S. S. Raghuvanshi. 1988. Comparison of B carrier and noncarrier population of diploid and autoteraploid *Trigonella foenum-graecum* L. *New Botanist* 15 (1): 19–22.

Khawar, K. M., S. Cocu, S. G. Onarici, C. Sancak, and S. Ozcan. 2002. Shoot regeneration from callus cultures of fenugreek. *Agroenvironment* 2002:26–29.

Khawar, K. M., S. Gulbitti Onarici, S. Cocu, S. Erisen, C. Saneak, and S. Ozcan. 2002. *In vitro* crown galls induced by *Agrobacterium tumefaciens* strain A281 (pTi B0542) in *Trigonella foenum-graecum* L. *Biologica Plantarum* 48 (3): 441–444.

Koli, N. R., and K. Ramakrishna. 2002. Frequency and spectrum of induced mutations and mutagenic effectiveness and efficiency in fenugreek. *Indian Journal of General Plant Breeding* 62 (4): 365–366.

Kumar, M., and B. M. Choudhary. 2003. Studies on genetic variability in fenugreek (*Trigonella foenum-graecum* L.). *Orissa Journal of Horticulture* 31 (1): 37–39.

Ladizinsky, G., and C. G. Vosa. 1986. Karyotype and C-banding in *Trigonella* section *foenum-graecum* (Fabaceae). *Plant System Evolution* 153 (1–2): 1–5.

Lakshmi, N., T. V. Rao, and T. Rao Venkateswara. 1984. Karyological and morphological investigations on some inbred strains of *Trigonella* L. *Genetica Iberian* 36 (3–4): 187–200.

Lakshmi, V., and S. K. Datta. 1989. Induced mutations of phylogenetic significance in fenugreek. *Legume Research* 12 (2): 91–97.

Leela, N. K. 2009. Flavors in tree and seed spices. In *Flavors, nutriceauticals and food colors from horticultural crops*, ed. A. Shamina, K. S. Krishnamurthy, K. N. Shiva, N. K. Leela, and B. Chempakam, 33–41. Calicut, Kerala (India): Indian Institute of Spices Research.

Leyel, C. F. 1987. *Elixirs of life.* London: Faber & Faber.

Li, X., M. J. Farn, L. B. Feng, X. Q. Shan, and Y. H. Feng. 1980. Analysis of the galactomannan gums in 24 seeds of Leguminosae. *Chinese Wu, Hsueh Pao* 22 (3): 302–304.

Loi, A., B. J. Nutt, R. McRobb, and M. A. Ewing. 2000. Potential new alternative annual pasture legumes for Australian Mediterranean farming system. *Option Mediterranean* 45:51–54.

Lust, J. B. 1974. *The herb book.* New York: Bantam Books Inc.

Madar, Z. 1984. Fenugreek (*Trigonella foenum-graecum*) as a means of reducing postpyramidal glucose level in diabetic rats. *Nutrition Report International* 29:1267–1273.

Madhuri, S., B. Manish, M. Sharon, and M. Bhaskare. 1998. Influence of habituation of *Trigonella foenum-graecum* callus on endogenous level of IAA and total alkaloid content in callus. *Indian Journal of Plant Physiology* 3:163–165.

Mahey, J., R. S. Raje, and D. L. Singhania. 2003. Studies on genetic variability and selection criteria in F_3 generation of a cross in fenugreek (*Trigonella foenum-graecum* L.). *Journal of Spices and Aromatic Crops* 12 (1): 19–28.

Makai, P. S., and S. Makai. 2004. Comparison of yield product of fenugreek (*Trigonella foenum-graecum*) varieties and determination of optimal germ number. *Acta Agronomica Ovariensis* 46 (1): 17–23.

Makai, P. S., S. Makai, and A. Kismanyoky. 2004. Comparative test of fenugreek of varieties. *Journal of Central European Agriculture* 5 (4): 259–262.

Malhotra, S. K. 2003. Crop improvement in seed spices crops. In *Breeding field crops*, ed. A. K. Sharma. Bikaner, India: Academic Publishers.

———. 2010. Genetic resources and their utilization in seed spices. Souvenir, *National Conference on Horticulture Biodiversity, Livelihood, Economic Development and Health Care*, University of Horticultural Sciences Campus, Bangalore, May 28–31, 2010, 83–97.

———. 2011a. Regeneration technologies in seed spices. In *Biotechnology in horticulture—Regeneration system*, vol. 1, ed. H. P. Singh, V. A. Parthsarthi, and Nirmalbabu. New Delhi: Westville Publ.

Malhotra, S. K. 2011b. Breading potential of indigenous germplasm of seed spices. In *Vegetable crops*: Genetic resources and Improvement. D. K. Singh and H. Chowdhuary (eds.). New Delhi: New India Publishing House. 477–497.

Malhotra, S. K., and M. K. Rana. 2008. Fenugreek. In *Scientific cultivation of vegetables*, ed. M. K. Rana, 345–361. Ludhiana: Kalyani Publishers.

Malhotra, S. K., and B. B. Vashishtha. 2007. Seed spices. In *Biodiversity of spices and aromatic crops*, ed. K. V. Peter and J. Abrahm. New Delhi: Daya Publishing House.

Manniche, L. 1989. *An ancient Egyptian herbal*. London: British Museum Pub. Ltd.

Mansour, E. H., and T. A. El-Adawy. 1994. Nutritional potential and functional properties of heat-treated and germinated fenugreek seeds. *Lebensmittel Wissenschaft und Technologie* 27:568–572.

Mansour, E. H., and A. H. Khalil. 2000. Evaluation of antioxidant properties of some plant extracts and their application to ground beef patties. *Food Chemistry* 69:135–141.

Marc, C. R., and G. Capraru. 2008. Influence of sodium phosphate (E 339) on mitotic division in *Trigonella foenum-graecum* L. *Sectiunea Genetica si Biologie Moleculara ToM* IX:67–70.

Marzougui, N., A. Ferchichi., F. Guarmi., and B. Mohamed. 2007. Morphological and chemical diversity among 38 Tunisian cultivars of *Trigonella foenum-graecum*. *Journal of Food Agriculture & Environment* 5 (3–4): 248–253.

Mathur, N. K., and G. M. Mathur. 2006. Fenugreek gum. *Science & Technology Entrepreneur* December, 2006:1–11.

McAnuff, M. A., F. O. Omoruyi, E. Y. S. A. Morrison, and H. N. Asemota. 2002. Plasma and liver lipid distribution in streptozotocin-induced rats fed sapogenin extract of the Jamaican bitter yam (*Dioscorea polygonoides*). *Nutrition Research* 22:1427–1434.

McCormick, K. M., R. M. Norton, and H. A. Eagles. 1998. Evaluation of a germplasm collection of fenugreek. In *Proceedings of 9th Australian Agro Conference*, 1998. Wagga Wagga, Australia.

———. 2006. Fenugreek has a role in southeastern Australian farming systems. *Proceedings "Groundbreaking Stuff" 13th Annual Agronomy Conference*, Perth, Australia.

McCormick, K. M., R. M. Norton, H. A. Eagles, and J. F. Kollmorgen. 2000. Fenugreek studies on a new crop for southeastern Australia. Biennial Report 1998–99. Joint Center for Crop Improvement, Private Bag 260, Horsham, Victoria, 3401, Australia, p. 15.

Meier, H., and J. S. G. Reid. 1977. Morphological aspects of the galactomannan formation in the endosperm of *Trigonella foenum-graecum*. *Planta* 133 (3): 243–248.

Mendhulkar, V. D., D. S. Gupta, and R. W. Raut. 2005. Mitotic depression in *Trigonella foenum-graecum* L. treated with sodium azide (SA). *Advances in Plant Science* 18 (2): 529–532.

Merkli, A., P. Christen, and I. Kapetanidis. 1997. Production of diosgenin by hairy root cultures of *Trigonella foenum-graecum* L. *Plant Cell Reports* 16 (9): 632–636.

Miller, J. I. 1969. *The spice trade of the Roman Empire 29 BC to AD 641*. Oxford, England: Clarendon Press.

Mir, Z., S. N. Acharya, P. S. Mir, W. G. Taylor, M. S. Zaman, G. J. Mears, and L. A. Goonewardene. 1997. Nutrient composition, *in vitro* gas production and digestibility of fenugreek (*Trigonella foenum-graecum*) and alfalfa forages. *Canadian Journal of Animal Science* 77 (1): 119–124.

Moschini, E. 1958. *Charatteristiche biologiche e colturali di Trigonella foenum-graecum L. edi Vicia sativa L. di diversa provenienza Esperienze e Ricerche*, pp. 10–11, Pisa.

Moyer, J. R., S. N. Acharya, Z. Mir, and R. C. Doran. 2003. Weed management in irrigated fenugreek grown for forage in rotation with other annual crops. *Canadian Journal of Plant Science* 83:181–188.

Multani, D. S. 1981. Tissue culture in diploid and autotetraploid strains of methi (*Trigonella foenum-graecum* L.). *Proceedings of Symposium Plant Cell Culture in Crop Improvement*, 435–439.

Murashige, T., and F. Skoog. 1962. A revised medium for rapid growth and bioassay with tobacco tissue culture. *Physiologia Plantarum* 15:473–479.

NHB. 2009. Indian Horticulture Database 2009. National Horticulture Board Gurgaon, Havana, India.

Nair, R. M., T. S. Dundas, M. Wallwork, D. C. Verlin, L. Warehouse, and K. Dowling. 2004. Breeding system in a population of *Trigonella balansae*. *Annals of Botany* 94 (6): 883–888.

Natural Database. 2010. Natural medicines comprehensive database of fenugreek. http://www.naturaldatabase. com/(S(we44ty55dms1i0552j50us2h))/nd/Search.aspx?cs=&s=ND&pt=9&Product=fenugreek&btnSea rch.x=10&btnSearch.y=6 (24.05.2010)

Oltraco, S., and Sabanci. 1996. Genetic resources of Turkey. In *Lathyrus genetic resources of Asia*, P. K. Arora, P. N. Mathur, and K. W. Riley, 77–87. Adham: Internationl Plant Genetic Resources Institute.

Oncina, R., J. M. Botia, J. A. Rio, and A. Ortuno. 2000. Bioproduction of diosgenin in callus cultures of *Trigonella foenum-graecum*. *Food Chemistry* 70 (4): 489–492.

Ortuno, A., R. Ornica, J. M. Botia, and J. A. del Rio. 2000. Bioproduction of diosgenin, a steroidal sapogenin, in plants and callus cultures of *Trigonella foenum-graecum* L. modulation by different plant growth regulators. *Recent Research Developments in Agriculture & Food Chemistry* 2:323–329.

Pathak, P., S. Srivastava, and S. Grover. 2000. Development of food products based on millets, legumes and fenugreek seeds and their suitability in diabetic diet. *International Journal of Food Science and Nutrition* 51:409–414.

Peraza, L. F., M. Rodriguez., C. Arias-Castro., J. M. Bessiere., and G. Calva-Calva. 2001. Sotolone production by hairy root cultures of *Trigonella foenum-graecum* L. in airlift with mesh bioreactors. *Journal of Agricultural and Food Chemistry* 49 (12): 6012–6019.

Petit, P., Y. Sauvaire, D. Hillaire-Buys, M. Manteghetti, Y. Baissac, R. Gross, and G. Ribes. 1995. Insulin stimulating effect of an amino acid, 4-hydroxy-isoleucine, purified from fenugreek seeds. *Diabetologia* 38 (SI): A101.

Petropoulos, G. A. 1973. Agronomic, genetic and chemical studies of *Trigonella foenum-graecum* L. PhD diss. Bath University, England.

———. 2002. *Fenugreek—The genus* Trigonella, 1–255. London: Taylor & Francis Inc.

Petropoulos, G. A., and P. Kouloumbis. 2002. Botany. In *Fenugreek: The genus* Trigonella, *medicinal and aromatic plants—Industrial profiles*, ed. G. A. Petropoulos. London: Taylor & Francis Inc.

Phitos, D., and J. Damboldt. 1985. Die Flora der Insel Kefallinia (Griechenland). *Botanika Chronika* 5 (1–2): 1–204.

Pilgaonkar, P. S., M. T. Rustomjee., A. S. Gandhi, and V. S. Bhumra. 2005. Fiber rich fraction of *Trigonella foenum-graecum* seeds and its use as a pharmaceutical excipient. Free US Patent No. 20050084549 (http://www.freepatentsonline.com/200500084549.html).

Polunin, O. 1988. *Flowers of Greece and the Balkans, a field guide*, 1. Oxford: Oxford University Press.

Pratap, P. S., M. K. Rana, and R. Mehra. 2002. Hisar Methi 350 fenugreek. *Indian Journal of Genetic Plant Breeding* 62 (1): 94.

Provorov, N. A., Y. D. Soskau, L. A. Lutova, O. A. Sokolova, and S. S. Bairamou. 1996. Investigation of fenugreek (*Trigonella foenum-graecum* L.) genotypes for fresh weight, seed productivity, symbiotic activity, callus formation and accumulation of steroids. *Euphytica* 88 (2): 129–138.

Provorov, N. A., and I. A. Tikhonovich. 2004. Genetic resource for improving nitrogen fixation in legume–shizobia symbiosis. *Genetic Resources and Crop Evolution* 50 (1): 89–99.

Pruthi, J. S. 2001. *Minor spices and condiments—Crop management and postharvest technology*, 228–241. New Delhi: ICAR.

Radwan, S. S., and C. K. Kokate. 1980. Production of higher levels of trigonelline by cell cultures of *Trigonella foenum-graecum* than by the differentiated plant. *Planta* 147:340–344.

Raghuraman, T. C., R. D. Sharma, and B. Sivakumar. 1994. Effect of fenugreek seeds on intravenous glucose disposition in non-insulin-dependent diabetic patients. *Phytotherapy Research* 8 (2): 83–86.

Raghuvanshi, S. S., and S. Joshi. 1964. *Trigonella foenum-graecum* B chromosome. *Current Science* 33:654.

Rajagopalan, M. S. 2001. Fenugreek a savory medicinal. *Supplement Industry Executive* 5 (6): 43–44.

Raje, R. S., D. Singh, and D. L. Singhania. 2003. Inheritance of giant mutant plant type in fenugreek (*Trigonella foenum-graecum* L.). *Journal of Spices and Aromatic Crops* 11 (2): 141–142.

Raje, R. S., D. L. Singhania, and D. Singh. 2003. Inheritance of powdery mildew resistance and growth habit in fenugreek (*Trigonella foenum-graecum* L.). *Journal of Spices and Aromatic Crops* 12 (20): 120–126.

Raju, J., D. Gupta, A. R. Rao, P. K. Yadav, and N. Z. Baquer. 2001. *Trigonella foenum-graecum* L. (fenugreek) seed powder improves glucose komocostasis in alloxan diabetic rat tissue by reversing the altered glycolytic, gluconegenic and lipogenic enzymes. *Molecular and Cellular Biochemistry* 224:45–51.

Ramesh, H. P., K. Yamaki, H. Ono, and T. Tushida. 2001. Two-dimensional NMR spectroscopic studies of fenugreek galactomanam without chemical fragmentation. *Carbohydrate Polymers* 45 (1): 69–77.

Ravi, R. K. 1987. Cytological studies in diploids and autotetraploids of *Trigonella foenum-graecum* L. *Annals of Biology* 3 (2): 64–67.

Reasat, M., J. Karapetyam, and A. Nasirzadeh. 2002. Karyotypic analysis of *Trigonella* genus of Fars Province. *Iranian Journal of Rangelands Forests Plant Breeding Genetic Research* 11 (1): 127–145.

Reda, A. A., F. A. Ahmed, S. A. Ghanem, and M. Suleiman. 2000. Factors affecting growth and trigonelline content of fenugreek calli. 2. Amino acids and casein hydrolysate. *Bulletin of National Research Center, Cairo* 25 (3): 269–279.

Reid, J. S. G., and H. Meier. 1970. Chemotaxonomic aspects of the reserve galactomannan in leguminous seeds. *Zeitschrift fur Pflanzenphysiologie* 62:89–92.

———. 1972. The function of the aleurone layer during galactomannan mobilization in germinating seeds of fenugreek, crimson clover and lucerne. *Planta* 106:44–60.

Riddoch, C. H., C. F. Mills, and G. G. Duthie. 1998. An evaluation of germinating beans as a source of vitamin C in refugee foods. *European Journal of Clinical Nutrition* 52:115–118.

Rosengarten, F. 1969. *The book of spices*. Wynnewood, PA: Livingston.

Rosser, A. 1985. The day of the yam. *Nursing Times* 81 (18): 47.

Rouk, H. F., and H. Mangesha. 1963. Fenugreek (*Trigonella foenum-graecum* L.)—Its relationship, geography and economic importance. Experimental Station Bulletin no. 20. Imper. Ethiopian College of Agric. & Mech. Arts.

Roy, R. P., and A. Singh. 1968. Cytomorphological studies of the colchicines-induced tetraploids *Trigonella foenum-graecum* L. *Genetica Iberian* 20 (1–2): 37–54.

Saha, A., and P. C. Kole. 2001. Genetic variability in fenugreek grown in subhumid lateritic belt of West Bengal. *Madras Agriculture Journal* 88 (4/6): 345–348.

Saini, N., K. Ramkrishna, A. Khar, and A. Sindhu. 1998. Effect of induced mutation on symbiotic nitrogen fixation in fenugreek. *Legume Research* 21 (2): 128–130.

Saleh, N., Z. El-Hawary, F. A. El-Shobabi, M. Abbassy, and S. R. Morcos. 1977. Vitamin content of fruits, vegetables in common use in Egypt. *Z. Ernobrungswiss* 16 (3): 158–162.

Saleh, N. A. 1996. *Breeding and cultural practices for fenugreek in Egypt*. Cairo, Egypt: National Research Center.

Sankara, R. D. S., and Y. G. Deosthale. 1981. Mineral composition of four Indian food legumes. *Journal of Food Science* 46:1962–1963.

Saraswat, K. S. 1984. Discovery of Emmer wheat and fenugreek from India. *Current Science* 53 (17): 925.

Sastry, E. V. D., B. L. Kumhar, and D. Singh. 2000. Association studies for seed yield and its attributes in M4 generation of fenugreek. *Proceedings of Centennial Conference on Spices and Aromatic Plants*, IISR, Calicut, September 20–23, 2000, 75–77.

Sauvaire, Y., J. C. Baccou, and P. Besancon. 1976. Nutritional value of the properties of fenugreek (*Trigonella foenum-graecum* L.). *Nutrition Report International* 14 (5): 527–537.

Sauvaire, Y., Y. Baissac, O. Leconte, P. Petit, and G. Ribes. 1996. Steroid saponins from fenugreek and some of their biological properties. *Advances in Experimental Medicine and Biology* 405:37–46.

Sauvaire, Y., P. Girardon, J. C. Baccou, and A. M. Risterucci. 1984. Change in the growth, proteins and free amino acid of developing seed and pod of fenugreek. *Phytochemistry* 23 (3): 479–486.

Sauvaire, Y., G. Ribes., J. C. Baccou, and M. M. Loubatieeres. 1991. Implications of steroidal sapogenins in the hypocholesterolemic effect of fenugreek. *Lipids*. 26(3): 191–197.

Schauenberg, P., and F. Paris. 1990. *Guide to medicinal plants*. Cambridge: Lutterworth Press.

Scherbukhin, V. D., and O. V. Anulov. 1990. Legume seed galactomannans. *Applied Biochemistry and Microbiology* 35:257–274.

Seghal, G., G. S. Chauhan, and B. K. Kumbhar. 2002. Physical and functional properties of mucilage from yellow mustard (*Sinapis alba* L.) and different varieties of fenugreek (*Trigonella foenum-graecum* L.) seeds. *Journal of Food Science Technology* 39:367–370.

Seidemann, J. 2005. *World spice plants, economic usages, botany and taxonomy*, 372–374. Berlin: Springer Verlag.

Sen, B., and S. Gupta. 1979. Differentiation in callus of leaf of two species of *Trigonella*. *Physiologia Plantarum* 45:425–428.

Serpukhova, V. I. 1934. Trudy, *Prikl. Bot. Genet. i selekcii Sen.*, 7 (1): 69–106 (Russian).

Settu, A., B. D. Ranjitha Kumari, and R. Jeya Mary. 1997. *In vitro* selection for salt tolerance in *Trigonella foenum-graecum* using callus and shoot tip cultures. In *Biotechnology of spices, medicinal and aromatic plants*, ed. S. Edison, K. V. Ramana, B. Sasikumar, K. Nirmal Babu, and J. Eapen Santhosh, 119–121. Calicut: Indian Society for Spices.

Shah, M. A., and P. S. Mir. 2004. Effect of dietary fenugreek seed on dairy cow pea performance and milk characteristics. *Canadian Journal of Animal Science* 84:725–729.

Shang, M., Y. Tezuka, S. Cai, J. Li, S. Kadota, W. Fan, and T. Namba. 1998. Studies on triterpenoids from common fenugreek. *Zhongcaoyao* 29:655–657.

Shanmugavelu, K. G., N. Kumar, and K. V. Peter. 2002. *Production techniques of spices and plantation crops*, 109–130, Jodhpur, India: Agrobios.

Sharma, B. D. 1984. Hypocholesterolemic activity of fenugreek an experimental study in rats. *Nutrition Reports International* 30:221–231.

Sharma, G. L., and R. Kamal. 1982. Diosgenin contents from seeds of *Trigonella foenum-graecum* L. collected from various geographical regions. *Indian Journal of Botany* 5 (1): 58–59.

Sharma, H. R., and G. S. Chauhan. 2000. Physicochemical and rheological quality characteristics of fenugreek (*Trigonella foenum-graecum* L.) supplemented wheat flour. *Journal of Food Science Technology* 37:91–94.

Sharma, R. D. 1986. An evaluation of hypocholesterolemic factor of fenugreek seeds in rats. *Nutrition Reports International* 33 (4): 669–677.

Sharma, R. D., A. Sarkar, D. K. Hazra, I. Misra, J. B. Singh, and B. B. Maheshwari. 1996. Toxicological evaluation fenugreek seeds: A long-term feeding experiment in diabetic patients. *Phytotherapy Research* 10 (6): 519–520.

Sharma, R. K., and E. V. D. Sastry. 2001. In *Seed spices—Production, quality and export*, ed. S. Agarwal, E. V. D Sastry, and R. K. Sharma. Jaipur: Pointer Publisher.

Sharon, M., and M. Bhaskare. 1998. Influence of habituation of *Trigonella foenum-graecum* callus on endogenous level of IAA and total alkaloid content in callus. *Indian Journal of Plant Physiology* 3 (2): 163–165.

Shekhawat, N. S., and A. W. Galston. 1983. Mesophyll protoplasts of fenugreek (*Trigonella foenum-graecum* L.): Isolation, culture and shoot regeneration. *Plant Cell Reports* 2 (3): 119–121.

Sigurbjornsson, B. 1983. Induced mutations. In *Crop breeding*, ed. D. R. Wood, 153–176. Madison: American Society of Agronomy and Crop Science Society of America.

Sigurbjornsson, B., and A. Micke. 1974. Philosophy and accomplishments of mutation breeding. In *Polyploidy and induced mutations in plant breeding. Proceedings of the International Atomic Energy Agency*, Vienna, 1972, Bari, 303–343.

Singh, A. 1973. Studies on the interspecific hybrids of *Trigonella corniculata* and *T. hamosa* and *T. cretica*. *Genetica* (Dordrecht) 44:264–269.

Singh, A., and R. P. Roy. 1970. Karyological studies in *Trigonella, Indigofera* and *Phaseolus*. *Nucleus* (Calcutta) 13:41–54.

Singh, A., and D. Singh. 1976. Karyotype studies in *Trigonella*. *Nucleus* (Calcutta) 19:13–16.

Singh, D., and A. Singh. 1974. A green trailing mutant of *Trigonella foenum-graecum* L. Methi. *Crop Improvement* 1 (1–2): 98–100.

———. 1976. Double trisomics in *Trigonella foenum-graecum* L. *Crop Improvement* 3 (1–2): 125–127.

Singh, J., A. K. Singh, and S. S. Raghuvanshi. 1986. Vigor and fertility spectra in autotetraploid fenugreek. *Indian Journal of Horticulture* 43 (3–4): 278–280.

———. 1992. Depolarization and fertility improvement in advanced autotetraploids of *Trigonella foenum-graecum* L. *Indian Journal of Genetic Plant Breeding* 52 (4): 385–389.

Singh, J., S. S. Raghuvanshi, and A. K. Singh. 1991. Performance studies in F6 lines of autotetraploid fenugreek. *Plant Breeding* 107 (3): 251–253.

Singh, R. J., K. S. Boora, and G. P. Lodhi. 1993. Inheritance of yield and its component characters in fenugreek. *Annals of Biology* 9 (1): 123–126.

Singh, J. V., J. P. Lodhi, R. N. Arora, B. S. Jhorar, C. Kishor, and N. K. Thakral. 1993. Association analysis for some quantitative traits in fenugreek. *International Journal Tropical Agriculture* 11(3): 182–186.

Singh, R. P., S. Shaswat, and S. Kapur. 2004. Free radical and oxidative stress in neurodegenerative diseases: relevance of dietary antioxidants. *Journal of Indian Academy of Clinical Medicine* 5 (3): 218–225.

Singh, S., V. K. Singh, and D. K. Singh. 1997. Molluscicidal activity of some common spice plants. *Biological Agriculture & Horticulture* 14 (3): 237–249.

Singhania, D. L., R. S. Raje, D. Singh, and S. S. Rajput. 2006. Fenugreek. *In Advances in spices research—History and achievements of spices research in India since independence*, ed. P. N. Ravindran, N. Babu, K. N. Shiva, and J. A. Kallupurackal, 757–783. Jodhpur, India: Agribios.

Sinskaja, E. 1961. Flora of the cultivated plants of the U.S.S.R. XIII. In *Perennial leguminous plants. Part I: Medicago, sweet clover, fenugreek.* Jerusalem: Israel Program for Scientific Translations.

Skaltsa, H. 2002. Chemical constituents. In *Fenugreek: The genus Trigonella, medicinal and aromatic plants—Industrial profile*, ed. G. A. Petropoulos, 132–161, London: Taylor & Francis Inc.

Slinkard, A. E., R. McVicar, C. Brenzil, et al. 2006. Fenugreek in Saskatchewan (online). Available at http://www.agriculture.gov.sk.ca/Default.aspx/DN=c6428c37-cab6-4e93-b862-e20a55af3586 (accessed June 11, 2008).

Small, E., P. Lassen, and B. S. Brookes. 1987. An expanded circumscription of *Medicago* based on explosive flower tripping. *Willdenowia* 16:415–437.

Smith, A. 1982. Selected markets for turmeric, coriander, cumin and fenugreek seed and curry powder, Tropical Product Institute, publication no. G165, London.

Spices Board. 2009. Major item-wise export and production of spices. http://www.indianspices.com/html/s0420sts.htm (accessed 6.11.2009).

Srichamroen, A., B. Ooraikul, T. Vasanthan, P. Chang, S. Acharya, and T. Basu. 2005. Compositional differences among five fenugreek experimental lines and the effect of seed fractionation on galactomannans extractability of a selected line. *International Journal Food Sciences and Nutrition.*

Srivastava, A., and S. S. Raghuvanshi. 1987. Buffering effect of B-chromosome system of *Trigonella foenum-graecum* against different soil types. *Theoretical and Applied Genetics* 75:807–810.

Stuart, M. 1986. *The encyclopedia of herbs and herbalism.* London: Orbis.

Suboh, S. M., Y. Y. Bilto, and T. A. Aburjai. 2004. Protective effects of selected medicinal plants against protein degradation, lipid peroxidation and deformability loss of oxidatively stressed human erythrocytes. *Phytotherapy Research* 18 (4): 280–284.

Subramanian, N. S. 1996. *Laboratory manual of plant taxonomy.* New Delhi: Vikas Pub.

Sur, P., M. A. Das, J. R. Gomes, N. P. Vedasiromoni, S. Sahu, R. M. Banerjee, R. M. Sharma, and D. K. Ganguly. 2001. *Trigonella foenum-graecum* (fenugreek) seed extract as an antineoplastic agent. *Phytotherapy Research* 15:257–259.

Tahiliani, P. and A. Kar. 2003. Mitigation of thyroxine induced hyperglycaemia by two plant extracts. *Phytotherapy Research* 17(3): 294–296.

Taranalli, A. D., and I. J. Kuppast. 1996. Study of wound healing activity of seeds of *Trigonella foenum-graecum* in rats. *Indian Journal of Pharmaceutical Science* 58:117–119.

Taylor, W. G., M. S. Zaman, Z. Mir, P. S. Mir, S. N. Acharya, G. J. Mears, and J. L. Elder. 1997. Analysis of steroidal sapogenins from amber fenugreek by capillary gas chromatography and combined gas chromatography mass spectrometry. *Journal of Agricultural and Food Chemistry* 45:753–759.

Taylor, W. G., H. L. Zulyniak, K. W. Richards, S. N. Acharya, S. Bittman, and J. L. Elder. 2002. Variation in diosgenin levels among 10 accessions of fenugreek seeds produced in western Canada. *Journal of Agricultural and Food Chemistry* 50:5994–5997.

Tayyaba, Z., S. Nazrul Hasnain, and K. Hasnain. 2001. Evaluation of the oral hypoglycemic effect of *Trigonella foenum-graecum* L. (methi) in normal mice. *Journal of Ethnopharmacology* 75 (2–3): 191–195.

Tayyaba, Z., I. A. Siddiqui, and S. Nazrul Hasnain. 2001. Nematicidal activity of *Trigonella foenum-graecum* L. *Phytotherapy Research* 15 (6): 538–540.

Thiel, R. J. 1997. Effects of naturopathic interventions on symptoms associated with seasonal allergic rhinitis. *American Naturopathic Medical Association Monitor* 1:4–9.

Thomas, J. E., S. K. Basu, and S. N. Acharya. 2006. Identification of *Trigonella* accessions which lack antimicrobial activity and are suitable for forage development. *Canadian Journal of Plant Science* 86:727–732.

Thompson, Coon, J. S. and E. Ernst. 2003. Herbs for serum cholesterol reduction: A systematic view. *Journal Farming Practies* 52(6): 468–78.

Tiran, D. 2003. The use of fenugreek for breast feeding women. *Comparative Therapeutic Nurse Midwifery* 9 (3): 155–156.

Toshniwal, R. P. 1984. Stability analysis in fenugreek. MSc (Ag.) thesis, Sukhadia University, Campus, Jobner, Rajasthan.

Tutin, T. G., and V. H. Heywood. 1964. *Flora European*, vols. I and II. Cambridge, England: Cambridge University Press.

USDA. 2001. Nutrient database for standard reference: Release 14. Washington, D.C.: USDA.

Vaitsis, T. 1985. Creation of a new variety of fenugreek named Ionia resistant to *Sclerotinia sclerotiorum*. Fodder and Pastures Research Institute, Larissa, Greece (unpublished data).

Varshney, I. P., and D. C. Jain. 1979. Study of glycosides from *Trigonella foenum-graecum* L. leaves. *National Academies of Science Letters* 2:391–392.

Varshney, I. P., and S. C. Sharma. 1996. Saponins XXXII *Trigonella foenum-graecum* seeds. *Journal of Indian Chemistry Society* 43:564–567.

Varshney, I. P., and A. R. Sood. 1969. Study of the sapogenins from *Trigonella corniculata*. *Journal of Indian Chemistry Society* 46:331–332.

Vasil'chenko, I. T. 1953. Bericht uber die Arten der Gattung. *Trigonella. Trudy Botany Institute Akademice Nauk*. U.S.S.R. 1, 10.

Vats, V., S. P. Yadav, and J. K. Grover. 2003. Effect of *Trigonella foenum-graecum* L. on glycogen content of tissues and the key enzymes of carbohydrate metabolism. *Journal of Ethnopharmacology* 85 (2–3): 237–242.

Vavilov, N. I. 1926. Centers of origin of cultivated plants. *Trends in Practical Botany & General Selection* 16:3–24.

Wanjari, K. B. 1976. Karyotype studies in *Trigonella foenum-graecum* L. from anther somatic cells. *College of Agriculture, Nagpur* 49:51–53.

Xu, Z. H., M. R. Davey, and E. C. Cocking. 1982. Callus formation from root protoplasts of *Glycine max*. *Zeitschrift fur Pflanzenphysiologie* 107:231–235.

———. 1985. Root protoplast isolation and culture in higher plants. *Scientia Sinica Series B* 28 (4): 386–393.

Yoshikawa, M., T. Murakani, et al. 1997. Medicinal food stuffs. IV. Fenugreek seed. (1): Structures of trigoneosides 1a, 1b, IIa, IIIa and IIIb, new furostanol saponins from the seeds of Indian *Trigonella foenum-graecum* L. *Chemical and Pharmaceutical Bulletin Tokyo* 45 (1): 81–87.

Zia, T., Siddiqui, I. A., and Nazrul-Hasnain. 2001. Nematicidal activity of *Trigonella foenum-graecum* L. *Phytotherapy Research* 15(6): 538–540.

Ziyyat, A., A. Legssyer, et al. 1997. Phytotherapy of hypertension and diabetes in oriental Morocco. *Journal of Ethnopharmacology* 58 (1): 45–54.

Zupancic, A., D. Baricevic, A. Umek, and A. Kristl. 2001. The importance of fertilizing on fenugreek yield and diosgenin content in the plant drug. *Rostlinna Vyroba* 47 (5): 218–224.

CHAPTER 25

Aloe vera

José Imery-Buiza

CONTENTS

25.1 INTRODUCTION

The medicinal properties and ornamental value of *Aloe vera* have been recognized by civilizations throughout the world and are described in ancient documents, including the Bible. Its therapeutic potential is demonstrated every day in the many laboratories, medical centers, and traditional hospitals that study and evaluate natural healing alternatives for diseases such as cancer, as well as gastrointestinal, skin, cardiovascular, respiratory, and metabolic disorders. Their investigations have found that the administration of *A. vera* extracts acts successfully as a treatment to alleviate these conditions, both by itself or as a coadjuvant of formal medicine effects. There are currently over 800 km^2 cultivated with *A. vera*, mainly in dry regions in the Americas, southern Europe, Africa, Arabia, India, China, and Australia; they produce the raw material for the food, pharmaceutical, and cosmetics industries.

Agricultural and industrial activities based on *A. vera* have a projected annual growth of nearly 8% with an overall market of about US\$200 billion. However, low genetic variability and the emergence of phytosanitary problems threaten crop productivity, generating the need to better understand ecological and agronomic aspects of this species and to search for new cultivars. Several research centers worldwide have oriented their efforts in this direction, carrying out basic studies in genetics, biotechnology, physiology, ecology, reproductive biology, plant pathology, entomology, and other scientific disciplines in order to obtain pathogen-tolerant experimental genotypes together with increases in crop yield, efficiency, and quality. This chapter reviews the results of these investigations and also considers the history, distribution, taxonomy, phylogeny, composition and therapeutic uses, main features, and agronomic problems of *A. vera*, as well as achievements and prospects in plant breeding of this important medicinal species.

25.2 COMPOSITION AND THERAPEUTIC USES

Aloe vera is one of the few succulent plants that, throughout history, has maintained an important place in the pharmacopoeias of various cultures of the world. It was used by the ancient Egyptians in mummification rituals, and, in China and India, it has been applied in traditional medicine since 400 BC. Dioscorides mentions its use for the treatment of almost all ailments, from insomnia to itching of the eyes (Rowley 1997; Batugal et al. 2004). *Aloe vera* leaves contain anthraquinones, saccharides, vitamins, amino acids, minerals, enzymes, fatty acids (Table 25.1), other emollient, healing, clotting, moisturizing, antiallergic, disinfectant, anti-inflammatory, astringent, choleretic, laxative, and other compounds (Yagi and Takeo 2003; Ramachandra and Srinivasa 2008).

Currently, the two most important components of *A. vera* leaves (latex and pulp) are studied in several research centers worldwide as regards their use by different ethnic groups and to determine the mechanisms of action of the compounds contained within them. Latex, which is yellow and has an unpleasant odor, is also called leaf exudate, bitter sap, blood, or juice and is rich in phenolic compounds—mainly chromones, anthraquinones, and anthrones with particular therapeutic effects (Table 25.2). Furthermore, a colorless mucilaginous material (gel) is stored within the leaf (hydroparenchyma, pulp, or glass) that contains glycoproteins, lectins, and large quantities of polysaccharides, including acemannan (rich in mannose)—one of the most important for the food, cosmetics, and drug industries (Table 25.3).

Other species in the *Aloe* genus are also used for therapeutic, ornamental, ecological, and superstitious purposes (Hodge 1953; Morton 1961; Reynolds 1966; Duke and DuCellier 1993; Rowley 1997; Newton 2004):

A. arborescens, A. ballyi, A. barberae, A. chrysostachys, A. dawei, A. ferox, A. kedongensis, A. perryi, A. rivae, A. ruspoliana, and *A. saponaria* are used in green spaces, gardens, streets, edges of yards, private property boundaries, and buffer zones to protect against erosion.

A. arborescens, A. ferox, and *A. perryi* are raw materials for the manufacture of laxatives.

A. confusa and *A. megalacantha* are used in dyes.

A. boylei, A. cooperi, A. kraussii, and *A. minima* are used as moisturizers and fresh vegetables.

A. ferox is used in jams and mixed with tea.

A. macrocarpa is used as a condiment.

A. marlothii and *A. saponaria* are used for soaps, dyed cloth, and skins.

A. vaotsanda is a source of wood for construction.

A. zebrina is used in the preparation of cakes and pastries.

A. dichotoma and *A. buettneri* are used in crafts and for hunting tools.

Table 25.1   List of Main Constituents of the Leaves of *Aloe vera*

Anthraquinones and anthrones

Aloe-emodin	Barbaloin	Ester of cinnamic acid
Aloetic acid	Crysophanic acid	Isobarbaloin
Aloin	Emodin	Resistanol
Anthranol		

Amino acids

Alanine	Hydroxyproline	Phenylalanine
Arginine	Isoleucine	Proline
Aspartic acid	Leucine	Threonine
Glutamic acid	Lysine	Tyrosine
Glycine	Methionine	Valine
Histidine		

Elements

Aluminum	Chlorine	Phosphorus
Barium	Chromium	Silicon
Boro	Iron	Sodium
Calcium	Magnesium	Strontium
Copper	Manganese	Zinc

Enzymes

Alkaline phosphatase	Cyclooxidase	Oxidase
Amylase	Cyclooxigenase	Phosphoenolpyruvate carboxylase
Carboxypeptidase	Lipase	Superoxide dismutase
Catalase		

Miscellaneous compounds

2(3*H*)-benzothiazolone	Linoleic acid	Palmitoleic acid
Arachidonic acid	Linolenic acid	Pentadecanoic acid
Beta-sitosterol	Margaric acid	Potassium sorbate
Campestrol	Methyldehydroabietic acid	Salicylic acid
Cholesterol	Methylhexadecanoic acid	Stearic acid
Dioctyl phthalate	Monooctyl phthalate	Triglycerides
Gibberillin	Myristic acid	Triterpenoid
Lauric acid	Octadecanoic acid	Uric acid
Lignins	Palmitic acid	

Proteins

Lectins	Lectin-like subtance

Saccharides and carbohydrates

Acetylated glucomannan	Galactan	l-Rhamnose
Acetylated mannan	Galactogalacturan	Mannan
Aldopentose	Galactoglucoarabinomannan	Mannose (more common)
Arabinogalactan	Galactose	Pectin substance
Arabinose	Glucogalactomannan	Xilose
Cellulose	Glucose	Xylan
Fucose		

Vitamins

B_1 (tiamine)	B_7 (choline)	C (ascorbic acid)
B_2 (riboflavine)	B_9 (folic acid)	E (α-tocopherol)
B_6 (pyridoxine)	Beta-carotene	

Sources: According to Yamaguchi, I., N. Mega, and H. Sanada. 1993. *Bioscience, Biotechnology & Biochemistry* 57:1350–1352; Femenia, A. et al. 1999. *Carbohydrate Polymers* 39:109–117; Kim, H. S., S. Kacew, and B. M. Lee. 1999. *Carcinogenesis* 20:1637–1640; Vogler, B. K., and E. Ernst. 1999. *British Journal of General Practice* 49:823–828; Larionova, M. et al. 2004. *Revista Cubana de Plantas Medicinales* 9:146; Hamman, J. H. 2008. *Molecules* 13:1599–1616.

Table 25.2   Effects of *Aloe vera* Exudate or Its Derivative Compounds

Effect	Author
Anti-inflammatory	Tian et al. (2003)
Antimicrobial activities (*Corynebacterium, Escherichia coli, Salmonella, Streptococcus*)	Levin et al. (1988); Cowan (1999); Hamman (2008)
Antioxidant	Hu, Xu, and Hu (2003); Patel et al. (2007); Hamman (2008)
Antitumor, anticancer	Pecere et al. (2000); Acevedo et al. (2004); Cárdenas, Quesada, and Medina (2006); Sheng et al. (2007)
Antiviral	Sydiskis et al. (1991); Rivero et al. (2002); Cheng et al. (2008)
Corrosion inhibitor	Prato et al. (2008)
Cytotoxic	Cui et al. (2008)
Decrease the fat content in blood	Boik (1995)
Diuretic	Zhou and Chen (1988)
Wound healing	Izhaki (2002)
Immunostimulant	Boik (1995)
Laxative	Capasso et al. (1998); Van den Gorkom, de Vries, and Kleibeuker (1999)

A. ballyi and *A. ruspoliana* are used as repellents of mammal pests (mice, rats, hyenas, etc.).
A. arborescens, *A. aristata*, *A. ecklonis*, and *A. saponaria* are used in magic rituals and as amulets.

There are many other examples of the ethnobotanical applications of *Aloe* species by African tribes.

The use of aloe plants as magical and superstitious symbols is widespread throughout the world, especially in Africa and in those places where people of African descent live or where components

Table 25.3   Effects of *Aloe vera* Gel

Effect	Author
Anticancer	Kim et al. (1999); Wasserman et al. (2002); Schmidt and Ernst (2004); Wang et al. (2004)
Antidiabetic	Buckle (2001); Shane (2001); Yeh and Eisenberg (2003); Miyuki et al. (2006); Noor et al. (2008)
Anti-inflammatory, wound healing, treatment of hemorrhoids and anal fissures	León et al. (1999); Beyra et al. (2004)
Antimicrobial activities (*Shigella, Streptococcus*)	Ferro et al. (2003)
Antioxidant	Hamman (2008)
Antiparasitic; antiasthmatic; antihepatitis; healing of pelvic inflammation, arthritis, candidiasis, digestive disorders, fatigue, acne, psoriasis, genital herpes, hypertension	Vogler and Ernst (1999); Castellanos et al. (2001); Beyra et al. (2004); Morais et al. (2005)
Antitoxic activity	Anshoo et al. (2005)
Antitumor, stimulation of proliferation of T- and B-cells	Leung et al. (2004)
Antitussive, expectorant	Bennett and Prance (2000)
Antiviral, immunostimulant	Djeraba and Quere (2000); Pugh et al. (2001)
Bacteriogenic	Prato et al. (2008)
Food	Vega et al. (2005)
Gastroprotective and treatment of gastroduodenal ulcers	Álvarez et al. (1996); Beyra et al. (2004)
Healing of bone trauma	González, Sotolongo, and Batista (2002)
Protecting the skin from burns and infections	Esteban et al. (2000); Beyra et al. (2004); Richardson et al. (2005)
Stimulation of the hematopoietic system	Naranjo et al. (2005)
Systemic and topical analgesics	Furones, Morón, and Piñedo (1996)

of their culture have been adopted (Reynolds 1966). In the Caribbean and other American countries, whole *A. vera* plants are placed above the entrance doors to houses as a warning to people with malicious intent and to keep adversities of all kinds at bay. Some people even put coins at the base of the plants in the hope of economic prosperity for their families.

25.3 ORIGIN

The species we know today as *Aloe vera* was originally recorded by Dioscorides and Pliny (during the first century) as being native to the Mediterranean and was later observed in large populations from the Cape Verde, Tenerife, Madeira, and the Canary Islands in the Atlantic Ocean, as well as Gibraltar, India, and China (Crosswhite and Crosswhite 1984; Haller 1990). Most authors agree, however, that this species is probably native to the Arabian peninsula, possibly introduced and naturalized by Phoenicians when they settled in the eastern Mediterranean and from there distributed to the rest of the world (Forster and Clifford 1986; Carter 1994; Rowley 1997; Newton 2004; Lyons 2006; Smith and Van Wyk 2008). Nevertheless, the overexploitation of natural populations of *A. vera* for over 3,000 years due to their direct use for medicinal purposes, as well as their introduction, cultivation, and marketing in different regions of the world, makes the origin of *A. vera* uncertain (Rowley 1997; Campestrini et al. 2006).

There are also large populations of *A. vera* in the Caribbean islands and continental countries in the New World (Hodge 1953; Duke and DuCellier 1993; Albornoz and Imery 2003). When the taxonomic synonym *Aloe barbadensis* Mill. is used, the meaning of the specific epithet assumes that the plant is originally from Barbados, but there is no other species of the *Aloe* genus growing in natural conditions on this Caribbean island, contrasting with the high genetic diversity shown by congeners from Africa and Arabia. Thus, it seems unlikely that it has evolved in this region.

25.4 ETYMOLOGY

In 1753, the Swedish naturalist Carl Linnaeus was the first to identify this plant as *Aloe perfoliata* var. *vera*. Later, Philip Miller found that a plant collected in Barbados in 1696 and cultivated in English gardens had not been identified and named it *Aloe barbadensis* in his book, *Gardener's Dictionary*, published on April 16, 1768. Miller did not know that the species he had named was the same as that previously identified by Linnaeus and that it was probably introduced to the Caribbean by European navigators during the fifteenth and sixteenth centuries. Ten days before the publication of Miller's book, another botanist, Nicolaas Laurens Burman, renamed Linneaus's plant as *Aloe vera* in his book, *Flora Indica*; thus, the scientific name finally accepted was *Aloe vera* (L.) Burm. f. (Newton 1979, 2004; Carter 1994; Vega et al. 2005; Lyons 2007).

Controversy over the identity of *Aloe vera* is not limited to Linnaeus's, Miller's, and Burman's studies. In 1819, Adrian Haworth named a very similar plant that was introduced into South China during the Middle Ages as *Aloe barbadensis* var. *chinensis*. Half a century later, in 1877, the English botanist J. G. Baker studied this species in greater depth during its adult stage and renamed it *Aloe chinensis*. However, as was mentioned in the case of Miller and specimens collected in the Caribbean, it is unlikely that the plants studied by Baker and Haworth were native to China (Hodge 1953; Hu 2003).

It is now well known that *Aloe* species show highly variable vegetative characteristics, depending on their fertility and the surrounding temperature, relative humidity, photoperiod, altitude, soil water availability, etc.; this leads to frequent identification errors. In this regard, Reynolds (1950) noted that many species of *Aloe* vary considerably in the size of the plants, length and width of leaves, presence of leaf spots, size of the inflorescence, lengths of flower and pedicels, etc. when they are grown under garden conditions. Later in this chapter, other attributes and genetic markers that allow for a more precise identification of *A. vera* and its discrimination from other *Aloe* species will be discussed.

Etymologically, the scientific name *Aloe vera* is composed of a binomial name: the generic term *aloe*, derived from the Arabic word *alloch*, which translates as "bitter," in combination with the specific epithet *vera*, which means genuine or real (referring to its status as the "true aloe"). This name was widely used by traders in ancient times to guarantee their buyers that their products contained the exudates from true aloe (*Aloe vera*) and not extracts from other species of *Aloe* blended with the juice of "agaloco" (*Aguilaria agallocha*), a resinous tree from the Euphorbiaceae (EINE 2006). Some vernacular names include the following (Hodge 1953; Hänsel et al. 1994; Gage 1996; Ritter 1998; WHO 1999; Hu 2003; Batugal et al. 2004; Canevaro 2004; Añez and Vásquez 2005; Patel et al. 2007; Smith and Van Wyk 2008):

Aalwyn	Ghai kunwar	Mediterranean aloe
Ahalim	Ghai kunwrar	Miracle plant
Allal	Gheekuar	Murr sbarr
Alloch	Ghikanvar	Musabar
Alloeh	Ghikuar	Musabbar
Aloe curaçao	Ghikumar	Obeiknowennoi sabur
Aloe vera	Ghikumari	Pita zabila
Aloes	Ghikwar	Rokai
Aloès	Ghirita kumari	Sabbara
Áloe	Ghiu kumari	Saber
Äloe	Ghrita kumari	Sabila
Aloès du Cape	Ghritakumari	Sábila
Aloès fèroce	Grahakanya	Sabilla
Aloes vrai	Gwar-patha	Sabr
Aloès vulgaire	Haang takhe	Saibr
Alovis	Hlaba	Savila
Azebre vegetal	Indian aloe	Savilla
Babosa	Isha irasu	Sebet sicutri
Babosa-medicinal	Jadaim	Semper vivum
Barbados aloe	Jadam	Shubiri
Barbadoes aloes	Kanyasara	Siang-tan
Barbadoes aloe	Korphad	Siba
Bergaalwyn	Kumari	Sibr
Bitteraalwyn	Kumaro	Star cactus
Bitter aloe	Kunvar pata	True aloe
Burn aloe	Kunwar	Tuna
Burn plant	Laloi	Umhlaba
Chinese aloe	Laluwe	Waan haang charakhe
Chirukattali	Lo-hoei	Wand of heaven
Curaçao aloe	Lo-hoi	Wan-hangchorakhe
Curaçao aloes	Lou-houey	West Indian aloe
Curaçao alos	Luchuy	Yaa dam
Dilang buwaya	Ludia byah	Yadam
Dilang halo	Luhui	Zábida
Echte aloe	Lu wei	Zábila
Erba babosa	Manjikattali	Zábira
Erva-babosa	Medicinal aloe	Zambila
First aid plant	Medicine plant	Závila

25.5 TAXONOMY AND EVOLUTION

Aloe vera is a monocotyledon belonging to the *Aloe* genus within the Aloaceae. This family is currently divided into five genera: *Aloe* L. (with over 550 species), *Astroloba* Uitewaal (7 species), *Chortolirion* Berger (1 species), *Gasteria* Duval (16 species), and *Haworthia* Duval (68 species) (Van Jaarsveld 1994; Smith and Van Wyk 2008). The taxonomic placement of this group of plants has been intensely debated. Originally, they were assigned to the tribe Aloineae in the Liliaceae, due to their lily-like flowers (Hutchinson 1959; Riley and Majumdar 1979). Later, they were relocated to the subfamily Alooideae within the Asphodelaceae (Dahlgren and Clifford 1982; Judd et al. 1999) and were left there, since phylogenetic studies based on chloroplast DNA (*rbc*L, *trn*L-F, and *mat*K) and repeated genomic sequences (ISSR) could not provide sufficient evidence for their separation (Chase et al. 2000; Treutlein et al. 2003). Later, based on an idea originally proposed by August Batsch in 1802, it was suggested that these succulent plants could be separated at a higher status (Smith and Steyn 2004).

By considering these plants as a monophyletic group with thick roots, succulent leaves arranged in rosettes, margins usually with teeth or spines, vascular bundles in a ring around the ground parenchyma, a cap of aloine-rich cells at the phloem pole, uniformity in chromosomes and chemical properties (mainly 1-methyl-8-hydroxianthraquinones from roots and anthrone-C-glycosides in the leaves), these plants were grouped in a new family called Aloaceae (Forster and Clifford 1986; Carter 1994; Smith and Van Wyk 1998, 2008; Viljoen, Van Wyk, and Newton 2001). For a time, this family was known as Aloeaceae, the taxon suggested by Cronquist (1981), which had a spelling error that was discussed for years until its eventual correction to the current term Aloaceae (Brummit 1992; Smith 1993; Glen and Hardy 2000).

In many publications, the name *Aloe barbadensis* has been used to refer to this species. When we consider the International Rules of Botanical Nomenclature, however, it is clear that this is a synonymy and that the correct name is *Aloe vera* (L.) Burm. f. (Newton 1979; Tucker, Duke, and Foster 1989). Other synonyms that have been recorded are *A. barbadensis* var. *chinensis* Haw., *A. chinensis* Bak., *A. elongata* Murray, *A. indica* Royle, *A. officinalis* Forssk., *A. perfoliata* L., *A. rubescens* DC, *A. vera* var. *littoralis* König ex Bak., *A. vera* var. *chinensis* Berger, and *A. vulgaris* Lam. (Hodge 1953; Hänsel et al. 1994). Some amateurs have even mistaken *A. vera* with *A. massawana* Reyn., *A. perryi* Bak., or *A. succotrina* Lam. due to the similarity of their leaf traits (Jacobsen 1954; Sapre 1975; Imery and Caldera 2002; Darokar et al. 2003; Lyons 2007). In order to clarify the complete identification of *Aloe* species and that of natural or experimental variants, further evaluations at different stages of plant development that combine several descriptors, including morphological, anatomical, pollinic, embryological, cytogenetic, biochemical, and molecular traits, are required (Valdés 1997; Imery-Buiza 2007a).

The greatest diversity of *Aloe* species is concentrated in southeast Africa and on the island of Madagascar (Treutlein et al. 2003). In South Africa, there are some 119 species, of which nearly 60% (71 species) are endemic. Other centers of origin are located in other countries of continental Africa (123 species), the Indian Ocean islands (86 species), and the Arabian peninsula (26 species). The highest level of endemism (100%) is observed on the island of Madagascar, which has 77 native species that have little phylogenetic connection with their continental congeners (Holland 1978; Newton 2004). In addition, there is a close relationship between the species from North Africa and those native to the Arabian peninsula; noteworthy among these is *A. officinalis* Forssk., an endemic species from Yemen very similar to *A. vera*.

Some scientists believe that the species now known as *A. vera* are ancient cultigens—the result of prehistoric selective breeding (Newton 1979) in the absence of a natural process of speciation in a given region. It is possible that *A. vera* is a product of hybridization between wild species or between wild and cultivated species selected in the same place for their potential usefulness, resulting in a low-fertility hybrid that has been maintained by vegetative propagation (Rowley 1997; Newton 2004; Imery-Buiza, Raymúndez, and Menéndez-Yuffa 2008). Thus, the high similarity between

A. officinalis and *A. vera* suggests that these species are related and that *A. officinalis* may be one of the progenitors of *A. vera* or at least a close ancestor undergoing a gradual process of evolution.

25.6 BOTANICAL AND ECOLOGICAL FEATURES

Aloe vera is a perennial herb, without apparent stems (stemless) in plants under 5 years old. In large populations more than 50 years old, there are adult plants with creeping rhizomes almost 40 cm long and 6–7 cm in diameter (Albornoz and Imery 2003). A small stem can be distinguished in cultivated plants whose leaves have been harvested from the base, leaving an uncovered section of 1.6 cm per year. In these plants, the stem is wrapped with thin ochre-colored layers, formed from the remains of the dried leaf ligules covering the compact sand-colored internodes. The beige lateral buds are aligned in tight, light brown knots below the ligule. From these the stem meristem generates stolons 6–11 mm in diameter and of variable length that grow and emerge from the soil in the form of suckers close to the mother plant. The main root may be distinguished by its darker color (brown to coffee), absence of sheaths, and emission of secondary roots of the same color at their base, which become lighter along their length and are yellowish at the apex (Figure 25.1).

Aloe vera plants show a high plasticity in the expression of leaf attributes, depending on their age and interactions with environmental factors (Yépez et al. 1993; Hernández-Cruz et al. 2002), which may lead to confusion in identification (Smith and Steyn 2004). In general, however, these plants have succulent narrow-lanceolate leaves, without fibers, that are upright; parallel venation; and soft, light green teeth that line the margins to the leaf apex. In young plants, the leaves emerge in the opposite direction and the blade is grass green with white spots irregularly distributed on both sides. In adult plants, the leaves are an apple-green color, arranged in rosettes without spots and covered by a waxy layer that gives them a grayish hue; they measure up to 70 cm long, 5–12 cm wide, and 1–3 cm thick. Plants that grow in water-stressed environments with nitrogen deficiency have yellowing to reddish leaves of smaller size and darker rigid teeth. The lack of other foliar nutrients (P, K, Ca, Mg, S, and Zn) induces marked changes in the shape, color, brightness, and size of the leaves (Figure 25.2). The lack of other foliar nutrients (P, K, Ca, Mg, S, and Zn) induces marked changes in the shape, color, brightness, and size of the leaves (Figure 25.2) (Rodríguez-Morales 1996; Fuentes-Carvajal, Véliz and Imery 2006). The interior of a mature leaf is shown in Figure 25.3, which clearly distinguished its components: cuticle, epidermal cells, stomatal apparatus, mesophyll, vascular bundles in a ring around ground hydroparenchyma, cap cells (rich in aloine) at the phloem pole, and a hydroparenchyma with large, thin-walled cells.

Flowers are grouped in indeterminate inflorescences with simple or compound racemes (maximum of two secondary axes) emerging laterally from the upper leaves. Each plant produces up to three inflorescences and a maximum of 380 flowers per raceme. The color of the six overlapping tepals depends on the state of maturity of the flower, from lemon yellow with apple-green lines before anthesis to yellow ochre along the whole perianth during flowering. The flower is pedicellate, tubular, gamotepal, zygomorphic, and bisexual and is directed toward the ground after anthesis. The androecium consists of six free stamens; cream-colored, flattened filaments; and oblong, yellow, dithecal, basifixed anthers with purple backs and longitudinal introrse dehiscence. Pollen grains are oval, smooth, and monosulcate (Sanghi and Sarna 2001). Pollen germination capacity (*in vitro*) is maintained for about 24 h after anthesis, then drops rapidly and declines completely at 130 h (Imery and Cárdenas 2006). The gynoecium has three zyncarpous carpels; an elongated, trilocular, superior ovary with two rows of anatropous ovules per locule; axial placentation; and an elongated, cylindrical open style, which terminates in a small, thin papillose stigma (Figure 25.4). Both sexual organs are exerts after anthesis.

Flowering occurs during the day, with little odor, abundant nectar, and ornithophily as the pollination syndrome. The highest volume of nectar (0.3–0.4 mL per flower), with a sucrose concentration

Figure 25.1 Stem and root in plants of *Aloe vera*. (A) Stem discovered by continuous harvest of leaves at the base of the plant. Ligule remains and suckers' emergence are observed. (B) Ligule (top), rhizome (middle), and main root with secondary roots (lower). (C) Details of the nodes (black arrows), internodes (white arrow), and lateral buds (circles) in the rhizome.

of 14–21%, occurs from the onset of anthesis until a day later. During November to May, about 20 flowers per raceme open per day. This profuse flowering positively affects the populations of *Apis mellifera*, and some beekeepers even suggest that the large amount of available nectar and pollen available from *A. vera* inflorescences induces bees to change their flight distance patterns and frequency of visits to other plant species (Figure 25.4A, Figure 25.5A) (Watt and Breyer 1962; Newton 2004; Velásquez-Arenas and Imery-Buiza 2008).

Figure 25.2 Nutrient deficiency symptoms in plants of *Aloe vera*. (A) Hydroponic culture solution (1X) and devoid of the elements nitrogen (–N), phosphorus (–P), potassium (–K), calcium (–Ca), magnesium (–Mg), sulfur (–S). (B) Plant grown on soil with zinc deficiency.

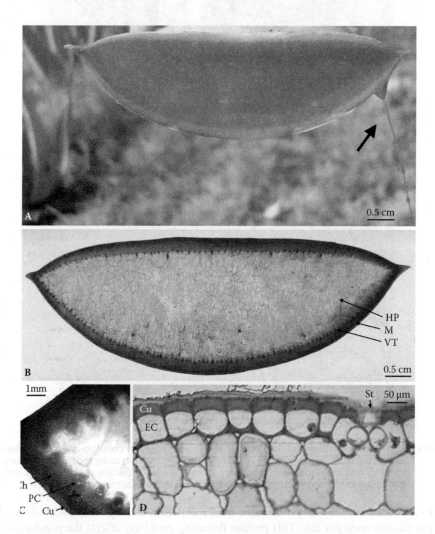

Figure 25.3 Cross section of a leaf of *Aloe vera*. (A) Recently cut leaf indicating accumulation of latex exudate (thick black arrow). (B) Leaf profile, indicating the hydroparenchyma (HP), mesophyll (M), and vascular tissue (VT). (C) Anatomical detail, noting the cuticle (Cu), cap of aloine cells at the phloem pole (PC), and chlorenchyma (Ch). (D) Details of the leaf surface: cuticle (Cu), epidermal cells (EC), and stomate (St).

Other diurnal visitors to the flowers are birds *Leucippus fallax*, *Amazilia tobaci*, and *Icterus nigrogularis* and hymenopteran insects in the *Trigona*, *Poliste*, *Aumenes*, and *Vespa* genera. Some less frequent nocturnal visitors are moths and ants (Figure 25.5). The anthers release pollen at least 1 day prior to stigma receptivity (a dichogamy condition that ranks as protandric). However, the flowers of *A. vera* are arranged in clusters, so protandry as a barrier to prevent autofertilization is

Figure 25.4   Flowers and fruits of *Aloe vera*. (A) Flowering plants grown in eastern Venezuela. (B) Inflorescence emergence. (C) Details of the immature raceme and its flower buds. (D) Flowering racemes of greenhouse plants. (E, F) Details of the anthers. (G) Long section of ovary showing aligned ovules within the locules (20–24 eggs per locule). (H) Cross section of ovary; three carpels, three locules, and anatropous ovules in axial placentation are visible. (I) Details of the papillose stigma. (J–L) Seed after crossing with pollen from *A. littoralis*.

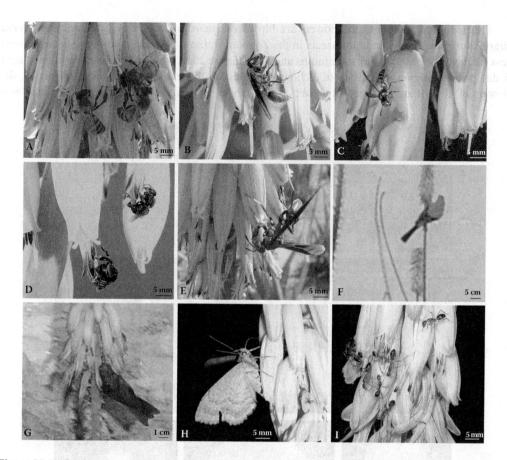

Figure 25.5 Some flower visitors of *Aloe vera*. Day visitors: (A) *Apis mellifera*, (B) *Eumenes* sp., (C) *Poliste* sp., (D) *Trigona* sp., (E) *Vespa* sp., (F) *Icterus nigrogularis*, (G) *Leucippus fallax*. Night visitors: (H) moths, (I) ants.

combined with a mechanism of self-incompatibility fairly widespread throughout the *Aloe* genus (Newton 2004; Imery-Buiza and Cequea-Ruíz 2008; Velásquez-Arenas and Imery-Buiza 2008). Flowers produce 253,000 grains of pollen and 60–66 ovules, equivalent to an average of 4,150 pollen grains per ovule, which indicates that plants of *A. vera* are obligate xenogamous (Cruden 1977) and that there is a marked trend toward cross-fertilization.

In lower and drier regions of the neotropics, where there are large plantations and naturalized populations of *A. vera*, fruits were not observed even after a lush flowering. In research centers and botanical gardens where several species of *Aloe* are conserved, however, the development of fruits and seeds after natural or manual crosses has been reported (Riley and Majumdar 1979; Imery-Buiza and Cequea-Ruíz 2008). Fruits are subglobular capsules—apple green during their development and changing gradually to toasted ochre from the apex toward the base. When fruits are dried, apical dehiscence occurs, releasing the seeds. Seeds are dark brown and flattened, with membranous wings that are copper colored toward the edges (Figure 25.5). The size of the fruit and seeds depends on the origin of the pollen and has been used as an indicator of crossability (Imery-Buiza 2005).

Aloe vera is a xerophytic plant with structural and physiological adaptations that allow it to survive in arid or semiarid regions with erratic rainfall. The succulence of the leaves is a xerophytic adaptation based on the presence of a specialized tissue (hydroparenchyma) where water is stored in large cells with thin walls (Figure 25.3). Mucilage contained in this tissue maintains the water status of the plant, due to little variation in its water potential even in drought conditions. *A. vera* plants show crassulacean acid metabolism (CAM) and open their stomata at night to convert atmospheric carbon dioxide into malic acid, which they use to perform photosynthesis during the day, when their stomata are closed to prevent loss of moisture (Heyes 1989; Salisbury and Ross 1991; Eller, Ruess, and Ferrari 1993; Díaz 2001). The CAM activity and the accumulation of large quantities of polysaccharides in the leaves are mechanisms of drought resistance and may aid the rehydration of other leaf cells with a lower water potential (Morse 1990; Goldstein, Andrade, and Nobel 1991; Clifford et al. 2002).

Aloe vera can be found anywhere in the world, but its extensive cultivation is limited to tropical and subtropical regions. In the Americas, it is commercially exploited in large plantations in the southern United States, Mexico, the Dominican Republic, Haiti, Puerto Rico, Cuba, the Bahamas, Barbados, Aruba, Bonaire, Curacao, Belize, Guatemala, Costa Rica, Colombia, Venezuela, Ecuador, Brazil, Peru, Bolivia, and down to northern Chile, Argentina, and Paraguay. In the Old World, it is cultivated throughout the Mediterranean and other dry areas of Spain, Italy, India, South Africa, the Arabian peninsula, China, Malaysia, Japan, Australia, Vietnam, Taiwan, and Nigeria (Haller 1990; WHO 1999; Newton 2004; EINE 2006).

25.7 PESTS AND DISEASES

Although *A. vera* is recognized as a species with few natural enemies and is used as a source of compounds with biological activity for the control of some organisms, there are reports of damage by arthropods, nematodes, fungi, and bacteria (Duke and DuCellier 1993; Saks and Barkai 1995; Simmonds 2004; Hernández, Bautista, and Velázquez 2007).

Fungi *Alternaria alternata, Alternaria* sp., *Botryodiplodia* sp., *Byssochlamys nivea, Cercosporidium* sp., *Colletotrichum* sp., *Coniothyrium concentricum, Corynespora* sp., *Curvularia* sp., *Drechslera spicifera* (= *Bipolaris spicifera*), *Exserohilum rostratum, Leptosphaeria nigrans, Macrophoma* sp., *Melanospora zamiae, Phyllosticta* sp., and *Physalospora* sp. cause leaf spots (Zhong et al. 1993; García 2000; IARI 2007; Kamalakannan et al. 2008; Farr and Rossman 2009). In the case of *Alternaria*, the fungus not only causes localized damage in *A. vera* leaves but also reduces the effectiveness of certain therapeutic compounds in infected plants (Pritam and Kale 2007). The fungal species *Uromyces* had been reported as responsible for rust disease in other *Aloe* species (McClymont 1989; Van Wyk and Smith 1996; Simmonds 2004), but there are now plantations of *A. vera* showing leaf damage caused by this pathogen (Montón et al. 2004).

Fungi *Fusarium oxysporum, Fusarium solani, Fusarium* sp., *Lasiodiplodia theobromae, Pythium ultimum, Phytophthora parasitica, Phytophthora* sp., and *Sclerotium rolfsii* produce stem rot (Averre and Reynolds 1964; Albarracín et al. 2001; Lugo and Medina 2001; Hirooka et al. 2007; Ji et al. 2007; Farr and Rossman 2009). Fungi *Fusarium oxysporum, Fusarium solani, Phytophthora parasitica, Phytophthora* sp., *Rhizoctonia solani, Rhizoctonia* sp., and *Sclerotium rolfsii* cause root rot (Duke and DuCellier 1993; Lugo and Medina 2001; Ji et al. 2007; Farr and Rossman 2009).

Bacteria *Pantoea agglomerans* produces aqueous spots that develop into necrotic spots, and *Erwinia chrysanthemi* causes watery rot followed by plant death (De Laat, Verhoeven, and Janse 1994; Lugo 1999; Lugo and Medina 2001; Trujillo et al. 2001; Mandal and Maiti 2005).

In Venezuela, the main *A. vera* pathogen is the bacterium *E. chrysanthemi*, which causes watery rot with an incidence of up to 4% in cultivated plants. In soils with poor drainage and excess water from continuous rainfall or badly managed irrigation, the incidence may reach 35%, exceeding the economic threshold level. The disease begins with chlorosis in older leaves and progresses to swelling in the base of the leaves, gas-bubble formation, loss of stiffness, overturning of the leaves, loss of leaf content, and plant death 8–10 days after the first symptom (Figure 25.6). Other economically important pathogens are the fungi *Alternaria* sp. (3%), *Macrophoma* sp. (1%), *Colletotrichum* sp. (0.7%), *Bipolaris* sp. (0.5%), *Rhizoctonia* sp. (0.5%), *Fusarium* sp., and *Sclerotium rolfsii* (0.4%) (Figure 25.7).

Figure 25.6   Symptoms of leaf rot caused by *Erwinia chrysanthemi* in plants of *Aloe vera*. (A) Early symptoms: chlorosis and swelling (arrow) at the base of the leaf (1–3 days). (B) Loss of stiffness and dumping the leaves (4–6 days). (C) Watery rot and death of the plant 8–10 days after seeing the first symptoms of disease.

The damage caused by pests of *A. vera* with an incidence less than 2% varies with their density, agronomic practices, and climatic conditions. There are reports of severe damage caused by the aphid *Aloephagus myersi* (Halbert 2002) and the nematodes *Meloidogyne incognita* and *Helicotylenchus dihystera* (Del Cid 2002; Martínez 2008). Some farmers from Chile and South Africa have reported that attacks by slugs (*Cantareus aspersus*) caused foliar damage.

In eastern Venezuela, the nematode *Meloidogyne* sp. parasitizes the roots and causes the weakening and eventual death of the plant (0.05%). The leaf-cutting ants *Acromyrmex octospinosus* cause defoliation of the plant (0.8%), mainly during very dry periods. The larvae of the lepidopteran

Figure 25.7 Some diseases caused by fungi in plants of *Aloe vera* grown in eastern Venezuela. (A) Leaf spot and necrosis caused by *Alternaria* sp. (B) Antragnosis by *Colletotrichum* sp. (C) Stem rot by *Fusarium* sp. (D) Root rot by *Rhizoctonia* sp.

Spodoptera sp. may damage the stem apices of plants grown over large areas (0.3%) and nurseries (1.2%) (Figure 25.8).

Other species of *Aloe* are attacked by mites (*Aceria aloinis, Eriophyes aloinis,* and *Tetranychus cinibarinus*), aphids (*Aphis craccivora, A. fabae, A. gossypii, Myzus persicae,* and *Toxoptera aurantii*), mealy bug (*Planococcus citri*), scales (*Coccus hesperidum, Duplachionaspis exalbida, D. humilis, Saissetia coffeae,* and *Separaspis capensis*), beetles (*Brachycerus* sp., *Mecistocersus aloes, Rhadinomerus illicitus,* and *Scyphophorus acupunctatus*), mirids (*Aloea australis* and *Helopeltis schoutedeni*), and thrips (*Frankliniella occidentalis, Heliothrips haemorrhoidalis,* and *Thrips tabaci*) (McClymont 1989; Duke and DuCellier 1993; Van Wyk and Smith 1996; Simmonds 2004; Fukada 2006; Kelly and Olsen 2006). The transmission of viruses and other biological agents causing diseases is a problem associated with infestation by these arthropods. Many of these small

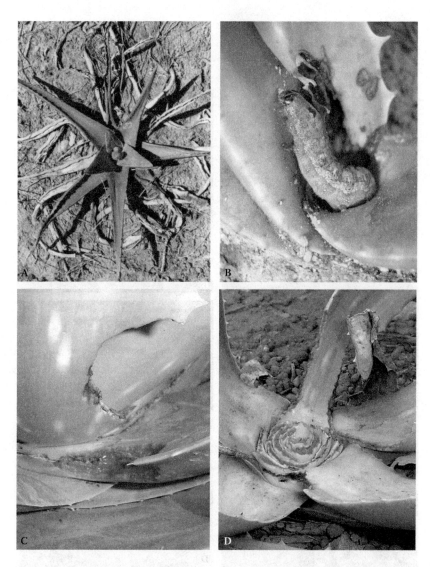

Figure 25.8   Damage caused by pests in some *Aloe vera* plants cultivated from eastern Venezuela. (A) Parasitism by *Meloidogyne* sp. (B) Damage to cauline apex by *Spodoptera* sp. (C, D) Defoliation caused by *Acromyrmex octospinosus*.

animals are endemic species from the Old World and have not yet been reported from neotropic plantations of *A. vera*; it is important, however, to maintain controls that prevent the introduction of these potential pests.

25.8 GENETICS

25.8.1 Karyotype

Most species in the Aloaceae present a bimodal karyotype with $2n = 2x = 14$ chromosomes, with the exception of some polyploid species with $2n = 21, 28, 35,$ or 42. The basic number $x = 7$ is composed of four large acrocentric/submetacentric chromosomes (L_1–L_4) from 12 to 18 μm and three small submetacentric chromosomes (S_1–S_3) from 4 to 6 µm (Snoad 1951; Brandham 1971; Riley and Majumdar 1979). There are differences in the length of chromosomal arms, the distribution of nucleolar organizers, meiotic configurations, and the quantity of DNA from one species to another (Brandham 1969a, 1969b, 1970, 1974a, 1974b, 1975, 1977a, 1977b, 1990; Brandham and Johnson 1977; Cavallini 1993; Takahashi et al. 1997; Adams et al. 2000; Imery and Caldera 2002; Bennett and Leitch 2005). However, they all show a marked similarity in the number and morphology of their chromosomes, due to the action of a karyotypic orthoselection process (Brandham and Doherty 1998).

The *A. vera* karyotype $2n = 2x = 14 = 8L + 6S$ has been studied after suspending mitosis in the root tips (Figure 25.9A–D). Among the eight large chromosomes (L), the first pair (L_1) may be

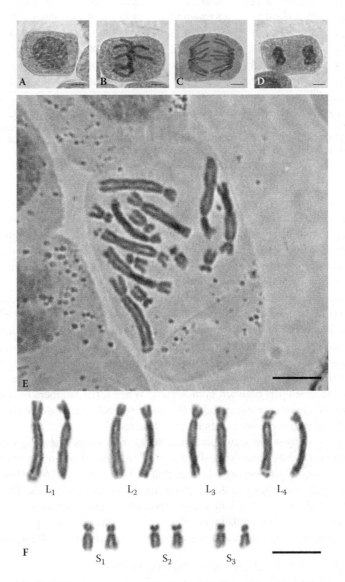

Figure 25.9 Mitosis and somatic chromosomes in diploid plants of *Aloe vera*. (A) Prophase. (B) Metaphase. (C) Anaphase. (D) Telophase. (E) Bimodal karyotype $2n = 2x = 14 = 8L + 6S$. (F) Karyogram with large chromosomes (L_1–L_4) and small chromosomes (S_1–S_3). Bar = 10 µm.

Table 25.4 Morphometric Characteristics of the Mitotic Chromosomes of *Aloe vera* (Naturalized Plant from Eastern Venezuela)

No.	p	q	L	RL	r	Cl_1	Cl_2
1	3.55 ± 0.07	10.40 ± 0.11	13.95 ± 0.09	10.35 ± 0.10	2.93 ± 0.08	sm	L_1
2	3.62 ± 0.04	10.72 ± 0.14	14.34 ± 0.16	10.64 ± 0.14	2.96 ± 0.04	sm	L_1
3	2.79 ± 0.02	10.66 ± 0.16	13.45 ± 0.16	9.98 ± 0.03	3.82 ± 0.06	st	L_2
4	2.78 ± 0.04	10.75 ± 0.09	13.52 ± 0.09	10.03 ± 0.06	3.87 ± 0.06	st	L_2
5	2.70 ± 0.05	10.17 ± 0.12	12.87 ± 0.15	9.55 ± 0.04	3.77 ± 0.06	st	L_3
6	2.65 ± 0.03	10.21 ± 0.12	13.86 ± 0.13	9.54 ± 0.04	3.85 ± 0.05	st	L_3
7	2.58 ± 0.04	10.63 ± 0.11	13.21 ± 0.13	9.80 ± 0.03	4.12 ± 0.06	st	L_4
8	2.38 ± 0.04	10.47 ± 0.12	12.85 ± 0.14	9.53 ± 0.03	4.41 ± 0.08	st	L_4
9	1.73 ± 0.03	3.25 ± 0.12	4.99 ± 0.12	3.70 ± 0.05	1.88 ± 0.07	sm	S_1
10	1.71 ± 0.03	3.31 ± 0.12	5.02 ± 0.13	3.72 ± 0.06	1.94 ± 0.07	sm	S_1
11	1.51 ± 0.03	2.98 ± 0.13	4.49 ± 0.15	3.33 ± 0.07	1.97 ± 0.18	sm	S_2
12	1.48 ± 0.03	2.94 ± 0.15	4.42 ± 0.16	3.28 ± 0.08	1.99 ± 0.11	sm	S_2
13	1.37 ± 0.04	3.08 ± 0.12	4.45 ± 0.14	3.30 ± 0.06	2.20 ± 0.11	sm	S_3
14	1.36 ± 0.04	3.03 ± 0.12	4.39 ± 0.13	3.25 ± 0.06	2.10 ± 0.04	sm	S_3

Notes: p: length of short arm (µm); q: length of long arm (µm); L: length of chromosome (µm); RL: relative length (%); r: arm ratio (q/p); Cl_1: chromosome classification according Levan et al. (1964); sm: centromere located in submedian position; st: subterminal position; Cl_2: chromosome classification according to Brandham (1971); L_1–L_4: large chromosomes; S_1–S_3: small chromosomes. Values indicate mean ± standard deviation from n = 30 cells (one cell per plant).

clearly distinguished by having the longest short arms (3.6 µm) and the fourth pair (L_4) for having the smallest short arms (2.5 µm); the second (L_2) and third (L_3) pairs have short arms of intermediate length (2.7 µm) and are hardly distinguishable with the naked eye. The small chromosomes (S) are less distinguishable from each other; the S_1 pair is slightly larger than the S_2 and S_3 pairs (Figure 25.9e). Table 25.4 summarizes the metric and morphological data of the somatic chromosomes of *A. vera* (using naturalized populations from eastern Venezuela).

Classical cytogenetic evaluations of *A. vera* require special care as regards sample size, accuracy of species identification, the sensitivity of chromatin to crushing during plate preparation, small differences in length between homologues, overlapping between sister chromatids, and the visualization of secondary constrictions. If large homologues (L_1–L_4) are measured after grouping them according to the characteristics of the short arms, discrepancies in the length of the long arms can be identified. These differences are not generally taken into account and may be diluted when calculating mean lengths between homologues. When large numbers of specimens (at least 30 plants from spaced clones) are examined in this way and each chromosome analyzed separately, small variations between large homologues may be detected.

These differences suggest the accumulation of structural changes that slightly alter the length of one or more chromosomes, signifying the origin of mutants of scientific interest. Differences between homologues could also endorse the idea that *A. vera* originated by hybridization between species with karyotypes that differ in the length of the larger chromosomes. These chromosomal changes are maintained due to the lack of sexual reproduction in this species.

The following karyological records reflect changes in chromosome morphology and a higher or lower probability of specifying the location of secondary constrictions in the *A. vera* karyotype:

$2n = 2x = 14 = 2L_{sat}sm + 4Lst + 2L_{sat}st + 4Sm + 2Ssm$ (Sapre 1978)

$K(2n) = 14 = 2A^{st}*4B^{st}2C^{st}6D^{sm}$ equivalent to $2n = 2x = 14 = 2L_{sat}st + 4Lst + 2L\ st + 6Ssm$ (Vij, Sharma, and Toor 1980)

$2n = 2x = 14 = 2Lsm + 4Lst + 2Lt + 2Sm + 4Ssm$ (Almasan et al. 1991)

$2n = 2x = 14 = 8Lsm + 6Ssm$ (Matos and Molina 1997)

$2n = 2x = 14 = 2L_{sat}sm + 4Lst + 2L_{sat}st + 6Ssm$ (Imery and Caldera 2002)

Some of the discrepancies found in the L_1 pair could be due to the fact that the arm ratio "r" (long arm length/short arm length), used when applying the Levan, Fredga, and Sandberg (1964) classification system, gives a value very close to the threshold between the *st* (subterminal centromere position) and *sm* (submedian centromere position) classes. A similar situation could apply for one or two pairs of small (S) with r values that generate an ambiguous location between the *sm* and *m* (median centromere position) classes. Moreover, the *t* class (terminal centromere position), assigned to the L_4 pair in a study by Almasan et al. (1991), provides evidence of structural chromosomal changes in specimens analyzed from the Philippines. Furthermore, the little-known triploid specimen ($2n = 3x = 21 = 12Lst + 9Sm$) found in a population from India (Abraham and Nagendra 1979) and possibly caused by the fusion of the $n = x = 4Lst + 3Sm$ (normal) and $n = 2x = 8Lst + 6Sm$ (not reduced) gametes is another example of the existence of *A. vera* cytotypes in different regions of the world.

Secondary constrictions are another species-specific attribute of the *A. vera* karyotype. When microscopic slide preparations of cells in prometaphase are obtained, four satellites of up to 1 μm long, separated by achromatic regions that indicate the position of the subterminal secondary constriction in the long arms (q) of chromosome L_1 and L_4, can easily be seen (Figure 25.9e, f). Occasionally, small satellites on the terminal short arms (p) of the S_1 pair are also observed. The use of rDNA probes (18S-5.8S-26S) in assays of *in situ* hybridization (FISH) confirms the association between secondary constrictions and nucleolar organizers (NORs) present at the qL_1, qL_4, and $_pS_1$ ends (Adams et al. 2000). This correspondence was also demonstrated by observing a maximum of six nucleoli in G_1 interphase cells (Duarte 2009) and Ag-NOR bands on most of the secondary constrictions (Imery-Buiza 2007a).

25.8.2 Constitutive Heterochromatin

The amount of DNA in the nucleus of somatic cells of *A. vera* has been estimated as being 2C = 33 pg (Zonneveld 2002). This large amount of genetic material is associated with histones and other proteins that form the chromatin of nuclei and chromosomes, of which only 5% is constitutive heterochromatin (Imery-Buiza 2007a; Velásquez 2009). When implementing a protocol based on denaturing ($Ba(OH)_2.8H_2O$, 2%), renaturing (NaCl, 0.3 M + $Na_3C_6H_5O_7.2H_2O$, 0.03 M, pH 7.4), and differential staining (Giemsa, 1.5%), constitutive heterochromatin (C-bands) located mainly in the centromeric regions of the six small chromosomes (S_1, S_2, and S_3 pairs) and in five large chromosomes (one L_1 and L_2 and L_4 pairs) is observed (Figure 25.10). In less condensed karyotypes, there is also a small C-band adjacent to the secondary constriction of L_4. In total, 11 centromeric C-bands ($1L_1 + 2L_2 + 2L_4 + 2S_1 + 2S_2 + 2S_3$) and a terminal C-band ($1L_4$) are revealed. In G_1 interphase nuclei, constitutive heterochromatin is present in 11 clearly distinguishable chromocenters oriented toward one end and in a smaller, more diffuse chromocenter toward the center of the nucleus.

25.8.3 Haploid Chromosomes

Another way to study the chromosomal characteristics of *Aloe* species is during the first pollen grain mitosis (Figure 25.11a–d), taking advantage of the ease of preparation and higher contrast between microscopic samples (Taylor 1925; Sapre 1975). In *A. vera*, the expected normal haploid set is $n = x = 7 = 1Lsm + 3Lst + 3Ssm$ (Figure 25.11e, f). Any deviation in the numbers and morphology of this distribution is indicative of chromosomal disorders during the process of nuclear division. The haploid complement $n = x = 7 = 3L + 1M + 3S$ in microspores of *A. vera* (Marshak 1934) shows a remarkable change in the size and shape of one of the large chromosomes, causing genetic defects in the pollen grains. Other deletions and numerical changes have been studied in our laboratory

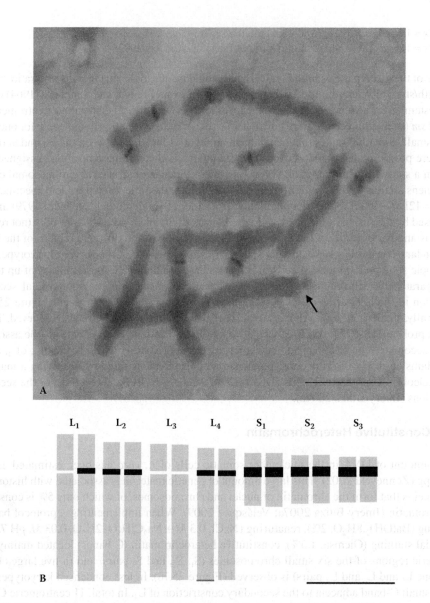

L_1 L_2 L_3 L_4 S_1 S_2 S_3

Figure 25.10   Constitutive heterochromatin in somatic chromosomes of *Aloe vera*. (A) Karyotype $2n = 2x = 14$ chromosomes with centromeric C-bands in $1L_1$, $2L_2$, $2L_4$, $2S_1$, $2S_2$ and $2S_3$, and a small terminal C-band in $1L_4$ (arrow). (B) Idiogram showing the C-banding patterns (in black). Bar = 10 μm.

(Figure 25.13o) to determine their relationships with pollen fertility and the origin of genetic variability from experimental hybrids.

25.8.4 Karyological Abnormalities

Karyological abnormalities are abundant in the processes of nuclear division preceding the formation of *A. vera* pollen. During the spread of pollen mother cells (PMCs), chromatidic bridges at anaphase and telophase, PMCs connected by thin chromatin bridge, PMCs without nuclei, and PMCs with one or more micronuclei of different sizes are commonly observed (Figure 25.13A–D). These alterations have been attributed to the action of endogenous genotoxic substances (mainly

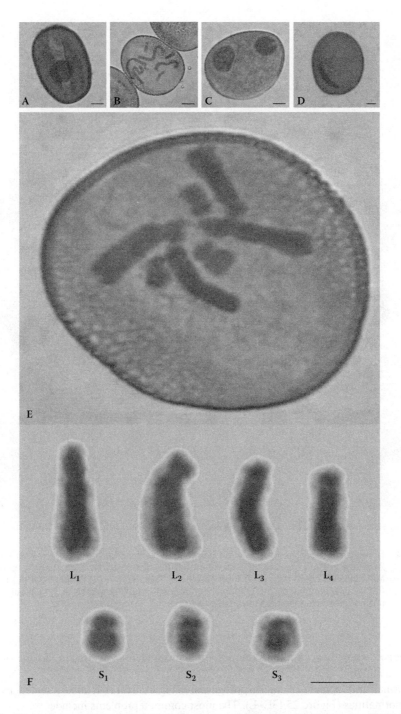

Figure 25.11 First pollen mitosis and haploid chromosomes of *Aloe vera*. (A) Microspore. (B) Prometaphase. (C) Immature pollen grain with two nuclei. (D) Mature pollen grain. (E) Bimodal haploid set $n = x = 7 = 4L + 3S$. (F) Karyogram with large chromosomes (L_1–L_4) and small chromosomes (S_1–S_3). Bar = 10 μm.

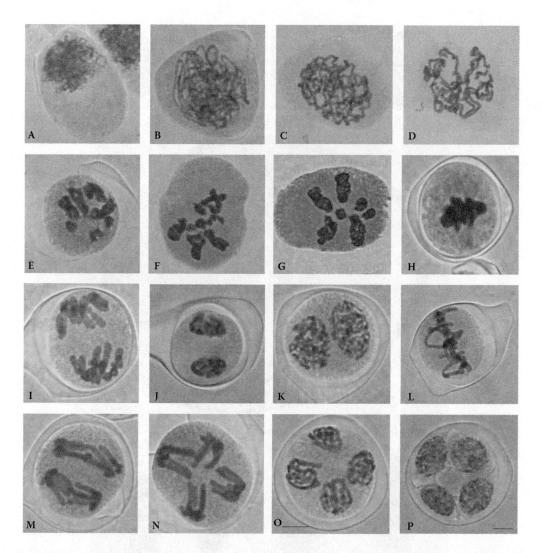

Figure 25.12 Normal microsporogenesis in *Aloe vera*. (A) Pollen mother cells (PMCs). (B) Leptotene. (C) Zygotene. (D) Pachytene. (E) Diplotene. (F) Diakinesis. (G) Metaphase-I, polar view. (H) Metaphase-I, equator view. (I) Anaphase-I. (J) Telophase-I. (K) Prophase-II. (L) Metaphase-II. (M) Parallel anaphase-II. (N) Opposite anaphase-II. (O) Telophase-II. (P) Cytokinesis with four microspores arranged as a tetrad. Bar = 10 μm.

anthraquinones) whose concentrations increase in plants under water stress and solar overexposure (Imery-Buiza 2007b).

The normal development of microsporogenesis (Figure 25.12) is also affected by numerous meiotic abnormalities (Figure 25.13E–L). The most common problems include

- agglutinations of bivalents at prophase-I and metaphase-I
- early migration of small chromosomes at metaphase-I
- one or two bridges between large chromosomes at anaphase-I and telophase-I that may persist until anaphase-II
- one or two acentric independent fragments or fragments linked (in the form of "V" or "U") to the bridges at anaphase-I

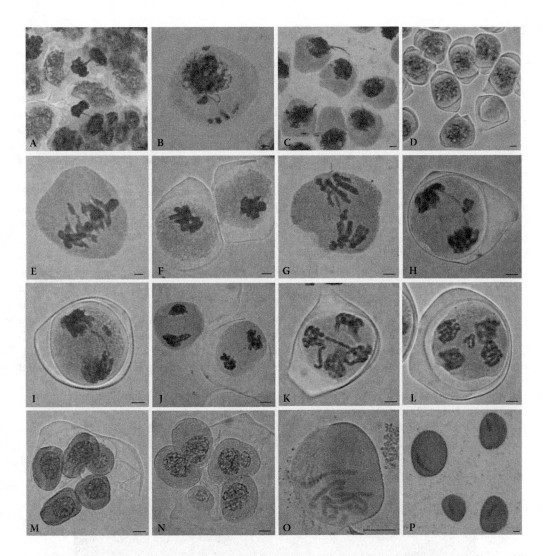

Figure 25.13 Karyological abnormalities during the formation of the pollen of *Aloe vera*. (A) Chromatidic bridges in telophase of mitotic proliferation of PMCs. (B) Micronuclei in PMC. (C) PMCs together by thin chromatin bridge. (D) PMCs without nucleus. (E) Early displacement of small chromosomes in metaphase-I. (F) Agglutination of bivalents at metaphase-I. (G) Acentric fragment in the form of "V" together with chromatidic bridge at anaphase-I. (H) Acentric fragment in the form of "U" together with chromatidic bridge at anaphase-I. (I) Acentric independent fragment and chromatidic bridge at anaphase-I. (J) One and two chromatidic bridges at telophase-I. (K) Chromatidic bridges at anaphase-II. (L) Acentric fragment at telophase-II. (M, N) One and two additional microspores. (O) Change in chromosome number (big extra chromosome), haploid abnormal set $n = x+1 = 8 = 5L + 3S$. (P) Pollen grains of different sizes. Bar = 10 μm.

one or two micronuclei at telophase-II and five or six microspores at the end of the process (Figure 25.13m, n) (Sapre 1975; Imery and Caldera 2002)

All of these chromosomal abnormalities cause gene duplications and deficiencies that reduce pollen fertility to less than 50%. Together with dichogamy and self-incompatibility, they explain the lack of fruits and seeds in naturalized populations that originated vegetatively from too few founder plants (Imery-Buiza and Cequea-Ruíz 2008).

25.8.5 Molecular Markers (RAPD-PCR)

The search for DNA markers in *A. vera* requires the standardization of DNA extraction protocols that remove as many of the polysaccharides and phenols present in the sample as possible. This has been achieved by incorporating polyvinylpyrrolidone and *n*-lauroylsarcosine in the initial incubation of mesophyll fragments in 3% CTAB extraction buffer, an additional cleaning step with chloroform/isoamyl alcohol, and high-salt treatment (NaCl, 2.5 M) between digestion with RNAnase and resuspension in TE buffer. By applying this modified CTAB protocol, isolates of DNA of up to 100 ng/µL with purities of about 1.8 and genomic fragments of high molecular weight have been achieved.

Random amplification polymorphic DNA (RAPD), using the polymerase chain reaction (PCR) and arbitrary primers OP05, OPA06, OPA07, OPA08, OPA09, OPA11, OPB04, OPB05, OPB07, and OPB10 (Operon Tech., United States), has revealed at least 50 informative bands from 340 to 2000 bp, of which $OPA05_{780}$, $OPA09_{440}$, $OPA11_{640}$, $OPA11_{750}$, $OPA11_{870}$, $OPA11_{1040}$, $OPB07_{580}$, $OPB07_{640}$, $OPB07_{670}$, $OPB07_{860}$, $OPB07_{920}$, and $OPB07_{1000}$ stand out as potential specific markers of *A. vera* (Imery-Buiza 2007a). The usefulness of the PCR-RAPD technique for generating species-specific markers has also been demonstrated by the accurate discrimination of commercial products of *A. vera* from other *Aloe* species (Shioda et al. 2003). In our laboratory, this simple technique has permitted us to amplify the characterization of many genotypes and identify associations between genetic markers. For example, in Figure 25.14, the products of amplification with the PCR-RAPD OPA11 primer, which produced polymorphic bands between *A. saponaria* (Figure 25.14a) and *A. vera* (Figure 25.14b) are shown. This gave us the opportunity to study the inheritance of several molecular markers in the progeny (Figure 25.14c–q).

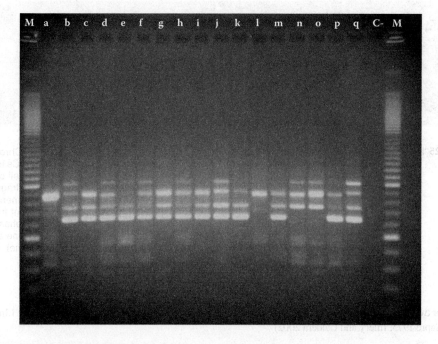

Figure 25.14 Agarose gel electrophoresis to separate fragments of random amplified polymorphic DNA (RAPD; OPA11 primer) in the genomes from (a) *Aloe saponaria*. (b) *A. vera*. (b) and (c–q) 15 experimental hybrids. M: molecular weight marker (100 bp); C–: negative control (without DNA).

25.9 PLANT BREEDING

25.9.1 Mutagenesis

The search for artificial mutants requires testing for dosimetry and sensitivity to mutagenic agents (IAEA 1995). In *A. vera*, the LD_{50} in young plants (15–20 cm) is produced by 6–8 Gy of gamma rays; for cut stems with at least five lateral buds (4–7 mm), the LD_{50} is produced by 0.5–1 Gy gamma rays, using a radioactive source of cobalt-60. At 30, 90, and 180 days after irradiation, the M_1V_1 lateral shoots and plants showed significant vegetative changes (Figure 25.15A–E). Among the descendants M_1V_2, a plant with seven tepals, distortions in the size and shape of the stamens and atrophy of the median segment to the style terminal were observed (Figure 25.15F, G). This mutant had a karyotype of $2n = 2x = 14 = 3Lsm + 5Lst + 1Mm + 5Ssm$, with a reduction in the length of the long arm (q) of one of the L_3 chromosomes and an increase in the length of the short arm (p) of an S_2, thus demonstrating the induction of a translocation qL_3/pS_2 (Figure 25.15H, J). Other M_1V_2 were evaluated to identify changes of phytochemical significance, associations between genetic markers, and chromosome mapping.

25.9.2 Polyploidy

In forage crops, roots, tubers, ornamental plants, or any other plant with agronomically important vegetative tissues, the artificial doubling of the chromosome number provides a rapid method of obtaining genetic improvements, due to an increase in the size of the somatic tissues (Briggs and Knowles 1967; Singh 2003). For polyploid cultivars, preliminary tests are performed with different inducing agents (colchicine, colcemid, chloral hydrate, chloroform, ether, and nitrous oxide), varying the experimental conditions (concentration, exposure time, and temperature) and tissue type and coadjuvant treatment (Blakeslee and Avery 1937; Sleper and Poehlman 2006).

In *A. vera*, the highest proportion of tetraploid plants ($2n = 4x = 28$) was achieved after immersing the rhizomes in a 0.15% colchicine solution for 24 h (Imery 2000; Imery and Cequea 2001) and also treating the seedlings *in vitro* with a 0.06% colchicine solution during 12 h (Wang et al. 2001). In both experiments, plants were more vigorous and had larger leaves as a result of the doubling of the chromosomes (Figure 25.16). Cytological studies confirmed the karyotype $2n = 4x = 28 = 4Lsm + 12Lst + 12Ssm$ (Figure 25.17) in artificial tetraploids, an increase in the size of the epidermal cells, and a decrease in stomatal density. From the agronomic point of view, these new polyploids improved foliar biomass production by over 30% compared to their diploid ancestors, and also showed increased tolerance to the bacterium *Erwinia chrysanthemi* in pathogenicity tests (Imery-Buiza 2007a). Currently, a first group of these polyploids is undergoing multiplication in order to be submitted for assessment for their official certification. A second group will be used as baseline genotypes for other areas of investigation.

25.9.3 Hybridization

Our germplasm bank is in a very dry tropical forest (Cumana, Venezuela) and houses 115 species of *Aloe* that were acquired according to international rules that regulate the traffic of species and plant health. Under these environmental conditions, the agronomic traits and other aspects of reproductive biology, cytogenetics, crossability, and tolerance to local bacteria and fungi of *A. ammophila*, *A. littoralis*, *A. saponaria*, *A. vacilans*, and *A. zebrina* qualifies them as potential donors. Other species have also been considered as suitable parents, but are less compatible with *A. vera* in conventional crosses.

Figure 25.15   Induced mutations by gamma rays on stem buds and young plants of *Aloe vera*. (A) Morphological changes in the suckers emerging 30 days after irradiation of buds. Morphological changes at 90 (B, C) and 180 (D, E) days after the irradiation of plants. Distortion in the size and shape of the stamens (F) and atrophy of the style (G) in mutant generation M_1V_2. (H) Karyotype of mutant M_1V_2 with a translocation qL_3/pS_2. (I) Karyotype in normal nonirradiated plants. (J) Comparative karyograms of normal plant (top) and mutants carrying the translocation qL_3/pS_2 (bottom).

Figure 25.16 Diploid (2*x*) and tetraploid (4*x*) plants of *Aloe vera* at (A) 4 months and (B) 18 months after the induction of polyploidy on stem buds. Both plants were emergents from rhizomes treated with colchicine. Left plant with diploid karyotype 2*n* = 2*x* = 14; right plant with tetraploid karyotype 2*n* = 4*x* = 28. (C) Details of the diploid (left) and tetraploid (right) leaves.

Interspecific hybridization reveals the heterozygosity accumulated by asexual propagation in *A. vera* and favors the formation of multiple genotypes in the progeny as a result of intra- and interchromosomic recombination in each parent. This genotypic variation provides a new alternative for the improvement of this species and counteracts the genetic homogeneity of traditional *A. vera* clones that represents a risk to survival when faced with potential environmental changes or the emergence of new pests and diseases (Natali, Castorena, and Cavallini 1990; Imery 2000). Moreover, combinations of polyploidy and interspecific hybridization have taken advantage of the contributions of both forms of improvement, generating a wide range of superior genotypes (Imery-Buiza et al. 2008).

Triploid hybrids between *A. vera* (4*x*) and *A. saponaria* (2*x*) showed variable karyotypes from 2*n* = 3*x* – 1 = 20 to 2*n* = 3*x* + 2 = 23, due to some deletions in individual L chromosomes and aneuploid variations in S chromosomes. Progeny with the expected karyotype 2*n* = 3*x* = 21 presented slight variations between homologues, due to differences in length between parental chromosomes. Furthermore, these studies revealed a new source of variation among the triploid hybrids caused by the inheritance of chromosomal abnormalities generated during tetraploid parent meiosis (Kumar and Tripathi 2002; Imery-Buiza 2007c, 2008). The absence of such variations among the diploid hybrids (*A. vera* × *A. littoralis*, *A. vera* × *A. saponaria*, and *A. vera* × *A. zebrina*) also demonstrated that polyploid *Aloe* genotypes are more tolerant of the deleterious effects of chromosomal deletions, previously suggested by Brandham (1976).

Preliminary agronomic evaluations catalogue the autotetraploids of *A. vera* and its triploid progeny with *A. saponaria* and *A. littoralis* as the most promising experimental materials so far achieved in our research center. One of the few disadvantages is that autotetraploids continue

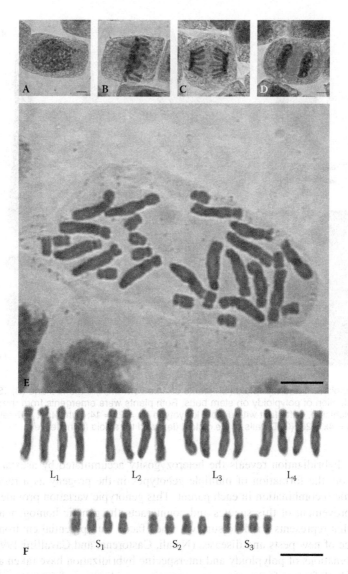

Figure 25.17 Mitosis and somatic chromosomes in tetraploid plants of *Aloe vera*. (A) Prophase. (B) Metaphase. (C) Anaphase. (D) Telophase. (E) Bimodal karyotype $2n = 4x = 28 = 16L + 12S$. (F) Karyogram with large chromosomes (L_1–L_4) and small chromosomes (S_1–S_3). Bar = 10 µm.

to be a genotype with a very small genetic base, whereas triploid and diploid hybrids have a wider genetic base and a large genotypic variability. Outstanding hybrids (10% of each F_1) were selected and incorporated into a clonal selection program. The rest of the hybrids were retained for further studies.

25.9.4 Genetic Transformation with *Agrobacterium*

A novel alternative for genetic improvement in *A. vera* is the application of transformation techniques using genetic engineering. Currently, *Agrobacterium tumefaciens C58C1* and *EHA105* strains are successfully used as vehicles in assays for the genetic transformation of *A. vera* explants, with 80% (*EHA105*) and 30% (*C58C1*) effectiveness, according to evaluations of the expression of *GUS* genes in selected transgenics in G418 medium (He et al. 2007). This model of transformation with *Agrobacterium* has also served to insert the *otsA* gene and regenerate transgenic *A. vera* plants that expressed nearly three times more trehalose synthase enzyme than their *A. vera* ancestors (Chen et al. 2007). In both cases, Southern blotting, PCR, and ELISA (enzyme-linked immunosorbent assay) analyses have demonstrated the effective integration and expression of genes in transformed plants. This new technology points the way toward a new era in the improvement of *A. vera* in which biotechnology and conventional techniques are combined to obtain transgenic cultivars with increased yield, physiological efficiency, tolerance to biotic factors, and other specific attributes.

25.10 MICROPROPAGATION

Conventional production of *A. vera* plants for new areas of cultivation generates no more than 10 suckers per year for each mother plant. When planning the establishment of large commercial plantations of 10,000–20,000 plants per hectare, this form of propagation for the acquisition of new material for planting is considered tedious, inefficient, and time consuming. Moreover, during the cutting of the suckers in the field, there is a risk of injury to both mother plants and propagules that increases the incidence and spread of disease. Plant tissue culture *in vitro* provides an alternative for meeting the demand for planting material of high quality, producing large numbers of pathogen-free homogenous plants.

Basically, the *in vitro* regeneration of *A. vera* involves the following:

collection of buds from stems
washing them in distilled water, ethanol, and sodium hypochlorite
removing excess leaf tissue
planting the meristems in MS culture medium (Murashige and Skoog 1962) supplemented with agar, sucrose, auxins, and cytokinins
growing them in a culture room (22–27°C, 80–85% relative humidity, 2000–2500 lux/16 h)
acclimation in a greenhouse (Natali et al. 1990; Araújo et al. 2002; Campestrini et al. 2006; Hosseini and Parsa 2007; Supe 2007; Singh and Sood 2009)

Other authors describe regeneration protocols from calluses or explants from leaves and inflorescences (Roy and Sarkar 1991; Cavallini et al. 1993; Richwine 1995; Garro-Monge, Gatica-Arias, and Valdez-Melara 2008).

In our laboratory, best results have been achieved by initially separating the rhizome from below the oldest green leaf without damaging the aerial part, which regenerates in a natural way. Cuttings 10–15 cm long (no roots or leaves) were washed, sealed at the ends with Bordeaux paste, and placed in sterile vermiculite in plastic germinators. After 20–30 days, lateral buds (8–10 mm) were collected and washed by immersion and continuous agitation in distilled water: NaClO (1.6%, 25 min), ethanol (70%, 3 min), NaClO (1.6%, 10 min), and distilled water (three times). The leaf tissue was then removed under a stereomicroscope in a sterile chamber and washed again (twice); finally, the meristems were planted in MS medium supplemented with agar (0.7%), sucrose (3%), 6-benzylaminopurine (5 mg/L), indole acetic acid (1 mg/L), and ascorbic acid (75 mg/L) (Figure 25.18). This technique has contributed to the *in vitro* conservation of experimental material of considerable scientific value and the massive propagation of outstanding genotypes.

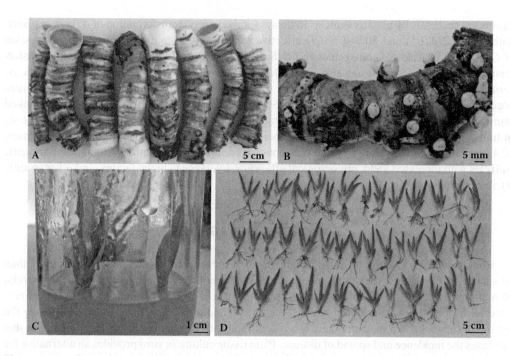

Figure 25.18   *In vitro* plants of *Aloe vera.* (A) Cuttings of adult plants. (B) Stem buds after cuttings on vermiculite by 20–30 days. (C) Regeneration of plants from buds cultured on MS medium supplemented with agar, sucrose, BA, IAA, and ascorbic acid. (D) Acclimated plants in greenhouse.

25.11 CONCLUSION

Aloe vera is a medicinal plant used since ancient times and cultivated in many warm regions of the world. A large amount of active ingredients and therapeutic properties are known in *A. vera*; however, research for genetic improvement is recent and limited by the form of vegetative propagation. The low genetic variability of *A. vera* has been overcome with the practice of hybridization assays using species that have high medicinal value. Moreover, the exploitation of polyploidy has played an important role in improving foliar biomass production and tolerance to diseases. The combination of these two breeding strategies has allowed us to explore an important source of genetic variation accumulated by the asexual reproduction of *A. vera* and donor species, in addition to contributions to the variability that have contributed in each parent recombination and chromosome mutations that occur throughout the formation of sex cells.

With polyploidy, hybridization, and even artificial mutations, a lot of experimental genotypes offer a promising future in the field of traditional breeding of *A. vera*. The application of new technologies has also offered alternative horizons for the improvement of *A. vera*, achieving successful propagation and maintenance of tissues in *in vitro* conditions and the production of transgenic plants via *Agrobacterium* transformation. This background envisions the integration of multidisciplinary groups to achieve significant advances in breeding, cultivation, and use of therapeutic benefits of *A. vera*.

ACKNOWLEDGMENTS

The author is grateful to Dr. Frances Osborn for review of the manuscript. Several investigations described in this chapter have been supported by Consejo de Investigación (CI-5-1001-0900/99 and CI-5-010101-1223/05 projects) of the Universidad de Oriente (Venezuela), for which the author is also grateful.

REFERENCES

Abraham, Z., and P. Nagendra. 1979. Occurrence of triploidy in *Aloe vera* Tourn. ex Linn. *Current Science* 48 (22): 1001–1002.

Acevedo, M., C. Russell, S. Patel, and R. Patel. 2004. Aloe-emodin modulates PKC isozymes, inhibits proliferation, and induces apoptosis in U-373MG glioma cells. *International Immunopharmacology* 4:1775–1784.

Adams, S. P., I. J. Leitch, M. D. Bennett, M. W. Chase, and A. R. Leitch. 2000. Ribosomal DNA evolution and phylogeny in *Aloe* (Asphodelaceae). *American Journal of Botany* 87:1578–1583.

Albarracín, N., M. Alcano, L. Vera, and M. Albarracín. 2001. Pudriciones basales en zábila (*Aloe vera* L.) en el estado Falcón. XVII Congreso Venezolano de Fitopatología, 155.

Albornoz, A. R., and J. Imery. 2003. Evaluación citogenética de ocho poblaciones de *Aloe vera* L. de la Península de Araya-Venezuela. *Ciencia (LUZ)* 11:5–13.

Almasan, A., A. Barrion, M. Casne, and G. De la Cruz. 1991. Karyotype analysis of *Aloe barbadensis* Mill. (Liliales: Liliaceae). *Philippine Agriculturist (Philippines)* 74:261–264.

Álvarez, A., I. Ramos, Y. Robaina, G. Pérez., M. Cuevas, and C. Carrillo. 1996. Efecto antiulceroso de fórmulas que contienen un extracto de *Aloe vera* L. (Sábila). *Revista Cubana de Plantas Medicinales* 1:31–36.

Añez, B., and J. Vásquez. 2005. Efecto de la densidad de población sobre el crecimiento y rendimiento de la zábila (*Aloe barbadensis* M.) *Revista de la Facultad de Agrononomía (LUZ)* 22:1–12.

Anshoo, G., S. Singh, A. S. Kulkarni, S. C. Pant, and R. Vijayaraghavan. 2005. Protective effect of *Aloe vera* L. gel against sulfur mustard-induced systematic toxicity and skin lesions. *Indian Journal of Pharmacology* 37:103–110.

Araújo, P. S., J. M. Oliveira, C. A. Neckel, C. Ianssen, A. C. Oltramari, R. D. Passos, E. Tiepo, D. B. Bach, and M. Maraschin. 2002. Micropropagação de babosa (*Aloe vera*—Liliaceae). *Biotecnologia Ciência & Desenvolvimento* 25:54–57.

Averre, C. W., and J. E. Reynolds. 1964. *Phythophthora* root and stem rot of aloe. *Proceedings of the Florida State Horticultural Society* 77:438–440.

Batugal, P. A., J. Kanniah, S. Y. Lee, and J. Oliver. 2004. *Medicinal plants research in Asia*, vol. I: *The framework and project work plans*. Serdang: IPGRI-APO.

Bennett, B. C., and G. T. Prance. 2000. Introduced plants in the indigenous pharmacopoeia of northern South America. *Economic Botany* 54:90–102.

Bennett, M. D., and I. J. Leitch. 2005. Nuclear DNA amounts in angiosperms: Progress, problems and prospects. *Annals of Botany* 95:45–90.

Beyra, A., M. C. León, E. Iglesias, D. Ferrándiz, R. Herrera, G. Volpato, D. Godínez, M. Guimarais, and R. Álvarez. 2004. Estudios etnobotánicos sobre plantas medicinales en la provincia de Camaguey (Cuba). *Anales del Jardín Botánico de Madrid* 61:185–204.

Blakeslee, A. F., and A. G. Avery. 1937. Methods of inducing doubling of chromosomes in plants. *Journal of Heredity* 18:393–411.

Boik, J. 1995. *Cancer and natural medicine.* Oregon: Medical Press.

Brandham, P. E. 1969a. Chromosome behavior in the Aloineae. I. The nature and significance of E-type bridges. *Chromosoma* 27:201–215.

———. 1969b. Chromosome behavior in the Aloineae. II. The frequency of interchange heterozygosity. *Chromosoma* 27:216–225.

———. 1970. Chromosome behavior in the Aloineae III. Correlations between spontaneous chromatid and subchromatid aberrations. *Chromosoma* 31:1–17.

———. 1971. The chromosome of the Liliaceae. III. Polyploidy and karyotype variation in the Aloineae. *Kew Bulletin* 25:381–399.

———. 1974a. Interchange and inversion polymorphism among populations of *Haworthia reinwardtii* var. *chalumnensis. Chromosoma* 47:85–108.

———. 1974b. The chromosomes of the Liliaceae. III. New cases of interchange hybridity in the Aloineae. *Kew Bulletin* 28:341–348.

———. 1975. Stabilized breakage of a duplicated chromosome segment in *Aloe. Chromosoma* 51:269–278.

———. 1976. *The frequency of spontaneous structural change.* In *Current chromosome research*, ed. K. Jones and P. Brandham. New York: North Holland Pub.

———. 1977a. The meiotic behavior of inversions in polyploid Aloineae. I. Paracentric inversions. *Chromosoma* 62:69–84.

———. 1977b. The meiotic behavior of inversions in polyploid Aloineae. II. Pericentric inversions. *Chromosoma* 62:85–91.

———. 1990. Meiotic crossing over between sites on opposite sides of the centromeres of homologues is frequent in hybrid Aloaceae. *Genome* 33:170–176.

Brandham, P. E., and M. J. Doherty. 1998. Genome size variation in the Aloaceae, an angiosperm family displaying karyotype orthoselection. *Annals of Botany* 82 (Suppl. A): 67–73.

Brandham, P. E., and M. A. Johnson. 1977. Population cytology of stomatal numerical chromosome variation in Aloineae (Liliaceae). *Plant Systematics & Evolution* 128:105–122.

Briggs, F., and P. Knowles. 1967. *Introduction to plant breeding*. California: Reinhold Publishing Corporation.

Brummit, R. 1992. *Vascular plant families and genera*. Kew, England: Royal Botanic Garden.

Buckle, J. 2001. Aromatherapy and diabetes. *Diabetes Spectrum* 14:124–126.

Campestrini, L. H., S. Kuhnen, P. M. Lemos, D. B. Bach, P. F. Dias, and M. Maraschin. 2006. Cloning protocol of *Aloe vera* as a study-case for "tailor-made" biotechnology to small farmers. *Journal of Technology Management & Innovation* 1:76–79.

Canevaro, S. 2004. *Aloe vera: El poder de su principio activo y las propiedades curativas de sus hojas*. Madrid: Susaeta Ediciones.

Capasso, F., F. Borrelli, R. Capasso, G. Di Carlo, A. A. Izzo, L. Pinto, N. Mascolo, S. Castaldo, and R. Longo. 1998. *Aloe* and its therapeutic use. *Phytotherapy Research* 12:S124–S127.

Cárdenas, C., A. R. Quesada, and M. A. Medina. 2006. Evaluation of the anti-angiogenic effect of aloe-emodin. *Cellular & Molecular Life Sciences* 63:3083–3089.

Carter, S. 1994. Aloaceae. In *Flora of tropical East Africa*, ed. R. M. Polhill, 1–57. Rotterdam: A. A. Balkema.

Castellanos, E., M. Rodríguez, T. Vásquez, and A. Sin. 2001. Efecto antiviral del extracto acuoso de *Aloe barbadensis* contra el virus de la hepatitis B. *Revista Cubana de Plantas Medicinales* 6:7–11.

Cavallini, A. 1993. Cytophotometric and biochemical analyses of DNA variations in the genus *Aloe* L. *International Journal of Plant Sciences* 154:169–174.

Cavallini, A., L. Natali, G. Cionini, O. Sassoli, and I. Castorena-Sánchez. 1993. *In vitro* culture of *Aloe barbadensis* Mill.: Quantitative DNA variations in regenerated plants. *Plant Science* 91:223–229.

Chase, M. W., A. De Bruijn, A. V. Cox, G. Reeves, P. J. Rudall, M. A. T. Johnson, and L. E. Eguiarte. 2000. Phylogenetics of Asphodelaceae (Asparagales): An analysis of plastid *rbcL* and *trnL-F* DNA sequences. *Annals of Botany* 86:935–951.

Chen, J., X. Zhang, Y. Xing, E. Cong, Y. Mao, H. Zhao, and J. Rong. 2007. *Agrobacterium*-mediated transformation of *Aloe* with trehalose synthase gene (*otsA*). *Acta Agronomica Sinica* 33:968–972.

Cheng, L., H. Wan, C. Ching, L. Shih, W. Lei, L. Ying, and L. Wei. 2008. Aloe-emodin is an interferon-inducing agent with antiviral activity against Japanese encephalitis virus and enterovirus 71. *International Journal of Antimicrobial Agents* 32:355–359.

Clifford, S. C., S. K. Arndt, M. Popp, and H. G. Jones. 2002. Mucilages and polysaccharides in *Ziziphus* species (Rhamnaceae): Localization, composition and physiological roles during drought stress. *Journal of Experimental Botany* 53:131–138.

Cowan, M. M. 1999. Plant products as antimicrobial agents. *Clinical Microbiology Reviews* 12:564–582.

Cronquist, A. 1981. *An integrated system of classification of flowering plant*, 2nd ed. New York: Columbia University Press.

Crosswhite, F. S., and C. D. Crosswhite. 1984. *Aloe vera*, plant symbolism and threshing floor: Light, life and good in our heritage. *Desert Plant* 6:46–50.

Cruden, R. 1977. Pollen-ovule ratio: A conservative indicator of breeding systems in flowering plants. *Evolution* 31:32–46.

Cui, X.-R., K. Takahashi, T. Shimamura, J. Koyanagi, F. Komada, and S. Saito. 2008. Preparation of 1,8-Di-*O*-alkylaloe-emodins and 15-amino-, 15-thiocyano-, and 15-selenocyanochrysophanol derivatives from aloe-emodin and studying their cytotoxic effects. *Chemical & Pharmaceutical Bulletin* 56:497–503.

Dahlgren, R., and H. T. Clifford. 1982. *The monocotyledons: A comparative study*. London: Academic Press.

Darokar, M., R. Rai, A. Gupta, A. Shasany, S. Rajkumar, V. Sunderasan, and S. Khanuja. 2003. Molecular assessment of germplasm diversity in *Aloe* spp. using RAPD and AFLP analysis. *Journal of Medicinal & Aromatic Plant Science* 25:354–361.

De Laat, P., J. Verhoeven, and J. Janse. 1994. Bacterial leaf rot of *Aloe vera* L. caused by *Erwinia chrysanthemi* biovar 3. *European Journal of Plant Pathology* 100:81–84.

Del Cid, J. 2002. Determinación de géneros y estudio de la distribución horizontal de los nematodos fitoparasíticos que afectan al cultivo de la sábila *Aloe vera* L. en el progreso. Thesis. Universidad de San Carlos de Guatemala.

Díaz, M. 2001. Ecología experimental y ecofisiología: Bases para el uso sostenible de los recursos naturals de las zonas áridas neo-tropicales. *Interciencia* 26:472–478.

Djeraba, A., and P. Quere. 2000. *In vivo* macrophage activation in chickens with acemannan, a complex carbohydrate extracted from *Aloe vera*. *International Journal of Immunopharmacology* 22:365–372.

Duarte, D. 2009. Ciclo del nucléolo en células meristemáticas y microsporogénicas de *Aloe vera* (L.) Burm. f. (Aloaceae). Thesis. Universidad de Oriente, Venezuela.

Duke, J., and J. DuCellier. 1993. *CRC handbook of alternative cash crops*. Boca Raton, FL: CRC Press.

Eller, B. M., B. R. Ruess, and S. Ferrari. 1993. Crassulacean acid metabolism (CAM) in the chlorenchyma and hydrenchyma of *Aloe* leaves and *Opuntia* cladodes: Field determinations in an *Aloe* habitat in southern Africa. *Botanica Helvetica* 103:201–205.

EINE (Equipo de Investigación Nueva Era). 2006. Aloe vera: *Una planta milagrosa*. Buenos Aires: Continente.

Esteban, A., J. M. Zapata, L. Casano, M. Martin, and B. Sabater. 2000. Peroxidase activity in *Aloe barbadensis* commercial gel: Probable role in skin protection. *Planta Medica* 66:724–727.

Farr, D. F., and A. Y. Rossman. 2009. Fungal databases, Systematic Mycology and Microbiology Laboratory, ARS, USDA. Retrieved June 25, 2009, from http://nt.ars-grin.gov/fungaldatabases/

Femenia, A., E. S. Sánchez, S. Simal, and C. Rossello. 1999. Compositional features of polysaccharides from *Aloe vera* (*Aloe barbadensis* Miller) plant tissues. *Carbohydrate Polymers* 39:109–117.

Ferro, V. A., F. Bradbury, P. Cameron, E. Shakir, S. R. Rahman, and W. H. Stimson. 2003. *In vitro* susceptibilities of *Shigella flexneri* and *Streptococcus pyogenes* to inner gel of *Aloe barbadensis* Miller. *Antimicrobial Agents & Chemotherapy* 47:1137–1139.

Forster, P. I., and H. T. Clifford. 1986. Aloeaceae. In *Flora of Australia*, ed. A. S. George, 66–77. Canberra: Australian Government Publishing Service.

Fuentes-Carvajal, A., J. A. Véliz, and J. Imery. 2006. Efecto de la deficiencia de macronutrientes en el desarrollo vegetativo de *Aloe vera*. *Interciencia* 31:116–122.

Fukada, M. 2006. The aloe mite: A new pest on Maui. *Malp News (Spring)*: 5. http://www.malp.org (accessed June 10, 2009).

Furones, J. A., F. Morón, and Z. Piñedo. 1996. Acción analgésica de un extracto acuoso liofilizado de *Aloe vera* L. en ratones. *Revista Cubana de Plantas Medicinales* 1:15–17.

Gage, D. 1996. Aloe vera: *Nature's soothing healer*. Rochester, MN: Healing Arts Press.

García, J. 2000. Determinación de enfermedades fungosas en el tejido aéreo de 30 especies medicinales en dos localidades del altiplano central. Thesis. Universidad de San Carlos de Guatemala.

Garro-Monge, G., A. M. Gatica-Arias, and M. Valdez-Melara. 2008. Somatic embryogenesis, plant regeneration and acemannan detection in aloe (*Aloe barbadensis* Mill.). *Agronomía Costarricense* 32:41–52.

Glen, H., and D. Hardy. 2000. Aloaceae. In *Flora of southern Africa*, ed. G. Germishuizen, 1–167. Pretoria: National Botanical Institute.

Goldstein, G., J. L. Andrade, and P. S. Nobel. 1991. Differences in water relations parameters for the chlorenchyma and the parenchyma of *Opuntia ficus-indica* under wet versus dry conditions. *Australian Journal of Plant Physiology* 18:95–107.

González, M., M. Sotolongo, and M. Batista. 2002. Extracto de *Aloe barbadensis* inyectable en fracturas experimentales. *Revista Cubana de Plantas Medicinales* 7:14–17.

Halbert, S. E. 2002. Ornamentals, foliage plant. Florida Department of Agriculture & Consumer Services. *Tri-ology* 41:1–11.

Haller, J. S. 1990. A drug for all seasons, medical and pharmacological history of aloe. *Bulletin of the New York Academy of Medicine* 66:647–659.

Hamman, J. H. 2008. Composition and applications of *Aloe vera* leaf gel. *Molecules* 13:1599–1616.

Hänsel, R., K. Keller, H. Rimpler, and G. Schneider. 1994. *Hagers Handbuch der Pharmazeutischen Praxis*, vol. 6, 5th ed. Berlin: Springer.

He, C., J. Zhang, J. Chen, X. Ye, L. Du, Y. Dong, and H. Zhao. 2007. Genetic transformation of *Aloe barbadensis* Miller by *Agrobacterium tumefaciens*. *Journal of Genetics & Genomics* 34:1053–1060.

Hernández, A. N., S. Bautista, and M. G. Velázquez. 2007. Prospectiva de extractos vegetales para controlar enfermedades postcosecha hortofrutícolas. *Revista Fitotecnia Mexicana* 30:119–123.

Hernández-Cruz, L. R., R. Rodríguez-García, R. D. Jasso, and J. L. Angulo-Sánchez. 2002. *Aloe vera* response to plastic mulch and nitrogen. In *Trends in new crops and new uses*, ed. J. Janick and A. Whipkey, 570–574. Alexandria: ASHS Press.

Heyes, J. A. 1989. Crassulacean acid metabolism in Zimbabwean succulents. *Excelsa* 14:14–20.

Hirooka, Y., T. Kobayashi, J. Takeuchi, T. Ono, Y. Ono, and K. T. Natsuaki. 2007. *Aloe* ring spot, a new disease of aloe caused by *Haematonectria haematococca* (Berk. & Broome) Samuels & Nirenberg (anamorph: *Fusarium* sp.). *Journal of General Plant Pathology* 73:330–335.

Hodge, W. H. 1953. The drug aloes of commerce, with special reference to the cape species. *Economic Botany* 7:99–129.

Holland, P. G. 1978. An evolutionary biogeography of the genus *Aloe*. *Journal of Biogeography* 5:213–226.

Hosseini, R., and M. Parsa. 2007. Micropropagation of *Aloe vera* L. grown in South Iran. *Pakistan Journal of Biological Sciences* 10:1134–1137.

Hu, S. 2003. *History of the introduction of exotic elements into traditional Chinese medicine.* Hong Kong: Commercial Press.

Hu, Y., J. Xu, and Q. Hu. 2003. Evaluation of antioxidant potential of *Aloe vera* (*Aloe barbadensis* Miller) extracts. *Journal of Agricultural and Food Chemistry* 51:7788–7791.

Hutchinson, J. 1959. *The families of flowering plants: Monocotyledons*, 2nd ed. Oxford, England: Clarendon Press.

Imery, J. 2000. Inducción de tetraploidía en *Aloe vera* (L.) Burm. f. (Aloaceae). Master's thesis. Universidad de Oriente, Venezuela.

———. 2002. Estudio cromosómico comparativo de cinco species de *Aloe* (Aloaceae). *Acta Botanica Venezuelica* 25:47–66.

Imery, J., and H. Cequea. 2001. Colchicine-induced autotetraploid in *Aloe vera* L. *Cytologia* 66:409–413.

Imery, J., and Y. Cárdenas. 2006. Durabilidad de la capacidad germinativa del polen en *Aloe vera* (L.) Burm. f. and *Aloe saponaria* Haw. *Revista Científica UDO Agrícola* 6:67–75.

Imery-Buiza, J. 2005. Cruzabilidad unidireccional de *Aloe vera* (L.) Burm. f. (Aloaceae). Memoria de la LV Convención Anual de la AsoVAC: 36.

———. 2008. Variaciones cromosómicas en microsporas de plantas tetraploides de *Aloe vera* (L.) Burm. f. (Aloaceae). *Memoria del XIII Congreso Latinoamericano de Genética:* 317.

———. 2007a. Caracterización genética de parentales e híbridos diploides [VS] y triploides [VVS] entre *Aloe vera* (L.) Burm. f. [2V, 4V] y *Aloe saponaria* Haw. [2S] (Aloaceae). Doctoral thesis. Universidad Central de Venezuela.

———. 2007b. Inestabilidad cariológica durante la formación de células madres del polen en *Aloe vera* (Aloaceae). *Revista de Biología Tropical* 55:805–813.

———. 2007c. Microsporogénesis en plantas tetraploides de *Aloe vera*. *Memoria del XVII Congreso Venezolano de Botánica:* 179–181.

Imery-Buiza, J., and H. Cequea-Ruíz. 2008. Autoincompatibilidad y protandría en poblaciones naturalizadas de *Aloe vera* de la península de Araya, Venezuela. *Polibotanica* 26:113–125.

Imery-Buiza, J., M. B. Raymúndez, and A. Menéndez-Yuffa. 2008. Karyotypic variability in experimental diploid and triploid hybrids of *Aloe vera* × *A. saponaria*. *Cytologia* 73:305–311.

IARI (Indian Agricultural Research Institute). 2007. *Annual report.* New Delhi: M/s Royal Offset Printers, A-89/1.

IAEA (International Atomic Energy Agency). 1995. *Manual on mutation breeding*, 2nd ed. Technical reports series no. 119.

Izhaki, I. 2002. Emodin—A secondary metabolite with multiple ecological functions in higher plants. *New Phytologist* 155:205–217.

Jacobsen, H. 1954. *A handbook of succulent plant.* London: Blandford Press.

Ji, G., L. Wei, Y. He, and Y. Wu. 2007. First report of aloe root and stem rot in China caused by *Fusarium solani*. *Plant Disease* 91:768.

Judd, W. S., C. S. Campbell, E. A. Kellogg, and P. F. Stevens. 1999. *Plant systematic: A phylogenetic approach.* Sunderland: Sinauer Associates, Inc.

Kamalakannan, A., C. Gopalakrishnan, R. Renuka, K. Kalpana, D. Lakshmi, and V. Valluvaparidasan. 2008. First report of *Alternaria alternata* causing leaf spot on *Aloe barbadensis* in India. *Australian Plant Disease Notes* 3:110–111.

Kelly, J., and M. Olsen. 2006. Problems and pest of agave, aloe, cactus and yucca. University of Arizona Cooperative Extension. AZ 1399.

Kim, H. S., S. Kacew, and B. M. Lee. 1999. *In vitro* chemopreventive effects of plant polysaccharides (*Aloe barbadensis* Miller, *Lentinus edodes*, *Ganoderma lucidum* and *Coriolus versicolor*). *Carcinogenesis* 20:1637–1640.

Kumar, G., and A. Tripathi. 2002. Microsporogenesis in a strong desynaptic mutant of tetraploid *Aloe barbadensis* Mill. *Nucleus* 45:143–146.

Larionova, M., R. Menéndez, O. V. Hernández, and V. Fusté. 2004. Estudio químico de los polisacáridos presentes en *Aloe vera* L. y *Aloe arborescens* Miller cultivados en Cuba. *Revista Cubana de Plantas Medicinales* 9:146.

León, J. E., V. P. Rosales, R. A. Rosales, and V. P. Hernández. 1999. Actividad antiinflamatoria y cicatrizante del ungüento rectal de *Aloe vera* L. (Sábila). *Revista Cubana de Plantas Medicinales* 3:106–109.

Leung, M. Y. H., C. Liu, L. F. Zhu, Y. Z. Hui, B.Yu, and K. P. Fung. 2004. Chemical and biological characterization of a polysaccharide biological response modifier from *Aloe vera* L. var *chinensis* (Haw.) Berg. *Glycobiology* 14:501–510.

Levan, A. K. Fredga, and A. A. Sandberg. 1964. Nomenclature for centromeric position on chromosomes. *Hereditas* 52:201–220.

Levin, H., R. Hazenfrantz, J. Friedman, and M. Perl. 1988. Partial purification and some properties of the antibacterial compounds from *Aloe vera*. *Phytotherapy Research* 1:1–3.

Lugo, Z. 1999. Zábila: Enfermedades y control. *Revista Fonaiap Divulga* 63:20–21.

Lugo, Z., and R. Medina. 2001. Diagnóstico preliminar de enfermedades en plantas de zábila (*Aloe vera* L.) en los municipios Colina, Federación, Miranda, Falcón y Sucre del estado Falcón. XVII Congreso Venezolano de Fitopatología, 4.

Lyons, G. 2006. The definitive *Aloe vera*, Vera? The jumping cholla, February 14. http://www.huntingtonbotanical.org/Desert/Cholla/feb06/feb06.htm (accessed June 10, 2009).

———. 2007. *Desert plants: A curator's introduction to the Huntington Desert Garden*. San Marino, CA: Huntington Library Press.

Mandal, K., and S. Maiti. 2005. Bacterial soft rot of aloe caused by *Pectobacterium chrysanthemi*: A new report from India. *Plant Pathology* 54:573.

Marshak, A. 1934. Chromosomes and compatibility in the Aloineae. *American Journal of Botany* 21:592–597.

Martínez, I. 2008. Nematofauna en plantas medicinales en la provincial de Las Tunas (Cuba). XLVIII Reunión Anual de la Sociedad Americana de Fitopatología, 19.

Matos, A., and J. Molina. 1997. Estudio citogenético en células radicales de *Aloe vera* L. *Revista de la Facultad de Agronomía (LUZ)* 14:173–182.

McClymont, D. S. 1989. Pest and diseases of aloes in Zimbabwe. *Excelsa* 14:100–105.

Miyuki, T., M. Eriko, I. Yousuke, H. Noriko, N. Kouji, Y. Muneo, T. Tomohiro, H. Hirotoshi, T. Mitunori, I. Masanori, and H. Riuuichi. 2006. Identification of five phytosterols from *Aloe vera* gel anti-diabetic compounds. *Biological & Pharmaceutical Bulletin* 29:1418–1422.

Montón, C., F. García, J. Armengol, L. A. Vicent, and J. García. 2004. Detección de *Uromyces aloes* sobre *Aloe vera*. *Phytoma* 163:22–25.

Morais, S. M., J. D. Pereira, A. R. Araújo, and E. Farias. 2005. Plantas medicinais usadas pelos índios Tapebas do Ceará. *Brazilian Journal of Pharmacognosy* 15:169–177.

Morse, S. R. 1990. Water balance in *Hemizonia luzulifolia*: The role of extracellular polysaccharides. *Plant, Cell & Environment* 13:39–48.

Morton, J. F. 1961. Folk uses and commercial exploitation of *Aloe* leaf pulp. Second Annual Meeting of the Society for Economic Botany, Massachusetts.

Murashige, T., and F. Skoog. 1962. A revised medium for rapid growth and bioassays with tobacco tissue cultures. *Physiological Plantarum* 15:473–497.

Naranjo, J., L. McCook, R. Menéndez, S. M. Martinez, M. Fernández, and M. R. Almarales. 2005. Evaluación biológica de un crudo de polisacáridos aislados del *Aloe barbadensis* Miller en ratones sometidos a mielosupresión. *Revista Cubana de Plantas Medicinales* 10:1–7.

Natali, L., I. Castorena, and A. Cavallini. 1990. *In vitro* culture of *Aloe barbadensis* Mill. micropropagation from vegetative meristems. *Plant Cell, Tissue and Organ Culture* 20:71–74.

Newton, L. E. 1979. In defense of the name *Aloe vera*. *Cactus and Succulent Journal of Great Britain* 41:29–30.

———. 2004. Aloes in habitat. In *Aloes: The genus* Aloe, ed. T. Reynolds, 3–14. Boca Raton, FL: CRC Press.

Noor, A., S. Gunasekaran, A. Soosai, and M. A. Vijayalakshmi. 2008. Antidiabetic activity of *Aloe vera* and histology of organs in streptozotobin-induced diabetic rats. *Current Science* 94:1070–1076.

Patel, R., R. Garg, S. Erande, and G. B. Maru. 2007. Chemopreventive herbal antioxidants: Current status and future perspectives. *Journal of Clinical Biochemistry & Nutrition* 40:82–91.

Pecere, T., M. Gazzola, C. Mucignat, C. Parolin, F. Vecchia, A. Cavaggioni, G. Basso, A. Diaspro, B. Salvato, M. Carli, and G. Palù. 2000. Aloe-emodin in a new type of anticancer agent with selective activity against neuroectodermal tumors. *Cancer Research* 60:2800–2804.

Prato, M. R., R. Ávila, C. Donquis, E. Medina, and R. Reyes. 2008. Antraquinonas en *Aloe vera barbadensis* de zonas semiáridas de Falcón, Venezuela, como inhibidores de la corrosión. *Multiciencias* 8:148–154.

Pritam, A., and P. G. Kale. 2007. Alteration in the antioxidant potential of *Aloe vera* due to fungal infection. *Plant Pathology Journal* 6:169–173.

Pugh, N., S. A. Ross, M. A. Elsohly, and D. S. Pasco. 2001. Characterization of aloeride, a new high-molecular-weight polysaccharide from *Aloe vera* with potent immunostimulatory activity. *Journal of Agricultural & Food Chemistry* 49:1030–1034.

Ramachandra, C. T., and P. Srinivasa. 2008. Processing of *Aloe vera* leaf gel: A review. *American Journal of Agricultural & Biological Science* 3:502–510.

Reynolds, G. 1950. *The aloes of South Africa.* Johannesburg: Aloes of South Africa Books Foundation.

———. 1966. *The aloes of tropical Africa and Madagascar.* Mbabane: Aloes Books Foundation.

Richardson, J., J. E. Smith, M. Mcintyre, R. Thomas, and K. Pilkington. 2005. *Aloe vera* for preventing radiation-induced skin reactions: A systematic literature review. *Clinical Oncology* 17:478–484.

Richwine, A. M. 1995. Establishment of *Aloe, Gasteria,* and *Haworthia* shoot cultures from inflorescence explants. *HortScience* 30:1443–1444.

Riley, H., and S. Majumdar. 1979. *The Aloineae: A biosystematic survey.* Lexington: University Press of Kentucky.

Ritter, L. 1998. Aloe Vera: *A mission discovered.* Triputic Laboratories. Dallas Rodríguez-Morales, A. 1996. Respuesta de la sábila (*Aloe vera*) a dosis crecientes de Nitrógeno, fósforo y potasio en Liberia, Guanacaste. Thesis. Universidad de Costa Rica.

Rivero, R., E. A. Rodríguez, R. Menéndez, J. A. Fernández, G. B. Alonso, and M. L. González. 2002. Obtención y caracterización preliminar de un extracto de *Aloe vera* L. con actividad antiviral. *Revista Cubana de Plantas Medicinales* 7:32–38.

Rodríguez-Morales, A. 1996. Respuesta de la sábila (*Aloe vera*) a dosis crecientes de nitrógeno, fósforo y potasio en Liberia, Guanacaste. Thesis. Universidad de Costa Rica.

Rowley, G. 1997. *A history of succulent plants.* Mill Valley, CA: Strawberry Press.

Roy, S. C., and A. Sarkar. 1991. *In vitro* regeneration and micropropagation of *Aloe vera* L. *Scientia Horticulturae* 47:107–113.

Saks, Y., and R. Barkai. 1995. *Aloe vera* gel activity against plant pathogenic fungi. *Postharvest Biological Technology* 6:159–165.

Salisbury, F., and C. Ross. 1991. *Plant physiology,* 3rd ed. Belmont: Wadsworth Publishing.

Sanghi, D., and N. J. Sarna. 2001. Palynological studies of some medicinal plants. *Journal of Phytology Research* 14:83–90.

Sapre, A. B. 1975. Meiosis and pollen mitosis in *Aloe barbadensis* Mill. (*A. perfoliata* var. *vera* L., *A. vera* Auth. Non Mill.) *Cytologia* 40:525–533.

———. 1978. Karyotype of *Aloe barbadensis* Mill.: A reinvestigation. *Cytologia* 43:237–241.

Schmidt, K., and E. Ernst. 2004. Assessing websites on complementary and alternative medicine for cancer. 2004. *Annals of Oncology* 15:733–742.

Shane, L. 2004. Biological complementary therapies: A focus on botanical products in diabetes. *Diabetes Spectrum* 13:199–208.

Sheng, C., L. Kai, C. Chun, F. Chia, and L. Chih. 2007. Aloe-emodin-induced apoptosis in human gastric carcinoma cells. *Food & Chemical Toxicology* 45:2292–2303.

Shioda, H., K. Satoh, F. Nagai, T. Okubo, T. Seto, T. Hamano, H. Kamimura, and I. Kano. 2003. Identification of *Aloe* species by random amplified polymorphic DNA (RAPD) analysis. *Shokuhin Eiseigaku Zasshi* 44:203–207.

Simmonds, M. 2004. Pest of aloes. In *Aloes: The genus Aloe,* ed. T. Reynolds, 367–379. Boca Raton, FL: CRC Press.

Singh, B., and N. Sood. 2009. Significance of explants preparation and sizing in *Aloe vera* L.—A highly efficient method for *in vitro* multiple shoot induction. *Scientia Horticulturae* 122:146–151.

Singh, R. J. 2004. *Plant cytogenetics*, 2nd ed. Boca Raton, FL: CRC Press.

Sleper, D. A., and J. M. Poehlman. 2006. *Breeding field crops*, 5th ed. Ames: Blackwell Publishing Professional.

Smith, G., and E. Steyn. 2004. Taxonomy of Aloaceae. In *Aloes: The genus Aloe*, ed. T. Reynolds, 15–36. Boca Raton, FL: CRC Press.

Smith, G., and B. Van Wyk. 1998. Asphodelaceae. In *Vascular plant genera of the world*, ed. K. Kubitzki, 130–140. Berlin: Springer–Verlag.

———. 2008. *Aloes in southern Africa*. Cape Town: Struik Publishing.

Smith, G. F. 1993. Familial orthography: Aloeaceae vs. Aloaceae. *Taxon* 42:87–90.

Snoad, B. 1951. Chromosome numbers of succulent plants. *Heredity* 5:279–283.

Supe, U. J. 2007. *In vitro* regeneration of *Aloe barbadensis*. *Biotechnology* 6:601–603.

Sydiskis, R. J., D. G. Owen, J. L. Lohr, K.-H. A. Rosler, and R. N. Blomster. 1991. Inactivation of enveloped viruses by anthraquinones extracted from plants. *Antimicrobial Agents & Chemotherapy* 35:2463–2466.

Takahashi, C., I. J. Leitch, A. Ryan, M. D. Bennett, and P. E. Brandham. 1997. The use of genomic *in situ* hybridization (GISH) to show transmission of recombinant chromosomes by a partially fertile bigeneric hybrid, *Gasteria lutzii* × *Aloe aristata* (Aloaceae), to its progeny. *Chromosoma* 105:342–348.

Tawfik, K. M., S. A. Sheteawi, and Z. A. El-Gawad. 2001. Growth and aloin production of *Aloe vera* and *Aloe eru* under different ecological conditions. *Egyptian Journal of Biology* 3:149–159.

Taylor, W. R. 1925. Cytological studies on *Gasteria* II. A comparison of the chromosomes of *Gasteria*, *Aloe* and *Haworthia*. *American Journal of Botany* 12:219–223.

Tian, J., J. Liu, J. Zhang, Z. Hu, and X. Chen. 2003. Fluorescence studies on the interactions of babaloin with bovine serum. *Chemical & Pharmaceutical Bulletin* 51:579–582.

Treutlein, J., G. F. Smith, B.-E. Van Wyk, and M. Wink. 2003. Phylogenetic relationships in Asphodelaceae (subfamily Alooideae) inferred from chloroplast DNA sequences (*rbc*L, *mat*K) and from genomic fingerprinting (ISSR). *Taxon* 52:193–207.

Trujillo, G., Y. Hernández, J. Imery, and J. Guevara. 2001. Aislamiento de bacterias fitopatógenas afectando *Aloe vera* en el Edo. Sucre. Memoria del XVII Congreso Venezolano de Fitopatología, 48.

Tucker, A. O., J. A. Duke, and S. Foster. 1989. Botanical nomenclature of medicinal plants. In *Herbs, spices and medicinal plants*, ed. L. E. Cracker and J. E. Simon, 169–242. Phoenix, AZ: Oryx Press.

Valdés, B. 1997. Caracteres taxonómicos: Citología y citogenética. In *Botánica*, ed. J. Izco et al., 133–154. Madrid: McGraw–Hill.

Van den Gorkom, B. A. P., E. G. de Vries, and J. H. Kleibeuker. 1999. Review article: Antranoid laxatives and their potential carcinogenic effects. *Alimentary Pharmacology & Therapeutics* 13:443–452.

Van Jaarsveld, E. 1994. *Gasterias of South Africa: A new revision of a major succulent group*. Pretoria: Fernwood Press.

Van Wyk, B. E., and G. F. Smith. 1996. *Guide to the aloes of South Africa*. Pretoria: Briza Publishing.

Vega, A., N. Ampuero, L. Díaz, and R. Lemus. 2005. El *Aloe vera* (*Aloe barbadensis* Miller) como componente de alimentos funcionales. *Revista Chilena de Nutrición* 32:208–214.

Velásquez, P. 2009. Heterocromatina constitutiva en siete especies del género *Aloe* L. (Aloaceae). Thesis. Universidad de Oriente, Venezuela.

Velásquez-Arenas, R., and J. Imery-Buiza. 2008. Fenología reproductiva y anatomía floral de las plantas *Aloe vera* y *Aloe saponaria* (Aloaceae) en Cumaná, Venezuela. *Revista de Biología Tropical* 56:1109–1125.

Vij, S. P., M. Sharma, and I. S. Toor. 1980. Cytogenetical investigations into some garden ornamentals II. The genus *Aloe* L. *Cytologia* 45:515–532.

Viljoen, A. M., B. E. Van Wyk, and L. E. Newton. 2001. The occurrence and taxonomic distribution of the anthrones aloin, aloinoside and microdontin. *Biochemical Systematics & Ecology* 29:53–67.

Vogler, B. K., and E. Ernst. 1999. *Aloe vera*: A systematic review of its clinical effectiveness. *British Journal of General Practice* 49:823–828.

Wang, L., X. Zheng, L. Li, and J. Gu. 2001. A preliminary study on the polyploid induction and variation of *Aloe vera*. *Acta Botanica Yunnanica* 23:493–496.

Wang, Z., J. Zhou, Z. Huang, A. Yang, Z. Liu, Y. Xia, Y. Zeng, and X. Zhu. 2004. *Aloe* polysaccharides mediated radioprotective effect through the inhibition of apoptosis. *Journal of Radiation Research* 45:447–454.

Wasserman, L., S. Avigad, E. Beery, J. Nordenberg, and E. Fenig. 2002. The effect of aloe emodin on the proliferation of a new merkel carcinoma cell line. *American Journal of Dermatopathology* 24:17–22.

Watt, J., and M. Breyer-Brandwijk. 1962. *The medicinal and poisonous plants of southern and eastern Africa.* Edinburgh: Livingstone.

WHO (World Health Organization). 1999. *Monographs on selected medicinal plants,* vol. 1. Geneva: World Health Organization.

Yagi, A., and S. Takeo. 2003. Anti-inflammatory constituents, aloesin and aloemannan in *Aloe* species and effects of tanshinon VI in *Salvia miltiorrhiza* on the heart. *Pharmaceutical Society of Japan* 123:517–532.

Yamaguchi, I., N. Mega, and H. Sanada. 1993. Components of the gel of *Aloe vera* (L.) Burm. f. *Bioscience, Biotechnology & Biochemistry* 57:1350–1352.

Yeh, G. Y., and D. M. Eisenberg. 2003. Systematic review of herbs and dietary supplements for glycemic control in diabetes. *Diabetes Care* 26:1277–1294.

Yépez, L. M., M. L. Díaz, E. Granadillo, and F. Chacín. 1993. Frecuencia óptima de riego y fertilización en *Aloe vera* L. *Turrialba* 43:261–267.

Zhong, Z., W. Ying, L. Yun, Z. Hong, and P. Wei. 1993. Studies of twenty-nine fungous disease on ornamental and floral plants. *Transactions of the Mycological Society of Republic of China* 8:69–76.

Zhou, X. M., and Q. H. Chen. 1988. Biochemical study of Chinese rhubarb XXII. Inhibitory effect of anthraquinone derivatives on sodium-potassium-ATPase of a rabbit renal medulla and their diuretic action. *Acta Pharmacologica Sinica* 23:17–20.

Zonneveld, B. J. M. 2002. Genome size analysis of selected species of *Aloe* (Aloaceae) reveals the most primitive species and results in some new combinations. *Bradleya* 20:5–12.

CHAPTER 26

Stevia rebaudiana—A Natural Substitute for Sugar

Vikas Jaitak, Kiran Kaul, V. K. Kaul, Virendra Singh, and Bikram Singh

CONTENTS

26.1 INTRODUCTION

Stevia rebaudiana Bertoni is a perennial plant, native to Paraguay, that is commonly known as sweet herb. The botanical identification of the plant was first given in 1889 (Bertoni 1899). The leaves of this plant have an incredible sweetness index and are used as herbal medicine for diabetes. Leaves are sources of natural sweetener due to the presence of nine steviol glycosides collectively known as steviosides, which are noncalorific and used in the food industry as sugar substitutes in ice creams and confectionery products.

Among the nine glycosides, stevioside and rebaudioside-A are the two important and major sweet constituents. Steviosides have many-fold advantages: they are nontoxic, heat stable, nonfermentive, flavor enhancing, and 100% natural. Since the restriction of artificial sweeteners in Japan, large quantities of *S. rebaudiana* leaves have been used for more than 50 years in many Japanese foods and in beverages. The sweet tasting products from the leaves are known as stevia sweeteners and have been commercialized in three basic forms in Japan: (1) stevia extract, (2) sugar transferred stevia extract, and (3) rebaudioside-A enriched stevia extract used in a variety of foods and beverages. More than 100 preparations containing stevia sweeteners are commercially available in Japan (Kinghorn 2002). Steviosides occupy 40% of the sweetener market in South Korea and 200–250 tons of steviosides are produced annually for export purposes, as well as for internal consumption as *soju*—a traditional Korean alcoholic beverage. In Japan and Brazil, *S. rebaudiana* is approved as a food additive and sugar substitute.

26.2 BOTANY

Stevia rebaudiana Bertoni is a herbaceous perennial plant of the Asteraceae family. The genus consists of herbs and shrubs (Gentry 1996). Morphological uniformity of flowers and capitula render the genus most distinctive within the tribe Eupatoreae (King and Robinson 1987). The plant grows up to 65 cm in height, with sessile, oppositely arranged lanceolate to oblanceolate, dark-green leaves with toothed margins. The leaves are sweet in taste. Trichome structures of the leaf surface are of two different sizes: large (4–5 µm) and small (2.5 µm) (Shaffert and Chebotar 1994). The plant has many branches and a root system with two types of roots. Fine roots (feeder roots) spread out on the surface of the soil while a thicker part (anchor root) grows deep into the soil. The flowers are small (7–15 mm), white, tubular, bisexual and arranged in an irregular cyme (Dwivedi 1999). The seed is an achene with feathery pappus (Lester 1999). Of the 280 species of the genus, only *S. rebaudiana* possesses sweet taste.

26.3 ORIGIN

Stevia rebaudiana is native to the valley of Rio Monday in the highlands of Paraguay, between 25 and 26°S latitude, where it grows in sandy soils near streams (Katayama et al. 1976). It was long ago known to the Guarani Indians of the Paraguayan highlands, who called it *caa-ehe*, meaning sweet herb (Lewis 1992). They especially used it in local green tea (*mate*—tea). Stevia was discovered in 1887 by South American scientist Antonio Bertoni. Approximately 80 wild species exist in North America and another 200 species are native to South America. The crop was first established in Japan (Sumida 1968). By the mid-1970s, standardized extracts and pure stevioside were utilized commercially in Japan for sweetening, flavoring in foods and beverages, and as a substitute for several synthetic sweeteners. Total market value of stevia sweeteners in Japan is around $23–35 million per year. The crop has been introduced in other countries, including Korea, Mexico, Indonesia, Tanzania, and Canada (Brandle and Rosa 1992; Megeji et al. 2005).

26.4 DISTRIBUTION

Stevia rebaudiana is distributed in southern United States and in Paraguay to Southern Brazil. It is presently under cultivation in Paraguay, Mexico, Central America, China, Malaysia, and South Korea. In Europe, it is reported to be cultivated in Spain, Belgium, and UK (Kinghorn 1992; Geuns 1998). In India, it is cultivated in Himachal Pradesh, Punjab, Haryana, Uttar Pradesh, Madhya Pradesh, West Bengal, Karnataka, and Tamil Nadu.

The Institute of Himalayan Bioresource Technology (IHBT) in Palampur, India, has catalyzed cultivation of *S. rebaudiana* in different agroclimatic zones; 1,250 acres have been covered under primary plantation and 3,000 acres through secondary plantation. This has yielded 5,000 tons of dry leaf biomass from primary plantation and 12,000 tons through secondary plantation. The total revenue generated from primary and secondary plantation has been $3.62 million, generating 1 million man-days.

26.5 IMPORTANCE

Stevia rebaudiana is distinguished by the presence of zero-calorie, high-potency sweeteners known as steviol glycosides, which are present in its leaf tissue (Ahmed and Dobberstein 1982a,b; Phillips 1987). These sweeteners are recommended for diabetes, obesity, high blood pressure, and dental caries; they are used for human consumption with no side effects (Kasai et al. 1981).This plant is of worldwide importance because its leaves have commercial importance and are used as non-nutritive sweeteners in Japan, Korea, China, and South America. Consumption of stevia extract in Japan and Korea is around 200 and 115 tons per year, respectively (Kinghorn, Wu, and Soejarto 2001). Its water extract has beneficial effects on human health, having hypoglycemic (Jeppesen et al. 2003), hypotensive (Chan et al. 2000) activity, and it is also a source of natural antioxidants (Xi, Sato, and Takeuchi 1998). It is a nontoxic sweetener and inhibits the formation of cavities and plaque in teeth (Elkins 1997). *Stevia rebaudiana* is reported to have cardiotonic properties (Machado, Chagas, and Reis 1986), improves gastrointestinal function (Alvarez 1986), and is effective against microbes (*Streptococcus mutans*, *Pseudomonas aeruginosa*, and *Proteus vulgaris*) (Yabu et al. 1997).

26.6 BIOLOGICAL ACTIVITIES

26.6.1 Anti-inflammatory and Immunomodulatory Activities

Boonkaewwan, Toskulkao, and Vongsakul (2006) reported anti-inflammatory and immuno-modulatory activities of stevioside and its aglycone, steviol. Stevioside at 1 mM significantly suppressed lipopolysaccharide (LPS)-induced release of tumor necrosis factor-alpha (TNF-α) and interleukin-beta (IL-β) and slightly suppressed nitric oxide release in THP-1 cells without exerting any direct toxic effect. Steviol at 100 μM did not suppress. Activation of IKKβ and transcription factor nuclear factor kappa-B (NF-κB) was suppressed by stevioside, as demonstrated by western blotting. Furthermore, only stevioside induced TNF-α, IL-β, and nitric oxide release in unstimulated THP-1 cells. Release of TNF-α could be partially neutralized by anti-TLR4 antibody. This study suggested that stevioside attenuates synthesis of inflammatory mediators in LPS-stimulated THP-1 cells by interfering with the IKKβ and NF-κB signaling pathway and stevioside-induced TNF-α secretion is partially mediated through TLR4.

26.6.2 Antimicrobial and Anticancer Activities

Ethyl acetate, acetone, chloroform, and water extracts of *S. rebaudiana* leaves were tested against different organisms (*Staphylococcus aureus*, *Salmonella typhi*, *Escherichia coli*, *Bacillus subtilis*, *Aeromonas hydrophila*, and *Vibrio cholerae*) for evaluation of antimicrobial activity, using an agar diffusion method. *Candida albicans*, *Cryptococcus neoformans*, *Trichophyton mentagrophytes*, and *Epidermophyton* species were selected for screening antiyeast and antifungal activities (Jayaraman, Manoharan, and Illanchezian 2008). The cytotoxic effects of different extracts on Vero and HEp2 cells were assayed using 3-(4,5-dimethylthiazol-2-yl)-2,5-diphenyltetrazolium bromide [MTT]. Among the four extracts tested, acetone and ethyl acetate extracts showed effective antibacterial activity. The acetone extract showed greater activity against Gram-positive than Gram-negative organisms. All the extracts were active against *Candida albicans* and *Epidermophyton* species. A 1:8 dilution of acetone extract was nontoxic to normal cells and had both anticancer and antiproliferative activities against cancerous cells.

26.6.3 Antioxidant Activity

Shukla et al. (2009) described antioxidant activity of ethanolic extract of *S. rebaudiana*. The DPPH activity (20, 40, 50, 100, and 200 µg/mL) did enhance in a dose-dependent manner, which was observed in the range of 36.93–68.76% as compared to ascorbic acid (64.26–82.58%). The IC_{50} values of ethanol extract and ascorbic acid in DPPH radical scavenging assay were obtained as 93.46 and 26.75 µg/mL, respectively. The ethanolic extract was also found to scavenge superoxide generated by EDTA/NBT system. Measurement of total phenolic content of the ethanol extract was achieved using Folin–Ciocalteau reagent containing 61.50 mg/g of phenolic content; this was found to be significantly higher when compared to reference standard gallic acid. The ethanol extract also inhibited the hydroxyl radical, nitric oxide, and superoxide anions with IC_{50} values of 93.46, 132.05, and 81.08 µg/mL, respectively. The IC_{50} values for the standard ascorbic acid were 26.75, 66.01, and 71.41 µg/mL, respectively. The results show that *S. rebaudiana* has a significant antioxidant activity.

26.6.4 Antirotavirus Activity

Anti-HRV (human rotavirus) activity of hot water extracts from *S. rebaudiana* was examined. It inhibited *in vitro* replication of all four serotypes of HRV (Kazuo et al. 2001). This inhibitory effect of *Stevia rebaudiana* extract (SE) was not reduced on prior exposure to HCl for 30 min at pH 2. Binding assay with radiolabeled purified viruses indicated that the inhibitory mechanism of SE is blockade of virus binding. The SE inhibited the binding of anti-VP7 monoclonal antibody to HRV-infected MA104 cells. The inhibitory components of SE were found to be heterogeneous anionic polysaccharides with different ion charges. The component analyses suggested that the purified fraction has highest inhibitory activity consisting of the anionic polysaccharide with molecular weight of 9800 and contains Ser and Ala as amino acids. Analysis of sugar residues suggests uronic acid or acids as sugar components. It did not contain amino and neutral sugars and sulfate residues. These findings suggested that SE may bind to 37 kD VP7 and interfere with the binding of VP7 to the cellular receptors by steric hindrance, which results in blockade of the virus attachment to cells.

26.6.5 Antifertility Activity

Kumar and Oommen (2008) report that oral intake of water-based sweet stevia extract and stevioside at doses of 500 mg/kg body weight and 800 mg/kg body weight, respectively, does not cause any significant female reproductive toxic effect in Swiss albino mice.

26.6.6 Antidiabetic Activity

Stevioside *in vitro* exerts a direct insulinotropic action in both isolated mouse islets and the clonal β-cell line, INS-118-19 (Jeppesen et al. 2000). The insulinotropic effect of stevioside, unlike classical sulfonylureas, is not associated with closure of ATP-sensitive potassium channels in β-cells, which has been associated with proarrhythmic effects on the heart. *In vitro*, like the incretin hormone glucagon-like peptide-1 (GLP-1), stevioside is glucose dependent. At normal glucose levels, no insulinotropic actions are observed. This indicates that stevioside is useful in the treatment of type 2 diabetes. Furthermore, long-term oral use of stevioside lowered blood pressure in the diabetic GK rat and caused enhanced first phase insulin response. In a long-term study on nondiabetic subjects with hypertension, a similar blood pressure lowering effect of stevioside has been demonstrated (Jeppesen et al. 2000). In a study in type 2 diabetic subjects, it was demonstrated that oral intake of stevioside causes a clear-cut reduction in glycemic response to a test meal (Jeppesen et al. 2000).

26.7 SOIL AND CLIMATE

Stevia rebaudiana is a subtropical plant and prefers a semihumid climate with an average temperature of 23.8°C and 139.7 cm rainfall. In its native Paraguay, *S. rebaudiana* grows in coarsely textured, infertile acidic sands and grasslands with shallow water tables. The plant prefers well-drained, lightly textured soil with organic matter and needs ample water so that the soil is consistently moist, but not wet. When stevia is cultivated on a commercial scale, it can be grown in well-drained red soil and sandy loam soil with a pH range of 6–7. The soil can be enriched with a basal dressing of 25 tons of well-rotted farmyard manure (FYM) per hectare. Saline soils should be avoided to cultivate this plant. Stevia can be successfully cultivated all over India except in areas that receive snowfall and where temperatures go below 5°C in winter. High summer temperatures do not affect this plant. In a subtropical climate, it grows well, and the concentration of steviosides in the leaves increases when the plants are grown under long-daylight conditions.

26.8 PROPAGATION AND CULTIVATION

Stevia rebaudiana can be propagated by cuttings as well as by seeds (Ramesh, Singh, and Megeji 2006). In many cultivars, seed germination rate is poor and crop raised through seedlings takes much time to establish. Clonal propagation is practiced for small-scale production but is not economically viable for large-scale production due to high labor inputs. Stevia is also propagated vegetatively through root stock division obtained from old plantation. Divided root stock is treated with GA_3 in 75–100 ppm solution for better sprouting in early spring.

The nursery is raised through seeds in the months of February and March. Seed germination takes place within 10–15 days. Seeds can be sown in plastic trays, pots, wooden boxes, or raised nursery beds of size 1.25 m × 1.0 m that have a good mixture of sand, soil, and organic manure. The nursery should be irrigated with a sprinkler daily, morning and evening, for the first 5 days and once a day during the next 5 days. Seedlings that are 2 months old in five- to seven-leaf stage and 8–10 cm height are ready for transplantation.

Saplings can be raised successfully by vegetative propagation using terminal cuttings of 10–15 cm height with four to six nodes. Stem cuttings should be taken from actively growing plants, avoiding extremely cold and high temperatures. Lower leaves are trimmed to facilitate planting. Cuttings are raised in pots or poly sleeves. In nursery beds, these are planted at 15 × 15 cm spacing, keeping one node inside the soil. Cuttings raised under partial shade and high humidity develop roots at a faster rate, within 8–10 days. Watering of plants is carried out daily through sprinklers for the first 10 days. Later, the crop is irrigated twice a week or as required. Bavistin (0.1%) and thiodan (0.25%) are sprayed to keep the plants free from diseases and insect pests.

Stevia crop is grown through transplantation as an annual or perennial crop. Rooted cuttings of 6–8 weeks old are transplanted in the field during March–April and July–August, avoiding periods of extreme hot and cold climate. Land should be prepared by repeatedly ploughing and harrowing. Irrigation and drainage channels should be laid out according to the layout of the field. Transplantation is carried out in furrows at a spacing of 45 × 45 cm and 45 × 30 cm in high- and medium-fertile soils, respectively. The crop is fertilized through organic manures and consumes nutrients moderately. A well rotted 25–30 ton/ha farmyard manure application at the time of field preparation is sufficient for crop growth.

Stevia cannot tolerate drought, so frequent irrigation is required. First irrigation is given immediately after transplantation and another is given after 2–3 days. Subsequent watering is carried out on a weekly basis till onset of monsoon. Actual numbers of irrigations are based on moisture-holding capacity of the soil and occurrence of natural precipitation. Weeds are removed manually. Hand weeding and hoeing twice are sufficient for control of weeds. Organic mulch can help in controlling germination and growth of weeds. Harvesting is carried out manually, leaving 8–10 cm stem height from the ground level. First harvest is taken 75–90 days after transplanting in the months of June and July. Subsequently, second harvest is taken 60–75 days after the first harvest in early September at the time of flower bud initiation. In case of late transplanting, a crop grown for a single cut, harvesting is carried out after 3–4 months of transplanting and continues till flowering begins as the maximum steviol glycosides are in leaves until the plant flowers.

Perennial crops may continue up to 4 years, once they are planted in the same field. Maximum leaf biomass is produced in the third or fourth year (Table 26.1). Flowering of the plant should be avoided and pinching of the apical bud is required to enhance bushy growth of the plant with side branches. Average fresh biomass yield of 20–30 ton/ha/year out of two harvests may be obtained; this gives dry herb yield of 4–5 tons. Because leaf is the commercial part, this has significance in terms of its yield. An average dried leaf yield of 17, 20, 23, and 2500 kg/ha separates from the total biomass yield during years 1, 2, 3, and 4, respectively. After harvesting, the whole plant is dried and the leaves are separated from stems for further processing. The stem has a very low concentration of steviol glycosides and is removed from the leaf biomass to minimize processing costs. Dried leaves are stored in airtight containers or in plastic bags and kept in a cool place. They can be directly marketed without further processing by the farmers. Stevia can also be grown under intercropping with wheat. It has been observed that stevioside content enhanced in comparison to other intercrops. Intercropping with wheat resulted in higher monetary benefits at 60 × 45 cm spacing (Ramesh, Singh, and Ahuja 2007).

26.9 PERFORMANCE OF *STEVIA REBAUDIANA* UNDER PALAMPUR AGROCLIMATIC CONDITIONS

Crop performance of two accessions of *S. rebaudiana* introduced in an experimental farm of the IHBT, Palampur, was satisfactory. The two accessions were least affected by biotic and abiotic stresses like high rainfall, frost, and infestation by insects and diseases. Stevioside content of the two accessions ranged between 6 and 8%. Accession 1 was superior in stevioside content and accession 2 was superior in leaf biomass. Variation in stevioside and rebaudioside-A concentrations in leaves was noted with the advancement of plant phenology and node position from apical bud. An increasing trend in stevioside concentration was observed in leaves from 1 to 10 node positions at vegetative stage. But at bud and flowering stages, a decreasing trend was observed due to higher accumulation of stevioside in young leaves at node position 1 (18.3%) in accession 1 and at node

Table 26.1 Economics of Cultivation of *Stevia rebaudiana*[a]

S. No.	Particulars	Year 1	Year 2	Year 3	Year 4
A.	**Establishment cost**	**1,297.31**	—	—	—
1.	Cost of 250 g seeds at $214.57/kg	53.64			
2.	Cost of raising seedlings on 500 m² lands for 75 days	214.57			
3.	Cost of uprooting nursery plants: 75,000 plants × $0.002/plant	16.10	—	—	—
4.	Transportation charges from nursery to plantation site	214.57	—	—	—
5.	Preparation of land: four ploughings, each followed by planking at $21.44/operation	85.79	—	—	—
6.	Preparation of plantation beds: 20 man-days	42.89	—	—	—
7.	FYM (35 tons) as basal dose at $ 7.88/ton	262.73	—	—	—
8.	Mixing of FYM in soil 10 @ man-days	21.44	—	—	—
9.	Transplanting of saplings @ 30 man-days	64.34	—	—	—
10.	Drying arrangements	321.71			
B.	**Variable costs**	**1,576.40**	**1,760.85**	**1,763.00**	**1,763.00**
11.	Gap filling, plant cost, and planting	—	34.31	34.31	34.31
12.	NPK fertilizers 100:60:40 ($21.44 + 25.46 + 6.43)	53.62	53.619	53.619	53.619
13.	Fertilizer application cost: 5 man-days	10.72	10.72	10.72	10.72
14.	Irrigation: 24 irrigations × $10.72/irrigation	257.48	257.48	257.48	257.48
15.	Hand weeding (two/year) @ 30 man-days/weeding	128.74	128.74	128.74	128.74
16.	Hoeing (two/year) @ 25 man-days	107.21	107.21	107.21	107.21
17.	Harvesting (two/year) @ 15 man-days	64.34	64.34	64.34	64.34
18.	Transportation of herbs from field to drying site	10.72	10.72	10.72	10.72
19.	Drying, postharvest handling, and forwarding charges	64.34	107.307	107.307	107.307
20.	Contingency cost	107.307	214.6	214.6	214.6
21.	Supervisory charges @ $64.34/month	772.78	772.78	772.78	772.78
C.	**Fixed costs**	**580.87**	**480.60**	**480.79**	**480.79**
22.	Rate of interest on establishment and variable costs @ 9%	258.88	158.613	158.613	158.613
23.	Rental value of land @ $321.71/annum	322.06	322.06	322.06	322.06
D.	**Total production cost (A + B + C)**	**3458.07**	**2242.9**	**2244.83**	**2244.83**
E.	**Gross income**	**3,648.46**	**4,292.30**	**4,936.15**	**5,365.38**
	Sale price of dried leaf @ $2.15/kg dried leaf. Yield of 1700 kg in year 1, 2000 kg in year 2, 2300 kg in year 3, and 2500 kg in year 4				
F.	**Net income**	**191.86**	**2,049.81**	**2,691.31**	**3,120.549**
G.	**Benefit cost ratio (BCR; gross income/ total production cost)**	**1.055**	**1.914**	**2.198**	**2.390**

Assumptions: Labor charges: $2.146 per man-day; total production cost per hectare for 4 years = $10,188.75; gross income per hectare during 4 years = $18,242.30; net income per hectare in 4 years = $8,051.17; average net income per hectare per year = $2,012.959; average benefit cost ratio (ABCR) = 1.889.

[a] Dollars per hectare per year.

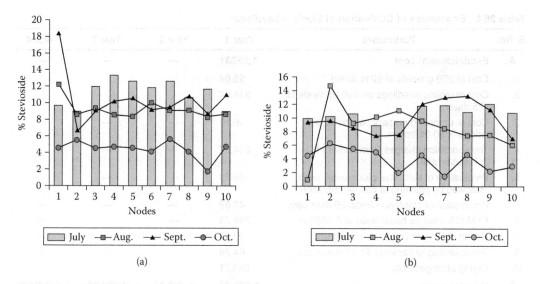

Figure 26.1 Differential distribution of stevioside between 1 and 10 nodes for (a) accession 1 and (b) accession 2.

position 2 (14.9%) in accession 2 (Figure 26.1). Likewise, maximum rebaudioside-A concentration (16%) in the leaves was attained during bud stage and minimum at seed-setting stage in accessions 1 and 2 (Figure 26.2).

Average stevioside concentration in leaves from 1 to 10 nodes was maximum at vegetative stage, as was rebaudioside-A at bud stage (Figure 26.3). The net photosynthetic rate was observed to decline with increasing age of fully expanded leaves. Monitoring of stevioside and rebaudioside-A concentrations in leaves during the whole growing cycle showed consistent increase up to the late vegetative stage, followed by gradual decline; the least concentration was at seed-setting stage (Figure 26.4). Such gradual increase in stevioside concentration in stevia leaves was reported earlier by Bondarev, Reshetnyak, and Nosov (2003). Also, similar trends were observed in steviol glycoside concentrations, total weekly rainfall, and relative humidity (Figure 26.4) under Palampur

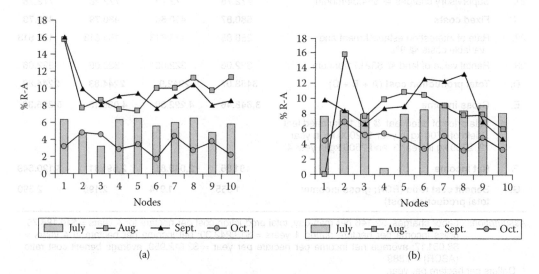

Figure 26.2 Differential distribution of rebaudioside-A between 1 and 10 nodes for (a) accession 1 and (b) accession 2.

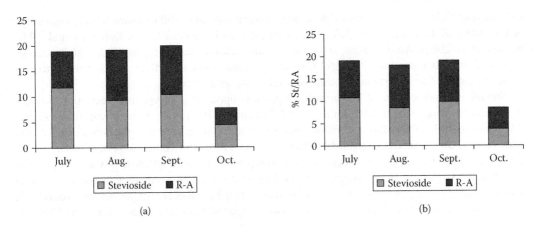

Figure 26.3 Ratio of stevioside to rebaudioside-A for (a) accession 1 and (b) accession 2.

agroclimatic conditions. Such results have been reported by Nepovim et al. (1998) under Czech Republic climatic conditions, indicating that rainfall alone is sufficient for adequate growth of the species in humid, wet areas (Hoyle 1992).

26.10 CONSTITUENTS OF *STEVIA REBAUDIANA*

26.10.1 Steviol Glycosides

Since precursors of steviol glycoside are synthesized in chloroplast, tissues devoid of chlorophyll contain no or trace amounts of these glycosides (Brandle and Rosa 1992; Bondarev et al. 2003). Leaves of *S. rebaudiana* contain nine steviol glycosides, out of which stevioside and rebaudioside-A are predominant. Stevioside is reported to be 300 times and rebaudioside-A 400 times sweeter than sucrose (Kasai et al. 1981). Taste of rebaudioside-A is better than that of stevioside, which gives a bitter aftertaste (DuBois and Stephenson 1985). Among nine steviol glycosides, diterpene steviol

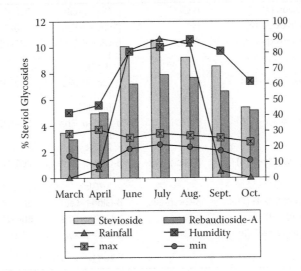

Figure 26.4 Effect of environmental factors on stevioside and rebaudioside-A content in *S. rebaudiana* during its cultivation cycle.

skeleton (*ent*-13-hydroxy kaur-16-en-19-oic acid) is common and exhibits characteristic organoleptic properties (Kohda et al. 1976; Sakamoto, Yamasaki, and Tanaka 1977a,b; Kobayahi et al. 1977; Brandle et al. 2002). Among these, stevioside (1) and rebaudioside-A (3) (see Table 26.2) are two major and important constituents useful for diabetes without any side effects in comparison to synthetic sweeteners like aspartame, saccharine, cyclamate, and sucralose.

Steviol glycoside content in leaves varies depending upon the genotype and production environment (Kinghorn and Soejarto 1985; Phillips 1987; Brandle and Rosa 1992; Brandle, Staratt, and Gizjen 1998; Starratt et al. 2002). Many reports on composition and content of steviol glycosides in leaves and other parts of *Stevia* have been published (Brandle et al. 1998; Nepovim et al. 1998; Bondarev et al. 2003). Maximum steviol glycoside concentration is reported in leaves, followed by very low percentages in inflorescence and stem and none in roots. However, information regarding their distribution in leaves at various node positions is limited. Differential distribution of two major steviol glycosides in leaves at various node positions (1–10) and at various stages of plant development is reported here for

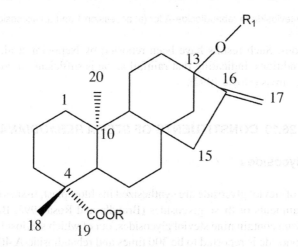

Table 26.2   Steviol Glycosides in *S. rebaudiana*

Diterpene Glycosides	R	R₁	RS
Stevioside (1)	β-Glc	β-Glc²–β-Glc¹	250–300
Steviolbioside (2)	H	β-Glc²–β-Glc¹	100–125
Rebaudioside-A (3)	β-Glc	β-Glc²–β-Glc¹ | β-³Glc	350–450
Rebaudioside-B (4)	H	β-Glc²–β-Glc¹ | β-³Glc	300–350
Rebaudioside-C (5)	β-Glc	β-Glc²–α-Rha¹ | β-³Glc	50–120
Rebaudioside-D (6)	β-Glc-β-Glc	β-Glc²–β-Glc¹ | β-³Glc	200–300
Rebaudioside-E (7)	β-Glc-β-Glc	β-Glc²–β-Glc¹	250–300
Dulcoside (8)	β-Glc	β-Glc²–α-Rha¹	50–120
Rebaudioside-F (9)	β-Glc	β-Glc²–β-Xyl¹ | β-³Glc	ND

Notes: Glc = β-d-glucopyranosyl; Rha = α-l-rhamnopyranosyl; Xyl = xylose; RS = relative sweetness; ND = No Data.

the first time. The dependence of steviol glycoside content on rainfall, relative humidity, temperature and day length has also been determined. Like other plant secondary metabolites, steviol glycosides function in a defensive capacity against herbivores, pests, and pathogens. Few investigations into the adaptive role of steviol glycosides have been conducted, and there is some evidence that steviosides have a deterrent effect on aphid feeding (Nanayakkara et al. 1987; Wink 2003).

26.10.2 Terpenes

In addition to sweet diterpene glycosides, several other diterpenes have been isolated from *S. rebaudiana*. These are jhanol (10), austroinulin (11), and 6-*O*-acetylaustroinulin (12) (Sholichin et al. 1980) (Figure 26.5). Jhanol (10), austroinulin (11), 6-*O*-acetylaustroinulin (12), and 7-*O*-acetyl-austroinulin (13) have been reported from stevia flowers (Darise et al. 1983). Eight additional diterpenes, called sterebins A–H (14-21), have been isolated and characterized from leaves (Oshima, Saito, and Hikino 1986, 1988). Triterpenes; amyrin acetate (22); three esters of lupeol (23), stigmasterol (24), stigmasterol-β-d-glucoside (25), and sitosterol (26); campesterol (27); and β-sitosterol-β-d-glucoside (28) have been isolated from leaves (Nabeta, Kasai, and Sugisawa 1974; Yasukawa et al. 1993; D'Agostino et al. 1984; Matsuo, Kanamori, and Sakamoto 1986) (Figure 26.6).

26.10.3 Flavonoids

Nineflavonoid glycosides—apigenin-4'-*O*-glucoside (29), kaempferol-3-*O*-rhamnoside (30), luteolin-7-*O*-glucoside (31), quercetrin-3-*O*-arabinoside (32), quercetin-3-*O*-glucoside (33), quercetin-3-*O*-rhamnoside (34), 5,7,3'-trihydroxy-3,6,4'trimethoxyflavone (35), apigenin-7-*O*-glucoside (36)—and quercetin-3-*O*-rutinoside (37) were identified in aqueous methanol extract of leaves (Matsuo Kanamori, and Sakawato 1986; Rajbhandari and Roberts 1983; Suzuki et al. 1976) from *S. rebaudiana* (Table 26.3).

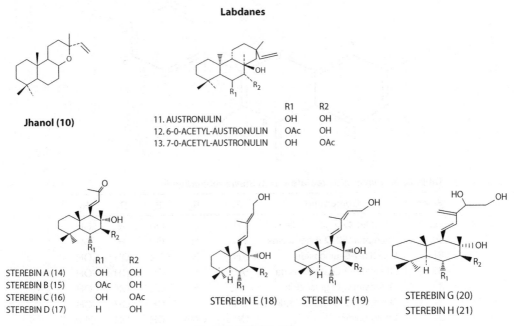

Figure 26.5   Structure of other diterpenes.

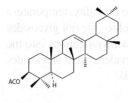
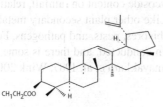
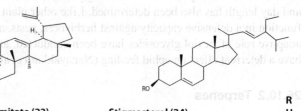

	R
β-Amyrin acetate (22)	
Lupeol-3-palmitate (23)	
Stigmasterol (24)	H
Stigmasterol-β-D-glucoside (25)	Glc

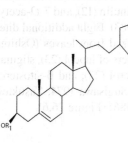

	R₁	R₂
β-Sitosterol (26)	H	CH₃
Campesterol (27)	H	H
β-Sitosterol-β-D-glucoside (28)	Glc	CH₃

Figure 26.6 Structure of triterpenes and sterols isolated from *Stevia rebaudiana*.

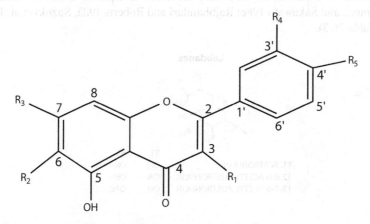

Table 26.3 Flavonoids Isolated from *Stevia rebaudiana*

Sr. no.	Compound	R₁	R₂	R₃	R₄	R₅
29.	Apigenin-4'-O-glucoside	H	H	OH	H	Glc
30.	Kaempferol-3-O-rhamnoside	Rha	H	OH	H	OH
31.	Luteolin-7-O-glucoside	H	H	Glc	OH	OH
32.	Quercetin-3-O-arabinoside	Ara	H	OH	OH	OH
33.	Quercetin-3-O-glucoside	Glc	H	OH	OH	OH
34.	Quercetin-3-O-rhamnoside	Rha	H	OH	OH	OH
35.	Centaureidin	OMe	OMe	OH	OH	OMe
36.	Apigenin-7-O-glucoside	H	H	Glc	H	OH
37.	Quercetin-3-O-rutinoside	Rut	H	OH	OH	OH

26.10.4 Volatile Oil Constituents

The major constituents identified in essential oil of S. *rebaudiana* are mainly sesquiterpenes; hydrocarbons β-caryophyllene, trans-β-farnesene, humulene, and ψ-cadinene; oxygenated sesquiterpenes nerolidol and β-caryophyllene oxide; and oxygenated monoterpenes linalool, terpinen-4-ol, and α-terpineol (Fujita, Taka, and Fujita 1977). Volatile constituents in hydrodistilled oil of dried leaves, spathulenol, caryophyllene oxide, and β-caryophyllene were identified (Martelli, Frattini, and Chialva 1985).

26.10.5 Miscellaneous Constituents

A number of common phytochemicals are identified in S. *rebaudiana*, such as pigments chlorophyll A and B and β-carotene (Cheng and Chang 1983) along with inorganic compounds comprising approximately 13% of the total extractables (i.e., potassium, calcium, iron, magnesium, phosphorus, sodium, and zinc). Tartaric acid is the major organic acid in chloroform extract of S. *rebaudiana* along with citric, formic, lactic, malic, and succinic acids (Cheng and Chang 1983). The phytohormone indole-3-acetonitrile and unspecified tannins have been reported from the seeds of S. *rebaudiana* (Randi and Felippe 1981; Brandle and Rosa 1992).

26.11 BIOSYNTHESIS OF STEVIOL GLYCOSIDES

Stevia rebaudiana leaves accumulate large quantities of steviol glycosides (Starratt et al. 2002; Phillips 1987). These are closely related to gibberellins and share part of their biosynthetic pathway. Initially, it was concluded that steviol was synthesized from kaurene via the mevalonate pathway (Mosettig et al. 1963; Hanson and White 1968; Ruddat, Heftmann, and Lang 1965). Like the synthesis of many diterpenes, it was later demonstrated, using *in vivo* labeling with [I-^{13}C] glucose and NMR spectroscopy, that the precursors of steviol are actually synthesized via the plastid localized methylerythritol 4-phosphate (MEP) pathway (Totte et al. 2000; Figure 26.7). Like all diterpenes, steviol is synthesized from GGDP, first by protonation-initiated cyclization to copalyl diphosphate (CDP) by CDP synthase (CPS). Next, kaurene (III) is produced from CDP by ionization-dependent cyclization catalyzed by kaurene synthase (KS). Kaurene is then oxidized in a three-step reaction to kaurenoic acid (IV) by kaurene oxidase (KO), a P450 monooxygenase that also functions in GA (gibberellin acid) biosynthesis (Bennett, Lieber, and Heftmann 1967; Helliwell et al. 1999; Kim, Sawa, and Shibata 1996). Steviol biosynthesis deviates from gibberellin biosynthesis with hydroxylation of kaurenoic acid by kaurenoic acid 13-hydroxylase (KAH) (Kim et al. 1996) to steviol (V) (Figure 26.7). Plant UDP-glycosyltransferases (UGTs) are a divergent group of enzymes that transfer a sugar residue from an activated donor to an acceptor molecule. The transfer of activated sugars like UDP-glucose to aglycone acceptor molecules helps to stabilize, detoxify, and solubilize metabolites and is often the end point of secondary metabolite pathways. Three of the four UGTs (UGT85C2, UGT74G1, and UGT76G1) involved in the synthesis of stevioside (VIII) and rebaudioside-A (IX) were subsequently identified and characterized (Brandle and Telmer 2007). Addition of C13-glucose to steviol is catalyzed by UGT85C2 and the C19-glucose by UGT74G1; finally, glucosylation of C3 of glucose at the C-13 position is catalyzed by UGT76G1. The UGT responsible for the synthesis of steviolbioside (VII) from steviolmonoside (VI) has not yet been identified. Brandle and Telmer (2007) reported that steviol glycoside synthesis is restricted to green tissues, with all the steps up to kaurene occurring in plastids. One of the two oxidation steps is located on the surface of the endoplasmic reticulum and glycosylation takes place in cytoplasm.

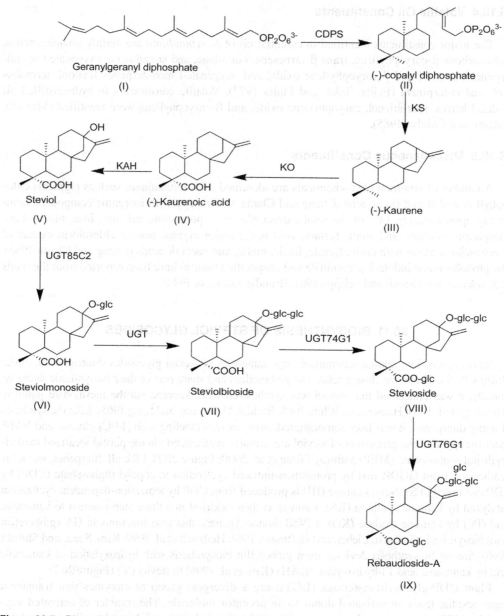

Figure 26.7 Biosynthetic pathway of the steviol glycosides. Abbreviations: CDPS = copalyl diphosphate synthase; KS = kaurene synthase; KO = kaurene oxidase; KAH = kaurenoic acid oxidase. UGTs are abbreviated to their numerical identifiers.

26.12 EXTRACTION AND QUANTIFICATION OF MAJOR STEVIOL GLYCOSIDES

Many patents on processing of *S. rebaudiana* leaves for the production of stevioside and rebaudioside-A are available, including 150 Japanese patents. Processing steps essentially are extraction, pretreatment, separation, purification, and refining of the end product. Most of the reported processes use coagulating agents and organic solvents. Some of the selected processes utilize chromatographic separation (Matsushita and Ikushige 1979) and chelating agents followed

Table 26.4   Yield of Stevioside and Rebaudioside-A with Different Extraction Methods

Sr. no.	Method	Time	Stevioside (%)	Rebaudioside-A (%)	Total (%)
1.	Conventional	12 h	6.54	1.20	7.74
2	Ultrasound	30 min	4.20	1.98	6.18
3.	Microwave	1 min	8.64	2.34	10.98

by solvent extraction (Kumar 1986). A process involving pretreatment of the extract with lime and use of ion exchange columns has been reported (Giovanetto 1990). A Japanese patent describes the production of steviol glycosides by supercritical fluid extraction using liquid carbon dioxide along with cosolvents such as methanol, ethanol, and acetone (Tan, Shibuta, and Tanaka 1988). Kumar et al. (2006) have filed a patent on the production of steviosides from *S. rebaudiana* using green technology. Pol et al. (2007) reported a pressurized fluid extraction (PFE) method for the extraction of stevioside from *S. rebaudiana* using methanol and water as solvents. In an earlier publication (Jaitak et al. 2009a,b), we reported comparison of percentage yield of major steviol glycosides (i.e., stevioside and rebaudioside-A) under microwave, ultrasound, and conventional techniques (Table 26.4). Figure 26.8 shows the high-performance liquid chromatography (HPLC) chromatogram of *S. rebaudiana* extract under microwave extraction and those of standards.

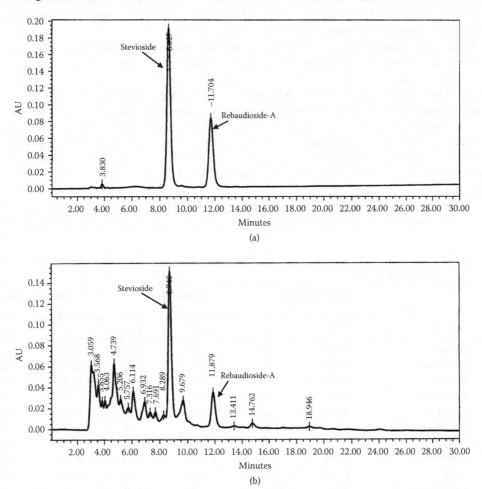

Figure 26.8   HPLC chromatogram of standards of (a) stevioside and rebaudioside-A; (b) *S. rebaudiana* extract by microwave irradiation.

Microwave-assisted extraction yielded 8.64% of stevioside and 2.34% of rebaudioside-A. Conventional and ultrasound techniques yielded 6.54 and 4.20% of stevioside and 1.20% and 1.98% of rebaudioside-A, respectively. This method is precise, saves considerable time and energy, and is suitable for quality control assessment of *S. rebaudiana* plant material used on a commercial scale by industry.

26.13 ANALYTICAL METHODS FOR THE QUANTIFICATION OF STEVIOL GLYCOSIDES IN *S. REBAUDIANA*

A range of analytical techniques have been reported in the literature for separation and quantification of sweet diterpene glycosides from the leaves of *S. rebaudiana*. Quantification of stevioside by a chemical method following enzymatic hydrolysis (Mizukami, Shiba, and Ohash 1982) and quantification of total glycosides by gas chromatography after acid hydrolysis of steviol glycosides have been reported (Sakaguchi and Kant 1982). Six steviol glycosides have been separated and identified by OPLC (overpressure thin-layer chromatography) method (Fullas et al. 1989). Quantification of two major steviol glycosides (i.e., stevioside and rebaudioside-A) by TLC has also been reported (Tanaka 1982). Dacome et al. (2005) have densitometrically quantified sugar and steviol derivatives. A number of HPLC methods have been reported for quantification of steviol glycosides using hydrophilic (OH) columns (Sholichin et al. 1980) and SECM (size exclusion chromatography method) (Yohei and Masataka 1978; Ahmed and Dobberstein 1982a,b).

Makapugay, Nanayakkara, and Kinghorn (1984) and Ahmed and Dobberstein (1982) quantified eight steviol glycosides using the HPLC method. Mauri et al. (1996) have used the capillary electrophoresis method to quantify two steviol glycosides, rebaudioside-A and steviolbioside, separated by semipreparative HPLC. We have recently reported a high-performance thin layer chromatography (HPTLC) method for the simultaneous quantification of major steviol glycosides (steviolbioside, stevioside, and rebaudioside-A) in *S. rebaudiana* from different locations in India (Jaitak et al. 2008). The method was found to be reproducible for quantitative analysis of steviol glycosides in leaves collected from 10 different locations. This method can be used as a quality control indicator for monitoring commercial production of stevioside and its allied molecules during different stages of processing of *S. rebaudiana* leaves on a commercial scale. Percentage variability of major steviol glycosides quantified by the HPTLC method is given in Table 26.5, and a three-dimensional overlay chromatogram of standard and sample is given in Figure 26.9.

Table 26.5 HPTLC Quantification of Glycosides Present in *S. rebaudiana* Grown in Different Locations in India

Sr. no.	Location	Steviolbioside		Stevioside		Rebaudioside-A	
		Average (n = 3) (%)	RSD (%)	Average (n = 3) (%)	RSD (%)	Average (n = 3) (%)	RSD (%)
1.	Sangrur	0.330	2.25	5.798	2.13	1.286	0.74
2.	Dharamsala	0.348	5.49	4.249	2.59	2.378	4.52
3.	Baroda	0.434	1.85	4.483	1.66	2.062	0.88
4.	Karimnagar	0.477	0.72	5.266	1.84	1.370	0.98
5.	Ghimtoli	0.881	0.98	3.348	1.03	1.302	0.20
6.	Chhatisgarh	0.340	3.42	4.846	1.68	1.380	1.69
7.	Nagpur	1.77	0.46	4.486	0.40	1.622	0.51
8.	Ahmedabad	0.399	1.42	5.384	0.23	2.062	2.85
9.	Gujrat	0.500	1.96	6.238	2.18	2.172	3.98
10.	IHBT (Palampur)	0.380	0.38	6.754	1.26	1.492	0.62

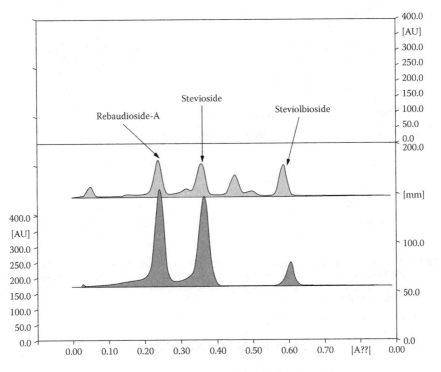

Figure 26.9 Three-dimensional overlay chromatogram of standard and sample.

26.14 ENZYMATIC MODIFICATION OF STEVIOSIDE
FOR REMOVAL OF BITTERNESS

Despite wide therapeutic applications of stevioside, its use is restricted for human consumption due to its bitter aftertaste and limits its application in food and pharmaceutical products. To overcome this problem, attention has been paid to producing debittered stevioside by modifying it through enzymatic biotransformations (Kasai et al. 1981; Fukunaga et al. 1989; Lobov et al. 1991; Yamamoto, Yoshikawa, and Okada 1994; Kusama et al. 1986; Abelyan et al. 2004; Ishikawa et al. 1990). However, yields of these biotransformed products are very low and, in certain cases, complex products are formed, thus removing bitterness partially. Authors have carried out enzymatic transglycosylation of stevioside using CGtase (cyclodextrin glucanotransferase) produced from *Bacillus firmus*, which yielded two products with a conversion yield up to 90% under microwave conditions (Jaitak et al. 2009b; Figure 26.10).

Table 26.6 shows the reaction conducted under conventional, ultrasound, and microwave conditions and yield comparison of products I and II. Comparative evaluation of enzymatic transglycosylation using three different techniques revealed that microwave-assisted transglycosylation of stevioside offers a faster alternative to the conventional method; it offers enhanced glycosylation yields of stevioside and is environmentally friendly. CGtase from *B. firmus* exhibited the specific property of a high transglycosylation rate under microwave and ultrasound techniques, catalyzing the production of only two modified glycosylated steviosides in comparison to an earlier reported mixture of nine glycosylated products. The procedure saves considerable time and energy and, from a commercial point of view, has potential for the industrial production of modified stevioside free from its bitter aftertaste.

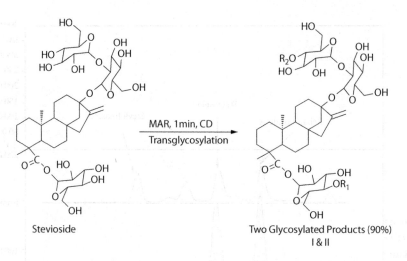

Figure 26.10 Enzymatic transglycosylation of stevioside under microwave-assisted conditions using CGtase and cyclodextrin as donor. R_1 = H; R_2= $-\alpha$-G^4 (I, 4'-O-α-d-glycosyl stevioside); R_1 = $-\alpha$-G'''^4– α-G''''^4; R_2 = H (II, 4''-O-α-d-maltosyl stevioside); MAR = microwave-assisted reaction; CD = β-cyclodextrin.

26.15 HEALTH AND SAFETY

Although *Stevia rebaudiana* has been used without any side effects for many years in Paraguay and other countries, in the United States, health and safety issues have been receiving considerable attention during the last 20 years, including claims and counterclaims regarding safety of steviol glycosides. The general safety of steviosides could be largely due to the fact that they do not hydrolyze or get absorbed in the digestive tract (Hutapea 1997). Many toxicological safety studies have been conducted for steviol glycosides in Japan (Johnson 1990). Some of these have confirmed the safety of stevia for diabetic use and prevention of cavities and plaque formation in teeth (Johnson 1990; Polyanskii, Rodionova, and Glagoleva 1997; Thamolwan and Narongsak 1997).

Studies have been conducted on carcinogenicity and mutagenicity by animal testing (Oliveira, Uehara and Valle 1988; Toyoda 1997). The safety of these glycosides has been confirmed by a wide range of studies on animals, chickens, and humans (Polyanskii et al. 1997; Melis and Sainati 1991; Melis 1995, 1997; White, Campbell, and Bernstein 1994; Sincholle and Marcorelles 1989; Smolyar 1993). Studies on food safety, including an extensive review of the literature undertaken prior to 1982 (Medon et al. 1982; Lee 1979), concluded that *S. rebaudiana* leaves and extracts are safe. Possible medicinal uses have been investigated often by using stevia extracts as intravenous

Table 26.6 Comparison of Enzymatic Glycosylation of Stevioside Using Conventional, Ultrasound- and Microwave-Assisted Techniques

Sr. no.	Technique	Time Duration	Conditions	P-I (%)	P-II (%)	Total Yield (%)
1.	Conventional	12 h	50°C, 200 rpm	46.2	23.8	70
2.	Ultrasound assisted	10 min	Pulse length 2 sec on and 1 sec off; duty 40%	50.4	24.6	75
3.	Microwave assisted	1 min	80 W at 50°C	66.0	24.0	90.0

infusion in rats; possible effects on glucose metabolism, diuresis, organ weight, and endocrine function have been studied (Oliveira, Uehara, and Valle 1988; Kinghorn 1987; Nunes and Pereira 1988; Suanarunsawat and Chaiyabutr 1996, 1997).

Stevia extract infusions have also shown some antiandrogenic activity in rats (Suanarunsawat and Chaiyabutr 1996). Beneficial effects of stevia extracts as antioxidants and to reduce blood pressure have also been reported (Chan et al. 1998; Xi Sato and Takeachi 1988). Steviol (a precursor in the biosynthesis of steviosides) can be produced from steviosides experimentally by using specific bacteria, but not *in situ* in the human body. Steviol exhibits some toxic and mutagenic activity (Tateo 1990).

Stevia sweeteners have a long history of safe use in foods and beverages in Japan and Paraguay. The safety of stevia sweeteners for human consumption has been established through rigorous research, including metabolic and pharmacokinetic studies. In June 2008, the Joint Expert Committee of Food Additives (JECFA) completed a review of the available scientific data as high-purity steviol glycosides and concluded that they are safe for use as general purpose sweeteners. Until recently, stevia sweeteners were only available as dietary supplements in the United States and were not permitted in foods and beverages. However, in December 2008, the U.S. Food and Drug Administration (FDA) approved and recognized steviol glycosides as GRAS (generally recognized as safe) as a general purpose sweetener. As a result, new food and beverage products containing these sweeteners have been introduced in the U.S. market.

26.16 MARKET IMPORTANCE OF *STEVIA REBAUDIANA*

Commercialization of *S. rebaudiana* leaves for sweetening and flavoring purposes has been quite rapid since its introduction in Japan. In recent years, about 200 metric tons of purified stevioside and other sweetener products have been produced from 2,000 metric tons of dried stevia leaves in the Japanese market (Kinghorn, Wu, and Soejarto 2001). Large-scale cultivation of *S. rebaudiana* leaves for the Japanese market is presently going on in China, especially in Fujian, Zhejiang, and Guangdong Provinces (Kinghorn et al. 2001). More land has been brought under stevia cultivation in China and the productivity of its leaves has gradually increased from 200 metric tons in 1982 to 1,300 metric tons (Kinghorn and Soejarto 1991).

Cultivation of *S. rebaudiana* for the Japanese market also takes place in Taiwan, Thailand, and Malaysia (Kinghorn and Soejarto 1991). About 115 metric tons of stevioside were consumed in Korea in 1995; the majority was used in sweetening of an alcoholic beverage, soju (Kinghorn et al. 2001). A refined extract of the leaves of *S. rebaudiana* (containing at least 60% steviosides) and pure stevioside (free from steviol and isosteviol) is approved in Brazil for sweetening chewing gum, diabetic foods and beverages, medicines, oral hygiene products, and soft drinks. Production of steviosides occurs in Parana Province in southern Brazil for the local market (Hanson and De Oliveira 1993; Oliveira Ferro Soejarto 1997). In the literature, many publications report cultivation of *S. rebaudiana* in Canada (Brandle, Stariatt and Gizjan 1998), the Czech Republic (Nepovim et al. 1998), India (Chalapathi et al. 1997), and Russia (Dzyuba and Vseross 1998).

26.17 CONCLUSION

An integrated approach is required, leading to research practices for desired end product from the crop. Among steviol glycosides, stevioside and rebaudioside-A have commercial importance, but the sweetness of rebaudioside-A is better than that of stevioside. The FDA has approved rebaudioside-A as safe for use in food products. Further research is needed in crop improvement studies through selective breeding for enhancing rebaudioside-A content and also to explore the possibilities of conversion of stevioside to rebaudioside-A on a commercial scale, using green chemistry. A wide scope exists for developing green processing technology for the production of a product rich in rebaudioside-A, which is preferred in the market. Demand for a modified, debittered, glycosylated product of stevioside is increasing day by day, but yields of these products are very low. More attention is needed to enhance the yield of modified biotransformed glycosylated products by using selective enzymatic reactions and identification of appropriate processing techniques that will produce a natural product of reliable quality.

ACKNOWLEDGMENTS

This work was carried out under CSIR mission mode project CMM-0014, which is duly acknowledged. Part of the work on biotransformation of stevioside was carried out under the project, Development of Enzymatic Biotransformation Process for Up-gradation of Stevioside, sponsored by the Department of Biotechnology, Government of India, New Delhi.

REFERENCES

Abelyan, V. A., A. M. Balyan, V. T. Ghochikyan, et al. 2004. Transglycosylation of stevioside by cyclodextrin glucanotransferases of various groups of microorganisms. *Applied Biochemistry and Microbiology* 40:129–134.

Ahmed, M. S., and R. H. Dobberstein. 1982a. *Stevia rebaudiana*. II. High-performance liquid chromatographic separation and quantitation of stevioside, rebaudioside-A and rebaudioside-C. *Journal of Chromatography* 236:523–526.

———. 1982b. *Stevia rebaudiana*. III. High-performance liquid chromatographic separation and quantitation of rebaudiosides B, D, E, dulcoside A, and steviolbioside. *Journal of Chromatography* 245:373–376.

Alvarez, M. 1986. *Stevia rebaudiana* (Bert) Bertoni: Toxicological aspects. Third Brazilian Seminar on *Stevia rebaudiana* (summaries), p. 427.

Bennett, R. D., E. R. Lieber, and E. Heftmann. 1967. Biosynthesis of steviol from (−)-kaurene. *Phytochemistry* 6:1107–1110.

Bertoni, M. S. 1899. El Caa-ehe (Eupatorium rebaudianum species nova). *Rivista de Agronomia* 1:35–37.

Bondarev, N. I., O. V. Reshetnyak, and A. M. Nosov. 2003. Effects of nutrient medium composition on development of *Stevia rebaudiana* shoots cultivated in the roller bioreactor and their production of SGs. *Plant Science* 165:845–850.

Boonkaewwan, C., C. Toskulkao, and M. Vongsakul. 2006. Anti-inflammatory and immunomodulatory activities of stevioside and its metabolite steviol on THP-1 cells. *Journal of Agricultural and Food Chemistry* 54:785–789.

Brandle, J. E., and N. Rosa. 1992. Heritability for yield, leaf-stem ratio and stevioside content estimated from a landrace cultivar of *Stevia rebaudiana*. *Canadian Journal of Plant Science* 72:1263–1266.

Brandle, J. E., A. N. Starratt, and M. Gijzen,1998. *Stevia rebaudiana*: Its agricultural, biological, and chemical properties. *Canadian Journal of Plant Sciences* 78:527–536.

Brandle, J. E., A. N. Starratt, C. W. Kirby, et al. 2002. Rebaudioside-F, a diterpene glycoside from *Stevia rebaudiana*. *Phytochemistry* 59:367–370.

Brandle, J. E., and P. G. Telmer. 2007. Steviol glycoside biosynthesis. *Phytochemistry* 68:1855–1863.

Chalapathi, M. V., S. Thimegowda, S. Sridhara, et al. 1997. Natural noncalorie sweetener stevia (*Stevia rebaudiana* Bertoni)—A future crop of India. *Crop Research* 14:347–350.

Chan, P., B. Tomlinson, Y. J. Chen, et al. 2000. A double-blind placebo-controlled study of the effectiveness and tolerability of oral stevioside in human hypertension. *Brazilian Journal of Clinical Pharmacology* 50:215–220.

Chan, P., D. Y. Xu, J. C. Liu, et al. 1998. The effect of stevioside on blood pressure and plasma catecholamines in spontaneously hypertensive rats. *Life Sciences* 63:1679–1684.

Cheng, T. F., and W. H. Chang. 1983. Studies on nonstevioside components of stevia extracts. *National Science Council Monthly, Taipei* 11:96–108.

Dacome, A. S., C. C. da Silva, E. M. Cecília, et al. 2005. Sweet diterpenic glycosides balance of a new cultivar of *Stevia rebaudiana* (Bert.) Bertoni: Isolation and quantitative distribution by chromatographic, spectroscopic, and electrophoretic methods. *Process Biochemistry* 40:3587–3594.

D'Agostino, M., F. De Simone, C. Pizza, et al. 1984. Sterols from *Stevia rebaudiana* Bertoni. *Bollettino-Societa Italiana di Biologia Sperimentable* 60:2237–2240.

Darise, M., H. Kohda, K. Mizutani, et al. 1983. Chemical constituents of flowers of *Stevia rebaudiana* Bertoni. *Agricultural and Biological Chemistry* 47:133–135.

DuBois, D. E., and R. A. Stephenson. 1985. Diterpenoid sweeteners. Synthesis and sensory evaluation of stevioside analogues with improved organoleptic properties. *Journal of Medicinal Chemistry* 28:93–98.

Dwivedi, R. S. 1999. Un-nurtured and untapped super sweet nonsacchariferous plant species in India. *Current Science* 76:1454–1461.

Dzyuba, O., and O. Vseross. 1998. *Stevia rebaudiana* (Bertoni) Hemsley—A new source of natural sweetener from Russia. *Rastitel'nye Resursy (Plant Resources)* 34:86–95.

Elkins, R. 1997. Stevia: *Nature's sweetener*, 6–7. Pleasant Grove, UT: Woodland Publishing.

Fujita, S., K. Taka, and Y. Fujita. 1977. Miscellaneous contributions to the essential oils of plants from various territories. XLI. On the components of the essential oil of *Stevia rebaudiana* Bertoni. *Yakugaku Zasshi* 97:692–694.

Fukunaga, Y., T. Miyata, N. Nakayasu, et al. 1989. Enzymatic transglucosylation products of stevioside. Separation and sweetness evaluation. *Agricultural and Biological Chemistry* 53:1603–1607.

Fullas, F., J. Kim, C. M. Compadre, et al. 1989. Separation of natural product sweetening agents using over-pressured layer chromatography. *Journal of Chromatography* 464:213–219.

Gentry, A. H. 1996. *A field guide of the families and genera of woody plants of northwest South America (Colombia, Ecuador, Peru) with supplementary notes on herbaceus taxa*, 895. Chicago: University of Chicago Press.

Geuns, J. M. C. 1998. *Stevia rebaudiana* Bertoni plants and dried leaves as novel food. Final version September 21, 1998, with addendum.

Giovanetto, R. H. 1990. Method for the recovery of steviosides from plant raw material. US patent no. 4892938.

Hanson, J. R., and B. H. De Oliveira. 1993. Stevioside and related sweet diterpenoid glycosides. *Natural Product Reports* 10:301–309.

Hanson, J. R., and A. F. White. 1968. Studies in terpenoid biosynthesis-II: The biosynthesis of steviol. *Phytochemistry* 7:595–597.

Helliwell, C. A., A. Poole, W. J. Peacock, et al. 1999. Arabidopsis *ent*-kaurene oxidase catalyzes three steps of gibberellin biosynthesis. *Plant Physiology* 119:507–510.

Hoyle, F. C. 1992. A review of four potential new crops for Australian agriculture, 34. Perth, Australia: Department of Agriculture.

Hutapea, A. M. 1997. Digestion of stevioside—A natural sweetener, by various digestive enzymes. *Journal of Clinical Biochemical Nutrition* 23:177–184.

Ishikawa, H., S. Kitahata, K. Ohtani, et al. 1991. Transfructosylation of rebaudioside-A with microbacterium β-fructosidase. *Chemical and Pharmaceutical Bulletin* 39:2043–2045.

Jaitak, V., Bandna, and V. K. Kaul. 2009a. An efficient microwave-assisted extraction process of stevioside and rebaudioside-A from *Stevia rebaudiana* (Bertoni). *Phytochemical Analysis* 20:240–245.

Jaitak, V., Bandna, V. K. Kaul, et al. 2009b. A simple and efficient enzymatic transglucosylation of stevioside by the enzyme from *Bacillus firmus*. *Biotechnology Letters* 31:1415–1420.

Jaitak, V., A. P. Gupta, V. K. Kaul, et al. 2008. Validated high-performance thin-layer chromatography method for steviol glycosides in *Stevia rebaudiana*. *Journal of Pharmaceutical and Biomedical Analysis* 47:790–794.

Jayaraman, S., M. S. Manoharan, and S. Illanchezian. 2008. In-vitro antimicrobial and antitumor activities of *Stevia rebaudiana* (Asteraceae) leaf extracts. *Tropical Journal of Pharmaceutical Research* 7:1143–1149.

Jeppesen, P. B., S. Gregersen, C. R. Poulsen, et al. 2000. Stevioside acts directly on pancreatic β-cells to secrete insulin: Actions independent of cyclic adenosine monophosphate and adenosine triphosphate-sensitive K⁺-channel activity. *Metabolism* 49:208–214.

Jeppesen, P. B., S. Gregersen, S. E. Rolfsen, et al. 2003. Antihyperglycemic and blood pressure-reducing effects of stevioside in the diabetic Goto-Kakizaki rat. *Metabolism* 52:372–378.

Johnson, E. D. R. 1990. Stevioside, naturally. Tuscon, AZ: The Calorie Control Council.

Kasai, R., N. Kaneda, O.Tanaka, et al. 1981. Sweet diterpene-glycosides of leaves of *Stevia rebaudiana* Bertoni. Synthesis and structure-sweetness relationship of rebaudiosides-A, -D, -E and their related glycosides. *Nippon Kagakukaishi* 1981:726–735.

Katayama, O., T. Sumida, H. Hayashi, et al. 1976. *The practical application of stevia and R&D data* (English translation), 747. ISU Company, Japan.

Kazuo, T., M. Masahiro, O. Kazutaka, et al. 2001. Analysis of antirotavirus activity of extract from *Stevia rebaudiana*. *Antiviral Research* 49:15–24.

Kim, K. K., Y. Sawa, and H. Shibata. 1996. Hydroxylation of *ent*-kaurenoic acid to steviol in *Stevia rebaudiana* Bertoni—Purification and partial characterization of the enzyme. *Archives of Biochemistry and Biophysics* 332:223–230.

King, R. M., and H. Robinson. 1987. *The genera of the eupatoreae (Asteraceae)*, 1–180. St. Louis: Missouri Botanical Garden Library.

Kinghorn, A. D. 1987. Biologically active compounds from plants with reputed medicinal and sweetening properties. *Journal of Natural Products* 50:1009–1024.

———. 1992. *Food ingredient safety review*: Stevia rebaudiana *leaves*. Herb Research Foundation, United States.

———. 2002. *The genus* Stevia: *Medicinal and aromatic plants—Industrial profiles*. Boca Raton, FL: CRC Press.

Kinghorn, A. D., and D. D. Soejarto. 1985. Current status of stevioside as a sweetening agent for human use. In *Economic and medicinal plant research*, ed. H. Wagner, H. Hikino, and N. R. Farnsworth. London: Academic Press.

———. 1991. Stevioside. In *Alternative sweetener*, 2nd ed. (revised and expanded), ed. L. O'Brien Nabors and R. C. Gelardi, 157–171. New York: Marcel Dekker, Inc.

Kinghorn, A. D., C. D. Wu, and D. D. Soejarto. 2001. Stevioside. In *Alternative sweetener*, 3rd ed. (revised and expanded), ed. L. O'Brien Nabors, 167–183. New York: Marcel Dekker, Inc.

Kobayahi, M., S. Horikawa, H. Isolde, et al. 1977. Dulcosides A and B. New diterpene glycosides from *Stevia rebaudiana*. *Phytochemistry* 16:1405–1408.

Kohda, H., R. Kasai, K. Yamasaki, et al. 1976. New sweet diterpene glucosides from *Stevia rebaudiana*. *Phytochemistry* 15:981–983.

Kumar, K., G. D. Kiran Babu, V. K. Kaul, et al. 2006. A process for the production of steviosides from *Stevia rebaudiana* Bertoni. US patent application no: 2006/0142555 A1.

Kumar, R. D., and O. V. Oommen. 2008. *Stevia rebaudiana* Bertoni does not produce female *reproductive* toxic effect: Study in Swiss albino mouse. *Journal of Endocrinology and Reproduction* 12:57–60.

Kumar, S. 1986. Method for recovery of stevioside. US patent no. 4599403.

Kusama, S., I. Kusakabe, Y. Nakamura, S. Eda, et al. 1986. Transglucosylation of stevioside by the enzyme system from *Streptomyces* sp. *Agricultural and Biological Chemistry* 50:2445–2451.

Lee, S. J. 1979. Studies on the new sweetening source plant stevia (*Stevia rebaudiana*) in Korea. I. A study on the safety of stevioside from *Stevia rebaudiana* as a new sweetening source. *Korean Journal of Food Science and Technology* 11:224–231.

Lester, T. 1999. *Stevia rebaudiana* (sweet honey leaf). *Australia New Crops Newsletter* no. 11.

Lewis, W. H. 1992. Early uses of *Stevia rebaudiana* leaves as sweetener in Paraguay. *Economic Botany* 46:336–337.

Lobov, S. V., R. Kasai, K. Ohtani, et al. 1991. Enzymatic production of sweet stevioside derivatives: Transglucosylation by glucosidases. *Agricultural and Biological Chemistry* 55:2959–2965.

Machado, E., A. M. Chagas, and D. S. Reis. 1986. *Stevia rebaudiana* (Bert.) Bertoni in the arterial pressure of the dog. Third Brazilian Seminar on *Stevia rebaudiana* (Bert.) Bertoni (summaries). Angelucci, E. (coordinator), p. 11.

Makapugay, H. C., N. P. D. Nanayakkara, and A. D. Kinghorn. 1984. Improved high-performance liquid chromatographic separation of *Stevia rebaudiana* sweet diterpene glycosides using linear gradient elution. *Journal of Chromatography* 283:390–395.

Martelli, A., C. Frattini, and F. Chialva. 1985. Unusual essential oils with aromatic properties. I. Volatile components of *Stevia rebaudiana* Bert. *Flavor and Fragrance Journal* 1:3–7.

Matsuo, T., H. Kanamori, and I. Sakamoto. 1986. Nonsweet glycosides in the leaves of *Stevia rebaudiana*. *Hiroshima-ken Eisei Kenkyusho Kenkyu Hokoku* 33:25–29.

Matsushita, S., and T. Ikushige. 1979. Separation of natural sweet component from natural extract. US patent no. 41717430.

Mauri, P., G. Catalano, C. Gardana, et al. 1996. Analysis of stevia glycosides by capillary electrophoresis. *Electrophoresis* 17:367–371.

Medon, P. J., J. M. Pezzuto, J. M. Hovanec-Brown, et al. 1982. Safety assessment of some *Stevia rebaudiana* sweet principles. *Federation Proceedings* 41:1568–1982.

Megeji, N. W., J. K. Kumar, V. Singh, et al. 2005. Introducing *Stevia rebaudiana*, a natural zero-calorie sweetener. *Current Science* 88:801–804.

Melis, M. S. 1995. Chronic administration of aqueous extract of *Stevia rebaudiana* in rats: Renal effects. *Journal of Ethnopharmacology* 47:129–134.

———. 1997. Effects of steviol on renal function and mean arterial pressure in rats. *Phytomedicine* 3:349–352.

Melis, M. S., and A. R. Sainati. 1991. Effect of calcium and verapamil on renal function of rats during treatment with stevioside. *Journal of Ethnopharmacology* 33:257–262.

Mizukami, H., K. Shiba, and H. Ohash. 1982. Enzymatic determination of stevioside in *Stevia rebaudiana*. *Phytochemistry* 21:1927–1930.

Mosettig, E., U. Beglinger, F. Dolder, et al. 1963. The absolute configuration of steviol and isosteviol. *Journal of American Chemical Society* 85:2305–2309.

Nabeta, K., T. Kasai, and H. Sugisawa. 1974. Phytosterol from the callus of *Stevia rebaudiana* Bertoni. *Agricultural and Biological Chemistry* 40:2103–2104.

Nanayakkara, N. P., J. A. Klocke, C. M. Compadre, et al. 1987. Characterization and feeding deterrent effects on the aphid, *Schizaphis graminum* of some derivatives of the sweet compounds, stevioside and rebaudioside A. *Journal of Natural Products* 50:434–441.

Nepovim, A., H. Drahosova, P. Velicek, et al. 1998. The effect of cultivation conditions on the content of stevioside in *Stevia rebaudiana* plants cultivated in Czech Republic. *Pharmacy and Pharmacology Letters* 8:19–21.

Nunes, B.d. A. P., and N. A. Pereira. 1988. Influence of the infusion of *Stevia rebaudiana* (Bert.) on the weight of sexual organs isolated from young mice. *Acta Amazonica* 18:1–2.

Oliveira-Filho, R.M., O.A. Uehara, and L.B.S. Valle. 1998. Endocrine parameters in rats following chronic treatment with concentrated extract of Stevia rebandiana. *Acta Amazonica* 18: 187–195.

Oliveira, Ferro, V. 1997. Faculty of Pharmaceutical Science University of Sao Paulo. Sao Paulo, Brazil. Private Communication.

Oshima, Y., J. I. Saito, and H. Hikino. 1986. Sterebins A, B, C, and D, bisnorditerpenoids of *Stevia rebaudiana* leaves. *Tetrahedron* 42:6443–6446.

———. 1988. Sterebins E, F, G, and H, diterpenoids of *Stevia rebaudiana* leaves. *Phytochemistry* 27:624–626.

Phillips, K. C. 1987. *Stevia*: Steps in developing a new sweetener. In *Developments in sweeteners 3*, ed. T. H. Grenby, 1–43. New York: Elsevier.

Pol, J., E. V. Ostra, P. Karasek, et al. 2007. Comparison of two different solvents employed for pressurised fluid extraction of stevioside from *Stevia rebaudiana*: Methanol versus water. *Analytical and Bioanalytical Chemistry* 388:1847–1857.

Polyanskii, K., N. S. Rodionova, and L. E. Glagoleva. 1997. Stevia in cultured milk deserts for medical and prophylactic purposes. *Molochnaya Promyshlennost* 5:511–515.

Rajbhandari, A., and M. F. Roberts. 1983. The flavonoids of *Stevia rebaudiana*. *Journal of Natural Products* 46:194–195.

Ramesh, K., V. Singh, and P. S. Ahuja. 2007. Production potential of *Stevia rebaudiana* under intercropping systems. *Archives of Agronomy and Soil Science* 53:443–458.

Ramesh, K., V. Singh, and N. W. Megeji. 2006. Cultivation of stevia [*Stevia rebaudiana* (Bert.) Bertoni]: A comprehensive review. *Advances in Agronomy* 89:137–177.

Randi, A. M., and G. M. Felippe. 1981. Substances promoting root growth from the achenes of *Stevia rebaudiana*. *Revista Brasileira Botanica* 4:49–51.

Ruddat, M., E. Heftmann, and A. Lang. 1965. Biosynthesis of steviol. *Archives of Biochemistry and Biophysics* 110:496–499.

Sakaguchi, M., and T. Kant. 1982. As pequisas japonesas com *Stevia rebaudiana* (Bert.) Bertoni e o estevio-sídeo. *Ciencia eCultura* 34:235–248.

Sakamoto, I., K. Yamasaki, and O. Tanaka. 1977a. Application of ¹³C NMR-spectroscopy to chemistry of natural glycoside rebaudioside-C, a new sweet diterpene glycoside of *Stevia rebaudiana*. *Chemical and Pharmaceutical Bulletin* 25:844–846.

———. 1977b. Application of ¹³C NMR-spectroscopy to chemistry of plant glycosides: Rebaudioside-D and -E, new sweet diterpene-glucosides of *Stevia rebaudiana* Bertoni. *Chemical and Pharmaceutical Bulletin* 25:3437–3439.

Shaffert, E. E., and A. A. Chebotar. 1994. Structure, topography and ontogeny of *Stevia rebaudiana* (Asteraceae) trichomes. *Botanicheskii Zhurnal St Petersburg* 79:38–48.

Sholichin, M., K. Yamasaki, R. Miyama, et al. 1980. Labdane-type diterpenes from *Stevia rebaudiana*. *Phytochemistry* 19:326–327.

Shukla, S., A. Mehta, V. K. Bajpai, et al. 2009. *In vitro antioxidant* activity and total phenolic content of ethanolic leaf extract of *Stevia rebaudiana* Bert. *Food and Chemical Toxicology* 47:2338–2345.

Sincholle, D., and P. Marcorelles. 1989. Study of the antiandrogenic activity of an extract of *Stevia rebaudiana* Bertoni. *Plantes Medicinales et Phytotherapie* 23 (4): 282–287.

Smolyar, V. I. 1993. Effect of saccharol glycosides on energy metabolism in animals with abnormal carbohydrate tolerance. *Voprosy Pitaniya* 1:38–40.

Staratt, A. N., C. W. Kirby, R. Pocs, et al. 2002. Rebaudioside-F, a diterpene glycoside from *Stevia rebaudiana*. *Phytochemistry* 59:367–370.

Suanarunsawat, T., and N. Chaiyabutr. 1996. The effect of intravenous infusion of stevioside on the urinary sodium. *Journal of Animal Physiology and Animal Nutrition* 76:141–150.

———. 1997. The effect of stevioside on glucose metabolism in rat. *Canadian Journal of Physiology and Pharmacology* 75:976–982.

Sumida, T. 1968. Reports on stevia introduced from Brazil as a new sweetness resource in Japan (English summary). *Journal of Central and Agricultural Experts Standard* 31:1–71.

Suzuki, H., T. Ikeda, T.Matsumoto, et al. 1976. Isolation and identification of rutin from cultured cells of *Stevia rebaudiana* Bertoni. *Agricultural and Biological Chemistry* 40:819–820.

Tan, S., Y. Shibuta, and O. Tanaka. 1988. Isolation of sweetener from *Stevia rebaudiana*. Japanese Kokai patent 63: 177,764.

Tanaka, O. 1982. Steviol-glycosides: New natural sweeteners. *Trends in Analytical Chemistry* 1:246–248.

Tateo, F. 1990. Technical and toxicological problems connected with the formulation of low-energy foods. 2. Mutagenic and fertility-modifying activity of extracts and constituents of *Stevia rebaudiana* Bertoni. [Italian]. *Revista della Societa Italiana di Scienza dell'Alimentazione* 19:19–22.

Thamolwan, S., and C. Narongsak. 1997. The effect of stevioside on glucose metabolism in rat. *Canadian Journal of Physiology and Pharmacology* 75:976–982.

Totte, N., L. Charon, M. Rohmer, et al. 2000. Biosynthesis of the diterpenoid steviol, an *ent*-kaurene derivative from *Stevia rebaudiana* Bertoni, via the methylerythritol phosphate pathway. *Tetrahedron Letters* 41:6407–6410.

Toyoda, K. 1997. Assessment of carcinogenecity of stevioside. *Food and Chemical Toxicology* 35:597–603.

White, J. R. K., R. K. Campbell, and R. Bernstein. 1994. Oral use of a topical preparation containing an extract of *Stevia rebaudiana* and the chrysanthemum flower in the management of hyperglycemia. *Diabetes Care* 17:940–941.

Wink, M. 2003. Evolution of secondary metabolites from an ecological and molecular phylogenetic perspective. *Phytochemistry* 64:3–19.

Xi, Y. Y., M. Sato, and M. Takeuchi. 1998. Antioxidant activity of Stevia rebaudiana. *Journal of Japanese Society Food Science Technology—Nippon Shokuhin Kagaku Kogaku Kaishi* 45:310–316.

Yabu, M., M. Takase, K. Toda, et al. 1997. Studies on stevioside, natural sweetener. Effect on the growth of some oral microorganisms. *Hiroshima Daikagu Zasshi* 8:12–17.

Yamamoto, K., K. Yoshikawa, and S. Okada. 1994. Effective production of glucosyl-stevioside by trans-α-1,6-transglucosylation of dextrin dextranase. *Bioscience Biotechnology and Biochemistry* 58:1657–1661.

Yasukawa, K., A. Yamaguchi, J. Arita, et al. 1993. Inhibitory effect of edible plant extracts on 12-*O*-tetradecanoyphorbol-13-acetate-induced ear edema in mice. *Phytotherapy Research* 7:185–189.

Yohei, H., and M. Masataka. 1978. High performance liquid chromatographic separation and quantification of stevia components on hydrophilic forebed column. *Journal of Chromatography* 161:403–405.

CHAPTER 27

Ipecac—*Carapichea ipecacuanha*

Carlos Roberto Carvalho and Luiz Orlando de Oliveira

CONTENTS

27.1 INTRODUCTION

Many eighteenth century Europeans were intrigued as to where the medicinal ipecac roots came from. Ipecac was introduced from Brazil to Europe in 1672 and preparations of ipecac roots became increasingly popular as a cure for amoebic dysentery, then a devastating disease in Europe. A physician named Helvetius, who received the sole right of vending the remedy from Louis XIV, the king of France, sold the secret to the French government for 1,000 Louis d'or coins and the composition became widely known by 1688. However, the origin of these mysterious roots remained speculative until 1800, when authentic specimens from Brazil were brought to Lisbon (Flückiger and Hanbury 1879).

Commercial harvesting of ipecac has occurred in Brazil since the eighteenth century, when the roots became a valuable trading good, and continued to be operational up to the mid-1950s. Harvesting at that time was careless and replanting a new crop did not follow the uprooting of native populations (Veloso 1947; Oliveira and Martins 1998). At the present time, the knowledge of ipecac as a medicinal plant has experienced an enormous shift in Brazil. Only the elderly are able to recognize ipecacs in their natural growth conditions and harvesting of wild plants has decreased sharply. However, other anthropic activities threaten ipecac. Currently, two of the most active forces of transformation are habitat fragmentation and habitat loss due to land conversion to agriculture (Oliveira and Martins 2002). This reality accelerated even further the demographic decline of wild populations of ipecac, and conservation measures are needed urgently.

To date, studies from several related disciplines are being used to characterize the genetic variability of wild populations of ipecac in Brazil. At the long term, conservationism (Skorupa and Assis 1998; Oliveira and Martins 2002), morphology (Assis and Giulietti 1999; Rossi, Oliveira, and Vieira 2005), genetics and physiology (Lima 2002), ecogeography (Martins et al. 2009), alkaloid profile (Garcia et al. 2005), cytology and reproduction (Souza et al. 2006), cytogenetics and cytometry (Rossi et al. 2008), and genetic diversity (Rossi et al. 2009) aim to provide conservationists and germplasm collectors with information about the spatial and temporal distribution of the existing genetic diversity of ipecac and assist the establishment of efficient *in situ* and *ex situ* conservation strategies.

27.2 NOMENCLATURE AND SYNONYMY

As cited by De Boer and Thulin (2005), the true ipecac displays the following synonymy: *Carapichea ipecacuanha* (Brot.) L. Andersson, Kew Bull. 57: 371. 2002 ≡ *Callicoca ipecacuanha* Brot., Trans. Linn. Soc. London 6: 137. 1802 ≡ *Psychotria ipecacuanha* (Brot.) Stokes, Bot. Mat. Med. 1: 365. 1812 ≡ *Cephaelis ipecacuanha* (Brot.) A. Rich, Bull. Fac. Med. 4: 92 1818 ≡ *Uragoga ipecacuanha* (Brot.) Baill., Hist. Plant. 7: 370. 1880 ≡ *Evea ipecacuanha* (Brot.) Standley, Contr. U.S. Natl. Herb. 18: 123. 1916. In some literature, the species is referred to as *Cephaelis acuminata*, which supposedly is the commercial source of the drug from Central America (Bruneton 1995; Itoh et al. 1999). However, the name *C. acuminata* has been considered a *nomen nudum* for ipecac—that is, a name that has been cited without a proper and complete description (Hatfield et al. 1981; Lorence 1999; De Boer and Thulin 2005). Herein, we use *Carapichea ipecacuanha* (Brot.) L. Andersson because this was the name Andersson (2002) used in a recent revision of the group. Ipecac, however, is most widely known for its previous Latin binomial *Psychotria ipecacuanha* (Brot.) Stokes.

27.3 BOTANY

Ipecac is a long-lived, perennial shrub that inhabits the deeply shaded understory of neotropical forests (Veloso 1947; Oliveira and Martins 1998). In natural growth conditions, ipecac is rarely found as a single individual. Instead, aerial stems of ipecac clump together to form circular or elliptical clusters with well-delimited borders (Figure 27.1). Cluster size may range from a few aerial stems to hundreds of aerial stems. An average aerial stem within a cluster averages 0.20 m tall and comprises about 11 nodes and up to eight totally expanded leaves (Oliveira and Martins 2002). Excavations revealed that nearby aerial stems within a cluster are attached frequently by horizontal, subterranean stems and that adventitious buds located in roots and subterranean stems play a key role in promoting horizontal spread of the cluster via vegetative propagation. Within a cluster, close

Figure 27.1   A cluster of ipecac at natural growth conditions.

neighbors are found frequently with overlapping or interconnected underground parts and show indications of natural fragmentation and stem decay (Rossi et al. 2009).

Ipecac clusters have characteristically low-frequency occurrence per unit area. The species still is locally common in the state of Mato Grosso, but has extremely low local density and is rarely encountered in the Atlantic range. Most of the extant populations we have located in the Atlantic range contain one to six clusters (L. O. Oliveira, unpubl. res.).

Ipecac is a distylous species and exhibits isomorphic clusters; that is, the clusters present only one of the two floral morphs (Rossi et al. 2005). Morphometric analyses revealed that anther length, stigma length, corolla diameter, and pollen grain diameter were consistently greater in short-styled flowers, regardless of the population investigated. Significant differences for floral traits in the short-styled morph were found among populations studied (Rossi et al. 2005).

The cylindrical branches emit out from the nodes, with 0.6–1.9 cm diameter, and the internodes are from 0.2 to 7.0 cm length (Figure 27.2A). The leaves are smooth and persistent in the superior part of the branches, oval, elliptical, and oblong (Figure 27.2B). The terminal inflorescence is wrapped by oval and acute bracts of greenish coloration; they present erect stalk or deflect with 1.2–3.5 cm length (Figure 27.2C). The flowers are relatively small (1 cm diameter), with a whitish to cream color, sessile hermaphrodites, and from 12 to 150 per inflorescence (Figure 27.2C, D). The fruit has an oval shape around 1.0 cm × 0.7 cm, with reddish to wine-red epicarp and containing two seeds (Figure 27.2E) (Oliveira and Martins 1998). The roots are annealed, with characteristic rounded ridges, about 0.6–1.7 cm in diameter, from 20 to 30 cm in length, whitish to reddish when fresh and gray when dry (Figure 27.3A–C). They are linked to a subterranean stem by a small, distinct filament. The roots have low odor when fresh, but taste bitter and cause nausea. A plant of 3 years of age can produce a primary root containing up to eight secondary roots. For micropropagation, the number of roots obtained can reach up to 15 secondary roots (Lameira 2002).

Figure 27.2   Details of specimen, flower, and fruits of ipecac. Bar = 1 cm.

27.4 ORIGINS AND GEOGRAPHICAL DISTRIBUTION

The contemporary populations of ipecac are confined to three clearly defined ranges (Skorupa and Assis 1998; Assis and Giulietti 1999; Rossi et al. 2009) (Figure 27.4):

the Atlantic range, in the central portion of the Mata Atlântica biome along the Brazilian coast, in the states of Bahia, Espirito Santo, Rio de Janeiro, and Minas Gerais

the Amazonian range, at the southwestern part of the Brazilian Amazonia biome, in the Brazilian states of Rondônia and Mato Grosso

the Central American range, the northern limit of ipecac, in Nicaragua, Costa Rica, and Panama

Collections from neighboring parts of Colombia (Departments of Choco, Antioquia, and Bolivar) indicate that the Central American range may extend farther south. The populations of the Amazonian

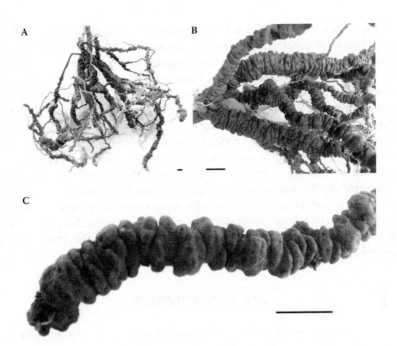

Figure 27.3 **(See color insert.)** The root system and the characteristic rounded ridges of an ipecac root. Bar = 1 cm.

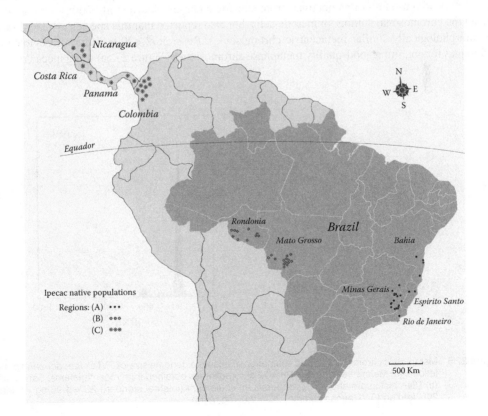

Figure 27.4 The geographical distribution of wild populations of ipecac.

range in Mato Grosso are at least 2500 km from the Central American range and 1600 km from the Atlantic range. These three geographic ranges form a mosaic whose pieces are separated on a continental-wide scale and each is surrounded by a vast landscape of apparently unsuitable environments.

Rossi et al. (2009) used intersimple sequence repeat (ISSR) markers to investigate whether the origin of the Brazilian disjunction of ipecacs could be attributed to human-mediated, long-distance dispersal events. A possible origin for the disjunct distribution of ipecac within Brazil was that indigenous peoples transported this valuable medicinal resource from one area to another (from the Amazonian range to the Atlantic range or vice versa). If this was the case, a signature of a recent source–founder relationship could be uncovered owing to the clonal and perennial nature of the species and the recent presence of humans on the American continent in geological time scale. The results suggested that human migrations likely neither fostered ipecac dissemination from an older range to previously unoccupied territories of a newer range nor reconnected formerly isolated populations of ipecac via translocations of plant material among ranges. Instead, the results favored the existence of a long-standing barrier preventing gene flow between the Atlantic populations and Amazonian populations (Rossi et al. 2009)

27.5 CYTOGENETICS

Cytogenetic studies performed in this species have reported $2n = 22$ chromosomes. Assis and Giulietti (1999) and Souza et al. (2006) obtained cytogenetic data in ipecac that contribute with taxonomy studies in this species. Initially using meiotic cells, Assis and Giulietti related that ipecac possesses $2n = 22$. The same chromosomic number was also mentioned by Rout, Samantaray, and Das (2000), who used root tips obtained from somatic embryos. Souza et al. (2006) not only confirmed the chromosome number in meiotic cells, but also reported that this species exhibited small and morphologically similar metacentric chromosomes. Rossi et al. (2008) improved cytological techniques for obtaining good-quality metaphase chromosomes (Figure 27.5a) using ipecac roots to

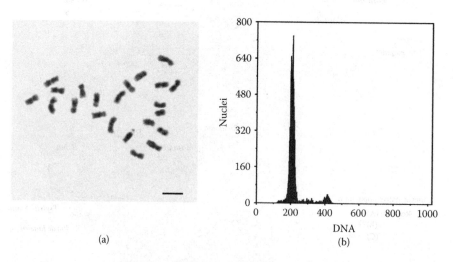

(a)   (b)

Figure 27.5 Metaphasic chromosomes and estimation of absolute genome size of TVI *C. ipecacuanha* populations. (a) Metaphase showing $2n = 22$ chromosomes obtained from root meristems. Bar = 5 μm. (b) DNA histogram showing *Lycopersicum sculentum* (internal standard 2C = 1.96 pg, channel 200, left) and TVI ipecac sample (2C = 2.05 pg, right).

classify the homologous sets of chromosomes and to construct a karyotype. They studied three ipecac populations: two (TVI and TLM) from the Atlantic and one (MOZ) from the Amazon forest.

However, metaphasic chromosomes obtained from root apical meristems of these populations
showed no karyotype differences, consisting of 11 chromosome pairs—four metacentric (4, 5, 8,
and 9) and seven submetacentric (1, 2, 3, 6, 7, 10, and 11), with lengths varying from 3.97 to 2.53 µm.
The secondary constriction was identified in the long arm of chromosome 6. Cytological characterization also revealed that, with regard to total length, each chromosome pair—2–3, 4–5, 6–7, 8–9,
and 10–11—is short- and long-arm size and chromosomic class, suggesting that this species might
be a tetraploid. By evaluating the nuclei 2C-DNA content of *C. ipecacuanha* by flow cytometry,
Rossi et al. (2008) identified two groups with distinct nuclei DNA amounts: one population with
2C = 1.24 pg and nine with 2C 2.05 pg (Figure 27.5b).

Reproductive and meiotic behavior studies were carried out on Brazilian accessions of ipecac
(*C. ipecacuanha*) by Souza et al. (2006). In these studies, they confirmed that chromosomes of
ipecac paired as 11 bivalents at metaphase I and showed that some irregular chromosome segregation in meiosis I and II associated with index and pollen viability. This suggests that the process of
sexual reproduction is not totally normal.

27.6 PHARMACOLOGICAL AND MEDICINAL ASPECTS

The pharmacological activities of the drug obtained from ipecac roots (called ipecac or ipecacuanha) as an expectorant, amebicide, and vomitive agent have been experimentally validated
(Bruneton 1995). The main therapeutic compounds—the isoquinoline alkaloids emetine and cephaeline—accumulate in the roots throughout the year (Garcia et al. 2005). Emetine acts mostly as an
expectorant and amebicide, whereas cephaeline is responsible for the emetic activity. Ipecac remains
in pharmacopoeias of many countries around the world, such as the Brazilian, French, Japanese, and
U.S. pharmacopoeias. Ipecac syrup is available as a nonprescription drug in some countries (e.g., in
the United States); in other countries, such as France, it is considered a controlled substance. When
taken for poisoning, ipecac may cause severe nausea, vomiting, and intestinal cramps. Cardiac and
skeletal muscle myopathies are frequent manifestations of chronic ipecac abuse by patients with an
eating disorder (Ho, Dweik, and Cohen 1998). Because of their low excretion rate (Bruneton 1995),
ipecac alkaloids remain in tissues for a long time.

Purified emetine is highly cytotoxic (Satou et al. 2002), inhibits protein (Li, Venkataraman,
and Ain 2006) and DNA syntheses (Burhans et al. 1991), and induces apoptosis in leukemia cells
(Möller et al. 2007). Although morphological data allow joining Central and South American
ipecacs into a single species (Assis and Giulietti 1999), alkaloid composition readily distinguishes
Panamanian from Brazilian ipecac. While cephaeline prevails in ipecac from Central America,
emetine is the major alkaloid in roots collected in Brazil (Gupta et al. 1986; Bruneton 1995;
Garcia et al. 2005).

Root samples collected in Brazil showed values for emetine and cephaeline contents that ranged
from 1.5 to 1.7% and from 0.6 to 0.7%, respectively (Costa 1978). Roots from Panamanian ipecac
yielded emetine and cephaeline contents that ranged from 0.49 to 1.26% and 0.65 to 1.19%, respectively
(Gupta et al. 1986), and from 0.42 to 1.60% and 0.76 to 1.82%, respectively (Hatfield et al. 1981).

Garcia et al. (2005) investigated the seasonal variation of emetine and cephaeline in four clusters
of ipecac at their natural growth conditions in the Atlantic range. High-performance liquid chromatography analyses showed cluster-to-cluster variation and indicated low temporal oscillation in
alkaloid content within a given cluster. The content of cephaeline remained very low in all clusters,
at all times. However, the pattern of emetine accumulation clearly distinguished two high-emetine

clusters from two low-emetine clusters. The presence of cluster-to-cluster variation suggested that different chemical profiles might exist at the within-population level and that performance in alkaloid accumulation cannot be readily predictable solely based on the geographical location of the cluster. Moreover, significant positive correlations between root size and emetine content indicated that root morphometry may be useful in distinguishing contrasting alkaloid profiles.

27.7 AGRICULTURE

The EMBRAPA (Empresa Brasileira de Pesquisa Agropecuária; www.embrapa.br/english), one of the most important institutions of agricultural research in Brazil, provides on its website suitable cultivar recommendations for several plants. Technical note no. 28 for ipecac, in Portuguese (Lameira 2002), can be downloaded from http://www.cpatu.embrapa.br/publicacoes_online/circular-tecnica-1/2002/cultivo-da-ipecacuanha-psychotria-ipecacuanha-brot-stokes/?searchterm=ipecacuanha. Part of this free technical note was translated into English and basically summarizes some suggestions for ipecac agriculture procedures. It was observed that ipecac can thrive only under conditions similar to its natural habitat, which is in regions of shaded tropical forest.

27.7.1 Agriculture Procedures

The cultivation of ipecac can be done at any time of year, in beds to avoid competition from the roots of other species and prepared with sandy soil or sandy-clay to facilitate the harvest of roots. Around 70,000 seedlings per hectare can be grown. The area between the beds should be covered with black to prevent the direct incidence of sunlight on ipecac, thus avoiding death of the plants during the development phases.

27.7.2 Propagation and Seedlings Preparation

The spread of ipecac by seed (sexual) is not recommended because of low germination; seeds germinate from 3 to 6 months after planting. Plants flower from seed after 2 years of cultivation. The species is a plant that spreads under certain ecophysiological conditions of humidity and high temperature and is not tolerant to intense solar radiation.

27.7.3 Harvest

The collection of ipecac should be done when the plant reaches its full development. In extraction, the plants in the forest have much competition; thus, the collection is performed at 3–4 years of age. The roots are in smaller numbers (four to six roots per plant on average) and, during the harvest, material is lost on the grounds of breach of the roots. Plants growing from these procedures may be harvested at 24 months. Harvesting is done throughout the year, preferably in the least dry season.

Micropropagation has served as a tool for rescuing endangered plants (Palomino et al. 1999). Studies on C. ipecacuanha to obtain whole plants in vitro have been successful (Jha, Sahu, and Mahato 1988; Ikeda et al. 1989; Jha and Jha 1989), but in situ germplasm preservation can also be an important strategy to prevent narrowing of the genetic base.

27.8 CONSERVATION

Given that ipecac reproduces via vegetative means, it is plausible that a cluster showing hundreds of aerial stems may actually represent a single genetic individual. A consequence of clonality is that conservation efforts should not rely solely on the number of clusters per unit area for defining conservation strategies for ipecac because the number of genetically unique individuals may be very small.

Ipecac is disappearing due to the destruction and degradation of its habitat owing to anthropogenic activities. The species inhabits only the understory and is highly sensitive to changes brought about by clearing, selective cutting, and incidental fires that allow long-term light penetration to the canopy floor; populations of ipecac decline rapidly when exposed to forest edge environments (Oliveira and Martins 2002). Although commercial harvesting has decreased substantially at the present time, human-led deforestation is reducing the number of suitable habitats and accelerating even further the demographic decline of wild populations of ipecac (Oliveira and Martins 2002). Ipecac is listed as endangered by the state of Minas Gerais (Mendonça and Lins 2000), the only Brazilian state within the geographical range of ipecac for which a plant conservation list has been published.

Currently, most of the genetic diversity of ipecac is not secure within protected areas. We are unaware of the presence of ipecac in any conservation unit within either the Amazonian range or the Atlantic range; Rio Doce State Park, located in the state of Minas Gerais, is the only exception. The finding that there is a strong genetic differentiation between ipecacs of the Amazonian range and ipecacs of the Atlantic range (Rossi et al. 2009) implies that genetic diversity of ipecac will be secured only when *in situ* strategies are aimed at conserving populations of both ranges.

Since 1995, several collecting expeditions have been organized with the aim of collecting germplasm of wild ipecac in the two Brazilian ranges of the species. These expeditions were part of an effort to conserve the genetic diversity of the species *ex situ*. The accessions were collected as living plants and are now part of a gene bank maintained at the Federal University of Viçosa at Viçosa, Minas Gerais, Brazil.

27.9 CONCLUSION

Commercial harvesting of ipecac roots has occurred in the Amazonian and Atlantic ranges since the eighteenth century and continued to be operational up to the mid-1950s. Overharvesting of wild plants in the past, combined with negligence in replanting, has led to a severe decline of native populations (Oliveira and Martins 1998).

England, the United States, and Canada have been highlighted among the countries that import ipecac, mainly for production of emetine hydrochloride. Basically, the Brazilian marketing of ipecac has been conducted by direct sale of dried roots and fluid extract obtained from roots. In the face of a progressive destruction of forests in the producing areas of the Brazilian states of Mato Grosso and Rondonia leading to the risk of species extinction, exports have decreased (Lameira 2002).

REFERENCES

Andersson, L. 2002. Re-establishment of Carapichea (Rubiaceae, Psychotrieae). *Kew Bulletin* 57:363–374.

Assis, M. C., and Giulietti, A. M. 1999. Morphological and anatomical differentiation in populations of "ipecacuanha"—*Psychotria ipecacuanha* (Brot.) Stokes (Rubiaceae). *Brazilian Journal of Botany* 22:205–216.

Bruneton, J. 1995. *Pharmacognosy, phytochemistry, medicinal plants*. Paris: Lavoisier.

Burhans, W. C., Vassiley, L. T., Wu, J. M., Nallaseth, F. S., and Depamphilis, M. L. 1991. Emetine allows identification of origins of mammalian DNA replication by imbalanced DNA synthesis, not through conservative nucleosome segregation. *EMBO Journal* 10:4351–4360.

Costa, A. F. 1978. *Farmacognosia.* Lisbon: Calouste Gulbekian.

De Boer, H. J., and Thulin, M. 2005. Lectotypification of *Callicocca ipecacuanha* Brot. and neotypification of *Cephaelis acuminata* H. Karst., with reference to the drug ipecac. *Taxon* 54:1080–1082.

Flückiger, F. A., and Hanbury, D. 1879. *Pharmacographia: A history of the principal drugs of vegetable origin met with in Great Britain and British India*, 2nd ed. London: Macmillan and Co.

Garcia, R. M. A., Oliveira, L. O., Moreira, M. A., and Barros, W. S. 2005. Variation in emetine and cephaeline contents in roots of wild ipecac (*Psychotria ipecacuanha*). *Biochemical Systematics and Ecology* 33:233–243.

Gupta, M. P., Cedeno, J. E., Soto, S. A. A., and Correa, D. M. A. 1986. Seasonal variation in the alkaloid content of Panamanian ipecac. *Fitoterapia* 57:147–151.

Hatfield, G. M., Arteaga, L., Dwyer, J. D., Arias, T. D., and Gupta, M. P. 1981. An investigation of Panamanian ipecac: Botanical source and alkaloid analysis. *Journal of Natural Products* 44:452–456.

Ho, P. C., Dweik, R., and Cohen, M. C. 1998. Rapidly reversible cardiomyopathy associated with chronic ipecac ingestion. *Clinical Cardiology* 21:780–783.

Ikeda, K., Teshima, D., Aoyamata, T., Satake, M., and Shimomura, K. 1989. Clonal propagation of *Cephaelis ipecacuanha*. *Plant Cell Reports* 7:288–291.

Itoh, A., Ikuta, Y., Baba, Y., Tanahashi, T., and Nagakura, N. 1999. Ipecac alkaloids from *Cephaelis acuminata*. *Phytochemistry* 52:1169–1176.

Jha, S., and Jha, T. 1989. Micropropagation of *Cephaelis ipecacuanha* A. Rich. *Plant Cell Reports* 8:437–439.

Jha, S., Sahu, N. P., and Mahato, S. B. 1988. Production of the alkaloids: Emetine and cephaeline in callus culture of *Cephaelis ipecacuanha*. *Planta Medica* 6:504–508.

Lameira, O. A. 2002. Cultivo da Ipecacuanha [*Psychotria ipecacuanha* (Brot.) Stokes] Ministério da Agricultura, pecuária e Abastecimento—Belém—PA—Brazil. *Circular Técnica* 28.

Li, W., Venkataraman, G. M., and Ain, K. B. 2006. Protein synthesis inhibitors, in synergy with 5-azacytidine, restore sodium/iodide symporter gene expression in human thyroid adenoma cell line, KAK-1, suggesting transactive transcriptional repressor. *Journal of Clinical Endocrinology and Metabolism* 92:1080–1087.

Lima, P. S. G. 2002. Divergência genética e efeito do nitrogênio total no crescimento *in vitro* de ipeca [Psychotria ipecacuanha (Brot) Stokes]. MSc diss., UFLA-Lavras Federal University.

Lorence, D. H. 1999. A nomenclator of Mexican and Central American Rubiaceae. *Monographs in Systematic Botany from the Missouri Botanical Garden* 73:1–177.

Martins, E. R., Oliveira, L. O., Maia, J. T. L. S., and Vieira, I. J. C. 2009. Estudo ecogeográfico da poaia [*Psychotria ipecacuanha* (Brot.) Stokes]. *Revista Brasileira de Plantas Medicinais* 11:24–32.

Mendonça, M. P., and Lins, V. 2000. Lista vermelha das espécies ameaçadas de extinção da flora de Minas Gerais. Belo Horizonte: Fundação Biodiversitas, Fundação Zoo-Botânica de Belo Horizonte-MG.

Möller, M., Herzer, K., Wenger, T., Herr, I., and Wink, M. 2007. The alkaloid emetine as a promising agent for the induction and enhancement of drug-induced apoptosis in leukemia cells. *Oncology Reports* 18:737–744.

Oliveira, L. O., and Martins, E. R. 1998. O desafio das plantas medicinais brasileiras: I—O caso da poaia (*Cephaelis ipecacuanha*). Campos de Goytacazes, RJ, UENF, 73 pp.

———. 2002. A quantitative assessment of genetic erosion in ipecac (*Psychotria ipecacuanha*). *Genetic Resources and Crop Evolution* 49:607–617.

Palomino, G., Doležel, J., Cid, R., Bruner, I., Méndez, I., and Rubluo, A. 1999. Nuclear genome stability of Mammillaria san-angelensis (Cactaceae) regenerants induced by auxins in long-term *in vitro* culture. *Plant Science* 141:191–200.

Rossi, A. A. B., Clarindo, W. R., Carvalho, C. R., and Oliveira, L. O. 2008. Karyotype and nuclear DNA content of *Psychotria ipecacuanha*: A medicinal species. *Cytologia* 73:53–60.

Rossi, A. A. B., Oliveira, L. O., Venturini, B. A., and Silva, R. S. 2009. Genetic diversity and geographic differentiation of disjunct Atlantic and Amazonian populations of *Psychotria ipecacuanha* (Rubiaceae). *Genetica* (The Hague) 136:57–67.

Rossi, A. A. B., Oliveira, L. O., and Vieira, M. F. 2005. Distyly and variation in floral traits in natural populations of *Psychotria ipecacuanha* (Brot.) Stokes (Rubiaceae). *Revista Brasileira de Botânica* 28:285–294.

Rout, G. R., Samantaray, S., and Das, P. 2000. *In vitro* somatic embryogenesis from callus cultures of *Cephaelis ipecacuanha* A. Richard. *Scientia Horticulturae* 86:71–79.

Satou, T., Akao, N., Matsuhashi, R., Koike, K., Fujita, K., and Nikaido, T. 2002. Inhibitory effect of isoquino-line alkaloids on movement of second-stage larvae of *Toxocara canis*. *Biological and Pharmaceutical Bulletin* 25:1651–1654.

Skorupa, L. A., and Assis, M. C. 1998. Collecting and conserving ipecac (*Psychotria ipecacuanha* Rubiaceae) germplasm in Brazil. *Economic Botany* 52:209–210.

Souza, M. M., Martins, E. R., Pereira, T. N., and Oliveira, L. O. 2006. Reproductive studies on ipecac (*Cephaelis ipecacuanha* (Brot.) A. Rich; Rubiaceae): Meiotic behavior and pollen viability. *Brazilian Journal of Biology* 66:29–41.

Veloso, H. P. 1947. As condições ecológicas da *Cephaelis ipecacuanha* Rich. *Memórias do Instituto Oswaldo Cruz* 45:361–372.

CHAPTER 28

Myrtus communis
Phytotherapy in the Mediterranean

Munir Ozturk, Salih Gucel, Ali Celik, Ernaz Altundag,
Tuba Mert, Eren Akcicek, and Sezgin Celik

CONTENTS

28.1 INTRODUCTION

With its extraordinary properties, plant life appeared millions of years ago on our planet. Nature has bestowed upon plants active molecules with natural affinities with the human body. As such, it is not surprising that the first humans found remedies for their aches by using plants (Plotkin 2000). This laid the foundation of ethnobotany in senso lato and phytotherapy in senso stricto. The fast-developing interdisciplinary field of ethnobotany is the study of the uses, conservation, and general economic as well as sociological importance of plants in human societies in the past and present (Baser, Honda, and Miki 1986; Ozturk and Ozcelik 1991; Baser 2002; Bonjar-Shaidi 2004).

Ethnobotany began when plants were classified differently in accordance with their uses, such as those with little or no use, those highly useful in several ways, pain- and illness-healing plants, and poisonous ones. It was during this period that the knowledge and manipulation of plant properties became associated with individuals. These individuals were the early medicine men or shamans, and we come across these people even now in some primitive societies. Many indigenous groups around the world are conversant with this knowledge. The knowledge of these people has a great potential value to humanity. This knowledge, however, is becoming extinct due to industrialization, building of dams, tourism, road building, construction of airports, warfare, missionary pressure, and migrations or through efforts to civilize the natives. The loss of this knowledge and of the natives themselves will be a grave hindrance to the progress of ethnobotany together with environmental conservation. The importance of ethnobotanical knowledge of the natives lies in their acquaintance with the properties of bioactive plants and their variants or ecotypes.

One of the subdivisions of ethnobotany is phytotherapy. In Greek, *phytos* means plants and therapy is the treatment. People nowadays are using the powerful curative properties of the plant world to stimulate the functioning of the organism and improve their physical, mental, and emotional well-being. The whole plant is the key to a more balanced treatment than the use of chemical extracts to cure patients. More and more people use natural methods. This field has given more than 80% of active substances to modern medicine.

This is a clear indication for the role of medicinal plants in the maintenance of health and treatment of diseases as therapeutic alternatives throughout the world still in the late twentieth and early twenty-first centuries (WHO 2002). It is a method that acts more slowly than synthetic chemical supplements, but every year millions of people go to hospitals due to the side effects of synthetic drugs and thousands die. Nature cures and herbal medicinal products are indispensable. Phytotherapy, in spite of its long history, does not possess a distinct scientific principle, but it offers new opportunities for herbal medicinal products. In phytotherapy, if we have wisdom to know and look back at our past, we can understand how to look forward to our future.

Turkey is one among the important gene centers of plant diversity in the world (Davis 1965–1985; Davis, Mill, and Tan 1988; Guner, Özhatay, and Ekim 2001). The number of flowering plant taxa distributed in the country is estimated to be around 10,000, which is very near to the number recorded for the entirety of Europe (Davis et al. 1988; Guner et al. 2001; Guvensen et al. 2006). Plant populations are suffering due to heavy industrialization causing environmental devastation. Furthermore, the knowledge and use of plants is fading at a fast pace. Of these plants, nearly 500 are used medicinally (Baytop 1984). Many, if not most, have local variants or ecotypes.

Myrtus communis belongs to the family Myrtaceae and is native to the Mediterranean region. It is commonly known as Mersin, Murt, and Hambeles in Turkey. The plant was held to be sacred to Venus and used as an emblem of love in wreaths and other decorations in Greco-Roman antiquity. It has been held as the emblem of honor and authority and was worn by Athenian judges in the exercise of their functions. Myrtle constituted the wreaths of the Grecian and Roman victors in the Olympian festivities. Scriptural allusions to it are abundant. In the Muslim tradition, it is regarded among the pure things carried by Adam from the Garden of Eden. A perusal of the literature reveals that few studies have been undertaken on this species in Turkey (Ahmad 1970; Dogan 1978; Akgul and Bayrak 1989; Ozek, Demirci, and Baser 2000; Cakir 2004; Sepici et al. 2004; Aydin and Ozcan 2007). An overview of the ecology of this newly reviving plant of the ancients and its role in the modern phytotherapy are discussed here.

28.2 MORPHOLOGY

Myrtus communis is a shrub that belongs to the Myrtaceae family. The plants are 40–250 cm tall, 50–200 cm wide, branched, and covered with a deep-grayish fissured bark; they are more than 5 m tall at some places (Figure 28.1a, b). The evergreen leaves are opposite, thick and lustrous, ovate, lanceolate, opposite, short petioled, closely pellucid-punctate, smooth, and shining with many small, translucent, oil-bearing glands. The solitary, axillary white flowers are about 2 cm long; they are borne on short stalks and have many stamens (Figure 28.2a). Fruit is a purplish-black, bluish-black, or white many-seeded fleshy berry, subglobular, containing four or five dull white reniform seeds. The average length, width, thickness, and diameter of fruits are around 15, 10, 8, and 12 mm, respectively (Figure 28.2b). The leaves, flowers, and berries are all fragrant. It is cultivated for ornamental purposes as well.

Figure 28.1(a)   A shrub form of *Myrtus communis.*

Figure 28.1(b)   A tree form of *Myrtus communis.*

Figure 28.2(a)   A flowering shoot of *Myrtus communis* L.

Figure 28.2(b)   A fruit-bearing shoot of *Myrtus communis* L.

28.3 ECOLOGY

Myrtus communis grows all along the coast in Turkey from the Syrian border up to Canakkale, on several islands, and at some places along the Black Sea coast (Demiriz 1956; Vardar and Ahmed 1971, 1973; Ahmed and Vardar 1973; Ozturk and Vardar 1974; Ozturk 1979). It is one of the characteristics of the floristically rich Mediterranean basin occupying macchias and at lower elevations among red pine forests. In the Taurus Mountains, it is found in pine forests and riversides just above sea level up to 500–600 m. The origin is still disputed. Early investigators consider it to be of Persian origin. Recent fossil studies in central Europe report the presence of *Myrtus* types of

Figure 28.3 Map showing the distribution of *Myrtus communis* in Turkey.

leaves, which have been named as *Myrtophyllum*, depicting the presence of *Myrtus* in this region along with other typical Mediterranean representatives like *Olea* and *Vitis*. As such, the origin of *M. communis* should lie in the Mediterranean region.

Fruits mature in autumn, dry or shrivel up, fall to the ground, and are carried by rain and wind away from their habitat. If fruits get embedded in a damp bed of alluvium, they germinate easily. Water is the causative factor for its recurrence alongside the streams. The minimum period of floating of an individual fruit is 4 days and the maximum is 20 days; the majority sink in 12 days. Birds help in the dispersal due to the plant's fleshy berries, as the seeds swallowed pass out of the gut undigested. It is a successful species with wide distribution and occupies many different ecological habitats at altitudes from 0 to 900 m, usually on the north-facing slopes. In Turkey, this hygrophilous species is found in the north, west, and south Anatolian regions, particularly in the states of Trabzon, Ordu, Samsun, Sinop, Kastamonu, Zonguldak, Bolu, Duzce, Bilecik, Istanbul, Canakkale, Bursa, Balikesir, Manisa, Izmir, Aydin, Denizli, Mugla, Antalya, Icel, Adana, Iskenderun, Osmaniye, and Antakya (Figure 28.3).

Myrtus communis is a prolific seed producer with a very high germination rate (80%) and the viability of the seeds is retained for over 2 years. Germination capacity increases with a decrease in the moisture content of seeds with maturity. The optimum germination takes place at 60–100% soil moisture, 9–15 hours duration of light, and 27°C constant and 30–22°C alternate temperatures (Figures 28.4 and 28.5; Table 28.1). It can be propagated easily by cuttings. This evergreen shrub or small tree adapts itself to many kinds of soils. The species is generally found on sandy-loam, sandy-clay-loam, clay-loam, loam, or silty-loam soils with pH varying from moderately acidic (5.35) to slightly alkaline (7.90); it is indifferent to the edaphic factor. However, the species occurs frequently on alkaline soils rich in calcium carbonate (0.66–40.96 meq.%) and calcium (18.43–31.0 meq.%) (Table 28.2).

The plants growing in moist and well-drained soils exhibit a luxuriant growth and are robust and rich in moisture content, calcium, and total nitrogen (Figure 28.6). On calcareous habitats, α-pinene and limonene are in greater amounts, while linalool, linalyl acetate, and *trans*-myrtanol acetate are higher in the plants of siliceous habitats. It is typically a plant that loves moderate temperatures, which is fully supported by its absence from areas like the central and eastern parts of Turkey and other regions of world where temperatures go down much below 0°C. Though typically a shady mesophyte, this species can grow in extremes of light conditions, but becomes bushy with profuse branching and short internodes in full sunlight as compared to plants in shade, which have longer internodes and big, lush, green leaves. Because they are not palatable, plants are not eaten by cattle and do not get trampled underfoot due to habit.

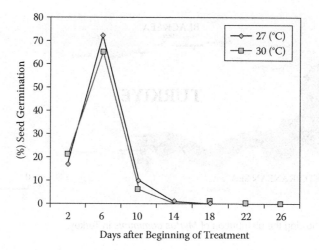

Figure 28.4 Germination percentage in *M. communis* at various times after exposure with light at intensity of 1800 lux.

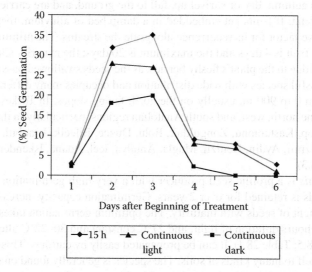

Figure 28.5 Effects of constant temperature on seed germination.

Table 28.1   Effects of Alternate Temperatures on Seed Germination

Alternate Temperature (°C)	Germination (%)	Minimum Days Required for Germination
30/12	88	5
30/22	100	3
30/27	100	2
27/6	58	5
27/12	85	4
27/22	90	3

Table 28.2 Physicochemical Analysis of the Soils

Moisture Content (%)	pH	Total Soluble Salts (%)	CaCO3 (%)	P (mm)	m.e.%			Organic Carbon (%)	Organic Matter (%)	Total N (%)
					K	Ca	Na			
22.66	7.77	0.41	23.11	2.96	6.55	26.72	0.12	0.69	1.19	0.12

Though it is a typical macchia element, the plant grows well in *Pinus brutia* forests together with *Arbutus unedo, Quercus infectoria*, and *Phillyrea* spp. The associates in macchias are *Phillyrea* spp. *Pistacia lentiscus*, and *Laurus nobilis* and around the periodically running streams *Nerium oleander* and *Laurus nobilis*. Along the coastal parts, xerophilous species like *Genista acantho-clada, Thymelaea tartonraira, Cistus parviflorus, Rhamnus alaternus*, and *Daphne gnidium* join this plant.

28.4 MEDICINAL IMPORTANCE

Myrtus communis is commonly called myrtle; its name goes back to the Greek *myrtos*, which was probably borrowed from a Semitic word. It has been used traditionally as an antiseptic and disinfectant drug and has therapeutic importance. The ancient popular use of myrtle is still rooted in coastal areas of the Mediterranean Sea (Sacchetti et al. 2007). It was one of the medicinal plants of the ancients and was practically obsolete in modern therapeutics until its revival in 1876 as a remedy for relaxation of parts with mucous and other profluvia. Myrtle has a slightly camphorous, fresh scent and a volatile secondary essential oil is found in most parts of the plant (Bradesi et al. 1997; Chalchat, Garry, and Michet 1998). Distillation of leaves and twigs yields a dextrorotatory, emerald-green volatile oil.

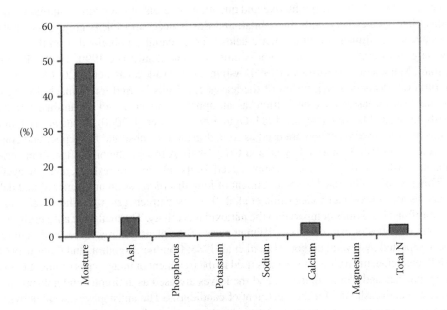

Figure 28.6 Shoot analysis data at flowering stage (% air dry weight).

- The oil is regarded as an active antiseptic and tonic consisting of a mixture of pinene, cineol, and dipentene with anti-inflammatory, anticatarrhal, antiseptic (urinary, pulmonary), astringent, balsamic, bactericidal, expectorant, regulator, and slightly sedative characteristics. It stimulates the gastric, renal, and pulmonic membranes, increasing their functions, and is reputed to possess decided antiseptic and deodorant powers. It is used against allergies, catarrh, catarrhal conjunctivitis, pharyngitis, chronic cough, colds, flu, infectious disease, tuberculosis, chest pains, and asthma; as a nerve sedative and stimulant to mucous membranes, bronchitis, cystitis and pyelitis, dry, and hollow cough, with tickling in chest; and for promoting digestion, treating urinary tract disorders, and preventing infections in wounds.
- Myrtle is also accepted by traditional healers in the Middle East as a curative agent in hemorrhoids and a useful herb for diabetes.
- The powder is sprinkled upon cotton with glycerin and applied to uterine ulcerations. Suppurative wounds and ulcers, intertrigo, and eczema are treated in the same manner, omitting the glycerin, while in cases with offensive discharges and threatened gangrene, a wine of myrtle is employed with the result of correcting the fetor and inducing granulation.
- An infusion of the leaves or diluted tincture relieves dysentery and gives excellent results when used as an injection in uterine prolapse, lax vaginal walls, and leucorrhea. In renal and cystic catarrh and colliquative sweating of phthisis, 15–40 grains of the powder are useful; 10–30 grains are needed to check wasting in menorrhagia. In doses of 2 minims every 2 or 3 hours, the oil (in capsules) is asserted to be prompt and curative in fetid bronchitis and pulmonary gangrene.
- An infusion is a wonderful eyewash for tired, irritated eyes and useful in treating allergic conjunctivitis. When consumed in the morning, it is a bit stimulating and helps to overcome mental fatigue accompanying allergies or chest colds. Its effects on hormonal imbalances of the thyroid and ovaries are also reported.
- Leaves contain terpinolene and are applied to scorpion stings. Tripolitan natives use these as abortifacients. The leaves of plants on calcareous habitats have higher percentages of α-pinene and limonene, whereas plants on siliceous soils have higher percentages of linalool, linalyl acetate, and *trans*-myrtanol acetate (Mansouri et al. 2001; Flamini et al. 2004).

Myrtle is known not only as an aromatic plant, but also for its pleasant smelling essential oil extract (present in numerous glands, especially in the leaves and mature fruits), which is rich in linear, cyclic, and bicyclic monoterpenes. Several compounds have been isolated from this plant (Pichon-Prum et al. 1989; Pichon-Prum, Joseph, and Raynaud 1998; Gauthier, Gourai, and Bellakhdar 1988; Martin et al. 1990, 1999). The qualitative and quantitative levels of the main compounds characterizing this plant species vary with the habitat, and this is necessary for its large-scale exploitation in pharmaceutical, alimentary, or cosmetic fields. The essential oil obtained from this species has been widely investigated. Its composition is quite variable (Lawrence 1990, 1993a, 1993b, 1996; Weyerstahl, Marschall, and Rustaiyan 1994; Asllani 2000; Yadegarinia et al. 2006).

In Turkey, a decoction or infusion of the leaves and fruits is used for stomach, hypoglycemic, cough, and oral diseases; for constipation; as an appetite enhancer and an antihemorrhagic; and externally for wound healing (Baytop 1984; Ogur 1994; Ozek et al. 2000). Fruits are rich in tannin. The fruits are very astringent and are used as a condiment as a substitute for pepper and considered a rich source of tannin (Canhoto, Lopes, and Cruz 1998). Although the oil composition varies with the habitat, myrtle is widely known and appreciated for its balsamic, refreshing, and astringent properties. This essential oil is used in the treatment of lung disorders, as an antibacterial and delousing agent, and as an antioxidant (Yadegarinia et al. 2006). The extracts generally show weak antibacterial and antifungal activities compared to the antioxidant activity. The analgesic antiperglycemic and antimycotic activities of extracts from different parts of the plant (mainly leaves and mature fruits) have been reported by several authors (Husni et al. 1988; Gauthier, Agoumi, and Gourai 1989).

In folk medicine, the fruit of this plant is used in the treatment of many types of infectious disease, including diarrhea and bloody diarrhea, and the leaves are used as antiseptic and anti-inflammatory agents, as a mouthwash, and for the treatment of candidiasis. The antistaphylococcal activity of the methanol crude extract of *M. communis* leaves has been reported previously (Mansouri et al. 2001).

The presence of antibacterial activity in different fractions and essential oil indicates that the extract possesses different compounds that have different activities. It is a rich source of polyphenols of therapeutic interest, such as myricitrin, hesperidin, hesperidinmethyl-chalcone, and esculin.

Out of the constituents characterizing myrtle oil, the most representative are myrtenol, myrtenyl acetate, limonene, linalool, pinene, and eucalyptol, as well as *p*-cymene, geraniol, nerol, and the phenylpropanoid derivative methyleugenol. These vary with habitat. Elfellah, Akhtar, and Khan (1984) reported no hypoglycemic activity on either acute or subacute administration, but potent activity was observed in streptozotocin-induced diabetic mice. In Turkish folk medicine, *M. communis* leaves and volatile oil obtained from the leaves are used to lower the blood glucose level in type 2 diabetic patients (Baytop 1999). The amount of volatile oil recommended for this treatment is 10 drops per day because of irritation and bleeding in mucous membranes and abortion in pregnancy (Baytop 1999). *Myrtus communis* oil may produce its hypoglycemic effect mainly by influencing the mechanism of intestinal transport of carbohydrates.

The fleshy berries are a rich source of resin, sugar, citric acid, malic acid, tannins and volatile oil eucalyptol, and γ-pinene. In past times, ripe fruits were used as food integrators because of their high vitamin content. The fruit decoction was used to bathe newborns with reddened skin, while the decoction of leaves and fruits was useful for washing sores. The decoction of the leaves is still used for vaginal lavage and enemas and against respiratory diseases. Seeds contain oleic, palmitic, and stearic acids. Methanolic extract of seeds and leaves shows antibacterial activity.

Myrtus communis represents an important natural source and possesses a good potential in the food and flavor industries. It is used by the food and cosmetic industries for its aromatic taste and scent. The oil from the leaves has been used in the cosmetics, sauces, confectionery, and beverage industries (Akgul and Bayrak 1989; Boelens and Jimenez 1992; Ogur 1994; Ozek et al. 2000). Moreover, at some areas in the Mediterranean, foods are flavored with leaves and mature fruits. Myrtle is both a top and middle note in perfumery; it blends well with most oils, including bay laurel, bay leaf, bergamot, clary sage, clove, ginger, lavender, lavandin, lime, rosemary, and spicy oils. The French distill an aromatic water from the leaves and flowers, which they call *eau d'ange*.

The leaves, berries, and twigs have been employed in flavoring food and wines. Herbal tea is prepared from the leaves. Leaf decoction is used for hair blackening and hair lotions are prepared from dry leaves and flowers together with some aromatic herbs. Egyptians used it in cosmetics and for skin ailments. It is good for oily skin, acne, and open pores. The oils extracted by steam distillation of fruits are used in both the flavor and fragrance industries (Dogan 1978).

The fruits are sold on a large scale in several provinces of Turkey and these serve as a food for young children. Each tree produces approximately 500 kg of fruit in the region of Antakya. The tree never fails to produce fruit and can be sustained without much care.

28.5 PRECAUTIONARY MEASURES

Myrtle should not be used with insulin or oral sulfonylureas because it may increase the blood glucose and thus should not be consumed orally as an antidiabetic agent. Large amounts taken orally can cause breathing and circulation problems, low blood pressure, and other complications. Myrtle may affect liver function.

REFERENCES

Ahmed, M. 1970. Autoecology and economic evaluation of *Myrtus communis* L. PhD thesis. Ege University, Izmir, Turkey, 173 pp.

Ahmed, M., and Y. Vardar. 1973. Distribution and plasticity of *M. communis*. *Phyton* (Austria) 15:145–150.

Akgul, A., and A. Bayrak. 1989. Essential oil content and composition of myrtle (*Myrtus communis* L.) leaves. *Doga TU Tar ve Or. D.* 13:143–147.

Asllani, U. 2000. Chemical composition of Albanian myrtle oil (*Myrtus communis* L.). *Journal of Essential Oil Research* 12:140–142.

Aydin, C., and M. Ozcan 2007. Determination of nutritional and physical properties of myrtle (*Myrtus communis* L.) fruits growing wild in Turkey. *Journal of Food Engineering* 79:453–458.

Baser, K. H. C. 2002. Aromatic biodiversity among the flowering plant taxa of Turkey. *Pure and Applied Chemistry* 74:527–545.

Baser, K. H. C., G. Honda, and W. Miki. 1986. Herb drugs and herbalists in Turkey. Institute for the Study of Languages and Cultures of Asia and Africa, Tokyo, no. 27.

Baytop, T. 1984. *Türkiyede Bitkiler ile Tedavi*. İstanbul Üniversitesi Yayınları, no. 3255, Eczacılık Fakültesi Yayın, 40:368–370.

———. 1999. *Treatment with plants in Turkey, past and present*. Nobel Tıp Kitapevleri.

Boelens, M. H., and R. Jimenez. 1992. The chemical composition of Spanish myrtle oils. *Journal of Essential Oil Research* 4:349–353.

Bonjar-Shahidi, G. H. 2004 Antibacterial screening of plants used in Iranian folkloric medicine. *Fitoterapia* 75:231–235.

Bradesi, T. P., J. Casanova, J. Costa, and A. F. Bernardini. 1997. Chemical composition of myrtle leaf oil from Corsica (France). *Journal of Essential Oil Research* 9:283–288.

Cakir, A. 2004. Essential oil and fatty acid composition of the fruits of *Hippophae rhamnoides* L. (sea buckthorn) and *Myrtus communis* L. from Turkey. *Biochemical Systematics and Ecology* 32:1–8.

Canhoto, J. M., M. L. Lopes, and G. S. Cruz. 1998. *In vitro* propagation of *Myrtus communis* through somatic embryogenesis and axillary shoot proliferation. In *1st International Meeting of Aromatic and Medicinal Mediterranean Plants*, Ansiao, Portugal.

Chalchat, J., R. P. Garry, and A. Michet. 1998. Essential oils of myrtle (*Myrtus communis* L.) of the Mediterranean littoral. *Journal of Essential Oil Research* 10:613–617.

Davis, P. H. 1965–1985. In *Flora of Turkey and the East Aegean Islands*, Volumes: 1–9. Edinburgh: Edinburgh University Press.

Davis, P. H., R. R. Mill, and K. Tan. 1988. *Flora of Turkey and the East Aegean Islands*, vol 10 (supplement 1). Edinburgh: Edinburgh University Press.

Demiriz, H. 1956. Ökologische Beobachtungen über das gemeinsame Auttreten von *Laurus nobilis* und *Myrtus communis* an Anatoliens Nord und Südküste. *Rev.FAc.Sci.de'l Univ.d'Istanbul* 21:237–266.

Dogan, A. 1978. *Myrtus communis* L. mersin bitkisinin ucucu yagi verimi, yagın fiziksel-kimyasal ozellikleri ve bilesimi uzerinde arastirmalar. *Ankara Universitesi Ziraat Fakultesi Yayin Numarasi* no. 678.

Elfellah, M. S., M. H. Akhtar, and M. T. Khan. 1984. The antidiabetic effect of the ethanol–water extracts of *Myrtus communis* using streptozotocin-induced hyperglycemia in mice. *Journal of Ethnopharmacology* 11:275–281.

Flamini, G., P. L. Cioni, I. Morelli, S. Maccioni, and R. Baldini. 2004. Phytochemical typologies in some populations of *Myrtus communis* L. on Caprione Promontory (east Liguria, Italy). *Food Chemistry* 85:599–604.

Gauthier, R., A. Agoumi, and M. Gourai. 1989. Activité d'extraits de *Myrtus communis* contre *Pediculus humanus capitis*. *Plantes Méd Phytothérapie* 28:95–108.

Gauthier, R., M. Gourai, and J. Bellakhdar. 1988. A propos de l'huile essentielle de *Myrtus communis* L. var. *Italica* et var. *Baetica* recolte au Maroc. *Review Morocco Pharmacy* 4:117–132.

Guner, A., N. Özhatay, and T. Ekim. 2001 *Flora of Turkey and the East Aegean Islands*, vol. 11. Edinburgh: Edinburgh University Press.

Guvensen, A., G. Gork, and M. Ozturk. 2006. An overview of the halophytes in Turkey. In *Sabkha ecosystems. Vol. II. West & Central Asia*, ed. Khan et al., 9–30. Dordrecht, the Netherlands: Springer.

Husni, A. A., H. M. Twais, K. Hanifa, and S. Ali. 1988. Pharmacological, phytochemical and antimicrobial studies on *Myrtus communis* L. Part 2: Glycemic and antimicrobial studies. *Journal of Biological Sciences Research* 19:41–52.

Lawrence, B. M. 1990. Progress in essential oils. Myrtle oil. *Perfumer and Flavorist* 15:65–66.

———. 1993a. *Essential oils*. Carol Stream, IL: Allured Co.

———. 1993b. Progress in essential oils. Myrtle oil. *Perfumer and Flavorist* 18:52–55.

———. 1996. Progress in essential oils. Myrtle oil. *Perfumer and Flavorist* 21:57–58.

Mansouri, S., A. Foroumadi, T. Ghaneie, and A. G. Najar. 2001. Antibacterial activity of the crude extracts and fractionated constituents of *M. communis*. *Pharmaceutical Biology* 39:399–401.

Martin, T., R. Begoña, V. Lucinda, F. Lidia, and D. Ana María. 1999. Polyphenolic compounds from pericarps of *Myrtus communis*. *Pharmaceutical Biology* 37:28–31.

Martin, T., L. Villaescusa, M. De Soto, A. Lucia, and A. M. Diaz. 1990. Determination of anthocyanic pigments in *Myrtus communis* berries. *Fitoterapia* 61:85.

Ogur, R. 1994 A research about myrtle tree (*Myrtus communis* L.). *Ecology Journal* 10:21–25.

Ozek, T., B. Demirci, and K. H. C. Baser. 2000. Chemical composition of Turkish myrtle oil. *Journal of Essential Oil Research* 12:541–544.

Ozturk, M., and H. Ozcelik. 1991. *Useful plants of East Anatolia*. Ankara: Siskav Press.

Ozturk, M. A. 1979. Preliminary observations on the edaphic and biotic relations of *M. communis* L. *Ege University Science Faculty Journal* B 3:137–142.

Ozturk, M. A., and Y. Vardar. 1974. Distribution and economical prospects of *M. communis* L. *Bitki* 1:100–107.

Pichon-Prum, N., M. J. Joseph, and J. Raynaud. 1998. Myricetin-3-d-(6"-*O*-galloyl-galactoside) de *Myrtus communis* L. (Myrtaceae). *Plants Medicines Phytotherapie* 26:86–92.

Pichon-Prum, N., J. Raynaud, L. Debourchieu, and M. J. Jy. 1989. Sur la presence d'un heteroside acyle de la quercetine dans les feuilles de *Myrtus communis* L. (Myrtaceae). *Pharmazie* 44:508–509.

Plotkin, M. J. 2000. *Medicine quest*. New York: Penguin Books Ltd.

Sacchetti, G., M. Muzzoli, G. A. Statti, F. Conforti, A. Bianchi, C. Agrimonti, M. Ballero, and F. Poli. 2007. Intraspecific biodiversity of Italian myrtle (*Myrtus communis*) through chemical markers profile and biological activities of leaf methanolic extracts. *Natural Product Research* 21:167–179.

Sepici, A., I. Gurbuz, C. Cevik, and E. Yesilada. 2004. Hypoglycemic effects of myrtle oil in normal and alloxan-diabetic rabbits. *Journal of Ethnopharmacology* 93:311–318.

Vardar, Y., and M. Ahmed. 1971. Water relations of *M. communis* seeds. Verh. Schweizer. *Naturforsch. Ges.* 150:70–75.

———. 1973. Some ecological aspects of *M. communis* L. *Jahrbücher für Systematik* 93:562–567.

Weyerstahl, P., H. Marschall, and A. Rustaiyan. 1994. Constituents of the essential oil of *Myrtus communis* L. from Iran. *Flavor and Fragrance Journal* 9:333–337.

WHO. 2002. *WHO traditional medicine strategy 2002–2005*. Geneva: WHO.

Yadegarinia, D., L. Gachkar, M. B. Rezaei, M. Taghizadeh, S. A. Astaneh, and I. Rasooli. 2006. Biochemical activities of Iranian *Mentha piperita* L. and *Myrtus communis* L. essential oils. *Phytochemistry* 67:1249–1255.

CHAPTER 29

Licorice (*Glycyrrhiza* species)

Marjan Nassiri-Asl and Hossein Hosseinzadeh

CONTENTS

29.1 INTRODUCTION

Licorice is a perennial herb native to regions of the Mediterranean, central and southern Russia, Asia Minor, and parts of Iran. It is now widely cultivated throughout Europe, the Middle East, and Asia (Blumenthal, Goldberg, and Brinckmann 2000). Licorice, described as "the grandfather of herbs," has been used medicinally since 500 BC (Ody 2000).

Licorice has a long history of medicinal uses in Europe and Asia. It is believed to be effective in treating peptic ulcer disease, constipation, cough, diabetes, cystitis, tuberculosis, wounds, kidney stones, lung ailments, and Addison's disease (Varshney, Jain, and Srivastava 1983; Dafni, Yaniv, and Palevitch 1984; Arseculeratne, Gunatilaka, and Panabokke 1985; Fujita et al. 1995; Yarnell 1997; Gray and Flatt 1997; Rajurkar and Pardeshi 1997; Armanini et al. 2002). It has also been used for its anabolic properties and its capacity to improve male sexual function (Sircar 1984; Nisteswar and Murthy 1989).

29.2 THE GENUS

The genus *Glycyrrhiza* L. [Fabaceae (Leguminosae)] consists of about 30 species, including *G. glabra* L. *G. uralensis* Fisch. *G. inflata* Batalin *G. aspera* Pall. *G. korshinskyi* Grig. and *G. eurycarpa*. *P.C. Li Glycyrrhiza glabra* also includes three varieties: (1) Persian and Turkish licorices are assigned to *G. glabra* var. *violacea*, (2) Russian licorice is classified as *G. glabra* var *gladulifera*, and (3) Spanish and Italian licorices are *G. glabra* var. *typica* (Nomura, Fukai, and Akiyama 2002). Licorice is also known by names such as liquorice, kanzoh, gancao, sweet root, and yasti-madhu (Blumenthal et al. 2000; Nomura et al. 2002).

29.3 GENETIC INFORMATION OF *GLYCYRRHIZA*

In the licorice species, 10 different genotypes (tg 1 through tg 9, and ADD) are recognized. It has been suggested that tg1 is the predominant genotype of the *Glycyrrhiza* species except for *G. uralensis*, *G. glabra*, or *G. inflata*. *G. glabra* and *G. inflata* have the tg2 genotype, *G. glabra* has the tg3 genotype, *G. inflata* has tg4 and tg5 genotypes, and *G. uralensis* has tg6 through tg9 genotypes. The ADD genotype is a hybrid between *G. uralensis* and *G. glabra* or *G. inflata* species (Kondo, Shiba, Nakamura, et al. 2007). Furthermore, it has been confirmed that glabridin, licochalcon A, and glycycoumarin function as species-specific constituents of *G. glabra*, *G. inflata*, and *G. uralensis*, respectively (Kondo Shiba, Nakamura, et al. 2007). The ADD genotype consists of the I-2 and I-3 genotypes of the internal transcribed spacer (ITS). The I-2 genotype was observed in the tg2 through tg5 genotypes, which are defined as the species-specific genotypes of *G. glabra* or *G. inflata*. The I-3 genotype was recognized in the tg6 through tg9 genotypes, which are defined as the species-specific genotypes of *G. uralensis* (Kondo, Shiba, Yamaji, et al. 2007).

29.4 CHEMICAL COMPOSITION

29.4.1 Active Components

Triterpenoid saponins (4–20%)—mostly glycyrrhizin, which is a mixture of potassium and calcium salts of glycyrrhizic acid (also known as glycyrrhizic or glycyrrhizinic acid), and a glycoside of glycyrrhetinic acid, which is 50 times as sweet as sugar—are in licorice root (Blumenthal et al. 2000). Other triterenes that have been found in licorice include liquiritic acid, glycyrretol, glabrolide, isoglaborlide, and licorice acid (Williamson 2003). Production of a high-concentration glycyrrhizin within a very short time period has been shown in controlled environments (Afreen, Zobayed, and Kozai 2005).

Flavonoids and chalcones, such as liquiritin, liquiritigenin, rhamnoliquiritin, neoliquiritin, isoliquiritin, isoliquiritigenin, neoisoliquiritin, licuraside, glabrolide, and licoflavonol, are found in licorice (Williamson 2003). The retrochalcones; licochalcone A, B, C, and D; and echinatin were isolated from the roots of *G. inflata*, and the minor flavonoids isotrifoliol and glisoflavanone were isolated from underground parts of *G. uralensis* (Hatano et al. 2000; Haraguchi 2001). 5,8-Dihydroxyflavone-7-*O*-β-d-glucuronide, glychionide A, and 5-hydroxy-8-methoxyl-flavone-7-*O*-β-d-glucuronide, glychionide B, were isolated from the roots of *G. glabra* (Li, Wang, and Deng 2005).

Glabridin, galbrene, glabrone, shinpterocarpin, licoisoflavones A and B, formononetin, glyzarin, and kumatakenin are isoflavonoid derivatives present in licorice (Williamson 2003). Also, hispaglabridin A and B, 4'*O*-methylglabridin and 3'-hydroxy-4'-*O*-methylglabridin (De Simone et al. 2001; Haraguchi 2001), and glabroisoflavanone A and B (Kinoshita, Tamura, and Mizutani 2005) have been found.

Other constituents include coumarins. Liqcoumarin, glabrocoumarone A and B, herniarin, umbelliferone, glycyrin, glycocoumarin, licofuranocoumarin, licopyranocoumarin, and glabrocoumarin are known as cumarins that present in *G. glabra* (De Simone et al. 2001; Haraguchi 2001; Williamson 2003; Kinoshita et al. 2005).

Based on phenolic constituents, licorice species were classified into three types: A, B, and C. Type A consists of roots and rhizomes of *G. uralensis* containing licopyranocoumarin, glycycoumarin, and/or licocoumarone, which are not found in *G. glabra* and *G. inflata*. Type B comprises *G. glabra*, which contains glabridin and glabrene that are not found in the samples of the other two species. Type C is *G. inflata*, which contains licochalcones A and B that are not found in the other two species (Hatano, Fukada, Miyase, et al. 1991).

Four dihydrostilbenes—dihydro-3,5-dihydroxy-4'-acetoxy-5'-isopentenylstilbene, dihydro-3,3', 4'-trihydroxy-5-*O*-isopentenyl-6-isopentenylstilbene, dihydro-3,5,3'-trihydroxy-4'-methoxystilbene, and dihydro-3,3'-dihydroxy-5β-d-*O*-glucopyranosyloxy-4'-methoxystilbene—were isolated from leaves of *G. glabra* grown in Sicily (Biondi, Rocco, and Ruberto 2005) (Figure 29.1).

29.4.2 Miscellaneous Compounds

Glycyrrhiza glabra extract contains fatty acids (C_2–C_{16}) and phenols (phenol, guaiacol), together with common saturated linear γ-lactones (C_6–C_{14}). A series of new 4-methyl-γ-lactones and 4-ethyl-γ-lactones in trace amounts has also been found in *G. glabra* (Näf and Jaquier 2006). Other compounds, such as asparagines, glucose, sucrose, starch, polysaccharides (arabinogalactants), and sterols (β-sitosterol, dihydrostigmasterol), are also present (Hayashi et al. 1998; Blumenthal et al. 2000).

29.5 PHARMACOLOGICAL EFFECTS

Anti-inflammatory properties of β-glycyhrritinic acid, the major metabolite of glycyrrhizin, have been shown in different experiments (Capasso et al. 1983; Amagaya et al. 1984; Inoue et al. 1989). Two mechanisms have been discussed for the anti-inflammatory effects of β-glycyhrritinic acid. First, it inhibits glucocorticoid metabolism and thus potentiates the anti-inflammatory response reported in skin and lung tissue after administration of β-glycyhrritinic acid (Teelucksingh et al. 1990; Schleimer 1991). Because it is a potent inhibitor of 11-β-hydroxysteroid hydroxygenase (Walker and Edwards 1991), it causes an accumulation of glucocorticoids with anti-inflammatory properties. Second, it inhibits the classical complement pathway (Kroes et al. 1997).

It was shown that glycyrrhizin has an anti-inflammatory effect by inhibiting the generation of reactive oxygen species (ROS) via neutrophils (Akamatsu et al. 1991; Wang and Nixon 2001). Also, glabridin could suppress the generation of ROS RAW 264.7 cells (Kang et al. 2005). Because glycyrrhizin reduces the development of acute inflammation, it could be used in mouse models of acute

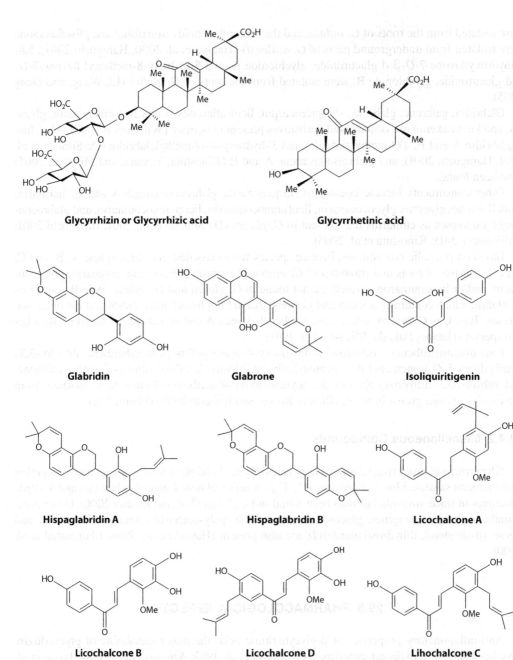

Figure 29.1 Chemical structures of some active components of licorice.

inflammation (carrageenan-induced pleurisy) to prevent the activation of nuclear factor (NF)-κB and STAT-3 (Menegazzi et al. 2008).

Glycyrrhiza glabra and glyderinine have also shown anti-inflammatory effects (Azimov, Zakirov, and Radzhapova 1988; Tokiwa et al. 2004). They reduced myocardial inflammatory edema in experimental myocardial damage (Zakirov, Aizimov, and Kurmukov 1999). In addition, glabridin and lichochalocone A have demonstrated anti-inflammatory properties in *in vivo* studies (Furuhashi et al. 2005; Kang et al. 2005).

Glycyrrhetinic acid did not inhibit prostaglandin biosynthesis catalyzed by cyclooxygenase 1 and 2 in an *in vitro* study (Perera et al. 2001). It seems that *G. radix* was involved in COX-2 inhibition (Kase et al. 1998). Moreover, glycyrrhizin and glycyrrhetinic acid could inhibit phospholipase A_2 (Kase et al. 1998). Some derivatives of glycyrrhetinic acid have also shown inhibitory effects against interleukin-1β (IL-1β)-induced prostaglandin E2 (PGE2) production in normal human dermal fibroblasts (NHDF) (Tsukahara et al. 2005).

Flavonoids with C5 aliphatic residues of licorice were effective against methicillin-resistant *Staphylococcus aureus* (MRSA) and restored the effects of oxacillin and β-lactam antibiotic against MRSA (Hatano et al. 2000a, 2005).

Antimicrobial activity of glabridin, glabrene, and licochalcone A were exhibited against *Helicobacter pylori in vitro* (Fukai et al. 2002a, 2002b).

Glycyrrhiza glabra extract has been found to have effective antibacterial properties against the following five bacteria: *Escherichia coli*, *Bacillus subtilis*, *Enterobacter aerogenes*, *Klebsiella pneunmoniae*, and *S. aureus* (Onkarappa, Shobha, and Chaya 2005).

Potent antibacterial activity of glycyrrhizol A and 6,8-diisoprenyl-5,7,4'-trihydroxyisoflavone from the root of *G. uralensis* was exhibited against *Streptococcus mutans* with minimum inhibitory concentrations of 1 and 2 µg/mL, respectively (He et al. 2006). Recently, 18-β-glycyrrhetinic acid, derived from the root of the *Glycyrrhiza* species, has reduced *in vitro* growth of *Candida albicans* strains that were isolated from patients with recurrent vulvovaginal candidiasis (RVVC) (Pellati et al. 2009).

Glycyrrhizic acid inhibits the replication of several viruses *in vitro* such as Epstein–Barr virus (EBV), herpes simplex virus, hepatitis A virus (HAV), hepatitis B virus (HBV), hepatitis C virus (HCV), human cytomegalovirus (CMV), human immunodeficiency virus (HIV), influenza virus, severe acute respiratory syndrome (SARS) coronavirus, and varicella zoster virus (VZV) (Pompei et al. 1979; Baba and Shigeta 1987; Ito et al. 1988; Crance et al. 1990; Numazaki, Umetsu, and Chiba 1994; Takahara, Watanabe, and Shiraki 1994; Sato et al. 1996; Utsunomiya et al. 1997; Van Rossum et al. 1998; Cinatl et al. 2003; Lin 2003). Some mechanisms have been suggested for the antiviral effects of glycyrrhizin (Van Rossum et al. 1998; Cohen 2005).

Glycocoumarine and licopyranocoumarin, two coumarins of *G. glabra*, may be able to inhibit giant cell formation in HIV-infected cell cultures without any cytotoxicity (Hatano et al. 1988; De Simone et al. 2001). Also, it was shown that lichochalcone A has anti-HIV properties (Hatano et al. 1988).

Curreli, Friedman-Kien, and Flore (2005) showed that glycyrrhizic acid induced apoptosis in primary effusion lymphoma (PEL) cells that were transformed by Kaposi's sarcoma-associated herpes virus (KSHV), and it terminated latent infection in B lymphocytes (Curreli et al. 2005).

Licochalcone A from Chinese licorice roots is known to possess antiplasmodial activity with IC_{50} values between 4.5 and 0.6 mg/mL (Chen et al. 1994; Jenett-Siems et al. 1999). It has altered the ultrastructure of the parasite mitochondria and inhibited parasitic function by selectively inhibiting fumarate reductase (FRD) in the respiratory chain of the parasite (Zhai et al. 1995; Chen et al. 2001).

Glycyrrhiza glabra was found to have significant antiplasmodial activity and selectivity for *Plasmodium falciparum* and *Plasmodium berghei* in *in vivo* and *in vitro* studies (Esmaeili et al. 2009).

Licochalcone A, B, C, and D and echinatin, isolated from *G. inflate*, were effective in preventing microsomal lipid peroxidation induced by Fe (III)-ADP/NADPH, and licochalcone B and D showed potent antioxidative and superoxide scavenging activities (Haraguchi et al. 1998). Hispaglabridin A showed potent antioxidative activity against peroxidation induced by Fe-ascorbate (Haraguchi 2001). In addition, other constituents of *G. glabra*, such as hispaglabridin A and B, 4'-*O*-methylglabridin, isoprenylchalcone derivative, and isoliquiritigenin, are antioxidants that protect against LDL oxidation (Vaya, Belinky, and Aviram 1997).

Moreover, glabridin, an isoflavan of *G. glabra*, was a potent antioxidant against LDL oxidation in *in vitro* and *in vivo* studies, inhibited lipid peroxidation in rat liver microsomes, and protected mitochondrial functions from oxidative stresses (Fuhrman et al. 1997; Vaya et al. 1997; Belinky et al. 1998; Haraguchi et al. 2000).

Consumption of licorice or glabridin by atherosclerotic apolipoprotein E-deficient (E^0) mice caused a significant reduction in LDL oxidation and in the development of atherosclerotic lesions (Fuhrman et al. 1997; Rosenblat et al. 1999).

In an *in vitro* study, glycyrrhizin was hepatoprotective, probably via the prevention of changes in cell membrane permeability (Nakamura, Fujii, and Ichihara 1985). Nevertheless, an *in vitro* study suggested that glycyrrhetinic acid is a better hepatoprotective drug than glycyrrhizin (Nose et al. 1994). This claim is based on the protective effects that glycyrrhetinic acid has demonstrated against carbon tetrachloride-induced hepatotoxicity and retrorsine-induced liver damage in rodents (Lin, Nnane, and Cheng 1999; Jeong et al. 2002).

Recently, glycyrrhizic acid has also shown protective effects against cytotoxicity induced by *tert*-butyl hydroperoxide (t-BHP) in cultured hepatocytes. Glycyrrhizic acid prevented intracellular GSH depletion and decreased ROS formation (Tripathi, Singh, and Kakkar 2009). Furthermore, in a hepatocyte model of cholestatic liver injury, glycyrrhizin exhibited proapoptotic properties, whereas glycyrrhetinic acid was a potent inhibitor of bile acid-induced apoptosis and necrosis (Gumpricht et al. 2005). In acute liver injury induced by carbon tetrachloride in mice, glycyrrhizin could down-regulate proinflammatory mediators (Lee et al. 2007).

The extract of *G. glabra* in cultured hepatocytes treated by acetaminophen or d-glactosamine increased their survival rate (Nakagiri, Oda, and Kamiya 2003). In addition, a combined regimen of *Salvia miltiorrhiza*, *Ligusticum chuanxiong*, and *G. glabra* (G) exerted antifibrotic effects in rats with dimethylnitrosamine-induced hepatic fibrosis (Lin et al. 2007).

The aqueous extract of licorice root has shown antiproliferative activity in MCF-7 breast cancer cells (Maggiolini et al. 2002; Dong et al. 2007). It inhibited the *in vivo* and *in vitro* proliferation of Ehrlich ascites tumor cells. It also inhibited angiogenesis in *in vivo* assay, as well as in peritoneal and chorioallantoic membrane assays (Sheela, Ramakrishna, and Salimath 2006). Also, the ethanol extract of *G. uralensis* root induced apoptosis and G1 cell cycle arrest in MCF-7 human breast cancer cells (Jo et al. 2005).

Glycyrrhetinic acid may also trigger proapoptotic pathways by changing mitochondrial permeability. This property may be useful for inducing apoptosis of tumor cells (Salvi et al. 2003; Fiore et al. 2004). Inhibitory effect of glycyrrhizic acid on aflatoxin B1-induced cytotoxicity was shown in human HepG2 cells (Chan, Chan, and Ho 2003). Recently, licochalcone E, a new retrochalcone from the roots of *G. inflata*, exhibited the most potent cytotoxic effect compared to the known antitumor agents licochalcone A and isoliquiritigenin (Yoon, Jung, and Cheon 2005).

There are many studies about the anticancer effects of several derivatives of the components of *G. glabra*. Isoliquiritigenin has shown antiproliferative activity in different cancer cell lines (Ma et al. 2001; Maggiolini et al. 2002; Kanazawa et al. 2003; Hsu et al. 2004; Ii et al. 2004; Jung et al. 2006). It also inhibited induction of aberrant crypt foci and colon carcinoma development in azoxymethane-treated ddY mice (Baba et al. 2002; Takahashi et al. 2004). In addition, it induced apoptotic cell death by inhibiting the NF-κB survival-signaling pathway in human hepatoma cell line (Hsu et al. 2005). Isoliquiritigenin reduced pulmonary metastasis in a pulmonary metastasis model of murine renal cell carcinoma cell line (Yamazaki et al. 2002).

Similarly, antitumor activities of lichochalcone A and the induction of apoptosis by modulating bcl-2 protein expression were shown in different cell lines (Rafi et al. 2000, 2002; Fu et al. 2004). Also, glabridin has antiproliferative effects in the human breast cell line (Tamir et al. 2000).

The aqueous extract of *G. glabra* has shown antidepressant effects as demonstrated using the forced-swim test and the tail-suspension test in mice (Dhingra and Sharma 2005). The ethanolic extract of *G. glabra* had anticonvulsant effects in pentylentetrazole (PTZ) and lithium–pilocarpine-induced convulsion models (Ambawade, Kasture, and Kasture 2002). Also, the aqueous extracts of *G. glabra* had anticonvulsant effects in the PTZ model in mice (Nassiri-Asl, Saroukhani, and Zamansoltani 2007).

The aqueous extract of *G. glabra* showed memory-enhancing effects in the plus-maze and passive-avoidance paradigms (Dhingra et al. 2004). Moreover, the chronic administration of *G. glabra* extract, in both low and high doses, improved performance of ovariectomized female rats in the passive-avoidance task (Fedotova, Krauz, and Papkovskaya 2005). Combined treatment with licorice root and vibration increased succinate dehydrogenase activity in different parts of the brain, improved brain energy supply, and ameliorated the effect of vibration (Oganisyan, Oganesyan, and Minasyan 2005). In addition, isoliquiritigenin has shown protective effects in cerebral ischemia-reperfusion injury in rats (Zhan and Yang 2006).

Glabridin inhibited serotonin reuptake (Ofir et al. 2003). It was shown that glabridin ameliorated cerebral injuries induced by middle cerebral artery occlusion (MCAO) in rats and staurosporine-induced damage in cultured cortical neurons of rats. This neuroprotective effect occurred through modulation of multiple pathways associated with apoptosis (Yu et al. 2008).

Carbenoxolone has shown anticonvulsant, sedative, and muscle relaxant properties in mice and in epilepsy-prone rats (Hosseinzadeh and Nassiri Asl 2003; Gareri et al. 2004). It also showed protective effects against acute ischemic reperfusion in skeletal muscle and hippocampi of rats (Hosseinzadeh, Asl, et al. 2005). However, carbenoxolone may decrease the learning performance of rats in a spatial memory task (Hosseinzadeh, Nassiri Asl, and Parvardeh 2005).

Licorice has shown antiplatelet aggregation effects (Yu et al. 2005). Glycyrrhizin has been known as a thrombin inhibitor in *in vitro* and *in vivo* studies and may represent a useful model in searching for new antithrombotic drugs (Francischetti, Monteiro, and Guimaraes 1997; Mendes-Silva et al. 2003). Also, *G. glabra* accelerates the metabolism of erythroid stem cells in bone marrow and increases animal resistance to stress (Adamyan et al. 2005).

Isoliquiritigenin has vasorelaxant properties (Yu and Kuo 1995). On the other hand, similar to the estrogen-like activities of glabridin in *in vivo* and *in vitro* studies, it was demonstrated that it could modulate vascular injury and atherogenesis. Thus, its use is suggested for the prevention of cardiovascular diseases in postmenopausal women (Somjen, Knol, et al. 2004).

In one study, licochalcone A inhibited rat vascular smooth muscle cell (rVSMC) proliferation. It also inhibited platelet-derived growth factor (PDGF)-induced rVSMC proliferation; PDGF-induced expression of cyclin A, cyclin D1, CDK2, and CDK4; and the phosphorylation of Rb, which arrests the cell cycle (Park et al. 2008).

Several immunomodulatory activities have been reported for glycyrrhizin and glycyrrhetinic acid (Kobayashi et al. 1993; Zhang et al. 1993; Kondo and Takano 1994; Raphael and Kuttan 2003). This immunomodulation has also been seen with licochalchone A (Barfod et al. 2002).

Glycyrrhizin selectively activated extrathymic T-cells in the liver and in human T-cell lines and glycyrrhizic acid enhanced Fas-mediated apoptosis without alteration of caspase-3–like activity (Kimura, Watanabe, and Abo 1992; Ishiwata et al. 1999). It improved the resistance to *Candida albicans* infection in mice infected with the LP-BM5 murine leukemia virus (MAIDS), stimulated macrophage-derived NO production, and upregulated iNOS expression through NF-κB transactivation in murine macrophages (Utsunomiya et al. 2000; Jeong and Kim 2002).

Inhibitory effects of glycyrrhizin on tumor necrosis factor (TNF)-α-induced IL-8 production have been shown in intestinal epithelial cells (Kang et al. 2005). In addition, there are several *in vitro* studies on the immunomodulatory effects of polysaccharide fractions obtained from shoots of *G. glabra* and hairy roots of *G. uralensis* (Nose et al. 1998).

Both anticomplementary activity and mitogenic activity have been shown for GR-2IIa and GR-2IIb, two isolated acidic polysaccharides from *G. uralensis* complementary activity (Zhao et al. 1991; Yamada et al. 1992; Kiyohara et al. 1996). The hemolytic activities of *G. uralensis* saponins and its adjuvant potentials against ovalbumin were established in mice (Sun and Pan 2006).

Glabridin showed antinephritic effects and radical scavenging activities in a mouse model of glomerular disease (Fukai et al. 2003). Glycyrrhizin may ameliorate gentamicin-induced acute renal failure in rats (Sohn, Kang, and Lee 2003). Also, the extract of G. radix could protect the kidneys against peroxynitrite (ONOO⁻)-induced oxidative stress by scavenging ONOO⁻ and/or its precursor, NO (Yokozawa et al. 2005).

Inhibitory effects of 69 compounds of *Glycyrrhiza* phenols have been reported on the growth of *Bacillus subtilis* H17 and M45. Some of the phenols, such as isoliquiritigenin, were positive in the rec-assay test (Fukai et al. 1998).

In one study, G. radix had a persistent antitussive effect in guinea pigs. This suggests that liquiritin apioside, the main antitussive component, plays an important role in the earlier phase, while liquiritigenin and liquiritin play an important role in the late phase (Kamei et al. 2005). This result agrees with previous studies showing the antitussive effects of licorice. Isoliquiritigenin relaxed guinea pig trachea in *in vitro* and *in vivo* studies through multiple intracellular actions (Liu et al. 2008).

In a mouse model of asthma, glycyrrhizin inhibited ovalbumin-induced airway constriction, airway hyperreactivity to methacholine, lung inflammation, and infiltration of eosinophils in the peribronchial and perivascular areas (Ram et al. 2006).

Glycyrrhitinic acid has been shown to inhibit gap junction channels. The inhibitory effects of 18-β-glycyrrhetinic acid on gap junction channels of arteriolar smooth muscle and endothelial cells, as well as pelvic, ureter, and mesenteric small renal arteries, have also been studied (Davidson and Baumgarten 1988; Yamamoto et al. 1998; Santicioli and Maggi 2000; Matchkov et al. 2004).

It seems that this herb acts on the metabolism of steroids through different mechanisms: G. glabra inhibited 11-β-hydroxysteroid dehydrogenase (11-β-HSD), 5-α-reductase, and 5-β-reductase (Latif, Conca, and Morris 1990; Fugh-Berman and Ernst 2001). Glycyrrhetinic acid inhibited 17-hydroxysteroid dehydrogenase (17-HSD) and 17-20 lyase (Armanini et al. 2003). Its incubation with microsomal fraction of testicular or ovarian tissue inhibited the conversion of androstenedione to testosterone and this study indicated that it inhibited the activity of 17-β-HSD (Sakamoto and Wakabayashi 1988). It also inhibited microsomal 3-β-HSD enzyme activity, causing a buildup of the intermediate, 5-α-dihydroaldosterone (DHAldo) (Latif et al. 1990).

Glycyrrhize glabra has exhibited estrogen-like activity (Armanini et al. 2002). It was shown that the configuration of these components could enable them to bind to and activate the estrogen receptor (Spignoli 2000). Recently, antiandrogenic activity of G. glabra in male rats was shown (Zamansoltani et al. 2009).

On the other hand, isoliquiritigenin, glabrene, and glabridin are phytoestrogens. Isoliquiritigenin and glabrene can bind to the human estrogen receptor with a higher affinity than glabridin. It seems that isoflavenes may serve as natural estrogen agonists in preventing the symptoms and diseases associated with estrogen deficiency (Tamir et al. 2000, 2001).

Isoliquiritigenin has both spasmogenic and spasmolytic roles that result from different mechanisms in regulating gastrointestinal motility, as shown in an *in vitro* study (Chen et al. 2009).

The growth of mouse osteoblastic (MC3T3-E1) cells and human cells was increased by glabridin (Somjen, Katzburg, et al. 2004; Choi 2005). The alcohol extract of licorice reduced the glucose levels of genetically diabetic KK-Aʸ mice (Kuroda et al. 2003).

Dermatological studies have shown that G. glabra and three flavonoids of licorice (licuraside, isoliquiritin, and licochalcone A) may play an important role as depigmenting agents because they inhibit tyrosinase (Fu et al. 2005; Adhikari et al. 2008). Similar results have been reported about glycyrrhisoflavone and glyasperin C (Kim et al. 2005). Licorice flavonoid oil in diet-induced obese rats decreased abdominal adipose tissue and hepatic and plasma triglycerides (Kamisoyama et al. 2008).

29.6 CLINICAL STUDIES

29.6.1 Gastrointestinal Effects

It was shown that oral licorice could heal ulcers as effectively as H2 blockers (Kassir 1985; Aly, Al-Alousi, and Salem 2005). Glycyrrhizinic acid, a major component of licorice, has antiulcer properties. By raising the local concentration of prostaglandins, which promote mucous secretion and cell proliferation in the stomach, glycyrrhizinic acid has led to healing of ulcers in experimental studies (Van Marle et al. 1981; Baker 1994).

Enoxolone and carbenoxolone, a hemisuccinate derivative of 18-β-glycyrrhetinic acid, are two synthetic derivatives of licorice that have been used in clinical therapy. This derivative has been used for the treatment of peptic ulcer disease and other GIT disorders, as well as skin, mouth, and throat disorders (Sweetman 2005). Carbenoxolone has been used to treat peptic ulcer disease and gastroesophageal reflux. It has also been used as a gel or mouthwash for symptomatic management of mouth ulceration (Sweetman 2005).

29.6.2 Anticancer Effects

Anticancer effects of licorice root have been identified by the National Cancer Institute (Craig 1999; Wang and Nixon 2001). As an ingredient in PC-SPES, a commercially available combination of eight herbs, it was used by patients with prostate cancer (DiPaola et al. 1998).

29.6.3 Antioxidative Effects

In topical formulations, *G. glabra* extract showed great antioxidant and free radical scavenging activity. It may be used to protect skin against free radical and ROS damage (Di Mambro and Fonseca 2005).

29.6.4 Antiviral and Hepatoprotective Effects

Glycyrrhizic acid is used intravenously for the treatment of chronic hepatitis B and C around the world, particularly in Asia. Stronger Neo-Minophagen C® (SNMC), which contains glycyrrhizic acid, was shown to decrease aminotrasferase levels in patients with chronic hepatitis in several double-blind studies (Van Rossum et al. 1999; Iino et al. 2001; Zhang and Wang 2002). It was suggested that glycyrrhizin has protective effects against the development of hepatocellular carcinoma (HCC) in patients with HCV-associated chronic hepatitis (Arase et al. 1997; Miyakawa and Iino 2001).

Licorice has been reported to have a direct hepatoprotective effect (Luper 1999; Leung, Ng, and Ho 2003). Glycyrrhizin is often used to treat patients with chronic liver damage who do not receive or respond to interferon (IFN) therapy (Okuno, Kojima, and Moriwaki 2001). SNMC, which contains 2 mg/mL of glycyrrhizin, has been used clinically as an antihepatitis agent (Shibata 2000).

29.6.5 Dermatological Studies

Glycyrrhiza glabra has been used in herbal medicine for skin eruptions, including dermatitis, eczema, pruritus, and cysts (Saeedi, Morteza-Semnani, and Ghoreishi 2003). Licorice extracts including glabridin have inhibited melanogenesis (Yokota et al. 1998; Petit and Pierard 2003). It was shown that glycyrrhetinic acid was effective in treatment of inflammatory dermatoses (Cohen and Heidary 2004). Deglycyrrhizinated licorice and carbenoxolone have been suggested to be useful for recurrent aphthus stomatitis (RAS) as immunomodulatory agents (Scully, Gorsky, and Lozado-Nur 2002).

Treatment with glycyrrhizin has been shown to be protective against UVB-irradiated melanoma (Rossi et al. 2005). Moreover, licorice extract and its active component, glycyrrhizic acid, have skin-whitening effects (Smith 1999). Licorice extract is used for hyperpigmentation disorders (Halder and Richards 2004).

Briganti, Camera, and Picardo (2003) demonstrated that liquiritin accelerates skin turnover. It was suggested that liquiritin causes depigmentation by two mechanisms. The first mechanism involves melanin dispersion by the pyran ring of the flavonoid nucleus, while the second mechanism involves accelerated epidermal renewal (Amer and Metwalli 2000). Glabridin protects against melanogenesis and inflammation by inhibiting tyrosinase activity of melanocytes. Therefore, it seems that hydroquinone will likely be replaced by licorice extract in a new preparation for dermal melasma (Piamphongsant 1998). However, in several cases, allergic dermatitis has developed in response to oil-soluble licorice extracts (Nishioka and Seguchi 1999).

29.6.6 Endocrinological Effects

Glycyrrhiza root was shown to decrease circulating levels of testosterone in men and women (Armanini, Bonanni, and Palermo 1999; Armanini et al. 2002, 2004; Rafi et al. 2002), but it did not significantly reduce salivary testosterone in men (Josephs et al. 2001). *Glycyrrhiza* also induced regular ovulation and pregnancy in infertile hyperandrogenic patients (Yaginuma et al. 1982).

In some traditional Chinese medicine preparations, the root of *G. glabra* is used to treat menopause-related symptoms. However, there are no clinical data regarding its safety or efficacy in the treatment of hot flashes (Santoro et al. 2004). Consuming licorice extract and glycyrrhetinic acid could decrease body fat mass in humans by inhibiting 11-β-HSD1 in fat cells (Armanini et al. 2005).

29.6.7 Respiratory Diseases

Licorice has been used as a cough-relieving medicinal herb since ancient times. Studies have suggested that the mucilage from licorice coats oral and throat mucosa, thus soothing irritability and relieving dry cough (Ody 2000; Puodziuniene et al. 2005).

29.6.8 Other Effects

The extract of *G. glabra* in combination with other herbs in ImmunoGuard® has been effective in the prophylactic management and treatment of patients with familial Mediterranean fever (FMF) (Amaryan et al. 2003). Double-blind, placebo-controlled pilot studies demonstrated that *Glycyrrhiza* herbal tincture ingested by human subjects stimulates activation and proliferation of various immune cells (Brush et al. 2006; Zwickey et al. 2007).

29.7 SIDE EFFECTS AND TOXICITY

The highest doses of licorice induced adverse effects in healthy subjects (Bernardi et al. 1994). Taking in large amounts of licorice may result in severe hypertension, hypokalemia, and other signs of mineralocorticoid excess. This hypertension is caused by decreased 11-β-HSD2 activity. Renal conversion of cortisol to cortisone is caused by 11-β-HSD2. The activity of 11-β-HSD-2 is blocked potently *in vivo* and *in vitro* by glycyrrhetinic acid (Monder et al. 1989; Ferrari et al. 2001; Palermo, Quinkler, and Stewart 2004). Thus, licorice leads to hyperactivation of renal mineralocorticoid receptors by the increased levels of cortisol, resulting in a state of apparent mineralocorticoid excess and suppression of the renin–angiotensin system (Stewart et al. 1990; Van Uum 2005). Chronic consumption of licorice showed paralysis and severe rhabdomyolysis in one patient (Van Den Bosch et al. 2005). It was reported that licorice is not teratogenic or genotoxic (Carmines, Lemus, and Gaworski 2005). In another study, licorice extract-mediated toxicity was observed in the liver of black molly fish (Radhakrishnan et al. 2005).

Hypokalemia and dysrhythmias were reported in one patient. This patient had ingested about 40–70 g of licorice candy every day for approximately 4 months (Eriiksson, Carlberg, and Hillorn 1999). After ingestion of licorice, transit visual loss has been reported in some cases (Dobbins and Saul 2000; Fraunfelder 2004).

Heavy glycyrrhizin exposure during pregnancy was significantly associated with lower gestational age. However, birth weight and maternal blood pressure did not change significantly (Strandberg et al. 2001).

29.8 PHARMACOKINETICS

After oral administration, glycyrrhizin is metabolized to glycyrrhetinic acid by intestinal bacteria that contain β-d-glucuronidase (Hattori et al. 1985). Intravenously administered glycyrrhizin is metabolized in the liver to 3-mono-glucuronide glycyrrhetinic acid by lysosomal β-d-glucuronidase. This metabolite is excreted with bile into the intestine, where it is metabolized by bacteria into glycyrrhetinic acid, which can then be reabsorbed (Akao et al. 1991).

Other components of the extract could affect the pharmacokinetics of glycyrrhizin and glycyrrhetinic acid (GA), a major metabolite of glycyrrhizin. After administration of aqueous licorice root extract (LE) to rats and humans, glycyrrhizin and GA levels were lower compared to those from glycyrrhizin alone. Also, the pharmacokinetic curves showed significant differences in the areas under the plasma-time curve (AUC), and Cmax and Tmax parameters. Data obtained from urine samples confirmed a reduced bioavailability of glycyrrhizin present in LE compared to pure glycyrrhizin. Interactions between the glycyrrhizin constituent and other components in LE during intestinal absorption were mentioned. Thus, the modified bioavailability could explain the adverse effects of the chronic oral administration of glycyrrhizin alone as opposed to LE (Cantelli-Forti et al. 1994).

However, it seems that the pharmacokinetics differ in other species. In another study, the AUCs of glycyrrhizin and GA after oral administration of LE were significantly higher than AUCs after administration of pure glycyrrhizin in rabbits. The bioavailability of glycyrrhizin and GA in rabbits was significantly better after administration of licorice than administration of pure glycyrrhizin. However, the presystemic metabolism of pure glycyrrhizin in rabbits is rather different from that found in rats, pigs, and humans (Hou et al. 2005). It has been shown that the pharmacokinetics of glycyrrhizin is nonlinear. After bolus intravenous administration at doses of 20, 50, or 100 mg/kg in rats, the decline in plasma glycyrrhizin concentration was generally biexponential at each dose, but the terminal disposition became much slower as the dose was increased.

In addition, the apparent total body clearance decreased significantly with increases in the dose. However, the apparent distribution volume after intravenous administration was unaffected by the dose (Tsai et al. 1992). Oral administration of different doses of 18-β-glycyrrhetinic acid (β-GRA) in healthy volunteers showed a biphasic decay of the plasma concentration-time curve at doses > 500 mg. The peak plasma concentration and the AUC increased with increasing β-GRA doses. Urinary elimination of β-GRA and its glucuronides over 24 h accounted for less than 1% of the dose administered. Data based on a single-dose kinetic analysis revealed that after multiple doses of 1.5 g β-GRA per day, 11-β-HSD was constantly inhibited, whereas at daily doses of 500 mg or less, such an inhibition occurred only transiently (Krahenbuhl et al. 1994).

The intravenous administration of glycyrrhizin in an animal model of liver disease (d-galac-tosamine-intoxicated [GAL] rat) significantly decreased the apparent volume of distribution (Vdss) and the total body clearance (CLtotal) compared to normal rats. When glycyrrhizin was adminis-tered orally, the AUC, mean residence time (MRT) and time to reach the maximum plasma con-centration (Tmax) for glycyrrhizin were higher, but the maximum plasma concentration (Cpmax) in GAL rats was lower than that in normal rats. However, the bioavailability of glycyrrhizin was not significantly changed. Also, the AUC for GA after oral administration of glycyrrhizin was higher in GAL rats than in normal rats, although there was no significant difference in MRT, Tmax, Cpmax, or GA bioavailability. However, the changes in the absorption rate and the reduction in the hepatic elimination rate in GAL rats could explain these differences (Wang et al. 1996). GA has a large volume of distribution and a long biological half-life (Tyler, Bradly, and Robbers 1988).

Using the *in situ* single-pass intestinal perfusion technique, Yang and colleagues (2008) demon-strated that in the absence of intestinal bacteria, glycyrrhizinate does not metabolize to GA. They suggested that the use of pharmaceutical carrier systems may allow effective oral administration of therapeutic levels of glycyrrhizinate without the side effects associated with GA.

In another study, liquiritin apioside showed a peak plasma concentration 15 min after adminis-tration in guinea pigs. The concentration gradually decreased and was almost undetectable 4 h after administration. Liquiritigenin, an aglycone of liquiritin apioside, appeared in the plasma 2 h after liquiritin apioside administration and remained for more than 6 h after administration. The plasma concentration of liquiritigenin remained unchanged for 15 min after administration and then gradu-ally increased for more than 6 h after administration (Kamei et al. 2005).

Glycyrrhizin, genistein, glycyrrhisoflavone, glicoricone, licofuranone, licopyranocoumarin, licocoumarone, and other licorice constituents were found to inhibit monoamine oxidase (MAO) *in vitro* (Hatano, Fukuda, Miyase, et al. 1991). However, the clinical significance of this is not known and these compounds are not found in all licorice species. The extracts of some licorice specimens of types A, B, and C inhibited 40–56% of xanthine oxidase activity. The extracts of some lico-rice specimens of types A and B also showed inhibitory effects on monoamine oxidase (44–64%) (Hatano, Fukada, Liu, et al. 1991).

29.9 DRUG INTERACTIONS

The extract of *G. uralensis* showed potent CYP3A4 inhibitory activity (Hu et al. 1999; Budzinski et al. 2000; Tsukamoto et al. 2005). Other components, such as (3R)-vestitol, 4-hydroxyguaiacol api-oglucoside, liquiritigenin 7,4′-diglucoside, and liquiritin apioside, showed potent CYP3A4 inhibi-tory activity (Tsukamoto et al. 2005). Glabridin was also found to inactivate the enzymatic activities of CYP 3A4 and 2B6 and was competitively inhibited by 2C9 (Kent et al. 2002).

On the other hand, prolonged intake of high doses of LE or glycyrrhizin may result in acceler-ated metabolism of coadministered drugs. Oral doses of LE or glycyrrhizin for 1, 4, or 10 consecu-tive days in mice were able to induce significant hepatic CYP3A- and, to a lesser extent, 2B1- and 1A2-dependent activities, as well as 6-β- (mainly associated with CYP3A), 2-α-, 6-α- (CYP2A1, 2B1), 7-α-, 16-α- (CYP2B9), and 16-β-testosterone hydroxylase (TH) activity. Thus, the induction of cytochrome P450-dependent activities by long-term ingestion of licorice may have clinical con-sequences for patients taking drugs metabolized by the same CYP enzymes (Paolini et al. 1998). Routine licorice consumers may be predisposed to CYP3A induction and the associated adverse effects. Consumption of licorice is contraindicated during pregnancy and for patients with liver disorders or hypokalemia.

The effects of licorice root on aldosterone may counteract antihypertensive actions of pre-scribed medications (Cassileth and Barazzuol 2001). A direct interaction of glycyrrhetinic acid absorption with sennosides and its derivatives has been found in humans (Mizuhara et al. 2005).

Oral administration of *G. glabra* increased the area under the curve (AUC) of prednisolone and decreased the total plasma clearance of it (Chen et al. 1991). It was shown that glycyrrhetinic acid potentiated the action of hydrocortisone in skin (Teelucksingh et al. 1990). It has also increased the excretions of acetaminophen-glucoronide conjugate in rats (Moon and Kim 1996).

Glycyrrhiza uralensis has increased metabolism of warfarin in rats (Mu et al. 2005). Oral contraceptives could increase sensitivity to the intake of glycyrrhizin. Hypertension, edema, and hypokalemia were reported in these women (Bernardi et al. 1994; De Klerk, Nieuwenhuis, and Beutle 1997). Recently, it was demonstrated that oral administration of glycyrrhizin and licorice could increase the AUC and MRT of methotrexate in rats (Lin et al. 2009).

29.10 FUTURE OUTLOOK

Licorice is used throughout the world as a traditional herbal remedy. The beneficial effects of licorice and its active constituents suggest a potential role for this plant in the treatment of different kinds of disease, such as cancer, atherosclerosis, immunodeficiency, hormone deficiency, and viral, skin, and respiratory diseases. However, further studies are necessary to confirm these effects.

REFERENCES

Adamyan, T. I., Gevorkyan, E. S., Minasyan, S. M., Oganesyan, K. R., and Kirakosyan, K. A. 2005. Effect of licorice root on peripheral blood indexes upon vibration exposure. *Bulletin of Experimental Biology and Medicine* 140:197–200.

Adhikari, A., Devkota, H. P., Takano, A., Masuda, K., Nakane, T. Basnet, P., and Skalko-Basnet, N. 2008. Screening of Nepalese crude drugs traditionally used to treat hyperpigmentation: *In vitro* tyrosinase inhibition. *International Journal of Cosmetic Science* 30:353–360.

Afreen, F., Zobayed, S. M. A., and Kozai, T. 2005. Spectral quality and UV-B stress stimulate glycyrrhizin concentration of *Glycyrrhiza uralensis* in hydroponic and pot system. *Plant Physiology and Biochemistry* 43:1074–1081.

Akamatsu, H., Komura, J., Asada, Y., and Niwa, Y. 1991. Mechanism of anti-inflammatory action of glycyrrhizin: Effects on neutrophil functions including reactive oxygen species generation. *Planta Medica* 57:119–121.

Akao, T., Akao, T., Hattori, M., Kanaoka, M., Yamamoto, K., Namaba, T., and Kobashi, K. 1991. Hydrolysis of glycyrrhizin to 18 β-glycyrrhetyl monoglucuronide by lysosomal β-d-glucuronidase of animal livers. *Biochemical Pharmacology* 41:1025–1029.

Aly, A. M., Al-Alousi, L., and Salem, H. A. 2005. Licorice: A possible anti-inflammatory and antiulcer drug. *AAPS Pharm SciTech* 6:E74–E82.

Amagaya, S., Sugishita, E., Ogihara, Y., Ogawa, S., Okada, K., and Aizawa, T. 1984. Comparative studies of the stereoisomers of glycyrrhetinic acid on anti-inflammatory activities. *Journal of Pharmacobiodynamics* 79:923–928.

Amaryan, G., Astvatsatryan, V., Gabrielyan, E., Panossian, A., Panosyan, V., and Wikman, G. 2003. Double-blind, placebo-controlled, randomized, pilot clinical trial of ImmunoGuard—A standardized fixed combination of *Andrographis paniculata* Nees, with *Eleutherococcus senticosus* Maxim, *Schizandra chinensis* Bail. and *Glycyrrhiza glabra* L. extracts in patients with familial Mediterranean fever. *Phytomedicine* 10:271–285.

Ambawade, S. D., Kasture, V. S., and Kasture, S. B. 2002. Anticonvulsant activity of roots and rhizomes of *G. glabra*. *Indian Journal of Pharmacology* 34:251–255.

Amer, M., and Metwalli, M. 2000. Topical liquiritin improves melasma. *International Journal of Dermatology* 39:299–301.

Arase, Y., Ikeda, K., Murashima, N., Chayama, K., Tsubota, A., Koida, I., Suzuki, Y. Saitoh, S., Kobayashi, M., and Kumada, H. 1997. The long-term efficacy of glycyrrhizin in chronic hepatitis C patients. *Cancer* 79:1494–1500.

Armanini, D., Bonannia, G., Mattarello, M. J., Fiore, C., Sartorato, P., and Palermo, M. 2003. Licorice consumption and serum testosterone in healthy man. *Experimental and Clinical Endocrinology and Diabetes* 111:341–343.

Armanini, D., Bonanni, G., and Palermo, M. 1999. Reduction of serum testosterone in men by licorice. *New England Journal of Medicine* 341:1158.

Armanini, D., Fiore, C., Mattarello, M. J., Bielenberg, J., and Palermo, M. 2002. History of the endocrine effects of licorice. *Experimental and Clinical Endocrinology and Diabetes* 110:257–261.

Armanini, D., Mattarello, M. J., Fiore, C., Bonanni, G., Scaroni, C., Sartorato, P., and Palermo, M. 2004. Licorice reduces serum testosterone in healthy women. *Steroids* 69:763–766.

Armanini, D., Nacamulli, D., Francini-Pesenti, F., Battagin, G., Ragazzi, E., and Fiore, C. 2005. Glycyrrhetinic acid, the active principle of licorice, can reduce the thickness of subcutaneous thigh fat through topical application. *Steroids* 70:538–542.

Arseculeratne, S. N., Gunatilaka, A. A. L., and Panabokke, R. G. 1985. Studies on medicinal plants of Srilanka. Part 14. Toxicity of medicinal herbs. *Journal of Ethnopharmacology* 13:323–335.

Azimov, M. M., Zakirov, U. B., and Radzhapova, Sh. D. 1988. Pharmacological study of the anti-inflammatory agent glyderinine. *Farmakologiia i Toksikologiia* 51:90–93.

Baba, M., Asano, R., Takigami, I., Takahashi, T., Ohumura, M., Okada, Y., Sugimoto, H., Arika, T., Nishino, H., and Okuyama, T. 2002. Studies on cancer chemoprevention by traditional folk medicines, XXV. Inhibitory effect of isoliquiritigenin on azoxymethane-induced murine colon aberrant crypt focus formation and carcinogenesis. *Biological Pharmaceutical Bulletin* 25:247–250.

Baba, M., and Shigeta, S. 1987. Antiviral activity of glycyrrhizin against varicella-zoster virus *in vitro*. *Antiviral Research* 7:99–107.

Baker, M. E. 1994. Licorice and enzymes other than 11-β-hydroxysteroid dehydrogenase: An evolutionary perspective. *Steroids* 59:136–141.

Barfod, L., Kemp, K., Hansen, M., and Kharazmi, A. 2002. Chalcones from Chinese licorice inhibit proliferation of T-cells and production of cytokines. *International Immunopharmacology* 2:545–555.

Belinky, P. A., Aviram, M., Furhman, B., Rosenblat, M., and Vaya, J. 1998. The antioxidative effects of the isoflavan glabridin on endogenous constituents of LDL during its oxidation. *Atherosclerosis* 137:49–61.

Bernardi, M., D'Intino, P. E., Trevisani, F., Cantelli-Forti, G., Raggi, M. A., Turchetto, E., and Gasbarrini, G. 1994. Effects of prolonged ingestion of graded doses of licorice by healthy volunteers. *Life Sciences* 55:863–872.

Biondi, D. M., Rocco, C., and Ruberto, G. 2005. Dihydrostilbene derivatives from *Glycyrrhiza glabra* leaves. *Journal of Natural Products* 68:1099–1102.

Blumenthal, M., Goldberg, A., and Brinckmann, J. 2000. *Herbal medicine: Expanded commission E monographs*, 233–236. Austin, TX: American Botanical Council.

Briganti, S., Camera, E., and Picardo, M. 2003. Chemical and instrumental approaches to treat hyperpigmentation. *Pigment Cell Research* 16:101–110.

Brush, J., Mendenhall, E., Guggenheim, A., Chan, T., Connelly, E., Soumyanath, A., Buresh, R., Barrett, R., and Zwickey, H. 2006. The effect of *Echinacea purpurea*, *Astragalus embranaceus* and *Glycyrrhiza glabra* on CD69 expression and immune cell activation in humans. *Phytotherapy Research* 20:687–695.

Budzinski, J. W., Foster, B. C., Vandenhoek, S., and Arnason, J. T. 2000. An *in vitro* evaluation of human cytochrome P450 3A4 inhibition by selected commercial herbal extracts and tinctures. *Phytomedicine* 7:273–282.

Cantelli-Forti, G., Maffei, F., Hrelia, P., Bugamelli, F., Bernardi, M., D'Intino, P., Maranesi, M., and Raggi, M. A. 1994. Interaction of liquorice on glycyrrhizin pharmacokinetics. *Environmental Health Perspectives* 102 (Suppl 9): 65–68.

Capasso, F., Mascolo, N., Autore, G., and Duraccio, M. R. 1983. Glycyrrhetinic acid, leucocytes and prostaglandins. *Journal of Pharmacy and Pharmacology* 35:332–335.

Carmines, E. L., Lemus, R., and Gaworski, C. L. 2005. Toxicologic evaluation of licorice extract as a cigarette ingredient. *Food and Chemistry Toxicology* 43:1303–1322.

Cassileth, B. R., and Barazzuol, J. D. 2001. Herbal products and other supplements problems of special relevance to surgery. *Journal of Pelvic Surgery* 7:21–26.

Chan, H. T., Chan, C., and Ho, J. W. 2003. Inhibition of glycyrrhizic acid on aflatoxin B1-induced cytotoxicity in hepatoma cells. *Toxicology* 188:211–217.

Chen, G., Zhu, L., Liu, Y., Zhou, Q., Chen, H., and Yang, J. 2009. Isoliquiritigenin, a flavonoid from licorice, plays a dual role in regulating gastrointestinal motility *in vitro* and *in vivo*. *Phytotherapy Research*
23:498–506.

Chen, M., Theander, T. G., Christensen, S. B., Hviid, L., Zhai, L., and Kharazmi, A. 1994. Licochalcone A, a
new antimalarial agent, inhibits *in vitro* growth of the human malaria parasite *Plasmodium falciparum*
and protects mice from *P. yoelii* infection. *Antimicrobial Agents and Chemotherapy* 38:1470–1475.

Chen, M., Zhai, L., Christensen, S. B., Theander, T. G., and Kharazmi, A. 2001. Inhibition of fumarate
reductase in *Leishmania* major and *L. donovani* by chalcones. *Antimicrobial Agents and Chemotherapy*
45:2023–2029.

Chen, M. F., Shimada, F., Kato, H., Yano, S., and Kanaoka, M. 1991. Effect of oral administration of glycyrrhizin on the pharmacokinetics of prednisolone. *Endocrinologica Japonica* 38:167–175.

Choi, E. M. 2005. The licorice root derived isoflavan glabridin increases the function of osteoblastic MC3T3-E1
cells. *Biochemical Pharmacology* 70:363–368.

Cinatl, J., Morgenstern, B., Bauer, G., Chandra, P., Rabenau, H., and Doerr, H. W. 2003. Glycyrrhizin, an active
component of licorice roots, and replication of SARS-associated corona virus. *Lancet* 361:2045–2046.

Cohen, D., and Heidary, N. 2004. Treatment of irritant and allergic contact dermatitis. *Dermatologic Therapy*
17:334–340.

Cohen, J. I. 2005. Licking latency with licorice. *Journal of Clinical Investigation* 115:591–593.

Craig, W. J. 1999. Health-promoting properties of common herbs. *American Journal of Clinical Nutrition*
70:491S–499S.

Crance, J. M., Biziagos, E., Passagot, J., Van Cuyck-Gandre, H., and Deloince, R. 1990. Inhibition of hepatitis
A virus replication *in vitro* by antiviral compounds. *Journal of Medical Virology* 31:155–160.

Curreli, F., Friedman-Kien, A. E., and Flore, O. 2005. Glycyrrhizic acid alters Kaposi sarcoma-associated herpes virus latency, triggering p53-mediated apoptosis in transformed B lymphocytes. *Journal of Clinical
Investigation* 115:642–652.

Dafni, A., Yaniv, Z., and Palevitch, D. 1984. Ethnobotanical survey of medicinal plants in northern Israel.
Journal of Ethnopharmacology 10:295–310.

Davidson, J. S., and Baumgarten, I. M. 1988. Glycyrrhetinic acid derivatives: A novel class of inhibitors of
gap-junctional intercellular communication. Structure–activity relationships. *Journal of Pharmacology
and Experimental Therapeutics* 246:1104–1107.

De Klerk, G. J., Nieuwenhuis, M. G., and Beutle, J. J. 1997. Hypokalemia and hypertension associated with use
of licorice flavored chewing gum. *British Medical Journal* 314:731–732.

De Simone, F., Aquino, R., De Tommasi, N., Mahmood, N., Piacente, S., and Pizza, C. 2001. Anti-HIV aromatic compounds from higher plants. In *Bioactive compounds from natural sources: Isolation, characterization and biological properties*, ed. C. Tringali, 325. New York: Taylor & Francis Inc.

Dhingra, D., Parle, M., and Kularni, S. K. 2004. Memory enhancing activity of *Glycyrrhiza glabra* in mice.
Journal of Ethnopharmacology 91:361–365.

Dhingra, D., and Sharma, D. 2006. Antidepressant-like activity of *Glycyrrhiza glabra* L. in mouse models of
immobility tests. *Progress in Neuro-Psychopharmacol and Biological Psychiatry* 30:449–454.

Di Mambro, V. M., and Fonseca, M. J. V. 2005. Assays of physical stability and antioxidant activity of a topical formulation added with different plant extracts. *Journal of Pharmaceutical and Biomedical Analysis* 37:287–295.

DiPaola, R. S., Zhang, H., Lambert, G. H., Meeker, R., Licitra, E., Rafi, M. M., Zhu, B. T., Spaulding, H., Goodin,
S., Toledano, M. B., Hait, W. N., and Gallo, M. A. 1998. Clinical and biologic activity of an estrogenic
herbal combination (PC-SPES) in prostate cancer. *New England Journal of Medicine* 339:785–791.

Dobbins, K. R. B., and Saul, R. F. 2000. Transient visual loss after licorice ingestion. *Journal of Neuro-
ophthalmology* 20:38–41.

Dong, S., Inoue, A., Zhu, Y., Tanji, M., and Kiyama, R. 2007. Activation of rapid signaling pathways and the
subsequent transcriptional regulation for the proliferation of breast cancer MCF-7 cells by treatment with
an extract of *Glycyrrhiza glabra* root. *Food and Chemical Toxicology* 45:2470–2478.

Eriiksson, J. W., Carlberg, B., and Hillorn, V. 1999. Life-threatening ventricular tachycardia due to licorice-
induced hypokalemia. *Journal of Internal Medicine* 245:307–310.

Esmaeili, S., Naghibi, F., Mosaddegh, M., Sahranavard, S., Ghafari, S., and Abdullah, N. R. 2009. Screening of
antiplasmodial properties among some traditionally used Iranian plants. *Journal of Ethnopharmacology*
121:400–404.

Fedotova, Y. U., Krauz, V. A., and Papkovskaya, A. A. 2005. The effect of dry cleared extract from licorice roots on the learning of ovariectomized rats. *Pharmaceutical Chemistry Journal* 39:422–424.

Ferrari, P., Sansonnens, A., Dick, B., and Frey, F. J. 2001. *In vivo* 11β-HSD-2 activity variability, salt-sensitivity, and effect of licorice. *Hypertension* 38:1330–1336.

Fiore, C., Salvi, M., Palermo, M., Sinigagliab, G., Armaninia, D., and Toninello, A. 2004. On the mechanism of mitochondrial permeability transition induction by glycyrrhetinic acid. *Biochimica et Biophysica Acta* 1658:195–201.

Francischetti, I. M., Monteiro, R. Q., and Guimaraes, J. A. 1997. Identification of glycyrrhizin as a thrombin inhibitor. *Biochemical and Biophysical Research Communications* 235:259–263.

Fraunfelder, F. W. 2004. Perspective ocular side effects from herbal medicines and nutritional supplements. *American Journal of Ophthalmology* 138:639–647.

Fu, B., Li, H., Wang, X., Lee, F. S. C., and Cui, S. 2005. Isolation and identification of flavonoids in licorice and a study of their inhibitory effects on tyrosinase. *Journal of Agricultural and Food Chemistry* 53:7408–7414.

Fu, Y., Hsieh, T. C., Guo, J., Kunicki, J., Lee, M. Y. W. T., Darzynkiewicz, Z., and Wu, J. M. 2004. Licochalcone-A, a novel flavonoid isolated from licorice root (*Glycyrrhiza glabra*), causes G2 and late-G1 arrests in androgen-independent PC-3 prostate cancer cells. *Biochemical and Biophysical Research Communication* 322:263–270.

Fugh-Berman, A., and Ernst, E. 2001. Herb–drug interactions: Review and assessment of report reliability. *British Journal of Clinical Pharmacology* 52:587–595.

Fuhrman, B., Buch, S., Vaya, J., Belinky, P. A., Coleman, R., Hayek, T., and Aviram, M. 1997. Licorice extract and its major polyphenol glabridin protect low-density lipoprotein against lipid peroxidation: *In vitro* and *ex vivo* studies in humans and in atherosclerotic apolipoprotein E-deficient mice. *American Journal of Clinical Nutrition* 66:267–275.

Fujita, T., Sezik, E., Tabata, M., Yesilada, E., Honda, G., and Takeda, Y. 1995. Traditional medicine in Turkey VII. Folk medicine in middle and regions. *Taan Economic Botany* 49:406–422. (From NAPRALERT)

Fukai, T., Cai, B.-S., Maruno, K., Miyakawa, Y., Konishi, M., and Nomura, T. 1998. An isoprenylated flavanone from *Glycyrrhiza glabra* and rec-assay of licorice phenols. *Phytochemistry* 49:2005–2013.

Fukai, T., Marumo, A., Kaitou, K., Kanda, T., Terada, S., and Nomura, T. 2002a. Anti-*Helicobacter pylori* flavonoids from licorice extract. *Life Sciences* 71:1449–1463.

———. 2002b. Antimicrobial activity of licorice flavonoids against methicillin-resistant *Staphylococcus aureus*. *Fitoterapia* 73:536–539.

Fukai, T., Satoh, K., Nomura, T., and Sakagami, T. 2003. Preliminary evaluation of antinephritis and radical scavenging activities of glabridin from *Glycyrrhiza glabra*. *Fitoterapia* 74:624–629.

Furuhashi, I., Iwata, S., Shibata, S., Sato, T., and Inoue, H. 2005. Inhibition by licochalcone A, a novel flavonoid isolated from licorice root, of IL-1β-induced PGE$_2$ production in human skin fibroblasts. *Journal of Pharmacy and Pharmacology* 57:1661–1666.

Gareri, P., Condorelli, D., Belluardo, D., Russo, E., Loiacono, A., Barresi, V., Trovato-Salinato, A., Mirone, M. B., Ibbadu, G. F., and De Sarro, G. 2004. Anticonvulsant effects of carbenoxolone in genetically epilepsy prone rats (GEPRs). *Neuropharmacology* 47:1205–1216.

Gray, A. M., and Flatt, P. R. 1997. Nature's own pharmacy: The diabetes perspective. *Proceedings of the Nutrition Society* 56:507–517.

Gumpricht, E., Dahl, R., Devereaux, M. W., and Sokol, R. J. 2005. Licorice compounds glycyrrhizin and 18β-glycyrrhetinic acid are potent modulators of bile acid-induced cytotoxicity in rat hepatocytes. *Journal of Biological Chemistry* 280:10556–10563.

Halder, R. M., and Richards, G. M. 2004. Topical agents used in the management of hyperpigmentation. *Skin Therapy Letter* 9:1–3.

Haraguchi, H. 2001. Antioxidative plant constituents. In *Bioactive compounds from natural sources: Isolation, characterization and biological properties*, ed. C. Tringali, 348–352. New York: Taylor & Francis Inc.

Haraguchi, H., Ishikawa, H., Mizutani, K., Tamura, Y., and Kinoshita, T. 1998. Antioxidative and superoxide scavenging activities of retrochalchones in *Glycyrrhiza inflata*. *Bioorganic and Medical Chemistry* 6:339–347.

Haraguchi, H., Yoshida, N., Ishikawa, H., Tamura, Y., Mizutani, K., and Kinoshita, T. 2000. Protection of mitochondrial functions against oxidative stresses by isoflavans from *Glycyrrhiza glabra*. *Journal of Pharmacy and Pharmacology* 52:219–223.

Hatano, T., Aga, Y., Shintani, Y., Ito, H., Okuda, T., and Yoshida, T. 2000. Minor flavonoids from licorice. *Phytochemistry* 55:959–963.

Hatano, T., Fukuda, T., Liu, Y. Z., Noro, T., and Okuda, T. 1991. Phenolic constituents of licorice. IV. Correlation of phenolic constituents and licorice specimens from various sources, and inhibitory effects of licorice extracts on xanthine oxidase and monoamine oxidase. *Yakugaku Zasshi* 111:311–321.

Hatano, T., Fukuda, T., Miyase, T. T., Noro, T., and Okuda, T. 1991. Phenolic constituents of licorice. III. Structures of glicoricone and licofuranone, and inhibitory effects of licorice constituents on monoamine oxidase. *Chemical and Pharmaceutical Bulletin* 39:1238–1243.

Hatano, T., Kusuda, M., Inada, K., Ogawa, T.-O., Shiota, S., Tsuchiya, T., and Yoshida, T. 2005. Effects of tannins and related polyphenols on methicillin-resistant *Staphylococcus aureus*. *Phytochemistry* 66:2047–2055.

Hatano, T., Shintani, Y., Aga, Y., Shiota, S., Tsuchiya, T., and Yoshida, T. 2000. Phenolic constituents of licorice. VIII. Structures of glicophenone and glicoisoflavanone, and effects of licorice phenolics on methicillin-resistant *Staphylococcus aureus*. *Chemical and Pharmaceutical Bulletin* 48:1286–1292.

Hatano, T., Yasuhara, T., Miyamoto, K., and Okuda, T. 1988. Anti-human immunodeficiency virus phenolics from licorice. *Chemical and Pharmaceutical Bulletin* 36:2286–2288.

Hattori, M., Sakamoto, T., Yamagishi, T., Sakamoto, K., Konishi, K., Kobashi, K., and Namba, T. 1985. Metabolism of glycyrrhizin by human intestinal flora. II. Isolation and characterization of human intestinal bacteria capable of metabolizing glycyrrhizin and related compounds. *Chemical and Pharmaceutical Bulletin* 33:210–217.

Hayashi, H., Hiraoka, N., Ikeshiro, Y., Yamamoto, H., and Yoshikawa, T. 1998. Seasonal variation of glycyrrhizin and isoliquiritigenin glycosides in the root of *Glycyrrhiza glabra* L. *Biological Pharmaceutical Bulletin* 21:987–989.

He, J., Chen, L., Heber, D., Shi, W., and Lu, Q.-Y. 2006. Antibacterial compounds from *Glycyrrhiza uralensis*. *Journal of Natural Products* 69:121–124.

Hosseinzadeh, H., Asl, M. N., Parvardeh, S., and Mansouri, S. M. T. 2005. The effects of carbenoxolone on spatial learning in the Morris water maze task in rats. *Medical Science Monitor* 11:88–94.

Hosseinzadeh, H., and Nassiri Asl. M. 2003. Anticonvulsant, sedative and muscle relaxant effects of carbenoxolone in mice. *BMC Pharmacology* 3:3.

Hosseinzadeh, H., Nassiri Asl, M., and Parvardeh, S. 2005. The effects of carbenoxolone, a semisynthetic derivative of glycyrrhizinic acid, on peripheral and central ischemia-reperfusion injuries in the skeletal muscle and hippocampus of rats. *Phytomedicine* 12:632–637.

Hou, Y. C., Hsiu, S. L., Ching, H., Lin, Y. T., Tsai, S. Y., Wen, K. C., and Chao, P. D. 2005. Profound difference of metabolic pharmacokinetics between pure glycyrrhizin and glycyrrhizin in licorice decoction. *Life Sciences* 76:1167–1176.

Hsu, Y. L., Kuo, P. L., Chiang, L. C., and Lin, C. C. 2004. Isoliquiritigenin inhibits the proliferation and induces the apoptosis of human non-small cell lung cancer A549 cells. *Clinical and Experimental Pharmacology and Physiology* 31:414–418.

Hsu, Y. L., Kuo, P. L., Lin, L. T., and Lin, C. C. 2005. Isoliquiritigenin inhibits cell proliferation and induces apoptosis in human hepatoma cells. *Planta Medica* 71:130–134.

Hu, W. Y., Li, Y. W., Hou, Y. N., He, K., Chen, J. F., But, P. P., and Zhu, X. Y. 1999. The induction of liver microsomal cytochrome P450 by *Glycyrrhiza uralensis* and glycyrrhetinic acid in mice. *Biomedical and Environmental Sciences* 12:10–14.

Ii, T., Satomi, Y., Katoh, D., Shimada, J., Baba, M., Okuyama, T., Nishino, H., and Kitamura, N. 2004. Induction of cell cycle arrest and p21[CIP1/WAF1] expression in human lung cancer cells by isoliquiritigenin. *Cancer Letters* 207:27–35.

Iino, S., Tango, T., Matsushima, T., Toda, G., Miyake, K., Hino, K., Kumada, H., Yasuda, K., Kuroki, T., Hirayama, C., and Suzuki, H. 2001. Therapeutic effects of stronger neo-minophagen C at different doses on chronic hepatitis and liver cirrhosis. *Hepatology Research* 19:31–40.

Inoue, H., Mori, T., Shibata, S., and Koshihara, Y. 1989. Modulation by glycyrrhetinic acid derivatives of TPA-induced mouse ear edema. *British Journal of Pharmacology* 96:204–210.

Ishiwata, S., Nakashita, K., Ozawa, Y., Niizeki, M., Noazki, S., Tomioka, Y., and Mizugaki, M. 1999. Fas-mediated apoptosis is enhanced by glycyrrhizin without alteration of caspase-3-like activity. *Biological Pharmaceutical Bulletin* 22:1163–1166.

Ito, M., Sato, A., Hirabayashi, K., Tanabe, F., Shigeta, S., Baba, M., De Clerq, E., Nakashima, H., and Yamamoto, N. 1988. Mechanism of inhibitory effect of glycyrrhizin on replication of human immunodeficiency virus (HIV). *Antiviral Research* 10:289–298.

Jenett-Siems, K., Mockenhaupt, F. P., Bienzle, U., Gupta, M. P., and Eich, E. 1999. *In vitro* antiplasmodial activity of Central American medicinal plants. *Tropical Medicine and International Health* 4:611–615.

Jeong, H. G., and Kim, J. Y. 2002. Induction of inducible nitric oxide synthase expression by 18β-glycyrrhetinic acid in macrophages. *FEBS Letters* 513:208–212.

Jeong, H. G., You, H. J., Park, S. J., Moon, A. R., Chung, Y. C., Kang, S. K., and Chun, H. K. 2002. Hepatoprotective effects of 18 β-glycyrrhetinic acid on carbon tetrachloride-induced liver injury: Inhibition of cytochrome P450 2E1 expression. *Pharmacological Research* 46:221–227.

Jo, E. H., Kim, S. H., Ra, J. C., Kim, S. R., Cho, S. D., Jung, J. W., Yang, S. R., Park, J. S, Hwang, J. W., Aruoma, O. I., et al. 2005. Chemopreventive properties of the ethanol extract of Chinese licorice (*Glycyrrhiza uralensis*) root: Induction of apoptosis and G1 cell cycle arrest in MCF-7 human breast cancer cells. *Cancer Letters* 230:239–247.

Josephs, R. A., Guinn, S., Harpe, M. L., and Askar, F. 2001. Licorice consumption and salivary testosterone concentrations. *Lancet* 358:1613–1614.

Jung, J. I., Lim, S. S., Choi, H. J., Cho, H. J., Shin, H.-K., Kim, E.-J., Chung, W.-Y., Park, K. K., and Park, J. H. Y. 2006. Isoliquiritigenin induces apoptosis by depolarizing mitochondrial membranes in prostate cancer cells. *Journal of Nutritional Biochemistry* 17:689–696.

Kamei, J., Saitoh, A., Asano, T., Nakamura, R., Ichiki, H., Iiduka, A., and Kubo, M. 2005. Pharmacokinetic and pharmacodynamic profiles of the antitussive principles of *Glycyrrhizae radix* (licorice), a main component of the kampo preparation bakumondo-to (mai-men-dong-tang). *European Journal of Pharmacology* 507:163–168.

Kamisoyama, H., Honda, K., Tominaga, Y., Yokota, S., and Hasegawa, S. 2008. Investigation of the antiobesity action of licorice flavonoid oil in diet-induced obese rats. *Bioscience Biotechnology Biochemistry* 72:3225–3231.

Kanazawa, M., Satomi, Y., Mizutani, Y., Ukimura, O., Kawauchi, A., Sakai, T., Baba, M., Okuyama, T., Nishino, H., and Miki, T. 2003. Isoliquiritigenin inhibits the growth of prostate cancer. *European Urology* 43:580–586.

Kang, O. H., Kim, J. A., Choi, Y. A., Park, H. J., Kim, D. K., An, Y. H., Choi, S. C., Yun, K. J., Nah, Y. H., Cai, X. F., et al. 2005. Inhibition of interleukin-8 production in the human colonic epithelial cell line HT-29 by 18-beta-glycyrrhetinic acid. *International Journal of Molecular Medicine* 15:981–985.

Kang, J. S., Yoon, Y. D., Cho, I. J., Han, M. H., Lee, C. W., Park, S. K., and Kim, H. M. 2005. Glabridin, an isoflavan from licorice root, inhibits inducible nitric-oxide synthase expression and improves survival of mice in experimental model of septic shock. *The Journal of Pharmacology and Experimental Therapeutics* 312: 1187–1194.

Kase, Y., Saitoh, K., Ishige, A., and Komatsu, Y. 1998. Mechanisms by which Hange-Shashin reduces prostaglandin E2 levels. *Biological Pharmaceutical Bulletin* 21:1277–1281.

Kassir, Z. A. 1985. Endoscopic controlled trial of four drug regimens in the treatment of chronic duodenal ulceration. *Irish Medical Journal* 78:153–156.

Kent, U. M., Aviram, M., Rosenbalt, M., and Hollenberg, P. F. 2002. The licorice root derived isoflavin glabridin inhibits the activities of human cytochrome P450S 3A4, 2B6 and 2C9. *Drug Metabolism and Disposition* 30:709–715.

Kim, H. J., Seo, S. H., Lee, B.-G., and Lee, Y. S. 2005. Identification of tyrosinase inhibitors from *Glycyrrhiza uralensis*. *Planta Medica* 71:785–787.

Kimura, M., Watanabe, H., and Abo, T. 1992. Selective activation of extrathymic T-cells in the liver by glycyrrhizin. *Biotherapy* 5:167–176.

Kinoshita, T., Tamura, Y., and Mizutani, K. 2005. The isolation and structure elucidation of minor isoflavonoids from licorice of *Glycyrrhiza glabra* origin. *Chemical and Pharmaceutical Bulletin* 53:847–849.

Kiyohara, H., Takemoto, N., Zhao, J. H., Kawamura, H., and Yamada, H. 1996. Pectic polysaccharides from roots of *Glycyrrhiza uralensis*: Possible contribution of neutral oligosaccharides in the galacturonase-resistant region to anti-complementary and mitogenic activities. *Planta Medica* 62:14–19.

Kobayashi, M., Schmitt, D. A., Utsunomiya, T., Pollard, R. B., and Suzuki, F. 1993. Inhibition of burn-associated suppressor cell generation by glycyrrhizin through the induction of contrasuppressor T-cells. *Immunology and Cell Biology* 71:181–189.

Kondo, K., Shiba, M., Nakamura, R., Morota, T., and Shoyama, Y. 2007. Constituent properties of licorices derived from *Glycyrrhiza uralensis*, *G. glabra*, or *G. inflata* identified by genetic information. *Biological Pharmaceutical Bulletin* 30:1271–1277.

Kondo, K., Shiba, M., Yamaji, H., Morota, T., Zhengmin, C., Huixia, P., and Shoyama, Y. 2007. Species identification of licorice using nrDNA and cpDNA genetic markers. *Biological Pharmaceutical Bulletin* 30:1497–1502.

Kondo, Y., and Takano, F. 1994. Nitric oxide production in mouse peritoneal macrophages enhanced with glycyrrhizin. *Biological Pharmaceutical Bulletin* 17:759–761.

Krahenbuhl, S., Hasler, F., Frey, B. M., Frey, F. J., Brenneisen, R., and Krapf, R. 1994. Kinetics and dynamics of orally administered 18 β-glycyrrhetinic acid in humans. *Journal of Clinical Endocrinology and Metabolism* 78:581–585.

Kroes, B. H., Beukelman, C. J., Van Den Berg, A. J. J., Wolbink, G. J., Van Dijk, H., and Labadie, R. P. 1997. Inhibition of human complement by beta-glycyrrhetinic acid. *Immunology* 90:115–120.

Kuroda, M., Mimaki, Y., Sashida, Y., Mae, T., Kishida, H., Nishiyama, T., Tsukagawa, M., Konishi, E., Takahashi, K., Kawada, T., et al. 2003. Phenolics with PPAR-γ ligand-binding activity obtained from licorice (*Glycyrrhiza uralensis* roots) and ameliorative effects of glycerin on genetically diabetic KK-A^y mice. *Bioorganic and Medicinal Chemistry Letters* 13:4267–4272.

Latif, S. A., Conca, T. J., and Morris, D. J. 1990. The effects of the licorice derivative, glycyrrhetinic acid, on hepatic 3-α- and 3-β-hydroxysteroid dehydrogenases and 5-α- and 5-β-reductase patways of metabolism of aldosterone in male rats. *Steroids* 55:52–58.

Lee, C. H., Park, S. W., Kim, Y. S., Kang, S. S., Kim, J. A., Lee, S. H., and Lee, S. M. 2007. Protective mechanism of glycyrrhizin on acute liver injury induced by carbon tetrachloride in mice. *Biological Pharmaceutical Bulletin* 30:1898–1904.

Leung, Y. K., Ng, T. B., and Ho, J. W. 2003. Transcriptional regulation of fosl-1 by licorice in rat clone 9 cells. *Life Sciences* 73:3109–3121.

Li, J. R., Wang, Y. Q., and Deng, Z. Z. 2005. Two new compounds from *Glycyrrhiza glabra*. *Journal of Asian Natural Products Research* 7:677–680.

Lin, G., Nnane, I. P., and Cheng, T. V. 1999. The effects of pretreatment with glycyrrhizin and glycyrrhetinic acid on the retrorsine-induced hepatotoxicity in rats. *Toxicon* 37:1259–1270.

Lin, J. C. 2003. Mechanism of action of glycyrrhizic acid in inhibition of Epstein–Barr virus replication *in vitro*. *Antiviral Research* 59:41–47.

Lin, S. P., Tsai, S. Y., Hou, Y. C., and Chao, P. D. 2009. Glycyrrhizin and licorice significantly affect the pharmacokinetics of methotrexate in rats. *Journal of Agricultural and Food Chemistry* 57:1854–1859.

Lin, Y. L., Hsu, Y. C., Chiu, Y. T., and Huang, Y. T. 2007. Antifibrotic effects of a herbal combination regimen on hepatic fibrotic rats. *Phytotherapy Research* 22:69–76.

Liu, B., Yang, J., Wen, Q., and Li, Y. 2008. Isoliquiritigenin, a flavonoid from licorice, relaxes guinea-pig tracheal smooth muscle *in vitro* and *in vivo*: Role of cGMP/PKG pathway. *European Journal of Pharmacology* 587:257–266.

Luper, S. 1999. A review of plants used in the treatment of liver disease: Part two. *Alternative Medicine Review* 4:178–188.

Ma, J., Fu, N. Y., Pang, D. B., Wu, W. Y., and Xu, A. L. 2001. Apoptosis induced by isoliquiritigenin in human gastric cancer MGC-803 cells. *Planta Medica* 67:754–757.

Maggiolini, M., Statti, G., Vivacqua, A., Gabriele, S., Rago, V., Loizzo, M., Menichini, F., and Amdo, S. 2002. Estrogenic and antiproliferative activities of isoliquiritigenin in MCF7 breast cancer cells. *Journal of Steroid and Biochemistry and Molecular Biology* 82:315–22.

Matchkov, V. V., Rahman, A., Peng, H., Nilsson, H., and Aalkjaer, C. 2004. Junctional and nonjunctional effects of heptanol and glycyrrhetinic acid derivates in rat mesenteric small arteries. *British Journal of Pharmacology* 142:961–972.

Mendes-Silva, W., Assafim, M., Ruta, B., Monteiro, R. Q., Guimaraes, J. A., and Zingali, R. B. 2003. Antithrombotic effect of glycyrrhizin, a plant-derived thrombin inhibitor. *Thrombosis Research* 112:93–98.

Menegazzi, M., Di Paola, R., Mazzon, E., Genovese, T., Crisafulli, C., Dal Bosco, M., Zou, Z., Suzuki, H., and Cuzzocrea, S. 2008. Glycyrrhizin attenuates the development of carrageenan-induced lung injury in mice. *Pharmacological Research* 58:22–31.

Miyakawa, Y., and Iino, S. 2001. Toward prevention of hepatocellular carcinoma developing in chronic hepatitis C. *Journal of Gastroenterology and Hepatology* 16:711–714.

Mizuhara, Y., Takizawa, Y., Ishihara, K., Asano, T., Kushida, H., Morota, T., Kase, Y., Takeda, S., Aburada, M., Nomura, M., and Yokogawa, K. 2005. The influence of the sennosides on absorption of glycyrrhetic acid in rats. *Biological Pharmaceutical Bulletin* 28:1897–1902.

Monder, C., Stewart, P. M., Lakshmi, V., Valentino, R., Burt, D., and Edwards, C. R. 1989. Licorice inhibits corticosteroid 11β-dehydrogenase of rat kidney and liver: *In vivo* and *in vitro* studies. *Endocrinology* 125:1046–1053.

Moon, A., and Kim, S. H. 1996. Effect of glycyrrhiza glabra roots and glycyrrizin on the glucoronidation in rats. *Planta Medica* 62:115–119.

Mu, Y., Zhang, J., Zhang, S., Zhou, H. H., Toma, D., Ren, S., Haung, L., Yaramus, M., Baum, A., Venkataramanan, R., and Xie, W. 2006. Traditional Chinese medicines wu wei zi (*Schisandra chinensis Baill*) and gan cao (*Glycyrrhiza uralensis* Fisch) activate PXR and increase warfarin clearance in rats. *Journal of Pharmacology and Experimental Therapeutics* 316:1369–1377.

Näf, R., and Jaquier, A. 2006. New lactones in licorice (*Glycyrrhiza glabra* L.). *Flavor and Fragrance Journal* 21:193–197.

Nakagiri, R., Oda, H., and Kamiya, T. 2003. Small-scale rat hepatocyte primary culture with applications for screening hepatoprotective substances. *Bioscience Biotechnology and Biochemistry* 67:1629–1635.

Nakamura, T., Fujii, T., and Ichihara, A. 1985. Enzyme leakage due to change of membrane permeability of primary cultured rat hepatocytes treated with various hepatotoxins and its prevention by glycyrrhizin. *Cell Biology and Toxicology* 1:285–295.

Nassiri-Asl, M., Saroukhani, S., and Zamansoltani, F. 2007. Anticonvulsant effects of *Glycyrrhiza glabra* in PTZ-induced seizure in mice. *International Journal of Pharmacology* 3:432–434.

Nishioka, K., and Seguchi, T. 1999. Contact allergy due to oil soluble licorice extracts in cosmetics products. *Contact Dermatitis* 40:56.

Nisteswar, K., and Murthy, V. K. 1989. Aphrodisiac effect of indigenous drugs—A myth or reality? *Probe* 28:89–92. (From NAPRALERT)

Nomura, T., Fukai, T., and Akiyama, T. 2002. Chemistry of phenolic compounds of licorice (*Glycyrrhiza* species) and their estrogenic and cytotoxic activities. *Pure and Applied Chemistry* 74:1199–1206.

Nose, M., Ito, M., Kamimura, K., Shimizu, M., and Ogihara, Y. 1994. A comparison of the antihepatotoxic activity between glycyrrhizin and glycyrrhetinic acid. *Planta Medica* 60:136–139.

Nose, M., Terawaki, K., Oguri, K., Ogihara, Y., Yoshimatsu, K., and Shimomura, K. 1998. Activation of macrophages by crude polysaccharide fractions obtained from shoots of *Glycyrrhiza glabra* and hairy roots of *Glycyrrhiza uralensis. Biological Pharmaceutical Bulletin* 21:1110–1112.

Numazaki, K., Umetsu, M., and Chiba, S. 1994. Effect of glycyrrhizin in children with liver dysfunction associated with cytomegalovirus infection. *Tohoku Journal of Experimental Medicine* 172:147–153.

Ody, P. 2000. *The complete guide to medicinal herbs*, 75. London: Dorling Kindersley.

Ofir, R., Tamir, S., Khatib, S., and Vaya, J. 2003. Inhibition of serotonin reuptake by licorice constituents. *Journal of Molecular Neuroscience* 20:135–140.

Oganisyan, A. O., Oganesyan, K. R., and Minasyan, S. M. 2005. Changes in succinate dehydrogenase activity in various parts of the brain during combined exposure to vibration and licorice root. *Neuroscience and Behavioral Physiology* 35:545–548.

Okuno, M., Kojima, S., and Moriwaki, H. 2001. Chemoprevention of hepatocellular carcinoma: Concept, progress and perspectives. *Journal of Gastroenterology and Hepatology* 16:1329–1335.

Onkarappa, R., Shobha, K. S., and Chaya, K. 2005. Efficacy of four medicinally important plant extracts (crude) against pathogenic bacteria. *Asian Journal of Microbiology Biotechnology and Environmental Sciences* 7:281–284.

Palermo, M., Quinkler, M., and Stewart, P. M. 2004. Apparent mineralocorticoid excess syndrome: An overview. *Arquivos Brasileiros de Endocrinologia Metabologia* 48:687–696.

Paolini, M., Pozzeti, L., Sapone, A., and Cantelli-Forti, G. 1998. Effect of licorice and glycyrrhizin on murine liver CYP-dependent monooxygenases. *Life Sciences* 62:571–582.

Park, J. H., Lim, H. J., Lee, K. S., Lee, S., Kwak, H. J., Cha, J. H., and Park, H. Y. 2008. Anti-proliferative effect of licochalcone A on vascular smooth muscle cells. *Biological Pharmaceutical Bulletin* 31:1996–2000.

Pellati, D., Fiore, C., Armanini, D., Rassu, M., and Bertoloni, G. 2009. *In vitro* effects of glycyrrhetinic acid on the growth of clinical isolates of *Candida albicans*. *Phototherapy Research* 23:572–574.

Perera, P., Ringbom, T., Huss, U., Vasange, M., and Bohlin, L. 2001. Search for natural products which affect cyclooxygenase-2. In *Bioactive compounds from natural sources: Isolation, characterization and biological properties*, ed. C. Tringali, 485, 462. New York: Taylor & Francis Inc.

Petit, L., and Pierard, E. 2003. Skin-lightening products revisited. *International Journal of Cosmetic Science* 25:109–181.

Piamphongsant, T. 1998. Treatment of melasma: A review with personal experience. *International Journal of Dermatology* 37:897–903.

Pompei, R., Flore, O., Marccialis, M. A., Pani, A., and Loddo, B. 1979. Glycyrrhizic acid inhibits virus growth and inactivates virus particles. *Nature* 281:689–690.

Puodziuniene, G., Janulis, V., Milasius, A., and Budnikas, V. 2005. Development of cough-relieving herbal teas. *Medicina* 41:500–505.

Radhakrishnan, N., Phil, M., Gnanamani, A., and Sadulla, S. 2005. Effect of licorice (*Glycyhrriza glabra* Linn.), a skin-whitening agent on black molly (*Poecilia latipinnaa*). *Journal of Applied Cosmetology* 23:149–158.

Rafi, M. M., Rosen, R. T., Vassil, A., Ho, C. T., Zhang, H., Ghai, G., Lambert, G., and DiPaola, R. S. 2000. Modulation of bcl-2 and cytotoxicity by licochalcone-A, a novel estrogenic flavonoid. *Anticancer Research* 20:2653–2658.

Rafi, M. M., Vastano, B. C., Zhu, N., Ho, C. T., Ghai, G., Rosen, R. T., Gallo, M. A., and DiPaola, R. S. 2002. Novel polyphenol molecule isolated from licorice root (*Glycyrrhiza glabra*) induces apoptosis, G2/M cell cycle arrest, and Bcl-2 phosphorylation in tumor cell lines. *Journal of Agricultural and Food Chemistry* 50:677–684.

Rajurkar, N. S., and Pardeshi, B. M. 1997. Analysis of some herbal plants from India used in the control of mellitus by NAA and AAS techniques. *Applied Radiation and Isotopes* 48:1059–1062.

Ram, A., Mabalirajan, U., Das, M., Bhattacharya, I., Dina, A. K., Gangal, S. V., and Ghosh, B. 2006. Glycyrrhizin alleviates experimental allergic asthma in mice. *International Immunopharmacology* 6:1468–1477.

Raphael, T. J., and Kuttan, G. 2003. Effect of naturally occurring triterpenoids glycyrrhizic acid, ursolic acid, oleanolic acid and nomilin on the immune system. *Phytomedicine* 10:483–489.

Rosenblat, M., Belinky, P., Vaya, J., Levy, R., Hayek, T., Coleman, R., Merchav, S., and Aviram, M. 1999. Macrophage enrichment with the isoflavan glabridin inhibits NADPH oxidase-induced cell-mediated oxidation of low density lipoprotein: A possible role for protein kinase C. *Journal of Biological Chemistry* 274:13790–13799.

Rossi, T., Benassi, L., Magnoni, C., Ruberto, A. I., Coppi, A., and Baggio, G. 2005. Effects of glycyrrhizin on UVB-irradiated melanoma cells. *In Vivo* 19:319–322.

Saeedi, M., Morteza-Semnani, K., and Ghoreishi, M. R. 2003. The treatment of atoptic dermatitis with licorice gel. *Journal of Dermatological Treatment* 14:153–157.

Sakamoto, K., and Wakabayashi, K. 1988. Inhibitory effect of glycyrrhetinic acid on testosterone production in rat gonads. *Endocrinologica Japonica* 35:333–342.

Salvi, M., Fiore, C., Armanini, D., and Toninello, A. 2003. Glycyrrhetinic acid-induced permeability transition in rat liver mitochondria. *Biochemical Pharmacology* 66:2375–2379.

Santicioli, P., and Maggi, C. A. 2000. Effect of 18 β-glycyrrhetinic acid on electromechanical coupling in the guinea-pig renal pelvis and ureter. *British Journal of Pharmacology* 129:163–169.

Santoro, N. F., Clarkson, T. B., Freedman, R. R., Fugh-Berman, A. J., Loprinzi, C. L., and Reame, N. K. 2004. Treatment of menopause-associated vasomotor symptoms: Position statement of menopause. *Menopause* 11:11–33.

Sato, H., Goto, W., Yamamura, J., Kurokawa, M., Kageyama, S., Takahara, T., Watanabe, A., and Shiraki, K. 1996. Therapeutic basis of glycyrrhizin on chronic hepatitis B. *Antiviral Research* 30:171–177.

Schleimer, R. P. 1991. Potential regulation of inflammation in the lung by local metabolism of hydrocortisone. *American Journal of Respiratory Cell and Molecular Biology* 4:166–173.

Scully, C., Gorsky, M., and Lozado-Nur, F. 2002. Aphtus ulcerations. *Dermatology Therapy* 15:185–205.

Sheela, M. L., Ramakrishna, M. K., and Salimath, B. P. 2006. Angiogenic and proliferative effects of the cytokine VEGF in Ehrlich ascites tumor cells is inhibited by *Glycyrrhiza glabra*. *International Immunopharmacology* 6:494–498.

Shibata, S. 2000. A drug over the millennia: Pharmacognosy, chemistry, and pharmacology of licorice. *Yakugaku Zasshi* 120:849–862.

Sircar, N. N. 1984. Pharmaco-therapeutics of dasemani drugs. *Ancient Science Life* 3:132–135. (From NAPRALERT)

Smith, W. P. 1999. The effects of topical L (+) lactic acid and ascorbic acid on skin whitening. *International Journal of Cosmetic Science* 21:33–40.

Sohn, E. J., Kang, D. J., and Lee, H. S. 2003. Protective effects of glycyrrhizin on gentamicin-induced acute renal failure in rats. *Pharmacology and Toxicology* 93:116–122.

Somjen, D., Katzburg, S., Vaya, J., Kaye, A. M., Hendel, D., Posner, G. H., and Tamir, S. 2004. Estrogenic activity of glabridin and glabrene from licorice roots on human osteoblasts and prepubertal rat skeletal tissues. *Journal of Steroid and Biochemistry and Molecular Biology* 91:241–246.

Somjen, D., Knol, E., Vaya, J., Stern, N., and Tamir, S. 2004. Estrogen-like activity of licorice root constituents: Glabridin and glabrene, in vascular tissues *in vitro* and *in vivo*. *Journal of Steroid and Biochemistry and Molecular Biology* 91:147–155.

Spignoli, G. 2000. Protective effects of dietary flavonoids on cardiovascular system and circulation. *European Bulletin of Drug Research* 8:1–8.

Stewart, P. M., Wallace, A. M., Atherden, S. M., Shearin, C. H., and Edwards, C. R. 1990. Mineralocorticoid activity of carbenoxolone: Contrasting effects of carbenoxolone and licorice on 11-β-hydroxysteroid dehydrogenase activity in man. *Clinical Science* 78:49–54.

Strandberg, T. E., Jarvenpaa, A. L., Vanhanen, H., and McKeigue, P. M. 2001. Birth outcome in relation to licorice consumption during pregnancy. *American Journal of Epidemiology* 153:1085–1088.

Sun, H. X., and Pan, H. J. 2006. Immunological adjuvant effect of *Glycyrrhiza uralensis* saponins on the immune responses to ovalbumin in mice. *Vaccine* 24:1914–1920.

Sweetman, S. C. 2005. *Martindale: The complete drug reference*, 1254–1255, 1264. London: Pharmaceutical Press.

Takahara, T., Watanabe, A., and Shiraki, K. 1994. Effects of glycyrrhizin on hepatitis B surface antigen: A biochemical and morphological study. *Journal of Hepatology* 21:601–609.

Takahashi, T., Takasuka, N., Iigo, M., Baba, M., Nishino, H., Tsuda, H., and Okuyama, T. 2004. Isoliquiritigenin, a flavonoid from licorice, reduces prostaglandin E_2 and nitric oxide, causes apoptosis, and suppresses aberrant crypt foci development. *Cancer Science* 95:448–453.

Tamir, S., Eizenberg, M., Somjen, D., Izrael, S., and Vaya, J. 2001. Estrogen-like activity of glabrene and other constituents isolated from licorice root. *Journal of Steroid and Biochemistry and Molecular Biology* 78:291–298.

Tamir, S., Eizenberg, M., Somjen, D., Stern, N., Shelach, R., Kaye, A., and Vaya, J. 2000. Estrogenic and antiproliferative properties of glabridin from licorice in human breast cancer cells. *Cancer Research* 60:5704–5709.

Teelucksingh, S., Mackie, A. D., Burt, D., McIntyre, M. A., Brett, L., and Edwards, C. R. 1990. Potentiation of hydrocortisone activity in skin by glycyrrhetinic acid. *Lancet* 335:1060–1063.

Tokiwa, T., Harada, K., Matsumura, T., and Tukiyama, T. 2004. Oriental medicinal herb, *Periploca sepium*, extract inhibits growth and IL-6 production of human synovial fibroblast-like cells. *Biological Pharmaceutical Bulletin* 27:1691–1693.

Tripathi, M., Singh, B. K., and Kakkar, P. 2009. Glycyrrhizic acid modulates t-BHP induced apoptosis in primary rat hepatocytes. *Food and Chemical Toxicology* 47:339–347.

Tsai, T. H., Liao, J. F., Shum, A. Y., and Chen, C. F. 1992. Pharmacokinetics of glycyrrhizin after intravenous administration to rats. *Journal of Pharmaceutical Sciences* 81:961–963.

Tsukahara, M., Nishino, T., Furuhashi, I., Inoue, H., Sato, T., and Matsumoto, H. 2005. Synthesis and inhibitory effect of novel glycyrrhetinic acid derivatives on IL-1β-induced prostaglandin E_2 production in normal human dermal fibroblasts. *Chemical and Pharmaceutical Bulletin* 53:1103–1110.

Tsukamoto, S., Aburatani, M., Yoshida, T., Yamashita, Y., El-Beih, A. A., and Ohta, T. 2005. CYP3A4 inhibitors isolated from licorice. *Biological Pharmaceutical Bulletin* 28:2000–2002.

Tyler, V. E., Bradly, L. R., and Robbers, J. E. 1988. *Pharmacognosy*, 9th ed., 68–69. Philadelphia: Lea and Febiger.

Utsunomiya, T., Kobayashi, M., Ito, M., Pollard, R. B., and Suzuki, F. 2000. Glycyrrhizin improves the resistance of MAIDS mice to opportunistic infection of *Candida albicans* through the modulation of MAIDS-associated type 2 T-cell responses. *Clinical Immunology* 95:145–155.

Utsunomiya, T., Kobayashi, M., Pollard, R. B., and Suzuki, F. 1997. Glycyrrhizin, an active component of licorice roots, reduces morbidity and mortality of mice infected with lethal doses of influenza virus. *Antimicrobial Agents Chemotherapy* 41:551–556.

Van Den Bosch, A. E., Van der Klooster, J. M., Zuidgeest, D. M. H., Ouwendijk, R. J. Th., and Dees, A. 2005. Severe hypokalemic paralysis and rhabdomyolysis due to ingestion of licorice. *Netherlands Journal of Medicine* 63:146–148.

Van Marle, J., Aarsen, P. N., Lind, A., and Van Weeren-Kramer, J. 1981. Deglycyrrhizinized licorice (DGL) and the renewal of rat stomach epithelium. *European Journal of Pharmacology* 72:219–225.

Van Rossum, T. G. J., Vulto, A. G., De Man, R. A., Brouwer, J. T., and Schalam, S. W. 1998. Review article: Glycyrrhizin as a potential treatment for chronic hepatitis C. *Alimentary and Pharmacological Therapy* 12:199–205.

Van Rossum, T. G. J., Vulto, A. G., Hop, W. C. J., Brouwer, J. T., Niesters, H. G. M., and Schalm, S. W. 1999. Intravenous glycyrrhizin for the treatment of chronic hepatitis C: A double-blind, randomized, placebo-controlled phase I/II trial. *Journal of Gastroenterology and Hepatology* 14:1093–1099.

Van Uum, S. H. 2005. Licorice and hypertension. *Netherlands Journal of Medicine* 63:119–120.

Varshney, I. P., Jain, D. C., and Srivastava, H. C. 1983. Study of saponins from *glycyrrhiza glabra* root. *International Journal of Crude Drug Research* 21:169–172. (From NAPRALERT)

Vaya, J., Belinky, P. A., and Aviram, M. 1997. Antioxidant constituents from licorice roots: Isolation, structure elucidation and antioxidative capacity toward LDL oxidation. *Free Radical Biological Medicine* 23:302–313.

Walker, B. R., and Edwards, C. R. 1991. 11-β-Hydroxysteroid dehydrogenase and enzyme-mediated receptor protection: Life after liquorice? *Clinical Endocrinology* 35:281–289.

Wang, Z., Okamoto, M., Kurosaki, Y., Nakayama, T., and Kimura, T. 1996. Pharmacokinetics of glycyrrhizin in rats with d-galactosamine-induced hepatic disease. *Biological Pharmaceutical Bulletin* 19:901–904.

Wang, Z. Y., and Nixon, D. W. 2001. Licorice and cancer. *Nutrition and Cancer* 39:1–11.

Williamson, E. M. 2003. Licorice. In *Potter's cyclopedia of herbal medicines*, 269–271. Saffron Walden, UK: C. W. Daniels.

Yaginuma, T., Izumi, R., Yasui, H., Arai, T., and Kawabata, M. 1982. Effect of traditional herbal medicine on serum testosterone levels and its induction of regular ovulation in hyperandrogenic and oligomenorrhagic women. *Nippon Sanka Fujinka Gakkai Zhasshi* 34:939–944.

Yamada, H., Kiyohara, H., Takemoto, N., Zhao, J. F., Kawamura, H., Komatsu, Y., Cyong, J. C., Aburada, M., and Hosoya, E. 1992. Mitogenic and complement activating activities of the herbal component of Juzen-Taiho-To. *Planta Medica* 58:166–170.

Yamamoto, Y., Fukuta, H., Nakahira, Y., and Suzuki, H. 1998. Blockade by 18β-glycyrrhetinic acid of intercellular electrical coupling in guinea-pig arterioles. *Journal of Physiology* 511:501–508.

Yamazaki, S., Morita, T., Endo, H., Hamamoto, T., Baba, M., Joichi, Y., Kaneko, S., Okada, Y., Okuyama, T., Nishino, H., and Tokue, A. 2002. Isoliquiritigenin suppresses pulmonary metastasis of mouse renal cell carcinoma. *Cancer Letter* 183:23–30.

Yang, J., Zhou, L., Wang, J., Wang, G., and Davey, A. K. 2008. The disposition of diammonium glycyrrhizinate and glycyrrhetinic acid in the isolated perfused rat intestine and liver. *Planta Medica* 74:1351–1356.

Yarnell, E. 1997. Botanical medicine for cystitis. *Alternative Complementary Therapy* 269–275. (From NAPRALERT)

Yokota, T., Nishio, H., Kubota, Y., and Mizoguchi, M. 1998. The inhibitory effect of glabridin from licorice extracts on melanogenesis and inflammation. *Pigment Cell Research* 11:355–361.

Yokozawa, T., Cho, E. J., Rhyu, D. Y., Shibahara, N., and Aoyagi, K. 2005. *Glycyrrhizae radix* attenuates peroxynitrite-induced renal oxidative damage through inhibition of protein nitration. *Free Radical Research* 39:203–211.

Yoon, G., Jung, Y. D., and Cheon, S. H. 2005. Cytotoxic allyl retrochalcone from the roots of Glycyrrhiza inflata. *Chemical and Pharmaceutical Bulletin* 53:694–695.

Yu, S. M., and Kuo, S. C. 1995. Vasorelaxant effect of isoliquiritigenin, a novel soluble guanylate cyclase activator, in rat aorta. *British Journal of Pharmacology* 114:1587–1594.

Yu, X. Q., Xue, C. C., Zhou, Z. W., Li, C. G., Du, Y. M., Liang, J., and Zhou, S. F. 2008. *In vitro* and *in vivo* neuroprotective effect and mechanisms of glabridin, a major active isoflavan from *Glycyrrhiza glabra* (licorice). *Life Sciences* 82:68–78.

Yu, Z., Ohtaki, Y., Kaid, K., Sasano, T., Shimauchi, H., Yokochie, T., Takada, T., Sugawara, S., Kumagai, K., and Endo, Y. 2005. Critical roles of platelets in lipopolysaccharide-induced lethality: Effects of glycyrrhizin and possible strategy for acute respiratory distress syndrome. *International Immunopharmacology* 5:571–580.

Zakirov, N. U., Aizimov, M. I., and Kurmukov, A. G. 1999. The cardioprotective action of 18-dehydroglycyrrhetic acid in experimental myocardial damage. *Eksperimental'naia i Klincheskaia Farmakologiia* 62:19–21.

Zamansoltani, F., Nassiri-Asl, M., Sarookhani, M. R., Jahani-Hashemi, H., and Zangivand, A. A. 2009. Antiandrogenic activities of *Glycyrrhiza glabra* in male rats. *International Journal of Andrology* 32:417–422.

Zhai, L., Blom, J., Chen, M., Christensen, S. B., and Kharazmi, A. 1995. The antileishmanial agent licochalcone A interferes with the function of parasite mitochondria. *Antimicrobial Agents Chemotherapy* 39:2742–2748.

Zhan, C., and Yang, J. 2006. Protective effects of isoliquiritigenin in transient middle cerebral artery occlusion-induced focal cerebral ischemia in rats. *Pharmacological Research* 53:303–309.

Zhang, L., and Wang, B. 2002. Randomized clinical trial with two doses (100 and 40 mL) of stronger neominophagen C in Chinese patients with chronic hepatitis B. *Hepatological Research* 24:220.

Zhang, Y. H., Isobe, K., Nagase, F., Lwin, T., Kato, M., Hamaguchi, M., Yokochi, T., and Nakashima, I. 1993. Glycyrrhizin as a promoter of the late signal transduction of interleukin-2 production by splenic lymphocytes. *Immunology* 79:528–534.

Zhao, J. F., Kioyahara, H., Yamada, H., Takemoto, N., and Kawamura, H. 1991. Heterogeneity and characterization of mitogenic and anti-complementary pectic polysaccharides from the roots of *Glycyrrhiza uralensis* Fisch et DC. *Carbohydrate Research* 219:149–172.

Zwickey, H., Brush, J., Iacullo, C. M., Connelly, E., Gregory, W. L., Soumyanath, A., and Buresh, R. 2007. The effect of *Echinacea purpurea*, *Astragalus membranaceus* and *Glycyrrhiza glabra* on CD25 expression in humans: A pilot study. *Phytotherapy Research* 21:1109–1112.

CHAPTER 30

Applications of Biotechnology and Molecular Markers in Botanical Drug Standardization and Quality Assurance

Kalpana Joshi, Preeti Chavan-Gautam, and Bhushan Patwardhan

CONTENTS

30.1 INTRODUCTION

Botanicals have a growing global market due to their demand as nutraceutical and herbal products. Herbal products are a very diverse category of plant products and extracts that are known as dietary supplements (United States), natural health products (Canada), phytomedicines (Europe), and traditional medicines (developing countries) (Foster, Arnason, and Briggs 2005). The world market for herbal medicine, including herbal products and raw materials, has been estimated to have an annual growth rate between 5 and 15%. Increased demand for medicinal plants for pharmaceuticals, phytochemicals, nutraceuticals, cosmetics, and other products is an opportunity sector for Indian trade and commerce (Singh, Singh, and Khanuja 2003).

Global trends in favor of botanicals have brought concerns over the quality, safety, and efficacy of these products. Various pharmacopoeias and regulatory agencies on botanicals, such as the WHO Guidelines on Good Agricultural and Collection Practices (GACP); Research and Evaluation of Traditional Medicine, US-FDA; European Scientific Cooperative on Phototherapy (ESCOP); AYUSH (Department of Ayurveda, Yoga, Unani, Siddha and Homeopathy; Government of India); Therapeutic Goods Administration (TGA), Australia; Complementary Healthcare Council of Australia; National Center for Complementary and Alternative Medicine; and Health Canada Natural Health Products Regulations, emphasize the need for establishing and verifying botanical identity of raw materials as a first step in ensuring quality, safety, and efficacy. In a recent review, Rader, Delmonte, and Trucksess (2007) reinforced the need for development of better tools for botanical authentication, extraction, and characterization of bioactive constituents and the detection of hazardous adulterants or contaminants because these have a bearing on quality and safety.

30.2 QUALITY CONTROL OF STARTING MATERIALS

In the context of botanicals, the term "standardization" generally means all the measures taken to ensure reproducible quality. Standardization is defined in the guidance documents published by the American Herbal Products Association: "Standardization refers to the body of information and controls necessary to produce material of reasonable consistency. This is achieved through minimizing the inherent variation of natural product composition through quality assurance practices applied to agricultural and manufacturing processes" (Eisner 2001).

Standardization of botanicals is a complex issue. Unlike modern pharmaceutical products that are usually single or combinations of purified compounds prepared from synthetic materials using reproducible manufacturing procedures, botanicals are usually whole herbs or their formulations or extracts consisting of several bioactive compounds and hence are difficult to standardize (Hon et al. 2003). Consistency in composition and biological activity of botanicals is further limited by problems in identifying plants, genetic variability, variable growing and harvesting conditions, differences in extracts, and lack of information about active pharmacologic principles (Marcus and Grollman 2002). In the absence of authentication as a first quality control step, problems may arise in the final product due to misidentification of the collected plant materials or adulteration or contamination with other species. Thus, authentication of the starting/raw material is important for ensuring reproducible quality, safety, and efficacy.

30.3 PROBLEMS IN QUALITY CONTROL OF STARTING/RAW MATERIALS

Several problems are encountered in authentication of starting material. Often, it is difficult to establish the identity of certain species that may be known by different binomial botanical names in different regions. Shankhapushpi, an important Medhya Rasayan in Ayurveda, is known by different botanical names—*Canscora nisatel*, *Evolvulus alsinoides*, and *Clitoria nisat*—in different regions of India. *Boerhaavia diffusa* is used widely as a quality-of-life enhancer in the traditional system of medicine. However, both *B. diffusa* and the plant *Trianthema portulacastrum* are known as Punarva, so both plants may be collected at the same time. Such wrong identification of medicinal plants could cause adverse effects (Dubey, Kumar, and Tripathi 2004).

Also, morphotaxonomic confusion exists regarding the identity of closely related species. Wild collection of plants for the use in botanicals can also affect quality. Plants collected in the wild may include nontargeted species either by accidental substitution or by intentional adulteration. For example, there is confusion regarding the identity and relationship of *P. amarus* and its close relatives *P. debilis*, *P. urinaria*, and *P. fraternus*, which grow together in the natural habitat. *Mucuna pruriens* is the best example of an unknown authentic plant with similarity in morphology. It is adulterated with other similar Papilionaceae seeds, such as *M. utilis* (sold as white variety) and *M. deeringiana* (sold as bigger variety). Apart from this, *M. cochinchinensis*, *Canavalia virosa*, and *C. ensiformis* are also sold in Indian markets. Substitution of *Periploca sepium* for *Eleutherococcus senticosus* has been widely documented and is regarded as responsible for the "hairy baby" case involving maternal/neonatal and rogenization (Awang 1997). Other examples include substitution of the bark of *Holarrhena antidysentrica* by *Wrightia tictoria* and *Saraca indica* by *Trema orientalis* (Prajapati et al. 2003).

Plantain (*Plantago ovata*) is often found to be contaminated by *Digitalis lanata* (Slifman et al. 1998). Certain rare and expensive medicinal plant species are often adulterated or substituted by morphologically similar, easily available, or less expensive species. For example, *Swertia chirata* is frequently adulterated or substituted by the cheaper *Andrographis paniculata*, and *Capsicum minimum* is substituted by *Capsicum annuum* (Sri Bhava Misra 1999).

In order to establish botanical identity, whole plants in flowering or fruiting stages are preferred. This is possible in the case of plants collected directly from the wild or field grown plants but is difficult in the case of plant materials purchased from the market. Moreover, confidence in botanical identification varies, depending in part on the preservation of the specimen and the level of experience of the individual analyst. Thus, the tools needed for authentication depend on the plant and process involved. These could be as simple as botanical/morphological identification or as elaborate as genetic or chemical profiling (Schilter et al. 2003; Khan 2006) or a combination of both.

30.4 CONVENTIONAL METHODS OF CRUDE DRUG IDENTIFICATION AND THEIR LIMITATIONS

Conventionally, macroscopic and microscopic examination and chemical profiling of crude herbal materials are carried out for authentication and quality control (Indian herbal pharmacopoeia, British herbal pharmacopoeia, WHO quality control methods). Macroscopic identity of medicinal plant materials is based on parameters like shape, size, color, texture, surface characteristics, fracture characteristics, odor, and taste, which are compared to standard reference material. Microscopy involves comparative microscopic inspection of broken as well as powdered test materials with the reference material.

Microscopic and macroscopic analyses are not a definite means of crude drug identification because these parameters are judged subjectively, and closely related species, substitutes, and adulterants may resemble the genuine material. Moreover, other factors, including environment, growth period, and storage conditions of the herbs, may affect the fine structure of the material. In chemical standardization, analysis of one or more specific chemical markers that can easily distinguish varieties is a preferred option. Such metabolites used as markers may or may not be therapeutically active but should ideally be neutral to environmental effects and management practices.

The routinely used chemoprofiling techniques are thin-layer chromatography (TLC) (Cui et al. 2005), high-performance liquid chromatography (HPLC) (Jin et al. 2006; Soares and Scarminio 2008), and gas chromatography (GC) (Di et al. 2004). Chromatographic fingerprints are used as reference standards to indicate the purity, identity, and stability of the herbal drug. Spectroscopic methods such as infrared (IR), nuclear magnetic resonance (NMR), and ultraviolet-visible (UV-vis) may also be used for identification (Ye, Yu, and Li 2005; Holmes et al. 2006; Pereda-Miranda et al. 2006; Chan et al. 2007). In addition, other techniques, such as volumetric analysis and gravimetric determinations, are frequently employed.

Increased sensitivity has been achieved by using hyphenated techniques, coupling HPLC or GC with other analytical systems such as mass spectrometry (MS). These include HPLC-MS, LC-MS, LC-MS-MS, GC-MS, and such (Chan et al. 2000; Weber et al. 2003; Li, Hu, and He 2007; Qian et al. 2007). Recently, capillary electrophoresis has been used to infer botanical sources and to assess the quality of herbal material (Pietta et al. 1994; Pietta, Mauri, and Bauer 1998; Obradovic et al. 2007) and also to detect and quantitate various phytochemicals (Xu et al. 2005). The limitation of using chemoprofiling data for confirming botanical identity is that the chemical composition of a plant varies with environment (geographic location, weather, soil conditions), physiology, genetic predisposition of a plant population toward certain metabolic pathways, plant part used, season, storage, and processing conditions (Joshi, Ranjekar, and Gupta 1999; McChesney 1999).

Moreover, closely related plant species may contain similar components from which a definite botanical identification may not be possible. Further, complications arise if commercial preparations have been "spiked" with other components (e.g., chemicals, pharmaceuticals) to mimic the authentic material or to produce a desired effect. Attempts to standardize crude materials, extracts, or other preparations chemically in the absence of botanical authentication and characterization could be misleading. Although chemical standardization is important, its utility is limited when the starting material is not well characterized botanically.

30.5 ALTERNATIVE APPROACHES TO BOTANICAL IDENTIFICATION

The limitations of routine pharmacognostic techniques have stimulated the search for alternative methods of medicinal plant characterization. Molecular markers such as protein and DNA have been explored for authentication and differentiation of plant species. Protein markers have been applied for species identification (Koga, Shoyama, and Nishioka 1991) and inter- and intraspecies differentiation (Sun et al. 1993; Koren and Zhuravlev 1998). The most commonly used protein markers are isozymes, which are variant forms of the same enzyme (Yu et al. 2005; Yurenkova et al. 2005). However, the use of protein markers is limited because protein patterns vary in different tissues, developmental stages, and environmental conditions and are subject to degradation on long-term storage of herbs. Also, distinguishable markers may not be easily identified in closely related species.

Molecular markers that reveal polymorphisms at the DNA level are called DNA markers. Recently, DNA-based markers have become versatile tools and have found applications in various fields like taxonomy, plant breeding, population genetics, genome mapping, genome evolution, marker-based gene tags, map-based cloning of agronomically important genes, and marker-assisted selection of desirable genotypes. DNA markers reveal genetic differences that are more informative, specific, and reliable than phenotypic differences such as morphology and chemical composition since the genetic composition is unique for each individual and is neutral to physiological and environmental factors. DNA marker technology can be very useful in identifying and differentiating closely related medicinal plant species. Advantages of DNA marker-based techniques are that DNA can be isolated from fresh or dried (Singh, Bandana, and Ahuja 1999; Warude et al. 2003) parts of a plant, such as leaves, stems, and roots, and that a small amount of sample is sufficient for analysis. Further, DNA markers are unaffected by environment. These techniques have been widely used for standardization of Traditional Chinese Medicine (TCM) and their application for quality control of Ayurvedic medicinal plants needs to be exploited.

30.6 TYPES OF DNA MARKERS USED IN PLANT GENETIC VARIATION ANALYSIS

Various types of DNA-based molecular techniques (Prince et al. 1995; Powell et al. 1996) are utilized to evaluate DNA polymorphism; they are generally classified as hybridization-based, polymerase chain reaction (PCR)-based (Mullis et al. 1986), and sequencing-based methods. Hybridization-based methods include restriction fragment length polymorphism (RFLP) (Botstein et al. 1980) and variable number tandem repeats (VNTR) (Nakamura, Julier, et al. 1987; Nakamura, Leppert, et al. 1987). PCR-based techniques include random amplified polymorphic DNA (RAPD) (Welsh and McClelland 1990; Williams et al. 1990), arbitrarily primed PCR (AP-PCR) (Welsh and McClelland 1991), DNA amplification fingerprinting (DAF) (Caetano-Anolles, Bassam, and Gresshof 1991; Caetano-Anolles and Bassam 1993), intersimple sequence repeats (ISSRs) (Zietkiewicz, Rafalski, and Labuda 1994), polymorphisms, and amplified fragment length polymorphism (AFLP) (Zabeau 1993; Vos et al. 1995). Sequencing-based strategies include analysis of single nucleotide polymorphisms (SNPs); polymorphisms in nuclear ribosomal DNA (rDNA) genes (J. S. Lee et al. 2006), such as ITS (internal transcribed spacer) 1, ITS2, and IGS; and plastid genes (Hori et al. 2006), such as *rbc*L, *mat*K, *trn*L–*trn*F, *ndh*F, and *psa*I–*acc*D.

DNA-based techniques have found a wide range of applications in commercially important plants such as food crops, horticultural crops, and, more recently, medicinal plants. The range of applications in medicinal plants include genotyping, assessment of genetic variation among closely related species and varieties, authentication of species, detection of adulteration or substitution (Table 30.1), medicinal plant breeding, and selection of desirable chemotypes (Joshi et al. 2004). Different DNA markers and their applications are reviewed here.

30.6.1 Restriction Fragment Length Polymorphism

The first DNA polymorphisms to be detected were differences in the length of DNA fragments after digestion with sequence-specific restriction endonucleases (i.e., RFLP). In RFLP analysis, restriction enzyme-digested genomic DNA is resolved by gel electrophoresis and then blotted onto a nitrocellulose membrane. The banding patterns are then visualized by hybridization with labeled probes. These probes are mostly species-specific single-locus probes of about 0.5–3.0 kb in size obtained from a cDNA library or a genomic library. RFLPs have their origin in the DNA rearrangements that occur due to evolutionary processes, point mutation within the restriction enzyme recognition site sequences, insertions or deletions within the fragments, and unequal crossing over. Presence and absence of fragments resulting from changes in restriction enzyme recognition sites are used for identifying species or population. RFLP is most suited to studies at the intraspecific level or among closely related taxa.

These markers were used for the first time in the construction of human genetic maps (Williams et al. 1990) and later adopted for plant genome mapping (Weber and Helentjaris 1989; Lind-Hallden, Hallden, and Sall 2002). RFLPs are codominant markers and can detect DNA fragments from all homologous chromosomes. Recently, they have been explored in medicinal plant studies. Some of the examples are cited next.

RFLP has been used to study interspecific genetic variation in *Glycyrrhiza* (Yamazaki et al. 1994), *Lupinus* (Yamazaki et al. 1993), *Epimedium* (Nakai, Shoyama, and Shiraishi 1996), and *Duboisia* (Mizukami, Ohbayashi, Kitamura et al. 1993) species. It has also been used for studying genetic diversity of *Glehnia littoralis* (Mizukami, Ohbayashi, Ohashi et al. 1993) and *Bupleurum falcatum* (Mizukami, Ohbayashi, Kitamura et al. 1993) genotypes. RFLP is less popular due to

Table 30.1   Detection of Adulterants/Contaminants of Some Commercially Important Medicinal Plants Using DNA-Based Molecular Techniques

Medicinal Plant	Adulterant/Contaminant	DNA-Based Method of Detection	Ref.
Alpinia galanga	A. conchigera A. suishaensis A. maclurei, A. polyantha	PCR and sequencing	Zhao et al. (2002)
Angelica sinensis	A. laxifoliata A. nitida.	Sequencing and PCR of rDNA ITS	Feng, Liu, and He (2010)
Crocus sativus	Carthamus tinctorius Hemerocallis fulva Hemerocallis citrina	Sequencing 5S-rRNA spacer domain	Ma et al. (2001)
Curcuma longa	Curcuma zedoaria (wild species)	RAPD	Sasikumar et al. (2005)
Dendrobium species	Nonmedicinal Dendrobium species and Pholidota	Sequencing	Lau et al. (2001)
Echinacea purpurea	Closely related Echinacea species	RAPD	Wolf, Zundorf, et al. (1999)
Hedyotis diffusa	H. corymbosa and other closely related species	Sequencing rDNA ITS	Li et al. (2010)
Ipomoea mauritiana	Pueraria tuberosa, Adenia hondala, Cycas circinalis	RAPD-SCAR	Devaiah, Balasubramani, and Venkatsubramanian (2010)
Panax ginseng	Mirabilis jalapa and Phytolacca acinosa	PCR-RFLP in ribosomal ITS1-5.8S-ITS2 region	Ngan et al. (1999)
Panax notoginseng	P. japonicus, Curcuma phaeocaulis, C. wenyujin, C. kwangsiensis	PCR direct sequencing matK and 18S rRNA genes	Cao, Liu, et al. (2001)
Panax ginseng	Other closely related Panax species	RAPD, AP-PCR	Shaw and Butt (1995)
Phyllanthus emblica	Closely related Phyllanthus species	SCAR	Dnyaneshwar et al. (2006)
Rheum species	Nonmedicinal Rheum species	PCR and sequencing	Yang et al. (2001)
Swertia mussotii	Closely related Swertia species	Sequencing rDNA internal transcribed spacer (ITS)	Xue et al. (2006)
Tribulus terrestris	Tribulus lanuginosus and T. subramanyamii	PCR and sequencing rDNA ITS	Balasubramani et al. (2010)
Zingiber officinale	Closely related Zingiber species	SCAR	Chavan et al. (2008)

certain limitations to its utility. First, the method requires a large amount of DNA for restriction digestion and Southern blotting. Second, the requirement of radioactive isotopes makes the analysis relatively expensive and hazardous. Third, the assay is time consuming and laborious and only one out of several markers may be polymorphic; this is highly inconvenient, especially for crosses between closely related species. PCR-RFLP is a modification of the original RFLP method wherein restriction site polymorphisms within PCR fragments (generated by either random or specific PCR primers) are studied by digesting with different restriction enzymes. This method is less cumbersome because the polymorphic patterns generated are easily analyzed by agarose gel electrophoresis.

30.6.2 Variable Number Tandem Repeats

Variable number tandem repeats loci (Nakamura, Leppert, et al. 1987) contain tandem repeats that vary in the number of repeat units between genotypes. The nucleotide sequence that is repeated is called the core sequence. VNTR loci can have many alleles that differ in length due to different numbers of the core sequence. VNTR loci are classified as microsatellites and minisatellites depending upon the length of the core sequence. In this technique, labeled probes for microsatellites and minisatellites are hybridized to filters containing DNA that has been digested with restriction enzymes.

Analysis of VNTR loci has been used to test genetic stability of *Hypericum perforatum* regenerated after cryopreservation (Urbanova, Kosuth, and Cellarova 2006). Direct amplification of minisatellites-region DNA (DAMD) (Heath, Iwama, and Devlin 1993) is a DNA fingerprinting method based on amplification of the regions rich in minisatellites at relatively high stringencies by using previously found VNTR core sequences as primers. It is employed in identification of species-specific sequences. The method has been used for authentication of *Panax ginseng* (Renshen) and *Panax quinquefolius* (Xiyangshen) (Ha et al. 2002) and characterization of varieties (Bhattacharya and Ranade 2001).

30.6.3 Random Amplified Polymorphic DNA

In this technique, a single 10-base oligonucleotide primer is used to amplify the genomic DNA. The primer binds to many different loci in the complex DNA template arbitrarily (in appropriate orientation). Amplification by PCR generates the products, which are analyzed by electrophoresis. A particular DNA fragment generated for one individual but not for another represents a DNA polymorphism and can be used as a genetic marker. This technique is widely used in crop plant characterization and has been extensively applied for medicinal plant species differentiation, authentication, and detection of adulterants. The RAPD technique is widely used because it is fast and inexpensive and requires no prior DNA sequence information.

RAPD analysis is currently a powerful tool to detect genetic variation. Variations in the RAPD profile among accessions of *Taxus wallichiana* (Shasany et al. 2000), *Azadirachta indica* (Farooqui, Ranade, and Sane 1998), *Codonopsis pilosula* (Fu et al. 1999), *Allium schoenoprasum* (Friesen and Blattner 1999), and *Andrographis paniculata* (Padmesh et al. 1999) have been reported. RAPD markers have been used to discriminate medicinal species *Scutellaria* (Hosokawa et al. 2000), *Melissa* (Wolf, Van Den Berg, et al. 1999), *Dendrobium* (M. Zhang et al. 2001), and *Echinacea* (Nieri et al. 2003). RAPD has been used to study the genetic diversity and relationships of

different *Scutellaria* germplasms (Shao et al. 2006)
farm races of *Morinda officinalis* (Ding, Xu, and Chu 2006)
Chrysanthemum morifolium accessions (Xu, Guo, and Wang 2006)
cultivated and natural populations of *Codonopsis pilosula* and *C. pilosula* var. *modesta* (Zhang et al. 2006)
wild and cultivated populations of *Paeonia lactiflora* (Zhou et al. 2002)
varieties of *Rehmannia glutinosa* (J. L. Cheng et al. 2002)
A. lancea and *A. chinensis* (Guo et al. 2001)
Chinese and Korean medicinal materials of *Niuxi achyranthis* root (Zheng, Guo, and Yan 2002)
Korean and Chinese *Astragali radix* (Na et al. 2004)
Echinacea (Kapteyn, Goldsbrough, and Simon 2002)
Cistanche species (Cui et al. 2004)

Further, RAPD markers have been used to authenticate *Bupleurum chinense* (Wang et al. 2003). Varietal characterization of Kenaf (*Hibiscus cannabinus*) (Z. Cheng et al. 2002) and *Capsicum annuum* using RAPD markers is reported (Hulya 2003). A RAPD primer that is selective for an elite strain, Aizu K-111, of *Panax ginseng* (including its cultured tissues) has been identified (Yukiko, Asaka, and Ichio 2001). Dried fruit samples of *Lycium barbarum* were differentiated from its related species using RAPD markers (K. Y. Zhang et al. 2001). The RAPD technique has also been used for determining the components of a Chinese herbal prescription, yu-ping-feng san. In this study, three herbs—*Astragalus membanaceus*, *Ledebouriella seseloides* and *Atractylodes macrocephala* Koidz—in the formulation have been detected using a single RAPD primer (Cheng et al. 1998).

The RAPD technique has been used to detect adulterations. It has been used to identify cortex *Magnoliae officinalis* (W. Liu et al. 2004) and different *Echinacea* (Wolf, Zundorf, et al. 1999) species and their adulterants. The RAPD fingerprint has been developed to support the chemotypic differences in oil quality of three different genotypes of *Pelargonium graveolens* (Shasany et al. 2002) and flavonoid composition of *Aconitum* species (Fico et al. 2003). An attempt has also been made to study variation in essential oil components and interspecific variations using the RAPD technique (Sangwan et al. 1999).

RAPD has been used to construct genetic linkage maps of *Eucalyptus grandis* and *E. urophylla* (Grattapaglia and Sederoff 1994) as well as for genetic mapping of Pacific yew (*Taxus bravifolia*). Phylogenetic relationships between seven species and three varieties of *Lycium* have been constructed using RAPD (Yin et al. 2005) as well as five medicinally important species of *Typhonium* (Araceae), including *T. venosum*, which was previously placed under the genus *Sauromatum*. RAPD has been used to test the genetic fidelity of the micropropagated medicinal plant *Chlorophytum arundinaceum* (Lattoo et al. 2006) and clones of *Tinospora cordifolia* (Rout 2006).

The advantages of RAPD are that it is a simple, fast technique and no prior sequence information is required. The application of RAPD in genotyping is limited by problems such as poor reproducibility, faint products, nonspecific amplification, and difficulty in scoring bands, which lead to inappropriate inferences. However, these limitations can be overcome by converting them into SCAR (sequence characterized amplified region) markers that are specific and reliable. Some variations of the RAPD technique include DAF, AP-PCR, cleaved amplified polymorphic sequences (CAPS), SCAR, and AFLP. Of these, SCAR and AFLP are the most widely applied due to their specificity and reproducibility.

30.6.4 DNA Amplification Fingerprinting

In this technique developed by Caetano-Anolles et al. (1991), single arbitrary primers as small as five bases are used to amplify DNA by PCR. Band patterns are analyzed using polyacrylamide gel electrophoresis and silver-staining. DAF requires careful optimization but is amenable to automation and fluorescent tagging of primers for quick and easy detection of amplified product. This technique has been useful in genetic typing and mapping studies. DAF has been used to identify Chinese traditional medicine, cortex *Magnolia officinalis* 'Houpo', and its counterfeits and substitutes (T. Wang et al. 2001).

30.6.5 Arbitrarily Primed Polymerase Chain Reaction

In AP-PCR, a single primer, 10–50 bases in length, is used for PCR wherein nonstringent annealing conditions are used in the first two cycles. The final products are structurally similar to RAPD. DNA fingerprinting and polymorphism in the Chinese drug "ku-di-dan" (herba elephantopi) and its substitutes were studied using AP-PCR and RAPD. The results were used for authentication of ku-di-dan and its substitutes (Cao, But, and Shaw 1996). Samples of *Astragalus membranaceus* from different provinces of China, such as Heilongjiang, Neimengu, and Shanxi, were differentiated using AP-PCR (Yip and Kwan 2006). This technique has also been used to authenticate *Panax* species (Shaw and Butt 1995) and *Taraxacum mongolicum* (Cao, But, and Shaw 1997), known as herba taraxaci in Chinese traditional medicine, and their adulterants as well as for identification of American and Oriental ginseng (Cheung et al. 1994) roots.

30.6.6 Cleaved Amplified Polymorphic Sequences

Here, polymorphic patterns are generated by restriction enzyme digestion of PCR products. These digestions are compared for their differential migration during electrophoresis. The PCR primer for this process can be synthesized based on the sequence information available in the data bank of genomic or cDNA sequences or cloned RAPD bands. These markers are codominant in nature. Recently, polymorphic CAPS-, SCAR-, and SNP-derived markers have been used for olive cultivar identification (Reale et al. 2006).

30.6.7 Sequence Characterized Amplified Region Markers

In this technique the RAPD marker termini are sequenced and longer primers (22–24 nucleotide bases) are designed for specific amplification of a particular locus (Paran and Michelmore 1993). The presence or absence of the band on amplification indicates variation in sequence. These markers are more reproducible and specific as compared to RAPD. Like RAPD, SCARs are usually dominant markers but can be made codominant by digestion with restriction enzymes. SCARs are useful in genetic mapping studies (codominant SCARs), map-based cloning, comparative mapping, or homology studies among related species. RAPD markers have been converted into SCAR markers for identification of medicinal plant species of *Panax* (J. Wang et al. 2001), *Echinacea* (Adinolfi et al. 2007), *Artemisia* (M. Y. Lee et al. 2006), *Phyllanthus emblica* (Dnyaneshwar et al. 2006), and *Zingiber officinale* (Chavan et al. 2008).

30.6.7.1 Case Study I: Development of SCAR Markers for Zingiber officinale Roscoe

Zingiber officinale (common or culinary ginger) is an official drug in Ayurvedic, Indian herbal, Chinese, Japanese, African, and British pharmacopoeias. The objective of the study was to develop DNA-based markers that can be applied for the identification and differentiation of the commercially important plant Z. *officinale* from the closely related species Z. *zerumbet* (pinecone, bitter, or shampoo ginger) and Z. *cassumunar* (cassumunar or plai [Thai] ginger). The rhizomes of the other two *Zingiber s*pecies used in the present study are morphologically similar to that of Z. *officinale* and can be used as its adulterants or contaminants.

Various methods, including macroscopy, microscopy, and chemoprofiling, have been reported for the quality control of crude ginger and its products. These methods are reported to have limitations in distinguishing Z. *officinale* from closely related species. Hence, newer complementary methods for correct identification of ginger are useful. In the present study, RAPD analysis was used to identify putative species-specific amplicons for Z. *officinale*. These were further cloned and sequenced to develop SCAR markers, which were tested in several non-*Zingiber* species commonly used in ginger-containing formulations. One of the markers, P3, was found to be specific for Z. *officinale* and was successfully applied for detection of Z. *officinale* from trikatu, a multicomponent formulation (Chavan et al. 2008).

30.6.7.2 Case Study II: Development of SCAR Markers for Phyllanthus emblica Linn.

Phyllanthus emblica (syn: *Emblica officinalis* [Indian gooseberry]; family: Euphorbiaceae) fruit is one of the top-selling botanicals and has diverse applications in the health care, food, and cosmetic industries. It has been well studied for immunomodulatory, anticancer, antioxidant, and antiulcer activities. An official drug of the Ayurvedic pharmacopoeia and Indian herbal pharmacopoeia, it forms a main ingredient of various multicomponent formulations. Correct genotype identification of the plant material therefore remains important for protection of the public health and industry. Chemoprofiling and morphological evaluation are routinely used for identification of the botanical. Chemical complexity and lack of therapeutic markers are some of the limitations associated with a chemical approach and subjective bias in morphological evaluation limits its use.

The present study was carried out to develop a DNA-based marker for identification of *Phyllanthus emblica* A putative marker (1.1 kb) specific for *P. emblica* was identified by random amplified polymorphic DNA (RAPD) technique. The SCAR marker was developed from the RAPD amplicon. It was found useful for identification of *P. emblica* in its commercial samples and triphalachurna, a multicomponent Ayurvedic formulation. The designed SCAR primer pair was used to amplify genomic DNA from the seven *Phyllanthus* species (including 11 *P. emblica* cultivars). A single, distinct, and brightly resolved band of 343 bp was obtained in DNA isolated from all the cultivars and no nonspecific amplification was observed in the other six *Phyllanthus* species (Dnyaneshwar et al. 2006).

30.6.8 Amplified Fragment Length Polymorphism

Amplified fragment length polymorphism is based on the amplification of subsets of genomic restriction fragments using PCR. Genomic DNA is digested simultaneously with a rare cutting and common cutting restriction enzyme. Adaptors are ligated to the ends of restriction fragments followed by amplification with adaptor-homologous primers. This technique is a combination of RFLP and PCR techniques and is useful in detection of polymorphism between closely related genotypes (Primrose and Twyman 2003). AFLP has the capacity to detect thousands of independent loci and can be used for DNA of any origin or complexity (Kumar 1999). The number of amplified fragments is controlled by the cleavage frequency of the rare cutting enzyme and the number of selective bases. Most AFLP fragments correspond to unique positions on the genome and hence can be exploited as landmarks in genetic and physical mapping.

The AFLP approach has been widely adopted by plant geneticists because it requires no previous sequence characterization of the target genome. In higher plants, AFLP may be the most effective way to generate high-density maps. The AFLP markers can also be used to detect corresponding cDNA clones and for fingerprinting of cloned DNA segments. The AFLP technique is reliable since stringent reaction conditions are used for primer annealing. Some examples are described next.

Genetic variation and relationships among *Withania* species (Negi, Singh, and Laksmikumaran 2000), populations of *Pinellia ternate* (Du, Ma, and Li 2006), and landraces of *Panax ginseng* (X. J. Ma et al. 2000) have been studied using AFLP. *Erythroxylum coca*, indigenous to the Andean region of South America, is grown historically as a source of homeopathic medicine. Two subspecies, *E. coca* var. *coca* and *E. coca* var. *ipadu*, are almost indistinguishable phenotypically; a related cocaine-bearing species also has two subspecies (*E. novogranatense* var. *novogranatense* and *E. novogranatense* var. *truxillense*) that are phenotypically similar, but morphologically distinguishable. AFLP analysis using a combination of five primer pairs proved optimal in differentiating the four taxa as well as a non-cocaine-bearing species, *E. aerolatum* (Johnson et al. 2003). AFLP has also been used for authentication of *Plectranthus* species (Passinho-Soares et al. 2006). AFLP markers efficiently and rapidly detecting genetic variations at the species as well as intraspecific level qualifies them as an efficient tool for estimating genetic similarity in plant species and for effective management of genetic resources.

AFLP analysis has also been found to be useful in predicting phytochemical markers in cultivated *Echinacea purpurea* germplasm and some related wild species (Baum et al. 2001). High reproducibility, rapid generation, and high frequency of identifiable polymorphisms make AFLP analysis a suitable technique for identifying polymorphisms; however, this technique is expensive because the bands are detected by silver-staining, fluorescent dye, or radioactivity (Mohan et al. 1997).

30.6.9 Intersimple Sequence Repeats

Intersimple sequence repeats are DNA fragments (100–3000 bp) located between adjacent, oppositely oriented microsatellites, amplified by PCR using microsatellite core sequences and a few selective nucleotides as a single primer (16–18 bp). An unlimited number of primers can be synthesized for various combinations of di-, tri-, tetra-, and pentanucleotides, etc. with an anchor made up of a few bases. About 10–60 fragments from multiple loci are generated simultaneously, separated by gel-electrophoresis, and scored as the presence or absence of fragments of particular sizes.

Because of the multilocus fingerprinting profiles obtained, ISSR analysis can be applied in studies involving genetic identity, parentage, and clone and strain identification. In addition, ISSRs are considered useful in gene mapping studies (Zietkiewicz et al. 1994; Gupta et al. 1994; Godwin, Aitken, and Smith 1997). These are the most dominant markers, though occasionally a few of them exhibit codominance. The initial studies focused on cultivated species, and demonstrated the hypervariable nature of ISSR markers. This technique can be exploited for a broad range of applications in plant species ranging from conservation biology to molecular ecology and systematics. This can be illustrated from the following examples:

Hemp (*Cannabis sativa*) is one of the most widely cultivated plants in the world. Cannabinoids are the active components found in this species. Chemical analysis of cannabinoid and ISSR fingerprinting of DNA were used to identify different samples of *C. sativa* for forensic purposes. ISSR fingerprinting could clearly differentiate between *Cannabis* samples that HPLC could not (Kojoma et al. 2002).

The genetic variation pattern revealed by ISSR markers was closely associated with that indicated by chemical constituents in the fruits of *V. rotundifolia* (Hu et al. 2007).

An ISSR-PCR system for molecular authentication and genetic variation in the *Dendrobium officinale* population has been developed (Shen et al. 2006).

Accessions of *Arabidopsis thaliana* (Barth, Melchinger, and Lubberstedt 2002) have been differentiated using CAPS and ISSR markers.

Recently, 12 different taxa recognized under "chirayat" complex were characterized using ISSR markers. Chirayat represents a group in which different species of *Swertia* and some other non-*Swertia* species are used as local substitutes/adulterants for the *Swertia chirayita* (Tamhankar et al. 2009).

30.6.10 Ribosomal and Plastid DNA-Based Analysis

Ribosomal genes, such as ITS1, ITS2, and IGS, and plastid genes, such as *rbc*L, *mat*K, *trn*L–*trn*F, *ndh*F, and *psa*I–*acc*D, have been extensively studied for detection of DNA sequence polymorphisms for identification of plants and have recently been explored as potential targets for plant DNA bar coding as part of CBOL (Consortium for the Barcode of Life), an international initiative devoted to developing DNA bar coding as a global standard for the identification of biological species.

Ribosomal DNA codes for the RNA component of the ribosome. The rDNA is a multigene family with nuclear copies in eukaryotes arranged in tandem arrays. They are organized in nucleolus organizer regions (NORs), potentially at more than one chromosomal location. Each unit within a single array consists of the genes coding for the small and large rRNA subunits (18S and 28S). The 5.8S nuclear rDNA gene lies embedded between these genes but separated by two internal transcribed spacers: ITS1 and ITS2. The external transcribed spacer (ETS) and the intergenic spacer (IGS) separate the large and small subunit rDNAs. The coding regions show little sequence divergence among closely related species, whereas the spacer regions exhibit higher rates of variability. Thus, the diversity of the spacer domain can be used as a molecular marker for species identification.

Nuclear ribosomal ITS region (including ITS1, ITS2, and 5.8S) sequence analysis has been used to identify and distinguish Fructus Schisandrae Sphenantherae (Gao et al. 2003), *Cannabis sativa* (Siniscalco 1999; Siniscalco, Caputo, and Cozzolino 1997), *Alpinia* species (Zhao et al. 2002), *Hypericum perforatum* (Crockett et al. 2004), *Dendrobium officinale* (Ding, Xu, et al. 2002; Ding, Wang, et al. 2002), *Cordyceps* species (Yue-Qin et al. 2002), *Artemisis iwayomogi* (Kim et al. 2004), and *Hedyotis diffusa* (Z. Liu et al. 2004) from their related species and adulterants. DNA sequence analysis of rDNA ITS and PCR-RFLP allowed effective and reliable differentiation of four medicinal *Codonopsis* species and their related adulterants, *Campanumoea javania* and *Platycodon grandiflorus*. An authentication procedure based upon the RFLP in the ribosomal ITS1-5.8S-ITS2 region is able to differentiate between *P. ginseng* and *P. quinquefolius* and discriminates the ginsengs from two common poisonous adulterants: *Mirabilis jalapa* and *Phytolacca acinosa* (Ngan et al. 1999). Molecular markers based on restriction site variation in the ITS region of nuclear ribosomal RNA genes, it was possible to identify A and B genomes of *Musa* (Nwakanma et al. 2003) and *Cannabis sativa* (Siniscalco 1998).

The 5S-rRNA coding sequence is highly conserved in higher eukaryotes, but the spacer sequence of the 5S-rRNA gene is variable among different species. The nucleotide sequence diversity of 5S-rRNA spacer domains has been used to authenticate and distinguish stigma of *Crocus sativus* and its adulterants *Carthamus tinctorius Hemerocallis fulva* and *Hemerocallis citrina* (Ma et al. 2001); *Angelica* species from China, Japan, and Korea (Zhao et al. 2003a); and *Astragalus* species and its closely related species *Hedysarum polybotrys* (X. Q. Ma et al. 2000). Sequence diversity of a 5S-rNA spacer was used to identify various species of *Fritillaria* (Cai et al. 1999) and also to authenticate Radix Adenophorae (Zhao et al. 2003b).

The most widely used nuclear sequences for phylogeny reconstruction at higher taxonomic levels in plants are 18S-ribosomal RNA genes. Based on 18S-rDNA sequence analysis, commercial ginseng samples were identified as belonging to one of three *Panax* species: *P. ginseng*, *P. japonicus*, and *P. quinquefolius* (Fushimi et al. 1996).

It has been found that 26S-rDNA sequences in seed plants evolve much more rapidly and are phylogenetically more informative as compared to nuclear ribosomal ITS and chloroplast *rbc*L sequences (Kuzoff et al. 1998). Polymorphisms of the D2 and D3 regions inside the 26S-rDNA gene of several *Fritillaria* species were identified by direct sequencing. Oligonucleotide probes specific for these polymorphisms were designed and printed on poly-lysine-coated slides to prepare the DNA chip for identification of different *Fritillaria* species (Tsoi et al. 2003). Recently, PCR markers based on the nuclear ribosomal DNA internal transcribed spacers have been used in finished food products (Urdiain et al. 2004) and dietary supplements (LeRoy et al. 2002) for authentication of additives and botanicals, respectively.

Chloroplast DNA (cpDNA) has been used extensively to infer plant phylogenies at different taxonomic levels. The *rbc*L gene, which encodes the large subunit of ribulose-1,5-bisphosphate carboxylase/oxygenase (RUBISCO), has been widely sequenced from numerous plant taxa. Phylogenies based on *rbc*L sequences were successfully obtained at the family level and also at higher levels. Phylogenetic relationships using *rbc*L sequences have also been inferred at lower taxonomic levels (inter- and intrageneric), indicating that *rbc*L can be used at the generic level.

However, in some instances, many relationships remain unclear. Although this gene works well for placing species to genera, it often does not have enough variation to separate closely related species (Gielly and Taberlet 1994). At present, a combination of *rbc*L and nuclear ribosomal DNA ITS or other noncoding regions of chloroplast genome, such as *trn*L (UAA) introns and the intergenic spacer between the *trn*L(UAA)3' exon and *trn*F (GAA) gene, may be a useful approach. Both are usually easily amplified and in combination will probably permit identification of most species. To identify and confirm the species of American (*Panax ginseng*) and Korean (*Panax quinquefolius*) ginseng samples, direct sequence analysis of the nuclear ribosomal ITS region and a portion of the chloroplast *rbc*L gene was performed. Sequence analysis of the *rbc*L gene was found to be less informative as compared to the ITS region (Mihalov, Marderosian, and Pierce 2000).

Coding regions of cpDNA such as *trn*K and *mat*K are frequently used to identify and distinguish between closely related species at the molecular level. Molecular authentication of *Atractylodes*-derived crude drugs (jutsu) was established based on PCR restriction fragment length polymorphism (RFLP) and direct sequencing of chloroplast *trn*K (Mizukami et al. 2000). The entire chloroplast *trn*K gene was sequenced for six medicinal *Curcuma* species from Sichuan Province of China. Sequence variability observed in these regions was used in identification of the *Curcuma* species at the DNA level (Cao and Komatsu 2003). DNA analysis of the *mat*K gene region of *R. palmatum*, *R. tanguticum*, *R. officinale*, and six other *Rheum* species was performed to clarify their phylogenetic relationship and further to develop a correct identification method for plants and drugs (Yang et al. 2004).

Several noncoding plastid regions (introns and intergenic spacers) appear to work well in some groups but are not variable enough in others to serve the purpose. Analysis of noncoding regions of cpDNA could extend the utility of the molecule at lower taxonomic levels. These zones tend to evolve more rapidly than do coding sequences, by the accumulation of insertions/deletions at a rate at least equal to that for nucleotide substitutions, and therefore can become very useful below the family level. Through sequencing, the *trn*L (UAA)/*trn*F (GAA) regions of chloroplast DNA of 13 species of *Rheum* (three medicinal rhubarb species and 10 adulterant ones), a molecular marker of the medicinal species was found (Yang et al. 2001). Intraspecific variation in *Cannabis sativa* was observed based on the intergenic spacer region of chloroplast DNA (Makino, Sekita, and Satake 2000). Based on sequencing of the intergenic spacer region between the *trn*L 3' exon and *trn*F exon (*trn*L–*trn*F IGS) and the *trn*L intron region, it was possible to identify and classify *Cinnamomum* species. Using a combination of coding and noncoding loci, the molecular phylogeny of *Nicotiana* species (Solanaceae) (Clarkson et al. 2004) has been inferred.

A potentially successful system for applying DNA bar codes to flowering plants is to use a combination of two or more genetic loci from the nuclear rDNA and the cpDNA. PCR direct sequencing was applied to determine the *mat*K and 18S rRNA sequences for six samples of *Pogostemon cablin* from different localities (Liu et al. 2002). Sequence variability of chloroplast *mat*K and 18S rRNA genes was used to identify *Panax notoginseng* and its adulterants (Fushimi et al. 2000; Cao, Liu, et al. 2001). In order to define the taxonomic position of a new *Panax* species discovered in Vietnam and include it in the molecular authentication of ginseng drugs, the 18S ribosomal RNA gene and *mat*K gene sequences of *P. vietnamensis* were determined and compared with previously reported *Panax* species (Komatsu et al. 2001). The botanical origins of Chinese and Japanese curcuma drugs were determined based on a comparison of their 18S-rRNA gene and *trn*K gene sequences (Cao, Sasaki, et al. 2001; Sasaki et al. 2002). The nuclear ITS region and the chloroplast intergenic spacer have been successfully used to study the phylogeny of two Chinese medicinal species: *Panax* (Lee and Wen 2004) and *Ephedra* (Long et al. 2004).

30.7 DNA MICROARRAYS IN PHARMACOGNOSY

Recently, microarrays have been applied for the DNA sequence-based identification of medicinal plants. To utilize DNA microarrays for identification and authentication of herbal material, it is necessary to identify a distinct DNA sequence that is unique to each species of medicinal plant. The DNA sequence information is then used to synthesize a corresponding probe on a silicon-based gene chip. These probes are capable of detecting complementary target DNA sequences if present in the test sample being analyzed. This DNA chip technology can provide a rapid, high-throughput tool for genotyping and plant species authentication (Tsoi et al. 2003). Recent advances in lab-on-chip technology have further enhanced the feasibility of automated, miniaturized, fast, and sensitive genetic authentication of species (Beer et al. 2007; Consolandi et al. 2006; Spaniolas et al. 2006; Qin et al. 2005).

Oligonucleotide probes specific for polymorphisms in the D2 and D3 regions of the 26S rDNA gene of several *Fritillaria* species were designed and printed on poly-lysine-coated slides to prepare a DNA chip. Differentiation of the various *Fritillaria* species was accomplished based on hybridization of fluorescence-labeled PCR products with the DNA chip. The results demonstrated the reliability of using DNA chips to identify different species of *Fritillaria* and that the DNA chip technology can provide a rapid, high-throughput tool for genotyping and plant species authentication (Tsoi et al. 2003). Similarly, using fluorescence-labeled ITS2 sequences as probes, distinctive signals were obtained for the five medicinal *Dendrobium* species listed in the Chinese pharmacopoeia.

The established microarray was able to detect the presence of *D. nobile* in a Chinese medicinal formulation containing nine herbal components (Zhang et al. 2003). A silicon-based DNA microarray using species-specific oligonucleotide probes is designed and fabricated to identify multiple toxic traditional Chinese medicinal plant species by parallel genotyping (Carles et al. 2005). Recently, a DNA microarray (PNX array) was developed for the identification of various *Panax* plants and drugs. The developed PNX array provided an objective and reliable method for the authentication of *Panax* plants and drugs as well as their derived health foods (Zhu, Fushimi, and Komatsu 2008).

Chip-based authentication of medicinal plants is a precise tool for quality control and safety monitoring of herbal pharmaceuticals and neutraceuticals and will significantly add to the medical potential and commercial profitability of herbal products. DNA microarray-based technology can provide an efficient, accurate, and quick means of testing the authenticity of hundreds of samples simultaneously; conventional chemical methodologies usually take several days for verification. This application of DNA microarrays will not only benefit the herbal drug industry but can also facilitate the identification of herbal products by regulatory authorities. An international initiative, the CBOL, is devoted to developing DNA bar coding as a tool for correct taxonomic identification of species, including medicinal plants. DNA microarrays can provide a suitable technology platform for such an initiative (Chase and Fay 2009; Thomas 2009). DNA microarrays have emerged as a powerful tool with application in pharmacodynamics, pharmacogenomics, toxicogenomics, and quality control of herbal drugs and extracts (Chavan, Joshi, and Patwardhan 2006).

30.8 CONCLUSION

Medicinal plant materials are generally processed for use as drugs, causing many morphological and anatomical characteristics as well as some chemical constituents to change and making it difficult to determine botanical origin based on anatomical and chemotaxonomical studies. In view of the limitations of current methods of quality control and standardization, DNA analysis can be suitably used for correct identification of medicinal plant species, especially for those where identity cannot be reliably established using conventional techniques. Chinese researchers have extensively applied the DNA marker techniques for authentication of herbs from traditional Chinese medicine. They have also been successful in applying this technique for standardization of semiprocessed herbal formulations.

Recently, there has been an increase in popularity of Ayurvedic medicine worldwide. In order to boost this growth, there is a need for stringent quality control measures. Several efforts have been made to assess the genetic diversity by phylogenetic analysis of Ayurveda-based medicinal plants. Although phylogenetic analysis is important in taxonomic classification of related species and assessment of diversity, DNA markers that can identify a species in a simple PCR reaction are required. There is a need for efforts toward development of DNA-based markers for identification of Ayurveda-based medicinal plants from crude, semiprocessed and processed materials. Such markers, along with conventional analytical techniques, can be successfully used for standardization of Ayurvedic medicine and will have industrial application.

This chapter brings to the fore the widespread application of DNA marker techniques in plant species identification and discrimination. A large number of DNA-based techniques have been explored and successfully applied. RAPD is one of the most extensively used techniques; however, very few have been converted into SCAR markers, which are comparatively more specific and reproducible. AFLP is an efficient tool for genotyping but is costly and cumbersome. Ribosomal and plastid DNA-based analyses have also been widely explored to study genetic variations and infer phylogenetic relationships among species. This type of analysis requires extensive sequencing of genotypes to reveal polymorphisms at specific loci that can then be used for species discrimination.

Although DNA analysis is currently considered to be cutting-edge technology, it has certain limitations due to which its use has been limited to academia. In order to establish a marker for identification of a particular species, DNA analysis of closely related species and/or varieties and common botanical contaminants and adulterants is necessary—a costly, time-consuming process. However, it is not impossible. Databases of DNA fingerprints and DNA sequences for a broad spectrum of plant species can be created and will be useful. Recent advances in DNA chip technology have made automated, high-throughout and sensitive genetic authentication of species possible.

Isolation of good-quality DNA suitable for analysis from semiprocessed or processed botanicals is also a challenge. Another important issue is that the DNA fingerprint will remain the same irrespective of the plant part used while the phytochemical content will vary with the plant part used, physiology, and environment. DNA fingerprinting ensures the presence of correct genotype but does not reveal content of the active principle or chemical constituents. Hence, DNA analysis and routine chemoprofiling techniques will have to be used hand in hand rather than in isolation.

Identification of quantitative trait loci (QTL) that are closely linked to a biologically active phytochemical will prove to be very useful. Several attempts have been made in recent years to correlate DNA markers with qualitative and quantitative variations in phytochemical composition among closely related species. Proper integration of molecular techniques and analytical tools will lead to the development of a comprehensive system of botanical characterization that can be conveniently applied at the industry level for quality control of botanicals.

REFERENCES

Adinolfi, B., Chicca, A., Martinotti, E., Breschi, M. C., and Nieri, P. 2007. Sequence characterized amplified region (SCAR) analysis on DNA from the three medicinal *Echinacea* species. *Fitoterapia* 78 (1): 43–45.

Awang, D. V. C. 1997. Quality control and good manufacturing practices: Safety and efficacy of commercial herbs. *Food and Drug Law Journal* 52:341–344.

Balasubramani, P. S., Murugan, R., Ravikumar, K., and Venkatasubramanian, P. 2010. Development of ITS sequence based molecular marker to distinguish, *Tribulus terrestris* L. (Zygophyllaceae) from its adulterants. *Fitoterapia* 81 (6): 503–508.

Barth, S., Melchinger, A. E., and Lubberstedt, T. 2002. Genetic diversity in *Arabidopsis thaliana* L. Heynh. investigated by cleaved amplified polymorphic sequence (CAPS) and intersimple sequence repeat (ISSR) markers. *Molecular Ecology* 11 (3): 495–505.

Baum, B. R., Mechanda, S., Livesey, J. F., Binns, S. E., and Arnason, J. T. 2001. Predicting quantitative phytochemical markers in single *Echinacea* plants or clones from their DNA fingerprints. *Phytochemistry* 56:543–549.

Beer, N. R., Hindson, B. J., Wheeler, E. K., Hall, S. B., Rose, K. A., Kennedy, I. M., and Colston, B. W. 2007. On-chip, real-time, single-copy polymerase chain reaction in picoliter droplets. *Analytical Chemistry* 79 (22): 8471–8475.

Bhattacharya, E., and Ranade, S. A. 2001. Molecular distinction amongst varieties of mulberry using RAPD and DAMD profiles. *BMC Plant Biology* 1:3.

Botstein, B., White, R. L., Skolnick, M., and Davis, R. W. 1980. Construction of a genetic linkage map in man using restriction fragment length polymorphisms. *American Journal of Human Genetics* 32:314–331.

Caetano-Anolles, G., and Bassam, B. J. 1993. DNA amplification fingerprinting using arbitrary oligonucleotide primers. *Applied Biochemistry and Biotechnology* 42 (2–3): 189–200.

Caetano-Anolles, G., Bassam, B. J., and Gresshof, P. M. 1991. High-resolution DNA amplification fingerprinting using very short arbitrary oligonucleotide primers. *Biotechnology* 9:553–557.

Cai, Z. H., Li, P., Dong, T. T. X., and Tsim, K. W. K. 1999. Molecular diversity of 5S-rRNA spacer domain in *Fritillaria* species revealed by PCR analysis. *Planta Medica* 65:360–364.

Cao, H., But, P. P., and Shaw, P. C. 1996. Authentication of the Chinese drug "ku-di-dan" (herba elephantopi) and its substitutes using random-primed polymerase. *Acta Pharmaceutica Sinica* 31 (7): 543–553.

———. 1997. Identification of herba taraxaci and its adulterants in Hong Kong market by DNA fingerprinting with random primed PCR. *China Journal of Chinese Materia Medica* 22 (4): 197–200.

Cao, H., and Komatsu, K. 2003. Molecular identification of six medicinal *Curcuma* plants produced in Sichuan: Evidence from plastid *trn*K gene sequences. *Acta Pharmaceutica Sinica* 38 (11): 871–875.

Cao, H., Liu, Y., Fushimi, H., and Komatsu, K. 2001. Identification of notoginseng (*Panax notoginseng*) and its adulterants using DNA sequencing. *Journal of Chinese Medicinal Materials* 24 (6): 398–402.

Cao, H., Sasaki, Y., Fushimi, H., and Komatsu, K. 2001. Molecular analysis of medicinally used Chinese and Japanese *Curcuma* based on 18S rRNA gene and trnK gene sequences. *Biological & Pharmaceutical Bulletin* 24 (12): 1389–1394.

Carles, M., Cheung, M. K., Moganti, S., Dong, T. T., Tsim, K. W., Ip, N. Y., et al. 2005. A DNA microarray for the authentication of toxic traditional Chinese medicinal plants. *Planta Medica* 71:580–584.

Chan, C. O., Chu, C. C., Mok, D. K., and Chau, F. T. 2007. Analysis of berberine and total alkaloid content in cortex phellodendri by near infrared spectroscopy (NIRS) compared with high-performance liquid chromatography coupled with ultra-visible spectrometric detection. *Analytica Chimica Acta* 592 (2): 121–131.

Chan, T. W., But, P. P., Cheng, S. W., Kwok, I. M., Lau, F. W., and Xu, H. X. 2000. Differentiation and authentication of *Panax ginseng*, *Panax quinquefolius*, and ginseng products by using HPLC/MS. *Analytical Chemistry* 72 (6): 1281–1287.

Chavan, P., Joshi, K., and Patwardhan, B. 2006. DNA microarrays in herbal drug research. *Evidence Based Complementary and Alternative Medicine* 3 (4): 447–457.

Chavan, P., Warude, D., Joshi, K., and Patwardhan, B. 2008. Development of SCAR marker as a complementary tool for identification of *Zingiber officinale* Roscoe from crude drug and multicomponent formulation. *Biotechnology & Applied Biochemistry* 50 (1): 61–69.

Chase, M. W., and Fay, M. F. 2009. Barcoding of plants and fungi. *Science* 325:682–683.

Cheng, J. L., Huang, L. Q., Shao, A. J., and Lin, S. F. 2002. RAPD analysis on different varieties of *Rehmannia glutinosa*. *China Journal of Chinese Materia Medica* 27 (7): 505–508.

Cheng, K. T., Tsay, H. S., Chen, C. F., and Chou, T. W. 1998. Determination of the components in a Chinese prescription, yu-ping-feng san, by RAPD analysis. *Planta Medica* 64 (6): 563–565.

Cheng, Z., Lu, B. R., Baldwin, B. S., Sameshima, K., and Chen, J. K. 2002. Comparative studies of genetic diversity in kenaf (*Hibiscus cannabinus* L.) varieties based on analysis of agronomic and RAPD data. *Hereditas* 136 (3): 231–239.

Cheung, K. S., Kwan, H. S., But, P. P., and Shaw, P. C. 1994. Pharmacognostical identification of American and Oriental ginseng roots by genomic fingerprinting using arbitrarily primed polymerase chain reaction (AP-PCR). *Journal of Ethnopharmacology* 42 (1): 67–69.

Clarkson, J. J., Knapp, S., Garcia, V. F., Olmstead, R. G., Leitch, A. R., and Chase, M. W. 2004. Phylogenetic relationships in *Nicotiana* (Solanaceae) inferred from multiple plastid DNA regions. *Phylogenetics & Evolution* 33 (1): 75–90.

Consolandi, C., Severgnini, M., Frosini, A., Caramenti, G., De Fazio, M., Ferrara, F., Zocco, A., Fischetti, A., Palmieri, M., and De Bellis, G. 2006. Polymerase chain reaction of 2-kb cyanobacterial gene and human anti-alpha1-chymotrypsin gene from genomic DNA on the In-Check single-use microfabricated silicon chip. *Analytical Biochemistry* 353 (2): 191–197.

Crockett, S. L., Douglas, A. W., Scheffler, B. E., and Khan, I. A. 2004. Genetic profiling of Hypericum (St. John's wort) species by nuclear ribosomal ITS sequence analysis. *Planta Medica* 70 (10): 929–935.

Cui, G. H., Chen, M., Huang, L. Q., Xiao, S. P., and Li, D. 2004. Study on genetic diversity of herba *Cistanche* by RAPD. *China Journal of Chinese Materia Medica* 29 (8): 727–730.

Cui, S., Fu, B., Lee, F. S., and Wang, X. 2005. Application of microemulsion thin layer chromatography for the fingerprinting of licorice (*Glycyrrhiza* spp.). *Journal of Chromatography B: Analytical Technologies in Biomedical and Life Sciences* 828 (1–2): 33–40.

Devaiah, K., Balasubramani, S. P., and Venkatsubramanian, P. 2011. Development of randomly amplified polymorphic DNA-based SCAR marker for identification of *Ipomoea mauritiana* Jacq (Convolvulaceae). *eCAM*.

Di, X., Shellie, R. A., Marriott, P. J., and Huie, C. W. 2004. Application of headspace solid-phase microextraction (HS-SPME) and comprehensive two-dimensional gas chromatography (GC × GC) for the chemical profiling of volatile oils in complex herbal mixtures. *Journal of Separation Science* 27 (5–6): 451–458.

Ding, P., Xu, J. Y., and Chu, T. L. 2006. RAPD analysis on germplasm resources of different farm races of *Morinda officinalis*. *Journal of Chinese Medicinal Materials* 29 (1): 1–3.

Ding, X., Xu, L., Wang, Z., Zhou, K., Xu, H., and Wang, Y. 2002. Authentication of stems of *Dendrobium officinale* by rDNA ITS region sequences. *Planta Medica* 68 (2): 191–192.

Ding, X. Y., Wang, Z. T., Xu, L. S., Xu, H., Zhou, K. Y., and Shi, G. X. 2002. Study on sequence difference and SNP phenomenon of rDNA ITS region in F type and H type population of *Dendrobium officinale*. *China Journal of Chinese Materia Medica* 27 (2): 85–89.

Du, J., Ma, X. J., and Li, X. D. 2006. AFLP fingerprinting of *Pinellia ternata* and its application. *China Journal of Chinese Materia Medica* 31 (1): 30–33.

Dubey, N. K., Kumar, R., and Tripathi, P. 2004. Global promotion of herbal medicine: India's opportunity. *Current Science* (India) 86 (1): 37–41.

Dnyaneshwar W., Preeti C., Kalpana J., Bhushan P. 2006. Development and application of RAPD-SCAR marker for identification of *Phyllanthus emblica* Linn. *Biological and Pharmaceutical Bulletin*, 29:2313–2316.

Eisner, S. 2001. *Guidance for manufacture and sale of bulk botanical extracts.* Silver Spring, MD: American Herbal Products Association.

Farooqui, N., Ranade, S. A., and Sane, P. V. 1998. RAPD profile variation amongst provenances of neem. *Biochemistry and Molecular Biology International* 45 (5): 931–939.

Feng, T., Liu, S., and He, X. J. 2010. Molecular authentication of the traditional Chinese medicinal plant Angelica sinensis based on internal transcribed spacer of nrDNA. *Electronic Journal of Biotechnology* 13(1).

Fico, G., Spada, A., Braca, A., Agradi, E., Morelli, I., and Tome, F. 2003. RAPD analysis and flavonoid composition of *Aconitum* as an aid for taxonomic discrimination. *Biochemical and Systematic Ecology* 31 (3): 293–301.

Foster, B. C., Arnason, J. T., and Briggs, C. J. 2005. Natural health products and drug disposition. *Annual Reviews—Pharmacology and Toxicology* 45:203–206.

Friesen, N., and Blattner, F. R. 1999. RAPD analysis reveals geographic differentiations within *Allium schoenoprasum* L. (Alliaceae). *Planta Medica* 65:157–160.

Fu, R. Z., Wang, J., Zhang, Y. B., Wang, Z. T., But, P. P., Li, N., and Shaw, P. C. 1999. Differentiation of medicinal *Codonopsis* species from adulterants by polymerase chain reaction fragment length polymorphism. *Planta Medica* 65:648–650.

Fushimi, H., Komatsu, K., Isobe, M., and Namba, T. 1996. 18S ribosomal RNA gene sequences of three *Panax* species and the corresponding ginseng drugs. *Biological and Pharmaceutical Bulletin* 19 (11): 1530–1532.

Fushimi, H., Komatsu, K., Namba, T., and Isobe, M. 2000. Genetic heterogeneity of ribosomal RNA gene and *mat*K gene in *Panax notoginseng*. *Planta Medica* 66 (7): 659–661.

Gao, J. P., Wang, Y. H., Qiao, C. F., and Chen, D. F. 2003. Ribosomal DNA ITS sequences analysis of the Chinese crude drug fructus schisandrae sphenantherae and fruts of *Schisandra viridis*. *China Journal of Chinese Materia Medica* 28 (8): 706–710.

Gielly, I., and Taberlet, P. 1994. The use of chloroplast DNA to resolve plant phylogenies: Noncoding versus *rbc*L sequences. *Molecular Biology and Evolution* 11 (5): 769–777.

Godwin, I. D., Aitken, E. A., and Smith, L. W. 1997. Application of intersimple sequence repeat (ISSR) markers to plant genetics. *Electrophoresis* 18 (9): 1524–1528.

Grattapaglia, D., and Sederoff, R. 1994. Genetic linkage maps of *Eucalyptus grandis* and *Eucalyptus urophylla* using a pseudo-test cross: Mapping strategy and RAPD markers. *Genetics* 137:1121–1137.

Guo, L. P., Huang, L. Q., Wang, M., Feng, X. F., Fu, G. F., and Yan, Y. N. 2001. A preliminary study on relationship between *Atractylodes lancea* and *A. chinensis* as analyzed by RAPD. *China Journal of Chinese Materia Medica* 26 (3): 156–158.

Gupta, M., Chyi, Y.-S., Romero-Severson, J., and Owen, J. L. 1994. Amplification of DNA markers from evolutionary diverse genomes using single primers of simple-sequence repeats. *Theoretical and Applied Genetics* 89:998–1006.

Ha, W. Y., Shaw, P. C., Liu, J., Yau, F. C., and Wang, J. 2002. Authentication of *Panax ginseng* and *Panax quinquefolius* using amplified fragment length polymorphism (AFLP) and directed amplification of minisatellite region DNA (DAMD). *Journal of Agricultural and Food Chemistry* 50:1871–1875.

Heath, D. D., Iwama, G. K., and Devlin, R. H. 1993. PCR primed with VNTR core sequences yields species specific patterns and hypervariable probes. *Nucleic Acids Research* 21:5782–5785.

Holmes, E., Tang, H., Wang, Y., and Seger, C. 2006. The assessment of plant metabolite profiles by NMR-based methodologies. *Planta Medica* 72 (9): 771–785.

Hon, C. C., Chow, Y. C., Zeng, F. Y., and Leung, F. C. C. 2003. Genetic authentication of ginseng and other traditional Chinese medicine. *Acta Pharmacologica Sinica* 24 (9): 841–846.

Hori, T. A., Hayashi, A., Sasanuma, T., and Kurita, S. 2006. Genetic variations in the chloroplast genome and phylogenetic clustering of Lycoris species. *Genes, Genetics & Systematics* 81 (4): 243–253.

Hosokawa, K., Minami, M., Kawahara, K., Nakamura, I., and Shibata, T. 2000. Discrimination among three species of medicinal *Scutellaria* plants using RAPD markers. *Planta Medica* 66 (3): 270–272.

Hu, Y., Zhang, Q., Xin, H., Qin, L. P., Lu, B. R., Rahman, K., and Zheng, H. 2007. Association between chemical and genetic variation of *Vitex rotundifolia* populations from different locations in China: Its implication for quality control of medicinal plants. *Biomedical Chromatography* 21 (9): 967–975.

Hulya, I. 2003. RAPD markers assisted varietal identification and genetic purity test in pepper, *Capsicum annuum*. *Scientia Horticulturae* 97 (3–4): 211–218.

Jin, W., Ge, R. L., Wei, Q. J., Bao, T. Y., Shi, H. M., and Tu, P. F. 2006. Development of high-performance liquid chromatographic fingerprint for the quality control of *Rheum tanguticum* Maxim. ex Balf. *Journal of Chromatography A* 1132 (1–2): 320–324.

Johnson, E. L., Saunders, J. A., Mischke, S., Helling, C. S., and Emche, S. D. 2003. Identification of *Erythroxylum* taxa by AFLP DNA analysis. *Phytochemistry* 64 (1): 187–197.

Joshi, K., Chavan, P., Warude, D., and Patwardhan, B. 2004. Molecular markers in herbal drug technology. *Current Science India* 87 (2): 159–165.

Joshi, S. P., Ranjekar, P. K., and Gupta, V. S. 1999. Molecular markers in plant genome analysis. *Current Science* 77:230–240.

Kapteyn, J., Goldsbrough, B., and Simon, E. 2002. Genetic relationships and diversity of commercially relevant *Echinacea* species. *Theoretical and Applied Genetics* 105 (2–3): 369–376.

Khan, I. A. 2006. Issues related to botanicals. *Life Science* 78:2033–2038.

Kim, S. Y., Chen, J. W., Liu, Z. Q., and Wang, Y. Z. 2004. Genetic identification of internal transcribed spacers sequence in rDNA of *Artemisia iwayomogi* Kitam. and other two *Artemisia* species. *Journal of Chinese Integrative Medicine* 2 (1): 58–61.

Koga, S., Shoyama, Y., and Nishioka, I. 1991. Studies on *Epimedium* species: Flavonol glycoside and isozymes. *Biochemical Systematics and Ecology* 19:315–318.

Kojoma, M., Iida, O., Makino, Y., Sekita, S., and Satake, M. 2002. DNA fingerprinting of *Cannabis sativa* using intersimple sequence repeat (ISSR) amplification. *Planta Medica* 68:60–63.

Komatsu, K., Zhu, S., Fushimi, H., Qui, T. K., Cai, S., and Kadota, S. 2001. Phylogenetic analysis based on 18S rRNA gene and matK gene sequences of *Panax vietnamensis* and five related species. *Planta Medica* 67 (5): 461–465.

Koren, O. G., and Zhuravlev, Y. N. 1998. Allozyme variations in two ginseng species *Panax ginseng* and *P.* quinqurfoliees: Advances in Gainseng Research. *Proceedings of 7th International Symposium on Ginseng*, Korea, September, 1998.

Kumar, L. S. 1999. DNA markers in plant improvement—Chromosome organization and evolution in a disomic polyploidy. *Biotechnology Advances* 17:143–182.

Kuzoff, R. K., Sweere, J. A., Soltis, D. E., Soltis, P. S., and Zimmer, E. A. 1998. The phylogenetic potential of entire 26S rDNA sequences in plants. *Molecular Biology and Evolution* 15 (3): 251–263.

Lattoo, S. K., Bamotra, S., Sapru Dhar, R., Khan, S., and Dhar, A. K. 2006. Rapid plant regeneration and analysis of genetic fidelity of *in vitro* derived plants of *Chlorophytum arundinaceum* Baker—An endangered medicinal herb. *Plant Cell Reports* 25 (6): 499–506.

Lau, D. T. W., Shaw, P. C., Wang, J., and But, P. P. H. 2001. Authentication of medicinal *Dendrobium* species by the internal transcribed spacer of ribosomal DNA. *Planta Medica* 67 (5): 456–460.

Lee, C., and Wen, J. 2004. Phylogeny of *Panax* using chloroplast *trnC–trnD* intergenic region and the utility of *trnC–trnD* in interspecific studies of plants. *Molecular Phylogenetics and Evolution* 31 (3): 894–903.

Lee, J. S., Lim, M. O., Cho, K. Y., Cho, J. H., Chang, S. Y., and Nam, D. H. 2006. Identification of medicinal mushroom species based on nuclear large subunit rDNA sequences. *Journal of Microbiology* 44 (1): 29–34.

Lee, M. Y., Doh, E. J., Park, C. H., Kim, Y. H., Kim, E. S., Ko, B. S., and Oh, S. E. 2006. Development of SCAR marker for discrimination of *Artemisia princeps* and *A. argyi* from other *Artemisia* herbs. *Biological and Pharmaceutical Bulletin* 29 (4): 629–633.

LeRoy, A., Potter, E., Woo, H. H., Heber, D., and Hirsch, A. M. 2002. Characterization and identification of alfalfa and red clover dietary supplements using a PCR-based method. *Journal of Agricultural and Food Chemistry* 50 (18): 5063–5069.

Li, M., Jiang, W., Hon, P. M., Cheng, L., Li, L. L., Zhou, J. R., Shaw, P. C., and But, P. P. H. 2010. Authentication of the anti-tumor herb Baihuasheshecao with bioactive marker compounds and molecular sequences. *Food Chemistry* 119 (3): 1239–1245.

Li, Y., Hu, Z., and He, L. 2007. An approach to develop binary chromatographic fingerprints of the total alkaloids from *Caulophyllum robustum* by high performance liquid chromatography/diode array detector and gas chromatography/mass spectrometry. *Journal of Pharmaceutical and Biomedical Analysis* 43 (5): 1667–1672.

Lind-Hallden, C., Hallden, C., and Sall, T. 2002. Genetic variation in *Arabidopsis suecica* and its parental species *A. arenosa* and *A. thaliana*. *Hereditas* 136 (1): 45.

Liu, W., Zhu, J., He, B., and Su, Y. 2004. Studies on random amplified polymorphic DNA fingerprinting of cortex *Magnoliae officinalis*. *Journal of Chinese Medicinal Materials* 27 (3): 164–169.

Liu, Y. P., Luo, J. P., Feng, Y. F., Guo, X. L., and Cao, H. 2002. DNA profiling of *Pogostemon cablin* chemotypes differing in essential oil composition. *Acta Pharmaceutica Sinica* 37 (4): 304–308.

Liu, Z., Hao, M., and Wang, J. 2004. Application of allele-specific primer in the identificatin of *Hedyotis diffusa*. *Journal of Chinese Medicinal Materials* 27 (7): 484–487.

Long, C., Kakiuchi, N., Takahashi, A., Komatsu, K., Cai, S., and Mikage, M. 2004. Phylogenetic analysis of the DNA sequence of the noncoding region of nuclear ribosomal DNA and chloroplast of *Ephedra* plants in China. *Planta Medica* 70 (11): 1080–1084.

Ma, X. J., Wang, X. Q., Xiao, P. G., and Hong, D. Y. 2000. A study on AFLP fingerprinting of land races of *Panax ginseng* L. *China Journal of Chinese Materia Medica* 25(12):707–10.

Ma, X. Q., Duan, J. A., Zhu, D. Y., Dong, T. T. X., and Tsim, K. W. K. 2000. Species identification of Radix Astragali (Huangqi) by DNA sequence of its 5S-rRNA spacer domain. *Phytochemistry* 54:363–368.

Ma, X. Q., Zhu, D. Y., Li, S. P., Dong, T. T. X., and Tsim, K. W. K. 2001. Authentic identification of stigma croci (stigma of *Crocus sativus*) from its adulterants by molecular genetic analysis. *Planta Medica* 67:183–186.

Makino, Y., Sekita, S., and Satake, M. 2000. Intraspecific variation in *Cannabis sativa* L. based on intergenic spacer region of chloroplast DNA. *Biological & Pharmaceutical Bulletin* 23 (6): 727–730.

Marcus, D. M., and Grollman, A. P. 2002. Botanical medicines–The need for new regulations. *New England Journal of Medicine* 347:2073–2076.

McChesney, J. D. 1999. Quality of botanical preparations: Environmental issues and methodology for detecting environmental contaminants. In *Botanical medicine—Efficacy, quality assurance and regulation*, ed. Eskinazi, D., Blumenthal, M., Farnsworth, N., and Riggins, C. W., 127–131. Larchmont, NY: Mary Ann Liebert, Inc.

Mihalov, J. J., Marderosian, A. D., and Pierce, J. C. 2000. DNA identification of commercial ginseng samples. *Journal of Agricultural and Food Chemistry* 48 (8): 3744–3752.

Mizukami, H., Ohbayashi, K., Kitamura, Y., and Ikenaga, T. 1993. Restriction fragment length polymorphisms (RFLPs) of medicinal plants and crude drugs. I. RFLP probes allow clear identification of *Duboisia* interspecific hybrid genotypes in both fresh and dried tissues. *Biological and Pharmaceutical Bulletin* 16 (4): 388–390.

Mizukami, H., Ohbayashi, K., and Ohashi, H. 1993. *Bupleurum falcatum* L. in northern Kyushu and Yamaguchi prefecture are genetically distinguished from other populations, based on DNA fingerprints. *Biological and Pharmaceutical Bulletin* 16 (7): 729–731.

Mizukami, H., Ohbayashi, K., Umetsu, K., and Hiraoka, N. 1993. Restriction fragment length polymorphisms of medicinal plants and crude drugs. II. Analysis of *Glehnia littoralis* of different geographical origin. *Biological and Pharmaceutical Bulletin* 16 (6): 611–612.

Mizukami, H., Okabe, Y., Kohda, H., and Hiraoka, N. 2000. Identification of the crude drug atractylodes rhizome (byaku-jutsu) and atractylodes lancea rhizome (so-jutsu) using chloroplast *trn*K sequence as a molecular marker. *Biological & Pharmaceutical Bulletin* 23 (5): 589–594.

Mohan, M., Nair, S., Bhagwat, A., Krisna, T. G., Yano, M., Bhatia, C. R., and Sasaki, T. 1997. Genome mapping, molecular markers and marker-assisted selection in crop plants. *Molecular Breeding* 3:87–103.

Mullis, K., Faloona, F., Scharf, S., Saiki, R., Horn, G., and Erilich, H. 1986. Specific enzymatic amplification of DNA *in vitro*: The polymerase chain reaction. *Cold Spring Harbor Symposia on Quantitative Biology* 51:263–273.

Na, H. J., Um, J. Y., Kim, S. C., Koh, K. H., Hwang, W. J., Lee, K. M., Kim, C. H., and Kim, H. M. 2004. Molecular discrimination of medicinal Astragali radix by RAPD analysis. *Immunopharmacology & Immunotoxicology* 26 (2): 265–272.

Nakai, R., Shoyama, Y., and Shiraishi, S. 1996. Genetic characterization of *Epimedium* species using random amplified polymorphic DNA (RAPD) and PCR restriction fragment length polymorphism (RFLP) diagnosis. *Biological and Pharmaceutical Bulletin* 19 (1): 67–70.

Nakamura, Y., Julier, C., Wolff, R., Holm, T., O'Connell, P., Lepert, M., and White, R. 1987. Characterization of human "midisatellite" sequence. *Nucleic Acids Research* 15:2537–2547.

Nakamura, Y., Leppert, M., O'Connell, P., Wolff, R., Holm, T., Culver, M., Martin, C., Fujimoto, E., Hoff, M., Kumlin, E., et al. 1987. Variable number of tandem repeat (VNTR) markers for human gene mapping. *Science* 235 (4796): 1616–1622.

Negi, M. S., Singh, A., and Laksmikumaran, M. 2000. Genetic variation and relationship among and within *Withania* species as revealed by AFLP marker. *Genome* 43:975–980.

Ngan, F., Shaw, P., But, P., and Wang, J. 1999. Molecular authentication of *Panax* species. *Phytochemistry* 50 (5): 787–791.

Nieri, P., Adinolfi, B., Morelli, I., Breschi, M. C., Simoni, G., and Martinotti, E. 2003. Genetic characterization of the three medicinal *Echinacea* species using RAPD analysis. *Planta Medica* 69 (7): 685–686.

Nwakanma, D. C., Pillay, M., Okoli, B. E., and Tenkouano, A. 2003. PCR-RFLP of the ribosomal DNA internal transcribed spacers (ITS) provides markers for the A and B genomes in *Musa* L. *Theoretical and Applied Genetics* 108 (1): 154–159.

Obradovic, M., Krajsek, S. S., Dermastia, M., and Kreft, S. 2007. A new method for the authentication of plant samples by analyzing fingerprint chromatograms. *Phytochemical Analysis* 18 (2): 123–132.

Padmesh, P., Sabu, K. K., Seeni, S., and Pushpangadan, P. 1999. The use of RAPD in assessing genetic variability in *Andrographis paniculata* Nees, a hepatoprotective drug. *Current Science* 76 (6): 833–835.

Paran, I., and Michelmore, R. W. 1993. Development of reliable PCR-based markers linked to downy mildew resistance genes in lettuce. *Theoretical & Applied Genetics* 85:985–993.

Passinho-Soares, H., Felix, D., Kaplan, M. A., Margis-Pinheiro, M., and Margis, R. 2006. Authentication of medicinal plant botanical identity by amplified fragmented length polymorphism dominant DNA marker: Inferences from the *Plectranthus* genus. *Planta Medica* 72 (10): 929–931.

Pereda-Miranda, R., Fragoso-Serrano, M., Escalante-Sanchez, E., Hernandez-Carlos, B., Linares, E., and Bye, R. 2006. Profiling of the resin glycoside content of Mexican jalap roots with purgative activity. *Journal of Natural Products* 69 (10): 1460–1466.

Pietta, P., Mauri, P., and Bauer, R. 1998. MEKC analysis of different *Echinacea* species. *Planta Medica* 64:649–652.

Pietta, P., Mauri, P., Bruno, A., and Merfort, I. 1994. MEKC as an improved method to detect falsifications in the flowers of *Arnica montana* and *A. chamissonis*. *Planta Medica* 60:369–372.

Powell, W., Morgante, M., Andre, C., Hanafey, M., Vogel, J., Tingey, S., and Rafalski, A. 1996. The comparison of RFLP, RAPD, AFLP, and SSR markers for germplasm analysis. *Molecular Breeding* 2:225–238.

Prajapati, N. D., Purohit, S. S., Sharma, A. K., and Kumar, T. A. 2003. *Handbook of medicinal plants*. Agrobios (India), Jodhpur.

Primrose, S. B., and Twyman, R. M. 2003. *Principles of genome analysis and genomics.* Blackwell Science, London.

Prince, J. P., Lackney, V. K., Angels, C., Blauth, J. R., and Kyle, M. M. 1995. A survey of DNA polymorphism within the genus *Capsicum* and the fingerprinting of pepper cultivars. *Genome* 38 (2): 224–231.

Qian, G. S., Wang, Q., Leung, K. S., Qin, Y., Zhao, Z., and Jiang, Z. H. 2007. Quality assessment of Rhizoma et Radix Notopterygii by HPTLC and HPLC fingerprinting and HPLC quantitative analysis. *Journal of Pharmaceutical and Biomedical Analysis* 44 (3): 812–817.

Qin, J., Leung, F. C., Fung, Y., Zhu, D., and Lin, B. 2005. Rapid authentication of ginseng species using microchip electrophoresis with laser-induced fluorescence detection. *Analytical & Bioanalytical Chemistry* 381 (4): 812–819.

Rader, J. I., Delmonte, P., and Trucksess, M. W. 2007. Recent studies on selected botanical dietary supplement ingredients. *Analytical Bioanalytical Chemistry* 389 (1): 27–35.

Reale, S., Doveri, S., Díaz, A., Angiolillo, A., Lucentini, L., Pilla, F., Martín, A., Donini, P., and Lee, D. 2006. SNP-based markers for discriminating olive (*Olea europaea* L.) cultivars. *Genome* 49 (9): 1193–1205.

Rout, G. R. 2006. Identification of *Tinospora cordifolia* (Willd.) Miers ex Hook. f. & Thomas using RAPD markers. *Zeitschrift für Naturforschung* [C] 61 (1–2): 118–122.

Sangwan, R. S., Sangwan, N. S., Jain, D. C., Kumar, S., and Ranade, S. A. 1999. RAPD profile based genetic characterization of chemotypic variants of *Artemisia annua* L. *Biochemistry and Molecular Biology International* 47 (6): 935–944.

Sasaki, Y., Fushimi, H., Cao, H., Cai, S. Q., and Komatsu, K. 2002. Sequence analysis of Chinese and Japanese *Curcuma* drugs on the 18S rRNA gene and *trn*K gene and the application of amplification-refractory mutation system analysis for their authentication. *Biological & Pharmaceutical Bulletin* 25 (12): 1593–1599.

Sasikumar, B., Syamkumar, S., Remyaa, T., and Zachariaha, J. 2005. PCR-based detection of adulteration in the market samples of turmeric powder. *Food Biotechnology* 18 (3): 299–306.

Schilter, B., Andersson, C., Anton, R., Constable, A., Kleiner, J., O'Brien, J., Renwick, A. G., Korver, O., Smit, F., and Walker, R. 2003. Guidance for the safety assessment of botanicals and botanical preparations for use in food and food supplements. *Food and Chemical Toxicology* 41:1625–1649.

Shao, A. J., Li, X., Huang, L. Q., Lin, S. F., and Chen, J. 2006. RAPD analysis of *Scutellaria baicalensis* from different germplasms. *China Journal of Chinese Materia Medica* 31 (6): 452–455.

Shasany, A. K., Aruna, V., Darokar, M. P., Kalra, A., Bahl, J. R., Bansal, R. P., and Khanuja, S. P. S. 2002. RAPD marking of three *Pelargonium graveolens* genotypes with chemotypic differences in oil quality. *Journal of Medicinal and Aromatic Plant Sciences* 24:729–732.

Shasany, A. K., Kukreja, A. K., Saikia, D., Darokar, M. P., Khanuja, S. P. S., and Kumar, S. 2000. Assessment of diversity among *Taxus wallichiana* accessions from northeast India using RAPD analysis. *PGR Newsletter* 121:27–31.

Shaw, P. C., and But, P. P. 1995. Authentication of *Panax* species and their adulterants by random-primed polymerase chain reaction. *Planta Medica* 61 (5): 466–469.

Shen, J., Ding, X. Y., Ding, G., Liu, D. Y., Tang, F., and He, J. 2006. Studies on population difference of *Dendrobium officinale* II establishment and optimization of the method of ISSR fingerprinting marker. *China Journal of Chinese Materia Medica* 31 (4): 291–294.

Singh, J., Singh, A. K., and Khanuja, S. P. S. 2003. Medicinal plants: India's opportunities. *Pharma Bioworld* 1:59–66.

Singh, M., Bandana, and Ahuja, P. S. 1999. Isolation and PCR amplification of genomic DNA from market samples of dry tea. *Plant Molecular Biology Reporter* 17:171–178.

Siniscalco, G. G., Caputo, P., and Cozzolino, S. 1997. Ribosomal DNA analysis as a tool for the identification of *Cannabis sativa* L. specimens of forensic interest. *Science & Justice* 37 (3): 171–174.

Siniscalco, G. S. 1998. Identification of *Cannabis sativa* L. (Cannabaceae) using restriction profiles of the internal transcribed spacer II (ITS2). *Science & Justice* 38 (4): 225–230.

———. 1999. Preliminary data on the usefulness of internal transcribed spacer I (ITS1) sequence in *Cannabis sativa* L. identification. *Journal of Forensic Science* 44 (3): 475–477.

Slifman, N. R., Obermeyer, W. R., Aloi, B. K., Musser, S. M., Correll, W. A., Jr., Cichowicz, S. M., Betz, J., and Love, L. A. 1998. Contamination of botanical dietary supplements by *Digitalis lanata*. *New England Journal of Medicine* 339:806–811.

Soares, P. K., and Scarminio, I. S. 2008. Multivariate chromatographic fingerprint preparation and authentication of plant material from the genus *Bauhinia*. *Phytochemical Analysis* 19 (1): 78–85.

Spaniolas, S., May, S. T., Bennett, M. J., and Tucker, G. A. 2006. Authentication of coffee by means of PCR-RFLP analysis and lab-on-a-chip capillary electrophoresis. *Journal of Agricultural and Food Chemistry* 54 (20): 7466–7470.

Sri Bhava Misra. 1999. *Bhavaprakash*, ed. Misra, B., and Vaisya, R. Chaukhambha Sanskrit Sansthan, Varanasi.

Sun, F., Cao, Y. Q., Liu, L. X., and Xu, S. M. 1993. Isozymatic patterns for Chinese ginseng and American medicine. *Chinese Traditional Herbal Drugs* 24:148–149.

Tamhankar, S., Ghate, V., Raut, A., and Rajput, B. 2009. Molecular profiling of "chirayat" complex using inter-simple sequence repeat (ISSR) markers. *Planta Medica* 75 (11): 1266–1270.

Thomas, C. 2009. Plant bar code soon to become reality. *Science* 325:526.

Tsoi, P. Y., Woo, H. S., Wong, M. S., Chen, S. L., Fong, W. F., Xiao, P. G., and Yang, M. S. 2003. Genotyping and species identification of *Fritillaria* by DNA chips. *Acta Pharmaceutica Sinica* 38 (3): 185–190.

Urbanova, M., Kosuth, J., and Cellarova, E. 2006. Genetic and biochemical analysis of *Hypericum perforatum* L. plants regenerated after cryopreservation. *Plant Cell Reports* 25 (2): 140–147.

Urdiain, M., Domenech-Sanchez, A., Alberti, S., Benedi, V. J., and Rossello, J. A. 2004. Identification of two additives, locust bean gum (E-410) and guar gum (E-412), in food products by DNA-based methods. *Food Additives and Contaminants* 21 (7): 619–625.

Vos, P., Hogers, R., Bleeker, M., Reijans, M., Van de Lee, T., Hornes, M., Fritjers, A., Pot, J., Peleman, J., Kuiper, M., and Zabeau, M. 1995. AFLP: A new technique for DNA fingerprinting. *Nucleic Acids Research* 23:4407–4414.

Wang, J., Ha, W. Y., Ngan, F. N., But, P. P., and Shaw, P. C. 2001. Application of sequence characterized amplified region (SCAR) analysis to authenticate *Panax* species and their adulterants. *Planta Medica* 67 (8): 781–783.

Wang, T., Su, Y., Zhu, J., Li, X., Zeng, O., and Xia, N. 2001. Studies on DNA amplification fingerprinting of cortex *Magnoliae officinalis*. *Journal of Chinese Medicinal Materials* 24 (10): 710–715.

Wang, X., Li, Y., Li, H., Zhang, Y., Zhao, L., and Yu, Y. 2003. RAPD analysis of genuineness on source of *Bupleurum chinense*. *Journal of Chinese Medicinal Materials* 26 (12): 855–856.

Warude, D., Chavan, P., Joshi, K., and Patwardhan, B. 2003. DNA isolation from fresh and dry plant samples with highly acidic tissue extracts. *Plant Molecular Biology Reporter* 21 (4): 467a–467f.

Weber, D., and Helentjaris, T. 1989. Mapping RFLP loci in maize using B–A translocations. *Genetics* 121:583-590.

Weber, H. A., Zart, M. K., Hodges, A. E., Molloy, H. M., O'Brien, B. M., Moody, L. A., Clark, A. P., Harris, R. K., Overstreet, J. D., and Smith, C. S. 2003. Chemical comparison of goldenseal (*Hydrastis canadensis* L.) root powder from three commercial suppliers. *Journal of Agricultural and Food Chemistry* 51 (25): 7352–7358.

Welsh, J., and McClelland, M. 1990. Fingerprinting genomes using PCR with arbitrary primers. *Nucleic Acids Research* 18:7213–7218.

———. 1991. Genomic fingerprints produced by PCR with consensus tRNA gene primers. *Nucleic Acids Research* 19:861–866.

Williams, J. G. K., Kubelik, A. R., Livak, J., Rafalski, J. A., and Tingey, S. V. 1990. DNA polymorphisms amplified by arbitrary primers are useful as genetic markers. *Nucleic Acids Research* 18 (22): 6531–6535.

Wolf, H. T., Van Den Berg, T., Czygan, F. C., Mosandl, A., Winckler, T., Zundorf, I., and Dingermann, T. 1999. Identification of *Melissa officinalis* subspecies by DNA fingerprinting. *Planta Medica* 65:83–85.

Wolf, H. T., Zundorf, I., Winckler, T., Bauer, R., and Dingermann, T. 1999. Characterization of *Echinacea* species and detection of possible adulterations by RAPD analysis. *Planta Medica* 65:773–774.

Xu, W. B., Guo, Q. S., and Wang, C. L. 2006. RAPD analysis for genetic diversity of *Chrysanthemum morifolium*. *China Journal of Chinese Materia Medica* 31 (1): 18–21.

Xu, X., Ye, H., Wang, W., and Chen, G. 2005. An improved method for the quantitation of flavonoids in Herba Leonuri by capillary electrophoresis. *Journal of Agricultural and Food Chemistry* 53 (15): 5853–5857.

Xue, C. Y., Li, D. Z., Lu, J. M., Yang, J. B., and Liu, J. Q. 2006. Molecular authentication of the traditional Tibetan medicinal plant *Swertia mussotii*. *Planta Medica* 72 (13): 1223–1226.

Yamazaki, M., Sato, A., Saito, K., and Murakoshi, I. 1993. Molecular phylogeny based on RFLP and its relation with alkaloid patterns in *Lupinus* plants. *Biological and Pharmaceutical Bulletin* 16 (11): 1182–1184.

Yamazaki, M., Sato, A., Shimomura, K., Saito, K., and Murakoshi, I. 1994. Genetic relationships among *Glycyrrhiza* plants determined by RAPD and RFLP analyses. *Biological and Pharmaceutical Bulletin* 17 (11): 1529–1531.

Yang, M., Zhang, D., Liu, J., and Zheng, J. 2001. A molecular marker that is specific to medicinal rhubarb based on chloroplast *trn*L/*trn*F sequences. *Planta Medica* 67 (8): 784–786.

Yang, Y. Y., Fushimi, H., Cai, S. Q., and Komatsu, K. 2004. Molecular analysis of rheum species used as rhei rhizoma based on the chloroplast *mat*K gene sequence and its application for identification. *Biological & Pharmaceutical Bulletin* 27 (3): 375–383.

Ye, X., Yu, H., and Li, P. 2005. Analysis of Chinese drug beimu and its fake species with clustering analysis and FTIR spectra. *Journal of Chinese Medicinal Materials* 28 (2): 89–91.

Yin, X. L., Fang, K. T., Liang, Y. Z., Wong, R. N., and Ha, A. W. 2005. Assessing phylogenetic relationships of Lycium samples using RAPD and entropy theory. *Acta Pharmacologica Sinica* 26 (10): 1217–1224.

Yip, P. Y., and Kwan, H. S. 2006. Molecular identification of *Astragalus* membranaceus at the species and locality levels. *Journal of Ethnopharmacology* 106 (2): 222–229.

Yu, C. Y., Hu, S. W., Zhao, H. X., Guo, A. G., and Sun, G. L. 2005. Genetic distances revealed by morphological characters, isozymes, proteins and RAPD markers and their relationships with hybrid performance in oilseed rape (*Brassica napus* L.). *Theoretical & Applied Genetics* 110 (3): 511–518.

Yue-Qin, C., Ning, W., Hui, Z., and Liang-Hu, Q. 2002. Differentiation of medicinal *Cordyceps* species by rDNA ITS sequence analysis. *Planta Medica* 68 (7): 635–639.

Yukiko, T. K., Asaka, I., and Ichio, I. 2001. A random amplified polymorphic DNA (RAPD) primer to assist the identification of a selected strain, Aizu K-111 of *Panax ginseng* and the sequence amplified. *Biological and Pharmaceutical Bulletin* 24 (10): 1210–1213.

Yurenkova, S. I., Kubrak, S. V., Titok, V. V., and Khotyljova, L. V. 2005. Flax species polymorphism for isozyme and metabolic markers. *Genetika* 41 (3): 334–340.

Zabeau, M. 1993. Selective restriction fragment amplification: A general method for DNA fingerprinting. European patent application publication no. 0534858A1.

Zhang, J. Q., Su, X., Wu, Q., Ding, S. S., and Sun, K. 2006. Analysis of RAPD on medicinal plants of *Codonopsis pilosula*. *Journal of Chinese Medicinal Materials* 29 (5): 417–420.

Zhang, K. Y., Leung, H. W., Yeung, H. W., and Wong, R. N. 2001. Differentiation of *Lycium barbarum* from its related *Lycium* species using random amplified polymorphic DNA. *Planta Medica* 67 (4): 379–381.

Zhang, M., Huang, H. R., Liao, S. M., and Gao, J. Y. 2001. Cluster analysis of *Dendrobium* by RAPD and design of specific primer for *Dendrobium candidum*. *China Journal of Chinese Materia Medica* 26 (7): 442–447.

Zhang, Y. B., Wang, J., Wang, Z. T., But, P. P., and Shaw, P. C. 2003. DNA microarray for identification of the herb of *Dendrobium* species from Chinese medicinal formulations. *Planta Medica* 69:1172–1174.

Zhao, K. J., Dong, T. T., Cui, X. M., Tu, P. F., and Tsim, K. W. 2003b. Genetic distinction of radix adenophorae from its adulterants by the DNA sequence of 5S-rRNA spacer domains. *American Journal of Chinese Medicine* 31 (6): 919–926.

Zhao, K. J., Dong, T. T., Tu, P. F., Song, Z. H., Lo, C. K., and Tsim, K. W. 2003a. Molecular genetic and chemical assessment of radix Angelica (Danggui) in China. *Journal of Agricultural and Food Chemistry* 51 (9): 2576–2583.

Zhao, Z. L., Wang, Z. T., Xu, L. S., and Zhou, K. Y. 2002. Studies on the molecular markers of rhizomes of some *Alpinia* species. *Planta Medica* 68 (6): 574–576.

Zheng, X. Z., Guo, B. L., and Yan, Y. N. 2002. Study on the genetic relationship between Chinese and Korean medicinal materials of niuxi with the method of RAPD. *China Journal of Chinese Materia Medica* 27 (6): 421–423.

Zhou, H. T., Hu, S. L., Guo, B. L., Feng, X. F., Yan, Y. N., and Li, J. S. 2002. A study on genetic variation between wild and cultivated populations of *Paeonia lactiflora* Pall. *Acta Pharmaceutica Sinica* 37 (5): 383–388.

Zhu, S., Fushimi, H., and Komatsu, K. 2008. Development of a DNA microarray for authentication of ginseng drugs based on 18S rRNA gene sequence. *Journal of Agricultural and Food Chemistry* 56 (11): 3953–3959.

Zietkiewicz, E., Rafalski, A., and Labuda, D. 1994. Genome fingerprinting by simple sequence repeat (SSR)-anchored polymerase chain reaction amplification. *Genomics* 20:176–183.

Appendix

BOOKS

Carlson, T.J.S., and L. Maffi.2004. Ethnobotany and Conservation of Biocultural Diversity. Advances in Economic Botany Volume 15. New York Botanical Garden Press.

Curtis, W. 1783. A catalogue of the British, medicinal, culinary, and agricultural plants, cultivated in the London Botanic Garden.

Duke, J. A. 2009. Duke's Handbook of Medicinal Plants of Latin America. CRC Press, Taylor&Francis Group, Boca Raton, Florida.

Duke, J.A. 2008. Duke's Handbook of Medicinal Plants of the Bible. CRC Press, Taylor&Francis Group.

Duke, J.A.2002. Handbooks of Medicinal Herbs (second edition). CRC Press, Boca Raton, Florida.

Ebadi, M.S. 2007. Pharmacodynamic basis of herbal medicine. CRC/Taylor & Francis, Boca Raton.

Foster, S., and R. L. Johnson. 2006. Desk Reference to Nature's Medicine. National Geographic Society, Washington, D.C.

Hardman, R. Medicinal and Aromatic Plants-Industrial Profiles. CRC Press, Taylor&Francis Group, Boca Ration, Florida.

Valerian: The Genus *Valeriana* (Peter Houghton; Editor)

Perilla: The Genus *Perilla* (He-ci Yu, Kenichi Kosuna, Megumi Haga; Editors)

Cannabis: The Genus *Cannabis* (David T Brown; Editor)

Poppy: The Genus *Papaver* (Jeno Bernath; Editor)

Neem: The Divine Tree, *Azadirachta indica* (H.S. Puri; Editor)

Ergot: The Genus *Claviceps* (Vladimir Kren, and Ladislav Cvak; Editor)

Caraway: The Genus *Carum* (Eva Nemeth; Editor)

Saffron: *Crocus Sativus* L. (Moshe Negbi, editor)

Tea Tree: The Genus *Melaleuca* (Ian Southwell, and Robert Lowe; Editors)

Basil: The Genus Ocimum (Raimo Hiltunen, and Yvonne Holm; Editors)

Fenugreek: The Genus *Trigonella* (Georgios A Petropoulos; Editor)

Ginkgo Biloba (Teris A vanBeek; Editor)

Black Pepper, *Piper Nigram* (P. N. Ravindran; Editor)

Sage: The Genus *Salvia* (Spiridon E. Kintzios; Editor)

Ginseng, the Genus *Panax* (William E Court; Editor)

Mistletoe: The Genus *Viscum* (Arndt Bussing; Editor)

Tea: Bioactivity and Therapeutic Potential (Yong-Su Zhen; Editor)

Artemisia (Colin W. Wright; Editor)

Stevia: The Genus *Stevia* (A. Douglas Kinghorn; Editor)

Vetiveria: The Genus *Vetiveria* (Massimo Maffei; Editor)

Narcissus and Daffodil: The Genus *Narcissus* (Gordon R Hanks; Editor)

Eucalyptus: The Genus *Eucalyptus* (J. J. W. Coppen; Editor)

Pueraria: The Genus *Pueraria* (Wing Ming Keung; Editor)

Thyme: The Genus *Thymus* (E. Stahl-Biskup and F. Sáez; Editors)

Oregano: The genera *Origanum* and *Lippia* (Spiridon E. Kintzios; Editor)

Citrus: The Genus *Citrus* (Giovanni Dugo and Angelo Di Giacomo; Editors)

Geranium and Pelargonium: History of Nomenclature, Usage and Cultivation (Maria Lis-Balchin; Editor)

Magnolia: The Genus *Magnolia* (Satyajit D. Sarker, Yuji Maruyama; Editors)

Lavender: The Genus *Lavandula* (Maria Lis-Balchin; Editor)

Cardamom: The Genus *Elettaria* (P. N. Ravindran, K.J. Madhusoodanan; Editors)

Hypericum: The genus *Hypericums* (Edzard Ernst; Editor)

Taxus: The Genus *Taxus* (Hideji Itokawa, Kuo-Hsiung Lee; Editors)

Capsicum: The genus *Capsicum* (Amit Krishna De; Editor)

Flax: The genus *Linum* (Alister D. Muir, Neil D. Westcott; Editors)

Urtica: The genus *Urtica* (Gulsel M. Kavalali; Editor)

Cinnamon and Cassia: The Genus *Cinnamomum* (P. N. Ravindran, K Nirmal-Babu, and M Shylaja; Editors)

Kava: From Ethnology to Pharmacology (Yadhu N. Singh; Editor)

Aloes: The genus *Aloe* (Tom Reynolds; Editor)

Echinacea: The genus *Echinacea* (Sandra Carol Miller, He-ci Yu; Editors)

Illicium, Pimpinella and Foeniculum (Manuel Miro Jodral; Editor)

Ginger: The Genus *Zingiber* (P. N. Ravindran, K. Nirmal Babu; Editors)

Chamomile: Industrial Profiles (Rolf Franke , Heinz Schilcher; Editors)

Pomegranates: Ancient Roots to Modern Medicine (David Heber, Risa N. Schulman , Navindra P. Seeram; Editors)

Mint: The Genus *Mentha* (Brian M. Lawrence; Editor)

Turmeric: The genus *Curcuma* (P. N. Ravindran, K. Nirmal Babu, and K. Sivaraman; Editors)

Essential Oil Bearing Grasses: The genus *Cymbopogon* (Anand Akhila; Editor)

Vanilla: (Eric Odoux, Michel Grisoni; Editors)

Sesame: The genus *Sesamum* (Dorothea Bedigian; Editor)

Iwu, M. M. 1993. Handbook of African Medicinal Plants. CRC Press, Boca Raton, FL.

Juliani, H. R., J. E. Simon, C.-T. Ho. 2009. African Natural Plant Products: New Discoveries and Challenges on Chemistry and Quality. American Chemical Society.

Karch, S. B. 1999. The consumer's guide to herbal medicine. Advanced Research Press, New York.

Li, T. S. C.2002. Chinese and related North American herbs: phytopharmacology and therapeutic values. CRC Press, Inc., Boca Raton, Florida.

Li., T. S. C. 2006. Taiwanese native medicinal plants: phytopharmacology and therapeutic values. CRC Press, Taylor & Francis Group. Boca Raton, Florida

Lindley, J. 1847. The Vegetable Kingdom; or, the Structure, Classification, and Uses of Plants. Bradbury & Evans, Whitefriars, London.

Lise, M. 1989. An Ancient Egyptian Herbal. University of Texas Press, Austin.

Mann, J. 1992. Murder, Magic, and Medicine. Oxford University Press, Inc., New York.

Marcello, S. The psychopharmacology of herbal medicine: plant drugs that alter mind, brain, and behavior. MIT Press, Cambridge, Massachusetts.

Raffauf, R. F. 1996. Plant Alkaloid-A guide to their discovery and distribution. The Haworth Press, Inc., New York.

Rosenthal, N. E. 1998. St. John's wort: the herbal way to feeling good. Harper Collins, New York.

Sneader, W. 2005. Drug Discovery A History. John Wiley & Sons Ltd., West Sussex, England.

Swain, T. 1972. Plants in the Development of Modern Medicine. Harvard University Press, Cambridge, Massachusetts.

Van Wyk, B.-E., M. Wink. 2004. Medicinal plants of the world. Timber Press, Portland, Oregon.

Wiart, C. 2006. Medicinal Plant of the Asia and the Pacific. CRC Press, Taylor&Francis Group, Boca Raton, Florida.

World Health Organization, 1998. Quality control methods for medicinal plant materials. World Health Organization, Geneva.

International Encyclopaedia of Medicinal Plants, 18 vols. (Vijay Verma, Meenu Sharma, Vishwanath Pandey and Satyavir Singh, 2008; ISBN: 9788126135455 / 8126135455), www.anmolpublications.com/

Journals

Journal of Medicinal Plants Research
(http://www.academicjournals.org/jmpr/About.htm)

Journal of Herbs, Spices & Medicinal Plants
(http://www.informaworld.com/smpp/title~ content= t792306868~db=all)

Medicinal Plants - International Journal of Phytomedicines and Related Industries
(http://www.indianjournals.com/ijor.aspx?target=ijor:mpijpri&type=home)

Fitoterapia
(http://www.elsevier.com/wps/find/journaldescription.cws_home/620051/description)

Journal of Ethnobiology and Ethnomedicine
*(*http://www.ethnobiomed.com/ *)*

Australian Journal of Medicinal Herbalism
*(*www.nhaa.org.au/*)*

British Journal Phytotherapy
*(*www.springerlink.com *)*

Herb Companion
*(*www.herbcompanion.com*)*

Herbalgram
*(*www.herbalgram.org*)*

International Immunopharmacology
www.elsevier.com

Journal of Ethnopharmacology
(www.elsevier.com/locate/jethpharm)

Journal of Herbs, Spices and Medicinal Plants
*(*www.haworthpress.com *)*

Journal of Medicinal and Aromatic Plant Science
(www.cimap.res.in/publications)

Journal of Natural Medicines
(http://www.springerlink.com/content/h1813g204613/)

Journal of Natural Products
*(*www.journalofnaturalproducts.com*)*

Medical Herbalism
*(*www.medherb.com*)*

Pharmaceutical Biology
*(*www.informaworld.com*)*

Phytomedicine
*(*www.phytomedicinejournal.com*)*

Phytotherapy Research
(www.wiley.com)

Planta Medica
(www.ga-online.org)

Review of Aromatic & Medicinal Plants
(www.cabdirect.org)

Journal of Medicinal Food Plants
(http://www.jmedfoodplants.com)

Bibliography

BOOKS:

Carlson, T. J. S., and L. Maffi. 2004. *Ethnobotany and Conservation of Biocultural Diversity. Advances in Economic Botany.* Volume 15. New York: Botanical Garden Press.

Duke, J. A. 2009. *Duke's Handbook of Medicinal Plants of Latin America.* CRC Press, Boca Raton, FL. Taylor & Francis Group.

Duke, J. A. 2008. *Duke's Handbook of Medicinal Plants of the Bible.* Boca Raton, FL: CRC Press; Taylor & Francis Group.

Duke, J. A. 2002. *Handbook of Medicinal Herbs* (Second Edition). Boca Raton, FL: CRC Press.

Ebadi, M. S. 2007. *Pharmacodynamic Basis of Herbal Medicine.* Boca Raton, FL: CRC Press; Taylor & Francis Group.

Foster, S., and R. L. Johnson. 2006. *Desk Reference to Nature's Medicine.* Washington, D.C.: National Geographic Society.

Medicinal and Aromatic Plants — Industrial Profiles. Series edited by Roland Hardman.

Individual volumes in this series provide both industry and academia with in-depth coverage of one major genus of industrial importance.

Volume 1: *Valerian,* edited by Peter J. Houghton
Volume 2: *Perilla,* edited by He-ci Yu, Kenichi Kosuna and Megumi Haga
Volume 3: *Poppy,* edited by Jenö Bernáth
Volume 4: *Cannabis,* edited by David T. Brown
Volume 5: *Neem,* edited by H.S. Puri
Volume 6: *Ergot,* edited by Vladimír Kˇren and Ladislav Cvak
Volume 7: *Caraway,* edited by Éva Németh
Volume 8: *Saffron,* edited by Moshe Negbi
Volume 9: *Tea Tree,* edited by Ian Southwell and Robert Lowe
Volume 10: *Basil,* edited by Raimo Hiltunen and Yvonne Holm
Volume 11: *Fenugreek,* edited by Georgios Petropoulos
Volume 12: *Ginkgo biloba,* edited by Teris A. Van Beek
Volume 13: *Black Pepper,* edited by P. N. Ravindran
Volume 14: *Sage,* edited by Spiridon E. Kintzios
Volume 15: *Ginseng,* edited by W.E. Court
Volume 16: *Mistletoe,* edited by Arndt Büssing
Volume 17: *Tea,* edited by Yong-su Zhen
Volume 18: *Artemisia,* edited by Colin W. Wright
Volume 19: *Stevia,* edited by A. Douglas Kinghorn
Volume 20: *Vetiveria,* edited by Massimo Maffei
Volume 21: *Narcissus and Daffodil,* edited by Gordon R. Hanks
Volume 22: *Eucalyptus,* edited by John J.W. Coppen
Volume 23: *Pueraria,* edited by Wing Ming Keung
Volume 24: *Thyme,* edited by E. Stahl-Biskup and F. Sáez
Volume 25: Oregano, edited by Spiridon E. Kintzios
Volume 26: *Citrus,* edited by Giovanni Dugo and Angelo Di Giacomo
Volume 27: *Geranium and Pelargonium,* edited by Maria Lis-Balchin
Volume 28: *Magnolia,* edited by Satyajit D. Sarker and Yuji Maruyama

Volume 29: *Lavender,* edited by Maria Lis-Balchin

Volume 30: *Cardamom,* edited by P. N. Ravindran and K. J. Madhusoodanan

Volume 31: *Hypericum,* edited by Edzard Ernst

Volume 32: *Taxus,* edited by H. Itokawa and K.H. Lee

Volume 33: *Capsicum,* edited by Amit Krish De

Volume 34: *Flax,* edited by Alister Muir and Niel Westcott

Volume 35: *Urtica,* edited by Gulsel Kavalali

Volume 36: *Cinnamon and Cassia,* edited by P. N. Ravindran, K. Nirmal Babu and M. Shylaja

Volume 37: *Kava,* edited by Yadhu N. Singh

Volume 38: *Aloes,* edited by Tom Reynolds

Volume 39: *Echinacea,* edited by Sandra Carol Miller Assistant Editor: He-ci Yu

Volume 40: *Illicium, Pimpinella and Foeniculum,* edited by Manuel Miró Jodral

Volume 41: *Ginger,* edited by P.N. Ravindran and K. Nirmal Babu

Volume 42: *Chamomile: Industrial Profiles,* edited by Rolf Franke and Heinz Schilcher

Volume 43: *Pomegranates: Ancient Roots to Modern Medicine,* edited by Navindra P. Seeram, Risa N. Schulman and David Heber

Volume 44: *Mint,* edited by Brian M. Lawrence

Volume 45: *Turmeric,* edited by P. N. Ravindran, K. Nirmal Babu, and K. Sivaraman

Volume 46: *Essential Oil-Bearing Grasses,* edited by Anand Akhila

Volume 47: *Vanilla,* edited by Eric Odoux and Michel Grisoni

Volume 48: *Sesame,* edited by Dorothea Bedigian

Iwu, M. M. 1993. *Handbook of African Medicinal Plants.* Boca Raton, FL: CRC Press.

Juliani, H. R., J. E. Simon, C.-T. Ho. 2009. *African Natural Plant Products: New Discoveries and Challenges on Chemistry and Quality.* Washington, D.C:American Chemical Society.

Karch, S.B. 1999. *The Consumer's Guide to Herbal Medicine.* New York: Advanced Research Press.

Li, T.S.C. 2002. *Chinese and Related North American Herbs: Phytopharmacology and Therapeutic Values.* Boca Raton, FL: CRC Press.

Li., T.S.C. 2006. Taiwanese *Native Medicinal Plants: Phytopharmacology and Therapeutic Values.* Boca Raton, FL: CRC Press; Taylor & Francis Group.

Lindley, J. 1847. *The Vegetable Kingdom; or, the Structure, Classification, and Uses of Plants.* Whitefriars, London: Bradbury & Evans.

Lise, M. 1989. *An Ancient Egyptian Herbal.* Austin, TX: University of Texas Press.

Mann, J. 1992. *Murder, Magic, and Medicine.* New York: Oxford University Press

Marcello, S. *The Psychopharmacology of Herbal Medicine: Plant Drugs That Alter Mind, Brain, and Behavior.* Cambridge, Massachusetts. MIT Press.

Raffauf, R. F. 1996. *Plant Alkaloid-A Guide to Their Discovery and Distribution.* New York: The Haworth Press, Inc.

Rosenthal, N.E. 1998. *St. John's Wort: The Herbal Way to Feeling Good.* New York: Harper Collins.

Sneader, W. 2005. *Drug Discovery A History.* West Sussex, England. John Wiley & Sons Ltd.

Swain, T. 1972. *Plants in the Development of Modern Medicine.* Cambridge, Massachusetts: Harvard University Press.

Van Wyk, B.-E., M. Wink. 2004. *Medicinal Plants of the World.* Portland, Oregon: Timber Press.

Wiart, C. 2006. *Medicinal Plants of the Asia and the Pacific.* Boca Raton, FL: CRC Press; Taylor & Francis Group.

World Health Organization, 1998. Quality control methods for medicinal plant materials. Geneva: World Health Organization.

Vijay Verma, V., Sharma, M., Pandey, V., and Singh, S. 2008; *International Encyclopaedia of Medicinal Plants,* 18 vols. Eastern Book Corporation, Delhi, India.

Journals

Australian Journal of Medicinal Herbalism (www.nhaa.org.au/)

British Journal Phytotherapy (www.springerlink.com)

Fitoterapia (http://www.elsevier.com/wps/find/journaldescription.cws_home/620051/description)

Herb Companion (www.herbcompanion.com)

Herbalgram (www.herbalgram.org)

International Immunopharmacology www.elsevier.com

Journal of Ethnobiology and Ethnomedicine (http://www.ethnobiomed.com/)

Journal of Ethnopharmacology (www.elsevier.com/locate/jethpharm)

*Journal of Herbs, Spices, and Medicinal Plants (*www.haworthpress.com)

Journal of Medicinal and Aromatic Plant Science (www.cimap.res.in/publications)

Journal of Medicinal Food Plants (http://www.jmedfoodplants.com)

Journal of Medicinal Plants Research (http://www.academicjournals.org/jmpr/About.htm)

Journal of Natural Medicines (http://www.springerlink.com/content/h1813g204613/)

Journal of Natural Products (www.journalofnaturalproducts.com)

Medical Herbalism (www.medherb.com)

Medicinal Plants - International Journal of Phytomedicines and Related Industries (http://www.
  indianjournals.com/ijor.aspx?target=ijor:mpijpri&type=home)

Pharmaceutical Biology (www.informaworld.com)

Phytomedicine (www.phytomedicinejournal.com)

Phytotherapy Research (www.wiley.com)

Planta Medica (www.ga-online.org)

Review of Aromatic & Medicinal Plants (www.cabdirect.org)

Index

Hybridization
  Aloe vera, 871, 873–874
  tea, 548
  turmeric crop improvement, 491, 494–495
Hybrids, fertility restoration, 732
Hydrangea, 5
Hydrastis canadensis
  cancer, 15
  plant-based drug discovery, 30
Hydrocharitaceae, 36
Hydrocotylaceae, 66
Hydrolytic enzymes, 557
Hydrophyllaceae, 36
Hydroxyl-isoleucine, 831
Hylocereus undatus, 110
Hymenophyllaceae, 35
Hyoscyamus niger
  chemical drugs originating from, 131
  clinical use/active principle, 67
  illicit use, 7
  medicinal usage, distribution, and plant part used in
      Turkmenistan, 215
  medicinal usage, name, and plant part used in Turkey,
      195
  plant-based drug discovery, 30
Hypecoideae, 358
Hypericaceae, 213
Hypericum sp., 201, 983
Hypericum helianthomoides, 213
Hypericum perforatum
  medicinal usage, name, and plant part used in Turkey,
      192
  plant-based drug discovery, 30
  ribosomal and plastid DNA-based analysis, 970
  variable number tandem repeats, 965
Hyperlipidemia, 532
Hypertension, *see* Blood pressure
Hypocholesterolemic properties, 5
Hypotensives, 26, *see also* Blood pressure
Hypothenemus hampei
  coffee, 590, 612
  data analysis, 615
  heterologous genes, 622
Hypoxidaceae, 36
Hyptis pectinata, 231

I

IAA, *see* Indole-3-acetic Acid (IAA)
IBPGR, *see* International Board for Plant Genetic
    Resources (IBPGR)
IBRG, *see* Indigenous Bioresources Research Group
    (IBRG)
Icacinaceae, 36
ICFR, *see* International Council for Fitogenetic
    Resources (ICFR)
ICMAP, *see* International Council for Medicinal and
    Aromatic Plants (ICMAP)
ICMR, *see* Indian Council of Medical Research (ICMR)

Icones Plantarum Medicinalium, 515
Icterus nigrogularis, 857
Identification/misidentification of herbs
  Ghana, 244
  Turkey, 186
IDRC, *see* International Development Research Center
    (IDRC)
IISR, *see* Indian Institute of Spices Research (IISR)
Ilacum, 652
Ilex sp., 150
Ilex paraguariensis, 30
Illicit production, 373–374, 376
Illicit use, 7
Illicium, pimpinella, and foeniculum, 984
Illustrated Reference of Botanical Nomenclature, 125
Immune function efficacy, 531
Immunomodulation, 15
Immunomodulatory activities, 887
Impatiens balsamina, 169
Importance, medicinal
  beverages, 5
  cosmeceutical products, 6–7
  fundamentals, 3–4
  illicit use, 7
  modern medicine, 4–5
  myrtle, 929–931
  quality control, 8
  spices, 5–6
  Stevia rebaudiana, 887, 903–904
Important Arts for the People's Welfare, 125
Imwonsipyukji, 520
INCB, *see* International Narcotics Control Board (INCB)
Index Kewensis, 144, 541
Index of Classics of the Sui Dynasty, 125
India
  ginger cultivars, 417, 426
  ginger germplasm conservation, 418
  popular traditional cultivars of turmeric, 479
  spices, 6
  turmeric germplasm conservation, 476–477
Indian Bdellium, 73
Indian Council of Medical Research (ICMR), 63
Indian gooseberry, 967–968
Indian Institute of Spices Research (IISR), 477
Indian traditional system of medicine
  Ayurvedic system, 55–58
  concepts, 56–61
  Council of Scientific and Industrial Research, 63
  ethnomedical botanical research, 61–62
  fundamentals, 54–55
  Golden Triangle partnership, 63–64
  health, sickness, and treatment, 56–57
  homeopathy, 61
  National Medicinal Plant Board, 64
  naturopathy, 60–61
  New Millennium Indian Technology Leadership
      Initiative, 63
  origin and history, 55–56, 58, 60–61
  Panchakarma, 57